	$\frac{\text{ft}}{\text{min}}$ (fpm)	$\frac{\text{cm}}{\text{s}}$	$\frac{\text{km}}{\text{hr}}$	$\frac{\text{ft}}{\text{sec}}$ (fps)	$\frac{\text{statute mile}}{\text{hr}}$ (mph)	Knots	$\frac{\text{m}}{\text{s}}$
$\frac{\text{ft}}{\text{min}}$ (fpm)							
$\frac{\text{cm}}{\text{s}}$	1.969 $\frac{\text{fpm}}{\text{cm/s}}$						
$\frac{\text{km}}{\text{hr}}$	54.68 $\frac{\text{fpm}}{\text{km/hr}}$	27.78 $\frac{\text{cm/s}}{\text{km/hr}}$					
$\frac{\text{ft}}{\text{sec}}$ (fps)	60 $\frac{\text{fpm}}{\text{fps}}$	30.48 $\frac{\text{cm/s}}{\text{fps}}$	1.097 $\frac{\text{km/hr}}{\text{fps}}$				
$\frac{\text{statute mile}}{\text{hr}}$ (mph)	88 $\frac{\text{fpm}}{\text{mph}}$	44.7 $\frac{\text{cm/s}}{\text{mph}}$	1.609 $\frac{\text{km/hr}}{\text{mph}}$	1.47 $\frac{\text{fps}}{\text{mph}}$			
Knots	101.3 $\frac{\text{fpm}}{\text{knot}}$	51.44 $\frac{\text{cm/s}}{\text{knot}}$	1.852 $\frac{\text{km/hr}}{\text{knot}}$	1.688 $\frac{\text{fps}}{\text{knot}}$	1.151 $\frac{\text{mph}}{\text{knot}}$		
$\frac{\text{m}}{\text{s}}$	196.9 $\frac{\text{fpm}}{\text{m/s}}$	100 $\frac{\text{cm/s}}{\text{m/s}}$	3.6 $\frac{\text{km/hr}}{\text{m/s}}$	3.281 $\frac{\text{fps}}{\text{m/s}}$	2.237 $\frac{\text{mph}}{\text{m/s}}$	1.944 $\frac{\text{knot}}{\text{m/s}}$	

Velocity conversion factors. Adapted from Table C.8.

Table C.11 Volume flow rate conversion factors.

$\frac{\text{m}^3}{\text{s}}$	$\frac{\text{ft}^3}{\text{sec}}$	$\frac{\text{Liter}}{\text{s}}$	$\frac{\text{ft}^3}{\text{min}}$	$\frac{\text{gal}}{\text{min}}$ (gpm)	$\frac{\text{cm}^3}{\text{s}}$	
	35.61 $\frac{\text{ft}^3/\text{sec}}{\text{m}^3/\text{s}}$	1000 $\frac{\text{l/s}}{\text{m}^3/\text{s}}$	2119 $\frac{\text{ft}^3/\text{min}}{\text{m}^3/\text{s}}$	15,850 $\frac{\text{gpm}}{\text{m}^3/\text{s}}$	60E6 $\frac{\text{cm}^3/\text{s}}{\text{m}^3/\text{s}}$	$\frac{\text{m}^3}{\text{s}}$
		28.32 $\frac{\text{l/s}}{\text{ft}^3/\text{sec}}$	60 $\frac{\text{ft}^3/\text{min}}{\text{ft}^3/\text{sec}}$	449 $\frac{\text{gpm}}{\text{ft}^3/\text{sec}}$	2.832E4 $\frac{\text{cm}^3/\text{s}}{\text{ft}^3/\text{sec}}$	$\frac{\text{ft}^3}{\text{sec}}$
			2.119 $\frac{\text{ft}^3/\text{min}}{\text{l/s}}$	15.85 $\frac{\text{gpm}}{\text{l/s}}$	1000 $\frac{\text{cm}^3/\text{s}}{\text{l/s}}$	$\frac{\text{liter}}{\text{s}}$
				7.481 $\frac{\text{gpm}}{\text{ft}^3/\text{min}}$	471.9 $\frac{\text{cm}^3/\text{s}}{\text{ft}^3/\text{min}}$	$\frac{\text{ft}^3}{\text{min}}$
					63.09 $\frac{\text{cm}^3/\text{s}}{\text{gpm}}$	$\frac{\text{gal}}{\text{min}}$ (gpm)
						$\frac{\text{cm}^3}{\text{s}}$

Pressure conversion factors. Adapted from Table C.7.

atm	Bar	psi	in. Hg (32°F)	in. H_2O (39.2°F)	mm Hg (0°C)	Pascal $\left(\frac{\text{N}}{\text{m}^2}\right)$	
	1.0132 $\frac{\text{bar}}{\text{atm}}$	14.696 $\frac{\text{psi}}{\text{atm}}$	29.92 $\frac{\text{in. Hg}}{\text{atm}}$	406.8 $\frac{\text{in. H}_2\text{O}}{\text{atm}}$	760 $\frac{\text{mm Hg}}{\text{atm}}$	1.013E5 $\frac{\text{Pa}}{\text{atm}}$	atm
		14.504 $\frac{\text{psi}}{\text{bar}}$	29.53 $\frac{\text{in. Hg}}{\text{bar}}$	401.5 $\frac{\text{in. H}_2\text{O}}{\text{bar}}$	750.1 $\frac{\text{mm Hg}}{\text{bar}}$	1.0E5 $\frac{\text{Pa}}{\text{bar}}$	Bar
			2.036 $\frac{\text{in. Hg}}{\text{psi}}$	27.68 $\frac{\text{in. H}_2\text{O}}{\text{psi}}$	51.71 $\frac{\text{mm Hg}}{\text{psi}}$	6.895E3 $\frac{\text{Pa}}{\text{psi}}$	psi
				13.60 $\frac{\text{in. H}_2\text{O}}{\text{in. Hg}}$	25.4 $\frac{\text{mm Hg}}{\text{in. Hg}}$	3386 $\frac{\text{Pa}}{\text{in. Hg}}$	in. Hg (32°F)
					1.868 $\frac{\text{mm Hg}}{\text{in. H}_2\text{O}}$	249.1 $\frac{\text{Pa}}{\text{in. H}_2\text{O}}$	in. H_2O (39.2°F)
						133.3 $\frac{\text{Pa}}{\text{mm Hg}}$	mm Hg (0°C)
							Pascal $\left(\frac{\text{N}}{\text{m}^2}\right)$

	$\frac{\text{ft}\cdot\text{lb}}{\text{min}}$	$\frac{\text{BTU}}{\text{hr}}$	$\frac{\text{ft}\cdot\text{lb}}{\text{sec}}$	$\frac{\text{gm}\cdot\text{cal}}{\text{s}}$	Horsepower (hp)	Kilowatt (kW)	$\frac{\text{BTU}}{\text{sec}}$
$\frac{\text{ft}\cdot\text{lb}}{\text{min}}$							
$\frac{\text{BTU}}{\text{hr}}$	12.96 $\frac{\text{ft}\cdot\text{lb/min}}{\text{BTU/hr}}$						
$\frac{\text{ft}\cdot\text{lb}}{\text{sec}}$	60 $\frac{\text{ft}\cdot\text{lb/min}}{\text{ft}\cdot\text{lb/sec}}$	4.629 $\frac{\text{BTU/hr}}{\text{ft}\cdot\text{lb/sec}}$					
$\frac{\text{gm}\cdot\text{cal}}{\text{s}}$	185.2 $\frac{\text{ft}\cdot\text{lb/min}}{\text{gm}\cdot\text{cal/s}}$	14.29 $\frac{\text{BTU/hr}}{\text{gm}\cdot\text{cal/s}}$	3.086 $\frac{\text{ft}\cdot\text{lb/sec}}{\text{gm}\cdot\text{cal/s}}$				
Horsepower (hp)	3.3E4 $\frac{\text{ft}\cdot\text{lb/min}}{\text{hp}}$	2542 $\frac{\text{BTU/hr}}{\text{hp}}$	550 $\frac{\text{ft}\cdot\text{lb/sec}}{\text{hp}}$	6.416E5 $\frac{\text{gm}\cdot\text{cal/s}}{\text{hp}}$			
Kilowatt (kW)	4.425E4 $\frac{\text{ft}\cdot\text{lb/min}}{\text{kW}}$	3413.4 $\frac{\text{BTU/hr}}{\text{kW}}$	737.6 $\frac{\text{ft}\cdot\text{lb/sec}}{\text{kW}}$	238.7 $\frac{\text{gm}\cdot\text{cal/s}}{\text{kW}}$	1.341 $\frac{\text{hp}}{\text{kW}}$		
$\frac{\text{BTU}}{\text{sec}}$	4.666E4 $\frac{\text{ft}\cdot\text{lb/min}}{\text{BTU/sec}}$	3600 $\frac{\text{BTU/hr}}{\text{BTU/sec}}$	778 $\frac{\text{ft}\cdot\text{lb/sec}}{\text{BTU/sec}}$	252.0 $\frac{\text{gm}\cdot\text{cal/s}}{\text{BTU/sec}}$	1.414 $\frac{\text{hp}}{\text{BTU/sec}}$	1.054 $\frac{\text{kW}}{\text{BTU/sec}}$	

Power conversion factors. Adapted from Table C.12.

Fundamentals of Fluid Mechanics

Fundamentals of FLUID MECHANICS

SECOND EDITION

Philip M. Gerhart
UNIVERSITY OF EVANSVILLE

Richard J. Gross
UNIVERSITY OF AKRON

John I. Hochstein
MEMPHIS STATE UNIVERSITY

ADDISON-WESLEY PUBLISHING COMPANY
Reading, Massachusetts ▪ Menlo Park, California ▪ New York
Don Mills, Ontario ▪ Wokingham, England ▪ Amsterdam ▪ Bonn ▪ Sydney
Singapore ▪ Tokyo ▪ Madrid ▪ San Juan ▪ Milan ▪ Paris

Faith Sherlock, Sponsoring Editor
Laurie L. McGuire, Associate Editor
Loren Hilgenhurst Stevens, Production Supervisor
Sheila J. Bendikian, Production Coordinator
Jerrold A. Moore, Copy Editor
Mark Ong, Text Designer
Peter M. Blaiwas, Cover Designer
Robert Forget, Art Editor
Nigel Andrews, Illustrator
Roy Logan, Manufacturing Supervisor

Library of Congress Cataloging-in-Publication Data

Gerhart, Philip M.
Fundamentals of fluid mechanics / Philip M. Gerhart, Richard J. Gross, John Hochstein. — 2nd ed.
p. cm.
ISBN 0-201-18358-7
1. Fluid mechanics. I. Gross, Richard J. II. Hochstein, John. III. Title.
TA357.G46 1992
620.1'06—dc20 91-15912
CIP

1 2 3 4 5 6 7 8 9 10—DO—959493929

Preface

This book was written to help prospective engineers learn fluid mechanics. Accordingly, it is primarily intended for use as a textbook in a sophomore- or junior-level course. The course may be taught by, and the students may be intending to become, civil engineers, mechanical engineers, or aerospace engineers. The assumption that the primary objective of the course is the preparation of engineers strongly influenced our writing. We emphasize applications of the science of fluid mechanics to the student's chosen profession. The variety of applications presented shows that the subject of fluid mechanics crosses the boundaries of all branches of engineering.

This book, like any other textbook, is one corner of a triangle in the learning process. The other corners are the instructor and the student. The only corner of this triangle that does not suffer the pressures of limited time is the textbook. We thus took pains to provide complete and careful discussions of complex phenomena and gave special attention to several points that always seem to trouble students. We are confident that the book's flexibility and completeness make it a valuable member of the learning triangle.

Objectives

The objectives of the book are

- to help students develop a physical awareness of the phenomena of fluid motion;
- to present the fundamental laws that govern all fluid behavior and impress on students that all fluid mechanics problems are solved by application of these fundamental laws;
- to develop practical methodologies for solving engineering flow problems, including the routine use of computer hardware and software;
- to give balanced emphasis to the differential and finite control volume approaches;
- to illustrate the extremely wide variety of fluid-related phenomena in everyday life and in modern technology;
- to prepare students to enter professional practice or to pursue graduate study; and
- to provide a valuable reference that can be used throughout an engineering career.

Pre- and Corequisites

We assumed that students have successfully completed courses in single- and multivariable calculus, statics, dynamics, and computer use and programming. We also assumed that most students have completed courses in mechanics of solids and ordinary differential equations or will be taking such courses concurrently with fluid mechanics. We did not assume either prior or concurrent study of thermodynamics. Any necessary information from thermodynamics is introduced in the text. Students familiar with thermodynamics can skim over this material.

Fluid mechanics can involve a great deal of mathematics. We have not hesitated to introduce and use appropriate mathematical formulations; however our priorities have always placed physical understanding and practical application ahead of mathematical formulation and manipulation.

Features

Style. Our writing style is casual and informal. Our intent was to create a friendly atmosphere for the reader.

Applications. We discuss both everyday life experiences that involve fluid mechanics principles (such as snowflakes carried in a wind gust or the aerodynamics of baseball pitching) and engineering applications of fluid mechanics. The applications are intended to show the importance of fluid mechanics in design, operation, maintenance, and other engineering functions. Applications are selected from many areas of civil, mechanical, and aerospace engineering.

Examples. The book contains 116 completely solved example problems, reflecting a wide variety of applications. Many illustrate the need to make assumptions or obtain data from outside sources. Several involve iterative solution or evaluation of several alternatives. Some illustrate the application of computers in fluid mechanics, from spreadsheets to computational fluid dynamics codes.

A degree of realism not often found in other texts is maintained. All examples are presented in a logical format. Careful attention is devoted to units in calculations. Discussions following each example comment on accuracy and generality of results, suggest alternative ways of solving the problem, and clarify tricky steps in the solution.

Problems. More than 1200 end-of-chapter problems are presented. More challenging problems are marked with the • symbol. Problems marked with the 🖳 symbol are specifically intended for computer solution, although we emphasize that personal computer hardware and software can be readily applied to almost every problem in this book.

Problems are drawn from all areas of fluid mechanics and vary in difficulty. Many are based on real-world applications. An effective problem-solving method is described in Chapter 1. Virtually every problem has been either tested in the classroom or independently checked by an experienced teacher of fluid mechanics.

Design. The primary practical application of fluid mechanics is to the design, operation, and maintenance of engineering systems. We present fluid mechanics within this context, describing the differences between design problems and analysis problems and including a few example problems that emphasize design.

For several years, we have assigned limited design projects in our own fluid mechanics courses. We included several successful design project problems in Appendix J.

Units. We used the International System (SI) of units, the English Engineering (EE) system of units, and the British Gravitational (BG) system of units. The SI units always appear in the text and in about 60 percent of the examples and problems. The EE units appear in about 30 percent of the examples and problems, and BG units appear in the remaining 10 percent.

As much as we might wish SI units to be used universally, U.S. industry still makes wide use of the older "foot, pound, inch" system. We believe that an engineering text written exclusively in either system does not adequately prepare students for the real world.

Computers. All kinds of computers, ranging from personal computers through powerful workstations to extremely large and powerful supercomputers, are available to today's engineers. In the past, the topic of computers in fluid mechanics evoked the image of large mainframe computers and huge, complicated codes for solving partial differential equation flow models. Applying a computer to a flow problem required a specialist. Current applications of computers to fluid mechanics still include such efforts, and we discuss this type of application in this text. A simple and powerful computational fluid dynamics code is available to instructors who adopt the book for their courses.

During the 1980s, use of computers in all areas of engineering took on a new meaning. Almost every practicing engineer and engineering student has access to a personal computer for everyday problem solving and information management. Spreadsheets and formula processing packages are easy to use and have greatly expanded the engineer's ability to make involved calculations and investigate various design options.

It is this everyday type of computing that we chose to emphasize in this text. Several chapters contain special sections and problems dedicated to using computers in fluid mechanics. Appendix F is dedicated to a review of basic numerical methods for problem solving, and floppy disks with basic numerical tools and fluid property data are available with the solutions manual to this text. We encourage instructors to make the disk available to students for copying. Occasionally, an example is solved by using a spreadsheet or a simple FORTRAN or BASIC program. We no longer avoid nonlinear algebraic equations or iterative solution techniques.

We readily sympathize with many of our colleagues who point out that the purpose of a first course in fluid mechanics is to teach fundamentals, not computing. We therefore, for the most part, packaged the material on computing separately.

Appendixes. Several appendixes are provided, giving data on fluid properties, geometric characteristics of areas, and pipe and tubing sizes in SI, BG, and EE units. An extensive list of conversion factors in a unique format is presented. Other items in the appendixes include compressible flow functions, a list of films, governing equations in cylindrical and spherical coordinates, an exhaustive list of symbols, a collection of design problems, a discussion of numerical methods, and a brief introduction to the computational fluid mechanics code. The appendixes are more extensive than customary, because we believe that they add to the long-term usefulness of the text. We find ourselves returning again and again to one or another of our old textbooks because of its extensive appendix.

Instructional Support. A comprehensive solutions manual is available to instructors from Addison-Wesley Publishing Company. Solutions to the problems are presented in the same format as the example problems. Virtually every solution has been tested in the classroom or independently checked by an experienced teacher of fluid mechanics. Additional information in the solutions manual includes a classification of all end-of-chapter problems by topic and separate tabulations of answers to even-numbered and odd-numbered problems.

Two IBM-compatible floppy disks are available to instructors. One has FORTRAN subroutines for numerical root finding, curve fitting, integration, and differential equation solution as well as fluid property data and compressible flow functions in computer readable format. This disk is provided to instructors with the solutions manual. A second disk provides a computational fluid dynamics code and is available to instructors on request.

Organization. The organization of the book is broadly typical of most fluid mechanics texts. We delayed introduction of most flow concepts until after the material on nonflowing fluids. Our experience indicates that introduction of Eulerian and Lagrangian descriptions, velocity fields, streamlines, and other ideas can be intimidating in an introductory chapter. A special chapter is dedicated to fundamental flow concepts. A simple outline of the book is as follows:

- Introduction (Chapter 1)
- Fluid Statics and Rigid Body Motion (Chapter 2)
- Fundamental Flow Concepts (Chapter 3)
- Basic Techniques of Flow Analysis (Chapters 4–6)
- Applications of Flow Analysis (Chapters 7–11)

Chapters 1–6 should be studied in order. Chapter 2 can be moved or even omitted. Chapters 7–11 can be taken in any order or omitted.

We tried to give balanced emphasis to the differential equation approach and the finite control volume approach. Because control volume analysis is more easily mastered by beginners, it is usually emphasized in an introductory course. Reflecting this tradition, the control volume material comprises a complete and self-contained chapter, which we placed ahead of the differential equation chapter. However, we paid careful attention to making the differential equation chapter equally complete and comprehensible. The balanced coverage of these topics is in keeping with our belief that both methods are important to the student's eventual mastery of fluid mechanics.

Changes from the First Edition

Readers acquainted with the first edition of *Fundamentals of Fluid Mechanics* should note the following changes in this second edition.

- The use of computers in fluid mechanics is woven throughout the book. Special consideration is given to the use of personal computers and associated software for routine problem solving.
- There are approximately 600 new problems, including about 50 computer problems. Many of the problems retained from the first edition were modified. Virtually all problems and solutions were tested in the classroom or independently checked by experienced teachers of fluid mechanics. The four-level classification system for problems was replaced with a less cumbersome two-level system that distinguishes challenging problems from others. Additionally, computer problems are clearly designated as such. Problems are grouped by topic.
- The order of presentation of the finite control volume method and the differential equation method was reversed. Chapter 4 on the finite control volume method was extensively rewritten and is more self-contained. Chapter 5 on the differential equation method also is more self-contained and contains significant material on methods and models for practical application of the differential equation method.
- Material on matching pumps and fans to systems and on selecting pumps and fans has been added to Chapter 7.
- About 10 percent of the example problems are either new or have been modified for improved clarity and realism or to more properly demonstrate a text concept.
- An appendix containing several fluid mechanics-based design projects has been added.
- A floppy disk with numerical analysis routines and appendix data is available. A computational fluid dynamics code is also available.

Course Planning

We included far more material than can be covered in a single semester (assumed to include 45 class periods) or quarter (assumed to include 40 class periods). By selecting and omitting material, instructors can construct courses specifically tailored to the needs of civil, mechanical, and aerospace engineers. There is sufficient material for a second-semester or second-quarter course in advanced or applied fluid mechanics. Some instructors may want to assign portions of the text that are not covered in class. Students may even read portions of the text that are not assigned (for one, the section on aerodynamics in sports looks interesting). We hope that students working on senior projects, as well as practicing engineers, will return to the book to find information not covered in their fluid mechanics course.

The following are suggested sections of the text to be included in courses for various purposes.

1. Single-semester or single-quarter introductory course for all types of engineers:

   Chapter 1

   Sections 2.1–2.4

   Chapter 3

   Sections 4.1–4.4.4, 4.5, 4.6

   Sections 5.1–5.3

   Chapter 6

   Sections 7.1–7.2.5, 7.3

   Sections 9.2.1–9.2.5

   Other material selected by instructor

2. Single-semester or single-quarter course for mechanical engineers:

   Chapter 1

   Sections 2.1–2.5

   Chapter 3

   Chapter 4

   Chapter 6

   Chapter 7 (omit 7.2.7)

   Sections 10.2–10.4

3. Single-semester or single-quarter course for civil engineers:

   Chapter 1

   Sections 2.1–2.5

   Chapter 3

   Chapter 4

   Chapter 6

   Chapter 7 (omit 7.2.7)

   Chapter 11

4. Single-semester or single-quarter course for aerospace engineers:

   Chapter 1

   Sections 2.1–2.3

   Chapter 3

   Sections 4.1–4.4.4, 4.5.3

Chapter 6

Chapter 8

Chapter 9

Sections 10.1–10.3, 10.6

5. Applied or advanced fluid mechanics (elective):

Chapter 3 (Review)

Chapter 5

Sections 7.2.7, 7.4, 7.5

Sections 8.3, 8.4

Chapter 9, Chapter 10, Chapter 11 } Various emphasis depending on instructor and discipline

Acknowledgements

This revision would not have been accomplished, or it would be significantly less than it is, without the help of a great number of people. Special thanks go to Don Fowley and Laurie McGuire of Addison-Wesley for believing in us and keeping us on track. Alice, Debbie, and Mary Lou once again displayed incredible love and patience, as did Terri, Mary Ann, Kathy, David, Billy, David, Ann, Marie, Daniel, Ann, and Andy. Several of them grew to be adults while this project was evolving.

On the technical side, we are grateful to the many instructors who provided thorough reviews of the manuscript and who subjected the problems and solutions to class testing and painstaking accuracy checks. Special thanks go to M. J. Marongiu. We also thank J. L. Vadnal, M. S. Korman, M. J. Marongiu, M. R. Miller, J. L. Mock, T. Johns, M. Talbert, D. Plummer, R. Loudermilk, N. Shuib, and N. Mohamad for supplying some of the new end-of-chapter problems and their solutions and B. Carpenter for her assistance in software development.

Finally, we thank all our professional colleagues who used the first edition of *Fundamentals of Fluid Mechanics*. They provided the motivation to prepare this second edition.

Evansville, Indiana
Akron, Ohio
Memphis, Tennessee

P.M.G.
R.J.G.
J.I.H.

MANUSCRIPT REVIEWERS

M. A. Alawady
Louisiana State University

N. Z. Azer
Kansas State University

David Bogard
University of Texas at Austin

Giles Brereton
University of Michigan

Colin Britcher
Old Dominion University

E. Eugene Callens, Jr.
Louisiana Tech University

Randall Charbeneau
University of Texas at Austin

Joseph C.-F. Chow
University of Illinois
at Chicago

Ruben Cohen
Rice University

Ronald A. Cox
University of Oklahoma

L. Michael Freeman
University of Alabama

Robert Fuessle
Bradley University

Richard Gardner
Washington University

Donald Gray
West Virginia University

B. K. Hodge
Mississippi State University

Charles V. Knight
University of Tennessee
at Chattanooga

Gunol Kocamustafaogullari
University of Wisconsin
at Milwaukee

Murray S. Korman
U.S. Naval Academy

Joe O. Ledbetter
University of Texas at Austin

Shen Lee
University of Missouri at Rolla

Gary Lehman
State University of New York
at Binghamton

Robert Medrow
University of Missouri at Rolla

Lee Robinson
Idaho State University

Leo T. Smith
University of Hartford

J. D. A. Walker
Lehigh University

PROBLEMS CLASSROOM TESTERS AND REVIEWERS

Tim Foutz
University of Georgia

Maurice J. Marongiu
Illinois Institute of Technology

Mardson Queiroz
Brigham Young University

J. I. Ramos
Carnegie Mellon University

Ting Wang
Clemson University

Daniel Woodbury
Brigham Young University

Contents

APPENDIXES 909

1 Introduction

1.1 FLUID MECHANICS IN ENGINEERING

Fluid mechanics is the branch of engineering science that is concerned with forces and energies generated by fluids at rest and in motion. The study of fluid mechanics involves applying the fundamental principles of mechanics and thermodynamics to develop physical understanding and analytic tools that engineers can use to design and evaluate equipment and processes involving fluids.*

You are probably reading this book because fluid mechanics is a required course in your curriculum. Most engineering curricula require fluid mechanics because its principles and methods find many technological applications in fields such as:

- Fluid transport
- Energy generation
- Environmental control
- Transportation

Fluid transport is movement of a fluid from one place to another so that the fluid may be used or processed. Examples include home and city water supply systems, cross-country oil, natural gas and agricultural chemical pipelines, and chemical plant piping. Engineers involved in fluid transport might design systems involving pumps, compressors, pipes, valves, and a host of other components. In addition to designing new systems, engineers may evaluate the adequacy of existing systems to meet new demands, or they may maintain or upgrade existing systems.

Very little useful energy is generated without fluid movement. Typical energy-conversion devices such as steam turbines, reciprocating engines, gas turbines, hydroelectric plants, and windmills involve many complicated flow processes. The thermodynamic cycles of these devices usually require fluid machinery such as pumps or compressors to do work on the fluid. Auxiliary equipment such as oil pumps, carburetors, fuel-injection systems, boiler draft fans, and cooling systems also involves fluid motion.

Environmental control involves fluid motion. Most building heating systems use a fluid to transport energy from a combustion process or other heat source to

* This discussion assumes that you already have some idea of what a fluid is. We present a formal definition shortly. The most common engineering fluids are air (a gas), water (a liquid), and steam (a vapor).

the heated spaces. In air-conditioning systems the circulating air is cooled by a flowing refrigerant. Similar processes occur in automobile engine cooling systems, in machine tool cooling systems, and in the cooling of electronic components by ambient air.

With the exception of space travel, all transportation takes place within a fluid medium (the atmosphere or a body of water). The relative motion between the fluid and the transportation device generates a force that opposes the desired motion. The application of fluid mechanics to vehicle design can minimize this force. The fluid often contributes in a positive way, such as by floating a ship or generating lift on airplane wings. In addition, ships and airplanes derive propulsive force from propellers or jet engines that interact with the surrounding fluid.

These examples are by no means exhaustive. Other engineering applications of fluid mechanics include the design of canals, harbors, and dams. The design of large structures must account for the effects of wind loading. In environmental engineering and biomedical engineering, engineers must deal with naturally occurring flow processes in the atmosphere and lakes, rivers, and seas or within the human body. The phenomena of fluid motion are central to the field of meteorology and weather forecasting.

Few engineers can function effectively without at least a rudimentary knowledge of fluid mechanics. Large numbers of engineers deal primarily with processes, devices, and systems in which a knowledge of fluid mechanics is essential to intelligent design, evaluation, maintenance, or decision making. You probably cannot foresee exactly what problems you will be called on to solve in your professional career; therefore you would be wise to obtain a firm grasp of the fundamentals of fluid mechanics.

These fundamentals include a knowledge of the nature of fluids and the properties used to describe them, the physical laws that govern fluid behavior, the ways in which these laws may be cast into mathematical form, and the various methodologies that may be used to solve engineering problems. We wrote this book to help you learn these fundamentals.

1.2 FLUIDS AND THE CONTINUUM HYPOTHESIS

All matter exists in one of two states: solid or fluid. You probably have a qualitative understanding of the difference between the two. A solid does not easily change its shape, but a fluid changes shape with relative ease. This concept of a fluid encompasses both liquids, which easily change shape but not volume, and gases, which easily change both shape and volume.

To develop a formal definition of a fluid, consider imaginary chunks of both a solid and a fluid (Fig. 1.1). The chunks are fixed along one edge, and a shear force is applied at the opposite edge.* A short time after application of the force, the solid assumes a deformed shape, which can be measured by the angle ϕ_1. If

* Note that it is easy to imagine actually carrying out this experiment with a solid such as steel or rubber but that it would be impossible to actually set a chunk of water on the table and carry out this experiment, because the water would flow all over the table. This phenomenon is exactly the nature of a fluid that we are seeking to define!

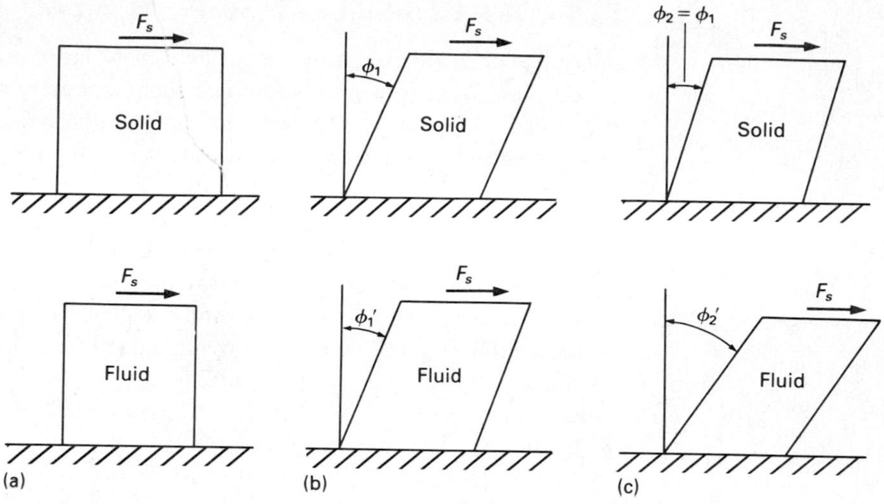

Figure 1.1 Response of samples of solid and fluid to applied shear force: Condition (a) corresponds to the instant of application of the force; condition (b) to a short time after the application of the force; and (c) to a later time.

we maintain this force and examine the solid at a later time, we find that the deformation is exactly the same, that is, $\phi_2 = \phi_1$. On application of a shear force, a solid assumes a certain deformed shape and retains that shape as long as the force is applied. For most solids, the magnitude of the deformation is proportional to the magnitude of the force, and the solid will return to the original undeformed shape if the force is removed, at least if the magnitude of the shear and deformation are below certain limits.

Consider the response of the fluid to the applied shear force. A short time after application of the force, the fluid assumes a deformed shape, as indicated by the angle ϕ'_1. At a later time, the deformation is greater, $\phi'_2 > \phi'_1$; in fact, the fluid continues to deform as long as the force is applied. If the force is removed, the fluid will not return to its undeformed shape but will retain whatever shape it had when the force was removed. Thus we can define a fluid:

> A fluid is a substance that deforms continuously under the action of an applied shear force or stress.

The process of continuous deformation is called *flowing*. A fluid is a substance that is able to flow.

Because a fluid will always flow if a shear stress is applied, it is not possible to analyze or discuss fluid behavior in terms of stress and deformation as is done in solid mechanics. It is necessary to consider the relation between stress and the *time rate* of deformation. A fluid deforms at a rate related to the applied stress. The fluid attains a state of "dynamic equilibrium" in which the applied stress is balanced by the resisting stress. Thus an alternative definition of a fluid is:

> A fluid is a substance that can resist shear only when moving.

The distinction between solid and fluid seems rather simple, and most substances can be easily classified; however, some substances can fool you. Cold tar seems to be solid; however, under the action of an applied shear it exhibits a slow rate of deformation and is thus a fluid. And what about toothpaste, which begins to flow only after a certain amount of stress is applied?

1.2.1 The Continuum Model

All substances are composed of an extremely large number of discrete particles called molecules. In a pure substance such as water, all molecules are identical; other substances such as air are mechanical mixtures of different types of molecules. Molecules interact with each other via collisions and intermolecular forces. The *phase* of a sample of matter—solid, liquid, or gas—is a consequence of the molecular spacing and intermolecular forces. The molecules of a solid are relatively close (spacing on the order of a molecular diameter) and exert large intermolecular forces; the molecules of a gas are far apart (spacing of an order of magnitude larger than the molecular diameter) and exert relatively weak intermolecular forces. Because intermolecular forces are weak, a gas easily changes its volume as well as its shape. The stronger intermolecular forces in a solid cause it to maintain both its volume and its shape. In a liquid, the intermolecular forces are sufficiently strong to maintain volume but not shape.

In principle it would be possible to describe a sample of fluid in terms of the dynamics of its individual molecules; in practice it is impossible because of the extremely large numbers of molecules involved. For most cases of practical interest, it is possible to ignore the molecular nature of matter and to assume that matter is continuous. This assumption is called the *continuum model* and may be applied to both solids and fluids. The continuum model assumes that molecular structure is so small relative to the dimensions involved in problems of practical interest that we may ignore it.

When using the continuum model, we describe a fluid in terms of its *properties*, which represent average characteristics of its molecular structure. As an example, we use the mass per unit volume or *density* rather than the number of molecules and the molecular mass. If matter were truly a continuum, the properties would be continuous functions of time and space. The velocity of a moving fluid is defined as the velocity of the center of mass of the collection of molecules near a given point.* Because the fluid properties and velocity are continuous functions, we can use calculus to analyze a continuum rather than applying discrete mathematics to each molecule.

For the continuum model to be valid, the smallest sample of matter of practical interest must contain a large number of molecules so that meaningful averages can be calculated. For air at sea-level conditions,† a volume of 10^{-9} mm^3 contains approximately 3×10^7 molecules. In most engineering problems, a volume of 10^{-9} mm^3 is very small, so the continuum model is valid. However, for certain cases, such as very-high-altitude flight, the molecular spacing becomes so large that a small volume contains only a few molecules, and the continuum model fails. For all the situations you will encounter in this book, the continuum model will be valid.

We should emphasize that, on adopting the continuum model, we seek the solution to a fluid mechanics problem in terms of the variation of a few fluid

---

* This fluid velocity is distinct from the random molecular velocity.

† By international agreement, sea-level conditions in the standard atmosphere correspond to a temperature of 15°C (59°F) and a pressure of 101,330 N/m^2 (14.696 lb/in^2).

properties and the fluid velocity throughout the fluid. These properties and velocities are continuous functions of space and time, defined for every *point* within the fluid. Our notion of a "point" must be expanded to include a very small volume of space that is nevertheless large enough to contain a large number of molecules. Within these limits, it is possible to have discontinuous "jumps" in fluid properties and velocities at the interface between two fluids or across a shock wave in a compressible fluid (discussed in Chapter 10).

1.3 FLUID PROPERTIES

Fluid properties characterize the state or condition of the fluid and macroscopically represent microscopic (molecular) structure and motion. You should already be familiar with properties such as pressure and temperature.

1.3.1 Density, Specific Volume, Specific Weight, Specific Gravity

The *density* of a fluid is its mass per unit volume. Density has a value at each point in a continuum and may vary from one point to another. Suppose that we pick an arbitrary volume $\Delta V\!\!\!\!-$ in a fluid. We let Δm represent the mass of fluid contained in this volume. The average density ($\bar{\rho}$) is

$$\bar{\rho} \equiv \frac{\Delta m}{\Delta V\!\!\!\!-}.$$

This average density might depend on the size and location of the chosen volume. For example, consider a glass half full of water. If the chosen volume coincides exactly with the volume of water in the bottom half of the glass, we would obtain a value of about 1000 kg/m^3 (62.4 lbm/ft^3) for the average density; however, if our chosen volume were the entire glass, the average density would be only half as large because the top half of the glass contains no water. An exact definition of the density must involve a limit. Our natural inclination would be to take the limit of $\bar{\rho}$ as $\Delta V\!\!\!\!-$ approaches zero; if we do, however, $\Delta V\!\!\!\!-$ will become so small that the molecular structure of the fluid will become apparent and the density will vary as indicated in Fig. 1.2. The irregularity at very small values of $\Delta V\!\!\!\!-$ is due to the molecular structure. As $\Delta V\!\!\!\!-$ becomes larger, the density varies continuously because a large number of molecules are included in $\Delta V\!\!\!\!-$. The smallest volume ($\delta V\!\!\!\!-$) for which the density variation is smooth is the limit of the continuum hypothesis. Thus we must define the fluid density by

$$\rho \equiv \lim_{\Delta V\!\!\!\!- \to \delta V\!\!\!\!-} \frac{\Delta m}{\Delta V\!\!\!\!-}.$$

The small volume $\delta V\!\!\!\!-$ represents the size of a typical "point" in the continuum. For fluids near atmospheric pressure and temperature, $\delta V\!\!\!\!-$ is of the order of 10^{-9} mm^3. We interpret all fluid properties as representing an average of the fluid molecular structure over this small volume.

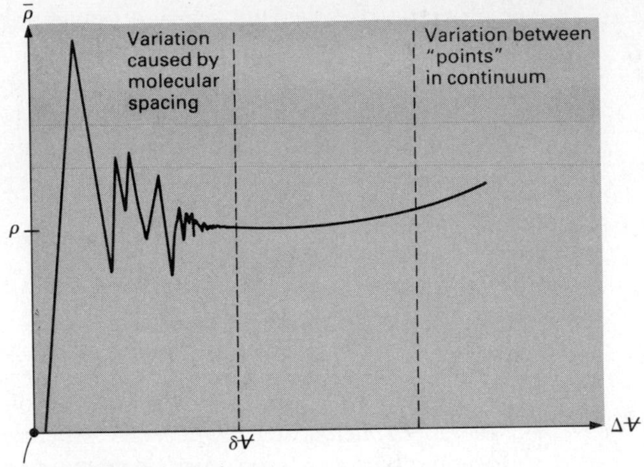

Figure 1.2 Dependence of (average) density on sample volume for small volumes.

Fluid density varies widely between fluids. At atmospheric conditions, the density of air is about 1.22 kg/m^3 (0.076 lbm/ft^3), the density of water is about 1000 kg/m^3 (62.4 lbm/ft^3), and the density of mercury is about 13,500 kg/m^3 (846 lbm/ft^3). For a particular fluid, density varies with temperature and pressure. This variation is quite strong for gases but fairly weak for liquids. If the density variation may be neglected, we say that a fluid is *incompressible*. Numerical values of density of several common fluids are presented in Appendix A, Tables A.3–A.6.

Several other fluid properties are directly related to density. The *specific volume* (v), defined by

$$v \equiv \frac{1}{\rho},$$

is of considerable use in thermodynamics but is seldom used in fluid mechanics.

The *specific weight* (γ) is the weight of the fluid per unit volume; thus

$$\gamma \equiv \rho g,$$

where g is the local acceleration of gravity.

The *specific gravity* of a fluid is the ratio of the density of the fluid to the density of a reference fluid. The defining equation is

$$S \equiv \frac{\rho}{\rho_{\mathrm{REF}}}.$$

For liquids, the reference fluid is pure water at 4°C and 101,330 N/m^2, $\rho_{\mathrm{REF}} =$ 1000 kg/m^3 (62.4 lbm/ft^3). Specific gravity of gases is usually based on dry air as the reference fluid.

Because specific volume, specific weight, and specific gravity are all directly related to density, they all are constant if the density is constant.

1.3.2 Pressure

Pressure is a fluid property of utmost importance. Most fluid mechanics problems involve prediction of fluid pressure or with the integrated effects of pressure over some surface or surfaces in contact with the fluid. Unlike density, which is usually one of the "known" quantities in a fluid mechanics problem, pressure is usually one of the "unknown" quantities to be determined by analysis or experiment. We define pressure as follows:

> Pressure is the normal compressive force per unit area acting on a real or imaginary surface in the fluid.

Consider a small surface area ΔA within a fluid, as shown in Fig. 1.3. A force ΔF_n acts normal to the surface. Tangential forces may also be present but are not relevant to the definition of pressure. The pressure is defined by

$$p \equiv \lim_{\Delta A \to \delta A} \frac{\Delta F_n}{\Delta A}.$$

The limiting value δA represents the lower bound of the continuum assumption, $\delta A \approx (\delta \forall)^{2/3}$. The pressure can vary from point to point in a fluid. Microscopically, pressure represents molecular momentum and intermolecular forces within the fluid. Pressure is positive for compression.

GAGE PRESS ≈ RELATIVE TO LOCAL ATM. PRESS

ABSOLUTE PRESS ≈ RELATIVE TO Ø PRESS.

We express pressure in units of force per unit area and measure it with respect to one of two datums. Pressure measured relative to local atmospheric pressure is called *gage pressure*. Pressure measured (or defined) relative to zero pressure is called *absolute pressure*. Gage and absolute pressures are related as follows:

Absolute pressure = Gage pressure + Atmospheric pressure in vicinity of gage.

Pressures below local atmospheric pressure are sometimes called *vacuum pressures*. A vacuum pressure is expressed as a positive number, so

$$\begin{aligned}\text{Vacuum pressure} &= \text{Atmospheric pressure} - \text{Absolute pressure}\\ &= -\text{Gage pressure.}\end{aligned}$$

Use of the subscripts "g" and "a" *with the units* of a pressure indicates whether the pressure is gage or absolute. Thus 10 psig means "ten pounds per square inch, gage" and 10 psia means "ten pounds per square inch, absolute." The subscripts "abs" and "gage" are also used for the same purpose.

To convert from gage to absolute pressure, you must know the atmospheric pressure at the time and location of measurement. Often this information is not

Figure 1.3 Illustration for definition of pressure.

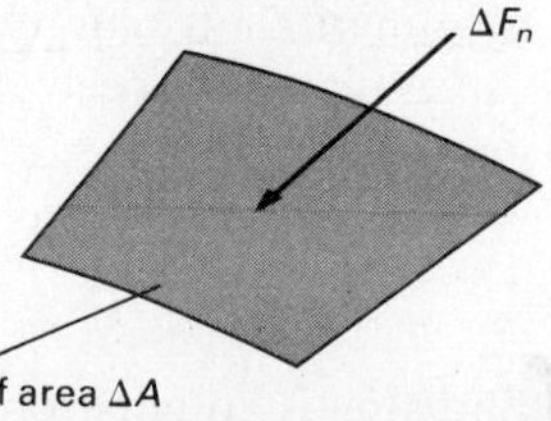

available. In such cases, we use a "standard" value of 101,330 Pa (14.696 lb/in^2, abs; 29.92 in. Hg, abs) for atmospheric pressure. The actual atmospheric pressure can be measured with an instrument called a barometer; hence atmospheric pressure is sometimes called barometric pressure.

In many situations that arise in fluid mechanics, we are more concerned with differences of pressure than with levels of pressure. Pressure differences are the same whether the pressures are considered as absolute or gage, so long as all pressures are based on a common datum. A pressure difference (say, between two locations in a pipeline) is not expressed in either "gage" or "absolute" units; these designations apply only to pressure levels.

1.3.3 Temperature and Energy-Related Properties

Temperature (T) is a property that is familiar to everyone but rather difficult to define with the same exactness as density or pressure. We normally associate temperature with the degree of "hotness" or "coldness," but this is hardly a precise definition. Temperature is often defined as that property whose difference gives rise to the transfer of heat. As heat is usually defined as flow of energy due to a difference of temperature, this definition is circular. We could develop an exact definition of temperature, using the second law of thermodynamics, but this definition would be of little use here. For our introductory purposes, we define the temperature of a fluid as a measure of (not "equal to") the energy contained in the molecular motions of the fluid.

We may express numerical values of temperature with respect to various scales. Some scales are absolute (temperature expressed as a value relative to the lowest possible temperature, "absolute zero"), and some scales are not absolute (temperatures expressed relative to an arbitrary datum). The most common non-absolute temperature scales are the Fahrenheit scale and the Celsius scale. The corresponding (i.e., equal except for selection of datum) absolute scales are the Rankine and Kelvin scales, respectively. The relations between these scales are given by

TEMP SCALES

$$T(\text{Rankine}) = T(\text{Fahrenheit}) + 459.67$$
$$T(\text{Kelvin}) = T(\text{Celsius}) + 273.15$$
$$T(\text{Rankine}) = 1.80\ T(\text{Kelvin}).$$

The *internal energy* (U) indicates the energy contained in random molecular motions and intermolecular forces. The *specific internal energy* ($\tilde{u}$) is the internal energy per unit mass. For single-phase fluids (liquids or gases), the internal energy is primarily a function of temperature.

A property closely related to internal energy is the *specific enthalpy* ($\tilde{h}$), defined by

$$\tilde{h} \equiv \tilde{u} + \frac{p}{\rho} = \tilde{u} + pv.$$

Enthalpy is an important property in the analysis of a compressible fluid because of the coupling between mechanical and thermal forms of energy.

We express the relationship between the internal energy or enthalpy of a fluid and the temperature of the fluid in terms of specific heat. The important specific heats are the *specific heat at constant volume* (c_v) and the *specific heat at constant pressure* (c_p), defined by

$$c_v \equiv \left.\frac{\partial \tilde{u}}{\partial T}\right)_v \quad \text{and} \quad c_p \equiv \left.\frac{\partial \tilde{h}}{\partial T}\right)_p .$$

For an incompressible fluid, all processes are constant specific volume and

$$\left.\frac{\partial (pv)}{\partial T}\right)_p = p \left.\frac{\partial v}{\partial T}\right)_p = 0,$$

so

$$c_p = c_v \qquad \text{(incompressible fluid).}$$

1.3.4 Property Relations and the Ideal Gas

Fluid properties are not independent. For a particular fluid, definite relations among the various properties exist. For most common fluids, specification of only two of the properties allows us to determine values for the other properties. Relations among the properties may be presented in graphical, tabular, or equation form. Tables A.3–A.6 in Appendix A are examples of a limited property relation that gives the density and other properties at particular conditions of temperature and pressure. The science of thermodynamics is especially concerned with the study of property relationships [1,2].*

At "low" pressure and "high" temperature, gas properties are closely approximated by the ideal gas law

$$p = \rho R T. \tag{1.1}$$

In this equation, temperature and pressure must be expressed in absolute units; R is the specific gas constant and is equal to the universal gas constant (R_o) divided by the molecular weight (M.W.) of the gas:

$$R = \frac{R_o}{\text{M.W.}} = \frac{8314}{\text{M.W.}}\ \text{N·m/kg·K} = \frac{1545}{\text{M.W.}}\ \text{ft·lb/lbm·°R}.$$

Liquids exhibit slight variation of density with temperature and pressure. No simple, exact equations are available for properties of liquids. For most practical purposes, we treat liquids as incompressible fluids.

For both ideal gases and incompressible fluids, the internal energy depends on only the temperature. Thus†

$$\tilde{u} = \tilde{u}\{T\} \quad \text{and} \quad c_v = \frac{d\tilde{u}}{dT} = c_v\{T\}.$$

* Numbers in brackets refer to references at the end of the chapter.

† The dashed parentheses { } mean "a function of."

In addition, for the ideal gas

$$\tilde{h} = \tilde{u} + \frac{p}{\rho} = \tilde{u}(T) + RT = \tilde{h}(T) \quad \text{and} \quad c_p = \frac{d\tilde{h}}{dT} = c_p(T).$$

Often, the assumption is that specific heats are constant. In this book, we assume that all gases may be modeled by the ideal gas law and constant specific heats.

EXAMPLE 1.1 **Illustrates the Ideal Gas Law and the Incompressible Fluid Assumption**

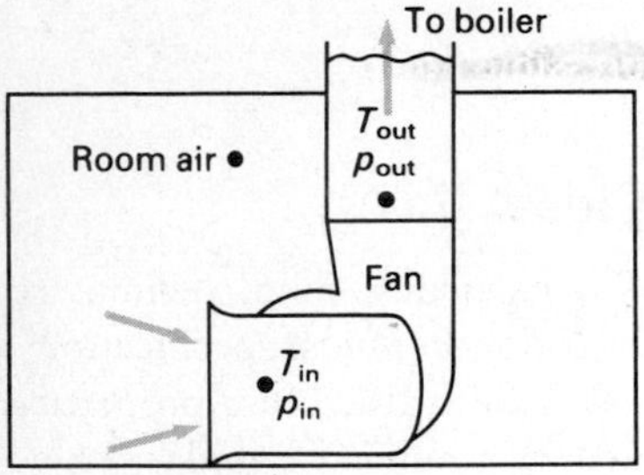

Figure E1.1 Fan setup for Example 1.1.

An engineer is designing a system to provide combustion air for a boiler, as shown in Fig. E1.1. A fan draws air from a large room through an inlet duct. The fan increases the air pressure and discharges it into an outlet duct. The pressure in the inlet duct is 1300 N/m^2, vacuum, and the temperature is 30.6°C. The pressure at the fan discharge is 1000 N/m^2, gage, and the temperature is 35.6°C. The barometric pressure is 100,000 N/m^2. Calculate the air density at the fan inlet and at the fan outlet. Comment on the possible use of the incompressible fluid model for the fan flow.

SOLUTION

Given

Fan inlet pressure = 1300 N/m^2, vacuum

Fan outlet pressure = 1000 N/m^2, gage

Fan inlet temperature = 30.6°C

Fan outlet temperature = 35.6°C

Barometric pressure = 100,000 N/m^2

Find

Density of the air at the fan inlet and outlet

Discuss the incompressible fluid model

Solution

Because the air pressure and temperature at the fan inlet are given, we may find the air density from the ideal gas law, Eq. (1.1):

$$\rho_{in} = \frac{p_{in}}{RT_{in}},$$

where the pressure and temperature must be expressed in absolute units. The absolute pressure is

$$\begin{aligned} p_{in} &= p_{atm} - p_{vac} \\ &= 100{,}000 \text{ N/m}^2 - 1300 \text{ N/m}^2 = 98{,}700 \text{ N/m}^2. \end{aligned}$$

From Table A.3 in Appendix A, $R = 287$ N·m/kg·K for air. Then

$$\rho_{in} = \frac{0.987 \times 10^5\ \text{N/m}^2}{(287\ \text{N}\cdot\text{m/kg}\cdot\text{K})(30.6 + 273.2)\ \text{K}}.$$

$$\rho_{in} = 1.13\ \text{kg/m}^3.$$ **ANSWER**

Similarly,

$$p_{out} = 100{,}000\ \text{N/m}^2 + 1000\ \text{N/m}^2$$

and

$$\rho_{out} = \frac{p_{out}}{RT_{out}} = \frac{(1.01 \times 10^5\ \text{N/m}^2)}{(287\ \text{N}\cdot\text{m/kg}\cdot\text{K})(35.6 + 273.2)\ \text{K}}.$$

$$\rho_{out} = 1.14\ \text{kg/m}^3.$$ **ANSWER**

The density difference as a percentage of the inlet density is

$$\Delta\rho(\%) = \frac{1.14 - 1.13}{1.13} \times 100 = 0.88\%,$$

so the incompressible fluid model should be relatively accurate.

Discussion

The data in this example come from a test on an actual power plant fan. Note that if we attempt to use the incompressible fluid model to estimate the outlet density as 1.13 kg/m³ and then use the measured outlet pressure to estimate the outlet temperature, we would calculate

$$T_{out} = \frac{p_{out}}{\rho R} = \frac{1.01 \times 10^5\ \text{N/m}^2}{(1.13\ \text{kg/m}^3)(287\ \text{N}\cdot\text{m/kg}\cdot\text{K})}$$
$$= 311.4\ \text{K} \quad \text{or} \quad 38.2^\circ\text{C}.$$

The temperature error of 2.6°C may be quite significant in energy-balance calculations; however, the density error of 0.01 kg/m³ is insignificant.

1.3.5 Viscosity

You learned earlier that a fluid is a substance that undergoes continuous deformation when subjected to a shear stress and that the shear stress is a function of the rate of deformation. For many common fluids, the shear stress is proportional to the rate of deformation. The constant of proportionality, called the *viscosity,* is a fluid property. To develop the defining equation for viscosity, we must first obtain an expression for the rate of deformation (i.e., the rate of shear strain) of a fluid particle. We do this by considering a flow in which the x-direction velocity of the fluid varies with y, as indicated in Fig. 1.4.

Because we assume that the fluid is continuous, the velocity is a continuous function (a smooth curve). We call a plot such as Fig. 1.4, with the fluid velocity at various values of y indicated by arrows whose tips are connected by a smooth curve, a *velocity profile.*

Imagine that we can isolate a small rectangular chunk of fluid from somewhere near the center of Fig. 1.4. The top of the chunk will move faster than the bottom,

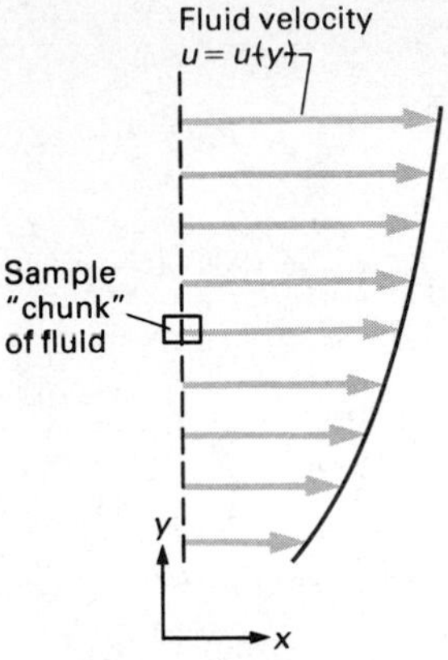

Figure 1.4 Typical fluid velocity variation.

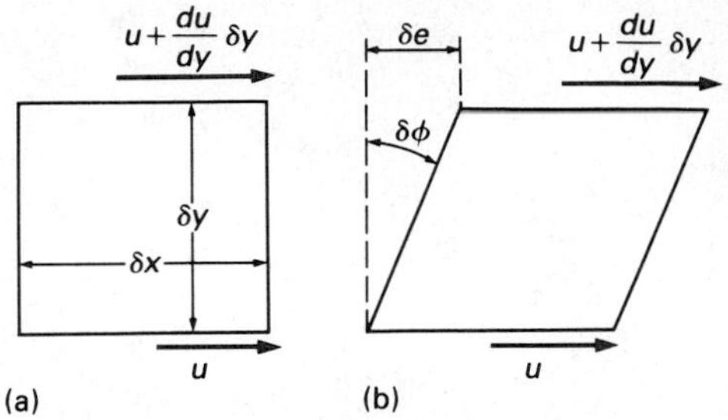

Figure 1.5 Fluid chunk (a) at time t and (b) at time $t + \delta t$.

so if it is square at time t, it will be sheared at time $t + \delta t$, as shown in Fig. 1.5. We measure the shear deformation by the angle $\delta\phi$, which we can relate to the fluid velocity as follows. As the chunk is small, the velocity of the top relative to the bottom is

$$u_{\text{rel}} \approx \left(u + \frac{du}{dy}\,\delta y\right) - u = \frac{du}{dy}\,\delta y.$$

The distance, δe, that the top moves relative to the bottom is

$$\delta e = u_{\text{rel}}\,\delta t = \frac{du}{dy}\,\delta y\,\delta t.$$

As $\delta\phi$ is small for small δt, we find that

$$\delta\phi \approx \tan(\delta\phi) = \frac{\delta e}{\delta y} = \frac{du}{dy}\,\delta t.$$

The shear strain rate is the rate of change* of $\delta\phi$, or

$$\frac{\delta\phi}{\delta t} = \frac{du}{dy}.$$

If the shear stress (τ) in the fluid is proportional to the strain rate, we can write

$$\tau = \mu\,\frac{du}{dy}. \tag{1.2}$$

The coefficient μ is the viscosity. Equation (1.2) is called *Newton's law of viscosity* because it was first suggested by Sir Isaac Newton. Fluids that obey this particular law and its generalization to multidimensional flow are called *Newtonian fluids*.

Not all fluids follow the Newtonian stress–strain relation. Some fluids, such

---

* If u also varies with x or t, we would have to write $\partial u/\partial y$.

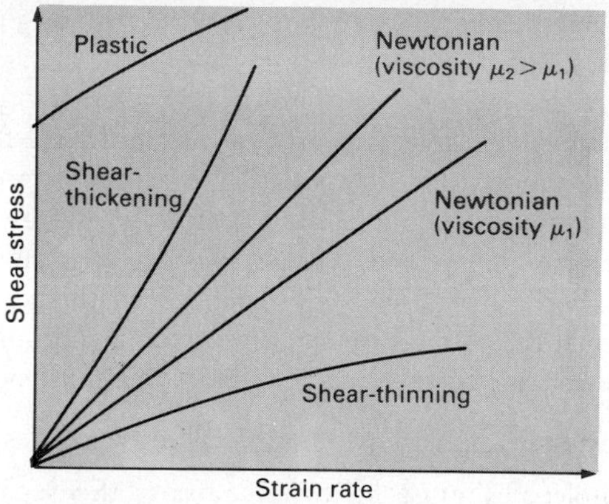

Figure 1.6 Various types of stress-strain rate relations.

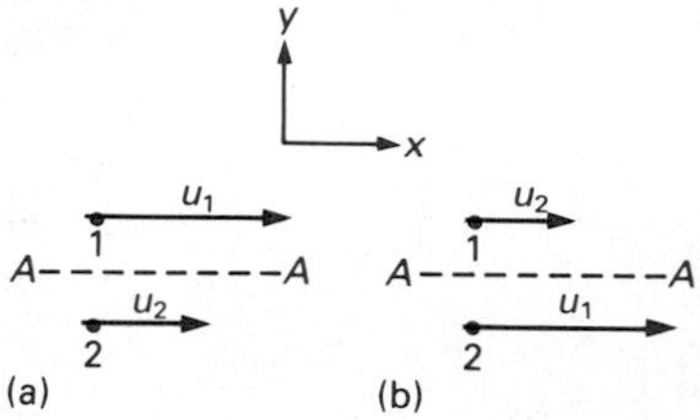

Figure 1.7 Representative gas molecules in a region of variable velocity.

as ketchup, are "shear-thinning"; that is, the coefficient of resistance decreases with increasing strain rate (it all comes out of the bottle at once!). Others, such as a mixture of corn starch and water, are "shear-thickening." Some fluids behave in a manner termed "plastic"; they do not begin to flow until a finite stress has been applied (the toothpaste mentioned in Section 1.2 is a plastic fluid). Fluids that do not follow the Newtonian relation are called *non-Newtonian.* We illustrate some possible types of stress–strain rate relations in Fig. 1.6. Viscosity of non-Newtonian fluids is a function of the strain rate.

Many common fluids such as air and other gases, water, and most simple solutions are Newtonian, as are most petroleum products. Solutions containing long-chain polymers, as well as slurries and suspensions, are usually non-Newtonian. Blood is a non-Newtonian fluid. In this book, we limit our study to Newtonian fluids.

The viscosity of Newtonian fluids varies considerably from fluid to fluid. For a particular fluid, viscosity varies rather strongly with temperature but only slightly with pressure. We usually neglect viscosity variation with pressure in practical calculations. Tables A.3–A.6 and Figs. A.1–A.4 in Appendix A contain viscosities of several common fluids.

We can demonstrate the mechanism of viscosity and its variation with temperature in a gas by presenting a simplified model of molecular behavior. Suppose the velocity distribution sketched in Fig. 1.4 represents a flowing gas. We further "magnify" the picture in Fig. 1.7 to show two gas molecules at different y locations. In Fig. 1.7(a), each molecule is moving to the right, with the fluid velocity appropriate to the molecule's particular location. Superimposed on its fluid velocity, each molecule also has a random molecular velocity and frequently collides with other gas molecules.* Suppose that the two molecules exchange places as a result of such collisions. We assume that, immediately after changing places, each

* For air at standard conditions, approximately 5×10^9 collisions per cubic meter occur each second. A typical molecule travels a distance of 3×10^{-8} m between collisions. This distance is called the *mean free path.*

molecule retains the fluid velocity appropriate to its original position. Figure 1.7(b) represents this condition.

A result of this interchange of position is a net change of momentum on either side of plane *A–A*. Because this process, as illustrated for two molecules, occurs for all of the gas molecules, an essentially continuous momentum exchange results from random molecular motions in the gas. If we adopt the continuum hypothesis, we must imagine that a shear stress exists along the plane *A–A* between the molecules. According to Newton's second law, this shear stress (force per unit area) must equal the rate of momentum exchange per unit area. Note that for this momentum exchange to take place, two actions must occur:

- The molecules must have different fluid velocities at different y locations.
- The molecules must be carried across the separating plane by random molecular motion.

In Newton's law of viscosity (Eq. 1.2), the velocity gradient, du/dy, represents the difference of fluid velocities, and the viscosity, μ, represents the molecular exchange process. As molecular agitation increases with temperature, we would expect the viscosity of a gas to increase with temperature, and this is exactly what happens!

The molecular momentum exchange model is quite accurate for gases; however, it does not explain the viscosity of liquids. Intermolecular forces contribute more to liquid viscosity than does molecular momentum transfer. Since intermolecular forces generally decrease with temperature, the viscosity of most liquids decreases with temperature.

Viscosity provides a way to quantify the adjectives *thick* and *thin* as applied to liquids. "Thick" liquids have high viscosity and do not flow easily; the opposite is true for "thin" liquids. Fluid density has little to do with "thickness." Motor oil is usually considered a "thick" liquid, especially when cold, but it is less dense than water ($S_{oil} \approx 0.8$).

The ratio of viscosity to density often appears in the equations describing fluid motion. This ratio is given the name *kinematic viscosity* and is usually denoted by the symbol ν:

$$\nu \equiv \frac{\mu}{\rho}.$$

The kinematic viscosity is so named because it contains only length and time dimensions. Tables A.3–A.6 in Appendix A give the kinematic viscosity of several common fluids as a function of temperature. When we need to distinguish the types of viscosity, we call μ the *dynamic viscosity* or *absolute viscosity.*

An effect usually associated with fluid viscosity is the *no-slip condition.* Whenever a fluid is in contact with a solid surface, the velocity of the fluid at the surface is equal to the velocity of the surface; that is, the fluid "sticks" to the surface and does not "slip" relative to it. This condition is true regardless of the type of fluid, type of surface, or surface roughness, so long as the continuum hypothesis is valid. Whenever a fluid flows over a surface, the variation of fluid velocity relative to the surface must be as sketched in Fig. 1.8. A velocity gradient always

ALWAYS TRUE WHEN CONTINUUM HYPOTHISES IS VALID

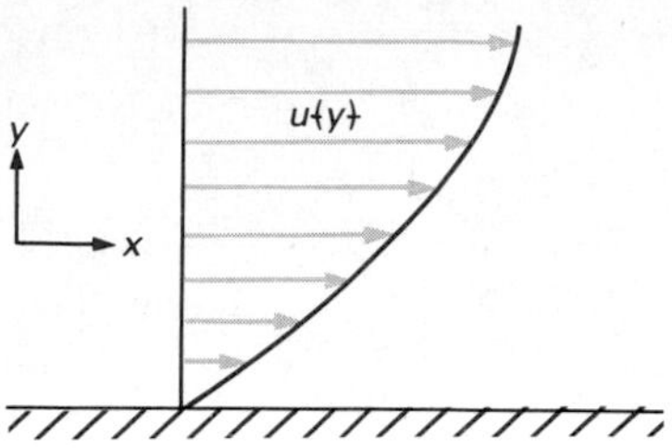

Figure 1.8 Variation of fluid velocity relative to a wall; note $u_{rel} = 0$ at the wall.

ASSUME INVISCID FLUID AWAY FROM SOLID BOUNDRIES

occurs near a surface when a fluid flows over the surface, and according to Newton's law of viscosity, shear stress always acts on the fluid in this region.

The equations of motion for flowing fluids become very complicated when shear stresses are included. For this reason, simplified analyses sometimes neglect shear stresses. An easy way to introduce the assumption of zero shear stress is to pretend that the fluid has zero viscosity. According to Eq. (1.2), $\mu = 0$ gives $\tau = 0$. A fluid with zero viscosity is called an *inviscid fluid*. Although no real fluid has zero viscosity, the viscosity of many fluids is small, so shear stresses in these fluids are small *if du/dy is not large*. This relation is often true, except in the immediate vicinity of solid boundaries (as in Fig. 1.8), so the assumption of an inviscid fluid is often useful for analyzing flow remote from solid boundaries.

The incompressible fluid and inviscid fluid models are combined in the so-called *ideal fluid*. The ideal fluid has $\mu \equiv 0$ and $\rho \equiv$ constant. Mathematicians and engineers have found many elegant solutions of the equations of motion of the ideal fluid. You must be careful to distinguish between the ideal fluid defined here and the ideal gas discussed in Section 1.3.4.

EXAMPLE 1.2 Illustrates the Calculation of Shear Stress from a Velocity Profile

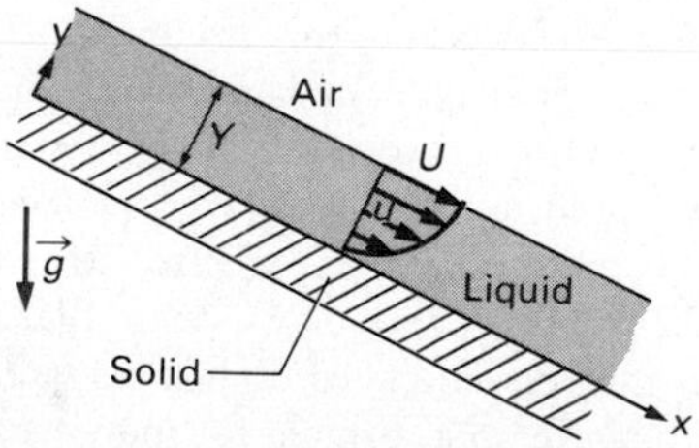

Figure E1.2 Liquid film flowing down an inclined surface.

Figure E1.2 shows a liquid film flowing down an inclined surface. The liquid is a Newtonian fluid with viscosity μ. The velocity profile is given by

$$u(y) = U\left[2\left(\frac{y}{Y}\right) - \left(\frac{y}{Y}\right)^2\right],$$

where U is a constant and Y is the thickness of the liquid layer. Find the shear stress at the fluid–solid interface ($y = 0$), at $y = Y/2$, and at the "free surface" ($y = Y$).

SOLUTION

Given

Velocity profile of a Newtonian fluid:

$$u(y) = U\left[2\left(\frac{y}{Y}\right) - \left(\frac{y}{Y}\right)^2\right].$$

Find

The shear stress at $y = 0$, $Y/2$, and Y

Solution

For a Newtonian fluid,

$$\tau = \mu \frac{du}{dy}.$$

Substituting the velocity u gives

$$\tau = \mu U \frac{d}{dy}\left[2\left(\frac{y}{Y}\right) - \left(\frac{y}{Y}\right)^2\right] = \mu \frac{U}{Y}\left[2 - 2\left(\frac{y}{Y}\right)\right].$$

The shear stresses at the various y locations are

$$\tau)_{y=0} = 2\mu \frac{U}{Y};$$ **ANSWER**

$$\tau)_{y=Y/2} = \frac{\mu U}{Y};$$ **ANSWER**

$$\tau)_{y=Y} = 0.$$ **ANSWER**

Discussion

The zero shear stress at the "free surface" is a common boundary condition for any type of liquid, as the air above the liquid exerts a negligible force on the liquid. For gasoline at 100°F with $Y = 0.05$ in. and $U = 13.4$ ft/sec, the shear stress at $y = 0$ would be 0.0025 lb/in^2, much smaller than the stresses normally encountered in solids.

EXAMPLE 1.3 Illustrates a Common Assumption for the Velocity Profile in Small Clearances and the Development of a "Working Formula"

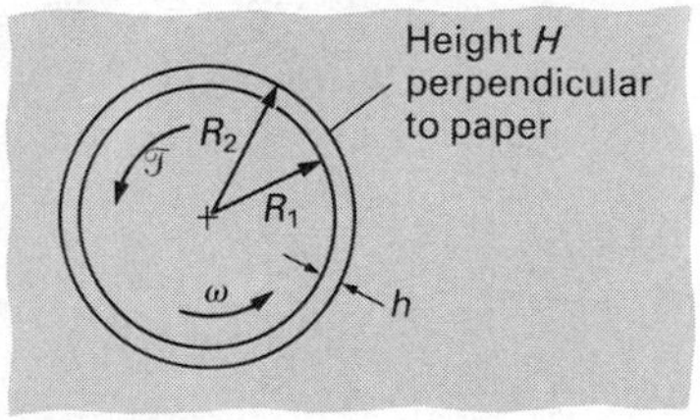

Figure E1.3a Concentric cylinder viscometer.

A concentric cylinder viscometer is a device used to measure absolute viscosity. Figures E1.3a–c illustrate the details of this viscometer. The fluid is contained between a fixed outer cylinder and an inner cylinder that is free to rotate. The application of a torque $\mathcal{T}$ causes the inner cylinder to rotate at a constant speed ω. The viscometer has height H, and the width of the gap h is quite small compared to either the radius R_1 or R_2. Because of the small gap, we may assume that the fluid velocity, V, in the tangential direction varies linearly across the gap. The fluid is Newtonian. Develop a formula for the viscosity μ in terms of the torque $\mathcal{T}$, speed ω, and the geometric parameters of the viscometer.

SOLUTION

Given

Concentric cylinder viscometer shown in Fig. E1.3a

Enlarged view of a portion of the fluid, Fig. E1.3b

Find

A formula relating μ to the applied torque $\mathcal{T}$ and rotational speed ω

F_1 = Applied force on fluid by inner cylinder

F_2 = Resisting force on fluid by fixed outer cylinder

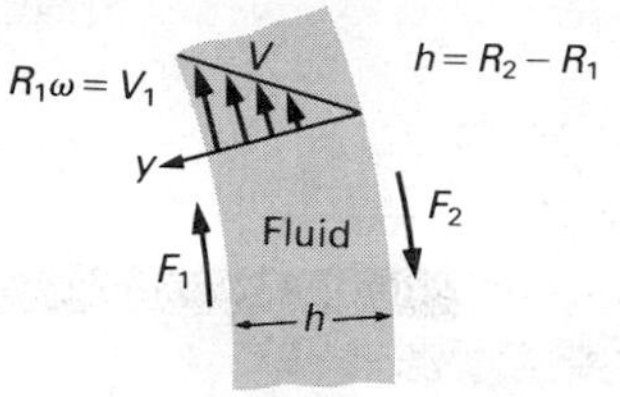

Figure E1.3b Enlarged view of portion of the fluid in concentric cylinder viscometer.

Solution

We see that μ appears in Newton's law of viscosity:

$$\tau = \mu \frac{du}{dy}, \quad \text{so} \quad \mu = \frac{\tau}{du/dy}.$$

To obtain the desired formula, we must express τ and du/dy in terms of the torque $\mathscr{T}$, speed ω, and other known quantities. Because we assume that the velocity varies linearly over the gap h,

$$\frac{du}{dy} = \frac{V_1}{h} = \frac{R_1\omega}{h}.$$

To express τ in terms of $\mathscr{T}$, we note that the force F_1 is the shear stress at the inner wall times the wall area, $2\pi R_1 H$. That is,

$$F_1 = 2\pi R_1 H \tau)_{y=h}, \quad \text{or} \quad \tau)_{y=h} = \frac{F_1}{2\pi R_1 H}.$$

However, the force F_1 times the radius R_1 equals the applied torque on the inner cylinder:

$$\mathscr{T} = F_1 R_1.$$

Substituting gives

$$\tau)_{y=h} = \frac{\mathscr{T}}{2\pi R_1^2 H}.$$

Substituting into Newton's law of viscosity yields

$$\mu = \frac{\tau)_{y=h}}{(du/dy)_{y=h}} = \left(\frac{\mathscr{T}}{2\pi R_1^2 H}\right)\left(\frac{h}{R_1\omega}\right);$$

$$\mu = \frac{\mathscr{T} h}{2\pi H \omega R_1^3}.$$

[Newtonian fluid, constant viscosity, constant speed, small gap ($h \ll R_1$)]. **ANSWER**

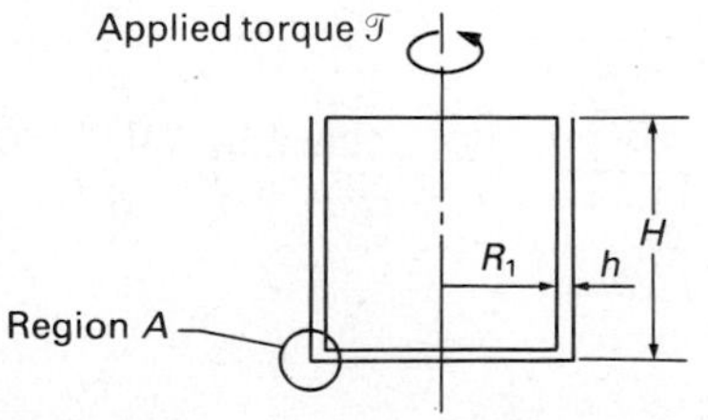

Figure E1.3c Side view of concentric cylinder viscometer.

Discussion

The conditions under which the "working formula" is valid are given alongside the formula in the answer. These limitations arise from the assumptions explicit in the problem statement; however, additional limitations exist that may not be quite so evident. Refer to Fig. E1.3c and note that the shear resists the applied torque at the bottom of the inner cylinder and on the side of the cylinder. Because we neglected the bottom torque, our equation is applicable only if $H \gg R_1$. To remove this restriction, we could express the resisting torque on the bottom of the inner cylinder in terms of μ and develop a more accurate "working formula." A further limitation arises because we neglected "end effects" on the gap velocity profile. Near the bottom of the viscometer (region A in Fig. E1.3c), the stationary bottom wall affects the gap velocity distribution. This effect is negligible if $h \ll H$. Our formula is therefore subject to the two additional restrictions $H \gg R_1$ and $H \gg h$.

1.3.6 Secondary Properties

In addition to density, pressure, temperature, and viscosity, several other fluid properties exist. We chose to designate them as secondary properties because the effects that they describe are often negligible.

Bulk Modulus of Elasticity. All real fluids exhibit some degree of compressibility; that is, the fluid density changes with pressure. The *bulk modulus of elasticity* is a fluid property that indicates the degree of compressibility.

Consider a fluid particle of volume $\forall$. If the particle is subjected to a pressure change dp, a volume change $d(\forall)$ results, and the bulk modulus of elasticity is

$$E_v = \frac{-dp}{(d\forall)/\forall} = -\forall\frac{dp}{d\forall}.$$

The negative sign is necessary because an increase in pressure results in a decrease in volume. Because the mass of the fluid particle is fixed, we can also express E_v in terms of the specific volume or the density:

$$E_v = -v\frac{dp}{dv} = \rho\frac{dp}{d\rho}. \tag{1.3}$$

The bulk modulus of elasticity is roughly equivalent to Young's modulus from solid mechanics. The units of E_v are those of pressure. For water at atmospheric conditions, E_v is approximately $2.1 \times 10^9\ \mathrm{N/m^2}$ ($3 \times 10^5\ \mathrm{lb/in^2}$).

If we consider a gas, the pressure–density relation also involves temperature; that is,

$$p = p\{\rho, T\};$$

thus

$$dp = \left.\frac{\partial p}{\partial \rho}\right)_T d\rho + \left.\frac{\partial p}{\partial T}\right)_\rho dT.$$

The exact relation between density change and pressure depends on temperature. If we assume constant temperature, we define the *isothermal bulk modulus* as

$$E_{v,T} \equiv \rho\left.\frac{\partial p}{\partial \rho}\right)_T.$$

For an ideal gas,

$$p = \rho RT, \quad \text{so} \quad E_{v,T} = \rho\frac{\partial(\rho RT)}{\partial \rho} = \rho RT = p.$$

The isothermal bulk modulus of a gas is equal to the pressure of the gas.

Example 1.1 implied that approximating a highly compressible fluid (in that case, air) as if it were incompressible may sometimes be possible. We can develop a criterion that allows us to check the validity of that approximation. From Eq.

(1.3), the relative change of density with pressure is

$$\frac{d\rho}{\rho} = \frac{dp}{E_v}.$$

If we accept a density change of 1% as insignificant, we can model a fluid as incompressible if

$$\frac{\Delta p}{E_v} < 0.01.$$

In Chapter 4 we show that pressure and fluid velocity can be related by

$$\Delta p = \tfrac{1}{2}\rho V^2.$$

Thus we can model a fluid as incompressible if

$$\frac{\rho V^2}{E_v} < 0.02.$$

For air at standard conditions, this quantity corresponds to a velocity of about 40 m/s (130 ft/sec).

Coefficient of Thermal Expansion. Changing the temperature of a fluid can also cause a change of density. Most fluids expand (that is, density decreases) as temperature increases. The *coefficient of thermal expansion* (α_T) measures this effect. Consider a fluid particle of volume $\forall$. Increasing the temperature by an amount dT causes an increase in the volume equal to $d(\forall)$. The coefficient of thermal expansion is

$$\alpha_T = \frac{d(\forall)/\forall}{dT} = \frac{1}{\forall}\frac{d\forall}{dT}.$$

As the mass of the fluid particle is constant, we can also express α_T in terms of specific volume or density:

$$\alpha_T = \frac{1}{v}\frac{dv}{dT} = -\frac{1}{\rho}\frac{d\rho}{dT}. \tag{1.4}$$

For water at atmospheric conditions, $\alpha_T \approx 1.5 \times 10^{-4}\,\mathrm{K}^{-1}$ ($9 \times 10^{-5}\,{}^\circ\mathrm{R}^{-1}$).

In general, the temperature–density relation also involves pressure, so we must write

$$d\rho = \left.\frac{\partial\rho}{\partial T}\right)_p dT + \left.\frac{\partial\rho}{\partial p}\right)_T dp.$$

If we specify that the process of thermal expansion takes place at constant pressure, then

$$\alpha_{T,p} = -\frac{1}{\rho}\left.\frac{\partial\rho}{\partial T}\right)_p.$$

For an ideal gas with

$$\rho = \frac{p}{RT},$$

we find that

$$\alpha_{T,p} = -\frac{1}{\rho}\frac{\partial}{\partial T}\left(\frac{p}{RT}\right) = \frac{1}{T}.$$

For air at sea-level conditions $\alpha_{T,p} \approx 3.5 \times 10^{-3}\ \mathrm{K}^{-1}$ ($1.9 \times 10^{-3}\ {}^\circ\mathrm{R}^{-1}$), 23 times the magnitude for water.

Surface Tension. *Surface tension* (σ) describes the forces at interfaces between a gas and a liquid (e.g., water–air), two liquids, or one or more liquids, a gas, and a solid (e.g., water–air–glass). Two familiar phenomena related to surface tension are the supporting of a small object of larger density (such as a steel sewing needle) by a water surface if the object is placed carefully on the surface and the beading of water into drops on the hood of a freshly waxed automobile.

In fluid mechanics, surface tension plays an important role in problems involving the formation of bubbles in liquids, the breakup of liquid jets into drops, and the determination of shapes of masses of liquid under conditions of "weightlessness" (zero g).

Consider the apparatus sketched in Fig. 1.9. A loop of wire is dipped in a soap solution so that a soap film is formed. This film, although only a few molecules thick, has two surfaces, one on each side. To move the slide (or even hold it fixed), we must apply a force *F*. This force is proportional to the length ℓ, so we write

$$F = 2\sigma\ell.$$

The constant of proportionality, σ, is the surface tension of the film. The factor 2 accounts for the two surfaces. Surface tension has the dimensions of force per unit length. Appendix A presents values of σ for some common liquids.

Thinking of *surface energy* rather than surface tension is sometimes more convenient. If we move the slide in Fig. 1.9 a distance Δx to the right, we generate more surface. The work required to generate this surface is

$$W = F(\Delta x) = 2\sigma\ell(\Delta x).$$

The product of Δx and ℓ is the amount of new surface created, ΔA, so

$$W = 2\sigma\,\Delta A.$$

This equation shows that we can interpret surface tension as the energy per unit area required to create or maintain a surface.

The relative magnitudes of the surface energies determine the shape of liquid surfaces that are in contact with solids and other fluids (usually air). Specific

Figure 1.9 Wire loop and soap film.

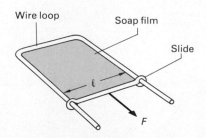

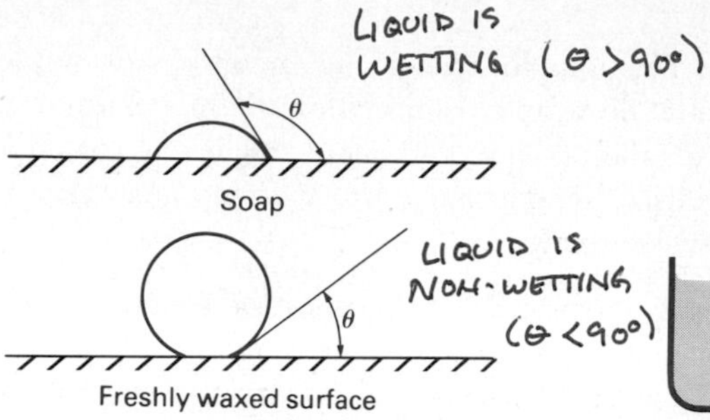

Figure 1.10 Drops of water on soap and freshly waxed surface.

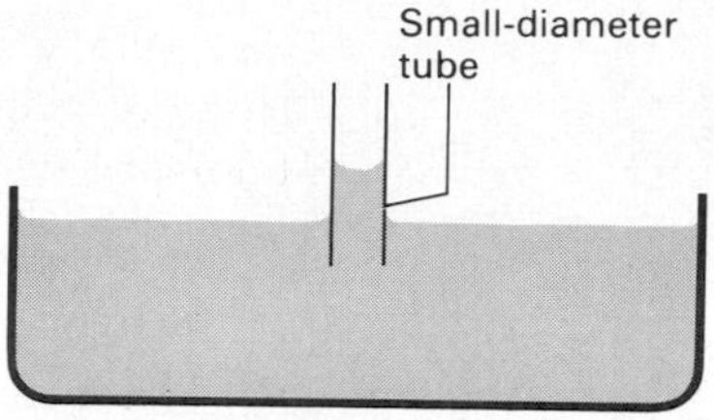

Figure 1.11 Illustration of capillary rise of liquid in small tube.

liquid–solid combinations are classified as wetting or nonwetting, depending on the contact angle between the liquid and solid surfaces. Figure 1.10 shows drops of water on soap and on a freshly waxed surface. If the contact angle θ is greater than 90°, the liquid is *wetting;* if it is less than 90°, the liquid is *nonwetting*. Wetting behavior is determined by the particular liquid–solid combination.

A final phenomenon associated with surface tension is *capillarity.* If a very-small-diameter glass tube is inserted in a pan of water, the water climbs the walls of the tube until the upward force due to surface tension is balanced by the downward pull of the weight of the fluid column (see Fig. 1.11). In a nonwetting situation, the liquid in the tube is depressed below the level of liquid in the pan. Fluids can be lifted to great heights by capillary action in very small spaces. Capillary action is the mechanism by which water seeps through cloth, moving through the gaps in the weave of the cloth. When cloth is "water-proofed," it is treated with chemicals that change the water–cloth surface from wetting to nonwetting; thus the water beads up on the surface rather than seeping through.

1.4 DIMENSIONS AND SYSTEMS OF UNITS

All engineering problems deal with physical entities that require quantitative description. We express the magnitude of the entity by a number and an associated unit of measurement. We may talk of a car with a mass of 1000 kilograms moving at a speed of 50 kilometers per hour or of a 2-inch-diameter pipe with a flow of 100 gallons per minute of water with a temperature of 150 degrees Fahrenheit. We recognize that "kilograms," "kilometers," "inches," "hours," "minutes," and "degrees Fahrenheit" are units of measurement for physical entities such as mass, length, time, and temperature. If we consider the situation carefully, we recognize two other facts. We know that we also could have expressed the pipe diameter as 0.16667 feet or 5.08 centimeters or 3.16×10^{-5} miles or any of several other combinations of numbers and units; however, the diameter of the pipe is a single entity that is independent of the choice of units used to express it. The pipe diameter is a characteristic of the pipe that can be expressed in a variety of units, so long as the units are of the same type.

We also recognize that the units of certain quantities are combinations of other units. The units for velocity are a combination of length and time units. The units for water flow are a combination of units of volume (gallons) and time; however, since a gallon contains 231 cubic inches, we could also express water flow in terms of units of length and time. We may generalize these observations by the following two statements:

- We may introduce the concept of a *dimension*, which is a generalization of the concept of units. The dimensions of a physical entity tell us what kind of units must be involved in a quantitative statement of magnitude of the entity. Diameter has the dimension of length, velocity has the dimension of length divided by time, and so on.
- We may select a set of independent *fundamental dimensions*. The dimensions of all physical entities may be expressed in terms of these fundamental dimensions.

The number of fundamental dimensions is surprisingly small, usually six: *mass, length, time, temperature, electric current,* and *luminous intensity*. The dimensions of all other physical entities may be derived from defining equations or from physical laws. Let's consider dimensions for velocity, energy, and force. As velocity is the rate of change of distance with time, it has dimensions of length divided by time. We use brackets [ ] to represent "the dimension of." The symbols $[M]$, $[L]$, $[T]$, and $[\theta]$ represent the fundamental dimensions of mass, length, time, and temperature. Using this notation, we write the previous statement as an equation,

$$[V] = \frac{[L]}{[T]} = \left[\frac{L}{T}\right],$$

which we read as, "The dimensions of velocity are length divided by time."

We may determine the dimensions of energy, $[E]$, by considering kinetic energy:

$$E_K = \tfrac{1}{2}mV^2.$$

The factor $\frac{1}{2}$ has no dimensions, so

$$[E] = [E_K] = [M][V]^2 = \left[\frac{ML^2}{T^2}\right].$$

Finally, we consider the dimension of force. Newton's second law ($\vec{F} = m\vec{a}$) shows that we can express force in dimensions of mass, length, and time:

$$[F] = \left[\frac{ML}{T^2}\right].$$

The selection of the number and types of fundamental dimensions is not unique. We could select force as a fundamental dimension rather than mass or select both force and mass as fundamental dimensions, increasing the number of fundamental dimensions from six to seven.*

---

* Increasing the number of "fundamental" dimensions beyond six is easy, but decreasing the number below six is very difficult.

Turning to the practical side of the question, various systems of units have been developed to enable us to express the dimensions of physical entities quantitatively. Although the number of fundamental dimensions is small, the number of units and unit systems is unnecessarily large. Unit systems developed in two ways. In the "scientific" method, a unit is defined for each fundamental dimension. The units of all physical entities are expressed as appropriate combinations of the units that have been assigned to the fundamental dimensions. Within this system, assigning specific names to certain combinations of units, such as the name "newton" for the combination kilogram·meter/second^2, is permissible.

In the "evolution" method, units were assigned to physical entities as the need arose. Units were often based on purely local standards or on specific physical phenomena. The "foot" was originally the length of the king's foot, and the Btu (British thermal unit) was defined as the amount of heat required to raise the temperature of one pound of water by one Fahrenheit degree. This method often resulted in the assigning of different units to physical entities with the same dimensions. For example, heat may be expressed in Btu and work in foot·pounds, although both heat and work are energy and have the dimensions $[ML^2/T^2]$, or $[FL]$.

Obviously, the "scientific" method is preferable from the standpoint of simplicity and precision; however, the need for workable and convenient unit systems in commerce led to widespread use of evolved systems of units long before any scientific systems were established. Today, all nations of the world have officially adopted the SI* system of units as standard; however, in the United States, considerable use of other units persists, both in commerce and in engineering.

We may change the units (but *not* the dimensions) of a particular entity by multiplying by a *conversion factor*. Conversion factors are derived from equivalence statements such as

$$1 \text{ ft} = 0.3048 \text{ m} \quad \text{and} \quad 1 \text{ Btu} = 778 \text{ ft}\cdot\text{lb}.$$

Rearranging these equations, we get

$$\frac{1 \text{ ft}}{0.3048 \text{ m}} = 1 \quad \text{and} \quad \frac{778 \text{ ft}\cdot\text{lb}}{1 \text{ Btu}} = 1.$$

The ratios on the left-hand sides are conversion factors.

Conversion factors are always equivalent to the pure number 1 and have no dimensions. Thus we may multiply any term in any equation by a conversion factor at any time without affecting the equation. Some famous conversion factors are sometimes assigned a symbol; for example, the Btu–ft·lb conversion is denoted by J. Whether represented by symbols or not, all conversion factors are equivalent to 1.

A major source of difficulty in most systems of units is the confusion between weight (a force) and mass and their units. We hope that as a result of your physics and elementary engineering mechanics courses, *you* know the difference between weight and mass; however, the distinction is by no means clear to everyone. If you go into the local supermarket to buy a "pound" of hamburger, are you buying weight or mass? If you fill out a professional résumé, do you list your "weight"

---

* SI stands for Système International or le Système International d' Unités.

as so many kilograms? We can illustrate the different approaches to the "weight–mass problem" by three unit systems: the SI system, the British gravitational (BG) system, and the English engineering (EE) system.

SI System. The SI system is a "scientific," or defined, system. The unit of mass is the kilogram (kg), the unit of length is the meter (m), the unit of time is the second (s), and the unit of temperature is the kelvin (K). The basis for the definition of these units is treated elsewhere [3]. The unit of force, the newton (N), is derived from Newton's second law:

$$1\ \mathrm{N} \equiv 1\ \mathrm{kg \cdot m/s^2}.$$

The weight* of a mass is given by

$$W = mg,$$

where g is the local acceleration of gravity. The standard acceleration of gravity on the earth is 9.807 $\mathrm{m/s^2}$, so a kilogram has a weight of 9.807 N under standard gravity.

The SI unit for energy is derived from

$$[E] = \left[\frac{ML^2}{T^2}\right] = [FL].$$

The energy unit is called a joule (J) and is defined by

$$1\ \mathrm{J} \equiv 1\ \mathrm{kg \cdot m^2/s^2} = 1\ \mathrm{N \cdot m}.$$

The unit of power is called a watt (W) and is defined by

$$1\ \mathrm{W} \equiv 1\ \mathrm{J/s} = 1\ \mathrm{N \cdot m/s} = 1\ \mathrm{kg \cdot m^2/s^3}.$$

British Gravitational System. The British gravitational (BG) system adopted force rather than mass as a fundamental dimension. The unit of length is the foot (ft). The foot is defined with respect to the meter: 1 ft = 0.3048 m. The unit of time is the second (sec).† The unit of temperature is either the degree Fahrenheit (°F) or the degree Rankine (°R). The unit of force is the pound (lb). If we rewrite Newton's second law,

$$m = \frac{F}{a},$$

the dimensions of mass become

$$[m] = \frac{[F]}{[a]} = \left[\frac{FT^2}{L}\right].$$

---

* In the SI system, the terms "force of gravity on" and "gravity force on" are preferred over "weight."

† Note that "second" is abbreviated "s" in SI units and "sec" in BG and EE units.

The unit of mass is the slug (slug), defined by

$$1 \text{ slug} = 1 \text{ lb}\cdot\text{sec}^2/\text{ft}.$$

As earth standard gravity is 32.174 ft/sec^2 in BG units, a mass of 1 slug weighs 32.174 lb under standard gravity. The unit of energy is the ft·lb, and the unit of power is the ft·lb/sec.

English Engineering System. The English engineering (EE) system is similar to the BG system in most respects; however, the mass–weight problem is solved by defining units for both mass and weight so that 1 unit of mass has a weight of 1 unit of force under conditions of standard gravity. The length, time, and temperature units are the same as in the BG system (i.e., ft, sec, °F, °R). The mass unit is the pound mass (lbm), and the force unit is the pound force (lb).* The pound force (in the EE system) and the pound (in the BG system) are equivalent units. *One pound mass has a weight of one pound force under conditions of standard gravity.* One pound mass has a weight other than one pound force at values of gravity other than "earth standard." Using Newton's second law,

$$W = mg,$$

we can calculate the weight of 1 lbm under conditions of standard gravity, which by definition is 1 lb:

$$1 \text{ lb} = 1 \text{ lbm} \times 32.174 \text{ ft/sec}^2.$$

Note, however, that the units are not the same on both sides of this equation and that 1 does not equal 32.174. We correct the problem by dividing the right-hand side by a conversion factor and write:

$$1 \text{ lb} = \frac{1 \text{ lbm} \times 32.174 \text{ ft/sec}^2}{32.174 \text{ ft}\cdot\text{lbm/lb}\cdot\text{sec}^2}.$$

We assign the conversion factor the symbol g_c, so

$$g_c = 32.174 \text{ ft}\cdot\text{lbm/lb}\cdot\text{sec}^2.$$

Because you can multiply any term in any equation by a conversion factor, you can multiply or divide by g_c any time you need to convert between units involving mass and units involving force.

You can also obtain other units of measurement by subdividing or combining the basic units. Table 1.1 lists dimensions, basic units, and other common units for several quantities of interest in fluid mechanics. Appendix C contains conversion factors between unit systems and between alternative units within a given system.

---

* In the EE system, the symbol lb_f is usually employed for the pound force. We use both the EE and BG systems in this book, so we use lbm for the pound mass and lb for the pound force.

TABLE 1.1 Various quantities, their dimensions, and their units.

Quantity	Dimensions*	SI Units	BG Units	EE Units	Other Common Units
Area	$[L]^2$	m^2	ft^2	ft^2	in^2
Volume	$[L]^3$	m^3	ft^3	ft^3	in^3 gallon (gal) liter (l)
Velocity	$[L/T]$	m/s	ft/sec	ft/sec	—
Acceleration	$[L/T^2]$	m/s^2	ft/sec^2	ft/sec^2	—
Pressure or stress	$[M/LT^2]$	Pa (pascal) ($1\ Pa = 1\ N/m^2$)	lb/ft^2	lb/ft^2	lb/in^2 (psi) kPa
Angular velocity	$[1/T]$	1/s (Hz)	rad/sec	rad/sec	revolutions per minute (rpm)
Energy, work, heat	$[ML^2/T^2]$	J	ft·lb	ft·lb	Btu
Power	$[ML^2/T^3]$	W	ft·lb/sec	ft·lb/sec	Btu/sec horsepower (hp)
Density	$[M/L^3]$	kg/m^3	$slug/ft^3$	lbm/ft^3	lbm/in^3
Viscosity	$[M/LT]$	Pa·s	slug/ft·sec	$lb \cdot sec/ft^2$	poise (P) (1 P = 1 Pa·s)
Specific heat, gas constant	$[L^2/T^2\theta]$	J/kg·K	ft·lb/slug·°R	ft·lb/lbm·°R	Btu/lbm·°R Btu/slug·°R
Modulus of elasticity	$[M/LT^2]$	Pa	lb/ft^2	lb/ft^2	psi kPa
Surface tension	$[M/T^2]$	N/m	lb/ft	lb/ft	lb/in.

* Based on selection of mass rather than force as fundamental.

EXAMPLE 1.4 Illustrates the Use of Conversion Factors

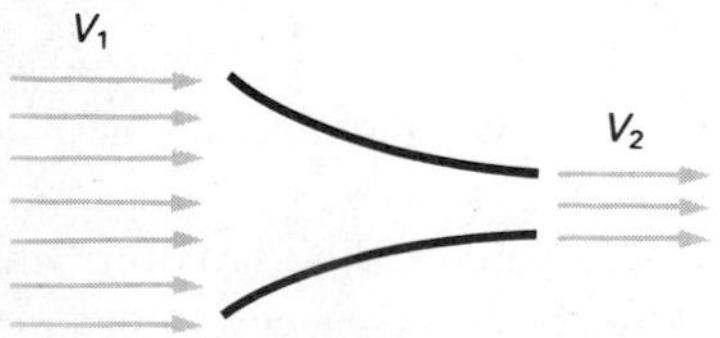

Figure E1.4 Nozzle.

Bernoulli's equation for an ideal fluid flowing in the nozzle shown in Fig. E1.4 is

$$\frac{p}{\rho} + \frac{V^2}{2} + gz = \text{Constant}.$$

The parameters and their values at a particular point *a* are

p = pressure, 101,400 N/m^2 (14.7 lb/in^2);

ρ = density, 1000 kg/m^3 (1.94 $slugs/ft^3$ or 62.4 lbm/ft^3);

V = velocity, 3 m/s (9.84 ft/sec);

g = local acceleration of gravity, 9.6 m/s^2 (31.5 ft/sec^2);

z = elevation above datum, 5 m (16.4 ft).

Calculate the value of the constant (C) in SI, BG, and EE units.

SOLUTION

Given

$$\frac{p}{\rho} + \frac{V^2}{2} + gz = \text{Constant},$$

with specific values for each term.

Find

Value of the constant in SI, BG, and EE units

Solution

For SI units:
Substituting terms, we get

$$\frac{101{,}400 \text{ N/m}^2}{1000 \text{ kg/m}^3} + \frac{3^2 \text{ (m/s)}^2}{2} + (9.6 \text{ m/s}^2)(5 \text{ m}) = \text{C}.$$

As

$$\text{N} = \text{kg}\cdot\text{m/s}^2,$$

the result is

$$101.4 \text{ m}^2/\text{s}^2 + 4.5 \text{ m}^2/\text{s}^2 + 48.0 \text{ m}^2/\text{s}^2,$$

so

$$C = 153.9 \text{ m}^2/\text{s}^2.$$ **ANSWER**

For BG units:
Substituting terms, we get

$$\frac{14.7 \text{ lb/in}^2}{1.94 \text{ slugs/ft}^3} + \frac{(9.84)^2 \text{ ft}^2/\text{sec}^2}{2} + (31.5 \text{ ft/sec}^2)(16.4 \text{ ft}) = C,$$

but

$$1 \text{ slug} = 1 \text{ lb}\cdot\text{sec}^2/\text{ft}.$$

Also, we must use the conversion factor

$$144 \text{ in}^2/\text{ft}^2$$

in the first term. Then

$$\frac{(14.7 \text{ lb/in}^2)(144 \text{ in}^2/\text{ft}^2)}{1.94 \text{ lb}\cdot\text{sec}^2/\text{ft}^4} + \frac{(9.84)^2 \text{ ft}^2/\text{sec}^2}{2} + (31.5 \text{ ft/sec}^2)(16.4 \text{ ft}) = C,$$

so

$$C = 1656 \text{ ft}^2/\text{sec}^2.$$ **ANSWER**

For EE units:

Substituting terms, we get

$$\frac{14.7 \text{ lb/in}^2}{62.4 \text{ lbm/ft}^3} + \frac{(9.84)^2}{2} \text{ ft}^2/\text{sec}^2 + (31.5 \text{ ft/sec}^2)(16.4 \text{ ft}) = C.$$

The first term needs two conversion factors,

$$\frac{(14.7\ \mathrm{lb/in^2})(144\ \mathrm{in^2/ft^2})(32.2\ \mathrm{ft\cdot lbm/lb\cdot sec^2})}{62.4\ \mathrm{lbm/ft^3}} + \frac{(9.84)^2}{2}\ \mathrm{ft^2/sec^2}$$
$$+ (31.5\ \mathrm{ft/sec^2})(16.4\ \mathrm{ft}) = C,$$

so

$$C = 1656\ \mathrm{ft^2/sec^2}.$$ **ANSWER**

Discussion

Note that as g_c is a conversion factor and is thus a dimensionless constant, its numerical value is 32.2 (rounded off from 32.174), even though the local acceleration of gravity is 31.5 ft/sec^2. *The value of g_c is independent of the local acceleration of gravity.*

Alternatively, we could choose the units of the constant as N·m/kg (ft·lb/slug or ft·lb/lbm) by selecting different positions for conversion factors. Note that we can convert the SI value of C to the English value:

$$(153.9\ \mathrm{m^2/s^2})(3.28\ \mathrm{ft/m})^2 = 1656\ \mathrm{ft^2/sec^2}.$$

1.5 SCOPE OF FLUID MECHANICS

We can put the study of fluid mechanics in perspective by developing a classification system for the various areas of fluid mechanics. Figure 1.12 summarizes the major subdivisions of fluid mechanics. As shown, the two main divisions are *fluid statics* and *fluid dynamics*.

1.5.1 Fluid Statics

We include in this category those cases in which a fluid undergoes "rigid body motion"; that is, the body of fluid moves without deformation. Pressure and gravity are the only forces acting on the fluid. No shear forces from fluid deformation

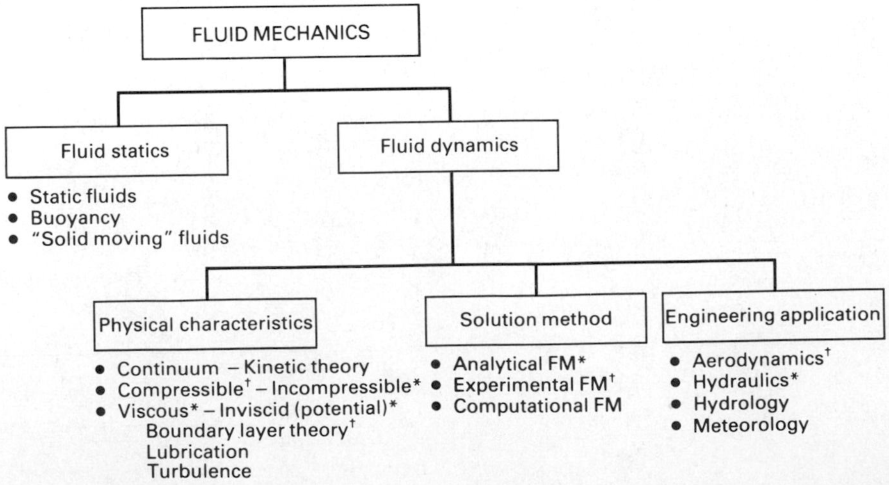

Figure 1.12 Specialty areas of fluid mechanics; areas marked with asterisk (*) are given special emphasis in this book; areas marked with dagger (†) are also covered.

are present. Fluid statics is an exact science; the only difficulties in practice arise from specific geometries of systems to be analyzed. Our discussion of fluid statics in Chapter 2 should prepare you to solve any fluid statics problems of practical interest. Examples of such problems include the calculation of forces on dams that impound large quantities of water, buoyant forces on objects submerged in liquids, stability of floating bodies such as ships, and pressure variation in a container of liquid that rotates as a rigid body.

1.5.2 Fluid Dynamics

Fluid dynamics deals with the mechanics of flowing fluids. In addition to gravitational forces (which are often negligible in fluid dynamics) and pressure forces, shear stress forces may be significant. Moreover, we must usually take fluid inertia into account. We can mathematically formulate most fluid dynamics problems using a few basic laws; however, exact solutions are available for only a relatively few simple cases.

Because of its complexity, we often subdivide fluid dynamics. We may base the division on the particular assumptions made about the *physical characteristics* of the fluid or flow problem, on the type of *method(s)* used to solve the problem, or on the type of *engineering application* involved. Following is a description of a few of these areas.

Physical Characteristics. Most fluid dynamics is based on the continuum hypothesis and may be called *continuum fluid dynamics;* however, one field, *kinetic theory,* considers the molecular nature of the fluid and applies statistical concepts together with the laws of particle dynamics to predict gas flows.

In *compressible fluid mechanics* (often called *gas dynamics*), fluid compressibility is of primary importance. The field of *incompressible flow* (often called *hydrodynamics*)* neglects fluid compressibility. The methods of gas dynamics are necessary in high-speed gas flows, whereas the methods of hydrodynamics are valid in low-speed gas flows as well as liquid flows.

Inviscid flow theory deals with an imaginary fluid with zero viscosity. This area is often called *potential flow theory* because, in the absence of viscosity, the velocity can be obtained from a scalar potential function. If we also introduce the incompressible fluid assumption, we obtain the *ideal fluid theory,* which is the most fully developed mathematical approach to flow prediction. Many exact solutions to the "ideal fluid" equations are known. These solutions are sometimes quite useful, but at other times they fail miserably.

Viscous flow analysis is subdivided into several specialties, including *boundary layer theory,* which assumes that viscous effects are confined to thin layers near walls (see Fig. 1.8); *lubrication,* which considers flow in small clearances; and *turbulence,* which seeks to develop models and calculation methods for turbulent flow.

---

* The prefix *hydro* comes from the Greek word for "water"; however, hydrodynamics can be applied to all types of fluids, even air.

Method for Solving Problems. We may classify methods for solving fluid dynamics problems as analytical, experimental, and computational. In *analytical fluid mechanics,* engineers use the tools of mathematics, modeling, and physical insight to develop solutions to engineering problems.

In *experimental fluid mechanics,* the tools include wind and water tunnels, as well as a considerable amount of instrumentation. In experimental fluid mechanics, the fluid "solves" the problem by taking a certain flow state; the challenge becomes measuring the flow and interpreting the results. This task is not as easy as you may think, especially as many experiments involve scale models rather than full-size apparatus. The experimental approach may be direct (such as measuring the response of an airplane to wind gusts or measuring the rate of flow through a steam turbine in a power plant) or indirect to provide basic information for use in mathematical models (such as studying the nature of turbulent flow near a rough wall). Several excellent references, [4–7], provide more information on experimental fluid mechanics.

Probably the fastest-growing area in fluid mechanics is *computational fluid mechanics.* In this field, engineers use the digital computer to solve approximate forms of differential equations of fluid motion. These approximations involve the use of finite-difference or finite-element representations of the equations. Engineers in this field must combine knowledge of the physics of fluid flow with skills in numerical analysis and computer programming. For more details, see references [8–11].

Engineering Application. The intended area of application provides the basis for the division among aerodynamics, hydraulics, hydrology, and meteorology. *Aerodynamics* deals with the evaluation of forces generated between surfaces and large bodies of fluid in which they are moving. The flow may be compressible or incompressible and may be assumed viscous or inviscid. The ultimate application may be the design of aircraft, automobiles, or submarines. *Hydraulics* concerns the flow of fluids in pipes and channels, in process or power-generating equipment, and in fluid machinery such as pumps or fans. The ultimate application may involve the design of a water supply system, chemical plant, power station, or hydraulic control system for an industrial machine.

Hydrology and *meteorology* involve flows that occur in the natural environment. Hydrology considers the motion of groundwater, seepage, drainage, precipitation runoff, and other fluid motions on or below the Earth's surface. Civil engineers make extensive use of the science of hydrology when designing water supply and waste treatment systems and devising schemes for flood control. Meteorology considers the very-large-scale fluid motions in the Earth's atmosphere. Understanding and predicting these motions are crucial to accurate weather forecasting and also to evaluating atmospheric conditions to help ensure safe aircraft flight.

Whichever specialty area of fluid mechanics interests you most, you will need to apply the fundamental principles of fluid mechanics to solve problems. Even if you do not become a fluid mechanics specialist, a knowledge of the fundamental principles of fluid mechanics will be very helpful, if not essential, in your career. As you read this book and study the subject of fluid mechanics, make sure that you grasp the basic ideas, definitions, and methodologies. A firm understanding of these fundamentals is essential whether you use the computer or wind tunnel and whether you design an airplane or select a pump. The purpose of this book is to help you learn the fundamentals rather than to "short-circuit" you directly to a particular specialty.

1.6 SOLVING FLUID MECHANICS PROBLEMS (A NOTE TO THE STUDENT)

If you want to learn the fundamentals of fluid mechanics (and obtain a good grade in your fluids course), you must do more than simply read this book. Your goal should be to obtain a working knowledge of the concepts and methodologies of fluid mechanics and to include this working knowledge in your professional "tool kit." You can gain a good working knowledge of the subject only by solving a variety of fluid mechanics problems.

1.6.1 Types of Problems

Generally speaking, you may be called on to solve two types of problems as part of your education or in your professional career: *analysis* problems and *design* problems.

Analysis problems are sometimes called "textbook" problems because of their extensive use in textbooks to help the student develop a basic understanding of a specific subject. Analysis problems have the following characteristics:

- The problem deals with a limited number of subject areas (usually only one).
- The given or available information is a nearly perfect match to the amount required; neither too much nor too little is provided, and the information is not contradictory.
- The solution is unique and can be found without the need for arbitrary choices by the analyst.

Analysis problems are not necessarily easy; in fact, they may be unsolvable because of limitations of time, resources, or available mathematical or computing power. Doing a lot of analysis problems is essential to learning a subject, so we present lots of them at the ends of chapters.

Design problems better represent the types of problems that occur in the practice of engineering. Characteristics of design problems include the following:

- Problems are often multidisciplinary, involving, say, fluid mechanics, electronics, stress analysis, and engineering economy.
- The information available is not exactly suited to the problem. Typically, too much information (possibly contradictory) is given about some aspects of the problem, and little or no information is given about other aspects. The engineer may have to dig out some information from other sources or proceed

with missing or incomplete information, making assumptions (sometimes crude) to fill in the gaps.

- The problem may have more than one solution, and several of the solutions may be equally satisfactory. The engineer may need to arbitrarily choose one of the possible solutions or some nontechnical criterion such as minimum cost or esthetic considerations may dictate the choice among several solutions. We call a problem with this characteristic an *open-ended problem.*

Design problems are not especially helpful for learning the fundamentals of a subject such as fluid mechanics, but knowing how to solve them is essential in professional practice. They also are useful in showing how a particular subject fits into the broader scope of engineering. We provide a few (relatively simple) design problems at the ends of chapters. In addition, Appendix J presents several more involved design problems that are suitable for junior- or senior-level engineering students.

1.6.2 A Systematic Approach to Problem Solving

You can greatly enhance your problem-solving ability by following a systematic approach. Here we suggest a six-step approach to problem solving, which is summarized in Fig. 1.13.

Step 1. Make a sketch to help you visualize the physical situation. On the sketch, indicate the given information in a descriptive manner so that you can recognize what you know about the problem. For example, indicate a flow rate by an arrow in the direction of the flow, with the magnitude indicated near the arrow. This sketch is analogous to the free-body diagram used in statics or dynamics. *Do not neglect this step.*

Step 2. List specifically the quantities you want to find. This task is more difficult for design problems because it requires an understanding of the physical phenomena involved.

Step 3. Identify the *primary** physical law or laws that apply to the particular problem. As an illustration of this step, recall the viscometer problem of Example 1.3: The applicable primary law is the stress–strain rate relation for fluids. After you identify the primary laws, select the appropriate mathematical forms of the laws. Be sure that any assumptions or restrictions introduced in development of the equations are valid for your particular problem.

Step 4. In many cases, the primary laws and equations are not sufficient to determine what you need to know. In that case, introducing *secondary*† equations

---

* The primary physical law is the one that most directly applies to the physical phenomenon. The quantity or quantities to be found usually appear explicitly in the mathematical forms of the primary laws.

† Secondary equations do not usually contain the quantity to be found but involve other variables that appear in the primary equations. The status of a particular law or equation as "primary" or "secondary" depends on the particular problem. The same law or equation may be primary in one problem and secondary (or irrelevant) in another problem.

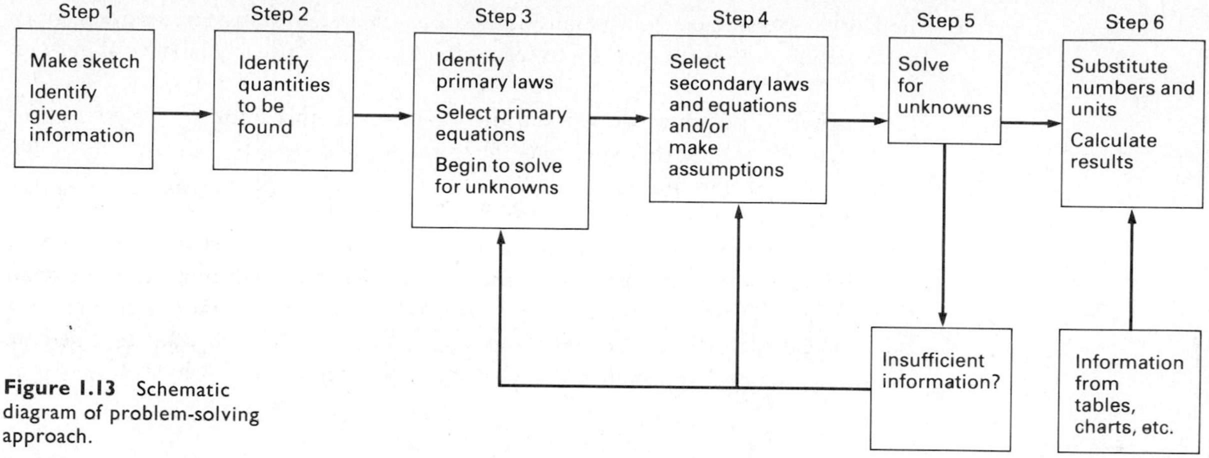

Figure 1.13 Schematic diagram of problem-solving approach.

and/or making some assumptions to fit the equations to the problem is necessary. The secondary equations may be based on other physical laws or on kinematic or geometric relations. Secondary equations in Example 1.3 are the equation relating torque, force, and radius; the equation relating force, stress, and area; the equation relating angular velocity, radius, and linear velocity; and the equation relating the surface area of a cylinder to radius and height. As in step 3, be sure that any inherent assumptions or restrictions in the secondary equations are valid.

If you must make assumptions, be careful that you do not "assume away the problem." Remember that any assumptions you make probably will affect the accuracy of your result. You should note the assumptions you introduce and also note any restrictions on their validity. The use of a linear velocity distribution for the fluid in the viscometer gap in Example 1.3 is an assumption. This velocity distribution will become a progressively worse approximation as the ratio of the gap width to cylinder radius increases.

Step 5. After you select the equations and introduce assumptions, manipulate the equations to solve for the desired unknown quantities in terms of given quantities. In carrying out this step, you usually come up against a roadblock. You might not have enough information to proceed past a certain point (as when the number of unknowns exceeds the number of equations). You must then return to step 4 (or even to step 3) and introduce more equations and/or assumptions. Initially, you may have to repeat the steps 3–4–5 cycle many times before proceeding to step 6. As you become experienced, you will do a better job of selecting equations and making proper assumptions in the first pass through steps 3 and 4, and need less feedback from step 5. Only with practice can you improve your selection–assumption process.

Step 6. After you carry the solution as far as you can in terms of symbols, introduce numbers and units (at the same time) into the problem and complete the solution. At this step, you may have to refer to tables or charts to find material (fluid) properties or conversion factors. You may also have to refer to charts to find other parameters.* Note any limitations of assumptions, accuracy of given information, or information taken from tables or charts. Clearly label your answer.

Insofar as possible, we worked the examples in this book by following this approach. We suggest that you study the examples and identify the problem-solving steps used in their solution.

Before leaving the topic of problem solving, we need to comment on the accuracy of numerical answers to engineering problems (see step 6). In general, you should not give a numerical answer that appears to be more accurate than the information used to obtain it (no matter how many decimal places you have on your calculator!). To illustrate this point, let's say that we want to calculate the force F caused by a uniform pressure p acting on an area A by

$$F = pA.$$

The pressure was measured as 91 psi with an uncertainty of ± 2 psi. The area is 61.0 in^2, with an uncertainty of ± 0.2 in^2.

Using the nominal values of pressure and area, we have

$$F = (91\ \text{lb/in}^2)(61\ \text{in}^2) = 5551.0\ \text{lb};$$

however, uncertainty considerations expand the range in which the answer could lie:

$$F_{max} = (91 + 2\ \text{lb/in}^2)(61.0 + 0.2\ \text{in}^2) = 5691.6\ \text{lb};$$

$$F_{min} = (91 - 2\ \text{lb/in}^2)(61.0 - 0.2\ \text{in}^2) = 5411.2\ \text{lb}.$$

Therefore we cannot justify writing our answer to more than two significant figures; or

$$F = 5500\ \text{lb}.$$

We usually shortcut this elaborate analysis by saying that the percent uncertainty in F is no less than the percent uncertainty in the most unreliable piece of given information. In this case, it is 2 psi out of 91 psi, or about 2.2% of 5551 (± 122). Therefore we immediately write

$$F = 5500\ \text{lb},$$

with a small uncertainty in the second 5. The idea is to retain only one significant figure with an uncertainty in it.

In addition to uncertainties in given data, numerical results have limited accuracy because of assumptions introduced in obtaining the results. We know of no quick and reliable way to evaluate the inaccuracy resulting from such assumptions. The best way would be to repeat the calculation without the assumptions, which is usually impractical. Otherwise, why did we need the assumptions in the first place? Another way would be to carry out a highly accurate experiment for comparison, which also is usually impractical. However, practice in doing prob-

* See, for example, the Moody chart, Fig. 7.11.

lems and experience will help give you a feel for the errors introduced by assumptions. Sometimes you can at least bracket the correct answer on the high and low sides by making conservative assumptions for both extremes.

1.7 COMPUTERS AND FLUID MECHANICS

The advent of the digital computer significantly enhanced our ability to analyze flow. Computational support ranges from calculators for "hand calculation" to supercomputers that are used to predict complicated flows from first principles. The cost associated with this support should be a factor in the choice to use it. Calculators are inexpensive and common, but supercomputers cost millions of dollars—and executing a single complicated analysis on one may cost hundreds of dollars. If this analysis saves thousands of dollars compared to performing costly experiments, the money is well spent.

Before we present an overview of how computers can be used to support flow analysis, a word of caution is in order. The use of computers does not lessen the importance of understanding the physics of fluid behavior. If anything, it makes an understanding of basic principles even more imperative. When analyses are limited to "hand calculations," a good analyst clearly states the assumptions required to reduce a complicated problem to one that is tractable with pencil and paper. If a problem is so complex that this approach is not possible, the analyst concludes that further study is required before the analysis can be performed. Today, with a good computer manual, a user can rather easily follow the instructions for assembling input to a large computer program and then execute that program to obtain an "answer." However, without an understanding of the principles of fluid mechanics—and the models imbedded in the program—use of the program may be inappropriate. Effectively, the assumptions being applied to the analysis are no longer easily visible for review but have been buried deep within computer code. Used without understanding, powerful computers and codes can generate impressive-looking results with little correspondence with reality! The responsibility for understanding the behavior of fluids still falls squarely on the shoulders of the engineer.

1.7.1 Numerical Methods as a Tool for Analysis

The main thrust of computer applications in this book is implementation of numerical methods to solve problems that would otherwise require tedious hand calculations. You can solve most textbook problems directly by using just pencil and paper and perhaps a simple calculator. This approach keeps exercises short and properly emphasizes understanding fundamental concepts. Unfortunately, it may also give you the mistaken notion that this is the norm for engineering analyses.

Real applications often do not produce such clean solutions. If you cannot develop an explicit equation for the variable of interest, you may have to use numerical methods to solve an implicit equation. Another common occurrence during an analysis is the appearance of an integral that is difficult or impossible to evaluate symbolically. A simple numerical integration algorithm may be all that is required to complete the analysis. Some problems require solution of a differential equation. At this stage in your education, analytic solution of a simple linear ordinary differential equation should be routine. However, a small change in problem statement can result in a nonlinear equation that cannot be solved without the assistance of a numerical method.

Comparison of the examples cited in the preceding paragraph with the general problem-solving process described in Section 1.6.2 reveals that numerical methods are usually used to assist with steps 5 and 6. As this is not a numerical methods textbook, we present the numerical methods without developing their theoretical underpinnings. In Appendix F we briefly present a few useful numerical methods. Included are simple, easily used algorithms for finding roots of equations, evaluating definite integrals, and solving ordinary differential equations. These algorithms are not necessarily the ultimate for their application; they are simply tools to enhance your problem-solving capability.

1.7.2 The Computer as a Tool for Design

Most design problems do not have a single correct answer. The design engineer may begin the solution process by developing a mathematical model of the physical system. Typically this model contains several parameters for which the engineer may select values. The range of possible parameter values is called the *design space*. The engineer searches the design space to find the best solution based on a set of performance criteria. This search can be a tedious process, even for a simple model with few parameters. The simplest way in which a computer can be used to support this process is to simply automate the calculation procedure. The engineer specifies a set of parameter values, and the computer calculates the corresponding system performance. Computer support becomes more desirable as the complexity of an analysis increases. The design process can be further automated if the computer model can predict system performance for several sets of parameter values. Even this approach may be inadequate for complicated problems. Implementing sophisticated optimization procedures to search for the best solution may be necessary. These optimization algorithms typically require substantial computational effort and therefore are best carried out on a computer. In general, no matter how simple or complicated a design problem may be, a computer permits a more thorough search of the design space and therefore the possibility of finding a better solution to the design problem.

1.7.3 Software for Problem Solving

Computer hardware provides the capability to perform the calculations required to support flow analyses. Computer software is required to put this capability to work on a particular problem. In general, four types of software are available to support fluid mechanics analyses.

First, task-specific software permits the user to make an analysis of a selected class of problems by simply entering the physical characteristics of the problem. For example, pipe-network programs are available to analyze the flow through a system of connected pipes. The user simply inputs the physical characteristics of the system, and the program performs all the calculations required to predict system performance. This type of software is widely used by practicing engineers but is of little value in learning the fundamentals of fluid mechanics.

Second, the most general purpose software available to support flow analysis is a programming language such as FORTRAN or BASIC. The engineer must formulate the sequence of calculations to be performed and then instruct the computer to perform them by issuing a series of commands in the form of a program. Certain tasks, such as solution of linear equation sets and root finding, appear in many analyses and can be viewed as a modular subtask. "Canned" programs to perform many of these modular tasks are available in a variety of languages. This type of software is available to instructors for distribution to their students. Using a package such as the IMSL Library® not only reduces programming time but can provide the user with state-of-the-art numerical performance.

Third, a new class of software given birth by the advent of personal computers, but now spreading to all computational platforms, is characterized by a high degree of interaction with the user. Software in this category is particularly useful in support of the design process because it enables the engineer to search the design space thoroughly but does not require a large software development effort. "Spreadsheets" such as Lotus 123®, Quattro®, and SuperCalc® use the paradigm of a large sheet of paper divided into rows and columns to form cells. The user can directly enter a value into a cell or can specify that a set of numerical operations be performed using data from other cells to compute a value to be assigned to that cell. The comprehensive nature of the computational functions available, and the flexibility with which they can be invoked, has led some to call spreadsheets programming languages in their own right. A group of products known as formula processors/equation solvers, typified by MathCad®, TK Solver!®, and Eureka®, also belong to this class of software. These products require the user to enter a mathematical statement of the problem of interest. The software then performs the numerical operations required to solve the problem and may provide graphic displays to assist the engineer in interpreting the results. We present solutions to some of the example problems in this book using this type of software, and you may find it useful in solving some of the end-of-chapter problems. Effort expended in learning how to use this type of software is rewarded by the simple and speedy solution of a wide range of problems.

The fourth type of software is the Computational Fluid Dynamics (CFD) code. CFD codes are powerful tools but they are typically very large and must be run on powerful computers. Although CFD is capable of solving very complex problems, simply running a CFD code will not provide you with any insight into the fundamental principles of fluid mechanics.

1.7.4 Computational Fluid Dynamics

Computer implementation of algorithms based on the fundamental physical principles governing fluid motion is known as Computational Fluid Dynamics (CFD). The crucial difference between CFD and the methods previously described is that those methods provide a solution to a numerical problem that arises after substantial analytic work had been performed on a specific fluid mechanics problem. The computational support is uniquely designed to solve the specific problem, which in many cases may be completed by hand or graphically but at the expense of accuracy or speed. In contrast, CFD strives for a single general formulation capable of solving a wide range of flow problems. Typical problems attacked by CFD cannot reasonably be solved by hand.

Practitioners of CFD combine knowledge of numerical methods with a knowledge of current computer technology to develop algorithms capable of solving general mathematical models of fluid flows. Usually, the mathematical model is in the form of a set of differential or integral equations. To solve these equations, CFD methods usually discretize the flow field into small but finite-size subregions by forming a computational mesh or grid. The most popular CFD methods are the finite-difference (FD), finite-volume (FV), and finite-element (FE) methods. Simple enumeration of all available CFD methods would be a formidable task, but any list would soon be obsolete as new algorithms are continually being developed.

The current capabilities of CFD are quite phenomenal, and some engineers believe that any flow problem can now be solved with the aid of computers. This is far from true. Two currently active areas of research are the modeling of turbulent flows and the modeling of hypersonic flows. Prominent researchers predict that an increase in speed of over two orders of magnitude compared to a CRAY-2 supercomputer is required before we can predict the details of even simple turbulent flows. Hypersonic flows are particularly challenging because of the breakdown of standard modeling assumptions, such as equilibrium chemistry. CFD is still a developing tool with many challenges and broad horizons. Of course, anyone interested in CFD must first have a firm grasp of the fundamental principles of fluid mechanics.

PROBLEMS

1. An engineer is using a positive displacement pump to achieve very low pressure levels in an air tank and wants to check the validity of the continuum assumption at the pump inlet. The air in the tank is at 10°C and 1.0×10^3 Pa. The inlet valve to the pump is 2.5 cm in diameter and opens a maximum of 0.30 cm. The continuum assumption is valid if the smallest dimension of the physical problem is at least 100 times the mean free path λ. The mean free path λ for gas molecules is given by kinetic theory as

$$\lambda = \frac{m}{\sqrt{2}\,\pi d^2 \rho},$$

where, for air,

m = Mass of an air molecule $= 4.8 \times 10^{-26}$ kg;

d = Diameter of an air molecule $= 3.7 \times 10^{-10}$ m;

ρ = Mass density of air.

Is the engineer justified in using the continuum assumption?

2. With the exception of the 410 bore, the gauge of a shotgun barrel indicates the number of round lead balls, each having the bore diameter of the barrel, that together weigh 1 lb. For example, a shotgun is called a 12-gauge shotgun if a $\frac{1}{12}$-lb lead ball fits the bore of the barrel. Find the diameter of a 12-gauge shotgun in inches and millimeters. Lead has a specific weight of 0.411 lb/in^3.

3. For air at 4°C (277K) and 101,330 N/m^2 as a reference, find the specific gravity of hydrogen at 25°C (298K) and 101,330 N/m^2.

4. Wine has a specific gravity of 1.15. A winery chemist decides to dilute wine with water to produce a specific gravity of 1.10. What percentage of the new volume is the added water?

5. At 4°C a mixture of automobile antifreeze (50% water and 50% ethylene glycol by volume) has a density of 1064 kg/m^3. If the water density is 1000 kg/m^3, find the density of the ethylene glycol.

6. A stick of butter at 35°F measures 1.25 in. × 1.25 in. × 4.65 in. and weighs 4 ounces. Find its specific weight.

7. A gallon of milk at 35°F weighs 8.60 lb. Find its specific weight. What additional information do you need in order to find the milk density?

8. A 12-in.-outside-diameter, $\frac{1}{2}$-in.-thick wall, 10-ft-long piston of a hydraulic lift for automobiles is made of steel and is used to raise automobiles weighing up to 3500 lb. What gage pressure is necessary at the bottom of the piston to support this load? Steel has a specific weight of 487 lb/ft^3.

9. Atmospheric pressure at sea level is 14.696 psia, or 760 mm Hg. Express this atmospheric pressure in feet of water and inches of mercury. Use Table C.7.

10. Assume that the atmospheric pressure in Denver, Colorado, the "mile high city," is 625 mm Hg. Consider a person who inhales one l of air per breath. How much less air mass is taken in per breath by the person in Denver than at sea level. Express your answer as a percentage reduction based on the sea level mass intake. Assume that the temperatures are the same at 20°C. For air, the specific gas constant is 287 N·m/kg·K.

11. A pressure gage indicates a pressure of 9.6 psig. The gage is located in an atmosphere where a barometer indicates the atmospheric pressure is 29.91 in. Hg. Find the absolute pressure corresponding to the pressure gage reading.

12. Even though women generally weigh less than men, why are the heel imprints of women's high-heel shoes more often found in soft floors than are the heel imprints of men's shoes?

13. A pressure gage indicates a pressure of 278 kPa where the atmospheric pressure is 100 kPa. What is the corresponding absolute pressure?

14. A vacuum gage indicates a pressure of 4.1 psi vacuum. The barometer indicates an atmospheric pressure of 29.923 in. Hg at the location of the vacuum gage. Find the absolute pressure corresponding to the vacuum gage reading.

15. A rigid tank initially contains air at 30 psia and 70°F. An air compressor adds 0.30 slug of air to the tank. The final conditions in the tank are 65 psia and 75°F. Calculate the volume of the tank.

16. A compressed air tank in a service station has a volume of 10 ft^3. It contains air at 70°F and 150 psia. How many tubeless tires can it fill to 44.7 psia at 70°F if each tire has a volume of 1.5 ft^3 and the compressed air tank is not refilled? The tank air temperature remains constants at 70°F because of heat transfer through the tank's large surface area.

17. The specific gravity of a gas is sometimes defined as the ratio of the density of the gas to the density of dry *air* at standard conditions. If standard conditions are taken at 4°C (39.2°F) and 101,330 N/m^2 absolute (14.696 lb/in^2), determine the specific gravity of dry air, carbon monoxide, and hydrogen at 20°C (68°F) and standard pressure.

18. A regulation basketball is initially flat and is then inflated to a pressure of approximately 24 lb/in^2 absolute. Consider the air temperature to be constant at 70°F. Find the mass of air required to inflate the basketball. The basketball's inside radius is 4.67 in.

19. Assume that the air volume in a small automobile tire is constant and equal to the volume between two concentric cylinders 13 cm high with diameters of 33 cm and 52 cm. The air in the tire is initially at 25°C and 202 kPa. Immediately after air is pumped into the tire, the temperature is 30°C and the pressure is 303 kPa. What mass of air was added to the tire? What would be the air pressure after the air has cooled to a temperature of 0°C?

20. A tank having a volume of 0.10 m^3 is filled with air at 25°C and 400 kPa absolute. The tank is connected to a tube of length L and inside diameter $d = 6.0$ cm, as shown in Fig. P1.20. A round wood plug is dropped into the tube and the valve separating the tank from the tube is opened. The air expands at constant temperature and pushes the wood plug through the tube. Find the length L so that the tank pressure drops to 200 kPa absolute as the wood plug leaves the tube.

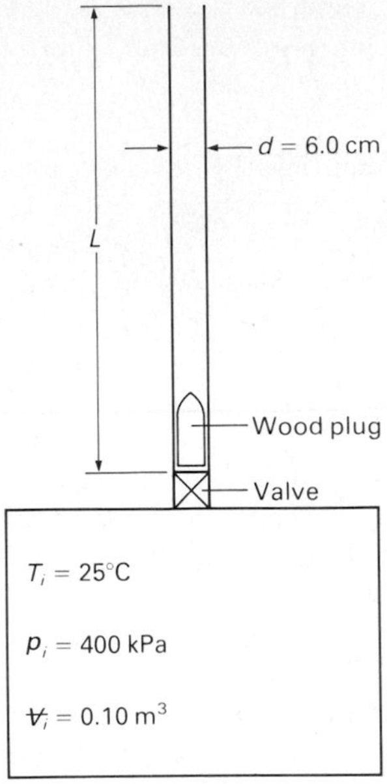

Figure P1.20

21. When air flows at high speed through a variable area passage, the Mach number, **M**, for an ideal gas is related to the area ratio, *AR*, by:

$$AR = \frac{0.5787(1 + 0.2\mathbf{M}^2)^3}{\mathbf{M}}.$$

For every value of *AR* (*AR* always greater than 1), there are two values of **M**, one greater than 1 and one less than 1, that satisfy this relationship. Determine both values of **M** for $AR = 1.3$.

22. Two important quantities in analysis of compressible flow are the area ratio, A/A^*, and the Mach number, **M**. For an ideal gas with $k = 1.4$, they are related by:

$$\frac{A}{A^*} = \frac{1}{\mathbf{M}} \frac{(1 + 0.2\mathbf{M}^2)^3}{1.728}.$$

Although the relationship is fine for computing A/A^*, analyses often require computation of **M**, given A/A^*, and this expression cannot be rearranged to be explicit in **M**. The following formula for a curve fit has been proposed for the specified range of A/A^*.

$$\mathbf{M} = a_1 + a_2\sqrt{\frac{A}{A^*} - 1} \quad \text{for } \mathbf{M} > 1;\ 1.0 < \frac{A}{A^*} < 2.9.$$

Use data from Table E.1 and determine appropriate values for a_1 and a_2.

23. Explain why house paint should be a shear thinning fluid.

24. A hydraulic lift in a service station has a 32.50-cm-diameter ram that slides in a 32.52-cm-diameter cylinder. The annular space is filled with SAE 10 oil at 20°C. The ram is traveling upward at the rate of 0.10 m/s. Find the frictional force when 3.0 m of the ram is engaged in the cylinder.

25. The oil between the journal bearing and the journal (shaft) shown in Fig. P1.25 is a Newtonian fluid. The "power loss" is the product of the angular velocity of the shaft and the torque between the oil and the shaft. Calculate the power loss.

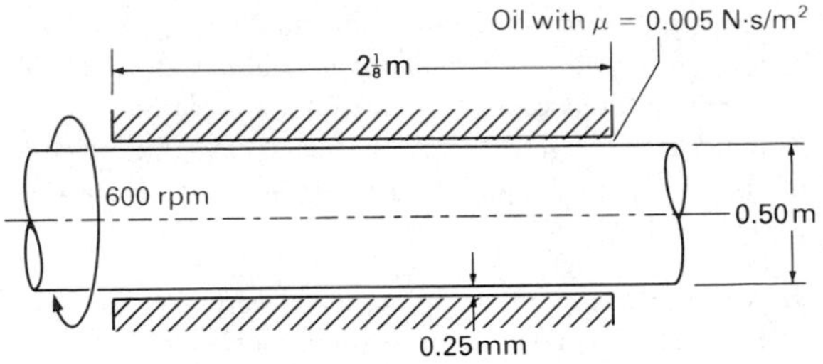

Figure P1.25

26. A plate measuring 25 cm × 25 cm is pulled horizontally through SAE 10 oil at $V = 0.100$ m/s, as shown in Fig. P1.26. The oil temperature is 40°C. Find the force *F*.

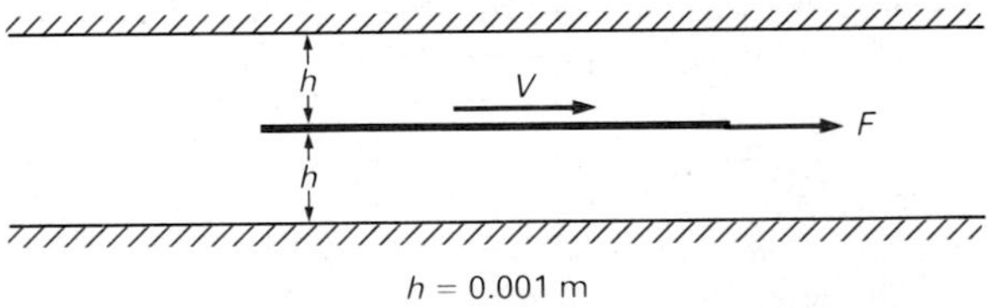

Figure P1.26

27. The plate in Problem 26 is still moving with a velocity of 0.100 m/s to the right, but the top and bottom walls are moving to the left at 0.050 m/s. Find the force *F*.

28. 20°C air flows over a large flat plate. The air near the surface forms a velocity profile that can be approximated by a half-sine wave, as shown in Fig. P1.28. Calculate the shear stress at the plate surface. If the plate is 1 m × 1 m and this stress acts uniformly over it, what is the force on the plate?

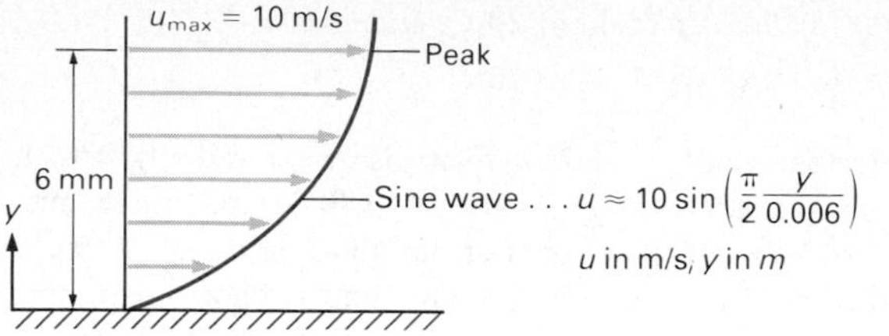

Figure P1.28

29. Three large plates are separated by thin layers of ethylene glycol and water, as shown in Fig. P1.29. The top plate moves to the right at 2 m/s. At what speed and in what direction must the bottom plate be moved to hold the center plate stationary?

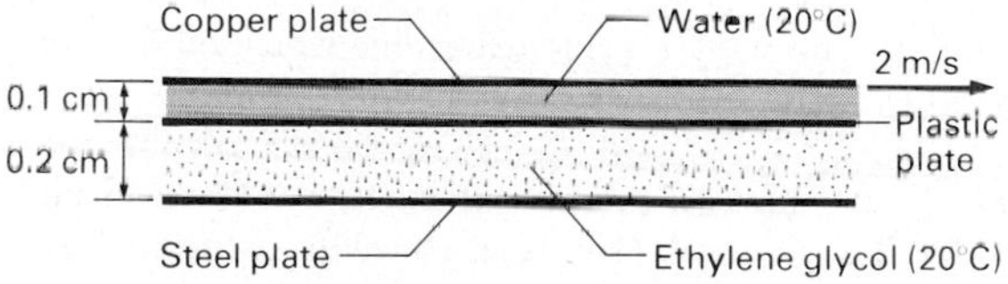

Figure P1.29

30. Oil (absolute viscosity = 0.0003 lb·sec/ft², density = 50 lbm/ft³) flows in the boundary layer, as shown in Fig. P1.30. The plate is 1 ft wide perpendicular to the paper. Calculate the shear stress at the plate surface.

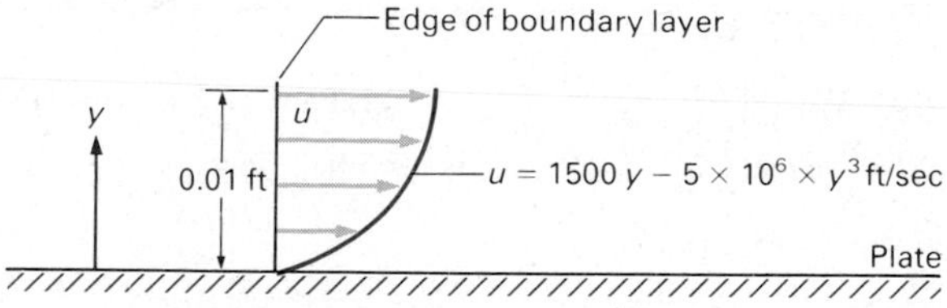

Figure P1.30

31. A large steel plate rests on a 30° incline. The incline has a 0.05-cm coating of 20°C lubricating oil over its entire surface area. The plate mass is 12 kg. At what velocity will the plate slide down the incline? (The plate is 1 m × 1 m × 1 cm.)

32. Data taken for a fluid are:

τ (lb/ft²)	0	0.010	0.021	0.029	0.041
$\frac{du}{dy}$ (1/sec)	0	50	100	150	200.

Is the fluid Newtonian? If so, find the fluid viscosity, μ.

33. Figure P1.33 shows a plate sliding down a plane inclined at an angle of 20° with the horizontal. The plate weighs 10 lb, measures 20 in. × 40 in., and has a velocity of 0.5 ft/sec. Estimate the thickness of the SAE 10W oil between the plate and the plane if the oil temperature is 50°F. Assume a linear oil velocity *across* the oil thickness and a Newtonian fluid.

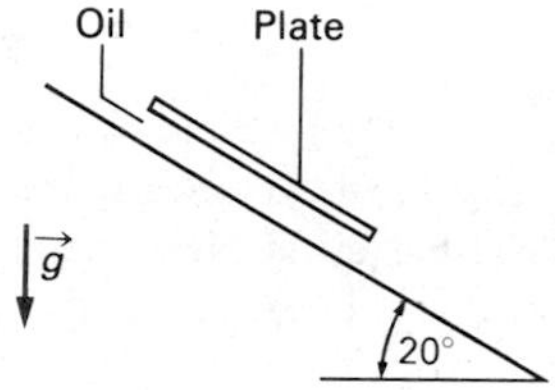

Figure P1.33

•**34.** Figure P1.34 shows a device known as a fluid drive. The driver rotates at an angular velocity ω_1, and the driven shaft rotates at a lower angular velocity ω_2. For steady-state conditions, ω_1 and ω_2 are constants. A torque $\mathscr{T}$ is transmitted by the driver to the driven shaft. Develop an expression for the slip $(\omega_1 - \omega_2)$ in terms of the applied torque $\mathscr{T}$, the narrow spacing h of the two discs, and the absolute viscosity μ of the fluid enclosed between the two discs, each of diameter D. Assume that the fluid is Newtonian. Then develop an expression for the efficiency, η, of the fluid drive in terms of ω_1, D, h, and $\mathscr{T}$. The efficiency is defined by

$$\eta = \frac{\omega_2 \mathscr{T}}{\omega_1 \mathscr{T}} = \frac{\omega_2}{\omega_1}.$$

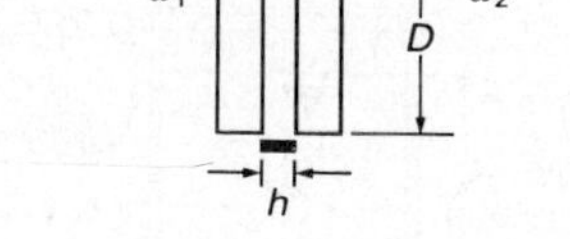

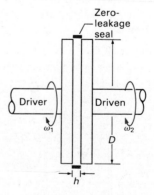

Figure P1.34

•**35.** Oil of viscosity μ fills the narrow gap of width t shown in Fig. P1.35. The truncated cone rotates at a constant rotational speed ω. Neglect surface tension and

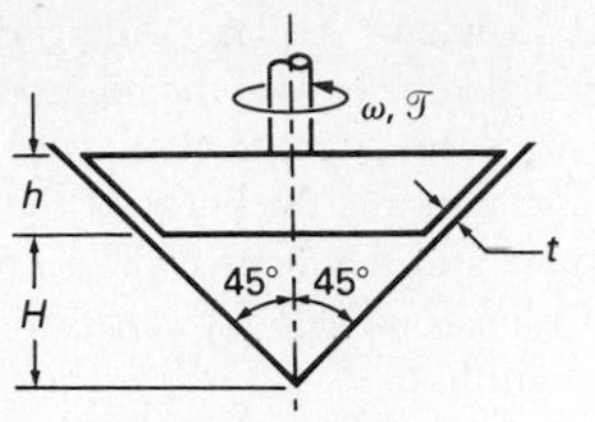

Figure P1.35

develop an expression for the fluid torque $\mathscr{T}$ as a function of the viscosity μ, the rotational speed ω, and the given dimensions. The oil is Newtonian. Assume that the oil velocity profile is linear across the narrow gap of thickness t.

• **36.** The circular disc of radius R shown in Fig. P1.36 is separated from a solid boundary by a distance h ($h \ll R$). The space is filled with a fluid of viscosity μ. The disc is rotated at angular velocity ω by a torque $\mathscr{T}$ applied to the shaft. Derive a relation between $\mathscr{T}$, μ, R, and ω.

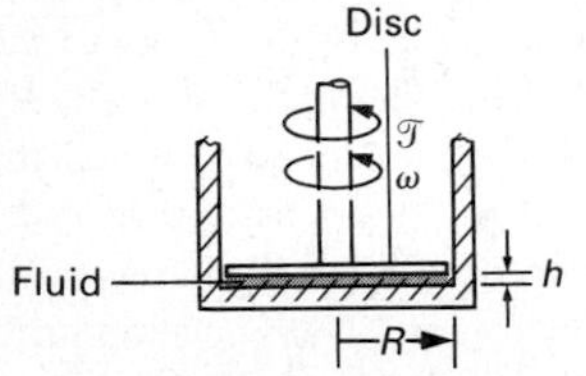

Figure P1.36

• **37.** The concentric cylinder viscometer shown in Fig. P1.37 uses a falling weight W to produce a constant rotational speed of the inner cylinder. Find the required weights if the inner cylinder rotates at 30 rpm and the fluid in the viscometer is

(a) tap water at 50°C,

(b) gasoline at 50°C ($\mu = 2.39 \times 10^{-3}$ N·s/m^2),

(c) kerosene at 20°C,

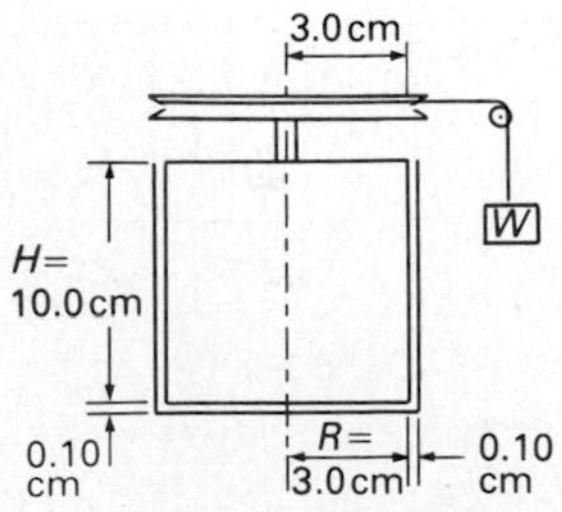

Figure P1.37

(d) SAE 10W oil at 20°C, and

(e) SAE 20W oil at 20°C.

Assume that the fluid velocity is linear radially *across* the vertical annular space of 0.10 cm thickness and *vertically* through the bottom fluid layer of 0.10 cm thickness at each radius r. The fluid is Newtonian, and there is no friction between the shaft and the bearings holding the shaft in place.

38. The viscosity, μ, of common liquids decreases as temperature increases. A plot of the common logarithm of viscosity ($\log_{10} \mu$) versus the reciprocal of the absolute temperature ($1/T$) is approximately a straight line. For

(a) an SAE 10W 30 engine oil and

(b) an SAE 10W engine oil,

plot $\log_{10} \mu$ versus $1/T$ and determine which oil is better for starting an engine in 0°C and 25°C weather.

39. A fluid flows in a plane channel, as shown in Fig. P1.39. The fluid viscosity is μ. Develop expressions for the shear stress in the fluid at the wall ($y = 0$) for both laminar and turbulent flow in terms of V_{max}, Y, and μ. Assume a Newtonian fluid.

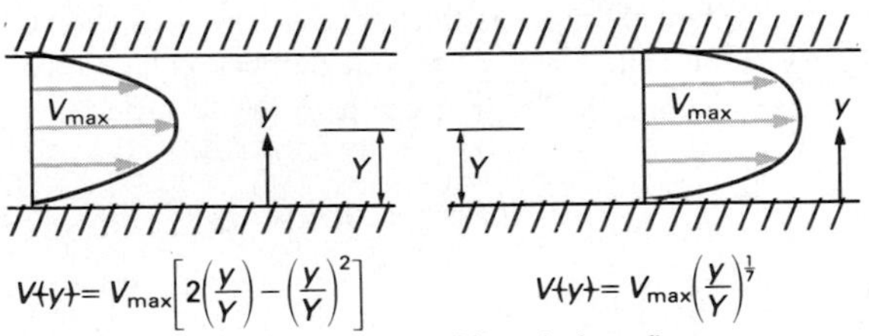

(a) Laminar flow; (b) turbulent flow.

Figure P1.39

40. In flow over bodies, the thin layer of fluid near the body that is slowed by frictional effects is known as the boundary layer. The following velocity profile was measured in flow over a flat plate (y measured normal to plate).

y (cm)	u (cm/s)	y (cm)	u (cm/s)
0.061	1.7	0.369	8.5
0.123	3.2	0.430	9.1
0.184	4.8	0.492	9.6
0.246	6.2	0.553	9.8
0.307	7.4	0.614	9.9

Note:

"No-slip" condition requires $u)_{y=0} = 0$.

Free-stream velocity, $U \approx 10$ cm/s.

Boundary layer thickness, $\delta \approx 0.614$ cm.

To obtain a simple approximation to these data, use the least squares criteria to compute the best parabolic approximation of the form:

$$\frac{u}{U} = a_0 + a_1\left(\frac{y}{\delta}\right) + a_2\left(\frac{y}{\delta}\right)^2.$$

41. The viscosity of liquid water is tabulated in Table A.5 from 0°C to 100°C for atmospheric pressure. The following equation is proposed to fit the data:

$$\ln\left(\frac{\mu_0}{\mu}\right) = a + b\left(\frac{T}{T_0}\right) + c\left(\frac{T}{T_0}\right)^2,$$

where T is the absolute temperature in K, $T_0 = 273$K, and $\mu_0 = \mu(T_0)$. Fit such an equation to the data of Table A.5. (A least-squares curve fit is preferred.)

42. Repeat Problem 41 using an equation of the form

$$\ln\left(\frac{\mu_0}{\mu}\right) = a + b\left(\frac{T}{T_0}\right).$$

Compare these values of μ calculated from the least squares curve fit with those of Problem 41.

43. The friction factor, f, is an important parameter in the analysis of flow through pipes and represents the resistance of the pipe to the flow of a viscous fluid through the pipe. Its value depends on the relative roughness, ε/D, and the Reynolds number, **R**. One of the most used relationships between these parameters is the Colebrook formula:

$$\frac{1}{\sqrt{f}} = -2.0\log\left(\frac{\varepsilon/D}{3.7} + \frac{2.51}{\mathbf{R}\sqrt{f}}\right).$$

What is the friction factor for flow at a Reynolds number of 1×10^4 through a pipe with a relative roughness of 0.0004?

44. The Sutherland formula is often used to compute an approximate viscosity for common gases,

$$\frac{\mu}{\mu_0} = \left(\frac{T}{T_0}\right)^{3/2}\left(\frac{T_0 + S}{T + S}\right),$$

where the subscript zero denotes a reference state and S, known as the Sutherland constant, is unique for each gas. This relationship is accurate to ± 2 percent for steam over a temperature range of 650–2700°R for the following constants: $T_0 = 750$°R, $\mu_0 = 1.144 \times 10^{-5}$ lbm/ft·sec, and $S = 1550$°R. What steam temperature corresponds to a viscosity of 2.312×10^{-5} lbm/ft·sec?

45. The velocity profile for turbulent flow near a solid boundary is frequently approximated by the law-of-the wall, law-of-the-wake profile,

$$\frac{u}{u_\tau} = 2.4\ln\left(\frac{yu_\tau}{\nu}\right) + 5.0 + 2.93\sin^2\left(\frac{\pi}{2}\frac{y}{\delta}\right),$$

where u_τ is the "friction velocity," δ is the boundary-layer thickness, and y is the distance measured perpendicular from the wall. Consider flow over a plate at a location where $u_\tau = 1.315$ cm/s, $\nu = 0.151$ cm²/s, and $\delta = 1.20$ cm. Compute the volume flow rate per unit width of the plate. Volume flow $\equiv Q \equiv \int_0^\delta u\,dy$. (Start integration at $y = 0.03$ cm to avoid singularity in logarithm at zero.)

46. Analysis of laminar boundary layer flow over a flat plate results in the following differential equation for a similarity function $f(\eta)$,

$$f''' + \tfrac{1}{2}ff'' = 0,$$

where $f(0) = 0$; $f'(0) = 0$; and $f''(\infty) = 1$. If $f''(0)$ were known instead of $f''(\infty)$, the equation could be numerically integrated using standard techniques. Use a value of $f''(0) = 0.33206$, show that $f'(10) \approx 1.0$, and determine the value of $f'(4.9)$.

47. The following shear stress measurements were made in a boundary layer:

y (cm)	τ (N/cm²)
0.0	6.528×10^{-8}
0.2	6.300×10^{-8}
0.4	5.649×10^{-8}
0.6	4.612×10^{-8}
0.8	3.262×10^{-8}
1.0	1.688×10^{-8}
1.2	0

where y is measured perpendicularly from the wall. Assume that this flow is laminar and determine the velocity profile for a fluid with $\mu = 0.151 \times 10^{-8}$ N·s/cm².

48. An important parameter in boundary layer flow is the displacement thickness, δ^*, defined as

$$\delta^* \equiv \int_0^\delta \left(1 - \frac{u}{U}\right) dy,$$

where u is the local velocity, U is the velocity at the outer edge of the boundary layer, y is measured perpendicular from the solid boundary, and δ is the boundary layer thickness. Compute the value of the displacement thickness of the boundary layer flow defined in Problem 45.

49. Automatic transmission fluid has specific gravities of 0.880 at 60°F, 0.821 at 210°F, and 0.789 at 300°F, all measured at atmospheric pressure. Find the fluid coefficient of thermal expansion at atmospheric pressure and 210°F.

50. The aerodynamic wing used on race cars is designed to give a "down-thrust." A race car travels at speeds up to 200 mph and in temperatures up to 110°F. The air flowing over the wing encounters a pressure drop of 1.0 lb/in^2 and a temperature drop of 20 F° (= 20 R°). Estimate the density change as a percentage of the undisturbed air density. Is the assumption of an incompressible fluid justified?

51. Mercury is used in many thermometers but water is not. Does mercury have a greater change in length than water for the same temperature change, making it a better thermometer fluid?

52. If the coefficient of thermal expansion is constant, how does the density of a liquid change with temperature for constant pressure?

53. A closed rigid container is filled with water at 15°C. Calculate the increase in pressure if the temperature is increased to 45°C.

54. Consider 1 m^3 each of water and air at 101 kPa and 20°C. What pressure change (in Pa) is required to produce the same volume change as a 40°C temperature change?

55. A method used to determine the surface tension of a liquid is to determine the force necessary to raise a wire ring through the air–liquid interface, as shown in Fig. P1.55. What is the value of the surface tension if a force of 0.015 N is required to raise a 4-cm-diameter ring? Consider the ring weightless, as a tensiometer (used to measure the surface tension) "zeroes" out the ring weight.

56. Calculate the pressure difference between the inside and outside of a spherical water droplet having a diameter of $\frac{1}{32}$ in. and a temperature of 50°F.

• **57.** A closed tank or vessel used to hold high-pressure fluids might be pressure tested. Let's consider a spherical vessel that is to be pressure-tested at 100°F and 400 psig. Would it be safer to pressure-test this vessel with water whose bulk modulus of elasticity is 3.27×10^5 lb/in^2 or with air whose bulk modulus of elasticity is 415 lb/in^2. The atmospheric pressure is 15 psia. [*Hint:* Should the vessel rupture during a pressure test, show that the water pressure would drop to atmospheric pressure over a smaller change in volume than would the air. Assume that each fluid expands isothermally from 415 psia to 15 psia.]

• **58.** The closed-loop, solar hot water system shown in Fig. P1.58 consists of type K copper tube, a tube bank in the solar collector, a tube bank in the heat exchanger, a pump to circulate water, and a surge tank. The surge tank prevents a dangerous pressure buildup as the water expands when heated. An enlarged view of the surge tank shows it to be a metal tank with a flexible bladder separating the water from an air space in the upper portion. The air in the upper portion can be pressurized to prevent boiling. In a particular system, there is 150 ft of $\frac{3}{4}$-in. and 100 ft of $\frac{1}{2}$-in. copper tube. The water and air are initially at 70°F and 20.0 lb/in^2 absolute pressure

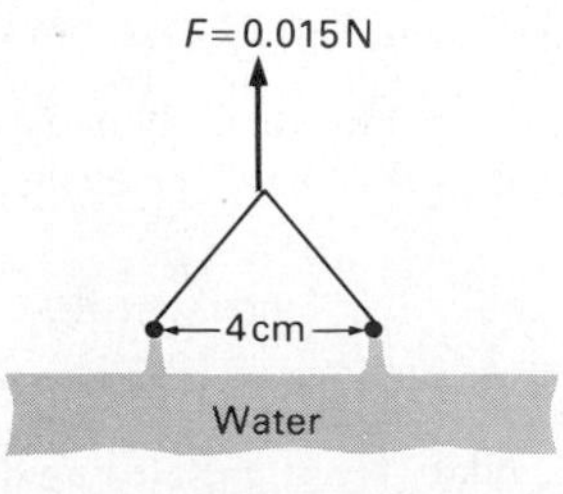

Figure P1.55

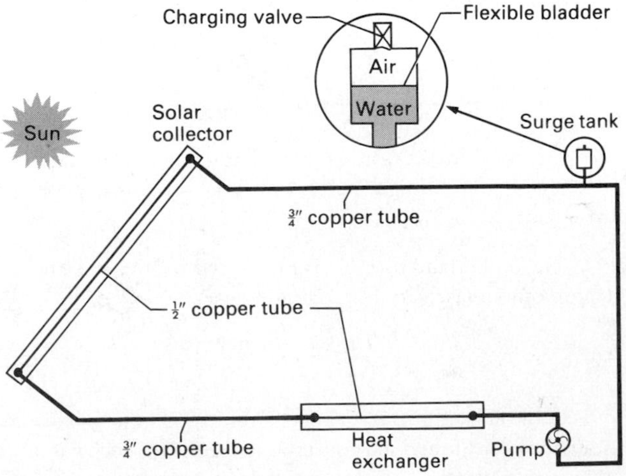

Figure P1.58

with no extension of the flexible bladder. Calculate the increase in volume of the water and in the air pressure if the average water temperature and the air temperature are increased to 120°F. The surge tank water volume and air volume are both initially 1.0 ft^3 and the pump is not operating. The increase in volume of the copper tube because of pressure and temperature changes is negligible.

• **59.** Explain how sweat soldering of copper pipe works from a fluid mechanics viewpoint.

• **60.** A glass tube is placed in a pan of water. Water rises in the tube to a height h above the level in the pan, as shown in Fig. P1.60. Show that the contact angle θ is given by

$$\theta = \cos^{-1}\left(\frac{\gamma h d}{4\sigma}\right),$$

where γ is the specific weight, d is the inside diameter of the tube, and σ is the surface tension. The pressure p at height h *above* the pan's water surface (and just below the glass tube's water surface) is given by

$$p = p_{\text{atm}} - \gamma h,$$

where p_{atm} is the atmospheric pressure.

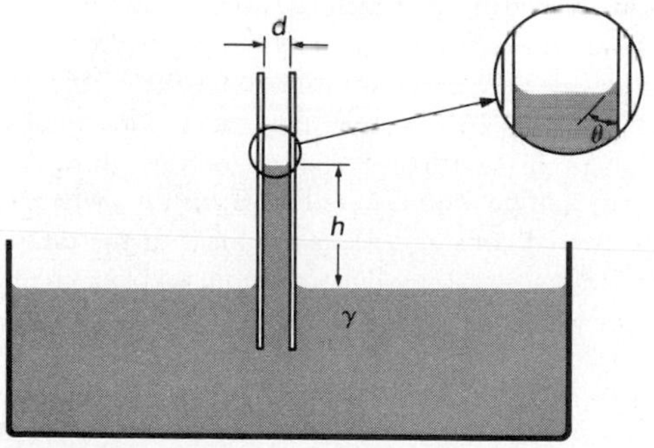

Figure P1.60

61. A soda straw with an inside diameter of 0.125 in. is inserted into a pan of water at 60°F. The water in the straw rises to a height of 0.150 in. above the water surface in the pan. Use the result of Problem 60 to find the angle θ shown in Fig. P1.60.

62. Use the surface tension data in Table A.5 to determine whether 10°C water or 50°C water would rise higher in a drinking straw. Is the difference significant? Assume that the contact angle θ does not change with temperature. Use the result of Problem 60.

63. The momentum flux (discussed in Chapter 4) is given by the product $\dot{m}V$, where $\dot{m}$ is mass flow rate and V is velocity. If mass flow rate is given in units of mass per unit time, show that the momentum flux can be expressed in units of force.

64. The air drag, $\mathscr{D}$, on a baseball thrown by a pitcher at a speed V in still air is given by

$$\mathscr{D} = \frac{C_D \rho S V^2}{2},$$

where

C_D = Dimensionless drag coefficient;

ρ = Air density; and

S = Area of the baseball as seen by the catcher (also called the projected area or the maximum cross-sectional area).

Show that the drag has units of force.

65. The Reynolds number, defined by

$$\text{R} = \frac{\rho V L}{\mu},$$

is a useful parameter in fluid mechanics. In this expression, ρ is the fluid density, μ is the fluid viscosity, V is the fluid velocity, and L is a relevant length of an object in a flow stream or of a channel in which the fluid is flowing. Show that the Reynolds number is dimensionless.

66. A commercial advertisement shows a pearl falling in a bottle of shampoo. If the diameter D of the pearl is quite small and the shampoo sufficiently viscous, the drag $\mathscr{D}$ on the pearl is given by Stokes's law,

$$\mathscr{D} = 3\pi\mu V D,$$

where V is the speed of the pearl and μ is the fluid viscosity. Show that the term on the right side of Stokes's law has units of force.

67. Air leaving a nozzle has absolute pressure 14.7 lb/in^2, temperature 50°F, and speed 100 ft/sec. Calculate

(a) the kinetic energy per unit mass of the air and

(b) the kinetic energy per unit volume of the air.

68. The universal gas constant R_0 is equal to 49,700 ft^2/(sec$^2\cdot$°R), or 8310 m^2/(s$^2\cdot$K). Show that these two magnitudes are equal.

69. The Mach number is a dimensionless ratio of the velocity of an object in a fluid to the speed of sound in the fluid. For an airplane flying at velocity V in air at absolute temperature T, the Mach number $\mathbf{M}$ is

$$\mathbf{M} = \frac{V}{\sqrt{kRT}},$$

where k is a dimensionless constant and R is the specific gas constant for air. Show that $\mathbf{M}$ is dimensionless.

70. The "power available in the wind" of velocity V through an area A is

$$\dot{W} = \tfrac{1}{2}\rho A V^3,$$

where ρ is the air density (0.075 lbm/ft^3). For an 18-mph wind, find the wind area A that will supply a power of 4 hp.

71. The axial velocity distribution for the flow of a viscous, Newtonian fluid in the circular pipe of inside radius R shown in Fig. P1.71 is

$$u = \frac{\Delta p_d}{4\mu L} R^2 \left[1 - \left(\frac{r}{R}\right)^2\right],$$

where the radius $r = 0$ at the centerline of the pipe, μ is the absolute viscosity, and Δp_d is the pressure drop in pipe length L. Find both the axial fluid velocity and the shear stress at $r = 0$ and at $r = R$.

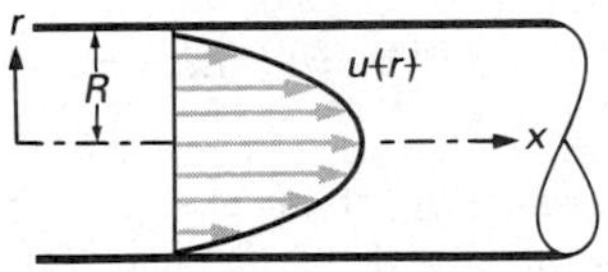

Figure P1.71

72. Air enters the converging nozzle shown in Fig. P1.72 at $T_1 = 70°F$ and $V_1 = 50$ ft/sec. At the exit of the nozzle, V_2 is given by

$$V_2 = \sqrt{V_1^2 + 2c_p(T_1 - T_2)},$$

where $c_p = 187$ ft·lb/lbm·°F and T_2 is the air temperature at the exit of the nozzle. Find the temperature T_2 for which $V_2 = 1000$ ft/sec.

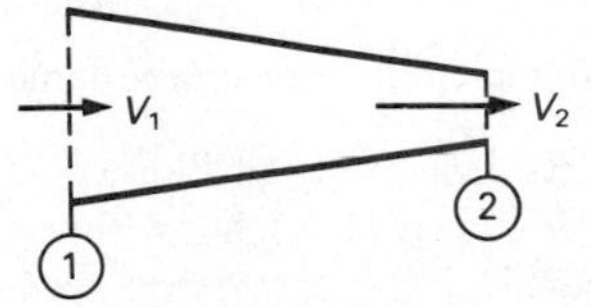

Figure P1.72

73. An electric motor has an output power of 20 hp and an efficiency of 0.80. Electricity costs \$0.07/kWh. Find the cost to run the electric motor for 24 hours.

74. A family has three teenagers, each of whom use a hair dryer for 10 min each morning. The dryer is rated at 1200 W and electricity costs \$0.08/kWh. Find the cost to operate the hair dryers for a 30-day month.

75. A bag of salt weighs 2.2 N at sea level. What is its mass in kilograms?

76. Sir Isaac Newton is reported to have said that the concept of gravity came to him while he was drinking tea in a garden and saw an apple fall. What is the weight in pounds of an apple that weighs 1.0 N?

77. A spring scale calibrated at sea level indicates the weight of a 10-lbm body as 8.0 lb. What is the local acceleration of gravity?

78. Consider a mass of 1.0 lbm. What is its weight at sea level and at an altitude of 10,000 ft above sea level?

79. The orbiter of the NASA space shuttle *Columbia* carries payloads of up to 29,000 kg (65,000 lbm) to orbits of up to 370 km (230 mi) altitude. Determine the weight of the payload at sea level and at the orbital altitude where the gravitational acceleration is 8.76 m/s^2 (28.743 ft/sec^2). Why is the payload said to be "weightless" in orbit? [You may want to use the earth's radius, 6380 km (3960 mi), to show that the gravitational acceleration of the earth is 8.76 m/s^2 at this altitude. This quantity can be found using $F = Gm_e m/p^2$, where G is a gravitational constant, m_e is the mass of the earth, and m is the mass of the object at radius r from the center of the earth.]

80. An equation for the frictional pressure loss Δp_L (inches H_2O) in a circular duct of inside diameter d (in.) and length L (ft) for air flowing with velocity V (ft/min) is

$$\Delta p_L = 0.027\left(\frac{L}{d^{1.22}}\right)\left(\frac{V}{V_0}\right)^{1.82},$$

where V_0 is a reference velocity equal to 1000 ft/min. Find the units of the "constant" 0.027.

81. How high above sea level must an object be raised for the object's weight to be decreased by 1.0 percent? Recall that the gravitational acceleration of an object decreases inversely with the square of the distance of the object from the center of the earth.

82. Show that each term in the following equation has units of lb/ft^3. Consider u a velocity, y a length, x a length, p a pressure, and μ an absolute viscosity.

$$0 = -\frac{\partial p}{\partial x} + \mu \frac{\partial^2 u}{\partial y^2}.$$

REFERENCES

1. Van Wylen, G. J., and R. E. Sonntag, *Fundamentals of Classical Thermodynamics,* Wiley, New York, 1986.
2. Holman, J. P., *Thermodynamics,* McGraw-Hill, New York, 1988.
3. ASME, "ASME Orientation and Guide for Use of SI (Metric) Units" (ASME Guide SI-1), American Society of Mechanical Engineers, New York.
4. Benedict, R. P., *Fundamentals of Temperature, Pressure and Flow Measurement* (2nd ed.), Wiley, New York, 1977.
5. Bradshaw, P., *Experimental Fluid Mechanics,* Macmillan, New York, 1964.
6. Bradshaw, P., *An Introduction to Turbulence and Its Measurement,* Pergamon Press, Oxford, 1971.
7. Dean, R. C., Jr., *Aerodynamic Measurements,* MIT Gas Turbine Lab Report, 1954, available from University Microfilms, Ann Arbor, MI.
8. Roache, P. J., *Computational Fluid Dynamics,* Hermosa Publishers, Albuquerque, 1976.
9. Chung, T. J., *Finite Element Analysis in Fluid Dynamics,* McGraw-Hill, New York, 1978.
10. Patankar, S. V., *Numerical Heat Transfer and Fluid Flow,* McGraw-Hill-Hemisphere, New York, 1980.
11. Anderson, D. A., J. C. Tannehill, and R. H. Pletcher, *Computational Fluid Mechanics and Heat Transfer,* McGraw-Hill-Hemisphere, 1987.

2 Mechanics of Nonflowing Fluids

In this chapter we discuss the mechanics of fluids that are not flowing; that is, the particles of the fluid are not experiencing any deformation. This condition is obviously satisfied for a static fluid. A second condition may also be classed as nonflow: "rigid body motion," in which the fluid moves without deformation. Practical applications of rigid body motion are not as numerous as those involving static fluids. Accordingly, in this chapter we concentrate mostly on static fluids.

2.1 PRESSURE AT A POINT: PASCAL'S LAW

If there is no deformation, there are no shear stresses acting on the fluid; the forces on any fluid particle are the result of gravity and pressure only.* We are mainly concerned here with calculating pressure distribution and evaluating the resultant pressure forces at interfaces between a fluid and a solid.

In a nonflowing fluid, the pressure is a scalar quantity; that is:

> The pressure at any point in a nonflowing fluid has a single value, independent of direction.

This statement is known as *Pascal's law*. To prove it, we consider the equilibrium of forces for the small fluid wedge shown in Fig. 2.1. For now, we assume that the wedge is static.

The forces on the fluid are the result of gravity and pressure, which we presume may be different for the y, z, and s directions (we neglect the x direction here for clarity). The pressure force on any face of the wedge is the product of the pressure and the area of the face. The wedge is at rest, so we can write

$$\sum F_y = p_y(\delta x\, \delta z) - p_s(\delta s\, \delta x) \sin\theta = 0,$$

and

$$\sum F_z = p_z(\delta x\, \delta y) - p_s(\delta s\, \delta x) \cos\theta - \gamma\left(\delta y\, \delta x\, \frac{\delta z}{2}\right) = 0.$$

But

$$\delta y = \delta s \cos\theta \quad \text{and} \quad \delta z = \delta s \sin\theta,$$

* We are neglecting surface tension and assuming that no electromagnetic forces act on the fluid.

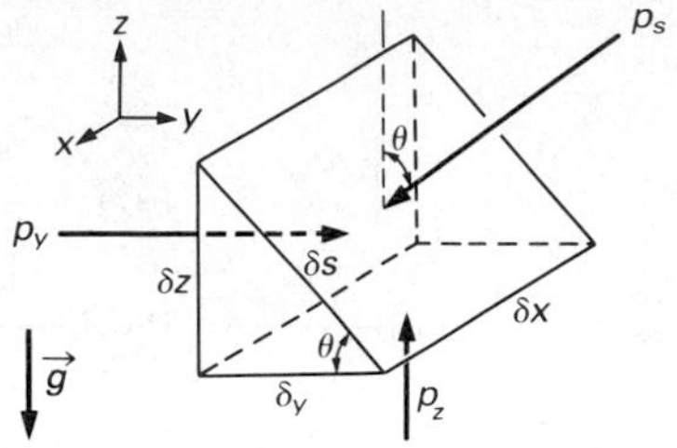

Figure 2.1 Wedge of fluid at rest.

so

$$p_y - p_s = 0 \quad \text{and} \quad p_z - p_s - \gamma \frac{\delta z}{2} = 0.$$

To evaluate this relation at a point, we take the limit as δx, δy, and δz approach zero, which results in

$$p_y = p_s \quad \text{and} \quad p_z = p_s.$$

As θ was arbitrary, these equations are valid for any angle. Note that the x and y axes are not unique; rotating the wedge 90° about the z axis would interchange the x and y axes, and we would conclude that $p_x = p_s$.

Imagine what would happen if the wedge were not at rest. If there were no shear stresses, the only change in the analysis would be to equate the force sums to the mass of the wedge times the appropriate acceleration:

$$\sum F_y = ma_y = \rho\left(\delta x\, \delta y \frac{\delta z}{2}\right)a_y \quad \text{and} \quad \sum F_z = ma_z = \rho\left(\delta x\, \delta y \frac{\delta z}{2}\right)a_z.$$

When we cancel common terms, the terms

$$\rho\left(\frac{\delta z}{2}\right)a_y \quad \text{and} \quad \rho\left(\frac{\delta z}{2}\right)a_z$$

remain. When we take the limit, these terms vanish. Therefore pressure is also independent of direction in a moving fluid *if no shear stresses are present.*

If shear stresses were present in the fluid, we would have to include them in our force balance. Forces due to shear stresses are proportional to surface area, so stress terms would not vanish as the fluid wedge is "shrunk" to a point. In a fluid with shear stresses, the pressure is not necessarily independent of direction. In such cases, an average pressure is defined by

$$p \equiv \tfrac{1}{3}(p_x + p_y + p_z).$$

The difference between the average pressure and the three "directional" pressures is seldom significant.

2.2 PRESSURE VARIATION IN A STATIC FLUID

Even though the pressure at a point is the same in all directions in a static fluid, the pressure may vary from point to point. To evaluate the pressure variation, let's

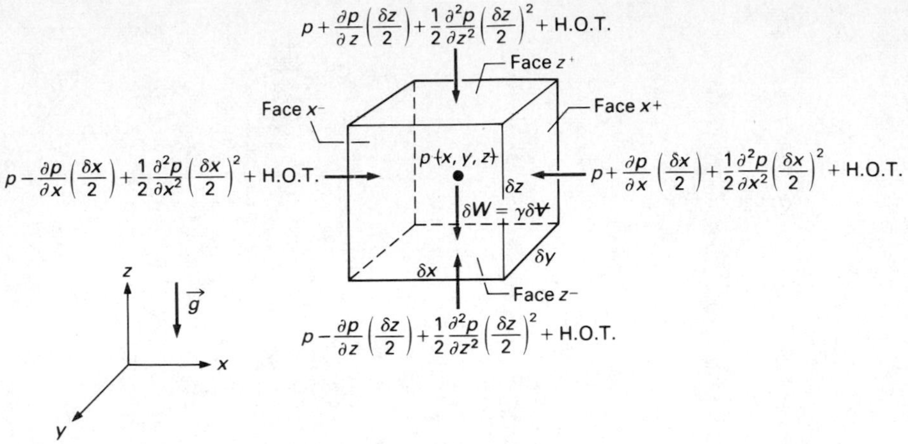

Figure 2.2 Cubic fluid element in a static fluid.

examine a cubical fluid element at rest (see Fig. 2.2). The sides of the fluid element are small but not necessarily infinitesimal. The volume of the fluid element is

$$\delta \forall = (\delta x)(\delta z)(\delta y).$$

The pressure at the center of the element is p. For simplicity, we assume that the pressure does not vary in the y direction. As the fluid is continuous, we can express the pressure at the faces of the element $\pm\delta x/2$ and $\pm\delta z/2$ from the center by Taylor series:

$$p_{z^+} = p + \frac{\partial p}{\partial z}\left(\frac{\delta z}{2}\right) + \frac{1}{2}\frac{\partial^2 p}{\partial z^2}\left(\frac{\delta z}{2}\right)^2 + \text{H.O.T.};$$

$$p_{z^-} = p - \frac{\partial p}{\partial z}\left(\frac{\delta z}{2}\right) + \frac{1}{2}\frac{\partial^2 p}{\partial z^2}\left(\frac{\delta z}{2}\right)^2 + \text{H.O.T.};$$

$$p_{x^+} = p + \frac{\partial p}{\partial x}\left(\frac{\delta x}{2}\right) + \frac{1}{2}\frac{\partial^2 p}{\partial x^2}\left(\frac{\delta x}{2}\right)^2 + \text{H.O.T.};$$

$$p_{x^-} = p - \frac{\partial p}{\partial x}\left(\frac{\delta x}{2}\right) + \frac{1}{2}\frac{\partial^2 p}{\partial x^2}\left(\frac{\delta x}{2}\right)^2 + \text{H.O.T.}$$

H.O.T. represents "higher order terms" involving δz and δx. The fluid element is at rest, so

$$\sum F_x = p_{x^-}(\delta z)(\delta y) - p_{x^+}(\delta z)(\delta y) = 0$$

and

$$\sum F_z = p_{z^-}(\delta x)(\delta y) - p_{z^+}(\delta x)(\delta y) - \gamma(\delta x)(\delta z)(\delta y) = 0.$$

Substituting the series expressions for the pressures gives

$$\left[p - \frac{\partial p}{\partial x}\left(\frac{\delta x}{2}\right) + \frac{1}{2}\frac{\partial^2 p}{\partial x^2}\left(\frac{\delta x}{2}\right)^2 - p - \frac{\partial p}{\partial x}\left(\frac{\delta x}{2}\right) - \frac{1}{2}\frac{\partial^2 p}{\partial x^2}\left(\frac{\delta x}{2}\right)^2 + \text{H.O.T.}\right](\delta z \delta y) = 0$$

and

$$\left[p - \frac{\partial p}{\partial z}\left(\frac{\delta z}{2}\right) + \frac{1}{2}\frac{\partial^2 p}{\partial z^2}\left(\frac{\delta z}{2}\right)^2 - p - \frac{\partial p}{\partial z}\left(\frac{\delta z}{2}\right) - \frac{1}{2}\frac{\partial^2 p}{\partial z^2}\left(\frac{\delta z}{2}\right)^2 + \text{H.O.T.} - \gamma(\delta z)\right](\delta x \delta y) = 0.$$

Combining and canceling terms where possible, we get

$$\left(-\frac{\partial p}{\partial x} + \text{H.O.T.}\right)(\delta x)(\delta z)(\delta y) = 0$$

and

$$\left(-\frac{\partial p}{\partial z} - \gamma + \text{H.O.T.}\right)(\delta x)(\delta z)(\delta y) = 0.$$

We now divide by $(\delta x)(\delta z)(\delta y)$, which incidentally is the volume of the fluid element, and then take the limit as δx, δy, and δz approach zero, reducing the fluid element to a point. The higher order terms contain positive powers of δx and δz, so they vanish as we pass to the limit. Our equations become

$$\frac{\partial p}{\partial x} = 0 \tag{2.1}$$

SHOWS PRESSURE IS CONSTANT IN A HORIZONTAL PLANE

and

$$\frac{\partial p}{\partial z} + \gamma = 0. \tag{2.2}$$

SHOWS THAT PRESSURE INCREASES WITH DEPTH

Before we discuss Eqs. (2.1) and (2.2), let's review the mathematical procedure used to obtain them. Because we assumed that the fluid was continuous, we could express the pressure near a point by a Taylor series. At the end of the derivation, we took the limit, shrinking the fluid element to a point. All terms involving derivatives higher than the first dropped out either by direct cancellation or because they included δx or δz to a positive power. When we use this type of derivation in the future, we do not include any terms beyond the first order, because we know that they will drop out in the limiting process.* Repeating the derivation of Eqs. (2.1) and (2.2), using only first-order terms in the series, would be good practice for you.

Equation (2.1) shows that the pressure does not vary in a horizontal plane (x could be any direction in the horizontal plane). Equation (2.2) shows that the pressure increases if we go "down" and decreases if we go "up." Because the

---

* We used this shortcut method to find the strain rate of a fluid element in Section 1.3.5. You might try repeating that derivation by retaining higher order velocity derivatives to see if you get the same result.

pressure changes in only one direction, we can replace the partial derivative with an ordinary derivative:

$$\frac{dp}{dz} = -\gamma. \tag{2.3}$$

Equation (2.3) is the basic equation of fluid statics.

2.2.1 Pressure Variation in a Constant-Density Fluid

If the specific weight of the fluid is constant (as in a liquid), we can easily integrate Eq. (2.3) to give

$$p(z) = -\gamma z + p_0, \tag{2.4}$$

where p_0 is the pressure at $z = 0$.

Consider a body of liquid with a free surface, as shown in Fig. 2.3. The free surface is at constant pressure, so it is horizontal. We introduce the *depth* of the liquid, h, measured downward from the free surface. As $z = -h$, we can write

$$p(h) = p_0 + \gamma h. \tag{2.5}$$

The pressure distribution implied by Eq. (2.5) is called a *hydrostatic pressure distribution*. The term γh is called the *hydrostatic pressure*.

Suppose that we have various containers of liquid as shown in Fig. 2.4. According to Eq. (2.5), the pressure at any point depends on the pressure at the free surface, the depth of the point, and the specific weight of the liquid, but *not* on the size or shape of the container. Thus the pressure at points A and B is the same, but the pressure at point C is different because C lies in a different fluid.

A practical use of this principle is the hydraulic lift system, illustrated in Fig. 2.5. Because elevation changes are usually small in such a system, the pressure at the piston essentially is equal to the pressure in the air tank. The weight of the automobile is supported by the pressure force on the piston. Only the magnitude of the pressure and the area of the piston have any effect on the weight-carrying capability of this system. The cross-sectional area of the air tank and the pressure transmission line can be any size.

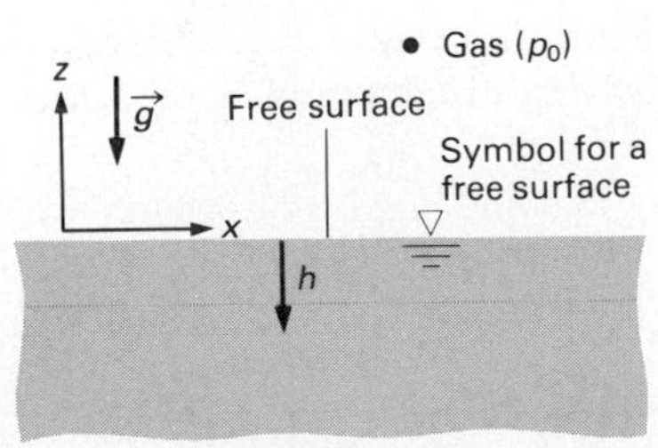

Figure 2.3 Body of liquid with a free surface.

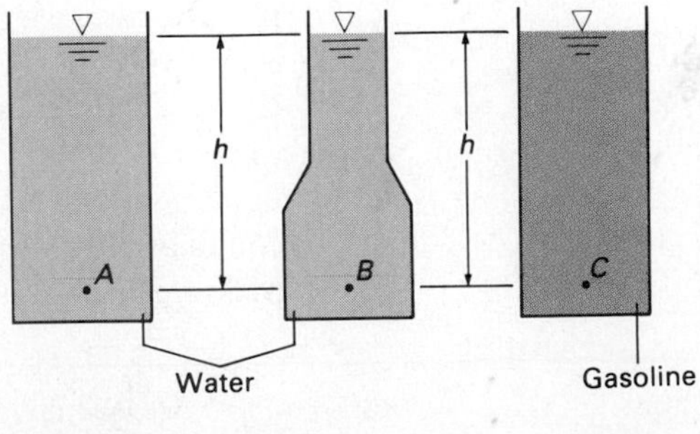

Figure 2.4 Liquids in different containers.

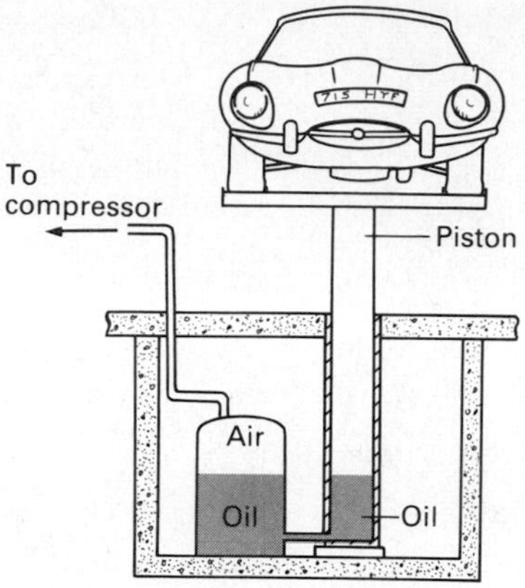

Figure 2.5 Hydraulic lift.

EXAMPLE 2.1 Illustrates the Calculation of Pressure in a Static Liquid

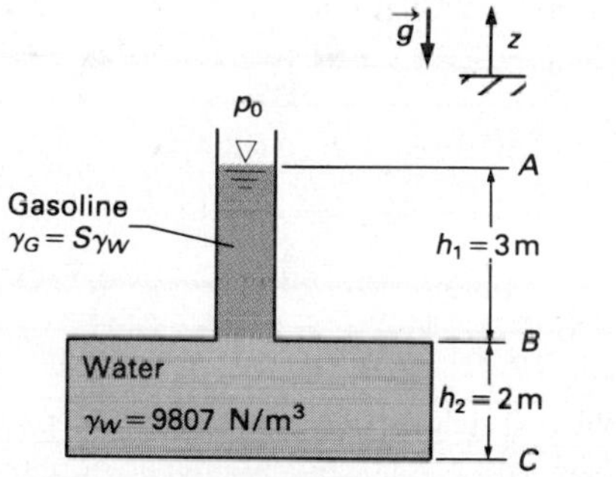

Figure E2.1a Tank of water with column of gasoline.

The tank of water shown in Fig. E2.1a has a 3-m column of gasoline ($S = 0.73$) above it. Find the pressure on the bottom of the tank. The atmospheric pressure is 101 kPa.

SOLUTION

Given

Tank of water with 3-m column of gasoline ($S = 0.73$) above it
Atmospheric pressure 101 kPa

Find

Pressure at bottom of tank

Solution

We use Eq. (2.5) to find the pressure at any point in a column of liquid when we know the pressure at another point and the difference in depth between the two points. We use the letters A, B, and C to represent the top of the gasoline, the interface between the gasoline and the water, and the bottom of the tank, respectively. Applying Eq. (2.5) separately to the gasoline and the water gives

$$p_B = p_A + \gamma_G h_1 \quad \text{and} \quad p_C = p_B + \gamma_w h_2.$$

Substituting the first equation into the second and setting $p_A = p_0$, we have

$$p_C = p_0 + \gamma_G h_1 + \gamma_w h_2.$$

The specific weight of gasoline is $\gamma_G = S\gamma_w$, so

$$p_C = p_0 + \gamma_w(Sh_1 + h_2).$$

Substituting numerical values, we obtain

$p_C = 101 \text{ kPa} + 9807 \text{ N/m}^3\,[0.73(3) + 2]\text{ m }(1 \text{ kPa}\cdot\text{m}^2/1000 \text{ N})$, or

$p_C = 142$ kPa.

ANSWER

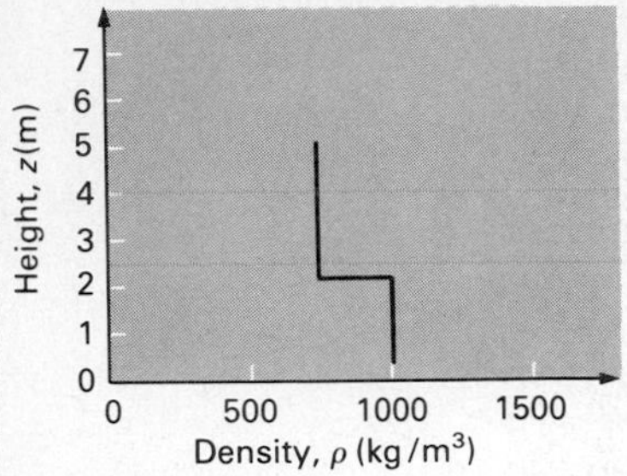

Figure E2.1b Variation of density with height.

Discussion

This example represents a simplified case of variable-density fluid. Figure E2.1b shows how the density varies with distance above the bottom of the tank. Using this density profile, we can integrate Eq. (2.3) to get

$$\int_{p_C}^{p_A} dp = -\int_{z_C}^{z_A} \gamma\, dz.$$

We must perform the integration in two parts because there are two distinct values of γ:

$$p_A - p_C = -\int_{z_C}^{z_B} \gamma_w\, dz - \int_{z_B}^{z_A} \gamma_G\, dz.$$

As $p_A = p_0$,

$$p_0 - p_C = -\gamma_w(z_B - z_C) - \gamma_G(z_A - z_B).$$

As $z_B - z_C = h_2$, $z_A - z_B = h_1$, and $\gamma_G = S\gamma_w$, we find that

$$p_C = p_0 + \gamma_w(Sh_1 + h_2),$$

which is the same result we obtained using Eq. (2.5) twice.

EXAMPLE 2.2 Illustrates That Pressure, Rather Than Force, is Transmitted through a Fluid and That Force is the Product of Pressure and Area

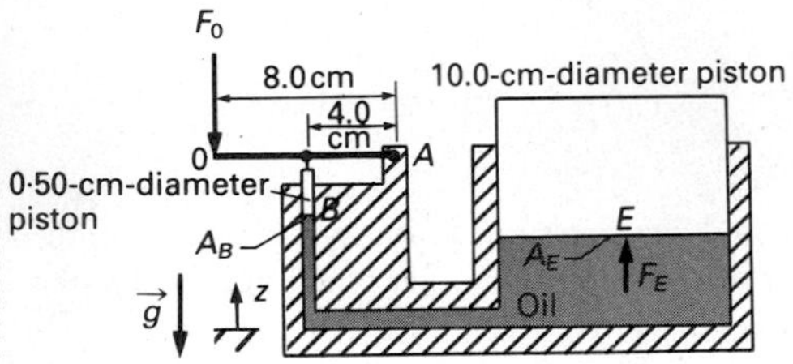

Figure E2.2a Hydraulic jack (or press).

Figure E2.2a shows a simplified sketch of a hydraulic jack. A force $F_0 = 100$ N is applied at a distance $\ell_0 = 8.0$ cm from the pivot point A. Find the force F_E at the bottom of the larger piston.

SOLUTION

Given

Hydraulic jack with force $F_0 = 100$ N applied at 8.0 cm from pivot point A. Piston diameters 0.50 cm and 10.0 cm, with the smaller piston 4.0 cm from pivot point A

Find

Force F_E at the bottom of the larger piston

Solution

As $p_E = F_E/A_E$, we must relate p_E to the applied force F_0. Referring to Fig. E2.2b, we take moments about the pivot point A. Neglecting the bar's weight, we have

$$\sum \mathcal{M}_A = -F_0\ell_0 + F_1\ell_1 = 0.$$

Solving for F_1, we obtain

$$F_1 = \frac{F_0\ell_0}{\ell_1} = \frac{(100 \text{ N})(8.0 \text{ cm})}{4.0 \text{ cm}} = 200 \text{ N}.$$

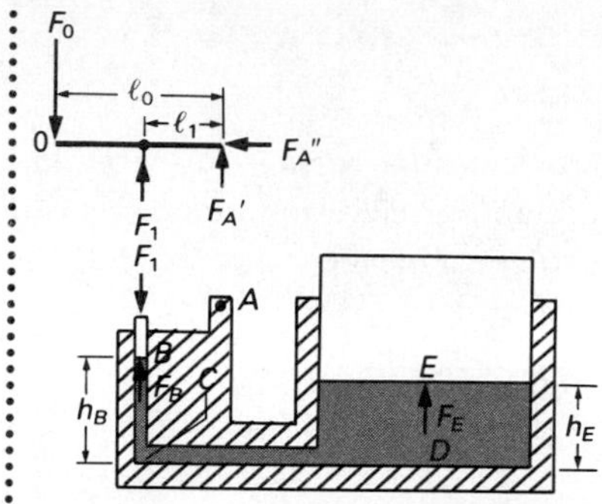

Figure E2.2b Forces on hydraulic jack.

If we neglect the weight of the smaller piston,

$$F_B = F_1 = 200 \text{ N}.$$

The pressure p_B is

$$p_B = \frac{F_B}{A_B} = \frac{200 \text{ N}}{(\pi/4)(0.5 \text{ cm})^2} = 1020 \text{ N/cm}^2 \ (=1.02 \times 10^4 \text{ kPa}).$$

Applying the hydrostatic pressure equation between levels B and C and between levels E and D gives

$$p_C = p_B + \gamma h_B \quad \text{and} \quad p_D = p_E + \gamma h_E.$$

Points C and D are at the same elevation in a static fluid, so

$$p_D = p_C.$$

Substituting for p_D and p_C gives

$$p_E + \gamma h_E = p_B + \gamma h_B \quad \text{or} \quad p_E = p_B + \gamma(h_B - h_E).$$

Assuming that points B and E are at the same elevation, we have

$$p_E = p_B = 1020 \text{ N/cm}^2.$$

The force F_E is

$$F_E = p_E A_E = (1020 \text{ N/cm}^2)\left(\frac{\pi}{4}\right)(10.0 \text{ cm})^2$$

$$F_E = 80{,}100 \text{ N}.$$

Only two significant figures are justified, so

$$F_E = 80{,}000 \text{ N}.$$ **ANSWER**

Discussion

Note that we could directly argue that $p_E = p_B$ if they are at the same elevation in a static fluid.* Then

$$p_E = \frac{F_E}{A_E} = \frac{F_B}{A_B} = p_B$$

and

$$F_E = \frac{A_E}{A_B} F_B = \frac{(\pi/4)(10.0 \text{ cm})^2}{(\pi/4)(0.50 \text{ cm})^2}(200 \text{ N})$$

$$= 80{,}000 \text{ N}$$

with no significant-figure problems. (Why did this happen?)

The pressure error introduced by the assumption that points B and E are at the same elevation is $\gamma(h_B - h_E)$. For an oil with $\gamma = 8400 \text{ N/m}^3$ and $(h_B - h_E) = 4$ cm, the force F_E is changed by a negligible amount:

$$\Delta F = \gamma(h_B - h_E)A_E$$

$$= (8400 \text{ N/m}^3)(0.04 \text{ m})\left(\frac{\pi}{4}\right)\left(\frac{10.0 \text{ cm} \times 1 \text{ m}}{100 \text{ cm}}\right)^2$$

$$= 2.6 \text{ N}.$$

* Remember that pressure, not force, is transmitted through the fluid.

2.2.2 Pressure Variation in a Variable-Density Fluid and the Standard Atmosphere

If the specific weight is variable, we use Eq. (2.3) to determine pressure distribution. We must relate the specific weight to the pressure and/or elevation before we can integrate the equation. A common case might involve an ideal gas (see Eq. 1.1) for which Eq. (2.3) becomes

$$\frac{dp}{dz} = -\rho g = -\frac{pg}{RT}.$$

Solving for pressure, we obtain

$$\int \frac{dp}{p} = \ln(p) = -\int \frac{g}{RT}\, dz + C. \tag{2.6}$$

To integrate on the right, we must know how the gas temperature varies with z.

A useful application of Eq. (2.6) is calculation of the variation of pressure with altitude in the Earth's atmosphere. Atmospheric temperature varies from day to day and season to season; however, calculations are usually based on the *U.S. Standard Atmosphere,* which has the temperature–altitude curve shown in Fig. 2.6(a). In the lower portion of the atmosphere, called the *troposphere,* temperature decreases linearly with altitude. At about 11 km (36,000 ft), temperature becomes constant in the region called the *stratosphere.* Temperature remains constant to an altitude of about 20 km (66,000 ft) and then begins to increase linearly. The outer region of the atmosphere is called the *ionosphere.* The ionosphere is extremely rarified and cannot be modeled as a continuum.

The temperature variation in the troposphere is

$$T = T_0 - Bz, \tag{2.7}$$

where T_0 is the surface temperature and B is a constant called the *lapse rate.* For the U.S. Standard Atmosphere,

$$T_0 = 15°\text{C} = 288.15 \text{ K} \qquad (59°\text{F or } 518.67°\text{R})$$

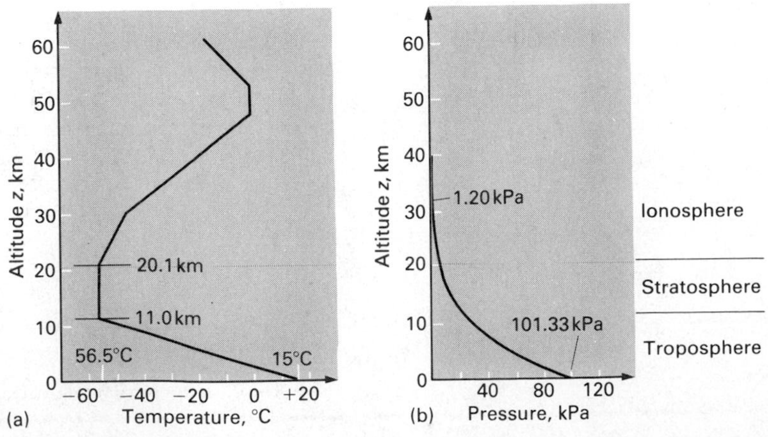

Figure 2.6 Temperature and pressure distribution in the U.S. Standard Atmosphere.

and

$$B = 0.00650 \text{ K/m} \qquad (0.003566°\text{R/ft}).$$

Substituting Eq. (2.7) into Eq. (2.6) and integrating, we obtain

$$p = p_0\left(1 - \frac{Bz}{T_0}\right)^{g/RB}. \tag{2.8}$$

The sea-level pressure is

$$p_0 = 101{,}330 \text{ Pa} \qquad (14.696 \text{ lb/in}^2).$$

In the stratosphere, the temperature is constant at

$$T = T_c = 216.7 \text{ K} = 390.1°\text{R}$$

and Eq. (2.6) becomes

$$\ln p = \int -\frac{g}{RT_c}\,dz + C.$$

Calling the pressure p_c at the lower edge of the stratosphere z_c, we get

$$p = p_c \exp\left[\frac{-g(z - z_c)}{RT_c}\right]. \tag{2.9}$$

For the U.S. Standard Atmosphere,

$$z_c = 11 \text{ km } (36{,}100 \text{ ft}) \quad \text{and} \quad p_c = 22{,}600 \text{ Pa } (3.28 \text{ lb/in}^2).$$

The rest of the atmosphere has either constant or linear temperature variation, so we can work out the pressure for the entire atmosphere. The results are shown in Fig. 2.6(b). The properties of the U.S. Standard Atmosphere are tabulated in Tables A.1 and A.2 in Appendix A.

2.3 MANOMETRY AND PRESSURE MEASUREMENT

2.3.1 Manometers

Recall the equation for pressure in a constant density static fluid, Eq. (2.5):

$$p - p_0 = \gamma h.$$

A simple and effective way to measure pressure is to measure the height of a column of liquid supported by the pressure. A device based on this principle is called a *manometer*. Figure 2.7 shows a simple manometer. The pressure at point 1 is

$$p_1 = p_{\text{atm}} + \gamma_L h.$$

Because 1 and 2 are at the same level,

$$p_1 = p_2,$$

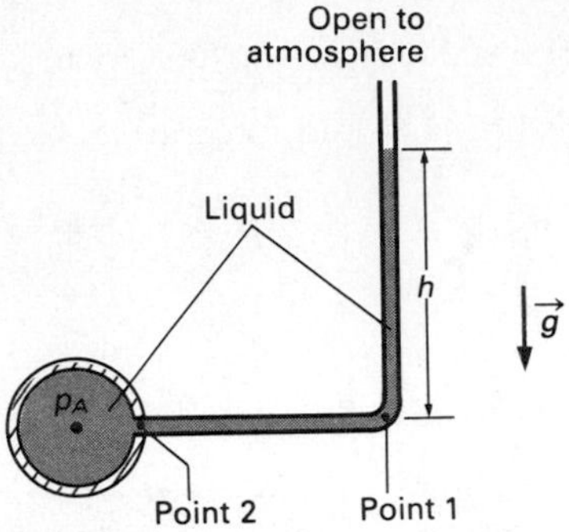

Figure 2.7 Simple manometer (piezometer).

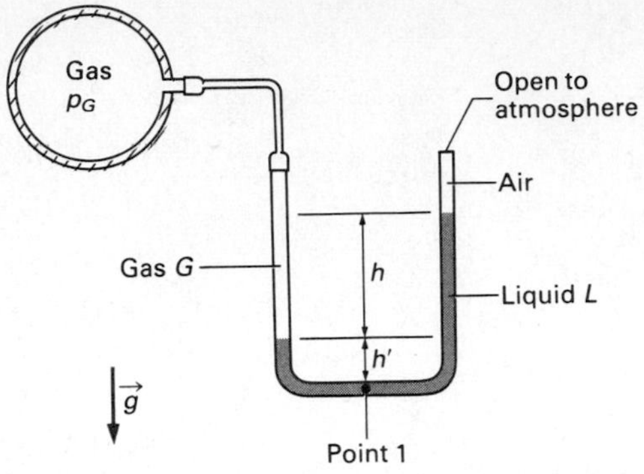

Figure 2.8 U-tube manometer.

and as 2 is in pipe A,

$$p_2 = p_A, \quad \text{so} \quad p_A - p_{\text{atm}} = \gamma_L h.$$

Measuring the height h and knowing the specific weight of the manometer fluid provides a simple and accurate pressure measurement. In this case, the fluid in the manometer is identical to the fluid whose pressure is being measured, and one end of the manometer is open to the atmosphere. This arrangement is called a *piezometer*. If the pressure in region A were very large and/or the fluid were very light (a gas, for example), a piezometer would be impractical because a very tall column of fluid would be required. An obvious solution to this problem is to use a denser liquid in the manometer.

One of the most common types of manometer is the U tube, as shown in Fig. 2.8. Because $p_1 = p_1$, we have

$$p_G + \gamma_L h' = p_{\text{atm}} + \gamma_L (h + h').$$

Canceling $\gamma_L h'$, we obtain

$$p_G - p_{\text{atm}} = \gamma_L h \tag{2.10}$$

as the equation of this manometer. Note that equal columns of height h' cancel in the equation. This condition is true in general, so we can just "jump" across the tube to the other side, ignoring the equal columns. In this development, we neglected the pressure due to the column of gas on the left of the tube and the column of air on the right. We illustrate the relative magnitudes of these pressures in Example 2.3.

Figure 2.9 shows a manometer with several different fluids. We can find an expression for the pressure difference $p_A - p_B$ by dividing the manometer at the high point "1" and the low point "2."

For the part of the manometer between "1" and A,

$$p_1 + \gamma_3 h_6 + \gamma_4 h_4 = p_A. \tag{2.11a}$$

For the part between "1" and "2,"

$$p_1 + \gamma_3 h_3 + \gamma_2 (h_2 + h_7) = p_2. \tag{2.11b}$$

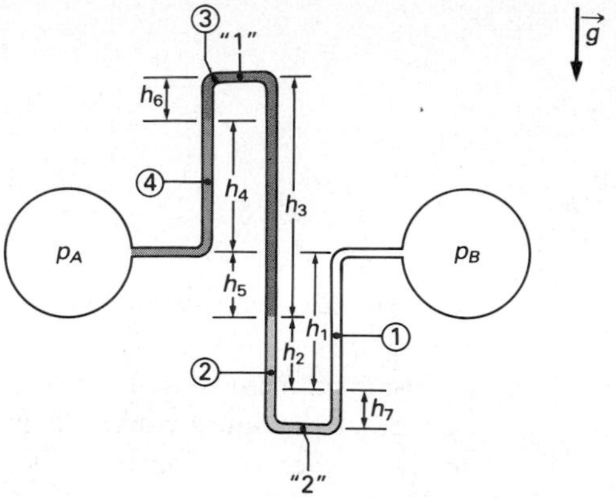

Figure 2.9 A complicated four-fluid manometer.

And for the part between B and "2,"

$$p_B + \gamma_1 h_1 + \gamma_2 h_7 = p_2. \tag{2.11c}$$

From Eqs. (2.11b) and (2.11c),

$$p_1 + \gamma_3 h_3 + \gamma_2 h_2 + \gamma_2 h_7 = p_B + \gamma_1 h_1 + \gamma_2 h_7,$$

so

$$p_1 = p_B + \gamma_1 h_1 - \gamma_3 h_3 - \gamma_2 h_2.$$

Substituting into Eq. (2.11a) and solving for $p_A - p_B$, we have

$$p_A - p_B = \gamma_1 h_1 - \gamma_2 h_2 - \gamma_3 (h_3 - h_6) + \gamma_4 h_4.$$

This development does not produce a general result; rather it illustrates the principles by which the equation for any manometer can be developed. The essence of the method is the application of the hydrostatic pressure equation and the balance of pressure at a point.

We can avoid the slight complexity of solving simultaneous equations by following a three-step *manometer rule*.

- Write the pressure at one end of the manometer.
- Proceed through the manometer, *adding* the hydrostatic pressure if you are going *down* and *subtracting* if going up (note that going up and down equal distances in the same fluid cancels).
- At any point, the algebraic sum of pressures is equal to the pressure at that point.

When we apply this procedure to the manometer of Fig. 2.9,

$$p_A - \gamma_4 h_4 - \gamma_3 h_6 + \gamma_3 h_3 + \gamma_2 h_2 - \gamma_1 h_1 = p_B,$$

so

$$p_A - p_B = \gamma_1 h_1 - \gamma_2 h_2 - \gamma_3 (h_3 - h_6) + \gamma_4 h_4. \tag{2.12}$$

We can change the *sensitivity* (change in reading h per unit change in pressure) and the maximum readable pressure by changing the liquid in the manometer. The

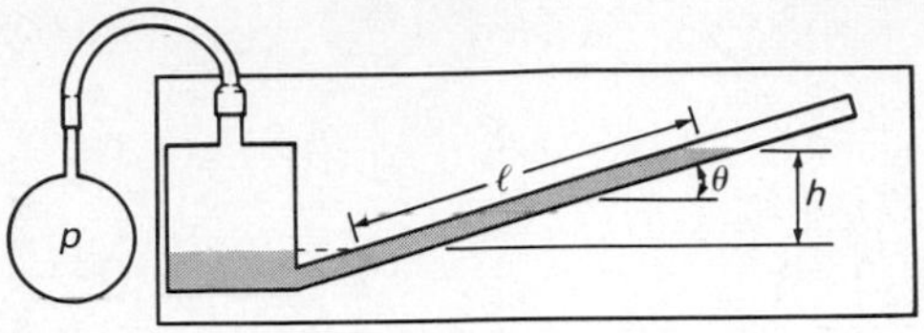

Figure 2.10 Inclined manometer.

most common manometer fluids are oil ($S \approx 0.8$), water ($S = 1.0$), and mercury ($S \approx 13.6$).

Using an inclined manometer, as shown in Fig. 2.10, increases sensitivity. The pressure of the liquid column is proportional to the vertical distance h; however, the scale deflection, ℓ, is magnified according to

$$\ell = \frac{h}{\sin\theta},$$

so the sensitivity increases as θ decreases.

The widespread use of manometers for measuring low pressures has led to the use of the "amount of length of fluid" (e.g., "inch of water" or "millimeter of mercury") as a unit of pressure. We convert these units to proper pressure units (force per unit area) by multiplying them by the specific weight of the manometric fluid. For example,

$$p\ (\text{lb/in}^2) = \gamma_{H_2O}\ (\text{lb/in}^3)p\ (\text{in. water})$$

EXAMPLE 2.3 Illustrates the Manometer Rule and Compares Pressures due to Columns of Gases and Liquids

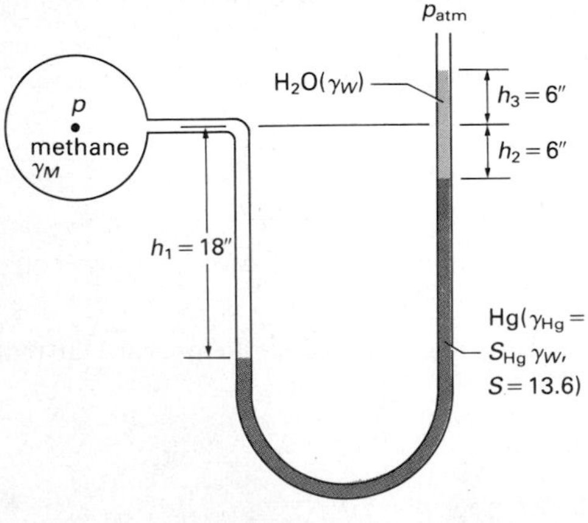

Figure E2.3 Gas pipe with manometer to measure methane pressure.

A gas pipe contains methane (CH_4) at a temperature of 70°F. A manometer is attached to the pipe as shown in Fig. E2.3. Calculate the pressure in the pipe. The atmospheric pressure, p_{atm}, is 14.696 lb/in^2. The gas constant R is 96.56 ft·lb/lbm·°R for methane.

SOLUTION

Given

Gas pipe containing methane at 70°F and connected manometer with mercury and water as measuring fluids
Figure E2.3

Find

The pressure in the pipe

Solution

Applying the manometer rule gives

$$p = p_{atm} + \gamma_w(h_3 + h_2) + S_{Hg}\gamma_w(h_1 - h_2) - \gamma_M h_1.$$

One purpose of this example is to compare the pressures from each column of fluid, so we need to compute each pressure individually.

The water pressure is

$$\gamma_w(h_3 + h_2) = \frac{(62.4\ \text{lb/ft}^3)(12.0\ \text{in.})}{1728\ \text{in}^3/\text{ft}^3}$$
$$= 0.433\ \text{lb/in}^2.$$

The mercury pressure is

$$S_{\text{Hg}}\gamma_w(h_1 - h_2) = \frac{13.6(62.4\ \text{lb/ft}^3)(12.0\ \text{in.})}{1728\ \text{in}^3/\text{ft}^3}$$
$$= 5.89\ \text{lb/in}^2.$$

The methane contribution is $\gamma_M h_1$; however, the specific weight of the methane is not given. Because the methane is at 70°F and its approximate pressure is 21.02 lb/in^2 (14.696 + 0.433 + 5.89), we may reasonably assume that the methane is an ideal gas. Using the ideal gas law to compute γ_M gives

$$\gamma_M h_1 = \rho_M g h_1 = \left(\frac{p}{RT}\right) g h_1.$$

The original equation then becomes

$$p = 21.02\ \text{lb/in}^2 - \frac{pgh_1}{RT}.$$

Solving for p, we have

$$p = \frac{21.02\ \text{lb/in}^2}{1 + gh_1/RT}.$$

Then

$$\frac{gh_1}{RT} = \frac{(32.2\ \text{ft/sec}^2)(18\ \text{in.})(1\ \text{ft/12 in.})}{(96.56\ \text{ft}\cdot\text{lb/lbm}\cdot{}^\circ\text{R})(459.67 + 70)^\circ\text{R}(32.2\ \text{ft}\cdot\text{lbm/lb}\cdot\text{sec}^2)}$$
$$= 2.93 \times 10^{-5},$$

a dimensionless number that is quite small compared to unity.

Therefore

$$p = 21.02\ \text{lb/in}^2.$$ **ANSWER**

We can now calculate the pressure from the column of methane:

$$\gamma_M h_1 = p\left(\frac{gh_1}{RT}\right) = (21.02\ \text{lb/in}^2)(2.93 \times 10^{-5})$$
$$= 0.000616\ \text{lb/in}^2.$$

Discussion

The relative contributions of the various fluids are

Mercury	5.89 lb/in^2
Water	0.433 lb/in^2
Methane	0.000616 lb/in^2

These numbers suggest that we can usually neglect the pressure contribution of a gas column compared to the pressure contribution of a liquid column, unless the gas column is very tall or we desire a high degree of accuracy.

The manometer principle is used in barometers (see Fig. 2.11) for measuring atmospheric pressure. When one end of the tube is closed off and the space is evacuated, the height of the fluid column measures the atmospheric pressure. Widespread use of this type of barometer with mercury as the manometric liquid resulted in adoption of the units of "inches of mercury" for atmospheric pressure among meteorologists.

2.3.2 Other Devices for Pressure Measurement

The manometer is a simple and accurate device for measuring pressure; however, its limits are the range of pressures it can measure (a pressure of 1000 lb/in^2 would require a mercury column more than 165 ft high!), its slow response to fluctuating pressures (because of inertia of the liquid column), and the difficulty of reading the liquid meniscus. Although alternative devices for pressure measurement generally are not based on the principles of fluid statics, we discuss them here for the sake of completeness.

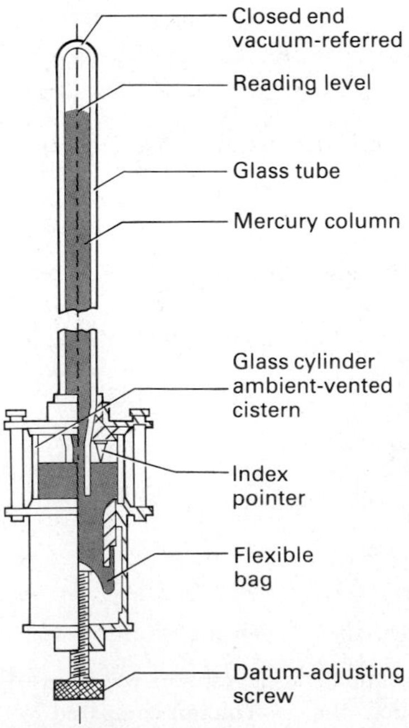

Figure 2.11 Typical (Fortin-type) barometer.

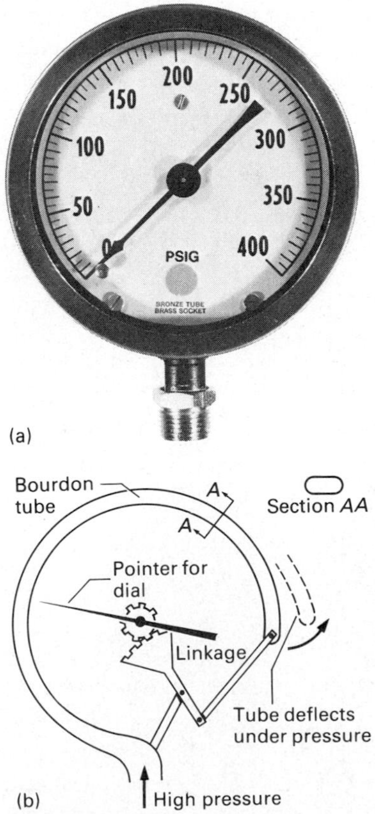

Figure 2.12 Bourdon tube pressure gage: (a) typical gage; (b) internal details.

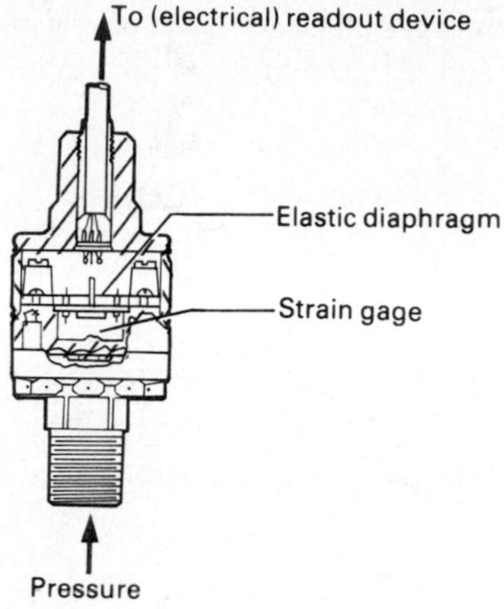

Figure 2.13 Strain-gage pressure transducer.

Most pressure-measuring devices operate on the principle that forces resulting from pressure produce deflection of an elastic element. Probably the most familiar and widely used pressure-measuring device is the Bourdon tube gage, illustrated in Fig. 2.12. When a pressure is applied to the open end of the curved, flattened tube, it tends to straighten out, much like the paper noisemaker popular at New Year's Eve parties. A linkage converts the motion of the tube to motion of a needle over the face of a dial. The dial can be calibrated to read the pressure in any units desired. By varying the size and rigidity of the tube, the manufacturer can construct a gage for pressures of almost any magnitude. Careful control of the tube material's quality and geometry, as well as careful calibration, ensures a reasonably accurate gage (to, say, ± 1.0 percent of the gage's full-scale reading).

Electric or electronic pressure transducer* applications cover a wide range of situations requiring accurate pressure measurement with rapid response, remote reading, and automatic recording. The most commonly used type of pressure transducer is the strain-gage type shown in Fig. 2.13. Application of pressure causes a deflection of the elastic diaphragm. This deflection is measured by the change of electrical resistance of a strain gage, a set of very fine wires whose electrical resistance changes when stretched. By calibration, the voltage output of a resistance-measuring bridge can be related to the applied pressure. The response of this type of gage to pressure changes is very rapid. The electrical output of the gage readily lends itself to automatic recording.

Similar characteristics are available with a piezoelectric pressure transducer, which uses a quartz crystal that generates a voltage when pressure is applied. Piezoelectric transducers have extremely rapid response to varying pressure because no fluid volume need be present in the transducer.

All pressure-measuring devices measure differences in pressure, not levels of pressure. Because one of the two pressures that an instrument senses is often local atmospheric pressure, we use the term "gage pressure" to describe pressure quoted

* *Transducer* is the name for a device that has one type of signal (say, pressure) as its input and another type (usually voltage) as its output.

relative to local atmospheric pressure, as discussed in Sec. 1.3.2. References [1] and [2] present more information about pressure measurement.

2.4 PRESSURE FORCES ON SURFACES

When a surface is in contact with a fluid, fluid pressure exerts a force on the surface. This force is distributed over the surface; however, it is often helpful in engineering calculations to replace the distributed force by a single resultant. To completely specify the resultant force, we must determine its magnitude, direction, and point of application. As

$$p = \frac{dF_n}{dA},$$

we can find the resultant force by integration:

$$F_n = \int p\, dA. \tag{2.13}$$

In general, we have to overcome two difficulties in performing this calculation.

- The pressure and area must be expressed in terms of a common variable to permit integration.
- If the surface is curved, the normal direction varies from point to point on the surface, and F_n is meaningless because no single "normal" direction characterizes the entire surface. In this case, the force at each point on the surface must be resolved into components and then each component integrated.

2.4.1 Forces on Plane Surfaces

We first consider a plane surface subjected to a *uniform* pressure p_0, as shown in Fig. 2.14. The net pressure force on the surface is

$$F = \int p\, dA = \int p_0\, dA = p_0 A. \tag{2.14}$$

Uniformly distributed pressures might occur in several different cases. Probably the most important case is that of a surface in contact with the atmosphere, where $p_0 = p_{\text{atm}}$. Another case involves a gas contained in a closed container or tank. A final case of interest is that of a horizontal plane surface submerged in a liquid, such as the bottom of the tank shown in Fig. 2.15. Because the entire bottom is at a uniform depth, the pressure on it is uniform and

$$F = p_0 A = (p_{\text{atm}} + \gamma_L h)A.$$

The force on the outside of the tank results from atmospheric pressure and is equal to $p_{\text{atm}}A$, so the net force from the liquid is

$$F_{\text{Liq}} = \gamma_L h A.$$

Note that in this case the force from the liquid is equal to the weight of the volume of liquid above the surface. However, the force would be the same for the tank shown in Fig. 2.16, although the volume of liquid contained in the tank is less.

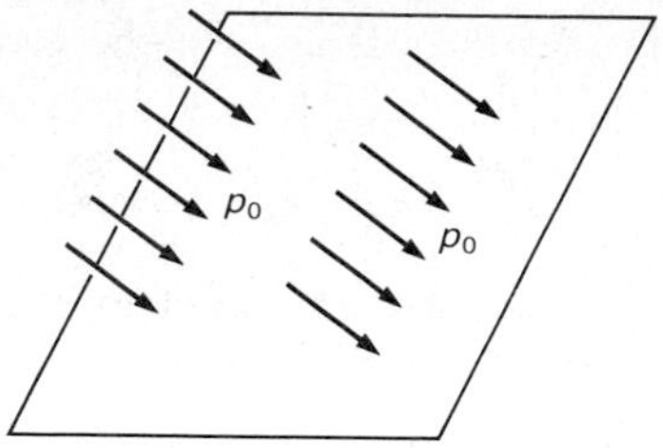

Figure 2.14 Plane surface subjected to a uniform pressure.

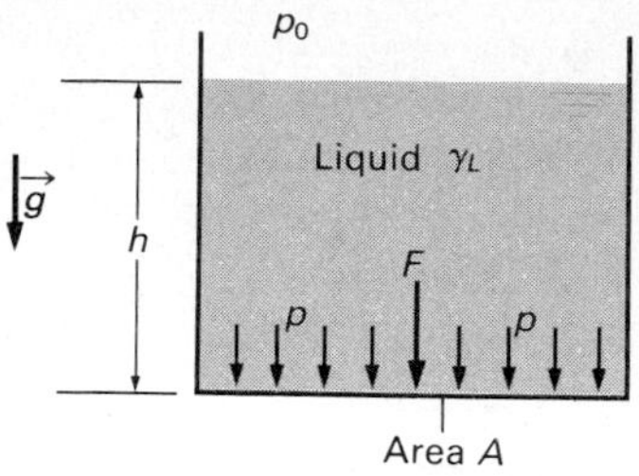

Figure 2.15 Tank of liquid.

Summarizing:

> The force on a plane surface caused by a uniform pressure is equal to the product of the pressure and the area. This force acts normal to the area.

Let's now evaluate forces on plane surfaces resulting from hydrostatic pressure distribution. Figure 2.17 shows an arbitrarily shaped inclined plane surface in contact with a static liquid. The force on a strip of area dA is

$$dF = p\,dA$$

and is normal to the surface.

The pressure on the strip is the sum of two terms representing the hydrostatic pressure and the pressure at the free surface. In many situations, the pressure at the free surface is atmospheric pressure, and the surface on which we wish to calculate the force is exposed to the atmosphere on one side. The net force on such a surface is the result of hydrostatic pressure only. Even when this condition is not true, we evaluate the force on any surface in two parts: the force resulting from γh and the force resulting from p_{atm}. The force on a submerged surface from p_{atm} is

$$F_{\text{atm}} = p_{\text{atm}}A.$$

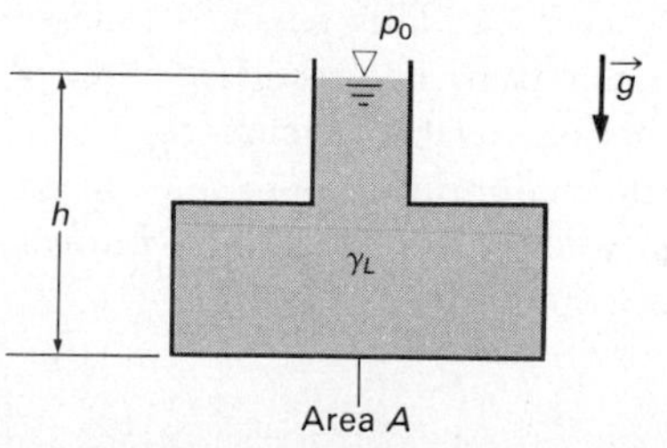

Figure 2.16 Another tank of liquid.

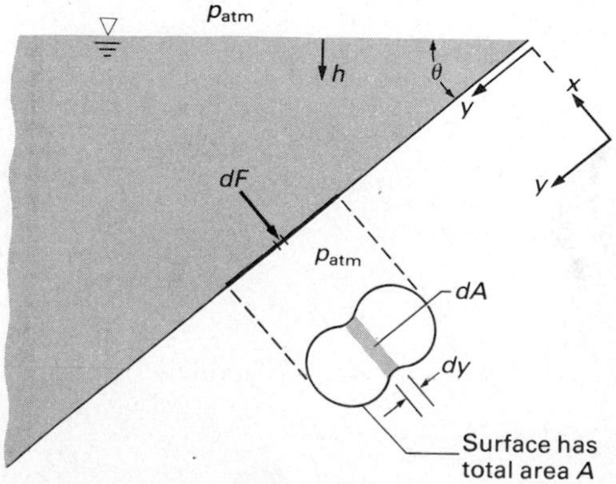

Figure 2.17 Arbitrary plane surface in contact with a liquid.

In the following analysis, we consider only the force caused by hydrostatic pressure, or

$$p = \gamma h,$$

where h is the depth from the free surface to the strip. We relate h to y, the distance from the free surface to the strip *in the plane of the area,* by

$$h = y \sin \theta, \tag{2.15}$$

so the force from hydrostatic pressure on the strip is

$$dF = \gamma y \sin \theta \, dA.$$

Integrating, and recognizing that γ and θ are constant,* we obtain

$$F = \gamma \sin \theta \int y \, dA. \tag{2.16}$$

We can replace the integral on the right by noting that the centroid of an area is defined by

$$y_c \equiv \frac{\int y \, dA}{A},$$

where y_c is the distance from the free surface to the centroid of area A, measured in the plane of area A. Substituting into Eq. (2.16), we have

$$F = \gamma y_c \sin \theta A.$$

Using Eq. (2.15), we obtain

$$h_c = y_c \sin \theta \tag{2.17}$$

and

$$F = \gamma h_c A. \tag{2.18}$$

The pressure at the centroid of the surface is given by

$$p_c = \gamma h_c, \tag{2.19}$$

so the net force on the surface is the product of the pressure at the centroid of the surface and the surface area. This result is independent of the shape of the surface or its angle of inclination. The resultant force is perpendicular to the surface and acts from the liquid toward the surface.

For plane surfaces, the major difficulty in calculating the net force is geometric, namely, locating the centroid. Appendix B gives information about the geometric characteristics of several common plane shapes.

* Note that θ is constant only if the surface is plane.

EXAMPLE 2.4 Illustrates the Calculation of Hydrostatic Force on a Plane Surface

A bathyscaph (Fig. E2.4) is a diving craft used for deep-sea observation; it consists of a thick-walled steel sphere attached to a large hull. The sphere protects the crew from the high pressure that exists at great depths. It has a flat, semicircular window with a 2.0-m diameter. In this case the top of the window is 10 m below the surface of the sea. Find the magnitude of the hydrostatic force on the outside of the observation window. The specific gravity of seawater is 1.028.

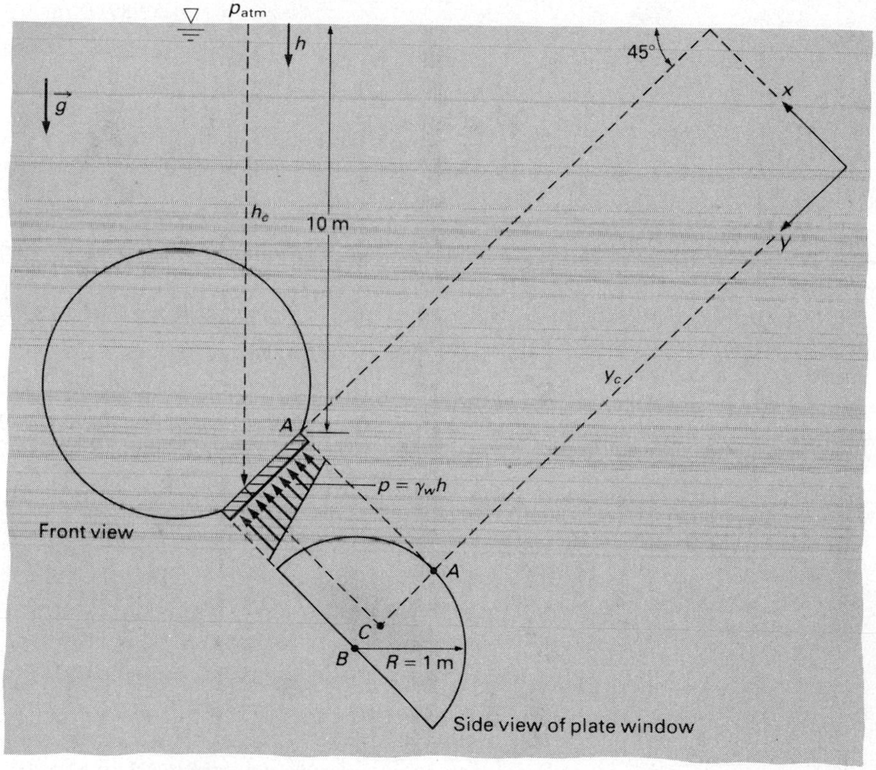

Figure E2.4 Bathyscaph with observation window.

SOLUTION

Given

Inclined semicircular window with 2.0-m diameter, making an angle of 45° with the horizontal

Top of window 10 m below sea level

Find

Magnitude of hydrostatic force on outside of window

Solution

Appendix B gives the location of the centroid of a semicircular area.

Using Fig. E2.4, we have

$$h_c - h_A = (1 - \tfrac{4}{3\pi})R \sin 45°.$$

Substituting numerical values gives

$$h_c - 10 \text{ m} = (1 - \tfrac{4}{3\pi})(1 \text{ m}) \sin 45° = 0.41 \text{ m},$$

or

$$h_c = 10 \text{ m} + 0.41 \text{ m} = 10.41 \text{ m}.$$

Assuming that seawater has a constant specific weight, we obtain

$$F = \gamma h_c A = S\gamma_w h_c A$$
$$= 1.028(9807 \text{ N/m}^3)(10.41 \text{ m})(\pi/2)(1.0 \text{ m})^2;$$
$$F = 1.65 \times 10^5 \text{ N}.$$ **ANSWER**

Discussion

Figure E2.4 shows the hydrostatic pressure distribution along a radius in a vertical plane. Pressure increases linearly with depth.

To complete the analysis of forces on plane surfaces, we must determine the point of application of the resultant force. This point is called the *center of pressure*. Because we can calculate the resultant force as the product of the area and the pressure at its centroid, we might be tempted to assume that the centroid is also the center of pressure. However, this assumption holds only for a *uniform* pressure distribution:

> If a surface is subjected to a *uniform* pressure, the center of pressure is at the centroid.*

In general, we must locate the center of pressure by equating the *moment* of the resultant force with the *moment* of the distributed (pressure) force. Consider a hydrostatic pressure variation, where the pressure is not uniform but increases with depth. The force on an element of area near the bottom of the surface has a greater moment than a force on an element near the top. Not only is the moment arm greater, but also the force itself is greater. Thus the center of pressure cannot be at the centroid of the area for a hydrostatic pressure distribution. Because the contribution to the moment is greater from locations in the deeper part of the liquid, the point of application of the resultant force must be shifted into the deeper portion. Hence we can state the general principle:

> The center of pressure of an area lies below the centroid of the area for hydrostatic pressure distribution.

To determine the center of pressure, let's use an inclined plane surface in a liquid, as shown in Fig. 2.18. Taking moments about the x and y axes through point 0, we obtain

$$Fy_p = \int yp\, dA \tag{2.20}$$

* This is exactly the case for the free-surface-pressure contribution to the force on a surface.

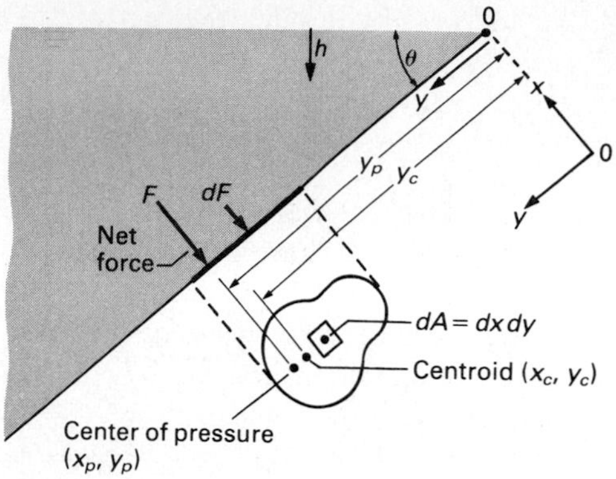

Figure 2.18 Arbitrary plane surface in contact with liquid; net force F acts at center of pressure.

and

$$Fx_p = \int xp\, dA. \tag{2.21}$$

We first consider y_p. The net force F is

$$F = \int p\, dA. \tag{2.22}$$

Substituting Eq. (2.22) into Eq. (2.20) gives

$$y_p = \frac{\int yp\, dA}{\int p\, dA}. \tag{2.23}$$

The pressure varies with y according to

$$p = \gamma y \sin\theta, \tag{2.24}$$

so

$$y_p = \frac{\int \gamma \sin\theta\, y^2\, dA}{\int \gamma \sin\theta\, y\, dA}.$$

As γ and $\sin\theta$ are constants, we cancel them and find that

$$y_p = \frac{\int y^2\, dA}{\int y\, dA} = \frac{I_x}{y_c A}. \tag{2.25}$$

I_x is the moment of inertia of the area about the x axis,

$$I_x = \int y^2\, dA.$$

Note that the x axis lies on the free surface. It is usually more convenient to express y_p in terms of the moment of inertia of the area about its centroidal axis. The parallel axis theorem for moments of inertia is

$$I_x = I_{xc} + y_c^2 A,$$

where I_{xc} is the moment of inertia of the area about an axis parallel to the x axis and passing through the centroid of the area. Substituting into Eq. (2.25) gives

$$y_p - y_c = \frac{I_{xc}}{y_c A}. \tag{2.26}$$

The center of pressure lies *below* the centroid a distance equal to $I_{xc}/y_c A$. Because I_{xc} and A are constant for a given area, the center of pressure moves closer to the centroid as the area moves deeper. Remember that we measure y_p from the free surface in the plane of the area.

We may calculate x_p by similar steps. From Eqs. (2.21) and (2.22), we have

$$x_p = \int \frac{x(\gamma y \sin\theta)\, dA}{\gamma y_c \sin\theta A}.$$

As γ and θ are constant,

$$x_p = \int \frac{xy\, dA}{y_c A} = \frac{I_{xy}}{y_c A}, \tag{2.27}$$

where I_{xy} is the product of inertia. Using the parallel axis theorem for the product of inertia results in

$$x_p - x_c = \frac{I_{xyc}}{y_c A}.$$

An easy way to compute x_p is to set the xy-coordinate system so that the y axis passes through the centroid of the area and $x = 0$ at the centroid. Then we can use the product of inertia about the centroid in Eq. (2.27). If the area is symmetric about either axis, the centroidal product of inertia is zero, and x_p coincides with the centroid. Moments of inertia of several plane shapes are tabulated in Appendix B.

EXAMPLE 2.5 Illustrates How the Center of Pressure Changes with Depth and How to Calculate Moments Due to Hydrostatic Pressure

Determine the locations of the center of pressure and the moments of the hydrostatic forces about the base of the semicircular observation window in Example 2.4 if the top of the window is 10 m, 100 m, and 10 km below the surface of the sea.

SOLUTION

Given

Inclined semicircular window with 2.0-m diameter making an angle of 45° with horizontal

Top of window 10 m, 100 m, and 10 km below sea level

Figure E2.5

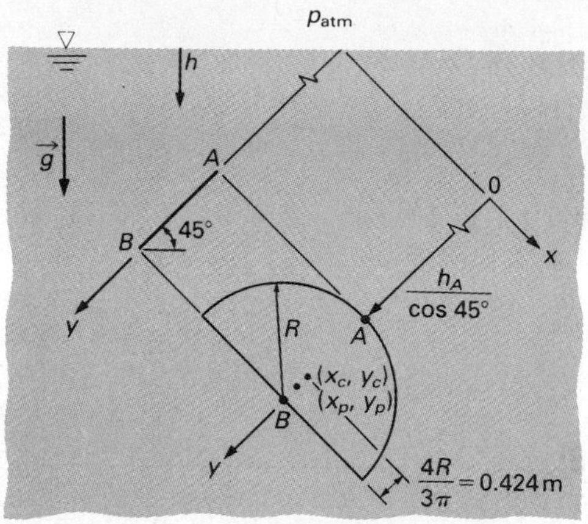

Figure E2.5 Centroid and center of pressure for a semicircular window.

Find

Centers of pressure and moments of hydrostatic forces about base of window

Solution

We first locate the centers of pressure (x_p, y_p). The y coordinate of the center of pressure is located relative to the centroid of the window, so we want to find $y_p - y_c$. Using Fig. E2.5, we obtain

$$y_c = \frac{h_A}{\cos 45°} + \left(R - \frac{4R}{3\pi}\right)$$

$$= 1.414h_A + 1.0 \text{ m}\left(1 - \frac{4}{3\pi}\right) = 1.414h_A + 0.576 \text{ m},$$

where h_A is in meters. For $h_A = 10$ m, using Eq. (2.26) and Table B.1 in Appendix B gives

$$y_c = 1.414(10 \text{ m}) + 0.576 \text{ m} = 14.7 \text{ m};$$

$$y_p - y_c = \frac{I_{xc}}{y_c A} = \frac{0.110R^4}{y_c(\pi R^2/2)} = \frac{0.110(1.0 \text{ m})^4}{(14.7 \text{ m})(\pi/2)(1.0 \text{ m})^2}, \quad \text{or}$$

$$y_p - y_c = 0.480 \text{ cm}.$$ **ANSWER**

For $h_A = 100$ m,

$$y_c = 1.414(100 \text{ m}) + 0.576 \text{ m} = 142 \text{ m};$$

$$y_p - y_c = \frac{I_{xc}}{y_c A} = \frac{0.110(1.0 \text{ m})^4}{(142 \text{ m})(\pi/2)(1.0 \text{ m})^2}, \quad \text{or}$$

$$y_p - y_c = 0.0490 \text{ cm}.$$ **ANSWER**

For $h_A = 10$ km,

$$y_c = 1.414(10{,}000 \text{ m}) + 0.576 \text{ m} = 14{,}140 \text{ m};$$

$$y_p - y_c = \frac{I_{xc}}{y_c A} = \frac{0.110(1.0 \text{ m})^4}{(14{,}140 \text{ m})(\pi/2)(1.0 \text{ m})^2}, \quad \text{or}$$

$$y_p - y_c = 0.00050 \text{ cm}.$$ **ANSWER**

The symmetry of the area and of the hydrostatic pressure about a radius in the vertical plane places the x coordinate of the center of pressure on the radius in the vertical plane (radius BA), so

$$x_p - x_c = \frac{I_{xyc}}{y_c A} = 0.$$ **ANSWER**

The moment of each resultant force about the base of the semicircular area is

$$\mathcal{M} = F\left[\frac{4R}{3\pi} - (y_p - y_c)\right]$$

$$= F\left[\frac{4(1.0 \text{ m})}{3\pi} - (y_p - y_c)\right]$$

$$= F[0.424 \text{ m} - (y_p - y_c)].$$

We use Eq. (2.18) to find the force F:

$$F = h_c A = \gamma y_c \sin 45^\circ\, A$$

$$= (10{,}080\ \text{N/m}^3)(y_c)\left(0.707\,\frac{\pi}{2}\right)(1.0\ \text{m})^2 = (11{,}190\ \text{N/m})\, y_c,$$

where y_c is in meters. For $h_A = 10$ m,

$$F = (11{,}190\ \text{N/m})(14.7\ \text{m}) = 1.65 \times 10^5\ \text{N}$$

and

$$\mathcal{M} = (1.65 \times 10^5\ \text{N})(0.424\ \text{m} - 0.0048\ \text{m}),\quad \text{or}$$

$$\mathcal{M} = 6.97 \times 10^4\ \text{N}\cdot\text{m}.$$ **ANSWER**

For $h_A = 100$ m,

$$F = (11{,}190\ \text{N/m})(142.2\ \text{m}) = 1.59 \times 10^6\ \text{N}$$

and

$$\mathcal{M} = (1.59 \times 10^6\ \text{N})(0.424\ \text{m} - 0.000490\ \text{m}),\quad \text{or}$$

$$\mathcal{M} = 6.73 \times 10^5\ \text{N}\cdot\text{m}.$$ **ANSWER**

For $h_A = 10$ km,

$$F = (11{,}190\ \text{N/m})(14{,}140\ \text{m}) = 1.58 \times 10^8\ \text{N}$$

and

$$\mathcal{M} = (1.58 \times 10^8\ \text{N})(0.424\ \text{m} - 0.0000050\ \text{m}),\quad \text{or}$$

$$\mathcal{M} = 6.699 \times 10^7\ \text{N}\cdot\text{m}.$$ **ANSWER**

Discussion

This example illustrates that the center of pressure moves closer to the centroid as a surface moves deeper; however, the center of pressure is always below the centroid. This example also illustrates a case where the area is symmetric with respect to the y axis and x_p coincides with x_c. Finally, it illustrates how we use the resultant force to calculate moments.

Note that the formulas developed in this section (particularly Eqs. 2.18, 2.26, and 2.27) apply only to forces caused by hydrostatic pressure on plane surfaces. However, the general principles of force and moment equality are applicable to any type of pressure distribution.

EXAMPLE 2.6 Illustrates How Resultant Forces are Related to Pressure Distribution for Other Than Hydrostatic Cases

One of the twin towers of the World Trade Center in New York City is 110 stories high. It measures 241 ft square and 1350 ft high. The pressure distribution on the windward side of this tower from a 15-mph wind is approximated by the equation

$$p = Az^4 + Bz,$$

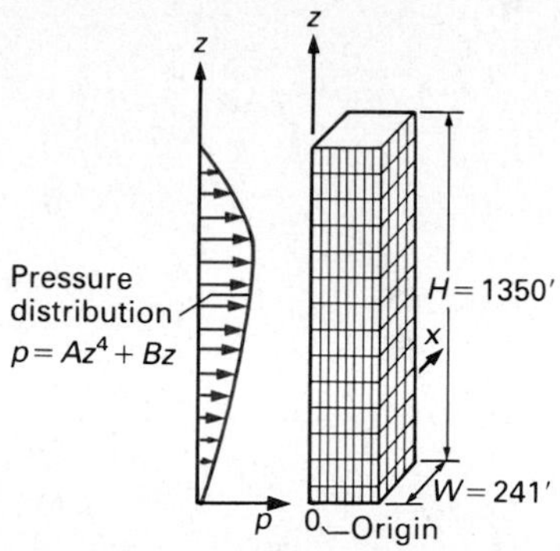

Figure E2.6a Pressure distribution on the windward side of a World Trade Center tower.

where

$A = -2.26 \times 10^{-11}$ lb/ft^6,

$B = 0.0556$ lb/ft^3, and

z is measured in feet above the ground.

Find the magnitude and location of the resultant wind force on the windward side of the building.

SOLUTION

Given

World Trade Center tower (see Fig. E2.6a)

Pressure distribution $p = Az^4 + Bz$

Find

Magnitude and location of resultant wind force

Solution

Figure E2.6b shows the resultant wind force on the building. If we assume that the pressure is uniform across the width, W, of the building, Eq. (2.13) gives the force as

$$F = \int p\, dA = \int_0^H pW\, dz.$$

Substituting for p gives

$$F = W \int_0^H (Az^4 + Bz)\, dz.$$

Integrating, we obtain

$$F = W\left(\frac{Az^5}{5} + \frac{Bz^2}{2}\right)_0^H$$

$$= WH^2\left(\frac{AH^3}{5} + \frac{B}{2}\right).$$

The numerical values give

$$F = (241 \text{ ft})(1350 \text{ ft})^2$$
$$\times \left[\frac{(-2.26 \times 10^{-11} \text{ lb/ft}^6)(1350 \text{ ft})^3}{5} + \frac{0.0556 \text{ lb/ft}^3}{2}\right];$$

$$F = 7.33 \times 10^6 \text{ lb}.$$ **ANSWER**

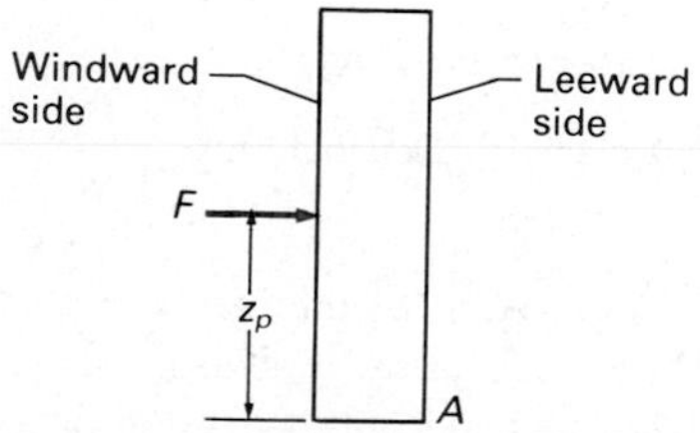

Figure E2.6b Wind force involved on a World Trade Center tower.

We find the vertical location of the resultant force by taking moments about the base of the building. Using Eq. (2.20), we have

$$Fz_p = \int zp\, dA = \int_0^H zpW\, dz.$$

Solving for z_p, we obtain

$$z_p = \frac{W \int_0^H z(Az^4 + Bz)\, dz}{F}.$$

Integrating gives

$$z_p = \frac{W}{F}\left(\frac{Az^6}{6} + \frac{Bz^3}{3}\right)_0^H = \frac{WH^3}{F}\left(\frac{AH^3}{6} + \frac{B}{3}\right).$$

The numerical values give

$$z_p = \frac{(241\ \text{ft})(1350\ \text{ft})^3}{(7.33 \times 10^6\ \text{lb})} \times \left[\frac{(-2.26 \times 10^{-11}\ \text{lb/ft}^6)(1350\ \text{ft})^3}{6} + \frac{0.0556\ \text{lb/ft}^3}{3}\right];$$

$$z_p = 750\ \text{ft}.$$ **ANSWER**

Because we assumed that the pressure is uniform across the width of the building, the x location of the resultant force is one half the distance from either side:

$$x_p = 120.5\ \text{ft}.$$ **ANSWER**

Discussion

When designing such a building, civil engineers must determine the necessary anchor force required to withstand wind loads. Moments taken about a lower edge of the building are equated to the moments of the anchoring force and the building's weight to find the necessary anchoring forces. In addition to the pressure on the windward side, a vacuum exists on the leeward side, resulting in a larger total wind force on the tower.

Suppose that we tried to determine the resultant force by calculating the pressure at the centroid of the windward side and multiplying this pressure by the area of the windward side. This approach yields

$$\begin{aligned} F &= p_{z=675}A \\ &= [(-2.26 \times 10^{-11}\ \text{lb/ft}^6)(675\ \text{ft})^4 \\ &\quad + (0.0556\ \text{lb/ft}^3)(675\ \text{ft})](241\ \text{ft})(1350\ \text{ft}) \\ &= 10.7 \times 10^6\ \text{lb}, \end{aligned}$$

which has about a 50% error. The reason for the incorrect answer is that the average pressure does not occur at the centroid of the area. The average pressure occurs at the centroid only when the pressure is uniform or increases linearly with distance.

EXAMPLE 2.7 Illustrates How Hydrostatic and Atmospheric Forces are Handled in the Same Problem

In the Apollo spacecraft, a 5.0 lb/in^2, abs, pure oxygen atmosphere was used. A similar atmosphere is proposed for use in the bathyscaph of Example 2.4. As part of the analysis to determine the structural integrity of the bathyscaph, we must determine both the magnitude and the location of the net force on hull's semicircular window. We consider the case where the top of the window is 100 m below the surface of the water. The atmospheric pressure above the water surface is 100 kPa.

SOLUTION

Given

Inclined semicircular window making an angle of 45° with the horizontal

Top of window 100 m below water surface

Atmospheric pressure above free surface of 100 kPa

Pressure inside bathyscaph of 5.0 psia

Find

Magnitude and location of net force acting on window

Solution

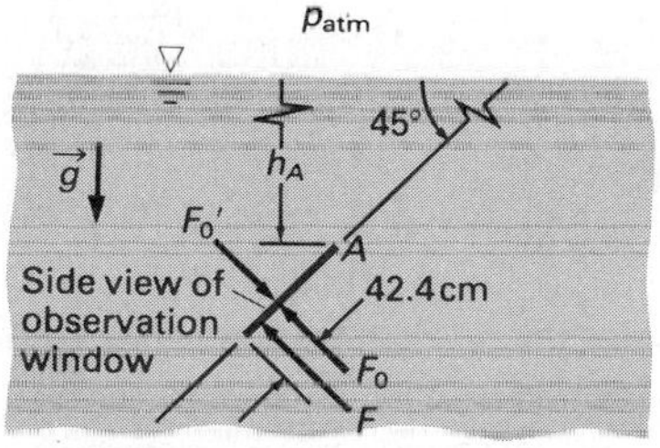

Figure E2.7a Forces on the bathyscaph observation window.

The three forces acting on the window are shown in Fig. E2.7a: The force F_0 exerted by the atmospheric pressure p_{atm} above the free surface acts on the outside at the centroid of the window; the force F_0' exerted by the atmospheric pressure p_0 inside the bathyscaph acts on the inside at the centroid of the window; and the force F exerted by the hydrostatic pressure acts on the outside of the hull and at the center of pressure. The net uniform pressure force is

$$F_0 - F_0' = p_{atm}A - p_0 A = (p_{atm} - p_0)A,$$

where

$$p_0 = (5.0 \text{ lb/in}^2)(6.895 \text{ kPa/lb/in}^2) = 34.5 \text{ kPa}.$$

Then

$$F_0 - F_0' = [(100 - 34.5)] \text{ kPa} \left(\frac{\pi}{2}\right)(1.0 \text{ m})^2(1000 \text{ N/m}^2\text{/kPa})$$
$$= 1.03 \times 10^5 \text{ N}.$$

The force $(F_0 - F_0')$ acts at the centriod of the window.

Using Fig. E2.7b and equating moments about a horizontal axis passing through the centroid to the moment of the resultant force F_R, we obtain*

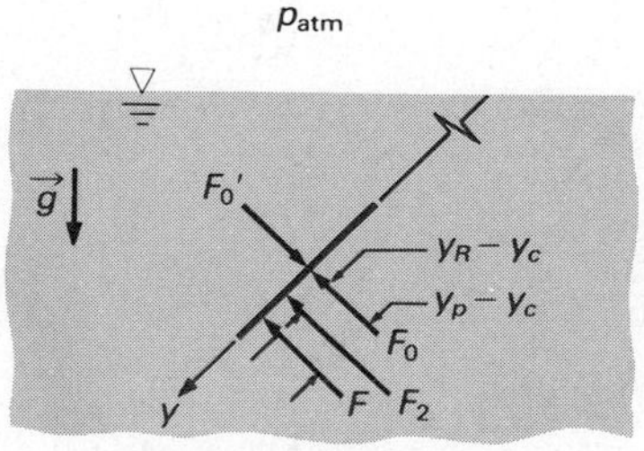

Figure E2.7b Forces and locations on the bathyscaph observation window.

$$F_R(y_R - y_c) = F(y_p - y_c).$$

Using

$$F_R = F + (F_0 - F_0')$$

and solving for $(y_R - y_c)$, we have

$$(y_R - y_c) = \frac{F(y_p - y_c)}{F + (F_0 - F_0')}.$$

* Note that F_0 and F_0' have no moment about the centroid.

The magnitudes of F and $(y_p - y_c)$ come from Example 2.5. For $h_A = 100$ m,

$$F = 1.59 \times 10^6 \text{ N};$$

$$y_p - y_c = 4.90 \times 10^{-4} \text{ m};$$

$$F_R = 1.59 \times 10^6 \text{ N} + 1.03 \times 10^5 \text{ N}, \quad \text{or}$$

$$F_R = 1.69 \times 10^6 \text{ N}.$$ **ANSWER**

$$y_R - y_c = \frac{(1.59 \times 10^6 \text{ N})(4.90 \times 10^{-4} \text{ m})}{1.69 \times 10^6 \text{ N}}, \quad \text{or}$$

$$y_R - y_c = 4.61 \times 10^{-4} \text{ m}.$$ **ANSWER**

The pressure is uniform along a horizontal element of the window and the semicircular window is symmetric about a radius in a vertical plane passing through the centroid. Thus we conclude that the x location, x_R, of the resultant force F_R is along the radius of symmetry, or

$$x_R = x_c.$$ **ANSWER**

Discussion

We took moments about the centroidal axis to obtain the y location of the resultant force as $y_R - y_c$. We could have used any other parallel horizontal axis; however, the force $(F_0 - F_0')$ would then have to be included.

2.4.2 Forces on Curved Surfaces

Next we consider a curved surface with hydrostatic pressure acting on it, as shown in Fig. 2.19. For simplicity, we use a two-dimensional curved surface, of which Fig. 2.19 shows only an edge view. Note that y is now a vertical coordinate.

As pressure always acts normal to area, the forces on each infinitesimal area have different directions. The force on each element of area may be broken into horizontal and vertical components. We can sum (integrate) the horizontal and vertical components separately.

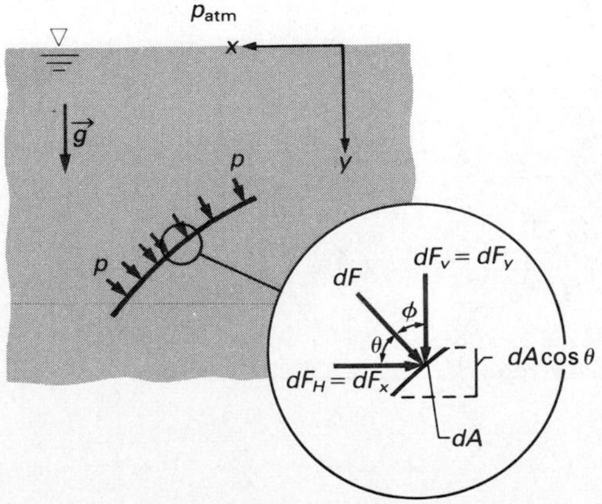

Figure 2.19 Curved surface in a liquid. Force on element of surface is broken into horizontal and vertical components.

First we consider the horizontal (x) component:

$$dF_x = dF \cos\theta = p\,dA \cos\theta.$$

But $dA \cos\theta$ is dA_x, the horizontal projection of dA, so we can write

$$F_x = \int p\,dA_x.$$

We can calculate the horizontal component of force on a curved area by calculating the force on the horizontal projection of the area:

$$F_x = \gamma h_{c,x} A_x. \tag{2.28}$$

By the same reasoning, we conclude that the line of action of the horizontal force passes through the center of pressure of the projected area (see Fig. 2.20a).

Next we consider the vertical component of a pressure force on a curved surface. The elemental vertical force is

$$dF_v = p\,dA \cos\phi.$$

For a surface submerged in a liquid with a free surface,

$$p = \gamma h, \quad \text{so} \quad F_v = \int dF_v = \int \gamma h \cos\phi\,dA.$$

Now $\int h \cos\phi\,dA$ is $\forall'$, the volume between the curved surface and the free surface (see Fig. 2.20b), so the vertical force is equal to the weight of fluid above the surface, or

$$F_v = \gamma \forall'. \tag{2.29}$$

The line of action of the vertical component of the force is through the centroid of the volume $\forall'$.

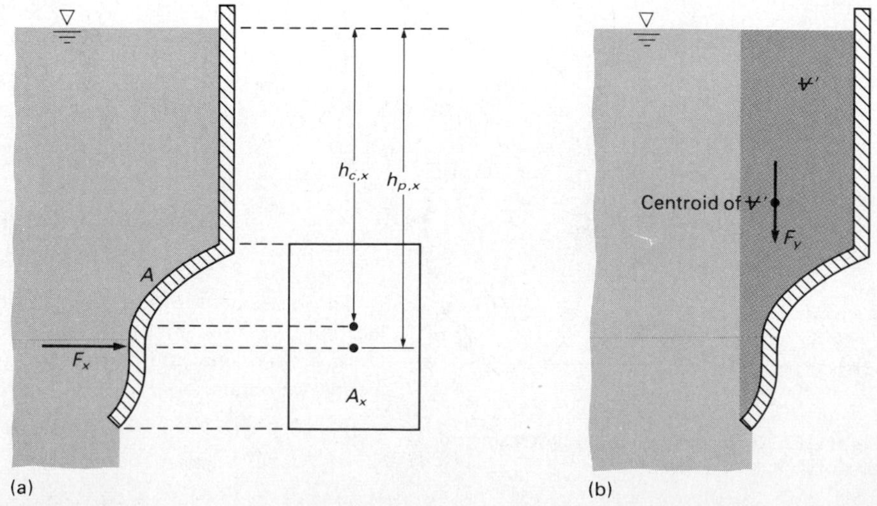

Figure 2.20 Curved surface: (a) horizontal projection of area; (b) volume above curved area.

EXAMPLE 2.8 Illustrates the Calculation of Forces on a Curved Surface

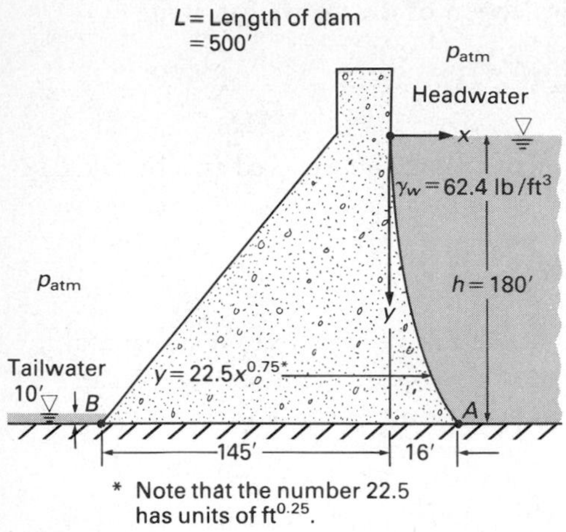

Figure E2.8a Solid gravity dam.

A gravity dam is designed so that the moment of the water forces tending to topple the dam about the toe (point B in Fig. E2.8a) is resisted primarily by the moment of the dam's weight. The concrete dam is 190 ft high and 500 ft long. It maintains a headwater height of 180 ft and a tailwater height of 10 ft. Find the magnitudes and locations of the lines of action of the horizontal and vertical forces acting on the upstream side of the dam.

SOLUTION

Given

Dam shown in Fig. E2.8a

Find

Forces F_x and F_y and their locations y_p and x_p, respectively, as shown in Fig. E2.8b

Solution

We find the horizontal force F_x from Eq. (2.28):

$$F_x = \gamma_w h_{c,x} A_x = (62.4\ \text{lb/ft}^3)(90\ \text{ft})(180\ \text{ft} \times 500\ \text{ft});$$

$$F_x = 5.05 \times 10^8\ \text{lb}.$$ **ANSWER**

We find the vertical force F_y from Eq. (2.29):

$$F_y = \gamma_w \mathcal{V}'.$$

We must obtain the volume ($\mathcal{V}'$) by integration. Referring to Fig. E2.8b, we get

$$F_y = \int dF_y = \gamma_w \int d\mathcal{V}' = \gamma_w L \int_0^{16} y\,dx,$$

where $L = 500$ ft. Substituting

$$y = 22.5x^{0.75}$$

gives

$$F_y = \gamma_w L \int_0^{16} 22.5x^{0.75}\,dx$$

$$= 22.5\gamma_w L\left(\frac{x^{1.75}}{1.75}\right)_0^{16} = 1646\ \text{ft}^2\ \gamma_w L$$

$$= (1646\ \text{ft}^2)(62.4\ \text{lb/ft}^3)(500\ \text{ft});$$

$$F_y = 5.13 \times 10^7\ \text{lb}.$$ **ANSWER**

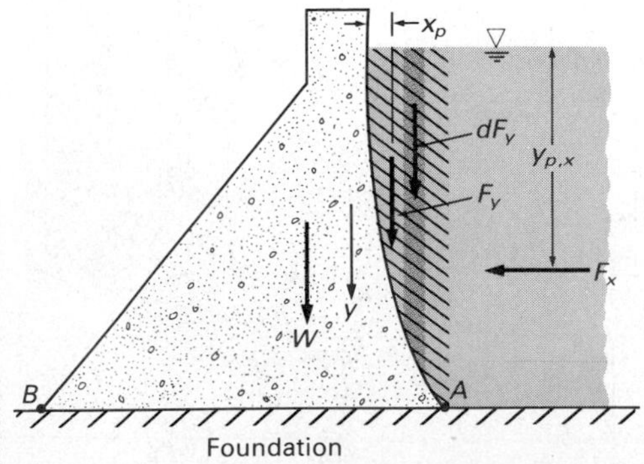

Figure E2.8b Forces acting on upstream side of dam.

The location $y_{p,x}$ of F_x is found in the same manner as the vertical location of a horizontal force on a vertical plane measuring $h \times L$. Using a z axis into the paper and Appendix B for the moment of inertia, we obtain

$$y_{p,x} = \frac{I_z}{y_c A} = \frac{\frac{1}{3}Lh^3}{h/2\ (hL)} = \frac{2h}{3} = \frac{2(180\ \text{ft})}{3};$$

$$y_{p,x} = 120 \text{ ft.}$$ **ANSWER**

The location of F_y is the centroid of the volume $\forall'$. We find the x coordinate of the centroid from

$$x_p = \frac{\int x\, d\forall'}{\int d\forall'} = \frac{\int_0^{16} x(Ly\, dx)}{\int_0^{16} Ly\, dx}.$$

Substituting $y = 22.5x^{0.75}$ gives

$$x_p = \frac{\int_0^{16} L(22.5x^{1.75})\, dx}{\int_0^{16} L(22.5x^{0.75})\, dx}.$$

Canceling constants and integrating, we obtain

$$x_p = \frac{1.75}{2.75}\left[\frac{(16 \text{ ft})^{2.75}}{(16 \text{ ft})^{1.75}}\right];$$

$$x_p = 10.2 \text{ ft.}$$ **ANSWER**

Discussion

Note that we did not find the z coordinates (into the paper) of F_x and F_y. Both are located halfway across the dam.

The magnitude of the resultant hydrostatic force F on the upstream side of the dam is

$$F = \sqrt{F_x^2 + F_y^2} = \sqrt{(5.05 \times 10^8 \text{ lb})^2 + (5.13 \times 10^7 \text{ lb})^2}$$
$$= 5.08 \times 10^8 \text{ lb}.$$

The line of action of F passes through the point

$$(x_p, y_p) = (10.2 \text{ ft}, 120 \text{ ft})$$

and makes an angle θ with the horizontal, where

$$\theta = \tan^{-1}\frac{F_y}{F_x}$$
$$= \tan^{-1}\frac{5.13 \times 10^7}{5.05 \times 10^8} = 5.8°.$$

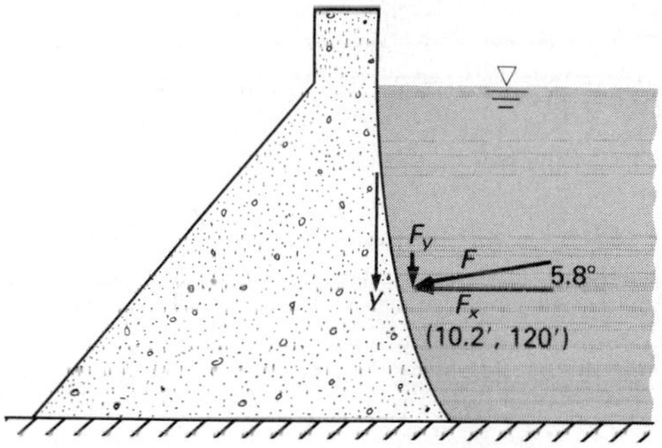

Figure E2.8c Resultant of upstream hydrostatic forces on dam.

Figure E2.8c shows these results. Note that the point (10.2, 120) is not on the surface of the dam. Note also that the constant 22.5 has units of $\text{ft}^{0.25}$ and that proper use of terms with fractional powers of units produced the correct units of force or distance.

We find pressure forces on three-dimensional curved surfaces the same way that we find pressure forces on two-dimensional curved surfaces. There are two horizontal components and one vertical component. Figure 2.21 illustrates the three components of pressure force on a three-dimensional curved surface.

In all our discussions and examples so far, the liquid has been above the curved surface. However, there may be no liquid directly above the curved surface, as in Fig. 2.22. We still calculate the pressure at each point on the curved surface from the specific weight of the liquid and the depth below the free surface; that is, the pressure is the same as if there were liquid above the surface. We calculate the force by integrating the pressure over the area. The net vertical force still has

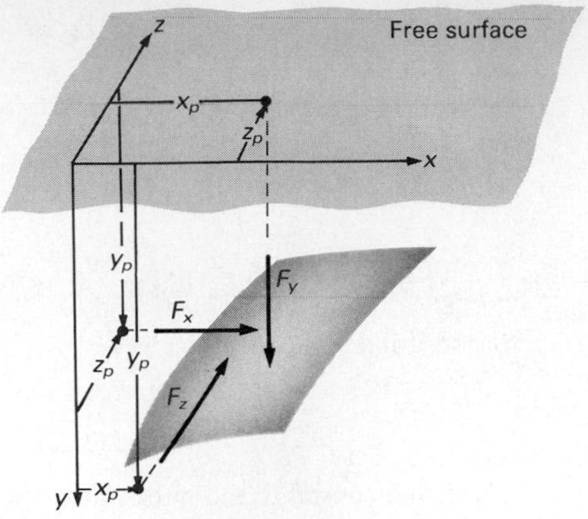

Figure 2.21 Components of a force on a three-dimensional curved surface. [*Note:* x_p for F_y may not equal x_p for F_z, etc.]

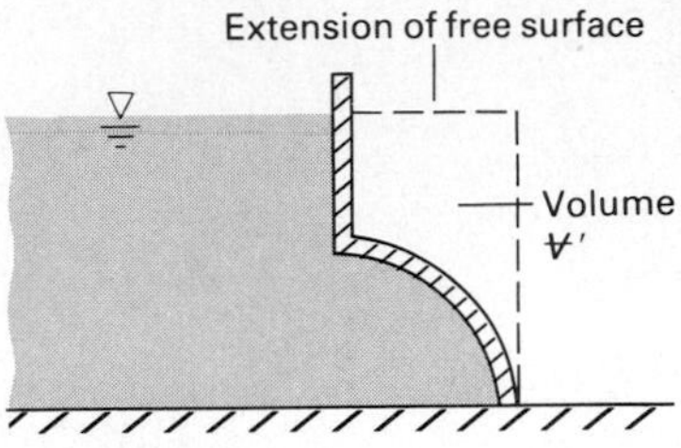

Figure 2.22 Curved surface with no liquid above it.

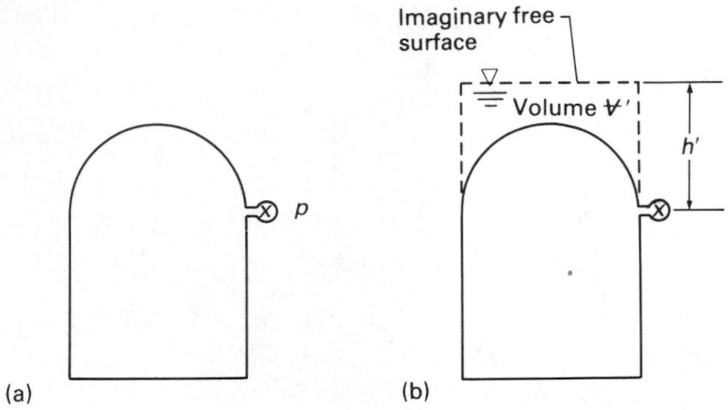

Figure 2.23 (a) Tank filled with pressurized fluid; (b) imaginary free surface above curved tank top.

a magnitude equal to the weight of liquid that would be contained in the volume $\forall'$ between the curved surface and an extension of the free surface, so we can still use Eq. (2.29). The only difference is that the vertical force will act *upward* rather than downward. You should always remember that pressure force acts *from* the liquid *toward* the contacting surface.

We can even use this approach to solve a problem involving no free surface at all, as with the pressurized container shown in Fig. 2.23(a). If we know the pressure at some level in the container, we can construct an imaginary free surface above this point (Fig. 2.23b). This "free surface" lies above the point of known pressure a distance h', where

$$h' = \frac{p}{\gamma}.$$

The net pressure force (from the fluid inside the tank) is equal to $\gamma\forall'$.* Note that this force is upward.

* Does γ have to be the specific weight of the fluid in the tank? What if h' does not lie above the top of the tank?

2.5 MECHANICS OF SUBMERGED AND FLOATING BODIES

2.5.1 Buoyancy

When a body is either fully or partially submerged in a fluid, a net upward force called the *buoyant force* acts on the body. This force is caused by the imbalance of pressures on the upper and lower surfaces.

Figure 2.24 shows a body submerged in a fluid. We want to calculate the net vertical force that pressure exerts on the body. The body is closed, so each "lower" area dA_1 has a corresponding "upper" area dA_2. The net pressure force on the two areas is

$$dF = p_1\, dA_{1,y} - p_2\, dA_{2,y}.$$

The vertical projections of dA_1 and dA_2 are equal, so

$$dF = (p_1 - p_2)\, dA_y = \gamma \ell\, dA_y.$$

Integrating, we obtain

$$F = \int \gamma \ell\, dA_y = \gamma \forall. \tag{2.30}$$

Equation (2.30) shows that:

> The buoyant force on a body submerged in a fluid is equal to the weight of the fluid displaced by the body.

We can also show that:

> The line of action of the buoyant force passes through the centroid of the displaced volume. This centroid is called the *center of buoyancy.*

These principles also apply if the body is only partly submerged. However, you must be careful to consider only the part of the body that is in the fluid.

This information about the magnitude and line of action of buoyant forces is called *Archimedes' first principle of buoyancy*, because it apparently was discovered by Archimedes in 220 B.C.

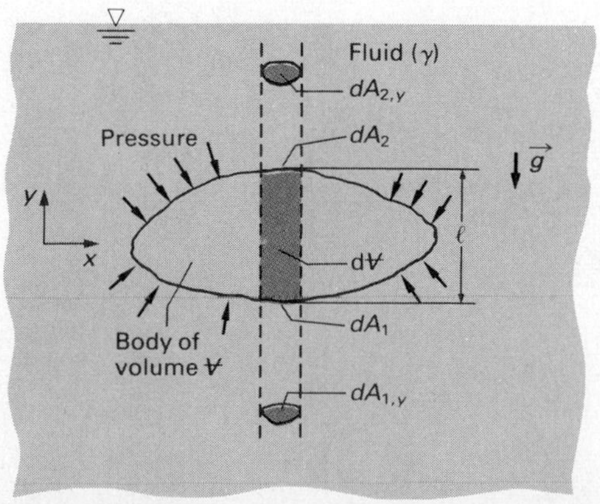

Figure 2.24 Arbitrary closed body submerged in fluid.

A body will float if its average density is less than the density of the fluid in which it is placed. For a floating body (see Fig. 2.25), the weight is

$$W = \gamma_f \forall_s, \tag{2.31}$$

where γ_f is the specific weight of the fluid and $\forall_s$ is the submerged volume. That is:

A floating body displaces a volume of fluid equivalent to its own weight.

This is *Archimedes' second principle of buoyancy*. When solving problems involving closed bodies in fluids, we can always replace the action of hydrostatic pressure by the buoyant force.

An application of the principles of buoyancy is the use of a *hydrometer* to measure the specific gravity of a liquid. Figure 2.26 illustrates a hydrometer floating in pure water ($S = 1.0$) and in another liquid ($S > 1.0$). If $S > 1.0$, the buoyant force per unit of submerged volume is greater, so less volume of submergence is necessary to balance the hydrometer's weight. In water,

$$W = \gamma_{H_2O} \forall_0, \tag{2.32}$$

but in the other fluid,

$$W = S\gamma_{H_2O}(\forall_0 - Ah).$$

Therefore

$$h = \frac{\forall_0}{A}\left(1 - \frac{1}{S}\right). \tag{2.33}$$

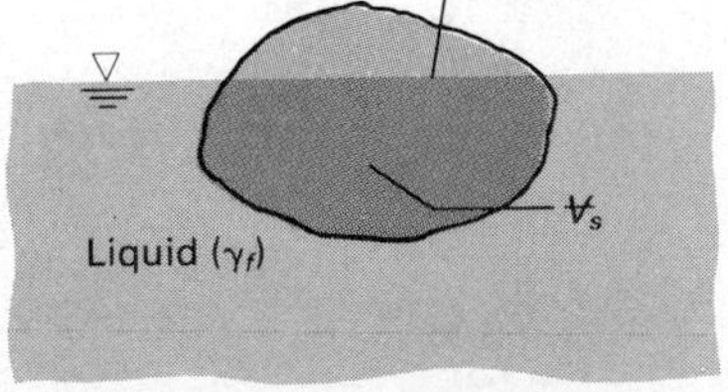

Figure 2.25 Floating body.

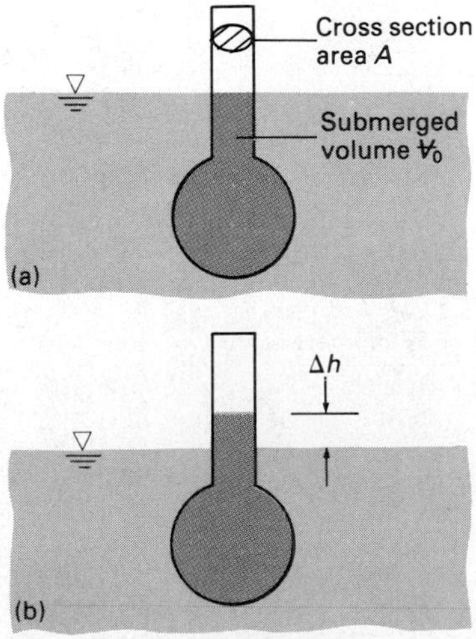

Figure 2.26 Hydrometer in two liquids: (a) water, $S = 1.0$; (b) other liquid, $S > 1.0$.

EXAMPLE 2.9 Illustrates Archimedes' Principles for a Floating Body

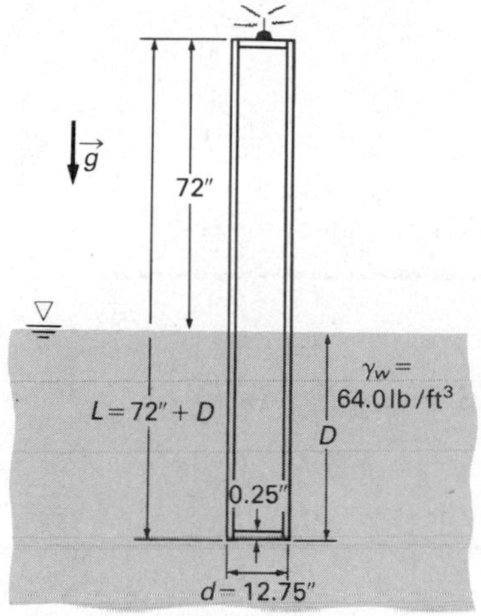

Figure E2.9 Spar buoy.

A *spar buoy* does not heave and roll when hit by ocean waves as much as other buoys do, and it is well suited for use in scientific experimental stations. Figure E2.9 shows a particular type of spar buoy, with a small radio sensor mounted on top. The buoy has a circular cross section and is supposed to float with 6 ft of its length protruding above the water. The other dimensions of the buoy are shown in the figure. Calculate the total height of the buoy if it is made from (a) aluminum and (b) steel. The specific weights of sea water, aluminum, and steel are 64 lb/ft³, 169 lb/ft³, and 490 lb/ft³, respectively.

SOLUTION

Given

Spar buoy, OD = 12.75 in., ID = 12.25 in., to float 6 ft out of water, constructed of (a) aluminum and (b) steel

Find

Total height of each buoy

Solution

The primary equation is Eq. (2.31), which states that the weight W of the floating buoy must equal the buoyant force $F_B(=\gamma_w \forall_s)$. The weight W of the buoy is

$$W = \gamma_B(\forall_{\text{sides}} + \forall_{\text{end plates}})$$

$$= \gamma_B\left[\frac{\pi}{4}[(12.75)^2 - (12.25)^2]L + 2\frac{\pi}{4}(12.25)^2(0.25)\right]\text{in}^3$$

$$\times\left(\frac{1\ \text{ft}^3}{1728\ \text{in}^3}\right)$$

$$= 0.00568\gamma_B L + 0.0341\gamma_B,$$

where γ_B is in lb/ft³ and L is in in., giving W in lb. The buoyant force F_B is

$$F_B = \gamma_w \forall_s = (64\ \text{lb/ft}^3)\left[\frac{\pi}{4}(12.75)^2(L-72)\right]\text{in}^3 \times \left(\frac{1\ \text{ft}^3}{1728\ \text{in}^3}\right)$$

$$= 4.73\,L - 340,$$

where L is in inches. Equating W and F_B, we have

$$0.00568\gamma_B L + 0.0341\gamma_B = 4.73L - 340,$$

or

$$L = \frac{340 + 0.0341\gamma_B}{4.73 - 0.00568\gamma_B}.$$

For aluminum,

$$L = \frac{340 + 0.0341(169)}{4.73 - 0.00568(169)};$$

$$L = 91.7\ \text{in.}$$ **ANSWER**

For steel,

$$L = \frac{340 + 0.0341(490)}{4.73 - 0.00568(490)};$$

$$L = 183 \text{ in.}$$ **ANSWER**

Discussion

Note that the aluminum buoy is submerged a depth of $91.7 - 72 = 19.7$ in. and the steel buoy is submerged a depth of $183 - 72 = 111$ in. We will use this information in a later problem.

2.5.2 Stability of Submerged and Floating Bodies

Buoyant force on a body always acts through the centroid of the displaced volume, whereas the weight always acts through the center of gravity. These characteristics can make a partially or totally submerged body either stable or unstable. An object is in stable equilibrium if a slight displacement generates forces or moments that restore the object to its original position. An object is in unstable equilibrium if a slight displacement generates forces or moments that further displace the object. An object is in neutral equilibrium if displacement generates no forces or moments. Stability is analogous to what happens upon slight displacement of a ball on the three surfaces shown in Fig. 2.27.

A completely or partially submerged (floating) body is in stable equilibrium if its center of gravity (G) lies below its center of buoyancy (B), as illustrated in Fig. 2.28. If the body is rotated, a moment is set up to right the body and return it to its original position with G directly below B. If the center of gravity of a totally submerged body is above the center of buoyancy, the body is in unstable equilibrium, because an unbalancing moment arises when the body is rotated, as illustrated in Fig. 2.29.

If the center of gravity of a floating body is above the center of buoyancy, the body may be stable *or* unstable, because the center of buoyancy shifts as the object is rotated. We can illustrate the reasons for this difference with two floating blocks of wood—a short, wide one (Fig. 2.30a and b) and a tall, narrow one (Fig. 2.30c and d). For both blocks, the center of buoyancy shifts to the right as some of the left side of the block moves above the surface. For the same angle of rotation, the center of buoyancy of the short, wide block moves farther to the right than does the center of buoyancy of the tall, narrow block. The result is that the new center of buoyancy (B') of the short, wide block now lies to the right of the center of gravity, whereas the new center of buoyancy of the tall, narrow block still lies to the left of the center of gravity. The short, wide block has a restoring moment and is stable, and the tall, narrow block has an upsetting moment and is unstable.

Figure 2.27 Ball-on-surface illustration of the concept of stability.

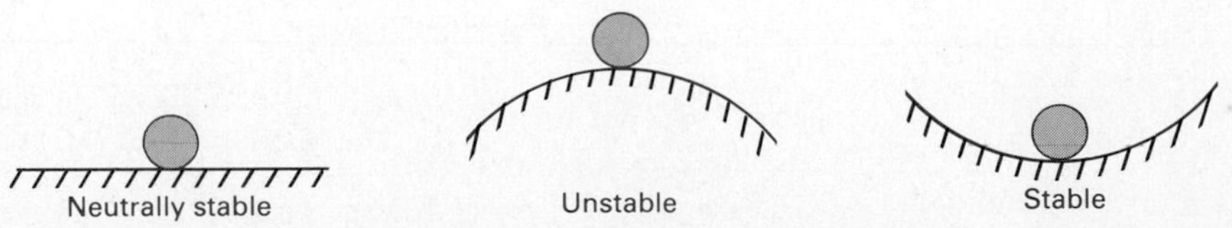

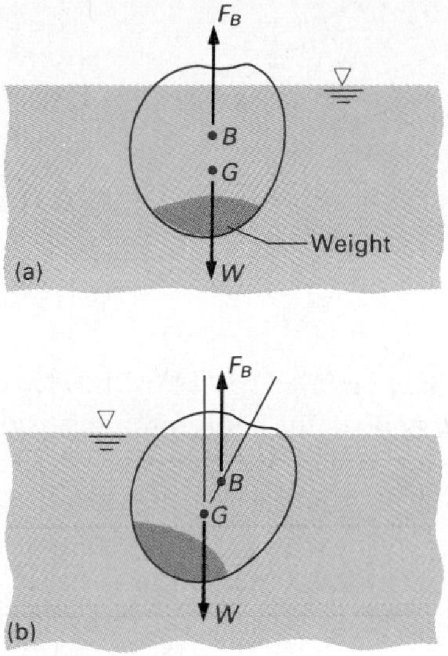

Figure 2.28 Floating body with center of gravity (G) below center of buoyancy (B): (a) equilibrium position; (b) tipped position.

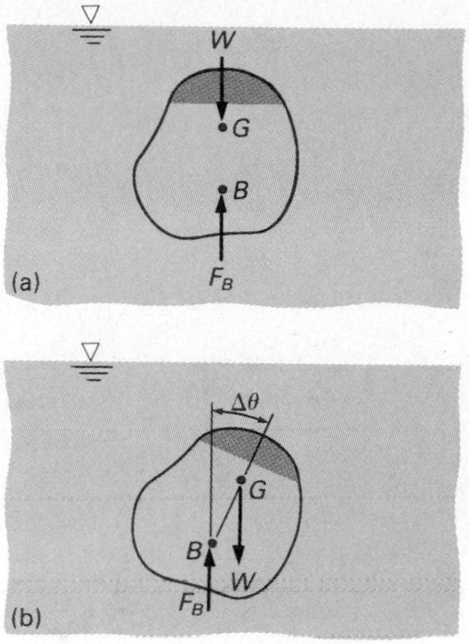

Figure 2.29 Submerged body with center of gravity (G) above center of buoyancy (B): (a) equilibrium position; (b) tipped position.

Figure 2.30 Comparison of stability of short, wide and tall, narrow blocks: (a) and (c) "equilibrium" position; (b) and (d) tipped position.

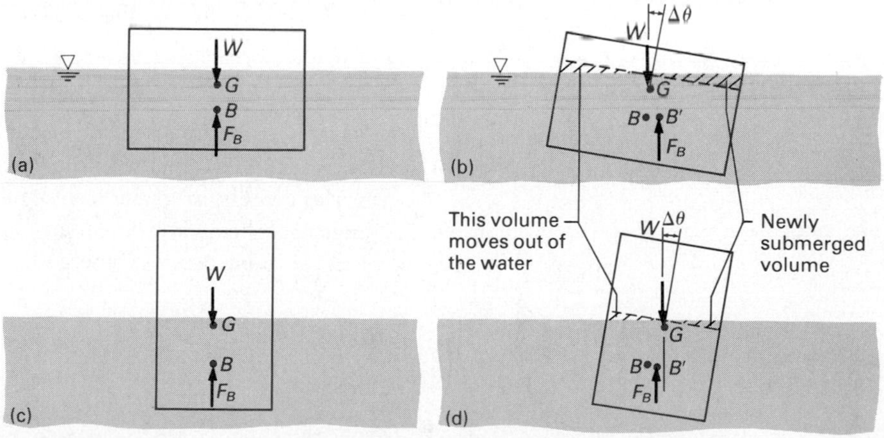

Engineers involved in the design of floating objects such as ships and barges must evaluate their stability. We can show how this is done for a simple situation with small angles of disturbance and no extraneous forces such as wind, using a ship floating in equilibrium (Fig. 2.31a). The center of gravity G lies above the center of buoyancy B. The weight acts downward through G, and the buoyant force acts upward through B.

Now suppose that the ship tips to the right by small angle $\Delta\theta$, the *angle of heel*. The center of gravity retains its position, but the center of buoyancy shifts

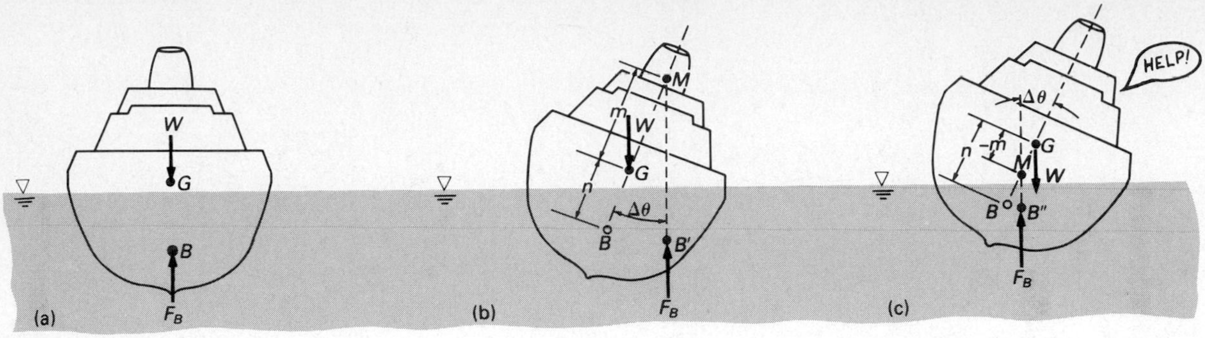

Figure 2.31 Ship with center of gravity (G) above the center of buoyancy (B): (a) equilibrium position; (b) tipped position, stable; (c) tipped position, unstable.

to B'. A new point, M, shown in Fig. 2.31(b), is the *metacenter*, which is the intersection of the line of action of the buoyant force and the line connecting the previous center of buoyancy B with the center of gravity.

Recall that the weight, W, and the buoyant force, F_B, are equal. These forces form a couple on the tipped ship. In Fig. 2.31(b), when the metacenter lies above G, the couple acts to right the ship and so it is stable. In Fig. 2.31(c), with the metacenter below G, the couple acts to tip the ship and the ship is unstable. Therefore:

> A floating object is stable if the metacenter lies above the center of gravity and unstable if the metacenter lies below the center of gravity.

The *metacentric height*, m, shown in Fig. 2.31, determines the location of the metacenter. The metacentric height is positive for the stable case and negative for the unstable case. We can develop an equation for m by equating the restoring/tipping couple to the change in buoyant moment generated by the change in submerged volume caused by the heel of the ship (see Fig. 2.32). The result is

$$m = I_x/\forall_s - n, \tag{2.34}$$

where I_x is the moment of inertia about the roll axis of the "plan view" area of the ship at the waterline, $\forall_s$ is the submerged volume, and n (shown in Fig. 2.31) is the distance between the centers of gravity and buoyancy. Equation (2.34) is valid only for small angles of heel. For large angles of heel, I_x could change appreciably as more of the ship becomes submerged.

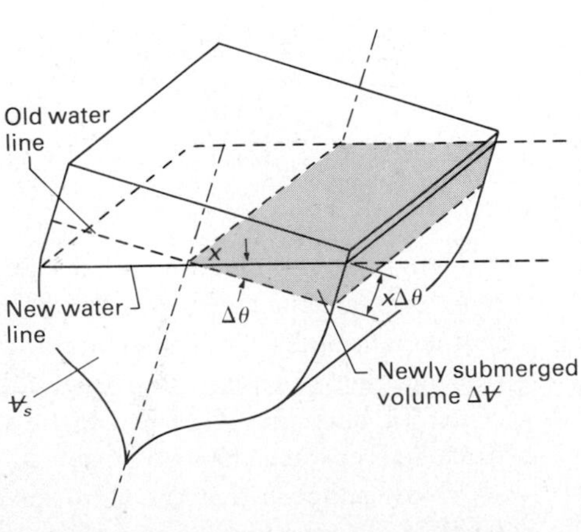

Figure 2.32 Volume ($\Delta\forall$) that becomes submerged when ship tips.

EXAMPLE 2.10 Illustrates How to Determine Whether a Floating Body is Stable

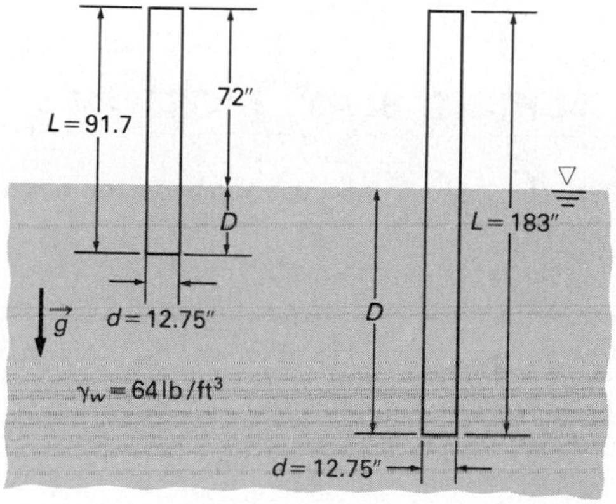

Figure E2.10 Aluminum and steel spar buoys.

Determine whether the steel and aluminum spar buoys sized as in Example 2.9 are stable in an upright position.

SOLUTION

Given

Two spar buoys shown in Fig. E2.10

Find

Whether the buoys are stable

Solution

Stability is determined by the sign of the metacentric height m, which we evaluate by using Eq. (2.34). Moment of inertia I_x for both buoys is that of a circle about its diameter. Using Appendix B, we find that

$$I_x = \frac{\pi d^4}{64} = \frac{\pi}{64}(12.75\text{ in})^4 = 1297\text{ in}^4.$$

To find the values of n, we need the depth of submergence D, which is

$$D = 91.7 - 72 = 19.7\text{ in.} \qquad \text{(for aluminum)};$$

$$D = 183 - 72 = 111\text{ in.} \qquad \text{(for steel)}.$$

Because the center of buoyancy lies at the centroid of the displaced volume and n represents the distance from the center of gravity (at the center of the spar buoy) to the center of buoyancy,

$$n = \frac{L}{2} - \frac{D}{2} = \frac{1}{2}(L - D);$$

$$n = \frac{1}{2}(91.7 - 19.7)\text{ in.} = 36\text{ in.} \qquad \text{(for aluminum)};$$

$$n = \frac{1}{2}(183 - 111)\text{ in.} = 36\text{ in.} \qquad \text{(for steel)}.$$

The volumes displaced are

$$\forall_s = \frac{\pi}{4}d^2 D;$$

$$\forall_s = \frac{\pi}{4}(12.75\text{ in.})^2(19.7\text{ in.}) = 2515\text{ in}^3 \qquad \text{(for aluminum)};$$

$$\forall_s = \frac{\pi}{4}(12.75\text{ in.})^2(111\text{ in.}) = 14{,}200\text{ in}^3 \qquad \text{(for steel)}.$$

Then

$$m = \frac{I_x}{\forall_s} - n = \frac{1297\text{ in}^4}{2515\text{ in}^3} - 36\text{ in.} = -35.5\text{ in.}$$

Aluminum buoy is unstable. **ANSWER**

And

$$m = \frac{1297 \text{ in}^4}{14{,}200 \text{ in}^3} - 36 \text{ in.} = -35.9 \text{ in.}$$

Steel buoy is unstable. **ANSWER**

Discussion

How can these buoys be made stable?

2.6 MECHANICS OF FLUIDS IN RIGID BODY MOTION

Thus far we have considered the mechanics of static fluids. With no motion of fluid relative to other fluid or to solid surfaces, there was no shear deformation and no shear stresses. This holds for another type of situation: "rigid body" motion in which the fluid moves as if it were a solid with no shear deformation.

The pressure at the center of the infinitesimal fluid element shown in Fig. 2.33 is p. The pressures on the z and y faces of the element are as shown. Pressure variation and acceleration in the x direction are omitted for clarity. The fluid element is experiencing a known acceleration $\vec{a}\ (= a_y\hat{j} + a_z\hat{k})$. There are no shear stresses. Applying Newton's second law in the y direction, we have

$$\sum F_y = ma_y.$$

The only y forces are from pressure, so

$$\left[p - \frac{\partial p}{\partial y}\left(\frac{\delta y}{2}\right)\right]\delta x\,\delta z - \left[p + \frac{\partial p}{\partial y}\left(\frac{\delta y}{2}\right)\right]\delta x\,\delta z = \rho(\delta x\,\delta y\,\delta z)a_y.$$

Canceling terms and shrinking the element to a point, we obtain

$$\frac{\partial p}{\partial y} = -\rho a_y. \tag{2.35}$$

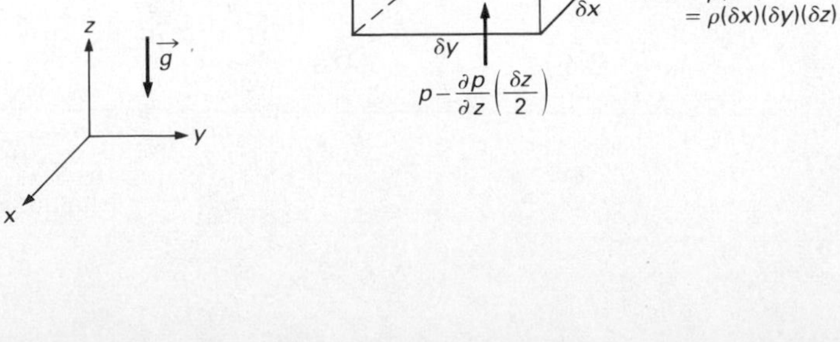

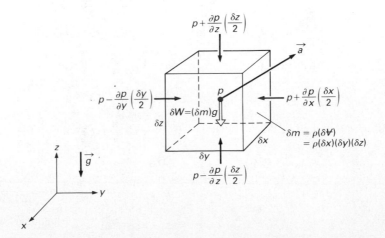

Figure 2.33 Fluid element experiencing a uniform acceleration $\vec{a}(= a_y\hat{j} + a_z\hat{k})$.

In the z direction, we have

$$\sum F_z = ma_z.$$

The z forces include gravity as well as pressure, so

$$\left[p - \frac{\partial p}{\partial z}\left(\frac{\delta z}{2}\right)\right]\delta x\,\delta y - \left[p + \frac{\partial p}{\partial z}\left(\frac{\delta z}{2}\right)\right]\delta x\,\delta y - \gamma\,(\delta x\,\delta y\,\delta z) = \rho(\delta x\,\delta y\,\delta z)a_z.$$

Canceling terms and shrinking the element, we obtain

$$\frac{\partial p}{\partial z} = -\gamma\left(1 + \frac{a_z}{g}\right). \tag{2.36}$$

As the x direction is similar to the y direction, we could include x variations by

$$\frac{\partial p}{\partial x} = -\rho a_x. \tag{2.37}$$

The total differential of pressure is

$$dp = \frac{\partial p}{\partial x}dx + \frac{\partial p}{\partial y}dy + \frac{\partial p}{\partial z}dz.$$

Substituting Eqs. (2.35), (2.36), and (2.37) gives

$$dp = -\rho a_x\,dx - \rho a_y\,dy - \gamma\left(1 + \frac{a_z}{g}\right)dz. \tag{2.38}$$

Integration of Eq. (2.38) yields the pressure distribution.

2.6.1 Uniform Linear Acceleration

One specific case of fluids in rigid body motion is that of a fluid moving in a plane, with a_x and a_z constant and a_y zero. For this case, Eq. (2.38) reduces to

$$dp = -\rho a_x\,dx - \gamma\left(1 + \frac{a_z}{g}\right)dz.$$

With a_x and a_z constant, we can integrate this equation, obtaining

$$p = -\rho a_x x - \gamma\left(1 + \frac{a_z}{g}\right)z + p_0, \tag{2.39}$$

where p_0 is the pressure at $x = z = 0$.

The shape of constant-pressure surfaces is interesting. If p is constant,

$$-\rho a_x x - \gamma\left(1 + \frac{a_z}{g}\right)z = \text{Constant},$$

so surfaces of constant pressure are planes with slope equal to $-a_x/(a_z + g)$.

A particularly simple and interesting case occurs if a_x is constant and a_y and a_z are zero (see Fig. 2.34). In this case, the free surface is a plane with slope $-a_x/g$, and the pressure distribution in the z direction is hydrostatic by Eq. (2.36).

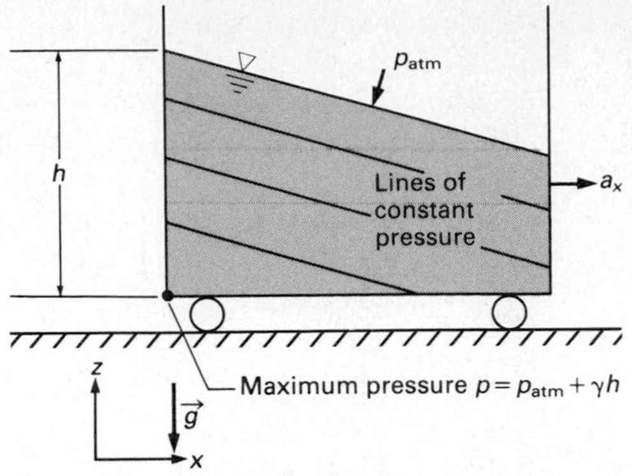

Figure 2.34 Container of liquid with constant acceleration in x direction.

2.6.2 Fluid Rotating about a Vertical Axis

A second specific case of fluids in rigid body motion is that of a container of liquid rotating at constant angular velocity (see Fig. 2.35). If the motion has been going on for some time, the fluid rotates as a rigid body. We could obtain the pressure distribution for this situation by integrating Eq. (2.38); however, it is more convenient to set up the pressure equations in polar coordinates before integrating.

For rigid body rotation, the velocity of each fluid particle is

$$V_r = V_z = 0 \quad \text{and} \quad V_\theta = r\omega.$$

We note from Eq. (2.38) that in rigid body motion the pressure gradient in any direction perpendicular to the gravity vector is proportional to acceleration in that direction. In this case,

$$a_z = a_\theta = 0 \quad \text{and} \quad a_r = -r\omega^2.$$

Thus

$$\frac{\partial p}{\partial \theta} = -\rho a_\theta = 0,$$

$$\frac{\partial p}{\partial z} = -\gamma\left(1 + \frac{a_z}{g}\right) = -\gamma \qquad \text{(hydrostatic pressure),}$$

and

$$\frac{\partial p}{\partial r} = -\rho a_r = \rho r\omega^2.$$

The total differential of pressure is

$$dp = \frac{\partial p}{\partial z}\,dz + \frac{\partial p}{\partial r}\,dr = -\gamma\,dz + \rho\omega^2 r\,dr.$$

Integrating, we obtain

$$p = \frac{\rho\omega^2}{2}r^2 - \gamma z + p_0. \tag{2.40}$$

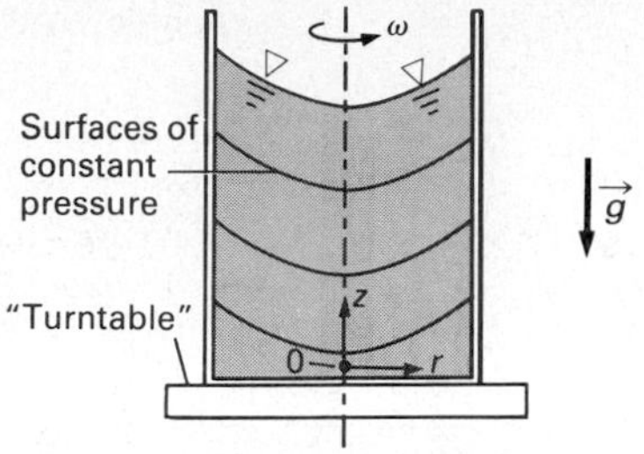

Figure 2.35 Tank of liquid rotating at constant angular velocity (ω).

If p is constant, Eq. (2.40) becomes

$$z = \frac{\omega^2}{2g} r^2 + \text{Constant.} \tag{2.41}$$

Surfaces of constant pressure, including the free surface, are paraboloids, as shown in Fig. 2.35.

EXAMPLE 2.11 Illustrates Rigid Body Motion of a Fluid Rotating at a Constant Angular Velocity

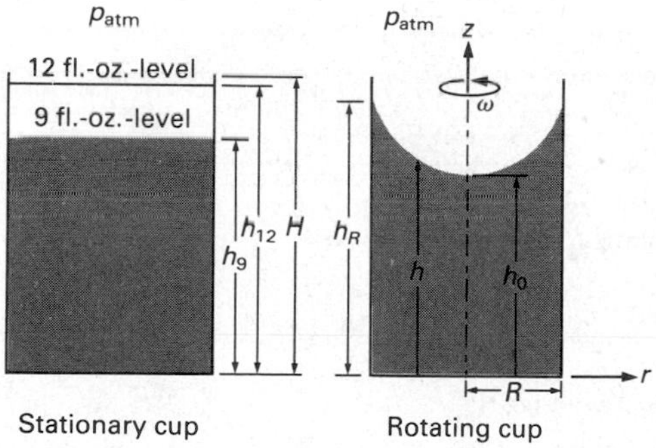

Figure E2.11a Stationary and rotating cup of soft drink.

A child buys a 12-fluid-ounce cup of a soft drink, drinks one fourth of it, and then places it directly in the center of a merry-go-round. The merry-go-round is then rotated at 15 rpm. Will any of the soft drink spill? The merry-go-round has a diameter of 10 ft; the cup has an inside diameter of 2.5 in.

SOLUTION

Given

12-oz cup with 9 fl oz of soft drink, rotating at 15 rpm

Inside diameter of cup 2.5 in.

Diameter of merry-go-round 10 ft

Figure E2.11a, which shows the cup stationary and rotating cup

Find

Height h_R of the soft drink when cup is rotating at 15 rpm

Compare with height of cup

Solution

We find the height H using the height h_{12}:

$$h_{12} = \frac{\text{Volume}}{\text{Area}} = \frac{(12\text{ fl oz})(1\text{ gal}/128\text{ fl oz})(231\text{ in}^3/\text{gal})}{(\pi/4)(2.5\text{ in.})^2}$$
$$= 4.41\text{ in.}$$

The difference between H and h_{12} is not given, so we must estimate it. Assuming that the top of a 12-oz cup is about $\frac{1}{2}$ in. above the 12-oz fluid level, we have

$$H \approx h_{12} + \tfrac{1}{2}\text{ in.}$$
$$\approx 4.91\text{ in.}$$

Next we find the height h_R from Eq. (2.40) by noting that $p = p_{atm}$ at $z = h$ and that $z = h_0$ at $r = 0$ at the free surface. Then

$$h = h_0 + \frac{\rho\omega^2 r^2}{2\gamma}.$$

Recognizing $\gamma = \rho g$, we find that

$$h = h_0 + \frac{\omega^2 r^2}{2g}.$$

Since $h = h_R$ at $r = R$, we have

$$h_R = h_0 + \frac{\omega^2 R^2}{2g}.$$

We can find h_0 by recognizing that the volume $\forall_1$ shown in Fig. E2.11b and given by

$$\forall_1 = \pi R^2 h_9$$

must equal the volume $\forall_2$, also in Fig. E2.11b, if no soft drink is spilled. We can calculate $\forall_2$ by

$$\forall_2 = \int_0^R 2\pi r h\, dr.$$

Substituting for h, we have

$$\forall_2 = 2\pi \int_0^R r\left(h_0 + \frac{\omega^2 r^2}{2g}\right) dr.$$

Integrating for constant h_0, ω, and g gives

$$\forall_2 = 2\pi\left(h_0 \frac{r^2}{2} + \frac{\omega^2 r^4}{8g}\right)_0^R = \pi R^2\left(h_0 + \frac{\omega^2 R^2}{4g}\right).$$

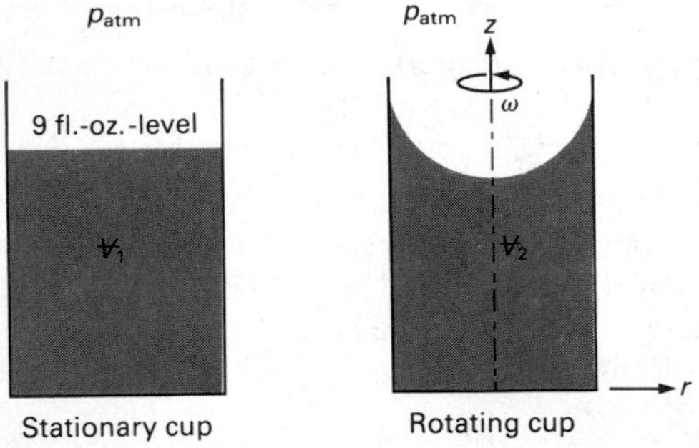

Figure E2.11b Volumes $\forall_1$ and $\forall_2$ of soft drink.

Equating $\forall_1$ and $\forall_2$, we have

$$\pi R^2 h_9 = \pi R^2\left(h_0 + \frac{\omega^2 R^2}{4g}\right) \quad \text{or} \quad h_0 = h_9 - \frac{\omega^2 R^2}{4g}.$$

Then

$$h_R = h_9 + \frac{\omega^2 R^2}{4g}.$$

The numerical values give

$$h_9 = \frac{\text{Volume}}{\text{Area}} = \frac{(9\text{ fl oz})(1.805\text{ in.}^3/\text{fl oz})}{(\pi/4)(2.5\text{ in.})^2} = 3.31\text{ in.}$$

and

$$h_R = 3.31\text{ in.} + \frac{(15\text{ rpm})^2(2\pi\text{ rad/rev})^2(1.25\text{ in.})^2}{4(386\text{ in./sec}^2)(60\text{ sec/min})^2}$$
$$= 3.31\text{ in.} + 0.002\text{ in.} = 3.31\text{ in.}$$

As $H > h_R$, we conclude:

No soft drink will spill from the cup. **ANSWER**

Discussion

The value of $\frac{1}{2}$ inch assumed for $(H - h_{12})$ was a realistic value but was not a significant factor in determining whether any soft drink was spilled.

You may have difficulty in recognizing that

$$\forall_2 = \int_0^R 2\pi r h \, dr$$

in a problem of this nature. If you do, a careful examination of Fig. E2.11b should be helpful.

PROBLEMS

1. Prove that the static pressure at a point is the same in all three directions in a static fluid.

2. The Empire State Building is 380 m high. Estimate the magnitude of the atmospheric pressure in pascals at the top of the building if the atmospheric pressure at the bottom is 760 mm Hg, abs., and the atmospheric temperature is (a) $-18°C$ and (b) 30°C. Assume constant air density equal to that at 760 mm Hg and the given temperature.

3. The deepest known spot in the oceans is the Challanger Deep in the Mariana Trench of the Pacific Ocean and is approximately 11,000 m below the surface. Assume that the salt water density is constant at 1025 kg/m^3 and determine the pressure at this depth.

4. A 30-ft-high downspout of a house is clogged at the bottom. Find the pressure at the bottom if the downspout is filled with 60°F rainwater.

5. Figure P2.5 shows a water tower used to maintain water pressure in a city's water lines. Find the height that the tower water level must be above the highest point, A, in the city for a water pressure of 400 kPa gage at this point. Also find the gage pressure, p_B.

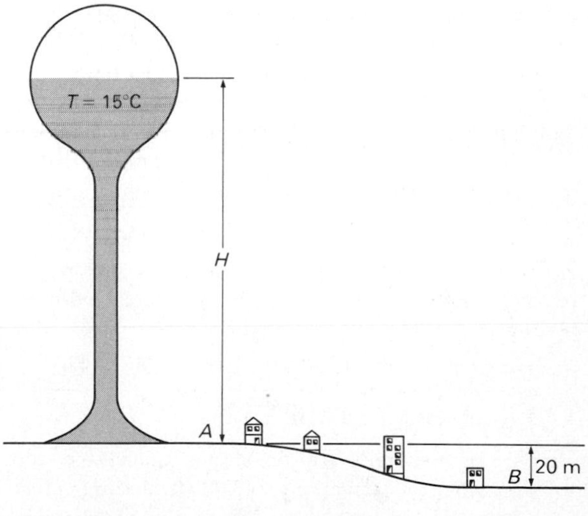

Figure P2.5

6. A submarine submerges by admitting seawater ($S =$ 1.03) into its ballast tanks. The amount of water admitted is controlled by air pressure, because seawater will cease to flow into the tank when the internal pressure (at the hull penetration) is equal to the hydrostatic pressure at the depth of the submarine. Consider a ballast tank with its outlet at the bottom. If the tank is filled to a depth of 2 ft and the air pressure in the tank is 40.6 psia, what is the submergence depth of the submarine?

•**7.** On a cold day in New York City, the atmospheric temperature was 0°C and the temperature inside the 380-m-high Empire State Building was 20°C. The air pressure inside the building at the building midheight is approximately equal to the outside atmospheric pressure at the same elevation, which is assumed to be 101 kPa. Assume that the air is an ideal gas, air density inside the building is constant (a good approximation), and air density outside the building is constant (another good approximation). Calculate the difference between the outside and inside air pressures at the bottom of the building.

8. A submarine submerges by admitting seawater ($S = 1.03$) into its ballast tanks. The amount of water admitted is controlled by air pressure, because seawater will cease to flow into the tank when the internal pressure (at the hull penetration) is equal to the hydrostatic pressure at the depth of the submarine. Consider a ballast tank, which can be modeled as a vertical half-cylinder ($R = 8$ ft, $L = 20$ ft) for which the air pressure control valve has failed shut. The failure occurred at the beginning of a dive from 60 ft to 1,000 ft. The tank was initially filled with seawater to a depth of 2 ft and the air was at a temperature of 40°F. As the weight of water in the tank is important in maintaining the boat's attitude, determine the weight of water in the tank as a function of depth during the dive. You may assume that tank internal pressure is always in equilibrium with the ocean's hydrostatic pressure and that the inlet pipe to the tank is at the bottom of the tank and penetrates the hull at the "depth" of the submarine. Compare your result to that of Problem 6.

9. Water boils at lower temperatures as pressure is reduced. Corresponding boiling temperatures and pressures are:

Temperature, °F	212	200	190	180
Pressure, psia	14.696	11.53	9.34	7.51

You are on a camping trip on Mount Rainier and stop for the night at an elevation of 10,000 ft. Assuming that you are in the Standard Atmosphere, what is the temperature of boiling water at this elevation?

10. Denver, Colorado, is called the "mile-high city" because its state capitol stands on land 1 mi above sea level. Assuming that the Standard Atmosphere exists, what is the pressure and temperature of the air in Denver? The temperature follows the lapse rate ($T = T_0 - Bz$).

11. The planet Venus has an average temperature of 427°C and a surface pressure of 2.0 MPa, abs. For a pure CO_2 atmosphere, a gravitational acceleration of 8.73 m/s^2, and an isothermal atmosphere, find the elevation where the atmospheric pressure is 101 kPa.

12. The deepest known spot in the oceans is the Challanger Deep in the Mariana Trench of the Pacific Ocean and is approximately 11,000 m below the surface. For a surface density of 1030 kg/m^3, a constant water temperature, and an isothermal bulk modulus of elasticity of 2.3×10^9 N/m^2, find the pressure at this depth.

13. The solar pond shown in Fig. P2.13 has a salt concentration that increases with water depth. The resulting water density increases as shown. Find the gage pressure at the bottom of the solar pond and compare it with the gage pressure if there were no salt in the pond.

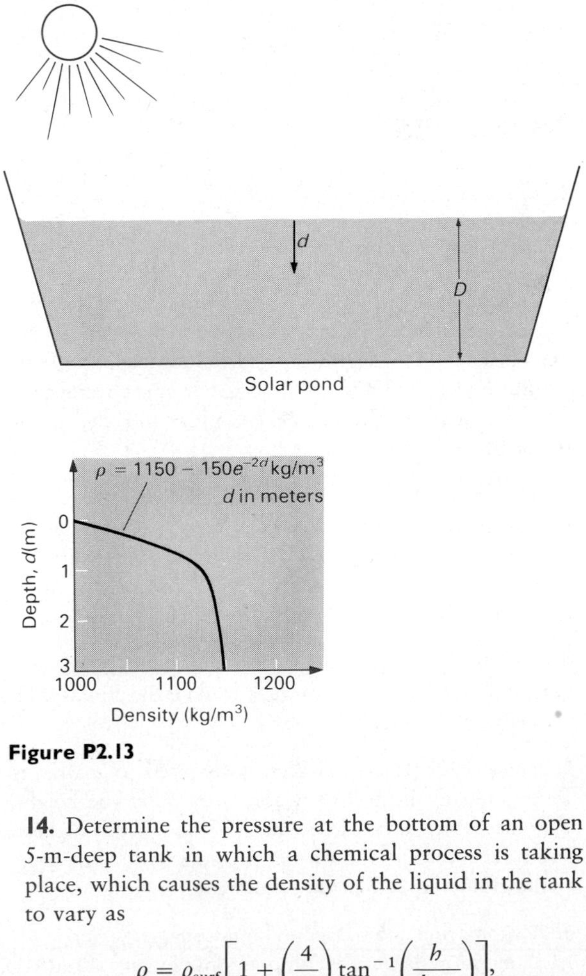

Figure P2.13

14. Determine the pressure at the bottom of an open 5-m-deep tank in which a chemical process is taking place, which causes the density of the liquid in the tank to vary as

$$\rho = \rho_{surf}\left[1 + \left(\frac{4}{\pi}\right)\tan^{-1}\left(\frac{h}{h_{bot}}\right)\right],$$

with $\rho_{surf} = 1700$ kg/m^3.

15. Determine the pressure at the bottom of an open 5-m-deep tank in which a chemical process is taking place that causes the density of the liquid in the tank to vary as

$$\rho = \rho_{\text{surf}} \sqrt{1 + \sin^2\left(\frac{h}{h_{\text{bot}}} \frac{\pi}{2}\right)},$$

where h is the distance from the free surface and $\rho_{\text{surf}} = 1700 \text{ kg/m}^3$.

16. A glass of liquid refreshment has a mixture, by volume, of 25 percent bourbon ($S \approx 0.78$) and 75 percent water. If the maximum suction that a person can develop is 100 mm Hg, what is the longest vertical length of a straw of $\frac{1}{4}$-in. inside diameter that this person can use and finish the drink?

17. A long glass tube is filled with light machine oil ($S = 0.834$), inverted, and then placed in a can of the same oil. Find the column height if the atmospheric pressure and temperature are 14.696 psia and 70°F, respectively. Assume that a vacuum exists above the oil at the top of the inverted tube.

18. Find the pressure p_T at the top of the water pipe shown in Fig. P2.18. The temperature is 10°C.

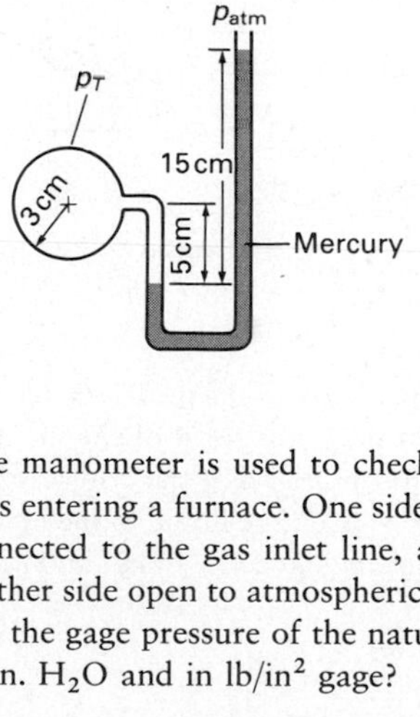

Figure P2.18

19. A U-tube manometer is used to check the pressure of natural gas entering a furnace. One side of the manometer is connected to the gas inlet line, and the water level in the other side open to atmospheric pressure rises 3 in. What is the gage pressure of the natural gas in the inlet line in in. H_2O and in lb/in^2 gage?

20. The container shown in Fig. P2.20 holds 60°F water and 60°F air as shown. Find the absolute pressures at locations A, B, and C.

21. A commercial manometer was not available, so a U-tube manometer was hastily constructed from two pieces of straight glass tubing with the same outside diameter and a piece of flexible tubing. When a pressure

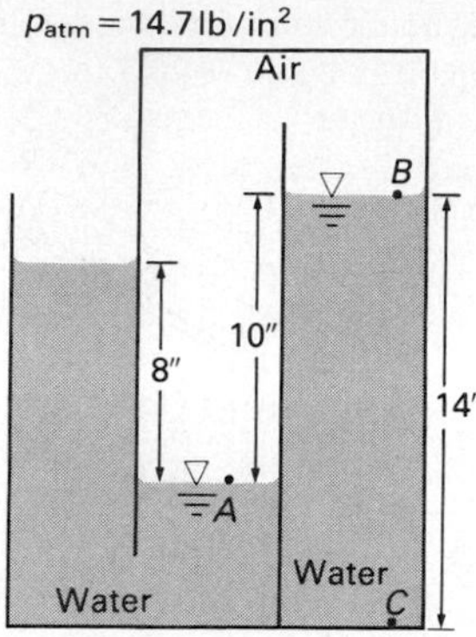

Figure P2.20

difference was applied to the manometer, the left water column rose twice as much as the right column fell. Explain the probable cause for the unequal change in the two water columns. Determine the applied pressure difference if the left column of 15°C water rose 10.0 cm.

22. The U-tube manometer shown in Fig. P2.22 has two fluids, water and oil ($S = 0.80$). Find the height difference between the free water surface and the free oil surface with no applied pressure difference.

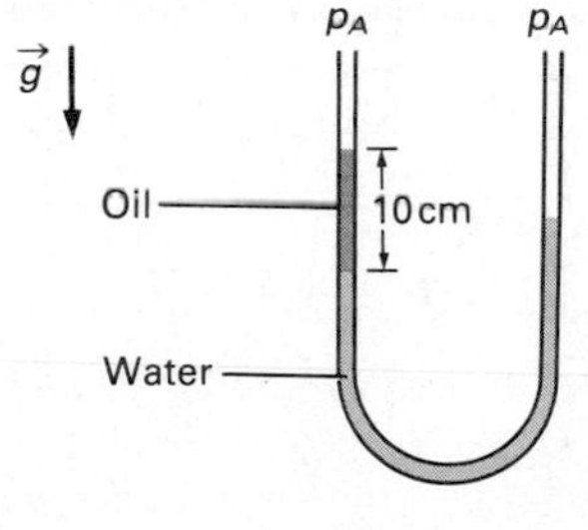

Figure P2.22

23. What is the specific gravity of the liquid in the left leg of the U-tube manometer shown in Fig. P2.23?

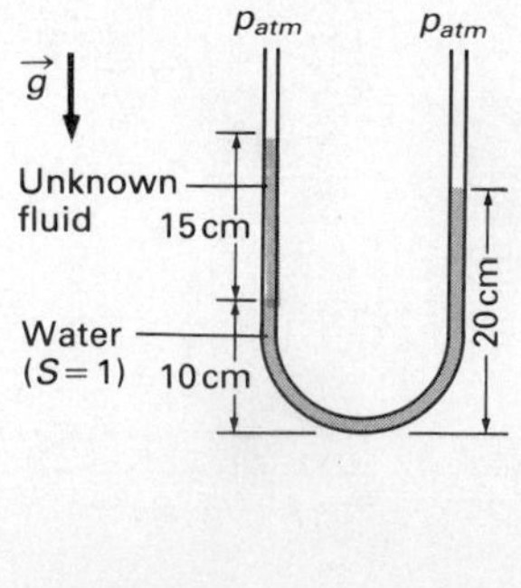

Figure P2.23

24. Determine the difference in pressure between points A and B in Fig. P2.24 if $h = 20.3$ in. and the mercury and water are at 50°F.

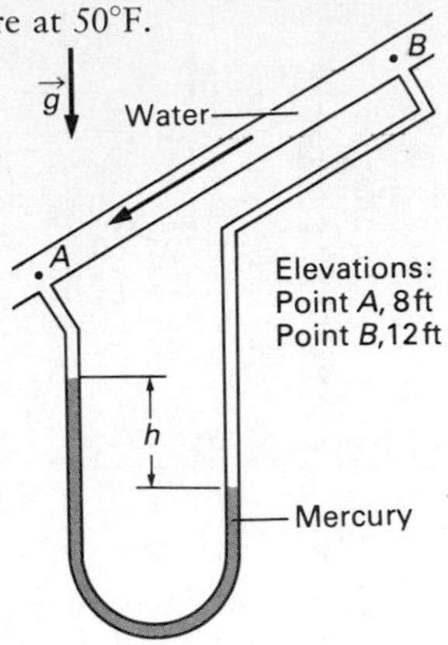

Figure P2.24

25. The difference in the two levels of the mercury manometer shown in Fig. P2.25 is 15 mm. Find the pressure difference between points A and B to which the manometer is connected in the stagnant water column. The fluids are at 10°C.

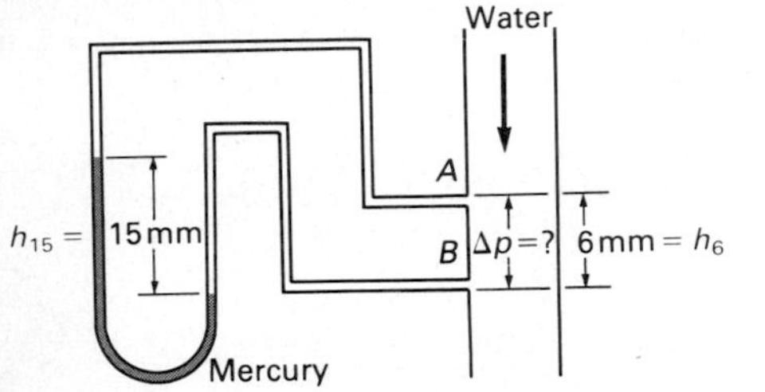

Figure P2.25

26. The manometer shown in Fig. P2.26 has an air bubble either in (a) the right horizontal line or (b) the left vertical leg. Find $h_1 - h_2$ for both cases if $p_A = p_B$.

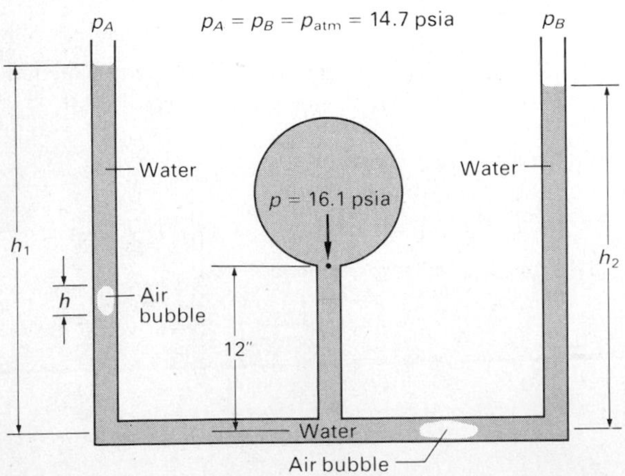

Figure P2.26

27. Find the percentage difference in the readings of the two identical U-tube manometers shown in Fig. P2.27. Manometer 90 uses 90°C water and manometer 30 uses 30°C water. Both have the same applied pressure difference. Does this percentage change with the magnitude of the applied pressure difference? Can the difference between the two readings be ignored?

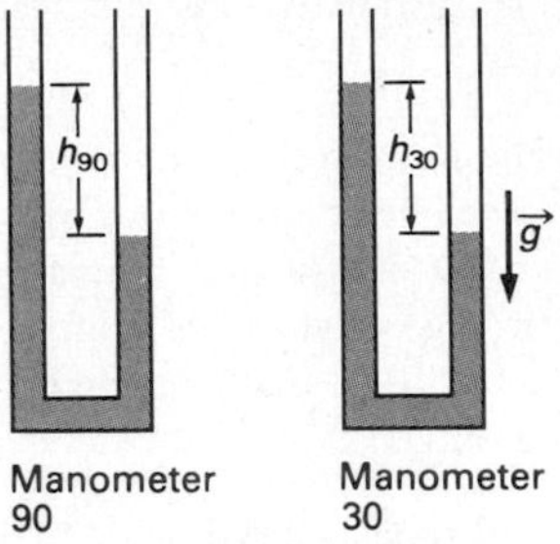

Figure P2.27

28. Consider the cistern manometer shown in Fig. P2.28. The scale is set up on the basis that the cistern area A_1 is infinite. However, A_1 is actually 50 times the internal cross-sectional area A_2 of the inclined tube. Find the percentage error (based on the scale reading) involved in using this scale.

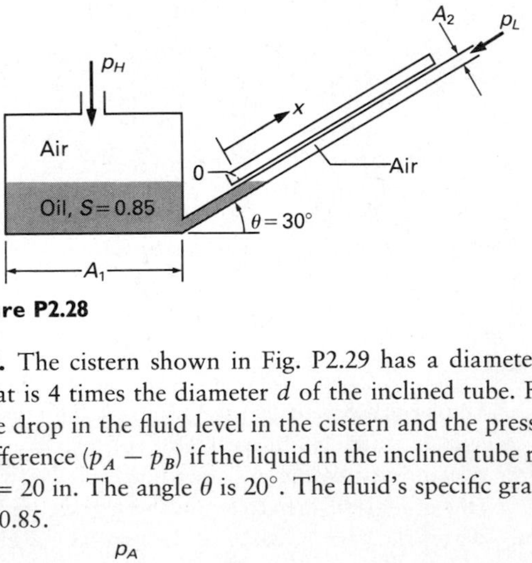

Figure P2.28

29. The cistern shown in Fig. P2.29 has a diameter D that is 4 times the diameter d of the inclined tube. Find the drop in the fluid level in the cistern and the pressure difference $(p_A - p_B)$ if the liquid in the inclined tube rises $\ell = 20$ in. The angle θ is 20°. The fluid's specific gravity is 0.85.

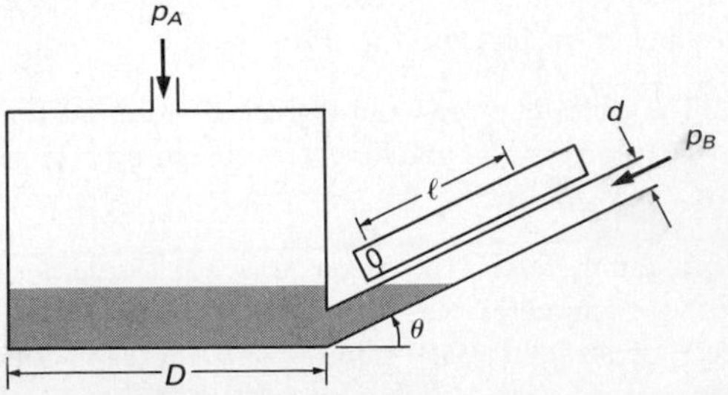

Figure P2.29

30. The sensitivity *Sen* of the manometer shown in Fig. P2.30 can be defined as

$$Sen = \frac{h}{p_R - p_L}.$$

Three manometer fluids with the listed specific gravities S are available:

Kerosene, $S = 0.82$;

SAE 10 oil, $S = 0.87$; and

Normal octane, $S = 0.71$.

Which fluid gives the highest sensitivity? The areas A_R and A_L are much larger than the cross-sectional area of the manometer tube, so $H \ll h$.

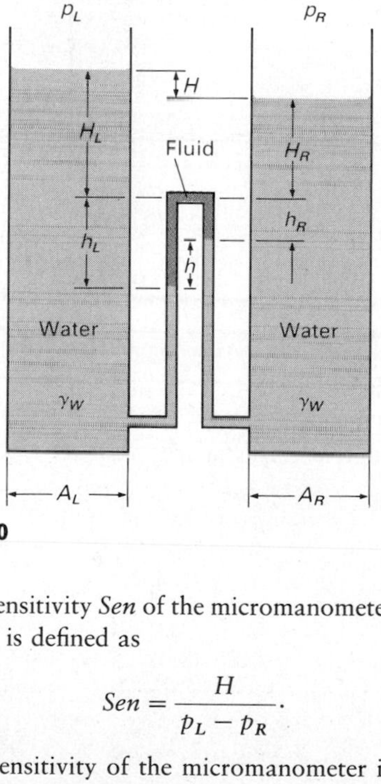

Figure P2.30

31. The sensitivity *Sen* of the micromanometer shown in Fig. P2.31 is defined as

$$Sen = \frac{H}{p_L - p_R}.$$

Find the sensitivity of the micromanometer in terms of the densities ρ_A and ρ_B. How can the sensitivity be increased?

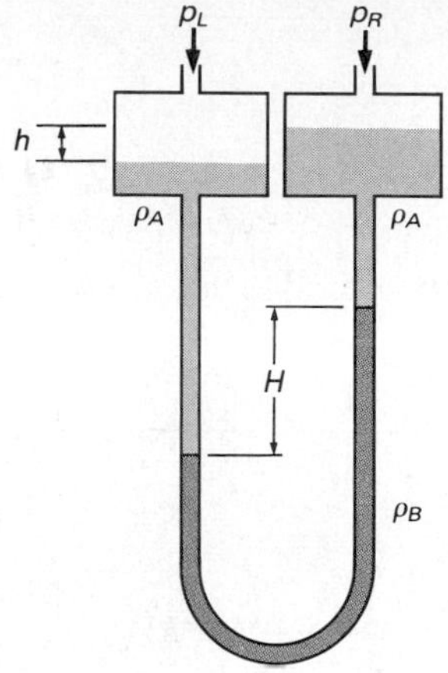

Figure P2.31

32. The U-tube manometer shown in Fig. P2.32 has legs that are 1.00 m long. When no pressure difference is applied across the manometer, each leg has 0.40 m of mercury. What is the maximum pressure difference that can be indicated by the manometer?

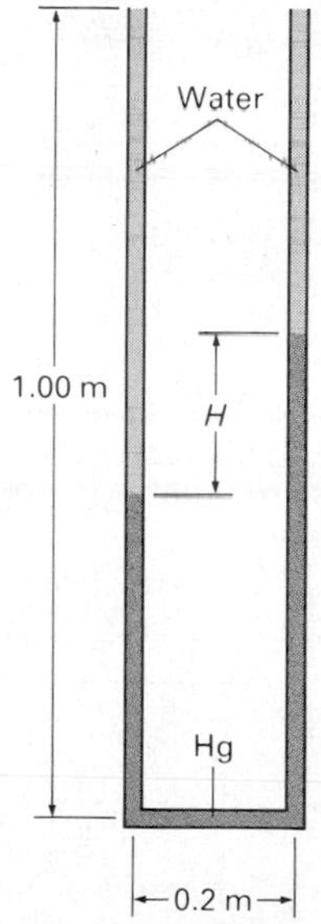

Figure P2.32

• 33. A tree has long tubes called *xylem tubes* to carry sap from the roots to the branches. The sap rises in the tubes partly because of surface tension and partly because of the higher pressure in the roots than in the branches. If sap has a surface tension of 0.0015 lb/ft and a density of 50 lbm/ft^3, what is the maximum height caused by surface tension to which the sap will rise in xylem tubes of the following diameters: 500 Å, 1000 Å, and 2000 Å? Assume a zero contact angle.

• 34. Figure P2.34 shows a circular bellows pressure gage used to measure a gage pressure p. Determine the spring constant k in N/m if k is defined as

$$k \equiv \frac{\text{Force } F \text{ on spring when bellows is subjected to gage pressure } p}{\text{Distance movable end of spring has moved from } p = 0 \text{ position}}$$

and if $x = 1.0$ cm for $p = 202{,}660$ N/m^2 gage and $p_{atm} = 101{,}330$ N/m^2, abs.

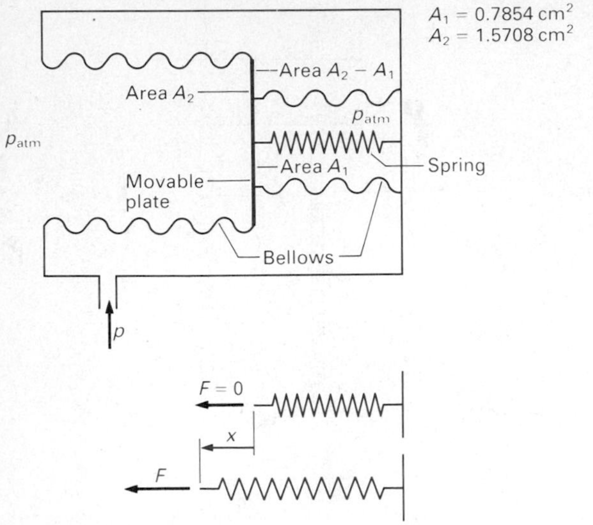

Figure P2.34

35. Figure P2.35 shows a circular bellows pressure gage used to measure an absolute pressure p. Determine the spring constant k in N/m if k is defined as

$$k \equiv \frac{\text{Force } F \text{ on spring when bellows is stretched to absolute pressure } p}{\text{Distance movable end of spring has moved from } p = 0 \text{ position}}$$

and if $x = 1.0$ cm for $p = 303{,}990\ \text{N/m}^2$, abs. and $p_{atm} = 101{,}330\ \text{N/m}^2$, abs.

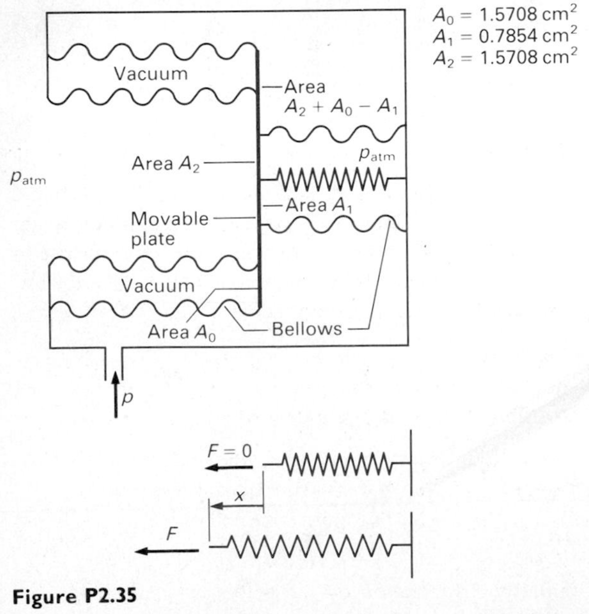

Figure P2.35

36. Find the center of pressure of an elliptical area of minor axis $2a$ and major axis $2b$ where axis $2a$ is vertical and axis $2b$ is horizontal. The center of the ellipse is a vertical distance h below the surface of the water ($h > a$). The fluid density is constant. Will the center of pressure of the ellipse change if the fluid is replaced by another constant-density fluid? Will the center of pressure of the ellipse change if the vertical axis is tilted back an angle α from the vertical about its horizontal axis? Explain.

37. Find the weight W needed to hold the wall shown in Fig. P2.37 upright. Wall is 10 m wide.

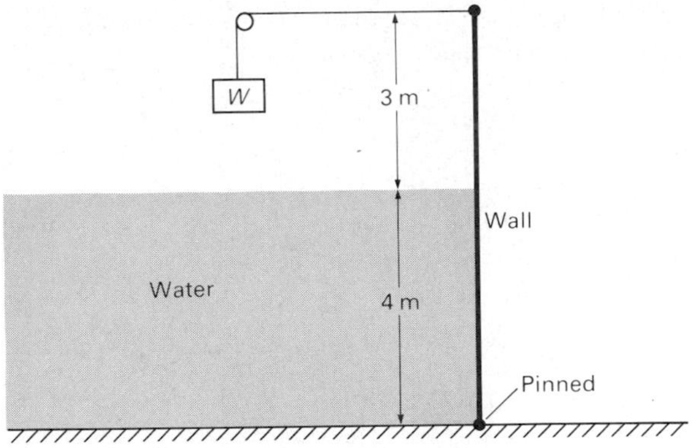

Figure P2.37

38. Determine the weight W needed to hold the wall shown in Fig. P2.38 in place. The wall weighs 2000 N/m^2 and is 10 m wide. The fluid is water.

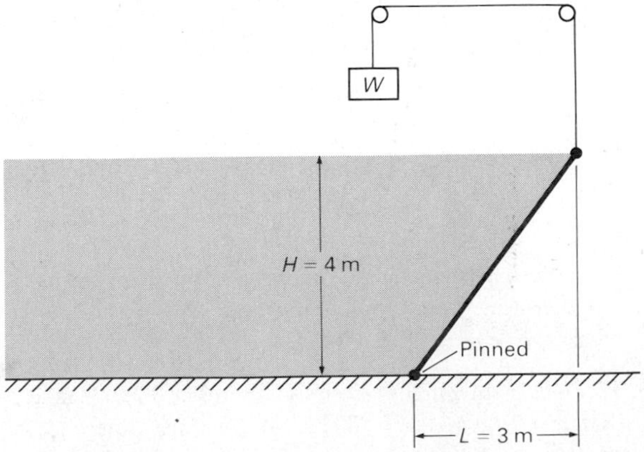

Figure P2.38

39. The hydraulic cylinder shown in Fig. P2.39, with a 4-in.-diameter piston, is advertised as being capable of providing a force of $F = 20$ tons. If the piston has a design pressure (the maximum pressure at which the cyl-

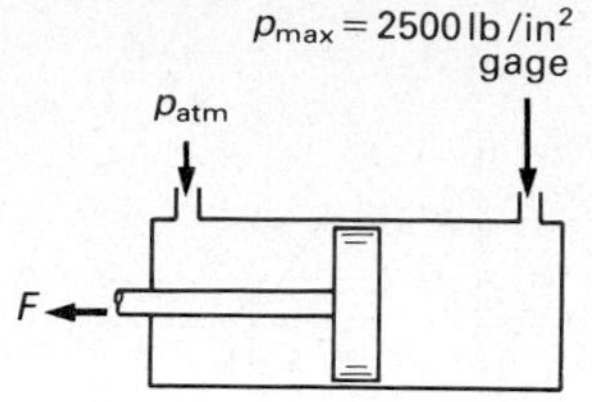

Figure P2.39

inder should safely operate) of 2500 lb/in^2, gage, can the cylinder safely provide the advertised force?

40. The container shown in Fig. P2.40 has square cross sections. Find the horizontal force on one of the four inclined surfaces, *ABCD*. Also find the net vertical force on the bottom, *EF*. Is the vertical force equal to the weight of the water in the container?

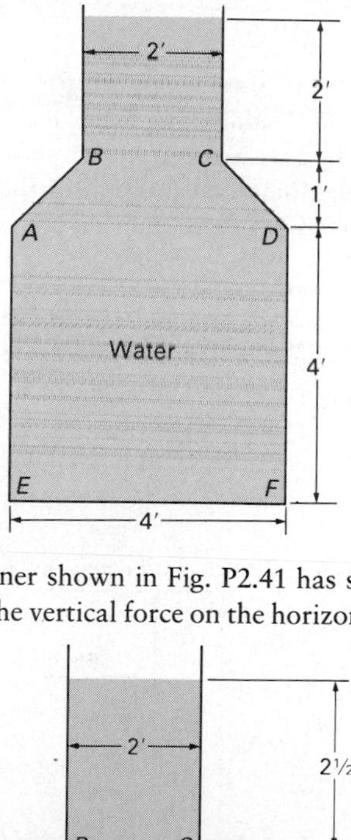

Figure P2.40

41. The container shown in Fig. P2.41 has square cross sections. Find the vertical force on the horizontal surface, *ABCD*.

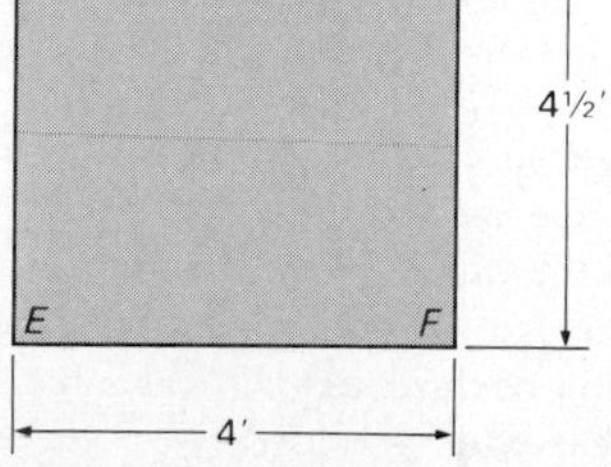

Figure P2.41

42. An automobile has just dropped into a river. The car door is approximately a rectangle, measures 36 in. wide and 40 in. high, and has hinges on a vertical side. The water level inside the car is up to the midheight of the door, and the air inside the car is at atmospheric pressure. Calculate the force required to open the door if the force is applied 24 in. from the hinge line. See Fig. P2.42. (The driver did not have the presence of mind to open the window to escape.)

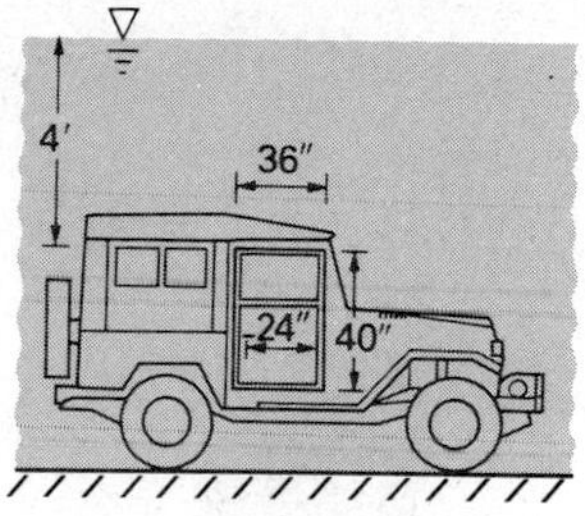

Figure P2.42

43. The gate shown in Fig. P2.43 is 10 ft wide and is exposed to freshwater on its left side. Find the magnitude of the net force on the gate and the location y^* of the pivot line below the top of the gate, so that there is no moment tending to open the gate for the water depth shown.

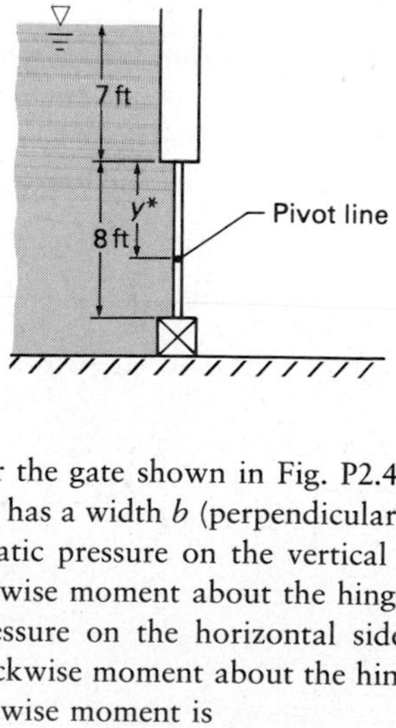

Figure P2.43

44. Consider the gate shown in Fig. P2.44. The gate is massless and has a width b (perpendicular to the paper). The hydrostatic pressure on the vertical side creates a counterclockwise moment about the hinge, and the hydrostatic pressure on the horizontal side (or bottom) creates a clockwise moment about the hinge. Show that the net clockwise moment is

$$\sum \mathcal{M} = \rho_w g h b\left(\frac{\ell^2}{2} - \frac{h^2}{6}\right).$$

45. Will the gate in Problem 44 ever open?

46. A tank contains 6 in. of oil ($S = 0.82$) above 6 in. of water ($S = 1.00$). Find the force on the bottom of the tank. See Fig. P2.46.

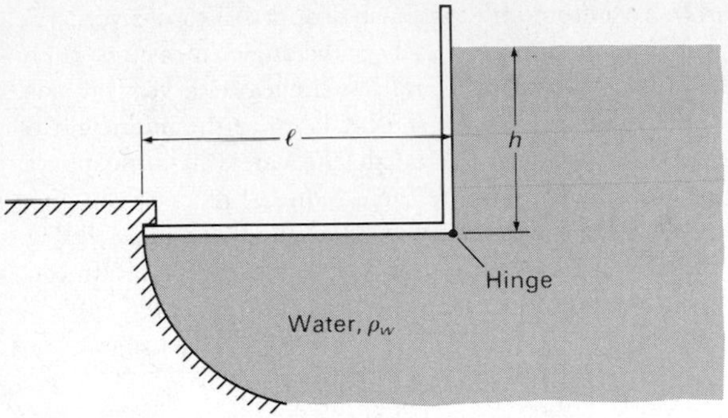

Figure P2.44

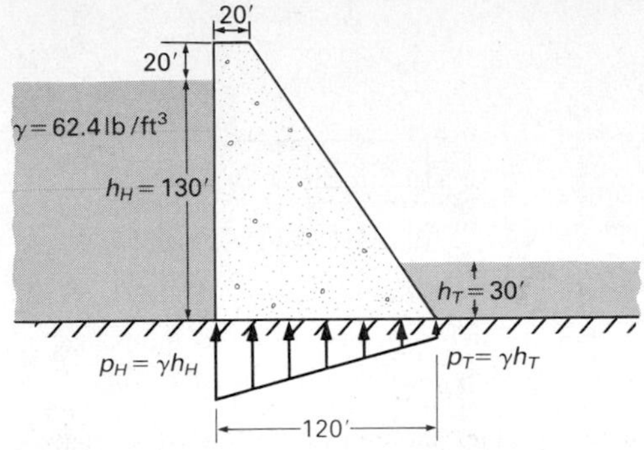

Figure P2.48

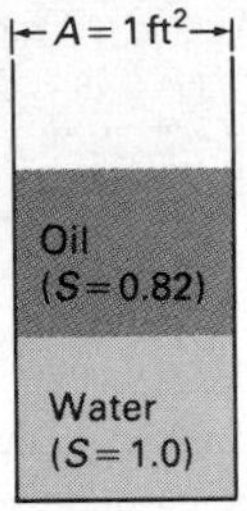

Figure P2.46

47. While building a high, tapered concrete wall, builders used the wooden forms shown in Fig. P2.47. If concrete has a specific gravity of about 2.5, find the total force on each of the three side sections (A, B, and C) of the wooden forms (neglect any restraining force of the two ends of the forms).

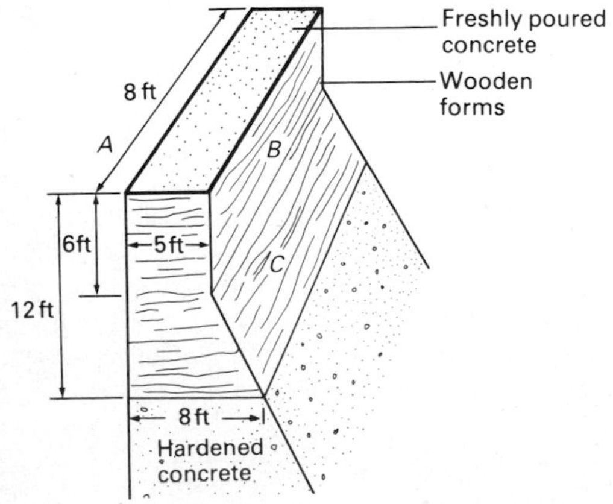

Figure P2.47

48. The concrete gravity dam shown in Figure P2.48 is 20 ft. wide at the top and 120 ft. wide at the base. Determine the magnitudes and locations of the hydrostatic forces acting on the left vertical wall of the dam and on the right inclined wall of the dam. Consider a unit length of the dam (b = 1 ft).

49. The concrete gravity dam shown in Problem 48 has a very pervious foundation. Water leaks under the dam to produce the pressure distribution shown and an uplift. Find the magnitude and location of the resultant force from this pressure distribution on the base of the dam. Consider a unit length of the dam (b = 1 ft).

50. Figure P2.50 is a representation of the Keswick gravity dam in California. Find the magnitudes and locations of the hydrostatic forces acting on the headwater vertical wall of the dam and on the tailwater inclined wall of the dam. Note that the slope given is the ratio of the run to the rise. Consider a unit length of the dam (b = 1 ft).

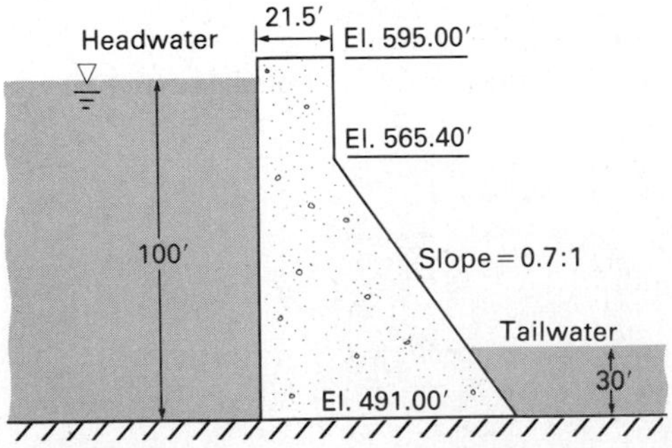

Figure P2.50

51. The Keswick dam in Problem 50 is made of concrete and has a specific weight of 150 lb/ft³. The hydrostatic forces and the weight of the dam produce a total vertical force of the dam on the foundation. Find the magnitude and location of this total vertical force. Consider a unit length of the dam (b = 1 ft).

52. The Keswick dam in Problem 50 is made of concrete and has a specific weight of 150 lb/ft^3. The coefficient of friction μ between the base of the dam and the foundation is 0.65. Is the dam likely to slide downstream? Consider a unit length of the dam ($b = 1$ ft).

53. Figure P2.53 is a representation of the Altus gravity dam in Oklahoma. Find the magnitudes and locations of the horizontal and vertical hydrostatic force components acting on the headwater wall of the dam and on the tailwater wall of the dam. Note that the slope given is the ratio of the run to the rise. Consider a unit length of the dam ($b = 1$ ft).

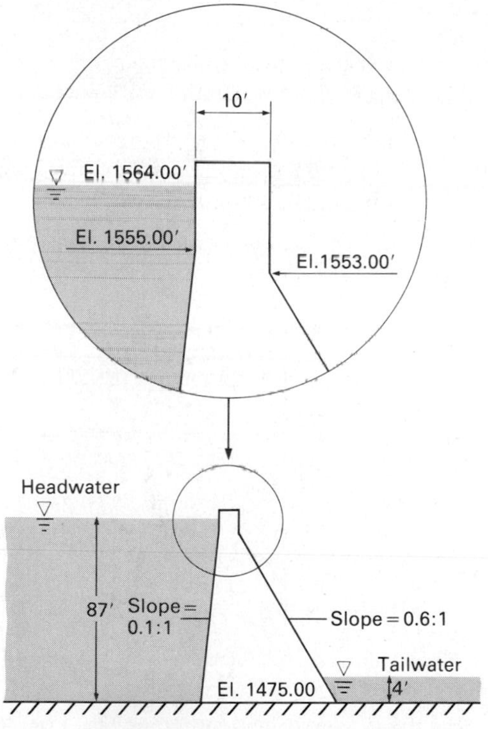

Figure P2.53

54. The Altus dam in Problem 53 is made of concrete with a density of 150 lbm/ft^3. The coefficient of friction μ between the base of the dam and the foundation is 0.65. Is the dam likely to slide downstream? Consider a unit length of the dam ($b = 1$ ft).

55. The dam shown in Fig. P2.55 is 200 ft long and is made of concrete with a specific gravity of 2.2. Find the magnitude and y coordinate of the line of action of the net horizontal force.

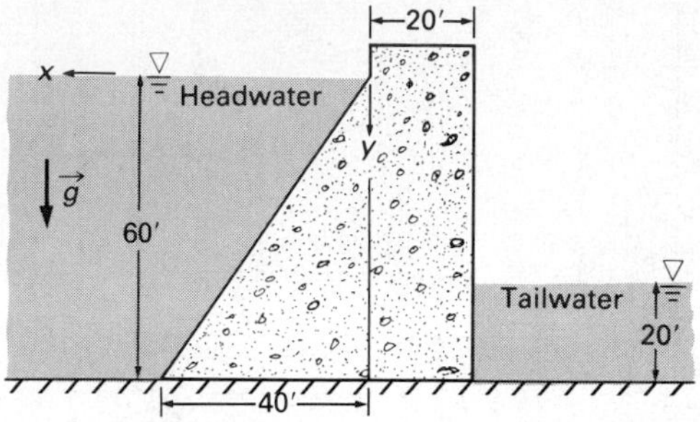

Figure P2.55

56. Repeat Problem 55 but find the magnitude and x coordinate of the line of action of the vertical force on the dam resulting from the water.

57. Determine the magnitude and direction of the force that must be applied to the bottom of the gate shown in Fig. P2.57 to keep the gate closed.

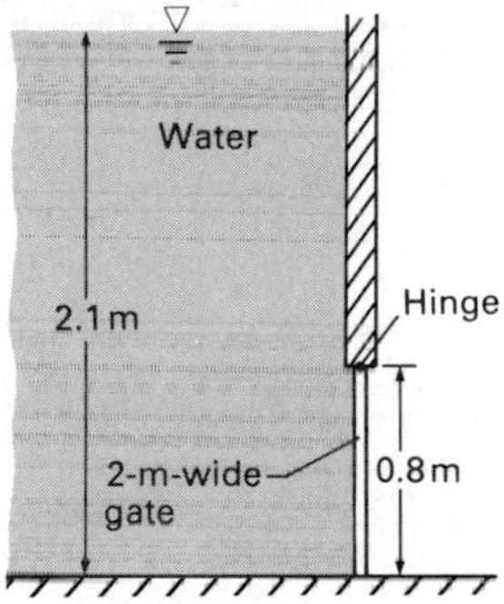

Figure P2.57

58. Find the maximum permissible water level so the gravity dam shown in Fig. P2.58 will not tip over. Assume that no hydrostatic pressure is exerted on the bottom of the dam. Concrete has a specific weight of 22,000 N/m^3.

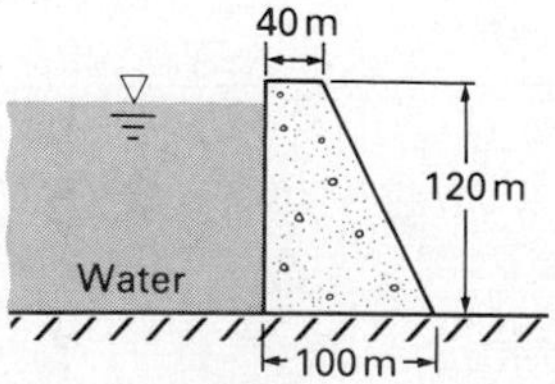

Figure P2.58

59. Find the magnitude and location of the net horizontal force on the gate shown in Fig. P2.59. The gate width is 5.0 m.

60. Find the total vertical force on the cylinder shown in Fig. P2.60.

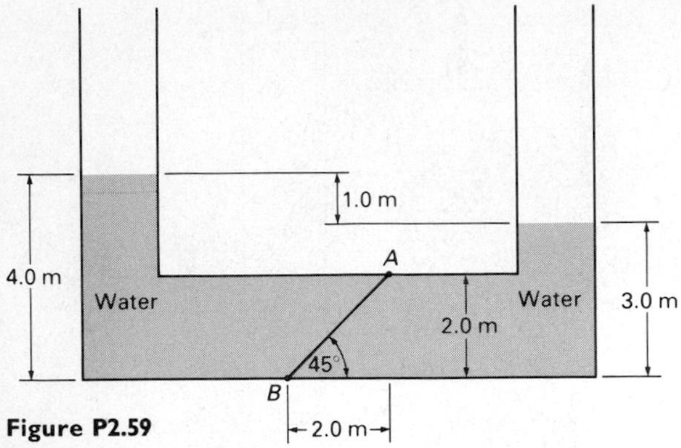

Figure P2.59

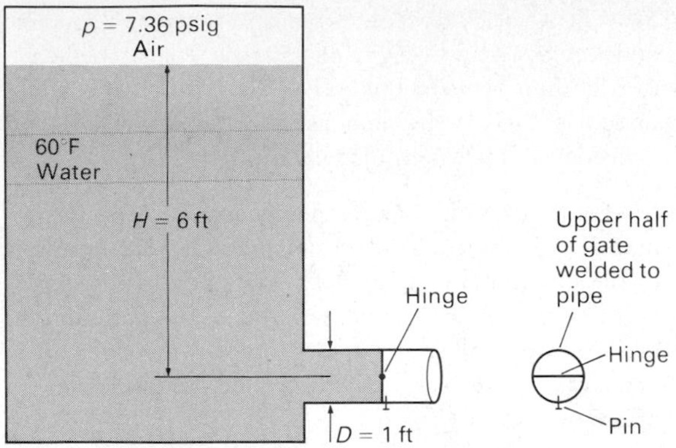

Figure P2.62

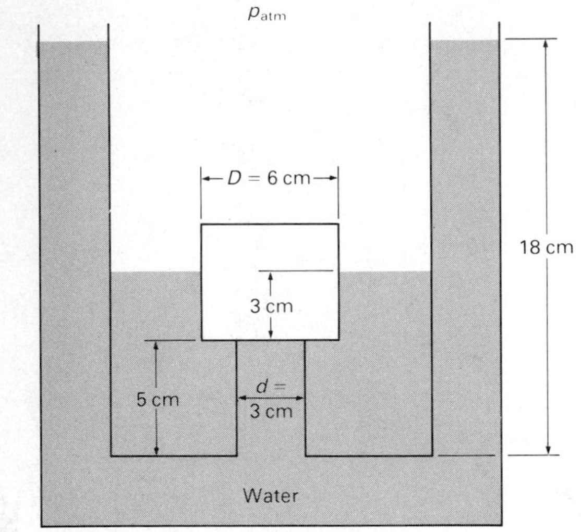

Figure P2.60

61. The triangular trough shown in Fig. P2.61 holds water at the depth shown. Find the force F in the support bar if the trough is 10 ft long.

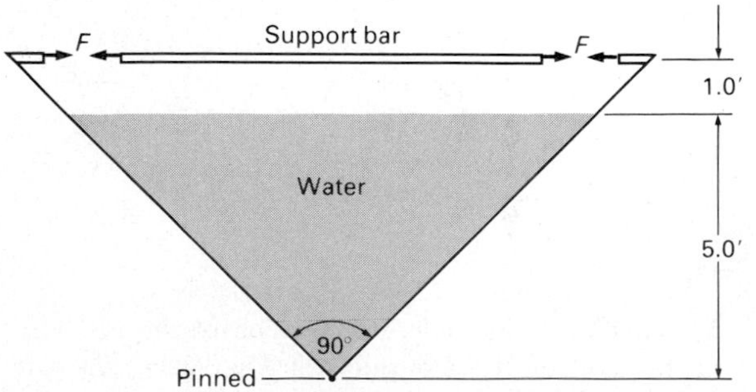

Figure P2.61

62. Pressurized air is trapped over a layer of water in a large tank, as shown in Fig. P2.62. The discharge pipe from the tank is closed by a semicircular gate, as shown. Calculate the shearing force in the restraining pin.

63. The gate shown in Fig. P2.63 is 1 m wide, is hinged at B, and opens automatically when the water becomes high enough. Is the gate open or closed for $H = 6$ m?

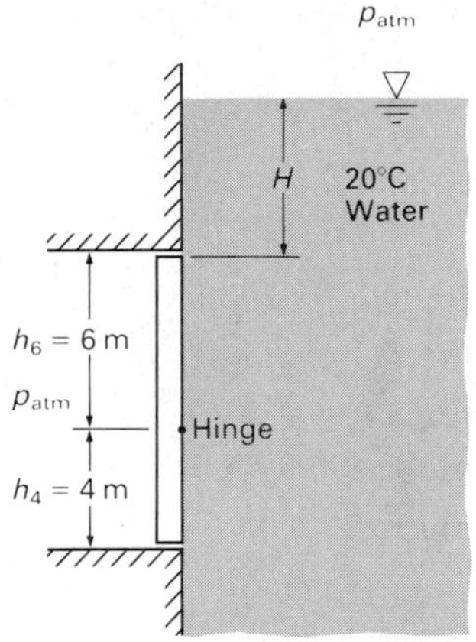

Figure P2.63

• **64.** Find the magnitude and location of the net vertical force on the gate in Problem 59.

• **65.** The concrete gravity dam shown in Fig. P2.48 has a very pervious foundation, and water leaks under the dam to produce the pressure distribution shown and an uplift. The dam is 20 ft wide at the top and 120 ft wide at the base. Will the dam topple over? The concrete has a specific weight of 140 lb/ft^3.

• **66.** The 300-lb gate in Fig. P2.66 is a 3-ft-diameter cylinder hinged on the left so that it can swing counterclockwise. Determine the force F required to just open or to just hold the gate closed. Indicate the direction of F as well as its magnitude. Do Archimedes's principles apply?

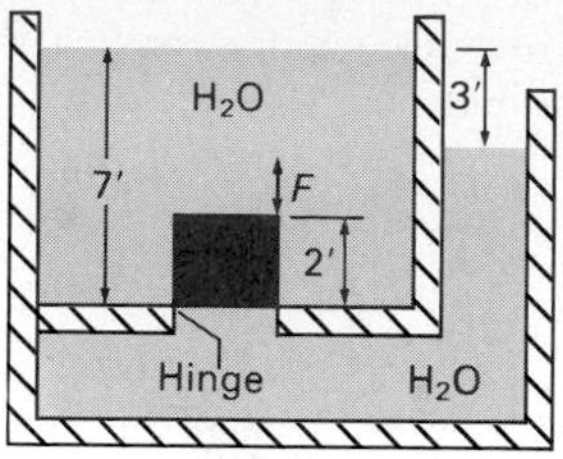

Figure P2.66

• **67.** Figure P2.50 is a representation of the Keswick gravity dam in California. The dam is made of concrete having a density of 150 lbm/ft^3. The hydrostatic uplift pressure on the base of the dam varies linearly from 67 percent of the headwater pressure at the upstream toe of the dam to 100 percent of the tailwater pressure at the downstream toe. Determine the hydrostatic uplift pressure at both the upstream toe and the downstream toe of the dam, the uplift force per unit length of the dam (perpendicular to the paper), and the location of the centers of pressure of the headwater and tailwater. Is this dam likely to overturn?

• **68.** Repeat Problem 67 for the Altus concrete dam shown in Fig. P2.53. Is this dam likely to overturn? There is no hydrostatic pressure on the base of the dam.

69. Wind blows against the large vertical wall shown in Fig. P2.69. Find the pressure p_0 at the bottom of the wall in psig. The pressure over the face of the wall is given by

$$p = p_0\left[1 - \left(\frac{y}{\ell}\right)^2\right],$$

and the pressure over the back of the wall is atmospheric. Both p and p_0 are measured in psig. Find the resultant force on the wall and its line of action.

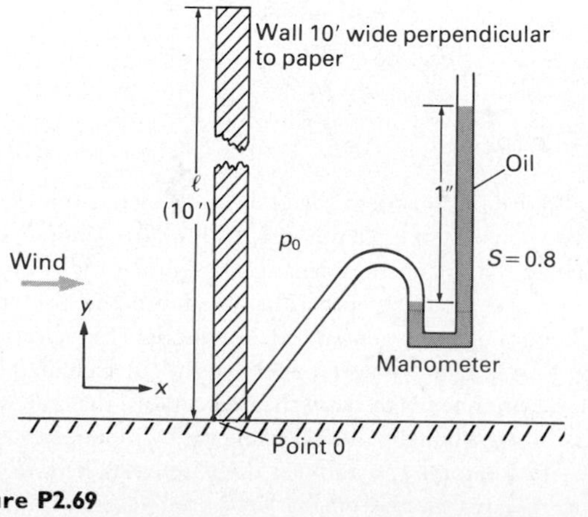

Figure P2.69

70. The submarine shown in Fig. P2.70 floats in seawater ($\gamma = 64.0$ lb/ft^3). The pressure inside the submarine is atmospheric. For shallow depths, a mercury manometer is used to indicate the depth. For the manometer levels shown, calculate the depth of the manometer tap in the submarine wall. Then find the minimum force required to open the submarine hatch at this depth. Neglect the weight of the hatch.

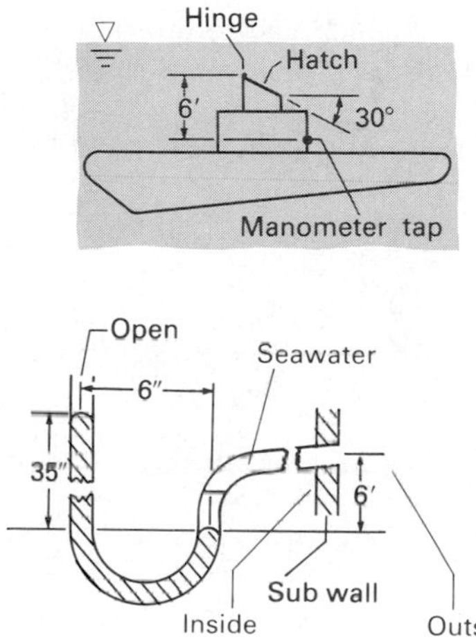

Figure P2.70

71. A little Dutch boy puts his thumb in a hole in a dike. If the sea level is 3 m above the boy's thumb, estimate the force in newtons that the boy must exert to keep the water from coming out the hole. The boy's thumb has a cross-sectional area of 3.0 cm^2.

72. Estimate the "surface tension" in a 0.82-ft-diameter balloon if the inside pressure exceeds the ambient pressure by 1.0 psi.

73. A small airplane has a mass of 700 kg and a wing area of 36 m^2. The average pressure on the lower wing surface is 70 kPa. What must be the average pressure on the upper wing surface for the plane to fly level? Use a gravitational acceleration of 9.81 m/s^2.

74. A yield point of 25,000 psi and a factor of safety of 2.0 are used in the design of a thin-walled cylindrical tank. The tank has an inside diameter of 18 in. and length of 24 in. The tank is used for storing compressed air up to pressures of 150 psig. Find the minimum permissible wall thickness.

75. Find the magnitude and location of the resultant force on the circular arc shown in Fig. P2.75. The arc is 25 m long.

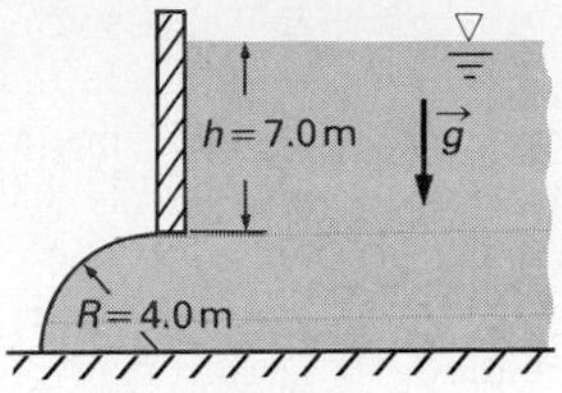

Figure P2.75

76. Find the magnitude and location of the resultant force on the hemispherical dome shown in Fig. P2.76.

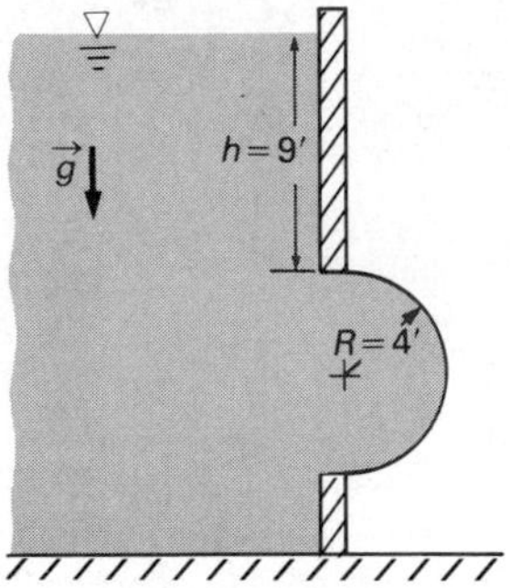

Figure P2.76

77. Figure P2.77 shows a cross section of a submerged tunnel used by automobiles to travel under a river. Find the magnitude and location of the resultant hydrostatic force on the circular roof of the tunnel. The tunnel is 4 mi long.

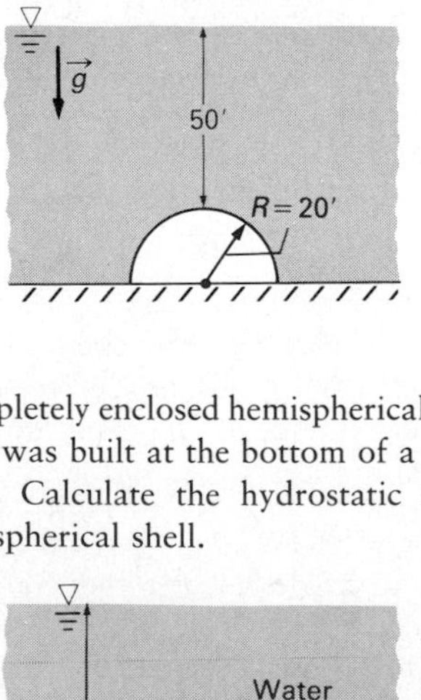

Figure P2.77

78. The completely enclosed hemispherical dome shown in Fig. P2.78 was built at the bottom of a deep lake for observations. Calculate the hydrostatic force on the dome's hemispherical shell.

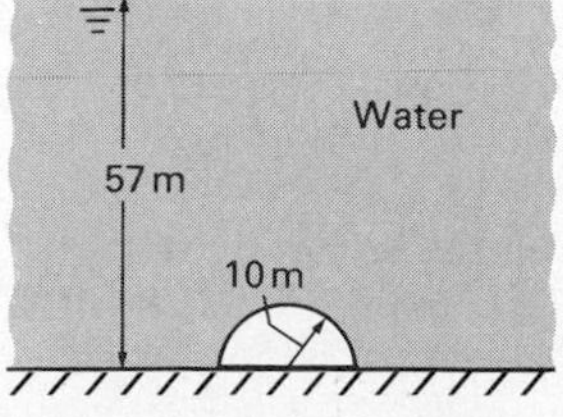

Figure P2.78

79. A step-in viewing window having the shape of a half-cylinder is built into the side of a large aquarium. See Fig. P2.79. Find the magnitude, direction, and location of the net horizontal forces on the viewing window.

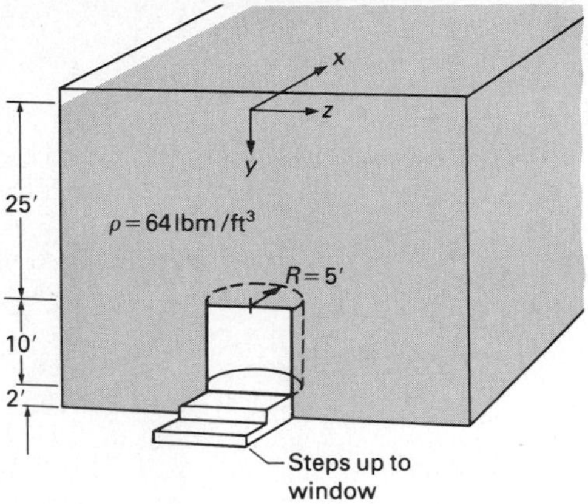

Figure P2.79

80. Find the magnitude, direction, and location of the net vertical force acting on the viewing window in Problem 79.

81. A cylindrical water tank has a diameter of 7 m and a height of 20 m. Find the forces F indicated in Fig. P2.81 if the tank is filled with water.

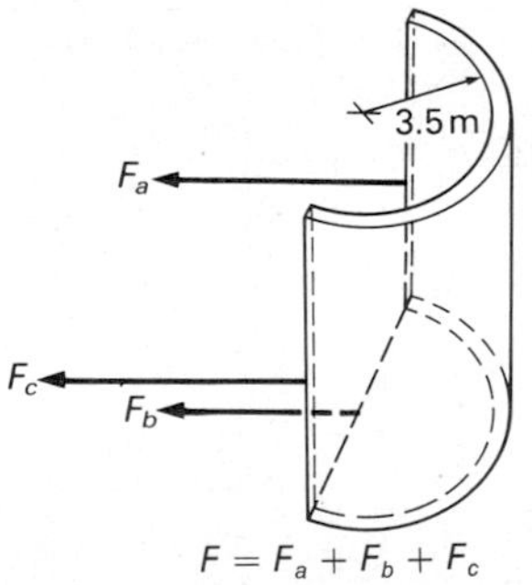

Figure P2.81

82. In drilling for oil in the Gulf of Mexico, some divers have to work at a depth of 1300 ft. (a) Assume that seawater has a constant density of 64 lb/ft^3 and compute the pressure at this depth. The divers breathe a mixture of helium and oxygen stored in cylinders, as shown in Fig. P2.82, at a pressure of 3000 psia. (b) Calculate the force, which tends to blow the end cap off, that the weld must resist while the diver is using the cylinder at 1300 ft. (c) After emptying a tank, a diver releases it. Will the tank rise or fall, and what is its initial acceleration?

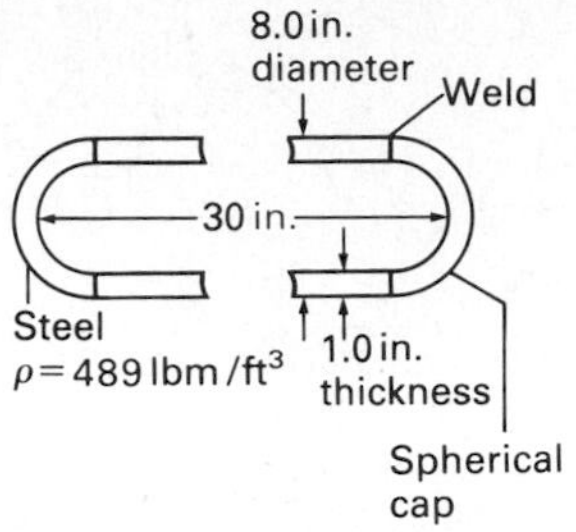

Figure P2.82

83. A 10-m-long log is stuck against a dam, as shown in Fig. P2.83. Find the magnitudes and locations of both the horizontal force and the vertical force of the water on the log in terms of the diameter D. The center of the log is at the same elevation as the top of the dam.

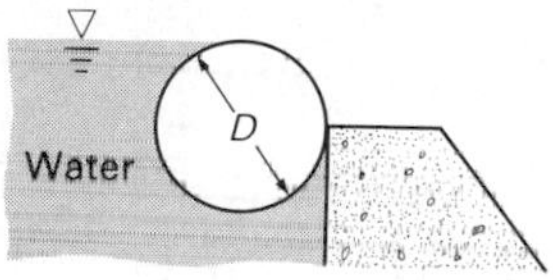

Figure P2.83

84. Find the net horizontal force on the 4.0-m-long log shown in Fig. P2.84.

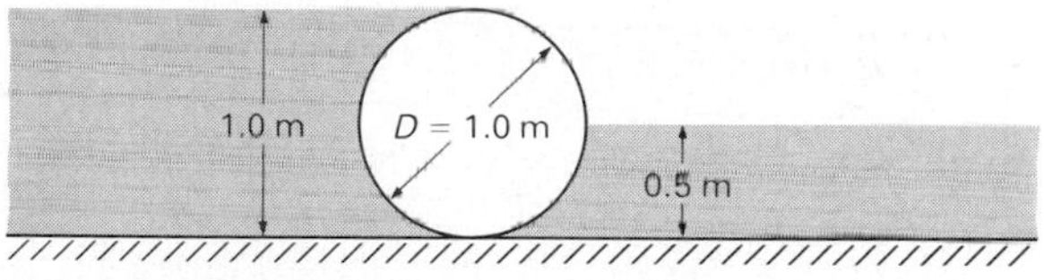

Figure P2.84

85. Consider the curved surface shown in Fig. P2.85(a) and (b). The two curved surfaces are identical. How are the vertical forces on the two surfaces alike? How are they different?

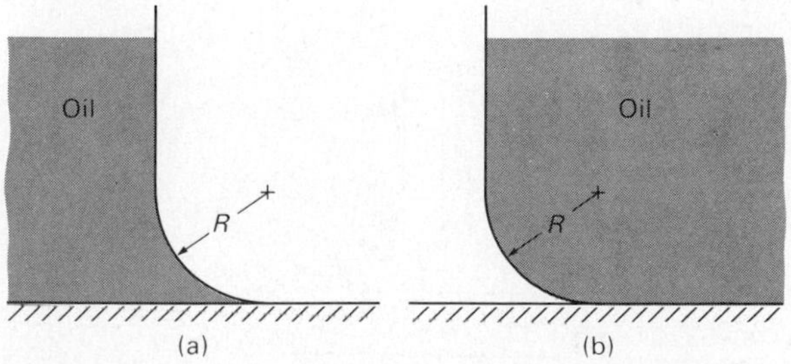

Figure P2.85

86. The container shown in Fig. P2.86 has circular cross sections. Find the vertical force on the inclined surface. Also find the net vertical force on the bottom, EF. Is the vertical force equal to the weight of the water in the container?

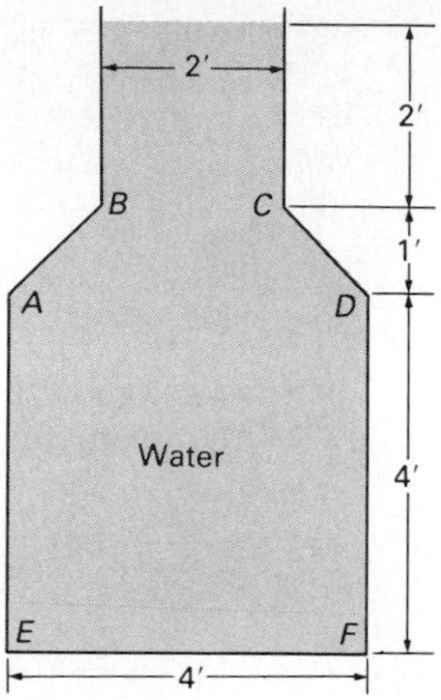

Figure P2.86

87. Why is extinguishing burning oil, grease, or gasoline with water dangerous?

88. A hydrometer is used to measure the specific gravity of liquids. Figure P2.88 shows a hydrometer with a scale etched on the surface. Would the higher specific gravities be near the top or the bottom of the hydrometer?

Figure P2.88

89. What volume fraction of an iceberg floats above seawater? Recall that an iceberg is frozen freshwater. For seawater, $S = 1.03$.

90. A barge is 40 ft wide by 120 ft long. The weight of the barge and its cargo is denoted by W. When in salt-free riverwater, it floats 0.25 ft deeper than when in seawater ($\gamma = 64$ lb/ft^3). Find the weight W.

91. A solid steel ($\gamma_{st} = 489$ lb/ft^3), 1.0-ft-diameter sphere rests on the bottom of a still lake. The lake is 36.0 ft deep. How much work is required to slowly raise the top of the sphere to the water surface?

92. A box open at the top measures 3.0 m × 4.0 m × 2.0 m high and has a mass of $M_0 = 1000$ kg. It is floating in 20°C water. How much mass must be added or removed so the box floats at half-depth?

93. Figure P2.93 shows a cone floating in freshwater. Find the mass of the cone necessary for it to float with half its height above the water; $R = 10$ cm and $H = 60$ cm. [*Hint:* The volume of the cone is $\pi R^2 H/3$.]

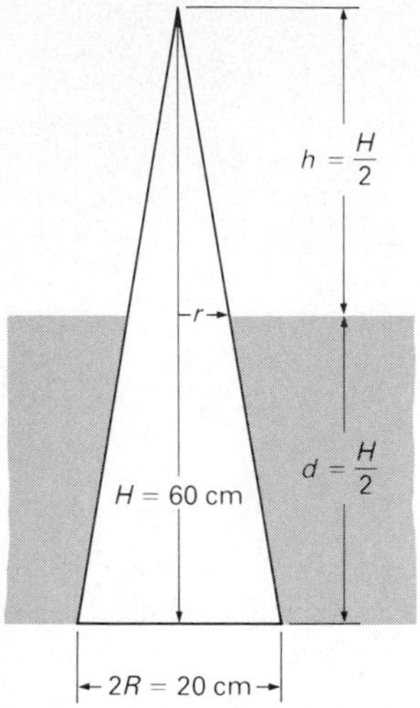

Figure P2.93

94. An object weighs 35 lb when totally submerged in water and 55 lb in air. Find the object's specific gravity and volume.

95. An empty, 100,000-lb barge has a flat bottom and vertical sides. The barge is 20 ft wide, 60 ft long, and 10 ft high. The sides are made of $\frac{3}{4}$-inch steel plate. Find the draft of the barge. (The draft is the depth of the bottom of the barge below the water surface.)

96. A spherical balloon filled with helium at 40°F and 20 psia has a 25-ft diameter. What load can it support in atmospheric air at 40°F and 14.696 psia? Neglect the balloon's weight.

97. Thirty cubic meters of wine having a specific gravity of 1.05 are fortified by removing 5 m^3 of wine and adding an equal volume of ethanol having a specific gravity of 0.85. How much deeper will a hydrometer float if it has a weight of 0.020 N and the stem protruding out of the wine has a diameter of 0.25 cm?

98. What is the minimum volume of a life preserver ($S = 0.2$) to support a 170-lb person in freshwater, with 80 percent of the person's body submerged in the water? The person's specific gravity is about 0.94.

99. Figure P2.99 shows the Baron's hot air balloon. The balloon, basket, burner, and butane weigh a total of 50 lb. The Baron weighs 150 lb. What is the required temperature inside the balloon to lift the balloon, basket, burner, butane, and Baron into 60°F air.

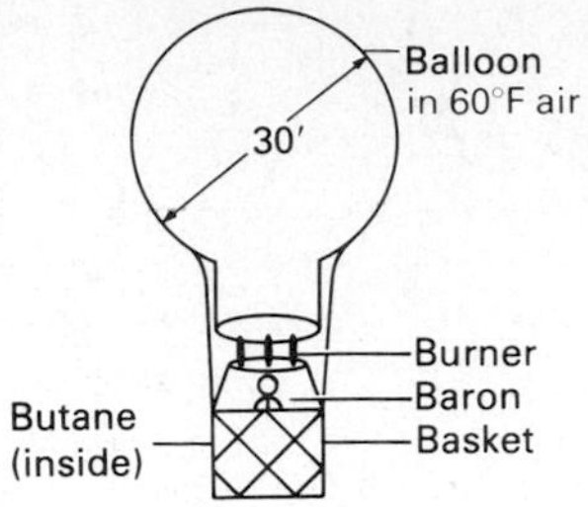

Figure P2.99

100. The block shown in Fig. P2.100 has a mass of 5.0 kg and measures 10 cm × 10 cm × 50 cm high. The massless support rod has length $\ell = 1.0$ m and diameter of of $d = 5.0$ cm. The volume of the pinned joints is negligible. Find the depth D. Assume that the block remains vertical.

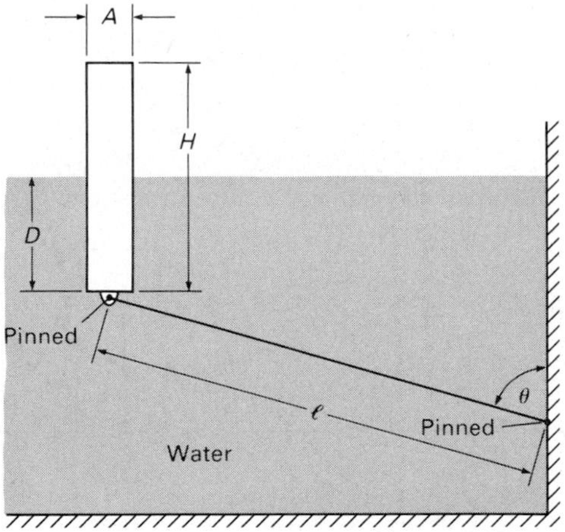

Figure P2.100

101. The wooden ($S = 0.5$) meter stick in Fig. P2.101 is floating with 13.0 cm out of freshwater. Determine the tension in the string if the meter stick's cross-sectional area is 2.0 cm^2.

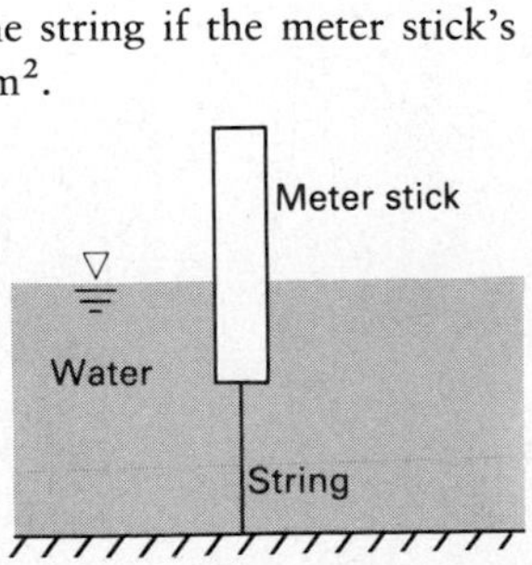

Figure P2.101

102. A child's balloon is a sphere 1 ft. in diameter. The balloon is filled with helium ($\rho = 0.014$ lbm/ft^3). The balloon material weighs 0.008 lbf/ft^2 of surface area. If the child releases the balloon, how high will it rise in the Standard Atmosphere. (Neglect expansion of the balloon as it rises.)

103. A flat-bottomed barge is to be designed to carry 10^6 kg of coal. The barge is 10 m wide and its sides and ends are 4 m high. How long should the barge be to carry the coal if the sidewalls are at least 1 m above the seawater? The barge is constructed of aluminum plate 7.5 cm thick. (The specific weights of seawater and aluminum are 1030 kg/m^3 and 2700 kg/m^3, respectively.)

104. General N and his companions are in a small speedboat and are being pursued by a large patrol boat. General N's boat reaches the lock shown in Fig. 2.104. Gate *A* is opened, and the speedboat rapidly enters the lock. Gate *A* is closed, the water level in the lock is raised to that outside gate *B*, gate *B* is opened, and the speedboat leaves the lock. The large patrol boat then enters the lock, the water level in the lock is raised to that outside gate *B*, and the patrol boat then leaves the lock. Is less water, the same amount of water, or more water added to the lock for General N's speedboat than for the patrol boat?

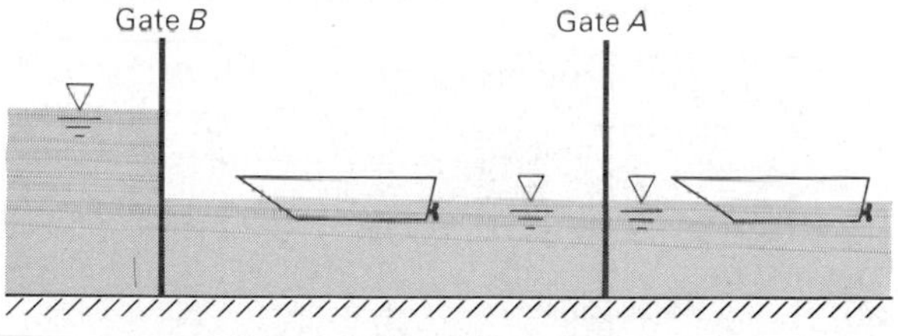

Figure P2.104

105. A 12-in. concrete block has a density of 138 lbm/ft^3 and weighs 55 lb. It is in a small rowboat in a small lake. The block is taken out of the boat and dropped in the lake. Will the lake-water level rise or fall?

• **106.** A steel can with a $\frac{1}{16}$-in. wall thickness has an inside diameter of 2.5 in. and height 4.0 in. The can is inverted, placed in a large pan of water, and allowed to come to an equilibrium position. See Fig. P2.106. Assume isothermal compression of the air in the can and determine the depth of submergence *D* of the can.

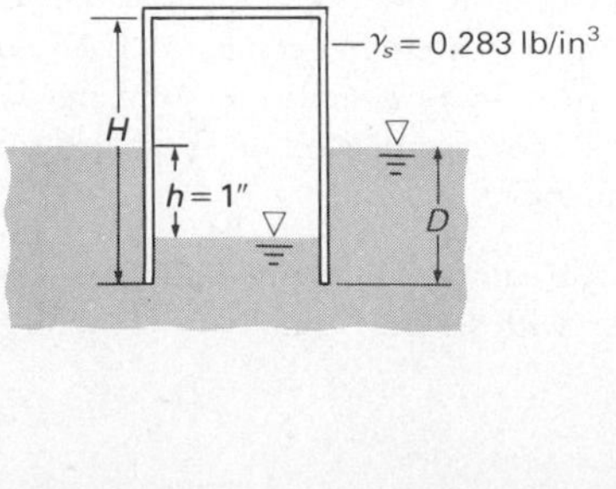

Figure P2.106

107. A floating 40-in.-thick piece of ice sinks 1 in. with a 500-lb polar bear in the center of the ice. What is the area of the ice in the plane of the water level? For seawater, $S = 1.03$.

• **108.** A submarine is modeled as a cylinder with a length of 300 ft, a diameter of 50 ft, and a conning tower as shown in Fig. P2.108. The submarine can dive a distance of 50 ft from the floating position in about 30 sec. Diving is accomplished by taking water into the ballast tank so the submarine will sink. When the submarine reaches the desired depth, some of the water in the ballast tank is discharged, leaving the submarine in "neutral buoyancy" (i.e., it will neither rise nor sink). For the conditions illustrated, find (a) the weight of the submarine and (b) the volume (or mass) of the water that must be in the ballast tank when the submarine is in neutral buoyancy. For seawater, $S = 1.03$.

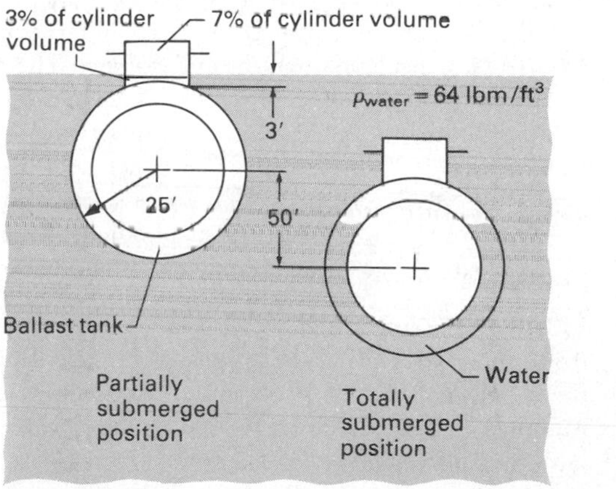

Figure P2.108

109. Secret Agent 007 is set adrift in a hollow steel sphere with an outside diameter of 2 m and a 0.5-cm-thick shell. The agent weighs 800 N. How deep will the sphere float in seawater? For steel, $S = 7.9$ and for seawater, $S = 1.03$.

110. Fir ($\gamma \approx 34$ lb/ft^3) and oak ($\gamma \approx 55$ lb/ft^3) trees felled by lumberjacks are floating down a river. Find the depth that each floats in the river.

• **111.** The float shown in Fig. P2.111 is used to maintain a constant water level in a tank found in most homes. The valve closes, decreasing the flow rate as the float rises, and the valve is completely closed (i.e., no flow)

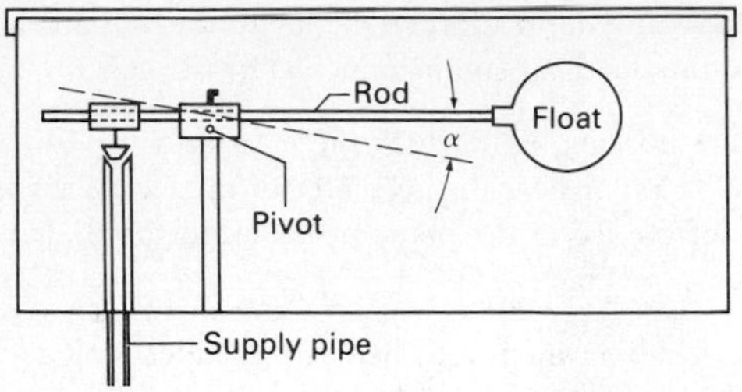

Figure P2.111

when the rod makes an angle $\alpha = 10°$ below the horizontal. The float and its support rod can be slid horizontally to move the float closer to or farther away from the pivot. If the water level at which a particular float stops the flow is too low, should the rod and float be moved to the right or left? Could the same result be obtained by bending the rod near the float upward or downward?

• **112.** A not-too-honest citizen is thinking of making bogus gold bars by first making a hollow iridium ($S = 22.5$) ingot and plating it with a thin layer of gold ($S = 19.3$) of negligible weight and volume. The bogus bar is to have a mass of 100 lbm. What must be the volumes of the bogus bar and of the air space inside the iridium so that an inspector would conclude it was real gold after weighing it in air and water to determine its density? Could lead ($S = 11.35$) or platinum ($S = 21.45$) be used instead of iridium? Would either be a good idea?

113. Find the angle θ at which a pinned wooden ($S = 0.50$) meter stick, as shown in Fig. P2.113, will float in the water. You must first obtain the weight and volume of a meter stick. Compare your calculated value of θ with that obtained by experiment.

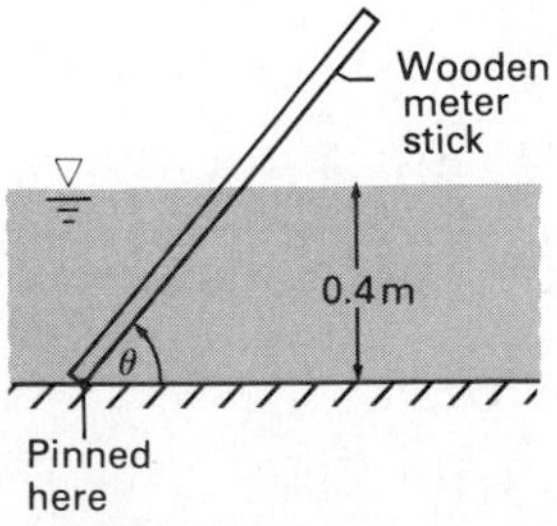

Figure P2.113

• **114.** A plastic straw has a 0.62-cm inside diameter, a 0.72-cm outside diameter, and a 20.5-cm length. It floats vertically in 20°C water with 1.0 cm protruding above the water, whereas it floats vertically in a soft drink with 2.0 cm protruding. The soft drink has been open to the air for a long period of time. Find the specific gravity of the soft drink and of the straw relative to 20°C water. How does this specific gravity of the soft drink compare with a measured density of 1030 kg/m³ ($S = 1.03$)? Explain why the straw in a glass of carbonated beverage will rise higher than the value predicted.

• **115.** A student wants to make a floating-ball hydrometer for an ethylene glycol-water mixture. See Fig. P2.115(a). The number of floating balls indicates the freezing temperature of the mixture when the mixture is at 15°C. Find the density of each of the four balls at 15°C from the following data. [Note that the lighter balls float when the mixture is warm (less dense) and the heavier balls start to float as the mixture gets colder (more dense).]

Number of Floating Balls	Freezing Temperature of Ethylene Glycol-Water Mixture
1	0°C
2	−10°C
3	−20°C
4	−30°C

The relation of the density of the ethylene glycol mixture to its freezing point is shown in Fig. P2.115(b).

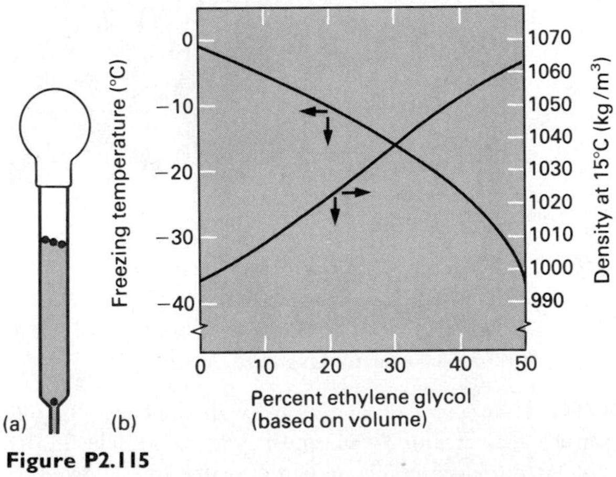

Figure P2.115

• **116.** Will the scale in Fig. P2.116 read the same or differently when the hollow aluminum ball is in the water? The ball has a circumference of 12 in. and a weight at sea level of 6.5 oz, and it is submerged halfway. Does the scale reading depend on the depth to which the ball is immersed?

• **117.** An empty 2-L plastic soda-pop bottle is filled to the top with water. A small (2.5-in. long) plastic "tiger,"

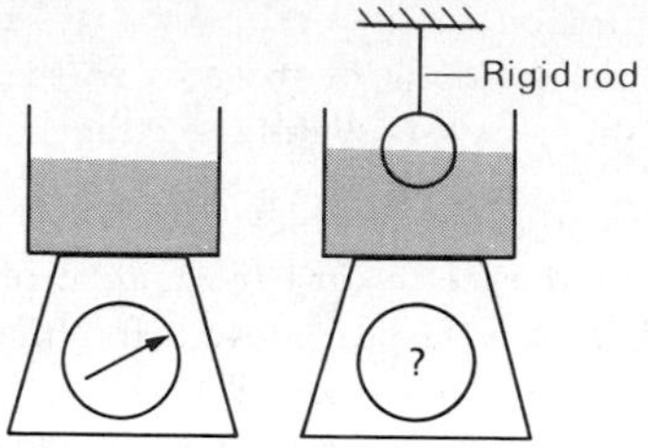

Figure P2.116

distributed in a breakfast cereal, is placed in the water and some water overflows from the bottle. The cap is then screwed on, pushing the tiger down a small amount. The tiger is then at the top of the water and no air bubbles are seen above the water. The plastic bottle is then squeezed on the sides and the tiger falls to the bottom of the water. When the pressure on the sides of the bottle is released, the tiger rises. Curiosity leads to an inspection of the tiger, which reveals that it has a small pin hole in the bottom and is hollow. Explain the behavior of the tiger.

• **118.** An empty 2-L plastic soda-pop bottle is filled to the top with water. A small glass test tube measuring about $2\frac{7}{8}$ in. long, $\frac{5}{16}$ in. inside diameter, and $\frac{3}{8}$ in. outside diameter is partially filled with water. It is then inverted and placed in the water, as shown in Fig. P2.118, so that the "top" (rounded part) of the test tube is even with the water level. Determine how much air must be inside the test tube. The glass specific weight is 170 lb/ft^3 and the water specific weight is 62.4 lb/ft^3. If the cap were placed on the soda-pop bottle and the sides of the bottle were squeezed, what would happen to the test tube?

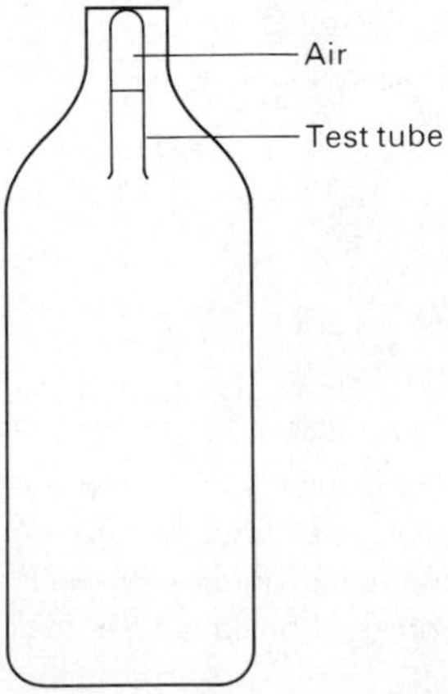

Figure P2.118

• **119.** A pipe is being installed underground, in a trench. Cement grout will be poured in and around the pipe to serve as a foundation. The grout will never be deeper than halfway up the pipe. Given the weight, W_p, of the pipe per foot in pounds, the radius (outer) of the pipe, r, in feet, and the specific weight of the grout, γ, in pounds per cubic foot, develop an expression for the distance from the centerline of the pipe to the top of the grout at which the pipe will "float."

120. A right circular cylinder of specific gravity S has radius R and height H. For what range of values of R/H will the cylinder float stably in water if the circular ends are horizontal?

121. A cube of specific gravity S has sides of length L. For what range of values of S will the cube float stably in water with a surface of the cube parallel to the water surface?

122. A drinking glass has a wall thickness of 3 mm, an inside diameter of 6.0 cm, and an inside depth of 18.0 cm. The glass has a specific gravity of 2.5. Is the drinking glass in stable or unstable equilibrium if placed upright and empty in a pan of 60°F water?

123. A solid cylindrical pine ($S = 0.50$) spar buoy has a cylindrical lead ($S = 11.3$) weight attached, as shown in Fig. P2.123. Determine the equilibrium position of the spar buoy in seawater (i.e., find d). Is this spar buoy stable or unstable? For seawater, $S = 1.03$.

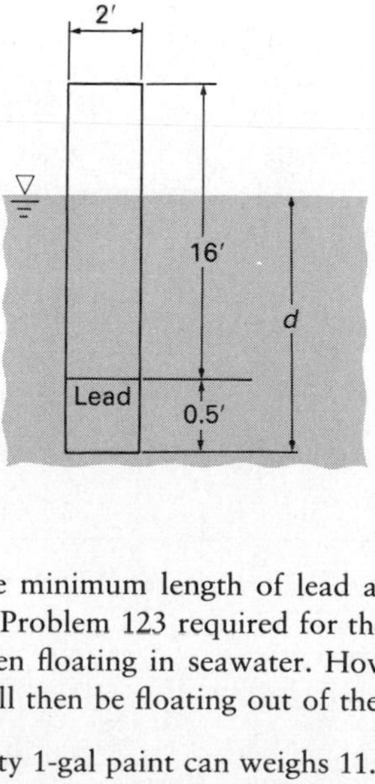

Figure P2.123

124. Find the minimum length of lead attached to the spar buoy in Problem 123 required for the spar buoy to be stable when floating in seawater. How much of the spar buoy will then be floating out of the water?

125. An empty 1-gal paint can weighs 11.5 oz and has a diameter of 6.55 in., a height of 7.37 in., and a center of gravity 3.8 in. above the bottom of the can. Will this paint can be stable while floating in water?

• **126.** The circular buoy shown in Fig. P2.126 is to float in freshwater to warn boaters. Will the buoy do its intended job? Is it necessary to connect a chain or rope from the bottom of the buoy to the bottom of the lake to stabilize the buoy?

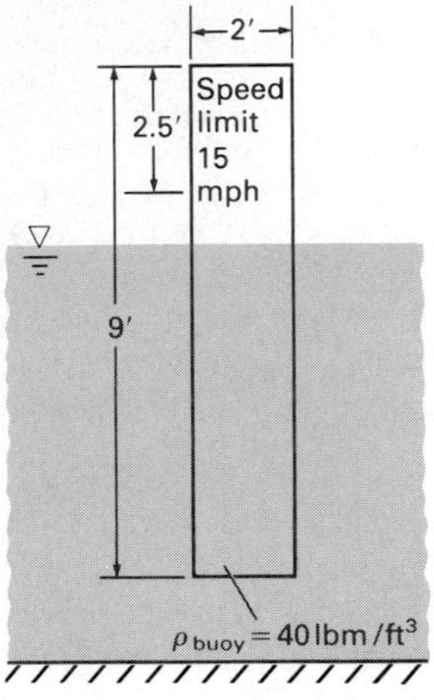

Figure P2.126

• **127.** A tall drinking glass has a 3.0-mm wall thickness, a 6.0-cm inside diameter, an 18.0-cm inside depth, and a specific gravity of 1.2. Compare the depths of submergence and the stability of the glass if it is placed upright and then upside down in 15°C water.

• **128.** Two young people are in a flat-bottomed boat and are being carried down a river by the current. The two people have a combined weight of 300 lb, and the boat weighs 20 lb per foot of width. A good estimate of the center of gravity is 1.7 feet above the bottom of the boat. As shown in Fig. P2.128, the boat has a rectangular cross section, its sides are 1.5 ft high, and it is 15 ft long. Find the width W of the boat so that it will not tip over if the boat lists slightly to either side.

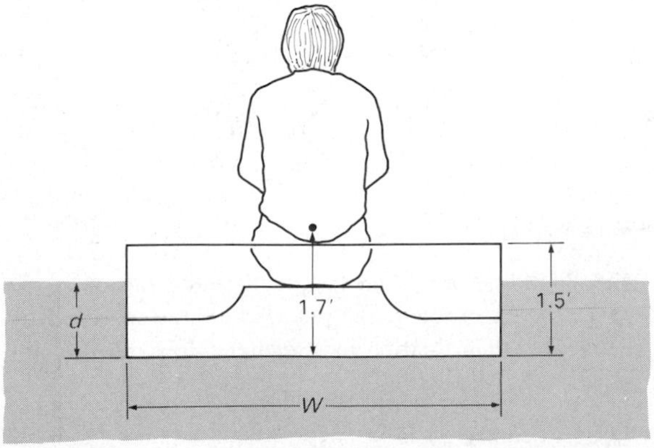

Figure P2.128

• **129.** The polar bear in Problem 107 slowly moves from the center of the ice toward the middle of one of the sides. How far can the bear move before the top surface of the ice becomes submerged? The ice has a square cross section with an area of 93.4 ft^2 in the plane of the water level. For seawater, $S = 1.03$; for ice, $S = 0.918$.

130. When an automobile brakes, the fuel gage indicates a fuller tank than when the automobile is traveling at a constant speed on a level road. Is the sensor for the fuel gage located near the front or rear of the fuel tank? Assume a constant deceleration.

131. A plastic glass has a square cross section measuring $2\frac{1}{2}$ in. on a side and is filled to within $\frac{1}{2}$ in. of the top with water. The glass is placed in a level spot in a car with two opposite sides parallel to the direction of travel. How fast can the driver of the car accelerate along a level road without spilling any of the water?

132. Repeat Problem 131 to find the maximum rate at which the car can decelerate.

133. The cart shown in Fig. P2.133 measures 10.0 cm long and 6.0 cm high and has rectangular cross sections. It is half-filled with water and accelerates down a 20° incline plane at $a = 1.0$ m/s^2. Find the height h.

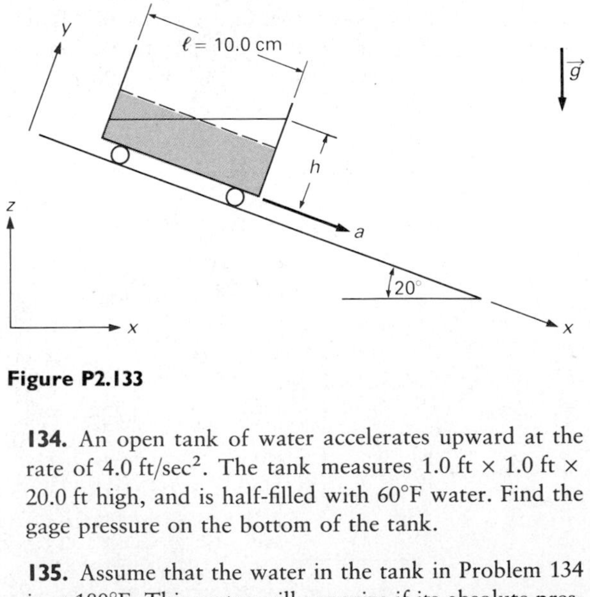

Figure P2.133

134. An open tank of water accelerates upward at the rate of 4.0 ft/sec^2. The tank measures 1.0 ft × 1.0 ft × 20.0 ft high, and is half-filled with 60°F water. Find the gage pressure on the bottom of the tank.

135. Assume that the water in the tank in Problem 134 is at 180°F. This water will vaporize if its absolute pressure drops to 7.52 psia. If the atmospheric pressure is 14.60 psia, calculate the tank acceleration (magnitude and direction) so the water at the bottom of the tank just starts to vaporize.

136. The U-tube manometer in Fig. P2.136 is used to measure the acceleration of the cart on which it sits. Develop an expression for the acceleration of the cart in terms of the liquid height h, the liquid density ρ, the local acceleration of gravity g, and the length ℓ.

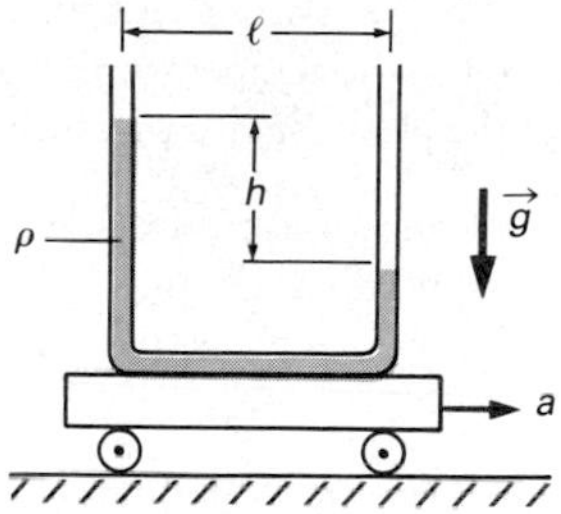

Figure P2.136

137. The cylinder in Fig. P2.137 accelerates to the left at the rate of 9.80 m/s^2. Find the tension in the string connecting the rod of circular cross section to the cylinder. The volume between the rod and the cylinder is completely filled with water at 10°C.

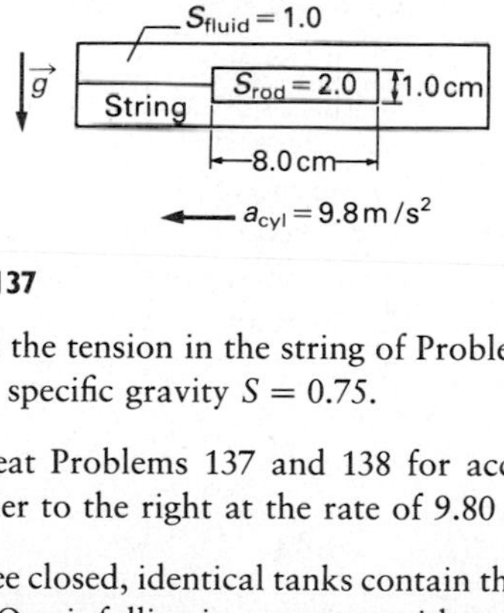

Figure P2.137

138. Find the tension in the string of Problem 137 if the rod has a specific gravity $S = 0.75$.

139. Repeat Problems 137 and 138 for acceleration of the cylinder to the right at the rate of 9.80 m/s^2.

140. Three closed, identical tanks contain the same mass of water. One is falling in a vacuum with an acceleration of 9.81 m/s^2, the second is falling but is retarded by air resistance and reaches a constant velocity of 100 m/s, and the third is falling but has an applied constant external force, so it accelerates downward at 12.1 m/s^2. Find the position of the water relative to the tank for each case. The local acceleration owing to gravity is 9.81 m/s^2.

• **141.** A cylinder of radius $R = 100$ cm is filled with oil ($S_o = 0.75$). As shown in Fig. P2.141, a 10° plastic sector

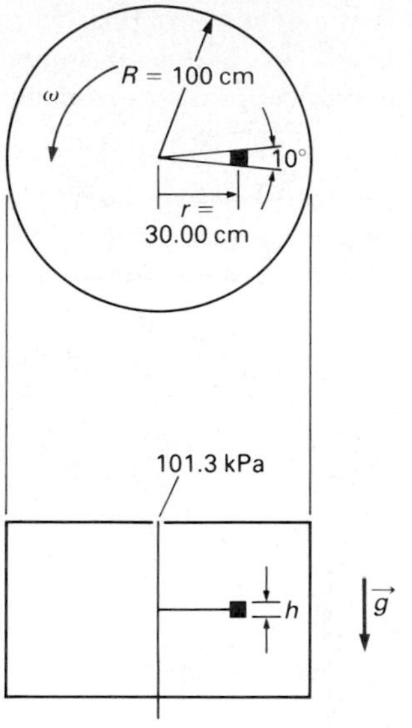

Figure P2.141

($S_c = 0.95$) with height h and thickness Δr of 5.0 cm is tied by a string to the center post passing vertically through the cylinder. A lid is put on the top of the cylinder. The cylinder is then rotated about its vertical axis at $\omega = 10$ Hz. Find the tension in the string if there is a hole at the top center of the cylinder so the pressure at that point is 101.3 kPa.

• **142.** Find the radial position of the plastic sector in Problem 141 if the specific gravity $S_c = 0.60$.

• **143.** A tank has a height of 5.0 cm and a square cross section measuring 5.0 cm on a side. The tank is one third full of water and is rotated in a horizontal plane with the bottom of the tank 100 cm from the center of rotation and two opposite sides parallel to the ground. What is the maximum rotational speed that the tank of water can be rotated with no water coming out of the tank?

• **144.** The U-tube in Fig. P2.144 rotates at 2.0 rev/sec. Find the absolute pressures at points C and B if the

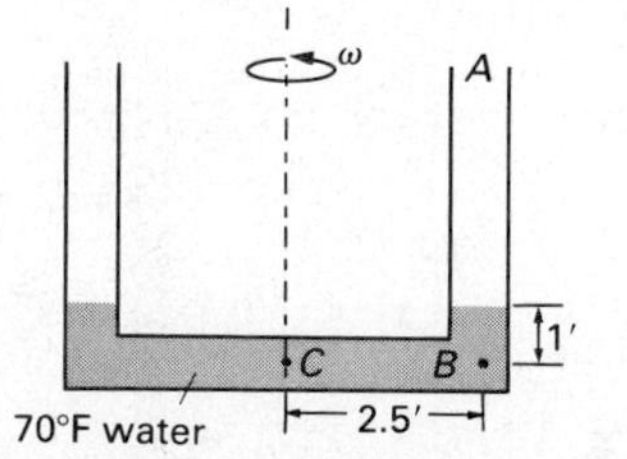

Figure P2.144

atmospheric pressure is 14.696 psia. Recall that 70°F water evaporates at an absolute pressure of 0.363 psia. Determine the absolute pressures at points C and B if the U-tube rotates at 2.0 rev/sec.

• **145.** Figure P2.145(a) shows the liquid phase of fluid *B* in a cylinder with the liquid under pressure from the weighted piston. The absolute pressure p_1 on the liquid is $p_1 = p_A + W/A$. For each pressure p_1, there is a temperature T_1, called the saturation temperature, at which the liquid begins to vaporize. As illustrated in Fig. P2.145(b), T_1 depends on the pressure p_1. Figure P2.145(b) also shows that for pressure p_1, fluid *B* is a liquid if $T < T_1$ and is a vapor if $T > T_1$. Note that if the cylinder in Fig. P2.145(a) were placed in a vacuum, p_A would be zero and the pressure p_1 on the liquid could be less than the atmospheric pressure (14.696 psia or 101,330 Pa, abs.) by using a lightweight piston. Now consider the barometer in Fig. P2.145(c), which has a mercury height of 29.90 in. The mercury temperature is 77°F. The space above the liquid mercury is filled with mercury vapor, whose pressure p_1 is called the saturation or vapor pressure for the temperature $T_1 = 77$°F and is 0.033 psia. Find the atmospheric pressure p_{atm}.

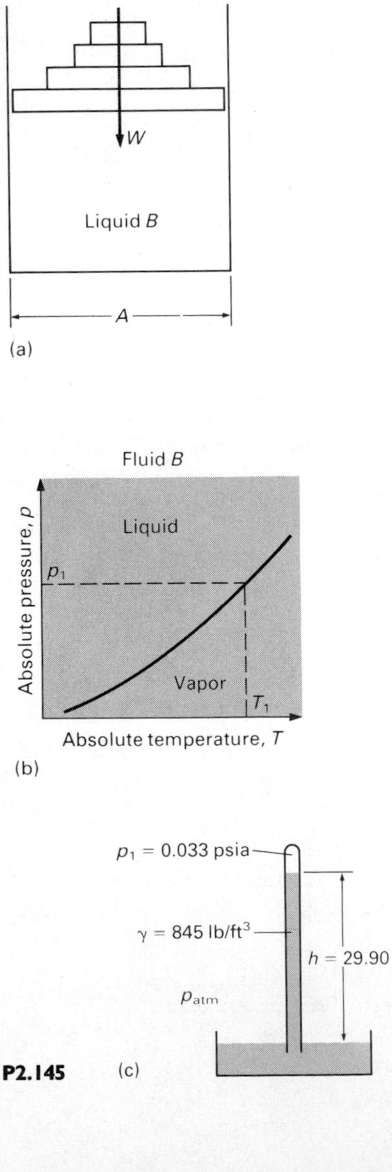

Figure P2.145

• **146.** Figure P2.146 shows a pump slowly taking water from a river. Water has the following saturation or vapor pressure as a function of the temperature (see Problem 145):

T (°F)	40	60	80	100	120	140
p (psia)	0.122	0.256	0.507	0.949	1.69	2.89

In a situation such as that shown in Fig. P2.146, the absolute pressure at the pump inlet could drop below the vapor pressure for the water temperature, and some of the liquid will evaporate or vaporize into vapor. Damage to the pump can occur when this vapor passes to other parts of the pump, where the pressure exceeds the vapor pressure for the water temperature. Figure P2.145(b) shows that the vapor will condense back into liquid. Because liquid water occupies less volume than an equal mass of water vapor, the condensation of water vapor into liquid creates small vacuum pockets that draw small pieces of material from the pump blades. The blade is said to erode and the pump is said to cavitate. Considering $p_{atm} = 14.60$ psia, find the height h at which the pump will cavitate.

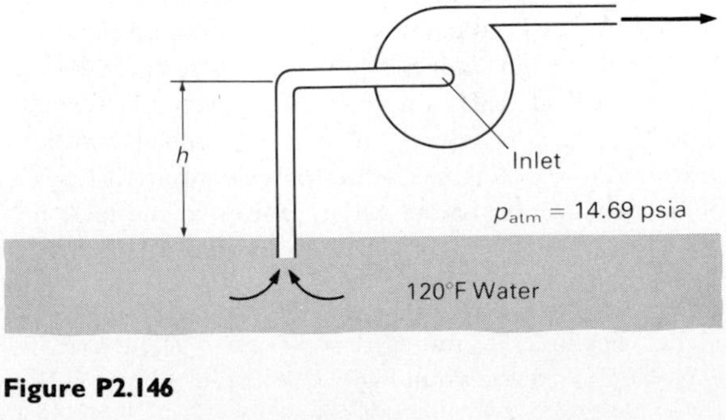

Figure P2.146

• **147.** Water at 25°C is flowing slowly down a vertical pipe and discharging to a river as shown in Fig. P2.147. If the saturation or vapor pressure for 25°C water is 3.17 kPa (see Problem 145), will the water at the top of the vertical pipe vaporize? The vertical pipe is filled with water.

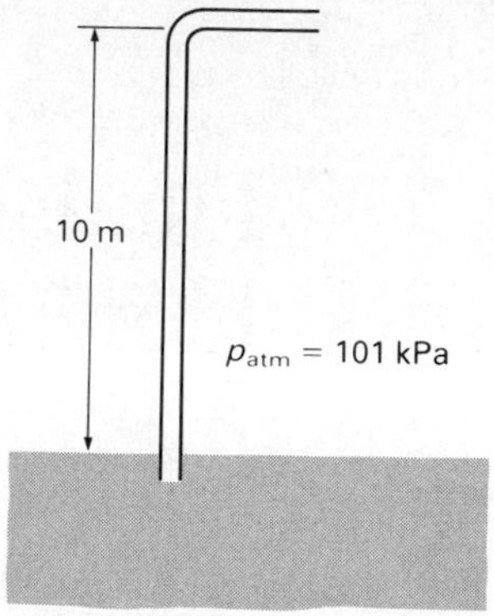

Figure P2.147

REFERENCES

1. Beckwith, T., N. Buck, and R. Marangoni, *Mechanical Measurements* (3rd ed.), Addison-Wesley, Reading, MA, 1981.

2. Benedict, R. P., *Fundamentals of Temperature, Pressure, and Flow Measurements* (2nd ed.), Wiley, New York, 1977.

3 Fundamental Concepts for Flow Analysis

Although there are many engineering applications of fluid statics and a few applications of rigid body motion, most fluid mechanics problems concern fluid flow. We covered most methods of analysis of nonflowing fluids in Chapter 2 but need several chapters to describe the basic methods of flow analysis.

In this chapter, we develop the fundamental concepts of flow analysis, including ways to describe fluid flow, natural laws that govern fluid flow, different approaches to formulating mathematical models of fluid flow, and methods that engineers use to solve flow problems. A clear understanding of these concepts is essential before you proceed to later chapters. We also suggest that you reread Sections 1.2 and 1.5 before beginning this chapter.

3.1 SOME TYPICAL FLOWS

Figures 3.1 and 3.2 illustrate typical flows of two broad types: internal or confined flows and external or unconfined flows. For internal flows, engineers are usually interested in the rate of fluid flow through the passage, changes of the fluid's energy during the flow, and, possibly, the force exerted on the confining walls. The fluid is entirely confined by the walls of the passage and, in the case of open channel flow, the free surface of the channel. The fluid involved may be gas or liquid. The flow may vary with time, as during the start-up of a pump, or may be invariant with time, as when a pump runs constantly.

In external flows, relative motion occurs between an object and a large mass of fluid. Engineers are usually interested in the forces exerted on the object by the fluid. They may also need to know the details of the fluid motion near the object's surface, so they can evaluate the rate of heat transfer to or from the surface. In this case, the fluid is not contained in a limited region; in fact, the fluid is often assumed to be infinite. The fluid may be gas or liquid or both, as with a sailboat. The flow may vary with time, such as during takeoff and climb of an airplane, or it may be invariant with time, as when an airplane cruises at constant speed and altitude in a still atmosphere.*

In spite of the apparent differences between these two types of flow, the fluid motion in both is governed by the same physical laws. Engineers may use different

* This flow is time invariant only in a coordinate system fixed to the airplane, a trick often used in analysis of external flows.

mathematical forms of these laws and different solution techniques, but the physical laws are the same. Initially, our approach to flow problems is as general as possible, so that different mathematical formulations do not mistakenly imply that there are several different "kinds" of fluid mechanics.

We introduced the division between internal and external flows primarily to show the widely different types of engineering flow problems. Engineers often model flows of one type by the opposite type. For example, engineers often study the flow over an airplane by placing a scale model of the airplane in a wind tunnel. If the model occupies only a small fraction of the wind tunnel's cross-sectional

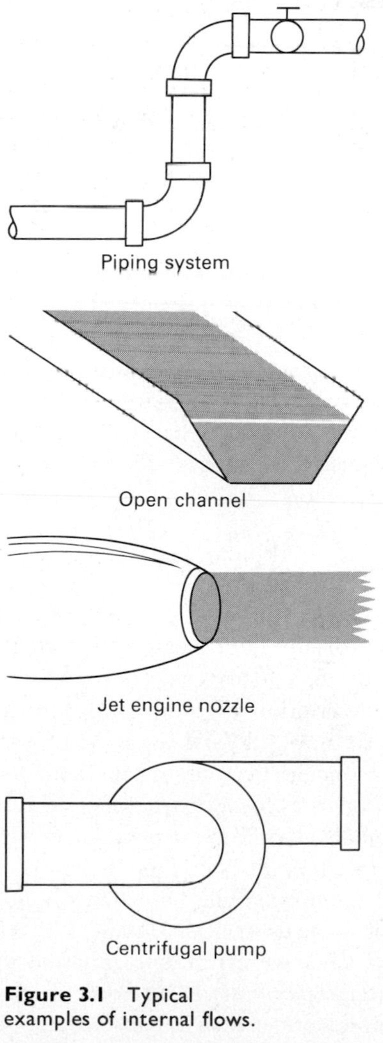

Figure 3.1 Typical examples of internal flows.

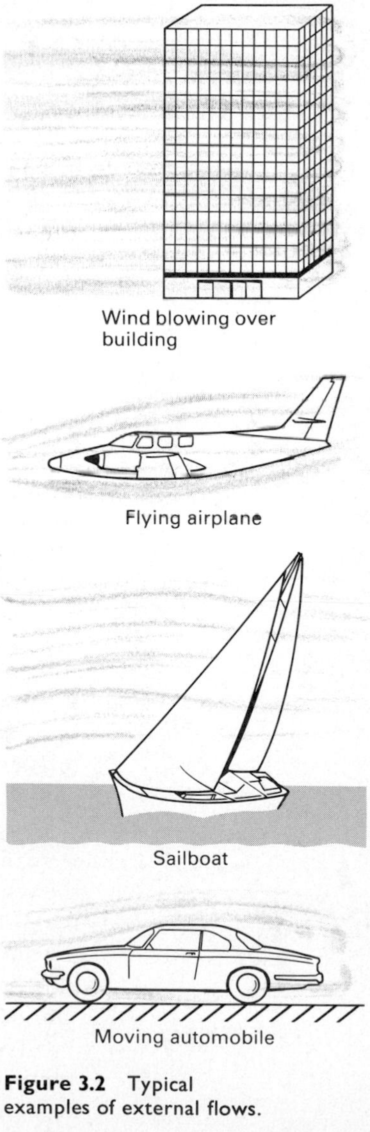

Figure 3.2 Typical examples of external flows.

area, the flow near the model is similar to the flow near the real airplane in the unconfined atmosphere. Some flows are a mixture of both types. For example, as an automobile moves along a road, the flow between the road surface and the underside of the auto is confined, while the flow over the sides and top is unconfined.

3.2 DESCRIBING FLUID FLOW

Before we can analyze fluid flow, we must be able to describe it. Also, we must be aware of the different types or *regimes* of fluid motion.

3.2.1 The Field Concept: Lagrangian Versus Eulerian Descriptions

Let's consider two of the flows illustrated in Figs. 3.1 and 3.2: the flow of water in a pipe and the flow of air around a moving automobile. In the pipe flow, the fluid is confined within the pipe; in the flow around the automobile, the entire atmosphere is the fluid of interest. In both cases, however, the fluid is continuous. A tiny mass, which we call a *fluid particle,* exists at each point in the fluid region.

A fluid particle is the tiny fluid mass that occupies a chosen point in the fluid.

There are infinitely many particles in any finite mass of fluid. Each particle may be characterized by its density, pressure, and other properties. If the fluid mass is moving, each particle also has some velocity. We define the velocity of a fluid particle as the velocity of the particle's center of mass. We use the symbol $\vec{V}$ for fluid velocity, because velocity is a vector. We may resolve the velocity vector into components along three arbitrarily selected directions. If we choose Cartesian coordinates (x, y, z), we use the symbols u, v, and w to represent the x, y, and z components, respectively. Thus

$$\vec{V} = u\hat{i} + v\hat{j} + w\hat{k}.$$

Because a fluid is deformable, the velocity of each fluid particle may be different. In addition, the velocity of any fluid particle may change over time. (According to Newton's laws of motion, velocity changes must be the result of forces on the particle.) A complete description of fluid motion requires specification of the velocity, pressure, and so on of all particles at all times. As a fluid is continuous, we can specify those characteristics only by mathematical functions that express the velocity and fluid properties for all particles at all times. Such a representation is called a *field representation,* and the dependent variables (such as velocity and pressure) are called *field variables.* The fluid region of interest (the water in the pipe or the air around the automobile) is called the *flow field.* See whether you can identify the other flow fields in Figs. 3.1 and 3.2.

To describe a flow field, we can adopt either of two approaches. The first approach, called the *Lagrangian description* (after the French mathematician J. L. deLagrange, 1736–1813), identifies each particular fluid particle and describes what happens to it over time. Mathematically we write fluid velocity as

$$\vec{V} = \vec{V}\{\text{particle identity}, t\}.^*$$

The independent variables are particle identity and time. The Lagrangian approach is widely used in solid mechanics, and in your study of dynamics, you undoubtedly calculated the velocities of a ball, a projectile, a particular block attached to a pulley system, and so on. A Lagrangian description is attractive if a small number of particles is involved, if all particles move as a rigid body, or if all particles are displaced only a small amount from their initial or equilibrium position. However, in a flowing fluid, identifying and keeping track of various particles is virtually impossible.[†] Further complications arise because a typical fluid particle often experiences a large displacement. For these reasons, a Lagrangian description is not very useful in fluid mechanics.

The second approach, called the *Eulerian description* (after the Swiss mathematician L. Euler, 1707–1783), focuses attention on a particular point (or region) in space and describes happenings at that point (or within and on the boundaries of the region) over time. The properties of a fluid particle depend on the location of the particle in space and time. Mathematically, we express the velocity field as

$$\vec{V} = \vec{V}\{x, y, z, t\}.$$

The independent variables are location in space, represented by the Cartesian coordinates (x, y, z) and time. We might talk about the fluid velocity at the outlet of the pipe 3 sec after the flow started or the air pressure 3 in. ahead of the automobile's hood ornament. At each instant of time, a different fluid particle probably occupies those positions, but that does not matter. Because identifying fixed points in space is usually easier than identifying individual pieces of fluid, the Eulerian description is most often used in fluid mechanics. Solving a fluid flow problem then requires determining the velocity, pressure, and so on as functions of space coordinates and time. We can then use the functions

$$\vec{V}\{x, y, z, t\} \quad \text{or} \quad p\{x, y, z, t\}$$

to find velocity or pressure anywhere within the field at any time simply by substituting values for x, y, z, and t.

The Eulerian description is especially suited to fluid mechanics problems because it does not depict what happens to any particular fluid particle. The engineering application of a flow analysis usually is concerned with the effects of the fluid motion on certain objects, such as the blades of a pump or the windows of a building. The pressure on the window, not the effect of the window on any particular fluid particle, is important to the building's designer.

F. M. White has illustrated the Eulerian and Lagrangian descriptions by applying them to traffic on a freeway [1]. Engineers who design the freeway are concerned with the number of cars that will use it, the traffic flow, and the number of cars that may be expected to enter or leave the freeway at each ramp. An

---

* Recall that dashed parentheses { } mean "is a function of."

† The only practical method identifies particles by their positions in space at a particular instant in time—for example, the initial or equilibrium position (x_0, y_0, z_0).

Eulerian description is perfect, because the freeway designer has no interest in any particular car. A police officer patrolling the freeway, however, is interested in identifying those particular cars that are breaking the law and in giving the drivers tickets. Only a Lagrangian approach would stand up in court!

EXAMPLE 3.1   **Illustrates a Velocity Field Given by an Eulerian Description and Some Characteristics of Fluid Flow**

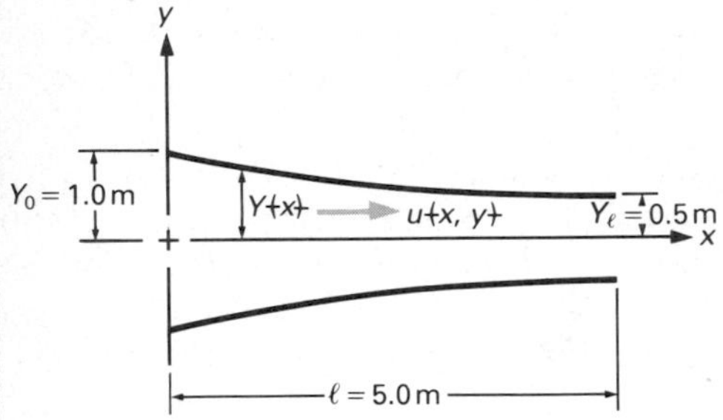

Figure E3.1a Two-dimensional channel.

A constant-density fluid flows in the converging, two-dimensional channel shown in Fig. E3.1a. The width perpendicular to the paper is quite large compared to the channel height. The velocity in the z direction is zero. The channel half-height, Y, and the fluid x velocity, u, are given by*

$$Y = \frac{Y_0}{1 + x/\ell} \quad \text{and} \quad u = u_0\left(1 + \frac{x}{\ell}\right)\left[1 - \left(\frac{y}{Y}\right)^2\right],$$

where x, y, Y, and ℓ are in meters, u is in m/s, $u_0 = 1.0$ m/s, and $Y_0 = 1.0$ m. Tabulate and plot the velocity distribution $u\{y\}$ at $x/\ell = 0, 0.5,$ and 1.0. Use y/Y values of $0, \pm 0.2, \pm 0.4, \pm 0.6, \pm 0.8,$ and ± 1.0.

SOLUTION

Given

$Y\{x\}$ and $u\{x, y\}$ as specified

Figure E3.1a

Find

u for specified values of x and y

Solution

Substitute numerical values of x/ℓ and y/Y into equation for $u\{x, y\}$.

The results are given in Table E3.1 and plotted in Fig. E3.1b.

ANSWER

Discussion

With the exception of the fluid in contact with the wall, the fluid velocity u increases as the channel height $2Y$ decreases. The fluid in contact with the wall has zero velocity, consistent with the no-slip condition.

This flow field also has a y component of velocity. This velocity, determined in Example 5.2, is given by:

$$v\{x, y\} = \frac{-u_0 y}{\ell} + \frac{u_0 y^3 (1 + x/\ell)^2}{\ell Y_0^2}.$$

Table E3.1 Tabulated values of the velocity $u\{x, y\}$ as a function of the length ratio x/ℓ, the height ratio y/Y, and the height y for the two-dimensional channel.

Length Ratio x/ℓ	Height Ratio y/Y	Height y (m)	Fluid Velocity u (m/s)
0.0	0.00	0.00	1.00
	±0.20	±0.20	0.96
	±0.40	±0.40	0.84
	±0.60	±0.60	0.64
	±0.80	±0.80	0.36
	±1.00	±1.00	0.00
0.5	0.00	0.00	1.50
	±0.20	±0.13	1.44
	±0.40	±0.27	1.26
	±0.60	±0.40	0.95
	±0.80	±0.53	0.54
	±1.00	±0.67	0.00
1.0	0.00	0.00	2.00
	±0.20	±0.10	1.92
	±0.40	±0.20	1.68
	±0.60	±0.30	1.28
	±0.80	±0.40	0.72
	±1.00	±0.50	0.00

---

* Note that u does not depend on time.

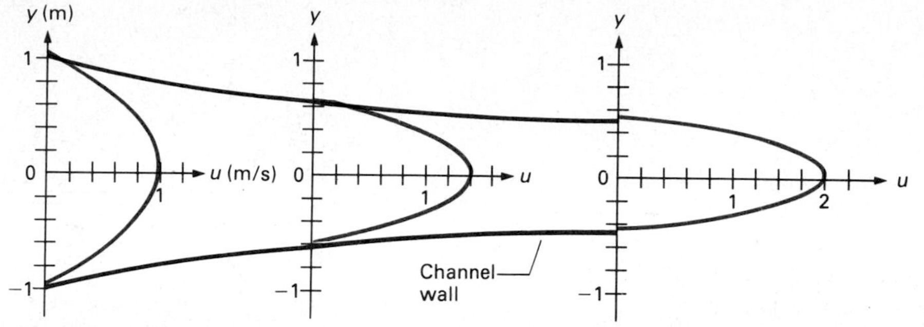

Figure E3.1b Plot of the velocity $u(x, y)$ as a function of the transverse coordinate y at various axial locations.

3.2.2 Visualizing the Velocity Field

To visualize the velocity field in a complex flow, we construct various flow lines on paper or in the laboratory.* These lines have the names streamline, pathline, streakline, and timeline. The first two are primarily analytic concepts; the last two are primarily laboratory concepts. The visual description of the flow given by these lines is loosely called the *flow pattern*.

A *streamline* is an imaginary line that is everywhere tangent to the fluid velocity vector. Figure 3.3 shows a few streamlines for the flow of an inviscid fluid over a circular cylinder. For any flow field, we can express the streamlines as a family of curves:

$$\Psi(x, y, z) = C(t) \tag{3.1}$$

We can (in principle) obtain the equations of the streamline curves from the condition that the streamline be tangent to the velocity vector; thus

$$\left.\frac{dy}{dx}\right)_{\Psi} = \frac{v}{u}; \tag{3.2a}$$

$$\left.\frac{dy}{dz}\right)_{\Psi} = \frac{v}{w}; \tag{3.2b}$$

$$\left.\frac{dz}{dx}\right)_{\Psi} = \frac{w}{u}. \tag{3.2c}$$

Because the fluid velocity (and therefore its components) may change with time, the streamlines also may change with time. If the velocity field does not change with time, the streamlines are fixed curves in space.

* Flow lines are illustrated very well in the film *Flow Visualization* [2].

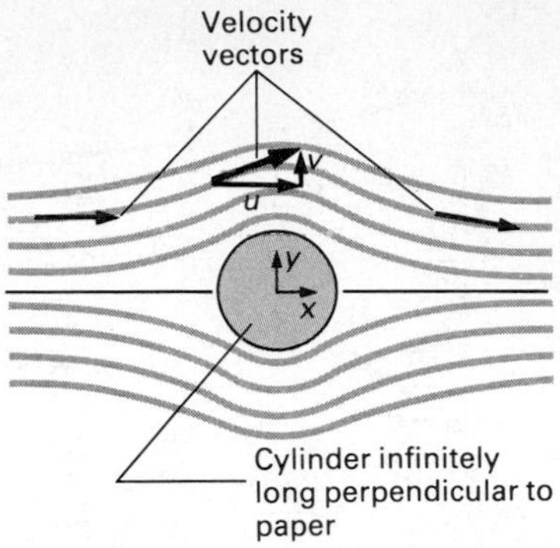

Figure 3.3 Typical streamlines for inviscid flow over a circular cylinder.

The streamline is an analytic concept. In the laboratory we have no way of marking curves that are instantaneously tangent to the fluid velocity, especially if the velocity varies with time.

A *pathline* is the curve marked out by the trajectory of a particular fluid particle as it moves through the flow field. Each fluid particle travels along its own pathline. The pathline is a Lagrangian concept; if we could analytically identify the pathline for all fluid particles, we could apply a Lagrangian method of description.

We can determine a pathline in the laboratory. We simply need to "mark" a fluid particle at some instant and then take a time-exposure photograph of its subsequent motion. We might mark a particle of water flowing in a channel by injecting dye with a hypodermic needle. Or we might electrolyze some of the water into hydrogen and oxygen bubbles by passing a pulse of current through a tiny wire in the water; the oxygen bubbles quickly dissolve, but the hydrogen bubbles remain, marking the fluid.

The equation of the pathline of a specific particle is

$$f(x, y, z) = C(t) \tag{3.3}$$

where

$$\frac{dx}{dt} = u_{\text{part}}(t); \tag{3.4a}$$

$$\frac{dy}{dt} = v_{\text{part}}(t); \tag{3.4b}$$

$$\frac{dz}{dt} = w_{\text{part}}(t). \tag{3.4c}$$

If the velocity field is not a function of time—and because any particle is always traveling in the direction of its own velocity, which is the same for all times at any point—a pathline and a streamline through the same point are identical. Stated another way, if the velocity is not a function of time, any fluid particle always travels along a unique streamline.

A *streakline* is a line made up of all particles that have passed a certain point. The streakline is essentially a laboratory concept. We can construct a streakline in the laboratory by continuously marking all fluid particles that pass a certain

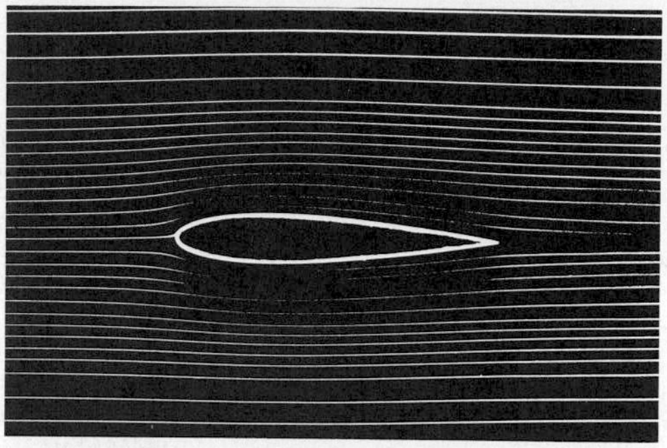

Figure 3.4 Smoke marks streaklines in flow over an airfoil in a wind tunnel (from the "NCFMF Book of Film Notes," 1974, The MIT Press with Education Development Center, Inc., Newton, Mass.).

point, say, by the continuous injection of dye in water or smoke in air (see Fig. 3.4). The marker material forms a streak downstream from the injection point. If the velocity field is not a function of time, all particles passing the injection point follow exactly the same path downstream and always move along the streamline passing through the point. A streakline is therefore identical to a pathline and a streamline for steady flow. As used in the laboratory, streaklines mark streamlines in steady flows, permitting the study of complicated flow patterns.

In a flow with a time-variant velocity, streamlines, pathlines, and streaklines are all different (see Fig. 3.5). We need all of them in order to visualize such a flow.

Streamlines, pathlines, and streaklines, although different in concept, reduce to identical lines in many cases. A timeline, however, is distinctly different. A *timeline* is a line of fluid particles that have been marked at a particular instant of time. The usual marking technique is to electrolyze a line of hydrogen bubbles in water by pulsing a current through a wire. Timelines are usually marked so that they are (initially) perpendicular to the direction of flow across the line. Each piece of a timeline subsequently travels at the velocity appropriate to its location at the time

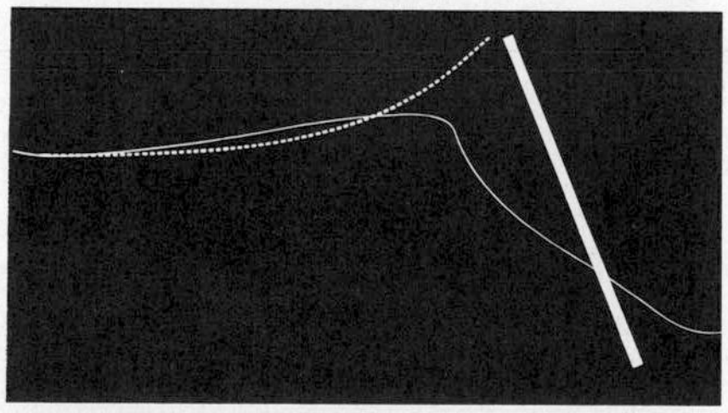

(a)

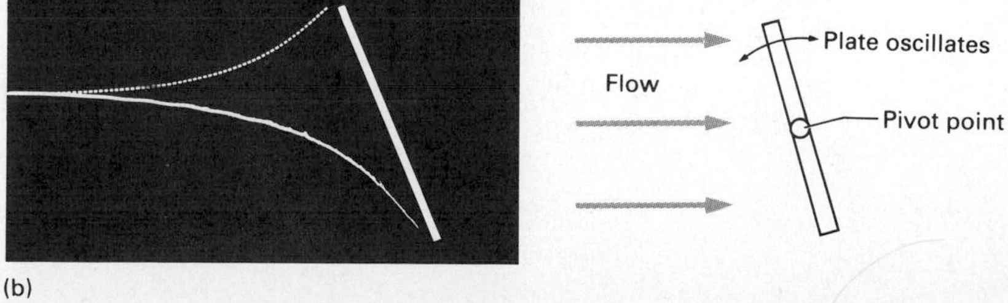

(b)

Figure 3.5 Difference between streamlines, streaklines, and pathlines in unsteady flow over an oscillating plate: (a) streamline (dotted) and pathline; (b) streamline and streakline. (From the "NCFMF Book of Film Notes," 1974, The MIT Press with Education Development Center, Inc., Newton, Mass.).

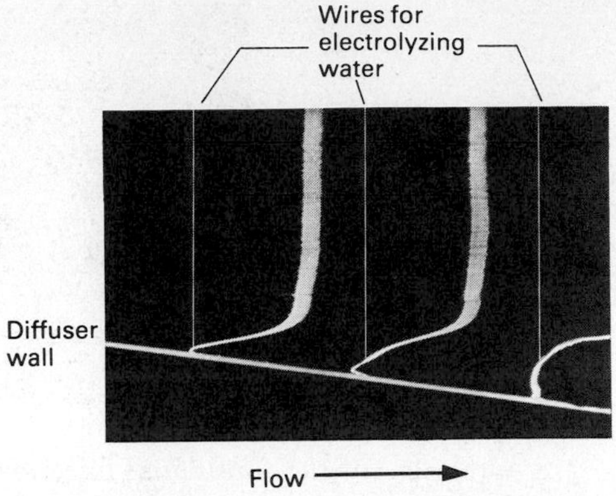

Figure 3.6 Timelines marked by hydrogen bubbles in water flow through a diffuser (from the "NCFMF Book of Film Notes," 1974, The MIT Press with Education Development Center, Inc., Newton, Mass.).

of marking. A photograph of the timeline taken at a later time shows how far the various elements of the line have traveled in the elapsed time (see Fig. 3.6). The distance traveled by each element is proportional to the velocity of the element, so the distorted timeline is a picture of the velocity profile at the instant and location of marking.

3.2.3 Classifying the Velocity Field

Dimensionality, directionality, and steadiness categorize the mathematical and physical nature of the velocity field. Recall that the fluid velocity is a *vector* function of four independent variables—three space coordinates and time. Using Cartesian coordinates, we write

$$\vec{V} = \vec{V}(x, y, z, t)$$

or, in terms of components,

$$\vec{V} = u(x, y, z, t)\hat{i} + v(x, y, z, t)\hat{j} + w(x, y, z, t)\hat{k}.$$

The *dimensionality* of a flow is the number of independent space coordinates necessary to specify the velocity.* From another point of view, dimensionality is the number of directions in which the velocity can vary. Dimensionality can vary from zero to three. A flow with zero dimensionality is called a *uniform flow*.

The *directionality* of a flow field is the number of nonzero velocity components present in the field, that is, the number of nonzero dependent variables in the velocity function. Directionality can vary from one† to three. It may seem that dimensionality and directionality are the same thing, but a few examples will show that they are really quite different.

One situation is the flow of a viscous fluid between two large parallel plates, as shown in Fig. 3.7. We assume that there is no flow perpendicular to the paper. The streamlines are straight, parallel lines. There is no velocity perpendicular to the streamlines, and the velocity profile does not change shape as the flow proceeds downstream. This flow has $v = 0$ and $w = 0$, and it is one-directional in x. The

---

* A slightly more general definition would involve all fluid properties rather than just the velocity.
† If directionality is zero, there is no flow!

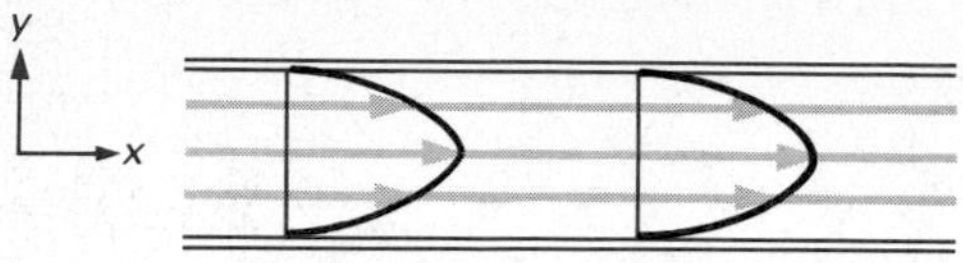

Figure 3.7 Flow of a viscous fluid between parallel plates: Flow is one-dimensional in y and one-directional in x.

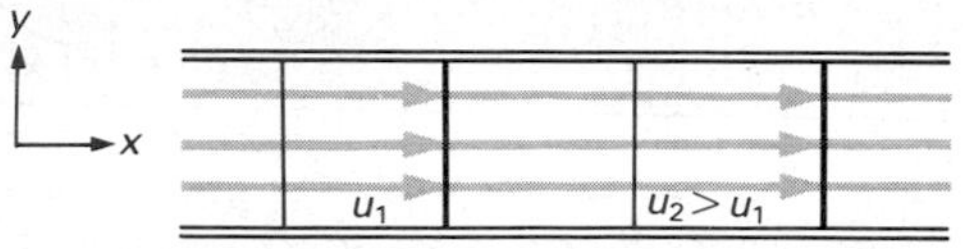

Figure 3.8 This flow is one-dimensional and one-directional in x.

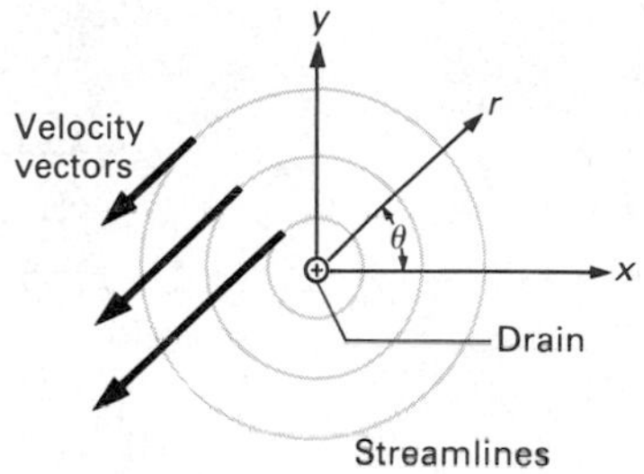

Figure 3.9 Approximate streamlines of swirling flow around a drain.

single velocity, u, varies in a direction perpendicular to the plates, and the flow is one-dimensional in y (but not in x).

Another situation is the flow shown in Fig. 3.8. The streamlines are straight, and $v = w = 0$. The velocity u varies in the x direction but not in the y direction. In this case, the single dimension and the single direction are aligned.

Then there is the flow over a circular cylinder as shown previously in Fig. 3.3. If the cylinder is very long perpendicular to the paper and if the fluid approaches the cylinder in a direction perpendicular to its axis, there is no velocity along the cylinder and the flow is identical in all planes perpendicular to the cylinder's axis. This flow is two-dimensional and two-directional. If we were to superimpose a uniform fluid velocity perpendicular to the paper on the entire flow field (equivalent to moving the cylinder parallel to its axis), the flow would become three-directional but would still be two-dimensional.*

When we perform an analysis, dimensionality determines whether we must solve (at least in principle) ordinary (one-dimensional) or partial (two- or three-dimensional) differential equations, and directionality determines the number of equations we must solve. Obviously, we want both dimensionality and directionality to be as low as possible to simplify the analysis. We have some control over dimensionality and directionality by our selection of a coordinate system, as the following illustration shows.

Consider the flow of water into a drain. Far from the drain, the water swirls around the drain in almost circular streamlines, with very little motion toward the drain. Figure 3.9 shows the approximate streamline pattern and the velocity vec-

* Unfortunately, the distinction between dimensionality and directionality is seldom clearly made. "Dimensionality" is usually used as a catch-all for both ideas. It is usually implied that a "two-dimensional" flow is also two-directional and that the two directions line up with the two dimensions. At least two of the preceding situations would violate one of these implications.

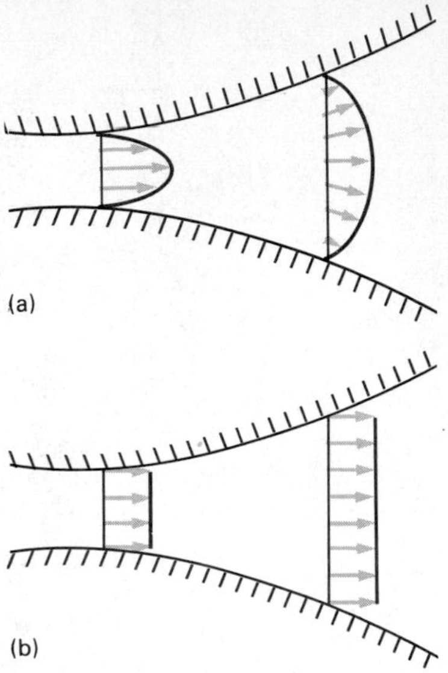

Figure 3.10 One-dimensional and one-directional flow approximation for diverging passage: (a) actual two-dimensional and two-directional flow; (b) approximate one-dimensional, one-directional flow.

tors. The swirling velocity increases as the flow moves toward the center. If we choose to analyze this flow in a Cartesian (x, y) coordinate system, the flow is two-dimensional and two-directional. If we choose a polar (r, θ) coordinate system, the flow is one-dimensional (in r) and one-directional (in θ).

We can often greatly simplify calculations with only a small sacrifice in accuracy by using average velocities to replace a velocity distribution in one or two directions. We can approximate the two-dimensional (and two-directional) flow in the diverging passage of Fig. 3.10(a) with the one-dimensional, one-directional flow of Fig. 3.10(b) by replacing the velocity distribution $u\{x, y\}$ with a uniform average velocity given by, say,*

$$\bar{u}\{x\} = \frac{1}{Y\{x\}} \int_0^{Y\{x\}} u\{x, y\}\, dy.$$

Thus by cleverly choosing a coordinate system and using average velocities, we can reduce some very complicated flows to rather simple problems.

The concept of flow steadiness or unsteadiness is closely associated with dimensionality in that it deals with one of the independent variables (time). A flow is *steady* if the velocity and other flow properties at all points in the field do not vary with time. A flow is *unsteady* if the velocity or any other property varies with time of any point.

Because time is an independent variable, steady flows (with one fewer variable) are easier to analyze than unsteady flows. Many flows that are unsteady with respect to one coordinate system are steady with respect to another coordinate

---

* There are several possibilities for defining an average, depending on exactly what you want to do in your analysis [3]. The point here is that an average velocity can be defined and that it reduces the dimensionality (and sometimes, the directionality) of the problem.

system. If we observe a passing boat from the bank of a river, we see that the flow of water around the boat is unsteady, even if the boat travels at constant speed. If we watch the water from the bow of the boat, we see a steady flow of water relative to the boat. In such a case, we can simply attach a coordinate system to the boat and perform a steady-flow analysis. Our analysis invariably makes use of Newton's second law of motion somewhere, so this trick works only if the coordinate system is moving at constant velocity. The trick fails at other times too. If the boat is approaching a pier, the motion is unsteady from both the boat and the pier, even if the boat is moving at constant velocity.

EXAMPLE 3.2 Illustrates the Difference Between Dimensionality and Directionality and Illustrates the Concept of Streamlines

Determine the dimensionality and the directionality of the velocity field of Example 3.1. Then sketch some typical streamlines.

SOLUTION

Given

Velocity field in Fig. E3.2a (previously illustrated in Fig. E3.1b)

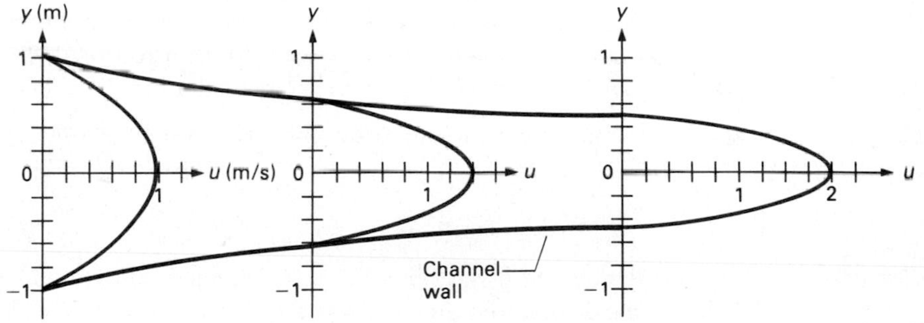

Figure E3.2a Plot of the velocity $u(x, y)$ as a function of the transverse coordinate y at various axial locations.

Find

Dimensionality and directionality of the flow

Several streamlines for the flow

Solution

The dimensionality of a velocity field is the number of independent space variables necessary to specify the velocity. The velocity u in Example 3.1 depends on both x and y, so the velocity field is two-dimensional.

The directionality of a velocity field is the number of nonzero velocity components in the field. We definitely have one nonzero velocity component, u. The problem statement of Example 3.1 says

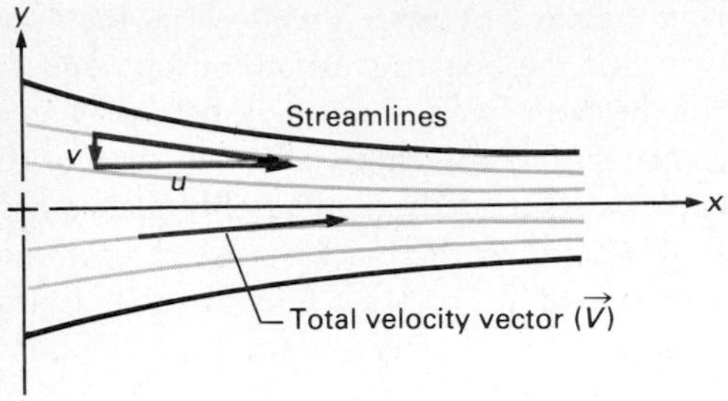

Figure E3.2b Typical streamlines for a flow field.

that the z velocity is zero. What about the y velocity component, v? The walls of the channel converge as the fluid flows from left to right, so they "push" the fluid toward the center of the channel. This configuration gives rise to a velocity v directed toward the center. As v is nonzero, we have two nonzero velocities, and the flow is two-directional.

The velocity field is two-dimensional, two-directional. **ANSWER**

A streamline is an imaginary line tangent to the fluid velocity vector. The total velocity of each particle is directed toward the center and toward the outlet of the channel.

The streamlines must be as shown in Fig. E3.2b. **ANSWER**

Discussion

Later we show how to determine v mathematically by knowing only $u(x, y)$. This flow is steady because the velocity field does not vary in time. Because the flow is steady, the streamlines shown in Fig. E3.2b correspond to both pathlines and streaklines. The velocity profiles shown in Fig. E3.2a correspond to timelines marked one second prior to the instant the "picture" is made.

3.2.4 Flow Regimes

Besides classification as steady or unsteady, uniform or one-, two-, or three-dimensional, and one-, two-, or three-directional, flows are divided into three flow regimes to describe the state of fluid motion. The three regimes are

- irrotational/inviscid flow (often called *ideal flow*);
- laminar flow; and
- turbulent flow.

The flow of water in a diverging channel, shown in Figs. 3.11 and 3.12, illustrates the distinctions among these three types of flow. The flow is from left to right and is made visible in Fig. 3.11 by hydrogen bubbles generated continuously by a wire stretched across the channel. In Fig. 3.12, velocity profiles are generated by three wires.

Probably the first thing to note about the flow is that the field divides into a rather large zone of nearly uniform flow in the center region of the channel and two thinner zones of nonuniform flow near the walls. First imagine a fluid particle from somewhere near the center of the uniform flow region (Fig. 3.13a). Because the top and bottom of the particle are moving in the same direction at the same speed, the particle is not being sheared or deformed, nor is it rotating (spinning) about its center.

Newton's law of viscosity connects shear stress with fluid shear deformation rate. For the nearly one-directional flow being considered, the equation (Eq. 1.2) is

$$\tau = \mu \frac{\partial u}{\partial y}.$$

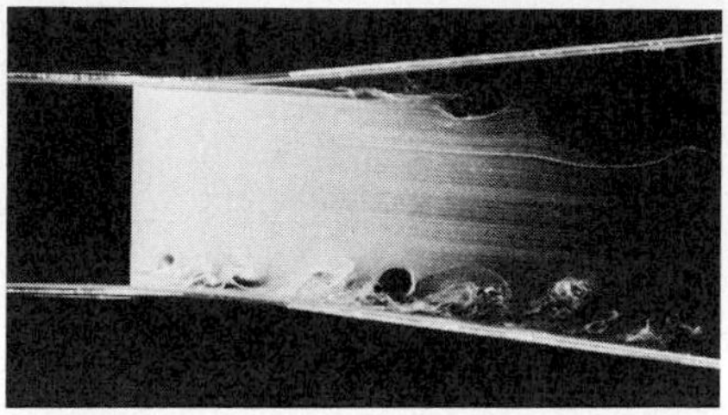

Figure 3.11 Water flowing in a diverging channel; flow is made visible by generating hydrogen bubbles from a wire spanning the channel (from the "NCFMF Book of Film Notes," 1974, The MIT Press with Education Development Center, Inc., Newton, Mass.).

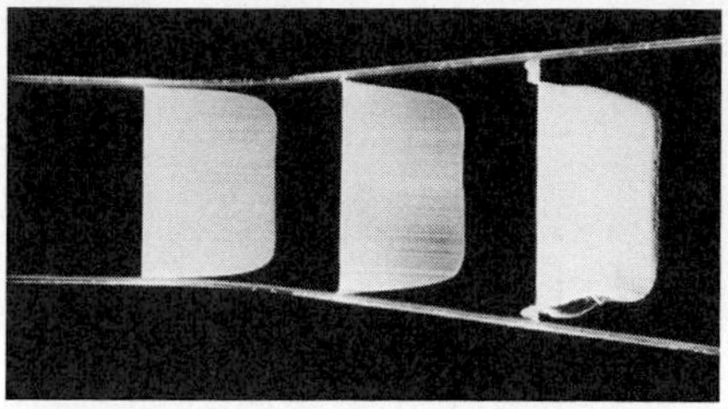

Figure 3.12 Velocity profiles at three locations in water flow in a diverging channel (from the "NCFMF Book of Film Notes," 1974, The MIT Press with Education Development Center, Inc., Newton, Mass.).

Many flows contain regions where the fluid shear deformation is small while, at the same time, the viscosity is not large. Mathematical models of these regions neglect the effects of shear stress. Equation (1.2) shows that a convenient way to do this is to set the fluid viscosity equal to zero; that is, to assume that the fluid has no viscosity. Regions of flow with small velocity gradients and negligible shear stress, like the center region of our channel, are called *irrotational* (to reflect the negligible particle spin), or *inviscid,* flow regions. Irrotational/inviscid flow is characterized by smooth streamlines and orderly fluid motion.

Figure 3.13(b) illustrates a fluid particle from one of the regions near the channel walls. Because the edge of the fluid particle nearer the center of the channel is moving faster than the edge nearer the wall, the particle is being deformed or sheared, and shear stress is significant in the particle's dynamics. Regions of flow where shear deformation is significant are called *shear flow* regions.

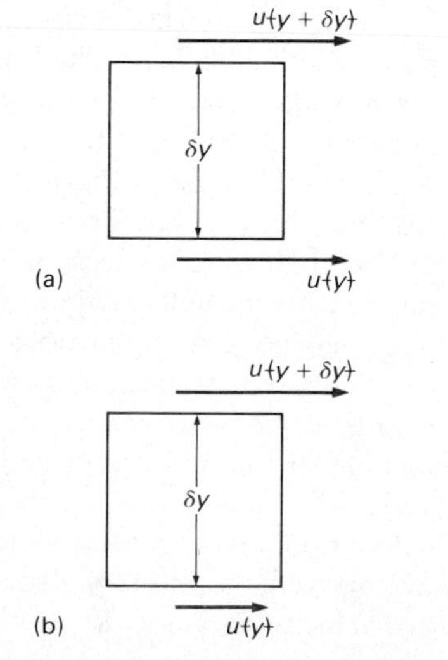

Figure 3.13 (a) Fluid particle near center of channel; (b) fluid particle near wall of channel.

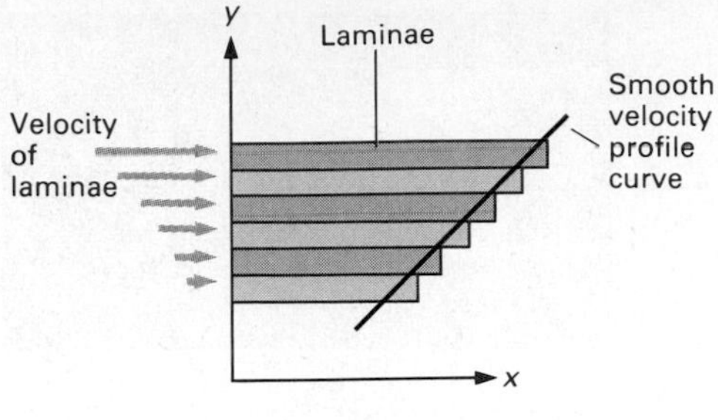

Figure 3.14 Laminar flow visualized as thin sheets (laminae) of fluid in relative sliding motion.

In many flows, such as the one illustrated in Figs. 3.11 and 3.12, shear flow occurs in thin regions near walls or other surfaces of discontinuity, while the remainder of the flow is nearly irrotational/inviscid. When this type of flow occurs, the thin region of shear flow is called a *boundary layer*. Note that not all flows can be divided into irrotational/inviscid and boundary layer regions; the flows illustrated in Figs. E3.1b and 3.7 are totally shear flow.

Examination of Figs. 3.11 and 3.12 reveals two types of shear flow. The shear flow nearer the entrance to the divergence is laminar, whereas that farther downstream is turbulent.

In *laminar flow,* the velocity gradient is not zero, the velocity profile is a smooth curve, and the fluid moves along smooth streamlines. The flow is called laminar because it looks like a series of thin sheets of fluid (laminae) sliding over one another (Fig. 3.14). In laminar flow, fluid particles move along fixed streamlines with very little mixing across streamlines. The concept of "friction" in the fluid is a good analogy for shear stress in laminar flow, even though the shear stress is actually the result of molecular momentum transfer and/or intermolecular forces (see Section 1.3.5 on viscosity).

Irrotational/inviscid and laminar flows contrast sharply to the disorderly fluid motion that we call turbulent flow. *Turbulent flow* is characterized by random fluctuations in fluid velocity and by intense mixing of the fluid. The disorderly pattern of bubbles near the lower wall of the channel of Fig. 3.11 is the result of the mixing action of the turbulent flow there. Turbulent flow is a familiar phenomenon in everyday life. Water usually flows from a kitchen faucet turbulently. A plume of smoke or steam from an industrial stack displays turbulent motion. You may have driven a car along a highway and watched "clouds" of snowflakes or dust billowing aimlessly about the highway; the snowflakes or dust serve as markers for turbulent wind gusts. In addition, turbulent flow occurs in a great many engineering applications.

Suppose that we insert a velocity-measuring instrument at a particular point near the lower wall of the channel of Fig. 3.11 and record the velocity as a function of time. If the instrument had a fast response time,* it would show a velocity-versus-time plot similar to that of Fig. 3.15. The flow is hopelessly unsteady; the velocity apparently changes randomly with time. We can, however, define a time-average velocity and then define a steady turbulent flow as one in which this average velocity does not vary with time at any point in the flow field.

---

* Response time characterizes how rapidly the instrument responds to changes in velocity.

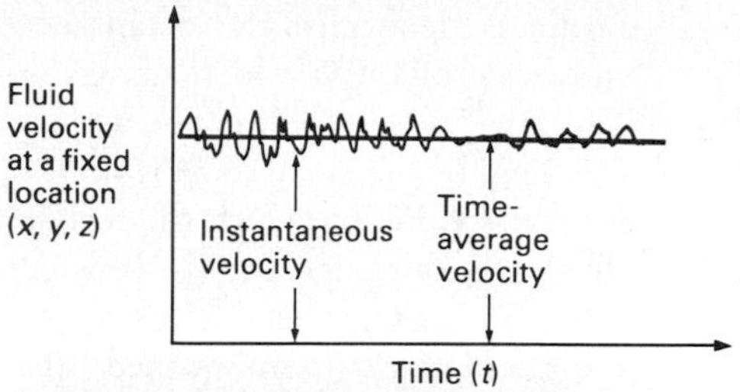

Figure 3.15 Velocity–time history for a fixed point in a turbulent flow.

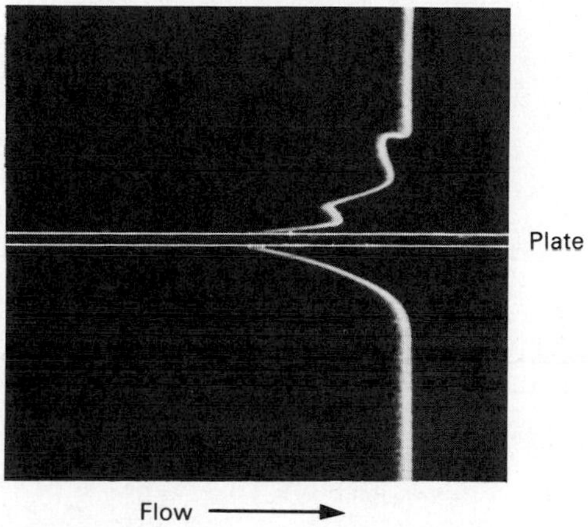

Figure 3.16 Instantaneous velocity profiles in laminar (lower) and turbulent (upper) flow over a solid plate (from the "NCFMF Book of Film Notes," 1974, The MIT Press with Education Development Center, Inc., Newton, Mass.).

The different types of flow generate different velocity profiles. Figure 3.16 shows two instantaneous velocity profiles for flow near a solid flat plate. These profiles are actually timelines generated shortly before the photograph was taken. The flow on the upper side of the plate is turbulent, and the flow on the lower side of the plate is laminar. If we generated instantaneous velocity profiles at another instant of time at the same location, the shape of the turbulent profile would be different, but the shape of the laminar profile would be the same. Figure 3.17 shows the superposition of a large number of velocity profiles. The laminar (lower)

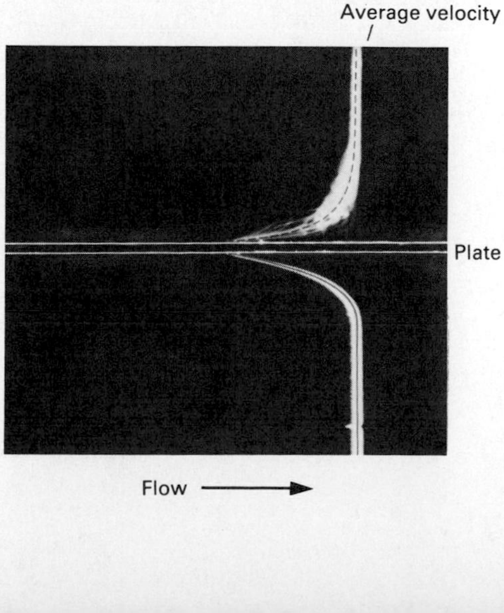

Figure 3.17 Superposition of a large number of laminar (lower) and turbulent (upper) instantaneous velocity profiles; lines indicate time average velocity profile (from the "NCFMF Book of Film Notes," 1974, The MIT Press with Education Development Center, Inc., Newton, Mass.).

profile is identical to the instantaneous profile of Fig. 3.17; however, the turbulent (upper) profile is a wide "smear" because of the different instantaneous profiles. Note that we can still define an average velocity profile for the turbulent flow.

Engineering analysis of turbulent flow is based on the idea of an average velocity distribution on which the turbulent fluctuations are superimposed. Figure 3.18 illustrates this concept. We consider the average velocity profile to be deterministic, that is, reproducible for the same boundary and initial conditions; however, the fluctuations (loosely called "the turbulence") are random, and we deal with them statistically. The details of turbulent flow are only partly understood. No theory is yet capable of predicting the details of the velocity–time trace of Figure 3.15.

The mixing action of turbulent flow is very important in engineering applications. Because of velocity fluctuations, small but macroscopic "lumps" of fluid (called *eddies*) are thrown about in the flow. Because these lumps carry mass, momentum, and energy, the rate of mixing of dissimilar fluids, "apparent" shear

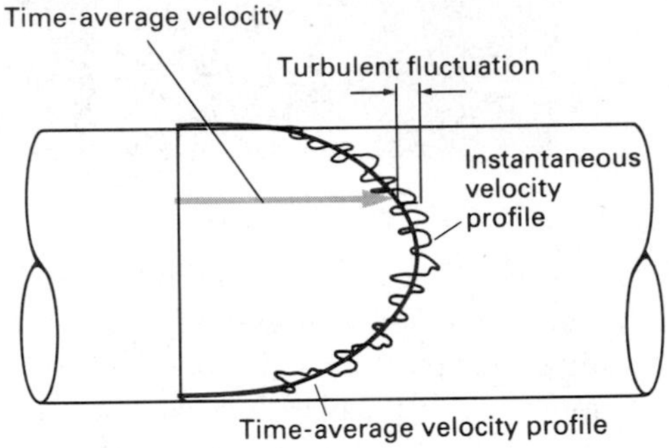

Figure 3.18 Instantaneous velocity profile in pipe flow; fluctuating velocity is superimposed on time-average velocity.

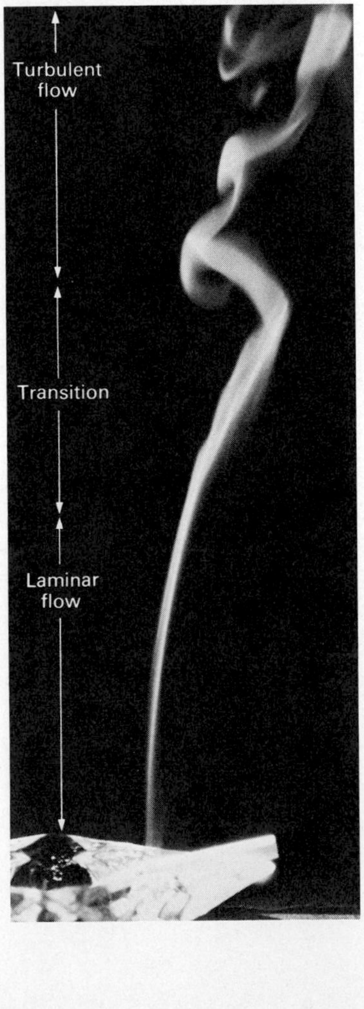

Figure 3.19 A plume of cigarette smoke illustrates laminar flow, transition, and turbulent flow (from the "NCFMF Book of Film Notes," 1974, The MIT Press with Education Development Center, Inc., Newton, Mass.).

stress (from momentum transfer), and "apparent" heat transfer are considerably larger than in laminar flow. This enhanced mixing is sometimes harmful, as in the increase of pressure drop in pipe flow, and sometimes helpful, as in the rapid mixing of cream into coffee after stirring (compared to the mixing of paint, usually a laminar process).

Laminar flow can change to turbulent flow in a process called *transition*. The plume of smoke rising from a cigarette* in Fig. 3.19 illustrates transition. The flow is initially laminar near the tip of the cigarette. Higher, the laminar flow becomes unstable. By mechanisms that are not fully understood, these instabilities grow and the flow becomes turbulent. The transition in the smoke plume occurs naturally without outside disturbances. Introducing disturbances (e.g., by roughening a surface in contact with the fluid) can promote transition. The flow near the lower wall in Fig. 3.11 and the flow over the top of the plate in Fig. 3.16 were made turbulent by roughening the wall upstream of the location shown in the figures.

Engineering approaches to turbulent flow are based on a combination of theory and experiment. Theory suggests how turbulent flows behave qualitatively and possible forms for mathematical description, but experimental data are necessary to make the results quantitative.

3.3 ANALYZING FLUID FLOW

We have concluded our discussion of the ways in which fluid motion may be described and classified. We can now begin to consider ways to analyze fluid flow.

3.3.1 The Fundamental Laws

Experience has shown that all fluid motion must be consistent with the following fundamental laws of nature.

- *The law of conservation of mass.* Mass can be neither created nor destroyed; it can only be transported or stored.
- *Newton's three laws of motion.*

  1. A mass remains in a state of equilibrium, that is, at rest or moving at constant velocity, unless acted on by an unbalanced force. (First law)
  2. The rate of change of momentum of a mass is proportional† to the net force acting on the mass. (Second law)
  3. Any force action has an equal (in magnitude) and opposite (in direction) force reaction. (Third law)

- *The first law of thermodynamics (law of conservation of energy).* Energy, like mass, can be neither created nor destroyed. Energy can be transported, changed in form, or stored.

---

* Illustrating transition is one of the few beneficial uses of a cigarette.

† The "constant of proportionality" involves only the conversion between units of force, mass, length, and time.

- *The second law of thermodynamics.* The second law deals with the availability of energy to perform useful work. The only possible natural processes are those that either decrease or, in the ideal case, maintain, the availability of the energy of the universe. The science of thermodynamics defines a material property called *entropy*, which quantifies the second law. The entropy of the universe must increase or, in the ideal case, remain constant in all natural processes.

- *The state postulate (law of property relations).* The various properties of a fluid are related. If a certain minimum number (usually two) of a fluid's properties are specified, the remainder of the properties can be determined* (see Section 1.3).

The important thing to remember about these laws is that they apply to all flows. They do not depend on the nature of the fluid, the geometry of the boundaries, or anything else. As far as we know, they have always been true and will continue to be true unless they are suspended by the Creator of the universe. Hence we can firmly base analysis of all flows on these laws.

In addition to these universal laws, several less fundamental "laws" apply in restricted circumstances. An example is Newton's law of viscosity (see Section 1.3.4):

> The shear stress in a fluid is proportional to the rate of deformation of the fluid.

This "law" is true only for some fluids and does not apply at all to solids. Such "laws" are better termed *constitutive relations*. We must use constitutive relations to solve most flow problems, but we must select them carefully to match the particular problem.

3.3.2 Mathematical Formulation: System Versus Control Volume

The fundamental laws are the basis of our understanding of fluid (and solid) motion. Besides understanding, an engineer needs predictive capability. To design a piece of equipment or a structure, an engineer must know in advance how it will affect and be affected by fluids that it contacts. An airplane's engine must be sufficiently powerful to overcome air resistance at the desired flight speed. A pipe must be large enough to convey water at a desired flow rate. A building must be strong enough to resist wind loads.

To obtain a predictive capability, we formulate the fundamental laws as mathematical expressions. We assign symbols to fluid velocity, pressure, and density, to the wind force on a building, and to the rate of flow in a pipe. We describe the proposed (or existing) equipment or structure (airplane, building, pipe) by certain geometric relations. Application of the fundamental laws permits us to write

---

* The statement of this law is universally valid; however, equations used to express it are of limited scope. The ideal gas law, Eq. (1.1), is less general than the statement that pressure depends on density and temperature.

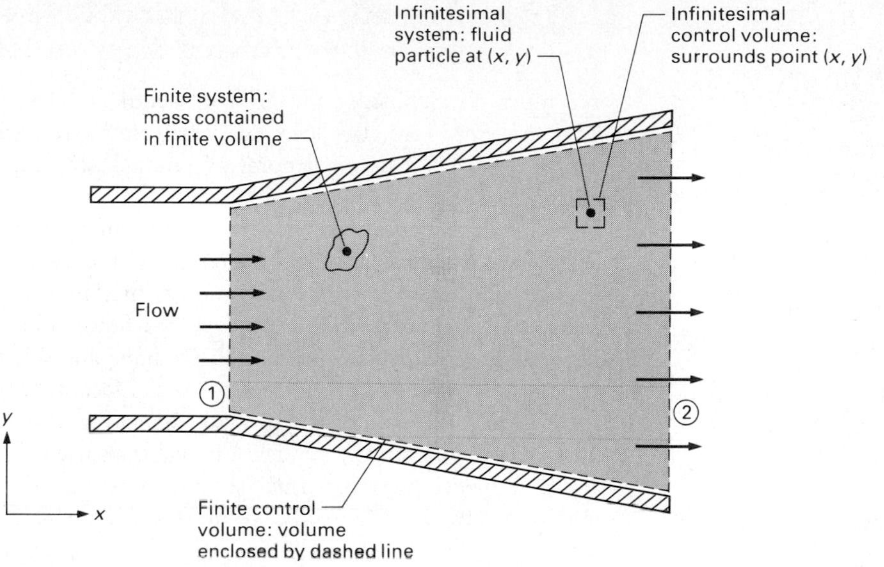

Figure 3.20 Illustration of systems and control volumes.

equations involving the symbols. By manipulating the symbols according to the rules of mathematics, we can "solve" the problem; that is, we can predict the velocity or pressure or wind force that we need to know.

To formulate the fundamental laws as mathematical models of flow, we must choose both a point of view and a level of detail. Here we consider the point of view; in the next section, we consider the level of detail.

We may apply the fundamental laws to either a system or a control volume.

A *system* is a specific fluid mass selected for analysis.

A system may be either infinitesimally small (a fluid particle) or finite (a lump of fluid).

A *control volume* is a specific region of space selected for analysis.

A control volume may be either infinitesimally small or finite; it may move or remain fixed in space. It is an imaginary volume, chosen by the analyst, and does not interfere with the flow. Figure 3.20 illustrates both infinitesimal and finite systems and control volumes that we might select for analysis of flow in a diverging channel such as that illustrated in Fig. 3.11.

The system point of view is related to a Lagrangian description of flow. Its advantage is that all the fundamental laws may be expressed directly in terms of a specific collection of mass. The control volume point of view is related to an Eulerian description of flow. Its advantage is that control volumes are easier to keep track of than systems. Unfortunately, Newton's laws of motion and the state postulate apply only to samples of matter, not to volumes. Thus we need to adopt the system point of view to formulate the fundamental laws but use the control volume point of view to apply them to flow problems! Fortunately, we can formally connect the two points of view by purely mathematical relationships, as we show after considering the choices of level of detail.

3.3.3 Mathematical Formulation: Differential Versus Finite Control Volume Approaches

We must now consider the level of detail at which we apply the fundamental laws and therefore, the level of detail of the resulting flow analysis. We must choose between a detailed point-by-point description and a global or "lumped" description.

The fundamental laws govern the fluid motion at every point in an entire flow field. The laws must also be satisfied in any finite region that contains many points and a finite fluid mass. We may express the laws either in a form that applies to each point or in a form that applies to a finite volume of space. The best choice depends on how much information we have and what we need to know.

Suppose that we are interested in the steady flow of air in a diverging duct such as that illustrated in Fig. 3.20. Plane 1 is the duct inlet and plane 2 is the duct outlet. For certain applications we may need to determine the fluid velocity and pressure at every point in the volume between plane 1 and plane 2. To determine the flow in such detail, we must apply the fundamental laws to each point in the field; that is, we apply the laws to each fluid particle or infinitesimal control volume. The result will be a set of differential equations with the fluid velocity and pressure as dependent variables and the location in the duct (x, y, z) as independent variables. The solution of these differential equations, together with appropriate boundary conditions representing the shape and size of the duct, the fluid velocity at the inlet, and so on, will be two functions, $\vec{V}\{x, y, z\}$ and $p\{x, y, z\}$, that can tell us the velocity and pressure at every point. We might call this a *local description* because it gives the detailed information at any locality. If the analysis calls for global information, such as the air-flow rate or the pressure force on the duct walls, we can obtain it by integrating the velocity or pressure distribution.*

Often, global information such as flow rate, force, and temperature change between inlet and outlet are all that an engineer needs to know. Determining (or at least estimating) these quantities directly by considering a finite control volume is often possible. Application of the fundamental laws instantaneously to the system that occupies a finite control volume yields control volume forms of the laws. For example, the principle of conservation of mass implies that nothing that results in a net creation or destruction of mass can occur within the control volume. Control volume equations deal with the relations between the fluid properties and the velocity at the control volume boundaries, without revealing the details of what is going on inside the volume. Control volume equations deal with a balance between the amount of a quantity that enters the control volume, the amount of the quantity that leaves the control volume, the amount of the quantity stored inside the control volume, and the net creation or destruction of the quantity within the control volume. In principle, control volume equations require integration of the fluid velocity and property distributions over the control surface; however, the

---

* We integrated pressure distributions (local information) to find resultant forces (global information) in Chapter 2.

Table 3.1 Comparison of differential and finite control volume formulations

	Advantages	Disadvantages
Differential formulation	1. Reveals all details of the flow 2. Forces fluid to obey fundamental laws at all points 3. Solves problem with minimum of input information (boundary conditions)	1. Produces differential equations that are often difficult or impossible to solve 2. Often requires that equations be solved by computer, which can be expensive for complex flows 3. May give more information than really needed
Finite control volume formulation	1. Simpler mathematics 2. Much less sensitive to approximations and assumptions; often yields quite useful approximate information with very crude assumptions 3. Pencil-and-paper method requiring about an hour's work 4. Often reveals only the information really needed	1. Does not reveal all details of flow; does not force fluid to obey fundamental laws at every point 2. Often yields *only* approximate answers 3. Requires more input information, such as velocity distribution at convenient boundaries 4. Often cannot give as much information as needed

engineer may often assume simplified distributions, such as the one illustrated in Fig. 3.10, and work with algebraic equations involving averaged velocity and fluid properties.

Both the finite control volume approach and the differential approach are important in fluid mechanics. Both have their advantages and disadvantages. The two approaches are compared in Table 3.1 and Fig. 3.21.

The differences between the differential and the finite control volume approaches sometimes lead students and engineers to conclude that there are two different kinds of fluid mechanics, as we mentioned in Section 3.1. We try to treat both approaches equally in this book and, in particular, to show how they may work together in the solution of engineering fluid mechanics problems.*

3.3.4 The Eulerian Derivative

The Eulerian derivative is a mathematical expression that connects the system and control volume points of view for a local or differential equation approach to flow analysis. It relates the rate of change of any property of a fluid particle (system) to the particle's location in the flow field.

---

* Students and engineers usually prefer the control volume approach because it involves less complicated mathematics. White [1] says that the approaches are equal but that the control volume approach is more equal.

(a)

Consider fluid particle or infinitesimal control volume

↓

Apply fundamental laws

↓

Differential equations for distributions, e.g., $V(x, y, z, t)$

↓

Simplify differential equations model or neglect certain effects

↓

Solve differential equations with boundary conditions for specific flow of interest

↓

Calculate quantities of interest (force, by integrating pressure; flow rate, by integrating velocity; streamlines, from velocities)

(b)

Consider finite control volume and enclosed system

↓

Apply fundamental laws

↓

Integral equations involving flows entering and leaving control volume

↓

Assume forms of velocity, pressure, and other property distributions

↓

Integrate to obtain algebraic working equations relating gross flow features (e.g., force, flow rate)

↓

Calculate global parameters by substituting known information into working equations

Figure 3.21 Flow chart for flow analysis by (a) differential approach; (b) finite control volume approach.

Imagine a specific fluid particle located at a specific point in a flow field (see Fig. 3.22). For generality, we assume a three-directional, three-dimensional, unsteady flow. We describe the flow in Cartesian coordinates. If we use an Eulerian description, the particle's properties and velocity depend on its location and time. Then for an arbitrary property b (b may be density, pressure, velocity, etc.), we can write

$$b_P = b(x_P, y_P, z_P, t),$$

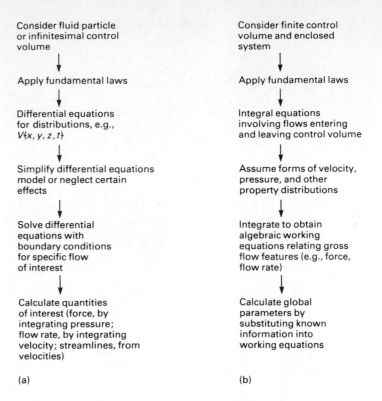

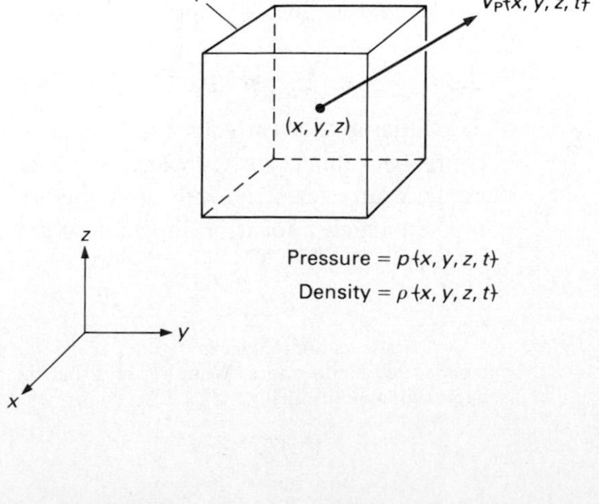

Figure 3.22 Particle P at point (x, y, z) at time t.

where the subscript P reminds us that we use the location of the particle.

The fundamental laws usually require rates of change of certain properties of matter (e.g., momentum, mass, energy, entropy). We determine the rate of change of b_P by using the rule for calculating total derivatives:

$$\frac{db_P}{dt} = \frac{\partial b}{\partial x_P}\frac{dx_P}{dt} + \frac{\partial b}{\partial y_P}\frac{dy_P}{dt} + \frac{\partial b}{\partial z_P}\frac{dz_P}{dt} + \frac{\partial b}{\partial t}.$$

However, x_P, y_P, and z_P are the position coordinates of the fluid particle, so their time derivatives are the particle's velocity components,

$$\frac{dx_P}{dt} = u_P, \quad \frac{dy_P}{dt} = v_P, \quad \text{and} \quad \frac{dz_P}{dt} = w_P.$$

Substituting, the rate of change of b is

$$\frac{db}{dt} = u\frac{\partial b}{\partial x} + v\frac{\partial b}{\partial y} + w\frac{\partial b}{\partial z} + \frac{\partial b}{\partial t}, \tag{3.5}$$

where we have now dropped the subscript P.

The complicated form of Eq. (3.5), requiring four terms to express a single total time derivative, is a consequence of selecting the Eulerian method of description in which fluid properties depend on both location and time. This type of derivative is variously called the *Eulerian derivative* (because it is necessary in an Eulerian description), the *material derivative* (because it expresses the rate of change of a property of a particle of the material), the *substantial derivative* (substituting the word *substance* for *material*), or the *total derivative.*

As the property b was arbitrary, we can drop it and write the derivative as a mathematical operator. To remind us of its special nature, we often write the operator as a D and rearrange it as follows:

$$\frac{D}{Dt} = \frac{\partial}{\partial t} + u\frac{\partial}{\partial x} + v\frac{\partial}{\partial y} + w\frac{\partial}{\partial z}. \tag{3.6}$$

A shorthand notation, using vectors, is

$$\frac{D}{Dt} = \frac{\partial}{\partial t} + \vec{V} \cdot \nabla, \tag{3.7}$$

where $\vec{V}$ is the fluid velocity vector and ∇ is the gradient operator.

Some comments on the physical meaning of Eqs. (3.5–3.7) are in order. The term db/dt (alternatively, D/Dt) is the total rate of change over time of the property b of the particle. That property may change because the particle moves to another location over time or because the value of the property at a specific location may change over time. For example, the temperature of a specific particle of air in the atmosphere may change as winds carry it to a region of different temperature, or if the particle remains at a fixed location, the temperaure may change from day to night or season to season. The rate of change resulting from changes of location is reflected in the terms containing the velocity. The rate of change resulting from unsteady effects at a fixed location is reflected in the partial derivative with time.

Remember the basic idea that, for any property of a fluid particle, we may write:

$$\begin{bmatrix}\text{Total rate of change}\\ \text{of property of a}\\ \text{fluid particle}\end{bmatrix} = \begin{bmatrix}\text{Rate of change}\\ \text{at a fixed location}\left(\dfrac{\partial}{\partial t}\right)\end{bmatrix} + \begin{bmatrix}\text{Velocity times}\\ \text{derivative with}\\ \text{respect to space}\\ \text{coordinates } (\vec{V}\cdot\nabla)\end{bmatrix}.$$

The first term on the right-hand side of this word equation is called the *local unsteady term*. The rightmost term is called the *convective rate of change*, because it reflects that the fluid is convected (carried) about in the field.

3.3.5 The Transport Theorem

The transport theorem is a mathematical expression that connects the system and control volume points of view for a global or finite control volume approach to flow analysis. It provides a way to identify a finite system and to evaluate the rate of change of any property or characteristic of that system by examining the flow through a control volume.

The fundamental laws deal with rates of change of certain properties of the system to which they are applied; for example, Newton's second law addresses the rate of change of system momentum, and the first law of thermodynamics addresses the rate of change of system energy. In order to apply these laws, we have to figure out how to identify a specific system in a moving, deforming fluid and how to calculate the rate of change of this system's properties. This is where a control volume comes in. Although fluid flows through a control volume continuously, a control volume contains a specific mass of fluid at any instant. We designate our system as *the fluid mass that instantaneously occupies the control*

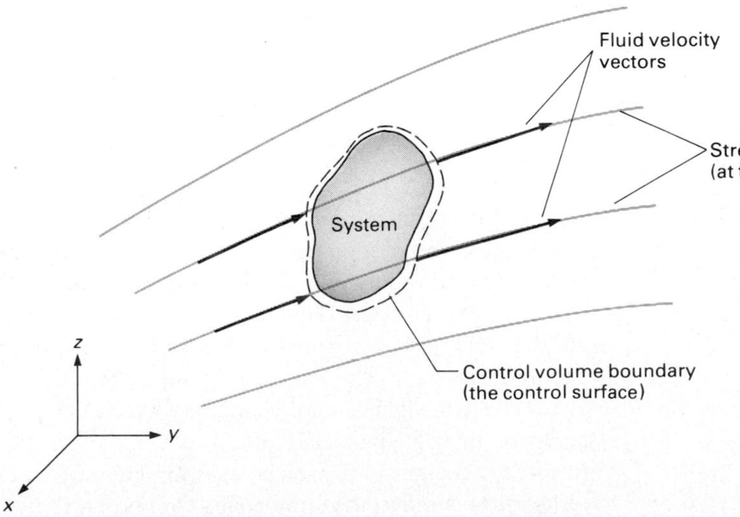

Figure 3.23 Control volume and enclosed system in a flow field.

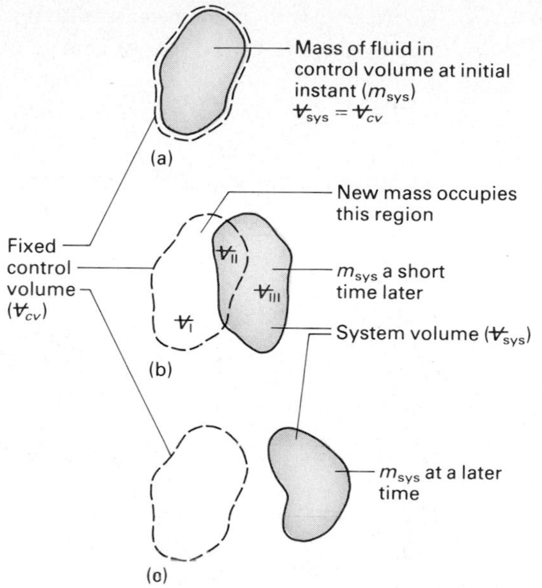

Figure 3.24 A system flows through a control volume.

volume. A different system occupies the control volume at each instant;* however, we may apply the fundamental laws to whatever system occupies the control volume at that instant. The control volume becomes a way to identify a specific system, even if only for a very short time.

Figure 3.23 shows a three-dimensional, three-directional flow field with a control volume superimposed. The mass contained in the control volume at the instant that the figure represents is chosen as a system. Figure 3.24 shows the passage of that system out of the control volume over time.

Figure 3.24(a) shows the situation at time t, when the system exactly fills the control volume. Figure 3.24(b) shows the situation at a later time, $t + \delta t$, when a portion of the system has left the control volume: V_I is the portion of the control volume that has been vacated by the system in time δt; V_{II} is the portion of the control volume that contained part of the system at time t and still contains part of the system at time $t + \delta t$; and V_{III} lies outside the control volume and contains the portion of the system that has left the control volume in time δt.

We can calculate the rate of change of an arbitrary property B of this system; B could be momentum, energy, or any other property. Property B may be nonuniformly distributed throughout the system, so we let b be the specific (per-unit-mass) property. If B represents system momentum ($\vec{M} = m\vec{V}$), then b is velocity ($\vec{V}$). If B represents system kinetic energy, then b is $V^2/2$. The total amount of B in the system is

$$B_{sys} = \int_{m_{sys}} b\, dm = \int_{V_{sys}} b\rho\, dV, \qquad (3.8)$$

* This idea is somewhat like taking motion pictures. We point the camera at a particular location (say, a busy street corner) and take pictures. We record frame-by-frame the pictures of people who occupy the location. When we run the film at regular speed, we see a continuous flow of people; but each frame shows a particular person occupying the location.

where m_{sys} is the system mass and $\forall_{sys}$ is the system volume.

By definition, the rate of change of B_{sys} is

$$\frac{dB_{sys}}{dt} \equiv \lim_{\delta t \to 0} \frac{B_{sys})_{t+\delta t} - B_{sys})_t}{\delta t}.$$

Substituting from Eq. (3.8), we have

$$\frac{dB_{sys}}{dt} = \lim_{\delta t \to 0} \frac{\int_{\forall_{sys}} \rho b \, d\forall)_{t+\delta t} - \int_{\forall_{sys}} \rho b \, d\forall)_t}{\delta t}. \tag{3.9}$$

At time t,

$$\forall_{sys})_t = \forall_{cv},$$

and at time $t + \delta t$,

$$\forall_{sys})_{t+\delta t} = \forall_{\mathrm{II}} + \forall_{\mathrm{III}} = \forall_{cv} - \forall_{\mathrm{I}} + \forall_{\mathrm{III}},$$

so

$$\frac{dB_{sys}}{dt} = \lim_{\delta t \to 0} \frac{\int_{\forall_{cv}} \rho b \, d\forall)_{t+\delta t} - \int_{\forall_{\mathrm{I}}} \rho b \, d\forall)_{t+\delta t} + \int_{\forall_{\mathrm{III}}} \rho b \, d\forall)_{t+\delta t} - \int_{\forall_{cv}} \rho b \, d\forall)_t}{\delta t}.$$

We can rewrite this expression as

$$\frac{dB_{sys}}{dt} = \underbrace{\lim_{\delta t \to 0} \frac{\int_{\forall_{cv}} \rho b \, d\forall)_{t+\delta t} - \int_{\forall_{cv}} \rho b \, d\forall)_t}{\delta t}}_{\text{Term Ⓐ}} + \underbrace{\lim_{\delta t \to 0} \frac{\int_{\forall_{\mathrm{III}}} \rho b \, d\forall)_{t+\delta t}}{\delta t}}_{\text{Term Ⓑ}}$$
$$- \underbrace{\lim_{\delta t \to 0} \frac{\int_{\forall_{\mathrm{I}}} \rho b \, d\forall)_{t+\delta t}}{\delta t}}_{\text{Term Ⓒ}}.$$

Each of the three terms has its own meaning. Term Ⓐ is similar to the right-hand side of Eq. (3.9), with one important difference: In term Ⓐ the integration is over the *control volume,* whereas in Eq. (3.9), the integration is over the *system volume.* Using the definition of a derivative, we find that

$$\text{Term Ⓐ} = \frac{d}{dt} \int_{\forall_{cv}} \rho b \, d\forall = \frac{d}{dt} (B_{cv}),$$

which represents the time rate of change of the amount of B contained in the control volume. Note that this amount of B could change because of changes in $\forall_{cv}$ (a moving or deforming control volume) and changes of ρ or b with time at points within the control volume. We sometimes call term Ⓐ the *rate of accumulation.*

The integral in the numerator of term Ⓑ represents the amount of B that flows out of the control volume between time t and time $t + \delta t$. Dividing by δt, we get the average rate of outflow, and passing to the limit, we get the instantaneous rate of outflow:

$$\text{Term Ⓑ} = \dot{B}_{out}.$$

The integral in term Ⓒ is the amount of B that flows into the control volume between time t and time $t + \delta t$. (Note that this B is not in the original system.) Dividing by δt, we get the average rate of inflow, and passing to the limit, we get the instantaneous rate of inflow:

$$\text{Term Ⓒ} = \dot{B}_{\text{in}}.$$

We sometimes call terms Ⓑ and Ⓒ the *flux terms*. Collecting terms Ⓐ, Ⓑ, and Ⓒ, we get

$$\frac{dB_{\text{sys}}}{dt} = \frac{d}{dt}\int_{\forall_{cv}} \rho b \, d\forall + \dot{B}_{\text{out}} - \dot{B}_{\text{in}}. \tag{3.10}$$

In words:

> The rate of change of any property of the system that occupies a control volume at any particular instant is equal to the instantaneous rate of accumulation of the property inside the control volume plus the difference between the instantaneous rates of outflow and inflow of the property. This latter term is the net rate of outflow of the property across the control surface.

Sometimes we can use Eq. (3.10) directly, but we often must relate $\dot{B}_{\text{out}}$ and $\dot{B}_{\text{in}}$ to the flow field. For $\dot{B}_{\text{out}}$, where B is a property of the fluid mass and the fluid is flowing across the control surface, each unit of mass leaving the control volume carries b units of B out with it. As b may not be uniform, we write

$$d\dot{B}_{\text{out}} = b(d\dot{m})_{\text{out}}$$

and

$$\dot{B}_{\text{out}} = \int d\dot{B}_{\text{out}} = \int b(d\dot{m})_{\text{out}}. \tag{3.11}$$

Similarly,

$$\dot{B}_{\text{in}} = \int d\dot{B}_{\text{in}} = \int b(d\dot{m})_{\text{in}}. \tag{3.12}$$

Now consider the rate of flow of mass out of the control volume. Figure 3.25 shows an enlargement of a small portion of the control surface. Mass flows out from the control volume through the area dA. Figure 3.25(a) shows the fluid velocity at dA at a particular time t; $\hat{n}$ is a unit vector drawn perpendicular to dA and

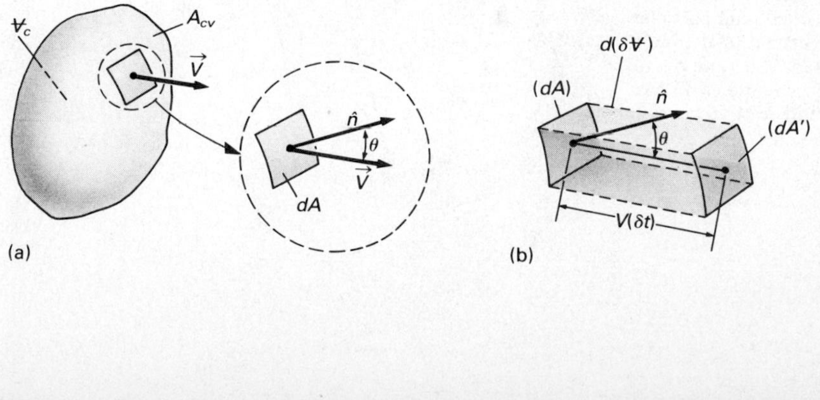

Figure 3.25 Details of outflow through a small piece of the control surface (dA): (a) velocity and unit outdrawn normal vector at surface; (b) fluid particles that were at dA at t are at (dA') at $t + \delta t$; fluid in volume $d(\delta\forall)$ has left control volume through dA.

pointing outward from the control volume. In general, the fluid velocity is not perpendicular to the control surface, so the angle θ between $\vec{V}$ and $\hat{n}$ is not zero. Figure 3.25(b) shows the fluid that has left the control volume through dA in time δt. The fluid that was at dA at time t has now moved to dA'. All the fluid in the volume $d(\delta \forall)$ between dA and dA' has flowed out of the control volume through dA in time δt. Thus can write

$$d(\delta m)_{\text{out}} = \rho d(\delta \forall). \tag{3.13}$$

From Fig. 3.25(b), we have

$$d(\delta \forall) = (V\,\delta t)\cos\theta\, dA. \tag{3.14}$$

To obtain an expression for the rate of outflow through dA, we substitute Eq. (3.14) into Eq. (3.13), divide by δt, and take the limit as $\delta t \to 0$. The result is

$$d\dot{m}_{\text{out}} = \rho V \cos\theta\, dA. \tag{3.15}$$

To identify the outflow area, note that fluid flows outward across the control surface at any point where $0° \leq \theta \leq 90°$. We can write in vector notation

$$d\dot{m}_{\text{out}} = \rho \vec{V} \cdot \hat{n}\, dA. \tag{3.16}$$

Remember that the magnitude of vector $\hat{n}$ is unity.

Next, consider the rate of mass flow into the control volume. Figure 3.26 illustrates a portion of the control surface through which mass enters the control volume. Figure 3.26(a) shows the velocity vector $\vec{V}$ and the unit outward normal vector $\hat{n}$ at the area dA at time t. Note that for inflow, the angle θ lies between 90° and 180°. The angle θ' is the complement of the angle θ. Figure 3.26(b) shows the situation at time $t + \delta t$. The fluid that was at dA at time t has now moved to dA', inside the control volume. All the fluid in volume $d(\delta \forall)$ flowed into the control volume across dA in time δt. We can write

$$d(\delta m)_{\text{in}} = \rho d(\delta \forall), \tag{3.17}$$

where

$$d(\delta \forall) = (V\,\delta t)\cos\theta'\, dA = -V\,\delta t \cos\theta\, dA.$$

Figure 3.26 Details of inflow through a small piece of control surface: (a) velocity and unit outdrawn normal vector at surface; (b) fluid particles that were at dA at time t are at (dA') at time $t + \delta t$; fluid in volume $d(\delta \forall)$ has entered the control volume.

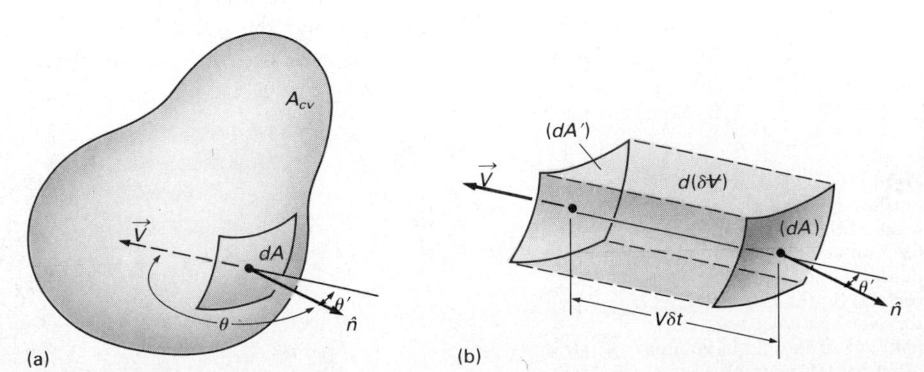

Dividing by δt and taking the limit as $\delta t \to 0$ gives

$$d\dot{m}_{\text{in}} = -\rho V \cos\theta \, dA. \tag{3.18}$$

In vector notation,

$$d\dot{m}_{\text{in}} = \rho \vec{V} \cdot \hat{n} \, dA. \tag{3.19}$$

Substituting the expressions for mass flow rate [Eqs. (3.15), (3.16), (3.18), and (3.19)] into Eqs. (3.11) and (3.12) gives

$$\dot{B}_{\text{out}} = \int_{A_{\text{out}}} \rho b V \cos\theta \, dA = \int_{A_{\text{out}}} \rho b (\vec{V} \cdot \hat{n}) \, dA. \tag{3.20}$$

and

$$\dot{B}_{\text{in}} = -\int_{A_{\text{in}}} \rho b V \cos\theta \, dA = \int_{A_{\text{in}}} \rho b (\vec{V} \cdot \hat{n}) \, dA. \tag{3.21}$$

Using Eqs. (3.20) and (3.21), we write Eq. (3.10) in the following forms:

$$\frac{dB_{\text{sys}}}{dt} = \frac{d}{dt} \int_{\mathcal{V}_{cv}} \rho b \, d\mathcal{V} + \int_{A_{\text{out}}} \rho b V \cos\theta \, dA + \int_{A_{\text{in}}} \rho b V \cos\theta \, dA, \tag{3.22a}$$

$$\frac{dB_{\text{sys}}}{dt} = \frac{d}{dt} \int_{\mathcal{V}_{cv}} \rho b \, d\mathcal{V} + \int_{A_{\text{out}}} \rho b V_n \, dA - \int_{A_{\text{in}}} \rho b V_n \, dA, \tag{3.22b}$$

and

$$\frac{dB_{\text{sys}}}{dt} = \frac{d}{dt} \int_{\mathcal{V}_{cv}} \rho b \, d\mathcal{V} + \oint_{A_{cv}} \rho b (\vec{V} \cdot \hat{n}) \, dA. \tag{3.22c}$$

In Eq. (3.22b), V_n represents the velocity component normal (perpendicular) to the control surface. We combine the two surface integrals in Eqs. (3.22a) and (3.22b) into a single integral over the entire surface, as in Eq. (3.22c), because any point on the surface has either inflow, outflow, or $\vec{V} \cdot \hat{n} = 0$.

Equation (3.22), in any of its forms, is the *transport theorem.** It allows us to relate the system and control volume points of view (alternatively, the Lagrangian and Eulerian methods of description) at a global level of detail. What the Eulerian derivative does at the differential level, the transport theorem does at the global level.

Engineering calculations seldom use directly the Eulerian derivative and the transport theorem. Their importance lies in applying the fundamental laws to flowing fluids, allowing us to develop the working equations that we do use to make calculations. The Eulerian derivative reappears in Chapter 5, when we present the differential approach to flow analysis. The transport theorem provides the point of departure for our development of the finite control volume approach in Chapter 4.

---

* Recall that B and b are "dummy" variables that represent any physical property of interest.

3.4 METHODS FOR SOLVING ENGINEERING FLOW PROBLEMS

Now that we have discussed methods for describing flow, examined different flow regimes, established the fundamental laws that govern fluid flow, and considered ways to apply the laws to analyze flow, we close this chapter by briefly examining ways that engineers use to solve flow problems. Broadly speaking, the choice is between mathematical methods and experimental methods.

Mathematical methods attempt to predict (calculate) solutions to flow problems using either pencil-and-paper or electronic computer methods. The flow is modeled by mathematical forms of the fundamental laws.

The most general mathematical approach is to use a differential equation (local) formulation that reveals all the details of the flow. We discuss the differential approach in Chapter 5, when we concentrate on formulating differential equation models for a variety of flows. Unfortunately, although we can set up the differential equations to describe almost any practically interesting flow, they usually cannot be solved in closed form, especially for turbulent flows or complicated geometries. Accordingly, mathematical methods usually are based on simplifications and approximations. Sometimes, simplifications are used just to obtain a less complicated differential equation model. If we attempted to predict the flow illustrated in Figs. 3.11 and 3.12, we would use different simplified equations for the irrotational/inviscid core and for the boundary layers.

Probably the most useful mathematical method is the finite control volume approach. Although the control volume forms of the fundamental laws are mathematically exact, engineers usually have to make some simplifying assumptions about the flow to make calculations. Using the control volume approach, engineers often obtain reasonably accurate estimates of key flow parameters with a few minutes' to a few hours' work. We present the control volume approach in Chapter 4, and the methods developed there lay the foundation for much of the remainder of this book.

With the flow modeling approach, we use information about fairly simple flows to build approximate solutions to more complex cases. For example, we might model the flow in a duct with a complicated cross section by the flow in an "equivalent" circular pipe, or we might model the flow in the shell of a heat exchanger by a network of circular pipes.

Using a computer to make calculations for any of the mathematical approaches saves time. One method requiring the use of a computer is computational fluid dynamics. Currently, this method offers the only practical way to achieve the generality of the differential equation formulation. We discussed CFD briefly in Chapter 1 and develop it further in Chapter 5.

Solving all engineering flow problems mathematically is not possible, or even desirable. Engineers often use experimental methods when mathematical methods are incapable of providing accurate or complete information. Experimental methods are necessary to verify calculations or to provide information for developing better mathematical models.

In experimental methods, engineers either set up and investigate a flow in a laboratory environment (Fig. 3.27) or make measurements in a real flow in the

Figure 3.27 Experimental fluid mechanics: An engineer is using a laser Doppler anemometer system to measure fluid velocities between the blades of a high-speed propeller. The propeller is mounted in a wind tunnel (photo courtesy of NASA Lewis Research Center).

field (Fig. 3.28). A laboratory investigation may involve either a scale model or the actual equipment. In experimental fluid mechanics, the fluid "solves" the problem and we have to extract the answers using instruments. Engineers apply experimental fluid mechanics on two levels. Probably the most obvious application is direct measurement of the quantities of interest. Examples include measuring the drag force on a scale model racing car chassis in a wind tunnel and direct

Figure 3.28 Solving a problem by measurements in the field: These engineers are measuring the flow rate through a large fan that exhausts flue gases from a boiler. The fan is driven by two 3000-hp electric motors. Gas velocity is measured at several points at the fan inlet plane and outlet plane. The flow rate is then calculated from the measured velocities.

measurement of flow rate in a city water line. Although such direct approaches are important, they are costly and sometimes impractical. If 100 alternative designs of a car chassis were proposed, a model of each would have to be tested. The water pipe in question may be buried under the brand new freeway!

A second application of experimental methods is research into basic phenomena. Detailed measurements of turbulent flow in a particular circular pipe can form the basis for a model of turbulent flow for all circular pipes. In fact, as no theory is capable of predicting turbulent flow, all information about turbulent flow is ultimately based on correlations of experimental data.

There are two key areas of concern in experimental fluid mechanics. The most obvious is that of measurement techniques. Our treatment of measurement techniques in this book is cursory; we discuss them only when they illustrate the application of some particular principle or equation of fluid mechanics (see the discussion of pressure measurement in Chapter 2, for example). Several excellent books deal with measurement techniques in general [4, 5] and experimental fluid mechanics in particular [6, 7, 8].

A less obvious but extremely important aspect of experimental fluid mechanics deals with efficient presentation and extrapolation of data. Experimental studies are often performed on scale models. How can we relate the scale-model data to full-scale performance?

A second question deals with the number of parameters that must be controlled in an experiment. Must we consider density, viscosity, pressure, velocity, and so on?

These problems are addressed by the technique of *dimensional analysis*, the key to "getting the most for your money" in experimental fluid mechanics. It is also the only way that the vast amount of experimental information available can be organized and passed on to other engineers.*

We discuss dimensional analysis in Chapter 6, completing the trio of approaches to flow analysis (finite control volume approach, differential equation approach, and experimental approach). The subsequent chapters illustrate applications of these approaches to several important flow situations.

PROBLEMS

1. The surface velocity of a river is measured at several locations x and can be reasonably represented by

$$V = V_0 + \Delta V(1 - e^{-ax}),$$

where V_0, ΔV, and a are constants. Find the Lagrangian description of the velocity of a fluid particle flowing along the surface if $x = 0$ at time $t = 0$.

2. A two-dimensional, unsteady velocity field is given by

$$u = 5x(1 + t) \quad \text{and} \quad v = 5y(-1 + t),$$

where u is the x-velocity component and v the y-velocity component. Find $x(t)$ and $y(t)$ if $x = x_0$ and $y = y_0$ at $t = 0$. Do the velocity components represent an Eulerian description or a Lagrangian description?

* Dimensional analysis also is useful in mathematical fluid mechanics, where it helps generalize and extend information generated in numerical solutions and suggests efficient forms for mathematical models.

3. A car accelerates from rest to a final constant velocity V_f and a police officer records the following velocities at various locations x along the highway:

$x = 0$	$V = 0$ mph
$x = 100$ ft	$V = 34.8$ mph
$x = 200$ ft	$V = 47.6$ mph
$x = 300$ ft	$V = 52.3$ mph
$x = 400$ ft	$V = 54.0$ mph
$x = 1000$ ft	$V = 55.0$ mph $= V_f$.

Find a mathematical Eulerian expression for the velocity V traveled by the car as a function of the final velocity V_f of the car and x if $t = 0$ at $V = 0$. [*Hint:* Try an exponential fit to the data.]

4. A test car is traveling along a level road at 88 km/hr. In order to study the acceleration characteristics of a newly installed engine, the car accelerates at its maximum possible rate. The test crew records the following velocities at various locations along the level road:

$x = 0$	$V = 88.5$ km/hr
$x = 0.1$ km	$V = 93.1$ km/hr
$x = 0.2$ km	$V = 98.3$ km/hr
$x = 0.3$ km	$V = 104.0$ km/hr
$x = 0.4$ km	$V = 110.3$ km/hr
$x = 0.5$ km	$V = 117.2$ km/hr
$x = 1.0$ km	$V = 164.5$ km/hr.

A preliminary study shows that these data follow an equation of the form $V = A(1 + e^{Bx})$, where A and B are positive constants. Find A and B and a Lagrangian expression $V = V(V_0, t)$, where V_0 is the car velocity at time $t = 0$ when $x = 0$.

5. A tornado has the following velocity components in polar coordinates:

$$V_r = -\frac{C_1}{r} \quad \text{and} \quad V_\theta = -\frac{C_2}{r}.$$

Note that the air is spiraling inward. Find an equation for the streamlines.

6. Streamlines are given in Cartesian coordinates by the equation.

$$\Psi = U\left(y - \frac{y}{x^2 + y^2}\right). \qquad x^2 + y^2 \geq 1.$$

Plot the streamlines for $\Psi = 0$, ± 0.0625, and describe the physical situation represented by this equation. The parameter U is an upstream uniform velocity of 1.0 ft/sec.

7. The velocity profiles shown in Fig. P3.7 represent the steady-state flow between two parallel, flat plates at several locations x. There is no velocity in the direction perpendicular to the paper. Identify the directionality and dimensionality of the flow at locations a, b, and c.

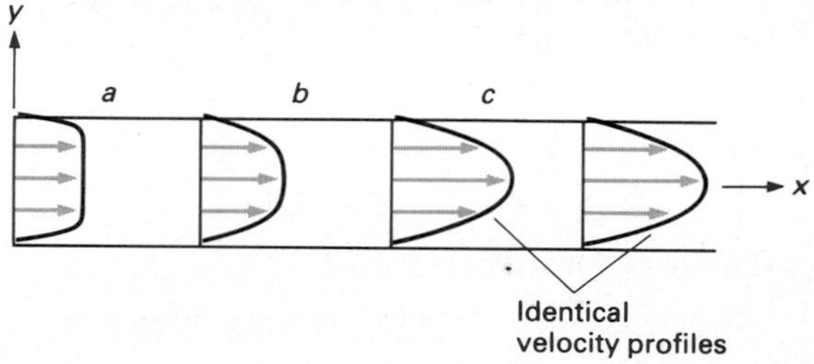

Figure P3.7

8. Air flows steadily through a circular, constant-diameter duct. The air is perfectly inviscid, so the velocity profile is flat across each flow area. However, the air density decreases as the air flows down the duct. Identify the directionality and dimensionality of the flow.

9. The jet pump in Fig. P3.9 has both a primary and a secondary flow at cross-section 1. The fluids mix and achieve the velocity profile shown at cross-section 2. Identify the directionality and dimensionality (a) at section 1, (b) at section 2, and (c) between sections 1 and 2. There is no velocity component perpendicular to the paper. Consider steady state.

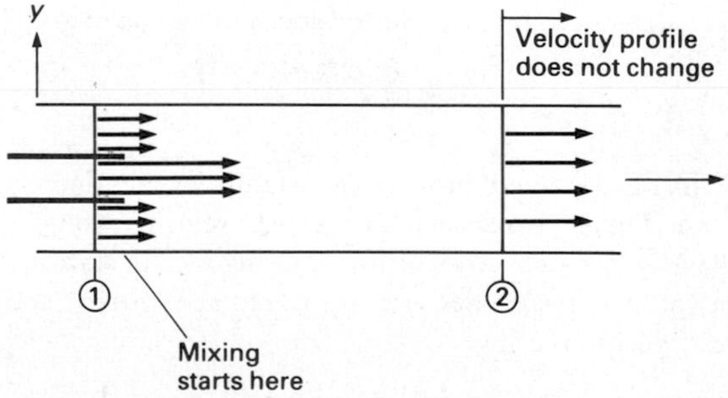

Figure P3.9

10. A small raindrop falling toward the Earth leaves a cloud at zero velocity and accelerates until it reaches a steady downward velocity of 25 m/s, when the raindrop's weight exactly equals the resistance of the air. Identify the directionality, dimensionality, and steadiness of the "flow." Ignore the rotational effects of the Earth and assume that the rain drop falls straight down.

11. Identify the flow in Problem 7 as inviscid or viscous.

12. Identify the flow in Problem 9 as inviscid or viscous.

13. Identify whether the falling rain particle in Problem 10 is a viscous phenomenon or an inviscid phenomenon.

14. List some beneficial and some harmful aspects of turbulence. Consider both engineering applications and everyday occurrences.

15. Classify the following as to laminar, turbulent, and/or almost inviscid:

(a) ketchup flowing out of a bottle;

(b) water flowing over a dam;

(c) boiling water (consider the liquid movement);

(d) a balloon slowly rising in the air;

(e) water coming out a faucet;

(f) air flow in lung alveoli;

(g) rainwater flowing down a street;

(h) sipping a milkshake through a straw; and

(i) sap oozing from a tree.

16. A faucet is turned on to full open and a gallon of 60°F water is collected in 20 sec. The copper pipe connected to the faucet has an inside diameter of about 0.50 in. The flow through this copper pipe will be laminar if the dimensionless Reynolds number, **R**, is less than approximately 2300. Using the value of absolute viscosity and density from Table A.6, determine if the flowing water in the copper pipe is laminar.

$$\mathbf{R} = \frac{\text{(Density)(Average fluid velocity)(Pipe diameter)}}{\text{Fluid absolute viscosity}}.$$

17. Air is delivered through a constant-diameter duct by a fan. The air is inviscid, so the fluid velocity profile is "flat" across each cross section. During the fan start-up, the following velocities were measured at the time t and axial positions x:

	$x=0$	$x=10$ m	$x=20$ m
$t=0$ s	$V=0$ m/s	$V=0$ m/s	$V=0$ m/s
$t=1.0$ s	$V=1.00$ m/s	$V=1.20$ m/s	$V=1.40$ m/s
$t=2.0$ s	$V=1.70$ m/s	$V=1.80$ m/s	$V=1.90$ m/s
$t=3.0$ s	$V=2.10$ m/s	$V=2.15$ m/s	$V=2.20$ m/s.

Calculate the local acceleration, the convective acceleration, and the total acceleration at $t = 1.0$ s and $x = 10$ m. What is the local acceleration after the fan has reached a steady air flow rate?

18. The following pressures for the air flow in Problem 17 were measured:

	$x=0$	$x=10$ m	$x=20$ m
$t=0$ s	$p=101$ kPa	$p=101$ kPa	$p=101$ kPa
$t=1.0$ s	$p=121$ kPa	$p=116$ kPa	$p=111$ kPa
$t=2.0$ s	$p=141$ kPa	$p=131$ kPa	$p=121$ kPa
$t=3.0$ s	$p=171$ kPa	$p=151$ kPa	$p=131$ kPa.

Find the local rate of change of pressure $\partial p/\partial t$ and the convective rate of change of pressure $V\,\partial p/\partial x$ at $t = 2.0$ s and $x = 10$ m.

19. Find the local acceleration $\partial u/\partial t$ and the convective acceleration $u\,\partial u/\partial x$ for $x = 2.5$ m and $y = 0.40$ m from the tabulated data in Example 3.1; $\ell = 5.0$ m. Compare your result(s) with that obtained directly from the given velocity profile.

20. Fluid flows through a pipe with a velocity of 2.0 ft/sec and is being heated, so the fluid temperature T at axial position x increases at a steady rate of 30.0°F/min. In addition, the fluid temperature is increasing in the axial direction at the rate of 2.0°F/ft. Find the value of the Eulerian derivative DT/Dt at position x.

21. Figure P3.21 illustrates a system and fixed control volume at time t and the system at a short time δt later. The system temperature is $T = 100$°F at time t and $\grave{T} =$ 103°F at time $t + \delta t$, where $\delta t = 0.1$ sec. The system mass, m, is 2.0 slugs, and 10 percent of it moves out of the control volume in $\delta t = 0.1$ sec. The energy per unit mass $\tilde{u}$ is $c_v T$, where $c_v = 32.0$ Btu/slug·°F. The energy U of the system at any time t is $m\tilde{u}$. Use the system or Lagrangian approach to evaluate DU/Dt of the system. Compare this result with the DU/Dt evaluated with that of the Eulerian time derivative and flux terms.

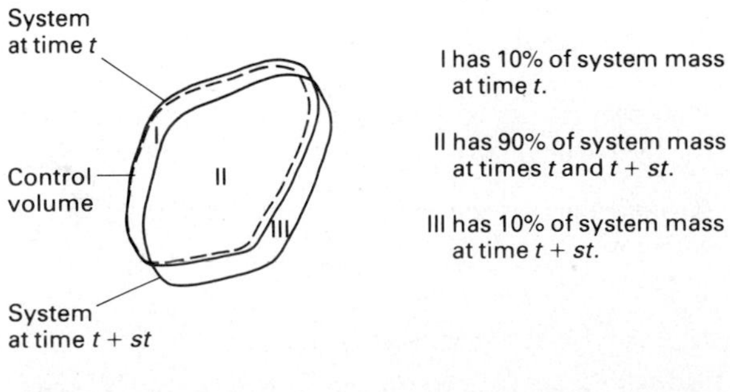

Figure P3.21

22. Figure P3.22 shows a fixed control volume. It has a volume $\forall_0 = 1.0\ \text{ft}^3$, a flow area $A = 1.0\ \text{ft}^2$, and a length $\ell_0 = 1.0$ ft. Position x represents the center of the control volume where the fluid velocity $V_0 = 1.0$ ft/sec and the density $\rho_0 = 1.800\ \text{slug/ft}^3$. Also, at position x the fluid density does not change locally with time but decreases in the axial direction at the linear rate of 0.25 slug/ft^4. Use the system or Lagrangian approach to evaluate dp/dt. Compare this result with that of the Eulerian time derivative and flux terms.

23. Find $D\forall/Dt$ for the system in Problem 22.

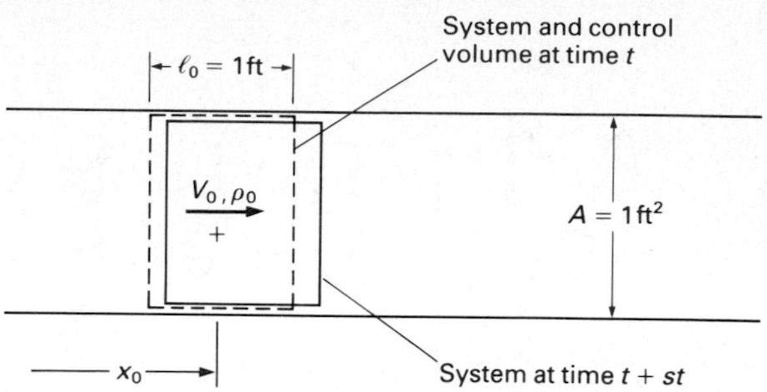

Figure P3.22

REFERENCES

1. White, F. M., *Fluid Mechanics,* McGraw-Hill, New York, 1979.

2. Kline, S. J., *Flow Visualization,* National Committee for Fluid Mechanics Films, distributed by Encyclopaedia Britannica Educational Corporation.

3. Gerhart, P. M., "Averaging Methods for Determining the Performance of Large Fans from Field Measurements," *Transactions of the American Society of Mechanical Engineers: Journal of Engineering for Power,* April 1981.

4. Beckwith, T., N. Buck, and R. Marangoni, *Mechanical Measurements* (3rd ed.), Addison-Wesley, Reading, MA, 1981.

5. Benedict, R. P., *Fundamentals of Temperature, Pressure, and Flow Measurements* (2nd ed.), Wiley, New York, 1977.

6. Bradshaw, P., *Experimental Fluid Mechanics,* Macmillan, New York, 1964.

7. Ower, E., and R. Pankhurst, *The Measurement of Air Flow* (5th ed.), Pergamon Press, Oxford, 1977.

8. Dean, R. C., Jr. (Ed.), *Aerodynamic Measurements,* MIT Gas Turbine Lab Report, 1954, available from University Microfilms, Ann Arbor, MI.

4 The Finite Control Volume Approach to Flow Analysis

In Chapter 3, we introduced three approaches that engineers use to solve flow problems. In this chapter, we develop the finite control volume approach. It ignores the details of a particular flow and gives relations between various gross features of the flow such as flow rate and force. This "lumped" or "global" feature is the finite control volume approach's greatest strength. Even with a relatively small amount of information, it can produce valuable results.

Introducing some simplifying assumptions (such as locally one-dimensional, uniform flow) is usually necessary to complete a flow analysis. The control volume approach is much less sensitive to such assumptions than the differential approach is. With a clever choice of control volume, careful use of given information, and some reasonable assumptions and approximations, we can almost always obtain at least a partial answer. In many cases, the approach produces the complete answer with a high degree of accuracy.

The finite control volume approach is a very important tool for flow analysis. Like any other single tool, it will not solve every flow problem, but it is frequently the best tool available. By carefully studying this chapter and by solving several practice problems, you can learn how to use this tool and come to appreciate its advantages and limitations.

4.1 OVERVIEW OF THE FINITE CONTROL VOLUME APPROACH

4.1.1 Methods for Developing Working Equations

The basic elements of the control volume method are the fundamental laws of nature described in Section 3.3.1 and the transport theorem derived in Section 3.3.5. The fundamental laws describe the fluid behavior in terms of a fixed mass of fluid, called a system. The transport theorem uses a control volume to instantaneously identify a specific system (the mass contained by the control volume) and relates rates of change of system properties to the flow field.

Each of the fundamental laws, when combined with the transport theorem, produces a specific equation. The law of conservation of mass produces the *continuity equation,* the law of conservation of energy produces the *energy equation,* and Newton's second law of motion produces the *momentum equation.* These three equations are the backbone of the finite control volume method. We may apply the general equations derived by this method to any control volume selected for a specific problem. The derivation need be done only once, and we do it in this chapter. We may then apply the working equations directly to specific problems.

The steps involved in using the finite control volume method are

- select a control volume appropriate to the flow;
- choose the necessary equations;
- use the known information and make reasonable assumptions to evaluate terms in the equations; and
- calculate the quantities of interest.

4.1.2 Selecting Appropriate Control Volumes

The choice of a control volume may greatly affect the complexity of an analysis. A control volume may be fixed and rigid or moving and deformable. A fixed, rigid control volume might always seem preferable, but it is not. The most important consideration is that the control volume be tailored to the problem. Figure 4.1 shows some control volumes especially tailored to flow problems.

Having decided on control volume type (fixed, moving, rigid, deformable), the analyst still must select the size, shape, and boundary location. In many problems, the choice is reasonably obvious. Sometimes, any of several choices will do. A good control volume selection has the following characteristics:

- The control volume fits the problem. (Try a rigid control volume for the piston and cylinder or a moving control volume for the pipe flow of Fig. 4.1 and see what happens!)

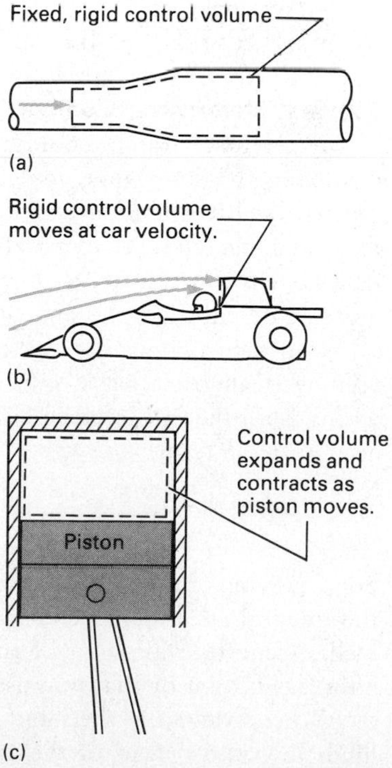
Fixed, rigid control volume
(a)
Rigid control volume moves at car velocity.
(b)
Control volume expands and contracts as piston moves.
Piston
(c)

Figure 4.1 Situations requiring different control volume choices: (a) flow in pipe with diffuser section; (b) flow into race car air scoop; (c) reciprocating air compressor.

- The quantities to be calculated (flows, forces, pressures, etc.) appear explicitly in the formulation. To do so usually requires that these quantities be defined at the boundaries of the control volume.
- The control surface should be perpendicular to the local fluid velocity.
- No unnecessary mathematical complexity is introduced by the choice.

Skill at selecting good control volumes can be developed only with practice. Studying the examples in this chapter with an eye toward the specific choice of control volume will also be valuable.

Evaluation of the terms in the transport theorem, and hence the corresponding terms in the working equations, may be affected by the type of control volume chosen. From Section 3.3.5, the transport theorem is

$$\frac{dB_{\text{sys}}}{dt} = \frac{d}{dt}\int_{\forall_{cv}} \rho b\, d\forall + \int_{A_{\text{out}}} \rho b V \cos\theta\, dA + \int_{A_{\text{in}}} \rho b V \cos\theta\, dA, \tag{4.1a}$$

or

$$\frac{dB_{\text{sys}}}{dt} = \frac{d}{dt}\int_{\forall_{cv}} \rho b d\forall + \int_{A_{\text{out}}} \rho b V_n dA - \int_{A_{\text{in}}} \rho b V_n dA, \tag{4.1b}$$

or

$$\frac{dB_{\text{sys}}}{dt} = \frac{d}{dt}\int_{\forall_{cv}} \rho b\, d\forall + \oint_{A_{cv}} \rho b(\vec{V}\cdot\hat{n})\, dA. \tag{4.1c}$$

Any of these forms of the equation may be applied to any control volume, fixed or moving, rigid or deformable. Similarly, the flow can be three-dimensional, three-directional, and unsteady. Most applications do not involve all these complexities. Let's examine some special cases involving combinations of these features.

Moving, Deforming Control Volume. You can apply the transport theorem in any reference frame (coordinate system) that you choose. If a control volume is moving and deforming, you generally must choose between a coordinate system that is fixed to some part of the control volume or a coordinate system that is fixed to some reference external to the control volume. Whichever type of coordinate system you choose, both the control volume geometry and the fluid velocity must be described in the same coordinate system.

Choosing a coordinate system that is fixed to at least one point in the control volume is almost always best, because this choice minimizes confusion in evaluating both the rate of accumulation term and the flux terms. Consider first the accumulation term

$$\frac{d}{dt}\int_{\forall_{cv}} \rho b\, d\forall.$$

For a moving or deforming control volume, $\forall_{cv}$ is time-dependent, so the limits of the integral are time-dependent. In addition, ρ and b may be time-dependent as well. Accurate evaluation of the term requires that you perform differentiation after integration or that you use Liebnitz's rule for differentiation.

When evaluating the flux terms, recall that the velocity measures the rate of fluid flow into or out of the control volume. If we choose a coordinate system fixed to the control volume, this velocity must be measured relative to the control surface. We use the symbol $\vec{V}_r$ to indicate velocity relative to the control surface. The flux terms are then written as

$$\dot{B}_{\text{out}} = \int \rho b V_r \cos\theta \, dA \quad \text{and} \quad \dot{B}_{\text{in}} = \int \rho b V_r \cos\theta' \, dA$$

or

$$\dot{B}_{\text{out}} - \dot{B}_{\text{in}} = \oint_{A_{cv}} \rho b(\vec{V}_r \cdot \hat{n}) \, dA$$

In this book, we assume that the transport theorem is applied in a coordinate system fixed to at least one point in the control volume. We, therefore, use $\vec{V}_r$ in the flux integrals and describe the control volume geometry in these coordinates. Note that the use of the relative velocity applies only to the mass flow; if the arbitrary property b contains any velocities (such as kinetic energy or momentum), these velocities may be referred to another coordinate system as necessary.

Fixed, Rigid Control Volume. If the control volume is fixed and rigid, then $\forall_{cv}$ does not change with time, and we can write

$$\frac{d}{dt}\int_{\forall_{cv}} \rho b \, d\forall = \frac{\partial}{\partial t}\int_{\forall_{cv}} \rho b \, d\forall = \int_{\forall_{cv}} \frac{\partial}{\partial t}(\rho b) \, d\forall;$$

that is, we may perform differentiation and integration in any order.

Steady-Flow, Fixed, Rigid Control Volume. In this case, the first term may be dropped altogether:

$$\frac{d}{dt}\int_{\forall_{cv}} \rho b \, d\forall \equiv 0.$$

4.2 THE CONTINUITY EQUATION

Combining the law of conservation of mass with the transport theorem yields one of the most useful equations in all of fluid mechanics: the *continuity equation*. Every flow analysis makes use of the continuity equation in some form.

4.2.1 Derivation of the Continuity Equation

For generality, we consider a three-dimensional, three-directional, unsteady flow field. A control volume is superimposed on the flow as shown in Fig. 4.2. As discussed in Section 3.3.5, we consider the system that instantaneously occupies the control volume. By definition, a system is a collection of mass of fixed identity, so

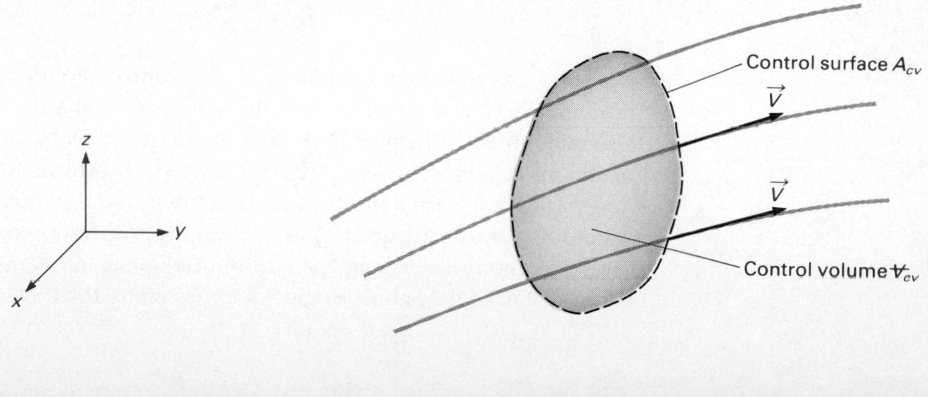

Figure 4.2 Flow field with control volume for continuity equation.

the law of conservation of mass is

$$\frac{dm_{\text{sys}}}{dt} = 0. \tag{4.2}$$

We evaluate the rate of change of system mass with the help of the transport theorem, using the vector form, Eq. (4.1c), for brevity. The general system property B is the system mass,

$$B_{\text{sys}} = m_{\text{sys}} = \int_{\forall_{\text{sys}}} \rho \, d\forall,$$

and the specific property b is

$$b = \frac{m_{\text{sys}}}{m_{\text{sys}}} = 1.$$

Substituting into the transport theorem and using Eq. (4.2), we get the continuity equation:

$$\frac{d}{dt}\int_{\forall_{cv}} \rho \, d\forall + \oint_{A_{cv}} \rho(\vec{V}_r \cdot \hat{n}) \, dA = 0. \tag{4.3a}$$

By Eqs. (4.1a) and (4.1b), we may also write the continuity equation as

$$\frac{d}{dt}\int_{\forall_{cv}} \rho \, d\forall + \int_{A_{\text{out}}} \rho V \cos\theta \, dA + \int_{A_{\text{in}}} \rho V \cos\theta \, dA = 0 \tag{4.3b}$$

and

$$\frac{d}{dt}\int_{\forall_{cv}} \rho \, d\forall + \int_{A_{\text{out}}} \rho V_n \, dA - \int_{A_{\text{in}}} \rho V_n \, dA = 0. \tag{4.3c}$$

Equations (4.3a–c) express the principle of conservation of mass for a control volume. Recalling that the first term of the transport theorem represents accumulation (storage) of a quantity within the control volume and that the remaining terms represent outflow and inflow of the quantity, we write the continuity equation in words as follows:

$$\begin{bmatrix}\text{Rate of accumulation}\\ \text{of mass inside}\\ \text{the control volume}\end{bmatrix} + \begin{bmatrix}\text{Rate of mass flow}\\ \text{leaving the}\\ \text{control volume}\end{bmatrix} - \begin{bmatrix}\text{Rate of mass flow}\\ \text{entering the}\\ \text{control volume}\end{bmatrix} = 0. \tag{4.4a}$$

In symbolic terms,

$$\frac{dm_{cv}}{dt} + \sum \dot{m}_{\text{out}} - \sum \dot{m}_{\text{in}} = 0, \tag{4.4b}$$

where m_{cv} is the mass contained in the control volume and $\dot{m}$ is the mass flow rate crossing the control surface. Equation (4.4) is a mass balance for the control volume. An often helpful approach to solving flow problems is to think in terms of the mass balance, beginning with Eq. (4.4), and then to express mass flows and mass storage in terms of the density, velocity, and control volume geometry, thus proceeding directly from Eq. (4.4) to an appropriate working equation. An alternative method of deriving the continuity equation begins with the hueristic mass balance, Eq. (4.4a), and substitutes expressions for flow rate and mass storage.

EXAMPLE 4.1 **Illustrates the Continuity Equation**

Show that the flow field in Example 3.1 satisfies the continuity equation.

SOLUTION

Given

Converging channel in Fig. E3.1a with

$$Y = \frac{Y_0}{1 + x/\ell} \quad \text{and} \quad u = u_0\left(1 + \frac{x}{\ell}\right)\left[1 - \left(\frac{y}{Y}\right)^2\right]$$

Constant density fluid, two-dimensional channel, and steady flow

Find

Whether the given velocity field satisfies Eq. (4.3a), or

$$\frac{d}{dt}\int_{\forall_{cv}} \rho \, d\forall + \oint_{A_{cv}} \rho(\vec{V}_r \cdot \hat{n}) \, dA \overset{?}{=} 0$$

Solution

The first term of the equation

$$\frac{d}{dt}\int_{\forall_{cv}} \rho \, d\forall = 0,$$

because the flow is steady. Because ρ is constant,

$$\oint_{A_{cv}} (\vec{V}_r \cdot \hat{n}) \, dA \overset{?}{=} 0.$$

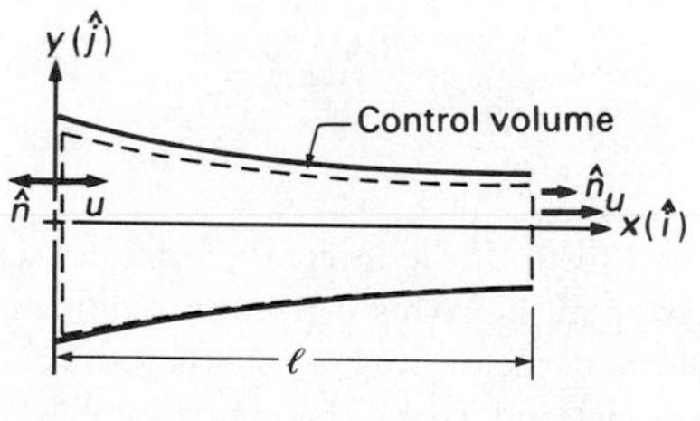

Figure E4.1 Control volume for converging two-dimensional channel.

For the control volume shown in Fig. E4.1, $\vec{V}_r \cdot \hat{n} = 0$ along the walls of the channel, because $u = v = 0$ at the wall. The flow is two-directional, so

$$\vec{V}_r = u\hat{i} + v\hat{j}.$$

At the outlet plane, the unit outward normal vector is $+\hat{i}$. At the inlet plane, the unit outward normal vector is $-\hat{i}$. Therefore

$$\oint_{A_{cv}} (\vec{V}_r \cdot \hat{n}) \, dA = \int_{A_{\text{out}}} (u\hat{i} + v\hat{j}) \cdot (\hat{i}) \, dA + \int_{A_{\text{in}}} (u\hat{i} + v\hat{j}) \cdot (-\hat{i}) \, dA$$
$$= \int_{A_{\text{out}}} u \, dA - \int_{A_{\text{in}}} u \, dA.$$

Substituting for u at the exit, we have

$$\int_{A_{\text{out}}} u \, dA = \int_{A_{\text{out}}} u_0\left(1 + \frac{\ell}{\ell}\right)\left[1 - \left(\frac{y}{Y_\ell}\right)^2\right] dA$$
$$= W u_0 \int_{-Y_\ell}^{+Y_\ell} 2\left[1 - \left(\frac{y}{Y_\ell}\right)^2\right] dy,$$

where Y_ℓ is the magnitude of the channel half-height at $x = \ell$ and W is the width of the channel (perpendicular to the paper). Integrating, we have

$$\int_{A_{out}} u\,dA = 2Wu_0\left[y - \frac{y^3}{3Y_\ell^2}\right]_{-Y_\ell}^{+Y_\ell} = \frac{8Wu_0Y_\ell}{3}.$$

As $Y_\ell = 0.5$ m and $u_0 = 1.0$ m/s,

$$\int_{A_{out}} u\,dA = \frac{4}{3}\,W.$$

Similarly,

$$\int_{A_{in}} u\,dA = \int_{A_{in}} u_0(1)\left[1 - \left(\frac{y}{Y_0}\right)^2\right]dA$$

$$= Wu_0\int_{-Y_0}^{+Y_0}\left[1 - \left(\frac{y}{Y_0}\right)^2\right]dy,$$

where Y_0 is the magnitude of the channel half-height at $x = 0$. Integrating, we obtain

$$\int_{A_{in}} u\,dA = Wu_0\left[y - \frac{y^3}{3Y_0^2}\right]_{-Y_0}^{+Y_0} = \frac{4Wu_0Y_0}{3}.$$

As $Y_0 = 1.0$ m and $u_0 = 1.0$ m/s,

$$\int_{A_{in}} u\,dA = \frac{4}{3}\,W.$$

Then

$$\oint_{A_{cv}} (\vec{V}\cdot\hat{n})\,dA = \frac{4}{3}\,W - \frac{4}{3}\,W = 0.$$

Thus

$$\frac{d}{dt}\int_{\forall_{cv}} \rho\,d\forall + \oint_{A_{cv}} \rho(\vec{V}\cdot\hat{n})\,dA = 0.$$ **ANSWER**

Discussion

Although the mathematical form of the integral continuity equation looks formidable, evaluation of the terms was not too difficult. Note that the unit outward normal vector was $+\hat{i}$ at the outlet and $-\hat{i}$ at the inlet. This vector is determined by the control volume geometry, not by the flow direction.

4.2.2 Flow Rate and Average Velocity

An important parameter for flow in a closed conduit (Fig. 4.3a) or an open channel (Fig. 4.3b), is the fluid flow rate. To evaluate the flow rate in any passage, we consider an imaginary surface that spans the passage. We usually select this surface so that it is perpendicular to the axis of the conduit or channel. We imagine a unit vector $\hat{n}$ perpendicular to the surface, pointing in the direction of flow ($0 < \theta < 90°$). The mass flow rate passing through the surface is

$$\dot{m} = \int \rho V \cos\theta\,dA \tag{4.5}$$

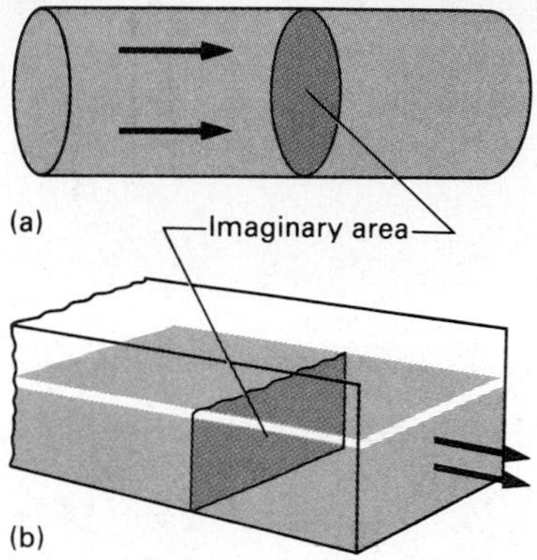

Figure 4.3 Areas for calculating flow rate in (a) a closed conduit and (b) an open channel.

or, in vector notation,

$$\dot{m} = \int \rho(\vec{V}_r \cdot \hat{n})\, dA. \tag{4.6}$$

A third way of writing the equation is

$$\dot{m} = \int \rho V_n\, dA, \tag{4.7}$$

where V_n is the component of the fluid velocity perpendicular to the area.

These equations for mass flow rate are correct for any choice of cross-sectional area. The analyst selects the area he or she likes best. The simplest choice is an area perpendicular to the duct axis or an area perpendicular to the velocity vector. In many cases, these choices reduce to the same area—but not always (see Example 4.2).

Sometimes engineers express flow rate in terms of volume rather than mass, as in rating pump discharge in gallons per minute or fan flow in cubic meters per minute. The volume flow rate, Q, crossing a surface may be computed by deleting the density from the mass flow:

$$Q = \int V \cos\theta\, dA, \tag{4.8}$$

$$Q = \int (\vec{V}_r \cdot \hat{n})\, dA, \tag{4.9}$$

or

$$Q = \int V_n\, dA. \tag{4.10}$$

If the fluid density is uniform over the area of integration,

$$\dot{m} = \rho Q. \tag{4.11}$$

If the density and velocity are both uniform over the cross-sectional area and if the area is perpendicular to the velocity, Eqs. (4.5)–(4.7) all reduce to

$$\dot{m} = \rho V A. \tag{4.12}$$

If the velocity is uniform and perpendicular to the area, Eqs. (4.8)–(4.10) all reduce to

$$Q = VA. \tag{4.13}$$

EXAMPLE 4.2 Illustrates Volume Flow Rate Calculation for Various Choices of Area

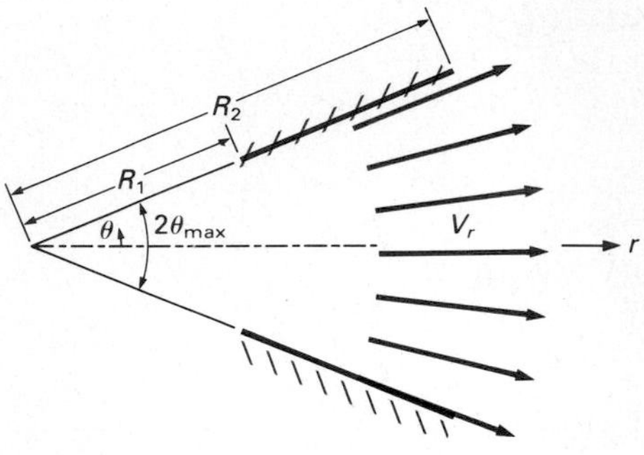

Figure E4.2a Plane diffuser.

Figure E4.2a shows an inviscid, constant-density fluid flowing steadily in a two-dimensional diffuser of constant width W (perpendicular to the page). The radial velocity $V_r(r, \theta)$ is given by

$$V_r(r, \theta) = \frac{C}{r},$$

where C is a constant, r is measured in meters, and V_r is measured in meters per second. The circumferential velocity $V_\theta = 0$. Calculate the volume flow rate for the two surfaces shown in Fig. E4.2b:

$$r = r_1 = \text{Constant, and}$$

$$r_1 \cos \theta_{max} = x_1 = \text{Constant.}$$

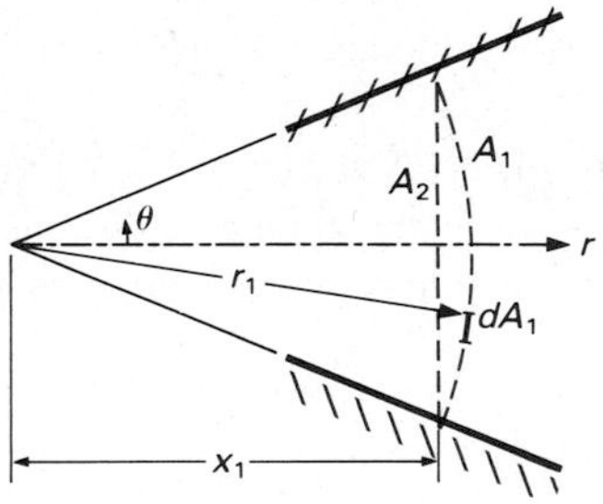

Figure E4.2b Plane diffuser with two surfaces to calculate volume flow rate.

SOLUTION

Given

Inviscid, constant-density fluid flowing through two-dimensional diffuser of width W (Fig. 4.2a)

Radial velocity V_r:

$$V_r(r, \theta) = \frac{C}{r}.$$

Find

Volume flow rate for two surfaces:

$r = r_1 =$ Constant

$x = x_1 =$ Constant

Solution

We use Eq. (4.10) to find the volume flow rate:

$$Q = \int V_n \, dA.$$

For the surface $r = r_1 =$ Constant, V_r is perpendicular to each differential area dA. Then

$$V_n = V_r \quad \text{and} \quad dA_1 = W r_1 \, d\theta.$$

Using the symmetry about the centerplane of the diffuser, we have

$$Q = 2 \int_0^{\theta_{max}} V_r (W r_1 \, d\theta).$$

Substituting for V_r gives

$$Q = 2\int_0^{\theta_{max}} \frac{C}{r_1} W r_1 \, d\theta = 2CW\theta\Big]_0^{\theta_{max}};$$

$$Q = 2WC\theta_{max}.$$ **ANSWER**

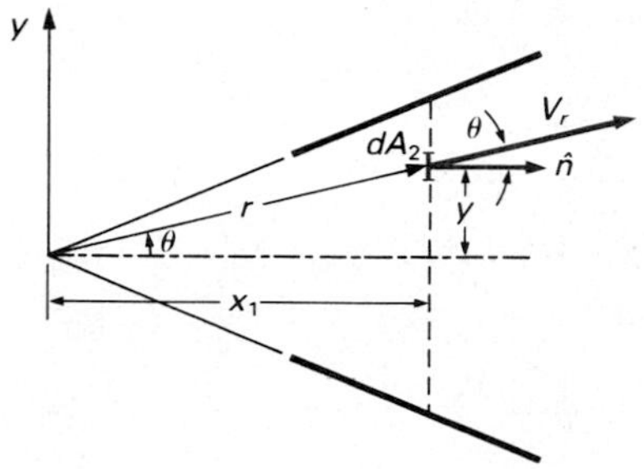

Figure E4.2c Relation of velocity and unit normal vectors.

For the surface $x = x_1$ = Constant, Fig. E4.2c shows that

$$V_n = V_r \cos\theta, \qquad dA_2 = W\,dy, \quad \text{and} \quad r = \sqrt{x_1^2 + y^2}.$$

Using the symmetry about the centerplane of the diffuser, we have

$$Q = 2\int_0^{y_{max}} V_r \cos\theta \,(W\,dy).$$

Substituting for V_r and r gives

$$Q = 2W\int_0^{y_{max}} \frac{C\cos\theta}{\sqrt{x_1^2 + y^2}}\, dy.$$

We calculate $\cos\theta$ from

$$\cos\theta = \frac{x_1}{r} = \frac{x_1}{\sqrt{x_1^2 + y^2}},$$

so

$$Q = 2WCx_1\int_0^{y_{max}} \frac{dy}{x_1^2 + y^2}.$$

Integrating, we have

$$Q = 2WCx_1\left[\frac{1}{x_1}\tan^{-1}\left(\frac{y}{x_1}\right)\right]_0^{y_{max}} = 2WC\tan^{-1}\left(\frac{y_{max}}{x_1}\right).$$

However, the angle whose tangent is y_{max}/x_1 is θ_{max}, so

$$Q = 2WC\theta_{max},$$ **ANSWER**

the same answer obtained for A_1.

Discussion

The units of r_1 were given as meters and V_r as meters per second. Therefore the units of C must be

$$[C] = [V_r \cdot r] = [\text{m/s}\cdot\text{m}] = [\text{m}^2/\text{s}].$$

The brackets indicate that we are considering only the units of the equation. Can you show that the units of Q are cubic meters per second?

Fluid velocity is seldom precisely uniform over the cross section of a pipe or channel. In Section 3.2.3, we discussed the desirability of approximating such flows, which are in fact multidimensional and multidirectional, as one-dimensional and one-directional by the use of average velocities. For approximate one-dimensional flow, Eq. (4.13) must be valid, so we define an average velocity ($\bar{V}$) by

$$Q = \int V_n \, dA = \bar{V}A.$$

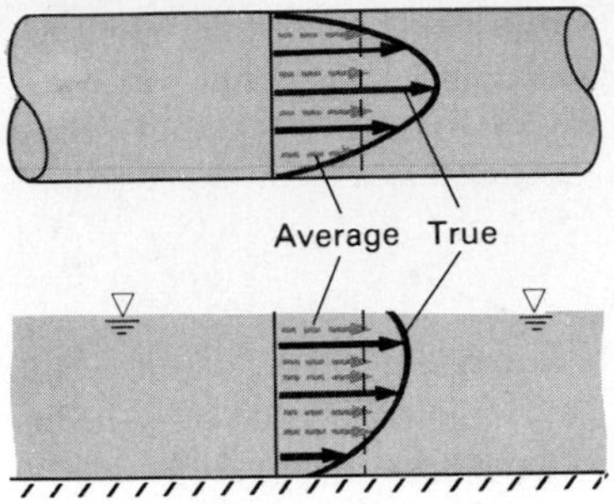

Figure 4.4 Comparison of true and average velocities for pipe and open channel flow.

Therefore

$$\bar{V} \equiv \frac{Q}{A} = \frac{1}{A}\int V_n\,dA, \tag{4.14}$$

where the overbar represents an average quantity. Although we do not normally use the overbar when dealing with one-dimensional flow, we use it here to distinguish the average velocity from the local velocity. Figure 4.4 shows true and average velocities for flow in a pipe or channel.

We also want the mass flow equation, Eq. (4.12), to apply to approximate one-dimensional flow. Thus

$$\dot{m} = \int \rho V_n\,dA = \bar{\rho}\bar{V}A. \tag{4.15}$$

To use this equation, we need an average density as well as an average velocity, which is no problem if the fluid is incompressible because $\bar{\rho} = \rho$. If the fluid's density varies from point to point in the cross section,* we find the average density by substituting Eq. (4.14) into Eq. (4.15),

$$\bar{\rho} = \frac{\int \rho V_n\,dA}{\int V_n\,dA} = \frac{\int \rho V_n\,dA}{\bar{V}A}. \tag{4.16}$$

Note that the proper average density is a "velocity-weighted" average. An average density computed by

$$\bar{\rho}' = \frac{\int \rho\,dA}{A}, \tag{4.17}$$

similar to the average velocity, is not the same as $\bar{\rho}$, and the mass flow rate cannot be correctly calculated by using $\bar{\rho}'$ in Eq. (4.15). In the remainder of this book, proper averages are implied when we use the one-dimensional flow approximation.

---

* This condition is not as rare as you might think. For example, air flowing in a pipe with a heated wall has a lower density at the hot wall than at the cooler center of the pipe.

EXAMPLE 4.3 Illustrates Average Velocity

Compute the average velocity $\bar{V}$ for the flow in Example 3.1.

SOLUTION

Given

Fluid velocity

$$u = u_0\left(1 + \frac{x}{\ell}\right)\left[1 - \left(\frac{y}{Y}\right)^2\right]$$

for the two-dimensional channel flow in Fig. E3.1a

Find

Average velocity $\bar{V}$

Solution

The problem is not adequately defined. Are we to calculate the average velocity at $x = 0, x = \ell$, or at all values of x? Because we know that u is a function of x and y, we calculate $\bar{V}(x)$, the average velocity as a function of x. We find the average velocity from Eq. (4.14):

$$\bar{V} = \frac{\int V_n\, dA}{A},$$

where V_n is the component of the velocity normal to dA. Because u is the x component of the total velocity and is normal to each area dA, as shown in Fig. E4.3,

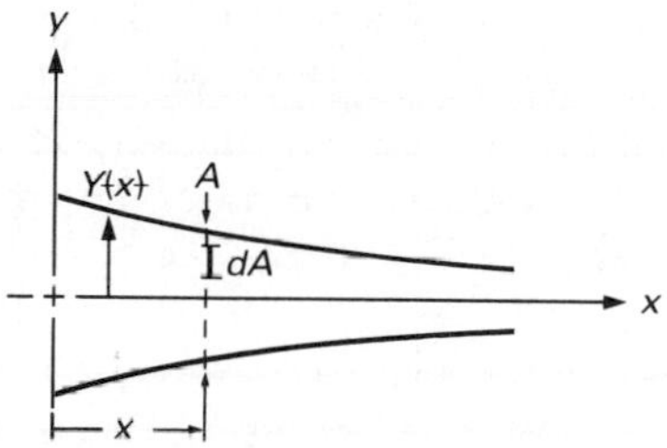

Figure E4.3 Flow area for calculating the average velocity $\bar{V}(x)$.

$$\bar{V}(x) = \frac{\int u(x, y)\, dA}{A}.$$

Using the channel width W, we obtain $dA = W\,dy$ and $A = 2WY$. Because $u(x, y)$ is symmetric to the centerplane of the channel,

$$\bar{V}(x) = \frac{2\int_0^Y u(x, y) W\, dy}{2WY} = \int_0^1 u_0\left(1 + \frac{x}{\ell}\right)\left[1 - \left(\frac{y}{Y}\right)^2\right] d\left(\frac{y}{Y}\right);$$

$$\bar{V}(x) = \frac{2}{3} u_0\left(1 + \frac{x}{\ell}\right).$$ **ANSWER**

Discussion

The first three columns of Table E4.3 give $\bar{V}$ and A as a function of the axial coordinate x/ℓ for a channel of width $W = 1.0$ m. The fourth column gives the product $\bar{V}A$ as a function of x/ℓ. Note that

$$(\bar{V}A)_{\text{in}(x=0)} = (\bar{V}A)_{\text{any }x}.$$

Why is this true?

The transverse velocity $v(x, y)$ is nonzero for this flow but does not influence the calculation of $\bar{V}$, because v is parallel to the cross-sectional area (perpendicular to $\hat{n}$).

Table E4.3 Values of $\bar{V}$, A, and $\bar{V}A$ as a function of x/ℓ for the flow in Example 3.1.

Length Ratio x/ℓ	Average Velocity $\bar{V}$ (m/s)	Area A (m^2)	$\bar{V}A$ (m^3/s)
0.0	0.667	2.000	1.334*
0.2	0.800	1.667	1.334
0.4	0.933	1.429	1.333
0.6	1.067	1.250	1.334
0.8	1.200	1.111	1.333
1.0	1.333	1.000	1.333

* Note the small error from rounding off $\bar{V}$ and A to four significant figures.

4.2.3 Simplified Forms of the Continuity Equation

We seldom use the continuity equation in its most general form, Eqs. (4.3a–c). In this section, we develop some simplified forms of the continuity equation and illustrate their application to a variety of flow problems. Although the simplified forms are usually easier and quicker to use, the more general form, Eqs. (4.3a–c), always produces identical results when correctly applied.

Steady, Uniform, One-Directional Flow. Consider the flow through the apparatus shown in Fig. 4.5. This apparatus may represent a water storage tank with multiple inlets and outlets, a steam turbine with steam extractions, or some other piece of equipment. We assume that the flow is steady (no accumulation of fluid inside the apparatus) and that the flow is uniform and one-directional at each inlet and outlet.

We select a control volume that encloses the entire apparatus (see Fig. 4.5) and write the integral continuity equation:

$$\frac{d}{dt}\int_{\forall_{cv}} \rho \, d\forall + \int_{A_{\text{out}}} \rho V \cos\theta \, dA + \int_{A_{\text{in}}} \rho V \cos\theta \, dA = 0.$$

The first term is zero for steady flow. Recall that for all the inflow surfaces, the third term is negative, so using the angle θ' ($\theta' = \pi - \theta$) instead of θ at the inlet (see Fig. 3.26), we get

$$\int_{A_{\text{in}}} \rho V \cos\theta' \, dA = \int_{A_{\text{out}}} \rho V \cos\theta \, dA.$$

Because mass flows in and out at a finite number of locations and the velocity and density are uniform at each location, we write the integrals as finite sums:

$$\sum (\rho V_n A)_{\text{in}} = \sum (\rho V_n A)_{\text{out}}. \tag{4.18}$$

Steady flow, finite number of uniform streams

An alternative form is

$$\sum \dot{m}_{\text{in}} = \sum \dot{m}_{\text{out}}. \tag{4.19}$$

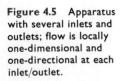

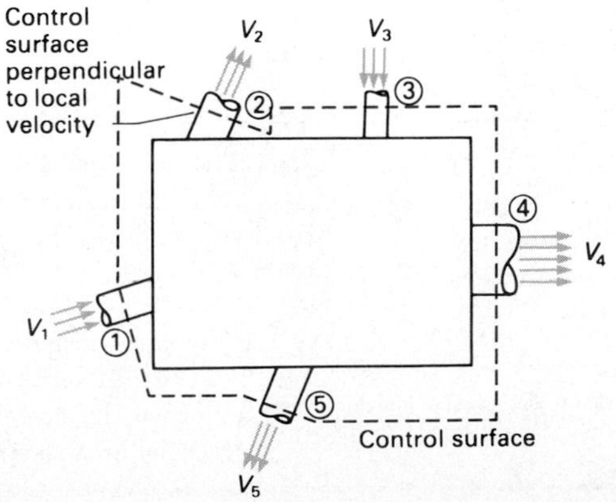

Figure 4.5 Apparatus with several inlets and outlets; flow is locally one-dimensional and one-directional at each inlet/outlet.

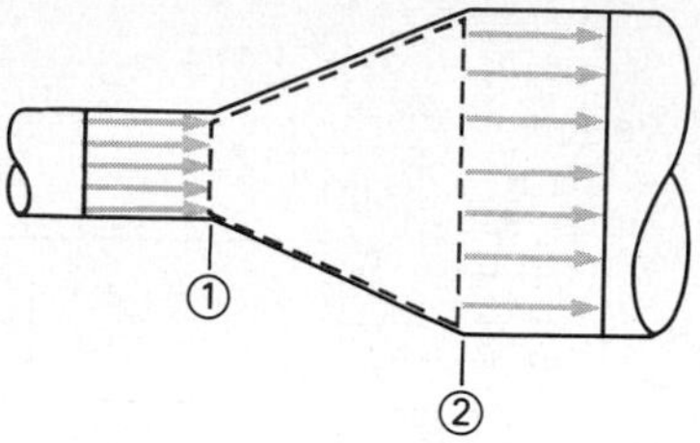

Figure 4.6 Conduit with one inlet and one outlet.

For the apparatus of Fig. 4.5, we have

$$\rho_1 V_1 A_1 + \rho_3 V_3 A_3 = \rho_2 V_2 A_2 + \rho_4 V_4 A_4 + \rho_5 V_5 A_5.$$

If the density of the fluid is the same at all points (a good assumption if the apparatus is a water tank but a poor one if the apparatus is a steam turbine), we may use volume flows instead of mass flows:

$$\sum (V_n A)_{\text{in}} = \sum (V_n A)_{\text{out}}, \tag{4.20}$$

Constant density, steady flow, finite number of uniform streams

or

$$\sum Q_{\text{in}} = \sum Q_{\text{out}}. \tag{4.21}$$

For a flow in a conduit or channel with only one inlet and one outlet, as shown in Fig. 4.6, we drop the summations, leaving

$$\rho_1 V_1 A_1 = \rho_2 V_2 A_2 \tag{4.22}$$

or, for constant density, $\rho_1 = \rho_2$,

$$V_1 A_1 = V_2 A_2. \tag{4.23}$$

Constant density only one inlet/one outlet

EXAMPLE 4.4 Illustrates the Steady, One-Dimensional Continuity Equation

Figure E4.4a shows a "tee" junction in a water pipeline. The main line of the tee is 6-in. (nominal) diameter type L copper pipe, and the branch is 4-in. (nominal) diameter type L copper pipe. A steady flow of water at 60°F enters the main line with an average velocity of 10 ft/sec and leaves the main line with an average velocity of 7 ft/sec. Find the water velocity, mass flow rate, and volume flow rate in the branch line.

SOLUTION

Given

Figure E4.4a. Water velocity entering tee is 10 ft/sec. Water velocity leaving tee is 7 ft/sec.

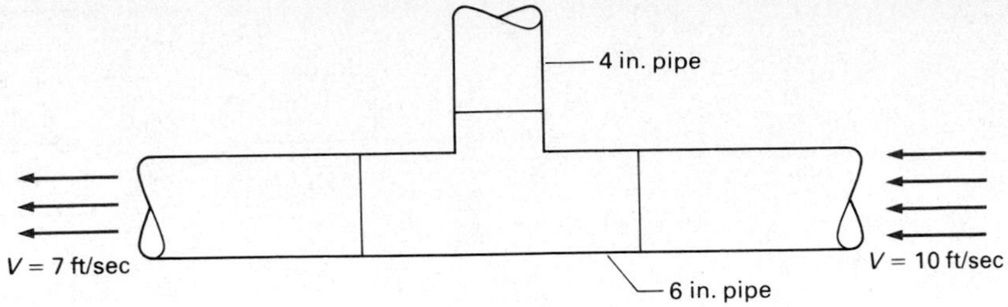

Figure E4.4a A "tee" junction in a pipeline.

Find

Water velocity and mass and volume flow rates in branch

Solution

We select a control volume as shown in Fig. E4.4b and apply the one-dimensional continuity equation. A proper form for this incompressible steady flow is Eq. (4.21):

$$\sum Q_{in} = \sum Q_{out},$$

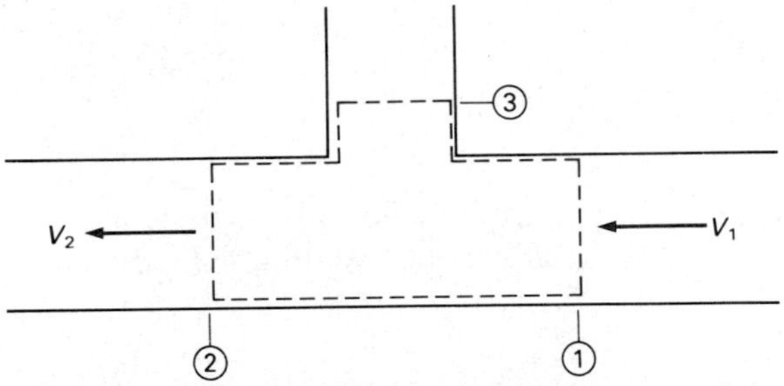

Figure E4.4b Control volume for flow in the "tee."

where the volume flow is the product of fluid velocity and cross-sectional area. We do not know whether the flow in the branch is entering or leaving the junction; however, as V_2 is smaller than V_1, we may reasonably assume that the branch flow is leaving the junction. The continuity equation becomes

$$V_1A_1 = V_2A_2 + V_3A_3.$$

Solving for V_3,

$$V_3 = \frac{V_1A_1}{A_3} - \frac{V_2A_2}{A_3}.$$

The areas are calculated from the pipe diameters. Appendix D gives dimensions of standard sizes of pipe and tubing. Note that we need the inside diameter of the pipes:

$$D_1 = D_2 = 5.845 \text{ in.} \quad \text{and} \quad D_3 = 3.905 \text{ in.}$$

The areas are

$$A_1 = A_2 = \frac{\pi}{4}(5.845 \text{ in.})^2 = 26.83 \text{ in}^2 = 0.1863 \text{ ft}^2,$$

and

$$A_3 = \frac{\pi}{4}(3.905 \text{ in.})^2 = 11.98 \text{ in}^2 = 0.0832 \text{ ft}^2.$$

The (average) velocity V_3 is

$$V_3 = (10 \text{ ft/sec})\left(\frac{0.1863 \text{ ft}^2}{0.0832 \text{ ft}^2}\right) - (7 \text{ ft/sec})\left(\frac{0.1863 \text{ ft}^2}{0.0832 \text{ ft}^2}\right);$$

$$V_3 = 6.72 \text{ ft/sec.}$$ **ANSWER**

The volume flow rate is

$$Q_3 = V_3 A_3 = (6.72 \text{ ft/sec})(0.0832 \text{ ft}^2);$$

$$Q_3 = 0.559 \text{ ft}^3/\text{sec.}$$ **ANSWER**

The mass flow rate is

$$\dot{m}_3 = \rho Q_3 = (62.4 \text{ lbm/ft}^3)(0.559 \text{ ft}^3/\text{sec});$$

$$\dot{m}_3 = 34.88 \text{ lbm/sec.}$$ **ANSWER**

Because V_3 calculated as a positive quantity, our assumption that the branch flow is leaving the tee is correct. Therefore $\dot{m}_3$ and Q_3 are flows leaving the junction.

Discussion

If the velocity V_3 had been negative, we would have concluded that the flow was entering the tee at the branch. In this problem, examining the data and concluding that the flow in the branch is leaving the control volume was relatively simple. But with several inlets and outlets, whether an unknown flow (velocity) is entering or leaving may not be obvious. In such cases, we simply assume a flow direction and let the sign of the calculated flow or velocity tell us whether our assumption is correct. (Also see the discussion for Example 4.6.)

EXAMPLE 4.5 Illustrates the Steady, One-Dimensional Continuity Equation

Figure E4.5a Hydraulic jump at the base of a dam.

A *hydraulic jump* is an open channel flow phenomenon whereby the depth of the flow stream abruptly increases. Engineers may design a dam spillway (a passage for excess water to run around or under the dam) so that a hydraulic jump occurs in the spillway to reduce the water velocity reentering the river. Figure E4.5a shows a hydraulic jump occurring at the base of a dam. Water flows at a steady rate with an average velocity $\bar{V}_1 = 11$ m/s. The depth increases from $y_1 = 2$ m to $y_2 = 6$ m. Find the average velocity downstream from the hydraulic jump.

SOLUTION

Given

Hydraulic jump shown in Fig. E4.5a

$\bar{V}_1 = 11$ m/s, $y_1 = 2$ m, and $y_2 = 6$ m

Find

Average velocity $\bar{V}_2$

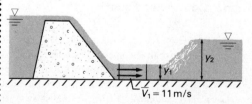

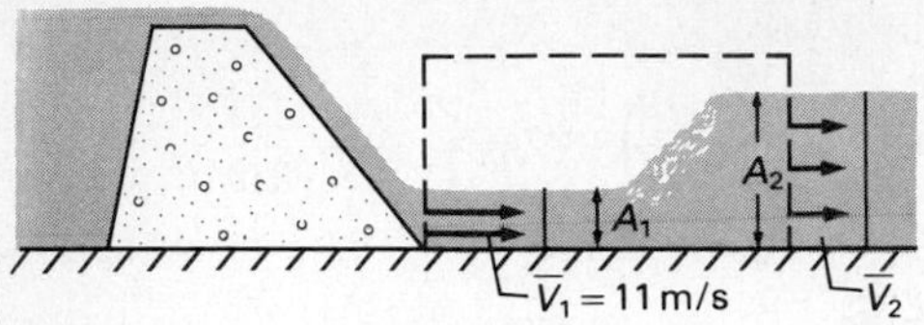

Figure E4.5b Control volume.

Solution

Apply the one-dimensional continuity equation to the control volume shown in Fig. E4.5b. For steady-flow conditions, Eq. (4.23) gives

$$\bar{V}_2 A_2 = \bar{V}_1 A_1.$$

Solving for $\bar{V}_2$, we have

$$\bar{V}_2 = \left(\frac{A_1}{A_2}\right)\bar{V}_1.$$

Each area is the product of the water depth and the width of the stream. We do not know the widths W_1 and W_2 of the channel; however, we expect the ratio W_1/W_2 to be close to 1.* Therefore

$$\frac{A_1}{A_2} = \frac{y_1 W_1}{y_2 W_2} \approx \frac{y_1}{y_2}$$

and

$$\bar{V}_2 = \left(\frac{y_1}{y_2}\right)\bar{V}_1 = \left(\frac{2 \text{ m}}{6 \text{ m}}\right)(11 \text{ m/s});$$

$$\bar{V}_2 = 3.67 \text{ m/s}.$$ **ANSWER**

Discussion

See the discussion for Example 4.6.

* If you were to observe a hydraulic jump at the base of a dam, it would occur within about 20 m of the base, and the width of the channel would not change appreciably over that distance.

EXAMPLE 4.6 Illustrates the Steady, One-Dimensional Continuity Equation

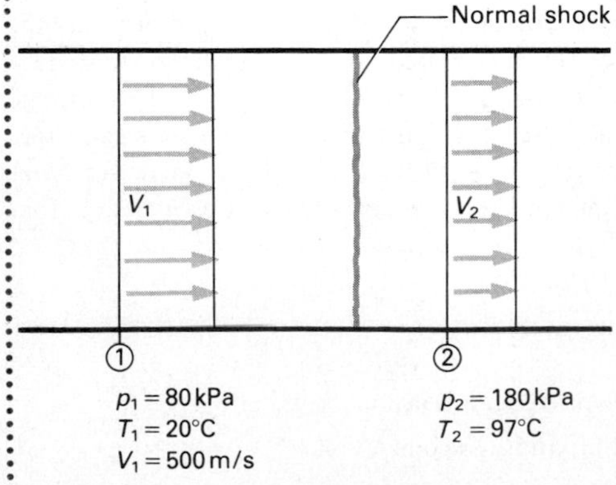

Figure E4.6a Normal shock in a constant-area duct.

A shock wave is a flow phenomenon that can occur when a compressible fluid flows at a speed greater than the speed of sound. There is an abrupt increase in the fluid pressure and temperature and a decrease in the fluid velocity across the shock. Shocks are so thin that we may assume they have zero thickness. Figure E4.6a shows air flowing through a normal shock (the shock is perpendicular to the fluid velocity) in a constant area duct. The pressure, temperature, and velocity upstream of the shock are $p_1 = 80$ kPa, $T_1 = 20$°C, and $V_1 = 500$ m/s. Downstream of the shock $p_2 = 180$ kPa and $T_2 = 97$°C. Find the downstream velocity V_2.

SOLUTION

Given

Figure E4.6a

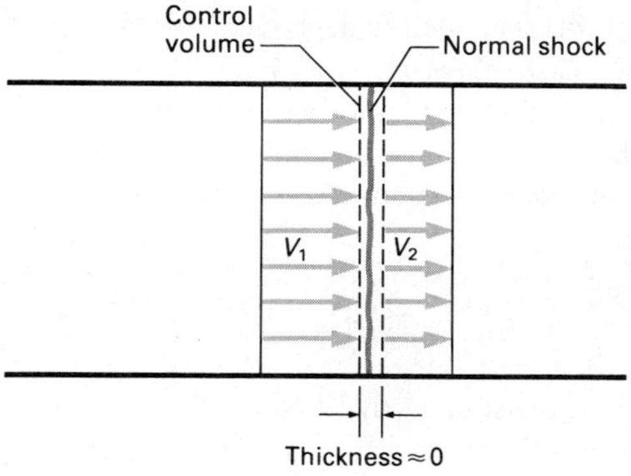

Figure E4.6b Control volume.

Find

Velocity V_2

Solution

Apply the one-dimensional continuity equation to the control volume shown in Fig. E4.6b. Because the shock is very thin, no mass can be stored inside the control volume, and the inlet mass flow must always equal the outlet mass flow. Therefore

$$\rho_2 V_2 A_2 = \rho_1 V_1 A_1.$$

We assume that the density is uniform over both area A_1 and area A_2. Noting that $A_1 = A_2$, we solve for V_2:

$$V_2 = \left(\frac{\rho_1}{\rho_2}\right) V_1.$$

We may assume that air is an ideal gas at the temperatures and pressures involved. The ideal gas law, Eq. (1.1), gives

$$\frac{\rho_1}{\rho_2} = \left(\frac{p_1}{RT_1}\right)\left(\frac{RT_2}{p_2}\right) = \left(\frac{p_1}{p_2}\right)\left(\frac{T_2}{T_1}\right).$$

Substituting into the continuity equation, we have

$$V_2 = \left(\frac{p_1}{p_2}\right)\left(\frac{T_2}{T_1}\right) V_1.$$

The numerical values give

$$V_2 = \left(\frac{80 \text{ kPa}}{180 \text{ kPa}}\right)\left[\frac{(273 + 97)\text{K}}{(273 + 20)\text{K}}\right](500 \text{ m/s});$$

$$V_2 = 280 \text{ m/s}.$$ **ANSWER**

Discussion

Examples 4.4, 4.5, and 4.6 illustrated applications of the steady, one-dimensional continuity equation. Examples 4.4 and 4.5 involved a liquid, so assuming constant density was realistic; Example 4.6 involved a gas with pressure and temperature changes large enough that density changes must be considered (cf. Example 1.1).

One-Dimensional, Nonsteady Flow. In nonsteady flow, mass accumulates in the control volume, so

$$\frac{d}{dt}(m_{cv}) = \frac{d}{dt}\int_{\mathcal{V}_{cv}} \rho \, d\mathcal{V} \neq 0.$$

For a finite number of inlets and outlets, the continuity equation is

$$\frac{d}{dt}\int_{\mathcal{V}_{cv}} \rho \, d\mathcal{V} = \sum \dot{m}_{\text{in}} - \sum \dot{m}_{\text{out}}. \tag{4.24}$$

We can evaluate the mass flow rates from density–velocity–area products.

To simplify the analysis of mass accumulation, we assume that the fluid density is uniform throughout the volume at any instant. Thus

$$\frac{d}{dt}\int_{\forall_{cv}} \rho \, d\forall = \frac{d}{dt}(\rho \forall_{cv}).$$

Either the volume or the density or both may be time-dependent. To investigate these possibilities, let's consider the two cases shown in Fig. 4.7.

In Fig. 4.7(a), air always fills the entire tank. As more air enters the tank, the density increases. The only sensible choice for a control volume is a volume that encloses the entire tank. Because this volume is constant, the continuity equation becomes

$$\forall_{cv}\frac{d\rho}{dt} = \dot{m}_{\text{in}}.$$

In Fig. 4.7(b), the water has constant density, so the level must rise. Suppose we choose a control volume that encloses the entire tank, as we did for the configuration in Fig. 4.7(a). At any instant, this control volume must be divided into two parts, one containing water and the other containing air. Then

$$\frac{d}{dt}(\rho \forall_{cv}) = \frac{d}{dt}(\rho_w \forall_w + \rho_a \forall_a) = \rho_w \frac{d\forall_w}{dt} + \rho_a \frac{d\forall_a}{dt}.$$

The air above the water is unconfined, so we assume that ρ_a is uniform and constant in time. Applying the continuity equation, we have

$$\rho_w \frac{d\forall_w}{dt} + \rho_a \frac{d\forall_a}{dt} = \dot{m}_{w,\text{in}} - \dot{m}_{a,\text{out}}.$$

The air and water don't mix, so we have

$$\rho_w \frac{d\forall_w}{dt} = \dot{m}_{w,\text{in}} \quad \text{and} \quad \rho_a \frac{d\forall_a}{dt} = -\dot{m}_{a,\text{out}}.$$

As $\forall_w = hA$ (see Fig. 4.7b), we get

$$\frac{dh}{dt} = \frac{\dot{m}_{w,\text{in}}}{\rho_w A} = \frac{Q_{w,\text{in}}}{A}.$$

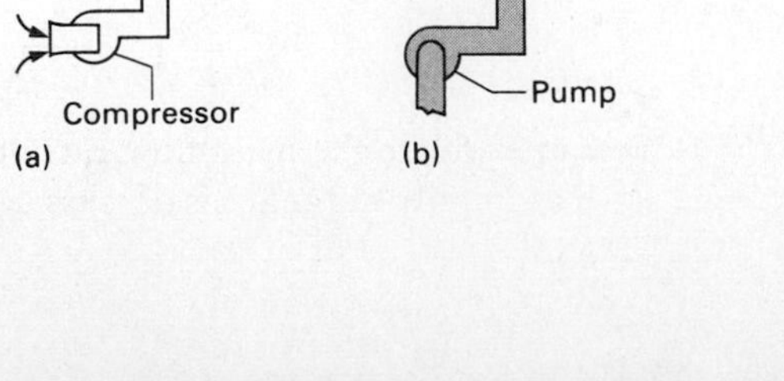

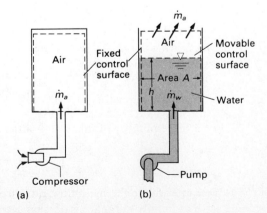

Figure 4.7 Two tank-filling operations: (a) a compressor fills a closed rigid tank with air; (b) a pump fills an open tank with water.

Considerable analysis was involved in arriving at this result.* Now let's choose a control volume that always coincides with the water in the tank. Then

$$\frac{d}{dt}(\rho_w \forall_{cv}) = \rho_w \frac{d\forall_{cv}}{dt}.$$

Because $\forall_{cv} = hA$, substituting into the continuity equation and solving for dh/dt gives

$$\frac{dh}{dt} = \frac{\dot{m}_{w,\text{in}}}{\rho_w A} = \frac{Q_{w,\text{in}}}{A}.$$

The choice of a deformable control volume saved us the trouble of considering the air in the tank. Always be alert to selection of control volumes that save work.

* Think how much more complicated the analysis would have been if the top of the tank had been closed, so that the air was compressed rather than allowed to escape.

EXAMPLE 4.7 Illustrates the Continuity Equation for Nonsteady Flow

Water is being added to a storage tank at the rate of 500 gal/min. At the same time water flows out the bottom through a 2.0-in. inside-diameter pipe with an average velocity of 60 ft/sec, as shown in Fig. E4.7a. The storage tank has an inside diameter of 10.0 ft. Find the rate at which the water level rises or falls.

SOLUTION

Given

Figure E4.7a

Storage tank with inside diameter 10.0 ft

$V = 60$ ft/sec in discharge pipe

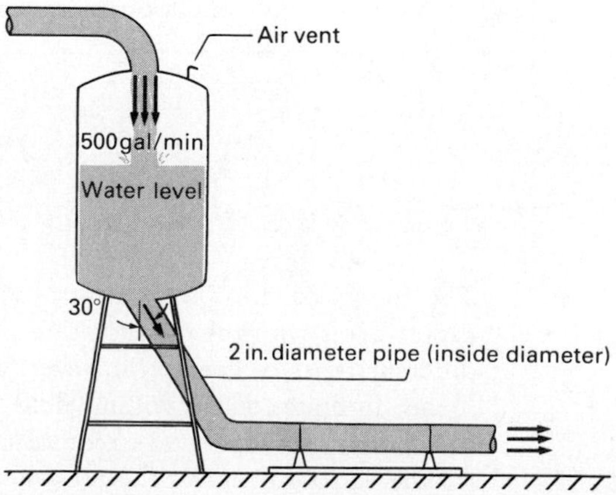

Figure E4.7a Water storage tank.

Find

Rate at which water level rises or falls

Solution

We choose the control volume shown in Fig. E4.7b, which coincides with the water in the tank. With constant water density, the integral continuity equation is

$$\rho_w \frac{d\mathcal{V}_{cv}}{dt} = \sum \dot{m}_{\text{in}} - \sum \dot{m}_{\text{out}} = \rho_w Q_{\text{in}} - \rho_w (VA)_{\text{out}}.$$

The volume is

$$\mathcal{V}_{cv} = \mathcal{V}_1 + \mathcal{V}_2 = \mathcal{V}_1 + A_T h.$$

Substituting into the continuity equation gives

$$\rho_w \frac{d}{dt}(\mathcal{V}_1 + A_T h) = \rho_w Q_{\text{in}} - \rho_w (VA)_{\text{out}}.$$

As ρ_w, $\mathcal{V}_1$, A_T, and A_{out} are constant,

$$A_T \frac{dh}{dt} = Q_{\text{in}} - (VA)_{\text{out}} \quad \text{or} \quad \frac{dh}{dt} = \frac{Q_{\text{in}} - (VA)_{\text{out}}}{A_T}.$$

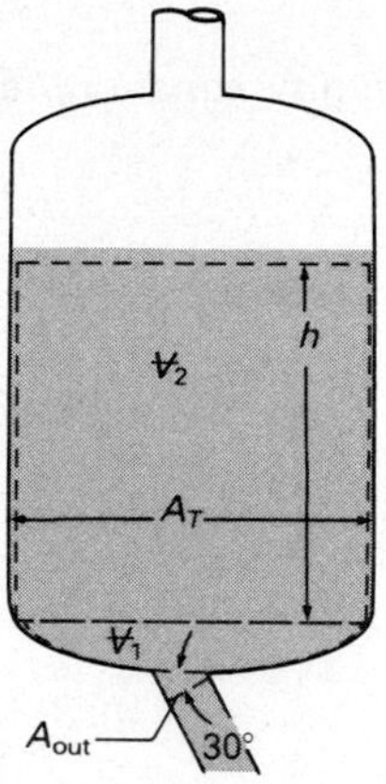

Figure E4.7b Control volume.

The numerical values give

$$\frac{dh}{dt} = \frac{(500 \text{ gal/min})(1 \text{ ft}^3/7.48 \text{ gal}) - (60 \text{ ft/sec})(\pi/4)(2/12 \text{ ft})^2(60 \text{ sec/min})}{(\pi/4)(10.0 \text{ ft})^2};$$

$$\frac{dh}{dt} = -0.149 \text{ ft/min}.$$ **ANSWER**

The water level is falling, as indicated by the negative sign.

Discussion

Note that we calculated mass flow rate two ways:

$$\dot{m} = \rho_w Q \quad \text{and} \quad \dot{m} = \rho_w VA.$$

Both equations are for a constant-density fluid.

4.3 THE ENERGY EQUATIONS

The energy equation for a control volume is similar to the continuity equation in that it deals with the conservation of a scalar quantity, energy. There are several different types of energy, however, and the energy equation has several different forms. In finite control volume fluid mechanics, an appropriate form of the energy equation is almost always required to solve a flow problem. The starting point for the various forms of the energy equation is the *general energy equation,* which we now derive.

4.3.1 Derivation of the General Energy Equation

We obtain the general energy equation by combining the transport theorem with the law of conservation of energy for a system. Because energy is conserved, we can write

$$\begin{bmatrix}\text{Rate of change of the}\\ \text{energy of a system}\end{bmatrix} = \begin{bmatrix}\text{Net rate of transfer of}\\ \text{energy to the system}\end{bmatrix}.$$

Before we begin the mathematics, let's examine the terms in this energy balance. Energy may be transferred to a system by two different processes, *heat* and *work*. Heat transfer (Q) is energy transfer across a system boundary because of any temperature difference between the system and its surroundings. Work (W) is energy transfer by the action of a force through a distance or any other energy transfer that could be replaced by action of a force through a distance.*

Intrinsic energy is energy stored in a system (that is, energy contained in the system's mass). There are five forms of intrinsic energy:

1. *kinetic energy,* the energy of motion;
2. *potential energy,* the energy owing to position in a force field (usually a gravitational field but sometimes an electric or magnetic field);
3. *internal energy,* the energy of molecular structure and motion;
4. *chemical energy,* the energy associated with the arrangement of atoms into molecules (released only in a chemical reaction); and
5. *nuclear energy,* the energy associated with the internal structure of atoms (released only by nuclear fission or fusion).

In many problems in fluid mechanics, chemical and nuclear reactions are absent; accordingly, we neglect chemical and nuclear energies in subsequent discussions.

The intrinsic energies are proportional to the amount of mass in the system and may be related to properties of the system (see Section 1.3). A fundamental postulate of thermodynamics is that internal energy is a system property and, as such, is related to other system properties, primarily mass and temperature.

We often deal with *specific energy,* which is energy per unit mass. Specific intrinsic energies are not proportional to system mass. Specific kinetic energy is one half the square of the fluid velocity ($V^2/2$). Specific potential energy is gravitational acceleration multiplied by elevation relative to a zero datum (gz). Specific internal energy† ($\tilde{u}$) is a property that depends only on other specific properties, primarily temperature.‡ We also define specific work, w, and specific heat transfer,

---

* The flow of electric current into a system may be considered work, because a mechanism could be devised (a frictionless electric motor) that completely converts the current flow to the action of a force through a distance (for example, using the motor to raise a weight). In fluid mechanics, we are almost always concerned with work that actually is the action of a force through a distance.

† The symbol u for internal energy per unit mass is almost universal in thermodynamics. In fluid mechanics, it might be confused with x-direction velocity. To avoid confusion, we use the slightly more cumbersome symbol $\tilde{u}$ to represent internal energy per unit mass.

‡ Similar to specific potential energy, a zero datum must be selected for internal energy. We specify that $\tilde{u} = 0$ at absolute zero temperature.

q, as work and heat transfer divided by the system mass; however, *specific work and specific heat transfer are not system properties.*

For later discussion, we classify various energies as either thermal or mechanical. *Thermal energies* are associated with temperature, molecular structure, and heat transfer. *Mechanical energies* are associated with force and motion. Neglecting chemical and nuclear energies, we classify the other forms as:

Mechanical Energies	**Thermal Energies**
Work	Heat
Kinetic energy	Internal energy
Potential energy	

Using these concepts, we amplify the energy balance for a system to read:

$$\begin{bmatrix}\text{Rate of}\\ \text{heat transfer}\\ \text{to system}\end{bmatrix} + \begin{bmatrix}\text{Rate of work}\\ \text{done on}\\ \text{system}\end{bmatrix} = \begin{bmatrix}\text{Rate of increase of intrinsic}\\ \text{(kinetic + potential + internal)}\\ \text{energy of system}\end{bmatrix}.$$

This equation makes no distinction between the thermal and mechanical energies of the system; an energy transfer may result in an increase of the system's mechanical energy, thermal energy, or both.

Introducing symbols and sign conventions customary in thermodynamics and fluid mechanics, we have

$$\frac{dQ}{dt} - \frac{dW}{dt} = \frac{dE_{sys}}{dt},$$

where

$$E_{sys} = \int_{\forall_{sys}} \rho\left(\tilde{u} + \frac{V^2}{2} + gz\right) d\forall;$$

heat transferred to a system is positive; and

work done by (that is, transferred from) a system is positive.

We obtain the control volume form of this equation by using the transport theorem, Eq. (4.1c), with $b = \tilde{u} + V^2/2 + gz$. The result is

$$\dot{Q} - \dot{W} = \frac{d}{dt}\int_{\forall_{cv}} \rho\left(\tilde{u} + \frac{V^2}{2} + gz\right) d\forall + \oint_{A_{cv}} \rho\left(\tilde{u} + \frac{V^2}{2} + gz\right)(\vec{V}_r \cdot \hat{n})\, dA. \tag{4.25}$$

The overdot $(\dot{\ })$ indicates rates of heat transfer and work. Rate of work is also called *power*.

If we neglect electrical and other equivalent forms of work, three types of work might be done on or by the fluid inside the control volume, as shown in Fig. 4.8.

- *Shaft work* (W_{shaft}) is transmitted by a rotating shaft, such as a pump drive shaft or a turbine output shaft that is "cut" by the control surface. This work

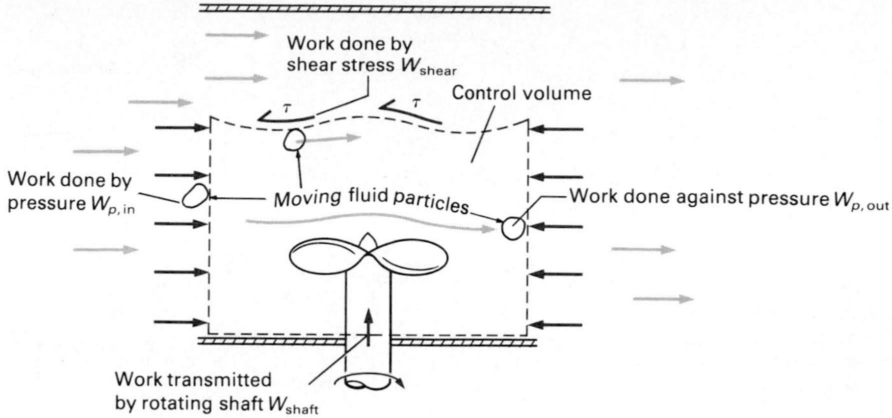

Figure 4.8 Various forms of work exchange with moving fluid.

is done by shear stresses in the "cut" shaft, so it is somewhat similar to shear work. Shaft work is sometimes called "pump work" (W_p) or "turbine work" (W_t) if these devices are present.

- *Shear work* (W_{shear}) is done by shear stresses in the fluid acting on the boundaries of the control volume.
- *Pressure work* ($W_{pressure}$) is done by fluid pressure acting on the boundaries of the control volume.

The total work is the sum of these three components:

$$W = W_{shaft} + W_{shear} + W_{pressure}.$$

Equation (4.25) requires the rate of work, which is simply

$$\dot{W} = \dot{W}_{shaft} + \dot{W}_{shear} + \dot{W}_{pressure}.$$

We can develop equations for calculating each of the work rates. We evaluate the shear work rate from the product of shear stress, area, and the fluid velocity component in the direction of the stress force:

$$\dot{W}_{shear} = \oint_{A_{cv}} \vec{V} \cdot d\vec{F}_{shear} = \oint_{A_{cv}} \vec{V} \cdot (\vec{\tau}\, dA), \tag{4.26}$$

where $\vec{\tau}$ is the shear stress vector, and is parallel to dA.

We often choose a control volume with control surfaces lying adjacent to solid boundaries where the fluid velocity is zero (Fig. 4.9a). In this case, there is no shear work even though there may be shear stress. If for some reason we choose a control surface that lies within a moving fluid (Fig. 4.9b), shear work occurs unless the fluid is inviscid.

The shaft work rate is equal to the shaft torque multiplied by its rotational speed:

$$\dot{W}_{shaft} = \mathcal{T}\omega. \tag{4.27}$$

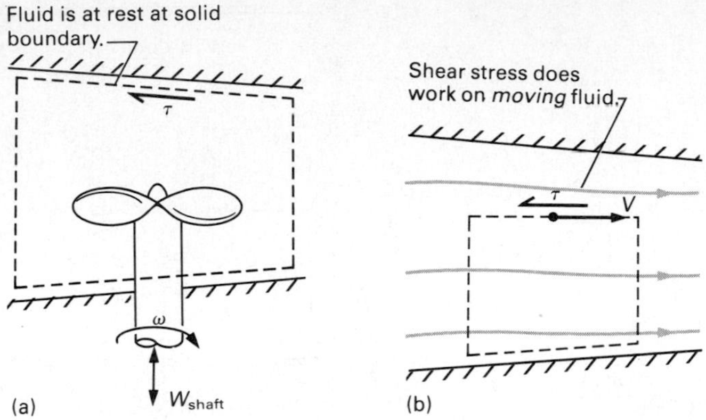

Figure 4.9 Control volumes (a) without and (b) with shear work and (a) with and (b) without shaft work.

We calculate the shaft torque by integrating the shear stress over the shaft cross section, so

$$\dot{W}_{\text{shaft}} = \omega \int_{A_{\text{shaft}}} \tau r \, dA. \tag{4.28}$$

Rather than calculating shaft work by Eq. (4.28), we usually leave it as a lumped term, $\dot{W}_{\text{shaft}}$.

We usually find it convenient to combine $\dot{W}_{\text{shaft}}$ and $\dot{W}_{\text{shear}}$ into a single work $\dot{W}_s$,

$$\dot{W}_s \equiv \dot{W}_{\text{shaft}} + \dot{W}_{\text{shear}}, \tag{4.29}$$

where $\dot{W}_s$ is work done on the fluid inside the control volume by forces other than those resulting from pressure.

We classify work done by pressure acting at the control volume boundary as either *flow work* or *work of (control volume) deformation*. We must always include flow work in the energy equation for moving fluids. Work of deformation occurs only if we select a deforming control volume.

We develop an expression for flow work by examining outflow and inflow for a control volume. Figure 4.10 shows a portion of a control surface where fluid is flowing outward. The fluid leaving the control volume must push the fluid outside the control volume out of the way. Fluid pressure at the control surface accomplishes this task.

At every point of outflow, the fluid leaving the control volume is doing work on the fluid outside. According to the sign convention, this is positive work. Similarly, at every point where fluid flows into the control volume, the fluid entering does work on the fluid inside as it pushes in. According to the sign convention, this is negative work.

Consider the detail of the outflow shown in Fig. 4.10. Because

$$\text{Work} = (\text{Force}) \times (\text{Distance moved in direction of force}),^*$$

* We neglect algebraic signs in this relation, because we already know that flow work is positive for outflow regions.

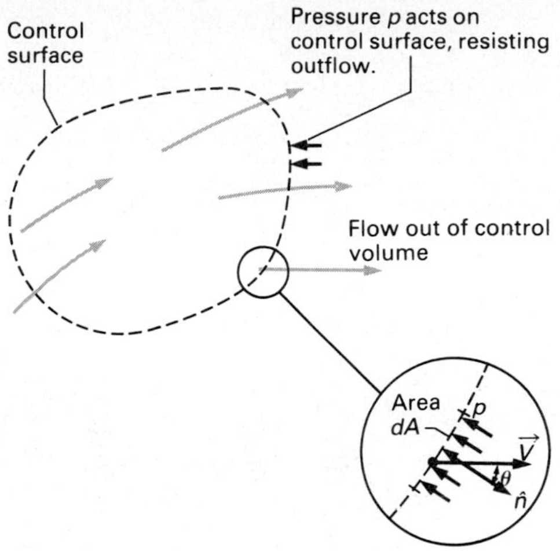

Figure 4.10 Detail of flow outward from control volume for calculation of pressure work.

then

$$d\dot{W}_{\text{flow}} = dF \times (\text{Rate of motion in direction of force}).$$

The pressure force is

$$dF = p\,dA,$$

and the rate of motion parallel to the force is

$$(\text{Rate of motion in direction of force}) = V_n = V\cos\theta = \vec{V}\cdot\hat{n}.$$

As we are calculating the work done by fluid leaving the control volume on fluid outside, we must use the velocity relative to the control surface ($\vec{V}_r$). The rate of flow work done at area dA is

$$d\dot{W}_{\text{flow}} = p(\vec{V}_r\cdot\hat{n})\,dA. \tag{4.30a}$$

Considering an inflow area, we calculate the flow work by the same expression, except that $(\vec{V}_r\cdot\hat{n})$ is negative, because the angle θ lies between 90° and 180°. This result is exactly what we want, because outflow work is positive and inflow work is negative. To obtain the total rate of flow work, all we need to do is integrate Eq. (4.30a) over the entire control surface:

$$\dot{W}_{\text{flow}} = \oint_{A_{cv}} p(\vec{V}_r\cdot\hat{n})\,dA. \tag{4.30b}$$

Work of deformation (W_D) occurs if the control volume is deforming. This work is the product of the pressure acting on the control surface and the displacement of the control surface. The rate of work is the product of pressure and the velocity of the control surface ($\vec{V}_c$). The total rate of deformation work is

$$\dot{W}_D = \oint_{A_{cv}} \vec{V}_c\cdot d\vec{F}_p = \oint_{A_{cv}} \vec{V}_c\cdot(p\hat{n}\,dA) = \oint_{A_{cv}} p(\vec{V}_c\cdot\hat{n})\,dA. \tag{4.31}$$

We now collect the various types of work,

$$\dot{W} = \dot{W}_{\text{shaft}} + \dot{W}_{\text{shear}} + \oint_{A_{cv}} p(\vec{V}_r\cdot\hat{n})\,dA + \oint_{A_{cv}} p(\vec{V}_c\cdot\hat{n})\,dA, \tag{4.32}$$

and substitute into Eq. (4.25) to get

$$\dot{Q} - \dot{W}_{\text{shaft}} - \dot{W}_{\text{shear}} - \oint_{A_{cv}} p(\vec{V}_r \cdot \hat{n})\, dA - \oint_{A_{cv}} p(\vec{V}_c \cdot \hat{n})\, dA$$
$$= \frac{d}{dt}\int_{\forall_{cv}} \rho\left(\tilde{u} + \frac{V^2}{2} + gz\right) d\forall + \oint_{A_{cv}} \rho\left(\tilde{u} + \frac{V^2}{2} + gz\right)(\vec{V}_r \cdot \hat{n})\, dA.$$

Note the similar form of the flow work integral on the left and the energy flux integral on the right. If we multiply and divide by the density ρ inside the flow work integral, move it to the right-hand side, and replace $\dot{W}_{\text{shaft}} + \dot{W}_{\text{shear}}$ by $\dot{W}_s$, we get

$$\dot{Q} - \dot{W}_s - \oint_{A_{cv}} p(\vec{V}_c \cdot \hat{n})\, dA = \frac{d}{dt}\int_{\forall_{cv}} \rho\left(\tilde{u} + \frac{V^2}{2} + gz\right) d\forall$$
$$+ \oint_{A_{cv}} \rho\left(\tilde{u} + \frac{p}{\rho} + \frac{V^2}{2} + gz\right)(\vec{V}_r \cdot \hat{n})\, dA. \quad (4.33)$$

Equation (4.33) is the *general energy equation.* Although central to the study of thermodynamics, it plays a surprisingly minor role in (incompressible) fluid mechanics. In thermodynamics, the sum $\tilde{u} + p/\rho$ is replaced by its equivalent, the fluid enthalpy $\tilde{h}$. Note that the enthalpy represents a combination of an intrinsic energy ($\tilde{u}$) and an energy transfer (flow work).

4.3.2 Some Simplified Forms of the General Energy Equation

Like the continuity equation, the general energy equation is seldom used in its most general form, Eq. (4.33). In this section, we examine some simplified forms and illustrate their use by examples.

Energy is a scalar, so velocities do not have to be measured with respect to an inertial reference frame. If you use a noninertial reference frame, you must be sure to include any work resulting from imaginary "forces," such as Coriolis or centrifugal forces.

Steady Flow, Fixed, Rigid Control Volume. Equation (4.33) applies for a moving, deforming control volume. If the control volume is fixed and rigid, then

$$\frac{d}{dt}\int_{\forall_{cv}} \rho\left(\tilde{u} + \frac{V^2}{2} + gz\right) d\forall = \int_{\forall_{cv}} \frac{\partial}{\partial t}\left[\rho\left(\tilde{u} + \frac{V^2}{2} + gz\right)\right] d\forall,$$

$$\oint_{A_{cv}} p(\vec{V}_c \cdot \hat{n})\, dA = 0,$$

and

$$\vec{V}_r = \vec{V}.$$

A great many applications involve steady flow through a fixed, rigid control volume. In this case, $\partial/\partial t = 0$, and the general energy equation reduces to

$$\dot{Q} - \dot{W}_s = \oint_{A_{cv}} \rho\left(\tilde{u} + \frac{p}{\rho} + \frac{V^2}{2} + gz\right)(\vec{V}\cdot\hat{n})\,dA. \tag{4.34}$$

Steady flow, fixed and rigid control volume

A common simplification is the assumption that the control volume has uniform inflow and outflow at a finite number of inlets and outlets (see Fig. 4.5). If we assume that the control volume is fixed and rigid, the general energy equation is

$$\begin{aligned}\dot{Q} - \dot{W}_s &= \frac{\partial}{\partial t}\int_{\forall_{cv}} \rho\left(\tilde{u} + \frac{V^2}{2} + gz\right)d\forall \\ &+ \sum_{\text{out}} (\rho A V_n)\left(\tilde{u} + \frac{p}{\rho} + \frac{V^2}{2} + gz\right) \\ &- \sum_{\text{in}} (\rho A V_n)\left(\tilde{u} + \frac{p}{\rho} + \frac{V^2}{2} + gz\right)\end{aligned} \tag{4.35a}$$

or

$$\begin{aligned}\dot{Q} - \dot{W}_s &= \frac{\partial}{\partial t}\int_{\forall_{cv}} \rho\left(\tilde{u} + \frac{V^2}{2} + gz\right)d\forall \\ &+ \sum_{\text{out}} \dot{m}\left(\tilde{u} + \frac{p}{\rho} + \frac{V^2}{2} + gz\right) \\ &- \sum_{\text{in}} \dot{m}\left(\tilde{u} + \frac{p}{\rho} + \frac{V^2}{2} + gz\right).\end{aligned} \tag{4.35b}$$

Fixed, rigid control volume with a finite number of uniform flow inlets and outlets

If the flow is steady, the first term on the right is zero. In Eq. (4.35), all terms under the summation signs are positive.* They are not vector equations, so the labels "in" and "out" serve to determine the algebraic signs.

* Unless z or $\tilde{u}$ is below an arbitrarily selected zero datum.

EXAMPLE 4.8 Illustrates the General Energy Equation

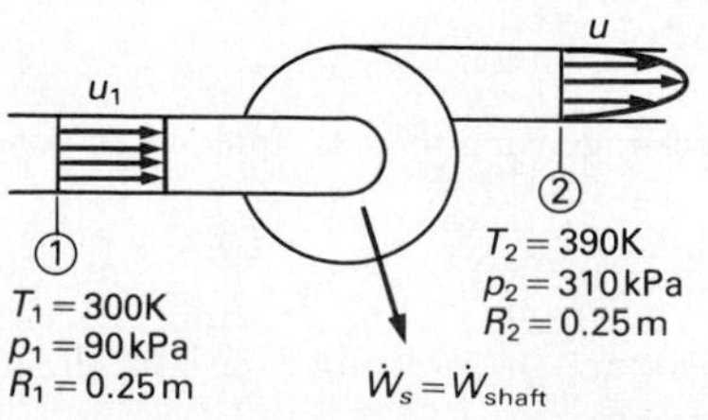

Figure E4.8a Air compressor with inlet and discharge pipes.

A compressor compresses 6 kg/s of air from inlet conditions of $T_1 =$ 300 K and $p_1 = 90$ kPa to discharge conditions of $T_2 = 390$ K and $p_2 = 310$ kPa (Fig. E4.8a). The air in the inlet pipe has a uniform velocity profile. The air in the discharge pipe has a parabolic velocity profile given by

$$u = u_{\max}\left[1 - \left(\frac{r}{R_2}\right)^2\right],$$

where R_2 is the inside radius of the discharge pipe. Elevation changes are negligible, and the internal energy change of the air is given by

$$\tilde{u}_2 - \tilde{u}_1 = c_v(T_2 - T_1), \qquad \text{with } c_v = 720 \text{ J/kg}\cdot\text{K}.$$

Assuming steady flow and negligible heat transfer, find the power required to drive the compressor.

SOLUTION

Given

Air compressor in Fig. E4.8a operating at a steady rate, with

$$\dot{m} = 6 \text{ kg/s}, \qquad R_1 = R_2 = 0.25 \text{ m},$$
$$T_1 = 300 \text{ K}, \qquad T_2 = 390 \text{ K},$$
$$p_1 = 90 \text{ kPa}, \quad \text{and} \quad p_2 = 310 \text{ kPa}$$

Uniform velocity profile at inlet and parabolic velocity profile at discharge

Internal energy change of air given by

$$\tilde{u}_2 - \tilde{u}_1 = c_v(T_2 - T_1) = (720 \text{ J/kg}\cdot\text{K})(T_2 - T_1)$$

Negligible heat transfer and elevation changes

Find

Power required to drive compressor

Solution

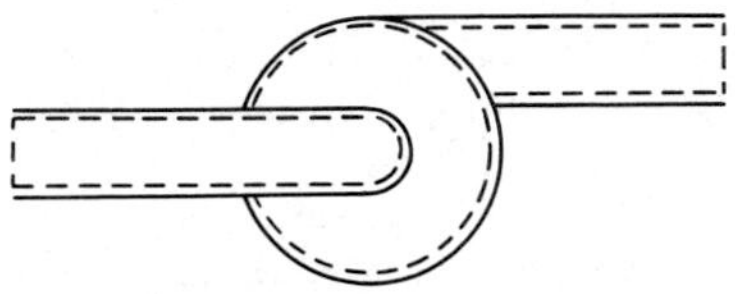

Figure E4.8b Control volume enclosing air in the compressor and piping.

We apply the general energy equation, Eq. (4.34), to a control volume enclosing the air in the compressor and piping (Fig. E4.8b):

$$\dot{Q} - \dot{W}_s = \oint_{A_{cv}} \rho\left(\tilde{u} + \frac{p}{\rho} + \frac{V^2}{2} + gz\right)(\vec{V}\cdot\hat{n})\,dA.$$

The rate of work $\dot{W}_s$ is

$$\dot{W}_s = \dot{W}_{\text{shaft}} + \dot{W}_{\text{shear}}.$$

The shear stress work is given by Eq. (4.26):

$$\dot{W}_{\text{shear}} = \oint_{A_{cv}} (\vec{\tau}\cdot\vec{V})\,dA,$$

where $\vec{V}$ is the fluid velocity at the control surface. The only shear stress along the control surface is between the fluid and the pipe wall or compressor outer casing, where $\vec{V} = 0$. Therefore

$$\dot{W}_{\text{shear}} = 0.$$

From the problem statement,

$$\dot{Q} = 0.$$

The negligible elevation change simplifies the integral to give

$$-\dot{W}_{\text{shaft}} = \oint_{A_{cv}} \rho\left(\tilde{u} + \frac{p}{\rho} + \frac{V^2}{2}\right)(\vec{V}\cdot\hat{n})\,dA.$$

Assuming uniform temperature and pressure over both the inlet and the exit and recognizing that

$$\int_{A_{\text{in}}} \rho(\vec{V}\cdot\hat{n})\,dA = -\dot{m}, \quad \int_{A_{\text{out}}} \rho(\vec{V}\cdot\hat{n})\,dA = \dot{m}, \quad \text{and} \quad V = |\vec{V}| = u$$

at the inlet and outlet, we have

$$-\dot{W}_{\text{shaft}} = \dot{m}\left[(\tilde{u}_2 - \tilde{u}_1) + \left(\frac{p_2}{\rho_2} - \frac{p_1}{\rho_1}\right)\right] - \dot{m}\,\frac{u_1^2}{2} + \int_{A_{\text{out}}} \rho\left(\frac{u^3}{2}\right) dA.$$

Substituting for the discharge velocity $u(r)$ gives

$$\int_{A_{\text{out}}} \rho\left(\frac{u^3}{2}\right) dA = 2\pi \int_0^{R_2} \frac{\rho u_{\max}^3}{2}\left[1 - \left(\frac{r}{R_2}\right)^2\right]^3 r\,dr.$$

We now assume that air is an ideal gas, so the uniform temperature and pressure assumption implies that the density is uniform over the exit area. Integration gives

$$\int_{A_{\text{out}}} \rho\,\frac{u^3}{2}\,dA = \frac{\pi \rho_2 u_{\max}^3 R_2^2}{8}.$$

We relate $u_{\max}$ to the average velocity by

$$\dot{m} = \rho_2 \bar{V}_2 A_2 = \int \rho_2 u_2\,dA = \int_0^{R_2} \rho_2 u_{\max}\left[1 - \left(\frac{r}{R_2}\right)^2\right](2\pi r\,dr).$$

Integrating gives

$$\rho_2 \bar{V}_2(\pi R_2^2) = \rho_2\left(\frac{u_{\max}}{2}\right)(\pi R_2^2).$$

Therefore

$$\bar{V}_2 = \frac{u_{\max}}{2} \quad \text{and} \quad \int_{A_{\text{out}}} \rho\,\frac{u^3}{2}\,dA = \frac{\pi \rho_2 u_{\max}^3 R_2^2}{8} = \rho_2 \bar{V}_2(\pi R_2^2)\bar{V}_2^2.$$

Because

$$\dot{m} = \rho_2 \bar{V}_2(\pi R_2^2)$$

and, by the ideal gas law, Eq. (1.1),

$$\frac{p}{\rho} = RT,$$

the energy equation becomes

$$-\dot{W}_{\text{shaft}} = \underbrace{\dot{m}c_v(T_2 - T_1)}_{\text{Internal energy}} + \underbrace{\dot{m}R(T_2 - T_1)}_{\text{Flow work}} + \underbrace{\dot{m}\left(\bar{V}_2^2 - \frac{u_1^2}{2}\right)}_{\text{Kinetic energy}}.$$

We calculate the velocities from Eqs. (4.15) and (1.1):

$$u_1 = \frac{\dot{m}}{\rho_1 A_1} = \frac{\dot{m}RT_1}{p_1 A_1} \quad \text{and} \quad \bar{V}_2 = \frac{\dot{m}}{\rho_2 A_2} = \frac{\dot{m}RT_2}{p_2 A_2}.$$

We obtain R, the gas constant, from Table A.3:

$$R = 287\ \text{N}\cdot\text{m/kg}\cdot\text{K}.$$

Also,

$$A_1 = A_2 = \pi R^2 = \pi(0.25\ \text{m})^2 = 0.196\ \text{m}^2,$$

so

$$u_1 = \frac{(6\ \text{kg/s})(287\ \text{N}\cdot\text{m/kg}\cdot\text{K})(300\ \text{K})}{(90{,}000\ \text{N/m}^2)(0.196\ \text{m}^2)} = 29.3\ \text{m/s},$$

and

$$\bar{V}_2 = \frac{(6\ \text{kg/s})(287\ \text{N}\cdot\text{m/kg}\cdot\text{K})(390\ \text{K})}{(310{,}000\ \text{N/m}^2)(0.196\ \text{m}^2)} = 11.1\ \text{m/s}.$$

Then

$$\begin{aligned}\dot{W}_{\text{shaft}} &= -(6\ \text{kg/s})(720\ \text{J/kg}\cdot\text{K})(390 - 300)\ \text{K} \\ &\quad - (6\ \text{kg/s})(287\ \text{N}\cdot\text{m/kg}\cdot\text{K})(390 - 300)\ \text{K} \\ &\quad - (6\ \text{kg/s})\left[(11.1\ \text{m/s})^2 - \frac{1}{2}(29.3\ \text{m/s})^2\right] \\ &= (\underbrace{-388{,}800}_{\text{Internal energy}} - \underbrace{155{,}000}_{\text{Flow work}} + \underbrace{1800}_{\text{Kinetic energy}})\ \text{J/s};\end{aligned}$$

$$\dot{W}_{\text{shaft}} = -542\ \text{kW}.$$ **ANSWER**

Discussion

Note that a large portion of the compressor input work appears as an increase in the thermal (internal) energy and the mechanical "pressure energy" of this compressible fluid. The kinetic energy change is much smaller. The negative sign means work is done on the air by the compressor.

The parabolic velocity profile at the compressor discharge would be appropriate for laminar flow at that location, a rather unlikely occurrence. We used the parabolic profile in this example purely for illustrative purposes.

Average Properties and Velocities. Usually, the uniform flow assumption is only an approximation, and we use average velocities and property values to represent the flow of energy at the various inlet and outlet planes. For extremely accurate calculations, we must carefully define the averages [1] so that they truly represent the associated energy flows. In most cases, appropriate average values of $\tilde{u}$, p, ρ, and z are readily apparent, because these properties are often closely uniform over the cross section. As the fluid velocity is usually nonuniform, representing the kinetic energy flux in terms of an average velocity is slightly more complicated.

For fluid flowing in a pipe with the velocity profile shown in Fig. 4.11, the flux of kinetic energy across plane 1 is

$$\dot{E}_{k,1} = \int_{A_1} \rho\left(\frac{V^2}{2}\right)(V_n\,dA),$$

and the average velocity at plane 1 is

$$\bar{V}_1 = \frac{\int_{A_1} V_n\,dA}{A_1}.$$

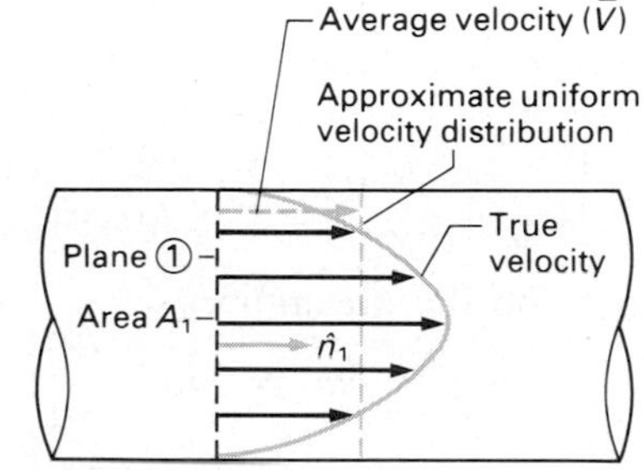

Figure 4.11 One-directional flow in a pipe with nonuniform velocity distribution.

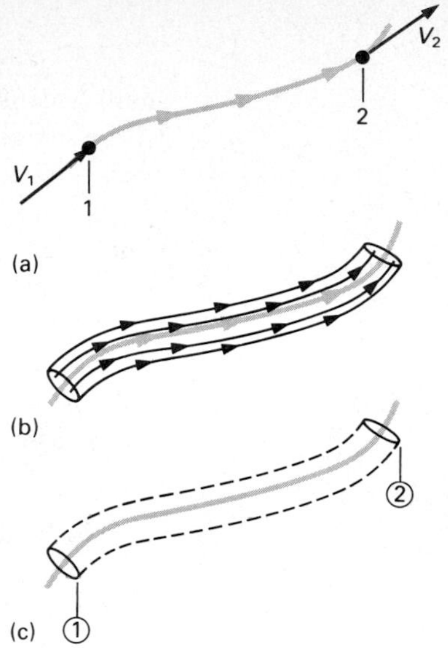

Figure 4.12 Flow along streamlines: (a) a single streamline; (b) streamtube with the streamline at the center; (c) control volume surrounding the streamtube.

However, because V is not uniform,

$$\dot{E}_{k,1} \neq \rho_1 \bar{V}_1 A_1 \left(\frac{\bar{V}_1^2}{2} \right).$$

The kinetic energy correction factor, α, is defined by

$$\alpha \equiv \frac{\int \frac{1}{2} V^3 \, dA}{\frac{1}{2} \bar{V}^3 A}. \tag{4.36}$$

The true kinetic energy flux across plane 1 is

$$\dot{E}_{k,1} = \alpha_1 (\tfrac{1}{2} \rho_1 \bar{V}_1^3) A_1. \tag{4.37}$$

With the kinetic energy correction factor, the general energy equation (for steady flow and a fixed, rigid control volume) is

$$\dot{Q} - \dot{W}_s = \sum_{\text{out}} \dot{m} \left[\bar{\tilde{u}} + \left(\frac{\bar{p}}{\rho} \right) + \alpha \frac{\bar{V}^2}{2} + \overline{gz} \right] - \sum_{\text{in}} \dot{m} \left[\bar{\tilde{u}} + \left(\frac{\bar{p}}{\rho} \right) + \alpha \frac{\bar{V}^2}{2} + \overline{gz} \right]. \tag{4.38}$$

For flow in a circular pipe, α ranges from 2.0 for fully developed laminar flow to about 1.05 for fully developed turbulent flow. The value of α for the flow at the compressor discharge in Example 4.8 is 2.0.

Energy Equation for a Streamline. An important special form of the general energy equation is the energy equation for flow along a single streamline. Figure 4.12(a) shows a single streamline in a steady flow, and Fig. 4.12(b) shows a small streamtube that surrounds the streamline. A *streamtube* is an imaginary tube whose surface consists entirely of streamlines. There is no velocity component perpendicular to the walls of a streamtube, so fluid enters or leaves only at the ends.

We select a control volume that is coincident with the streamtube and perpendicular to the central streamline at the ends (see Fig. 4.12). If we select the central streamline as a coordinate line, the flow is one-dimensional and one-directional. Because we have assumed that the flow is steady, the general energy equation for the streamtube control volume is

$$\delta\dot{Q} - \delta\dot{W}_s = \delta\dot{m}_2\left(\tilde{u}_2 + \frac{p_2}{\rho_2} + \frac{V_2^2}{2} + gz_2\right) - \delta\dot{m}_1\left(\tilde{u}_1 + \frac{p_1}{\rho_1} + \frac{V_1^2}{2} + gz_1\right).$$

No fluid crosses the sides of the streamtube, so

$$\delta\dot{m}_1 = \delta\dot{m}_2 = \delta\dot{m}.$$

Substituting $\delta\dot{m}$ for $\delta\dot{m}_1$ and $\delta\dot{m}_2$ and then dividing by $\delta\dot{m}$, we obtain

$$q - w_s = \tilde{u}_2 - \tilde{u}_1 + \frac{p_2}{\rho_2} - \frac{p_1}{\rho_1} + \frac{V_2^2}{2} - \frac{V_1^2}{2} + gz_2 - gz_1,$$

where

$$q = \frac{\delta\dot{Q}}{\delta\dot{m}}$$

is the heat transfer per unit mass of moving fluid, and

$$w_s = \frac{\delta\dot{W}_s}{\delta\dot{m}}$$

is the work per unit mass of moving fluid. If we take the limit of this expression, shrinking the streamtube and collapsing it around the central streamline, we get the steady-flow general energy equation along a streamline:

$$q - w_s = (\tilde{u}_2 - \tilde{u}_1) + \left(\frac{p_2}{\rho_2} - \frac{p_1}{\rho_1}\right) + \left(\frac{V_2^2}{2} - \frac{V_1^2}{2}\right) + (gz_2 - gz_1). \qquad (4.39)$$

Steady flow from point 1 to point 2 along a particular streamline

In steady flow, any particular mass particle always moves along a fixed streamline. We might interpret Eq. (4.39) as an energy balance for a specific particle that moves from position 1 to position 2 along a streamline. Then $(\tilde{u}_2 - \tilde{u}_1) + (V_2^2/2 - V_1^2/2) + (gz_2 - gz_1)$ is the net gain of the particle's instrinsic energy, $(-p_1/\rho_1)$ and (p_2/ρ_2) represent flow work done on the particle as it begins its trip from 1 to 2 and flow work done by the particle as it finishes its trip, and $(q - w_s)$ is the net energy transferred to the particle as it travels between 1 and 2. If we rewrite Eq. (4.39) as

$$\left(\tilde{u}_1 + \frac{p_1}{\rho_1} + \frac{V_1^2}{2} + gz_1\right) + (q - w_s) = \left(\tilde{u}_2 + \frac{p_2}{\rho_2} + \frac{V_2^2}{2} + gz_2\right), \qquad (4.40)$$

we might interpret it as follows:

$$\begin{bmatrix}\text{Initial energy}\\ \text{of fluid particle}\end{bmatrix} + \begin{bmatrix}\text{Net energy transferred}\\ \text{to fluid particle}\end{bmatrix} = \begin{bmatrix}\text{Final energy}\\ \text{of fluid particle}\end{bmatrix}. \tag{4.41}$$

This interpretation is helpful but not strictly correct, because it implies that p/ρ is intrinsic energy of the fluid rather than work done by pressure forces.

4.3.3 The Mechanical Energy Equation

The general energy equation is not as widely used in fluid mechanics as you might expect. Instead, the *mechanical energy equation,* which is a special form of the energy equation, is often used, especially for incompressible fluid flows. We examine the reasons for this after we develop the equation.

Let's consider steady flow along a single streamline. The fluid may be compressible or incompressible, and we consider both mechanical and thermal forms of energy. We rearrange Eq. (4.39) to obtain

$$-w_s - \left[(\tilde{u}_2 - \tilde{u}_1) + \int_1^2 p\,dv - q\right] - \int_1^2 \frac{dp}{\rho} = \frac{V_2^2}{2} - \frac{V_1^2}{2} + gz_2 - gz_1.$$

Besides grouping the terms differently, we have used

$$\frac{p_2}{\rho_2} - \frac{p_1}{\rho_1} = \int_1^2 d\left(\frac{p}{\rho}\right) = \int_1^2 \frac{dp}{\rho} + \int_1^2 p\,d\left(\frac{1}{\rho}\right) \quad \text{and} \quad v = \frac{1}{\rho}.$$

We take the integrals along the streamline. Their evaluation requires knowledge of the process experienced by a fluid particle as it travels from point 1 to point 2.

Note the terms in brackets [. . .]. By combining the first and second laws of thermodynamics [2, 3, 4], we may express this combination of terms as a single term, for which we use the symbol gh_L:

$$gh_L \equiv \left[\tilde{u}_2 - \tilde{u}_1 - q + \int_1^2 p\,dv.\right] \tag{4.42}$$

This single term has the following characteristics:

- gh_L is always positive or zero (a process with gh_L negative would violate the second law of thermodynamics);
- gh_L represents the loss of potential to perform useful work; and
- gh_L is zero only in an ideal process (of the many effects that can cause gh_L to be nonzero, the most common are friction and fluid viscous action).

We now write the energy equation for steady flow along a streamline as

$$-w_s - \int_1^2 \frac{dp}{\rho} - gh_L = \left(\frac{V_2^2}{2} + gz_2\right) - \left(\frac{V_1^2}{2} + gz_1\right). \tag{4.43}$$

Steady flow along a streamline

This equation is the (steady flow) *mechanical energy equation for a compressible fluid.* The terms on the right represent the change of the fluid's mechanical energy that occurs between the two ends of the streamline. The terms on the left represent the processes that occur along the streamline that bring about this change. They are, in order, work from shear stresses or machine parts, work from pressure forces, and loss of mechanical energy from (irreversible) conversion to thermal energy. Although w_s and dp/ρ can be zero, positive, or negative, gh_L cannot be negative and acts to decrease the fluid's mechanical energy. Thus gh_L is called the *mechanical energy loss* or, sometimes, simply the *energy loss* (which is incorrect because energy is never lost, just converted from a useful mechanical form to an unusable thermal form).

The symbol gh_L results from early hydraulic engineers' need to provide extra elevation in a water supply system to overcome this energy loss. Mechanical energy of fluid is sometimes expressed in terms of *head*, which is the elevation that would produce potential energy equal to the particular energy in question. Thus the *velocity head* is the elevation that would yield the same specific potential energy as the actual kinetic energy (velocity head $= (V^2/2)/g = V^2/2g$). In general, head is the specific energy divided by the gravitational acceleration. The term

$$h_L \equiv \frac{gh_L}{g},$$

which has the dimension of length, is called the *head loss*. (Remember that gh_L is one entity, not two.)

Incompressible Flow. Engineers use the mechanical energy equation most frequently in analyzing incompressible flows. If the fluid is incompressible, ρ does not vary and the pressure–density integral may be easily evaluated:

$$\int \frac{dp}{\rho} = \frac{p_2 - p_1}{\rho} = \frac{p_2}{\rho} - \frac{p_1}{\rho}.$$

Substituting into Eq. (4.43) and rearranging gives

$$\frac{p_1}{\rho} + \frac{V_1^2}{2} + gz_1 - w_s = \frac{p_2}{\rho} + \frac{V_2^2}{2} + gz_2 + gh_L. \tag{4.44}$$

Steady, incompressible flow along a streamline

Equation (4.44) is sometimes called simply "the energy equation" (for incompressible flow). Some engineers prefer to divide this equation by the gravitational acceleration, g, and work in terms of "head" rather than specific energy.

If we introduce the incompressible fluid model into Eq. (4.42), we get

$$\tilde{u}_2 - \tilde{u}_1 = q + gh_L. \tag{4.45}$$

Equation (4.45) is the *thermal energy equation for incompressible flow*. It shows us that the internal energy (and hence the temperature) of an incompressible fluid can be increased by two effects: heat transfer to the fluid and friction. Only one effect can cause an internal energy decrease, namely, heat transfer from the fluid, as gh_L cannot be negative.*

Equations (4.44) and (4.45) represent a splitting of the general energy equation into two separate parts for incompressible flow. These equations are almost always preferred to the general energy equation for incompressible flow (usually, only Eq. (4.44) is necessary). There are two reasons for this preference.

1. For incompressible fluids, we often do not care about thermal energies, so we lump them together. We do care about friction, pressure change, and work, so we keep them separate.

2. When thermal energy changes are present in an incompressible flow, for example, flow through a water heater, q and $\tilde{u}_2 - \tilde{u}_1$ are often orders of magnitude larger than the mechanical energy terms, and Eq. (4.45) is approximately true without gh_L. To investigate the mechanical aspects of such a flow (pressure drop, pump sizing, etc.), we must handle mechanical and thermal effects separately.

To use Eq. (4.44) requires evaluating the mechanical energy loss by considering the action of friction in the fluid.† The reason is that $(\tilde{u}_2 - \tilde{u}_1)$ and q are usually either so large that their difference is meaningless or so small that they cannot be calculated or measured with confidence.

---

* The relations among friction, heat, and internal energy—as well as the impossibility of negative friction loss—may be illustrated by a familiar example. On a cold day, you might rub the palms of your hands together to "warm" them. While you rub, friction converts the energy of motion into internal energy and the temperature of your hands rises. While you rub and certainly after you stop, heat is transferred from your hands to the surrounding air. In this case, friction loss is positive, internal energy change is positive because of the rubbing and negative because of the heat loss, and the heat transfer is negative. On a hot summer day, when undesirable heat is transferred to your hands from the surrounding air, finding a way to decrease the temperature (internal energy) of your hands with a consequent "negative friction" rubbing motion is impossible. Similarly, if a viscous oil is forced to flow through a long pipe, the oil will be heated by friction; however, cooling a pipe filled with stagnant oil will not cause the oil to flow backward!

† This is a very complicated problem. There are very few analytical solutions for head losses, but there is a fair amount of reliable measured data. This problem will be considered in Chapter 7.

EXAMPLE 4.9 **Illustrates the Mechanical Energy Equation for the Flow of an Incompressible Fluid**

A Boy Scout troop builds a camp shower with a 55-gal drum, as shown in Fig. E4.9. The scouts shower with 55°F water, but the troop leaders heat their water to 100°F. Frictionless flow occurs in both pipes. Find the flow rate to each shower stall when the water level in the drum is 3 ft above the pipe discharges.

The internal energy change of the water is given by

$$\tilde{u}_2 - \tilde{u}_1 = 1.0(T_2 - T_1),$$

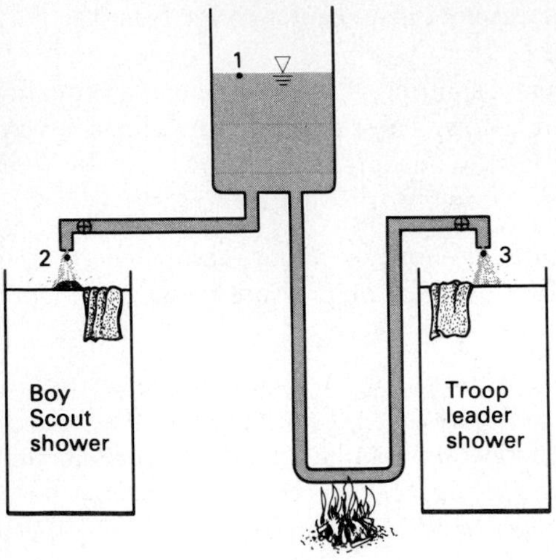

Figure E4.9 Shower installation.

where $\tilde{u}$ is in Btu/lbm and T is in °F. Find the heat added to each pound of water in each pipe. Neglect heating of the water by the sun or air.

SOLUTION

Given

55-gal drum with two discharge pipes
Frictionless flow
55°F water in scouts' line
100°F water in leaders' line

Find

Flow rate to each shower stall for water level 3 ft above pipe discharges
Heat added per pound of water in each pipe

Solution

We define point 1 as the water surface in the drum, point 2 as the scouts' end of the pipe, and point 3 as the leaders' end of the pipe. We write the mechanical energy equation for streamlines connecting points 1 and 2 and points 1 and 3:

$$\frac{p_1}{\rho} + \frac{V_1^2}{2} + gz_1 - w_s)_{1-2} = \frac{p_2}{\rho} + \frac{V_2^2}{2} + gz_2 + gh_L)_{1-2}$$

and

$$\frac{p_1}{\rho} + \frac{V_1^2}{2} + gz_1 - w_s)_{1-3} = \frac{p_3}{\rho} + \frac{V_3^2}{2} + gz_3 + gh_L)_{1-3}.$$

Neither pipe has a pump or turbine, so w_s is zero in both cases. Also, with the assumption of frictionless (inviscid) flow, the mechanical energy loss is zero in each pipe. The rate of fall of the water surface in the drum is small, so $V_1^2 \approx 0$. All three pressures are equal to atmospheric pressure, which we may take as zero. Finally, we may take the values of z_2 and z_3 as zero. Thus the mechanical energy equations reduce to

$$0 + 0 + gz_1 + 0 = 0 + \frac{V_2^2}{2} + 0 + 0$$

and

$$0 + 0 + gz_1 + 0 = 0 + \frac{V_3^2}{2} + 0 + 0,$$

so the discharge velocities are equal, that is,

$$V_2 = V_3 = \sqrt{2gz_1}.$$

The numerical values give

$$V_2 = V_3 = \sqrt{2(32.2 \text{ ft/sec}^2)(3 \text{ ft})};$$
$$V_2 = V_3 = 13.9 \text{ ft/sec}.$$ **ANSWER**

The thermal energy equation for incompressible flow, Eq. (4.45), is applicable to both flows:

$$q_{1-2} = \tilde{u}_2 - \tilde{u}_1 - gh_L)_{1-2} \quad \text{and} \quad q_{1-3} = \tilde{u}_3 - \tilde{u}_1 - gh_L)_{1-3}.$$

In both cases, $gh_L = 0$ by assumption. The internal energy changes are

$$\tilde{u}_2 - \tilde{u}_1 = 1.0(T_2 - T_1) \quad \text{and} \quad \tilde{u}_3 - \tilde{u}_1 = 1.0(T_3 - T_1).$$

Substitution gives

$$q_{1-2} = 1.0(T_2 - T_1) \quad \text{and} \quad q_{1-3} = 1.0(T_3 - T_1).$$

The numerical values are

$$q_{1-2} = (1.0\ \text{Btu/lbm °F})(55 - 55)\text{°F},$$

or

$$q_{1-2} = 0,$$ **ANSWER**

and

$$q_{1-3} = (1.0\ \text{Btu/lbm °F})(100 - 55)\text{°F},$$

or

$$q_{1-3} = 45\ \text{Btu/lbm}.$$ **ANSWER**

Discussion

This example illustrates that the mechanical energy equation and the thermal energy equation are applicable to the same problem *if* the flow is incompressible and frictionless. Why? Note that the heat transfer in the leaders' shower pipe had no effect on the mechanical energy equation.

Control Volume and Average Property Forms of the Mechanical and Thermal Energy Equations. The (incompressible flow) mechanical and thermal energy equations, Eqs. (4.44) and (4.45), apply to flow between two points on a single streamline. We may also need forms of these equations for a finite control volume. Let's first consider a control volume with a single inlet and a single outlet (such as a pipe or nozzle). We may express the mechanical energy equation in terms of average fluid properties and velocities over the inlet and outlet planes and the mass flow rate as

$$\dot{m}\left[\frac{\bar{p}_1}{\rho} + \alpha_1\left(\frac{\bar{V}_1^2}{2}\right) + \overline{gz}_1\right] - \dot{W}_s = \dot{m}\left[\frac{\bar{p}_2}{\rho} + \alpha_2\left(\frac{\bar{V}_2^2}{2}\right) + \overline{gz}_2\right] + \dot{m}(\overline{gh_L}), \tag{4.46}$$

where $\overline{gh_L}$ is the mass flow averaged mechanical energy loss,

$$\overline{gh_L} = \frac{\int_{\substack{\text{All}\\\text{streamlines}}} (gh_L)\, d\dot{m}}{\dot{m}} = \frac{\dot{L}}{\dot{m}},$$

and $\dot{L}$ is the total rate of mechanical energy loss within the control volume, more properly calculated by

$$\dot{L} = \int_{\Psi_{cv}} (\text{Loss per unit volume})\, d\Psi. \tag{4.47}$$

Mechanical energy losses are not generally distributed uniformly throughout the control volume.

Dividing Eq. (4.46) by the mass flow rate gives the *specific mechanical energy equation:*

$$\frac{\bar{p}_1}{\rho} + \alpha_1\left(\frac{\bar{V}_1^2}{2}\right) + \overline{gz}_1 - w_s = \frac{\bar{p}_2}{\rho} + \alpha_2\left(\frac{\bar{V}_2^2}{2}\right) + \overline{gz}_2 + \overline{gh}_L. \tag{4.48}$$

By subtracting the mechanical energy equation from the general energy equation, we find that

$$\dot{m}(\bar{\bar{u}}_2 - \bar{\bar{u}}_1) = \dot{Q} + \dot{L}$$

and

$$(\bar{\bar{u}}_2 - \bar{\bar{u}}_1) = q - \overline{gh}_L. \tag{4.49}$$

Equations (4.48) and (4.49) are similar to the streamline forms, Eqs. (4.44) and (4.45). These latest equations apply on average and not to any particular streamline.

Finally, we extend the mechanical energy equation to a control volume with multiple inlets and outlets, such as that shown in Fig. 4.13:

$$\sum_{\text{in}} \dot{m}\left(\frac{\bar{p}}{\rho} + \alpha\frac{\bar{V}^2}{2} + \overline{gz}\right) - \dot{W}_s = \sum_{\text{out}} \dot{m}\left(\frac{\bar{p}}{\rho} + \alpha\frac{\bar{V}^2}{2} + \overline{gz}\right) + \dot{L}. \tag{4.50}$$

As indicated by Eq. (4.47), $\dot{L}$ must be evaluated from an integral throughout the entire control volume or estimated from an empirical or hueristic model (see Example 4.11).

Recall that our most useful forms of the mechanical and thermal energy equations, Eqs. (4.44)–(4.50), are valid only for steady, incompressible flow. Unsteady

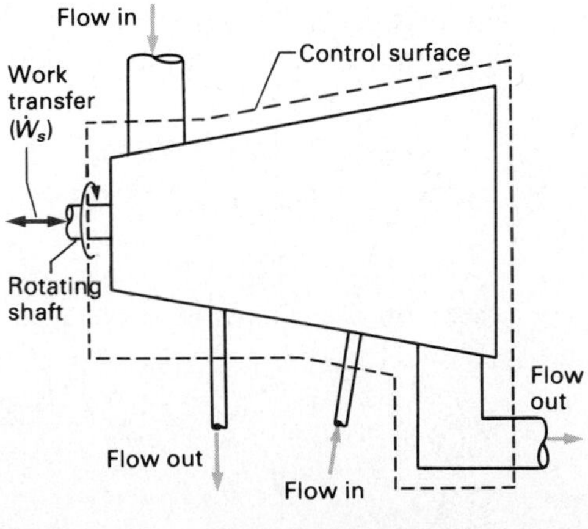

Figure 4.13 Control volume with multiple inlets and outlets.

flow analysis requires the addition of terms to represent storage of energy within the control volume. Compressible flow analysis is usually based on the general energy equation rather than the mechanical energy equation.

EXAMPLE 4.10 Illustrates Energy Equations for the Flow of an Incompressible, Viscous Fluid

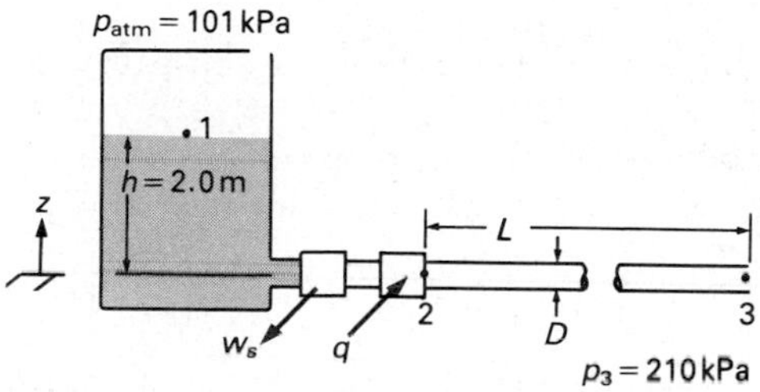

Figure E4.10 Flow from storage tank through pump, heater, and pipeline.

A storage tank contains fuel oil at 0°C. The fuel oil is pumped from the storage tank through a steam heater and through a 50-m-long pipe at an average velocity of 2.5 m/s, as shown in Fig. E4.10. The pipe's inside diameter is 4.0 cm, and the discharge pressure is 210 kPa. The fuel oil is heated to 70°C in the steam heater and has an internal energy change given by

$$\tilde{u}_2 - \tilde{u}_1 = 2000(T_2 - T_1),$$

where $\tilde{u}$ is measured in J/kg and T in °C. The mechanical energy loss in the pipe is expressed by

$$\overline{gh_L} = f\left(\frac{L}{D}\right)\frac{\overline{V}^2}{2},$$

where $f = 0.02$ and $\overline{V}$ is the average velocity. The fuel oil density is 850 kg/m³. Assume that the tank, pump, and pipeline are all thermally insulated. Find the pump work per unit mass of flowing fluid, the temperature rise of the oil in the pipe, the heater discharge pressure, and the heat transfer to the fuel oil in the heater. Neglect mechanical energy loss in the pump and heater.

SOLUTION

Given

Fuel oil pumped from storage tank through steam heater and through pipe with discharge pressure of 210 kPa

Pipe's inside diameter 4.0 cm and length 50 m

Oil internal energy related to temperature by

$$\tilde{u}_2 - \tilde{u}_1 = 2000(T_2 - T_1),$$

with

$$T_2 = 70°\text{C} \quad \text{and} \quad T_1 = 0°\text{C}.$$

Mechanical energy loss of

$$\overline{gh_L} = 0.02\left(\frac{L}{D}\right)\left(\frac{\overline{V}^2}{2}\right)$$

in pipe, with

$$\overline{V} = 2.5 \text{ m/s} \quad \text{and} \quad \rho = 850 \text{ kg/m}^3 = \text{Constant}.$$

Pipeline thermally insulated and $\overline{gh_L} \approx 0$ in pump and heater.

Find

Pump work per unit mass

Temperature rise of oil in pipe

Pressure at heater outlet

Heat transfer to fuel oil in heater

Solution

Because we consider the density constant, we apply the mechanical energy equation, Eq. (4.46), between points 1 and 3:

$$\frac{\bar{p}_1}{\rho} + \alpha_1\left(\frac{\bar{V}_1^2}{2}\right) + \overline{gz_1} - w_s = \frac{\bar{p}_3}{\rho} + \alpha_3\left(\frac{\bar{V}_3^2}{2}\right) + \overline{gz_3} + \overline{gh_L})_{1-3}.$$

We assume that the oil level in the tank drops slowly, so $\bar{V}_1^2 \approx 0$. We next substitute the given expression for $\overline{gh_L}$ and solve for w_s:

$$w_s = \frac{\bar{p}_1 - \bar{p}_3}{\rho} - \alpha_3\left(\frac{\bar{V}_3^2}{2}\right) + g(\bar{z}_1 - \bar{z}_3) - f\left(\frac{L}{D}\right)\left(\frac{\bar{V}^2}{2}\right).$$

From the continuity equation, the average velocity in the pipe is constant, so $\bar{V}_3 = \bar{V}$. We also assume that $\alpha_3 \approx 1.0$. Then because $\bar{z}_1 - \bar{z}_3 = h$, substituting gives

$$w_s = \frac{\bar{p}_1 - \bar{p}_3}{\rho} + gh - \left[f\left(\frac{L}{D}\right) + 1\right]\left(\frac{\bar{V}^2}{2}\right).$$

The numerical values give

$$\begin{aligned} w_s = {} & \frac{(101 - 210)\text{ kPa }(1000\text{ N/m}^2\text{/kPa})}{850\text{ kg/m}^3} \\ & + (9.807\text{ m/s}^2)(2.0\text{ m})(1\text{ N}\cdot\text{s}^2\text{/kg}\cdot\text{m}) \\ & - \left[0.02\left(\frac{50\text{ m}}{0.04\text{ m}}\right) + 1\right] \\ & \times \frac{(2.5\text{ m/s})^2}{2}(1\text{ N}\cdot\text{s}^2\text{/kg}\cdot\text{m}); \end{aligned}$$

$$w_s = -189.9\text{ J/kg.}$$ **ANSWER**

We find the temperature rise of the oil in the pipeline by applying Eq. (4.49) between points 2 and 3:

$$\bar{\bar{u}}_3 - \bar{\bar{u}}_2 = q + \overline{gh_L})_{2-3}.$$

The problem statement implies that $q = 0$. Substituting for $(\bar{\bar{u}}_3 - \bar{\bar{u}}_2)$ and $\overline{gh_L}$ gives

$$2000(\bar{T}_3 - \bar{T}_2) = f\left(\frac{L}{D}\right)\left(\frac{\bar{V}^2}{2}\right) \quad \text{or} \quad \bar{T}_3 - \bar{T}_2 = \frac{f(L/D)\bar{V}^2}{2(2000)}.$$

The numerical values give

$$\bar{T}_3 - \bar{T}_2 = \frac{0.02(50\text{ m})}{(2000\text{ J/kg}\cdot{}^\circ\text{C})(0.04\text{ m})}\left[\frac{(2.5\text{ m/s})^2}{2}\right],$$

$$\bar{T}_3 - \bar{T}_2 = 0.039^\circ\text{C}.$$ **ANSWER**

We find the heater discharge pressure by applying the mechanical energy equation between points 2 and 3:

$$\frac{\bar{p}_2}{\rho} + \alpha_2\left(\frac{\bar{V}_2^2}{2}\right) + g\bar{z}_2 - w_s = \frac{\bar{p}_3}{\rho} + \alpha_3\left(\frac{\bar{V}_3^2}{2}\right) + g\bar{z}_3 + \overline{gh_L})_{2-3}.$$

With no shaft work or elevation change,

$$w_s = 0 \quad \text{and} \quad \bar{z}_2 = \bar{z}_3.$$

The continuity equation gives

$$\bar{V}_2 = \bar{V}_3 = \bar{V}.$$

Assuming that $\alpha_2 = \alpha_3$ and solving for $\bar{p}_2$, we obtain

$$\bar{p}_2 = \bar{p}_3 + \rho\overline{gh_L})_{2-3}.$$

Substituting for $\overline{gh_L}$, we have

$$\bar{p}_2 = \bar{p}_3 + \rho f\left(\frac{L}{D}\right)\left(\frac{\bar{V}^2}{2}\right).$$

The numerical values give

$$\bar{p}_2 = 210 \text{ kPa} + 0.02(850 \text{ kg/m}^3)\left(\frac{50 \text{ m}}{0.04 \text{ m}}\right) \times \frac{(2.5 \text{ m/s})^2}{2}\left(\frac{1 \text{ kPa}\cdot\text{m}\cdot\text{s}^2}{1000 \text{ kg}}\right),$$

or

$$\bar{p}_2 = 276 \text{ kPa}.$$ **ANSWER**

We find the heat transferred to the oil in the steam heater by applying the thermal energy equation for incompressible flow between points 1 and 2:

$$\bar{\tilde{u}}_2 - \bar{\tilde{u}}_1 = q + \overline{gh_L})_{1-2}.$$

By assumption, $\overline{gh_L})_{1-2} = 0$, so

$$q = \bar{\tilde{u}}_2 - \bar{\tilde{u}}_1 = 2000(\bar{T}_2 - \bar{T}_1).$$

Substituting numerical values, we obtain

$$q = (2000 \text{ J/kg}\cdot{}^\circ\text{C})(70^\circ\text{C} - 0^\circ\text{C}) = 140 \text{ kJ/kg}.$$

Because

$$\dot{m} = \rho\bar{V}A = 850 \text{ kg/m}^3(2.5 \text{ m/s})\left(\frac{\pi}{4}\right)(0.04)^2 \text{ m}^2 = 2.67 \text{ kg/s},$$

then

$$\dot{Q} = \dot{m}q = 2.67 \text{ kg/s}(140 \text{ kJ/kg}),$$

or

$$\dot{Q} = 374 \text{ kW}.$$ **ANSWER**

Discussion

Note that the pressure drop in the pipe ($\bar{p}_2 - \bar{p}_3 = 66$ kPa) is significant, whereas the temperature rise ($\bar{T}_3 - \bar{T}_2 = 0.039^\circ$C) is small. In many incompressible fluid flows, the relative temperature change is several orders of magnitude smaller than the relative pressure change. Compare this result with the temperature and pressure changes of the compressible fluid across the compressor in Example 4.9. Do you expect relative temperature and pressure changes of the same or significantly different orders of magnitude for the flow of a compressible fluid in a pipeline?

The kinetic energy change in this example is quite small compared to the changes of other quantities. What are some types of physical problems in which the kinetic energy change is a significant factor?

Also note that the heat transfer was several orders of magnitude larger than the work input or the potential energy of the oil in the reservoir. Our neglect of the mechanical energy loss in the heater and pump had a negligible effect on the heat transfer calculation.

EXAMPLE 4.11 Illustrates the Mechanical Energy Equation for Incompressible Flow and Various Methods for Calculating Mechanical Energy Loss

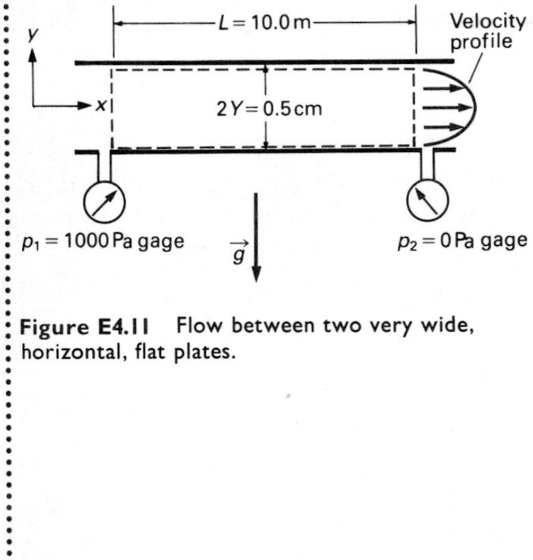

Figure E4.11 Flow between two very wide, horizontal, flat plates.

An incompressible viscous fluid flows between two horizontal parallel plates as shown in Fig. E4.11. Pressure gages are placed at two locations. The plates are spaced 0.5 cm apart and are very wide perpendicular to the page. Flow is laminar and steady. We show in Chapter 5 that if the plates are very long and wide, the fluid velocity profile does not change in the x-direction or perpendicular to the page and is given by

$$u = -\frac{Y^2}{2\mu}\left(\frac{\Delta p}{L}\right)\left[1 - \left(\frac{y}{Y}\right)^2\right],$$

where Δp is the pressure change that occurs in length L.

Calculate the average mechanical energy loss $\overline{gh_L}$ between the two pressure gages. Then show that the mechanical energy loss also satisfies the equations

$$\overline{gh_L} = \frac{\int_{V_{cv}} \mu(\partial u/\partial y)^2 \, dV}{\dot{m}} \tag{E4.1}$$

and

$$\overline{gh_L} = f(\mathrm{R})\left(\frac{L}{Y}\right)\left(\frac{\bar{V}^2}{2}\right), \tag{E4.2}$$

where

$$\mathrm{R} = \frac{\rho \bar{V} Y}{\mu}.$$

The fluid density is 850 kg/m^3.

SOLUTION

Given

Flow between parallel plates in Fig. E4.11

Velocity profile equation

Fluid density 850 kg/m^3

Find

Average mechanical energy loss $\overline{gh_L}$ and verify Eqs. (E4.1) and (E4.2)

Solution

We select a control volume enclosing the fluid between the plates and the inlet and outlet planes. Because the fluid is incompressible, we use the specific mechanical energy equation, Eq. (4.48):

$$\frac{\bar{p}_1}{\rho} + \alpha_1\left(\frac{\bar{V}_1^2}{2}\right) + \overline{gz_1} - w_s = \frac{\bar{p}_2}{\rho} + \alpha_2\left(\frac{\bar{V}_2^2}{2}\right) + \overline{gz_2} + \overline{gh_L}.$$

The plates are horizontal, so

$$\overline{gz_1} = \overline{gz_2}.$$

There are no pumps or stresses doing work on the fluid at the boundary of the control volume, so $w_s = 0$. From the continuity equation,

$$\bar{V}_1(2Y) = \bar{V}_2(2Y), \quad \text{so} \quad \bar{V}_1 = \bar{V}_2.$$

We evaluate the kinetic energy correction factors, α_1 and α_2, by substituting the velocity profile into the defining equation, Eq. (4.36), and simplifying to get

$$\alpha = \frac{1}{A}\int\left(\frac{u}{\bar{V}}\right)^3 dA.$$

We leave the integration as an exercise for you. Actually, it is unnecessary, because the velocity profile is identical at 1 and 2. Thus $\alpha_1 = \alpha_2$. With these simplifications, the energy equation gives

$$\overline{gh_L} = \frac{\bar{p}_1 - \bar{p}_2}{\rho} = \frac{p_1 - p_2}{\rho}. \tag{E4.3}$$

Substituting numbers, we have

$$\overline{gh_L} = \frac{1000 \text{ N/m}^2}{850 \text{ kg/m}^3};$$

$$\overline{gh_L} = 1.18 \text{ J/kg}.$$ **ANSWER**

We demonstrate the validity of Eq. (E4.1) by showing that the expression on the right-hand side of Eq. (E4.1) reduces to $(p_2 - p_1)/\rho$. Differentiating the given velocity profile, we obtain

$$\frac{\partial u}{\partial y} = \frac{y}{\mu}\left(\frac{\Delta p}{L}\right).$$

Substituting into the right-hand side of Eq. (E4.1) and denoting the plate width by W, we get

$$\frac{\int_{\forall_{cv}} \mu(\partial u/\partial y)^2 \, d\forall}{\dot{m}} = \frac{2\mu \int_0^Y [(y/\mu)(\Delta p/L)]^2 WL\, dy}{\rho\bar{V}(2Y)W}$$

$$= \frac{L(\Delta p/L)^2}{\rho\mu\bar{V}Y}\int_0^Y y^2\, dy = \frac{Y^2 L}{3\rho\mu\bar{V}}\left(\frac{\Delta p}{L}\right)^2.$$

The average velocity is

$$\bar{V} = \frac{2\int_0^Y u\,dy}{2Y} = \frac{-2\int_0^Y (Y^2/2\mu)(\Delta p/L)[1-(y/Y)^2]\,dy}{2Y}.$$

Integrating and simplifying give

$$\bar{V} = -\frac{1}{3}\left(\frac{Y^2}{\mu}\right)\left(\frac{\Delta p}{L}\right) = \frac{2}{3}u_{\max}.$$

Substituting the average velocity, we obtain

$$\frac{\int_{\forall_{cv}} \mu(\partial u/\partial y)^2\,d\forall}{\dot{m}} = -\frac{1}{\rho}\left(\frac{\Delta p}{L}\right)L = \frac{p_1 - p_2}{\rho}.$$

Thus

$$\frac{\int_{\forall_{cv}} \mu(\partial u/\partial y)^2\,d\forall}{\dot{m}} = \frac{p_1 - p_2}{\rho} = \overline{gh_L}.$$

Equation (E4.1) is verified. **ANSWER**

To verify Eq. (E4.2), we first note from the average velocity equation that

$$-\frac{\Delta p}{L} = \frac{3\mu\bar{V}}{Y^2}.$$

Multiplying the right-hand side by $2\bar{V}/2\bar{V}$ and both sides by L/ρ, we obtain

$$-\frac{\Delta p}{L}\left(\frac{L}{\rho}\right) = \frac{6\mu}{\rho\bar{V}Y}\left(\frac{L}{Y}\right)\left(\frac{\bar{V}^2}{2}\right).$$

But

$$-\frac{\Delta p}{L}(L) = p_1 - p_2 \quad \text{and} \quad \frac{\rho\bar{V}Y}{\mu} = \mathbf{R},$$

so

$$\frac{p_1 - p_2}{\rho} = \overline{gh_L} = \left(\frac{6}{\mathbf{R}}\right)\left(\frac{L}{Y}\right)\left(\frac{\bar{V}^2}{2}\right).$$

As $6/\mathbf{R}$ is clearly a function of $\mathbf{R}$,

Eq. (E4.2) is verified. **ANSWER**

Discussion

The easiest calculation of $\overline{gh_L}$ was from the incompressible energy equation itself; however, we could make this calculation only because we knew all the other terms. Usually, we need to evaluate $\overline{gh_L}$ from independent information so that we can use the incompressible energy equation to calculate pressure or work.

We did not derive Eqs. (E4.1) and (E4.2); we only verified them for a particular case. Equations (E4.1) and (E4.2) are, in fact, more general than Eq. (E4.3). Equation (E4.1) may be used to calculate mechanical energy loss for laminar flow if the velocity distribution is known. Equation (E4.2) is used to calculate mechanical energy loss in any fully developed pipe or channel flow, laminar or turbulent; however, $f\{\mathbf{R}\}$ must usually be obtained by experiment.

4.3.4 Bernoulli's Equation

One of the most famous and useful equations in all of fluid mechanics is Bernoulli's equation. This equation may be developed as a special form of the energy equation, as we do here. Or it may be developed by integrating Newton's second law of motion for a fluid particle in a steady, inviscid, incompressible flow, as we do in Chapter 5.

We begin by considering a single streamline in a steady, incompressible flow. The flow may be internal or external, as shown in Fig. 4.14. We apply the mechanical energy equation, Eq. (4.44), between two points on the streamline:

$$\frac{p_1}{\rho} + \frac{V_1^2}{2} + gz_1 - w_s = \frac{p_2}{\rho} + \frac{V_2^2}{2} + gz_2 + gh_L.$$

We now assume that the flow is frictionless (inviscid) and that there is no shaft or shear work. Then

$$\frac{p_1}{\rho} + \frac{V_1^2}{2} + gz_1 = \frac{p_2}{\rho} + \frac{V_2^2}{2} + gz_2. \tag{4.51}$$

Steady, incompressible, frictionless flow
along a single streamline, no work

As points 1 and 2 are arbitrary, the sum of the three terms is the same for any point on the streamline. Thus an alternate form of the equation is

$$\frac{p}{\rho} + \frac{V^2}{2} + gz = \text{Constant.} \tag{4.52}$$

Steady, incompressible, frictionless flow
along a single streamline, no work

Equations (4.51) and (4.52) are alternative forms of Bernoulli's equation. We have listed all of the assumptions used in its derivation along with the equation. Before using Bernoulli's equation in an engineering analysis (or in a homework problem), satisfy yourself that the assumptions used in deriving the equation are valid for the application.

Bernoulli's equation is probably the simplest form of the energy equation. It tells us that the fluid's mechanical energy, measured by gravitational potential

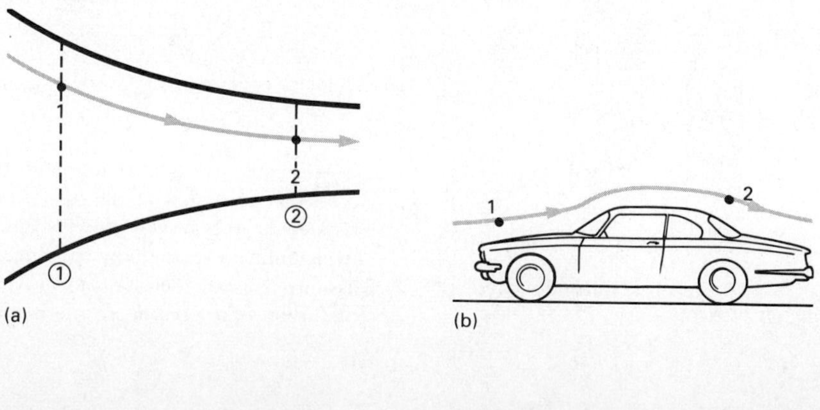

Figure 4.14 Two points 1 and 2 on the same streamline: (a) flow in a nozzle; (b) flow over an automobile.

energy, kinetic energy, and pressure,* is constant if there is no external work or shear stress. The energy may be rearranged between the three forms, but the total remains constant. Note also that if the flow is incompressible, heat transfer has no effect on Bernoulli's equation (see Example 4.9).

Although the assumptions in Bernoulli's equation are seldom strictly true, they often are good approximations. Liquid and low-speed gas flows are nearly incompressible. Viscous (frictional) effects are usually negligible in flows with rapid acceleration, such as nozzles, in flows with no confining walls and negligible mixing, such as liquid jets in air, and in flows outside viscous boundary layers. If the flow field varies slowly with time, we may model it as quasi-steady. (See the discussion of Example 4.12.) In Chapter 5, we show that sometimes the restriction to a single streamline may be relaxed, but for now, we respect that restriction.

The following examples illustrate the use of Bernoulli's equation in several cases where its assumptions are approximately true. The discussion of Example 4.13 suggests a common practice in fluid mechanics: modifying the predictions of Bernoulli's equation with experimentally derived information that accounts for small friction and/or compressibility effects.

* Our earlier development of the general energy and mechanical energy equations showed that pressure appears in the energy equation because of flow work, not because it is intrinsic fluid energy. This condition is also true in Bernoulli's equation. Nevertheless, thinking of p/ρ as "potential energy" because of pressure in the same way that gz is potential energy because of gravity is convenient.

EXAMPLE 4.12 Illustrates Bernoulli's Equation

A storage tank has an inside diameter of 30 in. and a 1.0-in.-diameter nozzle in its side. At a particular instant, the water level is 4 ft above the center of the hole. Find the velocity of the water leaving the hole at this instant.

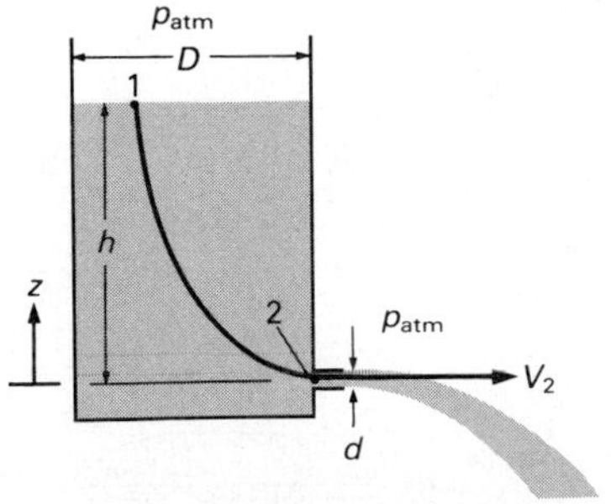

Figure E4.12 Storage tank and nozzle with exiting jet of water.

SOLUTION

Given

Figure E4.12

$D = 30$ in., $d = 1.0$ in., and $h = 4.0$ ft

Find

Velocity of jet leaving nozzle, V_2, in Fig. E4.12

Solution

Because the cross-sectional area of the tank is much greater than the cross-sectional area of the nozzle, we expect the water level in the tank to drop relatively slowly. We obtain an approximate solution by assuming a steady-flow condition at each instant of time. We also assume constant density and gravity and inviscid fluid. No pumps or turbines are present, so there is no work. All the assumptions in

Bernoulli's equation are valid, so we write the equation between point 1 on the surface and point 2 at the jet exit to get

$$\frac{p_1}{\rho} + \frac{V_1^2}{2} + gz_1 = \frac{p_2}{\rho} + \frac{V_2^2}{2} + gz_2.$$

We want to solve for V_2. Because the water level drops slowly, $V_1 \approx 0$ (see the Discussion). We establish a reference level for elevation by setting $z_2 = 0$, so $z_1 = h$. The tank is open to the atmosphere, so $p_1 = p_{atm}$, and the nozzle discharges to the atmosphere, so $p_2 = p_{atm}$. (Do you think p_2 is $(p_{atm} + \gamma h)$? Do you understand why it isn't?) We now solve for V_2 and get

$$V^2 = 2\left(\frac{p_2 - p_1}{\rho} + gh\right),$$

or, as $p_1 = p_2$,

$$V_2 = \sqrt{2gh}.$$

The numerical values give

$$V_2 = \sqrt{2(32.2 \text{ ft/sec}^2)(4 \text{ ft})};$$

$$V_2 = 16.0 \text{ ft/sec}.$$ **ANSWER**

Discussion

Note that the fluid density does not appear in the resulting equation for V_2; therefore this analysis suggests that V_2 is independent of the type of fluid. We may calculate the mass flow rate from the nozzle by

$$\dot{m} = \rho A_2 V_2,$$

if we assume that V_2 is uniform over area A_2 (all streamlines in area A_2 originate at the free surface in the tank and all streamlines have $z_2 \approx 0$). For 50°F water, the numerical values give

$$\dot{m} = (62.4 \text{ lbm/ft}^3)\left(\frac{\pi}{4}\right)\left[1.0 \text{ in.}\left(\frac{1 \text{ ft}}{12 \text{ in.}}\right)\right]^2 (16.0 \text{ ft/sec})$$
$$= 5.45 \text{ lbm/sec}.$$

Let's examine the assumptions and conditions on Bernoulli's equation. The fluid is water, so the flow is very nearly incompressible. There are no pumps or turbines, so no useful work is done. We cannot—and, in fact, do not need to—locate exactly a unique streamline that extends from 1 to 2, but we can certainly imagine that such a streamline exists. The only questionable assumptions are inviscid flow and steady flow.

Some viscous effects occur in the flow, mostly in the vicinity of the hole. Experiments show that in this type of flow, the major effect is on the area of the water jet rather than its velocity. Our velocity calculation is reasonably accurate.

Flow steadiness is related to the rate that the water level in the tank drops, that is, to the velocity V_1. Using our calculated jet velocity of 16 ft/sec, the continuity equation implies that the water surface must be falling at $V_1 = (d/D)^2 V_2 = 0.018$ ft/sec. This seems rather small, but note that the Bernoulli equation deals with the *square* of the velocity, so we are neglecting only 0.018^2 ($= 0.0003$) compared to 16^2 ($= 256$). Thus we are certainly justified in modeling the flow as steady, especially if we use the instantaneous value of h whenever we make a calculation. The use of instantaneous values of slowly varying parameters together with steady-state versions of the governing equations is called a *quasi-steady* flow model.

EXAMPLE 4.13 Illustrates the Simultaneous Application of the Continuity and Bernoulli Equations

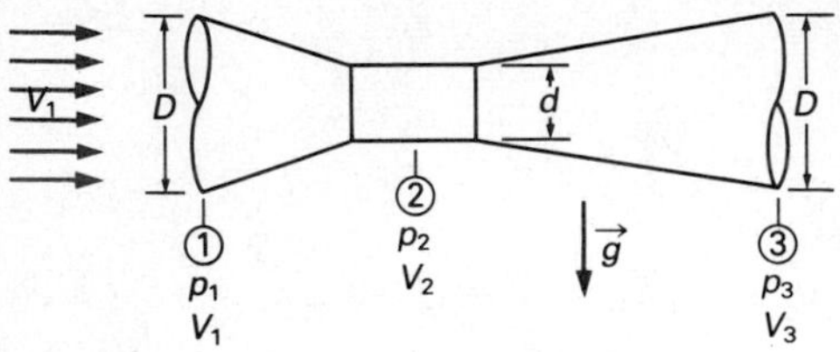

Figure E4.13a Horizontal Venturi with entering uniform velocity.

Water at 10°C enters the horizontal Venturi tube in Fig. E4.13a with a uniform and steady velocity of 2.0 m/s and an inlet pressure of 150 kPa. Find the pressures at the throat, 2, where $d = 3.0$ cm, and at the exit, 3, where $D = 6.0$ cm.

SOLUTION

Given

Figure E4.13a

$D = 6.0$ cm and $d = 3.0$ cm

$V_1 = 2.0$ m/s and $p_1 = 150$ kPa

Find

p_2 and p_3

Solution

We assume constant density and uniform velocity over planes 1 and 2. Applying the continuity equation between these two planes, we obtain

$$V_1 A_1 = V_2 A_2 \quad \text{or} \quad V_2 = \left(\frac{A_1}{A_2}\right) V_1 = \left(\frac{D}{d}\right)^2 V_1.$$

Assuming an inviscid fluid and applying Bernoulli's equation to a streamline connecting cross sections 1 and 2, we have

$$\frac{p_1}{\rho} + \frac{V_1^2}{2} + gz_1 = \frac{p_2}{\rho} + \frac{V_2^2}{2} + gz_2.$$

Assuming that $z_1 \approx z_2$ and solving for p_2, we get

$$p_2 = p_1 + \frac{\rho}{2}(V_1^2 - V_2^2).$$

Substituting for V_2 gives

$$p_2 = p_1 + \frac{\rho}{2}[1 - (D/d)^4]V_1^2.$$

Using numerical values from Table A.5, we obtain

$$p_2 = 150 \text{ kPa} + \left(\frac{10^3 \text{ kg/m}^3}{2}\right)$$

$$\times \left[1 - \left(\frac{6.0 \text{ cm}}{3.0 \text{ cm}}\right)^4\right](2.0 \text{ m/s})^2(\text{N}\cdot\text{s}^2/\text{kg}\cdot\text{m})$$

$$= 150 \text{ kPa} - 30{,}000 \text{ N/m}^2;$$

$$p_2 = 120 \text{ kPa}.$$ **ANSWER**

Similarly, applying the continuity equation, Eq. (4.23), between planes 1 and 3 gives $V_1A_1 = V_3A_3$. As $A_3 = A_1$, then $V_3 = V_1$. Bernoulli's equation between planes 1 and 3 gives

$$\frac{p_1}{\rho} + \frac{V_1^2}{2} + gz_1 = \frac{p_3}{\rho} + \frac{V_3^2}{2} + gz_3.$$

Assuming that $z_1 \approx z_3$ and using $V_3 = V_1$, we have

$$\frac{p_1}{\rho} = \frac{p_3}{\rho},$$

or

$$p_3 = p_1 = 150 \text{ kPa}.$$ **ANSWER**

Discussion

The water has constant density. The continuity equation shows that the velocity increases in the converging section and decreases in the diverging section. Bernoulli's equation then shows that the static pressure decreases in the converging section and increases in the diverging section. The diverging section is a diffuser. A *diffuser* increases the static pressure by decreasing the fluid kinetic energy. The converging section acts as a nozzle. A *nozzle* increases the fluid velocity (kinetic energy) by decreasing the static pressure.

Because we assumed that the water is inviscid, the static pressure drop $(p_1 - p_2)$ is fully recovered in the diffuser by decreasing the fluid velocity to V_1. However, full pressure recovery would not occur in a real Venturi tube. Viscous effects would produce a net pressure drop between sections 1 and 3 (see Fig. E4.13b).

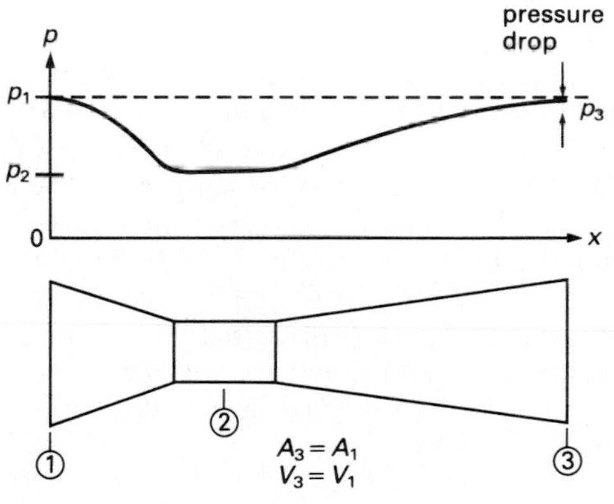

Figure E4.13b Typical static pressure distribution in a real Venturi.

Venturi tubes such as the one illustrated in this example are often used to measure velocity or flow rate in a pipeline. If we combine the continuity and Bernoulli equations to find the throat velocity, V_2, we get

$$V_2 = \sqrt{\frac{2(p_1 - p_2)}{\rho[1 - (d/D)^4]}}.$$

Measurement of the two diameters and the two pressures allows us to determine the velocity and, from it and the throat diameter, the flow rate. The measured velocity and flow rate are somewhat inaccurate because of small friction effects, which Bernoulli's equation neglects. Accounting for such effects by introducing a multiplying coefficient, C_v, to adjust the theoretical value is common practice. That is,

$$V_2 = C_v\sqrt{\frac{2(p_1 - p_2)}{\rho[1 - (d/D)^4]}},$$

where the magnitude of C_v is found by experiment. We discuss Venturi tubes and other flow measuring devices further in Chapter 7.

EXAMPLE 4.14 **Illustrates Bernoulli's Equation, the Concept of Streamlines, and Velocities as Vectors**

A city has a fire truck whose pump and hose can deliver 1000 gal/min with a nozzle velocity of 120 ft/sec. The tallest building in the city is 100 ft high. The firefighters hold the nozzle at an angle of 75° from the ground. Find the minimum distance the firefighters must stand from the building to put out a fire on the roof without the aid of a ladder. The firefighters hold the hose 5 ft above the ground. Assume that the water velocity is not reduced by air resistance.

SOLUTION

Given

Fire hose delivers 1000 gal/min

Nozzle velocity 120 ft/sec

Tallest building 100 ft high

Nozzle inclined at 75° angle and held 5 ft above ground

Find

Minimum distance firefighters must stand from the building to put out fire on roof

Solution

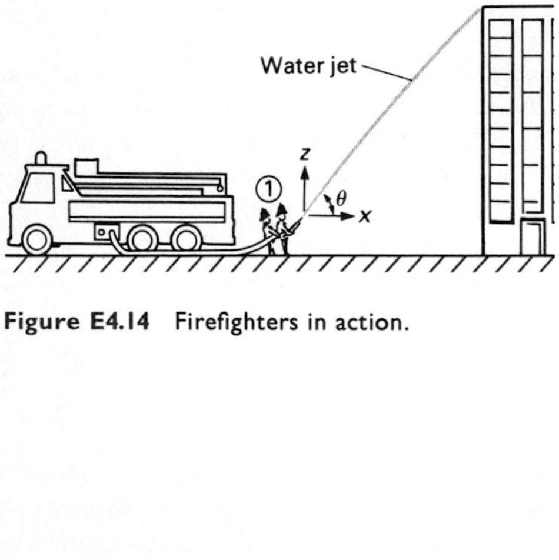

Figure E4.14 Firefighters in action.

Figure E4.14 shows a sketch of the physical situation. We define a coordinate system with its origin at the nozzle and hence seek the horizontal (x) location of the building relative to the nozzle. If we assume that the water jet just clears the corner of the building as shown, the location of the building relative to the jet is the value of x that corresponds to $z = 95$ ft (100 ft height − 5 ft initial elevation). We assume steady flow, so the trajectory of the water jet's centerline is simultaneously a streakline, a pathline, and a streamline. The slope of a streamline is equal to the slope of the velocity vector, so the water jet trajectory is described by

$$\frac{dx}{dz} = \frac{V_x}{V_z}.$$

Bernoulli's equation can help us find the water velocities, because the flow is steady and incompressible and there is no work. Of course, we assume that the flow is frictionless. Between the origin (point 1) and any other point on the streamline, Bernoulli's equation is

$$\frac{p_1}{\rho} + \frac{V_1^2}{2} + gz_1 = \frac{p}{\rho} + \frac{V^2}{2} + gz.$$

At any point, $p = p_1 = p_{\text{atm}}$. Taking $z_1 = 0$ and solving for V^2 give

$$V^2 = V_1^2 - 2gz.$$

With negligible air resistance, there is no force on the fluid in the x

(horizontal) direction. Thus

$$V_x = V_1 \cos\theta.$$

We write Bernoulli's equation as

$$V_x^2 + V_z^2 = V_1^2(\cos^2\theta + \sin^2\theta) - 2gz.$$

As $V_x^2 = V_1^2 \cos^2\theta$, we get

$$V_z^2 = V_1^2 \sin^2\theta - 2gz \quad \text{or} \quad V_z = \sqrt{V_1^2 \sin^2\theta - 2gz}.$$

Substituting the velocities, we get the jet trajectory equation:

$$\frac{dx}{dz} = \frac{V_1 \cos\theta}{(V_1^2 \sin^2\theta - 2gz)^{1/2}}.$$

Multiplying by dz and integrating* gives

$$x = V_1 \cos\theta \frac{(V_1 \sin\theta - \sqrt{V_1^2 \sin^2\theta - 2gz})}{g}.$$

Rearrangement gives

$$x = \frac{V_1^2 \sin 2\theta}{2g}\left(1 - \sqrt{1 - \frac{2gz}{V_1^2 \sin^2\theta}}\right).$$

Substituting $z = 95$ ft, $\theta = 75°$, and $V_1 = 120$ ft/sec gives the minimum distance that the firefighters must stand from the building:

$$x = \frac{(120 \text{ ft/sec})^2 (\sin 150°)}{2(32.2 \text{ ft/sec}^2)}\left[1 - \sqrt{1 - \frac{2(32.2 \text{ ft/sec}^2)(95 \text{ ft})}{(120 \text{ ft/sec})^2 (\sin^2 75°)}}\right],$$

or

$$x = 29.3 \text{ ft.}$$ **ANSWER**

Discussion

If a nozzle velocity of less than approximately 81 ft/sec were used, the quantity $(1 - 2gz/V_1^2 \sin^2\theta)$ would then be negative. What would this result mean?

Two interesting extensions of this problem are to

1. find the maximum height to which the water jet could rise; and
2. show that the maximum horizontal travel of the water jet (assuming no intervening building) is obtained when the nozzle is held at 45°.

* We leave the integration for you to do. It is not as difficult as it might appear, because everything except z is a constant.

Alternative Forms for Bernoulli's Equation. Bernoulli's equation is a special form of the energy equation, and each term in Eqs. (4.51) and (4.52) has the dimensions of energy per unit mass. Other forms of Bernoulli's equation are the pressure form, obtained by multiplying the equation by density, ρ,

$$p + \tfrac{1}{2}\rho V^2 + \gamma z = \text{Constant}, \tag{4.53}$$

and the head form, obtained by dividing the equation by the acceleration of gravity,

$$\frac{p}{\gamma} + \frac{V^2}{2g} + z = \text{Constant.} \tag{4.54}$$

In the pressure form, each term has the dimensions and units of pressure. The terms often have names that imply that they actually are pressures. True fluid pressure, p, is called *static pressure,* because it is the pressure that would be measured by an instrument that is static with respect to the fluid. Of course, if the instrument were static with respect to the fluid, it would have to move along with the fluid. The term $\frac{1}{2}\rho V^2$, which is actually kinetic energy per unit volume, is called *velocity pressure* or, sometimes, *dynamic pressure*. If a fluid particle is brought to rest ($V = 0$) without shear stress or shaft work, the pressure of the particle rises by an amount equal to $\frac{1}{2}\rho V^2$. The kinetic energy will have been converted into pressure. The term γz, which is actually the potential energy per unit volume, is called *hydrostatic pressure,* by obvious analogy to Eq. (2.5). Note that γz is not actually a pressure but could be converted to pressure if the fluid were brought to zero elevation. The constant on the right-hand side of Eq. (4.53) is called *total pressure* (p_T) or, sometimes, *stagnation pressure* (p_0). Total pressure is the maximum pressure that could be attained by a fluid particle unless mechanical work were done on it. To attain this pressure, the particle would have to be brought to rest and to zero elevation so that the kinetic energy (velocity pressure) and potential energy (hydrostatic pressure) could be converted into true pressure.*

In the head form of Bernoulli's equation, Eq. (4.54), each term has the dimensions and units of length—p/γ is called the *pressure head* or *static head,* $V^2/2g$ is called the *velocity head,* and z is called the *elevation head.* The sum of static and elevation heads ($p/\gamma + z$) is called the *piezometric head,* because this magnitude is the height to which fluid would rise in a piezometer attached to a pipe with flowing fluid (Fig. 2.7). The constant on the right-hand side of Eq. (4.54) is called the *total head* (h_T). The concept of head is quite useful when we consider flow systems where gravitational potential energy drives the flow. Head is energy per unit weight.

Pitot and Pitot-Static Tubes. Pitot and Pitot-static tubes are used for measuring total pressure, static pressure, and velocity. Figure 4.15 shows a Pitot tube, which is basically a circular tube bent into an "L" shape with one leg pointed directly into a flow stream. The other leg is capped by a pressure measuring device. The figure also shows a streamline that connects the free stream with the stagnation point (a *stagnation point* is a point where the velocity is zero) at the nose of the tube. Applying Bernoulli's equation between points in the free stream and at the nose of the tube and taking $z = 0$ at the tube centerline, we get

$$p_1 + \tfrac{1}{2}\rho V_1^2 = p_2 = p_T.$$

Because the fluid in the tube is not moving, we may use the hydrostatic pressure equation to compute the pressure at the tube nose from the gage measurement and

* Recall that this entire discussion is applicable only if the fluid density is constant.

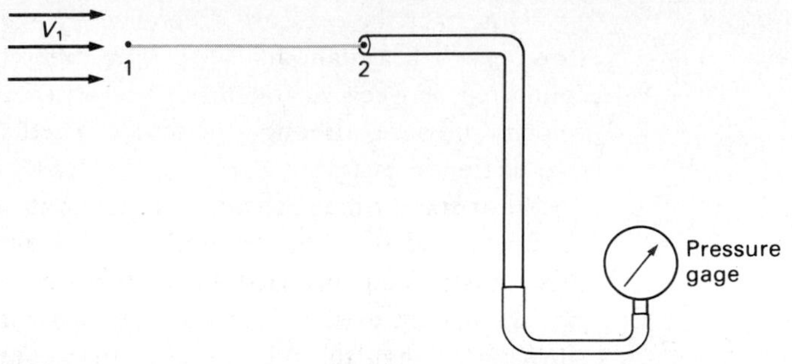

Figure 4.15 Pitot tube in a flow stream.

the system's geometry. In practice, the difference between the pressure at the gage and at the tube nose is often negligible. Thus the Pitot tube measures the total pressure of the flow stream.

A Pitot-static tube measures the static pressure as well as the total pressure and, indirectly, the fluid velocity. Figure 4.16 shows a Pitot-static tube. A second tube—concentric with the first, sealed off from it, and connected to a second pressure measuring device—has been added to the basic Pitot tube. Several small holes are drilled in the outer tube several tube diameters downstream from the nose to sense the static pressure of the fluid. Several holes are necessary so that the pressure sensed by the gage at B is an average that more accurately represents the free-stream static pressure. Thus

$$p_A = p_T \quad \text{and} \quad p_B = p_1,$$

so, from Bernoulli's equation,

$$V_1 = \sqrt{\frac{2(p_A - p_B)}{\rho}}. \tag{4.55}$$

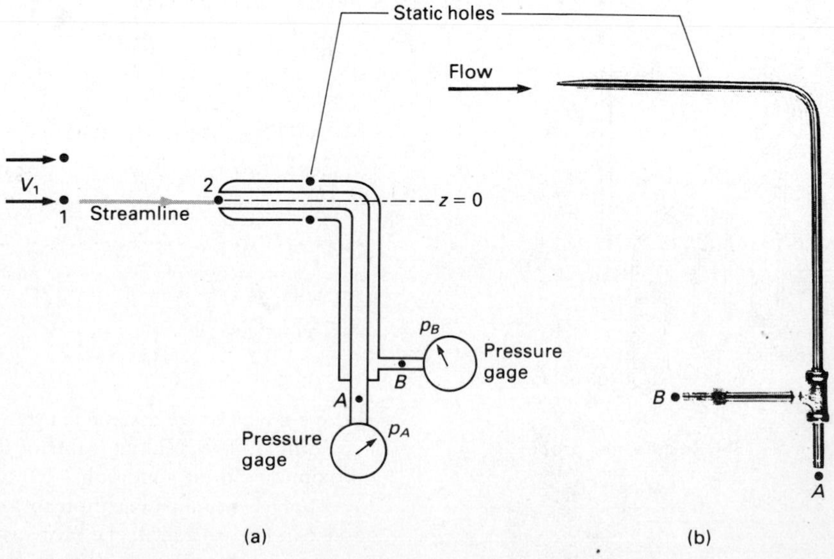

Figure 4.16 Pitot-static tube: (a) schematic drawing; (b) actual tube.

Note that the pressure difference $p_A - p_B$ is equal to the fluid velocity pressure $(\frac{1}{2}\rho V^2)$. If a manometer or other differential pressure measuring device were connected between A and B, we could say that the device directly measures the velocity pressure, although the device is actually measuring the difference between two distinct pressures.

Pitot-static tubes provide very accurate measurements of fluid velocity. The major sources of error are fluid compressibility and misalignment between the tube and the fluid velocity vector. The effect of compressibility is limited mostly to the equation used to calculate velocity from the measured total and static pressures rather than the tube's ability to accurately measure these pressures. Tubes with spherical or ellipisoidal noses can minimize the effects of misalignment, up to angles of about 20°. Specially designed tubes that use the Pitot-static principle are used to align the total pressure hole with the velocity and to measure the velocity direction. The following example examines one of these tubes.

EXAMPLE 4.15 Illustrates Bernoulli's Equation for Pitot Tubes

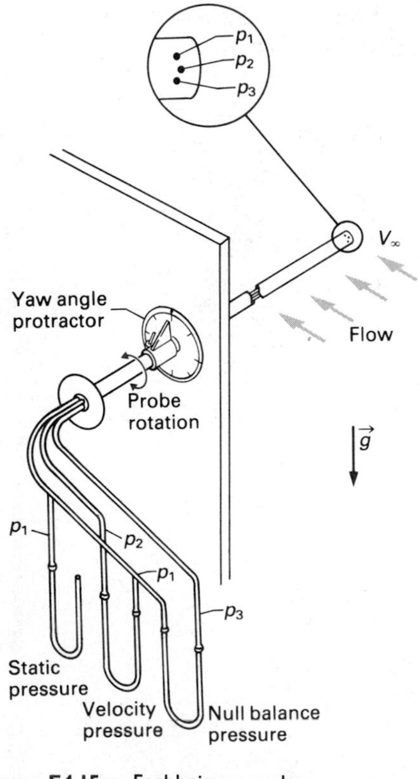

Figure E4.15a Fechheimer probe.

One difficulty with using the Pitot-static tube is the possible misalignment of the stagnation hole with the streamlines. This difficulty may be partially overcome by using the Fechheimer probe shown in Fig. E4.15a. It has a cylindrical body, a leading stagnation hole, and two off-center holes serving for alignment in one plane and as a static pressure reference. The Fechheimer probe is properly aligned in one plane when $p_1 = p_3$. Find θ_0 in Fig. E4.15b so that $p_1 = p_3 = p_\infty$. The velocity over the cylindrical surface is given by

$$V_\theta = 2V_\infty \sin\theta \qquad (0 \le \theta \le 180°).$$

Then calculate V_∞ for air that is flowing, with $p_1 = p_3 = 0$ (gage), and $p_2 - p_1 = 5.0$ cm of water.

SOLUTION

Given

Fechheimer probe in Fig. E4.15a

Velocity distribution over cylindrical body: $V_\theta = 2V_\infty \sin\theta$

Air flowing with $p_1 = p_3 = 0$ and $p_2 - p_1 = 5$ cm H_2O

Find

Angle θ_0 so that $p_1 = p_3 = p_\infty$

Air velocity V_∞

Solution

In this analysis, we assume constant density and gravity and an inviscid fluid. The resulting value of θ_0 is valid only for situations approximating these conditions.

Apply Bernoulli's equation to a streamline connecting points ∞ and 2 in Fig. E4.15b. If the probe is properly aligned, half the parti-

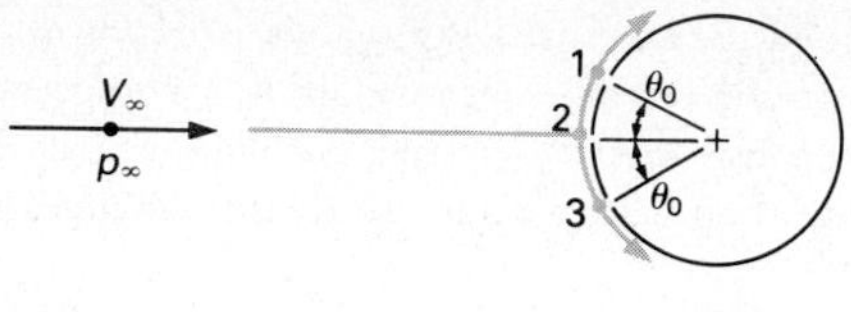

Figure E4.15b Flow over cylindrical body of Fechheimer probe.

cles moving along the ∞–2 streamline pass over point 1 and the other half pass over point 3. Applying Bernoulli's equation along the streamline from ∞ to 1 and along the streamline from ∞ to 3 gives

$$\frac{p_\infty}{\rho} + \frac{V_\infty^2}{2} + gz_\infty = \frac{p_1}{\rho} + \frac{V_1^2}{2} + gz_1$$

and

$$\frac{p_\infty}{\rho} + \frac{V_\infty^2}{2} + gz_\infty = \frac{p_3}{\rho} + \frac{V_3^2}{2} + gz_3,$$

where $z_\infty \approx z_1 \approx z_3$. Solving for $p_1 - p_\infty$ and $p_3 - p_\infty$ gives

$$p_1 - p_\infty = 0 = \frac{\rho}{2}(V_\infty^2 - V_1^2) \quad \text{and} \quad p_3 - p_\infty = 0 = \frac{\rho}{2}(V_\infty^2 - V_3^2).$$

Substituting for V_1 or V_3 gives the same result:

$$0 = V_\infty^2(1 - 4\sin^2\theta_0).$$

Solving for $\sin\theta_0$ gives

$$\sin\theta_0 = \tfrac{1}{2},$$

so

$$\theta_0 = 30°.$$ **ANSWER**

If we assume that the "static holes" 1 and 3 are correctly placed so that p_1 and p_3 are the true free-stream static pressure when the probe is properly aligned, the velocity is

$$V_\infty = \sqrt{\frac{2(p_2 - p_1)}{\rho}}.$$

As p_1 is atmospheric (zero gage), we use Table A.1 for the air density, if we assume standard sea level conditions. Thus

$$\rho = 1.225 \text{ kg/m}^3.$$

The pressure difference is

$$p_2 - p_1 = \gamma h = 9810 \text{ N/m}^3 \text{ (0.05 m)} = 490.5 \text{ N/m}^2,$$

so the air velocity is

$$V_\infty = \sqrt{\frac{2(490.5)\text{ N/m}^2}{1.225 \text{ kg/m}^3}},$$

or

$$V_\infty = 28.3 \text{ m/s}.$$ **ANSWER**

Discussion

The probe is properly aligned with the flow when the null balance pressure manometer gives a pressure difference $p_1 - p_3 =$ zero. The yaw angle protractor measures flow direction.

The given velocity distribution for the flow over the cylindrical probe body is the theoretical one for an inviscid fluid. The actual location of the free-stream pressure is at about 39°.

You may be concerned that we used Bernoulli's equation to calculate the velocity of an air stream, because air is a compressible fluid. For the pressure change involved in this problem, the air density changes by no more than 0.35 percent, so density changes may be neglected.

4.3.5 Summary and Comparison of Various Forms of the Energy Equation

In our presentation of the energy equation, we developed several different forms for it. You may have trouble deciding which form of the energy equation to use and occasionally may use too many energy equations. This section is intended to help you select the correct equation.

Figure 4.17 shows a "typical" flow system. Suppose that the flow is steady and that you want to apply an energy equation to a streamline from point 1 to point 2. You must choose among the following three equations.

- *General energy equation; Eq. (4.40):*

$$\tilde{u}_1 + \frac{p_1}{\rho_1} + \frac{V_1^2}{2} + gz_1 + q - w_s = \tilde{u}_2 + \frac{p_2}{\rho_2} + \frac{V_2^2}{2} + gz_2.$$

- *Mechanical energy equation; Eq. (4.44):*

$$\frac{p_1}{\rho} + \frac{V_1^2}{2} + gz_1 - w_s = \frac{p_2}{\rho} + \frac{V_2^2}{2} + gz_2 + gh_L.$$

- *Bernoulli's equation; Eq. (4.51):*

$$\frac{p_1}{\rho} + \frac{V_1^2}{2} + gz_1 = \frac{p_2}{\rho} + \frac{V_2^2}{2} + gz_2.$$

You should choose the simplest equation that accurately represents the actual flow. You may use only *one* of the three equations, because either the assumptions and restrictions on a particular equation make it inconsistent with the others or the equations actually are identical.

Table 4.1 summarizes the assumptions, restrictions, and conditions on these equations. Suppose first that the flow in a pipe is compressible. You cannot use the mechanical energy or Bernoulli's equations; the general energy equation is the only possible choice. Next assume that the flow is incompressible but has shear stresses.

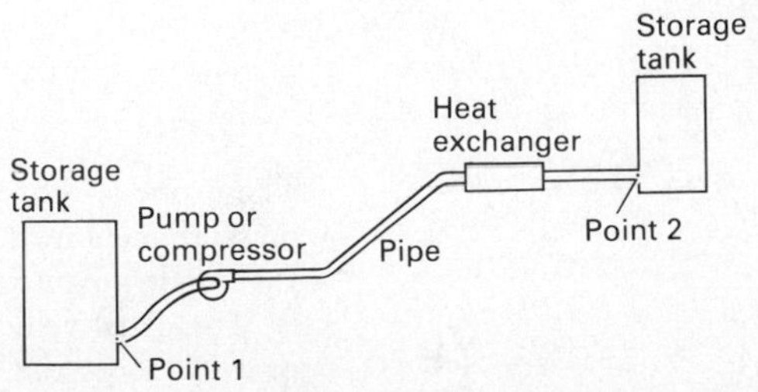

Figure 4.17 "Typical" flow system.

Table 4.1 Assumptions and conditions for the Bernoulli, general energy, and mechanical energy equations

	General Energy	Mechanical Energy	Bernoulli
Steady/unsteady	Steady only	Steady only	Steady only
Compressibility	Compressible or incompressible	Incompressible	Incompressible
Friction	Allowed	Allowed	None
Heat transfer	Allowed	Allowed	Allowed
Shaft work	Allowed	Allowed	None

Bernoulli's equation does not apply, but the general energy and mechanical energy equations do; however, if the flow is incompressible, $\rho_2 = \rho_1$ and, from Eq. (4.45),

$$gh_L = (\tilde{u}_2 - \tilde{u}_1) - q,$$

so the general energy and mechanical energy equations are actually the same. The mechanical energy equation is generally simpler because one quantity (gh_L) replaces three (q, $\tilde{u}_2$, $\tilde{u}_1$). Finally, suppose that the flow is incompressible, with no friction and no work. You would know immediately that Bernoulli's equation applies. A check of Table 4.1 shows that the other two equations are also valid; however, if there is no friction, then

$$gh_L = 0 \quad \text{and} \quad \tilde{u}_2 - \tilde{u}_1 = q.$$

Because $w_s = 0$, the general energy and mechanical energy equations reduce to Bernoulli's equation automatically.

A crucial point in choosing the proper form of the energy equation is the fluid's compressibility. Obviously, liquids usually behave as if they were incompressible. Sometimes you can treat a flowing gas as incompressible. The choice of an incompressible fluid model depends as much on the specific flow situation as it does on the fluid involved.

The most obvious criterion for incompressible flow is that the largest possible density change in the flow field be negligibly small. Examples 1.1 and 4.15 suggested this direct approach. Section 1.3.6 considered the coupling between density and pressure, and by borrowing results that we have now shown rely on Bernoulli's equation, suggested that you may neglect density variation if the dynamic pressure of the flow is considerably smaller than the bulk modulus of elasticity of the fluid.

An alternative criterion for the incompressible fluid model involves the effect of compressibility on the energy equation. If compressibility is significant, mechanical and thermal energies may be of comparable magnitude, and you must use the general energy equation. But, if mechanical and thermal energies are of different orders of magnitude, the larger form of energy dominates the general energy equation. To examine the smaller form of energy, you have to utilize separate mechanical and thermal energy equations. You may do so only for incompressible flow, using Eq. (4.44), and if needed, Eq. (4.45). Thus you should assume that the flow is incompressible if the ratio

$$\frac{\text{Typical mechanical energy}}{\text{Typical thermal energy}}$$

is very small. The typical mechanical energy is ($V^2/2$), and the typical thermal energy is ($\tilde{u}_1$). For air flowing at 50 m/s, $V^2/2$ is 1250 m^2/s^2. If the air temperature is 20°C, its internal energy (u) is approximately 210,000 m^2/s^2, so the ratio

$$\frac{V^2/2}{\tilde{u}} \approx 0.006$$

is quite small. Thus you may treat the air as if it were incompressible. At 150 m/s, however, compressibility may be significant.

The information in Table 4.1 regarding compressibility, work, friction, and heat transfer applies to the control volume and mass flow averaged forms of the energy equation [Eqs. (4.33)–(4.35), (4.38), (4.46), and (4.47)–(4.50)] as well as the streamline forms. Equations (4.33) and (4.35) may be applied to unsteady flow; all others listed apply only to steady flow. Each equation has certain restrictions and conditions and should be used only if they are valid. For example, if the fluid velocity and pressure vary across streamlines at a given cross-sectional plane, the mass averaged or control volume forms of an equation are more appropriate than the streamline form.

You still must never mix energy equations. You cannot use both Bernoulli's equation and the finite control volume general energy equation for the same region of flow; either the equations will not give independent information or one of them (Bernoulli in this case) is invalid. The streamline and mass averaged forms of an equation are either identical (if the properties and velocity are truly uniform) or inconsistent. The finite control volume and mass flow averaged forms are redundant, because one may be obtained from the other simply by multiplying or dividing by the mass flow rate.

In summary:

> One and only one form of the energy equation may be used for a particular region of flow.

4.4 THE MOMENTUM EQUATIONS

The two momentum equations express Newton's second law of motion for a fluid flowing through a finite control volume: The *linear momentum equation* relates fluid linear momentum to the forces that act on the fluid, and the *angular momentum equation* relates fluid angular momentum to the moment or torque acting on the fluid. Of the two, the linear momentum equation is the most frequently used and is often referred to simply as "the momentum equation." The following material concentrates on the linear momentum equation, although most of it applies to the angular momentum equation if "torque" is substituted for "force."

Together with the continuity and energy equations, the momentum equation is one of the cornerstones of the control volume approach. The following are the main differences between the momentum equation and the continuity and energy equations:

- The momentum equation deals with vector quantities (force and momentum), whereas the continuity and energy equations deal with scalar quantities.
- The continuity and energy equations deal with quantities that are conserved (mass and energy), whereas the momentum equation deals with a balance between entities that are not conserved (force and momentum).

The momentum equation is usually used to calculate force interactions between a moving fluid and solid objects in contact with it. Almost all control volume flow analyses rely on the continuity equation and most rely on some form of the energy equation, as a general rule, but only those analyses that require evaluation of forces require use of the momentum equation.

Unlike the energy equation, which is most conveniently expressed in several alternative forms with each form applicable to a certain type of problem, the momentum equation is best applied by starting with the general form and evaluating the terms as appropriate for the specific problem at hand. In this section, we first derive the linear momentum equation. Next we discuss evaluation of its terms and illustrate its use with fixed, rigid control volumes. Then we consider the linear momentum equation for moving and deforming control volumes. Finally, we consider briefly the angular momentum equation.

4.4.1 Derivation of the Linear Momentum Equation

We obtain the linear momentum equation by combining the transport theorem with Newton's second law of motion for a system. Figure 4.18 shows a flow field that includes a solid object in contact with a moving fluid and a control volume of

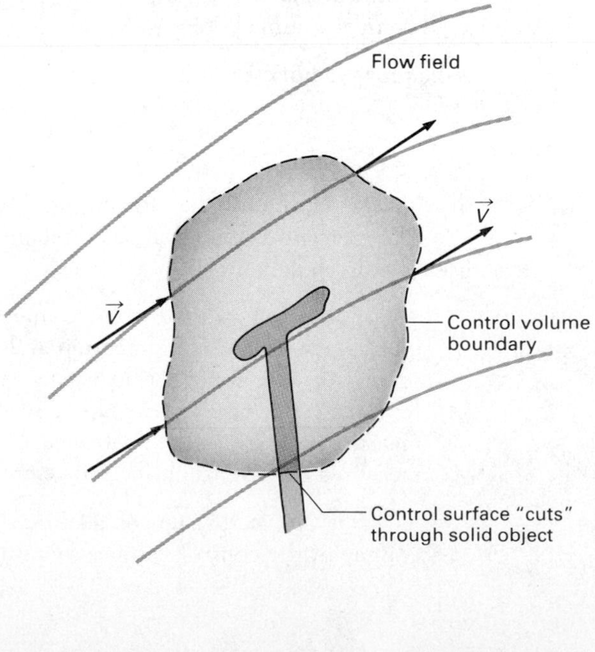

Figure 4.18 Flow field with solid object and superimposed control volume.

arbitrary shape superimposed on the flow field. Note that the control surface "cuts" through the solid object so that part of the object lies inside the control volume. As always, we consider the mass inside the control volume (including the portion of the solid object) as a system. Newton's second law for this system is

$$\frac{d\vec{M}_{\text{sys}}}{dt} = \sum \vec{F}_{\text{sys}},$$

where $\vec{M}_{\text{sys}}$ is the system momentum,

$$\vec{M}_{\text{sys}} = \int_{\forall_{\text{sys}}} \rho \vec{V}\, d\forall,$$

and $\sum \vec{F}_{\text{sys}}$ is the sum of the forces acting on the system.

To use the transport theorem, we first note that the property B is the momentum, $\vec{M}$, and that the per-unit-mass property, b, is the velocity, $\vec{V}$. Using Eq. (4.1c) gives

$$\frac{d}{dt}\int_{\forall_{cv}} \rho \vec{V}\, d\forall + \oint_{A_{cv}} \rho \vec{V}(\vec{V}_r \cdot \hat{n})\, dA = \sum \vec{F} \tag{4.56a}$$

and using Eq. (4.1b) gives

$$\frac{d}{dt}\int_{\forall_{cv}} \rho \vec{V}\, d\forall + \int_{A_{\text{out}}} \rho \vec{V}(V_n)\, dA - \int_{A_{\text{in}}} \rho \vec{V}(V_n)\, dA = \sum \vec{F}. \tag{4.56b}$$

Either of these equations may be called the linear momentum equation.

The following verbal expression is the linear momentum theorem:

$$\begin{bmatrix}\text{Rate of}\\ \text{accumulation}\\ \text{of momentum}\\ \text{inside the}\\ \text{control volume}\end{bmatrix} + \begin{bmatrix}\text{Rate of}\\ \text{momentum}\\ \text{flow leaving}\\ \text{the control}\\ \text{volume}\end{bmatrix} - \begin{bmatrix}\text{Rate of}\\ \text{momentum}\\ \text{flow entering}\\ \text{the control}\\ \text{volume}\end{bmatrix} = \begin{bmatrix}\text{Sum of}\\ \text{forces on the}\\ \text{mass inside}\\ \text{the control}\\ \text{volume}\end{bmatrix}.$$

A helpful symbolic form is

$$\frac{d}{dt}(\vec{M}_{cv}) + \dot{\vec{M}}_{\text{out}} - \dot{\vec{M}}_{\text{in}} = \sum \vec{F}, \tag{4.57}$$

in which $\dot{\vec{M}}$ represents the momentum flow.

The momentum equation has three important characteristics that anyone who uses it must understand.

- The velocity $\vec{V}$ in the volume integral and the first velocity $\vec{V}$ in the area integral represent the momentum and must be measured relative to an inertial reference. The second velocity in the area integral (written as $\vec{V}_r$ or V_n) represents the flow rate passing through the control surface and must be measured relative to the control surface. The quantities $\vec{V}_r$ and V_n need not be measured relative to an inertial reference.

- $\sum \vec{F}$ is the vector sum of all forces acting on the mass within the control volume. If the control volume contains any solid material, as in Fig. 4.18, the forces on this mass are included. The term $\sum \vec{F}$ is sometimes called "the sum of forces acting on the control volume," omitting the words "mass within." Note that forces from all sources (gravitational, pressure, magnetic, etc.) are included in $\sum \vec{F}$.

• The momentum equation is a vector equation. As such, it represents three separate component (x, y, z or r, θ, x, etc.) equations. The vector nature of the equation is embodied in $\sum \vec{F}$ and $\vec{V}$ (but not in $\vec{V}_r$ or V_n).

4.4.2 Evaluating the Terms in the Linear Momentum Equation

To apply the linear momentum equation to a particular flow problem, we must evaluate the terms in Eq. (4.57). This equation takes many different forms, depending on the specific application. Here, we examine various forms of the primary terms in the equation. We discuss separately the evaluation of the momentum accumulation term, the momentum flux term(s), and the sum-of-forces term.

Momentum Accumulation. The momentum accumulation term,

$$\frac{d}{dt}(\vec{M}_{cv}) = \frac{d}{dt}\int_{\forall_{cv}} \rho\vec{V}\,d\forall,$$

may be important if mass accumulates or depletes in the control volume or if the control volume accelerates. Accelerating control volumes introduce the special problem of noninertial coordinates. Analysis of accelerating control volumes is considered in a later section.

If the control volume is fixed and rigid, the accumulation term is

$$\frac{d}{dt}(\vec{M}_{cv}) = \frac{\partial}{\partial t}\int_{\forall_{cv}} \rho\vec{V}\,d\forall = \int_{\forall_{cv}} \frac{\partial}{\partial t}(\rho\vec{V})\,d\forall.$$

If the flow is steady, the momentum accumulation term vanishes, or

$$\frac{d}{dt}(\vec{M}_{cv}) = 0. \qquad \text{Steady flow}$$

The condition of steady flow is probably the most widely used simplification of the linear momentum equation.

Momentum Flux. The momentum flux terms are the two area integrals in Eq. (4.56b). Recall that V_n is the fluid velocity component perpendicular to the control surface and measured relative to the control surface. Although V_n has no algebraic sign, each portion of the control surface with flow crossing it must be properly identified as "inflow" or "outflow." The velocity $\vec{V}$ is measured relative to an inertial reference. If the control volume is either fixed or moving at constant velocity, it is considered to be an inertial reference, and both $\vec{V}$ and V_n are usually measured with respect to the control volume. In this case, V_n is one component of $\vec{V}$.

If the velocity is nonuniformly distributed over a portion of the control surface, analytical or numerical integration is necessary to evaluate the momentum flux.

Figure 4.19 illustrates a practical example of such a flow. The momentum equation may be used to calculate the force on the mounting struts. Note in this case that the fluid velocity vector, $\vec{V}$, has only one component and that it is equal to V_n.

In many practical cases, engineers avoid integration of velocity profiles by taking advantage of certain characteristics of a particular problem or by making some simplifying assumptions. We discuss these special cases, but in doing so, we hope you do not conclude that avoiding integration in evaluating momentum flux is always possible.

We often choose a control volume with inflow and outflow at only a finite number of locations and assume that the flow is locally uniform and one-directional at these locations (Fig. 4.20). In this case, the momentum flux may be calculated by

$$\dot{\vec{M}}_{\text{out}} - \dot{\vec{M}}_{\text{in}} = \sum_{\text{out}} (\rho V_n A)\vec{V} - \sum_{\text{in}} (\rho V_n A)\vec{V}. \tag{4.58a}$$

Locally uniform flow

The summations are taken over the number of outlets and inlets, respectively. At any outlet or inlet,

$$\dot{m} = \rho V_n A,$$

so we may also write Eq. (4.58a) as

$$\dot{\vec{M}}_{\text{out}} - \dot{\vec{M}}_{\text{in}} = \sum_{\text{out}} \dot{m}\vec{V} - \sum_{\text{in}} \dot{m}\vec{V}. \tag{4.58b}$$

Locally uniform flow

We evaluate the momentum flux—a vector—as separate components. For the flow situation shown in Fig. 4.20, the three component momentum fluxes are

$$\dot{M}_{x,\text{out}} - \dot{M}_{x,\text{in}} = \sum_{\text{out}} (\rho V_n A)u - \sum_{\text{in}} (\rho V_n A)u, \tag{4.59a}$$

$$\dot{M}_{y,\text{out}} - \dot{M}_{y,\text{in}} = \sum_{\text{out}} (\rho V_n A)v - \sum_{\text{in}} (\rho V_n A)v, \tag{4.59b}$$

$$\dot{M}_{z,\text{out}} - \dot{M}_{z,\text{in}} = \sum_{\text{out}} (\rho V_n A)w - \sum_{\text{in}} (\rho V_n A)w. \tag{4.59c}$$

Locally uniform flow

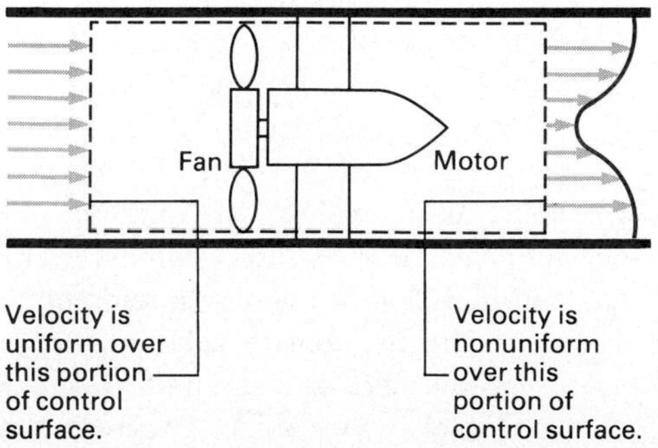

Fan
Motor
Velocity is uniform over this portion of control surface.
Velocity is nonuniform over this portion of control surface.

Figure 4.19 Control volume for analysis of flow through a fan in a duct, illustrating nonuniform velocity over part of the control surface.

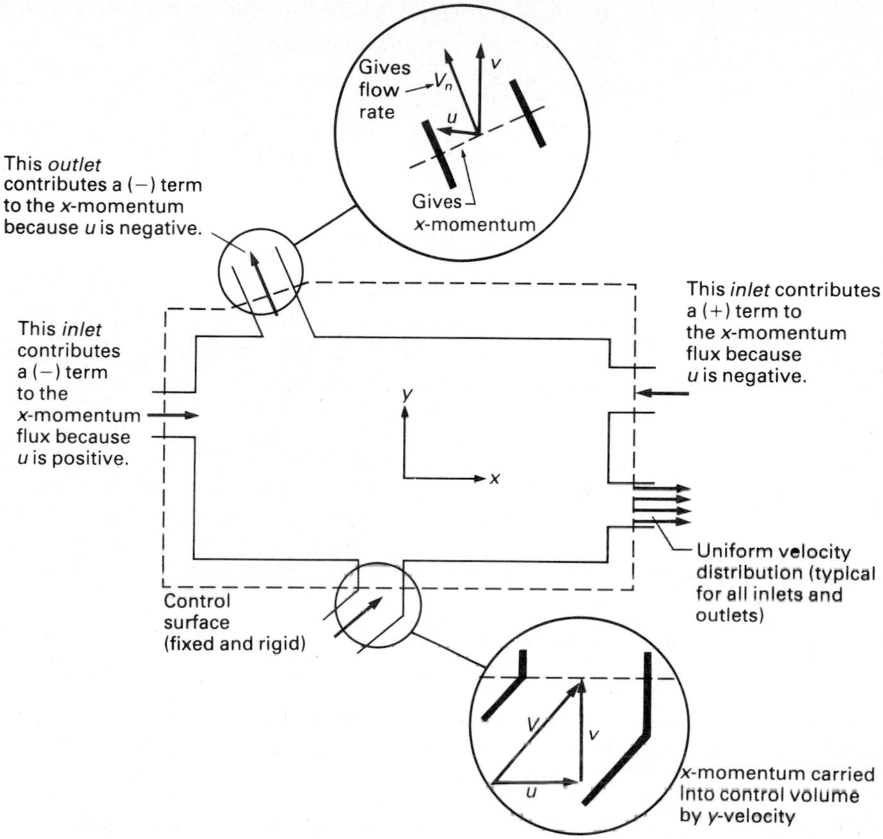

Gives
flow
rate
V_n
v
u
Gives
x-momentum
This *outlet* contributes a (−) term to the x-momentum because u is negative.
This *inlet* contributes a (+) term to the x-momentum flux because u is negative.
This *inlet* contributes a (−) term to the x-momentum flux because u is positive.
y
x
Uniform velocity distribution (typical for all inlets and outlets)
Control surface (fixed and rigid)
V
v
u
x-momentum carried into control volume by y-velocity

Figure 4.20 Control volume with a finite number of inlets and outlets, each with locally uniform and one-directional flow.

Even if the flow is not locally uniform, we may evaluate the momentum flux in terms of components. The x-momentum flux is

$$\dot{M}_{x,\text{out}} - \dot{M}_{x,\text{in}} = \int_{A_{\text{out}}} \rho u(V_n)\,dA - \int_{A_{\text{in}}} \rho u(V_n)\,dA, \tag{4.60}$$

and similar expressions give the y- and z-momentum fluxes.

When using these component equations to calculate momentum flux, you must obey the following rules:

- Calculating the mass flow part of the momentum flux ($\rho V_n A$) requires use of the relative normal velocity (V_n) in all three component equations. The velocity component (u, v, or w) is used only for the momentum per unit mass part of the momentum flux.

- Positive directions must be assigned to x, y, and z. Velocities (u, v, w) are entered as positive numbers if they are in the positive direction and as negative numbers if they are opposite this direction. As a result, any of the momentum flux terms may have a positive or negative sign. The sign is not determined only by the label "in" or "out" (see Fig. 4.20).

- The *components* of the momentum flux are combined with *components* of the momentum accumulation (if accumulation is not zero) and equated to the *components* of the sum of forces; for example,

$$\frac{d}{dt}(M_{x,cv}) + \dot{M}_{x,\text{out}} - \dot{M}_{x,\text{in}} = \sum F_x. \tag{4.61}$$

- The force terms must obey the same sign convention as that for the momentum flux and accumulation terms.

In many practical cases, the fluid velocity is locally one-directional but not uniform. Figure 4.11 illustrates this situation for flow in a circular pipe. At plane 1, the flow is one-directional (parallel to the pipe axis) but nonuniform. The momentum flux at plane 1 is

$$\dot{\vec{M}}_1 = \int \rho\vec{V}[V\cos(0)]\,dA = \left(\int \rho V^2\,dA\right)\hat{n}_1.$$

To avoid a tedious integration, replacing the nonuniformly distributed velocity with the average velocity would be desirable:

$$\bar{V} = \frac{\int V_n\,dA}{A}.$$

Unfortunately, this approximation does not give the correct momentum flux, because the real V is nonuniform. In general $\dot{\vec{M}}_1 > (\rho\bar{V}^2A)\hat{n}_1$.

We evaluate the true momentum flux by using a *momentum correction factor* (β), defined by

$$\beta \equiv \frac{\int \rho V^2\,dA}{\rho\bar{V}^2A}. \tag{4.62}$$

We then calculate the momentum flux across an arbitrary plane by*

$$\dot{\vec{M}} = \beta(\rho\bar{V}^2A)\hat{n}_v. \tag{4.63}$$

For a control volume with a finite number of inlets and outlets with locally one-directional flow,

$$\dot{\vec{M}}_{\text{out}} - \dot{\vec{M}}_{\text{in}} = \sum_{\text{out}} \beta(\rho\bar{V}^2A)\hat{n}_v - \sum_{\text{in}} \beta(\rho\bar{V}^2A)\hat{n}_v. \tag{4.64}$$

Steady flow, finite number of inlets/outlets
Locally, one-directional

The vector nature of the momentum flux is indicated by the unit vector $\hat{n}_v$, which is parallel to the local velocity vector.

For fully developed flow in a circular pipe, numerical values of β range from 1.33 for laminar flow to as low as 1.02 for turbulent flow. As β is often approximately equal to 1.0, momentum correction factors are often ignored in engineering calculations.

---

* We have only a single plane, not an entire control surface, so we do not know whether $\dot{M}$ is "in" or "out." The unit vector $\hat{n}_v$ points in the same direction as V.

EXAMPLE 4.16 Illustrates Momentum Flux and the Momentum Correction Factor

Figure E4.16 shows a constant-density fluid flowing at a steady rate in the entrance region of a circular pipe of inside diameter $2R$. The velocity profile is uniform across the inlet and parabolic at a cross section 60 diameters downstream from the inlet. Find the momentum flux and momentum correction factor at each cross section.

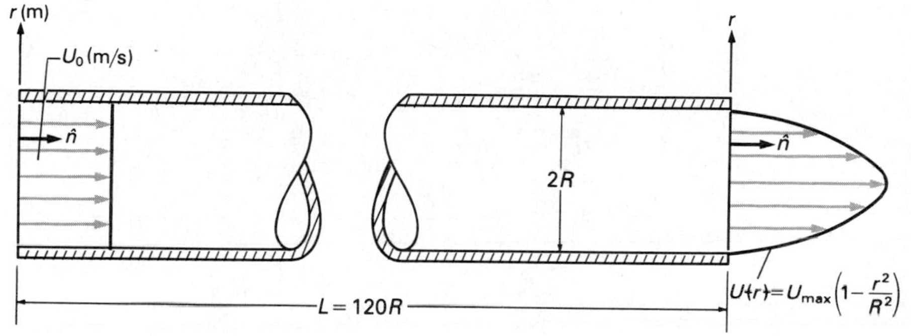

Figure E4.16 Velocity distribution at two cross sections in a circular pipe.

SOLUTION

Given

Constant-density fluid flowing at a steady rate

Velocity profile uniform at inlet and parabolic 60 diameters downstream

Find

Momentum flux and momentum correction factor at each cross section

Solution

The momentum flux is

$$\dot{\vec{M}} = \int \rho \vec{V}(V_n)\, dA.$$

Density is constant and $\hat{n}$ is taken as positive to the right (Fig. E4.16). At the inlet,

$$\vec{V} = V\hat{n}, \qquad V = V_n = U_0 \quad \text{and} \quad \dot{\vec{M}} = \int \rho U_0^2 \hat{n}\, dA.$$

Because ρ, U_0, and $\hat{n}$ are constant,

$$\dot{\vec{M}}_{\text{inlet}} = \rho U_0^2 A \hat{n}.$$

Because the flow area is circular,

$$\dot{\vec{M}}_{\text{inlet}} = \pi R^2 \rho U_0^2 \hat{n}.$$ **ANSWER**

The momentum correction factor β for a constant-density fluid is defined by Eq. (4.62). At the inlet,

$$\beta \equiv \frac{\int \rho V^2\, dA}{\rho \bar{V}^2 A} = \frac{\int \rho U_0^2\, dA}{\rho U_0^2 A},$$

so

$$\beta_{\text{inlet}} = 1. \qquad \textbf{ANSWER}$$

The momentum flux 60 diameters downstream is

$$\dot{\vec{M}} = \int \rho u^2 \hat{n}\, dA.$$

With

$$u = u_{\max}\left[1 - \left(\frac{r}{R}\right)^2\right],$$

we find $u_{\max}$ from the continuity equation:

$$\rho U_0 \pi R^2 = \int \rho u\, dA = 2\pi\rho \int_0^R u_{\max}\left(1 - \frac{r^2}{R^2}\right) r\, dr.$$

Integrating gives

$$\rho U_0 \pi R^2 = \frac{\pi \rho u_{\max} R^2}{2}.$$

Solving for $u_{\max}$ gives

$$u_{\max} = 2U_0 \quad \text{so} \quad u = 2U_0\left[1 - \left(\frac{r}{R}\right)^2\right].$$

The momentum flux then is

$$\dot{\vec{M}} = \rho \int_0^R (2U_0)^2 \left(1 - \frac{r^2}{R^2}\right)^2 2\pi r\, dr\, \hat{n}.$$

Integrating gives

$$\dot{\vec{M}} = \tfrac{4}{3}\pi R^2 U_0^2 \hat{n}. \qquad \textbf{ANSWER}$$

The momentum correction factor is

$$\beta \equiv \frac{\int \rho V^2\, dA}{\rho \bar{V}^2 A},$$

where, for our case, $\bar{V} = U_0$ and $\int \rho V^2\, dA$ is the magnitude of the momentum flux. Substituting gives

$$\beta = \frac{\frac{4}{3}\rho\pi R^2 U_0^2}{\rho U_0^2 \pi R^2},$$

so

$$\beta_{60D} = \tfrac{4}{3}. \qquad \textbf{ANSWER}$$

Discussion

Both locations have the same average velocity and $\beta_{60D} > \beta_{\text{inlet}}$, so $\dot{\vec{M}}_{60D} > \dot{\vec{M}}_{\text{inlet}}$. Equation (4.57) shows that the net force acting on the fluid between the inlet and outlet must be positive.

Sum-of-Forces Term ($\sum \vec{F}$). The term $\sum \vec{F}$ represents the sum of all forces on the mass inside the control volume. "All forces" means all types of forces and all

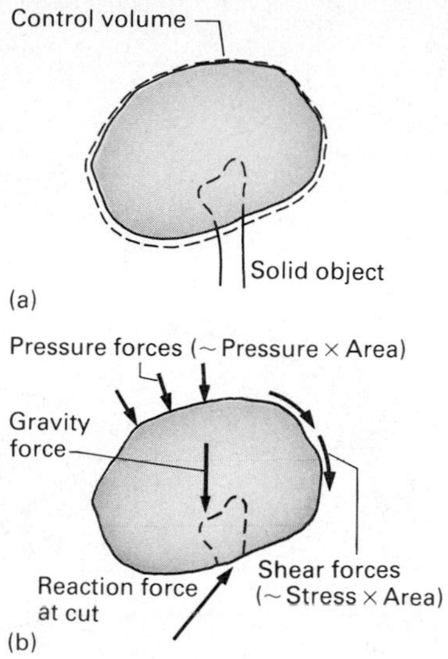

Figure 4.21 Forces acting on mass within a control volume: (a) mass in a control volume; (b) free body of mass in control volume.

locations where force may act. Evaluation of $\sum \vec{F}$ or of a specific part of $\sum \vec{F}$ is the most common objective of a control volume momentum analysis.

Considering the mass within a control volume as a free body, as in Fig. 4.21, we split the force into three types:

$$\sum \vec{F} = \sum \vec{F}_{\text{gravity}} + \sum \vec{F}_{\text{pressure}} + \sum \vec{F}_{\text{shear}}. \tag{4.65}$$

Gravity is called a *body force* because it acts on the mass as a whole. Pressure and shear are called *surface forces* because they act on the mass at the control surface where it interfaces with material outside. All surface forces inside the control volume (say, between the fluid and the portion of the solid object enclosed in the control volume) cancel because, by Newton's third law, they occur in equal and opposite pairs.

The surface forces include those where the control surface "slices" through solid objects that protrude into the fluid. These reaction forces are usually left in a single lump, so we write

$$\sum \vec{F} = \sum \vec{F}_{\text{pressure}} + \sum \vec{F}_{\text{shear (in fluid)}} + \sum \vec{F}_{\text{gravity}} + \vec{F}_R. \tag{4.66}$$

The pressure force can be evaluated by

$$\sum \vec{F}_{\text{pressure}} = -\oint_{A_{cv}} p\hat{n}\, dA. \tag{4.67}$$

The direction of the pressure force on an element dA is opposite to the outdrawn unit normal vector $\hat{n}$, because pressure is perpendicular to area and is positive for compression. Evaluation of the pressure integral is often easier than it looks. We may drop any constant uniform pressure from consideration, because the integral of a constant over a closed surface is zero—as illustrated in Fig. 4.22. Figure 4.22(a) shows part of a water jet being deflected by a turning vane. A classic fluid mechanics problem involves calculating the force that the jet exerts on the vane.

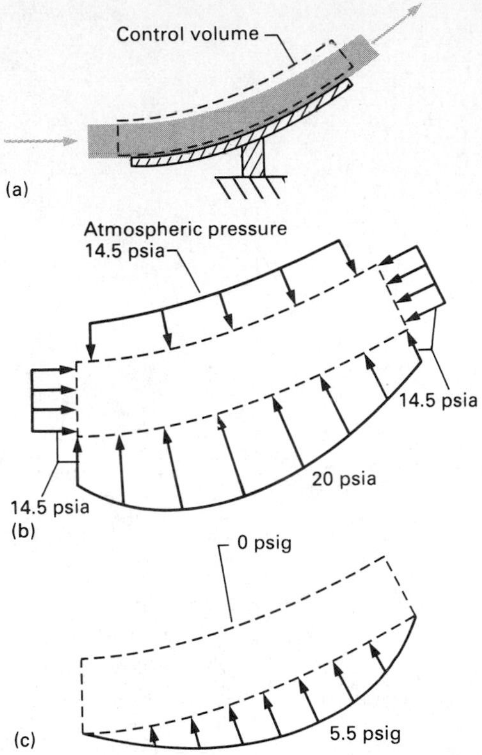

Figure 4.22 Removing part of the pressure integral: (a) jet deflection by a fixed vane; (b) pressure distribution on control volume; (c) pressure distribution with atmospheric pressure removed.

Figure 4.22(b) shows the pressure distribution around the selected control volume. We evaluate the pressure force as follows:

$$\vec{F}_{\text{pressure}} = -\oint p\hat{n}\,dA = -\oint (p - 14.5)\hat{n}\,dA - \oint (14.5)\hat{n}\,dA$$
$$= -\oint (p - 14.5)\hat{n}\,dA = -\oint p_{\text{gage}}\hat{n}\,dA,$$

where p_{gage} is zero over much of the control surface, as shown in Fig. 4.22(c), greatly simplifying the pressure force integral. Note that any uniform component of pressure may be removed in this way; it need not be atmospheric pressure.

We evaluate the shear forces on the fluid by

$$\sum \vec{F}_{\text{shear (in fluid)}} = \oint_{A_{cv,\text{ fluid}}} \vec{\tau}\,dA, \tag{4.68}$$

where $A_{cv,\text{ fluid}}$ is the portion of the control surface that is in contact with fluid and $\vec{\tau}$, the stress vector, is defined as the shear force per unit area. Because it is dependent on the specific orientation of the elemental area,* $\vec{\tau}$ is more complicated than the pressure. In problems involving the linear momentum equation, we often evaluate the shear stress force as a single lump rather than by integration. This

* If you are familiar with the concept of the stress tensor, you may know that τ is the dot product of the stress tensor and the unit normal vector $\hat{n}$.

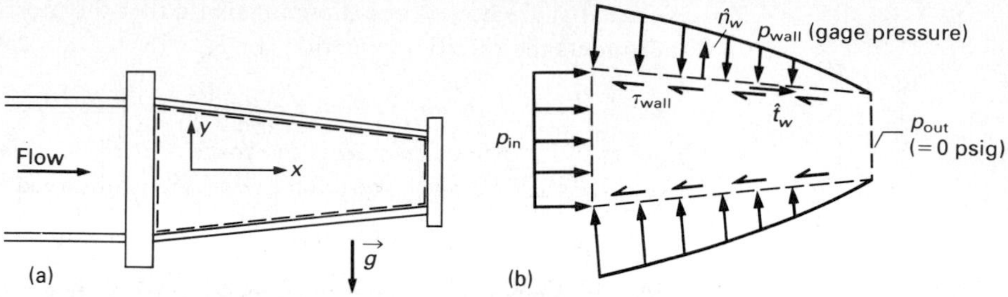

Figure 4.23 Pressure and shear stress on fluid flowing in a fire hose nozzle; (a) nozzle and control volume; (b) pressure and shear stress on fluid.

trick of leaving part of the force as a lump is also commonly applied to the pressure integral, splitting it into parts that are evaluated by integration and parts that are left as lumps. The various lump forces from the pressure and shear stress integrals may be combined into a single force. Often, calculation of this force is the objective of an analysis.

To illustrate these ideas, let's consider the force on the fire hose nozzle shown in Fig. 4.23. We select a coordinate system and control volume as shown. The sum of forces is

$$\sum \vec{F} = \sum \vec{F}_{\text{pressure}} + \sum \vec{F}_{\text{shear}} + \sum \vec{F}_{\text{gravity}}.$$

The gravity force is simply the weight, $-W\hat{j}$, of the water in the nozzle. The pressure force is

$$\sum \vec{F}_{\text{pressure}} = \left(\int_{A_{\text{in}}} p\, dA\right)\hat{i} - \left(\int_{A_{\text{out}}} p\, dA\right)\hat{i} - \int_{A_w} p\hat{n}_w\, dA.$$

The shear force is the result of shear stress at the nozzle walls; that is

$$\int_{A_w} \vec{\tau}\, dA = -\int_{A_w} \tau_w \hat{t}_w\, dA,$$

where $\hat{t}_w$ is a unit vector tangent to the wall, pointing in the flow direction.

Substituting the expressions for pressure, shear, and gravity forces, we have

$$\sum \vec{F} = \left(\int_{A_{\text{in}}} p\, dA\right)\hat{i} - \left(\int_{A_{\text{out}}} p\, dA\right)\hat{i} - W\hat{j} - \int_{A_w} \tau\hat{t}_w\, dA - \int_{A_w} p\hat{n}_w\, dA.$$

Evaluation of the last two integrals on the right requires detailed information on wall pressure and shear stress. The sum of these two integrals is the net force that the nozzle wall exerts on the fluid:

$$-\int_{A_w} \tau\hat{t}_w\, dA - \int_{A_w} p\hat{n}_w\, dA = \vec{F}_w.$$

Substituting this result, we find that the sum of forces is

$$\sum \vec{F} = \left(\int_{A_{\text{in}}} p\, dA - \int_{A_{\text{out}}} p\, dA\right)\hat{i} - W\hat{j} + \vec{F}_w.$$

In a typical problem, we would know the inlet and outlet pressures and areas and the weight, so we may move these terms to the momentum side of the linear momentum equation and calculate the unknown wall force as a single lump. If we assume that the flow is one-directional and that the pressure is uniform at the inlet and outlet, the resulting equation for F_w is

$$\vec{F}_w = (\beta_{out}\rho_{out}\bar{V}^2_{out}A_{out} - \beta_{in}\rho_{in}\bar{V}^2_{in}A_{in} + p_{out}A_{out} - p_{in}A_{in})\hat{i} + W\hat{j}.$$

Note that $\vec{F}_w$ thus calculated is the *force that the nozzle wall exerts on the fluid.* The force that the fluid exerts on the nozzle is obtained from Newton's third law:

$$\vec{F}(\text{fluid on nozzle}) = -\vec{F}_w(\text{nozzle on fluid}).$$

A final point to consider in connection with the force term is its vector nature: $\sum \vec{F}$ has x, y, and z components, $\sum F_x$, $\sum F_y$, and $\sum F_z$. These components have signs; they are positive if the force acts in the positive coordinate direction and negative if the force acts in the negative coordinate direction. If you are solving the linear momentum equation for an unknown force and do not know its direction (sign), simply assume that it is positive when writing the momentum equation. If the calculated force is positive, you made the right assumption. If the calculated force is negative, it acts in the negative coordinate direction.

EXAMPLE 4.17 Illustrates Calculation of the Force Terms for the Linear Momentum Equation

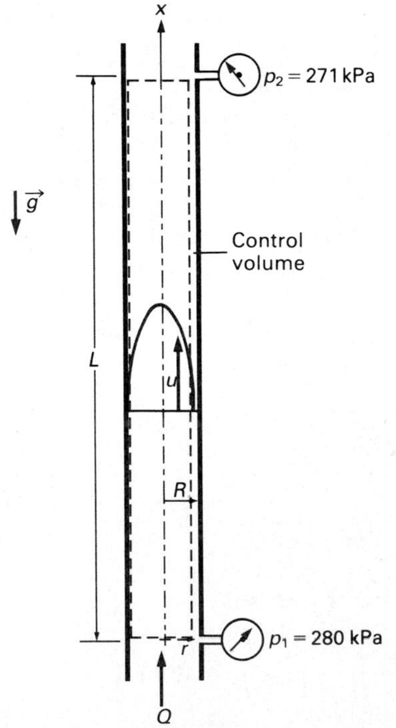

Figure E4.17 Flow of oil upward through a vertical pipe.

Lubricating oil at 20°C flows upward through a vertical pipe with an inside diameter of 2.0 cm and a length of 1.0 m. The axial velocity profile is

$$u = u_{max}\left[1 - \left(\frac{r}{R}\right)^2\right],$$

where R is the pipe radius and $u_{max} = 1.0$ m/s and is identical at each cross section. For the control volume shown in Fig. E4.17, find the net force acting on the fluid.

SOLUTION

Given

Oil at 20°C flowing upward through a vertical pipe

Pipe inside diameter 2.0 cm and length 1.0 m

Axial velocity given by

$$u = u_{max}\left[1 - \left(\frac{r}{R}\right)^2\right],$$

where R is the pipe radius and $u_{max} = 1.0$ m/s

Steady, fully developed flow

Figure E4.17

Find

Net force on fluid in control volume

Solution

In this problem, we are concerned with the force in the axial direction, $\sum F_x$. This force consists of

- pressure force $(F_x)_p$;
- shear force $(F_x)_s$; and
- gravity force $(F_x)_g$.

The pressure force is

$$(F_x)_p = \int_{A_{in}} p\,dA - \int_{A_{out}} p\,dA.$$

We assume that the pressure is uniform over each horizontal plane, so the pressure force is

$$(F_x)_p = \frac{(p_1 - p_2)(\pi)D^2}{4}.$$

The numerical values give

$$(F_x)_p = (280{,}000 - 271{,}000)\ \text{N/m}^2\ (\pi)\left[\frac{(2.0\ \text{cm})^2}{4(100\ \text{cm/m})^2}\right] = 2.83\ \text{N}.$$

The shear force is

$$(F_x)_s = \int_{A_w} \tau_w\,dA.$$

Because the oil is a Newtonian fluid,*

$$\tau_w = \mu\left(\frac{du}{dr}\right)_{r=R}.$$

Substituting for u gives

$$\tau_w = \mu u_{max}\left(\frac{-2r}{R^2}\right)_{r=R} \quad \text{or} \quad \tau_w = \frac{-2\mu u_{max}}{R}.$$

The negative sign means that the stress at the pipe wall exerts a downward force on the fluid in the control volume. For this case, the shear stress does not vary from point to point on the pipe wall, so the total shear stress force is $(F_x)_s = \tau_w A_w = \tau_w(\pi D)L = \tau_w(2\pi R)L$.

Substituting the expression for τ_w gives

$$(F_x)_s = -4\pi\mu u_{max}L.$$

Table A.5 gives $\mu = 1.31 \times 10^{-2}\ \text{N}\cdot\text{s/m}^2$. Then

$$(F_x)_s = -4\pi(1.31 \times 10^{-2}\ \text{N}\cdot\text{s/m}^2)(1.0\ \text{m/s})(1.0\ \text{m}) = -0.16\ \text{N}.$$

The gravity force is

$$(F_x)_g = -\int_{V_{cv}} \gamma\,dV = -\gamma V_{cv},$$

where γ is the specific weight and $V_{cv} = \pi R^2 L$. Thus

$$(F_x)_g = -\gamma\pi R^2 L.$$

* The shear stress on the fluid due to the wall is the negative of the shear stress on the wall due to the fluid.

Data from Table A.5 give

$$\gamma = \rho g = (871\ \text{kg/m}^3)(9.807\ \text{m/s}^2) = 8542\ \text{N/m}^3$$

and

$$(F_x)_g = \frac{-(8542\ \text{N/m}^3)(\pi)(1.0\ \text{cm})^2(1.0\ \text{m})}{(100\ \text{cm/m})^2} = -2.68\ \text{N}.$$

The net axial force $\sum F_x$ is

$$\sum F_x = (F_x)_p + (F_x)_s + (F_x)_g = (2.83\ \text{N}) + (-0.16\ \text{N}) + (-2.68\ \text{N});$$

$$\sum F_x \approx -0.01\ \text{N}.$$ **ANSWER**

Discussion

The velocity profile does not change between the inlet and the outlet of the control volume. Thus there is no momentum change, and the net axial force should be zero, by the momentum theorem. The small value that we calculated is the result of round-off error or errors in the data. Note that pressure was measured only to the nearest kPa.

4.4.3 Applying the Linear Momentum Equation

We have considered several specialized forms of the various terms in the linear momentum equation. We have deliberately not substituted all these specialized terms into the general equation to obtain specialized equations. We believe that each problem should be formulated by first writing the general expression for the linear momentum equation, Eq. (4.56) or Eq. (4.57), and then introducing appropriate specialized forms of the terms as needed. By carefully studying the following examples and working several problems, you can become adept at applying the linear momentum equation.

EXAMPLE 4.18 **Illustrates the Linear Momentum Equation and the Choice of Alternative Control Volumes**

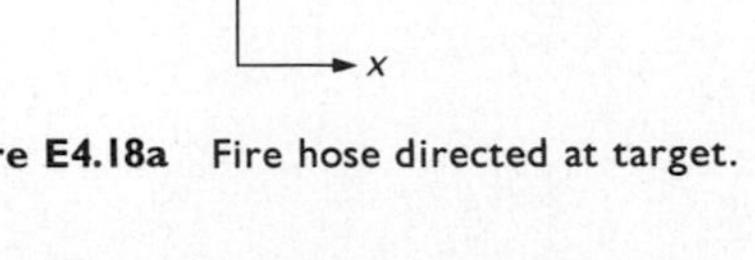

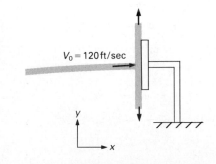

Figure E4.18a Fire hose directed at target.

The firefighters from Example 4.14 are training by directing their fire hose against the 4-ft × 6-ft rectangular target shown in Fig. E4.18a. The water leaves the fire hose at a rate of 1000 gal/min and strikes the target with a velocity of 120 ft/sec. Find the horizontal force required to anchor the target to the ground. Use each of the control volumes shown in Fig. E4.18b.

SOLUTION

Given

1000-gal/min water stream with 120-ft/sec velocity

Water stream hitting a 4-ft × 6-ft rectangular target

Figures E4.18a and E4.18b

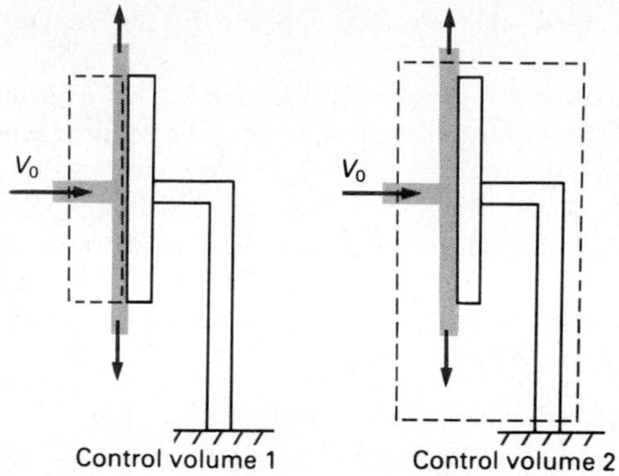

Figure E4.18b Control volumes used to find the force of the supporting pipe on the target.

Find

Anchoring force F_p for supporting pipe using each of the control volumes in Fig. E4.18b.

Solution

We first consider control volume 1. As shown in Fig. E4.18c, F_w is the horizontal force exerted on the mass in the control volume (i.e., the water) by the target. Intuition suggests that this force would act to the left; intuition, however, is not always correct, so we show F_w acting in the positive coordinate direction. We let algebraic signs tell us the direction of forces in this analysis.

Figure E4.18d is a free-body diagram of the target. The force that the water exerts on the target is equal in magnitude and opposite in direction to the force on the water. Thus *if* our assumed direction is correct, the force on the target must act as shown in Fig. E4.18d. The anchoring force on the pipe, F_p, is shown in the positive direction. Applying Newton's first law to the target, we have

$$\sum F_x = F_p + (-F_w) = 0 \quad \text{so} \quad F_p = F_w.$$

We find F_w by applying the momentum equation to the control volume. We assume steady flow, so the momentum equation is

$$\dot{M}_{x,\text{out}} - \dot{M}_{x,\text{in}} = \sum F_x.$$

Equation (4.59a) gives

$$\dot{M}_{x,\text{out}} - \dot{M}_{x,\text{in}} = \sum (\rho V_n A)_{\text{out}} u_{\text{out}} - (\rho V_n A)_{\text{in}} u_{\text{in}}.$$

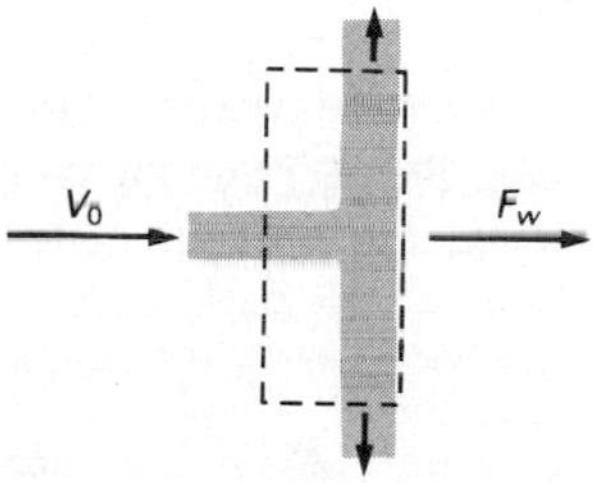

Figure E4.18c Flows and horizontal force on control volume 1.

Assuming 60°F water, we get $\rho = 62.4\ \text{lbm/ft}^3$ from Table A.6. Equation (4.14) gives

$$A_{\text{in}} = \frac{Q}{(V_n)_{\text{in}}} = \frac{(1000\ \text{gal/min})(1\ \text{min}/60\ \text{sec})}{(120\ \text{ft/sec})(7.48\ \text{gal/ft}^3)} = 0.0186\ \text{ft}^2,$$

where $(V_n)_{\text{in}} = u_{\text{in}} = V_0$. We know neither the exit area A_{out} nor the exit velocity $(V_n)_{\text{out}}$ but do not need to find them because $u_{\text{out}} = 0$. Therefore

$$\dot{M}_{x,\text{out}} - \dot{M}_{x,\text{in}} = 0 - \rho V_0 A_{\text{in}}(V_0) = -\rho A_{\text{in}} V_0^2.$$

Because we assumed that F_w acts on the control volume in the positive x direction,

$$\sum F_x = +F_w.$$

As atmospheric pressure acts uniformly around the control volume and target, there is no resultant pressure force, except for the part already included in F_w. The linear momentum equation for control volume 1 gives

$$F_w = -\rho A_{\text{in}} V_0^2 \quad \text{and} \quad F_p = -\rho A_{\text{in}} V_0^2.$$

The numerical values give

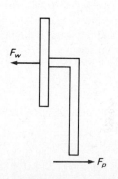

Figure E4.18d Free-body diagram of target.

$$F_P = \frac{-(62.4\ \text{lbm/ft}^3)(0.0186\ \text{ft}^2)(120\ \text{ft/sec})^2}{32.2\ \text{ft}\cdot\text{lbm/lb}\cdot\text{sec}^2};$$

$$F_P = -519\ \text{lb}.$$ **ANSWER**

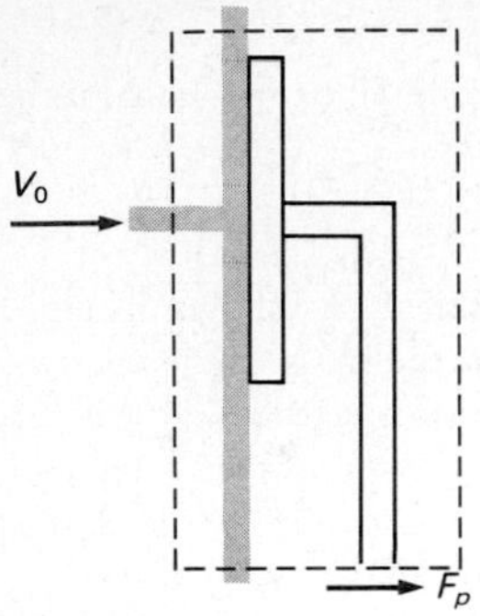

Figure E4.18e Force and flows for control volume 2.

The negative sign means that F_P acts opposite to the assumed direction.

We now repeat the analysis using control volume 2. Figure E4.18e shows the forces and flows for this control volume. The desired pipe anchoring force now appears explicitly where the control volume slices through the pipe. We show the force exerted on the pipe by the ground as acting in the positive direction, although by now we are fairly sure that it acts in the other direction. We apply the momentum equation to control volume 2, using the same assumptions and deductions as for control volume 1 to get

$$\dot{M}_{x,\text{out}} - \dot{M}_{x,\text{in}} = 0 - \rho A_{\text{in}} V_0(V_0) \quad \text{and} \quad \sum F_x = F_P.$$

The linear momentum equation gives

$$F_P = -\rho A_{\text{in}} V_0^2.$$

This result is the same as before. The numerical result then is

$$F_P = -519 \text{ lb.}$$ **ANSWER**

Discussion

Both choices of control volume yielded the same result, so either choice is correct. Control volume 2 required less work because the force that we wanted to calculate appeared explicitly in the momentum equation. The reason is that the control volume boundary was located at the junction of the pipe and the ground. For control volume 2, the intermediate forces that we had to consider for control volume 1 (i.e., the forces between the fluid and the target) did not appear in the calculation, because they were "buried" inside the control volume.

EXAMPLE 4.19 Illustrates the Linear Momentum Equation with Nonuniform Pressure Distribution

Figure E4.19a shows a transition section in a small canal. Upstream from the transition, the canal is 5 m wide, the water is 4 m deep, and the water velocity is 0.75 m/s. Leaving the transition and downstream from it, the canal is 1.5 m wide and the water is 3.6 m deep. Find the force on the walls of the transition section.

SOLUTION

Given

Water with $V_1 = 0.75$ m/s flowing through transition, as shown in Fig. E4.19a

Find

Force on transition walls

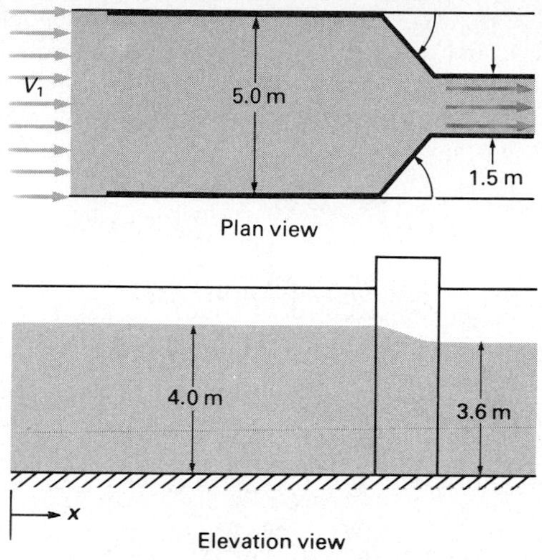

Figure E4.19a Water flow in a canal transition.

Solution

The control volume and relevant details are shown in Fig. E4.19b.

The linear momentum equation, written for the flow direction, is

$$\dot{M}_{x,\text{out}} - \dot{M}_{x,\text{in}} = \sum F_x.$$

We assume uniform velocity over both inlet and outlet, so the momentum flux terms are

$$\dot{M}_{x,\text{out}} - \dot{M}_{x,\text{in}} = \dot{m}_{\text{out}} V_{x,\text{out}} - \dot{m}_{\text{in}} V_{x,\text{in}}.$$

For steady flow, the continuity equation gives

$$\dot{m}_{\text{in}} = \dot{m}_{\text{out}} = \dot{m},$$

and for constant density,

$$V_{\text{in}} A_{\text{in}} = V_{\text{out}} A_{\text{out}},$$

so

$$V_1 A_1 = V_2 A_2 \quad \text{and} \quad V_2 = \left(\frac{A_1}{A_2}\right) V_1.$$

The momentum flux terms become

$$\dot{M}_{x,\text{out}} - \dot{M}_{x,\text{in}} = \dot{m} V_1 \left(\frac{A_1}{A_2} - 1\right) = \rho V_1^2 (W_1 h_1) \left(\frac{W_1 h_1}{W_2 h_2} - 1\right).$$

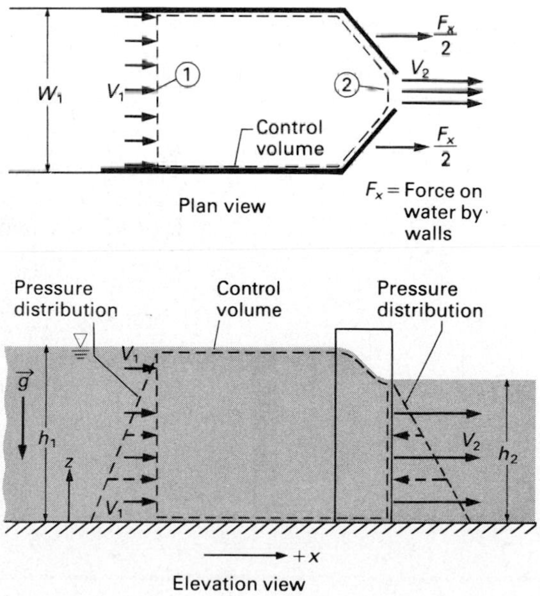

Figure E4.19b Control volume.

The force term includes the force F_x from the walls and the force from pressure at the inlet and outlet of the control volume. To evaluate the pressure force on the fluid, we must know the pressure distribution. The pressure along the free surface is atmospheric, which we may conveniently take as zero. To evaluate the pressure distribution over the inlet, plane 1, and the outlet, plane 2, we must borrow a result from Chapter 5 and treat it as an assumption. If the streamlines are straight, the pressure variation across streamlines is hydrostatic. Therefore

$$p_1 = \gamma(h_1 - z) \quad \text{and} \quad p_2 = \gamma(h_2 - z).$$

The net pressure force is

$$F_{\text{pressure}} = \int_0^{h_1} p_1 W_1 \, dz - \int_0^{h_2} p_2 W_2 \, dz.$$

Substituting the pressure distributions and integrating gives

$$F_{\text{pressure}} = \frac{\gamma h_1^2 W_1}{2} - \frac{\gamma h_2^2 W_2}{2}.$$

The sum of forces is

$$\sum F_x = F_x + \frac{\gamma h_1^2 W_1}{2} - \frac{\gamma h_2^2 W_2}{2}.$$

Equating this expression to the momentum flux and solving for F_x gives

$$F_x = \rho V_1^2 W_1 h_1 \left(\frac{h_1 W_1}{h_2 W_2} - 1 \right) + \frac{\gamma}{2} (h_2^2 W_2 - h_1^2 W_1).$$

Substituting numerical values, we obtain

$$F_x = (1000 \text{ kg/m}^3)(0.75 \text{ m/s})^2 (5 \text{ m})(4 \text{ m}) \left[\frac{(4)(5)}{(3.6)(1.5)} - 1 \right]$$

$$+ \frac{9810 \text{ N/m}^3}{2} [(3.6 \text{ m})^2 (1.5 \text{ m}) - (4 \text{ m})^2 (5 \text{ m})] = -267 \text{ kN}.$$

The negative sign means that the force acts to the left, which is logical because the transition walls are holding the water back. The force of the water on the walls is equal and opposite, or

$$F_{\text{H}_2\text{O on walls}} = 267 \text{ KN} \qquad \text{(to the right).}$$ **ANSWER**

Discussion

The pressure distribution is hydrostatic, so we could have borrowed results from Chapter 2 to calculate the pressure force on the inlet and outlet areas:

$$F_{\text{pressure}} = \gamma(\text{Area}) \left(\frac{\text{Depth}}{2} \right).$$

EXAMPLE 4.20 Illustrates the Vector Nature of the Linear Momentum Equation

A carbon steel, 14-in., schedule 30, 90° elbow is to be butt welded to a pipe carrying water at 4000 gal/min. Find the force required in the weld to support the elbow. The elbow weighs 150 lb. Figure E4.20a shows the pipe and elbow configuration.

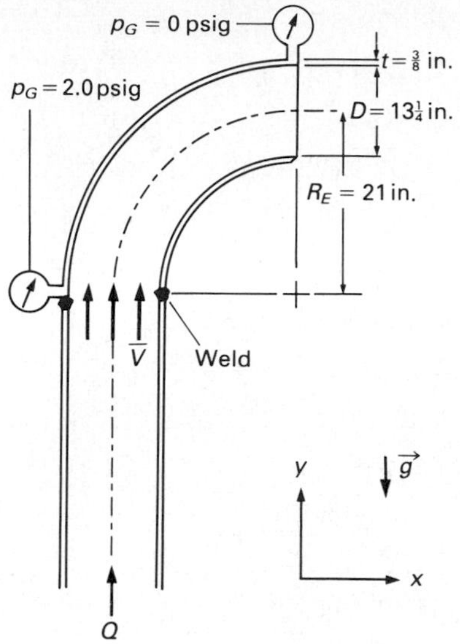

Figure E4.20a Water flowing through 90° elbow butt welded to 14-in. pipe.

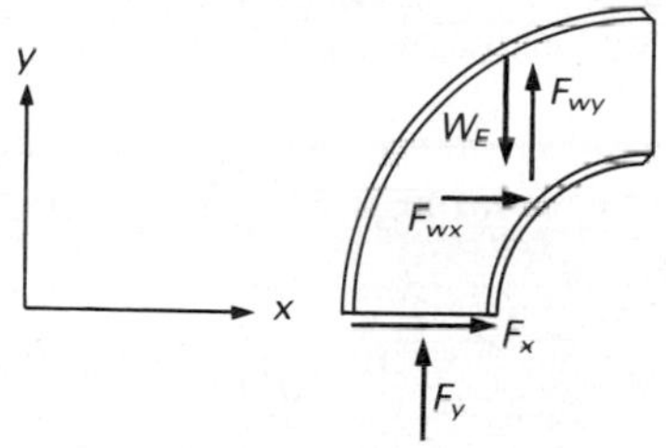

Figure E4.20b Forces on 90° elbow.

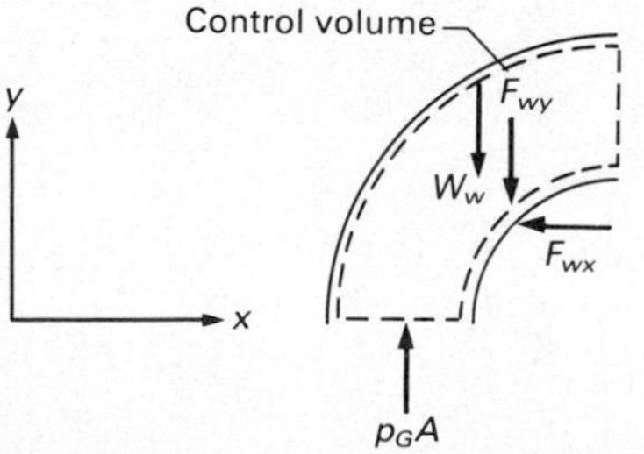

Figure E4.20c Forces on water in control volume.

SOLUTION

Given

Figure E4.20a

Carbon steel, 14-in., schedule 30, 90° elbow for butt welding

Water flow rate 4000 gal/min

Elbow weight 150 lb

Find

Force required in butt weld to support elbow

Solution

Figure E4.20b is a free-body diagram of the elbow. The following forces are involved:

F_y = Vertical component of the force of the weld on the elbow.

F_x = Horizontal component of the force of the weld on the elbow.

W_E = Force of gravity on the elbow.

F_{wy} = Vertical component of the force of the water on the elbow.

F_{wx} = Horizontal component of the force of the water on the elbow.

Applying Newton's first law in the horizontal (x) direction gives

$$F_x = -F_{wx}.$$

Applying Newton's first law in the vertical (y) direction gives

$$F_y = W_E - F_{wy}.$$

We evaluate F_{wy} and F_{wx} by applying the linear momentum equation to a control volume enclosing the water in the elbow (Fig. E4.20c). For steady flow, the x and y components of Eq. (4.57) are

$$\dot{M}_{x,\text{out}} - \dot{M}_{x,\text{in}} = \sum F_x \quad \text{and} \quad \dot{M}_{y,\text{out}} - \dot{M}_{y,\text{in}} = \sum F_y.$$

Using Newton's third law, we obtain the force terms:

$$\sum F_x = -F_{wx} \quad \text{and} \quad \sum F_y = p_G A - W_w - F_{wy},$$

where p_G is the gage pressure and W_w is the force of gravity on the water in the elbow.

We now assume that the velocity profile is uniform at the inlet and outlet ($\beta_{\text{in}} = \beta_{\text{out}} = 1$). With this assumption, Eqs. (4.59a) and (4.59b) give

$$\dot{M}_{x,\text{out}} - \dot{M}_{x,\text{in}} = (\rho V_n A u)_{\text{out}} - (\rho V_n A u)_{\text{in}}$$

and

$$\dot{M}_{y,\text{out}} - \dot{M}_{y,\text{in}} = (\rho V_n A v)_{\text{out}} - (\rho V_n A v)_{\text{in}}.$$

The continuity equation gives

$$(\rho V_n A)_{\text{in}} = (\rho V_n A)_{\text{out}}.$$

Assuming constant water density and noting that

$$A_{\text{in}} = A_{\text{out}} = A,$$

we find that

$$(V_n)_{\text{in}} = (V_n)_{\text{out}} = \bar{V}.$$

The momentum flux terms then are

$$\dot{M}_{x,\text{out}} - \dot{M}_{x,\text{in}} = (\rho\bar{V}A)(\bar{V}) - (\rho\bar{V}A)(0) = \rho\bar{V}^2A$$

and

$$\dot{M}_{y,\text{out}} - \dot{M}_{y,\text{in}} = (\rho\bar{V}A)(0) - (\rho\bar{V}A)(\bar{V}) = -\rho\bar{V}^2A.$$

Substituting into the x-momentum equation gives

$$F_{wx} = -\rho\bar{V}^2A \quad \text{so} \quad F_x = \rho\bar{V}^2A.$$

The numerical values give

$$A = \frac{\pi D^2}{4} = \frac{\pi(13.25 \text{ in.})^2}{4(144 \text{ in}^2/\text{ft}^2)} = 0.958 \text{ ft}^2$$

and

$$\bar{V} = \frac{Q}{A} = \frac{(4000 \text{ gal/min})(1 \text{ ft}^3/7.48 \text{ gal})(1 \text{ min}/60 \text{ sec})}{0.958 \text{ ft}^2} = 9.31 \text{ ft/sec}.$$

Assuming 60°F water and using Table A.6, $\rho = 62.3$ lbm/ft^3. Then

$$F_x = \frac{(62.3 \text{ lbm/ft}^3)(9.31 \text{ ft/sec})^2(0.958 \text{ ft}^2)}{32.2 \text{ ft}\cdot\text{lbm/lb}\cdot\text{sec}^2},$$

or

$$F_x = 161 \text{ lb.}$$ **ANSWER**

Because it is positive, F_x acts to the right.

Substituting the y-momentum flux term and the y forces into the y-linear momentum equation gives

$$-\rho\bar{V}^2A = p_GA - W_w - F_{wy} \quad \text{or} \quad F_{wy} = \rho\bar{V}^2A + p_GA - W_w,$$

and

$$F_y = W_E - \rho\bar{V}^2A - p_GA + W_w.$$

We obtain the approximate weight of the water in the elbow by estimating the inside volume of the elbow and multiplying it by the specific weight of the water. The water volume is approximately

$$\forall_w \approx \frac{\pi D^2}{4}(L) \approx \frac{\pi D^2}{4}\left(\frac{\pi R_E}{2}\right),$$

where L is the length of the centerline and is calculated from the radius of curvature of the elbow. Then

$$W_w = \gamma\forall_w \approx (62.3 \text{ lb/ft}^3)\left(\frac{\pi^2}{8}\right)\left[\frac{(13.25 \text{ in.})^2(21 \text{ in.})}{1728 \text{ in}^3/\text{ft}^3}\right] \approx 164 \text{ lb.}$$

The y-momentum flux term is

$$\rho\bar{V}^2A = \frac{(62.3 \text{ lbm/ft}^3)(9.31 \text{ ft/sec})^2(0.958 \text{ ft}^2)}{(32.2 \text{ ft}\cdot\text{lbm/lb}\cdot\text{sec}^2)} = 161 \text{ lb.}$$

We now compute F_y:

$$F_y = 150\text{ lb} - 161\text{ lb} - \frac{(2.0\text{ lb/in}^2)(0.958\text{ ft}^2)}{(1\text{ ft}^2/144\text{ in}^2)} + 164\text{ lb},$$

or

$$F_y = -123\text{ lb}.$$ **ANSWER**

Because it is negative, F_y acts downward.

Discussion

Note that a force acts in the direction assumed if calculated to be positive and acts opposite to the direction assumed if calculated to be negative. For convenience, always assume an unknown force to be in the positive coordinate direction.

Also note that the linear momentum equation is a vector equation, has three components, and may be written in any of the three directions. You might try writing the linear momentum equation in the z direction and showing that $F_z = 0$. Another interesting exercise would be to include realistic values for the momentum correction factor (say, $\beta \approx 1.05$ for turbulent flow) in the analysis.

EXAMPLE 4.21 Illustrates the Generality of the Linear Momentum Equation

Figure E4.21a shows a 400-m-long mechanized walkway in an airport. The walkway moves at a speed of 4 km/hr and transports people and their luggage from one location to another. Some 2000 persons per hour use the walkway at peak traffic times. Estimate the force that must be supplied to the walkway by the five drive cylinders. Neglect friction.

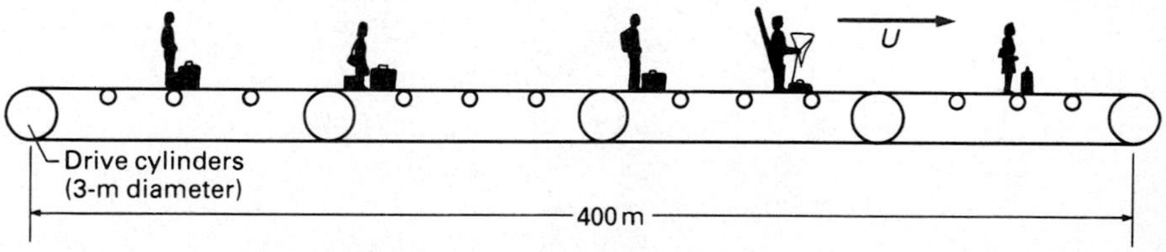

Figure E4.21a Mechanized walkway at an airport.

SOLUTION

Given

400-m mechanized walkway moving at 4 km/hr

Peak traffic rate 2000 persons per hour

Figure E4.21a

Find

Force that must be supplied to walkway by five drive cylinders

Solution

We assume that people enter and leave the walkway at the same rate, so that a steady-flow situation exists during the peak traffic period.

We apply the linear momentum equation to a control volume enclosing the walkway and write it in the direction of travel (the x direction; see Fig. E4.21b):

$$\dot{M}_{x,\text{out}} - \dot{M}_{x,\text{in}} = \sum F_x,$$

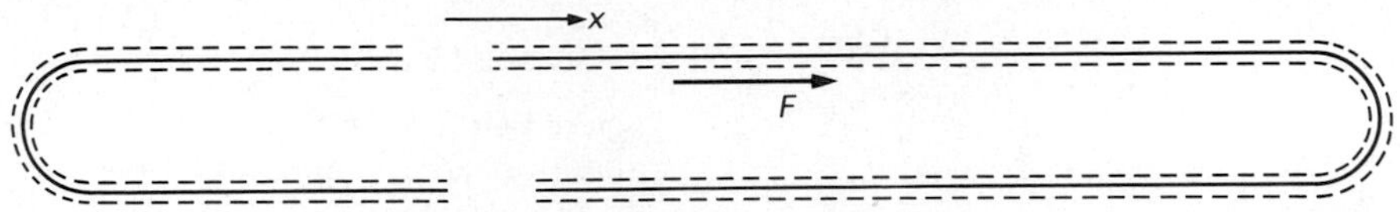

Figure E4.21b Control volume for mechanized walkway. (Drawings not to scale.)

but

$$\dot{M}_{x,\text{out}} - \dot{M}_{x,\text{in}} = \sum_{\text{out}} \dot{m}u - \sum_{\text{in}} \dot{m}u.$$

We assume that the people and luggage moving onto the walkway have no x-velocity component and that any people and luggage leaving the walkway have an x-velocity component equal to that of the walkway. Therefore

$$u_{\text{out}} = U \quad \text{and} \quad u_{\text{in}} = 0.$$

We take $\dot{m}_{\text{in}}$ and $\dot{m}_{\text{out}}$ to represent the rate of mass entering and leaving the control volume, specifically the mass of people and their luggage moving onto or off the walkway in unit time. Assuming an average mass of 75 kg per person (with luggage), we get

$$\dot{m}_{\text{out}} = \dot{m}_{\text{in}} \approx (75\ \text{kg/person})(2000\ \text{persons/hr})(1\ \text{hr}/3600\ \text{s})$$
$$= 41.67\ \text{kg/s}.$$

The linear momentum theorem gives

$$\dot{m}(U) - \dot{m}(0) = F \quad \text{or} \quad F = \dot{m}U.$$

The numerical values give

$$F = (41.67\ \text{kg/s})(4\ \text{km/hr})(10^3\ \text{m/km})(1\ \text{hr}/3600\ \text{s})(1\ \text{N s}^2/\text{kg m});$$
$$F = 46.3\ \text{N}.$$ **ANSWER**

Discussion

Although this was not a fluid mechanics problem, the principles of fluid mechanics (developed from Newton's second law) applied. This situation is not uncommon; you will encounter similar situations in your professional career.

The calculated force is small. The actual force would be considerably larger because of friction in the system and aerodynamic drag on the people being transported. The calculated force represents the force required to accelerate the passengers to the walkway's speed.

We were not given one critical item of information needed to solve this problem: the average mass of a person with luggage. The assumed value of 75 kg seems reasonable but may be in error by ± 10 percent. The calculated force also contains this error. Engineers are often obliged to assume reasonable values for unavailable information. In such cases, they must use the results with caution.

4.4.4 Linear Momentum Equation for Moving and Deforming Control Volumes

The general form of the linear momentum equation, Eq. (4.56) or Eq. (4.57), is applicable to moving or deforming control volumes if the velocities are measured with respect to an inertial reference. We consider separately the constant-velocity control volume, deforming control volume, and accelerating control volume.

Constant-Velocity Control Volume. The most common moving control volume application involves a control volume with constant velocity. Suppose that an engineer needs to analyze the flow through an airplane's jet engines. The logical control volume choice is one that encloses the engine and travels along with it (Fig. 4.24). If the airplane is cruising at constant velocity, the control volume also moves with constant velocity. One way to handle this problem is to add a uniform velocity equal and opposite to the airplane velocity at every point in the field, which brings the airplane, and hence the control volume, to rest. This method is equivalent to attaching a coordinate system to the control volume (airplane). In such a coordinate system, the control volume is fixed, so the methods of the last section can be employed. The forces acting on the material in the control volume are unaffected by this transformation. This approach is called a *Galilean transformation.*

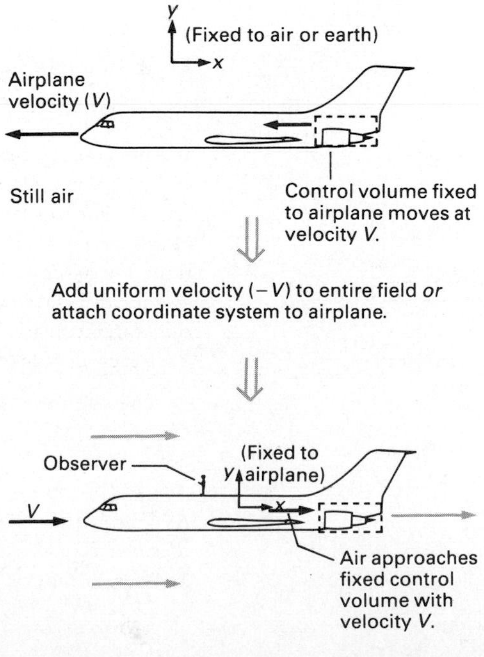

Figure 4.24 Transforming moving control volume to fixed control volume.

Another way to approach this problem is to imagine an observer riding on the airplane. This observer sees the air approaching the airplane with a velocity that is equal in magnitude and opposite in direction to the airplane's velocity. The observer naturally chooses a coordinate system that is fixed to the airplane, draws a control volume around the engine, and proceeds to analyze the flow through a fixed control volume. If the airplane is not accelerating, the observer may use Eq. (4.56), or any of its simplified forms, with the velocities defined with respect to the airplane-based coordinates. The forces acting on the material in the control volume are unaffected by a Galilean transformation or by the motion of the observer.

EXAMPLE 4.22 Illustrates Application of the Linear Momentum Equation to a Control Volume Moving with Constant Velocity

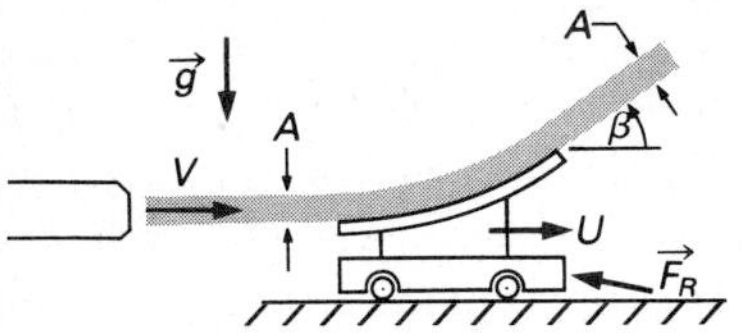

Figure E4.22a Water impinging on a moving vane.

The curved vane in Fig. E4.22a moves with a constant absolute velocity U along a horizontal frictionless surface. Water flows from the nozzle with a constant absolute velocity V and impinges on the vane. Develop an expression for the force $\vec{F}_R$ resisting the motion of the vane. Neglect the force of gravity.

SOLUTION

Given

Curved vane in Fig. E4.22a

Absolute velocity U along horizontal frictionless surface

Water with absolute horizontal velocity V impinging on vane

Find

Expression for force $\vec{F}_R$ resisting motion of vane

Solution

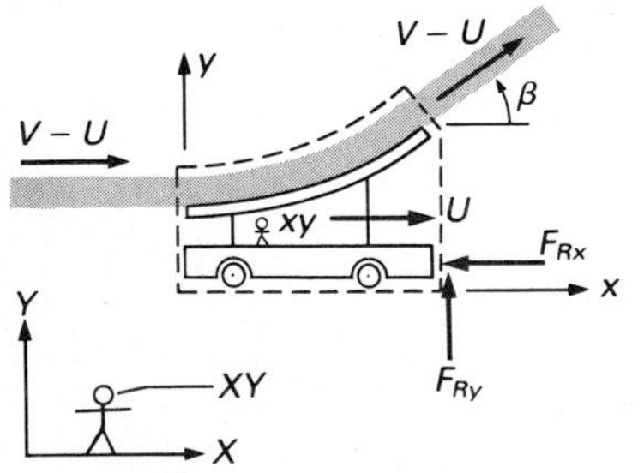

Figure E4.22b Control volume moving with vane and observer *xy*.

We imagine an observer riding on the vane and choose a coordinate system (x, y) and control volume fixed to the vane, as shown in Fig. E4.22b. For this control volume, we may remove the uniform atmospheric pressure acting on the entire control surface from consideration. We assume that the velocity V is greater than the velocity U, so the vane does not outrun the water. In addition, we assume that the water density is constant and that fluid velocities are uniform over each flow area.

The appropriate principle is the linear momentum equation:

$$\frac{d}{dt}(\vec{M}_{cv}) + \dot{\vec{M}}_{\text{out}} - \dot{\vec{M}}_{\text{in}} = \sum \vec{F}.$$

As V and U are constant and the vane does not outrun the water, the observer sees the identical flow situation inside the control volume at every instant of time and concludes that

$$\frac{d}{dt}(\vec{M}_{cv}) = 0.$$

The observer then uses Eqs. (4.59a) and (4.59b) to obtain

$$\sum_{\text{out}} (\rho V_n A)u - \sum_{\text{in}} (\rho V_n A)u = -F_{Rx} \quad \text{and}$$

$$\sum_{\text{out}} (\rho V_n A)v - \sum_{\text{in}} (\rho V_n A)v = F_{Ry},$$

where ρ, A_{in}, and A_{out} are given information.

If we neglect fluid friction between the water jet and the vane, the vane does not change the speed of the fluid relative to the vane; the vane changes only the direction of flow. To the observer on the vane, the water enters with speed $(V - U)$, parallel to the x direction, and leaves with speed $(V - U)$, at angle β. The velocities for the momentum equations are

$$V_{n,\text{in}} = V - U, \qquad V_{n,\text{out}} = V - U,$$

$$u_{\text{in}} = V - U, \qquad v_{\text{in}} = 0,$$

$$u_{\text{out}} = (V - U)\cos\beta, \quad \text{and} \quad v_{\text{out}} = (V - U)\sin\beta.$$

The forces are independent of the observer (and the coordinate system). The x and y components, again denoted F_{Rx} and F_{Ry}, are shown in Fig. E4.22(b).

Substituting these expressions into the x component of the linear momentum theorem gives

$$\rho(V - U)A(V - U)\cos\beta - \rho(V - U)A(V - U) = -F_{Rx},$$

or

$$F_{Rx} = \rho A(V - U)^2(1 - \cos\beta). \qquad \textbf{ANSWER}$$

Substituting the appropriate expressions into the y component, the linear momentum equation gives

$$\rho(V - U)A(V - U)\sin\beta - \rho(V - U)A(0) = F_{Ry},$$

or

$$F_{Ry} = \rho A(V - U)^2 \sin\beta. \qquad \textbf{ANSWER}$$

Discussion

If we apply Bernoulli's equation to the streamline at the jet surface and note that this streamline has every point at atmospheric pressure, we can prove the assertion that the jet's relative speed is unaffected by passage over the vane, in the absence of friction.

If there were no resisting force, the vane would accelerate to the right until $U = V$. The resisting force may be caused by friction or an external load attached to the vane. The water jet-vane system generates power in the amount of:

$$\dot{W} = F_{Rx}U = \rho AU(V - U)^2(1 - \cos\beta).$$

Deforming Control Volumes. The most common application of a deforming control volume involves situations where the control volume must grow or shrink to completely contain the fluid. The control volume inside the piston–cylinder shown in Fig. 4.1 is a typical example. Mass and momentum often accumulate within deforming control volumes, so the first term of the linear momentum equation,

$$\frac{d}{dt}(\vec{M}_{cv}) = \frac{d}{dt}\int_{\forall_{cv}} \rho\vec{V}\,d\forall,$$

must be carefully evaluated.

EXAMPLE 4.23 Illustrates a Deforming Control Volume

Redo Example 4.22, using a control volume that is fixed to the nozzle at one end and to the moving cart at the other end so that the control volume stretches as the cart moves. Find only F_{Rx}.

SOLUTION

Given

Situation identical to that in Example 4.22

Control volume fixed at one end and moving at other

Find

Force F_{Rx}

Solution

The specified control volume is shown in Fig. E4.23. We choose a coordinate system fixed to the nozzle and the ground. The water enters this control volume at the nozzle with velocity V in the x direction. In Example 4.22, we found that the speed of the water relative to the vane is unchanged by passage over the vane. The velocity of the water leaving the vane and control volume is the vector sum of the vane speed, U, in the x direction and speed $(V - U)$ at angle β.

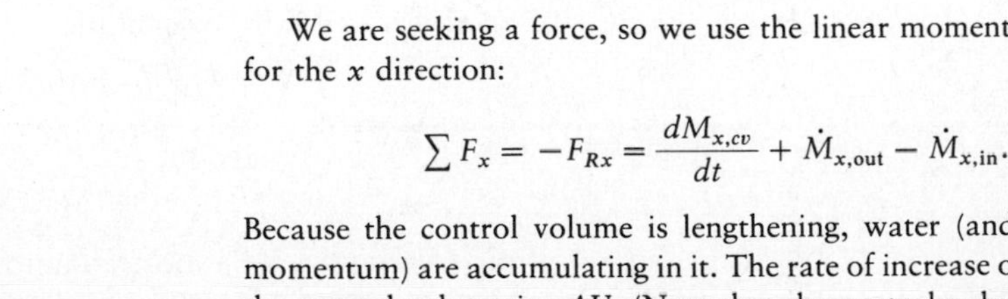

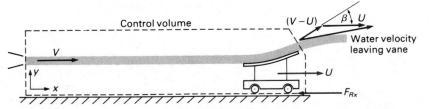

Figure E4.23 Control volume fixed to both nozzle and vane.

We are seeking a force, so we use the linear momentum equation for the x direction:

$$\sum F_x = -F_{Rx} = \frac{dM_{x,cv}}{dt} + \dot{M}_{x,\text{out}} - \dot{M}_{x,\text{in}}.$$

Because the control volume is lengthening, water (and the water's momentum) are accumulating in it. The rate of increase of mass inside the control volume is ρAU. (Note that the control volume lengthens at the cart's speed, not at the jet's speed.) The water's momentum per unit mass is V, so

$$\frac{dM_{x,cv}}{dt} = \rho AUV.$$

The velocities for evaluating the momentum fluxes are

$$V_{n,\text{in}} = V, \qquad V_{n,\text{out}} = V - U,$$
$$u_{\text{in}} = V, \quad \text{and} \quad u_{\text{out}} = U + (V - U)\cos\beta.$$

Substituting into the momentum equation gives

$$-F_{Rx} = \rho AUV + \rho A(V - U)[U + (V - U)\cos\beta] - \rho AV^2.$$

Multiplying by -1 and simplifying yield

$$F_{Rx} = \rho A(V - U)^2(1 - \cos\beta).$$

Discussion

Fortunately, we got the same result as in Example 4.22. You may form your own opinion, but we believe that the simpler analysis used the control volume moving with the cart rather than the deforming control volume. In certain cases, however, choosing a deforming control volume simplifies the analysis.

Accelerating Control Volumes. Applying the linear momentum equation to accelerating control volumes sometimes is necessary. An engineer who designs pumps often uses a control volume fixed to a rotating impeller. When analyzing atmospheric flows for weather prediction, a meteorologist usually must account for the fact that a control volume fixed to the surface of the earth has a (small) acceleration with respect to the fixed stars because of the Earth's rotation. Momentum analysis of fluid motion through an accelerating control volume is complicated by the fact that a coordinate system fixed with respect to the control volume is noninertial.

An engineer may select a control volume that is fixed in a noninertial coordinate system, because the flows into and out of the control volume may be expressed conveniently in this system. Figure 4.25 shows a control volume that is fixed with respect to a noninertial reference frame, xyz. The noninertial frame moves with respect to an inertial reference frame, XYZ. We can apply Newton's second law only in the XYZ frame. For the system that instantaneously occupies the control volume,

$$\frac{d}{dt}(\vec{M}_{\text{sys}})_{XYZ} = \sum \vec{F}.$$

To derive an expression for the rate of change of system momentum, we first consider a single mass particle within the control volume. The momentum of this particle is

$$\delta\vec{M}_{XYZ} = (\delta m)\vec{V}_{XYZ}.$$

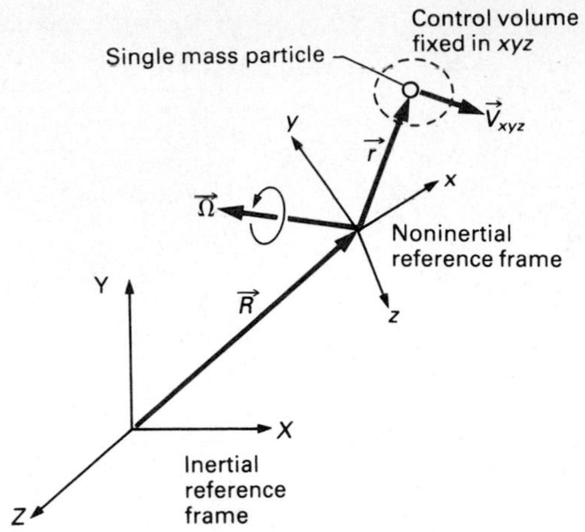

Figure 4.25 Control volume in noninertial reference frame.

From elementary dynamics, we have

$$\vec{V}_{XYZ} = \vec{V}_{xyz} + \frac{d\vec{R}}{dt} + \vec{\Omega} \times \vec{r}. \tag{4.69}$$

The rate of change of the particle's momentum is

$$\frac{d}{dt}(\delta \vec{M}_{XYZ}) = \frac{d}{dt}(\delta m \vec{V}_{XYZ}) = \delta m\left(\frac{d\vec{V}_{XYZ}}{dt}\right).$$

Borrowing another result from elementary dynamics, we find that

$$\frac{d\vec{V}_{XYZ}}{dt} = \frac{d\vec{V}_{xyz}}{dt} + \frac{d^2\vec{R}}{dt^2} + \frac{d\vec{\Omega}}{dt} \times \vec{r} + 2\vec{\Omega} \times \vec{V}_{xyz} + \vec{\Omega} \times \vec{\Omega} \times \vec{r}. \tag{4.70}$$

We obtain the rate of change of system momentum by integrating over all particles that make up the system:

$$\begin{aligned} \frac{d}{dt}\vec{M}_{XYZ} &= \int_{\forall_{\text{sys}}} \frac{d\vec{V}_{XYZ}}{dt}(dm) = \int_{\forall_{\text{sys}}} \rho \frac{d\vec{V}_{XYZ}}{dt}(d\forall) \\ &= \int_{\forall_{\text{sys}}} \rho\left(\frac{d\vec{V}_{xyz}}{dt} + \frac{d^2\vec{R}}{dt^2} + \frac{d\vec{\Omega}}{dt} \times \vec{r} + 2\vec{\Omega} \times \vec{V}_{xyz} + \vec{\Omega} \times \vec{\Omega} \times \vec{r}\right) d\forall. \end{aligned}$$

The integral

$$\int_{\forall_{\text{sys}}} \rho\left(\frac{d\vec{V}_{xyz}}{dt}\right) d\forall$$

represents the rate of change of the system's xyz momentum. An observer in the xyz frame would calculate this rate of change by using the transport theorem for a fixed control volume:

$$\int_{\forall_{\text{sys}}} \rho\left(\frac{d\vec{V}_{xyz}}{dt}\right) d\forall = \frac{\partial}{\partial t}\int_{\forall_{cv}} \rho \vec{V}_{xyz}\, d\forall + \oint_{A_{cv}} \rho \vec{V}_{xyz}(\vec{V}_{xyz} \cdot \hat{n})\, dA.$$

The remainder of the terms that make up the rate of change of the system's XYZ momentum are not affected by the flow as viewed from the xyz frame. The system of interest instantaneously fills the control volume, allowing integration of these terms over the control volume $\forall_{cv}$. Newton's second law for a control volume that is fixed with respect to a noninertial reference frame thus becomes

$$\frac{\partial}{\partial t}\int_{\forall_{cv}} \rho\vec{V}_{xyz}\,d\forall + \oint_{A_{cv}} \rho\vec{V}_{xyz}(\vec{V}_{xyz}\cdot\hat{n})\,dA$$
$$+\int_{\forall_{cv}} \rho\left[\frac{d^2\vec{R}}{dt^2} + \left(\frac{d\vec{\Omega}}{dt}\times\vec{r}\right) + (2\vec{\Omega}\times\vec{V}_{xyz}) + (\vec{\Omega}\times\vec{\Omega}\times\vec{r})\right]d\forall = \sum\vec{F}. \quad (4.71)$$

This equation is the linear momentum equation for a noninertial rigid control volume. The first two terms on the left are the "ordinary" momentum accumulation and momentum flux terms, evaluated by an observer in the noninertial (moving) reference frame. To an observer in this frame, the additional momentum terms appear to be forces. Often these terms are transferred to the force side of the equation and treated as additional forces. An observer in the moving reference frame would use the following form of the equation:

$$\frac{\partial}{\partial t}\int_{\forall_{cv}} \rho\vec{V}\,d\forall + \oint_{A_{cv}} \rho\vec{V}(\vec{V}_r\cdot\hat{n})\,dA = \sum\vec{F}$$
$$-\int_{\forall_{cv}} \rho\left[\frac{d^2\vec{R}}{dt} + \left(\frac{d\vec{\Omega}}{dt}\times\vec{r}\right) + (2\vec{\Omega}\times\vec{V}) + (\vec{\Omega}\times\vec{\Omega}\times\vec{r})\right]d\forall. \quad (4.72)$$

The last two terms in the brackets under the integral on the right contribute the Coriolis "force" and centrifugal "force". The xyz subscript was omitted from the velocity, $\vec{V}$, because this equation is written from the viewpoint of an observer in the xyz system.

4.4.5 The Angular Momentum Equation

The angular momentum equation is an alternative formulation of Newton's second law of motion and often is more convenient to use than the linear momentum equation for systems exhibiting rotary motion or when moments of forces are required. We derive this equation and briefly discuss it. Much of the methodology used for the linear momentum equation may be easily adapted to the angular momentum equation.

Our starting point is the angular momentum law for a system:

$$\sum(\mathcal{M}_0)_s = \frac{d}{dt}\int_{\forall_{sys}} \rho(\vec{r}_0\times\vec{V})\,d\forall, \quad (4.73)$$

where

$$\sum(\mathcal{M}_0)_s = \sum(\vec{r}_0\times\vec{F})_{sys}$$

is the sum of moments about an arbitrary point 0 of the forces acting on the system and r_0 is the position vector measured from the inertial point 0.

Equation (4.73) does not represent an independent fundamental law of nature; it is derived from Newton's second law of motion. Let's consider a single mass particle as a system. Figure 4.26 shows the particle, its velocity $\vec{V}$, the net force $\vec{F}$ acting on the

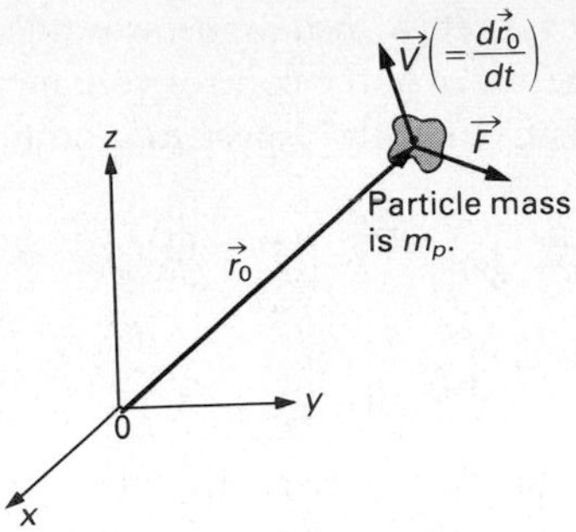

Figure 4.26 Mass particle with velocity $\vec{V}$ at position $\vec{r}_0$; force $\vec{F}$ is acting on the particle.

particle, and the vector from an arbitrary point 0 to the particle, r_0. Newton's second law for the particle is

$$\vec{F} = \frac{d}{dt}(m_p\vec{V}).$$

Taking the vector (cross) product of $\vec{r}_0$ with this equation, we have

$$\vec{r}_0 \times \vec{F} = \vec{r}_0 \times \frac{d}{dt}(m_p\vec{V}). \tag{4.74}$$

From the definition of the moment of a force,

$$\vec{\mathscr{M}}_0 \equiv \vec{r}_0 \times \vec{F}. \tag{4.75}$$

Consider

$$\frac{d}{dt}[\vec{r}_0 \times (m_p\vec{V})].$$

According to the product rule for derivatives,

$$\frac{d}{dt}[\vec{r}_0 \times (m_p\vec{V})] = \frac{d\vec{r}_0}{dt} \times (m_p\vec{V}) + \vec{r}_0 \times \frac{d}{dt}(m_p\vec{V}).$$

The derivative of the position vector $\vec{r}_0$ is the particle velocity $\vec{V}$, so

$$\frac{d}{dt}[\vec{r}_0 \times (m_p\vec{V})] = \vec{V} \times (m_p\vec{V}) + \vec{r}_0 \times \frac{d}{dt}(m_p\vec{V}).$$

But,

$$\vec{V} \times (m_p\vec{V}) = m_p(\vec{V} \times \vec{V}) = 0,$$

because the vector product of identical vectors is zero. Then

$$\frac{d}{dt}[\vec{r}_0 \times (m_p\vec{V})] = \vec{r}_0 \times \frac{d}{dt}(m_p\vec{V}).$$

When we combine this result and Eq. (4.75), Eq. (4.74) becomes

$$\vec{\mathscr{M}}_0 = \frac{d}{dt}[\vec{r}_0 \times (m_p\vec{V})]. \tag{4.76}$$

If the single particle is part of a finite system, Eq. (4.76) applies to each particle of the system. If the system is continuous, Eq. (4.76) applies to each differential piece of the system with

$$m_p = \int dm = \int \rho \, d\forall.$$

Integrating the individual equations for each differential mass yields Eq. (4.73).

We obtain the angular momentum equation by applying the transport theorem with $b = \vec{r}_0 \times \vec{V}$:

$$\sum \vec{\mathcal{M}}_0 = \frac{d}{dt} \int_{\forall_{cv}} \rho(\vec{r}_0 \times \vec{V}) \, d\forall + \oint_{A_{cv}} \rho(\vec{r}_0 \times \vec{V})(\vec{V}_r \cdot \hat{n}) \, dA. \tag{4.77}$$

In Eq. (4.77), $\vec{V}_r$ is the velocity relative to the control surface, and $\vec{V}$ must be measured with respect to an inertial coordinate system. $\sum \vec{\mathcal{M}}_0$ is the vector sum of the moments of all forces acting on the mass inside the control volume. Extension to a noninertial coordinate system requires addition of the term

$$-\int_{\forall_{cv}} \rho\left(\vec{r}_0 \times \left[\frac{d^2\vec{R}}{dt^2} + \frac{d\vec{\Omega}}{dt} \times \vec{r}_0 + 2\vec{\Omega} \times \vec{V} + \vec{\Omega} \times \vec{\Omega} \times \vec{r}_0\right]\right) d\forall$$

to the left-hand side of Eq. (4.77). (See Fig. 4.25.)

Equation (4.77) in general form is applicable to nonsteady, multidimensional flow through a moving, deformable control volume. Simplified forms are applicable to several common cases, similar to simplified forms of the other control volume equations. For example, the moment of momentum equation for steady flow through a control volume with a finite number of inlets and outlets, each with locally one-directional and uniform flow, is

$$\begin{aligned} \sum \vec{\mathcal{M}}_0 &= \sum_{\text{out}} (\rho V_n A)(\vec{r}_0 \times \vec{V}) - \sum_{\text{in}} (\rho V_n A)(\vec{r}_0 \times \vec{V}) \\ &= \sum_{\text{out}} \dot{m}(\vec{r}_0 \times \vec{V}) - \sum_{\text{in}} \dot{m}(\vec{r}_0 \times \vec{V}). \end{aligned} \tag{4.78}$$

Steady, locally uniform flow

Like the linear momentum equation, the angular momentum equation is a vector equation, so it has three separate component equations.

EXAMPLE 4.24 Illustrates the Angular Momentum Equation

Figure E4.24a shows an isometric view of a short section of pipe that protrudes from a wall. The pipe diameter is 20 cm, and both pipe bends are 90°. Water enters the pipe at the base and exits at the open end with a speed of 10 m/s. Calculate the torsional moment and the bending moment at the base of the pipe. Neglect the weight of the water and pipe.

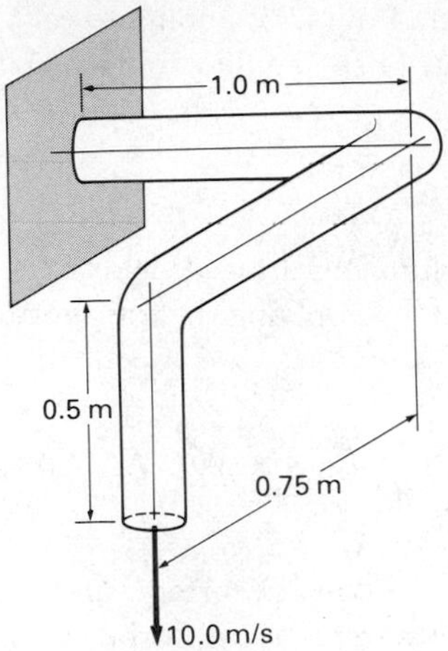

Figure E4.24a Isometric view of pipe section.

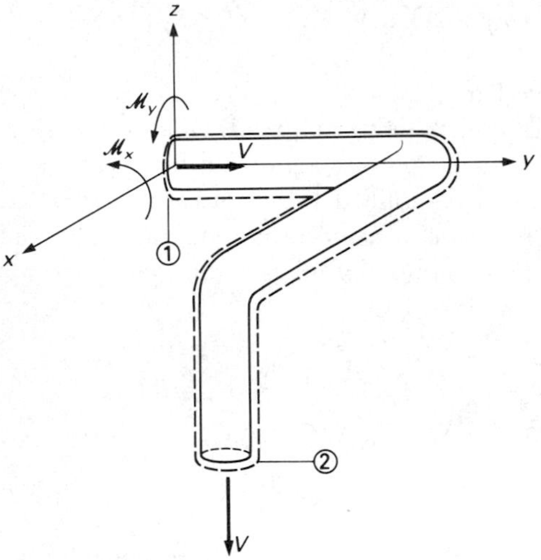

Figure E4.24b Coordinate system and control volume.

SOLUTION

Given

20 cm pipe with two 90° bends as shown

Water flowing through pipe at 10 m/s

Find

Torsional moment and bending moment at base of pipe

Solution

We choose the coordinate system and control volume shown in Fig. E4.24b. The control volume "slices" through the pipe at the base. The moments shown are the moments acting on the material in the control volume because of the anchoring at the wall. The term $\mathcal{M}_x$ is the bending moment, and the term $\mathcal{M}_y$ is the torsional moment. These moments are shown acting in the positive direction for the chosen coordinate system. We assume locally uniform and incompressible flow. The angular momentum equation, Eq. (4.78), gives

$$\mathcal{M}_x\hat{i} + \mathcal{M}_y\hat{j} = (\rho VA)_2(\vec{r}_2 \times \vec{V}_2) - (\rho VA)_1(\vec{r}_1 \times \vec{V}_1).$$

The continuity equation gives

$$(\rho AV)_2 = (\rho AV)_1 = \rho AV.$$

Also,

$$\vec{r}_1 = 0, \qquad \vec{r}_2 = 0.75\hat{i} + 1.0\hat{j} - 0.5\hat{k},$$
$$\vec{V}_1 = 10\hat{j}, \quad \text{and} \quad \vec{V}_2 = -10\hat{k},$$

and so

$$\begin{aligned}\mathcal{M}_x\hat{i} + \mathcal{M}_y\hat{j} &= (1000\ \text{kg/m}^3)\pi(0.1\ \text{m})^2(10\ \text{m/s}) \\ &\quad \times (0.75\hat{i}\ \text{m} + 1.0\hat{j}\ \text{m} - 0.5\hat{k}\ \text{m}) \times (-10\hat{k}\ \text{m/s}) \\ &= 2356\hat{j}\ \text{N}\cdot\text{m} - 3142\hat{i}\ \text{N}\cdot\text{m}.\end{aligned}$$

Equating components, we get

$$\mathcal{M}_x = -3142\ \text{N}\cdot\text{m} \quad \text{and} \quad \mathcal{M}_y = 2356\ \text{N}\cdot\text{m}.$$ **ANSWER**

Discussion

What do the algebraic signs on the answers mean? You should convince yourself from physical reasoning that the calculated moments have the correct directions. Recall that we calculated the restraining moment exerted on the pipe by the base. Using vector notation permitted us to do both parts of the problem simultaneously and helped us to avoid errors in geometry and algebraic signs.

4.4.6 An Application of the Angular Momentum Equation: Turbomachinery

The angular momentum equation is useful in the analysis of turbomachinery. A *turbomachine* is a device that uses a moving rotor, carrying a set of *blades* or *vanes,* to transfer work to or from a moving stream of fluid. If the work is done on the fluid by the rotor, the machine is called a *pump* or *compressor;* if the fluid delivers work to the rotor, the machine is called a *turbine*. Figure 4.27 illustrates typical turbomachines.

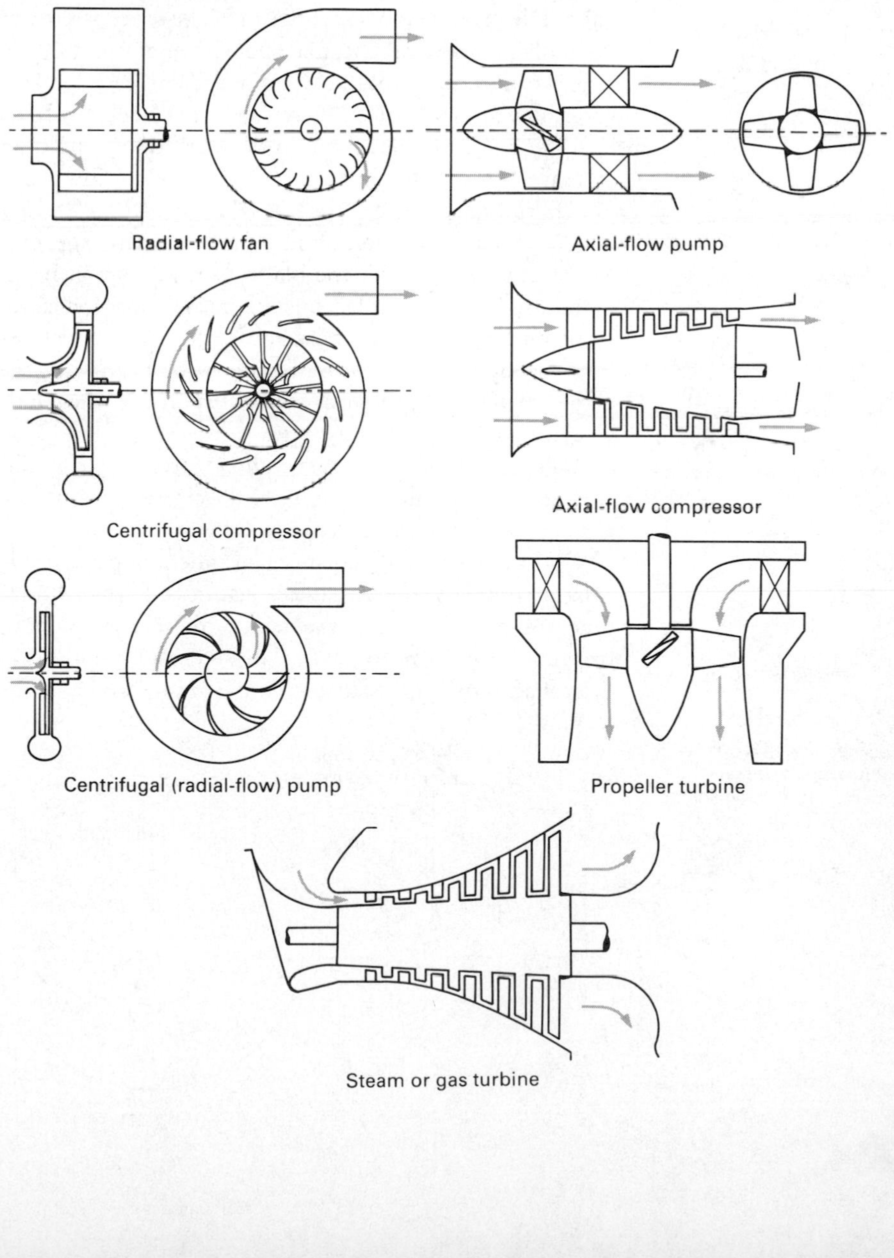

Figure 4.27 Sketches of typical turbomachines.

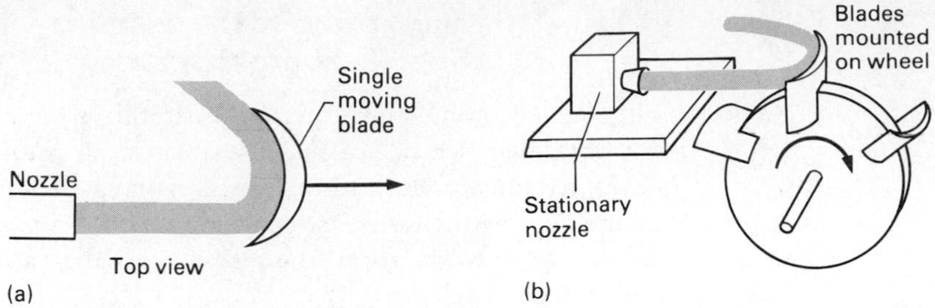

Figure 4.28 A simple turbine made of several moving vanes for fluid jet deflection.

The principle of turbomachinery operation is not difficult. In Examples 4.22 and 4.23, we considered the force generated when a jet of fluid is deflected by a moving vane. The product of vane force and vane speed is the rate of work done on the vane. A simple turbine could be made up of several vanes (Fig. 4.28). Instead of moving in a straight path, the vanes move in a circular path because they are connected to the rotor. If we run the rotor "backward," driving it with a motor and pushing the blades into the fluid, the blade would do work on the fluid and the fluid's energy would increase. This is the principle of pump operation.

Turbomachines are classified as *axial flow, radial flow,* or *mixed flow,* depending on the direction of fluid motion with respect to the rotor's axis of rotation as the fluid passes over the blades (see Fig. 4.27). In an axial-flow rotor, the fluid maintains an essentially constant radial position as it flows from rotor inlet to rotor outlet. In a radial-flow rotor, the fluid moves primarily radially from rotor inlet to outlet, although the fluid may be moving in the axial direction at the machine inlet or outlet. In a mixed-flow rotor, the fluid has both axial and radial velocity components as it passes through the rotor. For a given diameter and rotational speed, radial- and mixed-flow rotors are capable of exchanging more work with the moving fluid than axial-flow rotors, because the "centrifugal forces" associated with the radius change do work.

To apply the angular momentum equation to a turbomachine flow, let's consider a control volume that lies just outside the rotor (Fig. 4.29). The control volume is fixed. We can use the angular momentum equation to compute the moment (torque) about any axis. The most useful computation is the torque about the driveshaft axis (z), because it contains the driving torque. For moments about the

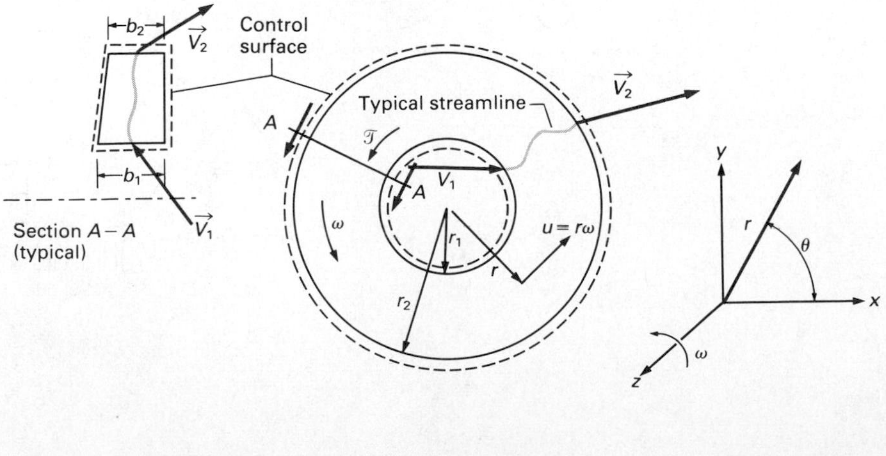

Figure 4.29 Control volume enclosing a generalized turbomachine rotor.

z axis, Eq. (4.77) becomes

$$\sum \mathcal{M}_z = \left[\frac{d}{dt} \int_{\mathcal{V}_{cv}} \rho(\vec{r} \times \vec{\mathcal{V}})\, dV \right] \cdot \hat{k} + \left[\oint_{A_{cv}} \rho(\vec{r} \times \vec{V})(\vec{V}_r \cdot \hat{n})\, dA \right] \cdot \hat{k}, \tag{4.79}$$

where $\hat{k}$ is the unit vector in the z direction. Taking the dot product with $\hat{k}$ produces the z component on the right side. We now introduce the following assumptions:

- The only moment acting on the material in the control volume is the driveshaft torque $\mathcal{T}_z$. We neglect moments resulting from the weight of the material in the control volume and the pressure and shear forces in the fluid at the control surface.*
- Angular momentum does not accumulate inside the control volume.
- The flow is uniform at the inlet (radius r_1) and at the outlet (radius r_2).

With these assumptions, the angular momentum equation becomes

$$\mathcal{T}_z = \dot{m}_2(r_2 V_{\theta 2}) - \dot{m}_1(r_1 V_{\theta 1}),$$

where V_θ is the tangential component of the fluid velocity. The mass flow rates, $\dot{m}_2$ and $\dot{m}_1$, come from integration of the product of density and the normal component of velocity over the control volume inlet and outlet areas, the rV_θ terms come from $(\vec{r} \times \vec{V}) \cdot \hat{k}$, and $\mathcal{T}_z$ is the torque required to drive the rotor.

Application of the continuity equation gives

$$\dot{m}_1 = \dot{m}_2 = \dot{m},$$

so we can write

$$\mathcal{T}_z = \dot{m}(r_2 V_{\theta 2} - r_1 V_{\theta 1}). \tag{4.80}$$

The power transferred to the fluid by the rotor (that is, the rate at which the rotor does work on the fluid) is equal to the product of the driveshaft torque and the rotational speed ω:

$$-\dot{W}_s = \mathcal{T}_z \omega = \dot{m}\omega(r_2 V_{\theta 2} - r_1 V_{\theta 1}).$$

The minus sign is added to $\dot{W}_s$ according to the customary sign convention for work. As the control volume lies just outside the rotor, r_1 and r_2 are the radii of the rotor at inlet and outlet respectively.† The product of these rotor radii with the rotor angular velocity equals the *linear* velocity of the rotor at points 1 and 2. In turbomachinery analysis, U is customarily used to represent the linear speed of the rotor at a point and V_u to represent the tangential ("U-wise") component of the fluid velocity, so we write

$$-\dot{W}_s = \dot{m}(U_2 V_{u2} - U_1 V_{u1}). \tag{4.81}$$

* Note that moments due to pressure would be zero for a circular control volume.

† Note carefully that 1 is rotor *inlet* and 2 is rotor *outlet*. In this discussion, 1 is the smaller radius and 2 is larger; in some machines, the inlet and outlet may be reversed, with fluid entering at the larger radius and exiting at the smaller.

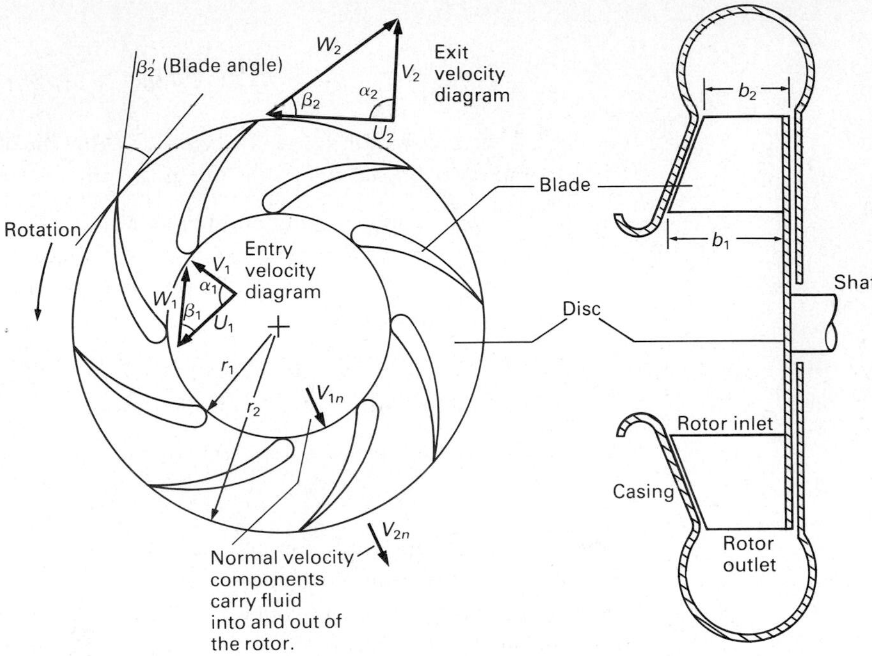

Figure 4.30 Details of flow entering and leaving a radial-flow rotor.

Equation (4.81) is called *Euler's pump and turbine equation.* It applies to turbines (for which U_2V_{u2} must be less than U_1V_{u1}, so that $\dot{W}_s$ is positive) as well as to pumps (for which $U_2V_{u2} > U_1V_{u1}$).

Drawing *velocity diagrams* to represent the fluid and rotor velocities at various points in the machine facilitates analysis and design of turbomachines. The fluid velocity can be viewed from two coordinate systems, one fixed to the rotor and one fixed to the machine casing. The velocity as viewed from a coordinate system fixed to the rotor is called the *relative velocity* and is assigned the symbol $\vec{W}$. Because the relative coordinate system moves at the rotor velocity $\vec{U}$ (written here as a vector), the relation between the absolute velocity, $\vec{V}$, and the relative velocity is

$$\vec{V} = \vec{U} + \vec{W}. \tag{4.82}$$

Velocity diagrams reflect this vector relation. Figure 4.30 shows velocity diagrams for flow entering and leaving a blade in a radial-flow turbomachine. In terms of the geometry of the velocity diagram, the power absorbed by the rotor and transferred to the fluid is

$$-\dot{W}_s = \dot{m}(U_2V_2 \cos \alpha_2 - U_1V_1 \cos \alpha_1). \tag{4.83}$$

The mass flow rate through the rotor may be calculated as the flow across the inlet or outlet:*

* The following equation is applicable only to radial-flow rotors because of the expression used to calculate the areas ($2\pi rb$). A similar equation can be developed for axial-flow rotors where the flow area is an annulus.

$$\dot{m} = \rho_1 V_1 \sin \alpha_1 (2\pi r_1 b_1) = \rho_2 V_2 \sin \alpha_2 (2\pi r_2 b_2). \quad (4.84)$$

The angles in the velocity diagrams are related to the angles of the blades; a particularly simple assumption is

$$\beta_2 \text{ (velocity diagram angle)} \approx \beta_2' \text{ (blade angle)}. \quad (4.85)$$

Simple preliminary analysis and design of turbomachines can be based on Eqs. (4.83), (4.84), and (4.85), together with consideration of the geometry of the velocity diagrams.

EXAMPLE 4.25 Illustrates the Analysis of a Simple Turbomachine

W is fluid velocity relative to point of entry or exit.

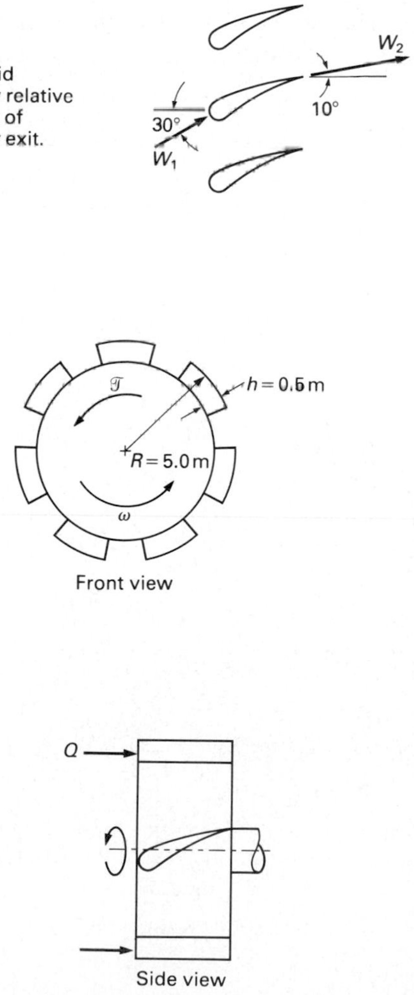

Figure E4.25a Sketch of hydraulic axial flow turbine.

The axial-flow hydraulic turbine sketched in Fig. E4.25a has a water flow rate of 75 m^3/s, an outer radius $R = 5.0$ m, and a blade height $h = 0.5$ m. Assume uniform properties and velocities over both the inlet and the exit. The water temperature is 20°C, and the turbine rotates at 60 rpm. The relative velocities W_1 and W_2 make angles of 30° and 10°, respectively, with the normal to the flow area. Find the output torque and power developed by the turbine.

SOLUTION

Given

Turbine in Fig. 4.25a

Volume flow rate 75 m^3/s

Outer radius 5.0 m

Blade height 0.5 m

Rotational speed 60 rpm

Fluid 20°C water

Find

Output torque and power developed by turbine

Solution

We use Eq. (4.80) to find the torque:

$$\mathcal{T} = \dot{m}(r_2 V_{u2} - r_1 V_{u1}).$$

The mass flow rate is

$$\dot{m} = \rho Q = (998 \text{ kg/m}^3)(75 \text{ m}^3/\text{s}) = 74{,}850 \text{ kg/s}.$$

For an axial-flow machine where the blade height h is small compared to R,

$$r_2 = r_1 \approx R - \tfrac{1}{2} h = 5.0 \text{ m} - \tfrac{1}{2}(0.5 \text{ m}) = 4.75 \text{ m}.$$

We find the tangential components of the absolute velocity from the velocity triangles in Fig. E4.25b, where V represents the fluid absolute velocity, W the fluid velocity relative to the turbine wheel, and U the

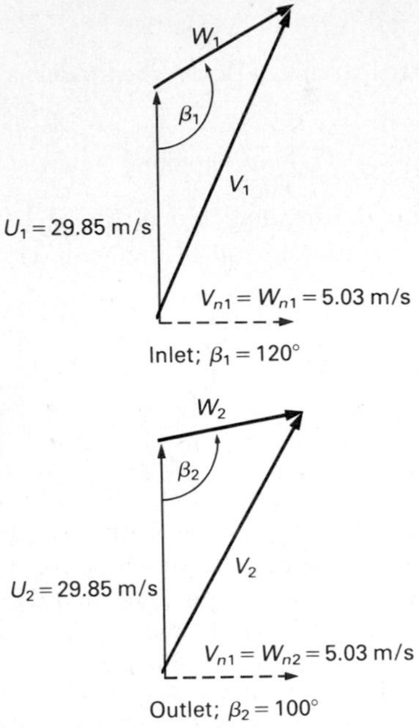

Figure E4.25b Velocity diagrams for axial-flow turbine.

wheel peripheral velocity. Also,

$$U_1 = r_1\omega = (4.75 \text{ m})\left(60 \frac{\text{rev}}{\text{min}}\right)\left(2\pi \frac{\text{rad}}{\text{rev}}\right)\left(\frac{1 \text{ min}}{60 \text{ s}}\right) = 29.85 \text{ m/s},$$

$$U_2 = r_2\omega \approx r_1\omega = 29.85 \text{ m/s},$$

$$V_{n1} = \frac{Q}{A_1} = \frac{75 \text{ m}^3/\text{s}}{\pi[(5.0 \text{ m})^2 - (4.5 \text{ m})^2]} = 5.03 \text{ m/s},$$

and

$$V_{n2} = \frac{Q}{A_2} = 5.03 \text{ m/s},$$

where V_{n1} and V_{n2} represent the normal components of the absolute velocities V_1 and V_2 (or of the relative velocities W_1 and W_2), respectively.

Referring again to Fig. E4.25b, we calculate the tangential components of the absolute velocities as

$$V_{u1} = U_1 + W_1 \cos(\pi - \beta_1) = U_1 - W_1 \cos \beta_1$$

and

$$V_{u2} = U_2 + W_2 \cos(\pi - \beta_2) = U_2 - W_2 \cos \beta_2.$$

As

$$W_{n1} = W_1 \sin(\pi - \beta_1) = W_1 \sin \beta_1$$

and

$$W_{n2} = W_2 \sin(\pi - \beta_2) = W_2 \sin \beta_2,$$

we have

$$V_{u1} = U_1 - \frac{W_{n1}}{\tan \beta_1} \quad \text{and} \quad V_{u2} = U_2 - \frac{W_{n2}}{\tan \beta_2}.$$

The numerical values give

$$V_{u1} = 29.85 \text{ m/s} - \frac{5.03 \text{ m/s}}{\tan 120°} = 32.75 \text{ m/s}$$

and

$$V_{u2} = 29.85 \text{ m/s} - \frac{5.03 \text{ m/s}}{\tan 100°} = 30.74 \text{ m/s}.$$

The torque is

$$\mathscr{T} = (74{,}850 \text{ kg/s})[(4.75 \text{ m})(30.74 \text{ m/s}) - (4.75 \text{ m})(32.75 \text{ m/s})]$$
$$= -7.15 \times 10^5 \text{ N}\cdot\text{m}.$$

The significance of the negative sign is that the torque is in the direction opposite that assumed to be positive ($\mathscr{T}$ is the load torque that resists rotation of the turbine). The magnitude is

$$\mathscr{T} = 7.15 \times 10^5 \text{ N}\cdot\text{m}.$$ **ANSWER**

The output power is

$$\dot{W} = \mathscr{T}\omega = (7.15 \times 10^5 \text{ N}\cdot\text{m})(60 \text{ rev/min})(2\pi \text{ rad/rev})(1 \text{ min}/60 \text{ s});$$
$$\dot{W}_s = 4.49 \times 10^6 \text{ N}\cdot\text{m/s},$$

or

$$\dot{W}_s = 4490 \text{ kw}.$$ **ANSWER**

Discussion

The water entering the rotor has a significant velocity component in the direction of rotor motion ($V_{u1} \neq 0$; see Fig. E4.25b). A set of nozzles or deflector vanes must be provided to direct the flow as it enters the turbine.

More about Turbomachinery. More extensive considerations of flow in turbomachines are beyond the scope of this book. Here, we can only mention in passing that the major effort in this field is directed along two paths:

- Relating the flow over the blades to the blade shapes. Crudely, we may think of this analysis as an improvement to the "perfectly guided fluid" assumption embodied in Eq. (4.85) and the one-dimensional flow assumption embodied in Eqs. (4.83) and (4.84). This requires analysis of fluid motion in a coordinate system fixed to the rotor.

- Calculating the "losses" associated with the flow through the turbomachine. Although Eq. (4.83) applied to a pump, for example, accurately calculates the work transferred between the fluid and the rotor (within the limits of the one-dimensional flow assumption), not all this work results in an increase in the fluid's mechanical energy. If flow conditions are poor, well over half the work done on the fluid may be converted into "losses."

The fluid mechanics of turbomachinery is currently a very active area of research. If you are interested in further reading in this area, consult references [5–8].

4.5 PUTTING CONTINUITY, ENERGY, AND MOMENTUM TO WORK

Practical applications of the control volume approach involve the appropriate forms of the continuity, energy, and momentum equations to model relatively complex flow situations. In this section, we present three topics associated with practical engineering application of control volume flow analysis.

4.5.1 Simultaneous Application of Continuity, Energy, and Momentum

In Sections 4.2–4.4, we developed the continuity, energy, and momentum equations in relative isolation. Most practical flow problems require the simultaneous use of two or all three of these equations.* When tackling such a problem, you may not

* You may have noted that most of the examples dealing with energy and momentum also required the continuity equation.

realize at the beginning that all three equations are needed. You should always remember that continuity, energy, and momentum are three *independent* equations and that all flows must obey the physical laws they represent. It is never incorrect to use all three equations in a problem; however, using one of the equations may not be necessary, or you may lack sufficient information to do so effectively. Of course, you should never simultaneously use two forms of the same principle (e.g., Bernoulli's and the general energy equations) for the same control volume.

EXAMPLE 4.26 Illustrates Simultaneous Application of the Continuity, Energy, and Linear Momentum Equations

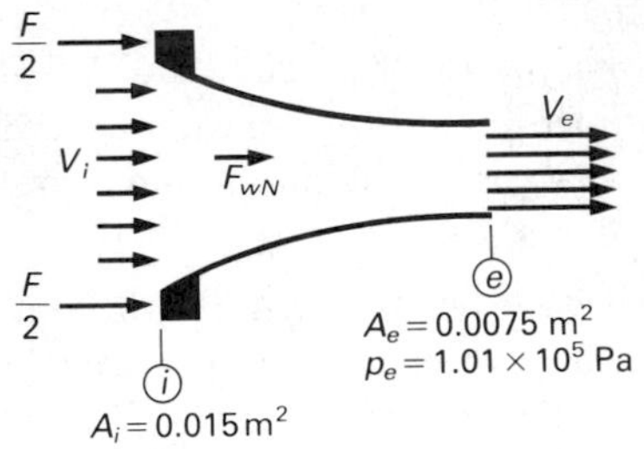

Figure E4.26a Nozzle.

Water at 20°C flows through the nozzle in Fig. E4.26a at a steady rate of 12.0 m^3/min. Find the force required to hold the nozzle. Neglect gravitational force.

SOLUTION

Given

Nozzle in Fig. E4.26a

Water flow rate 12.0 m^3/min

Neglect the gravitational force

Find

Force required to hold nozzle

Solution

Apply Newton's first law to the nozzle:

$$2\left(\frac{F}{2}\right) + F_{wN} = 0,$$

where F_{wN} is the force of the water on the nozzle. Then

$$F = -F_{wN} = F_{Nw},$$

which says that we can find the force required to hold the nozzle by finding the force of the nozzle on the water (F_{Nw}). Figure E4.26b shows a control volume enclosing the fluid, with F_{Nw} assumed to be positive. Its actual direction is determined by the algebraic sign of the result.

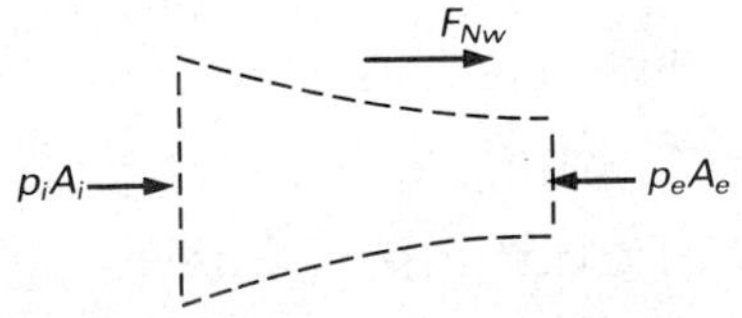

Figure E4.26b Control volume for water in nozzle.

We assume that the fluid density and velocity are uniform over each flow area, so the linear momentum equation for the x direction is

$$\dot{M}_{x,\text{out}} - \dot{M}_{x,\text{in}} = \sum F_x,$$

where

$$\dot{M}_{x,\text{out}} - \dot{M}_{x,\text{in}} = \rho_e V_e^2 A_e - \rho_i V_i^2 A_i,$$

and

$$\sum F_x = p_i A_i + F_{Nw},$$

where p_i is the gage pressure, because we may drop any constant pressure that acts around the entire control volume. Thus

$$F_{Nw} = \rho_e V_e^2 A_e - \rho_i V_i^2 A_i - p_i A_i.$$

The fluid is water, we may assume that the flow is incompressible, and so we know that ρ_e and ρ_i are equal. We also know A_e and A_i. The preceding equation contains the unknowns F_{Nw}, V_e, V_i, and p_i. We need more equations! The continuity equation for this steady incompressible flow gives

$$V_e A_e = V_i A_i = Q \qquad (= 12 \text{ m}^3/\text{min}),$$

so

$$V_e = \frac{Q}{A_e} \quad \text{and} \quad V_i = \frac{Q}{A_i}.$$

The energy equation can also be applied. The nozzle flow is steady and incompressible and no work is done. We assume that the flow is frictionless and apply Bernoulli's equation to a "typical" streamline between the inlet and outlet planes to get

$$\frac{p_i}{\rho} + \frac{V_i^2}{2} + gz_i = \frac{p_e}{\rho} + \frac{V_e^2}{2} + gz_e.$$

Substituting for the velocities, assuming that $z_e = z_i$, and noting that $p_e = 0$ because we are using gage pressures, we obtain

$$p_i = \frac{\rho Q^2}{2}\left(\frac{1}{A_e^2} - \frac{1}{A_i^2}\right).$$

Finally, incorporating the velocity and pressure expressions into the momentum equation gives

$$F_{Nw} = \rho Q^2\left(\frac{1}{A_e} - \frac{1}{A_i}\right) - \frac{\rho Q^2 A_i}{2}\left(\frac{1}{A_e^2} - \frac{1}{A_i^2}\right)$$
$$= -\frac{\rho Q^2 (A_i - A_e)^2}{2A_e^2 A_i}.$$

Substituting numerical values and units gives

$$F_{Nw} = \left(998 \frac{\text{kg}}{\text{m}^3}\right)\left(12.0 \frac{\text{m}^3}{\text{min}}\right)^2\left(\frac{1 \text{ min}}{60 \text{ s}}\right)^2 \frac{(0.015 \text{ m}^2 - 0.0075 \text{ m}^2)^2}{2(0.0075 \text{ m}^2)^2(0.015 \text{ m}^2)};$$

$$F_{Nw} = -1331 \text{ N}.$$ **ANSWER**

We assumed that the positive direction was to the right but the numerical value is negative, so F_{Nw} acts to the left.

Discussion

Effectively, we were able to calculate three quantities—a velocity, a pressure, and a force—from the three fundamental equations. Our problem-solving method followed the outline suggested in Section 1.6: We began with an equation involving the desired quantity (F_{Nw}) and introduced other equations as they were needed. In addition to the three fundamental equations, we had to make several assumptions. Some of them, such as incompressible flow, are very accurate. Others, such as uniform, frictionless flow, are probably satisfactory for nozzle flow but would be questionable if the flow passage were a diffuser.

4.5.2 Using the Control Volume Approach for "Engineering Design" Problems

The two most prevalent obstacles to successful application of the control volume approach to engineering design problems are incomplete information and the need for tedious computation. Computers have made many previously tedious calculations commonplace. The next section will consider computer-aided control volume analysis in detail.

Accurate control volume analysis requires information about velocity, shear stress, and pressure distributions; models for mechanical energy loss; data on fluid properties; and considerable information about flow geometry. Frequently, however, some of this information is not available, and the engineer must seek it out or make some reasonable assumptions in order to proceed. This is particularly true in design problems, most of which have several possible solutions, depending on the assumptions and choices made by the design engineer. The systematic approach to problem solving presented in Section 1.6 is especially helpful to avoid making too many arbitrary assumptions. The following example illustrates a fairly involved "engineering design" type of analysis.

EXAMPLE 4.27 **Illustrates an "Engineering Design" Problem**

Your company is designing a snowblower similar to that shown in Fig. E4.27a. The primary design requirement is that the machine must clear a 16-in.-wide path in 6-in.-deep snow. The vanes are driven by a small gasoline engine. The drive pulley and belt system reduce the vane rotational speed to one fourth of the engine speed. The company currently manufactures two gasoline engines, models 100A and 100B. The performance characteristics of the two engines at wide-open throttle (WOT) conditions are shown in Fig. E4.27b. The company prefers to use the smaller and cheaper model 100A in the snowblower. Is that possible? If not, is model 100B acceptable?

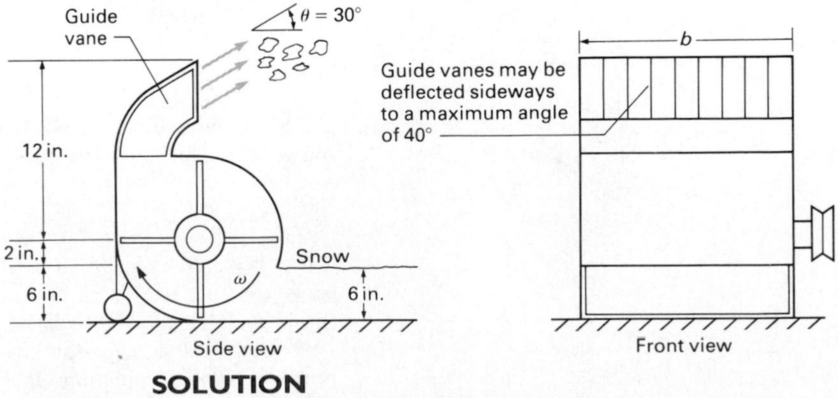

Figure E4.27a Side and front views of proposed snowblower.

SOLUTION

Given

Snowblower and gasoline engine characteristics shown in Fig.E4.27a and E4.27b.

Snowblower vanes' rotational speed is $\frac{1}{4}$ engine speed

Snowblower must clear a path 16 in. wide and 6 in. deep

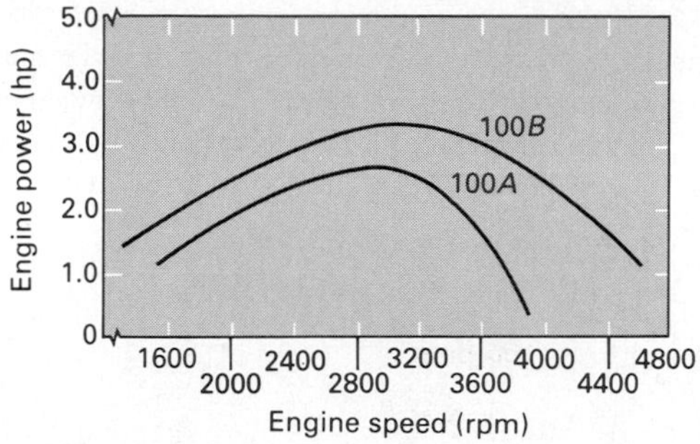

Figure E4.27b Performance characteristics of available engines.

Find

Is engine model 100A acceptable?

If not, is engine model 100B acceptable?

Solution

This is a complicated problem, and you may not immediately see a method of solution. We begin by making a sketch of the physical situation and identifying one or more "design criteria." We need to translate the design criteria into a mathematical statement in order to use an analysis to answer the relevant question.

We are interested in the ability of the engines to drive the snowblower, so our primary design criterion is:

> The engine selected must produce sufficient power to drive the snowblower.

We must therefore estimate the power required to drive the snowblower and compare it to the engine's power. To simplify the problem, we neglect energy loss in the drive belt and pulleys, so that the power required from the engine is equal to the power input to the snowblower rotor. Our mathematical problem is:

> Estimate the power required to drive the snowblower rotor.

Note that, essentially, our first design criterion and the corresponding mathematical problem are based on the principle of energy conservation.

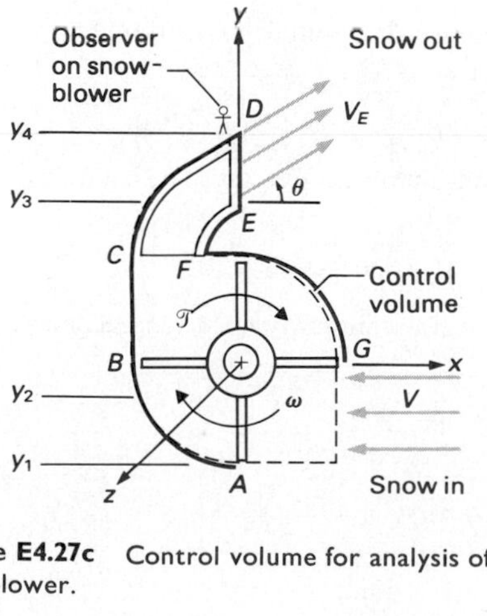

Figure E4.27c Control volume for analysis of snowblower.

Figure E4.27c shows a control volume for analyzing the flow of snow through the rotor and vanes. The input power at the rotor drive shaft is the product of the shaft torque and the rotational speed; that is,

$$\dot{W}_s = \mathcal{T}\omega.$$

This expression tells us that we need relations and/or equations to find $\mathcal{T}$ and ω. An appropriate equation for $\mathcal{T}$ is Eq. (4.77). To apply it, we choose a coordinate system and an observer fixed on the snowblower, as shown in Fig. E4.27c. Equation (4.77) gives

$$\sum \mathcal{M}_0 = \frac{d}{dt}\int_{\forall_{cv}} \rho(\vec{r} \times \vec{V})\, d\forall + \oint_{A_{cv}} \rho(\vec{r} \times \vec{V})(V_n)\, dA.$$

The observer sees a steady-flow condition, so we write

$$\frac{d}{dt}\int_{\forall_{cv}} \rho(\vec{r} \times \vec{V})\, d\forall = 0.$$

We then specialize the angular momentum theorem to the z component:

$$\sum \mathcal{M}_z = \oint_{A_{cv}} \rho(\vec{r} \times \vec{V})_z(V_n)\, dA.$$

Neglecting friction between the snow and the casing and pressure forces on the material inside the control volume, we get

$$\sum \mathscr{M}_z = -\mathscr{T}.$$

We now assume that the velocity is uniform over both the inlet and outlet areas. The inlet velocity V is normal to the inlet area, whereas the outlet velocity V_E makes an angle of 30° with the normal to the area. The snow leaving the blower is assumed not to be deflected sideways. At the inlet,

$$V_n = -V \quad \text{and} \quad (r_0 \times \vec{V})_z \, dA = -yVb\,dy,$$

where b is the vane width. At the outlet,

$$V_n = V_E \cos\theta$$

and

$$(r_0 \times \vec{V})_z \, dA = -yV_E \sin(\pi/2 - \theta)\, b\,dy = -yV_E \cos\theta\, b\,dy.$$

Substituting into the angular momentum theorem gives

$$-\mathscr{T} = \int_{y_1}^{y_2} \rho(-yVb\,dy)(-V) + \int_{y_3}^{y_4} \rho(-yV_E\cos\theta\,dy)(V_E\cos\theta).$$

Assuming that density is constant and that both V and V_E are uniform over their respective areas, we integrate and obtain

$$-\mathscr{T} = \rho V^2 b \left.\frac{y^2}{2}\right|_{y_1}^{y_2} - \rho V_E^2 \cos^2\theta\, b \left.\frac{y^2}{2}\right|_{y_3}^{y_4}$$

or

$$-\mathscr{T} = \frac{\rho V^2 b}{2}(y_2^2 - y_1^2) - \frac{\rho V_E^2 \cos^2\theta\, b}{2}(y_4^2 - y_3^2).$$

We next apply the integral continuity equation to the control volume. For constant density,

$$Vb(y_2 - y_1) = V_E \cos\theta\, b(y_4 - y_3),$$

where V_E and y_3 are both unknown. We assume that V_E is equal to the vane tip velocity, given by

$$V_E = V_{\text{tip}} = R\omega,$$

where $R = -y_1$. The integral continuity equation becomes

$$Vb(y_2 - y_1) = R\omega \cos\theta\, b(y_4 - y_3),$$

or

$$y_4 - y_3 = \frac{V}{R\omega\cos\theta}(y_2 - y_1) \quad \text{and} \quad y_3^2 = \left[y_4 - \frac{V(y_2 - y_1)}{R\omega\cos\theta}\right]^2.$$

Then

$$y_4^2 - y_3^2 = y_4^2 - \left[y_4^2 - \frac{2y_4V(y_2 - y_1)}{R\omega\cos\theta} + \left(\frac{V(y_2 - y_1)}{R\omega\cos\theta}\right)^2\right]$$

$$= \frac{2Vy_4(y_2 - y_1)}{R\omega\cos\theta} - \left(\frac{V(y_2 - y_1)}{R\omega\cos\theta}\right)^2.$$

The required torque $\mathscr{T}$ is then

$$-\mathscr{T} = \frac{\rho V^2 b}{2}(y_2^2 - y_1^2) - \frac{\rho R^2\omega^2 \cos^2\theta\, b}{2}\left[\frac{2Vy_4(y_2 - y_1)}{R\omega\cos\theta} - \left(\frac{V(y_2 - y_1)}{R\omega\cos\theta}\right)^2\right].$$

Simplifying yields

$$\mathscr{T} = \rho b V(y_2 - y_1)(y_4 R\omega\cos\theta + y_1 V).$$

Because $R = -y_1$,

$$\mathscr{T} = \rho b V R(y_2 + R)(y_4\omega\cos\theta - V).$$

The power input to the vanes is

$$\dot{W}_z = \mathscr{T}\omega \quad \text{or} \quad \dot{W}_z = \rho b V R\omega(y_2 + R)(y_4\omega\cos\theta - V).$$

The known numerical values are

$b = 16$ in. $= 1.33$ ft;

$R = 8$ in. $= 0.667$ ft;

$y_2 = -2$ in. $= -0.167$ ft;

$y_4 = 12.0$ in. $= 1.0$ ft; and

$\cos\theta = \cos 30^\circ = 0.866$.

We also need a value for the density of snow. A quick check of the Appendixes at the back of this book reveals no snow properties; in fact, they are rather hard to come by. During a recent winter in Akron, Ohio, we conducted some experiments to determine the density of snow. We found that it varied from 20 lbm/ft^3 for wet snow to 6 lbm/ft^3 for dry, fluffy snow. Because the snowblower must operate in all kinds of snow, the best (conservative) approach is to use the higher value.

Substituting all known numerical values gives the input power as

$$\dot{W}_s = \frac{(20\ \text{lbm/ft}^3)(1.33\ \text{ft})V(0.667\ \text{ft})\omega(-0.167\ \text{ft} + 0.667\ \text{ft})}{(32.2\ \text{ft}\cdot\text{lbm/lb}\cdot\text{sec}^2)} \times [(1.0\ \text{ft})\omega(0.866) - V]$$
$$= 0.275V\omega(0.866\omega - V)\ \text{ft}\cdot\text{lb/sec}.$$

For use in this equation, V must be in feet per second and ω in radians per second.

The power required to rotate the vanes depends on the vane rotational speed and the velocity at which the snowblower is being pushed. The pushing velocity is unspecified, so we must assume a reasonable value. We assume that $V = 2$ ft/sec, so the vane power requirement becomes

$$\dot{W}_s = 0.275(2.0)\omega(0.866\omega - 2.0)\ \text{ft}\cdot\text{lb/sec}$$
$$= (0.476\omega^2 - 1.10\omega)\ \text{ft}\cdot\text{lb/sec}(1\ \text{sec}\cdot\text{hp}/550\ \text{ft}\cdot\text{lb})$$
$$= (8.66\omega^2 - 20\omega) \times 10^{-4}\ \text{hp},$$

where ω is still in rad/sec. Converting ω to rpm gives

$$\dot{W}_s = [8.66\omega^2(2\pi \text{ rad/rev})^2(1 \text{ min/60 sec})^2 - 20\omega(2\pi \text{ rad/rev})(1 \text{ min/60 sec})] \times 10^{-4} \text{ hp}$$
$$= (0.095\omega^2 - 2.09\omega) \times 10^{-4} \text{ hp}.$$

Based on our previous assumption of negligible power loss in the pulley system, the vane input power must equal the engine output power. The preceding equation then represents the power that must be supplied by an engine to rotate the vanes at a rotational speed ω. This equation is plotted in Fig. E4.27d as curve C, as are the actual power outputs of the two engines. Curve C and the curve for engine 100A do not intersect. Thus a snowblower equipped with engine 100A cannot remove 6 in. of wet snow while being pushed at 2.0 ft/sec with the engine operating at WOT. The conclusion is:

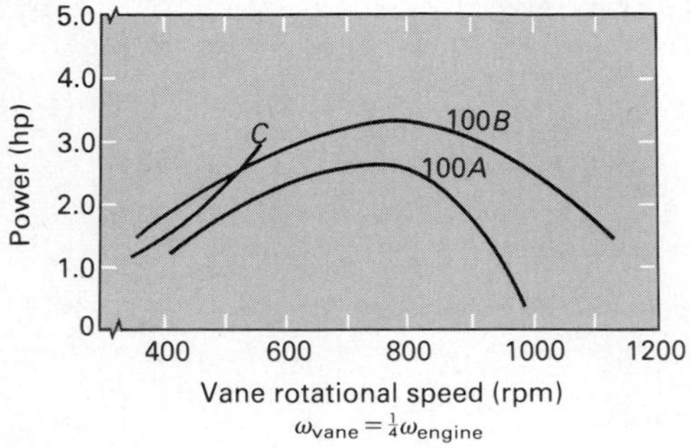

Figure E4.27d Engine output power and required vane power as a function of vane rotational speed.

Engine 100A does not meet the design specifications. **ANSWER**

Note that curve C and the curve for engine 100B intersect at about 515 rpm. This means that the vanes would rotate at 515 rpm if engine 100B is installed in the snowblower and operated at wide-open throttle in 6-in.-deep wet snow. Although the snowblower would work with engine 100B, we cannot be sure that it is acceptable. To be useful, the snowblower must throw the snow a substantial distance. We now propose a second design criterion:

The snowblower must throw the snow at least 20 feet.

We translate this criterion into a requirement for the minimum acceptable rotational speed. The 20-foot throw requirement implies that the velocity of the snow leaving the vanes (V_E) must be large enough to carry the snow 20 ft. We have already assumed that the exit velocity of the snow is equal to the vane tip velocity; thus the minimum acceptable vane rotational speed is

$$\omega_{\text{min}} = \frac{V_{E,\text{min}}}{R},$$

where $V_{E,\text{min}}$ is the velocity required to carry the snow 20 ft.

Figure E4.27e depicts the trajectory of the snow. If we neglect air drag on the particles and clumps of snow, their trajectory is the well-known parabola derived in elementary physics, in engineering dynamics, and in Example 4.14 in this book. Using the x' and y' coordinates indicated in Fig. E4.27e and noting that V is the snowblower velocity and V_E is the speed relative to the snowblower, we obtain

$$x' = \frac{(V_E \cos\theta + V)^2}{g}\left(V_E \sin\theta - \sqrt{V_E^2 \sin^2\theta - 2gy'}\right).$$

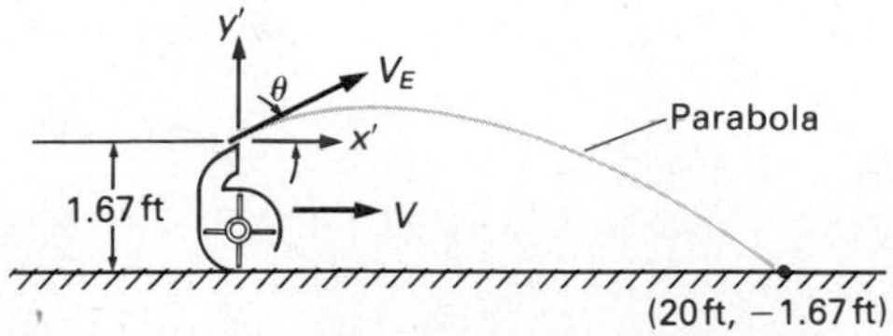

Figure E4.27e Trajectory of snow leaving snowblower.

Substituting

$$x' = 20 \text{ ft}, \quad y' = -1.67 \text{ ft}, \quad \theta = 30°, \quad \text{and} \quad g = 32.2 \text{ ft/sec}^2$$

and simplifying give

$$-11.45V_E^2 - 29.78V_E + 6447 = 0.$$

Using the quadratic formula and discarding the unrealistic negative value, we get

$$V_E = 22.46 \text{ ft/sec}.$$

This quantity is $V_{E,\min}$, because we assumed a 20-ft throw. The minimum acceptable rotational speed is

$$\omega_{\min} = \frac{V_{E,\min}}{R} = \frac{22.46 \text{ ft/sec}}{0.667 \text{ ft}} = 33.67 \text{ rad/sec}.$$

Converting to rpm yields

$$\omega_{\min} = 37.6 \text{ rad/sec}(1 \text{ rev}/2\pi \text{ rad})(60 \text{ sec/min}) \approx 321 \text{ rpm}.$$

The wide-open throttle value of ω with engine 100B is greater than this minimum acceptable value, so

Engine model 100B meets the design specifications. **ANSWER**

Discussion

Because the vane's rotational speed at wide-open throttle is larger than the minimum acceptable speed, snow would be thrown farther than 20 ft if the engine is operated at WOT. In practice, the snowblower engine would be operated at a lower throttle setting. Commercially available snowblowers often are equipped with a governing device that automatically adjusts the throttle position to maintain a fixed rotational speed.

Our assumptions in this example fall into three categories:

1. neglect of items considered minor;
2. conservative assumptions; and
3. unavailable information that must be estimated or researched.

Examples of the first type of assumptions and the reasons for these assumptions follow:

- We neglected the frictional force between the snow and the casing, because we expected it to be small. Including it would have required a more detailed analysis, which would have been quite cumbersome.
- Uniform inlet and exit velocity and exit velocity equal to vane tip velocity seemed reasonable. All snow entering the machine does so at the speed at which a person pushes the machine. Uniform exit velocity would have to be checked, but, as it leaves the machine, the snow would have a velocity close to that of the vane tip. The snow is last in contact with the vane tips as it leaves the vanes.
- Negligible power loss in the pulley system was reasonable. Pulley systems transmit power at an efficiency of about 95 percent, which we could have incorporated into the analysis without complicating it too much.

Examples of the second type of assumptions and the reasons why these assumptions are conservative follow:

- We used the maximum snow density, that of wet snow, because the greatest snow density gives the maximum power requirement. (See the equation for $\dot{W}_s$.)

We assumed that snow leaving the snowblower would not be deflected sideways. If it were deflected an angle ϕ (sideways and upward such that $\phi > \theta$), the power equation would be

$$\dot{W}_s = \rho b V R \omega (y_2 + R)(y_4 \omega \cos \phi - V).$$

Because

$$y_4 = y_3 + \left(\frac{V}{R\omega \cos \phi}\right)(y_2 - y_1),$$

then

$$\dot{W}_s = \rho b V R \omega (y_2 + R)\left[y_3 \omega \cos \phi + \frac{V(y_2 - y_1)}{R} - V\right].$$

The power decreases as ϕ increases; hence we have considered the case of the maximum power requirement by using $\phi = \theta$, as 0 is the minimum value of ϕ.

• Neglect of the pressure forces of the casing on the snow is partly justified, because the pressure along the circular surface AB in Fig. E4.27c is directed toward the center of rotation of the vanes. Moments were taken about this line of rotation, so the moment of the pressure force along AB is zero. The pressure forces along surfaces BC, CD, and EF are not directed toward the center of rotation. However, the pressure along CD is expected to exceed the pressure along EF (pressure increases as we move away from the center of curvature). In addition, area CD exceeds EF, and the clockwise moment $\mathcal{M}_{CD}$ of the pressure forces along CD exceeds the counterclockwise moment $\mathcal{M}_{EF}$ of the pressure force along EF. Taking moments gives

$$\sum \mathcal{M}_z = -\mathcal{T} - \mathcal{M}_{CD} + \mathcal{M}_{EF}.$$

Solving for $\mathcal{T}$ in the angular momentum equation gives a smaller value if we include these two pressure forces. Hence neglect of these two pressure forces (and of that along BC) gives larger torque and power estimates.

The third type of assumptions involves quantities that were not given and had to be postulated if we were to solve the problem. The density of wet snow was the best available, the 20-ft throw seems reasonable for a snowblower, and the 2.0-ft/sec velocity of pushing the snowblower is a reasonable estimate.

We should note however that reducing either the pushing speed, V, or the snow density will give a lower power estimate and could lead us to conclude that engine 100A is acceptable. Perhaps the potential customers would accept a slower speed or a narrower "cut" (<16 in.) in heavy snow if the price of the product were low enough.

4.5.3 Computer Support for Finite Control Volume Analyses

An important characteristic of the finite control volume method is that we can reduce a complicated flow to a rather simple analytical problem by carefully choosing a control volume and by using simplifying assumptions, such as locally uniform flow. In many cases, the mathematical model is so simple that a "pencil-and-paper" solution is feasible (possibly with the support of a hand-held calculator). The examples presented so far in this chapter illustrate this type of problem. As a general rule, problems that can be formulated in terms of a small number of explicit algebraic equations, or one or two simple ordinary differential equations, can be solved by "hand."

Control volume based analyses in support of engineering analysis or design often exhibit the following features:

- Nonlinear and implicit algebraic equations are present.
- One or more nonlinear ordinary differential equations appear.
- One or more definite integrals must be evaluated, and an integration formula is not immediately obvious.
- A large number of tedious or iterative calculations is required (especially true in design problems).

We can easily overcome such difficulties by using appropriate numerical methods with computer support.

Appendix F presents a brief introduction to several numerical tools that are helpful in solving the kinds of problems that arise in finite control volume analysis. We selected these particular tools for their simplicity in concept and application. In general, the equations associated with finite control volume analysis are mathematically well-behaved and amenable to solution by these techniques. Because use of these tools typically requires a large number of arithmetic operations, the aid of a programmable calculator or computer is indispensible.

Simply applying a particular numerical technique to a problem is not sufficient to guarantee a correct solution. First, you must be certain that you have implemented the algorithm correctly. The best way to test for correctness is to use the intended method to solve a similar problem for which you know an analytical solution. If you cannot find one, you should perform a careful pencil-and-paper check of the numerical results. Execution of several successful analyses may be required before you place confidence in the method. Even after you are certain that you have correctly implemented the algorithm, many methods involve parameters that you must arbitrarily select for a particular application. These parameters usually present a trade-off between accuracy and computational expense. You may need to select values that are "small enough" or "large enough" that they do not significantly detract from the accuracy of your solution. The time-step size to be used in the numerical solution of a time-dependent ordinary differential equation is one example of a user-specified parameter. A study of the dependence of the solution on these parameters is called a *convergence study*. Even simple problems may require a convergence study before a solution is complete.

To illustrate the interaction between a specific problem and computational support, we present a sequence of closely related examples, the first of which is easily solved using pencil and paper.

EXAMPLE 4.28 Starting Point for a Sequence of Examples Illustrating Numerical/Computer Support for Control Volume Analysis

A tank of liquid chemical fire extinguisher hangs from the ceiling above a factory floor. A hose connects the bottom of the tank to a nozzle 12 ft below. The tank is rectangular in cross section with dimensions of (5 ft × 5 ft × 5 ft) and the nozzle has a 2-in. diameter. If the tank is full initially, how long will the extinguisher supply last after the nozzle is opened? Assume that mechanical energy losses through the hose and nozzle are negligible.

SOLUTION

Given

Tank dimensions 5 ft × 5 ft × 5 ft

2-in. diameter outlet nozzle

Nozzle 12 ft below tank bottom

Figure E4.28

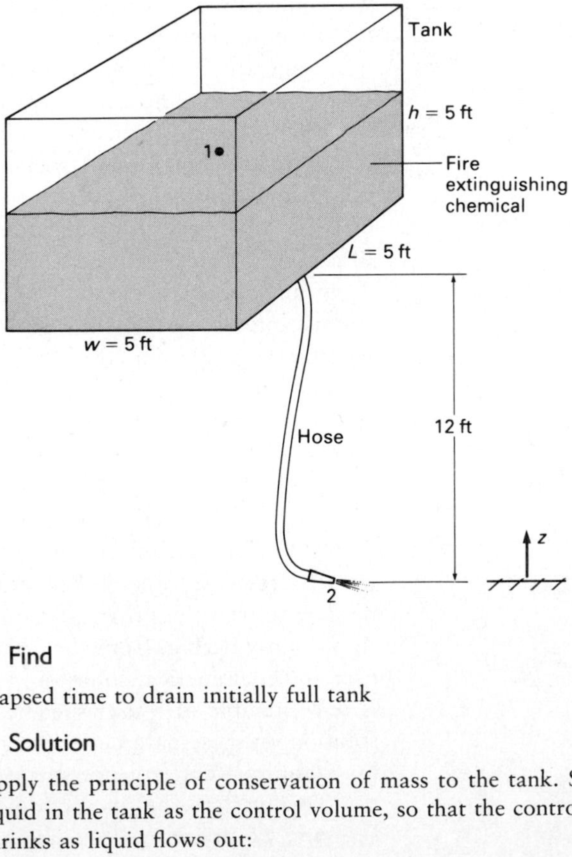

Figure E4.28 Fire extinguisher tank with hose.

Find

Elapsed time to drain initially full tank

Solution

Apply the principle of conservation of mass to the tank. Select the liquid in the tank as the control volume, so that the control volume shrinks as liquid flows out:

$$\frac{dm}{dt} = \rho \frac{d\forall}{dt} = \dot{m}_{\text{in}} - \dot{m}_{\text{out}} = -\rho Q_{\text{out}}$$

and

$$\frac{d\forall}{dt} = A \frac{dz_1}{dt} = wL \frac{dz_1}{dt} = -Q_{\text{out}} = -V_2 A_2.$$

Assuming inviscid flow, we write the Bernoulli equation between point 1 on the free surface in the tank and point 2 at the nozzle exit:

$$\frac{p_1}{\rho} + \frac{V_1^2}{2} + gz_1 = \frac{p_2}{\rho} + \frac{V_2^2}{2} + gz_2.$$

The pressure difference is zero, because both locations are at atmospheric pressure. The flow area at point 1 is much larger than the flow area at point 2, so the kinetic energy at point 1 is negligible compared to the kinetic energy at point 2. The zero elevation datum is selected at point 2. Solving for the velocity at the nozzle exit gives

$$V_2 = \sqrt{2g(z_1 - z_2)} = \sqrt{2gz_1}.$$

Substituting this result into the continuity equation, we have

$$\frac{dz_1}{dt} = \frac{1}{wL}(-V_2 A_2) = -\sqrt{2gz_1}\,\frac{A_2}{wL}.$$

We now rearrange and integrate from the initial values, t_i, z_i, to the instantaneous values, t, z_1, to get

$$\int_{z_i}^{z_1} \frac{dz_1}{\sqrt{z_1}} = 2(\sqrt{z_1} - \sqrt{z_i}) = \int_{t_i}^{t} -\sqrt{2g}\,\frac{A_2}{wL}\,dt$$

$$= -\sqrt{2g}\,\frac{A_2}{wL}(t - t_i).$$

Starting the draining process with the tank full at $t_i = 0$, $z_i = 12$ ft + 5 ft = 17 ft, and draining until the tank is empty, $t = t_f$, $z_1 = 12$ ft, provides the solution:

$$t = 188 \text{ sec.}$$ **ANSWER**

Discussion

The combination of simple physics and simple geometry permitted development of an explicit expression for the desired unknown.

A Change in Configuration. The solution of Example 4.28 did not require computational support. However, a change in the problem statement can have a significant impact on the difficulty of the problem. For example, a less expensive way to build the holding tank is by wrapping a single sheet of metal around two end plates with a curved cross section, requiring fewer welds to fabricate the tank.

EXAMPLE 4.29 Illustrates a Problem That Apparently Requires Numerical/Computer Solution

The tank in Example 4.28 is to be replaced with a less expensive tank with the same height and total volume, but with a cross section given by $y = x^2$, as shown in Fig. E4.29a. If the tank is full initially, how long can it supply extinguisher to fight a fire?

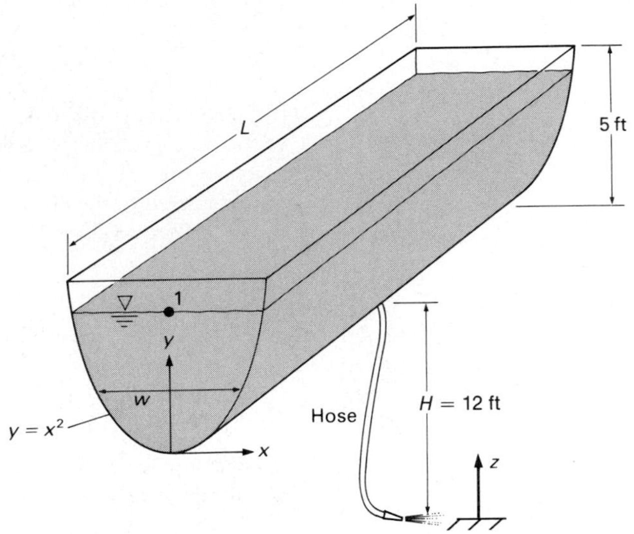

Figure E4.29a Modified fire extinguisher tank.

SOLUTION

Given

Figure E4.29a

125-ft^3 tank

2-in. diameter outlet nozzle

Nozzle 12 ft below tank bottom

Find

Elapsed time to drain initially full tank

Solution

The basic physics is unchanged from that of Example 4.28, so the solution procedure is the same. The crucial difference is that the volume in the tank is no longer a linear function of fill height. The expression for conservation of mass in the tank is now slightly different:

$$\frac{d\mathcal{V}}{dt} = wL\frac{dz_1}{dt} = (2x_1)L\frac{dz_1}{dt} = 2\sqrt{y_1}L\frac{dz_1}{dt} = -V_2A_2.$$

Application of the Bernoulli equation is identical to that of Example 4.28, yielding the same expression for the velocity at the nozzle exit:

$$V_2 = \sqrt{2gz_1}.$$

Substituting this expression into the continuity equation yields

$$2\sqrt{y_1}\,L\,\frac{dz_1}{dt} = -\sqrt{2gz_1}\,A_2.$$

Because $y_1 = z_1 - H$, we may rewrite the differential equation as

$$\frac{\sqrt{z_1 - H}}{\sqrt{z_1}}\frac{dz_1}{dt} = \sqrt{1 - \frac{H}{z_1}}\,\frac{dz_1}{dt} = \frac{-\sqrt{2g}}{2}\frac{A_2}{L}.$$

Inspecting this equation reveals two unknowns: z and L. We find the tank length easily by imposing the constraint that the tank volume is equal to 125 ft^3. Thus

$$L = \frac{V\!\!\!\!-}{A} = \frac{V\!\!\!\!-}{\frac{4}{3}(y_{max})^{3/2}} = \frac{(125)}{\frac{4}{3}(5)^{3/2}} = 8.39 \text{ ft}.$$

The problem is now reduced to one (differential) equation and one unknown. Because an analytic integration of this differential equation is not immediately obvious, we solve the problem using a numerical technique. From the many techniques suitable for solving this type of equation, we select the simplest, Euler's method, described in Appendix F. To apply the method, we rearrange the differential equation to obtain an explicit expression for the first derivative:

$$\frac{dz_1}{dt} = \frac{\sqrt{z_1}}{\sqrt{z_1 - H}}\cdot\frac{-\sqrt{2g}}{2}\frac{A_2}{L} = -\sqrt{\frac{2g}{1 - H/z_1}}\,\frac{A_2}{2L}.$$

We use Euler's method to estimate the change in free surface height over a specified time interval, Δt:

$$z_1(t + \Delta t) = z_1(t) + \left.\frac{dz_1}{dt}\right)_t \Delta t.$$

Knowing the initial conditions, we move the solution forward through time in increments of Δt. Before proceeding, however, we must select an appropriate value for Δt. Fortunately, we can look to Example 4.28 for guidance in selecting an appropriate time-step size. Draining the rectangular tank took 188 seconds, so a reasonable starting value for Δt is 10 seconds. This increment should result in approximately 20 steps, which we can review carefully without too much effort.

In order to proceed, we must select a means of implementing Euler's method on a computer. The proliferation of personal computers—and software for them—has greatly increased the number of options available. One approach is to use a traditional programming language, such as FORTRAN or BASIC, to write a program that performs the necessary calculations. Another approach is to use a spreadsheet program, such as Lotus 123® or Quattro®, in which we assign the appropriate operations to the cells in the spreadsheet. A third alternative is to use a "math-package," such as MathCad® or Mathematica®, in which the software can directly "solve" the differential equation. There is no "correct" choice. The programming language gives maximum flexibility and modularity; the spreadsheet offers instant organizational structure and graphic display of results;

and the math-package requires the least input to accomplish the task. In the final analysis, the choice of software to support a flow analysis is really a matter of personal preference and availability. The only requirement is correct implementation of an appropriate numerical method.

With the exception of the math-package, the first task in preparing the software is to organize mentally the steps necessary to implement the numerical method for the problem being solved. In order to maximize generality, we organize the steps before we identify a particular implementation language. The steps required to solve the draining problem are as follows:

- Initialize variables describing geometry and initial conditions.
- Increment the time level.
- Compute values at the new time level.
- Check for termination condition (tank empty)
  — if not empty, return to step 2 and repeat process
  — if empty, terminate calculation.

We implement these steps using FORTRAN, a "driver" program, and the Euler method subroutine presented in Appendix F. The idea is to isolate the fundamentals of the method in the subroutine so that it can be easily used for many applications. Problem specifics are contained in the driver and must change from problem to problem. The driver program in Fig. E4.29b is appropriate for the tank draining problem.

We first test our method by solving Example 4.28 with the driver and subroutine. Selecting a time step of 10 sec produces an elapsed time of 187 sec to drain the rectangular tank. Decreasing the time-step size to 5 sec reproduces the value given by the analytic solution. We now have confidence in the correctness of the subroutine and a good indication of a suitable time-step size.

To solve the current problem, we select for the first run a time step of 5 sec and compute an elapsed time to drain the tank of 189 sec.* To study the convergence of this solution, we rerun the program with time-step sizes of 1, 0.5, 0.1, and 0.05 sec. The predicted drain times from these runs are 186, 186, 185, and 185 sec, respectively. Therefore:

The entire supply of extinguisher in a full tank is depleted in 185 seconds. **ANSWER**

Discussion

Changing the holding tank of the previous example to one in which the cross-sectional area is a function of fill height did not change the basic physics but did make the solution of the governing equations more difficult. We verified the correctness of the numerical tool that we used to solve the problem by using it to solve a similar problem for which we already had an analytic solution.

* The value of 189 sec is obtained by linear interpolation between 185 sec and 190 sec.

```
          program ex4Cb
c.......... initialize problem
          data hosel/12./, i/0/, t/0./, tankh/5./
          z         = hosel + tankh
          zstop     = hosel
          c.......... inquire for time-step size
          write (*, 6)
          read (*, 7) dt
          write (*, 3)
          write (*, 2) i, t, z
c.......... if-loop to advance solution until tank is empty
        5 i    =    i + l
          call Euler (t, dt, z, ztpdt)
          t = t + dt
          write (*, 2) i, t, ztpdt
          if ( ztpdt .gt. zstop ) then
                z = ztpdt
                go to 5
          endif
c.......... terminate calculation
          tend = t – (dt/(z–ztpdt))*(zstop–ztpdt)
          write (*, l) dt, tend
          stop
c
        1 format (t2, 79 ('*')/t2, '**',' Using Euler's Method with a '
         &   , 'time-step size of', e10.3,' seconds;', t79, '**'/t2, '**'
         &   , ' The tank is completely drained in', e10.3,' seconds.', t79
         &   , '**' /t2, 79 ( '*' ) )
        2 format (t2, i8, t15, ell.4, t31, ell.4)
        3 format (t2, 'STEP NO.', t20, 't', t36, 'z')
        6 format (t2, 'Enter time-step size:')
        7 format (f10.3)
          end
        function yprime (x, y)
c       function subprogram to compute first derivative of y.
        data dnozzl/.1667/, g/32.2/, hosel/12./, pi/3.14159/, tankw/5./
c.......... example c4a
c         tankl     = 5.
c         const     = sqrt (2. *g) * (0.25*pi*dnozzl**2) / (tankw*tankl)
c         yprime    = const*sqrt(y)
c
c.......... example c4b
          tankl     = 8.39
          const     = –sqrt (2. *g) * ( 0.25*pi*dnozzl**2) / (2. *tankl)
          yprime    = const / sqrt ( l – hosel/y )
          return
          end

c****************************************************************
c**   Driver program to solve Examples 4.28 and 4.29 from GGH text.   **
c**       Variable names are self-evident.                              **
c**       Problem is setup and solved in EE units (lb, ft, s)           **
c**                                          Written:  GGH/26JAN90      **
c**                                    last modified:                   **
c****************************************************************
```

Figure E4.29b "Driver" program for tank draining problem.

Although the subroutine and external function used to solve this problem were simple, combining them with a "driver" is a good practice. By isolating the solution procedure in a subroutine that is not modified after testing against a known solution, you have confidence that it is not a source of errors.

We performed a convergence study to ensure that the solution is independent of our choice of parameters for the numerical tool. Although reducing the time-step size from 0.5 to 0.1 sec did change the answer in the third significant place, whether this is significant compared to other approximations introduced during the solution is questionable.

As a final note, we should point out that performing an analytic integration of the governing equation by using a substitution; $z = H\cosh^2\xi$, is in fact possible. The resulting equation is

$$t_f = \sqrt{\frac{2}{g}}\frac{LH}{(\pi/4)d^2}\left[\frac{1}{4}[\sinh(\xi_f) - \sinh(\xi_i)] + \frac{\xi_i - \xi_f}{2}\right].$$

Inserting the appropriate values for this problem predicts an elapsed drain time of 185 sec.

Utility of Computer Solution. By no means should you think that the programming time invested in the preceding example was wasted. First of all, for many engineers obtaining the numerical solution would consume less time and effort than searching for the appropriate trick to integrate the differential equation. Secondly, the numerical tool is now capable of computing the drain time for any tank cross-sectional distribution that the user can specify. In fact, this is one of the great strengths of properly constructed numerical tools: A range of problems can be solved with only minor modification of the tool. Also, another "twist" to the problem reveals the utility of a general computer program for solving the tank draining problem.

EXAMPLE 4.30 Illustrates Computer Application to a "Design" Problem and Use of Spreadsheets for Problem Solving

A new requirement has been imposed on our fire extinguisher system by the fire-safety inspector: The flow of extinguisher must last at least 5 min. The question now becomes one of nozzle sizing. What is the maximum nozzle size that can be used for this application?

SOLUTION

Given

125-ft^3 tank shown in Fig. 4.29a

Nozzle 12 ft below tank bottom

Minimum of 5 min to drain tank

Find

Maximum permissible nozzle diameter

Solution

The basic physics, and with the exception of the nozzle size, the geometry of Example 4.29 is unchanged. Therefore the governing differential equation is the same:

$$\frac{dz_1}{dt} = -\sqrt{\frac{2g}{1 - H/z_1}}\,\frac{A_2}{2L}.$$

For this problem, we use a spreadsheet to implement Euler's method. Figure E4.30a shows a Lotus 123® spreadsheet for solving the tank draining problem. To check for correctness of the implementation, we first use the spreadsheet to recompute the solution to Example 4.29a.

Anticipating that various values will be used for the nozzle diameter and time-step size, we enter these variables in cells C3 and C5. The nozzle area is calculated in cell F5 by the formula @PI*C3*C3/4/144. Column headings for t, z, and dz/dt are entered

	A	B	C	D	E	F	G	H
		(Solution of dz/dt = – sqrt [2g/(1–h/z)] * A/2L)						
1		Diameter	2 in.	Area	0.021816 ft^2			
		Time step	5 sec					
2			t	z	dz/dt			
3			0.00	17.00	–0.0192			
4			5.00	16.90	–0.0194			
5			10.00	16.81	–0.0195			
6			15.00	16.71	–0.0196			
7			20.00	16.61	–0.0198			
8			25.00	16.51	–0.0199			
9			30.00	16.41	–0.0201			
10			35.00	16.31	–0.0203			
11			40.00	16.21	–0.0205			
12			45.00	16.11	–0.0207			
13			130.00	14.17	–0.0267			
14			135.00	14.03	–0.0274			
15			140.00	13.89	–0.0282			
16			145.00	13.75	–0.0292			
17			150.00	13.61	–0.0303			
18			155.00	13.46	–0.0317			
19			160.00	13.30	–0.0334			
20			165.00	13.13	–0.0355			
21			170.00	12.95	–0.0385			
22			175.00	12.76	–0.0427			
23			180.00	12.55	–0.0499			
24			185.00	12.30	–0.0671			
25			190.00	11.96	ERR			

Figure E4.30a Lotus 123® spreadsheet for tank draining.

in cells C7, D7, and E7. The initial conditions for $t(0)$ and z (17[ft]) are entered in cells C8 and D8. The value of dz/dt at $t = 0$ is calculated in cell E8 by the formula

−@SQRT(2*32.17/(1 − 12/D8))*$F$3/2/8.39.

The values for g, h, and L are entered as numerical constants, because they will not vary in our study.

We are now ready to advance the solution through the first time step. We calculate the new value of t by adding the time step to the last available value of t. To do this, the formula +C8+$C$5 is entered in cell C9. To calculate the next value of z, we simply multiply the most current value of dz/dt by the time step and add the result to the most current value of z. The formula +C8+E8*$C$5 does this in cell D9. The value of dz/dt at this time can be calculated in cell E9 by COPYing cell D8 into cell D9. The spreadsheet automatically adjusts the cell references to produce

−@SQRT(2*32.17/(1 − 12/D9))*$F$3/2/8.39.

With this work, the basic spreadsheet is set up. We now advance the calculation through time by COPYing the range of cells C9 . . E9 directly below itself. To calculate the tank draining time, we need only to keep COPYing until the magnitude of z falls to 12.0 or less. This occurs in row 46, for which the display is

190.00 11.96 ERR.

The ERR appears in the dz/dt column because $(1 - 12/z)$ inside the SQRT has become negative. We avoid this slight inconvenience by using

−@SQRT(2*32.17/(1 − 12/@MAX(Cxx,12.001)))*$F$3/2/8.39

in the dz/dt cells and

@MIN(12,+Cxx+Exx*$C$5)

in the z cells. This approach provides a constant value of $z = 12$ after the tank is empty. Specifying a time-step size of 5 sec reproduces the solution obtained in Example 4.29(a).

In the current problem, we know the drain time (300 sec) and must use an iterative approach to find the maximum permissible nozzle size. The basic procedure is to guess a nozzle size and calculate the resulting drain time. We use a time-step size of 5 sec to expedite execution until we are near the correct diameter. Knowing that the 2-in. diameter nozzle drains in approximately 3 min, we proceed with the following sequence of guesses and compute the corresponding drain times. (The range of rows in the spreadsheet will have to be expanded downward to allow enough time for the tank to drain.)

Diameter	Drain Time
1.00	745
1.50	334
1.75	246
1.62	286

1.56	309
1.59	297
1.58	301

The last three appear to be near the answer, so we switch to smaller time-step sizes to obtain more accurate results. Again, the range of spreadsheet rows has to be increased to capture the entire draining process.

Time Step	Diameter	Drain Time
1.0	1.58	298
1.0	1.57	302
0.5	1.57	301
0.2	1.57	300

We are now confident that our solution is accurate.

$$D = 1.57 \text{ in.}$$ **ANSWER**

Discussion

Although the physics and basic geometry are unchanged from Example 4.29, an iterative procedure is required because a different variable was the unknown. With computer support, the iteration was relatively painless.

The sequence of "guesses" follows a numerical method known as *bisection* or *interval halving*. Once an upper and lower bound on the independent variable is established, the interval between them is halved to search for the answer. This procedure is described in more detail as a root-finding method in Appendix F. A more sophisticated search procedure, such as the Newton–Rhapson procedure, converges more quickly but requires more calculations to obtain a next guess.

The choice of a spreadsheet implementation does not significantly alter the nature of the solution. One advantage offered by most spreadsheet programs is the option of graphic display of results. For instance, the fill height as a function of time is shown in Fig. E4.30b.

Rather than having a human in the loop, the iterative search procedure may be implemented as part of the software. The bisection algorithm is a simple way to automate the search; however, more sophisticated root finders may become attractive, because the machine performs the calculations required to obtain the next guess. The benefit of an automated search algorithm also becomes clear if the analysis must be repeated for a variety of tank shapes in a variety of installations.

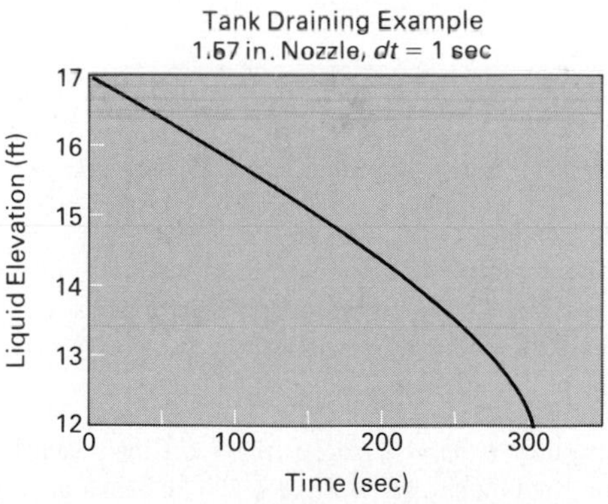

Figure E.4.30b Plot of spreadsheet results.

Reviewing the Analysis. We review the results from an engineering perspective. The (quasi-steady) mathematical model of the physical system neglects kinetic energy at the free surface and mechanical energy losses in the system. Although we can extend the computed answer to more significant digits, to do so is inappropriate because of the modeling simplifications. The power of good computational tools often tricks engineers into thinking that their results are more accurate than the model on which the results are based. Neglecting the kinetic energy at the free surface is a minor assumption. Neglecting the losses has a major impact on the predicted drain time, causing an overprediction of flow velocity and an underprediction of drain time. Therefore our analysis is conservative: In the physical system, the drain time for the selected nozzle size exceeds the minimum requirement.

Concluding Remarks. As you can see, there is definitely a place for computers in support of finite control volume analysis. Again, you should recall the caution expressed in the discussion of computational tools in Chapter 1: There is no substitute for good engineering judgment! Review the plausibility of your solution. Presenting the results of a long and sophisticated analysis—only to have a listener point out that the solution violates a fundamental principle and therefore must be in error—can be extremely embarrassing. Use computer-based techniques to support your engineering efforts but do not blindly rely on them to provide "the answer."

PROBLEMS

1. Water having a density of 1000 kg/m^3 and flowing out of a 2.0-cm-inside-diameter pipe is collected in a weigh tank at the rate of 0.981 N/s. Find the average water velocity in the pipe.

2. A liquid flows through an annular area having an inside radius of 2.0 in. and an outside radius of 5.0 in. The flow rate was measured to be 15 gal/min. Find the average liquid velocity in the annular area.

3. A rectangular tank measuring 50 cm long × 25 cm wide × 15 cm high is being filled with a liquid by a tube having an inside diameter of 5.0 cm and a flow rate of 100 cc/s. Calculate the average fluid velocity in the tube and the rate at which the liquid depth increases with time.

4. Figure P4.4 shows the impeller of a centrifugal pump. The width $b = 5.0$ cm. Liquid enters with a velocity W_1 *relative* to the impeller and at an angle $\theta_1 = 30°$ with a normal to the inlet area $2\pi r_1 b$. Liquid leaves with a velocity W_2 *relative* to the impeller and at an angle $\theta_2 = 45°$ with a normal to the outlet area $2\pi r_2 b$. The numerical values are $W_1 = 5.0$ m/s, $W_2 = 3.06$ m/s, $r_1 = 0.75$ m, and $r_2 = 1.50$ m. Determine the inlet and outlet volume flow rates.

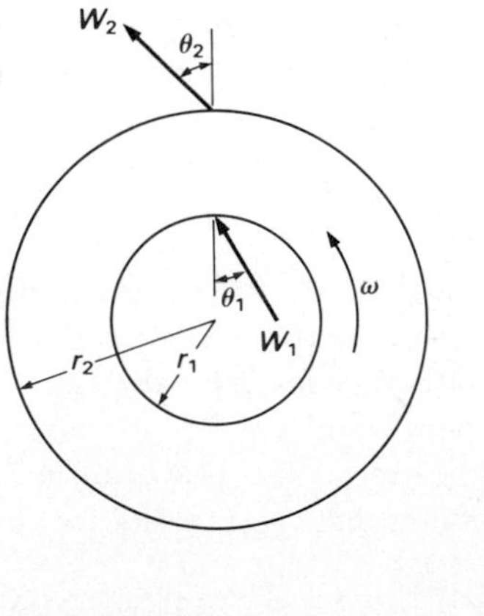

Figure P4.4

5. Rain is falling vertically with a velocity of 3 ft/sec. There are 200 $\frac{1}{8}$-in.-diameter water droplets per cubic foot. How long will it take for a small vessel measuring 1 in. on a side and 4 in. high to fill up with water?

6. A hypodermic syringe is being used to inject oil into a machine part. The inside diameter of the glass tube is 1.0 cm, and the plunger moves at the rate of 0.5 cm/s. How long will it take to discharge 12 cc of oil?

7. A stationary tube is used to inject oil into bearings moving on the belt shown in Fig. P4.7. The bearings have a diameter of 3.0 cm, and the belt moves at a speed of

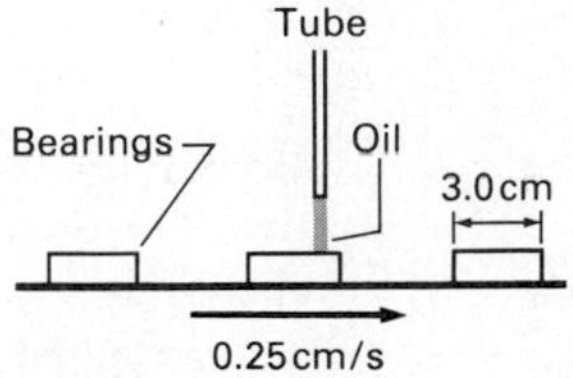

Figure P4.7

0.25 cm/s. What must be the minimum flow rate out of the tube to provide each bearing with at least 3.0 cc of oil?

8. Rainwater is flowing at a steady rate down a street having an incline of 30° with the horizontal. A scrap of paper on the surface of the water moves at the rate of $U = 1$ ft/sec. The street is 20 ft wide and the water depth $h = \frac{1}{2}$ in. Estimate the flow rate of the rainwater for the following water velocity profiles:

(a) $u = U$;

(b) $u = Uy/h$;

(c) $u = U\left(2\frac{y}{h} - \frac{y^2}{h^2}\right)$.

9. A river is approximately 30 ft wide and has an average depth of 7 ft. The surface water flows at an estimated speed of $U = 6$ ft/sec. Estimate the river flow rate if the fluid velocity varies linearly with depth.

10. A constant-density fluid flows through a converging section having an area A given by

$$A = \frac{A_0}{1 + (x/\ell)},$$

where A_0 is the area at $x = 0$. Determine the velocity and acceleration of the fluid in Eulerian form and then the velocity and acceleration of a fluid particle in Lagrangian form. The velocity is V_0 at $x = 0$ when $t = 0$.

11. Kerosene has an average velocity of 1.2 m/s through a 0.025-m-inside-diameter pipe. Calculate the volume flow rate.

12. A 2-in. schedule 40 pipe has an average liquid velocity of 0.5 ft/sec. The liquid density is 1.7 slugs/ft^3. Find the mass flow rate.

13. Type L copper tube is connected by a tee. Oil enters by two legs of the tee and leaves by the third leg. The two inlet flow rates are 10.0 gal/min and 6.0 gal/min. Find the average velocity in the outlet tube, which has a nominal diameter of 1.0 in.

14. The weight W in Fig. P4.14 is falling at the rate of 3.0 cm/s. Find the liquid flow rate Q leaving the discharge tube.

15. Air enters the axial compressor shown in Fig. P4.15 at 101 kPa absolute, 27°C, and an average axial velocity component of 1.5 m/s. Find the air mass flow rate through the compressor.

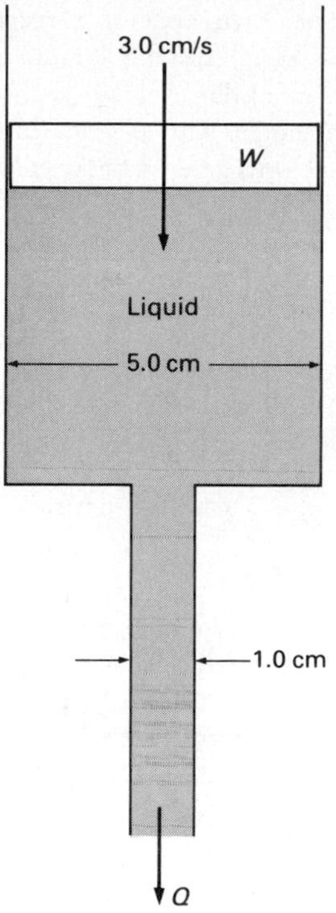

Figure P4.14

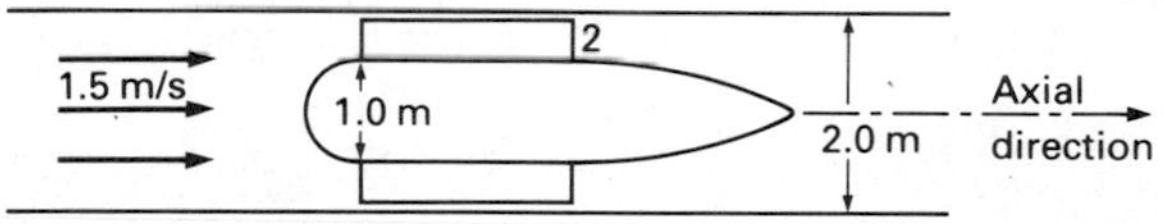

Figure P4.15

16. The absolute pressure and temperature at point 2 of the axial compressor in Fig. P4.15 are 202 kPa and 87°C. The air mass flow rate is 5.53 kg/s. Calculate the average axial velocity at point 2.

17. Laminar flow in a circular pipe has the following velocity profile:

$$u = u_{max}\left[1 - \left(\frac{r}{R}\right)^2\right],$$

where r is a radial coordinate measured from the pipe centerline, u_{max} is the maximum fluid velocity and occurs at $r = 0$, and R is the inside radius of the pipe. Find the volume flow rate passing through the pipe in terms of R and u_{max}.

18. The cross-sectional area of a rectangular duct is divided into 16 equal rectangular areas, as shown in Fig. P4.18. The axial fluid velocity measured in feet per second in each smaller area is given in the figure. Estimate the volume flow rate and average axial velocity.

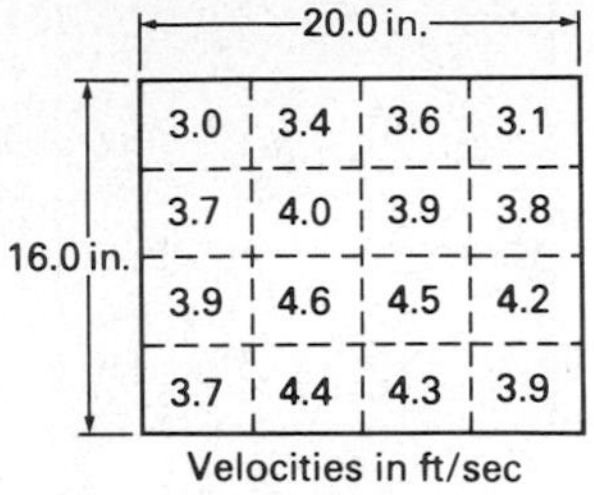

Figure P4.18

19. A thin, cylindrical annulus has a spiral flow with a velocity of 4.0 ft/sec, which makes an angle of 30° with the cylinder's axial direction. The inside annulus radius $r_1 = 0.50$ ft and the outside annulus radius $r_2 = 0.60$ ft. Find the volume flow rate.

•**20.** A student has 7 min to go from one building to another through a moderate rain. Would the student get wetter by running or by walking? Assume that the rain is falling straight down.

21. The following velocity profile was measured for a turbulent flow through a 20-cm-diameter pipe.

r (cm)	*V* (m/s)
0.0	50.0
1.0	49.3
2.0	48.4
3.0	47.5
4.0	46.5
5.0	45.3
6.0	43.9
7.0	42.1
8.0	39.7
9.0	36.0
10.0	0.0

What is the volume flow rate through the pipe?

22. A woman is emptying her aquarium at a steady rate with a small pump. The water is pumped to a 12-in.-diameter cylindrical bucket, and its depth is increasing at the rate of 4.0 in. per minute. Find the rate at which the aquarium water level is dropping if the aquarium measures 24 in. (wide) × 36 in. (long) × 18 in. (high).

23. A thin, cylindrical annulus has a spiral flow with a velocity of $V = 4.0$ ft/sec., which makes an angle of 30° with the cylinder's axial direction, as shown in Fig. P4.23. The inside annulus radius $r_1 = 0.50$ ft and the outside annulus radius $r_2 = 0.60$ ft. Straightening vanes redirect the flow, so the velocity becomes totally axial. Calculate this new axial velocity.

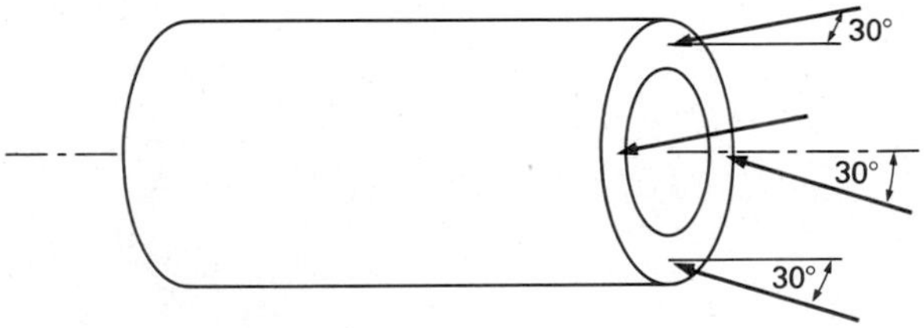

Figure P4.23

24. A rectangular tank measuring 25 ft long × 20 ft wide × 15 ft high is being filled with water at the rate of 500 gal/min. Water is leaking out the bottom at a rate of 20 gal/min. Find the rate at which the water level rises.

25. A rectangular tank measuring 50 cm × 25 cm × 15 cm is being filled with liquid by a tube having a constant flow rate of 100 cc/s. How long will it take to fill the tank?

26. Find the time required to fill the tank in Problem 24 if the tank is initially empty.

27. A rectangular tank measuring 10 m × 8 m × 5 m high is being filled with oil at the rate of 400 L/min. Oil drains from a hole at the midheight at the rate of 100 L/min whenever the oil level is above the midheight. Find the amount of time required to fill the tank.

28. The jet pump shown in Fig. P4.28 has a primary stream of flow area $A_1 = 0.1$ m^2 and velocity $V_1 = 6.0$ m/s, and a secondary stream of flow area $A_2 = 0.9$ m^2 and velocity $V_2 = 1.0$ m/s. The two streams mix and enter a duct of flow area $A = 1.0$ m^2. For constant density, find the downstream average velocity.

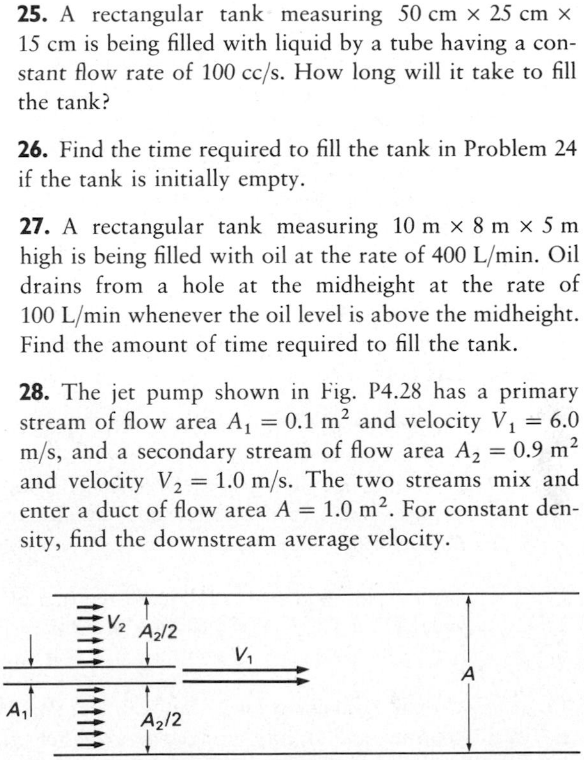

Figure P4.28

29. A 0.5-cm-outlet-diameter nozzle is connected to a 3.0-cm-inside-diameter pipe. The constant density fluid flowing in the pipe has a velocity of 1.0 m/s. What is the nozzle outlet velocity?

30. An ideal gas at 198 kPa absolute, 400K, and velocity 10 m/s is flowing in a 1.0-cm-inside-diameter pipe. The gas leaves through an outlet diameter of 0.5 cm at 101 kPa absolute and 300K. Determine the outlet velocity.

31. In the vortex tube shown in Fig. P4.31, air enters at 202 kPa absolute and 300K. Hot air leaves at 150 kPa absolute and 350K, whereas cold air leaves at 101 kPa absolute and 250K. The hot air mass flow rate, $\dot{m}_H$, equals the cold air mass flow rate, $\dot{m}_C$. Find the ratio of the hot air exit area to cold air exit area for equal exit velocities.

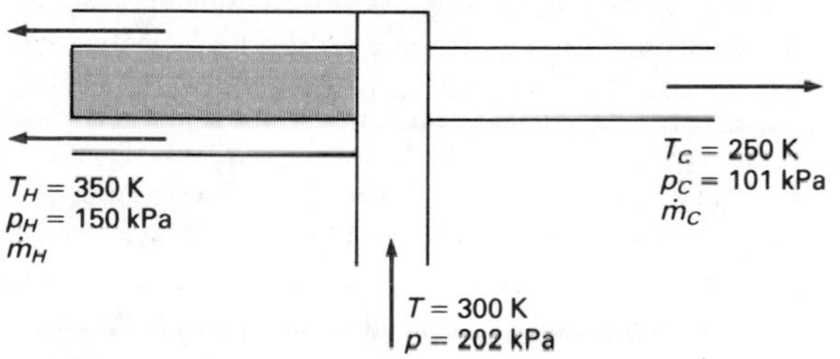

Figure P4.31

32. The velocity field in a nozzle is both one-dimensional and one-directional. The axial velocity is

$$u = U_0 e^{x/4\ell},$$

where $x = 0$ represents the inlet and the inlet area is 0.25 m^2. For constant-density flow, find the exit area where $x = \ell$.

33. Write an equation for the nozzle flow area as a function of the coordinate x for the flow of a constant-density fluid in Problem 32.

34. Brine flows through a pipe of inside radius $R = 1.5$ in. with an axial velocity of

$$u = U_0\left[1 - \left(\frac{r}{R}\right)^2\right],$$

where $U_0 = 1.5$ ft/sec. The salt concentration is

$$c = c_0\left(\frac{r}{R}\right),$$

where $c_0 = 0.00155$ slug/ft^3. Find the mass flow rate of salt through the pipe. Is it equal to the average brine velocity times the average brine concentration times the pipe flow area? Explain.

35. Oil flows into a tank at the rate of 10 m^3/s and leaves through a 0.5-m^2 hole in the side of the tank with a velocity of

$$V = \sqrt{19.6h},$$

where V is in meters per second and h, the vertical distance from the centerline of the hole to the free surface of the oil, is in meters. Find the steady-state value of h.

36. Show that $\bar{\rho}$ in Eq. (4.16) equals $\bar{\rho}'$ in Eq. (4.17) for constant density.

37. Show that $\bar{\rho}$ in Eq. (4.16) does not necessarily equal $\bar{\rho}'$ in Eq. (4.17) for a variable-density fluid. Use

$$u = U_0\left[1 - \left(\frac{r}{R}\right)^2\right] \quad \text{and} \quad \rho = \rho_0\left[1 + \left(\frac{r}{R}\right)^2\right],$$

where R is a pipe radius, U_0 is the centerline velocity, and ρ_0 is the centerline density.

38. A tank has horizontal cross-sectional area A and height H. Water flows into the tank at a steady rate Q_i and leaves at a rate of

$$Q_e = A_e\sqrt{2gh},$$

where A_e is the exit area of a nozzle in the side of the tank, g is the local acceleration of gravity, and h is the water height above the centerline of the nozzle. Develop the differential equation for the water height h as a function of time.

39. Civil engineering students are participating in a concrete canoe race. The canoe, shown in Fig. P4.39, springs a leak in its bottom. The velocity of the water through the hole is 10 ft/sec. The students set up a hand pump to bail out the canoe. Find the rate at which water is filling or emptying the canoe. Assume that the water temperature is 60°F.

40. In a furnace, 18 slugs of air react with every slug of methane (CH_4) for combustion. The products of combustion leave at 300°F and 14.7 psia and have a specific gas constant of 1700 ft·lb/slug·°R. Find the velocity of the products of combustion through a 6-in-diameter duct if the furnace burns 0.373 slug of fuel per hour.

41. Oil passes through the porous base of the thrust bearing shown in Fig. P4.41 at the rate of $q = 0.12$ L/cm^2·min. Find the velocity of the oil leaving the outside area of height $h = 0.1$ cm and diameter $D = 5.0$ cm.

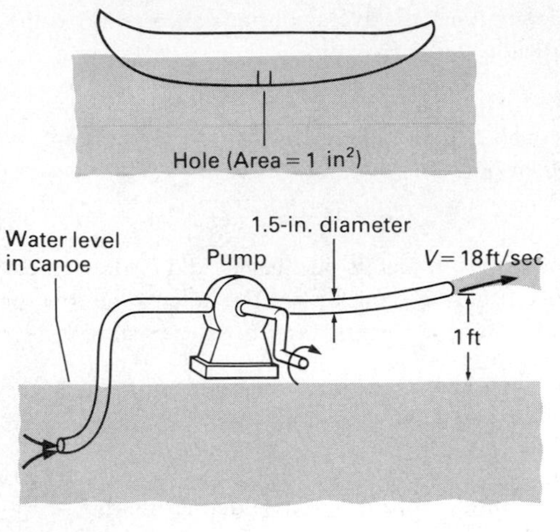

Figure P4.39

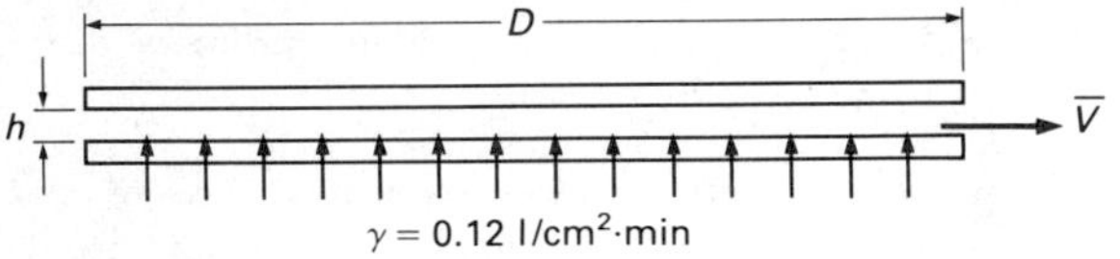

Figure P4.41

42. The rocket engine shown in Fig. P4.42 has a liquid fuel flow rate of 10 slugs/min and a liquid oxygen flow rate of 25 slugs/min. The products of combustion leave the exhaust nozzle at $p_e = 12$ psia, $T_e = 1500°$R, and a velocity of 5000 ft/sec. The specific gas constant for the products is 1650 ft·lb/slug·°R. Find the exit area A_e.

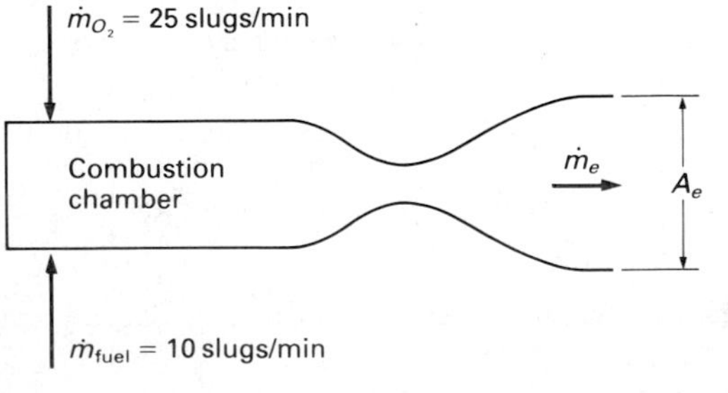

Figure P4.42

43. In the vortex tube shown in Fig. P4.43, air enters at 29.4 psia and 540°R. Warmer air leaves at 21.8 psia and 630°R, and cold air leaves at 14.7 psia and 495°R. The cold air mass flow rate, $\dot{m}_C$, is twice the hot air mass flow rate, $\dot{m}_H$. Find the ratio of the hot air exit area to cold air exit area for equal exit velocities.

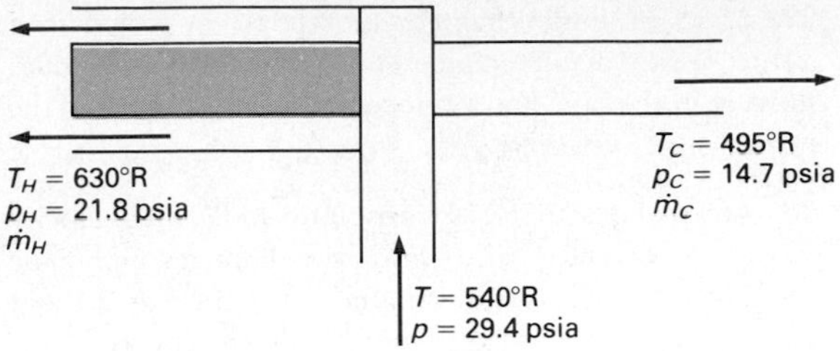

Figure P4.43

44. A gas with a density of 0.00333 slug/ft³ is flowing in a circular duct of 24-in. diameter with a velocity of 4.0 ft/sec. As the gas flows down the duct, the density decreases to 0.00111 slug/ft³. Calculate the new average gas velocity.

45. Repeat Problem 44 for a duct diameter that decreases from 24 in. to 18 in. Find the new average gas velocity at the 18 in. diameter section, where the density is 0.00111 slug/ft³.

46. Oil is drained from the upper rectangular tank to the lower square tank shown in Fig. P4.46. The oil depth d_S is increasing at the rate of 1.0 ft/min. Determine the time rate of change of the oil level in the rectangular tank (i.e., dd_R/dt).

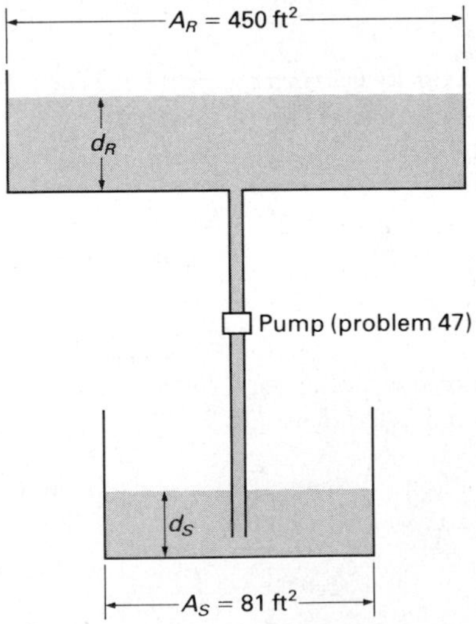

Figure P4.46

47. Oil is pumped from the lower square tank to the upper rectangular tank shown in Fig. P4.46 at the rate of 25 ft³/min. Find the time rate of change of the two oil depths, d_S and d_R.

48. Rainwater flows down a 4.0-m-wide rectangular open channel. The water depth is 1.5 m, and the average velocity is 3.2 m/s. A kilometer downstream the water depth has decreased to 1.2 m. Find the new average velocity. Assume a constant channel width.

49. At the downstream point in Problem 48, the channel width has decreased to 3.0 m. All other values remain the same. Find the average downstream velocity.

50. The device shown in Fig. P4.50 has a water inlet velocity of $V_1 = 100$ m/s and pumps water from the well at 0.1 m^3/s. Determine the downstream velocity, V_2. Assume constant water density.

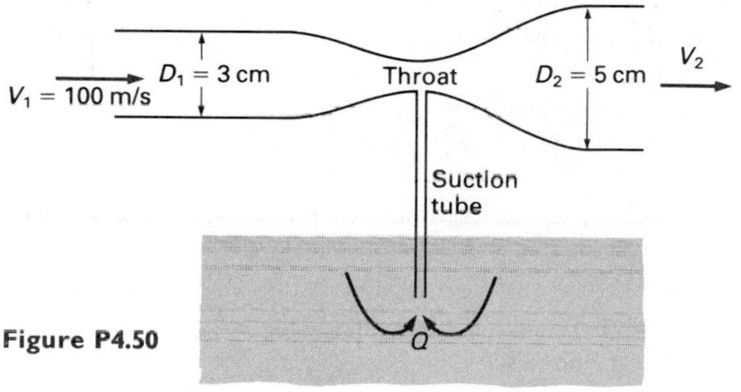

Figure P4.50

51. Find the minimum throat area in Problem 50 so that the maximum water velocity does not exceed 150 m/s.

52. Figure P4.52 shows a two-reservoir water supply system. The water level in reservoir 1 drops at the rate of 0.01 m/min, and the water level in reservoir 2 drops at the rate of 0.015 m/min. Calculate the average velocity V_3 in the 0.50-m-diameter pipe.

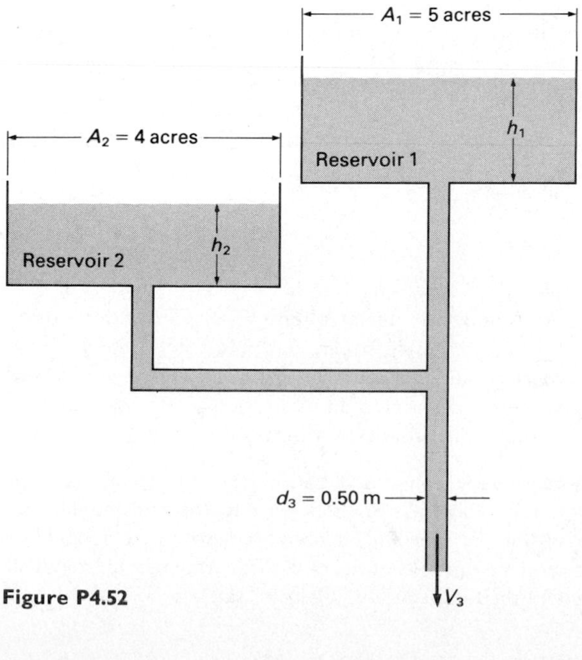

Figure P4.52

53. Rain is falling straight down on a street 12 m wide and 200 m long. The rain's water droplets are perfect spheres with a diameter of 0.3 cm. There are 4×10^4 droplets per cubic meter, and the droplets are falling with a velocity of 0.5 m/s. The water drains from the street into a 20-cm-inside-diameter pipe. Find the pipe's water velocity if the pipe is running full.

54. Liquid alcohol is drained from the rectangular tank to the 10-ft-long cylindrical tank shown in Fig. P4.54. The liquid depth d_R in the rectangular tank is dropping at the rate of 4.0 in./min. Calculate the rate at which the liquid depth d_c is increasing in the cylindrical tank for the depth d_c indicated.

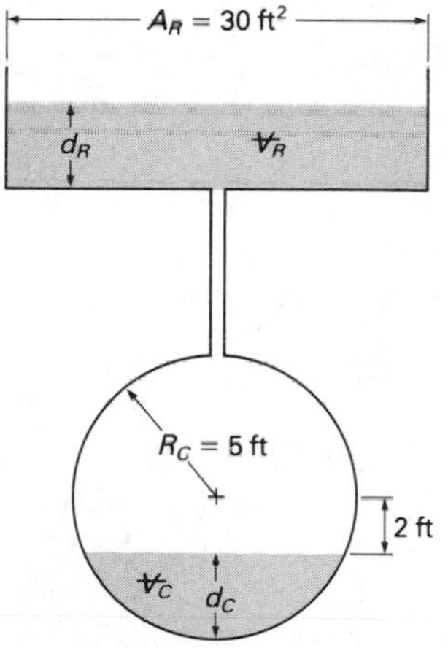

Figure P4.54

55. The hull of a ship is often subdivided into watertight compartments to prevent the ship's sinking from a single puncture of the hull. Consider a rectangular compartment with a floor area of 200 ft^2 and a 2-ft-wide doorway with the door sill 1 ft above the floor of the compartment, as shown in Fig. P4.55. If water flows through a

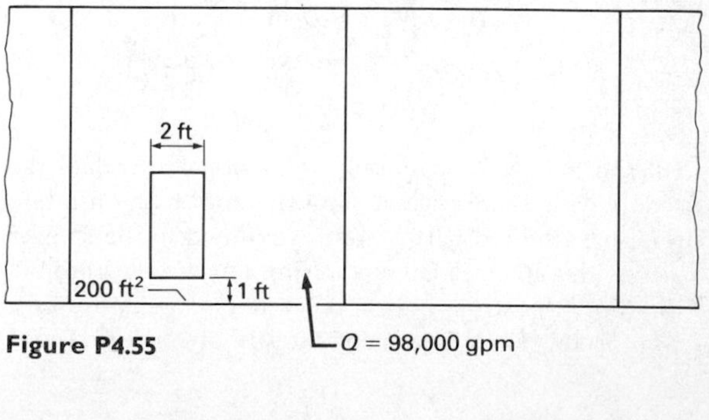

Figure P4.55

puncture in the hull into the compartment at a rate of 98,000 gal/min, how long will it take for water to begin to flow through the doorway?

56. Air at 20°C and atmospheric pressure flows into a pipe, as shown in Fig. P4.56. The pipe wall is perforated so that air flows out through it. The pipe diameter is 1 m, the pipe length is 4 m, and air enters the pipe with a velocity of 45 m/s. The holes in the pipe wall are 6 mm in diameter and there are 800 holes/m^2 of wall area. The air velocity at each hole is 10 m/s. Compute the mass flow rate of air entering the pipe at the left end, the volume flow rate of air leaving the pipe at the right end, and the air velocity at the pipe exit.

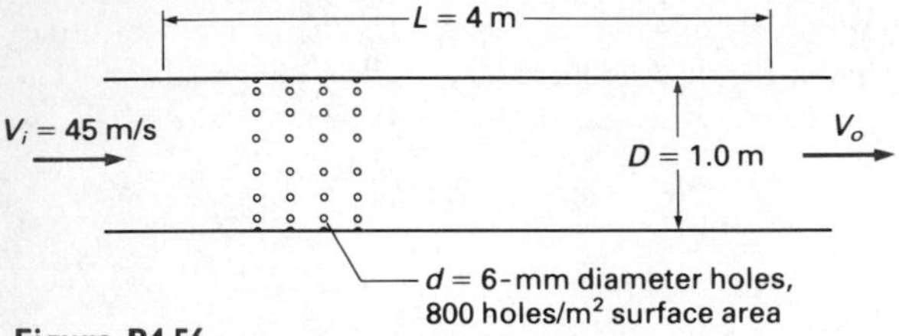

Figure P4.56

57. Oil for lubricating the thrust bearing shown in Fig. P4.57 flows into the space between the bearing surfaces through a circular inlet pipe with velocity

$$u = U_0\left[1 - \left(\frac{r}{R}\right)^2\right],$$

where $R = 1.5$ mm. The oil has a specific gravity $S = 0.86$ and flows in the inlet pipe at a rate of 0.006 kg/s. Compute the *average* velocity V_1 of the oil in the inlet pipe and the average velocity V_2 at the outlet (plane 2) and the maximum velocity in the oil inlet pipe (U_0). Assume radial flow.

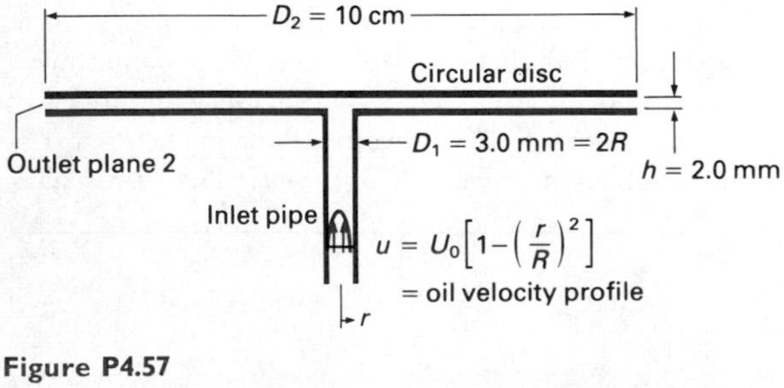

Figure P4.57

58. On a Boy Scout outing, the troop leader had the scouts design a hot-water shower. From high on a hill, the boys directed part of a stream down to the shower area with a 10-cm (I.D.) pipe. Using a massive fire around the pipe, the scouts raised the water temperature to a comfortable level. Determine the size of conduit needed to slow the water from 6.0 m/s to 2.0 m/s. The water density is constant.

59. A storage tank has a volume $\mathcal{V}_A = 10$ m^3. Water is entering with an average velocity of $V_1 = 50$ cm/s through a 10.0-cm-inside-diameter pipe. At the bottom of the tank, water is withdrawn at a constant rate of 500 cc/min. How long will it take to fill the tank if it is initially empty. Assume constant water density.

60. Water is flowing into a 6 m × 6 m compartment of a ship at 17.0 m^3/s. A 25-cm-diameter hose is being used to pump the water out at an average velocity of 20 m/s. Is the water level rising in the compartment?

61. Molten plastic at a temperature of 510°F is augered through an extruder barrel by a screw occupying $\frac{3}{5}$ of the barrel's volume (Fig. P4.61). The extruder is 16 ft long and has an inner diameter of 8 in. The barrel is connected to an adapter having a volume of 0.48 ft^3. The adapter is then connected to a die of equal volume. The plastic exiting the die is immediately rolled into sheets. The line is producing 4-ft widths of material at a rate of 30 ft/min and a gauge thickness of 187 mil. What is the axial velocity, V_1, of the plastic in the barrel? Assume that the plastic density is constant as it solidifies from a liquid (in the extruder) into a solid sheet.

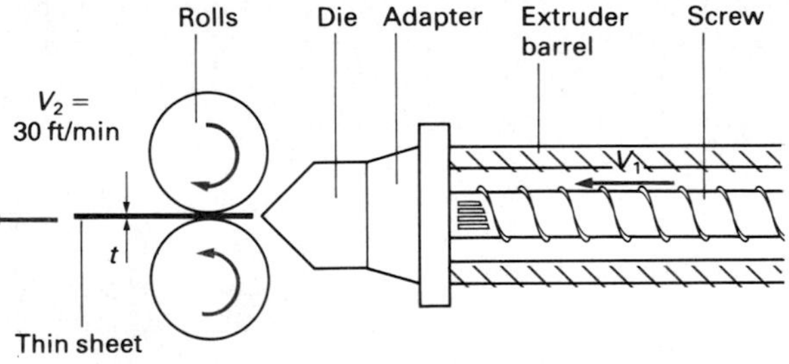

Figure P4.61

• **62.** A tank contains a volume $\mathcal{V}_0$ of pure water at time $t = 0$. An entering stream of volume flow rate Q_i has a constant salt concentration c_i, and the salt is well mixed in the tank. Develop an expression for the salt concentration c in the tank as a function of time.

• **63.** A tank contains 300 m^3 of pure water. Brine containing 3 kg/m^3 of salt flows into the tank at the rate of 3 m^3/s. The tank contents are well mixed, and the mixture runs out of the tank at the rate of 2 m^3/s. When will the tank contain 100 kg of salt?

•**64.** Oil flows into an initially empty tank of 10.0 m^2 cross section at the rate of 10 m^3/s and leaves through a 0.50-m^2 hole in the side of the tank with a velocity

$$V = \sqrt{19.6h},$$

where V is in m/s and h is in m. Find $h(t)$, where h is the vertical distance from the centerline of the hole to the free surface of the oil.

•**65.** Find the depth $Y(x)$ of rainwater flowing steadily down a hill having a slope of 30° with the horizontal, as shown in Figure P4.65. The rain is falling vertically with a velocity of 3.0 ft/sec, and there are 250 $\frac{1}{8}$-in.-diameter water droplets per cubic foot of air. The water flowing down the hill has the velocity profile

$$u = U\frac{y}{Y}$$

where u is the fluid velocity parallel to the hill and Y is the water depth perpendicular to the hill at a distance x from the top. Assume that $Y = 0$ at the top, where $x = 0$ and $U = 1.0$ ft/sec.

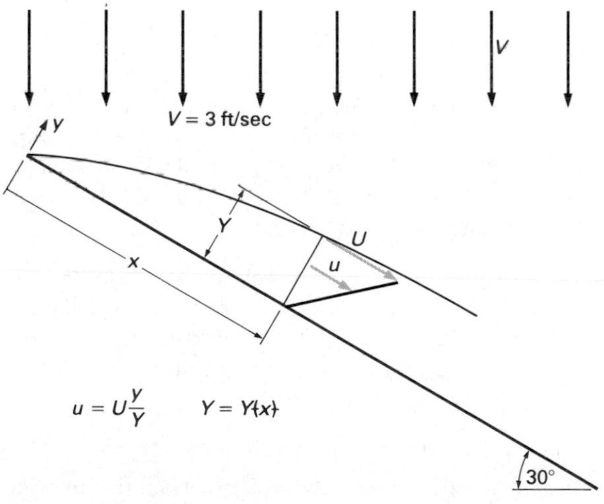

Figure P4.65

•**66.** Figure P4.66 shows a coal pulverizer in a power plant. Lump coal and hot air enter the pulverizer in separate streams. The coal is ground to a very fine powder and mixed with the air. The air–coal mixture leaves the pulverizer as a single stream. The coal powder is so fine and so well mixed with the air that the exiting stream may be treated as a continuous fluid. Using the information given in Fig. P4.66, calculate the mass flow rate of air entering the pulverizer. The mass ratio of coal to air is 1:1. If the temperature and pressure of the exiting stream are the same as the entering air, what is the approximate density of the coal–air mixture? Calculate the velocity of the coal–air mixture in the delivery pipe. The density of coal is 50 lbm/ft³.

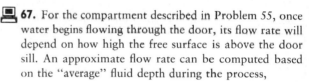

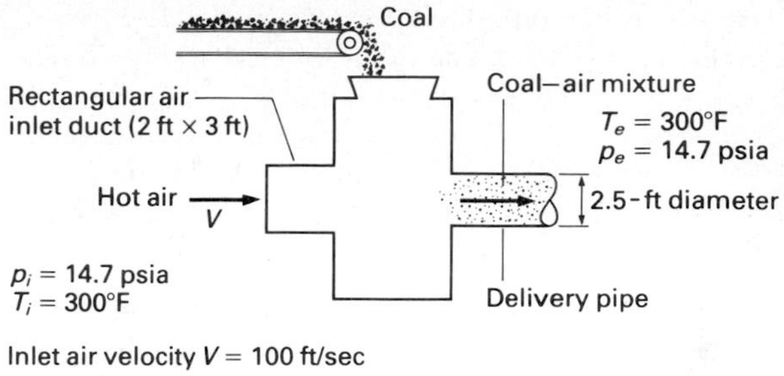

Figure P4.66

67. For the compartment described in Problem 55, once water begins flowing through the door, its flow rate will depend on how high the free surface is above the door sill. An approximate flow rate can be computed based on the "average" fluid depth during the process,

$$Q = 0.5w(H_{avg})^{1.5}\sqrt{2g},$$

where w = door width and H_{avg} = average height of fluid (above sill). If an automatic door-closing mechanism activates when the seawater reaches a depth of 5 ft in the compartment ($H_{avg} = 2$ ft), how long does an occupant have to safely leave the compartment?

68. For the compartment described in Problem 55, once water begins flowing through the door, its flow rate can be computed by

$$Q = 0.5w(H)^{1.5}\sqrt{2g},$$

where w = door width and H = instantaneous height of fluid (above sill). If an automatic door-closing mechanism activates when the seawater reaches a depth of 5 ft in the compartment, how long does an occupant have to safely leave the compartment after the hull is punctured?

69. For the compartment described in Problem 55, the flow rate from the sea into the compartment depends on the depth of the puncture below the surface of the sea according to

$$Q_{in} = A\sqrt{2gd},$$

where A = cross-sectional area of the puncture and d = depth of the puncture below sea level. A valid expression for flow rate through the compartment door is given in Problem 68. An automatic door-closing mechanism activates when the sea water reaches a depth of 5 ft in the compartment. For a ship with a draught (distance from sea level to bottom to keel) of 30 ft, determine the relationship (table or graph) between the depth of the hull puncture and the elapsed time before the door-closing mechanism is activated for a circular puncture 3 ft in diameter.

70. A storage tank of variable cross-sectional area full of liquid is to be drained for repair. Analysis of the fluid mechanics of the drain process produced the following relationship between liquid depth z and elapsed time from opening of the drain valve:

$$\frac{dz}{dt} = -C_1 A \sqrt{\frac{z + C_2}{z}},$$

where A is the exit area, C_1 and C_2 are geometry-dependent constants with the following values: $C_1 = 0.50\ (\text{ft}\cdot\text{s})^{-1}$ and $C_2 = 12$ ft. If the tank is filled with liquid to a depth of 5 ft, how long will it take to drain the tank through a drain valve with an exit inside diameter of 1.6 in.?

71. Calculate the kinetic energy correction factor for each of the following velocity profiles for a circular pipe:

(a) $u = u_{\max}\left(1 - \frac{r}{R}\right)$;

(b) $u = u_{\max}\left(1 - \frac{r^2}{R^2}\right)$;

(c) $u = u_{\max}\left(1 - \frac{r}{R}\right)^{1/9}$.

R is the pipe radius and r is the radial coordinate.

72. The cross-sectional area of a rectangular duct is divided into 16 equal rectangular areas, as shown in Fig. P4.72. The axial fluid velocity measured in feet per second in each smaller area is shown. Estimate the kinetic energy correction factor.

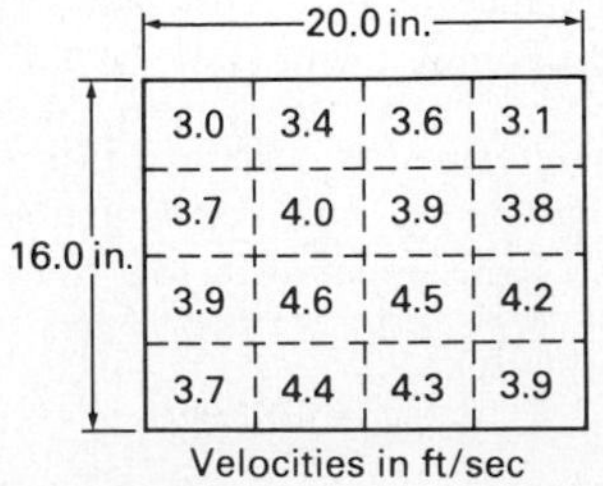

Figure P4.72

73. The fluid axial velocities shown in Fig. P4.73 are the average velocities measured in ft/s in each annular area of a duct. Find the kinetic energy correction factor for the flowing fluid.

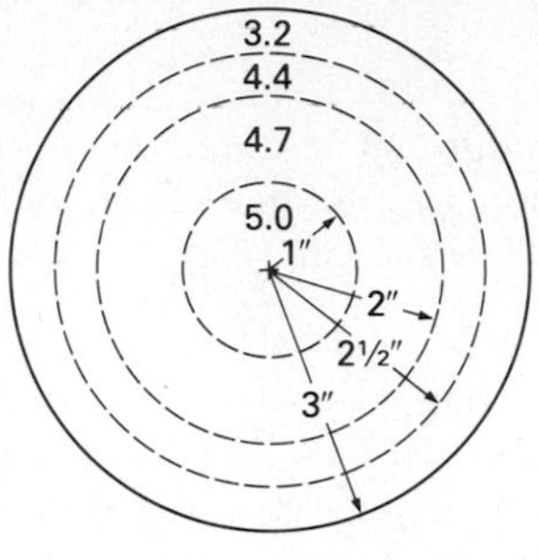

Figure P4.73

74. A motor-driven fan has a 25°C air flow rate of 200 m^3/min and a pressure rise of 20 cm of water. The fan inlet and outlet velocities are equal. The motor has an output power of 10.0 kW and draws a steady current of 60 amps at 220 volts. Find the fan and motor efficiencies for constant air density. The efficiency η is defined as

$$\eta = \frac{\text{Power output}}{\text{Power input}}.$$

75. A fluid flows through a nozzle without heat loss or gain. The inlet enthalpy $(\tilde{u}_1 + p_1/\rho_1)$ is 1000 Btu/slug, and the exit enthalpy $(\tilde{u}_2 + p_2/\rho_2)$ is 500 Btu/slug. The inlet velocity is 10 ft/sec. Determine the exit velocity. Neglect potential energy changes.

76. Air enters a converging nozzle at 40 psia, 150°F, and a velocity of 30 ft/sec. The inlet area is 0.25 ft^2, the exit pressure is 35 psia, and the exit velocity is 150 ft/sec. Find the exit area and the exit temperature if the air internal energy change is given by

$$\tilde{u}_2 - \tilde{u}_1 = (5.5\ \text{Btu/slug}\cdot{}^\circ\text{F})(T_2 - T_1),$$

and the nozzle is thermally insulated from the ambient air.

77. A diffuser is used to reduce the velocity of a flow stream. A particular air diffuser has inlet conditions of 250 m/s, 202 kPa, and 300°C and exit velocity of 50 m/s. The diffuser is thermally insulated from the ambient air, and the internal energy change of the air is given by

$$\tilde{u}_2 - \tilde{u}_1 = (750\ \text{J/kg}\cdot{}^\circ\text{C})(T_2 - T_1).$$

Find the exit temperature. Elevation changes are negligible.

78. The flow rate through the pump shown in Fig. P4.78 is 2 ft^3/sec; $d_1 = d_2 = 4$ in. and $d_3 = 0.5$ in. Calculate the pump head in feet. Neglect mechanical energy losses.

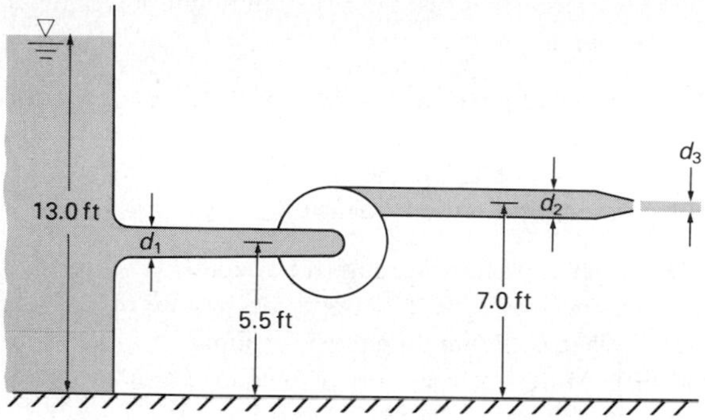

Figure P4.78

79. Figure P4.79 shows a subsonic wind tunnel that consists of a converging frictionless nozzle, test section, diffuser, fan, and connecting ductwork. The test section has a flow area of 1.0 m^2 and a maximum velocity of 120 km/hr of 10°C air. The ductwork has a flow area of 9.0 m^2. The mechanical energy loss in the wind tunnel is given by $KV^2/2$, where $K = 15$ and V is the velocity in the ductwork. Find the pressure drop across the nozzle and the power input to the air by the fan.

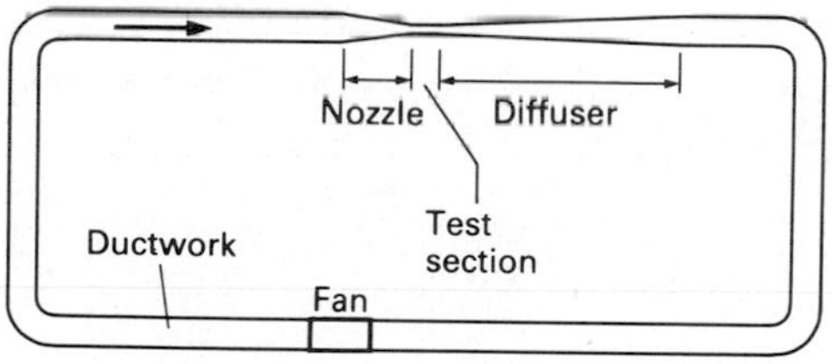

Figure P4.79

80. The hydraulic turbine shown in Fig. P4.80 has an efficiency of 90 percent (ratio of output power to available power). The 10°C water flow rate is 10,000 m^3/min.

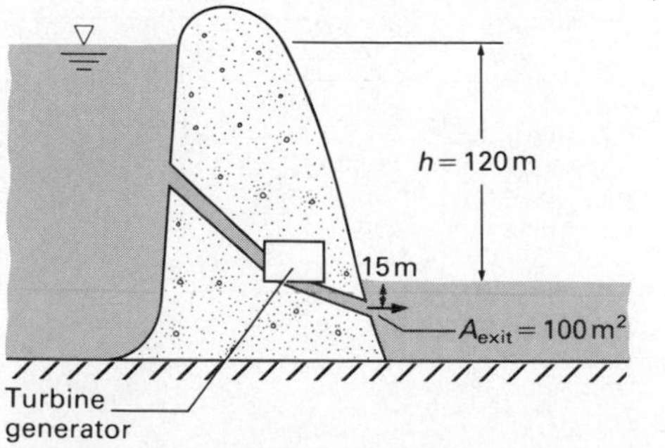

Figure P4.80

Determine the turbine output power in kilowatts for frictionless pipe flow.

81. A horizontal oil ($S = 0.80$) pipeline has a flow rate of 1000 gal/min and a pressure drop of 3.0 psi per 100 ft. Find the power input to the oil by a pump installed every 3 mi. The pump has no losses.

82. The sump pump shown in Fig. P4.82 has an input current of 20 amps and an input voltage of 110 volts. What is the maximum possible height h that the 60°F water can be raised if the pump has a capacity of 100 gal/min? Assume that the pump is 100 percent efficient; that is, the entire pump input power is transferred to the fluid.

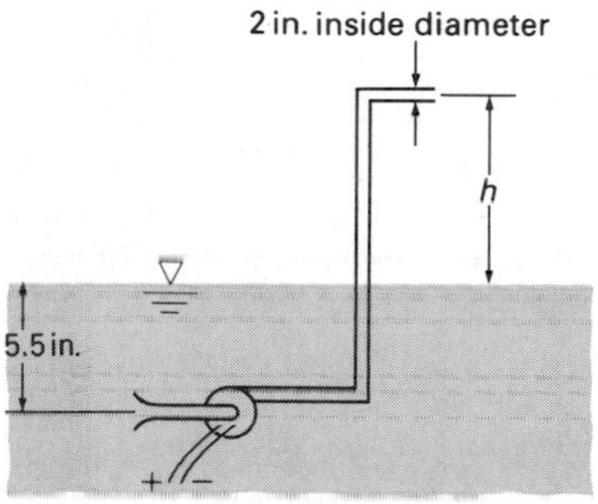

Figure P4.82

83. A pump delivers 60°F water at 2000 gpm through a vertical pipeline 300 ft high. Find the minimum horsepower input to the pump. Assume that atmospheric pressure exists at both the pump inlet and the pipeline outlet and that kinetic energy changes are negligible.

84. Figure P4.84 shows the mixing of two streams. The shear stress between each fluid and its adjacent walls is negligible. Why can't Bernoulli's equation be applied between points in stream 1 and the mixed stream or between points in stream 2 and the mixed stream?

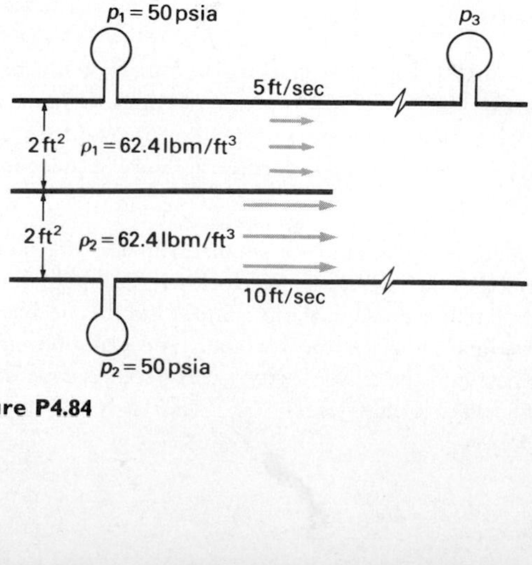

Figure P4.84

85. Figure P4.85 shows an open-circuit wind tunnel. The two manometers indicate the pressure in the test section and at the fan discharge. Assume that air is incompressible, the flow is steady, and elevation changes are negligible. Calculate the power input to the fan in horsepower. The mechanical energy loss downstream from the test section is given by $KV^2/2$, where $K = 4.0$ and V is the test section velocity. The air temperature is 40°F.

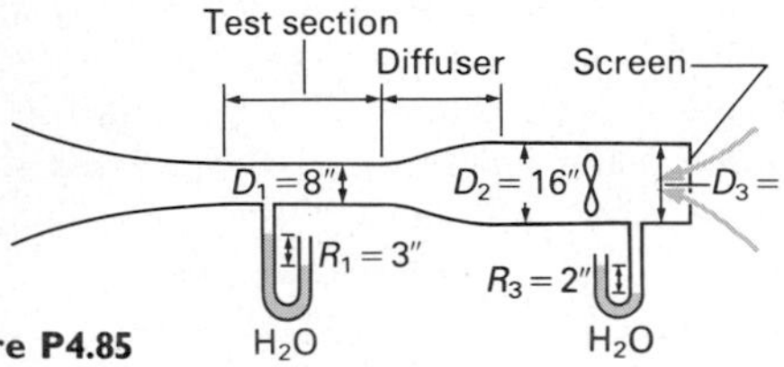

Figure P4.85

86. A diver is using a flexible hose to suck up a mixture of water and sand having a specific gravity of 1.5. The inside diameter of the hose is 10 cm. How much of the mixture can be picked up per minute by a 1.5-kW pump? The end of the flexible hose (connected to the pump inlet) is 6.0 m below the water surface, and the open end of the hose connected to the pump discharge is at the water surface and atmospheric pressure. Neglect all friction losses.

87. A centrifugal pump has an input power of 75 kW and a flow rate of 10.0 m^3/min. Neglect elevation changes. What is the maximum pressure rise that can be generated by the pump at this flow rate?

88. Water at 20°C flows through a hydraulic turbine at the rate of 10 m^3/s. The inlet pressure and velocity are 250 kPa gage and 10 m/s, respectively, and the discharge pressure and velocity are −90 kPa gage and 1.0 m/s, respectively. Determine the maximum power delivered to the turbine by the water, assuming negligible elevation changes.

89. A small pump has a shaft work input of 300 N·m/kg, a pressure rise of 200 kPa, and negligible kinetic and potential energy changes. Steady-state conditions exist and the water density is constant. Find the mechanical energy loss.

90. Water at 60°F enters a small turbine at 50 psia and 50 ft/sec. The turbine shaft work is 3000 ft·lb/slug of water passing through the turbine. Find the mechanical energy loss, gh_L, in the turbine. The exit pressure is 13.0 psia and the exit velocity is 20 ft/sec. Assume negligible elevation changes.

91. A pump delivers 310 slugs/hr of 60°F water from inlet conditions of 15 psia and zero elevation to exit conditions of 15 psia and 100 ft elevation with negligible kinetic energy changes. Find the minimum power input to the pump.

92. An exhaust fan in a wall of a building has a pressure rise of 3.0 in. of water and a volume flow rate of 5000 ft^3/min. Calculate the minimum power input to the fan. Assume constant air density.

93. Water is pumped through a horizontal pipeline with a pressure drop of 202 kPa and a flow rate of 100 m^3/min. What is the minimum power input to a pump installed in this pipeline in order to eliminate the net pressure drop? See Fig. P4.93.

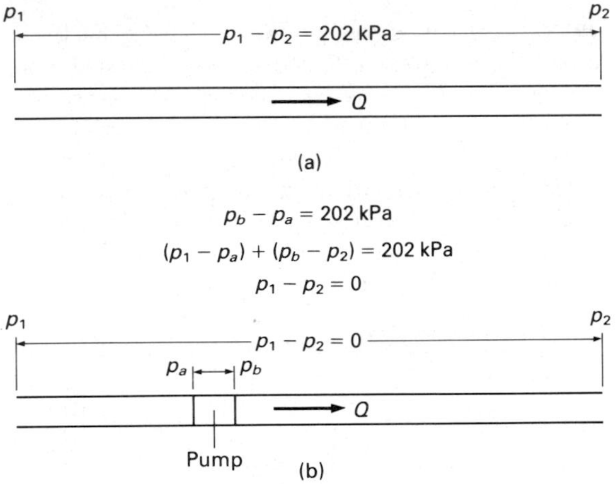

Figure P4.93

94. Determine the volume flow rate and minimum power input to the water pump in Fig. P4.94.

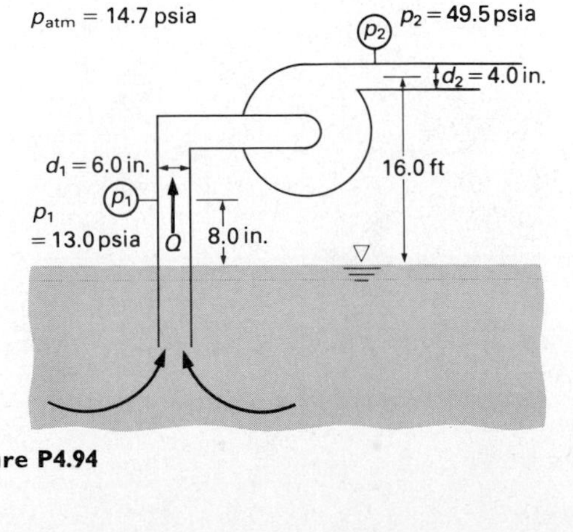

Figure P4.94

95. Combustion products at 300°C flow up a stack, as shown in Fig. P4.95. The gases have constant density, a molecular weight of 29.0, and a flow rate of 20,000 kg/min. Compute the pressure p_1 at the bottom of the chimney if the mechanical energy loss is given by

$$gh_L = 0.28V^2,$$

where V is the velocity.

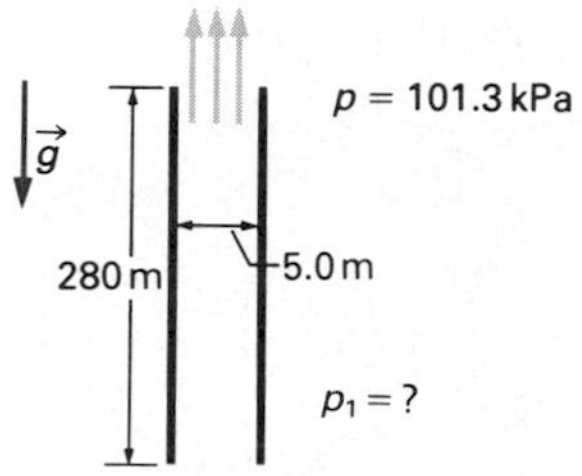

Figure P4.95

96. Oil ($S = 0.85$) at 40°F flows through a long, horizontal pipeline and has a pressure drop of 150 psi. Find the internal energy change per unit mass of fluid if the pipe is thermally insulated.

97. Oil ($S = 0.85$) at 20°C flows through a 3000-m-long, thermally insulated, horizontal pipeline. The mechanical energy loss is given by

$$gh_L = 0.0075\left(\frac{L}{D}\right)\frac{\bar{V}^2}{2},$$

where D is the pipe inside diameter and is 25 cm. Pressure gages indicate that the pressure drop is 150 kPa. Calculate the average fluid velocity $\bar{V}$.

98. Liquid water at 40°F flows down a vertical, thermally insulated, 2-in. schedule 40 steel pipe. The temperature change of the water is related to its internal energy change by

$$\tilde{u}_2 - \tilde{u}_1 = 32.2 \text{ Btu/slug}\cdot{}^\circ\text{F}(T_2 - T_1).$$

What is the temperature change of the water per 100 ft of drop if the pressure drop ($p_1 - p_2$) per 100 ft is 12.0 psi?

99. Figure P4.99 shows a test rig for evaluating the loss coefficient, K, for a valve. Mechanical energy losses in valves are modeled by the equation:

$$gh_L = \text{K}\left(\frac{V^2}{2}\right),$$

where gh_L is the mechanical energy loss and V is the flow velocity entering the valve. In a particular test, the pressure gage reads 40 kPa, gage, and the 1.5-m^3 catch tank fills in 2 min 55 sec. Calculate the loss coefficient for a water temperature of 20°C.

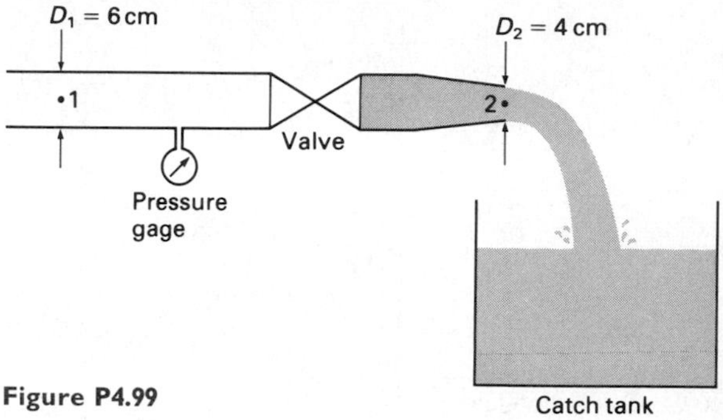

Figure P4.99

100. Lubricating oil at 20°C flows through the system shown in Fig. P4.100 at 0.1 m^3/s. The energy losses in the short 40-cm inlet pipe are negligible. The energy losses in the 20-cm discharge pipe are given by

$$gh_L = 2.5V^2,$$

where V is the oil velocity in the 20-cm pipe. Calculate the pump input power to the fluid.

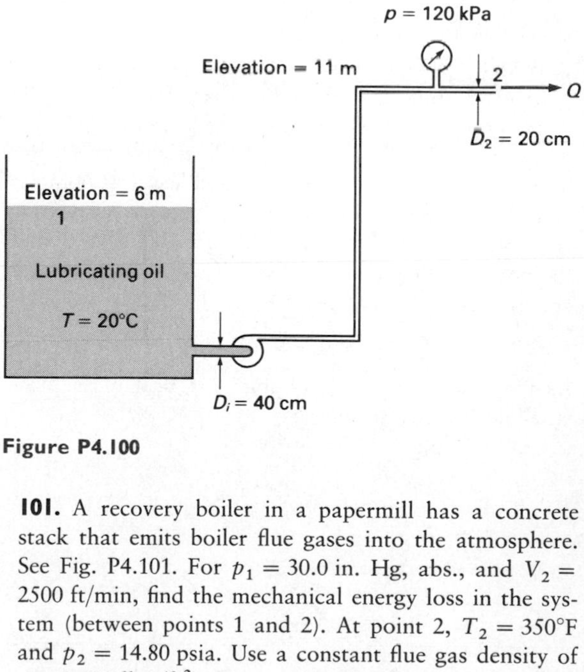

Figure P4.100

101. A recovery boiler in a papermill has a concrete stack that emits boiler flue gases into the atmosphere. See Fig. P4.101. For $p_1 = 30.0$ in. Hg, abs., and $V_2 = 2500$ ft/min, find the mechanical energy loss in the system (between points 1 and 2). At point 2, $T_2 = 350$°F and $p_2 = 14.80$ psia. Use a constant flue gas density of $\rho = 0.055$ lbm/ft^3.

102. A pump located in the basement of a six-story building pumps 60°F water to the sixth floor. When the water reaches the sixth floor, it must have a pressure

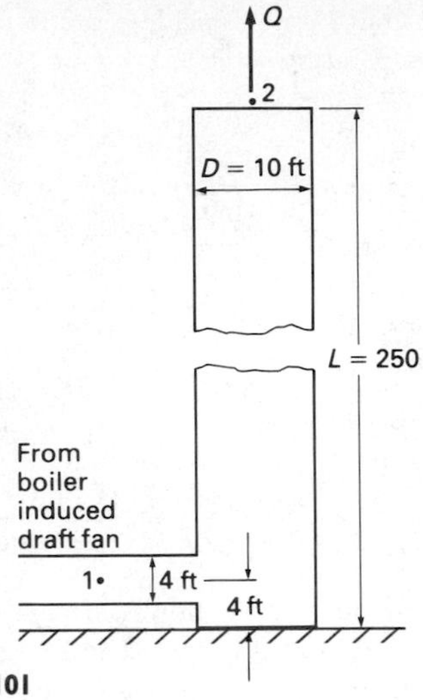

Figure P4.101

of 50 psig. The pressure at the pump inlet is 45 psig. For a 60°F water flow rate of 100 gpm, find the power transferred from the pump to the fluid. Assume that the energy loss in the pump discharge piping is negligible. The sixth floor piping is 87 ft above the pump and the entire piping system has the same diameter.

103. Figure P4.103 shows a lateral pipe fitting. This particular fitting has a mainline diameter of 4 in. The diameter of the lateral is 3 in. and the lateral angle is 45°. Water at 60°F is flowing in the fitting. Flow enters the fitting from the left end (point 1) and leaves at points 2 and 3. Measurements show that the pressure at point 1 is 34 psig, the flow rate at point 2 is 1.0 ft³/sec., and the flow rate at point 1 is 1.63 ft³/sec. The mechanical energy loss in the fitting can be modeled as

$$gh_L = 0.1V_1^2$$

for the flow from point 1 to point 2. Calculate the pressure and velocity at point 2.

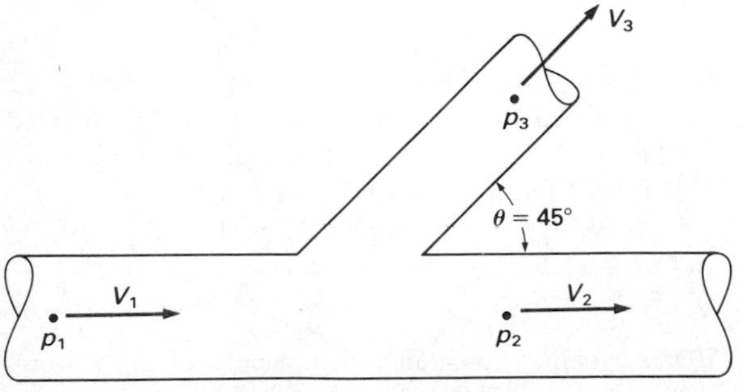

Figure P4.103

104. Figure P4.104 shows a pump testing setup. Water is drawn from a sump and pumped through a pipe containing a valve. The water is discharged into a catch tank sitting on a scale. During a test run, 800 lb of water is collected in the catch tank in 15 sec. The pump power input to the fluid during this period is 700 ft·lb/sec. Calculate the water velocity in the pipe and the mechanical energy loss (ft·lb/lb$_m$) in the pipe and valve.

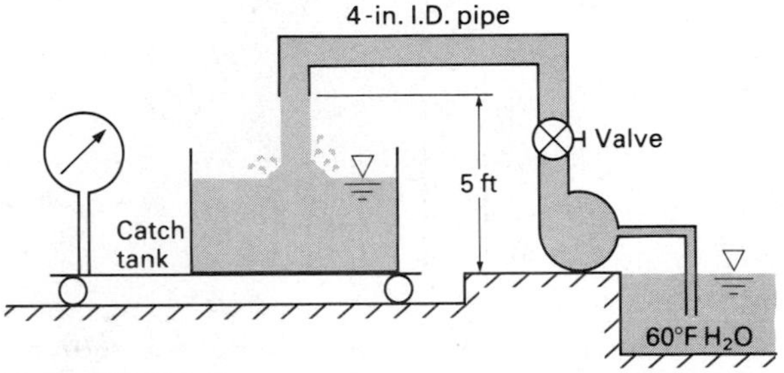

Figure P4.104

• **105.** Fuel oil is heated as it flows upward in a 0.2-m-diamater vertical tube. The tube is 25 m long. At the entrance, the velocity is uniform at 6.0 m/s, the pressure is 280 kPa, and the fluid density is 850 kg/m³. At the exit, the velocity is uniform at 7.0 m/s, and the density is uniform. The mechanical energy loss is given by

$$d(gh_L) = \frac{0.02V^2}{2D}\,dx,$$

where V is the average velocity at each cross section and D is the tube diameter. The fluid velocity changes linearly with distance along the tube. Determine the pressure drop along the tube. What is the change in the fluid's internal energy other than from heat transfer ($\Delta\tilde{u} - q$)?

• **106.** Liquid water at 40°F flows upward through a vertical, thermally insulated, 2-in. schedule 40 steel pipe at the rate of 100 gal/min. The water temperature change is related to its internal energy change by the relation in Problem 98. Find the temperature change of the water for a pressure drop ($p_1 - p_2$) of 12.0 psi and

$$gh_L = 0.0065\left(\frac{L}{D}\right)V^2.$$

107. Air enters a vacuum pump at 7.0 kPa, 15°C, and 30 m/s through an inlet area of 0.004 m². The air leaves the vacuum pump at 101 kPa and a negligible velocity. The input power to the pump is 0.16 kW, and the pump is thermally insulated from its surroundings. What is the air discharge temperature if the elevation changes are negligible?

• **108.** A 10-ft-diameter water storage tank has 75 slugs of water below 1.0 slug of air at 60 psia and 70°F. A 1-in.-diameter plug is pulled from the bottom of the tank, and the flowing water is used to produce work. Find the maximum work that can be developed by the water from the tank. The air temperature is constant at 70°F.

• **109.** The pump shown in Fig. P4.109 has a constant input power of 7.46 kW and no energy losses. Determine the time needed to fill the storage tank. Neglect energy losses and kinetic energy.

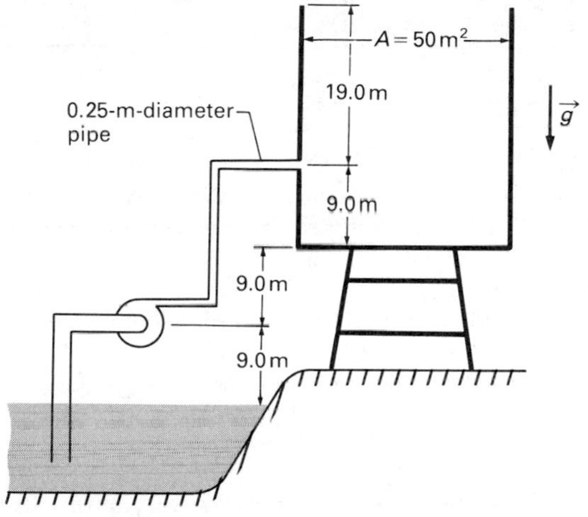

Figure P4.109

• **110.** A reciprocating pump has a constant volume flow rate of $Q = 100$ gal/min and a static pressure rise Δp. The pump has a constant efficiency (output power $Q\,\Delta p$ divided by the input power) of 0.75. Find the energy input required to fill the 20,000-gal storage tank through pipeline *A* shown in Fig. P4.110. The fluid has a specific gravity of 0.78. Assume inviscid flow and neglect kinetic energy.

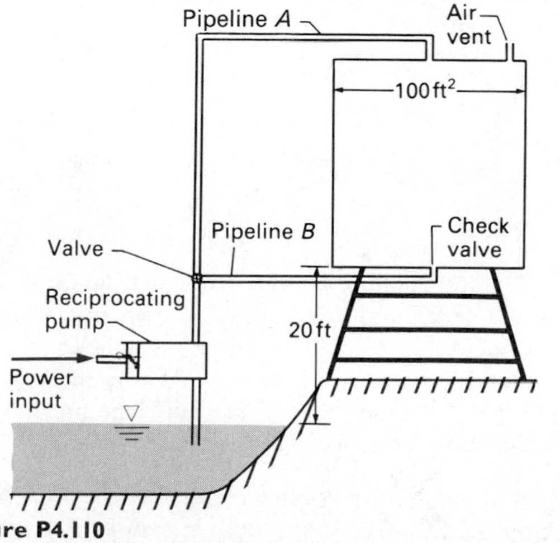

Figure P4.110

• **111.** Repeat Problem 110 for pipeline *B*. The check valve permits fluid to enter the storage tank from pipeline *B* but stops any fluid from flowing out of the tank. Compare your result with that of Problem 110 and state your conclusions.

• **112.** Water is pumped steadily through the apparatus shown in Fig. P4.112. The pipe area and gage pressure are shown for both outlet sections 1 and 2. Assume that the 40°F water is frictionless and incompressible. Compute the horsepower input to the pump. The total volume flow rate $Q_T = 1.0$ ft³/sec. $D_a = 4.0$ in.

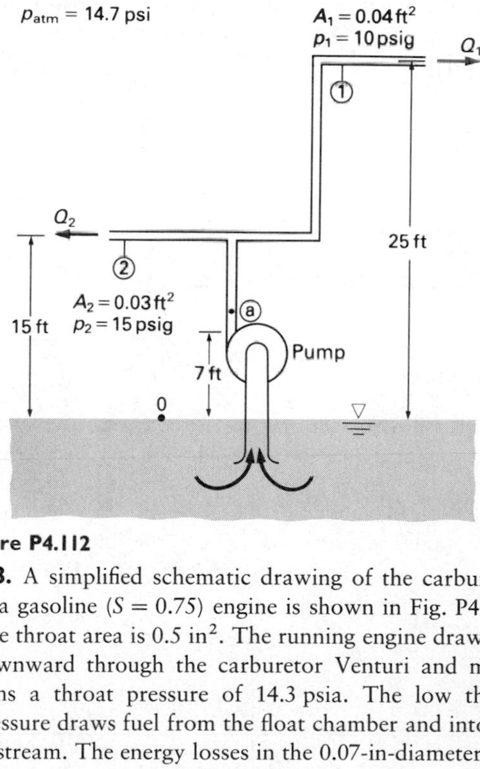

Figure P4.112

• **113.** A simplified schematic drawing of the carburetor of a gasoline ($S = 0.75$) engine is shown in Fig. P4.113. The throat area is 0.5 in². The running engine draws air downward through the carburetor Venturi and maintains a throat pressure of 14.3 psia. The low throat pressure draws fuel from the float chamber and into the airstream. The energy losses in the 0.07-in-diameter fuel metering line and valve are given by

$$gh_L = \frac{KV^2}{2g},$$

where K = 6.0 and V is the fuel velocity in the metering line. Assume that the air is an ideal fluid having a constant density $\rho_A = 0.075$ lbm/ft³. The atmospheric pressure is 14.7 psia. Calculate the air-to-fuel ratio ($\dot{m}_a/\dot{m}_f$).

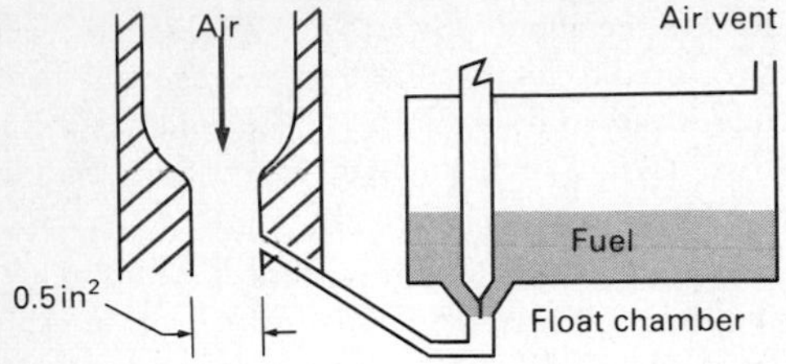

Figure P4.113

114. The pump shown in Fig. P4.109 has a constant input power of 10 hp and no energy losses. Find the time needed to fill the storage tank. Assume negligible energy losses.

115. Figure P4.115 shows a tube for siphoning water from an aquarium. Determine the rate at which the water leaves the aquarium for the conditions shown. Is there an advantage to having the large-diameter section? The water flow is inviscid.

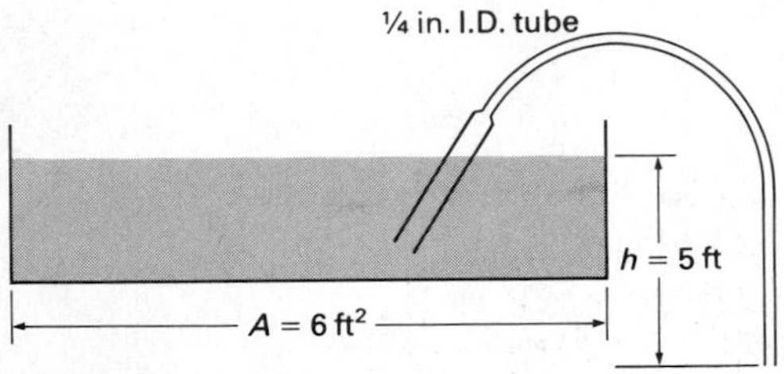

Figure P4.115

116. Air flows radially outward between the two parallel circular plates shown in Fig. P4.116. The pressure at the outer radius $R_o = 5.0$ cm is atmospheric. Find the pressure at the inner radius $R_i = 0.5$ cm if the air density is constant, the air flow is inviscid, the volume flow rate is 4.0 L/s, and the plate spacing is $h = 0.125$ cm.

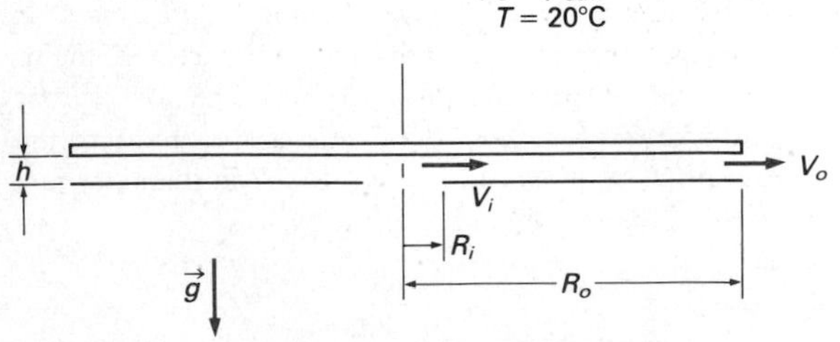

Figure P4.116

117. Repeat Problem 116 for air flowing radially inward.

118. A hole is to be drilled in the side of the tank shown in Fig. P4.118 so the liquid will travel the farthest horizontal distance L. Determine the height h. Assume inviscid flow.

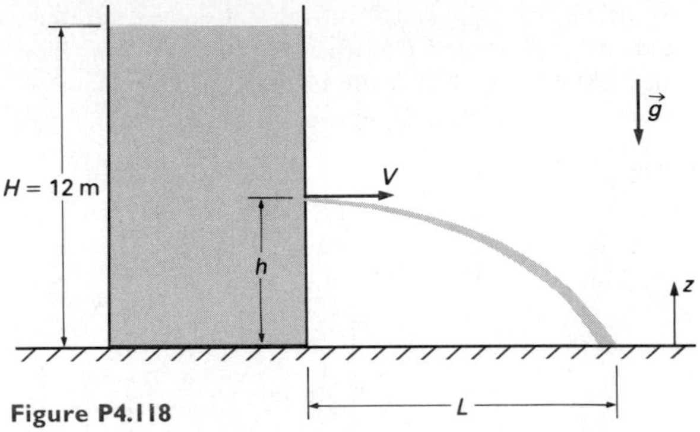

Figure P4.118

119. A liquid stream directed vertically upward leaves a nozzle with a steady velocity V_0 and cross-sectional area A_0. Find the velocity V and cross-sectional area A as a function of the vertical position y.

120. The tank shown in Fig. P4.120 contains air at atmospheric pressure above the water surface. The velocity of the water flowing from the tank is 22.5 ft/sec. Determine the water level (h).

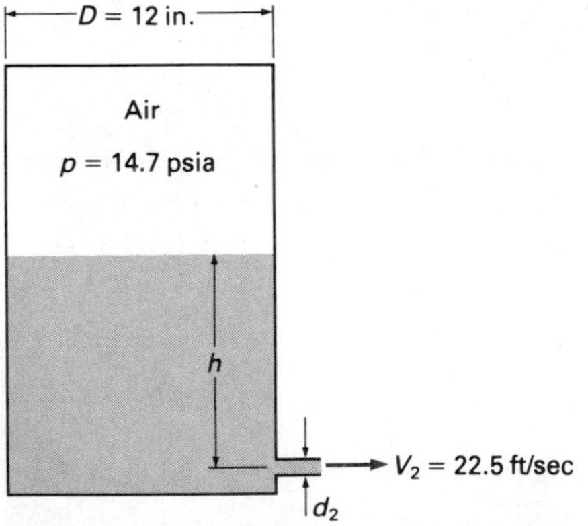

Figure P4.120

121. A gutter running along the side of a house is 6 in. wide and 40 ft long. During a hard downpour, water is 1 in. deep in the gutter. The gutter has only one downspout, and it is 3 in. in diameter. What is the velocity of the water entering the downspout? The pressure at the downspout entrance is atmospheric pressure.

122. A 15-mph wind is blowing against a closed door that opens outward. If the door measures 3 ft × 7 ft, estimate the force required to start opening the door. Assume that the air pressure on the inside of the door is equal to the local atmospheric pressure, that the air temperature is 60°F, and that the door handle is 3 ft from the hinge line.

123. The axial velocity of flow through a nozzle is given by

$$u = \frac{U_0}{[1 - (x/2\ell)]^2},$$

where U_0 is the entrance velocity at $x = 0$ and ℓ is the length of the nozzle. The flow is approximately incompressible and inviscid. Gravity acts perpendicular to the axial (x) direction. Calculate the pressure drop across the nozzle (i.e., from $x = 0$ to $x = \ell$) for $U_0 = 5$ m/s and $\ell = 3.0$ m. The fluid is water, the density is 1000 kg/m^3, and the flow is one-dimensional.

124. The pressure and average velocity at point A in the pipe shown in Fig. P4.124 are 16.0 psia and 4.0 ft/sec, respectively. Find the height h and the pressure and average velocity at point B. Fluid fills the 1-in.-diameter discharge pipe.

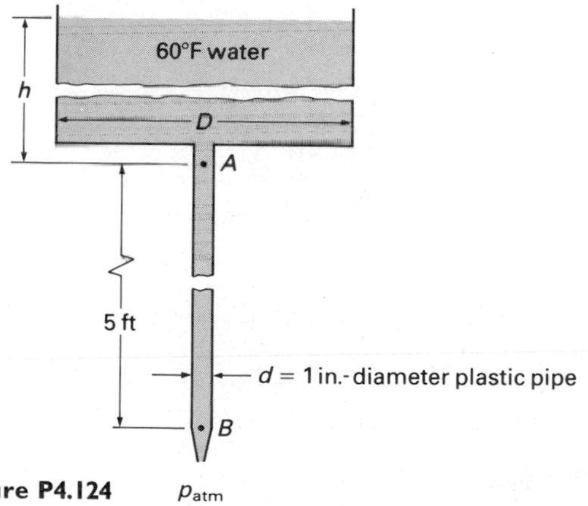

Figure P4.124

125. Fortified wine containing five parts wine with a specific gravity of 1.05 and one part ethanol with a specific gravity of 0.85 flows through a Venturi having an inlet diameter of 2.0 cm and a throat diameter of 0.50 cm. Determine the pressure drop between the inlet and throat of the Venturi for a flow rate of 0.001 m^3/s.

126. What is the minimum height for an oil ($S = 0.75$) manometer to measure airplane speeds up to 30 m/s at altitudes up to 1500 m? The manometer is connected to a Pitot-static tube. See Fig. P4.126.

127. Figure P4.127 shows a duct for testing a centrifugal fan. Air is drawn from the atmosphere (p_{atm} = 14.7 psia, T_{atm} = 70°F). The inlet box is 4 ft × 2 ft. At section 1, the duct is 2.5 ft square. At section 2, the duct is circular and has a diameter of 3 ft. A water manometer in the inlet box measures a static pressure of −2.0 in. of water. Calculate the volume flow rate of air into the fan and the average fluid velocity at both sections 1 and 2. Assume constant density.

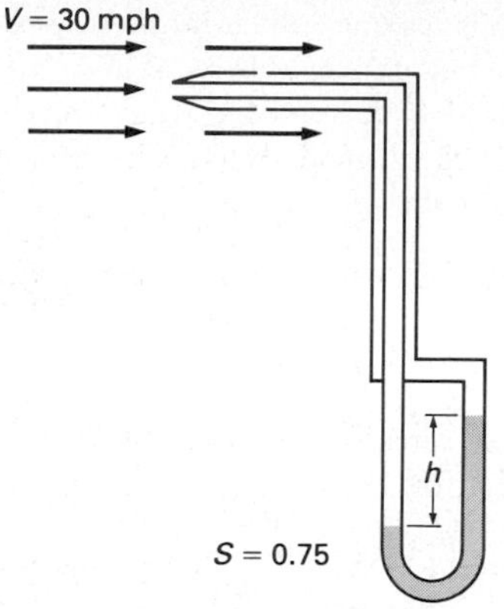

Figure P4.126

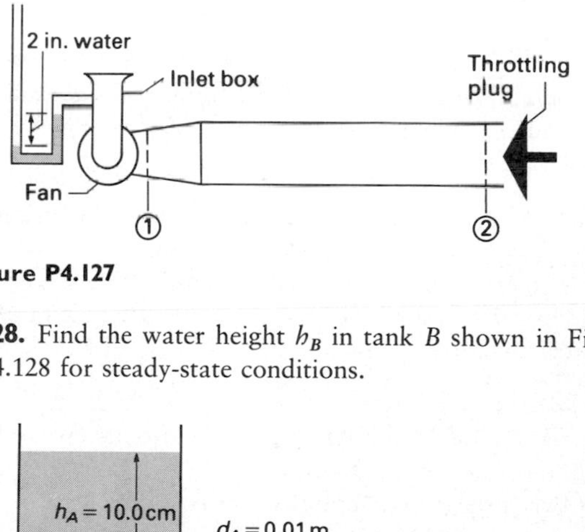

Figure P4.127

128. Find the water height h_B in tank B shown in Fig. P4.128 for steady-state conditions.

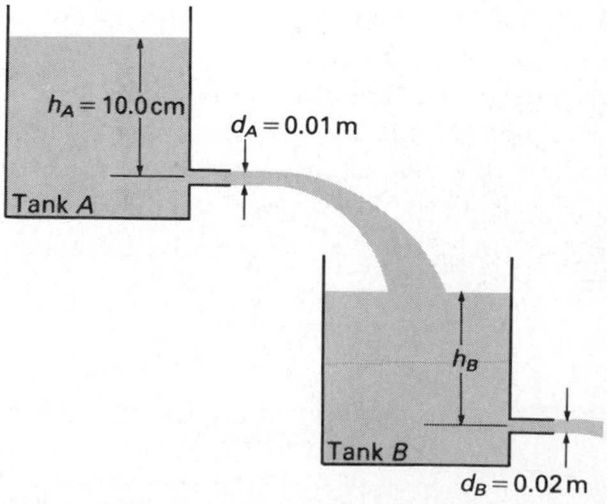

Figure P4.128

129. Water is leaving a hole in the side of a tank with a horizontal velocity of 20 ft/sec. The hole is 25 ft above the ground. How far will the water travel horizontally before it hits the ground? What is its velocity just before the water hits the ground?

130. For the Pitot-static tube shown in Fig. 4.16, $p_A - p_B = 1.4$ psi, and the flowing fluid is water. Find the fluid velocity V_1.

131. At what rate does oil ($S = 0.85$) flow from the tank shown in Fig. P4.131?

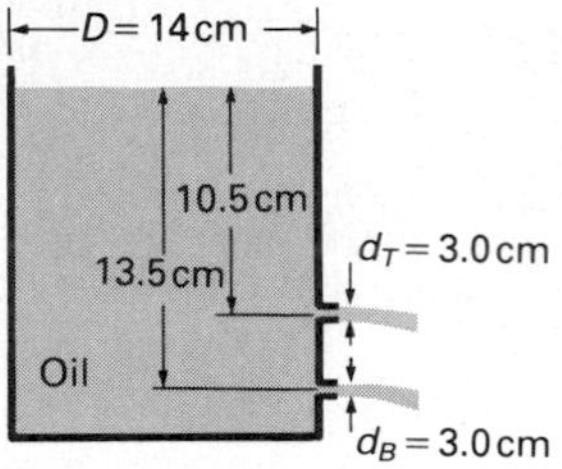

Figure P4.131

132. A fire hose has a nozzle outlet velocity of 30 mph. What is the maximum height the water can reach?

133. Water leaves a pump at 200 kPa and a velocity of 12 m/s. It then enters a diffuser to increase its pressure to 250 kPa. What must be the ratio of the outlet area to the inlet area of the diffuser?

134. A sump pump is submerged in 60°F ordinary water that vaporizes at a pressure of 0.256 psia. The pump inlet has an inside diameter of 2.067 in. and is 15 ft below the water surface. Find the maximum possible flow rate before cavitation (vaporization) occurs.

135. The cap is removed from the fire hydrant shown in Fig. P4.135, and water shoots a horizontal distance of 70 ft before hitting the ground. The vertical distance between the ground and the bottom of the 6-in.-inside-diameter pipe is 14 in. Estimate the distance ℓ and find the volume flow rate as the water leaves the hydrant. Assume that the velocity over the hydrant exit is uniform.

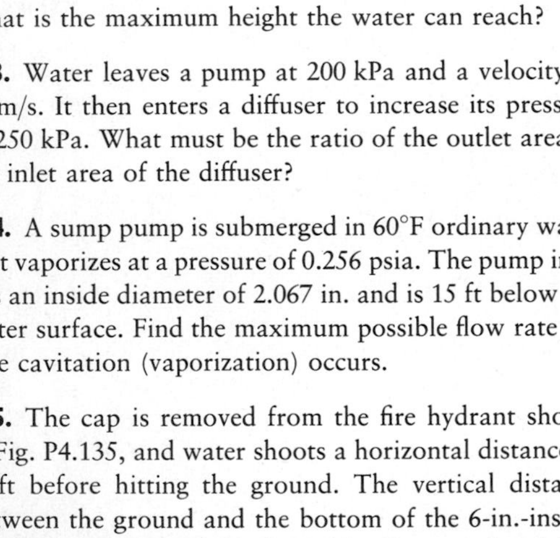

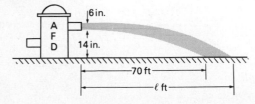

Figure P4.135

136. Someone siphoned 15 gal of gasoline from a gas tank in the middle of the night. The gas tank measures 12 in. wide × 24 in. long × 18 in. high and was full when the thief started. If the siphoning tube has an inside diameter of $\frac{1}{2}$ in., find the minimum amount of time needed to siphon the 15 gal from the tank. Assume that at any instant of time the steady-state equations are adequate to predict the gasoline velocity in the siphon tube. Also assume that the end of the siphon tube outside the gas tank is at the same level as the bottom of the tank. You may consider the gasoline to be inviscid.

137. The water clock (clepsydra) shown in Fig. P4.137 is an ancient device for measuring time by the falling water level in a large glass container. The water slowly drains out through a small hole in the bottom. Determine the approximate shape $R(z)$ of a container of circular cross section required for the water level to fall at a constant rate of 5 cm/hr if the drain hole is 1.5 mm in diameter. What size container (h and R_0) permits the clock to run for 24 hr without refilling?

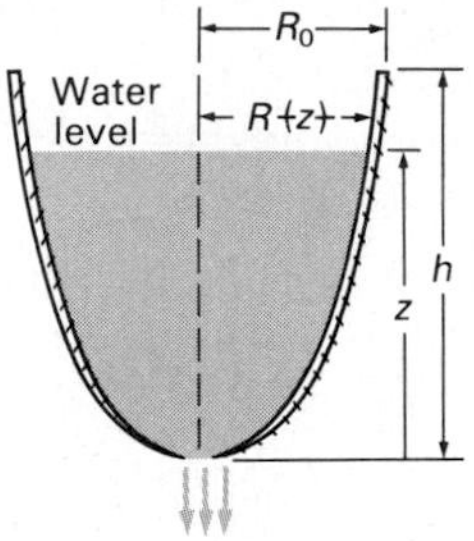

Figure P4.137

138. Air flows over the airfoil shown in Fig. P4.138. Sensors give the pressures shown at points A, B, C, and D. Find the air velocities just above points A, B, C, and D. The air density is 0.0020 slug/ft³.

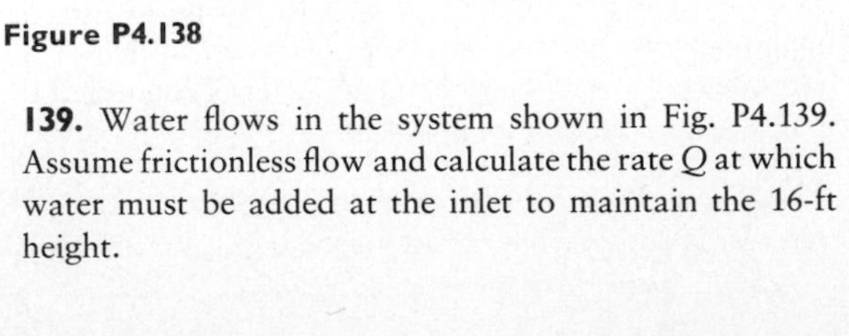

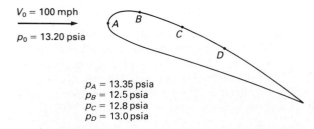

Figure P4.138

139. Water flows in the system shown in Fig. P4.139. Assume frictionless flow and calculate the rate Q at which water must be added at the inlet to maintain the 16-ft height.

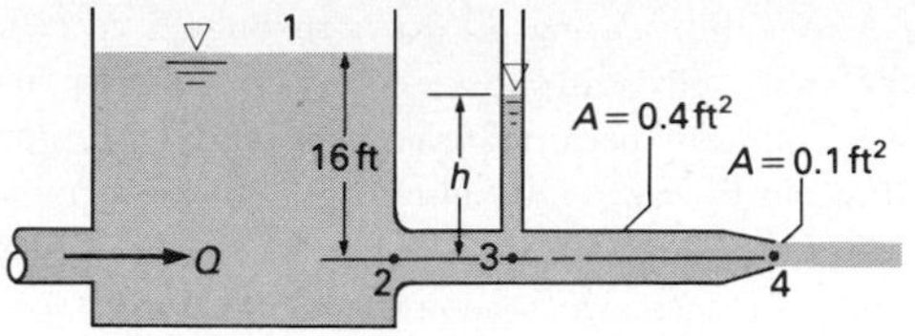

Figure P4.139

140. Calculate the height h in feet of water in the static-pressure tube in Problem 139.

141. A water jet flows up the vane shown in Fig. P4.141. The inlet area A_0 and velocity V_0 are $\frac{1}{12}$ ft^2 and 12 ft/sec, respectively. Assume inviscid flow and uniform velocity over both the inlet and outlet. Determine the exit area A_e and exit velocity V_e if the jet width (perpendicular to the paper) remains constant at $\frac{1}{6}$ ft.

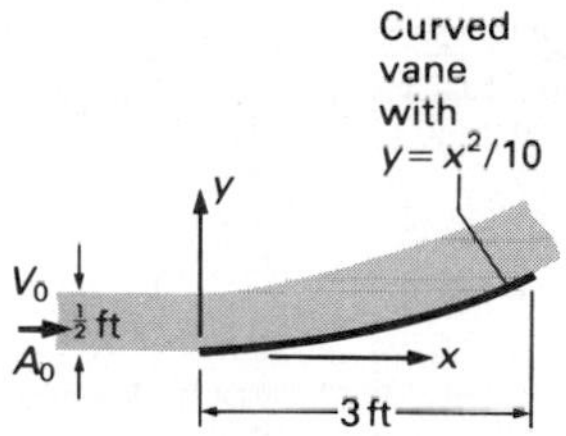

Figure P4.141

142. A liquid jet has a velocity of 100 ft/sec and makes an angle of 45° with the horizontal. Find the maximum possible height that the fluid can attain.

143. A U-tube manometer indicates the difference in the stagnation pressures across a pump. The water flow rate through the pump is 100 gal/min, the inlet pipe is $2\frac{1}{2}$-in. nominal diameter, type L copper pipe, the discharge pipe is a 4-in. nominal diameter, type L copper pipe, and the pump has a static pressure rise of 45 psi. Find the difference in the mercury levels of the U-tube manometer. There is no elevation difference across the pump.

144. An airship flies at a speed of 30 km/hr into a 15 km/hr head wind. Find the stagnation pressure at the nose of the airship if the air temperature is 20°C and the ambient static pressure is 100 kPa.

145. The water from a fire hose reaches a peak of 90 ft above the nozzle with a horizontal velocity of 20 ft/sec. Determine the velocity of the water leaving the nozzle.

146. Which of the two circular tanks shown in Fig. P4.146 will be completely drained first? Assume that the flow is inviscid.

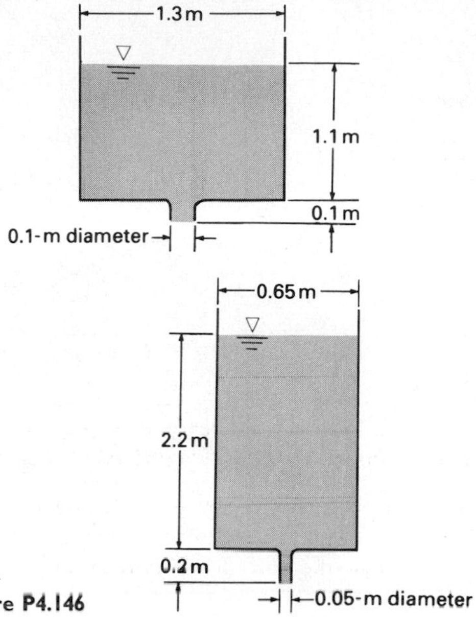

Figure P4.146

147. Find the water mass flow rate at the nozzle outlet *o* shown in Fig. P4.147. The water density is 1.9 slugs/ft^3, and the flow is inviscid.

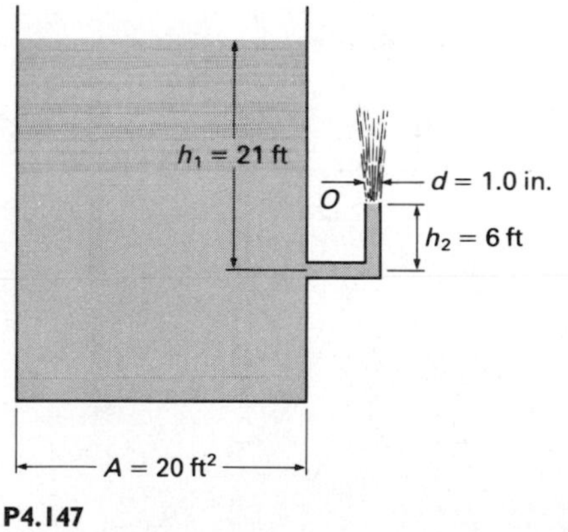

Figure P4.147

148. Calculate the maximum height to which the exiting water stream in Problem 147 would rise if the nozzle exit velocity is 31.1 ft/sec.

149. Air approaches a wind machine at $V_1 = 20$ mph, as shown in Fig. P4.149. The wind machine blade has a diameter of 10 ft and, while rotating, creates a swept area of 78.5 ft^2. The air stream passing through the 78.5-ft^2 swept area has an area of 19.6 ft^2 at a distance 20 ft upstream from the blade, where the pressure is the atmospheric pressure of 14.6 psia. Find the air velocity

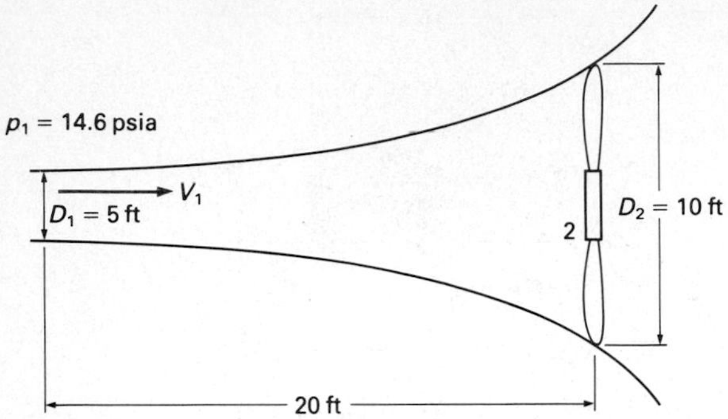

Figure P4.149

V_2 and static pressure difference $p_2 - p_1$. Assume a constant air density of 0.00233 slug/ft^3 and inviscid flow.

150. Water is drained from the tank shown in Fig. P4.150 by the 4-cm-inside-diameter hose. The flow is inviscid. What are the water flow rate and the hose outlet velocity?

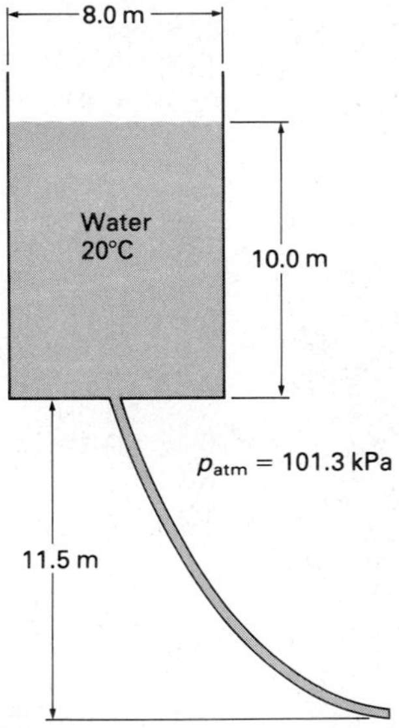

Figure P4.150

151. The tank in Problem 150 is capped, and the air space above the 20°C water is pressurized to increase the discharge flow rate to 0.03 m^3/s. Assume constant water density and inviscid flow and find the pressure in the air space.

152. The tank in Problem 150 is capped, and the air pressure above the water is adjusted so that the discharge flow rate is 0.02 m^3/s. Assume constant water density and inviscid flow. Find the air gage pressure p.

153. A 68°F lubricating oil flows through a converging channel with circular cross sections as shown in Fig. P4.153. The static pressure at point 1, where the diameter is 5.0-in., is p_1 = 80 psia. The total pressure at point 2, where the diameter is 2.0-in., is p_{02} = 82 psia. Calculate the flow rate through the converging section.

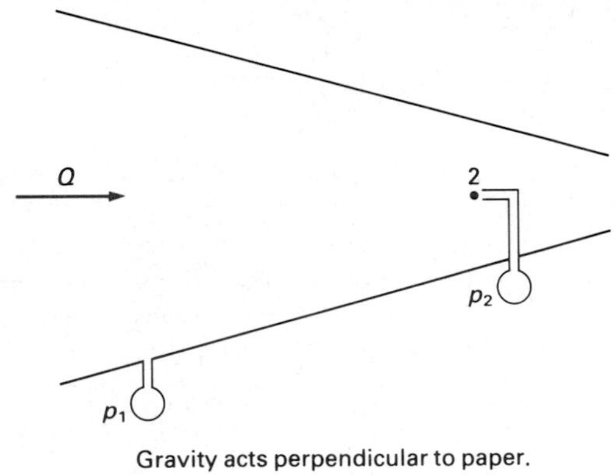

Figure P4.153

154. What is the static pressure at plane 2 in Problem 153 if the flow rate is Q = 2.52 ft^3/sec?

155. The water leaving the lawn sprinkler hose shown in Fig. P4.155 reaches the heights shown. Each of the 25 nozzles has an exit area of 1.5 cm^2. Find the volume flow rate of water entering the sprinkler.

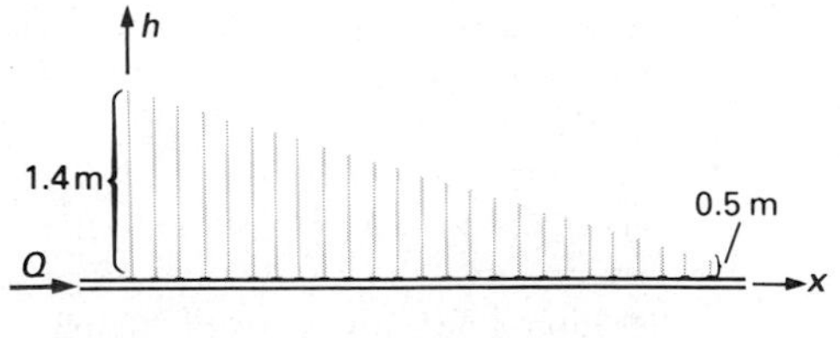

Figure P4.155

156. Consider the two tanks shown in Fig. P4.156. Tank A initially contains salt water with 35 grams of salt per liter of solution and h_A = 10.0 m. Tank B initially contains pure water with h_B = 10.0 m. Both tanks have a diameter of 8.0 m. Find the salt concentration of tank B in grams per liter of solution at 10 s after the initial time t = 0. Assume that the contents of both tanks are well mixed, so the salt is uniformly distributed throughout each tank. Neglect the volume of salt water below the outlet of Tank B.

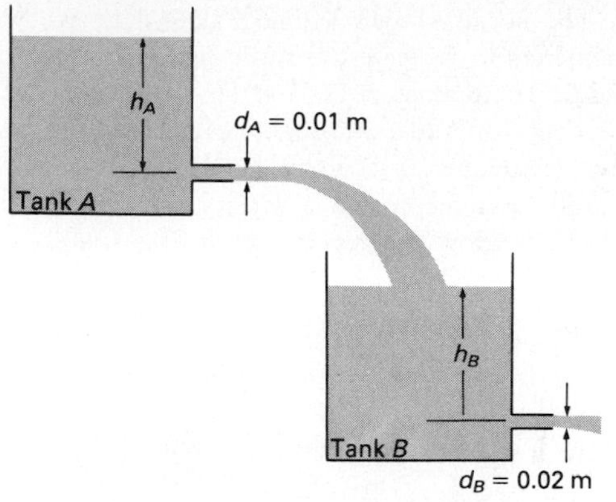

Figure P4.156

157. The "wye" fitting shown in Fig. P4.157 lies in a horizontal plane. The fitting splits the inlet flow into two equal parts. At section 1, the water velocity is 12 ft/sec and the pressure is 20 psig. Calculate the water pressure at sections 2 and 3. Assume inviscid flow and a water temperature of 60°F.

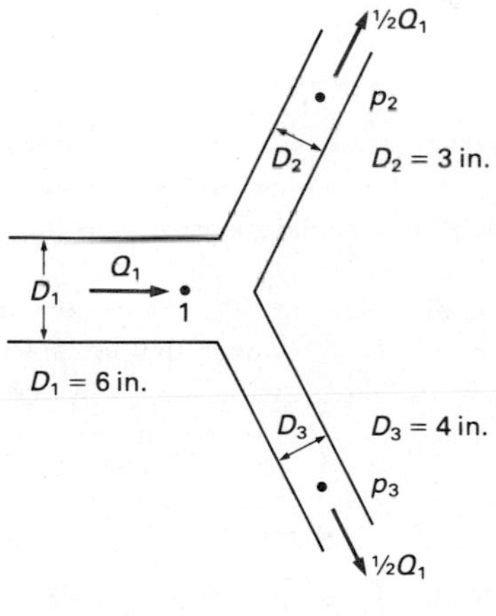

Figure P4.157

158. Standard atmospheric air ($p = 14.7$, psia, $T =$ 70°F) flows in a channel that is divided by a splitter plate. Pitot tubes are located in each half of the channel, as shown in Fig. P4.158. The channel is 1 ft wide perpendicular to the paper. Calculate the total air mass flow rate in the channel.

159. If the diameter of a jar in Fig. P4.159 is 10 cm, estimate the shortest time to empty the contents of the jar if the throat pressure remains constant at −2.354 kPa, gage (i.e., a vacuum of 2.354 kPa).

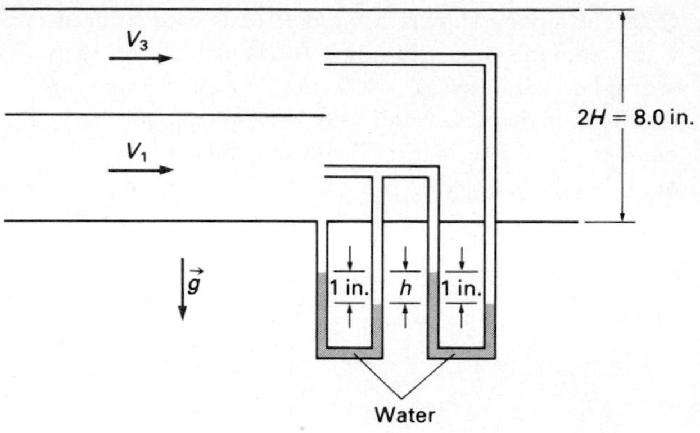

Figure P4.158

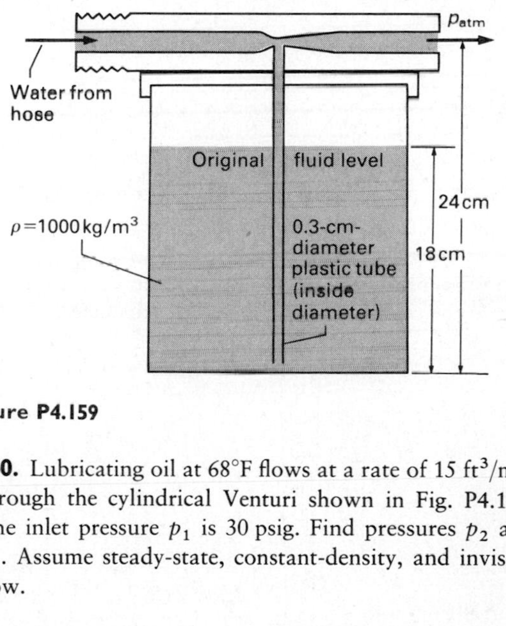

Figure P4.159

160. Lubricating oil at 68°F flows at a rate of 15 ft³/min through the cylindrical Venturi shown in Fig. P4.160. The inlet pressure p_1 is 30 psig. Find pressures p_2 and p_3. Assume steady-state, constant-density, and inviscid flow.

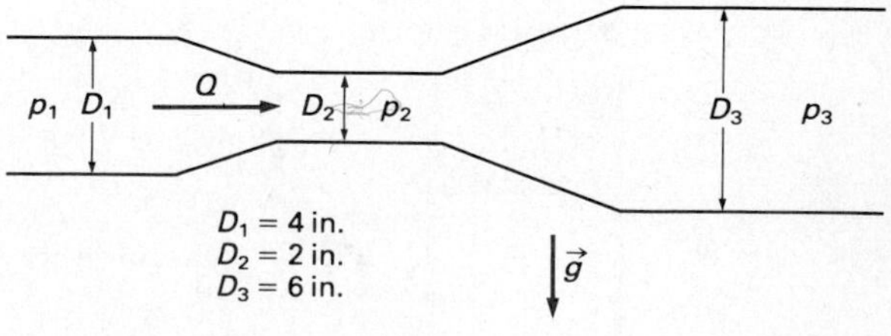

Figure P4.160

161. The 68°F lubricating oil in Problem 160 produces a pressure difference of $p_1 - p_2 = 0.05$ psi. Find the oil flow rate and the pressure difference $p_1 - p_3$.

162. The siphon shown in Fig. P4.162 is used to drain water from the tank whose area A_0 is much greater than the siphon exit area $A_e = 4.0\ \text{cm}^2$. How much time is required for the tank water level to drop from the 1.0-m mark to the 0.5-m mark? The flow is inviscid, and the fluid density is constant.

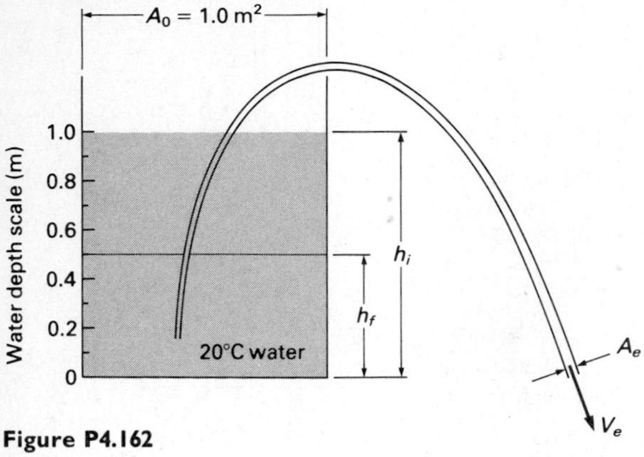

Figure P4.162

163. The syringe shown in Fig. P4.163 has an inside diameter of 1.0 cm. The syringe is held vertically, and the plunger is pushed upward at the rate of 4.0 cm/s. Find the maximum height to which the 20°C water will rise if the outlet measures 2.0 mm in diameter.

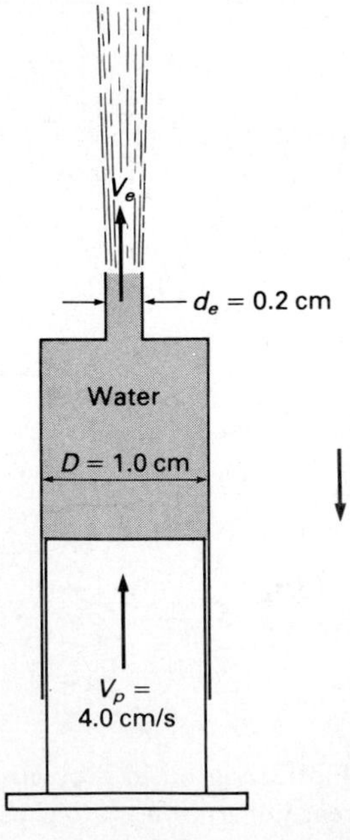

Figure P4.163

164. The nozzle shown in Fig. P4.164 has two water manometers to indicate the static pressures at sections 1 and 2. The diameters D_1 and D_2 are 8 in. and 2 in., respectively. Air flows through the nozzle, and the air and water temperatures are 60°F. Find the air volume flow rate. Assume constant density and inviscid flow. Neglect elevation changes.

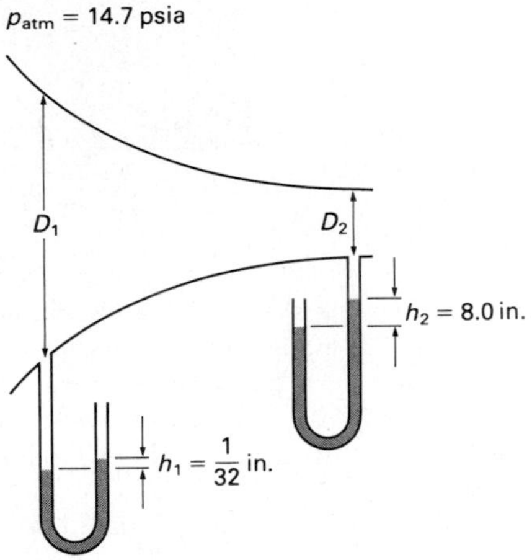

Figure P4.164

165. The storage tank shown in Fig. P4.165 has an inside diameter of 20 ft and a height of 28 ft. Water is supplied to this tank at a rate of $Q_i = 20\ \text{ft}^3/\text{min}$ through a $d =$ 4.0-in.-inside-diameter pipe. At one instant, the level of water in the tank is 18 ft above the pipe inlet. Find the pressure p_i in the pipe just before the water enters the tank. The water temperature is 50°F.

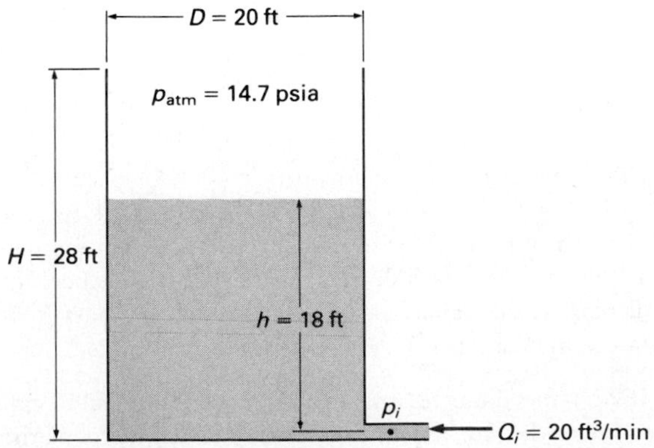

Figure P4.165

166. The piston shown in Fig. P4.166 is forcing 70°F water out the exit at 12 ft/sec. The exit pressure has been measured as $p_e = 40$ psig. Determine the force on the piston for a piston diameter $D_p = 3.0$ in.

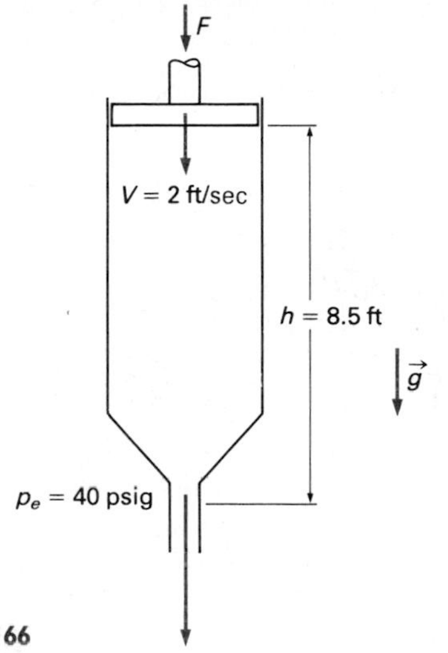

Figure P4.166

167. Figure P4.167 shows two tall towers. Air at 10°C is blowing toward the two towers at $V_0 = 30$ km/hr. If the two towers are 10 m apart and half the air flow approaching the two towers passes between them, find the minimum air pressure between the two towers. Assume constant air density, inviscid flow, and steady-state conditions. The atmospheric pressure is 101 kPa.

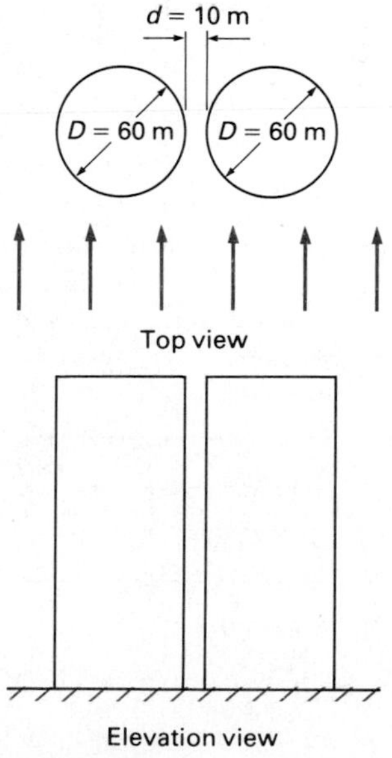

Figure P4.167

• 168. A cylinder of radius $r = 2.0$ cm is rotating about a horizontal axis at a speed of 125 rad/s. Air at 20°C is blowing over the cylinder, perpendicular to the axis of rotation, at 120 km/hr as shown in Fig. P4.168. Assume steady-state and constant air density conditions. Also assume that the air at the top of the cylinder has been sped up by the cylinder's rotation and that the air at the bottom of the cylinder has been slowed down by the cylinder. As a result, the air at the top is moving at twice the upstream velocity V_0 plus the cylinder surface velocity and the air at the bottom is moving at twice the upstream velocity V_0 minus the cylinder surface velocity. Calculate the pressure p_T at the top of the cylinder and the pressure p_B at the bottom of the cylinder.

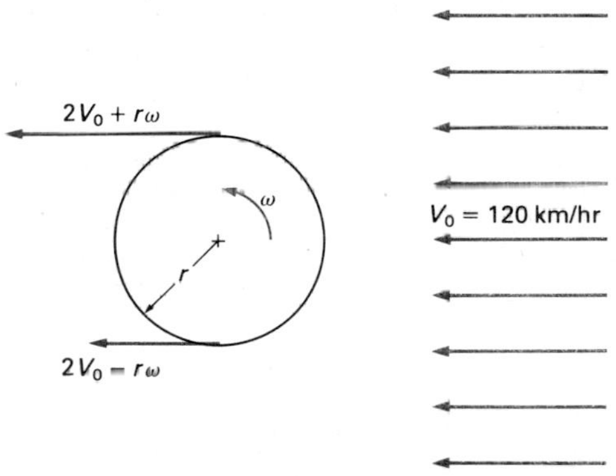

Figure P4.168

• 169. A commercial spray jar used to spray lawns and gardens is shown in Fig. P4.159. The inlet diameter $D_1 = 2.0$ cm, the throat diameter $D_t = 0.4$ cm, and the inlet water pressure $p_1 = 150$ kPa, gage. Find the required (minimum) water flow rate from the hose when the last drop of water is taken from the jar. The water temperature is 20°C.

• 170. Water drains from the higher of two large tanks through the Venturi in the discharge line, as shown in Fig. P4.170. Determine the minimum height H so that

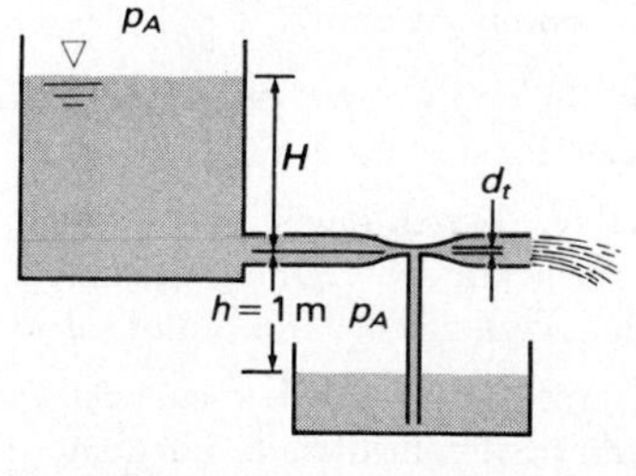

Figure P4.170

water will start to flow from the lower tank into the Venturi throat. The throat diameter $d_t = 0.50$ cm, and the water temperature is 20°C.

171. A water supply tank for a small town is constructed in the shape of an oblate spheroid (a shape formed by spinning an ellipse about its minor axis). The minor axis of the tank is 30 ft, the major axis is 60 ft, and the bottom of the tank is 75 ft above ground level. How long does it take to drain a half-full tank through a ground-level pipe with a 6-in. I.D.?

172. The conical sugar dispenser shown in Fig. P4.172 is used in a restaurant to refill individual sugar containers. If the bottom cap accidently jams open, how long will it take for a completely full dispenser to empty? (You may assume that the granular sugar behaves like an inviscid fluid.)

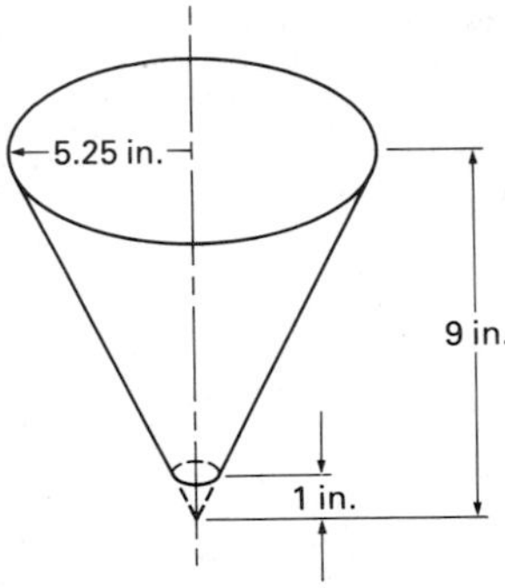

Figure P4.172

173. Air flows through a horizontal, converging nozzle. The inlet velocity, inlet pressure, inlet temperature, and outlet pressure are known. For each case listed below, indicate whether the (a) general energy equation, (b) mechanical energy equation, and/or (c) Bernoulli's equation can be used to find the desired quantity. [*Note:* The conservation of mass principle may also be needed.] Steady-state conditions apply.

(1) Constant and known mass density. Frictionless flow. Heat transfer. Find outlet velocity.

(2) Variable, but known mass density variation because of heating of fluid. Frictionless flow. Inlet and outlet areas given. Find outlet velocity.

(3) Constant and known mass density. Friction. Nozzle thermally insulated. Inlet and outlet areas known. Find fluid temperature change.

174. Find the vertical and horizontal components of force required to hold the tank shown in Fig. P4.174 in place. The tank contains 100 m³ of 20°C water and weighs 10^5 N.

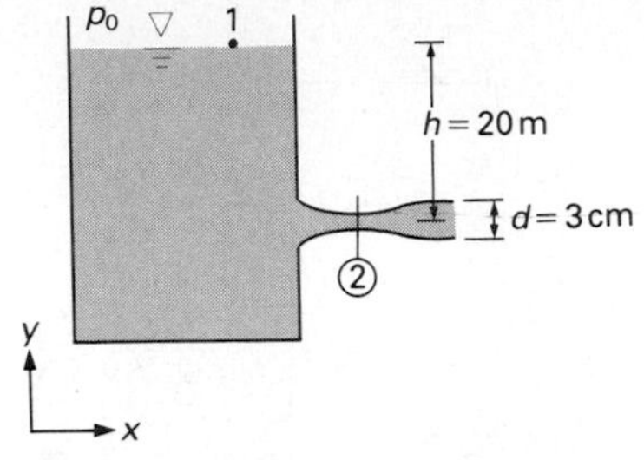

Figure P4.174

175. Two constant-density streams mix as shown in Fig. P4.175. Find the pressure drop $(p_1 - p_3)$ caused by mixing. Neglect friction between the fluids and the walls.

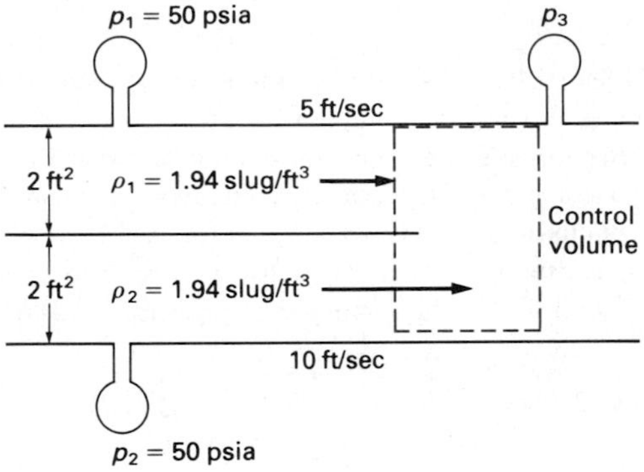

Figure P4.175

176. Calculate the pressure change $(p_2 - p_1)$ for the jet pump shown in Fig. P4.176. The fluid is 20°C water. Assume negligible friction at the walls and uniform pressure over each flow area.

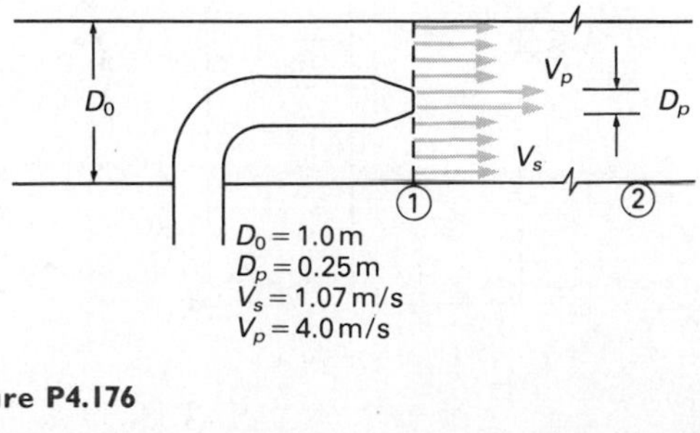

Figure P4.176

177. Find the x and y components of the force required to hold the bend shown in Fig. P4.177 in place. Pressures p_1 and p_2 are gage pressures.

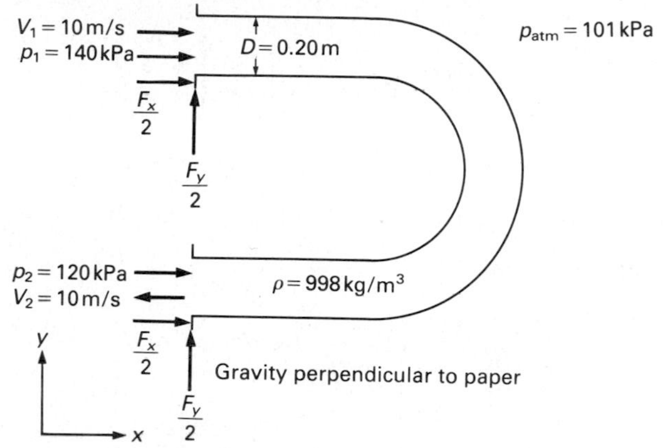

Figure P4.177

178. Fortified wine having a specific gravity of 1.03 flows through a 2-in., 90° standard elbow with a pressure drop of 0.50 psi and a flow rate of 40 gal/min. Determine all the horizontal forces on the elbow if it is in a horizontal plane and the inlet pressure is 12 psig.

179. Water flows steadily between fixed vanes, as shown in Fig. P4.179. Find the x and y components of the water's force on the vanes. The total volume flow rate is 100 m³/s, pressure $p_1 = 150$ kPa, and pressure $p_2 = 101$ kPa.

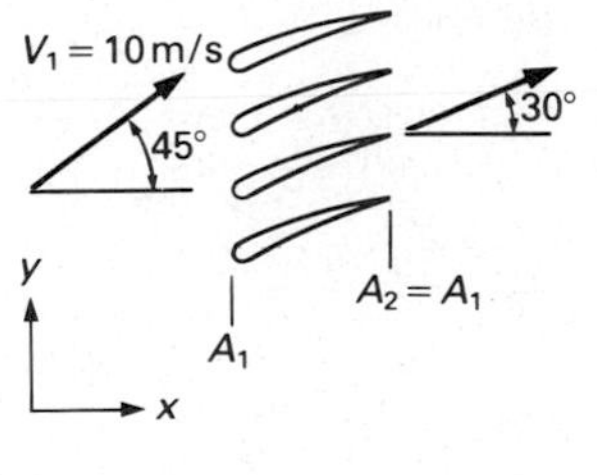

Figure P4.179

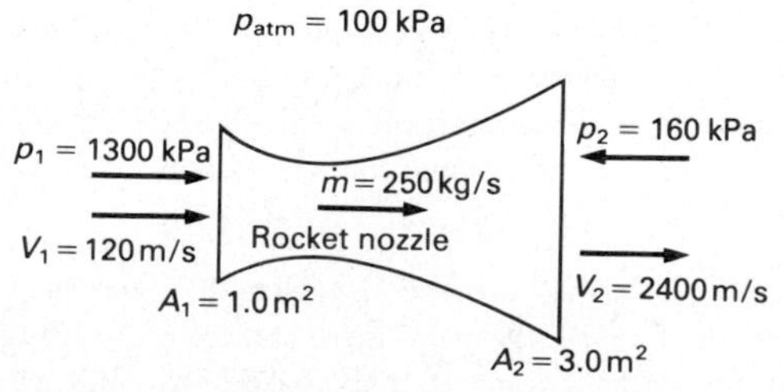

Figure P4.180

180. The nozzle of a rocket has the velocities and absolute pressures shown in Fig. P4.180. Find the force F required to hold the nozzle in place. Neglect gravity.

181. Water at 60°F and at a flow rate of 80 gal/min flows through a 6-in. schedule 40 pipe, through a reducer, and then through a 4-in. schedule 40 pipe. The pressure at the reducer inlet is 40 psig and at the reducer outlet is 34 psig. Find the force of the water on the reducer. Assume that atmospheric pressure is 14.7 psia.

182. What fraction of the mass flow rate $\dot{m}$ is diverted down the plate shown in Fig. P4.182? Neglect friction between the water and the plate and the effect of gravity.

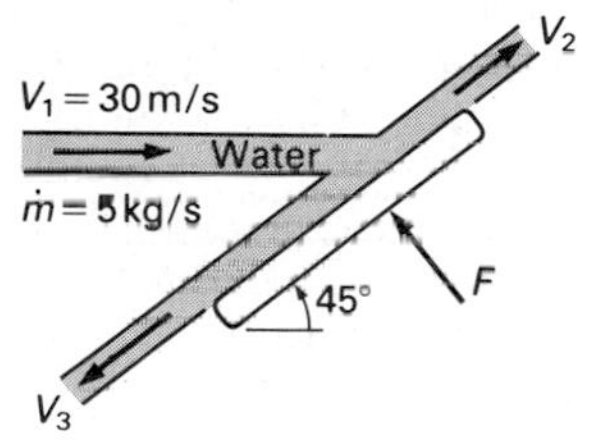

Figure P4.182

183. Find the magnitude of the force F required to hold the plate in Fig. P4.182 stationary.

184. Figure P4.184 shows coal being dropped from a hopper onto a conveyor belt at a constant rate of 25 ft³/sec. The coal has a specific gravity ranging from 1.12 to 1.50. The belt has a speed of 5.0 ft/sec and a loaded length of 15.0 ft. Estimate the torque required to turn the drive pulley of the conveyor belt. The drive pulley diameter is 2.0 ft. Assume that there is no friction between the belt and the other rollers.

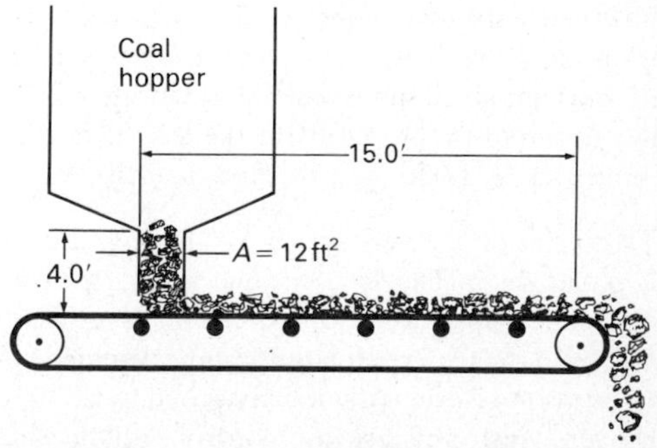

Figure P4.184

185. The stationary jet engine shown in Fig. P4.185 produces a thrust of 200 kN. Find the mass flow rate through the engine.

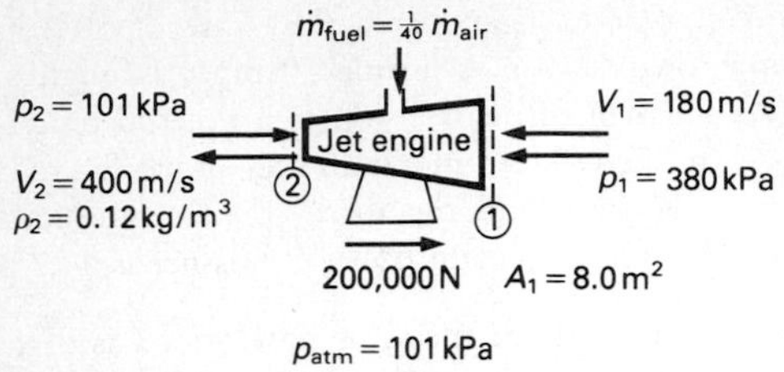

Figure P4.185

186. Air flows through a horizontal, 90° miter bend at the rate of 40 m^3/s. The duct cross section is rectangular, measuring 0.67 m × 0.50 m. The air is at 15°C and 101 kPa. Calculate the horizontal forces on the bend resulting from the turning of the air. Atmospheric pressure acts inside the duct.

187. A 20°C water jet is aimed at a hole in the wall, as shown in Fig. P4.187. Twenty-five percent of the water escapes through the hole, and the remainder is turned 90° to the water jet. Find the force exerted on the wall by the jet.

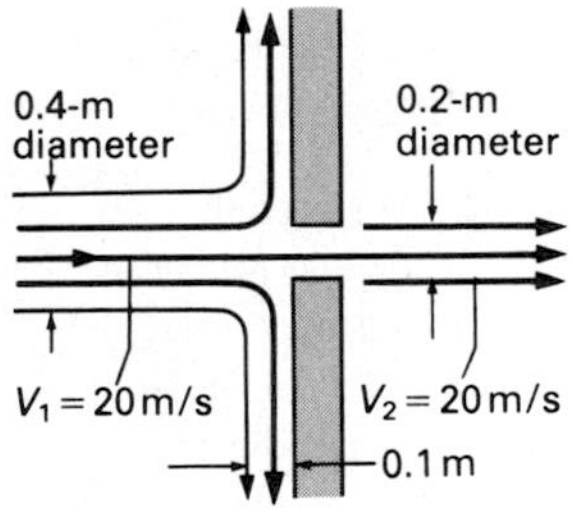

Figure P4.187

188. Water flows from the tank shown in Fig. P4.188. Find the spring scale reading as a function of the water height h if the spring scale is set at zero when the tank is not on the beam. The balance beam remains approximately horizontal. Assume that the time rate of change of momentum of the water inside the tank is negligible. The tank weighs 1.0 lb, and the flow is inviscid.

189. Water at 60°F flows over a 100-ft-long, 25-ft-high dam, as shown in Fig. P4.189. The water velocity V_1 some distance upstream from the dam is uniform at 10 ft/sec, and the water velocity V_2 some distance downstream is also uniform. Assume inviscid flow and a constant river width and use the control volume shown. Calculate the force exerted on the dam by the water.

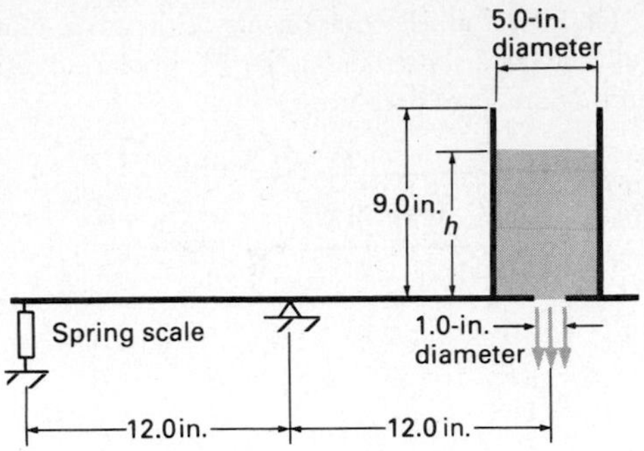

Figure P4.188

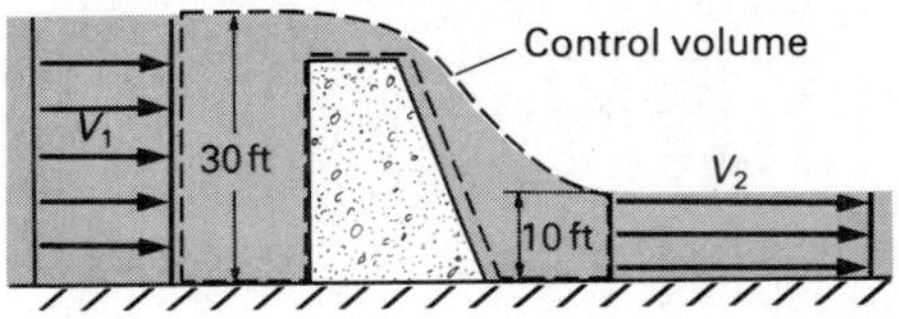

Figure P4.189

190. When the pump shown in Fig. P4.190 is stopped, there is no flow through the system and the spring force is zero. With the pump running, a 6-in.-diameter jet leaves the pipe, and the spring force is 420 lb. The water surface elevation in the tank is constant. Determine the water flow rate and the power consumed by the pump. Assume quasi-steady flow.

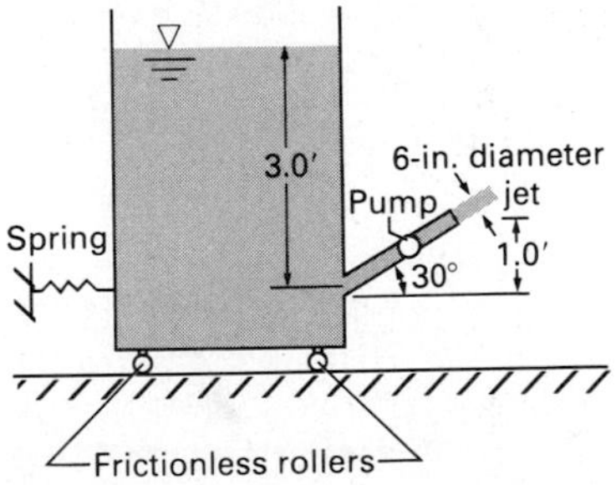

Figure P4.190

191. A 4-in. schedule 40 steel pipe ends in a tee. The flow rate is 2.0 ft^3/sec of 77°F lubricating oil ($\rho = 1.58$ slugs/ft^3). Find the x-direction force on the tee. See Fig. P4.191. Neglect gravity and assume that the flow divides evenly; $p_1 = 20$ psig, $p_2 = 0$ psig, and $p_3 = 0$ psig.

192. Water flows up the 2-cm-inside-diameter pipe shown in Fig. P4.192 at a flow rate of $Q = 1.0$ m^3/s. Find the bending moment at the base of the pipe caused by the flow of the water, which is inviscid.

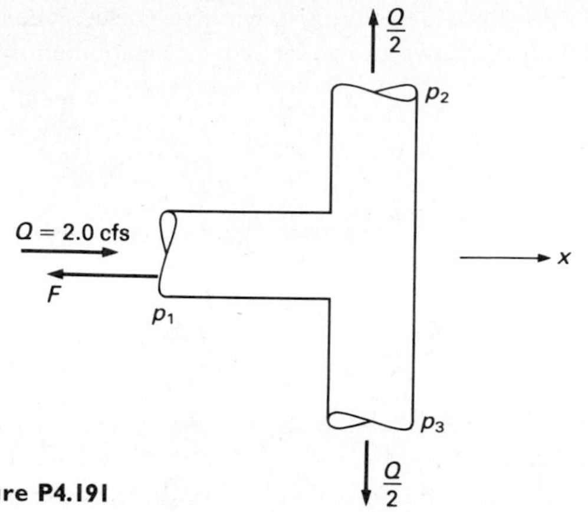

Figure P4.191

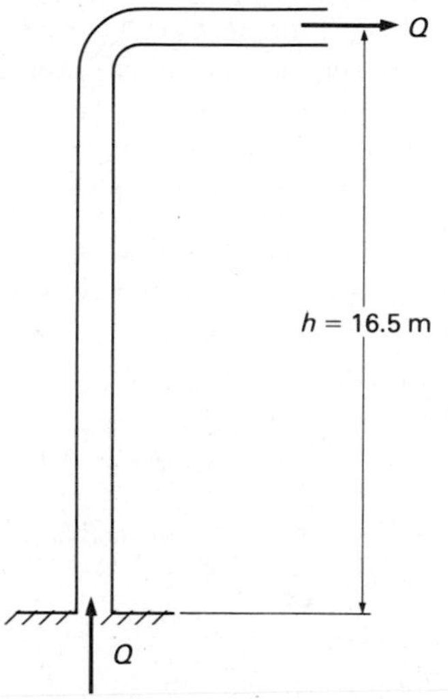

Figure P4.192

193. Air at $T_1 = 300\text{K}$, $p_1 = 303$ kPa, and $V_1 = 0.5$ m/s enters the Venturi shown in Fig. P4.193. The air leaves at $T_2 = 220\text{K}$ and $p_2 = 101$ kPa; $A_1 = 0.6\ \text{m}^2$ and $A_2 = 1.0\ \text{m}^2$. Calculate the horizontal force required to hold the Venturi stationary.

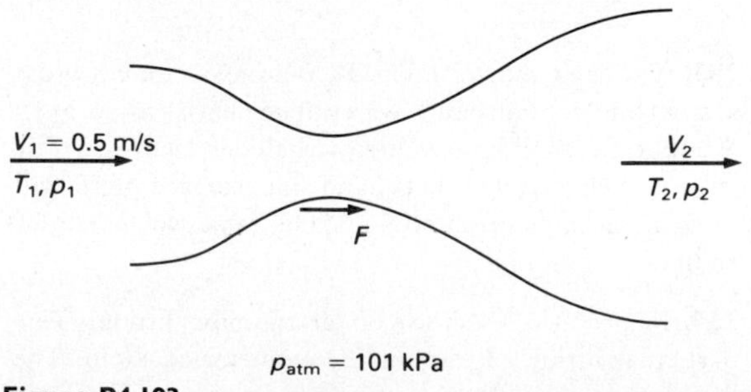

Figure P4.193

194. Find the horizontal and vertical forces to hold stationary the nozzle shown in Fig. P4.194. The fluid flowing through it is 10°C liquid water; $A_1 = 1.0\ \text{m}^2$, $A_2 = 0.25\ \text{m}^2$, $V_1 = 20$ m/s, $p_2 = p_{\text{atm}}$, and $p_1 = p_{\text{atm}} +$ 30 kPa. Neglect gravity.

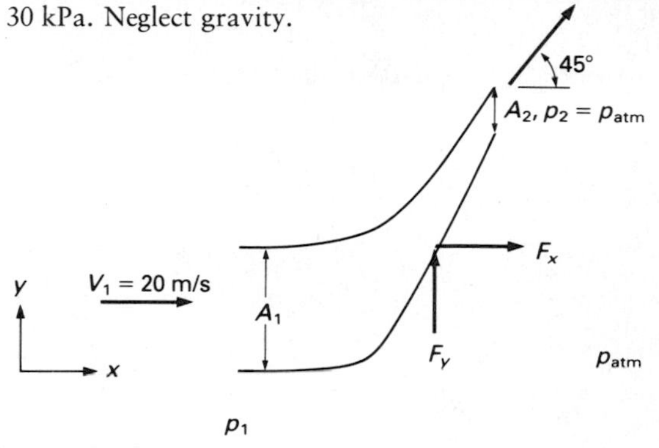

Figure P4.194

195. Water at 60°F leaves the nozzle shown in Fig. P4.195 with a velocity of 5 ft/sec. The nozzle exit area is 0.725 in^2. The water falls a distance of 60 ft before hitting the flat plate. What force does the water stream exert on the flat plate?

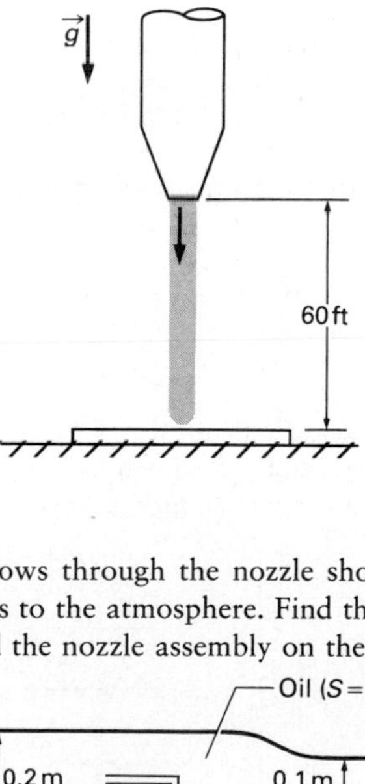

Figure P4.195

196. Water flows through the nozzle shown in P4.196 and discharges to the atmosphere. Find the force F necessary to hold the nozzle assembly on the support.

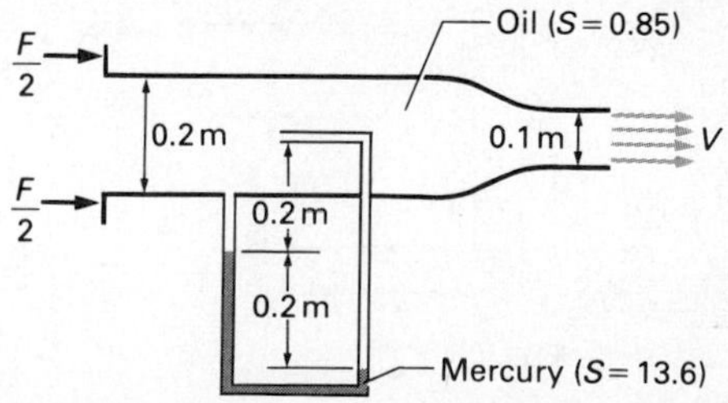

Figure P4.196

197. How much force must firefighters exert to hold the nozzle shown in Fig. P4.197? Assume that no force is exerted on the nozzle by the hose, which has a 6-in. inside diameter, and that the water is at 60°F. Neglect gravity.

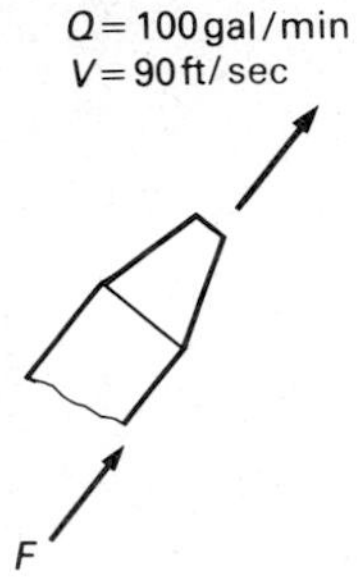

Figure P4.197

198. Find the forces F_x and F_y required to hold steady the tank shown in Fig. P4.198. The pipe diameter D is that required for a constant water level. Assume inviscid flow and neglect the weight of the tank.

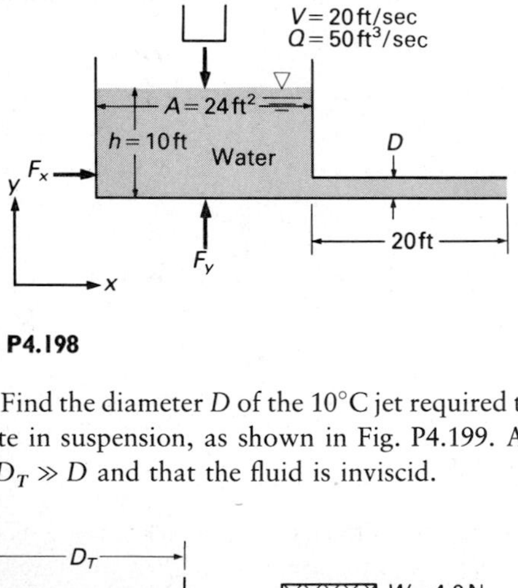

Figure P4.198

199. Find the diameter D of the 10°C jet required to hold a plate in suspension, as shown in Fig. P4.199. Assume that $D_T \gg D$ and that the fluid is inviscid.

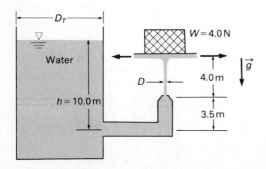

Figure P4.199

200. A 0.10-m-diameter, 20°C water stream hits a conical deflector as shown in Fig. P4.200. Calculate the force F required to hold the deflector stationary.

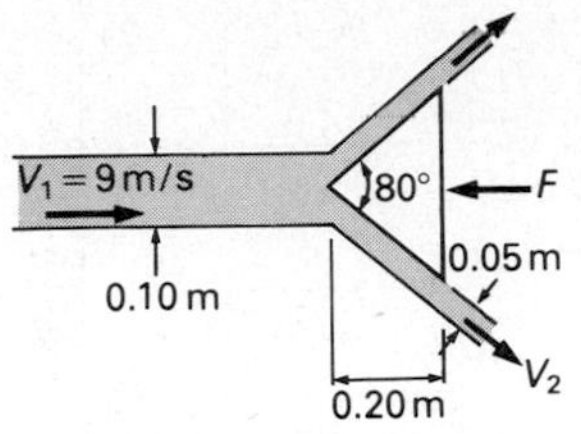

Figure P4.200

201. Air enters a stationary jet engine at a uniform velocity of 200 m/s and leaves at a uniform velocity of 600 m/s. The fuel mass flow rate is $\frac{1}{40}$ of the air mass flow rate into the engine. The inlet area is 0.25 m^2, the exit area is 0.10 m^2, the inlet air density is 0.75 kg/m^3, and the exit pressure is equal to the atmospheric pressure. How much thrust does the engine develop? The fuel enters the engine in a direction perpendicular to the air flow through the engine.

202. Find the force components F_x and F_y required to hold the box shown in Fig. P4.202 stationary. The fluid is oil and has specific gravity of 0.85. Neglect gravity effects. Atmospheric pressure acts around the entire box. Steady flow conditions prevail.

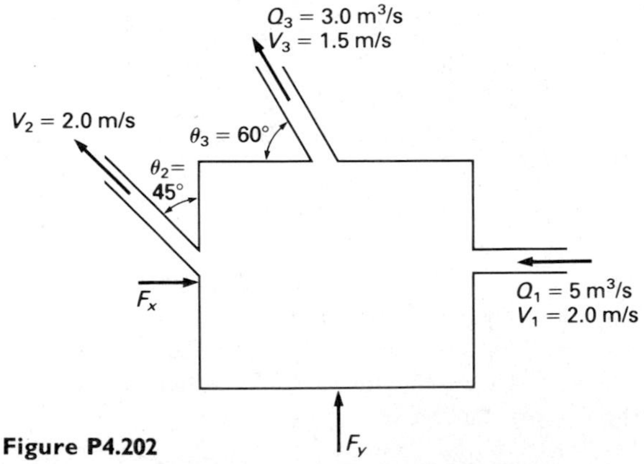

Figure P4.202

203. The tank shown in Fig. P4.203 has a circular cross section and contains 60°F water. The depth H = 6.25 ft. What is the minimum value of the coefficient of static friction between the tank and the surface that will prevent the tank from sliding? The tank weighs 25 lbs empty.

204. Figure P4.204 shows a lateral pipe fitting. This particular fitting has a mainline diameter of 4.0 in. The diameter of the lateral is 3.0 in., and the lateral angle is

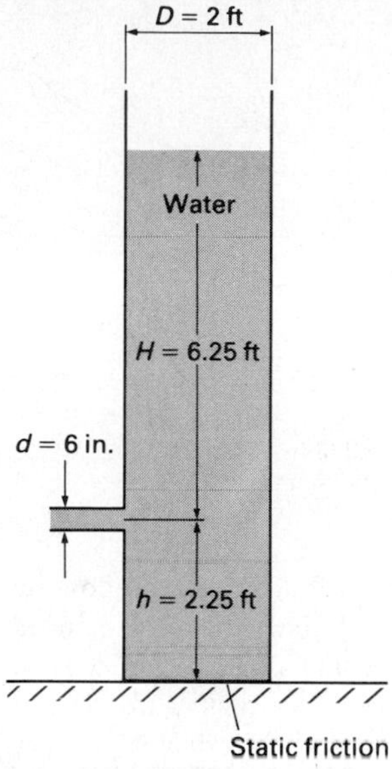

Figure P4.203

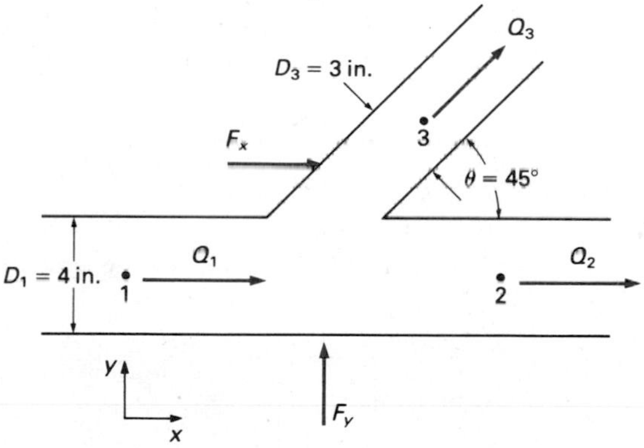

Figure P4.204

45°; 60°F water is flowing in the lateral. Measurements show that the pressure at point 1 is 34.0 psig, the pressure at point 2 is 35.0 psig, the pressure at point 3 is 33.5 psig, and the flow rate at point 2 is 1.0 ft^3/sec. Determine the horizontal and vertical force components (F_x and F_y) required to hold the lateral fitting stationary. Neglect gravity. $Q_1 = 1.63$ ft^3/sec.

205. Water flows in the conical diffuser shown in Fig. P4.205. Calculate the water velocity leaving the diffuser, the pressure at the diffuser exit, and the force on the diffuser walls.

206. A pipe delivers water into a catch tank sitting on a scale, as shown in Fig. P4.206. At one point during a test, the water velocity in the pipe reaches 3 m/s and the

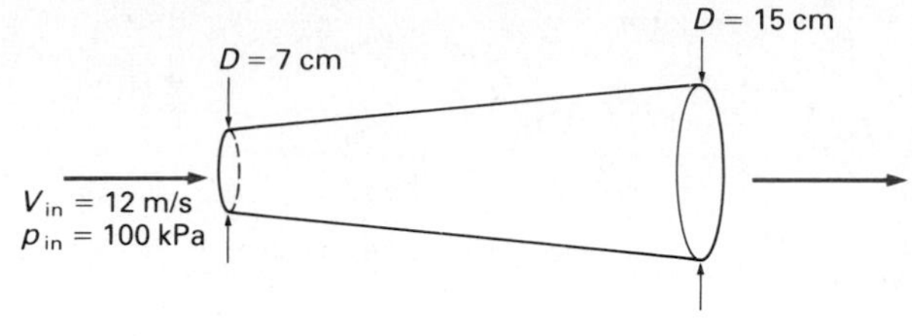

Figure P4.205

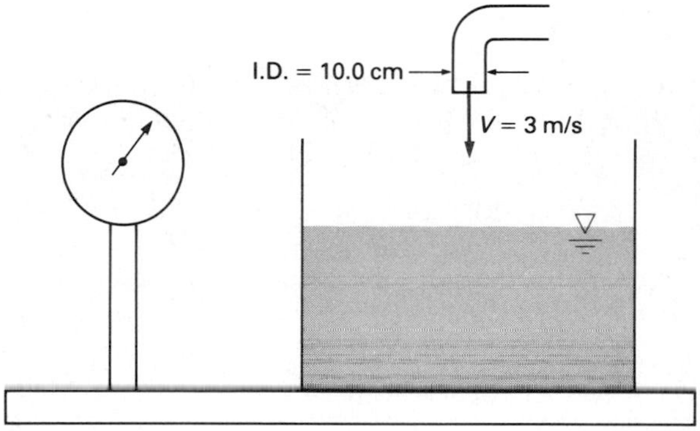

Figure P4.206

scale reads 200 N. How much water is in the catch tank at that instant? The scale is "zeroed" when the catch tank is empty, and the water temperature is 20°C.

207. The water flow in the penstock elbow shown in Fig. P4.207 is 5 ft^3/s. Calculate the necessary restraining force at the penstock elbow anchors. Neglect gravity.

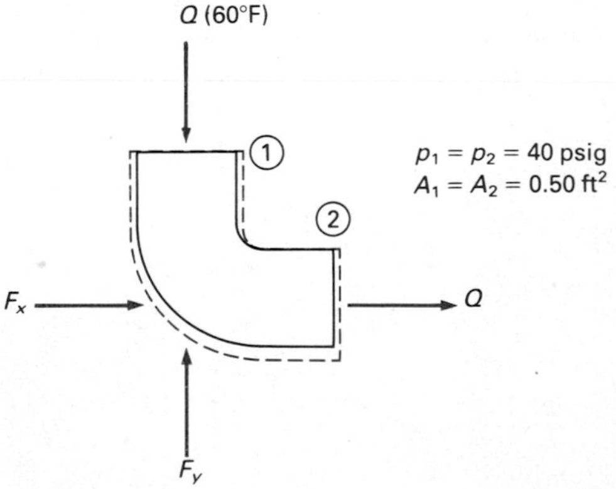

Figure P4.207

208. Figure P4.208 shows the configuration of the center (tail-mounted) jet engine on an airliner. The airliner is cruising at altitude, and the velocities shown are relative to an observer on board. Calculate the thrust force that the engine exerts on the airplane.

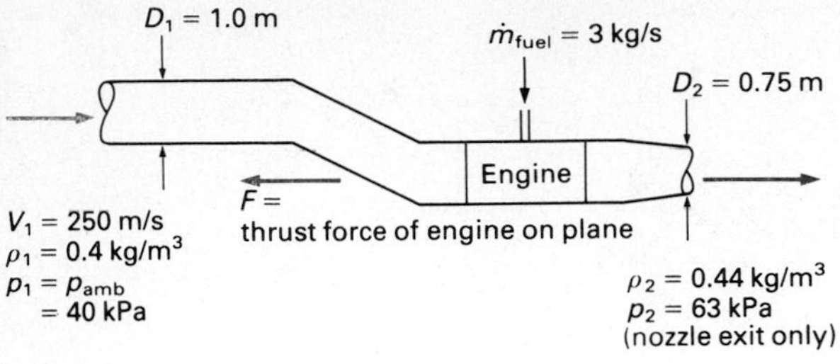

Figure P4.208

209. The plate shown in Fig. P4.209 is 0.5 m wide perpendicular to the paper. Calculate the velocity of the water jet required to hold the plate upright.

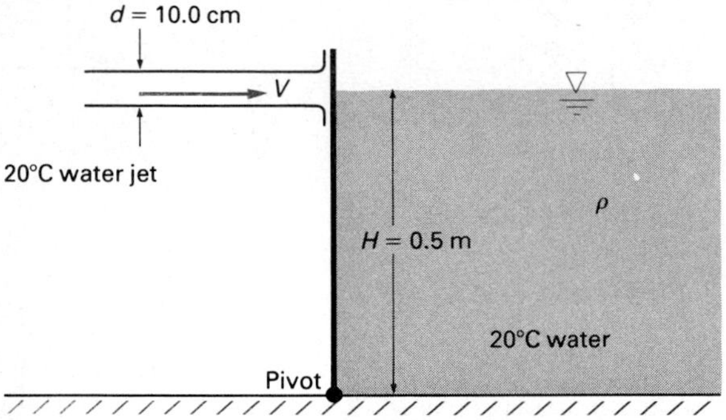

Figure P4.209

210. The nozzle shown in Fig. P4.210 has a water flow rate of 0.495 ft^3/s. Find the force F to keep the nozzle stationary. Assume inviscid flow, uniform velocity over each area, and constant fluid density at 60°F.

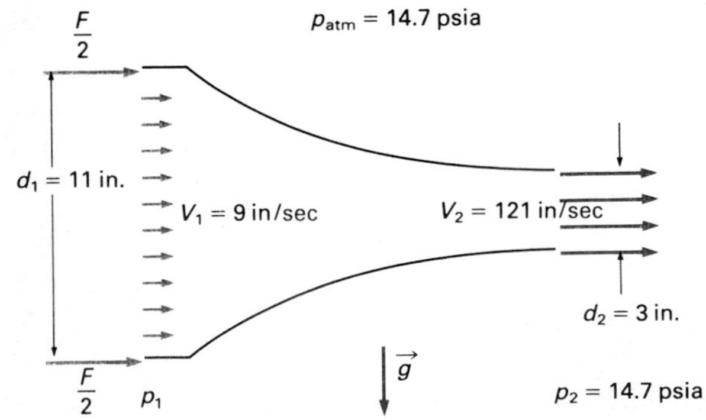

Figure P4.210

211. The diffuser shown in Fig. P4.211 has a water flow rate of 14.14 m^3/s. Calculate the force F required to keep the diffuser stationary. Assume that the fluid density is constant at 20°C.

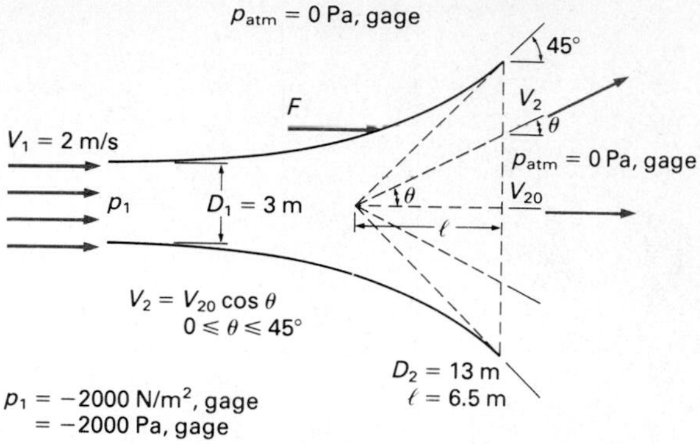

Figure P4.211

212. Air flows through the axial flow turbine shown in Fig. P4.212. Find the axial force F_x to hold the turbine stationary. Assume constant air density, with $p_1 = p_2$ and uniform air velocity over the flow area $\pi(R_0^2 - R_i^2)$ with $R_i = 0.5$m and $R_0 = 0.7$ m.

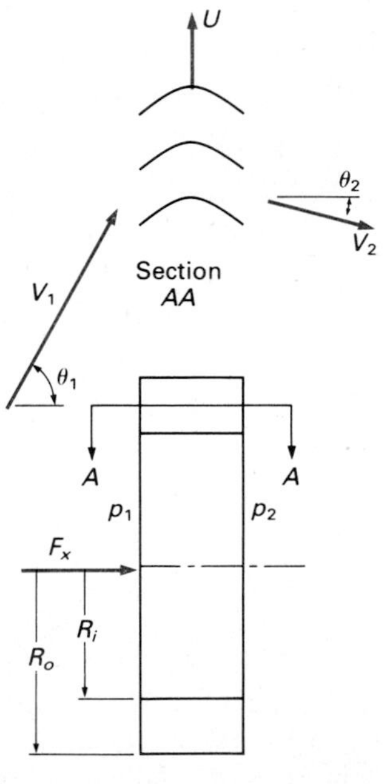

Figure P4.212

213. Air at $p_1 = 287$ kPa, $T_1 = 500$K, and $\rho_1 = 2.0$ kg/m^3 enters the turbine shown in Fig. P4.213 and leaves at $p_2 = 143.5$ kPa, $T_2 = 400$K, and $\rho_2 = 1.25$ kg/m^3. The flow area is $\pi(R_0^2 - R_i^2)$ with $R_i = 0.5$ m and $R_0 = 0.7$ m. The inlet absolute velocity $V_1 = 200$ m/s. Assume that the air velocities are uniform over their respective flow areas and that the air is an ideal gas. Find the axial force F_x to hold the turbine stationary.

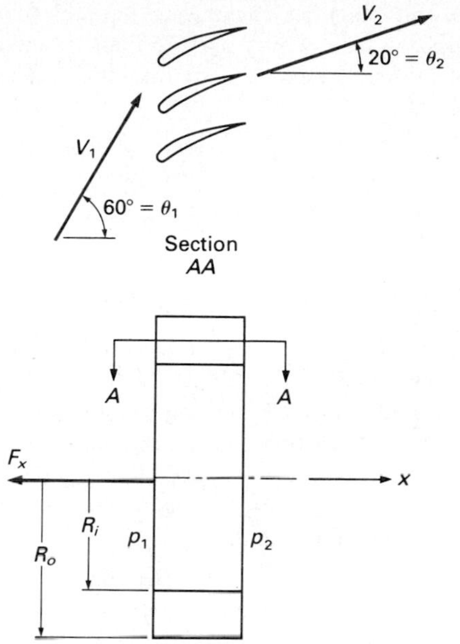

Figure P4.213

214. The fan shown in Fig. P4.214 has an inlet air flow rate of 5000 cfm, an inlet flow diameter of 2.0 ft, and an exit flow area of 3.0 ft^2. Determine the horizontal force on the bolts holding the fan to its support. Assume constant air density at 60°F.

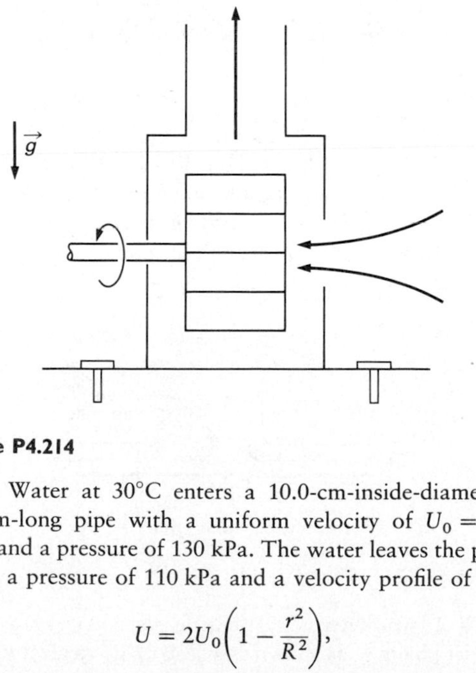

Figure P4.214

215. Water at 30°C enters a 10.0-cm-inside-diameter, 2.0-m-long pipe with a uniform velocity of $U_0 = 3.0$ m/s and a pressure of 130 kPa. The water leaves the pipe with a pressure of 110 kPa and a velocity profile of

$$U = 2U_0\left(1 - \frac{r^2}{R^2}\right),$$

where R is the pipe inside radius. Find the frictional force of the pipe on the water. Assume constant density.

•**216.** A 60°F water jet leaves the nozzle shown in Fig. P4.216 with velocity $V_0 = 10$ ft/sec through area $A_0 = 0.5$ ft^2. The jet is directed upward to support the wedge of weight $W = 10$ lb. The wedge deflects the jet stream equally to the right and left and is long enough (perpendicular to the page) that end effects are negligible. The water density is constant. Write the linear momentum equation for the control volume shown to find the height h. Neglect the weight of any water in the control volume and any change in the height of the water as it flows through the control volume.

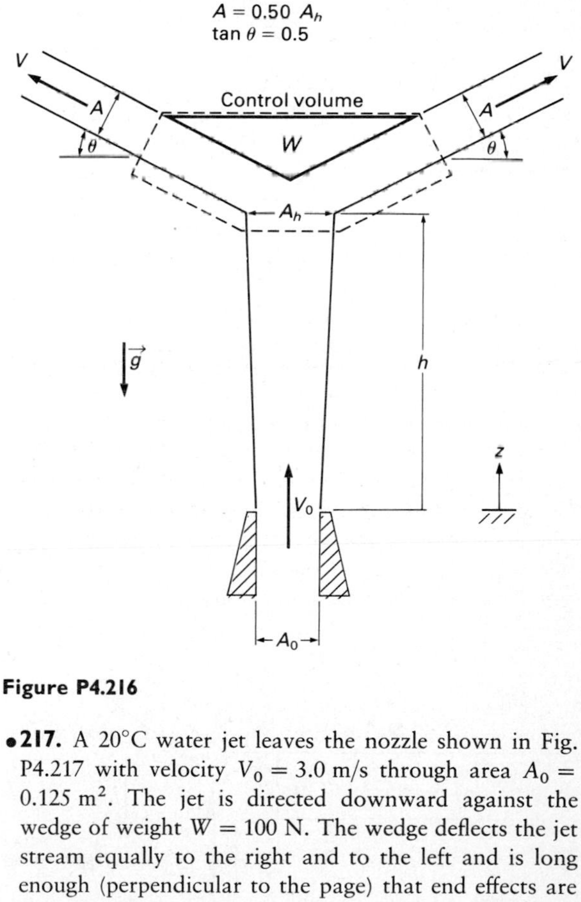

Figure P4.216

•**217.** A 20°C water jet leaves the nozzle shown in Fig. P4.217 with velocity $V_0 = 3.0$ m/s through area $A_0 = 0.125$ m^2. The jet is directed downward against the wedge of weight $W = 100$ N. The wedge deflects the jet stream equally to the right and to the left and is long enough (perpendicular to the page) that end effects are negligible. The water density is constant. Write the linear momentum equation for the control volume shown to find the force F for $h = 1.0$ m. Neglect the weight of any water in the control volume and change in the height of the water as it flows through the control volume.

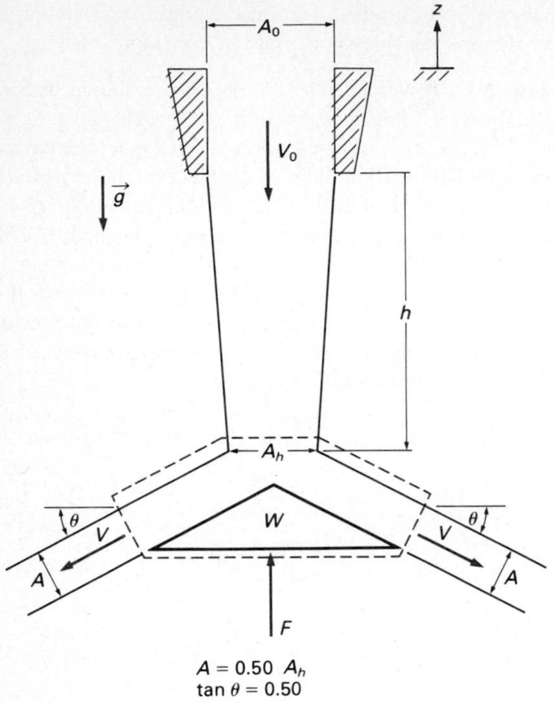

Figure P4.217

• **218.** Figure P4.218 shows a sharp-edged splitter plate used to control the flow of a liquid jet W units wide by H_1 units high. Write expressions for the deflection angle θ and the force F of the jet on the splitter plate as a function of the fluid density ρ, H_1, W, V, and plate insertion h. This force F has no components parallel to the plate. Assume that the jet flow is inviscid, that the jet width W remains constant, $H_2/H_3 = H_2'/H_3' = h/(H_1 - h)$, and constant fluid density.

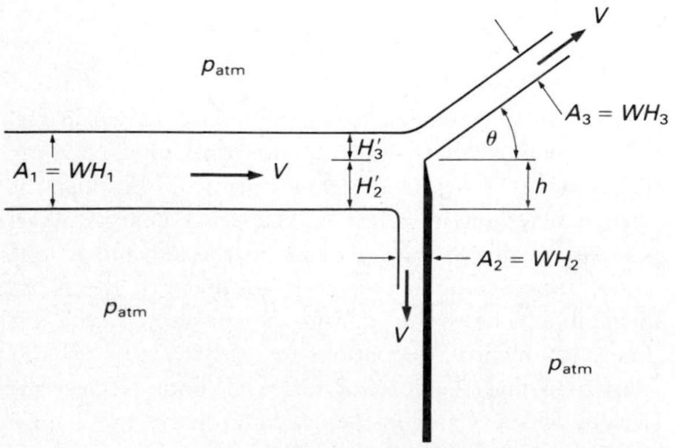

Figure P4.218

• **219.** A stationary jet engine has an inlet velocity of 400 ft/sec, an inlet density of 0.075 lbm/ft^3, and an inlet area of 1.0 ft^2. The exit velocity from the circular exit is given by

$$u = (3300 \text{ ft/sec})\left(1 - \frac{r}{R}\right)^{1/7},$$

with radius $R = 1.0$ ft. The exit density is 0.0040 lbm/ft^3. Determine the thrust generated by the engine (atmospheric pressure exists around the engine).

• **220.** The centrifugal pump shown in Fig. P4.220 delivers 60°F water at the rate of 10 ft^3/sec. The water is an ideal fluid, and the pump, piping, and internal fluid weigh 500 lb. Each bolt holding the pump to the floor can withstand a shear force of 500 lb and a tensile force of 1000 lb. Find the minimum number of bolts required to hold the pump to the floor.

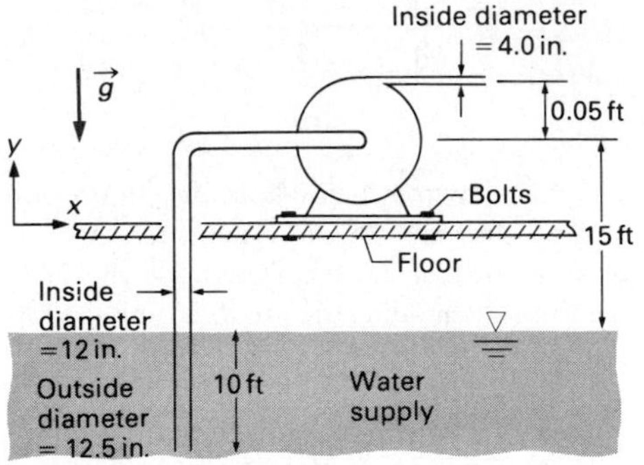

Figure P4.220

• **221.** What spacing h is required to lift the disc shown in Fig. P4.221? The disc weighs 10.0 N. Assume that the 20°C air density is constant. Note that the air pressure in the circular space of height h above the disc is less than the atmospheric pressure and that the air velocity decreases with radius.

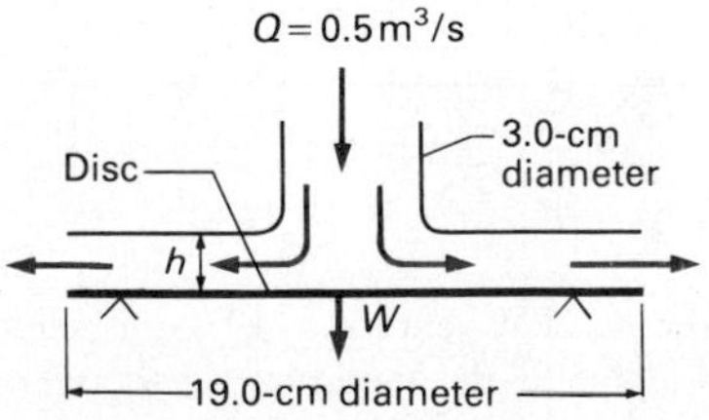

Figure P4.221

• **222.** A two-dimensional object is placed in a 2.0-m-wide water channel, as shown in Fig. P4.222. Neglect friction between the walls and the water; $p_0 - p_1 = 10$ kPa. Find the drag per unit length of the object.

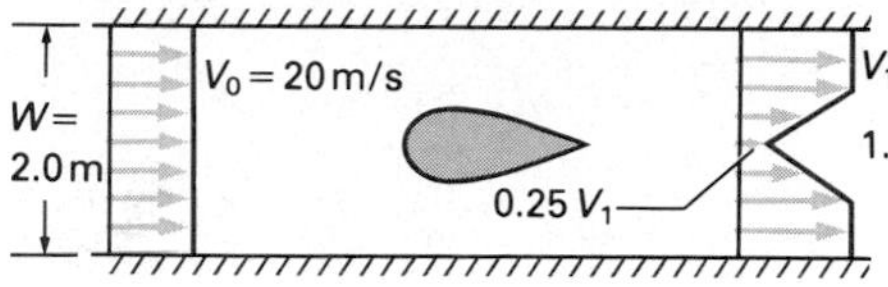

Figure P4.222

•**223.** Figure P4.223 shows a two-dimensional object that has been placed in an infinite stream and the resulting velocity profiles. Calculate the drag force per unit length of the fluid on the object in terms of the fluid density ρ, velocity V_0, and h.

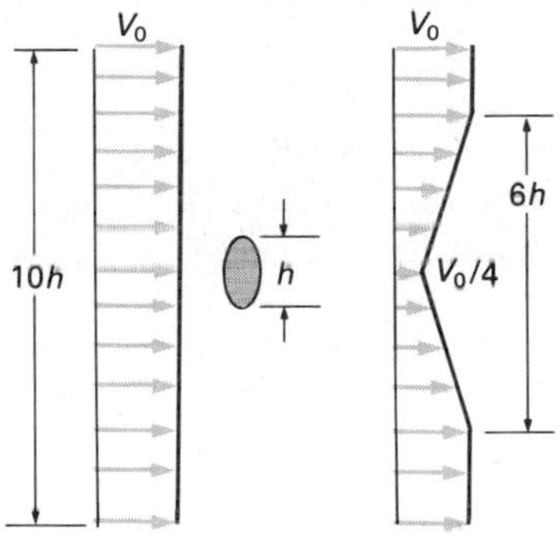

Figure P4.223

•**224.** Water at 20°C flows over the block shown in Fig. P4.224. The block is $W = 5.0$ m wide (perpendicular to the page), the flow is inviscid, and the water velocity is uniform over each flow area. The flow is two-directional (x, z), so there is no velocity component perpendicular to the page. Because of the nature of this flow, there are two possible flow heights h_2. Find the horizontal force exerted by the water on the block for each flow height.

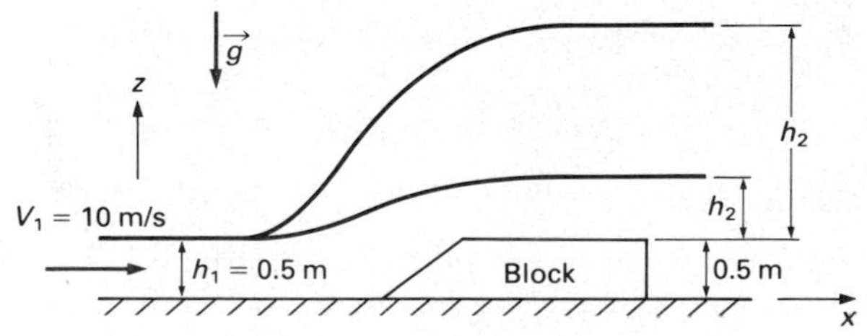

Figure P4.224

•**225.** Show that the magnitude of the pressure pulse Δp that travels upstream at the speed of sound c due to the rapid closure of a valve in a liquid piping system is given by

$$\Delta p = \rho V c,$$

where ρ is the fluid density, V is the fluid velocity before closure, and c is the speed of sound in the liquid. [*Hint:* Apply the linear momentum equation to a control volume of differential length.]

226. Water (60°F) flows through an annular space formed by inserting a 1-in.-radius solid cylinder into a 1.5-in.-radius tube. The following axial velocities were measured in the annular space.

$(r - r_i)/(r_o - r_i)$	u (ft/s)
0.1	12
0.2	23
0.3	28
0.4	33
0.5	34
0.6	31
0.7	28
0.8	21
0.9	10

Assume that the no-slip condition ($u = 0$) exists at the solid boundaries. What are the rates of mass, momentum, and kinetic energy flow through the annular space?

227. Analyzing the annular flow of Problem 226 yields an expression relating pressure drop to axial velocity:

$$u = -\frac{1}{4\mu}\frac{dp}{dx}\left[r_0^2 - r^2 + (r_0^2 - r_i^2)\frac{\ln(r_0/r)}{\ln(r_i/r_0)}\right].$$

For this profile and a water temperature of 60°F, compute the pressure gradient using a least-squares fit of the velocity data.

228. Assuming a pump-motor efficiency of 60 percent and the relationship between pressure drop and velocity presented in Problem 227, how much electrical power is required to generate the flow of Problem 226 through 50 ft of pipe?

229. A fireboat is traveling at 35 mph and is discharging water toward the rear at the rate of 390 gal/min through a 6-in.-diameter nozzle and at a 45° angle above the horizontal. How much faster would the fireboat travel if the water were directed at a fixed, solid wall?

230. A commercial airliner with four jet engines is traveling horizontally at a constant speed of 600 mph in still air. Each engine has an inlet area of 1.0 ft², an exit area of 0.30 ft^2, an air–fuel ratio of 35:1 on a mass basis, an exhaust velocity of 1500 ft/sec relative to the plane, and an exhaust pressure equal to the local atmospheric pressure. The ambient temperature and pressure are −65°F and 4.0 psia, respectively. Find the drag on the airplane.

231. A gun boat weighs 100,000 lb and is traveling at 30 knots. How much less engine horsepower is required to maintain this constant speed if the boat has two guns shooting 60 rounds per minute to the rear with a muzzle velocity of 3000 ft/sec? Each round weighs 0.125 lb.

232. A moving jet engine draws in air at 15 kg/s with an inlet relative velocity of 150 m/s and an exit relative velocity of 700 m/s. Find the engine thrust if atmospheric pressure exists around the engine.

233. Figure P4.233 shows water with a constant absolute velocity V striking a cart moving with velocity U. Find the cart's velocity U so that the power developed by the cart ($\dot{W} = UF_{Rx}$) is a maximum for fixed values of V, A, ρ, and β; F_{Rx} is the horizontal component of $\vec{F}_R$.

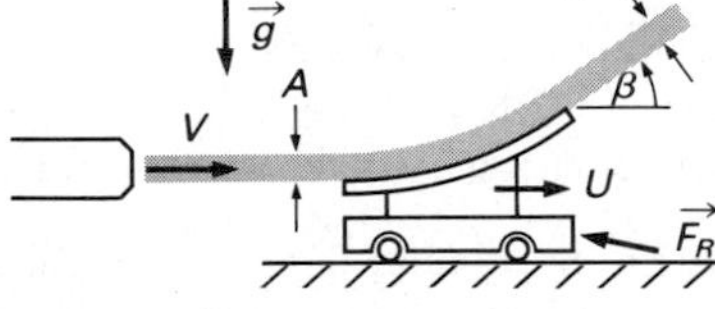

Figure P4.233

234. A sailboat travels 20 km/hr with a tailwind of 30 km/hr. An estimate of the air flow pattern is shown in Fig. P4.234. Find the force of the air on the sailboat by using an observer on the sailboat to write the linear momentum equation.

235. Repeat Problem 234, but use an observer fixed on the shore to write the linear momemtum equation.

236. The sailboat in Problem 234 is moving at an absolute (relative to the shore) and constant velocity of 20 km/hr in a river that has an absolute and constant velocity of V_R in the same direction as the sailboat's velocity V_B. If $V_R = 5$ km/hr, find the velocities of the air entering and leaving a control volume moving with the sail for an observer moving with the river.

237. The propeller-driven swamp boat in Fig. P4.237 travels at a constant velocity of 20 km/hr in stationary air. The 25°C air jet leaving the propeller has a diameter of 2.0 m and a velocity of 100 km/hr relative to the boat

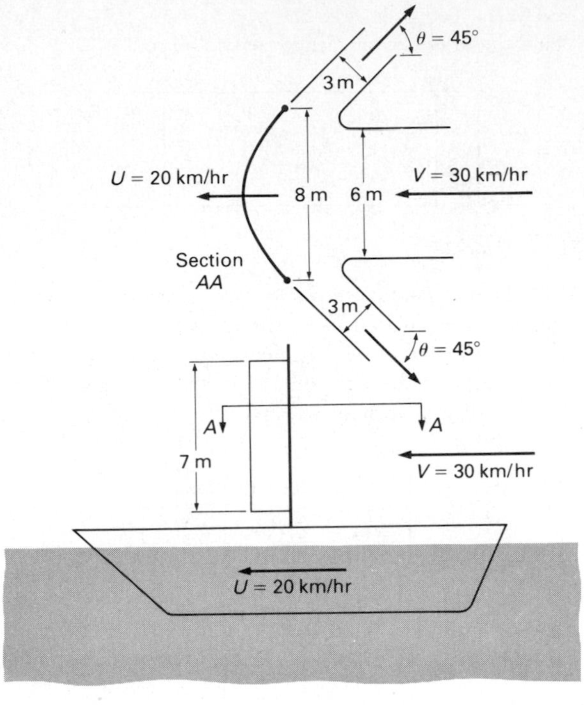

Figure P4.234

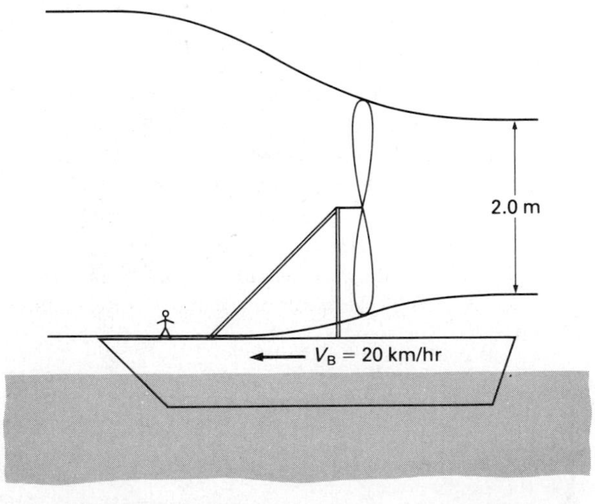

Figure P4.237

at the rear of the boat. Find the forward thrust of the air on the propeller. Use an observer on the boat to write the linear momentum equation. Assume constant air density.

238. Repeat Problem 237, but use an observer on the shore to write the linear momentum equation.

239. The propeller-driven swamp boat in Problem 237 travels at an absolute and constant velocity of $V_B = 20$ km/hr in a river that has an absolute and constant velocity of $V_R = 10$ km/hr in the same direction as the swamp boat. Air at 25°C leaving the rear of the boat has a velocity $W = 100$ km/hr relative to the boat. Find the velocity of the air leaving the rear of the boat relative to an observer moving with the river.

240. Water at 20°C leaves the nozzle with a jet velocity $V_j = 5.0$ m/s through an area of $A_j = 4.0$ cm². The jet impinges on a conical disc that moves toward the jet with a velocity of $V_d = 2.0$ m/s. The jet hits the disc and deflects uniformly and circumferentially. The deflected jet thickness $t = 1.0$ cm at the disc radius $R = 3$ cm as shown in Fig. P4.240. Find the force of the jet on the disc.

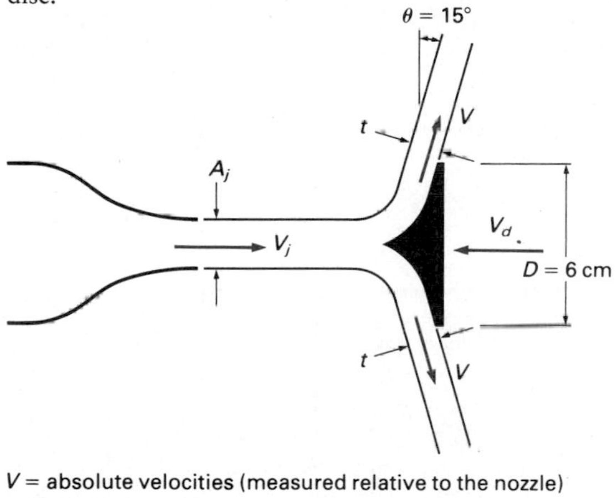

Figure P4.240

•241. The cart in Fig. P4.241 holds a tank of 20°C water, which is pumped out at a rate of $Q(t)$ m³/s. The tank has a constant cross-sectional area A and an original height h_0. Determine the flow rate $Q(t)$ required for the cart to have a constant velocity of $V_c = 0.25$ m/s. The cart rolls with no friction or air resistance. The water density is constant, and the pump discharge area is 0.025 m².

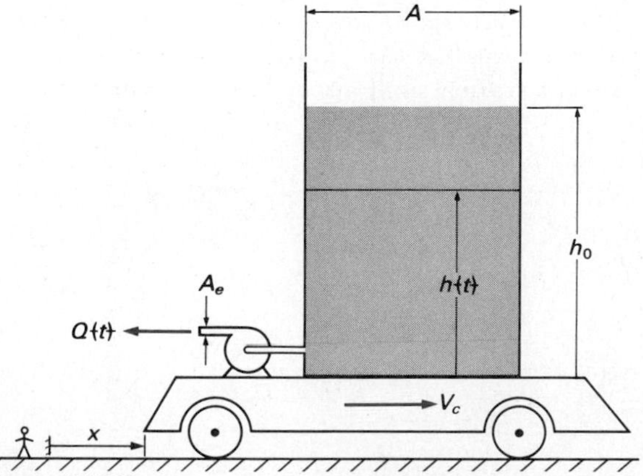

Figure P4.241

•242. Repeat Problem 241 for a cart that has a constant drag force of CV_c, where $C = 4.0$ N·s/m.

•243. A 2-ft-wide snow shovel is pushed with a constant velocity of $V = 2$ ft/sec along a level street with a horizontal force $P_x(t)$ as shown in Fig. P4.243. The snow depth $d = 4.0$ in., the snow density $\rho_s = 25$ lbm/ft³, and the snow shovel's frictional force $F = 0.4W$, where W is the weight of the snow on the shovel and the shovel itself. The shovel weighs $W_{sh} = 4$ lb. Calculate $P(t)$ when the shovel is moving with velocity V at the time it "hits" the snow of depth d at time $t = 0$.

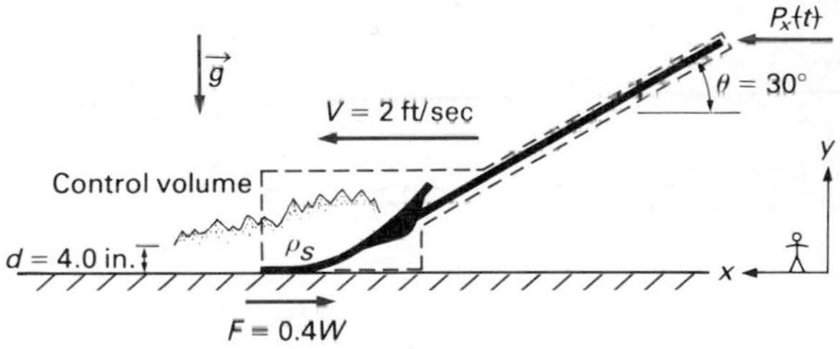

Figure P4.243

•244. The tank shown in Fig. P4.244 has a mass of $m_C = 0.01$ slug with $A = 12$ in², H = 13.5 in., $h_0 = 2.5$ in., and $A_e = 0.5$ in². The tank is filled with water ($\rho = 0.00112$ slug/in³) and is pushed along a level surface at

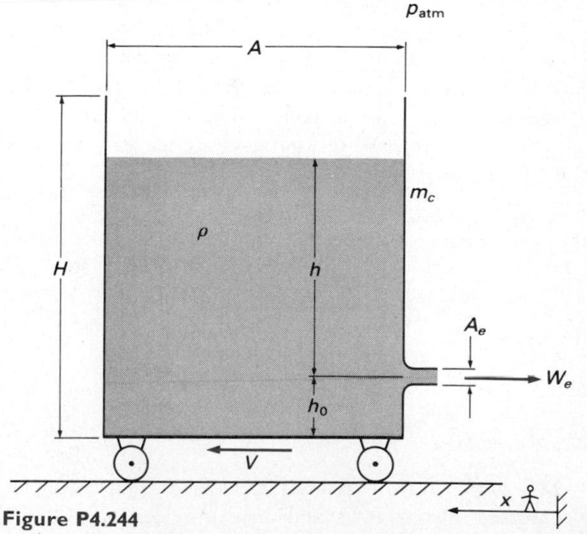

Figure P4.244

a constant velocity of 1.0 in/sec. At a time designated as $t = 0$, a cork is removed from the nozzle at the rear of the tank and the pushing force is removed. Find the new force F as a function of time that must be used to maintain the constant velocity.

• **245.** The locomotive in Fig. P4.245 is plowing at a steady rate through 0.2-m-deep snow while traveling at 20 km/hr. The snow density is 320 kg/m^3. The angle α that the front of the locomotive makes with the direction of travel is 120°. Find the power required to remove the snow. Neither the area nor the velocity of the snow as it is pushed off the side of the track is known. Assume that the snow density is constant and find the required power for various ratios of $k = \rho_1 A_1/\rho_2 A_2$.

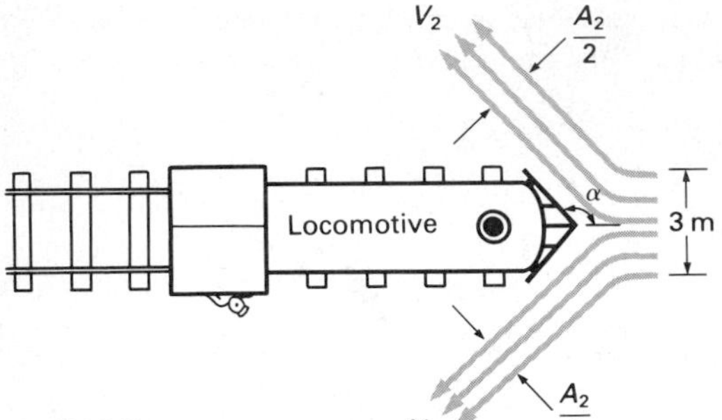

Figure P4.245

• **246.** Students leading in a concrete canoe race are dealing with a leak in their canoe. They are removing 60°F water from the canoe with a hand pump. In an attempt to slow the canoe behind them, they direct the pump's discharge water jet against the bow of the trailing canoe. This jet's water velocity is 18 ft/sec relative to the trailing canoe. See Fig. P4.246. Compute the force exerted on the trailing canoe by the water jet. Will this action increase or decrease the velocity of the leading canoe? Express your answer by assuming that $V_e = k(V_j + V_c)$, where V_j and V_c are the jet and canoe velocities, respectively.

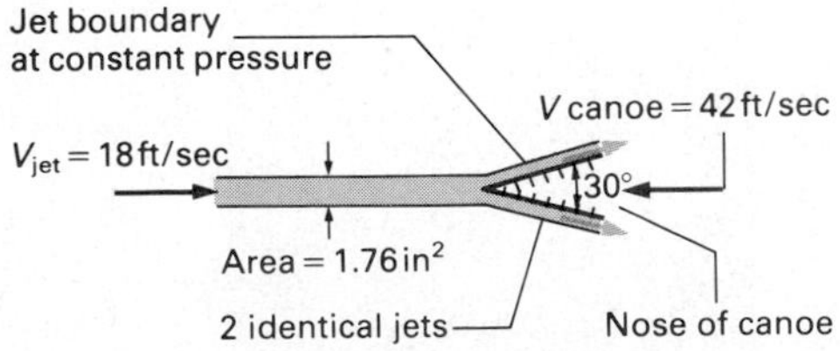

Figure P4.246

247. Water at 10°C from a fire hose having a volume flow rate of 0.20 m^3/s and an exit velocity of 1.0 m/s strikes a cart having a circular area of 0.20 m^2, as shown in Fig. P4.247. If the cart's mass is 10 kg and its initial velocity is zero, what is its initial acceleration?

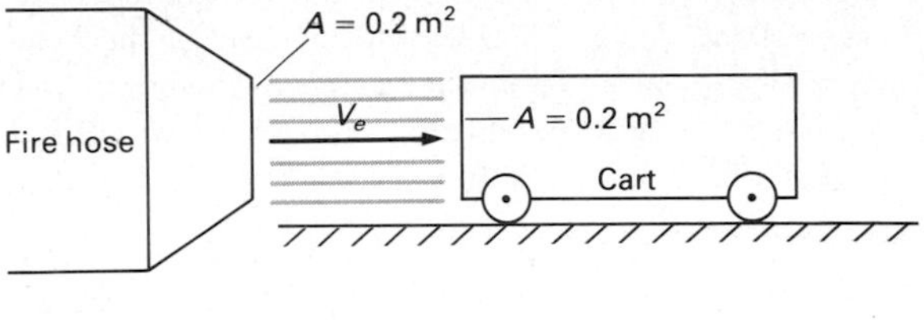

Figure P4.247

248. The cart shown in Fig. P4.248 holds a tank of 20°C water, which is pumped out at a rate of $Q\{h\}$ m^3/s. The tank has a constant cross-sectional area $A = 0.025$ m^2 and an original height $h_0 = 0.25$ m. The cart, tank, and pump have a mass of 10 kg and experience no drag or friction force. The water density is constant and the pump discharge area $A_e = 0.001$ m^2. Find the pump flow rate $Q\{h\}$ required for the cart to accelerate at 0.01 m/s^2.

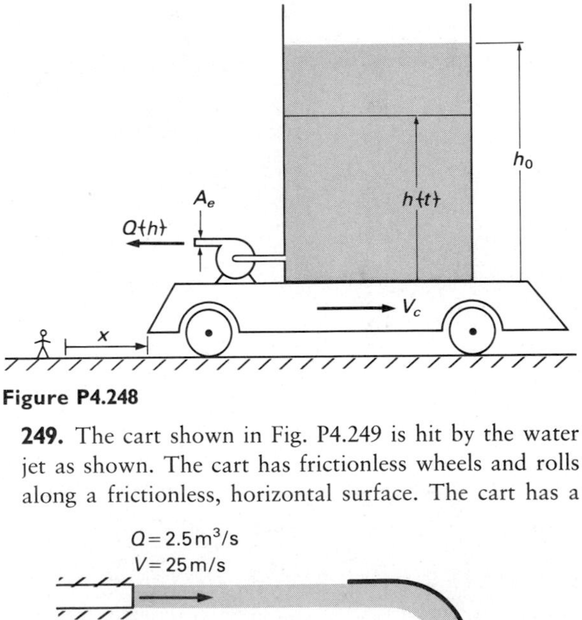

Figure P4.248

249. The cart shown in Fig. P4.249 is hit by the water jet as shown. The cart has frictionless wheels and rolls along a frictionless, horizontal surface. The cart has a

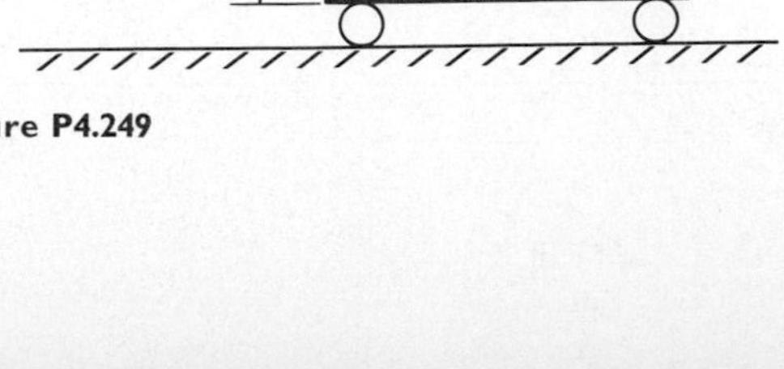

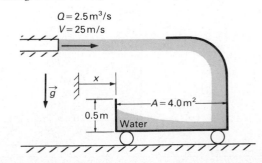

Figure P4.249

mass of 4.0 kg and can hold up to 2.0 m^3 of water. Find the differential equation for the position x as a function of time, if the cart initially has no water and a velocity of zero when first hit by the water jet. Assume that the entire water mass in the cart has the cart velocity V_C.

• **250.** Find the acceleration of the cart shown in Fig. P4.250 as a function of the water height in the cart, which varies with time. The initial total mass is m_0, and the fluid density is ρ_0. Assume frictionless bearings, a frictionless surface, constant fluid density, uniform velocity over area A_N, all the fluid in the cart has the cart velocity, no air drag, and $A \gg A_N$.

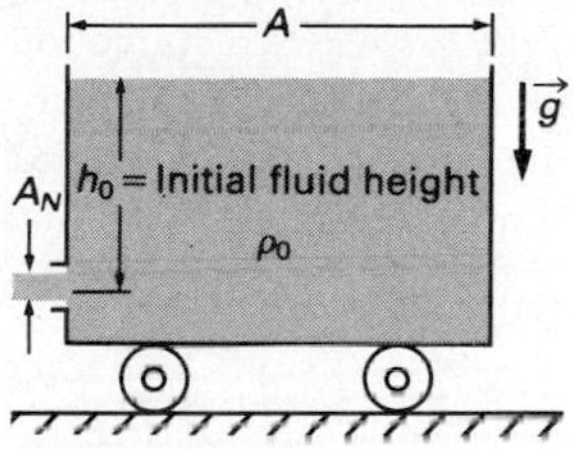

Figure P4.250

• **251.** A jet of water ($\rho = 1000$ kg/m^3) hits a cart, as shown in Fig. P4.251. The cart has a mass of 5.0 kg, the velocity $V = 1.0$ m/s, the area $A = 0.10$ m^2, $\beta = 45°$, and $U = 0$ at time $t = 0$. The air resistance on the cart can be expressed as

$$F_R = KU,$$

where $K = 20.0$ N·s/m. Write the differential equation for the cart velocity $U(t)$.

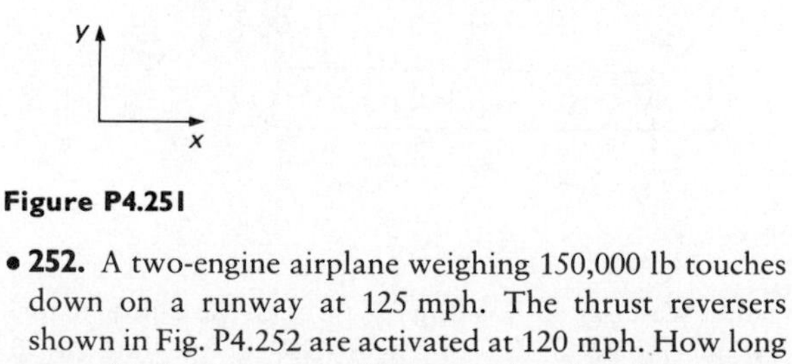

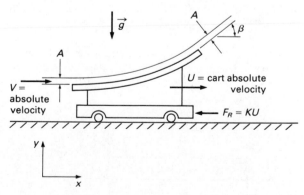

Figure P4.251

• **252.** A two-engine airplane weighing 150,000 lb touches down on a runway at 125 mph. The thrust reversers shown in Fig. P4.252 are activated at 120 mph. How long must the thrust reversers be operated to slow the airplane

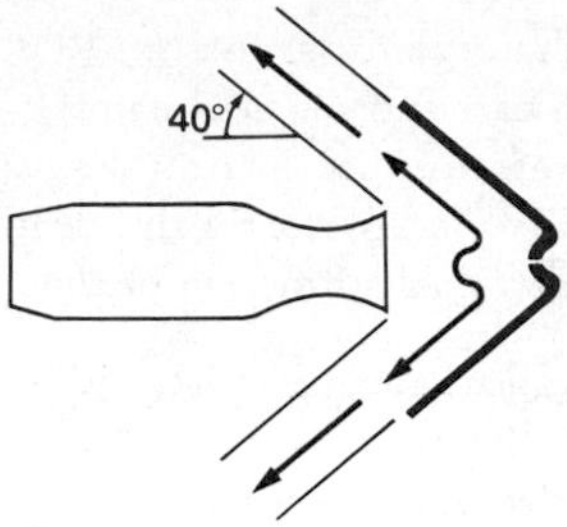

Figure P4.252

to 30 mph? The mass flow rate from each engine is 200 lbm/sec, with a density of 0.02 lbm/ft^3 and a relative velocity of 2500 ft/sec. Neglect air drag on the airplane and assume that no other braking forces are applied to the airplane. Also assume that the entire mass inside the airplane has the airplane's velocity and that the airplane travels in a straight path. The mass of the fuel is negligible compared to the total mass. The fuel flow rate through the engine is negligible compared to the air flow rate through the engine.

• **253.** An unmanned space vehicle has a velocity of 7500 m/s and a mass of 1500 kg and is moving in outer space with negligible gravitational attraction. A rocket on the engine fires at a rate of 5 kg/s, with an exit velocity of 1000 m/s for 2.0 s to accelerate the space vehicle. What is the new velocity of the vehicle? Assume that the entire mass *inside* the vehicle has the rocket velocity.

• **254.** The rocket-powered sled shown in Fig. P4.254 is traveling at 600 mph and is slowed by lowering a scoop on the sled into water. Find the distance in which the sled can be slowed to 300 mph. The sled weighs 4000 lb, and the scoop is 6 ft wide. Neglect air resistance. The water that hits the scoop is deflected sideways 90°.

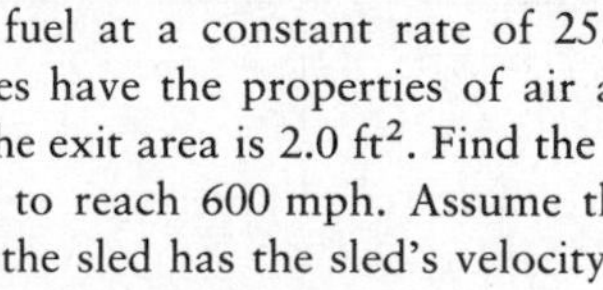

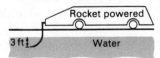

Figure P4.254

• **255.** The sled in Problem 254 is brought up to speed by burning fuel at a constant rate of 25.0 lb/sec. The exhaust gases have the properties of air at 1600 F and 14.7 psia. The exit area is 2.0 ft^2. Find the time required for the sled to reach 600 mph. Assume that the entire mass inside the sled has the sled's velocity.

• **256.** A 4-ton cart carrying 8 tons of coal ($S = 1.25$) is rolling freely down a 10° incline at a constant speed of

1.0 mph. A chute with an area of 10 ft^2 opens in the bottom of the cart and the coal empties out of the cart at a constant rate in 20 sec. The air drag $\mathscr{D}$ is CV^2, where $C = 0.030\ \text{lb}\cdot\text{hr}^2/\text{mi}^2$. Write the differential equation for the cart velocity V as a function of time.

•**257.** A semitrailer having a weight W_S and carrying coal weighing W_C is stopped on an incline making an angle θ with the horizontal. At time $t = 0$, the engine is started and a constant force F is applied to the road through the tires. At the same time, the tailgate of the trailer comes loose and coal falls straight downward (relative to the trailer) at the constant rate of $\dot{W}_C$ with velocity V_C. The air drag on the trailer is equal to CV^2, where C is a constant. Write the differential equation for the semitrailer velocity V as a function of time as the semitrailer moves up the incline. Assume that all the coal inside the trailer has the trailer's velocity.

258. Emergency supplies weighing 40,000 lb are to be dropped into a remote drought-stricken area from a B-52 (weight without payload of 270,000 lb), in a single pass over a dry river bed within 30 seconds to prevent scattering. The unloading pass begins in level flight at 3000 ft when the pilot fixes the engine thrust to match the drag at the initial air-speed of 200 knots. The lift and drag on the plane can be approximated by $\mathscr{L} = K_L|V|^2$ and $\mathscr{D} = K_D|V|^2$. At the beginning of the pass, the pilot fixes the wing surfaces so that $K_L = 2.89\ \text{lb}\cdot\text{sec}^2/\text{ft}^2$ and $K_D = 0.24\ \text{lb}\cdot\text{sec}^2/\text{ft}^2$. Assume that the plane's fuselage always aligns with the velocity vector. At the end of the unloading pass, what is the speed of the plane, the angle of flight with respect to horizontal, and the altitude of the plane?

•**259.** The balloon in Fig. P4.259 is filled with air to an absolute pressure of $p_0 = 105$ kPa while the outside atmospheric pressure $p_{\text{atm}} = 101$ kPa absolute. The balloon diameter, denoted by D, is D_0 at time $t = 0$ when a plug is removed from the tube of diameter d. Because $d \ll D$ and the air pressure inside the balloon is not much greater than the atmospheric pressure, the air leaks out slowly and the air density may be considered constant. The mass of the balloon is negligible. Develop an expression for the force of the support on the balloon as a function of the diameter D, air density ρ, diameter d, gravitational acceleration g, and time t. The balloon remains a sphere at all times.

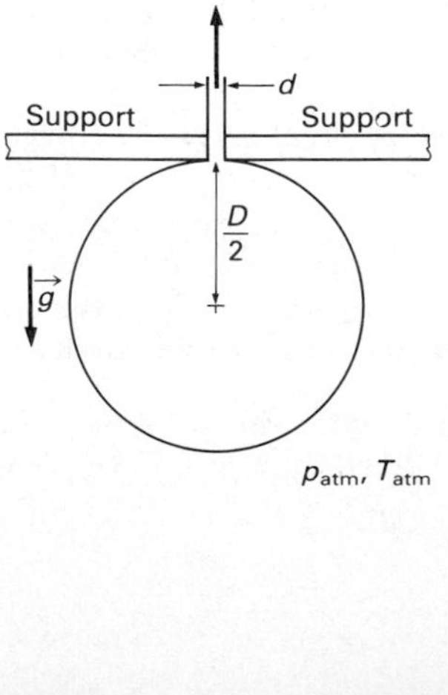

Figure P4.259

•**260.** Repeat Problem 259 for constant balloon surface tension σ and air density ρ related to the air pressure by $\rho = cp$, where c is a constant.

•**261.** Calculate the force $\vec{F}(t)$ required to hold steady the tank shown in Fig. P4.261. The pipe diameter D is 10.0 cm. Assume inviscid flow and neglect the weight of the tank. The initial height H is 0.95 m.

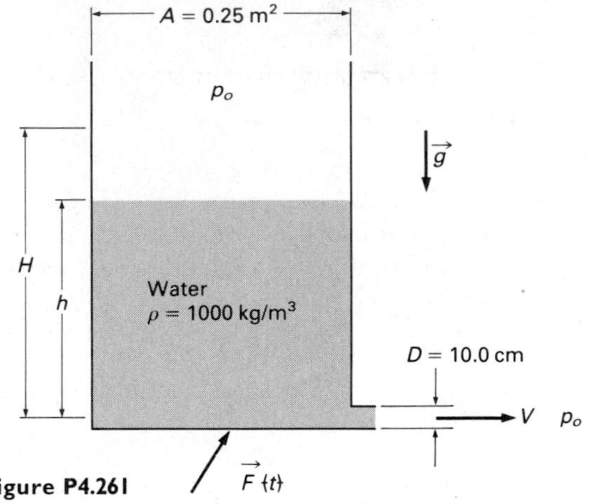

Figure P4.261

•**262.** Figure P4.262 shows a simplified sketch of a dishwasher water supply manifold. Find the resisting torque for a water temperature of 140°F. $Q = 0.25$ gal/min and $\omega = 3$ rpm.

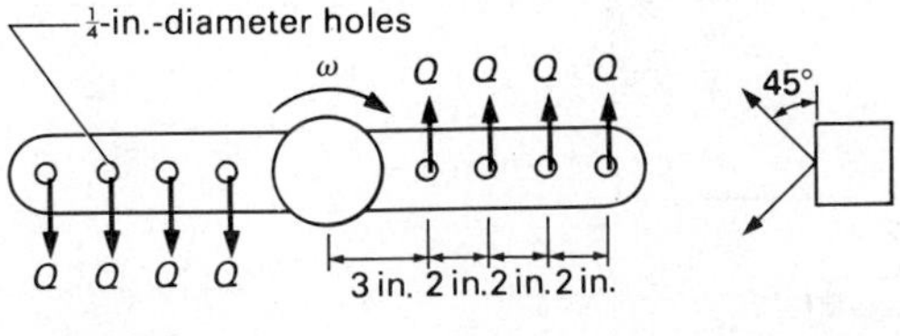

Figure P4.262

263. Water at 60°F is flowing through the 2-in schedule 40 steel pipe shown in Fig. P4.263 at the rate of 90 gal/min. Determine the torque developed at the base

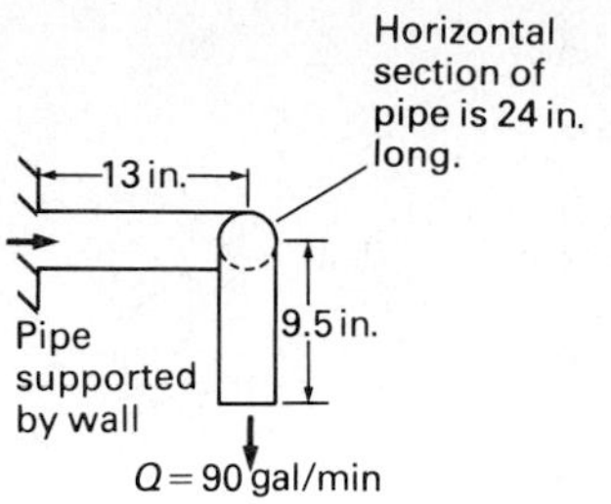

Figure P4.263

where the pipe is supported. Neglect the pipe and water weights. Steady-state conditions apply.

264. Water enters a centrifugal pump with an absolute velocity $V_1 = 10$ m/s in the radial direction and leaves with an absolute velocity V_2, which makes an angle of $\theta_2 = 60°$ with the radial direction, as shown in Fig. P4.264. The impeller width (perpendicular to the paper) is $b = 0.125$ m, $R_1 = 0.125$ m, and $R_2 = 0.35$ m. Find the input torque $\mathcal{T}$ required to drive the pump if there are no friction losses.

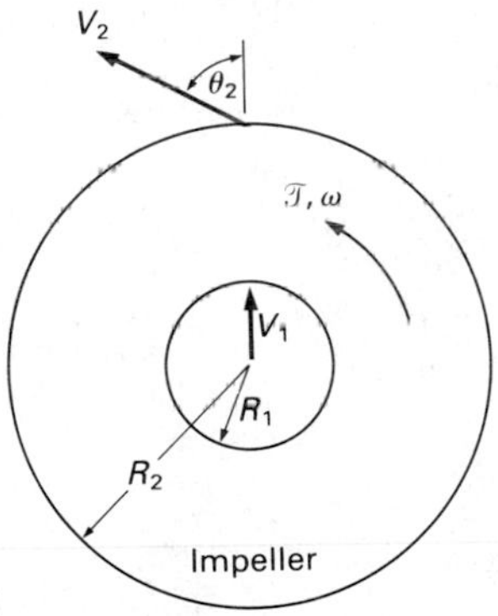

Figure P4.264

265. A lossless motor drives the fan shown in Fig. P4.265 at 40 Hz. The power input to the motor is 40 amps at 440 volts. For the geometry shown, what is the discharge flow rate of air through the fan? Assume that the tangential component of the velocity leaving the impeller is equal to that of the impeller at that point. The exit air temperature $T_2 = 15°$C.

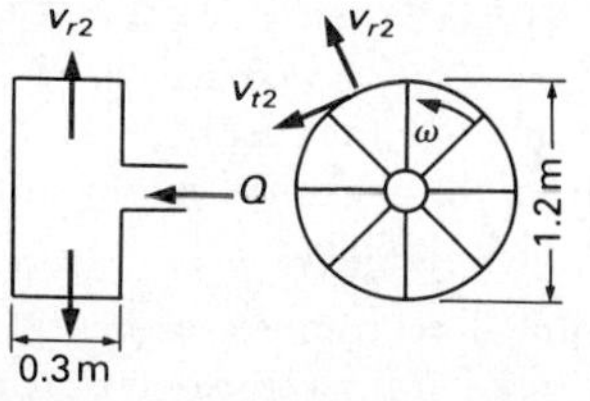

Figure P4.265

266. Calculate the torque required to drive the pump shown in Fig. P4. 266 at 30 Hz and to deliver 20°C water at 3.0 m^3/min.

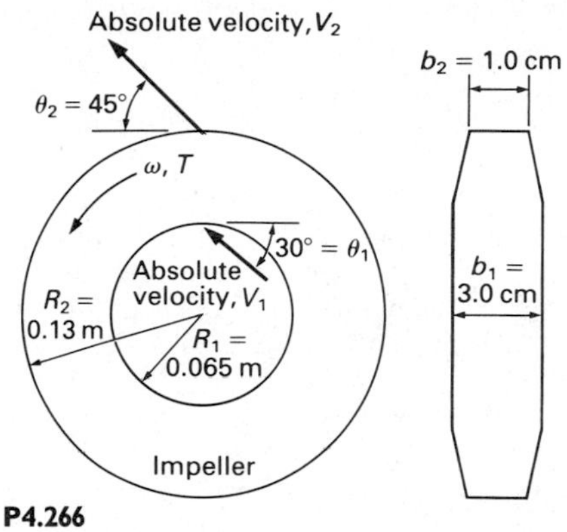

Figure P4.266

267. The circular water sprinkler shown in Fig. P4.267 has a total flow rate of 0.32 m^3/min and four discharge holes, each having a discharge area of 1.0 cm^2. The bearing frictional torque amounts to $\mathcal{T}_R = (2\pi \text{ N·m/Hz})\omega$, where ω is the rotational speed in Hz. Find the rotational speed ω. The fluid enters the sprinkler at the center in the z direction. Note that θ_2 locates V_2 relative to the hole, while ϕ_2 locates V_2 relative to the radius.

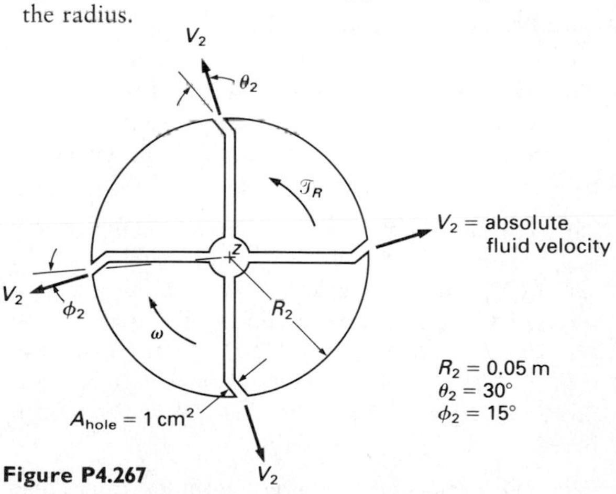

Figure P4.267

268. The frictionless converging-stationary nozzle of the hydraulic turbine shown in Fig. P4.268 has an inlet pressure $p_0 = 480$ kPa, a negligible inlet velocity V_0, and an exit pressure $p_1 = 101$ kPa. The velocity V_1 is used to drive the axial flow turbine, which has a rotational speed of 185.7 rpm, an outside radius of $R = 1.20$ m, and a blade height of $h = 0.40$ m. Determine velocities V_1 and V_2 and the power transmitted by the fluid to the turbine. Assume that the fluid density is constant and that

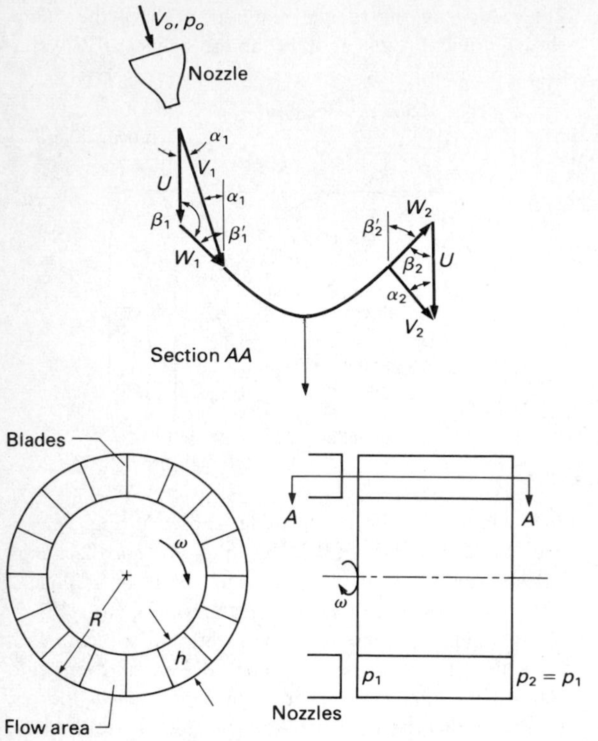

Figure P4.268

the velocities (W) of the fluid relative to the blade are directed as shown in Fig. P4.268. Assume that the velocities are uniform over the flow inlet and outlet areas and use an average blade velocity at the blade midheight ($r = 1.00$ m) and a water temperature of 20°C.

269. The power output for the turbine in Problem 268 is approximately 5.0 MW. Repeat this problem for the velocity V_1 making an angle $\alpha_1 = 25°$ and a turbine rotor speed of 127 rpm. Values remaining the same are $p_0 =$ 480 kPa, V_0 negligible, $p_1 = 101$ kPa, $R = 1.20$ m, $h =$ 0.40 m, $\beta_1' = \beta_2' = 45°$, and 10°C water. Compare the power output with that of Problem 268 and state your observations.

270. The formula for the power transmitted from a fluid to the rotor of a turbomachine is given in the text as

$$\dot{W}_S = -\dot{m}(U_2V_2\cos\alpha_2 - U_1V_1\cos\alpha_1),$$

where the subscript 2 represents the outlet and the subscript 1 represents the inlet. Consider an inlet or outlet where the velocity diagram involving U, V, and W is broken into components, as in Fig. P4.270. Show that

$$\dot{W}_S = \dot{m}\left[\frac{V_1^2 - V_2^2}{2} + \frac{U_1^2 - U_2^2}{2} - \frac{W_1^2 - W_2^2}{2}\right].$$

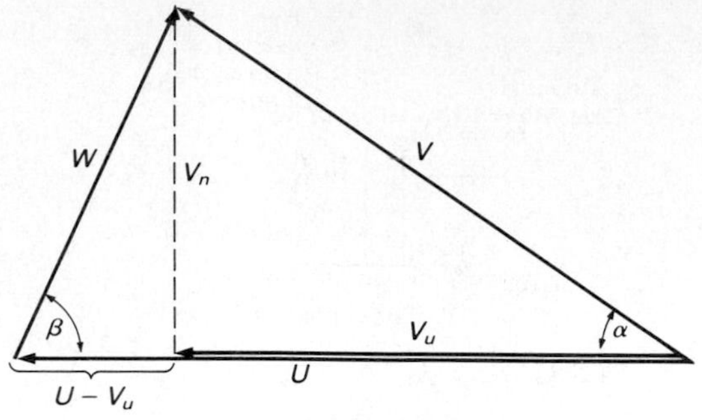

V_u = component of V in U direction
$= V\cos\alpha$

V_n = component of V in direction normal to U
$= V\sin\alpha = W\sin\beta$

Figure P4.270

271. An axial flow compressor shown in Fig. P4.271 has the inlet and outlet velocity diagrams shown. Calculate the work per unit mass by using the two equations in Problem 270. Quantities are $U_1 = U_2 = U = 762$ ft/sec, $V_1 = 440$ ft/sec, $W_1 = 880$ ft/sec, $\alpha_1 = 90°$, $V_2 = 545$ ft/sec, $W_2 = 622$ ft/sec, and $\alpha_2 = 53.8°$.

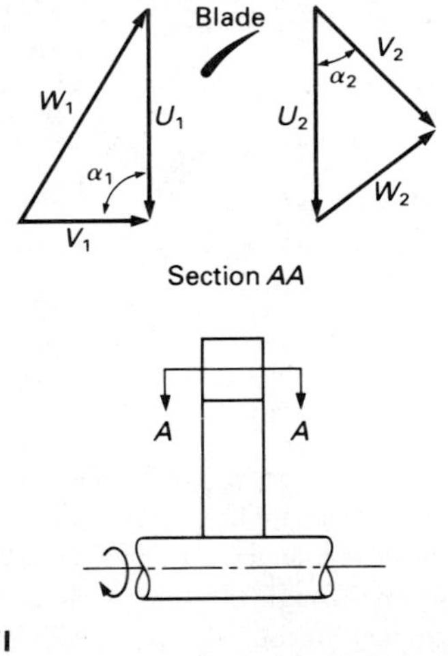

Figure P4.271

272. The axial flow gas turbine shown in Fig. P4.272 has a mean blade radius of $R = 5.0$ in., a rotational speed of 15,000 rpm, a mass flow rate of 10.0 lbm/sec, $W_1 = 972$ ft/sec, $V_1 = 1560$ ft/sec, $\alpha_1 = 20°$, $W_2 = 874$ ft/sec, $V_2 = 483$ ft/sec, and $\alpha_2 = 99.4°$. Find the power transmitted from the gas to the turbine blade using both of the equations in Problem 270.

273. A centrifugal air compressor has a rotor inner diameter of $D_1 = 2.0$ in., a rotor outer diameter of $D_2 =$ 6.5 in., a rotor depth of 10 in., and a rotor rotational

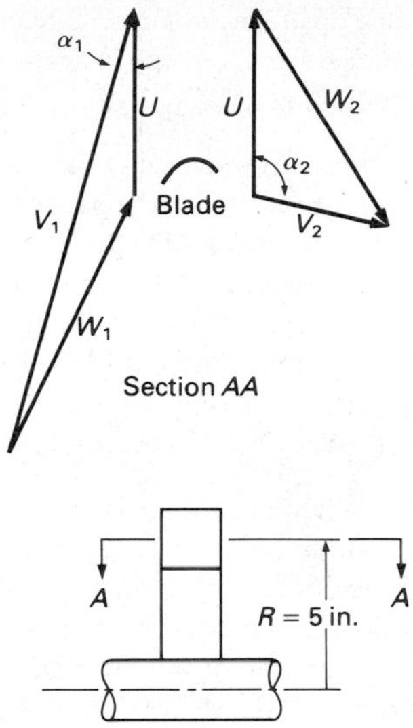

Figure P4.272

speed of 3600 rpm. The fluid relative velocities (W) are purely radial. The compressor delivers an air mass flow rate of 1.0 lbm/sec with $T_1 = 70°F$, $p_1 = 14.7$ psia, $T_2 = 240°F$, and $p_2 = 33.1$ psia. Find the power transferred by the rotor to the air and the inlet relative velocity W_1.

274. The hydraulic turbine shown in Fig. P4.274 has a 10°C water flow rate of 36.4 m^3/s, an inlet radius of $R_1 = 1.0$ m, an outlet radius $R_2 = 0.50$ m, a blade depth (perpendicular to paper) $b = 0.50$ m, and a rotational speed of $N = 360$ rpm; $V_1 = 50$ m/s, $V_2 = 30$ m/s, $\alpha_1 = 13.4°$ and $\alpha_2 = 40°$. Calculate the power transferred by the fluid to the rotor, the inlet relative velocity W_1, and the direction W_1 makes with the radius at the inlet.

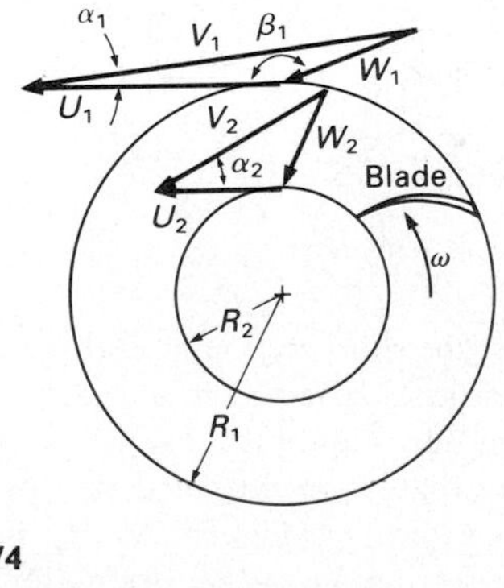

Figure P4.274

•275. The axial flow steam turbine rotor shown in Fig. P4.275 has a blade outer radius $R_0 = 2.40$ ft, a blade inner radius $R_i = 2.00$ ft, a steam inlet pressure $p_1 =$ 200 psia, a steam inlet density $\rho_1 = 0.296$ lbm/ft^3, and an inlet absolute velocity $V_1 = 1000$ ft/sec making an angle of 70° with the axial direction. The steam outlet pressure $p_2 = 50$ psia and outlet density is 0.1014 lbm/ft^3. $\beta_2 = 40°$. The rotor rotates at 3600 rpm. Using a blade velocity at the blade midheight ($R = 2.20$ ft), find the power transferred from the steam to the rotor.

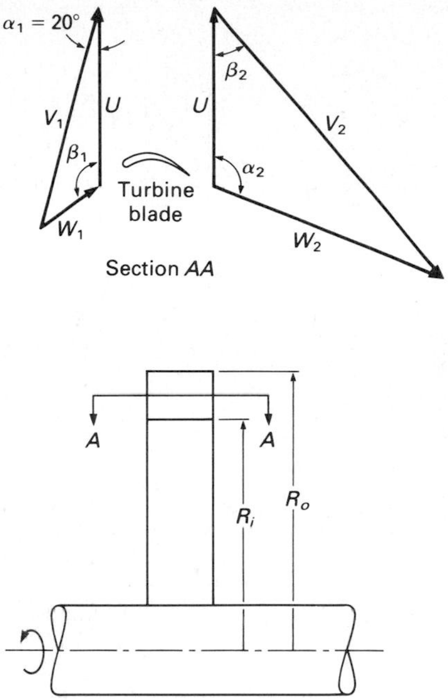

Figure P4.275

•276. The axial flow steam turbine design in Problem 275 has been changed to that shown in Fig. P4.276. The steam inlet conditions, rotor speed, blade angles, and steam outlet pressure and density are unchanged. Use a blade velocity at the blade midheight for the inlet ($R_1 = 2.2$ ft) and outlet ($R_2 = 2.0$ ft), and assume no flow in the radial direction at the outlet [only axial and circumferential (i.e., tangential) velocity components]. Find the power transferred from the steam to the rotor.

•277. A centrifugal fan has a power input of 25 kW, an inner radius of $R_1 = 0.5$ m, an outer radius of $R_2 = 1.0$ m, and delivers 100 kg/s of air. There are no friction losses, the air inlet absolute velocity has no tangential component, and the outlet absolute velocity has a tangential component equal to the blade velocity at the outer radius R_2. The rotor depth (i.e., blade height) is 1.0 m. What is the required rotational speed of the rotor?

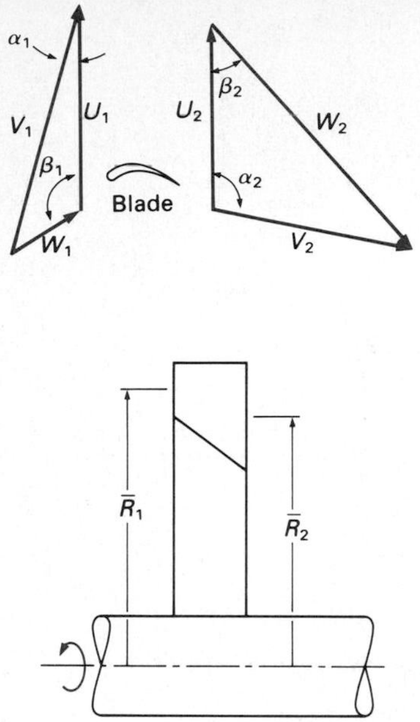

Figure P4.276

•**278.** A windmill has an approach velocity $V_1 = 24$ mph (35.3 ft/sec) and a blade diameter of 100 ft. The blade rotates at 15 rpm. Figure P4.278 shows a blade cross-sectional profile and the velocity diagrams for a short section of the blade at a radius of $r = 25.0$ ft. The outlet relative velocity W_2 is tangent to the blade at the outlet so $\beta_2 = \beta_2' = 30°$. The air density is constant as it flows over the blade. Assume that the flow area for the mass flow rate that interacts with this short section of the blade is the same upstream and downstream of the blade (i.e., $A_1 = A_2$). Find the velocity diagram downstream of the blade by finding U, W_2, V_2, and α_2.

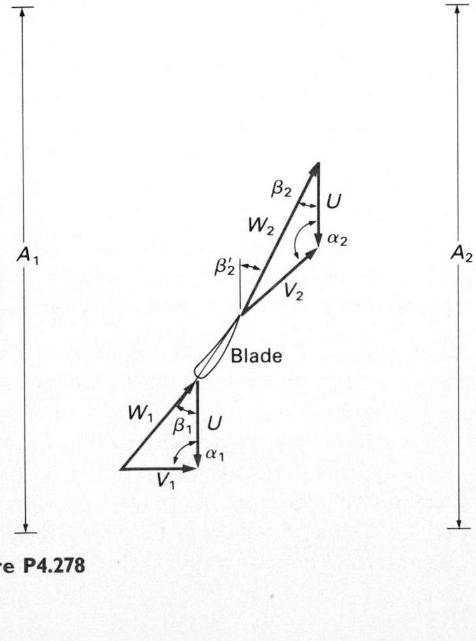

Figure P4.278

•**279.** Consider the short section in Problem 278 as being one foot long (perpendicular to the paper in Fig. P4.278). This short section produces a torque of 1350 ft·lb. Find the area $A_1 = A_2$ in Fig. P4.278 that interacts with this windmill blade section. Note the assumption that all the air interacting with this windmill blade section has the velocity pattern shown in Fig. P4.278. Use an air temperature and pressure of 60°F and 14.4 psia, respectively.

•**280.** A propeller-driven airplane is traveling with a velocity $V_1 = 200$ mph (294 ft/sec). The propeller diameter is 10 ft and it rotates at 3000 rpm. Figure P4.280 shows a propeller cross-sectional profile and the velocity diagrams for a short section of the propeller at a radius of 3.0 ft. The outlet relative velocity W_2 is assumed tangent to the propeller at the outlet, so $\beta_2' = \beta_2 = 50°$. The air density is constant as it flows over the propeller. Assume that the flow area for the mass flow rate interacting with this short section of the propeller is the same upstream and downstream (inlet and outlet) from the propeller (i.e., $A_1 = A_2$). Produce the velocity diagram downstream from the propeller by finding U, W_2, V_2, and α_2.

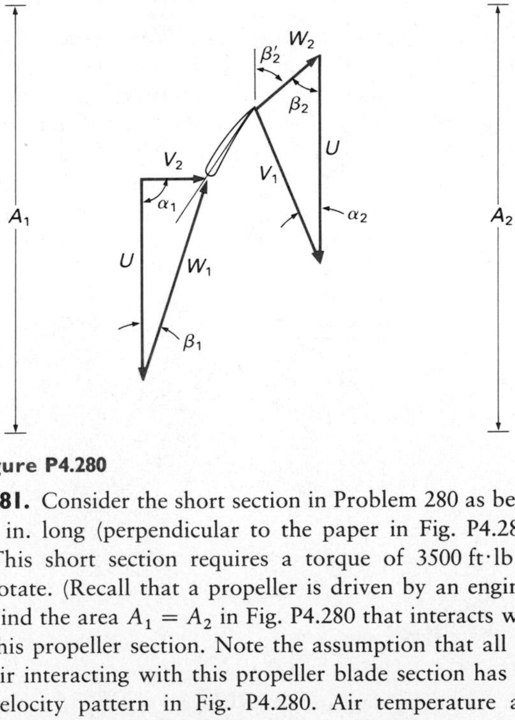

Figure P4.280

•**281.** Consider the short section in Problem 280 as being 6 in. long (perpendicular to the paper in Fig. P4.280). This short section requires a torque of 3500 ft·lb to rotate. (Recall that a propeller is driven by an engine.) Find the area $A_1 = A_2$ in Fig. P4.280 that interacts with this propeller section. Note the assumption that all the air interacting with this propeller blade section has the velocity pattern in Fig. P4.280. Air temperature and pressure are 60°F and 14.4 psia, respectively.

REFERENCES

1. Gerhart, P. M., "Averaging Methods for Determining the Performance of Large Fans from Field Tests," *ASME Journal of Engineering for Power,* October 1981.

2. Van Wylen, G., and R. Sonntag, *Fundamentals of Classical Thermodynamics* (3rd ed.), John Wiley & Sons, New York, 1986.

3. Holman, J., *Thermodynamics* (3rd ed.), McGraw-Hill, New York, 1988.

4. Moran, M., and H. Shapiro, *Fundamentals of Engineering Thermodynamics,* John Wiley & Sons, New York, 1988.

5. Shepherd, D. G., *Principles of Turbomachinery,* Macmillan, New York, 1956.

6. Dixon, S. L., *Fluid Mechanics; Thermodynamics of Turbomachinery* (3rd ed.), Pergamon Press, Oxford, 1981.

7. Stepanoff, A. J., *Centrifugal and Axial Flow Pumps,* John Wiley & Sons, New York, 1948.

8. Wilson, D. G., *The Design of High Efficiency Gas Turbines and Turbomachines,* Cambridge, Mass., MIT Press, 1984.

5 The Differential Approach to Flow Analysis

In this chapter, we consider the differential approach to flow analysis. We concentrate on developing the governing differential equations and appropriate boundary conditions and initial conditions and on simplified flow models that yield more "solvable" differential equations.

5.1 PRELIMINARY CONCEPTS

5.1.1 Overview of the Differential Approach

The differential approach to flow analysis begins with formulation of the law of conservation of mass and Newton's second law of motion as differential equations. Because these laws are valid at every point in a flow field, the resulting differential equations are capable of describing every point in the field. Deriving the differential equations is a relatively straightforward exercise, and they are quite general. By the time we finish this chapter, most of the relevant equations will be at hand.

Another necessary task is to prescribe boundary conditions and initial conditions for the equations. Boundary conditions are derived from the physical constraints that exist at the boundaries of the flow (for example, the velocity of a water particle in contact with a pipe wall must have zero velocity relative to the pipe). Initial conditions describe the flow field at the initial instant of time (for example, the velocity of all of the water in a pipe is zero at the instant that a valve is opened). Boundary conditions and initial conditions are unique to a given problem. The main reason is variations in geometry: Flow over a streamlined airfoil is different from flow over a square rod and both are different from flow through a circular pipe.

The (general) differential equations, together with the (problem-dependent) boundary and initial conditions, are a complete mathematical model of a particular flow. If we can solve the differential equations and incorporate the correct boundary and initial conditions, we are rewarded with information about the flow at every point in the field for all time (or at least until the conditions change, as when the valve is closed again). We can calculate flow rates from the known velocities and forces from the known pressure and shear stress distributions.

There are two pitfalls in the differential approach. First, each different flow geometry must be solved separately; only the differential equations are general. Second, finding an analytic solution to the differential equations is often difficult or impossible. Complicated flow geometries (such as the cooling air passages inside a personal computer cabinet or the steam flow path through a turbine), fluid shear stress, and fluid turbulence complicate the model significantly. In such cases, we often turn to simplified forms of the equations to obtain approximate solutions. Alternatively, we may tackle accurate approximations to the differential equations with the methods of computational fluid dynamics. Advances in digital computer hardware and software have made the differential approach a feasible alternative for many important engineering flow problems.

5.1.2 Alternatives for Formulating the Differential Equations

The most general flow field would be three-directional and three-dimensional. We can certainly derive the differential equations for such a general case; however, we follow a simpler path in this chapter. Usually, we consider a two-directional, two-dimensional flow for detailed developments and then simply state the appropriate three-dimensional forms of the equations.

Two choices must be made before we can begin. First, we must choose a coordinate system for describing the flow and expressing the differential equations. Common choices are Cartesian coordinates, cylindrical coordinates, and streamline coordinates. The equations may be expressed in any coordinate system that we choose. The geometry of the specific problem usually dictates the best coordinate system; for example, flow in a circular pipe is best analyzed in cylindrical coordinates, whereas flow in a rectangular duct is best analyzed in Cartesian coordinates. We do most of our work in this chapter in Cartesian coordinates.

The second choice that we must make is between applying the fundamental laws to a fluid particle (system) or applying the control volume equations to a tiny control volume. Both methods give identical differential equations, so the choice is relevant only when deriving the equations, not when applying them to a flow problem. In this chapter, we use the control volume method for conservation of mass and the particle method for Newton's second law.

5.1.3 Kinematics of a Fluid Particle

Kinematics is the study of space–time relationships. In order to understand a complicated fluid motion, we must consider quantities such as fluid velocity, acceleration, rotation, and deformation. Fluid kinematics and dynamics interact strongly, because force is related to acceleration and fluid shear stress is related to the rate of shear deformation.

Consider a small cubical fluid particle in a two-directional, two-dimensional, unsteady flow. As the particle moves from point to point, its motion is a combination of translation, rotation, stretching, and shear deformation. Figure 5.1 illustrates all these components of motion. We can express them in terms of the velocity field. Figure 5.2 shows the particle located at point (x, y) at time t. The fluid velocity at the center of the particle (in an Eulerian description) is

$$\vec{V} = \vec{V}(x, y, t) = u(x, y, t)\hat{i} + v(x, y, t)\hat{j}.$$

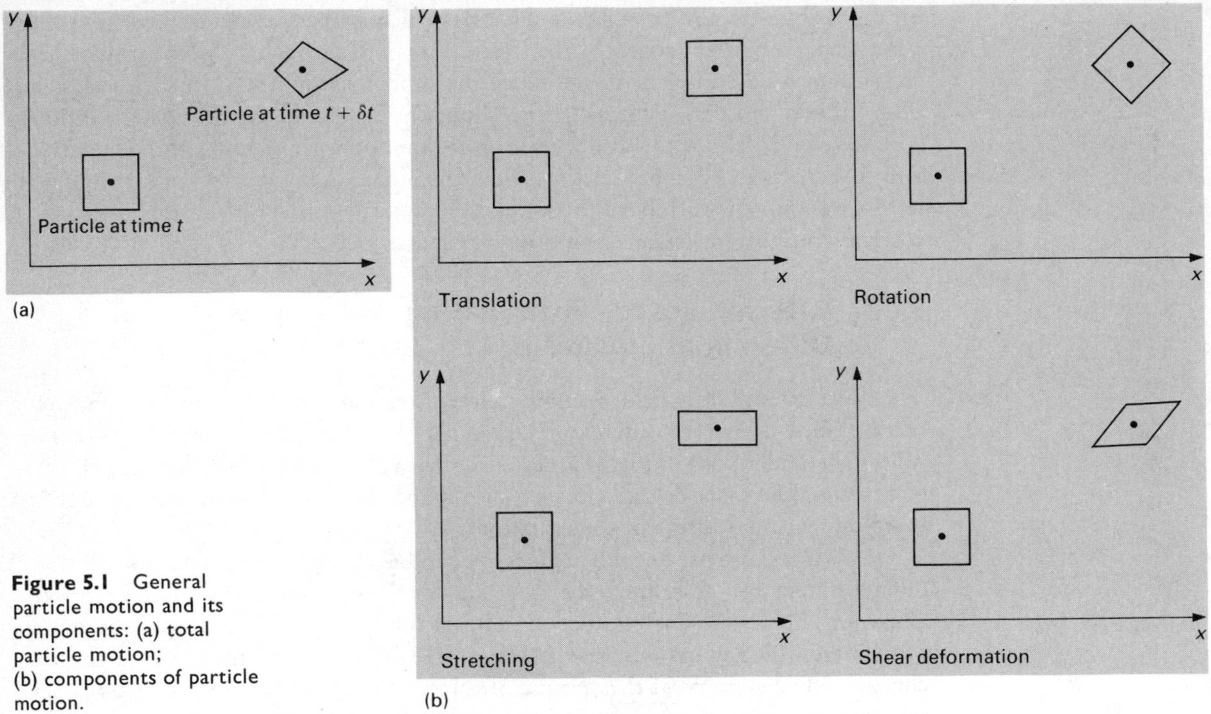

Figure 5.1 General particle motion and its components: (a) total particle motion; (b) components of particle motion.

The velocity is continuously variable so the velocities at the faces of the fluid particle may be different from the velocity at the center. Figure 5.2 shows these velocities, as approximated by the first term of a Taylor series.

Velocity. The translation of the fluid particle is given by the particle velocity, $\vec{V}$.

Acceleration. The acceleration of the particle is the total time rate of change of its velocity. We introduce the subscript p to remind us that we are considering a

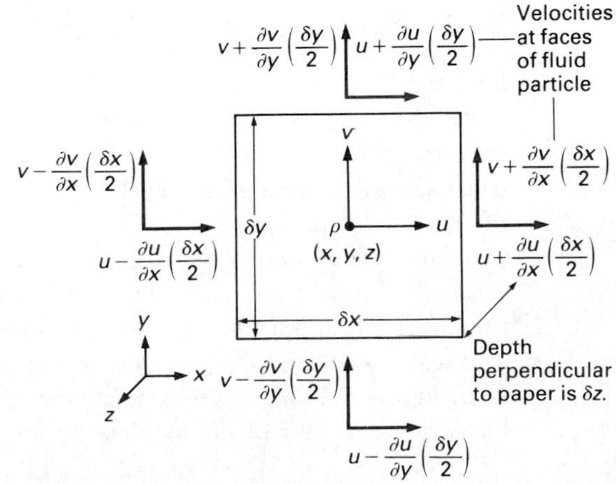

Figure 5.2 Fluid particle in two-dimensional, two-directional velocity field; particle center is at point (x, y) with velocity (u, v) at time t; the particle's z coordinate is arbitrary.

specific fluid particle, and write

$$\vec{V}_p = \vec{V}_p(x_p, y_p, t).$$

The acceleration is

$$\vec{a}_p = \frac{d}{dt}(\vec{V}_p) = \frac{d}{dt}(\vec{V}_p(x_p, y_p, t)).$$

We write the velocity and acceleration in x and y components:

$$a_{p,x} = \frac{d}{dt}(u_p(x_p, y_p, t)) \quad \text{and} \quad a_{p,y} = \frac{d}{dt}(v_p(x_p, y_p, t)).$$

Considering $a_{p,x}$, we have

$$a_{p,x} = \frac{\partial u_p}{\partial x_p}\left(\frac{dx_p}{dt}\right) + \frac{\partial u_p}{\partial y_p}\left(\frac{dy_p}{dt}\right) + \frac{\partial u_p}{\partial t}.$$

However,

$$\frac{dx_p}{dt} = u_p \quad \text{and} \quad \frac{dy_p}{dt} = v_p,$$

so

$$a_{p,x} = \frac{\partial u_p}{\partial t} + u_p \frac{\partial u_p}{\partial x_p} + v_p \frac{\partial u_p}{\partial y_p}.$$

At the instant in question, the fluid particle is at $x_p = x$, $y_p = y$, with $u_p = u$ and $v_p = v$, so

$$a_x = \frac{\partial u}{\partial t} + u \frac{\partial u}{\partial x} + v \frac{\partial u}{\partial y}.$$

By a similar development,

$$a_y = \frac{\partial v}{\partial t} + u \frac{\partial v}{\partial x} + v \frac{\partial v}{\partial y}.$$

Note that the derivatives on the right-hand side of both acceleration equations are the Eulerian derivatives from Eq. (3.10) (in two dimensions), so

$$a_x = \frac{Du}{Dt} \quad \text{and} \quad a_y = \frac{Dv}{Dt}.$$

We can extend these expressions to three dimensions and three directions:

$$a_x = \frac{Du}{Dt} = \frac{\partial u}{\partial t} + u \frac{\partial u}{\partial x} + v \frac{\partial u}{\partial y} + w \frac{\partial u}{\partial z}; \tag{5.1}$$

$$a_y = \frac{Dv}{Dt} = \frac{\partial v}{\partial t} + u \frac{\partial v}{\partial x} + v \frac{\partial v}{\partial y} + w \frac{\partial v}{\partial z}; \tag{5.2}$$

$$a_z = \frac{Dw}{Dt} = \frac{\partial w}{\partial t} + u \frac{\partial w}{\partial x} + v \frac{\partial w}{\partial y} + w \frac{\partial w}{\partial z}. \tag{5.3}$$

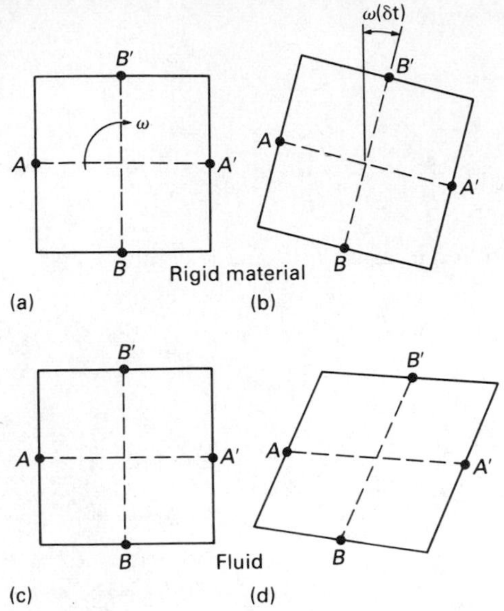

Figure 5.3 Comparison of rotation of rectangular particles of rigid material at (a) time t and (b) time $t + \delta t$; and of fluid at (c) time t and (d) time $t + \delta t$.

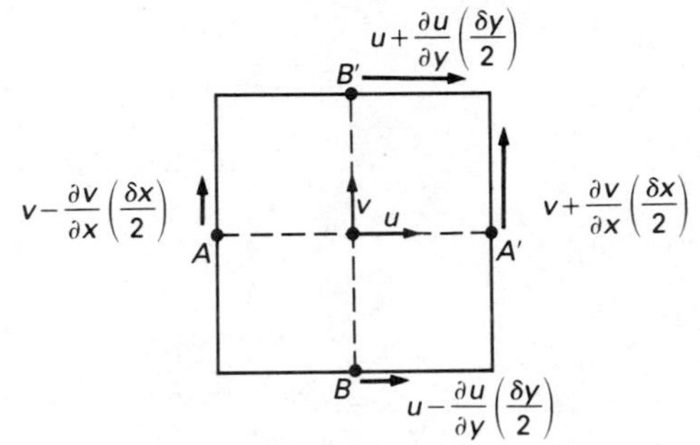

Figure 5.4 Rectangular fluid particle with two instantaneous perpendicular lines AA' and BB'; velocities perpendicular to AA' and BB' are also shown.

Or we can write them in a single equation using vector notation:

$$\vec{a} = \frac{D\vec{V}}{Dt} = \frac{\partial \vec{V}}{\partial t} + (\vec{V} \cdot \nabla)\vec{V}. \tag{5.4}$$

Angular Velocity and Vorticity. A purely mathematical procedure yields the acceleration of a fluid particle. However, we cannot obtain the angular velocity quite so rigorously. To understand why, consider a particle of a rigid material (such as steel), as shown in Fig. 5.3(a). Two perpendicular lines are marked on the particle. Because the particle is rigid, both lines rotate at the same angular velocity. If the angular velocity is not zero, a short time later the lines appear as in (b). Next, consider a rectangular fluid particle with two perpendicular lines marked on it (Fig. 5.3c). The fluid is deformable, so a short time later the fluid and the lines appear as shown in part (d). Because of deformation, the two lines may have rotated different amounts. To obtain a unique value for the angular velocity of the fluid particle, we must introduce a definition:

> The instantaneous angular velocity (ω) of a fluid particle is the average of the instantaneous angular velocities of two mutually perpendicular lines on the fluid particle.

Figure 5.4 shows a fluid particle with two lines AA' and BB'. By definition,

$$\omega \equiv \tfrac{1}{2}(\omega_{AA'} + \omega_{BB'}), \tag{5.5}$$

where

$$\omega_{AA'} = \frac{v_{A'} - v_A}{\delta x} = \frac{[v + (\partial v/\partial x)(\delta x/2)] - [v - (\partial v/\partial x)(\delta x/2)]}{\delta x} = \frac{\partial v}{\partial x} \tag{5.6}$$

and

$$\omega_{BB'} = -\frac{u_{B'} - u_B}{\delta y} = -\frac{[u + (\partial u/\partial y)(\delta y/2)] - [u - (\partial u/\partial y)(\delta y/2)]}{\delta y} = -\frac{\partial u}{\partial y}, \tag{5.7}$$

so

$$\omega_z = \frac{1}{2}\left(\frac{\partial v}{\partial x} - \frac{\partial u}{\partial y}\right). \tag{5.8}$$

We added the subscript z to ω to remind us that this is actually the angular velocity about the z axis (perpendicular to the paper). The $\frac{1}{2}$ is somewhat inconvenient, so instead of the angular velocity ω_z, we often use the vorticity ζ_z, defined by

$$\zeta_z \equiv 2\omega_z. \tag{5.9}$$

In three-directional, three-dimensional flow, the angular velocity and vorticity have three components:

$$\zeta_x = 2\omega_x = \frac{\partial w}{\partial y} - \frac{\partial v}{\partial z}; \tag{5.10}$$

$$\zeta_y = 2\omega_y = \frac{\partial u}{\partial z} - \frac{\partial w}{\partial x}; \tag{5.11}$$

$$\zeta_z = 2\omega_z = \frac{\partial v}{\partial x} - \frac{\partial u}{\partial y}. \tag{5.12}$$

In vector notation,

$$\vec{\zeta} = 2\vec{\omega} = \nabla \times \vec{V}. \tag{5.13}$$

A flow in which angular velocity and vorticity are zero is called an *irrotational flow*. In irrotational three-dimensional flow, all three components of $\vec{\omega}$ and $\vec{\zeta}$ must equal zero.

Rate of Volumetric Strain (Stretching). A fluid particle can be "stretched" and "squeezed," as illustrated in Fig. 5.5. This process can cause a change in the volume of the fluid particle. The rate of change of volume, divided by the volume itself, is called the *rate of volumetric strain:*

$$\left(\frac{1}{\delta \forall}\right) d\left(\frac{\delta \forall}{dt}\right).$$

The volume of this fluid particle is (refer also to Fig. 5.2)

$$\delta \forall = (\delta x)(\delta y)(\delta z).$$

In two-directional, two-dimensional flow, the particle stretches or shrinks in both

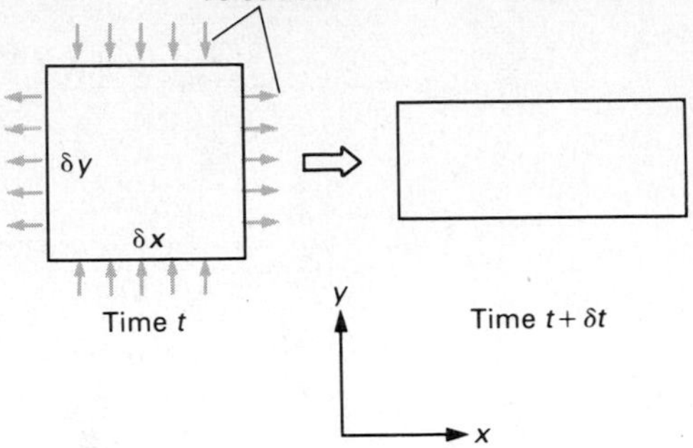

Figure 5.5 Fluid particle stretched in x direction and squeezed in y direction.

the x and y directions, so

$$\frac{d(\delta V\!\!\!/)}{dt} = (\delta z)(\delta y)\frac{d(\delta x)}{dt} + (\delta z)(\delta x)\frac{d(\delta y)}{dt}.$$

We have δx as the distance between the left and right faces of the fluid particle; $d(\delta x)/dt$ is the relative x velocity between the two faces:

$$\frac{d(\delta x)}{dt} = u)_{\text{right face}} - u)_{\text{left face}}$$

or

$$\frac{d(\delta x)}{dt} = \left[u + \frac{\partial u}{\partial x}\left(\frac{\partial x}{2}\right)\right] - \left[u - \frac{\partial u}{\partial x}\left(\frac{\partial x}{2}\right)\right].$$

Simplifying, we get

$$\frac{d(\delta x)}{dt} = \frac{\partial u}{\partial x}(\delta x).$$

By a similar argument,

$$\frac{d(\delta y)}{dt} = \frac{\partial v}{\partial y}(\delta y),$$

so the rate of volumetric strain is

$$\frac{1}{\delta V\!\!\!/}\frac{d(\delta V\!\!\!/)}{dt} = \frac{\partial u}{\partial x} + \frac{\partial v}{\partial y}.$$

For a three-dimensional, three-directional flow, we have

$$\frac{1}{\delta V\!\!\!/}\frac{d(\delta V\!\!\!/)}{dt} = \frac{\partial u}{\partial x} + \frac{\partial v}{\partial y} + \frac{\partial w}{\partial z}. \tag{5.14}$$

In vector notation,

$$\frac{1}{\delta V\!\!\!/}\frac{d(\delta V\!\!\!/)}{dt} = \nabla \cdot \vec{V}. \tag{5.15}$$

Rate of Shear Deformation (Shear Strain Rate). In addition to rotation and volumetric strain, a fluid particle in a flow field also experiences shear deformation. Next to acceleration, the rate of shear deformation is the most important kinematic property because the shear stresses in the fluid are related to this strain rate (see Sections 1.2 and 1.3.5).

Our earlier discussions of shear deformation in fluids (see Figs. 1.1, 1.4, and 1.5) used a somewhat simplified picture, with the upper and lower edges of a rectangular fluid particle always remaining parallel to the x coordinate. A more general definition of strain rate can be developed with the aid of a sketch. Figure 5.6 shows an initially rectangular fluid particle with two perpendicular lines on it and the deformed fluid particle at a later time. The average shear deformation, ϕ, is defined by

$$\phi \equiv \tfrac{1}{2}(\phi_1 + \phi_2),$$

with ϕ_1 and ϕ_2 positive as shown. The rate of shear deformation, $\dot{\phi}$, is

$$\dot{\phi} = \frac{d\phi}{dt} = \frac{1}{2}\left(\frac{d\phi_1}{dt} + \frac{d\phi_2}{dt}\right). \tag{5.16}$$

Borrowing results from the discussion of angular velocity, we have

$$\frac{d\phi_1}{dt} = \omega_{AA'} \quad \text{and} \quad \frac{d\phi_2}{dt} = -\omega_{BB'}.$$

When we use Eqs. (5.6) and (5.7), Eq. (5.16) becomes

$$\dot{\phi} = \frac{1}{2}\left(\frac{\partial u}{\partial y} + \frac{\partial v}{\partial x}\right). \tag{5.17}$$

This expression can be extended to a three-dimensional fluid particle, where there are six component rates of shear strain, but we do not do that here.

Note that in Figs. 1.4 and 1.5, we assumed that $\partial v/\partial x = 0$. In flows of the type sketched in Figs. 1.4 and 1.5, it is usually true that $\partial u/\partial y \gg \partial v/\partial x$, so we assume $\partial v/\partial x$ to be zero.

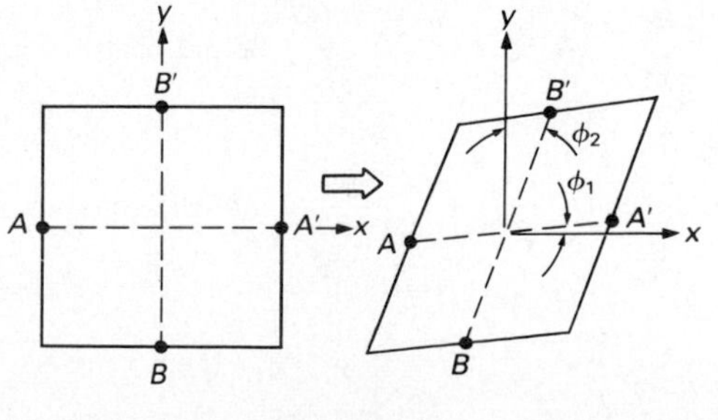

Figure 5.6 Deformation of fluid particles; deformation angles ϕ_1 and ϕ_2 are positive as shown.

EXAMPLE 5.1 Illustrates the Kinematics of a Two-Dimensional, Two-Directional Flow

The flow considered in Examples 3.1, 3.3, and 4.1 has an x component of velocity given by

$$u = u_0\left(1 + \frac{x}{\ell}\right)\left[1 - \left(\frac{y}{Y}\right)^2\right].$$

In Example 5.2, we find that the y component of velocity is given by

$$v = u_0\left[\frac{y^3}{\ell Y_0^2}\left(1 + \frac{x}{\ell}\right)^2 - \frac{y}{\ell}\right].$$

Calculate the linear acceleration, rotation, vorticity, rate of volumetric strain, and rate of shear deformation for the flow.

SOLUTION

Given

Velocity distribution

Fig. E5.1

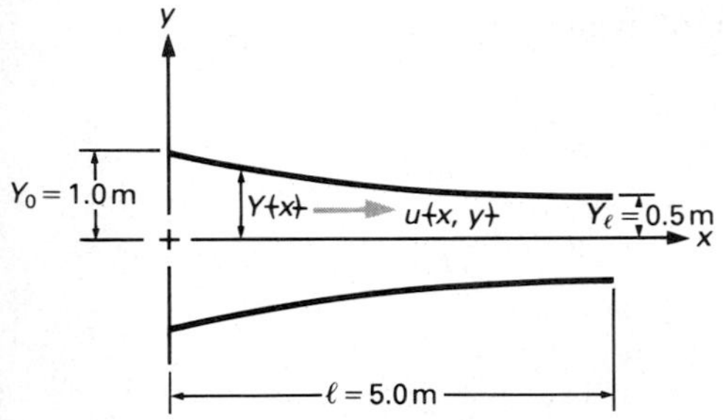

Figure E5.1 Two-dimensional channel.

Find

Linear acceleration

Rotation

Vorticity

Rate of volumetric strain

Rate of shear deformation

Solution

There are two linear accelerations, a_x and a_y. Equations (5.1) and (5.2) give

$$a_x = \frac{Du}{Dt} = \frac{\partial u}{\partial t} + u\frac{\partial u}{\partial x} + v\frac{\partial u}{\partial y};$$

$$a_y = \frac{Dv}{Dt} = \frac{\partial v}{\partial t} + u\frac{\partial v}{\partial x} + v\frac{\partial v}{\partial y}.$$

By inspection,

$$\frac{\partial u}{\partial t} = \frac{\partial v}{\partial t} = 0.$$

Before continuing, we substitute

$$Y = \frac{Y_0}{[1 + (x/\ell)]}$$

into the equation for u, which gives

$$u = u_0\left(1 + \frac{x}{\ell}\right)\left[1 - \frac{y^2}{Y_0^2}\left(1 + \frac{x}{\ell}\right)^2\right]$$

$$= u_0\left[\left(1 + \frac{x}{\ell}\right) - \frac{y^2}{Y_0^2}\left(1 + \frac{x}{\ell}\right)^3\right].$$

Then

$$\frac{\partial u}{\partial x} = u_0\left[\frac{1}{\ell} - \frac{3y^2}{\ell Y_0^2}\left(1 + \frac{x}{\ell}\right)^2\right];$$

$$\frac{\partial u}{\partial y} = u_0\left[-\frac{2y}{Y_0^2}\left(1 + \frac{x}{\ell}\right)^3\right];$$

$$\frac{\partial v}{\partial x} = u_0\left[\frac{2y^3}{\ell^2 Y_0^2}\left(1 + \frac{x}{\ell}\right)\right];$$

$$\frac{\partial v}{\partial y} = u_0\left[\frac{3y^2}{\ell Y_0^2}\left(1 + \frac{x}{\ell}\right)^2 - \frac{1}{\ell}\right].$$

Substituting and simplifying, we obtain

$$a_x = \frac{u_0^2}{\ell}\left[\left(1 + \frac{x}{\ell}\right) - 2\left(\frac{y}{Y_0}\right)^2\left(1 + \frac{x}{\ell}\right)^3 + \left(\frac{y}{Y_0}\right)^4\left(1 + \frac{x}{\ell}\right)^5\right]$$

and

$$a_y = \frac{u_0^2}{\ell^2}\left[y - 2\frac{y^3}{Y_0^2}\left(1 + \frac{x}{\ell}\right)^2 + \frac{y^5}{Y_0^4}\left(1 + \frac{x}{\ell}\right)^4\right].$$ **ANSWER**

The angular velocity is

$$\omega_z = \frac{1}{2}\left(\frac{\partial v}{\partial x} - \frac{\partial u}{\partial y}\right).$$

Substituting for $\partial v/\partial x$ and $\partial u/\partial y$, we have

$$\omega_z = \frac{y u_0}{Y_0^2}\left(1 + \frac{x}{\ell}\right)\left[\frac{y^2}{\ell^2} + \left(1 + \frac{x}{\ell}\right)^2\right].$$ **ANSWER**

The vorticity is

$$\zeta_2 = 2\omega_2,$$

so

$$\zeta_z = \frac{2y u_0}{Y_0^2}\left(1 + \frac{x}{\ell}\right)\left[\frac{y^2}{\ell^2} + \left(1 + \frac{x}{\ell}\right)^2\right].$$ **ANSWER**

The rate of volumetric strain is

$$\frac{1}{\delta V}\frac{d(\delta V)}{dt} = \left(\frac{\partial u}{\partial x} + \frac{\partial v}{\partial y}\right).$$

Substituting for $\partial u/\partial x$ and $\partial v/\partial y$, we have

$$\frac{1}{\delta V}\frac{d(\delta V)}{dt} = 0.$$ **ANSWER**

The rate of shear deformation is

$$\dot{\phi} = \frac{1}{2}\left(\frac{\partial u}{\partial y} + \frac{\partial v}{\partial x}\right).$$

Substituting for $\partial u/\partial y$ and $\partial v/\partial x$, we obtain

$$\dot{\phi} = \frac{y u_0}{Y_0^2}\left(1 + \frac{x}{\ell}\right)\left[-\left(1 + \frac{x}{\ell}\right)^2 + \frac{y^2}{\ell^2}\right].$$ **ANSWER**

Discussion

It would be informative if we could observe the rotation, stretching, and deformation of a fluid particle as it moved through a flow field. Even though we can mathematically separate each quantity, the rotation, stretching, and deformation occur simultaneously and one cannot be isolated from the others. Calculating a few numerical values of $\partial u/\partial y$ and $\partial v/\partial x$ and comparing them would be instructive for you.

5.1.4 Velocity and Acceleration in Streamline Coordinates

In Chapter 3, we defined a streamline as a line that is everywhere tangent to the fluid velocity vector. In Chapter 4, we found that applying the energy equation between two points on the same streamline often is convenient. When we did that, we were actually using the streamline as a coordinate line for describing the flow. Sometimes, solving the differential equations that describe fluid motion is easier if streamline coordinates are used. In this section, we describe streamline coordinates and derive expressions for fluid velocity and acceleration in streamline coordinates. As usual, we concentrate on two-dimensional, two-directional (planar) flow for convenience.

In a streamline coordinate system, the coordinates are the flow steamlines and a set of lines normal (perpendicular) to them. Figure 5.7 illustrates a streamline coordinate system for two-dimensional, two-directional flow. The coordinate lines are the streamlines (s) and the normal lines (n). The n lines are perpendicular to the streamlines and point toward their center of curvature. Note that although the s lines and n lines are always perpendicular to each other, the s and n directions are variable because the streamlines are not straight.

The situation illustrated in Fig. 5.7 presumes that the streamlines always lie in a single plane, which is true only for two-dimensional, two-directional flow. In three-dimensional, three-directional flow, the streamlines are curves in space

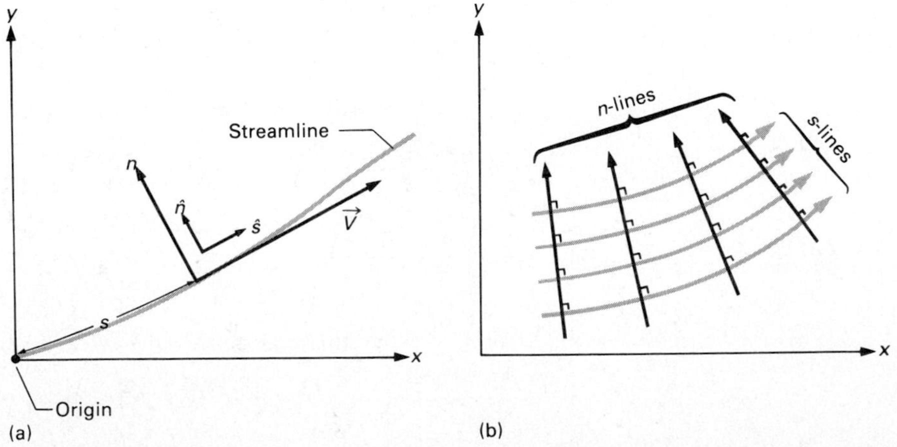

Figure 5.7 Streamline coordinate system: (a) coordinates for a single streamline; (b) a grid of s and n lines in a plane.

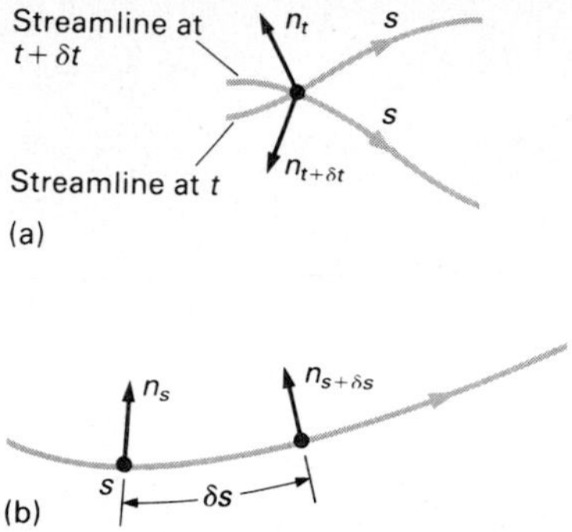

Figure 5.8 Changes of the n direction: (a) streamline changes with time; $n_{t+\delta t}$ is different from n_t: (b) changes with movement along a streamline; $n_{s+\delta s}$ is different from n_s.

and there are two normal vectors, $\hat{n}$ and $\hat{m}$; $\hat{n}$ and $\hat{m}$ are perpendicular to $\hat{s}$ and to each other.

The major advantage of the *sn*-coordinate system is that the velocity at any point is always parallel to the s direction. We write the velocity vector

$$\vec{V} = V_s\hat{s} + V_n\hat{n} = V_s\hat{s},$$

because $V_n \equiv 0$. As the fluid velocity is always parallel to the (local) s direction, any fluid particle is instantaneously moving along the s line passing through it. If the flow is steady, any fluid particle always moves along the same s line.

The s-direction acceleration is

$$a_s = \frac{DV_s}{Dt} = \frac{\partial V_s}{\partial t} + \frac{\partial V_s}{\partial s}\left(\frac{ds}{dt}\right) + \frac{\partial V_s}{\partial n}\left(\frac{dn}{dt}\right).$$

We write the velocity components* as

$$V_s = V_s\{s, n, t\} \quad \text{and} \quad V_n = V_n\{s, n, t\}.$$

Also

$$\frac{ds}{dt} = V_s \quad \text{and} \quad \frac{dn}{dt} = V_n = 0,$$

so

$$a_s = \frac{\partial V_s}{\partial t} + V_s\frac{\partial V_s}{\partial s}. \tag{5.18}$$

Next consider the n direction. Note that, although $V_n = 0$ at any instant, it does not follow that a_n is zero, because the n direction can change with time or with movement along a streamline (see Fig. 5.8). The n-direction acceleration is

$$a_n = \frac{DV_n}{Dt} = \frac{\partial V_n}{\partial t} + \frac{\partial V_n}{\partial s}\left(\frac{ds}{dt}\right) + \frac{\partial V_n}{\partial n}\left(\frac{dn}{dt}\right) = \frac{\partial V_n}{\partial t} + V_s\frac{\partial V_n}{\partial s}. \tag{5.19}$$

---

* You may think that we need not consider V_n at all, because it is instantaneously zero; that would be a mistake, as you will soon see.

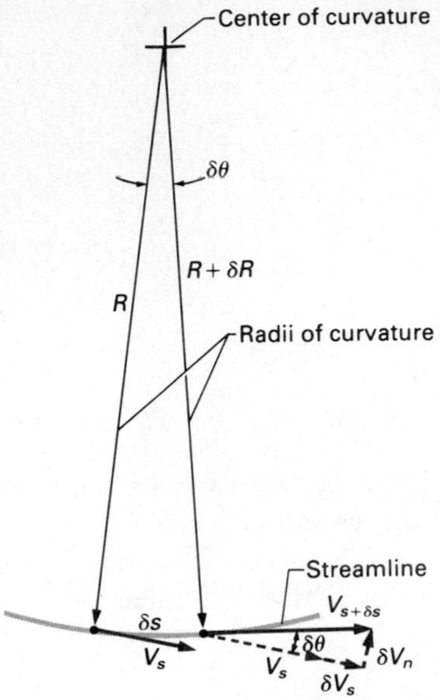

Figure 5.9 Changes of velocity along a streamline.

We can further simplify Eq. (5.19) by examining the geometry of Fig. 5.9, which shows the fluid velocity at two points along a streamline in *steady* flow. The change of normal velocity, δV_n, because of movement along the streamline from s to $s + \delta s$ is

$$\delta V_n \approx V_s \tan(\delta\theta) \approx V_s(\delta\theta).$$

We may also write

$$\delta V_n \approx \frac{\partial V_n}{\partial s}(\delta s).$$

From the geometry of Fig. 5.9, we have

$$\delta\theta \approx \frac{\delta s}{R},$$

where R is the local radius of curvature of the streamline. Combining the three preceding equations gives

$$\frac{\partial V_n}{\partial s} \approx \frac{V_s}{R}.$$

If we use this equation, the second term in Eq. (5.19) becomes

$$V_s\left(\frac{\partial V_n}{\partial s}\right) \approx V_s\left(\frac{V_s}{R}\right) = \frac{V_s^2}{R}.$$

If we take the limit of this expression as $\delta s \to 0$, the normal acceleration becomes

$$a_n = \frac{\partial V_n}{\partial t} + \frac{V_s^2}{R}. \tag{5.20}$$

5.2 THE DIFFERENTIAL CONTINUITY EQUATION

5.2.1 Derivation of the Differential Continuity Equation

To formulate a differential equation representing the principle of conservation of mass, we consider a two-dimensional, two-directional, unsteady flow field. Figure 5.10 shows a point in the field. A control volume of dimensions δx, δy, δz is superimposed around the point. The fluid velocities at the faces of the control volume are shown. Figure 5.10 is much like Fig. 5.2; however, the dotted lines refer to the boundaries of a fixed control volume rather than the boundaries of a specific fluid particle in this case.

We assume that the velocities are uniform across the faces of the small control volume and that the fluid density at the center of the control volume is an appropriate average. The principle of conservation of mass implies that

$$\frac{d}{dt}(m_{\text{cv}}) + \sum \dot{m}_{\text{out}} - \sum \dot{m}_{\text{in}} = 0.$$

The volume, $\delta \forall$, of the control volume is constant, so

$$\frac{d}{dt}(m_{\text{cv}}) = \frac{d}{dt}(\rho\, \delta \forall) = \delta \forall \frac{d\rho}{dt}$$

$$= (\delta x)(\delta y)(\delta z)\frac{d\rho}{dt}.$$

Because the control volume is fixed in space, the time derivative of density is taken at a fixed point and should be written $\partial \rho / \partial t$, so

$$\frac{d}{dt}(\rho\, \delta \forall) = (\delta x)(\delta y)(\delta z)\frac{\partial \rho}{\partial t}.$$

Mass flows into the control volume through the left and bottom faces and out through the right and top faces, and only the velocity component perpendicular to an area carries mass through the area. Thus the mass flow rates are

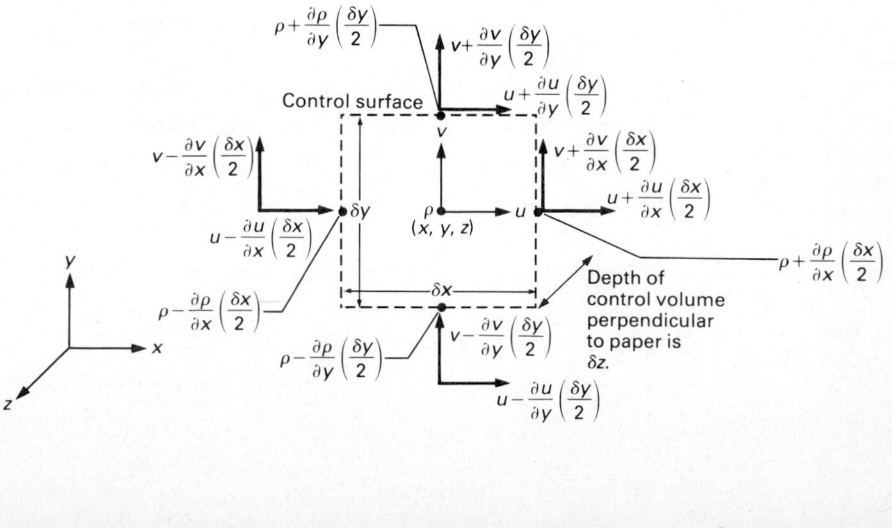

Figure 5.10 Rectangular control volume enclosing point (x, y) in a two-dimensional flow field; velocities and densities at control volume faces included.

$$\sum \dot{m}_{\text{in}} = \left[\rho - \frac{\partial \rho}{\partial x}\left(\frac{\delta x}{2}\right)\right]\left[u - \frac{\partial u}{\partial x}\left(\frac{\delta x}{2}\right)\right](\delta y)(\delta z)$$
$$+ \left[\left(\rho - \frac{\partial \rho}{\partial y}\left(\frac{\delta y}{2}\right)\right]\left[v - \frac{\partial v}{\partial y}\left(\frac{\delta y}{2}\right)\right](\delta x)(\delta z)$$

and

$$\sum \dot{m}_{\text{out}} = \left[\rho + \frac{\partial \rho}{\partial x}\left(\frac{\delta x}{2}\right)\right]\left[u + \frac{\partial u}{\partial x}\left(\frac{\delta x}{2}\right)\right](\delta y)(\delta z)$$
$$+ \left[\rho + \frac{\partial \rho}{\partial y}\left(\frac{\delta y}{2}\right)\right]\left[v + \frac{\partial v}{\partial y}\left(\frac{\delta y}{2}\right)\right](\delta x)(\delta z).$$

Substituting into the continuity equation and dropping higher order terms give

$$(\delta x)(\delta y)(\delta z)\left(\frac{\partial \rho}{\partial t}\right) + (\delta x)(\delta y)(\delta z)\left[\rho \frac{\partial u}{\partial x} + u \frac{\partial \rho}{\partial x} + \rho \frac{\partial v}{\partial y} + v \frac{\partial \rho}{\partial y}\right] = 0.$$

We now divide by $(\delta x)(\delta y)(\delta z)$ and take the limit as δx, δy, and δz approach zero, shrinking the control volume to a point. The equation becomes

$$\frac{\partial \rho}{\partial t} + u \frac{\partial \rho}{\partial x} + \rho \frac{\partial u}{\partial x} + v \frac{\partial \rho}{\partial y} + \rho \frac{\partial v}{\partial y} = 0.$$

Combining terms in accordance with the product rule for derivatives, we get

$$\frac{\partial \rho}{\partial t} + \frac{\partial(\rho u)}{\partial x} + \frac{\partial(\rho v)}{\partial y} = 0,$$

which is the differential continuity equation for two-dimensional, two-directional flow. For three-dimensional, three-directional flow, the continuity equation is

$$\frac{\partial \rho}{\partial t} + \frac{\partial(\rho u)}{\partial x} + \frac{\partial(\rho v)}{\partial y} + \frac{\partial(\rho w)}{\partial z} = 0. \tag{5.21}$$

We write the continuity equation in vector notation as

$$\frac{\partial \rho}{\partial t} + \nabla \cdot (\rho \vec{V}) = 0. \tag{5.22}$$

Equation (5.22) may be derived in or transformed to other coordinate systems. Appendix H gives the differential continuity equation in cylindrical and spherical coordinates.

Two (almost) obvious special forms of the differential continuity equation are the equation for steady, compressible flow,

$$\frac{\partial(\rho u)}{\partial x} + \frac{\partial(\rho v)}{\partial y} + \frac{\partial(\rho w)}{\partial z} = 0, \tag{5.23}$$

Steady, compressible flow

and the equation for incompressible flow (steady or unsteady),

$$\frac{\partial u}{\partial x} + \frac{\partial v}{\partial y} + \frac{\partial w}{\partial z} = 0. \tag{5.24}$$

Steady or unsteady incompressible flow

The differential continuity equation is one of a set of differential equations used to predict a flow field in detail. The derivation of the equation shows that it expresses the mass balance at a point in a flow field, but the mathematical format of the resulting differential equation certainly does not call the principle of mass conservation directly to mind. The following discussion shows how the equation can be understood from a different point of view.

The transport theorem shows that the control volume expression of any fundamental law is equivalent to expressing that law for the specific mass (system) that occupies the control volume at a particular instant. Therefore the continuity equation, Eq. (5.22), also expresses the principle of mass conservation for the fluid particle that is located inside the control volume. Using the product rule for derivatives, we write the continuity equation as

$$\frac{\partial \rho}{\partial t} + \vec{V} \cdot \nabla \rho + \rho(\nabla \cdot \vec{V}) = 0.$$

But

$$\frac{\partial \rho}{\partial t} + \vec{V} \cdot \nabla \rho = \frac{D\rho}{Dt},$$

which is the instantaneous rate of change of the particle's density, and

$$\nabla \cdot \vec{V} = \frac{1}{\delta \forall} d(\delta \forall)/dt = \frac{1}{\delta \forall} \frac{D(\delta \forall)}{Dt},$$

which is the instantaneous rate of volumetric strain of the particle, by Eq. (5.15). Multiplying by the particle volume, $\delta \forall$, gives

$$\delta \forall \left(\frac{D\rho}{Dt}\right) + \frac{\rho D(\delta \forall)}{Dt} = \frac{D(\rho\, \delta \forall)}{Dt} = 0.$$

This expression is equivalent to the statement that the particle mass, $\rho\, \delta \forall$, is invariant with time. Viewed simply, the continuity equation states that because mass is conserved, any change in the density of a fluid particle must be compensated by a corresponding change in the particle's volume.

EXAMPLE 5.2 Illustrates the Differential Continuity Equation

Find the transverse velocity $v(x, y)$ for the fluid flowing in the converging channel of Example 3.1. (See also Examples 3.3, 4.1, and 5.1.)

SOLUTION

Given

Constant density fluid flowing in a converging channel with a half-height of

$$Y = \frac{Y_0}{[1 + (x/\ell)]}$$

Longitudinal velocity, given by

$$u = u_0\left(1 + \frac{x}{\ell}\right)\left[1 - \left(\frac{y}{Y}\right)^2\right]$$

with $u_0 = 1.0$ m/s

Fig. E5.1

Find

$v\{x, y\}$

Solution

The appropriate form of the differential continuity equation is Eq. (5.24), with $\partial w/\partial z = 0$. Then we rearrange the remaining two terms:

$$\frac{\partial v}{\partial y} = -\frac{\partial u}{\partial x}.$$

The equations for $u\{x, y\}$ and $Y\{x\}$ give

$$u = u_0\left(1 + \frac{x}{\ell}\right)\left[1 - \frac{y^2}{Y_0^2}\left(1 + \frac{x}{\ell}\right)^2\right].$$

Substituting this equation for u into the differential continuity equation gives

$$\frac{\partial v}{\partial y} = -\frac{\partial}{\partial x}\left\{u_0\left(1 + \frac{x}{\ell}\right)\left[1 - \frac{y^2}{Y_0^2}\left(1 + \frac{x}{\ell}\right)^2\right]\right\}$$

$$= -\frac{u_0}{\ell}\left[1 - \frac{3y^2}{Y_0^2}\left(1 + \frac{x}{\ell}\right)^2\right].$$

Multiplying by dy and integrating, we obtain

$$v = -\int \frac{u_0}{\ell}\,dy + \int \frac{3y^2 u_0}{\ell Y_0^2}\left(1 + \frac{x}{\ell}\right)^2 dy$$

$$= -\frac{u_0 y}{\ell} + \frac{u_0 y^3}{\ell Y_0^2}\left(1 + \frac{x}{\ell}\right)^2 + C\{x\},$$

where $C\{x\}$ is a "constant" of integration. We are integrating with respect to y, so $C\{x\}$ cannot be a function of y but could possibly be a function of the other coordinate, x. The fluid immediately adjacent to the wall has zero velocity; therefore $v = 0$ at $y = Y$ for all x.

We can use either boundary to evaluate $C\{x\}$. Using the upper boundary where $Y = Y_0/[1 + (x/\ell)]$ gives

$$0 = -\frac{u_0 Y_0}{\ell[1 + (x/\ell)]} + \frac{u_0 Y_0^3}{\ell Y_0^2[1 + (x/\ell)]} + C\{x\} \quad \text{and} \quad C\{x\} = 0.$$

Then

$$v\{x, y\} = -\frac{y u_0}{\ell} + \frac{y^3 u_0}{\ell Y_0^2}\left(1 + \frac{x}{\ell}\right)^2.$$

A slightly more convenient form is

$$v\{x, y\} = -u_0\frac{y}{\ell}\{1 - (y/Y_0)^2[1 + (x/\ell)]^2\}.$$ **ANSWER**

Table E5.2 Tabulated values of the velocity $v\{x, y\}$ for the two-dimensional channel of Example 5.2

Longitudinal Coordinate x (m)	Height* y (m)	Transverse Velocity v (m/s)*
0.0	0.00	0.000
0.0	±0.20	∓0.038
0.0	±0.40	∓0.067
0.0	±0.60	∓0.077
0.0	±0.80	∓0.058
0.0	±1.00	0.000
2.5	0.00	0.000
2.5	±0.13	∓0.024
2.5	±0.27	∓0.045
2.5	±0.40	∓0.051
2.5	±0.67	∓0.000
5.0	0.00	0.000
5.0	±0.10	∓0.019
5.0	±0.20	∓0.034
5.0	±0.30	∓0.078
5.0	±0.40	∓0.029
5.0	±0.50	0.000

* Top sign for top half of channel, lower sign for lower half of channel.

Discussion

The velocity $v(x, y)$ is tabulated in Table E5.2. It is zero at the wall (satisfying the boundary condition), zero along the center of the channel, negative for positive y, and positive for negative y. This pattern indicates that fluid is moving toward the center of the channel, a quite reasonable result since we would expect the converging walls to "push" the fluid toward the center. See Example 3.3.

The velocity u was given in units of m/s, and the velocity v would have the same units. This statement can be verified by noting that u_0 appears in the equation for v.

5.2.2 Stream Function for Two-Dimensional, Two-Directional Flow

Suppose that we were interested in predicting the details of a two-directional, two-dimensional flow. Our minimum objective would be prediction of two velocity components (u and v, if we choose Cartesian coordinates) and the pressure, p. To achieve the objective would require the solution of three differential equations for the three unknowns; the differential continuity equation would be one of the necessary equations. Obviously, it would be to our advantage if we could reduce the number of equations and the number of unknowns. The stream function allows us to do just that.

Suppose that we could find *one* function $\Psi(x, y)$ from which we could obtain *both* velocity components according to:

$$u = \frac{\partial \Psi}{\partial y} \tag{5.25a}$$

and

$$v = -\frac{\partial \Psi}{\partial x}. \tag{5.25b}$$

If we could do so, we could replace both u and v with Ψ in our calculations.

The obvious question is: "Does such a function exist?" This question requires both a mathematical and a physical answer. The rules of multivariable calculus require that Ψ satisfy the condition

$$\frac{\partial}{\partial x}\left(\frac{\partial \Psi}{\partial y}\right) - \frac{\partial}{\partial y}\left(\frac{\partial \Psi}{\partial x}\right) = 0,$$

if it is a continuous function with continuous derivatives.

The physical requirement is that the flow field derived from Ψ must satisfy the continuity equation. Engineers simply are not interested in any "flow fields" that do not satisfy conservation of mass! Assuming incompressible flow, the continuity equation is

$$\frac{\partial u}{\partial x} + \frac{\partial v}{\partial y} = 0.$$

Substituting Equations (5.25a) and (5.25b), we get

$$\frac{\partial}{\partial x}\left(\frac{\partial \Psi}{\partial y}\right)+\frac{\partial}{\partial y}\left(-\frac{\partial \Psi}{\partial x}\right)=0,$$

which is identical to the mathematical requirement. We conclude that it is possible to derive realistic velocities from a function Ψ and that these velocities satisfy the continuity equation for incompressible, two-directional, two-dimensional flow. For this type of flow, we can reformulate the problem in terms of Ψ and drop continuity from the set of equations.

The function Ψ is called the *stream function* or the *streamline function*. From Eq. (3.2a), a streamline in two-dimensional, two-directional flow must satisfy the equation

$$\left(\frac{dy}{dx}\right)_{\Psi}=\frac{v}{u}.$$

On a line where Ψ = Constant, we write

$$d\Psi=\frac{\partial \Psi}{\partial x}\,dx+\frac{\partial \Psi}{\partial y}\,dy=0.$$

Using Eqs. (5.25a) and (5.25b), we get

$$\frac{dy}{dx}=-\frac{\partial \Psi/\partial x}{\partial \Psi/\partial y}=\frac{v}{u},$$

which is the equation for a streamline. Therefore the stream function defined by Eqs. (5.25a) and (5.25b) is identical to the streamline function defined by Eq. (3.1).

EXAMPLE 5.3 Illustrates the Stream Function

Consider the stream function given by $\Psi = 4xy$. Find the corresponding fluid velocity components and show that they satisfy the differential continuity equation. Then sketch a few streamlines and suggest any practical applications of the resulting flow field.

SOLUTION

Given

Stream function given by $\Psi = 4xy$

Find

Velocity components

That velocities satisfy continuity

Sketch streamlines

Practical example of flow pattern

Solution

The velocity components are given by Eqs. (5.25a) and (5.25b):

$$u = \frac{\partial \Psi}{\partial y} = 4x$$ **ANSWER**

and

$$v = -\frac{\partial \Psi}{\partial x} = -4y.$$ **ANSWER**

The continuity equation for incompressible two-directional, two-dimensional flow is

$$\frac{\partial u}{\partial x} + \frac{\partial v}{\partial y} = 0.$$

Substituting the velocity components yields

$$\frac{\partial}{\partial x}(4x) + \frac{\partial}{\partial y}(-4y) = 4 - 4 = 0.$$

Continuity is satisfied. **ANSWER**

We may choose between two methods for sketching some streamlines. We may choose a few points in the xy plane [for example (0, 0), (1, 1), (1, 2), (2, 1)], calculate u and v at those points, and sketch the velocity vector at the points. This approach gives us some idea of the streamline shapes, as streamlines are tangent to the velocity vectors.

An alternative method is to set Ψ equal to various constants (e.g., $\Psi = 4xy = 0$, $\Psi = 4xy = 1$, $\Psi = 4xy = 2$, etc.) and plot the resulting curves. Applying either method, we conclude that the streamlines for this flow are hyperbolas. Sketching only the upper half-plane gives

the streamlines shown in Fig. E5.3. **ANSWER**

Upon examination, we could conclude that these streamlines might model the flow near the stagnation point on the nose of a blunt body

Figure E5.3 Some streamlines from $\Psi = 4xy$.

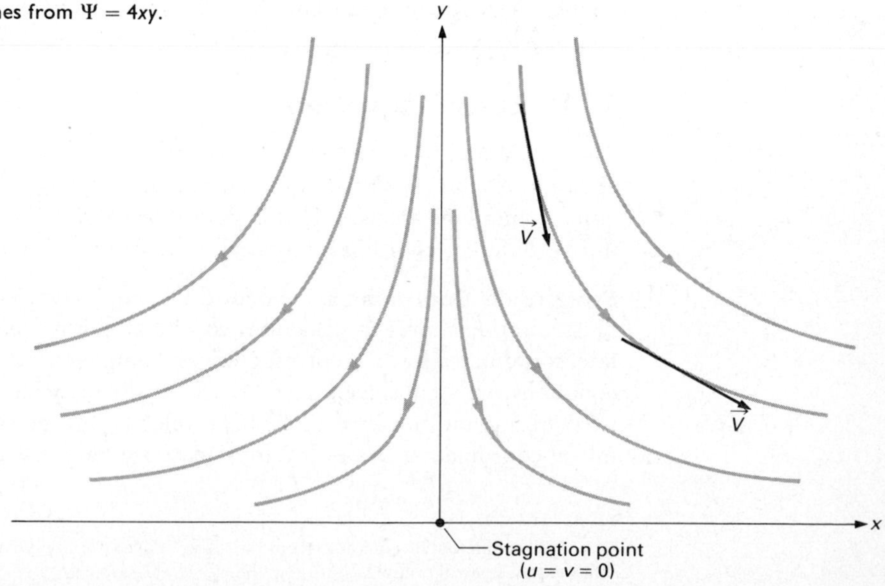

in a streaming flow. If we considered only the upper right quarter-plane, the streamlines might model flow in a 90° corner.

Discussion

In this example, a stream function was given, and we were asked to find the velocity field and discover a realistic physical system that would produce a flow with the corresponding streamlines. In a more realistic problem (and a considerably more difficult one), the physical system would be given, and we would be required to find the stream function. The problem given in this example is an example of an *inverse problem* of fluid mechanics: "Given a flow pattern, find the corresponding physical system geometry."

5.3 DYNAMICS OF INVISCID FLOW

We now apply Newton's second law of motion to develop additional differential equations to describe fluid motion. For an infinitesimal fluid particle,* the law is

$$\begin{pmatrix}\text{Mass of}\\ \text{fluid particle}\end{pmatrix} \times \begin{pmatrix}\text{Acceleration of}\\ \text{fluid particle}\end{pmatrix} = \begin{pmatrix}\text{Sum of forces acting}\\ \text{on fluid particle}\end{pmatrix}. \qquad (5.26)$$

In general, three types of forces may act on a fluid particle: body forces (such as weight), pressure forces, and shear stress forces. Shear stress forces depend on both the viscosity and the fluid velocity field. Their inclusion in the mathematical model leads to a rather complicated set of differential equations. We begin our investigation of differential forms of Newton's law by considering flows in which the effects of shear stress can be neglected. Doing so implies that we neglect the effects of both viscosity and turbulence. As an imaginary fluid with no viscosity would have neither shear stress nor turbulence, we call a flow with negligible shear stresses an *inviscid flow*.

5.3.1 Euler's Equations

The differential form of Newton's second law for an inviscid flow is called Euler's equation if a single vector equation is used or Euler's equations if separate component equations are used. In this section, we derive and discuss Euler's equation(s) in two different coordinates: streamline coordinates and Cartesian coordinates.

Streamline Coordinates. Because the fluid velocity and acceleration are expressed rather simply in streamline coordinates, the equations of motion in streamline coordinates have a correspondingly simple form. In order to develop these equations, let's consider Fig. 5.11, which shows a rectangular fluid particle at a particular point on a streamline. The velocity and pressure fields are expressed in the *sn*-coordinate system. A Cartesian coordinate system is also shown for refer-

* The differential equations could also be derived by applying the momentum equation to an infinitesimal control volume.

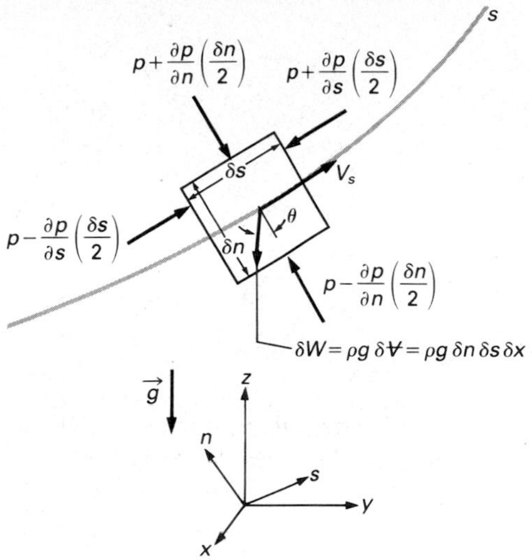

Figure 5.11 A rectangular fluid particle in streamline coordinates; pressures and gravity force acting on the particle are shown. Depth perpendicular to paper is δx.

ence. Note that the z coordinate points upward. We assume that the flow is two-dimensional and two-directional and so must develop separate equations of motion for the s and n directions.

First, we consider the s (streamwise) direction. Newton's second law gives

$$\sum \delta F_s = (\delta m_p)a_s, \tag{5.27}$$

where $\sum \delta F_s$ represents the sum of the streamwise components of all forces acting on the particle and δm_p is the mass of the particle,

$$\delta m_p = \rho\, \delta \forall = \rho(\delta s)(\delta n)(\delta x). \tag{5.28}$$

Using Eq. (5.18), we find that

$$(\delta m_p)a_s = \rho\left[\frac{\partial V_s}{\partial t} + V_s\frac{\partial V_s}{\partial s}\right](\delta s)(\delta n)(\delta x). \tag{5.29}$$

Now we consider the forces on the particle. As we decided to neglect shear stress, the only forces are from pressure and gravity. With the aid of Fig. 5.11, we find

$$\sum \delta F_s = \underbrace{\left[p - \frac{\partial p}{\partial s}\left(\frac{\delta s}{2}\right)\right](\delta n)(\delta x) - \left[p + \frac{\partial p}{\partial s}\left(\frac{\delta s}{2}\right)\right](\delta n)(\delta x)}_{\text{net pressure force}} - \underbrace{\delta W \sin\theta}_{\text{gravity force}}.$$

The particle's weight is

$$\delta W = g(\delta m_p) = g\rho(\delta s)(\delta n)(\delta x),$$

and the force term can be simplified so that

$$\sum \delta F_s = \left(-\frac{\partial p}{\partial s} - \rho g \sin\theta\right)(\delta n)(\delta s)(\delta x). \tag{5.30}$$

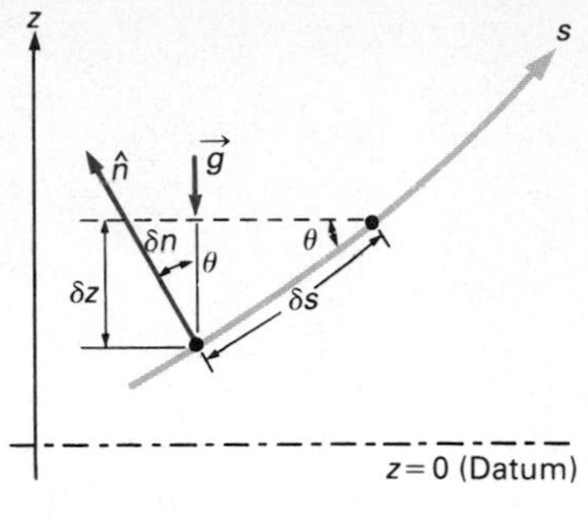

Figure 5.12 Illustration of relation between movement along a streamline (δs) and vertical displacement (δz).

We now substitute Eqs. (5.29) and (5.30) into Eq. (5.27), divide by $(\delta n)(\delta s)(\delta x)$, and take the limit as δn, δs, and δx approach zero. The result is

$$-\frac{\partial p}{\partial s} - \rho g \sin\theta = \rho\left[\frac{\partial V_s}{\partial t} + V_s\frac{\partial V_s}{\partial s}\right]. \tag{5.31}$$

An alternative form of Eq. (5.31) is obtained by dividing by ρ and writing the acceleration terms on the left and the force terms on the right:

$$\frac{\partial V_s}{\partial t} + V_s\frac{\partial V_s}{\partial s} = -\frac{1}{\rho}\frac{\partial p}{\partial s} - g\sin\theta. \tag{5.32}$$

To prepare for integration of this equation, we replace the second term on the right by an expression involving a derivative with respect to s. We now consider the relation between the s and z directions shown in Fig. 5.12. Movement along the streamline a distance δs results in an upward movement δz, where

$$\delta z \approx \delta s \sin\theta.$$

Dividing by δs and taking the limit as $\delta s \to 0$, we have

$$\frac{\partial z}{\partial s} = \sin\theta.$$

Substituting into Eq. (5.32), we get

$$\frac{\partial V_s}{\partial t} + V_s\frac{\partial V_s}{\partial s} + \frac{1}{\rho}\frac{\partial p}{\partial s} + g\frac{\partial z}{\partial s} = 0. \tag{5.33}$$

This is Euler's equation for the streamwise direction. It is valid if shear stress is negligible.

We now consider the equation of motion for the n direction. Newton's second law for the n direction for the particle of Fig. 5.11 is

$$\sum \delta F_n = (\delta m_p)a_n. \tag{5.34}$$

The mass of the particle is given by Eq. (5.28), and the n-direction acceleration is given by Eq. (5.20). The net pressure force in the n direction is

$$\delta F_{n,\text{pressure}} = \left[\left(p - \frac{\partial p}{\partial n}\frac{\delta n}{2}\right) - \left(p + \frac{\partial p}{\partial n}\frac{\delta n}{2}\right)\right](\delta s)(\delta x) = -\frac{\partial p}{\partial n}(\delta s)(\delta n)(\delta x),$$

the n component of the gravity force is

$$\delta W_n = -\delta W \cos\theta = -\rho g \cos\theta\,(\delta s)(\delta n)(\delta x),$$

and there is no shear force (by assumption). Next, we substitute the expressions for δm_p, a_n, and the forces into Eq. (5.34), divide by $(\delta s)(\delta n)(\delta x)$, and take the limit to obtain

$$-\frac{\partial p}{\partial n} - \rho g \cos\theta = \rho\left(\frac{\partial V_n}{\partial t} + \frac{V_s^2}{R}\right).$$

An alternative form is

$$\frac{\partial V_n}{\partial t} + \frac{V_s^2}{R} = -\frac{1}{\rho}\frac{\partial p}{\partial n} - g\cos\theta. \tag{5.35}$$

We can replace $\cos\theta$ by noting from Fig. 5.12 that

$$\frac{\partial z}{\partial n} = \lim_{\substack{\delta n \to 0 \\ s\ \text{fixed}}} \frac{\delta z}{\delta n} = \cos\theta.$$

We then write Eq. (5.35) as

$$\frac{\partial V_n}{\partial t} + \frac{V_s^2}{R} + \frac{1}{\rho}\frac{\partial p}{\partial n} + g\frac{\partial z}{\partial n} = 0. \tag{5.36}$$

Equation (5.36) represents Newton's second law in a direction normal to a streamline in the absence of shear stresses.

For steady, inviscid flow, Euler's equations in streamline coordinates are

$$V_s\frac{\partial V_s}{\partial s} + \frac{1}{\rho}\frac{\partial p}{\partial s} + g\frac{\partial z}{\partial s} = 0 \tag{5.37}$$

and

$$\frac{V_s^2}{R} + \frac{1}{\rho}\frac{\partial p}{\partial n} + g\frac{\partial z}{\partial n} = 0. \tag{5.38}$$

Steady, inviscid flow, streamline coordinates

EXAMPLE 5.4 **Illustrates Euler's Equations in Streamline Coordinates**

Figure E5.4a shows an ideal fluid (zero viscosity and constant density) flowing through a planar converging nozzle that lies in a horizontal plane. Compare the pressures at points 1 and 2, at 3 and 4, and at 5 and 6.

SOLUTION

Given

Figure E5.4a with points as shown

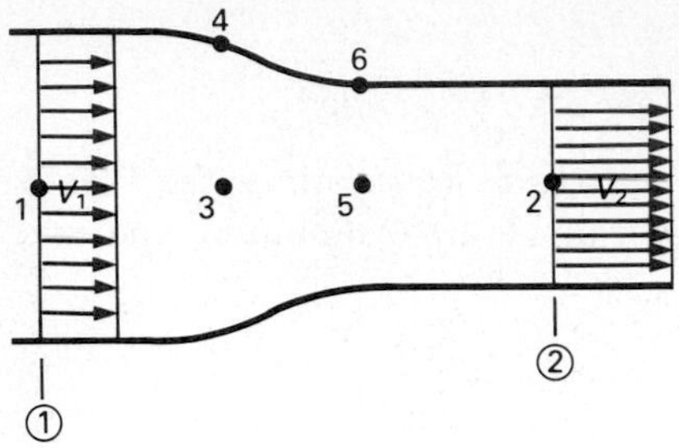

Figure E5.4a Planar nozzle.

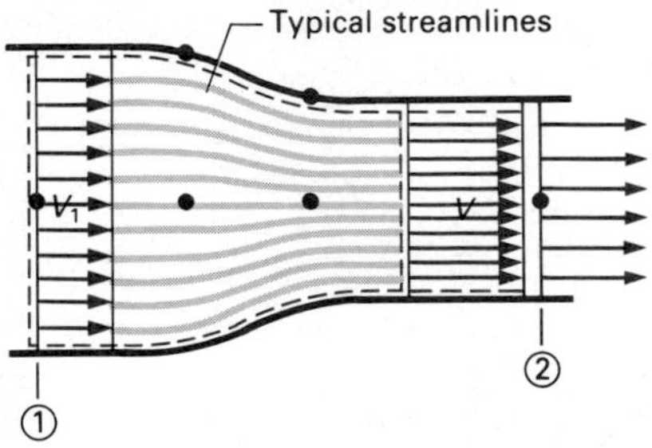

Figure E5.4b Planar nozzle with control volume and typical streamlines.

Find

Compare pressures at points shown

Solution

We estimate the shape of the streamlines from the shape of the nozzle walls. Figure E5.4b shows some typical streamlines. It also shows a control volume for applying the integral continuity equation. From the continuity equation for steady, incompressible flow, we conclude that the velocity increases from 1 to 3 to 5 to 2. Therefore, along the line 1-3-5-2,

$$\frac{\partial V}{\partial s} > 0.$$

From Eq. (5.37) and the specification that the nozzle lies in a horizontal plane,

$$\frac{\partial p}{\partial s} = -\rho V \frac{\partial V}{\partial s}.$$

As $\partial V/\partial s > 0$, we conclude that $\partial p/\partial s < 0$ and that the pressure falls along line 1–3–5–2. Therefore

$$p_1 > p_3 > p_5 > p_2. \qquad \textbf{ANSWER}$$

From Eq. (5.38),

$$\frac{\partial p}{\partial n} = -\rho \frac{V^2}{R}.$$

Recall that n points toward the center of curvature. Both V^2 and R are positive, so the pressure increases outward from the center of curvature. From Fig. E5.4b, we conclude that

$$p_4 > p_3 \quad \text{and} \quad p_5 > p_6. \qquad \textbf{ANSWER}$$

Discussion

Although consideration of Euler's equations allowed us to comment on the relative magnitudes of the pressures, it did not permit us to calculate their values. The equations must be integrated before we can calculate any numerical values for pressure.

Cartesian Coordinates. Euler's equations in streamline coordinates have one serious limitation: We must know, or at least be able to estimate, the streamline pattern. Having the equations in coordinates that we can establish without any prior knowledge of the flow field would be very useful. So we derive Euler's equations in Cartesian coordinates. As usual, we do a detailed development for a two-dimensional, two-directional flow.

Figure 5.13 shows a fluid particle located at point (x, y) in a flow described in Cartesian coordinates. For generality, we assume that the gravity vector is not aligned with any of the coordinate axes. Figure 5.13 also shows the pressures and the gravity force acting on the fluid particle. By assumption, there are no shear

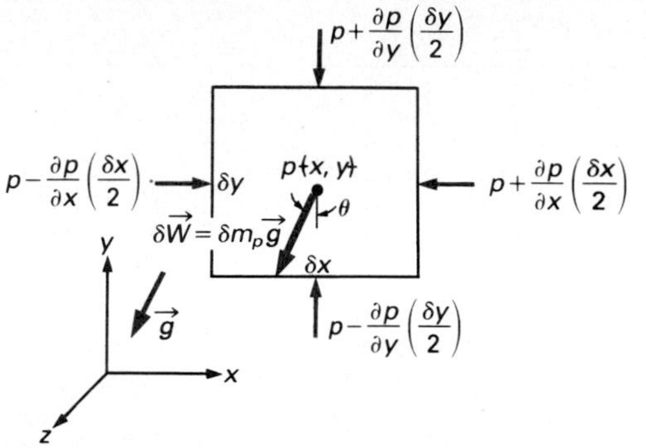

Figure 5.13 Rectangular fluid particle in two-dimensional, two-directional flow showing pressures and gravity force; depth of particle (perpendicular to paper) is δz.

stresses, so Newton's second law for the x direction is

$$\delta m_p a_x = \sum \delta F_x = \delta F_{x,\text{pressure}} + \delta F_{x,\text{gravity}}.$$

The mass of the particle is

$$\delta m_p = \rho\, \delta \forall = \rho(\delta x)(\delta y)(\delta z).$$

The x component of the particle's acceleration is given by Eq. (5.1), with $w = 0$ and $\partial u/\partial z = 0$ for two-dimensional, two-directional flow:

$$a_x = \frac{\partial u}{\partial t} + u\frac{\partial u}{\partial x} + v\frac{\partial u}{\partial y}.$$

The net pressure force in the x direction is

$$\begin{aligned}\delta F_{x,\text{pressure}} &= \left[p - \frac{\partial p}{\partial x}\left(\frac{\delta x}{2}\right)\right](\delta y)(\delta z) - \left[p + \frac{\partial p}{\partial x}\left(\frac{\delta x}{2}\right)\right](\delta y)(\delta z)\\ &= -\frac{\partial p}{\partial x}(\delta x)(\delta y)(\delta z).\end{aligned}$$

The x component of the gravity force is

$$\delta F_{x,\text{gravity}} = \delta W_x = (\delta m_p) g_x = \rho(\delta x)(\delta y)(\delta z) g_x,$$

where g_x is the x component of the gravity vector.

Substituting the forces and acceleration into Newton's second law gives

$$\rho(\delta x)(\delta y)(\delta z)\left(\frac{\partial u}{\partial t} + u\frac{\partial u}{\partial x} + v\frac{\partial u}{\partial y}\right) = -\frac{\partial p}{\partial x}(\delta x)(\delta y)(\delta z) + \rho g_x(\delta x)(\delta y)(\delta z).$$

We divide by $\rho(\delta x)(\delta y)(\delta z)$ and take the limit, shrinking the fluid particle to the point (x, y):

$$\frac{\partial u}{\partial t} + u\frac{\partial u}{\partial x} + v\frac{\partial u}{\partial y} = -\frac{1}{\rho}\frac{\partial p}{\partial x} + g_x. \tag{5.39}$$

Next, we consider Newton's second law for the y direction:

$$\delta m_p a_y = \delta F_{y,\text{pressure}} + \delta F_{y,\text{gravity}},$$

with δm_p the same as before. We have a_y, given by Eq. (5.2) with $w = 0$ and

$\partial v/\partial z = 0$:

$$a_y = \frac{\partial v}{\partial t} + u\frac{\partial v}{\partial x} + v\frac{\partial v}{\partial y}.$$

The net pressure force is

$$\delta F_{y,\text{pressure}} = -\frac{\partial p}{\partial y}(\delta y)(\delta x)(\delta z),$$

and the gravity force is

$$\delta F_{y,\text{gravity}} = \rho(\delta x)(\delta y)(\delta z)g_y.$$

Substituting into Newton's second law, we have

$$\rho(\delta x)(\delta y)(\delta z)\left(\frac{\partial v}{\partial t} + u\frac{\partial v}{\partial x} + v\frac{\partial v}{\partial y}\right) = -\frac{\partial p}{\partial y}(\delta x)(\delta y)(\delta z) + \rho g_y(\delta x)(\delta y)(\delta z).$$

Dividing by $\rho(\delta x)(\delta y)(\delta z)$ and taking the limit, we get

$$\frac{\partial v}{\partial t} + u\frac{\partial v}{\partial x} + v\frac{\partial v}{\partial y} = -\frac{1}{\rho}\frac{\partial p}{\partial y} + g_y. \tag{5.40}$$

Equations (5.39) and (5.40) are Euler's equations for two-dimensional, two-directional flow. The equations for three-dimensional, three-directional flow are similar. The x and y equations must contain the acceleration terms $w\,\partial u/\partial z$ and $w\,\partial v/\partial z$, respectively. The z-direction equation is similar to the equations for the x and y directions, with a_z appearing on the left and the z-direction pressure gradient and gravity vector component appearing on the right. Euler's equations for three-dimensional, three-directional, unsteady flow are

$$\frac{\partial u}{\partial t} + u\frac{\partial u}{\partial x} + v\frac{\partial u}{\partial y} + w\frac{\partial u}{\partial z} = -\frac{1}{\rho}\frac{\partial p}{\partial x} + g_x; \tag{5.41}$$

$$\frac{\partial v}{\partial t} + u\frac{\partial v}{\partial x} + v\frac{\partial v}{\partial y} + w\frac{\partial v}{\partial z} = -\frac{1}{\rho}\frac{\partial p}{\partial y} + g_y; \tag{5.42}$$

$$\frac{\partial w}{\partial t} + u\frac{\partial w}{\partial x} + v\frac{\partial w}{\partial y} + w\frac{\partial w}{\partial z} = -\frac{1}{\rho}\frac{\partial p}{\partial z} + g_z. \tag{5.43}$$

The equations for special cases (steady, two-dimensional flow) can be obtained from these equations by deleting the appropriate terms.

Commenting on the significance of these equations might be helpful at this point. If we want to compute the details of a three-dimensional, three-directional flow, we need equations for computing distributions of the three velocity components (u, v, w) and the pressure, p. To calculate four dependent variables, we need four differential equations. The three Euler equations, Eqs. (5.41), (5.42), and (5.43), together with the differential continuity equation, Eq. (5.22), provide the necessary equations, at least if a relationship between density and pressure is known and we are willing to assume inviscid flow.

Euler's equations may be expressed in coordinates other than streamline and Cartesian. The equations in other coordinates may be derived by applying Newton's second law to a representative fluid particle in that coordinate system or by mathematical transformation between coordinate systems. Finally, as Euler's equations represent Newton's second law, they may be written as a single vector equation:

$$\frac{D\vec{V}}{Dt} = \frac{\partial \vec{V}}{\partial t} + (\vec{V} \cdot \nabla)\vec{V} = -\frac{\nabla p}{\rho} + \vec{g}. \tag{5.44}$$

Euler's equations in cylindrical and spherical coordinates may be obtained from Appendix H. These forms are particularly useful when a flow has cylindrical or spherical symmetry, because many terms drop out.

5.3.2 Integration of Euler's Equations

Differential equation models of fluid flow are generally useful only if the differential equations can be integrated. Most often, the equations can be integrated only in the context of a specific problem geometry. In a few special cases, however, a general integral can be found. In this section, we find integrated forms of Euler's equations. Of course, because we consider Euler's equations, we limit consideration to inviscid flow.

Integration Along a Streamline in Steady Flow. Consider the streamwise Euler equation in streamline coordinates. We limit consideration to steady flow. The equation is

$$V \frac{\partial V}{\partial s} + \frac{1}{\rho}\frac{\partial p}{\partial s} + g\frac{\partial z}{\partial s} = 0.$$

The subscript s has been dropped from V, because the velocity is totally in the s direction. We multiply by ds to get

$$V \frac{\partial V}{\partial s}\,ds + \frac{1}{\rho}\frac{\partial p}{\partial s}\,ds + g\frac{\partial z}{\partial s}\,ds = 0. \tag{5.45}$$

In general, the total differential of any parameter of the flow field (say pressure p) is given by

$$dp = \frac{\partial p}{\partial s}\,ds + \frac{\partial p}{\partial n}\,dn,$$

because p is a function of both s and n. If we restrict ourselves to remaining on the same streamline,

$$dn = 0 \quad \text{and} \quad dp = \frac{\partial p}{\partial s}\,ds.$$

Similar relations hold for other properties.

With the restriction of staying on the same streamline, Eq. (5.45) becomes

$$V\,dV + \frac{dp}{\rho} + g\,dz = 0. \tag{5.46}$$

Integrating

$$\frac{V^2}{2} + \int \frac{dp}{\rho} + \int g\,dz = C \qquad \text{(a constant)}.$$

In most flows of engineering interest, the acceleration of gravity does not vary throughout the field. Thus we write

$$\int \frac{dp}{\rho} + \frac{V^2}{2} + gz = \text{Constant.} \tag{5.47}$$

Steady, inviscid flow along a streamline

Equation (5.47) is *Bernoulli's equation for compressible flow*. If we also assume constant density (incompressible flow), we get

$$\frac{p}{\rho} + \frac{V^2}{2} + gz = \text{Constant.} \tag{5.48}$$

Steady, inviscid, incompressible flow along a streamline

Equation (5.48) is, of course, the familiar Bernoulli's equation that we derived from energy considerations as Eq. (4.52). Note that the list of assumptions and restrictions on the equation is slightly different here than for Eq. (4.52).

We have already examined the usefulness of Bernoulli's equation in Section 4.3.4, so we need not repeat that material here. Examining the connection between the *energy* interpretation of Bernoulli's equation and the *dynamic* interpretation of this chapter is instructive, however. As developed in Section 4.3.4, Bernoulli's equation is a special form of the mechanical energy equation. When Eq. (5.37), which involves force and acceleration, is multiplied by ds to produce Eq. (5.45), the result is an energy equation. This result can be understood by examining the three terms in Eq. (5.46).

- $-(dp/\rho)$ is the work per unit mass done by pressure on a particle of fluid as it moves a distance ds. This can be verified by noting that $-(\partial p/\partial s)$ is the net pressure force per unit volume; $-(1/\rho)(\partial p/\partial s)$ is the net pressure force per unit mass; and (Force per unit mass) × (Displacement) = (Work per unit mass).
- $V\,dV$ $[= d(V^2/2)]$ is the change of kinetic energy per unit mass of a fluid particle as it moves a distance ds.
- $g\,dz$ is usually interpreted as the change of gravitational potential energy per unit mass of a fluid particle as it moves a distance ds; $-g\,dz$ can also be interpreted as the work done on the fluid particle by the gravity force.

The equivalence between work done by the gravitational force and the change of gravitational potential energy is sometimes extended to pressure, and dp/ρ is interpreted as "change of potential energy caused by pressure."

Integration Normal to Streamlines. Although Bernoulli's equation is very useful, our derivation restricts the equation to flow along a single streamline. To relate pressure, velocity, and elevation changes between streamlines, we must consider the equation of motion for the normal direction. If we assume steady flow, the appropriate equation is Eq. (5.38). Following the procedure used in deriving

Bernoulli's equation, we multiply by dn and integrate in the direction normal to the streamlines:

$$\int \frac{V^2}{R}\,dn + \int \frac{dp}{\rho} + \int g\,dz = \text{Constant}. \tag{5.49}$$

The pressure and elevation integrals may be evaluated in the same way as in Bernoulli's equation, if we assume constant density and gravity; however, evaluating the first integral is not possible, because, in general, we do not know how V and R vary in the n direction. The normal equation can be integrated only for specific cases in which V and R are known as functions of n. An important case is when the flow streamlines are straight and parallel, so that R, the radius of curvature, is infinite, and

$$\int \frac{V^2}{R}\,dn = 0.$$

If ρ and g are constant,

$$\frac{p}{\rho} + gz = \text{Constant}$$

from streamline to streamline, if the steamlines are straight and parallel.

For other cases, $\int (V^2/R)\,dn$ can be evaluated only if the velocity and streamline shapes are known. In many cases, this information is not known; however, Eqs. (5.38) and (5.49) can still help us make a qualitative estimate of the pressure distribution. You should remember that the normal equation must be applied across streamlines and that applying Bernoulli's equation across streamlines may lead to errors.

EXAMPLE 5.5 **Illustrates Proper and Improper Use of Bernoulli's Equation and the Normal Equation**

Figure E5.5 shows two simple flow fields called a *free vortex* and a *forced vortex*. The circumferential velocities in a free vortex and a forced vortex, respectively, are

$$V = \frac{C}{r} \quad \text{and} \quad V = r\omega.$$

We take $C = 4.0\ \text{m}^2/\text{s}$ and $\omega = 1.0$ rad/s; r is in m, and V is in m/s.

Apply Bernoulli's equation and Euler's normal equation between point A, where $r_A = 2.0$ m, and point B, where $r_B = 8.0$ m, to calculate the pressure difference $(p_B - p_A)$ for both types of vortex. Both flows involve 20°C water and lie in a horizontal plane.

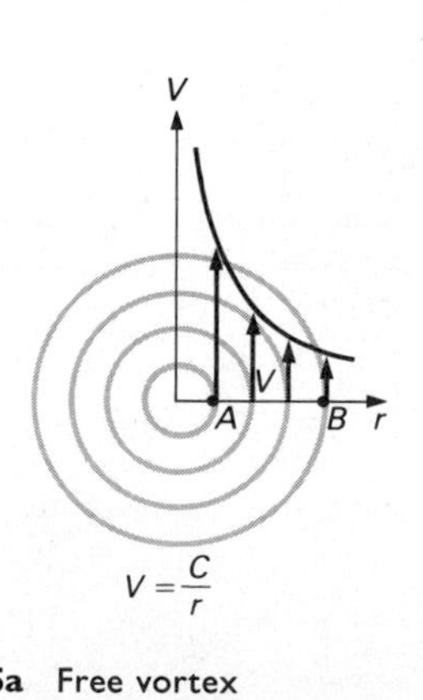

Figure E5.5a Free vortex

SOLUTION

Given

Figure E5.5a and E5.5b

Free vortex with $V = (4.0\ \text{m}^2/\text{s})/r$

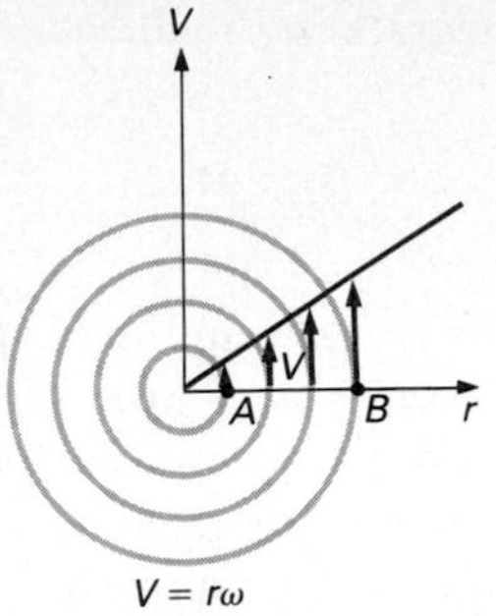

Figure E5.5b Forced vortex with typical velocity distributions.

Forced vortex with $V = (1.0 \text{ rad/s})r$, where V is in m/s and r is in m
20°C water
Flows lying in a horizontal plane

Find

Pressure difference $(p_B - p_A)$ for both flows, where $r_A = 2.0$ m and $r_B = 8.0$ m

Solution

Bernoulli's equation is

$$\frac{V_1^2}{2} + \frac{p_1}{\rho} + gz_1 = \frac{V_2^2}{2} + \frac{p_2}{\rho} + gz_2.$$

In Fig. E5.5, we identify point A as 1 and point B as 2 and note that $z_A = z_B$, so

$$p_B - p_A = \frac{\rho}{2}(V_A^2 - V_B^2).$$

For the free vortex,

$$p_B - p_A = \frac{\rho C^2}{2}\left(\frac{1}{r_A^2} - \frac{1}{r_B^2}\right).$$

Table A.5 gives $\rho = 998$ kg/m^3. Then

$$p_B - p_A = \frac{(998 \text{ kg/m}^3)(4.0 \text{ m}^2/\text{s})^2}{2}\left[\frac{1}{(2.0 \text{ m})^2} - \frac{1}{(8.0 \text{ m})^2}\right].$$

Thus

$$p_B - p_A = 1.87 \text{ kPa}. \qquad \textbf{ANSWER}$$

Bernoulli's equation, free vortex

For the forced vortex,

$$p_B - p_A = \frac{\rho\omega^2}{2}(r_A^2 - r_B^2)$$

$$= \frac{(998 \text{ kg/m}^3)(1.0 \text{ rad/s})^2}{2}[(2.0 \text{ m})^2 - (8.0 \text{ m})^2], \text{ or}$$

$$p_B - p_A = -31.9 \text{ kPa}. \qquad \textbf{ANSWER}$$

Bernoulli's equation, forced vortex

Next, we apply Euler's normal equation between points A and B of both vortexes. Integrating Eq. (5.49) with $dz = 0$ and constant density gives

$$p_B - p_A = -\rho \int_A^B \frac{V^2}{R}\, dn.$$

Because the normal coordinate points toward the center of curvature, $dn = -dr$. Also, $R = r$, so

$$p_B - p_A = \rho \int_A^B \frac{V^2}{r}\, dr.$$

For the free vortex, the normal equation is

$$p_B - p_A = \rho \int_A^B \frac{C^2}{r^3}\, dr$$

$$= \frac{\rho C^2}{2}\left(\frac{1}{r_A^2} - \frac{1}{r_B^2}\right)$$

$$= \frac{(998\ \text{kg/m}^3)(4.0\ \text{m}^2/\text{s})^2}{2}\left[\frac{1}{(2.0\ \text{m})^2} - \frac{1}{(8.0\ \text{m})^2}\right]$$

or

$$p_B - p_A = 1.87\ \text{kPa}.$$ **ANSWER**

Normal equation, free vortex

This answer is the same as that obtained from Bernoulli's equation.

For the forced vortex, the normal equation is

$$p_B - p_A = \rho \int_A^B \frac{r^2\omega^2}{r}\, dr$$

$$= \frac{\rho\omega^2}{2}(r_B^2 - r_A^2)$$

$$= \frac{(998\ \text{kg/m}^3)(1.0\ \text{rad/s})^2}{2}\left[(8.0\ \text{m})^2 - (2.0\ \text{m})^2\right]$$

or

$$p_B - p_A = 29.9\ \text{kPa}.$$ **ANSWER**

Normal equation, forced vortex

This answer is *not* the same as that obtained from Bernoulli's equation.

Discussion

Euler's normal equation was properly applied in both cases and gave us the correct answers. Bernoulli's equation was not properly applied, because we violated one of the assumptions used in development of the equation, namely, that the equation must be applied along a streamline. We applied Bernoulli's equation normal to streamlines, so we should not expect the correct answers from it. Surprisingly, the calculations showed that Bernoulli's equation did give us the correct answer for the free vortex, even though we violated one of the assumptions implicit in the equation. We explain this surprising result in the next section.

Misapplication of Bernoulli's equation in this example illustrates an important point:

> Applying an equation to a situation in which all the assumptions used in the development of that equation are not satisfied usually leads to incorrect answers.

Also note that Bernoulli's equation erroneously indicated that the pressure decreased with radius for the forced vortex. Recall that a mass particle moves in a straight line unless acted on by a force; in this case each particle tends to move in the tangential direction. The particle can follow a circular path only if the pressure at its larger radius exceeds the pressure at its inner radius and forces the particle to continually change its direction to form the circular path.

Integration for Steady, Irrotational Flow. If a steady, inviscid flow field is also irrotational, we can integrate Euler's equations in any direction and apply Bernoulli's equation between any two points, whether they are on the same streamline or not. To derive this important result, we consider a two-directional, two-dimensional flow and choose Cartesian coordinates to describe it. Figure 5.14 shows the flow field and the coordinate system. Anticipating the result, we choose the symbol ξ for the coordinate normal to the xy plane and use z to denote the elevation, measured opposite the gravity vector. For this flow, Euler's equations, from Eqs. (5.41) and (5.42), are

$$u\frac{\partial u}{\partial x} + v\frac{\partial u}{\partial y} = -\frac{1}{\rho}\frac{\partial p}{\partial x} + g_x$$

and

$$u\frac{\partial v}{\partial x} + v\frac{\partial v}{\partial y} = -\frac{1}{\rho}\frac{\partial p}{\partial y} + g_y.$$

First, we consider the gravity term. Figure 5.15 shows the geometric relationship between the x and y axes and the z direction, as well as the x and y components of the gravity vector. From Fig. 5.15,

$$g_x = -|g|\cos\theta = -g\cos\theta \quad \text{and} \quad g_y = -|g|\sin\theta = -g\sin\theta.$$

If the fluid particle moves from x to $x + \delta x$, it moves "up" an amount δz, where

$$\delta z = \delta x \cos\theta, \quad \text{so} \quad \cos\theta = \frac{\delta z}{\delta x} = \frac{\partial z}{\partial x}.$$

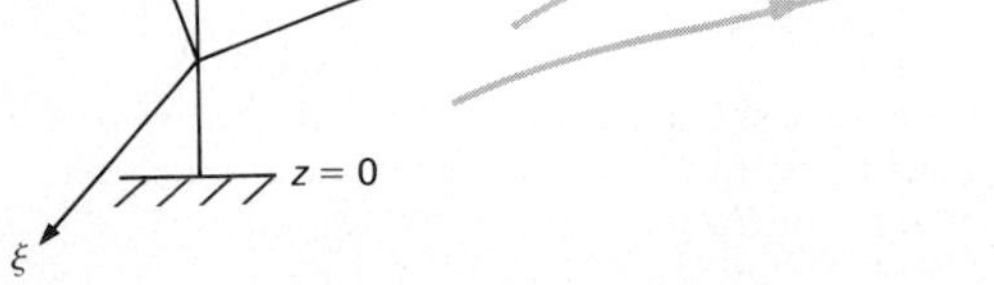

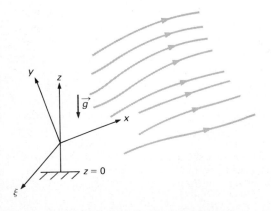

Figure 5.14 Flow in $xy\xi$ coordinate system; z is elevation.

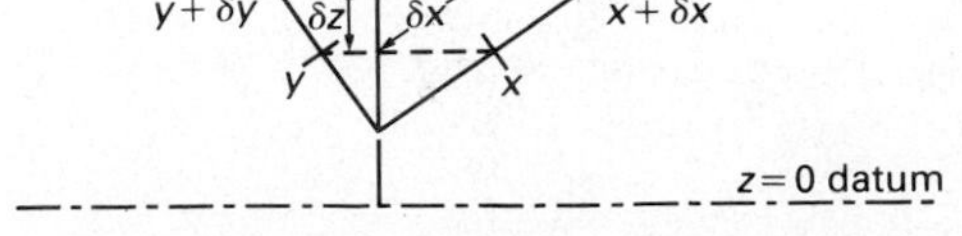

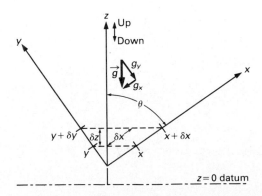

Figure 5.15 xy coordinate system with z coordinate (elevation) pointing upward; movement from x to $(x + \delta x)$ or from y to $(y + \delta y)$ results in elevation change δz.

Similarly,

$$\sin\theta = \frac{\partial z}{\partial y}.$$

Therefore

$$g_x = -g\frac{\partial z}{\partial x} \quad \text{and} \quad g_y = -g\frac{\partial z}{\partial y}.$$

Substituting the g components into the right-hand side of the Euler equations gives

$$u\frac{\partial u}{\partial x} + v\frac{\partial u}{\partial y} = -\frac{1}{\rho}\frac{\partial p}{\partial x} - g\frac{\partial z}{\partial x} \tag{5.50}$$

and

$$u\frac{\partial v}{\partial x} + v\frac{\partial v}{\partial y} = -\frac{1}{\rho}\frac{\partial p}{\partial y} - g\frac{\partial z}{\partial y}. \tag{5.51}$$

Next, we introduce the condition of irrotationality. An irrotational flow has angular velocity ($\vec{\omega}$) and vorticity ($\vec{\zeta} = 2\vec{\omega}$) equal to zero. From Eq. (5.12),

$$\zeta_\xi = \frac{\partial v}{\partial x} - \frac{\partial u}{\partial y} = 0.$$

Thus

$$\frac{\partial u}{\partial y} = \frac{\partial v}{\partial x}.$$

If we use this relationship, Eq. (5.50) becomes

$$u\frac{\partial u}{\partial x} + v\frac{\partial v}{\partial x} = \frac{\partial}{\partial x}\left(\frac{u^2}{2} + \frac{v^2}{2}\right) = -\frac{1}{\rho}\frac{\partial p}{\partial x} - g\frac{\partial z}{\partial x},$$

and Eq. (5.51) becomes

$$u\frac{\partial u}{\partial y} + v\frac{\partial v}{\partial y} = \frac{\partial}{\partial y}\left(\frac{u^2}{2} + \frac{v^2}{2}\right) = -\frac{1}{\rho}\frac{\partial p}{\partial y} - g\frac{\partial z}{\partial y}.$$

The magnitude of the velocity vector $\vec{V}$ is given by

$$|\vec{V}|^2 = V^2 = u^2 + v^2,$$

so substituting and rearranging (constant g is assumed), we obtain

$$\frac{\partial}{\partial x}\left(\frac{V^2}{2} + gz\right) + \frac{1}{\rho}\frac{\partial p}{\partial x} = 0 \tag{5.52}$$

and

$$\frac{\partial}{\partial y}\left(\frac{V^2}{2} + gz\right) + \frac{1}{\rho}\frac{\partial p}{\partial y} = 0. \tag{5.53}$$

Multiplying Eq. (5.52) by dx and Eq. (5.53) by dy, adding them and using the fact that, for any quantity,

$$d(\;) = \frac{\partial(\;)}{\partial x}(dx) + \frac{\partial(\;)}{\partial y}(dy),$$

we get

$$d\left(\frac{V^2}{2}\right) + d(gz) + \frac{dp}{\rho} = 0.$$

Integrating gives

$$\frac{V^2}{2} + gz + \int \frac{dp}{\rho} = \text{Constant.} \tag{5.54}$$

If ρ is a constant, we get

$$\frac{V^2}{2} + \frac{p}{\rho} + gz = \text{Constant.} \tag{5.55}$$

Equation (5.55) is identical to Bernoulli's equation, but there are some important differences in application. Bernoulli's equation applies only along a streamline but contains no restrictions about irrotationality. Equation (5.55) applies between any two points in a flow, even across streamlines, if the flow is irrotational. Because Eq. (5.55) is identical in form to Bernoulli's equation, we do not treat it as a different equation but simply relax one of the restrictions on Bernoulli's equation:

Bernoulli's equation may be applied across streamlines if the flow is irrotational.

Although we carried out this development for two-dimensional, two-directional flow, it is also valid for three-dimensional, three-directional flow, so long as the flow is irrotational (all three components of vorticity equal to zero).

We can now explain the surprising result of Example 5.5. A free vortex flow field is irrotational, so Bernoulli's equation could indeed be accurately applied between streamlines. A forced vortex flow field is rotational (in fact, every fluid particle has the same vorticity), so Bernoulli's equation gives incorrect results if applied between streamlines.*

5.4 DYNAMICS OF VISCOUS FLOW

The Euler and Bernoulli equations considered in Section 5.3 are helpful tools for flow analysis, but their usefulness is limited. From the beginning of their derivation, we excluded shear stress forces in the fluid. Although the resulting inviscid flow model is useful for certain types of flow, in other cases it is completely invalid.

* Note that we have not demonstrated that these flows are indeed irrotational and rotational, respectively. We have only stated the fact. We suggest that you calculate the vorticity for both flow fields and prove the statements for yourself.

In this section, we consider the differential equations that describe flow with shear stress and/or turbulence. Introducing these effects complicates the analysis considerably. Whereas Bernoulli's equation is a powerful and quite general integral of the inviscid flow equations, the viscous flow equations have no general integrals. The velocity–pressure relationship must be worked out individually for each flow geometry. In inviscid flow, we found the streamline coordinate system to be quite useful, but in viscous flow, there is little advantage in using streamline coordinates.* In most viscous flow analyses, simple coordinate systems—e.g., Cartesian or cylindrical—are used.

We might use either of two methods in considering the differential approach to viscous flow analysis. When faced with a specific problem, we could apply Newton's second law to a fluid particle that represents that specific situation. We could introduce simplifying assumptions (such as steady flow or neglect of one component of the pressure gradient) as the derivation is done. We illustrate this method in Chapters 7, 9, and 10. The alternative approach is to derive the general equations and then simplify them as appropriate when a specific problem is considered. The method utilized in this chapter approximates the latter approach. In order to simplify the discussion with minimal loss of generality, we concentrate on two-directional, two-dimensional, incompressible flow. Appropriate results for three-dimensional, three-directional, compressible flow are stated but not derived. For similar reasons, we limit consideration to Cartesian coordinates. Equations for cylindrical and spherical coordinates are given in Appendix H.

5.4.1 The Cauchy Equations

The Cauchy equations are the differential equations of motion expressed in terms of stress. Consider a fluid particle located at point (x, y) in a two-dimensional, two-directional flow field. The particle experiences forces from viscous stress as well as gravity and pressure. Newton's second law, for the x and y directions, respectively, gives

$$(\delta m_p)a_x = \delta F_{x,\text{pressure}} + \delta F_{x,\text{gravity}} + \delta F_{x,\text{stress}} \tag{5.56}$$

and

$$(\delta m_p)a_y = \delta F_{y,\text{pressure}} + \delta F_{y,\text{gravity}} + \delta F_{y,\text{stress}}. \tag{5.57}$$

With the exception of stress forces, all terms in these equations have already been found during considerations of inviscid flow. The already-derived terms are

$$\delta m_p = \rho(\delta x)(\delta y)(\delta z);$$

$$a_x = \frac{\partial u}{\partial t} + u\frac{\partial u}{\partial x} + v\frac{\partial u}{\partial y};$$

$$a_y = \frac{\partial v}{\partial t} + u\frac{\partial v}{\partial x} + v\frac{\partial v}{\partial y};$$

* The reason is that, in viscous flow, each equation contains at least one second-order partial derivative with respect to the perpendicular direction; i.e., the s-component equation contains at least $\partial^2 V/\partial n^2$ in addition to the several s-direction derivatives that appear in the inviscid flow version of the equation.

$$\delta F_{x,\text{pressure}} = -\frac{\partial p}{\partial x}(\delta x)(\delta y)(\delta z);$$

$$\delta F_{y,\text{pressure}} = -\frac{\partial p}{\partial y}(\delta x)(\delta y)(\delta z);$$

$$\delta F_{x,\text{gravity}} = \rho g_x(\delta x)(\delta y)(\delta z);$$

$$\delta F_{y,\text{gravity}} = \rho g_y(\delta x)(\delta y)(\delta z).$$

In order to determine the net stress force acting on a fluid particle, we consider Fig. 5.16, which shows a rectangular fluid particle at point (x, y) in a two-dimensional, two-directional flow field. The force from gravity and the fluid velocities are omitted for clarity. In general, two types of stress may act on the fluid: tangential (shear) stress and normal stress. Up to this point, we have assumed that the normal stress in fluids is completely represented by pressure. In Section 2.1, we showed that the normal stress (pressure) indeed is equal in all directions for a static fluid or a moving inviscid fluid, but we also noted that the normal stress might be different in different directions for a viscous fluid. This possible difference in normal stresses results from volumetric strain of the fluid particle. Figure 5.16 shows the total normal stress in two parts: the pressure p (now defined as the *average* of the normal stresses at a point) and the normal viscous stresses σ_x and σ_y. The values p, σ_x, and σ_y are at the point (x, y). Values at the faces of the rectangular fluid particle are approximated by the first terms of a Taylor series.

The shear stresses τ_{xy} and τ_{yx} also are shown. The first subscript on the shear stress identifies the coordinate direction perpendicular to the face of the particle on which the particular shear stress acts. The second subscript identifies the direction in which the particular shear stress acts. In three-dimensional flow, an x face would have stresses τ_{xy} and τ_{xz} acting on it.

The forces on the particle from stresses other than pressure are

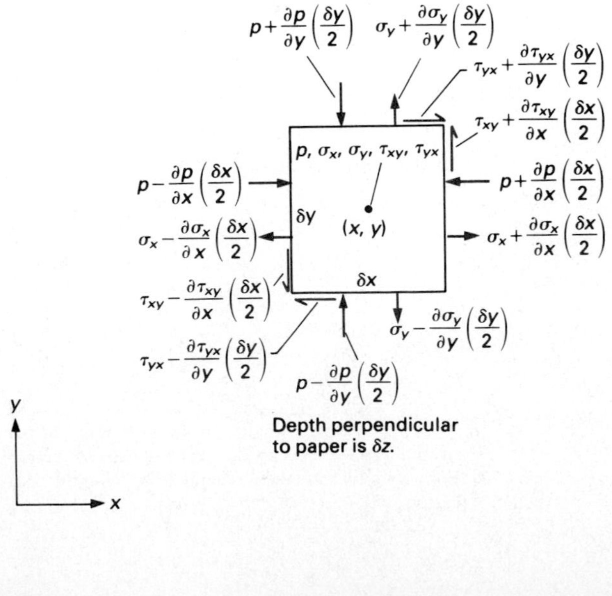

Figure 5.16 Pressures and stresses acting on a rectangular fluid particle.

$$\delta F_{x,\text{stress}} = \left[\sigma_x + \frac{\partial \sigma_x}{\partial x}\left(\frac{\delta x}{2}\right)\right](\delta y)(\delta z) - \left[\sigma_x - \frac{\partial \sigma_x}{\partial x}\left(\frac{\delta x}{2}\right)\right](\delta y)(\delta z)$$
$$+ \left[\tau_{yx} + \frac{\partial \tau_{yx}}{\partial y}\left(\frac{\delta y}{2}\right)\right](\delta x)(\delta z) - \left[\tau_{yx} - \frac{\partial \tau_{yx}}{\partial y}\left(\frac{\delta y}{2}\right)\right](\delta x)(\delta z)$$

and

$$\delta F_{y,\text{stress}} = \left[\sigma_y + \frac{\partial \sigma_y}{\partial y}\left(\frac{\delta y}{2}\right)\right](\delta x)(\delta z) - \left[\sigma_y - \frac{\partial \sigma_y}{\partial y}\left(\frac{\delta y}{2}\right)\right](\delta x)(\delta z)$$
$$+ \left[\tau_{xy} + \frac{\partial \tau_{xy}}{\partial x}\left(\frac{\delta x}{2}\right)\right](\delta y)(\delta z) - \left[\tau_{xy} - \frac{\partial \tau_{xy}}{\partial x}\left(\frac{\delta x}{2}\right)\right](\delta y)(\delta z).$$

Simplifying yields

$$\delta F_{x,\text{stress}} = \left(\frac{\partial \sigma_x}{\partial x} + \frac{\partial \tau_{yx}}{\partial y}\right)(\delta x)(\delta y)(\delta z)$$

and

$$\delta F_{y,\text{stress}} = \left(\frac{\partial \sigma_y}{\partial y} + \frac{\partial \tau_{xy}}{\partial x}\right)(\delta x)(\delta y)(\delta z).$$

Next, we substitute the particle's mass and acceleration and all the forces into Eqs. (5.56) and (5.57), divide by $(\delta x)(\delta y)(\delta z)$, and take the limit, shrinking the particle to the point (x, y). The results are

$$\rho\left(\frac{\partial u}{\partial t} + u\frac{\partial u}{\partial x} + v\frac{\partial u}{\partial y}\right) = -\frac{\partial p}{\partial x} + \rho g_x + \frac{\partial \sigma_x}{\partial x} + \frac{\partial \tau_{yx}}{\partial y} \qquad (5.58)$$

and

$$\rho\left(\frac{\partial v}{\partial t} + u\frac{\partial v}{\partial x} + v\frac{\partial v}{\partial y}\right) = -\frac{\partial p}{\partial y} + \rho g_y + \frac{\partial \sigma_y}{\partial y} + \frac{\partial \tau_{xy}}{\partial x}. \qquad (5.59)$$

These are the Cauchy equations for two-dimensional, two-directional flow. Compared to the Euler equations, each has two extra terms on the right to represent shear and normal viscous stresses.

Two-directional, two-dimensional viscous flow must satisfy the Cauchy equations and the differential continuity equation. These three equations, however, contain seven dependent variables (u, v, p, τ_{xy}, τ_{yx}, σ_x, and σ_y). If we consider the rotational equilibrium of a fluid particle,* $\tau_{xy} = \tau_{yx}$. Even so, we still have six dependent variables and only three equations. To obtain a closed set of equations, we must relate the fluid stresses to the velocity field.

5.4.2 The Navier–Stokes Equations

Stress in fluids is related to rate of strain. The simplest possible model of this relationship is a linear proportionality. A fluid that follows the simple proportionality

* This is left as an exercise for you.

relationship is called a Newtonian fluid. In Section 1.3.5, we showed that a fluid property—viscosity—is used to relate stress and strain rate. The defining equation that we used for viscosity, Eq. (1.2), was based on essentially one-dimensional, one-directional flow in which the strain rate had only one component. The viscosity concept can be extended to multidimensional, multidirectional flows, with shear stress proportional to shear strain rate and normal viscous stress proportional to normal (volumetric) strain rate.

The strain rates are related to the velocity field by Eqs. (5.14) and (5.17). By considering the *principal axes*—a set of coordinates in which only normal stresses and strains appear—we can show that the same definition of viscosity applies to both shear and normal stresses, at least for an incompressible fluid. References [1] and [2] present a complete derivation of the stress–strain rate relationship for a Newtonian fluid. The resulting stress–strain rate relationships are

$$\tau_{xy} = \tau_{yx} = 2\mu\dot{\phi} = \mu\left(\frac{\partial u}{\partial y} + \frac{\partial v}{\partial x}\right), \tag{5.60}$$

$$\sigma_x = 2\mu \frac{\partial u}{\partial x}, \tag{5.61}$$

and

$$\sigma_y = 2\mu \frac{\partial v}{\partial y}. \tag{5.62}$$

These relations must be altered slightly for a compressible fluid.

If we use Eqs. (5.60) and (5.61), the last two terms on the right-hand side of Eq. (5.58) become

$$\frac{\partial \sigma_x}{\partial x} + \frac{\partial \tau_{yx}}{\partial y} = \frac{\partial}{\partial x}\left(2\mu \frac{\partial u}{\partial x}\right) + \frac{\partial}{\partial y}\left[\mu\left(\frac{\partial u}{\partial y} + \frac{\partial v}{\partial x}\right)\right].$$

If we assume constant viscosity,

$$\frac{\partial \sigma_x}{\partial x} + \frac{\partial \tau_{yx}}{\partial y} = 2\mu \frac{\partial}{\partial x}\left(\frac{\partial u}{\partial x}\right) + \mu \frac{\partial}{\partial y}\left(\frac{\partial u}{\partial y}\right) + \mu \frac{\partial}{\partial y}\left(\frac{\partial v}{\partial x}\right).$$

Interchanging the order of differentiation of the last term, we obtain

$$\frac{\partial \sigma_x}{\partial x} + \frac{\partial \tau_{yx}}{\partial y} = \mu\left(\frac{\partial^2 u}{\partial x^2} + \frac{\partial^2 u}{\partial y^2}\right) + \mu \frac{\partial}{\partial x}\left(\frac{\partial u}{\partial x} + \frac{\partial v}{\partial y}\right).$$

The second term in parentheses is zero by the continuity equation. Substituting the remaining terms into Eq. (5.58), we get

$$\rho\left(\frac{\partial u}{\partial t} + u\frac{\partial u}{\partial x} + v\frac{\partial u}{\partial y}\right) = -\frac{\partial p}{\partial x} + \rho g_x + \mu\left(\frac{\partial^2 u}{\partial x^2} + \frac{\partial^2 u}{\partial y^2}\right).$$

Similarly, Eq. (5.59) becomes

$$\rho\left(\frac{\partial v}{\partial t} + u\frac{\partial v}{\partial x} + v\frac{\partial v}{\partial y}\right) = -\frac{\partial p}{\partial y} + \rho g_y + \mu\left(\frac{\partial^2 v}{\partial x^2} + \frac{\partial^2 v}{\partial y^2}\right).$$

If we extend the equations to three-dimensional, three-directional flow, they become

$$\rho\left(\frac{\partial u}{\partial t} + u\frac{\partial u}{\partial x} + v\frac{\partial u}{\partial y} + w\frac{\partial u}{\partial z}\right) = -\frac{\partial p}{\partial x} + \rho g_x + \mu\left(\frac{\partial^2 u}{\partial x^2} + \frac{\partial^2 u}{\partial y^2} + \frac{\partial^2 u}{\partial z^2}\right); \tag{5.63}$$

$$\rho\left(\frac{\partial v}{\partial t} + u\frac{\partial v}{\partial x} + v\frac{\partial v}{\partial y} + w\frac{\partial v}{\partial z}\right) = -\frac{\partial p}{\partial y} + \rho g_y + \mu\left(\frac{\partial^2 v}{\partial x^2} + \frac{\partial^2 v}{\partial y^2} + \frac{\partial^2 v}{\partial z^2}\right); \tag{5.64}$$

$$\rho\left(\frac{\partial w}{\partial t} + u\frac{\partial w}{\partial x} + v\frac{\partial w}{\partial y} + w\frac{\partial w}{\partial z}\right) = -\frac{\partial p}{\partial z} + \rho g_z + \mu\left(\frac{\partial^2 w}{\partial x^2} + \frac{\partial^2 w}{\partial y^2} + \frac{\partial^2 w}{\partial z^2}\right). \tag{5.65}$$

These are the *Navier–Stokes equations* for an incompressible, constant viscosity fluid. Together with the differential continuity equation, Eq. (5.24), they describe the motion of a viscous, incompressible fluid. Note that the four equations have only four unknowns (u, v, w, and p) if ρ and μ are known. The equations can be derived in or transformed to other coordinates, such as cylindrical or spherical (see Appendix H).

The Navier–Stokes equations are coupled, nonlinear, partial differential equations and are among the most well-known and difficult equations in all of engineering science. In their compressible fluid form, the only important restriction on the equations themselves is their limitation to Newtonian fluids. In principle, the equations apply to turbulent flow as well as laminar flow (a special form for turbulent flow is derived in Section 5.4.3). Except for a few special and highly simplified flows, the Navier–Stokes equations have defied solution from the time of their initial derivation (in about 1825) until the present. Today, powerful electronic computers have made accurate, although technically approximate, solutions possible for a respectable range of flows.

Many simplifications of the Navier–Stokes equations are possible. If we let $\mu = 0$, we obtain the Euler equations. For very slow motion, we might neglect the acceleration terms. If the velocity is zero everywhere, the hydrostatic pressure formula results! The equations have an easily solvable form for fully developed pipe flow. Our emphasis in this textbook is on specific but highly practical special cases rather than on general solutions of the Navier–Stokes equations.

5.4.3 The Reynolds–Navier–Stokes Equations for Turbulent Flow

We introduced the phenomenon of turbulent flow in Section 3.2.4, but we have said little about it since. Turbulent fluid motion is both complex and common. Certainly, turbulent flow obeys the fundamental laws of mass and energy conservation and Newton's laws of motion. The control volume approach of Chapter 4 and the continuity and Navier–Stokes equations of this chapter are applicable to turbulent flow, but significant practical problems are involved. In the control volume approach, the problems involved are the need for an accurate model of shear stress forces when required for the momentum equation and the need for accurate models for the mechanical energy loss. These needs are addressed in large part in Chapters 7, 8, and 10.

The Navier–Stokes equations as presented in Eqs. (5.63)–(5.65) are limited in a practical sense to laminar flow. In laminar flow, even an unsteady one, the fluid moves along smooth streamlines, and the velocity at any point is either fixed in time or varies in a regular manner. In turbulent flow, the fluid particles are violently mixed and the fluid velocity at a point varies randomly with time, as shown in Fig. 5.17.

Solution of the unsteady Navier–Stokes equations in fine enough detail to capture the random fluctuations of turbulence is practically impossible. Instead, engineers work with time averages in turbulent flow. The time average of the (single point) velocity trace given in Fig. 5.17 is

$$\langle V \rangle \equiv \frac{1}{T} \int_0^T V\,dt \tag{5.66}$$

Note that $\langle V \rangle$ is a *time* average of the instantaneous velocity at a point and is different from $\bar{V}$, the *area* average. The $\langle\ \rangle$ notation is cumbersome and is used only when distinguishing between instantaneous and time-average velocities (or other properties) for turbulent flow is necessary. The instantaneous velocity is related to the time average according to

$$V = \langle V \rangle + V', \tag{5.67}$$

where V' is the fluctuating component of the velocity. By definition,

$$\langle V' \rangle = 0. \tag{5.68}$$

This relation also holds for the velocity components in two- or three-directional

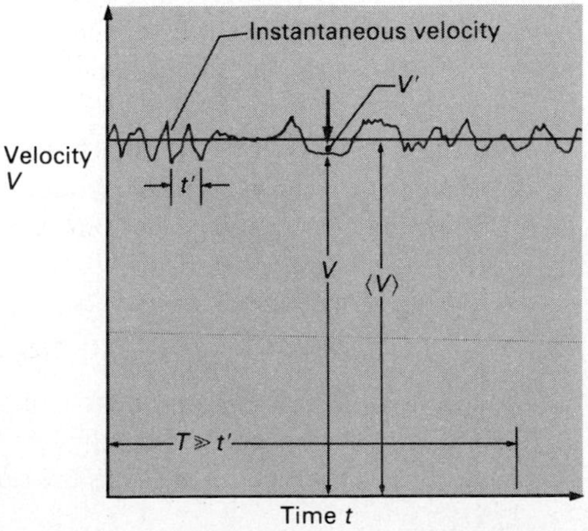

Figure 5.17 Illustration of the definition of time-average velocity in turbulent flow. Velocity V is measured at a single point.

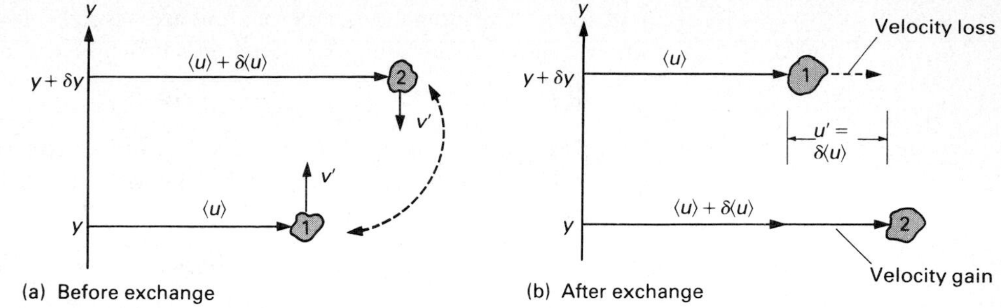

Figure 5.18 Two lumps of fluid with different mean velocities exchange places in a turbulent flow.

flow:

$$u = \langle u \rangle + u'; \tag{5.69a}$$

$$v = \langle v \rangle + v'; \tag{5.69b}$$

$$w = \langle w \rangle + w'. \tag{5.69c}$$

Most engineering considerations of turbulent flow are concerned only with evaluation of the time-average velocity. The time-average velocity characterizes the net flow of mass, momentum, and energy. When we refer to the fluid velocity in turbulent flow, we usually mean the time-average velocity and simply write V for local time-average velocity and $\bar{V}$ for area-and-time-average velocity.

If we limit consideration to the time-average flow, we must account for additional *apparent stresses** in the fluid caused by turbulent motion. We can demonstrate the existence of these apparent stresses by simple physical arguments. In turbulent flow, macroscopic "lumps" of fluid are thrown about by the turbulent motion. Figure 5.18 illustrates two typical fluid lumps in a turbulent flow. The time-average flow is one-directional ($\langle u \rangle$) and one-dimensional (y). The turbulent fluctuation velocities are two-directional (u', v'). The time-average velocities of the two lumps are $\langle u \rangle$ and $\langle u \rangle + \delta\langle u \rangle$.

Suppose that the two lumps exchange places because of turbulent mixing. The slower lump arrives at its new location with velocity $\langle u \rangle$, and the faster lump arrives at its new location with velocity $\langle u \rangle + \delta\langle u \rangle$. Now, neither lump has the velocity appropriate for its location. At $y + \delta y$, the velocity perturbation is $u' \approx -\delta\langle u \rangle$, and at y, the velocity perturbation is $u' \approx +\delta\langle u \rangle$. The time-average velocities at y and $y + \delta y$ are still $\langle u \rangle$ and $\langle u \rangle + \delta\langle u \rangle$. The interchange of the two lumps has resulted in an increase of momentum at location y and a decrease in momentum at $y + \delta y$. As this type of interchange occurs continuously in turbulent flow, there is a net rate of momentum transfer between y and $y + \delta y$ because of the velocity fluctuations. From the time-average point of view, this momentum transfer is equivalent to a force or—on a per unit area basis—a stress.† The momentum represented by u' is transported between y and $y + \delta y$ by the velocity

* Apparent stresses are sometimes called *turbulent stresses* or *Reynolds stresses*.

† The strong parallel between this discussion and that of viscosity in a gas presented in Section 1.3.5 is no accident. The salient difference is that viscosity is caused by molecular momentum transport and is a fluid property, whereas turbulent momentum transport is a *flow* characteristic rather than a *fluid* characteristic.

perturbation v'. The momentum flux (per unit area) is $\rho u'v'$. If we treat this momentum flux as a stress, according to d'Alembert's principle, its instantaneous value is

$$\tau_{\text{app}} = -\rho u'v',$$

where τ_{app} is the instantaneous apparent stress resulting from turbulent momentum transport.* The time average of τ_{app} is the apparent stress that we must include in the mechanics of the time-average flow:

$$\langle \tau_{\text{app}} \rangle = -\rho \langle u'v' \rangle. \tag{5.70}$$

We can demonstrate the concept of apparent stress in a more formal way from the instantaneous Navier–Stokes equations. We begin with Eqs. (5.63)–(5.65) and substitute the sum of the time-averaged and fluctuating velocity components, as given by Eqs. (5.69a–c), for the instantaneous velocity components. We then time average the Navier–Stokes equations, using the procedure implied in Eq. (5.66). We omit the details of this lengthy manipulation, except to note that the principle embodied in Eq. (5.68) greatly simplifies the result. After the dust settles, we are left with the following equations for the time-averaged velocity and pressure:†

$$\begin{aligned}\rho\left(\frac{\partial\langle u\rangle}{\partial t} + \langle u\rangle\frac{\partial\langle u\rangle}{\partial x} + \langle v\rangle\frac{\partial\langle u\rangle}{\partial y} + \langle w\rangle\frac{\partial\langle u\rangle}{\partial z}\right) &= -\frac{\partial\langle p\rangle}{\partial x} + \rho g_x + \frac{\partial}{\partial x}\left(\mu\frac{\partial\langle u\rangle}{\partial x} - \rho\langle u'^2\rangle\right)\\ &+\frac{\partial}{\partial y}\left(\mu\frac{\partial\langle u\rangle}{\partial y} - \rho\langle u'v'\rangle\right) + \frac{\partial}{\partial z}\left(\mu\frac{\partial\langle u\rangle}{\partial z} - \rho\langle u'w'\rangle\right),\end{aligned} \tag{5.71}$$

$$\begin{aligned}\rho\left(\frac{\partial\langle v\rangle}{\partial t} + \langle u\rangle\frac{\partial\langle v\rangle}{\partial x} + \langle v\rangle\frac{\partial\langle v\rangle}{\partial y} + \langle w\rangle\frac{\partial\langle v\rangle}{\partial z}\right) &= -\frac{\partial\langle p\rangle}{\partial y} + \rho g_y + \frac{\partial}{\partial x}\left(\mu\frac{\partial\langle v\rangle}{\partial x} - \rho\langle u'v'\rangle\right)\\ &+\frac{\partial}{\partial y}\left(\mu\frac{\partial\langle v\rangle}{\partial y} - \rho\langle v'^2\rangle\right) + \frac{\partial}{\partial z}\left(\mu\frac{\partial\langle v\rangle}{\partial z} - \rho\langle v'w'\rangle\right),\end{aligned} \tag{5.72}$$

and

$$\begin{aligned}\rho\left(\frac{\partial\langle w\rangle}{\partial t} + \langle u\rangle\frac{\partial\langle w\rangle}{\partial x} + \langle v\rangle\frac{\partial\langle w\rangle}{\partial y} + \langle w\rangle\frac{\partial\langle w\rangle}{\partial z}\right) &= -\frac{\partial\langle p\rangle}{\partial z} + \rho g_z + \frac{\partial}{\partial x}\left(\mu\frac{\partial\langle w\rangle}{\partial x} - \rho\langle u'w'\rangle\right)\\ &+\frac{\partial}{\partial y}\left(\mu\frac{\partial\langle w\rangle}{\partial y} - \rho\langle v'w'\rangle\right) + \frac{\partial}{\partial z}\left(\mu\frac{\partial\langle w\rangle}{\partial z} - \rho\langle w'^2\rangle\right).\end{aligned} \tag{5.73}$$

These very formidable equations are called the *Reynolds–Navier–Stokes equations*. Note that the time averages of the cross-products and squares of the

---

* Note that positive values of v' (upward movement) generate negative values of u'.

† Note that $\langle u'^2\rangle = \langle u'u'\rangle \neq \langle u'\rangle\langle u'\rangle$ and that $\langle u'v'\rangle \neq \langle u'\rangle\langle v'\rangle$.

fluctuating velocities have been grouped with the shear stress and normal viscous stress terms, highlighting their interpretation as apparent stresses from turbulence.

As difficult as these equations may seem, the true situation is even worse than a quick glance reveals. These equations replace Eqs. (5.63)–(5.65) when we consider turbulent flow, so we still have only four equations but have just added six more unknown quantities (the six apparent stresses). Practically speaking, the only way out of this dilemma is to introduce ad hoc, semiempirical models of the turbulent stresses, relating them to other flow parameters, such as the velocity gradients.

Clearly, much experimental and analytical work is required to develop a workable differential equation model of turbulent flow. Turbulence modeling is an active area of research, and new developments are always appearing. At present, however, most engineering calculations of turbulent flow are made with the control volume method with correlations of experimental data for shear stress and mechanical energy loss.

EXAMPLE 5.6 Illustrates a Simple Model for Turbulent Stress

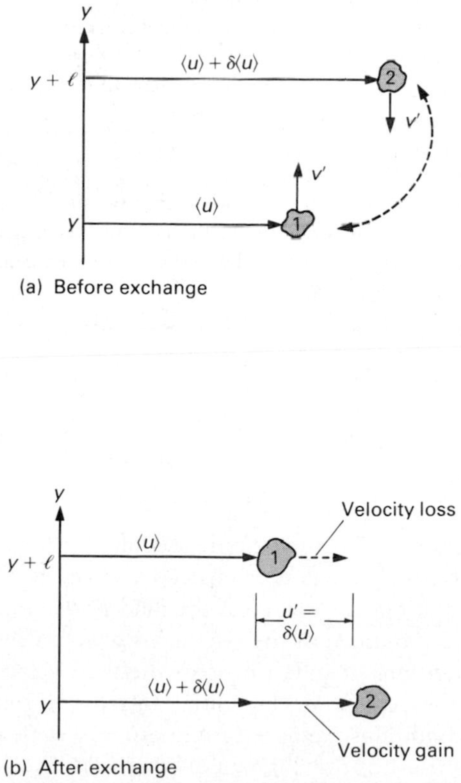

Figure E5.6 Two lumps of fluid with different mean velocities exchange places in a turbulent flow.

Let's consider the momentum interchange model of turbulent flow illustrated in Fig. 5.18. Suppose that the average distance separating two fluid lumps that exchange places is ℓ. Develop an expression for the apparent shear stress caused by turbulent motion in terms of ℓ and other relevant parameters of the flow.

SOLUTION

Given

Momentum interchange model of Fig. E5.6

Distance traveled by fluid lumps is ℓ

Find

A model for turbulent shear stress

Solution

The apparent stress due to turbulence is given by Eq. (5.70):

$$\langle \tau_{app} \rangle = -\rho \langle u'v' \rangle.$$

We can estimate the x-velocity perturbation for the momentum interchange model as

$$u' \approx \delta\langle u \rangle \approx \frac{\partial \langle u \rangle}{\partial y}\,\delta y,$$

but

$$\delta y = \ell,$$

so

$$u' \approx \ell\,\frac{\partial \langle u \rangle}{\partial y}.$$

We assume that the y-velocity perturbation is approximately equal in magnitude but opposite in direction to the x-velocity perturbation, so

$$v' \approx -u' \approx -\ell \frac{\partial \langle u \rangle}{\partial y}.$$

The apparent stress is

$$\langle \tau_{\text{app}} \rangle = -\rho \langle u'v' \rangle = -\rho(-\ell^2)\left(\frac{\partial \langle u \rangle}{\partial y}\right)^2 = \rho \ell^2 \left(\frac{\partial \langle u \rangle}{\partial y}\right)^2.$$

We can assign the correct algebraic sign to the stress by noting that a positive (mean flow) velocity gradient yields a positive stress. We therefore insert a set of absolute value signs and let the velocity gradient determine the correct algebraic sign. The final model is

$$\langle \tau_{\text{app}} \rangle = \rho \ell^2 \left| \frac{\partial \langle u \rangle}{\partial y} \right| \left(\frac{\partial \langle u \rangle}{\partial y} \right).$$ **ANSWER**

Discussion

The turbulent stress model allows us to represent the stress in terms of the time mean flow field, if a relationship for ℓ can be prescribed. The term ℓ is called the *mixing length,* and the model derived is called the mixing length model. In practice, a relationship for ℓ is obtained by reasonable assumption and adjusted by comparison of results with experimental data. For turbulent flow near a solid surface such as a pipe wall or flat plate, the relationship $\ell = \kappa y$, with κ an empirical constant, is often suggested.

The mixing length model is by no means the only possible model for turbulent stresses; in fact, it is over 60 years old and new models are being developed continuously. It is, however, a venerable and useful, although simple, idea.

5.5 BOUNDARY AND INITIAL CONDITIONS FOR FLOW PROBLEMS

If we intend to solve differential equations to predict the details of a particular flow, we must specify boundary conditions. If the flow is unsteady, we must also specify initial conditions. Boundary conditions are values of the fluid velocity and properties at the boundaries (walls, inlets, outlets) of the region for which we are trying to calculate the flow. Initial conditions are the values of the fluid velocity and properties at all points within the region at the initial instant of time. Mathematically, boundary and initial conditions are necessary in order to evaluate any constants of integration in the solutions of the differential equations.

In this discussion, we limit our consideration to incompressible fluids with constant viscosity. We need boundary and initial conditions for pressure and velocity. For a turbulent flow, boundary and initial conditions for the time-average velocity and pressure are required.

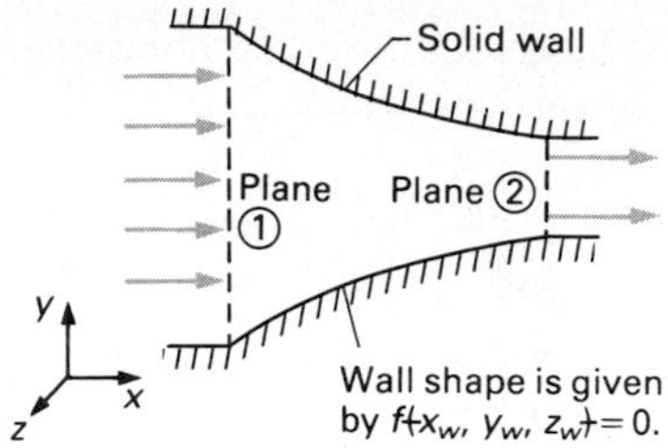

Figure 5.19 Flow through a nozzle.

Consider the problem of predicting the velocity and pressure field for flow in the nozzle shown in Fig. 5.19. Suppose that we are interested only in the flow in the region between the inlet and the outlet, planes 1 and 2. The flow is governed by the continuity and Navier–Stokes equations. If the flow is unsteady, the initial conditions are

$$\vec{V}(x, y, z, t = 0) = \vec{V}_i(x, y, z)$$

and

$$p(x, y, z, t = 0) = p_i(x, y, z),$$

where $\vec{V}_i$ and p_i are known functions. We customarily set $t = 0$ at the beginning of the calculation. Initial conditions are usually relatively easy to determine. If we were interested in evaluating the start-up of flow from rest, we would probably make $\vec{V}_i = 0$ everywhere; p_i might be uniform or there might be a hydrostatic pressure distribution.

Next, we consider the boundary conditions. The boundaries of the flow field are the inlet (plane 1), the outlet (plane 2), and the side walls of the nozzle. The mathematical character of the governing equations is such that we must specify values of the velocity at all points on the boundary. At the solid walls, there can be no component of velocity perpendicular to the wall:

$$V_{\text{perp to wall}} = V_n = 0.$$

Also, if the fluid is viscous, there is no slip at the wall; that is,

$$V_{\text{tan to wall}} = V_t = 0.$$

Together these conditions give

$$\vec{V}(x_w, y_w, z_w, t) = 0,$$

where x_w, y_w, and z_w are the coordinates of the nozzle wall and satisfy some relation that describes the nozzle geometry; that is,

$$f(x_w, y_w, z_w) = 0.$$

At the inlet and outlet planes, we must specify the pressure as well as the velocity:

$$\vec{V}_{\text{at plane 1}} = \vec{V}_1(x_1, y_1, z_1, t) \qquad p_{\text{at plane 1}} = p_1(x_1, y_1, z_1, t)$$

and

$$\vec{V}_{\text{at plane 2}} = \vec{V}_2(x_2, y_2, z_2, t) \qquad p_{\text{at plane 2}} = p_2(x_2, y_2, z_2, t).$$

These conditions are not specific to our particular nozzle; in fact, we can generalize them to all flows.

- At a solid boundary, the normal component of velocity relative to the boundary is zero. (5.74)
- At a solid boundary, the tangential component of velocity relative to the boundary is zero. (5.75)
- At an inlet, the fluid pressure and velocity distributions must be known as functions of time. (5.76)
- At an outlet, the fluid pressure and velocity distributions must be known as functions of time. (5.77)

Notice that we inserted the phrase *relative to the boundary* in the first two conditions. This phrase accounts for moving solid boundaries, as well as fixed solid boundaries.

Some special situations must be mentioned. First, suppose that we are interested in predicting the flow around an airplane cruising at constant speed V_∞ through the atmosphere, as shown in Fig. 5.20(a). We can consider this as a steady flow, if we attach a coordinate system to the airplane. In this system, the airplane is at rest and the air approaches the airplane at constant speed, V_∞, as shown in Fig. 5.20(b). For this flow, the solid surfaces of the airplane are obviously one boundary, but where are the "inlet" and "outlet" boundaries? These boundaries are far upstream and far downstream. Mathematically, we put them at $\pm\infty$. There also are boundaries far above and far below the airplane. Mathematically, we handle these conditions by writing

$$\lim_{x,y,z\to\infty} \vec{V}(x, y, z) = V_\infty \hat{i} \quad \text{and} \quad \lim_{x,y,z\to\infty} p(x, y, z) = p_\infty.$$

Next, suppose that we decide to analyze a particular flow using the inviscid fluid model. The governing equations are now the Euler equations rather than the Navier–Stokes equations. This approach immediately leads to a mathematical problem. The Euler equations are of lower order than the Navier–Stokes equations, because the Euler equations do not contain any second derivatives. Thus enforcing all the boundary conditions, Eqs. (5.74)–(5.77), is impossible. We must drop one of these conditions. The only sensible condition to drop is Eq. (5.75), the no-slip condition, because the non-slip of fluid at a solid boundary is associated with viscosity. For an inviscid fluid, only the conditions expressed by Eqs. (5.74), (5.76), and (5.77) are applied. The tangential (slip) velocity at a solid surface is calculated as part of the solution and is not zero.

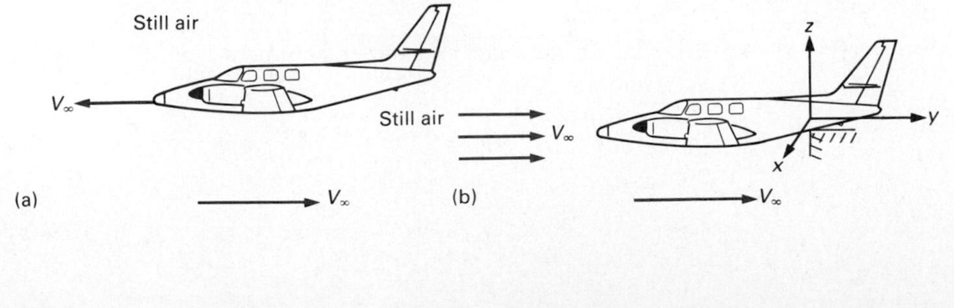

Figure 5.20 Flow around an airplane: (a) actual case: airplane moves at constant speed V_∞ through still air; (b) flow from coordinate system fixed to airplane: air approaches airplane at constant speed V_∞.

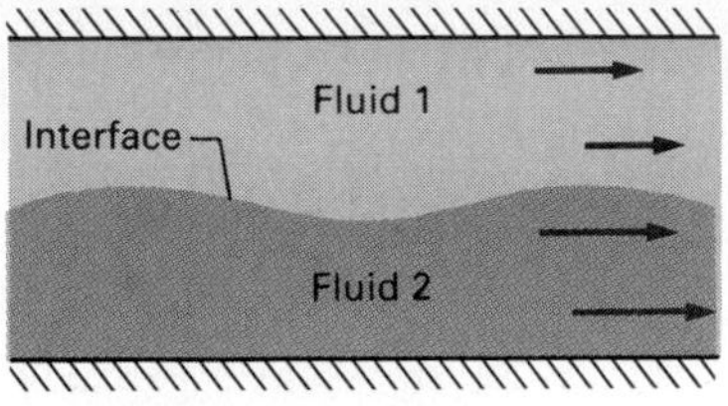

Figure 5.21 Flow of two imiscible fluids that form an interface between them.

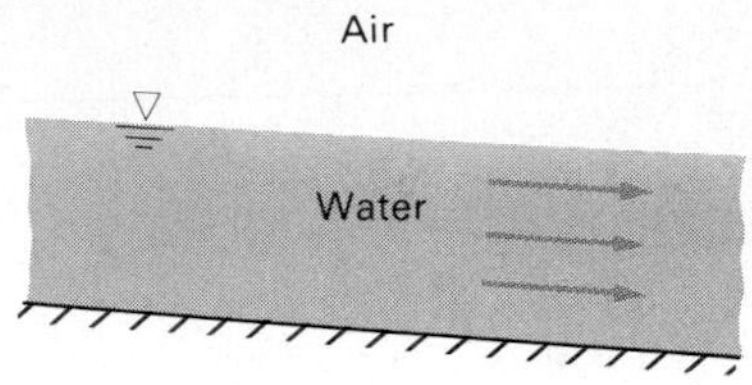

Figure 5.22 Flow in an open channel with an air–water interface.

If there is an internal boundary in the flow field, we must pay special attention to boundary conditions. Consider the flow of two immiscible fluids that form an interface between them (Fig. 5.21). This situation may represent a layer of oil over water or a free surface with air over water. The flow fields of the two fluids must be patched together at the interface by means of boundary conditions, which must express the conditions of continuity of mass and balance of forces at the interface. The necessary conditions are:

- The velocities of each fluid normal to the interface at the interface must be equal. (5.78)
- The shear stresses in each fluid must be equal at the interface. (5.79)
- The pressures in each fluid at the interface must be equal except for surface tension effects. (5.80)

These conditions can be very complicated, so we consider only two special cases here. The first is an air–water interface in an open channel (Fig. 5.22). In this case, we usually simplify the conditions to

$$\tau = 0 \tag{5.79a}$$

and

$$p_{\text{liquid}} = p_{\text{air}} \tag{5.80a}$$

at the interface.

The second special case involves the issue of a jet into a large quiescent body of fluid (see Fig. 5.23). This situation may represent water issuing from a hose into still air, air escaping from a nozzle into the atmosphere, or water issuing from a pipe submerged in a lake or ocean. Let's consider the pressure of the fluid in the jet immediately at the plane of issue. The condition in Eq. (5.80) states that the pressure just inside the jet must equal the pressure just outside the jet; that is, the pressure inside the jet must equal the pressure of the quiescent body of fluid, except for surface tension effects. There is an exception to this condition. If the fluid in the jet is compressible and moving at a speed greater than the speed of sound, it is possible for the jet pressure to adjust to the exhaust pressure by means of compression or expansion waves.

In most cases encountered in engineering practice, surface tension effects are negligible, and the fluid moves at a speed less than the speed of sound, so we

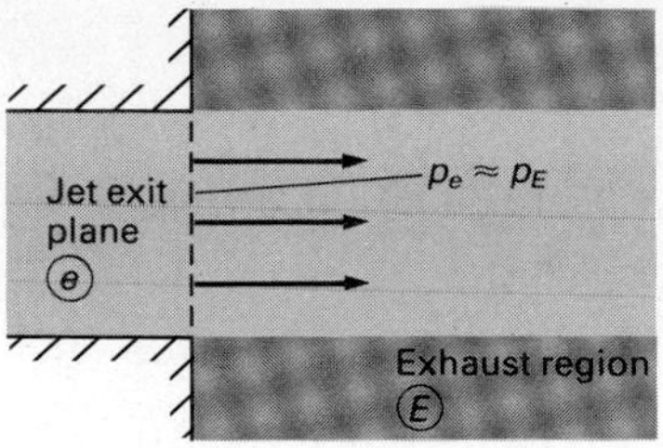

Figure 5.23 Jet of fluid issuing into a large quiescent "exhaust region."

usually are justified in simply setting the pressure just inside the jet equal to the pressure of the exhaust region. If the streamlines in the exiting jet are straight, the pressure is uniform (or hydrostatic) across the entire jet and equal to the exhaust region pressure. Note that this condition applies irrespective of whether the jet and exhaust region fluids are similar or dissimilar (i.e., air–air or water–air). We formally write this jet exit condition as

$$p_e = p_E. \tag{5.80b}$$

When we introduced the need for boundary and initial conditions, our justification was primarily mathematical; however, if we consider the facts expressed by the boundary conditions, we see that these conditions are necessary to make the mathematical solutions to the differential equations representative of the true physical behavior of the fluid. Boundary conditions perform another very important task for us. If we consider, for example, the Navier–Stokes equations, we note that they are exactly the same for all flow geometries. Only the boundary conditions distinguish one flow from another. Specifying proper boundary conditions for a flow problem is as important as using the right governing equations.

EXAMPLE 5.7 Illustrates Boundary Conditions

Write the necessary boundary conditions for the steady, two-directional, two-dimensional flow of a constant-density, Newtonian fluid flowing in the entrance region between the two wide, parallel, flat plates shown in Fig. E5.7.

SOLUTION

Given

Two wide, parallel, flat plates

Fig. E5.7

Constant-density, Newtonian fluid

Find

Necessary boundary conditions to find $u\{x, y\}$ and $v\{x, y\}$.

Solution

Boundary conditions are necessary for both u and v at the inlet, 1, and the outlet, 2, at the upper solid boundary, and at the lower solid

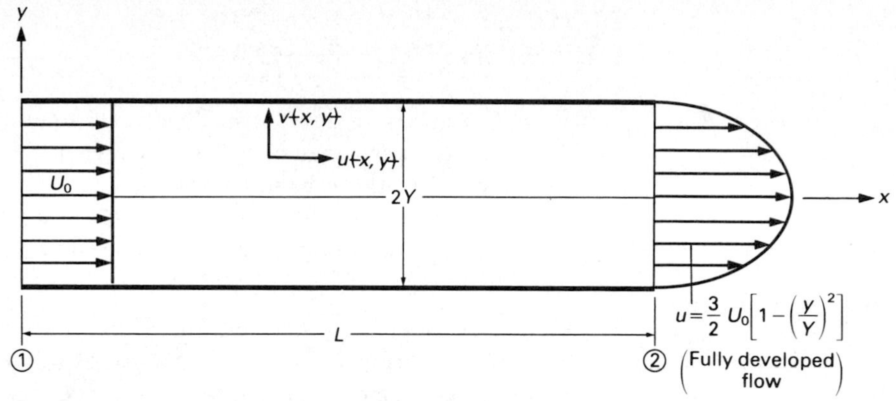

Figure E5.7 Entrance region between two wide, parallel, flat plates.

boundary. See Eqs. (5.74)–(5.77). At the inlet,

$$u(0, y) = U_0;$$
$$v(0, y) = 0;$$
$$p(0, y) = p_0 - \gamma y, \qquad \textbf{ANSWER}$$

where p_0 is the static pressure at $y = 0$.

At the outlet,

$$u(L, y) = \frac{3}{2} U_0 \left[1 - \left(\frac{y}{Y}\right)^2\right];$$
$$v(L, y) = 0;$$
$$p(L, y) = p_L - \gamma y, \qquad \textbf{ANSWER}$$

where p_L is the static pressure at $y = 0$.

At the upper solid boundary,

$$u(x, Y) = 0;$$
$$v(x, Y) = 0. \qquad \textbf{ANSWER}$$

At the lower solid boundary,

$$u(x, -Y) = 0;$$
$$v(x, -Y) = 0. \qquad \textbf{ANSWER}$$

Discussion

If the fluid were inviscid, the boundary conditions at the upper and lower boundaries would be $v(x, Y) = 0$ and $v(x, -Y) = 0$. There would be no boundary conditions on u at the upper and lower boundaries.

The requirement that the pressure p_L be known at $x = L$ is rather stringent. This requirement is often replaced by the fully developed flow condition

$$\frac{\partial p}{\partial x} = \text{Constant} \qquad (\text{for } x \geq L).$$

5.6 MODELS AND METHODS FOR USING THE DIFFERENTIAL APPROACH

Differential equations for flow are fine, but unless solutions can be found, the equations are of little use to engineers. Also, mathematical solutions must accurately represent the real world to be of any use.

In this section, we briefly examine some flows and flow models that permit us to solve the differential equations and obtain useful and accurate information. We examine three cases:

- Fully developed internal flow, in which the physical nature of the flow dictates that many terms in the Navier–Stokes equations vanish and the resulting exact differential equations can be readily solved.
- The potential flow and boundary layer model, in which the physical nature of the flow allows us to divide it into two distinct regions. In each region, certain simplifying assumptions can be used to reduce the Navier–Stokes equations to a more easily solved approximate form.
- Computational fluid dynamics, in which the partial differential equations are approximated by dividing the flow field into a large number of small but finite regions. This approach results in a large number of coupled algebraic equations that require computer solution.

In order to keep the discussion as simple as possible, while still capturing the main ideas, we consider only incompressible flow and concentrate on inviscid and laminar flow. Similar ideas may be used for turbulent flow, but calculations are more difficult and more approximate, owing to the need for models of turbulence.

5.6.1 Steady, Fully Developed Internal Flow

We first consider a viscous fluid flowing inside a long conduit (pipe or channel), as illustrated in Fig. 5.24. The conduit is straight and of constant cross-sectional area, and the flow is steady. Far downstream from the location where the fluid enters the conduit, the velocity profile appears as shown in Fig. 5.24, with zero velocity at the wall and maximum velocity near the centerline. We define *fully developed flow* as the condition when the velocity profile does not change as the flow proceeds downstream.* Mathematically, this definition translates to:

$$\frac{\partial u}{\partial x} = 0.$$

Definition of fully developed flow

Note that the velocity varies in the cross-stream direction.

Applying the continuity equation, we get

$$\frac{\partial u}{\partial x} + \frac{\partial v}{\partial y} = 0 + \frac{\partial v}{\partial y} = \frac{\partial v}{\partial y} = 0$$

* There are other, subtly different, definitions for fully developed flow, but this definition is sufficient for our purposes.

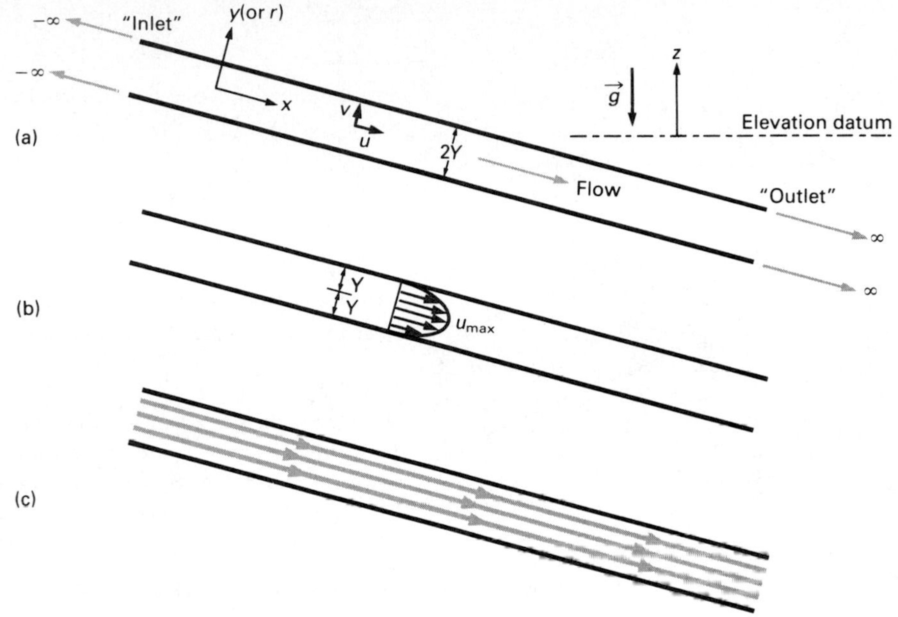

Figure 5.24 Viscous flow between two large (infinite) parallel plates: (a) basic geometry and coordinate system; (b) expected velocity profile (u velocity); (c) typical streamlines.

for Cartesian coordinates and

$$\frac{\partial u}{\partial x} + \frac{1}{r}\frac{\partial (rv)}{\partial r} = 0 + \frac{1}{r}\frac{\partial (rv)}{\partial r} = \frac{1}{r}\frac{\partial (rv)}{\partial r} = 0$$

for cylindrical coordinates (see Appendix H). Because the cross-stream velocity, v, is zero at the conduit wall, it is zero everywhere. The velocity vector is parallel to the conduit walls and centerline so the streamlines are all straight and parallel.

Because v and $\partial u/\partial x$ vanish everywhere, all of the acceleration terms and all but one of the viscous stress terms vanish from the Navier–Stokes equations for fully developed flow. This condition generally reduces the equations to an easily solved form.

EXAMPLE 5.8 **Illustrates Simplification and Solution of the Navier–Stokes Equations for Steady, Fully Developed Flow**

Steady, fully developed flow occurs between two very large parallel flat plates, as shown in Fig. E5.8. The plates are inclined at angle θ with respect to the horizontal and are separated by distance $2Y$. The extent of the plates in the x and ξ directions is very large compared to Y. Begin with the general Navier–Stokes equations, Eqs. (5.63)–(5.65), and an appropriate set of boundary conditions. Simplify the equations and solve them to obtain an expression for the velocity profile. Finally, obtain a relationship between the average velocity and the pressure and elevation.

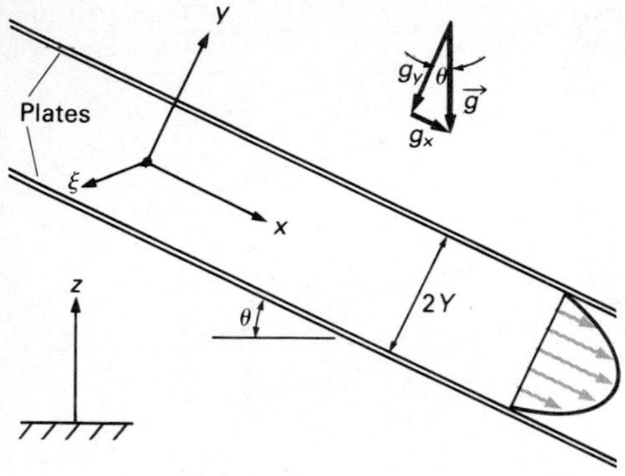

Figure E5.8 Flow between parallel plates.

SOLUTION

Given

Steady, fully developed flow between parallel plates
Navier–Stokes equations

Find

Boundary conditions
Simplified equations
Velocity profile
Average velocity–pressure/elevation relationship

Solution

We choose the Cartesian coordinate system shown in Fig. E5.8. Because it is customary to use z for elevation, we use ξ for the coordinate perpendicular to the xy plane. In this coordinate system, the Navier–Stokes equations are

$$\rho\left(\frac{\partial u}{\partial t} + u\frac{\partial u}{\partial x} + v\frac{\partial u}{\partial y} + w\frac{\partial u}{\partial \xi}\right) = -\frac{\partial p}{\partial x} + \rho g_x + \mu\left(\frac{\partial^2 u}{\partial x^2} + \frac{\partial^2 u}{\partial y^2} + \frac{\partial^2 u}{\partial \xi^2}\right);$$

$$\rho\left(\frac{\partial v}{\partial t} + u\frac{\partial v}{\partial x} + v\frac{\partial v}{\partial y} + w\frac{\partial v}{\partial \xi}\right) = -\frac{\partial p}{\partial y} + \rho g_y + \mu\left(\frac{\partial^2 v}{\partial x^2} + \frac{\partial^2 v}{\partial y^2} + \frac{\partial^2 v}{\partial \xi^2}\right);$$

$$\rho\left(\frac{\partial w}{\partial t} + u\frac{\partial w}{\partial x} + v\frac{\partial w}{\partial y} + w\frac{\partial w}{\partial \xi}\right) = -\frac{\partial p}{\partial \xi} + \rho g_\xi + \mu\left(\frac{\partial^2 w}{\partial x^2} + \frac{\partial^2 w}{\partial y^2} + \frac{\partial^2 w}{\partial \xi^2}\right).$$

The boundary conditions are the no-slip and impermeability conditions at the plate surfaces:

$$u = w = 0 \quad \text{at} \quad y = \pm Y; \qquad v = 0 \quad \text{at} \quad y = \pm Y.$$ **ANSWER**

For fully developed flow, we do not need boundary conditions at the channel inlet and outlet. (The mathematical model is based on the assumption that there is no inlet or outlet!)

We now use the problem specifications and boundary conditions to simplify the equations. For steady flow,

$$\frac{\partial u}{\partial t} = \frac{\partial v}{\partial t} = \frac{\partial w}{\partial t} = 0.$$

Next, we assume that the x and ξ directions have been selected so that the flow is aligned with the x direction and that there is no flow in the ξ direction. Then we have

$$w \equiv 0$$ **ANSWER**

and

$$\frac{\partial(\text{any quantity})}{\partial \xi} = 0.$$

We now consider the differential continuity equation:

$$\frac{\partial u}{\partial x} + \frac{\partial v}{\partial y} + \frac{\partial w}{\partial \xi} = 0.$$

With fully developed flow,

$$\frac{\partial u}{\partial x} \equiv 0,$$

and the continuity equation becomes

$$0 + \frac{\partial v}{\partial y} + 0 = 0.$$

From the boundary conditions, v is zero at the surface of the plates and so it is zero everywhere:

$$v \equiv 0. \qquad \textbf{ANSWER}$$

Substituting all our velocity results up to this point into the Navier–Stokes equations gives

$$\rho\left[0 + u(0) + 0\left(\frac{\partial u}{\partial y}\right) + (0)(0)\right] = -\frac{\partial p}{\partial x} + \rho g_x + \mu\left[0 + \frac{\partial^2 u}{\partial y^2} + 0\right],$$

$$\rho[0 + u(0) + (0)(0) + (0)(0)] = -\frac{\partial p}{\partial y} + \rho g_y + \mu[(0) + (0) + (0)],$$

and

$$\rho[0 + u(0) + (0)(0) + (0)(0)] = -\frac{\partial p}{\partial \xi} + \rho g_\xi + \mu[(0) + (0) + (0)].$$

So

$$0 = -\frac{\partial p}{\partial x} + \rho g_x + \mu \frac{d^2 u}{dy^2};$$

$$0 = -\frac{\partial p}{\partial y} + \rho g_y;$$

$$0 = -\frac{\partial p}{\partial \xi} + \rho g_\xi.$$

Note that $\partial^2 u/\partial y^2$ has been changed to $d^2 u/dy^2$, because u depends only on y.

The channel orientation in Fig. E5.8 shows that

$$g_\xi = 0, \quad g_y = -g \cos\theta, \quad \text{and} \quad g_x = g \sin\theta.$$

Referring to Fig. E5.8 gives

$$\cos\theta = \frac{\partial z}{\partial y}, \quad \sin\theta = -\frac{\partial z}{\partial x}, \quad \text{and} \quad \frac{\partial z}{\partial \xi} = 0.$$

Then

$$g_y = -g\left(\frac{\partial z}{\partial y}\right) \quad \text{and} \quad g_x = -g\left(\frac{\partial z}{\partial x}\right).$$

Introducing the specific weight ($\gamma = \rho g$),

$$0 = -\frac{\partial p}{\partial y} - \gamma \frac{\partial z}{\partial y} = -\frac{\partial}{\partial y}(p + \gamma z)$$

and

$$0 = -\frac{\partial p}{\partial \xi} - \gamma \frac{\partial z}{\partial \xi} = -\frac{\partial}{\partial \xi}(p + \gamma z),$$

so

$$-\frac{\partial p}{\partial x} - \gamma \frac{\partial z}{\partial x} = -\frac{\partial (p + \gamma z)}{\partial x} = -\frac{d}{dx}(p + \gamma z).$$

Finally, all that remains of the three Navier–Stokes equations is

$$0 = -\frac{d}{dx}(p + \gamma z) + \mu\left(\frac{d^2 u}{dy^2}\right).$$ **ANSWER**

To solve this equation, we rewrite it as

$$\mu \frac{d^2 u}{dy^2} = \frac{d}{dx}(p + \gamma z).$$

Each term of this equation must be equal to a constant, because the left term is a function only of y and the right term is a function only of x. If the equality is to be true for all x's and y's, the terms must be constant. To remind us of this, we let

$$\tilde{p} = \frac{d}{dx}(p + \gamma z).$$

To obtain the velocity distribution, we must solve

$$\frac{d^2 u}{dy^2} = \frac{\tilde{p}}{\mu}.$$

Integrating twice, we obtain

$$u = \frac{\tilde{p}}{2\mu}(y^2) + C_1 y + C_2.$$

We evaluate the constants of integration from the boundary conditions:

$$u = 0 \qquad \text{at } y = \pm Y,$$

giving

$$C_1 = 0, \quad \text{and} \quad C_2 = -\frac{\tilde{p}Y^2}{2\mu}.$$

The velocity distribution is given by

$$u = \left(-\frac{\tilde{p}Y^2}{2\mu}\right)\left[1 - \left(\frac{y}{Y}\right)^2\right].$$ **ANSWER**

The maximum velocity is

$$u_{\max} = -\frac{\tilde{p}Y^2}{2\mu} \qquad (\text{at } y = 0).$$

The average velocity, $\bar{V}$, is obtained by integration:

$$\bar{V} = \frac{1}{2Y}\int_{-Y}^{Y} u\,dy = -\frac{2}{3}\frac{\tilde{p}Y^2}{2\mu} = \frac{2}{3}u_{\max}.$$

We reintroduce the definition of $\tilde{p}$ and get

$$u = \left(\frac{-Y^2}{2\mu}\right)\left(\frac{d(p+\gamma z)}{dx}\right)\left[1-\left(\frac{y}{Y}\right)^2\right], \tag{5.81}$$

or

$$u_{\max} = \left(\frac{-Y^2}{2\mu}\right)\frac{d(p+\gamma z)}{dx} \tag{5.82}$$

and

$$\bar{V} = \frac{2}{3}u_{\max} = \left(\frac{-Y^2}{3\mu}\right)\frac{d(p+\gamma z)}{dx}. \tag{5.83}$$

ANSWER

Discussion

These results have general usefulness, so we assigned equation numbers to them. Equation (5.82) was used in Example 4.11 (note that $\tilde{p} = \Delta(p+\gamma z)/L$). That example illustrates the relationship between the mechanical energy loss and the detailed velocity distribution. Flow near the inlet to the parallel plate channel would not be fully developed and the analysis of this example would not be applicable. Example 5.7 examined entrance region flow. Note that the fully developed velocity profile that we derived here was given as the outlet boundary condition in that problem.

The same methodology that we used in this example could be used to find the velocity profile and average velocity–pressure/elevation relationship for fully developed flow in a circular pipe. We solve that problem using a slightly different approach in Chapter 7.

5.6.2 The Potential Flow and Boundary Layer Model

Fully developed flow, considered in Section 5.6.1, admits an exact solution of the Navier–Stokes equations. All the terms that are dropped from the full equations are truly zero if the flow is fully developed. The potential flow and boundary layer model, on the other hand, is an approximate model. The Navier–Stokes equations are simplified by neglecting certain terms that are small when compared with other terms in the equations.

The most significant simplification of the Navier–Stokes equations results when we neglect shear stress and consider inviscid flow. The resulting differential equations are Euler's equations; if the flow is also irrotational, we have a general solution available in Bernoulli's equation. Because many fluids (most importantly, air and water) have relatively small values of viscosity, seemingly we could almost always use an inviscid flow model with success. Unfortunately, this simply is not true. Inviscid flow models alone cannot calculate energy losses in internal flows and cannot predict drag force in streaming flow over an object. Flow patterns predicted by neglecting shear stresses are often grossly in error.

We can see the reason for the failure of the inviscid flow model alone by examining Figs. 3.6, 3.11, 3.12, 3.16, and 3.17. No matter how small the viscosity, there is always a region near a wall where the velocity falls off to zero. If the viscosity is small, this region is thin and the velocity gradient (e.g., $\partial u/\partial y$) in it is large. A large velocity gradient means a large strain rate and significant vorticity. As a result, the shear stress, which is the product of the viscosity and the velocity gradient, is always significant somewhere in the flow. This region near flow boundaries where shear and vorticity cannot be neglected is called a *boundary layer*. Often, the boundary layer is thin and it has a relatively minor effect on the overall flow. Sometimes, the boundary layer becomes very thick and disturbs the entire flow pattern. (Compare Figs. 3.4 and 8.27.)

The potential flow and boundary layer model is useful for flows with thin boundary layers. The basis of the model is the *boundary layer hypothesis,* first proposed by L. Prandtl in 1904:

> In a flow with large velocity and small viscosity, the effects of fluid viscosity and rotation (shear stress, turbulence, and vorticity) are confined to thin regions near solid surfaces or surfaces of velocity discontinuity. Outside these thin regions, the flow is nearly inviscid and irrotational.

Figure 5.25 illustrates this concept for both internal and external flows. Note that boundary layers typically grow thicker in the downstream direction, so in an internal flow, the boundary layer may grow to fill the entire channel.

The potential flow/boundary layer model divides the flow field into two distinct regions. In each region, certain small terms in the equations are dropped to obtain a mathematically simpler model. The entire flow field is then obtained by "patching" the two regions together.

We analyze the outer region by assuming that all shear stress forces are zero and that the flow is irrotational. As a first approximation, the boundary layer is

Figure 5.25 Division of flow into potential flow and boundary layer regions: (a) boundary layer and wake in external flow; (b) boundary layer and ideal flow regions in developing flow.

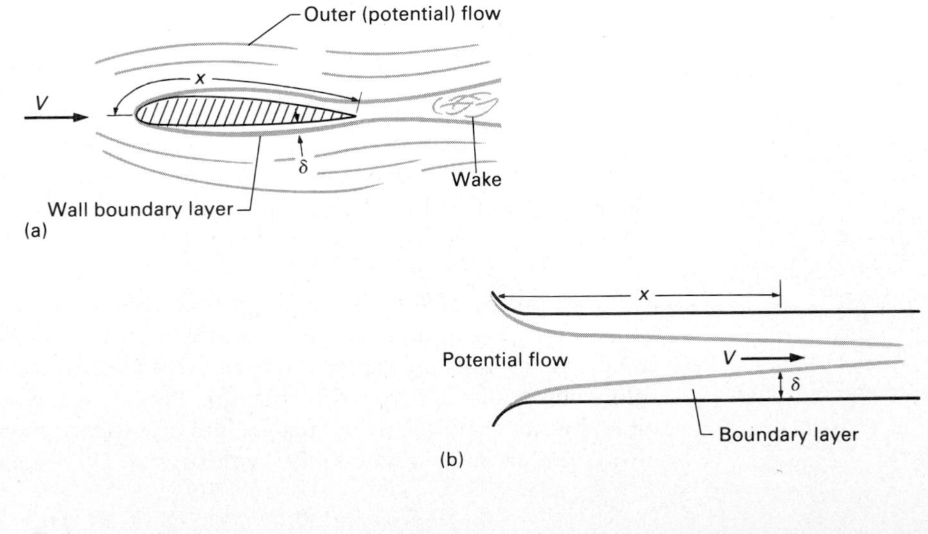

assumed to be so thin that its presence is neglected. The velocity field for an irrotational flow may be derived from the gradient of a scalar *velocity potential function;* therefore the outer region is called a potential flow. The potential flow analysis yields the streamline pattern in a region that extends from "near" the surface to the outer boundaries of the flow field and the pressure distribution for the entire flow, including that on the surface itself.

We model the boundary layer region by assuming that it is so thin that the pressure is constant across the boundary layer and equal to the surface pressure calculated from the potential flow analysis. If the boundary layer is very thin, the only significant viscous stress is the shear stress parallel to the main flow direction. The governing equations for the boundary layer region reduce to the continuity equation and a simplified version of the momentum equation for the direction parallel to the surface/main flow. Although the differential equations that describe the boundary layer remain nonlinear partial differential equations, they are of a form much simpler to solve than the full Navier–Stokes equations.

Complete calculation of a flow field using the potential flow/boundary layer model is illustrated schematically in Fig. 5.26. We begin the calculations with the potential flow, neglecting the presence of the boundary layer. Then, using the surface pressure distribution from the potential flow calculation, we calculate the boundary layer flow. Usually, the calculations end at this point; however, a more accurate analysis may be obtained by repeating the potential flow calculation, taking into effect the presence of the boundary layer, then recalculating the boundary layer, and so on until convergence.

Development of methods for calculating potential flow and boundary layers has proceeded along separate paths. Neither task is as simple as calculating fully developed flow, but each is simpler than tackling the full Navier–Stokes equations for the entire flow field. We consider methods for potential flow calculation and

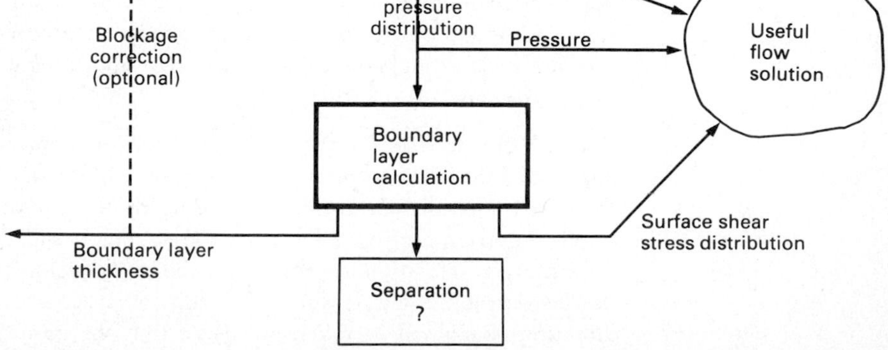

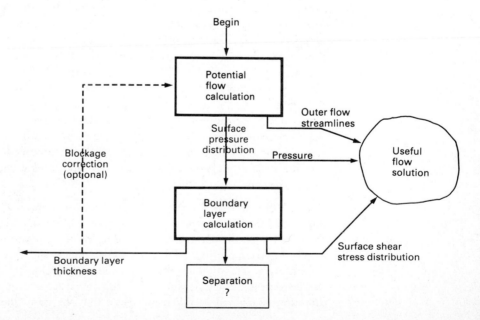

Figure 5.26 Schematic chart of calculations in the potential flow/boundary layer model.

boundary layer calculation in Chapter 9. In Chapter 8, we will use the boundary layer idea qualitatively to explain important phenomena in external flow.

5.6.3 Computational Fluid Dynamics

The difficulty of solving the set of nonlinear partial differential equations that describe the general motion of a fluid has been well documented in the preceding sections. We demonstrated the direct solution of the simple yet practical problem of fully developed flow. We also discussed simplified models suitable for the solution of classes of problems with special characteristics. Despite these, and other more sophisticated attempts at analytical solutions, analytical prediction of many important flows is beyond our understanding and mathematics. The advent of high-speed digital computers has provided a new tool with which to attack these flow problems: computational fluid dynamics (CFD).

As discussed in Chapter 1, a variety of CFD methods have been developed. The most prevalent of these methods are the finite difference method, the finite element method, and the finite volume method. This section is a brief introduction to CFD, so we selected a single method for exposition in sufficient depth to enable you to actually solve some flow problems: the *finite volume method*. This approach is perhaps the most accessible, because it relies heavily on concepts already developed in the study of the finite control volume method. In fact, understanding the reasoning behind the finite volume approach will actually reinforce many of the concepts introduced in this and the preceding chapters.

The first step in most CFD methods is to divide the flow field into a finite number of subregions. The idea behind the finite volume approach is to apply the continuity and momentum principles to each of these small but finite volumes within the flow field. In order to make sufficient progress to actually solve some problems, we consider only constant property flows and seek solutions in terms of *primitive variables* (u, v, and p). We use a uniform mesh in which each volume is the same size as all the other volumes and limit our consideration to unsteady flow. As we have concentrated almost exclusively on steady flows up to this point considering unsteady flow may seem strange. However, computing a flow by "marching" forward through time has certain computational advantages. A steady flow can be computed by tracing its development from rest over a time period long enough for the flow to become constant in time. Just as we divide the flow field into a number of small volumes, we divide time into a number of small steps. Although our choices are restrictive, the methods developed under these restrictions are extendable to less restrictive conditions without changing the underlying concepts.

Finite Volume Formulation. Figure 5.27(a) depicts a section of a flow field discretized into a uniform mesh of volumes, called *cells*. Figure 5.27(b) displays the staggered-grid convention we use to develop the equations. Pressure is defined as a cell-centered quantity and velocities are defined as face-centered quantities. Computing every velocity and pressure in the mesh yields an approximate solution to the flow problem. We assume that the velocities are in the positive coordinate direction. Care will be taken to ensure that the modeling equations behave cor-

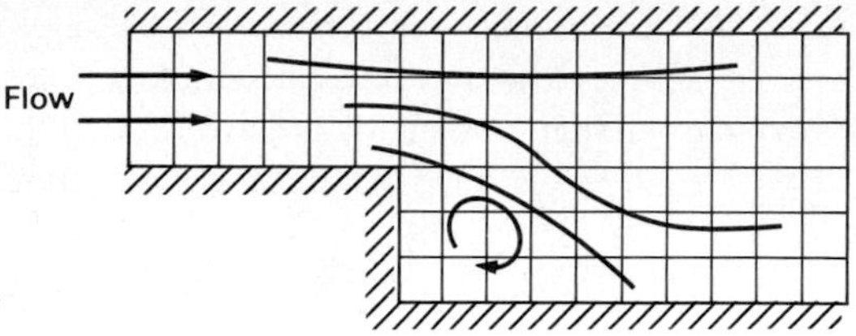

(a)

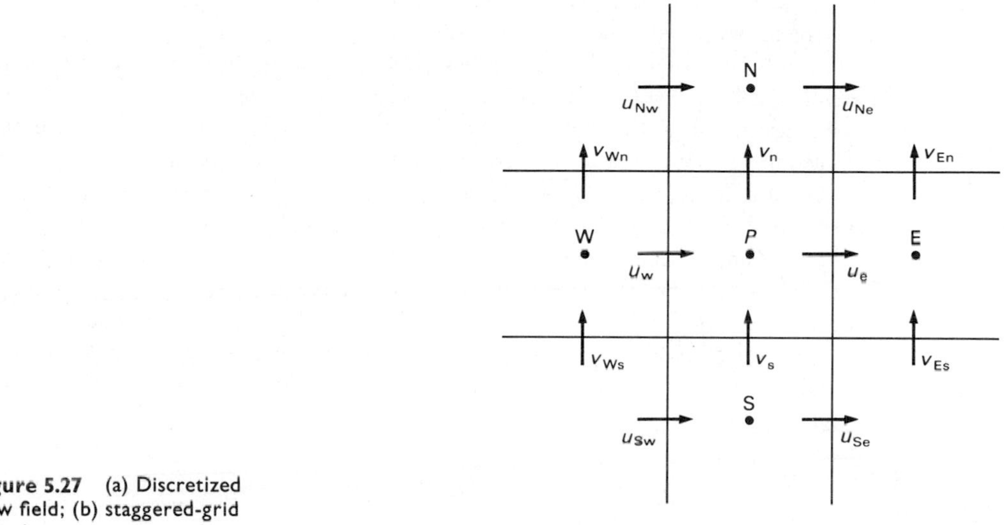

(b)

Figure 5.27 (a) Discretized flow field; (b) staggered-grid notation.

rectly if the flow is in the negative coordinate direction. We adopt a compass notation to identify the neighboring cells. Capital letters signify a cell center, whereas lowercase letters signify a cell face. Combined lowercase subscripts, such as ne, indicate the point at the intersection of two faces. Upper and lower case combinations indicate faces of neighboring cells, e.g., Ne indicates the "east" face of the "North" cell.

The continuity equation, applied to a control volume corresponding to a computational cell, can be expressed as

$$\begin{bmatrix}\text{Rate of change of}\\ \text{mass in the cell}\end{bmatrix} = \begin{bmatrix}\text{Net mass flow rate}\\ \text{into the cell}\end{bmatrix}.$$

The net flow rate into the cell is given by

$$\begin{bmatrix}\text{Net mass flow}\\ \text{rate into cell}\end{bmatrix} = \dot{m}_w + \dot{m}_s - \dot{m}_e - \dot{m}_n.$$

As the velocities are defined at the cell faces, this expression is readily rewritten in terms of cell geometry, fluid density, and velocities:

$$\begin{bmatrix}\text{Net mass flow}\\ \text{rate into cell}\end{bmatrix} = (\rho u A)_w + (\rho v A)_s - (\rho u A)_e - (\rho v A)_n. \tag{5.84}$$

Turning our attention toward the rate-of-change term, we see that we must start to identify the time level at which our variables are defined. We approximate the rate of change of a quantity with time as the difference in that quantity at two different time levels divided by the elapsed time. That is,

$$\frac{d(m_{\text{cell}})}{dt} = \frac{(m_{\text{cell}})_{t+\delta t} - (m_{\text{cell}})_t}{\delta t} = \frac{(\rho \forall)_{t+\delta t} - (\rho \forall)_t}{\delta t}. \tag{5.85}$$

Before we combine Eqs. (5.84) and (5.85), we should identify the time level at which the flow rates in Eq. (5.84) are to be evaluated. A superscript, $(\ )^n$, is used to identify the time level. Time-level n indicates that n steps have occurred since the initial time. If the flow rates are evaluated at the old time level (time-level n), the only unknown is the new time-level mass in the cell and we can solve for it explicitly. Hence this choice is known as an *explicit formulation*. If we choose to evaluate the flow rates at the new time level (time-level $n + 1$), the equation will contain several unknowns. Such a formulation is known as an *implicit formulation*. Although the system of equations generated by an implicit formulation for the entire grid of cells may be well posed (an equal number of equations and unknowns) and therefore have a unique solution, each individual equation cannot be solved directly for a new time-level unknown, as with the explicit formulation. Each formulation has its merits. The explicit is simpler in concept but suffers from limitations on allowable time-step size. The implicit does not suffer from a time-step size limitation but requires more computations per step. The explicit formulation for the continuity equation is:

$$\frac{(\rho \forall)^{n+1} - (\rho \forall)^n}{\delta t} = [(\rho u A)_{\text{w}} + (\rho v A)_{\text{s}} - (\rho u A)_{\text{e}} - (\rho v A)_{\text{n}}]^n. \tag{5.86}$$

To obtain an implicit formulation, the superscript n on the right-hand side is replaced by $n + 1$.

Application of the momentum equation requires careful consideration of the staggered-grid and its implications. We begin with a word equation:

$$\begin{bmatrix} \text{Sum of forces} \\ \text{acting on mass in} \\ \text{the control volume} \end{bmatrix} = \begin{bmatrix} \text{Rate of change} \\ \text{of momentum in} \\ \text{control volume} \end{bmatrix} + \begin{bmatrix} \text{Net rate of flow} \\ \text{of momentum out} \\ \text{of control volume} \end{bmatrix}.$$

When we apply the momentum principle to a control volume, the momentum storage or depletion is evaluated within the control volume, but the pressure force, shear force, and momentum flows are evaluated at the control surface. If the control volume is selected to be the computational cell, the pressure force must be computed at the cell faces even though the pressure is a cell-centered quantity. Similarly, computation of the shear forces is based on inconveniently located variables.

Consider an alternative control volume for the x-momentum equation centered on the east face of the computational cell as shown in Fig. 5.28(a). The net pressure force acting in the x direction is readily expressed in terms of the variables associated with the staggered-grid convention:

$$\begin{bmatrix} \text{Net pressure force} \\ \text{in } x \text{ direction} \end{bmatrix} = (pA)_{\text{P}} - (pA)_{\text{E}}. \tag{5.87}$$

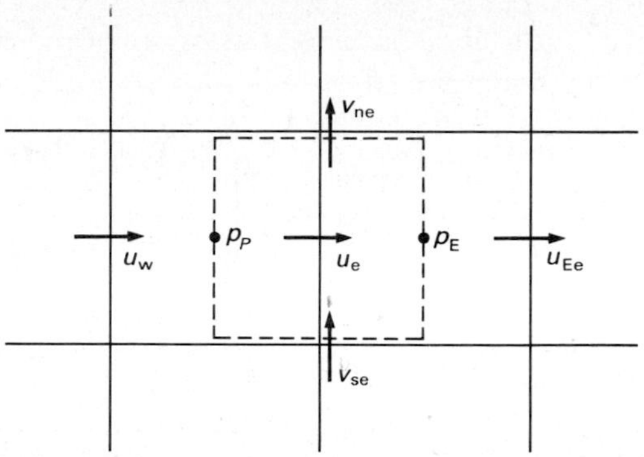

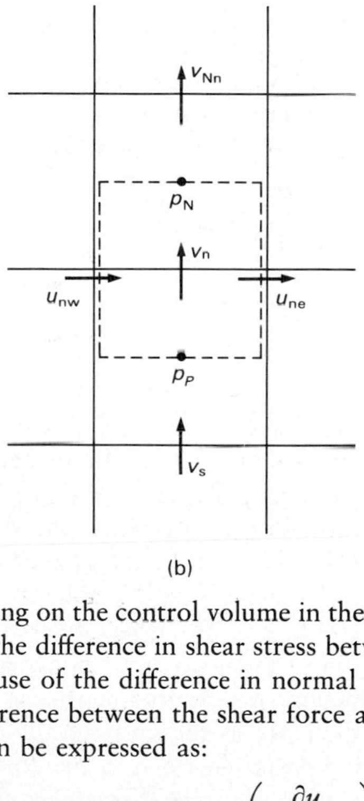

Figure 5.28 (a) The *x*-momentum control volume; (b) the *y*-momentum control volume.

The net viscous force acting on the control volume in the x direction has two components: one because of the difference in shear stress between the north and south faces and the other because of the difference in normal viscous forces at the east and west faces. The difference between the shear force at the top and the bottom of the control volume can be expressed as:

$$\begin{bmatrix}\text{Net shear force} \\ \text{in } x \text{ direction}\end{bmatrix} = (\tau A)_{\text{ne}} - (\tau A)_{\text{se}} = \left(\mu \frac{\partial u}{\partial y} A\right)_{\text{ne}} - \left(\mu \frac{\partial u}{\partial y} A\right)_{\text{se}}$$
$$= \mu\left[\frac{u_{\text{Ne}} - u_{\text{e}}}{\delta y_{\text{ne}}}\right] A_{\text{ne}} - \mu\left[\frac{u_{\text{e}} - u_{\text{Se}}}{\delta y_{\text{se}}}\right] A_{\text{se}}.$$

Note that these are the best available estimates, because the velocities are required at locations aligned with the center of the control volume. For a constant viscosity fluid and a uniform mesh, the net shear force in the x direction simplifies to:

$$F_{\text{shear},x} = \mu\left[\frac{u_{\text{Ne}} - 2u_{\text{e}} + u_{\text{Se}}}{\delta y}\right] \cdot A_y \qquad (\text{Note: } A_{\text{ne}} = A_{\text{se}} = A_y). \qquad (5.88\text{a})$$

An expression for the normal viscous stress can be developed in an analogous manner:

$$F_{\text{normal},x} = \mu\left[\frac{u_{\text{Ee}} - 2u_{\text{e}} + u_{\text{w}}}{\delta x}\right] \cdot A_x \qquad (\text{Note: } A_{\text{E}} = A_{\text{P}} = A_x). \qquad (5.88b)$$

The gravity force on the fluid in the control volume can be expressed as:

$$\left[\begin{array}{c}\text{Gravity force on mass}\\ \text{in control volume}\end{array}\right] = F_{g,x} = \rho \forall g_x. \qquad (5.89)$$

The rate of change of x momentum in the control volume depicted in Fig. 5.28(a) is easily written as the difference in momentum between two time levels divided by the elapsed time:

$$\left[\begin{array}{c}\text{Rate of change of}\\ x\text{ momentum in control volume}\end{array}\right] = \frac{d(M_x)}{dt} = \frac{(\rho \forall u_{\text{e}})^{n+1} - (\rho \forall u_{\text{e}})^{n}}{\delta t}. \qquad (5.90)$$

The net rate of flow of x momentum out of the control volume can be expressed as:

$$\begin{aligned}\left[\begin{array}{c}\text{Net rate of flow of}\\ x\text{ momentum out}\\ \text{of control volume}\end{array}\right] &= \dot{M}_{x_{\text{E}}} - \dot{M}_{x_{\text{P}}} + \dot{M}_{x_{\text{ne}}} - \dot{M}_{x_{\text{se}}}\\ &= \dot{m}_{\text{E}}u_{\text{E}} - \dot{m}_{\text{P}}u_{\text{P}} + \dot{m}_{\text{ne}}u_{\text{ne}} - \dot{m}_{\text{se}}u_{\text{se}}\\ &= (\rho uA)_{\text{E}}u_{\text{E}} - (\rho uA)_{\text{P}}u_{\text{P}} + (\rho vA)_{\text{ne}}u_{\text{ne}} - (\rho vA)_{\text{se}}u_{\text{se}}. \qquad (5.91)\end{aligned}$$

The "shifted" control volume of Fig. 5.28(a) allowed accurate and convenient expressions for the stress, pressure, gravity, and momentum storage terms, but computation of the flow-of-momentum term requires variables at locations other than where they are defined. One idea is to use linear interpolation to obtain values at the desired locations, and there are sound mathematical arguments for doing so. Unfortunately, the algorithm produced by this idea is numerically unstable. To obtain a simple stable formulation, we use a technique called *upwind differencing* or *donor-cell differencing*. In this formulation, we assume that convection of momentum far outweighs viscous diffusion of momentum in the flow direction. We approximate the momentum-per-unit-mass at the control volume boundary by the nearest value "upwind" from the boundary. For instance, u_{P} is approximated by u_{w} if the flow is in the positive direction but by u_{e} if the flow is in the negative direction. You should note that, although the velocity associated with the momentum is assigned an "upwind" value, the velocity associated with mass flow rate is still computed from a linear interpolation. The upwind formulation has two shortcomings: It burdens a computer code with the logic required to select the upwind values, and it introduces "artificial viscosity," which reduces the accuracy of the solution.

A variable dependent on the flow direction is identified as a double subscripted variable with a slash between the subscripts. For instance, $u_{\text{w/e}}$ indicates either the west or east face velocity, depending on the flow direction.

The flow-of-momentum term can be written as:

$$\begin{bmatrix}\text{Net rate of flow of}\\ x\text{ momentum out}\\ \text{of control volume}\end{bmatrix} = \sum \dot{M}_x$$

$$= (\rho A_x)\left(\frac{u_e + u_{Ee}}{2}\right)u_{e/Ee} - (\rho A_x)\left(\frac{u_w + u_e}{2}\right)u_{w/e}$$

$$+ (\rho A_y)\left(\frac{v_n + v_{En}}{2}\right)u_{e/Ne} - (\rho A_y)\left(\frac{v_s + v_{Es}}{2}\right)u_{e/Se}. \quad (5.92)$$

The flow-of-momentum term, as well as all other terms in the x-momentum equation have now been expressed solely in terms of variables evaluated at the location where they are defined in the staggered-grid convention. We collect all the terms and write the x-momentum equation as:

$$\frac{(\rho \forall u_e)^{n+1} - (\rho \forall u_e)^n}{\delta t}$$

$$= (\rho A_x)\left(\frac{u_w + u_e}{2}\right)u_{w/e} + (\rho A_y)\left(\frac{v_s + v_{Es}}{2}\right)u_{e/Se} - (\rho A_x)\left(\frac{u_e + u_{Ee}}{2}\right)u_{e/Ee}$$

$$- (\rho A_y)\left(\frac{v_n + v_{En}}{2}\right)u_{e/Ne} + (p_P - p_E)A_x + \rho \forall g_x$$

$$+ \mu\left[\left(\frac{u_{Ee} - 2u_e + u_w}{\delta x}\right)A_x + \left(\frac{u_{Ne} - 2u_e + u_{Se}}{\delta y}\right)A_y\right] \quad (5.93)$$

where the momentum-flow terms have been transferred to the right-hand side of the equation.

As we are considering only incompressible flow, the amount of mass in the control volume cannot change with time. We therefore divide continuity and the x-momentum equation by the product $\rho\forall$ to obtain the following expressions:

$$0 = [(uA)_w + (vA)_s - (uA)_e - (vA)_w]; \quad (5.94)$$

$$\frac{(u_e)^{n+1} - (u_e)^n}{\delta t} = -\left[\frac{\frac{1}{2}(u_e + u_{Ee})u_{e/Ee} - \frac{1}{2}(u_w + u_e)u_{w/e}}{\delta x}\right]$$

$$- \left[\frac{\frac{1}{2}(v_n + v_{En})u_{e/Ne} - \frac{1}{2}(v_s + v_{Es})u_{Se/e}}{\delta y}\right] - \frac{1}{\rho}\frac{[p_E - p_P]}{\delta x}$$

$$+ \nu\left[\frac{u_{Ee} - 2u_e + u_w}{(\delta x)^2} + \frac{u_{Ne} - 2u_e + u_{Se}}{(\delta y)^2}\right] + g_x. \quad (5.95)$$

We can develop the y-momentum equation by considering the control volume depicted in Fig. 5.28(b) and using the same reasoning as for the x-momentum equation. The resulting equation is:

$$\frac{(v_n)^{n+1} - (v_n)^n}{\delta t} = -\left[\frac{\frac{1}{2}(u_{Ne} + u_e)v_{n/En} - \frac{1}{2}(u_{Nw} + u_w)v_{Wn/n}}{\delta x}\right]$$

$$- \left[\frac{\frac{1}{2}(v_n + v_{Nn})v_{n/Nn} - \frac{1}{2}(v_s + v_n)v_{s/n}}{\delta y}\right] - \frac{1}{\rho}\frac{[p_N - p_P]}{\delta y}$$

$$+ \nu\left[\frac{v_{En} - 2v_n + v_{Wn}}{(\delta x)^2} + \frac{v_{Nn} - 2v_n + v_s}{(\delta y)^2}\right] + g_y. \quad (5.96)$$

Equations 5.94–5.96 can be written for each cell in the mesh. When combined with equations for cells at mesh boundaries, they form a set of algebraic equations involving the velocity components and the pressure for the entire mesh. Of course, we must still decide whether to evaluate the terms on the right-hand sides at time-level n or time-level $n + 1$; that is, we must choose between explicit and implicit formulations.

Boundary Conditions. There are two basic approaches to imposing boundary conditions at the edges of a computational mesh. The first is to install logic in the code to test for the mesh edge and then use special coding for cells with a face at the mesh edge. The second is to add a layer of "ghost" cells around the outside of the mesh. At the beginning of a computational step, the ghost-cell variables are set to values so that, when all the regular cells are computed using the standard coding, the correct boundary conditions are imposed at the flow boundary. Although both approaches have merit and both are widely used, we present the ghost-cell technique. One argument in favor of this approach is that the resulting code is more easily "vectorized" and therefore likely to perform better on today's supercomputers. We present the formulation for five common boundary conditions: a free-slip rigid boundary, a no-slip rigid boundary, a specified flow (velocity) boundary, a continuative boundary, and a specified velocity rigid boundary.

A free-slip rigid boundary is one that does not change position with time and at which the viscous stress is zero. It has one of the simplest boundary conditions to impose and can be used to model flow at a nondeforming free surface. Figure 5.29 shows a free-slip boundary aligned with the y direction at the east edge of a column of cells. The impenetrable condition is easily imposed by specifying u_e to be zero. If the value of v_{En} is set to equal v_n and the value of v_{Es} is set to equal v_s, the rate of change of v with respect to x is also zero. When the momentum

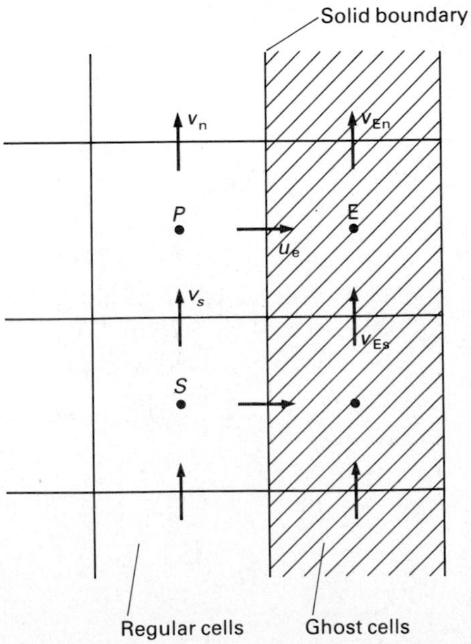

Figure 5.29 Cell configuration for implementing boundary conditions.

equation computes the viscous stress at the east edge of the computational cell, it will also be zero, producing a free-slip boundary.

A no-slip boundary condition requires that the code compute a zero velocity at the location of the boundary. Again, referring to Figure 5.29, we specify a value of zero for u_e. The v velocity in a ghost cell is set equal in magnitude to the v velocity in the neighboring regular cell but is assigned the opposite sign. If the regular cell and the ghost cell are the same size, and the differencing procedure for the momentum equation is based on a linear variation in velocity between points at which the velocity is defined, this formulation results in a zero value of the tangential velocity at the boundary.

A continuative boundary approximates the fully developed flow condition. Referring to Figure 5.29, we simply set $u_e = u_w$ for the ghost cell. A specified flow boundary condition is used when there is a known flow condition at a mesh edge. One example of such an application is a tank inlet with a known inflow rate. Using the ghost-cell technique, we set the velocity normal to the cell face at which the inflow occurs equal to the known value. Referring to Fig. 5.29, we set u_e to a positive value for a specified outflow rate or to a negative value for a specified inflow rate.

A specified velocity rigid boundary is a generalization of the no-slip boundary. Instead of requiring that the tangential velocity be zero at the boundary, the ghost-cell velocities are set to values that produce the specified value at the boundary. The ghost-cell value is computed by linear interpolation between the regular cell velocity and the ghost-cell velocity. The following equation applies to the configuration depicted in Fig. 5.29:

$$v_{\mathrm{En}} = 2(v_{\mathrm{boundary}}) - v_{\mathrm{n}}.$$

A quick check shows that the no-slip boundary condition is recovered if the boundary velocity is set to zero.

Solution Algorithm. The problem of modeling incompressible constant property flow through the discretized region has now been reduced to solution of simultaneous algebraic equations for the primitive variables u, v, and p. Although the entire array of techniques available for algebraic problems can now be brought to bear, solution of this set of equations is far from trivial. Direct approaches, such as matrix inversion techniques, encounter difficulties because of the nonlinearity of the equations and the large number of cells required to model most flow problems. Iterative techniques can be used to solve nonlinear problems, but the user may encounter convergence difficulties unless the method and the system are well understood. The entire problem is complicated by the fact that pressure does not appear as a variable in the continuity equation. Although this lack may not seem troubling at first—as only a few unknowns typically appear in each equation of any system of equations—it is a significant feature that severely restricts the numerical tools that can be used successfully to solve this particular problem. We focus the presentation in this textbook on one solution technique from the class of solution procedures known as *pressure correction algorithms*.

Because the solution procedure is applicable to difference formulations other than those presented in the preceding section, we develop it in a general way, independently of our specific finite volume formulation. We start by specializing the governing differential equations, Eqs. (5.24), (5.63), and (5.64), for plane two-dimensional, unsteady, incompressible flow:*

$$\frac{\partial u}{\partial x} + \frac{\partial v}{\partial y} = 0;$$

$$\frac{\partial u}{\partial t} + u\frac{\partial u}{\partial x} + v\frac{\partial u}{\partial y} = -\frac{1}{\rho}\frac{\partial p}{\partial x} + \nu\left(\frac{\partial^2 u}{\partial x^2} + \frac{\partial^2 u}{\partial y^2}\right) + g_x;$$

$$\frac{\partial v}{\partial t} + u\frac{\partial v}{\partial x} + v\frac{\partial v}{\partial y} = -\frac{1}{\rho}\frac{\partial p}{\partial y} + \nu\left(\frac{\partial^2 v}{\partial x^2} + \frac{\partial^2 v}{\partial y^2}\right) + g_y.$$

Using the simple differencing for the time-terms that was introduced in the previous section, we rewrite the momentum equations:

$$\frac{u^{n+1} - u^n}{\delta t} = -\frac{1}{\rho}\frac{\partial p}{\partial x} + \nu\left(\frac{\partial^2 u}{\partial x^2} + \frac{\partial^2 u}{\partial y^2}\right) + g_x - \left(u\frac{\partial u}{\partial x} + v\frac{\partial u}{\partial y}\right);$$

$$\frac{v^{n+1} - v^n}{\delta t} = -\frac{1}{\rho}\frac{\partial p}{\partial y} + \nu\left(\frac{\partial^2 v}{\partial x^2} + \frac{\partial^2 v}{\partial y^2}\right) + g_y - \left(u\frac{\partial v}{\partial x} + v\frac{\partial v}{\partial y}\right).$$

Solving for the new time-level velocities yields:

$$u^{n+1} = u^n - \frac{\delta t}{\rho}\frac{\partial p}{\partial x} + \delta t\left[\nu\left(\frac{\partial^2 u}{\partial x^2} + \frac{\partial^2 u}{\partial y^2}\right) + g_x - \left(u\frac{\partial u}{\partial x} + v\frac{\partial u}{\partial y}\right)\right]; \tag{5.97}$$

$$v^{n+1} = v^n - \frac{\delta t}{\rho}\frac{\partial p}{\partial y} + \delta t\left[\nu\left(\frac{\partial^2 v}{\partial x^2} + \frac{\partial^2 v}{\partial y^2}\right) + g_y - \left(u\frac{\partial v}{\partial x} + v\frac{\partial v}{\partial y}\right)\right]. \tag{5.98}$$

If we evaluate all of the right-hand-side terms at time-level n, we obtain an explicit formulation with two equations and three unknowns (u, v, p). The third equation is continuity. We enforce continuity at the new time level to ensure that the velocity field computed at each time level satisfies continuity. That is,

$$\left(\frac{\partial u}{\partial x}\right)^{n+1} + \left(\frac{\partial v}{\partial y}\right)^{n+1} = 0. \tag{5.99}$$

You can now see the difficulty mentioned in the introduction to this section: Pressure does not appear as a variable in the continuity equation, so we apparently have no way to compute p^{n+1}! The heart of a solution procedure based on a pressure correction algorithm is to recast the continuity equation so that pressure does appear as a variable in the equation that represents conservation of mass.

If we solve the momentum equations, Eqs. (5.97) and (5.98), using a fully explicit formulation, the new time-level velocities (u^{n+1}, v^{n+1}) do not necessarily satisfy continuity, because we would still be using the old time-level pressure field. We therefore seek to use continuity to "correct" the pressure field from old time-level values to new time-level values, so that the velocities produced by the momen-

---

* Our finite-volume formulation is one method for computing an approximate solution to these differential equations.

tum equations satisfy continuity at the new time level. We rewrite the momentum equations using an implicit evaluation of the pressure gradient terms:

$$u^{n+1} = u^n - \frac{\delta t}{\rho}\left(\frac{\partial p}{\partial x}\right)^{n+1} + \delta t\left[\nu\left(\frac{\partial^2 u}{\partial x^2} + \frac{\partial^2 u}{\partial y^2}\right) + g_x - \left(u\frac{\partial u}{\partial x} + v\frac{\partial u}{\partial y}\right)\right]^n; \qquad (5.100)$$

$$v^{n+1} = v^n - \frac{\delta t}{\rho}\left(\frac{\partial p}{\partial y}\right)^{n+1} + \delta t\left[\nu\left(\frac{\partial^2 v}{\partial x^2} + \frac{\partial^2 v}{\partial y^2}\right) + g_y - \left(u\frac{\partial v}{\partial x} + v\frac{\partial v}{\partial y}\right)\right]^n. \qquad (5.101)$$

Recalling the goal of rewriting continuity as an equation for "correcting" the pressure from the old time level to the new time level, we split the pressure into an old time-level component plus a "correction" ($\Delta p \equiv p^{n+1} - p^n$). Thus

$$\left(\frac{\partial p}{\partial x}\right)^{n+1} = \frac{\partial(p^{n+1} - p^n + p^n)}{\partial x} = \frac{\partial(p^{n+1} - p^n)}{\partial x} + \left(\frac{\partial p}{\partial x}\right)^n = \left(\frac{\partial \Delta p}{\partial x}\right) + \left(\frac{\partial p}{\partial x}\right)^n \qquad (5.102)$$

and

$$\left(\frac{\partial p}{\partial y}\right)^{n+1} = \left(\frac{\partial \Delta p}{\partial y}\right) + \left(\frac{\partial p}{\partial y}\right)^n. \qquad (5.103)$$

Using this formulation for the pressure gradient, we define the following nomenclature:

$$Q_x^n = \frac{\delta t}{\rho}\left[-\frac{\partial p}{\partial x} + \mu\left(\frac{\partial^2 u}{\partial x^2} + \frac{\partial^2 u}{\partial y^2}\right) + \rho g_x - \rho\left(u\frac{\partial u}{\partial x} + v\frac{\partial u}{\partial y}\right)\right]^n; \qquad (5.104)$$

$$Q_y^n = \frac{\delta t}{\rho}\left[-\frac{\partial p}{\partial y} + \mu\left(\frac{\partial^2 v}{\partial v^2} + \frac{\partial^2 u}{\partial y^2}\right) + \rho g_y - \rho\left(u\frac{\partial v}{\partial x} + v\frac{\partial v}{\partial y}\right)\right]^n. \qquad (5.105)$$

We then rewrite the momentum equations:

$$u^{n+1} = u^n - \frac{\delta t}{\rho}\frac{\partial \Delta p}{\partial x} + Q_x^n; \qquad (5.106)$$

$$v^{n+1} = v^n - \frac{\delta t}{\rho}\frac{\partial \Delta p}{\partial y} + Q_y^n. \qquad (5.107)$$

We now substitute these expressions for the new time-level velocities into the continuity equation:

$$0 = \left(\frac{\partial u}{\partial x} + \frac{\partial v}{\partial y}\right)^{n+1} = \left(\frac{\partial u}{\partial x} + \frac{\partial v}{\partial y}\right)^n - \frac{\delta t}{\rho}\left(\frac{\partial^2 \Delta p}{\partial x^2} + \frac{\partial^2 \Delta p}{\partial y^2}\right) + \left(\frac{\partial Q_x}{\partial x} + \frac{\partial Q_y}{\partial y}\right)^n. \qquad (5.108)$$

The first term on the right-hand side of Eq. (5.108) is simply continuity expressed at the old time level and is therefore equal to zero. As the Q terms are evaluated at the old time level, their values are known. The principle of conservation of mass, applied at time-level $n + 1$, can now be written:

$$\left(\frac{\partial^2 \Delta p}{\partial x^2} + \frac{\partial^2 \Delta p}{\partial y^2}\right) = \nabla^2(\Delta p) = -\frac{\rho}{\delta t}\left(\frac{\partial Q_x}{\partial x} + \frac{\partial Q_y}{\partial y}\right)^n. \tag{5.109}$$

We have now succeeded in transforming the continuity equation into a "pressure correction" equation. The resulting equation is in a standard form known as *Poisson's equation.* Poisson's equation appears in the modeling of many physical processes, and many procedures have been developed for solving it. Substituting a solution for Δp into the momentum equations brings the velocity field up to the new time level.

An iterative procedure for solving the pressure correction form of the continuity equation is less demanding of computational resources than is a direct Poisson solver. The procedure begins by computing a first approximation to the new time-level velocities. A numerical superscript in a circle denotes an iteration level. The first estimate is computed from a fully explicit formulation of the momentum equations, with $\Delta p^{①} = 0$:

$$u^{①} = u^n + Q_x^n; \tag{5.110}$$

$$v^{①} = v^n + Q_y^n. \tag{5.111}$$

To obtain an improved estimate of the new time-level velocities, we must compute a pressure correction. We begin by noting that the explicit part of the momentum equation is now embedded in our first estimate velocity field:

$$u^{②} = u^n - \frac{\delta t}{\rho}\left(\frac{\partial \Delta p}{\partial x}\right)^{②} + Q_x^n = u^{①} - \frac{\delta t}{\rho}\left(\frac{\partial \Delta p}{\partial x}\right)^{②}; \tag{5.112}$$

$$v^{②} = v^n - \frac{\delta t}{\rho}\left(\frac{\partial \Delta p}{\partial y}\right)^{②} + Q_y^n = v^{①} - \frac{\delta t}{\rho}\left(\frac{\partial \Delta p}{\partial y}\right)^{②}. \tag{5.113}$$

To solve for the pressure correction for the second iteration, we substitute these expressions into continuity:

$$\frac{\partial u^{②}}{\partial x} + \frac{\partial v^{②}}{\partial y} = \frac{\partial u^{①}}{\partial x} - \frac{\delta t}{\rho}\frac{\partial^2 \Delta p^{②}}{\partial x^2} + \frac{\partial v^{①}}{\partial y} - \frac{\delta t}{\rho}\frac{\partial^2 \Delta p^{②}}{\partial y^2} = 0. \tag{5.114}$$

We collect terms computed from first iteration values and rearrange the equation into a standard Poisson form. We use the symbol D to represent the divergence of the velocity field

$$D^{①} = \frac{\partial u^{①}}{\partial x} + \frac{\partial v^{①}}{\partial y}, \tag{5.115}$$

so Eq. (5.114) becomes

$$D^{①} - \frac{\delta t}{\rho}\left(\frac{\partial^2 \Delta p^{②}}{\partial x^2} + \frac{\partial^2 \Delta p^{②}}{\partial y^2}\right) = 0, \tag{5.116}$$

or

$$\nabla^2(\Delta p^{②}) = \frac{\rho}{\delta t} D^{①}. \tag{5.117}$$

A simple centered-difference approximation of the second derivatives of Δp appearing in Eq. (5.117) is both efficient and sufficiently accurate:

$$\begin{aligned}\nabla^2(\Delta p^{②}) &= \left(\frac{\partial^2 \Delta p}{\partial x^2} + \frac{\partial^2 \Delta p}{\partial y^2}\right)^{②} \\ &\approx \left(\frac{\Delta p_E - 2\,\Delta p_P + \Delta p_W}{(\Delta x)^2} + \frac{\Delta p_N - 2\,\Delta p_P + \Delta p_S}{(\Delta y)^2}\right)^{②}.\end{aligned} \tag{5.118}$$

Substituting and rearranging give:

$$-2\left(\frac{1}{(\Delta x)^2} + \frac{1}{(\Delta y)^2}\right)\Delta p_P^{②} + \left(\frac{\Delta p_E + \Delta p_W}{(\Delta x)^2}\right)^{②} + \left(\frac{\Delta p_N + \Delta p_S}{(\Delta y)^2}\right)^{②} = \frac{\rho}{\Delta t} D^{①}. \tag{5.119}$$

We now have to solve this system for $\Delta p^{②}$. If we retain all the terms, we must use a numerical procedure for solving simultaneous equations. Instead, we note that with each iteration, we will be converging on the new time-level pressures and that the pressure corrections will therefore decrease toward zero. We therefore neglect the second two terms on the left-hand side of Eq. (5.119) and solve explicitly for Δp_p:

$$\Delta p_p^{②} = \frac{\rho}{\Delta t} \frac{-D^{①}}{2[1/(\Delta x)^2 + 1/(\Delta y)^2]}. \tag{5.120}$$

Equation (5.120) shows that the new pressure correction is proportional to the value of D from the most recent iteration. The divergence of the velocity field, D, can be interpreted as either the net volumetric strain rate or the net rate of volumetric flow out of the cell. Continuity tells us that the net flow rate out of the cell must be zero; otherwise, mass is being created or destroyed within the cell.

We now compute second iterate approximations to the pressure and velocity fields:

$$p^{②} = p^{①} + \Delta p^{②}; \tag{5.121}$$

$$u^{②} = u^{①} - \frac{\delta t}{\rho}\left(\frac{\partial \Delta p}{\partial x}\right)^{②}; \tag{5.122}$$

$$v^{②} = v^{①} - \frac{\delta t}{\rho}\left(\frac{\partial \Delta p}{\partial y}\right)^{②}. \tag{5.123}$$

These second iterate velocities do not exactly satisfy continuity, but the total divergence in the field will have decreased compared to the first iterate approximations.

We now compute third iterate approximations in exactly the same manner as we computed the second iterate values:

$$\Delta p^{③} = \frac{\rho}{\Delta t} \frac{-D^{②}}{2[1/(\Delta x)^2 + 1/(\Delta y)^2]} \tag{5.124}$$

and

$$p^{③} = p^{②} + \Delta p^{③} \tag{5.125}$$

$$u^{③} = u^{②} - \frac{\delta t}{\rho}\left(\frac{\partial \Delta p}{\partial x}\right)^{③}; \tag{5.126}$$

$$v^{③} = v^{②} - \frac{\delta t}{\rho}\left(\frac{\partial \Delta p}{\partial y}\right)^{③}. \tag{5.127}$$

The iteration pattern is now clear. With each iteration, the divergence in the field will decrease and so will the subsequent pressure correction. Typically the iteration procedure is repeated until the maximum divergence in each computational cell has decreased to less than a user-specified tolerance:

$$|D^{ⓚ}| < \text{tolerance}. \tag{5.128}$$

When this condition has been satisfied, the computations for this time step are complete. The final velocity and pressure fields may be recorded as output, (i.e., $u^{n+1} = u^{ⓚ}$, $v^{n+1} = v^{ⓚ}$, and $p^{n+1} = p^{ⓚ}$). If the solution is to continue, the known values are designated as old time-level values, and the entire process is repeated for the next time step.

Time-Step Size Limitations. Virtually all numerical methods require the user to select an interval size of one sort or another. Discretization of the flow field required specification of cell size. Solution of time-dependent models requires specification of a time-step size. All such choices affect the accuracy of the solution and may also significantly affect the functioning of the numerical method. For some methods, if the interval selected is outside an acceptable range of values, the numerical method becomes *unstable* and fails to converge to a meaningful solution. The technique described in the preceding sections does have stability limits that must be respected in order to obtain a reliable solution. We omit the details of the analysis used to determine the conditions for stability and simply present the results along with heuristic explanations.

The first stability criterion is related to how fast mass is convected through a computational cell. To obtain a stable solution, the time-step size must be below the following limit:

$$\delta t \leq \frac{1}{|\delta x/u| + |\delta y/v|}. \tag{5.129}$$

This criterion is often referred to as the Courant–Friedrichs–Lewy (CFL) condition and may be interpreted as limiting the distance that a fluid particle travels in a single time step to no more than a single computational cell length. If this condition is violated, information can be transported such a great distance in a single time step that it completely skips over a computational cell, never affecting the variables associated with that cell.

A second stability criterion is related to momentum transfer by viscosity. The condition is:

$$\delta t \leq \frac{1}{2\nu}\left|\frac{\delta x^2 \delta y^2}{\delta x^2 + \delta y^2}\right|. \tag{5.130}$$

This criterion may be interpreted as limiting the transport of momentum by the action of viscosity to no more than a single cell in a single time step. The smaller of the two limiting time steps determined by Eqs. (5.129) and (5.130) must be enforced as an upper limit on δt to ensure stability.

Using an appropriate time step produces a stable solution in the previously developed difference formulation and solution procedure. You should note the interaction between mesh size and time-step size. If we refine the mesh, using a larger number of smaller volumes in order to obtain a more accurate computation, we must also decrease the time-step size in order to maintain stability. Both the larger number of volumes and the smaller time step combine to increase the number of computer operations required to obtain a solution.

Driven Cavity Flow. Finally, we present a computational model of a driven cavity flow as an application of the methodology presented in the preceding sections. This geometrically simple flow is often used as a test case for evaluating new algorithms. The geometry, depicted in Fig. 5.30, is a square space defined by stationary walls on three sides and a moving wall on the fourth side. Clearly, the boundaries of the flow are easily aligned with a rectangular mesh, and the boundary conditions developed in the preceding section are sufficient for modeling this flow. If we set the density, viscosity, and cavity length to values of unity, the Reynolds number for this flow ($R = \rho VL/\mu$) has the same value as the plate velocity, which is 100 for this example. A brief description of the code used to model this flow is provided in Appendix I.

We begin our analyses by using a coarse mesh to model the flow. The advantage of such a preliminary run is to verify that the geometry is correctly defined and that the code is functioning properly. Figure 5.31 shows a uniform 5 × 5 mesh and the velocity field and streamline pattern predicted by the code for this mesh. Although the basic flow field looks reasonable, it is missing many flow details. We now begin a convergence study using finer and finer meshes until, when the solution no longer changes significantly with further refinement, we have obtained a "mesh independent" solution. Figure 5.32 presents a uniform 20 × 20 mesh and the associated flow field prediction. Figure 5.33 presents a uniform 40 × 40 mesh and the

Figure 5.30 Driven cavity.

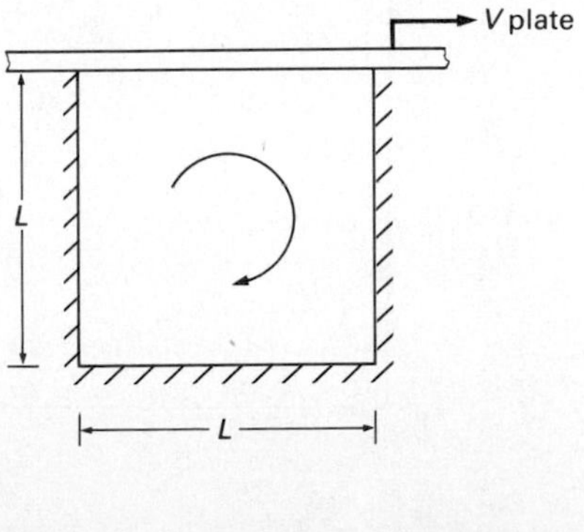

Figure 5.31 (a) 5 × 5 mesh.
(b) Velocity field computed with 5 × 5 mesh.
(c) Streamline pattern computed with 5 × 5 mesh.

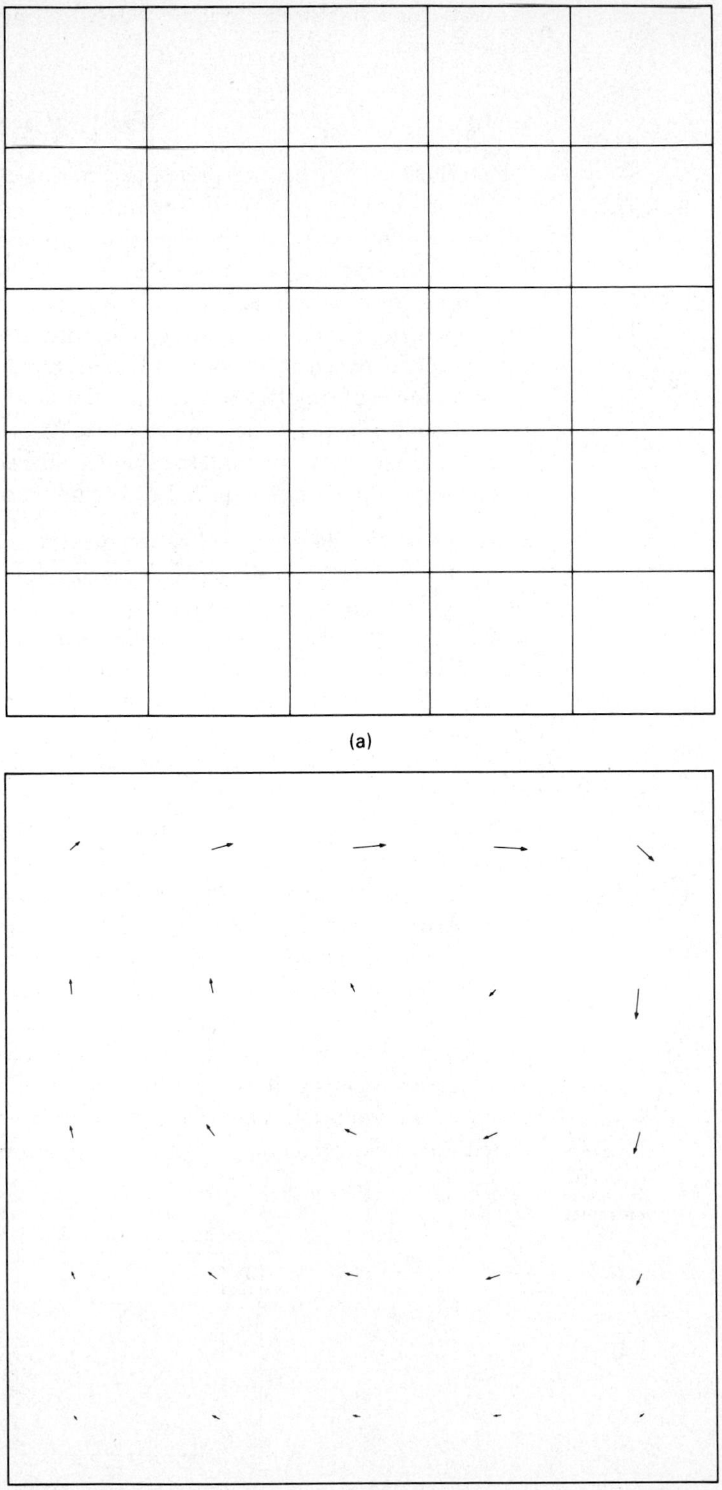

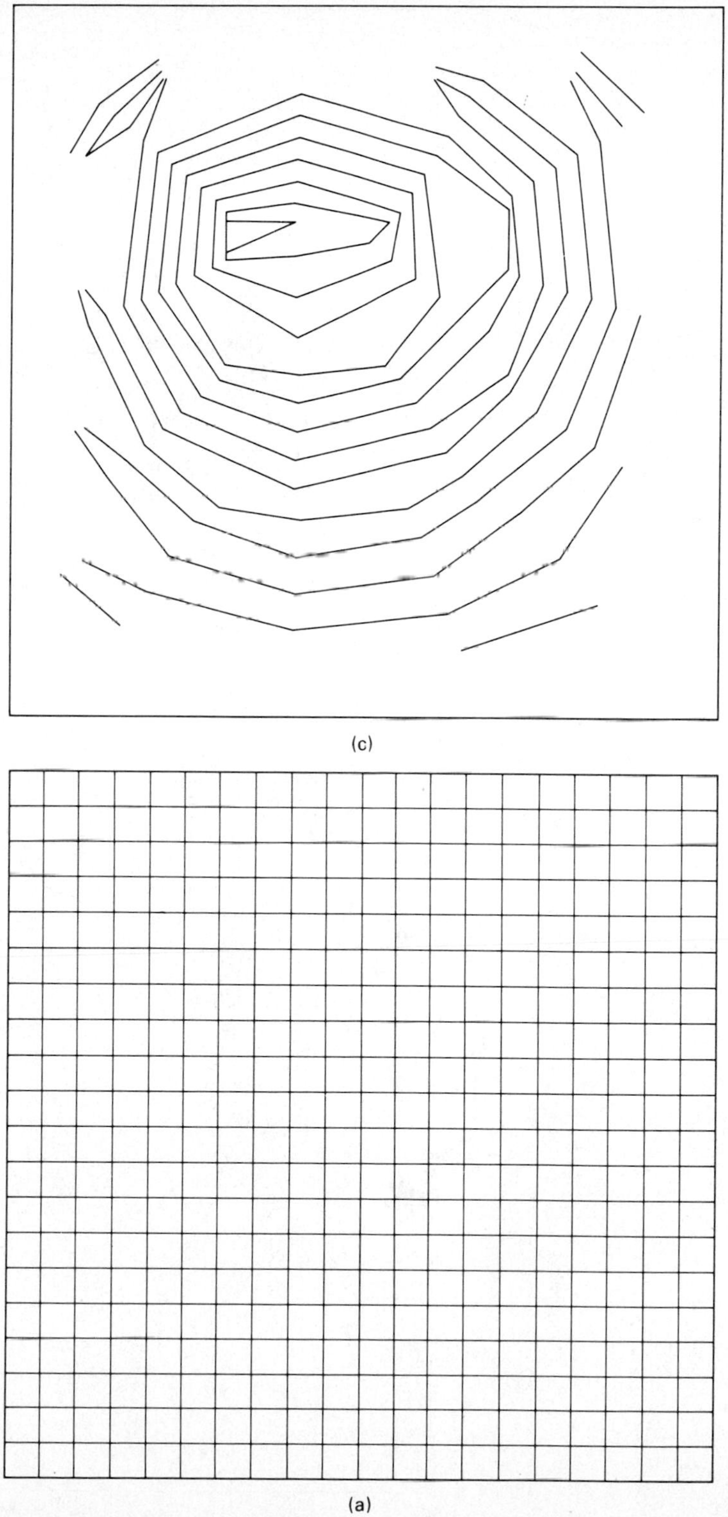

Figure 5.32 (a) 20 × 20 mesh. (b) Velocity field computed with 20 × 20 mesh. (c) Streamline pattern computed with 20 × 20 mesh.

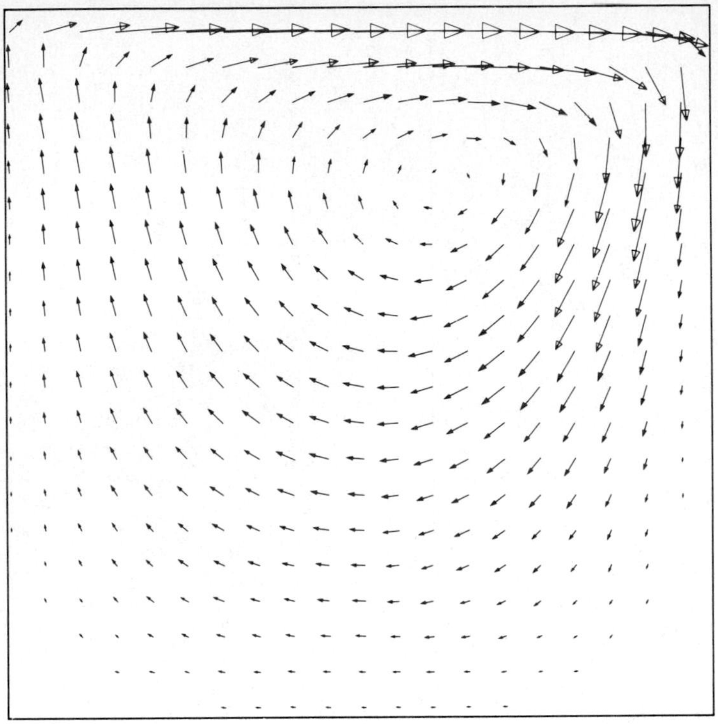

(b)

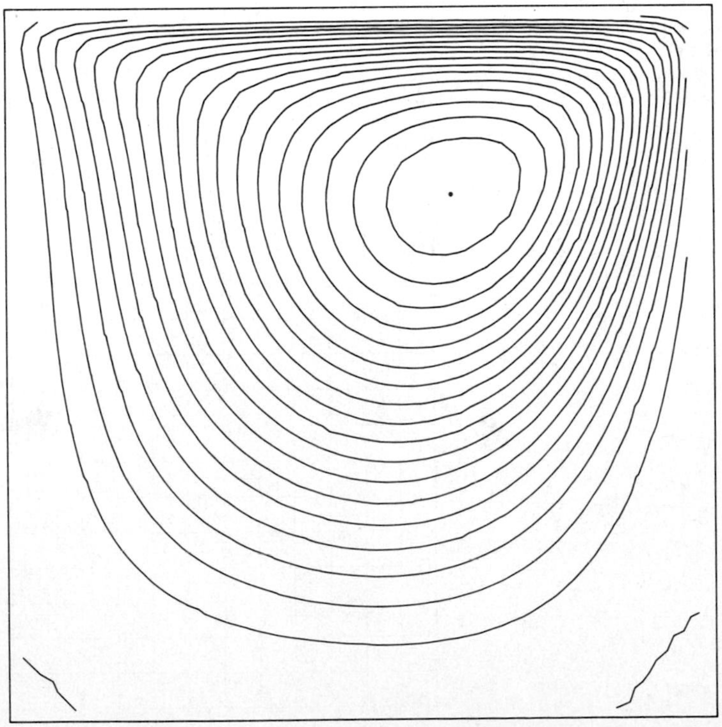

(c)

Figure 5.33 (a) 40 × 40 mesh. (b) Velocity field computed with 40 × 40 mesh. (c) Streamline pattern computed with 40 × 40 mesh.

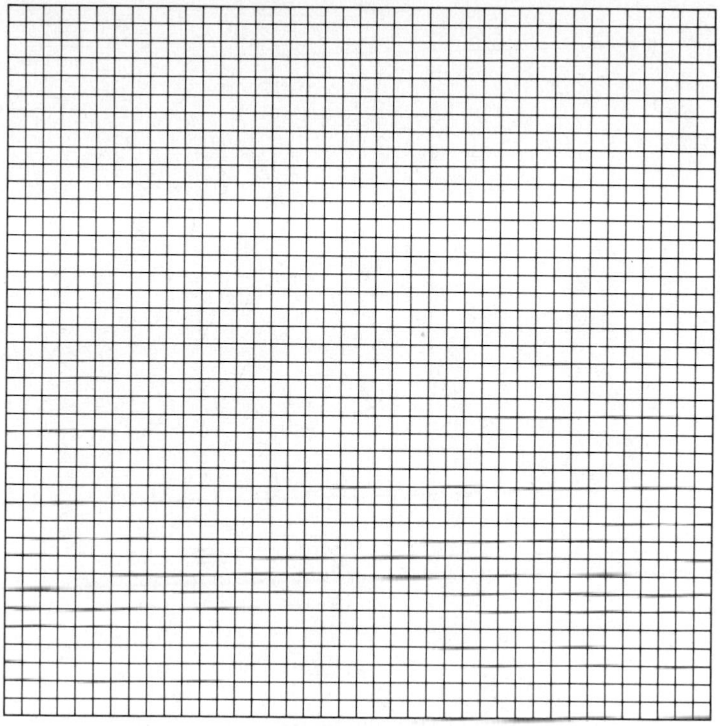

(a)

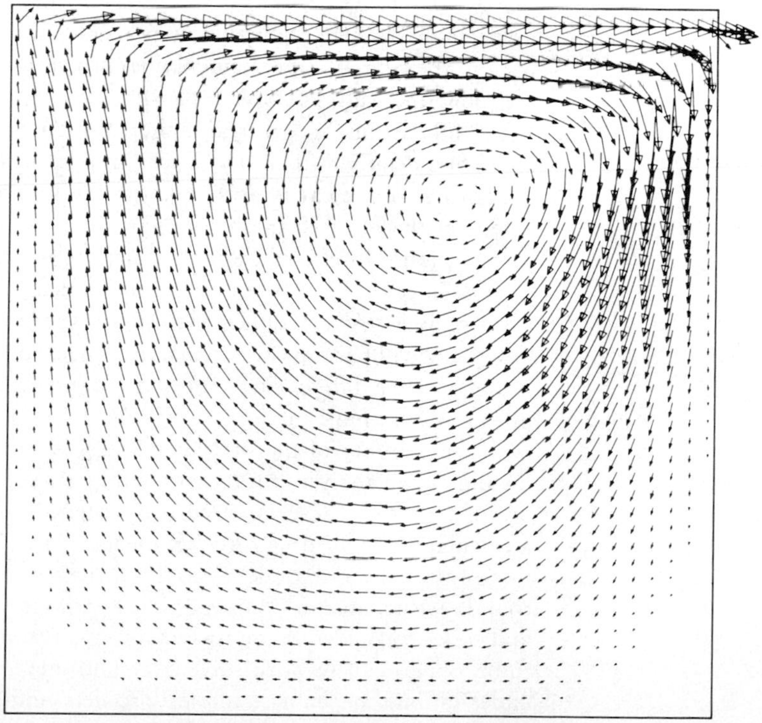

(b)

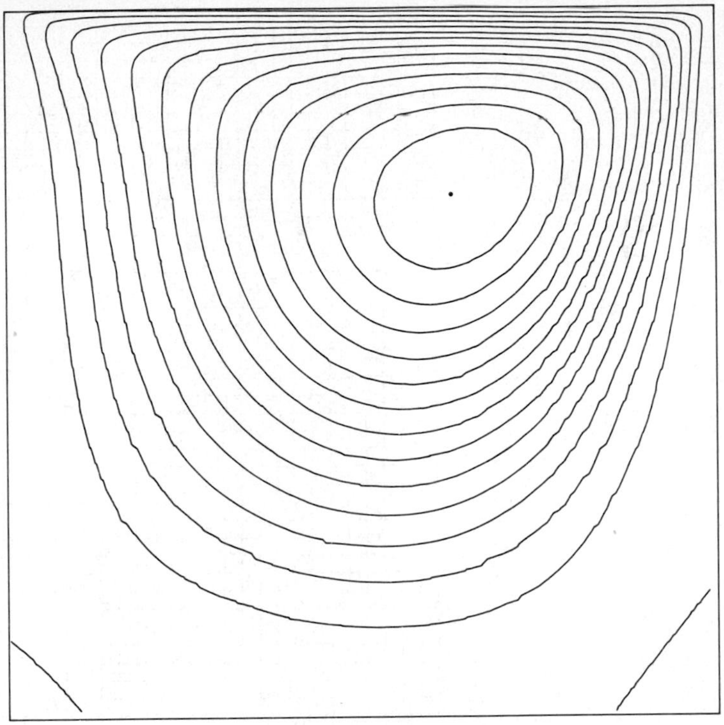

(c)

associated flow field prediction. It appears that quadrupling the number of volumes used in the computational model does not significantly alter the flow solution, and we judge the analysis to have converged.

While the main features of the solution are correct, we have in fact missed some fine details in the flow. There should be small counter-rotating vortexes in the corners of the cavity away from the sliding plate. Not only were they not captured by the 40×40 mesh, they are unlikely to be captured using the present formulation no matter how fine a mesh is used. The culprit is the "artificial viscosity" introduced by the upwind differencing in the momentum equation. This additional viscosity is sufficient to suppress the weak corner vortexes to the extent that they do not appear in our numerical solution. Capturing these vortexes requires use of a better differencing procedure for evaluating the flow-of-momentum terms in the momentum equations.

The inability of the preceding analysis to capture the finest details does not invalidate the analysis but rather emphasizes the need for care and judgment in the use of CFD. The absence of corner vortexes will probably not significantly affect a calculation of the drag force on the flat plate but would certainly effect a simulation of the collection of debris in the corners of the cavity. We have come full-circle to the caveat stated in Chapter 1. Used without understanding, powerful computers and codes can generate impressive-looking results, which may be inappropriate for the problem you want to solve. A sound understanding of the fundamental principles of fluid mechanics, and the problem under study, is required to ensure that accurate solutions are obtained from a CFD analysis.

PROBLEMS

1. The velocity components of u and v of a two-dimensional flow are given by

$$u = ax + \frac{bx}{x^2y^2} \quad \text{and} \quad v = ay + \frac{by}{x^2y^2},$$

where a and b are constants. Calculate the vorticity, the rate of volumetric strain, and the rate of shear deformation.

2. Is the flow of Problem 1 irrotational? Justify your answer.

3. Find the accelerations a_x and a_y of the flow in Problem 1.

4. The vorticity at any point in a flow is given by the curl of the velocity vector $\vec{V}$,

$$\vec{\zeta} = \nabla \times \vec{V},$$

where ∇ is the gradient operator. Show that this expression reduces to Eq. (5.12) for two-dimensional flow in Cartesian coordinates.

5. The vorticity at any point in a flow field is given by the curl of the velocity vector,

$$\vec{\zeta} = \nabla \times \vec{V},$$

where ∇ is the gradient operator. In two-dimensional cylindrical coordinates (r, θ),

$$\nabla = \hat{r}\frac{1}{r}\left(\frac{\partial(r\;)}{\partial r}\right) + \hat{\theta}\frac{1}{r}\left(\frac{\partial}{\partial\theta}\right) \text{ and } \vec{V} = \hat{r}V_r + \hat{\theta}V_\theta,$$

where $\hat{r}$ and $\hat{\theta}$ are the respective unit vectors. Show that

$$\zeta_x = \frac{1}{r}\left(\frac{\partial}{\partial r}\right)(rV_\theta) - \frac{1}{r}\left(\frac{\partial V_r}{\partial\theta}\right).$$

6. Consider the flow field

$$V_r = \frac{a}{r^2}\cos\theta, \qquad V_\theta = \frac{a}{r^2}\sin\theta.$$

Use the result of Problem 5 to determine the vorticity ζ_x. Is this flow irrotational?

7. Consider the velocity field $V_\theta = br$, where b is a constant. Use the result of Problem 5 to determine the vorticity. Is this flow irrotational?

8. Consider the velocity field $V_\theta = b/r$, where b is a constant. Use the result of Problem 5 to determine the vorticity. Is this flow irrotational?

•**9.** Show that the flow field described by the velocity components

$$u = \frac{-2xyz}{(x^2+y^2)^2}, \qquad v = \frac{(x^2-y^2)z}{(x^2+y^2)^2},$$

and

$$w = \frac{y}{x^2+y^2}$$

is a possible incompressible fluid flow. What is the vorticity?

•**10.** Find the volumetric strain for the flow field described in Problem 9.

•**11.** Find the rate of shear deformation for the flow field described in Problem 9.

•**12.** Determine the acceleration a_x of a fluid particle as a function of x, y, and z for the flow field described in Problem 9.

13. The stream function in polar coordinates, for an inviscid, constant density, two-dimensional, and two-directional flow is $\Psi = A\theta - Br$, where $A = 4.0$ ft^2/sec·rad and $B = 1.0$ ft/sec. Plot the streamline for $\Psi = 2.0$ ft^2/sec with $\theta > 0.50$ rad.

14. For the two-dimensional and two-directional flow over the cylinder shown in Fig. P5.14, the radial velocity V_r and the circumferential velocity V_θ are given by

$$V_r = -V_\infty \cos\theta\left(1 - \frac{R^2}{r^2}\right)$$

and

$$V_\theta = V_\infty \sin\theta\left(1 + \frac{R^2}{r^2}\right)$$

for $r \geq R$, where V_∞ is the fluid velocity far upstream from the cylinder. The fluid density is constant, and the flow is two dimensional so a stream function Ψ exists. In cylindrical coordinates, Ψ is related to V_r and V_θ by

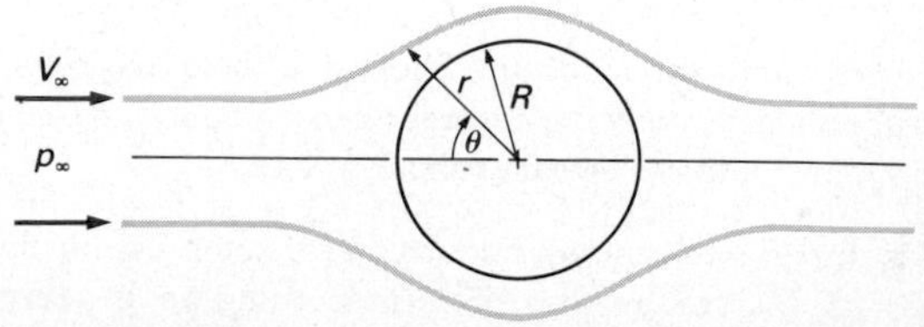

Figure P5.14

$$V_r = \frac{1}{r}\frac{\partial \Psi}{\partial \theta} \quad \text{and} \quad V_\theta = -\frac{\partial \Psi}{\partial r}.$$

Find $\Psi(r, \theta)$.

15. For the two-dimensional and two-directional flow over the cylinder shown in Fig. P5.15, the radial velocity V_r and circumferential velocity V_θ are given by

$$V_r = -V_\infty \cos\theta \left(1 - \frac{R^2}{r^2}\right)$$

and

$$V_\theta = V_\infty \sin\theta \left(1 + \frac{R^2}{r^2}\right) + \frac{2RV_\infty}{r}$$

for $r \geq R$, where V_∞ is the fluid velocity far upstream from the cylinder. The fluid density is constant, and the flow is two dimensional so a stream function Ψ exists. In cylindrical coordinates, Ψ is related to V_r and V_θ by

$$V_r = \frac{1}{r}\frac{\partial \Psi}{\partial \theta} \quad \text{and} \quad V_\theta = -\frac{\partial \Psi}{\partial r}.$$

Find $\Psi(r, \theta)$.

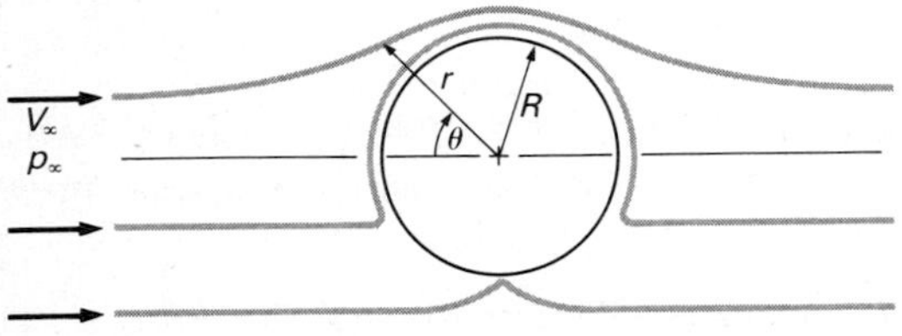

Figure P5.15

16. A two-dimensional, two-directional flow has the stream function

$$\Psi(x, y) = -V_\infty y\left(1 - \frac{R^2}{x^2 + y^2}\right) + C,$$

$$x^2 + y^2 \geq R^2,$$

where C is a constant. Find $u(x, y)$ and $v(x, y)$.

17. A two-dimensional, two-directional flow has the velocity components

$$u = ax + \frac{b}{xy^2} \quad \text{and} \quad v = -\left(ay + \frac{b}{x^2 y}\right),$$

where a and b are constant. Show that these two velocity components satisfy the constant density continuity equation, and find the stream function $\Psi(x, y)$.

18. Integrate Bernoulli's equation for compressible flow, Eq. (5.47), for an ideal gas undergoing an isothermal process along a streamline.

19. Integrate Bernoulli's equation for compressible flow, Eq. (5.47), for a fluid whose pressure p and density ρ obey the expression $p = C\rho^n$ along a streamline, where C and n are constants.

20. Consider the two-dimensional and two-directional air flow over the cylinder shown in Fig. P5.20. The radial velocity V_r and the circumferential velocity V_θ are given by

$$V_r = -V_\infty \cos\theta \left(1 - \frac{R^2}{r^2}\right)$$

and

$$V_\theta = V_\infty \sin\theta \left(1 + \frac{R^2}{r^2}\right),$$

where V_∞ is the velocity far upstream from the cylinder of radius R and p_∞ is the static pressure far upstream from the cylinder; $r \geq R$ and θ is the angle as shown. Assume constant density, steady flow, negligible gravity, and inviscid flow. Plot the pressure coefficient $[(p - p_\infty/\frac{1}{2}\rho V_\infty^2)]$ on the surface of the cylinder $(r = R)$.

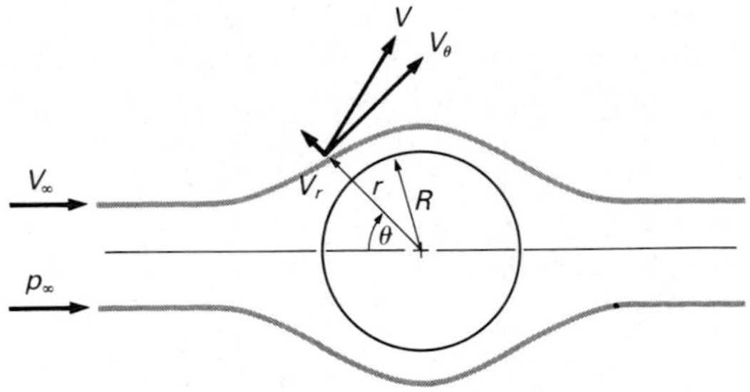

Figure P5.20

21. Consider the two-dimensional and two-directional air flow over the cylinder shown in Fig. P5.21. The radial velocity V_r and the circumferential velocity V_θ are given by

$$V_r = -V_\infty \cos\theta \left(1 - \frac{R^2}{r^2}\right)$$

and

$$V_\theta = V_\infty \sin\theta \left(1 + \frac{R^2}{r^2}\right) + \frac{2RV_\infty}{r},$$

where V_∞ is the velocity far upstream from the cylinder of radius R and p_∞ is the static pressure far upstream from the cylinder; $r \geq R$ and θ is the angle as shown. Assume constant density, steady flow, negligible gravity, and inviscid flow. Plot the pressure coefficient (see Problem 20) on the surface of the cylinder.

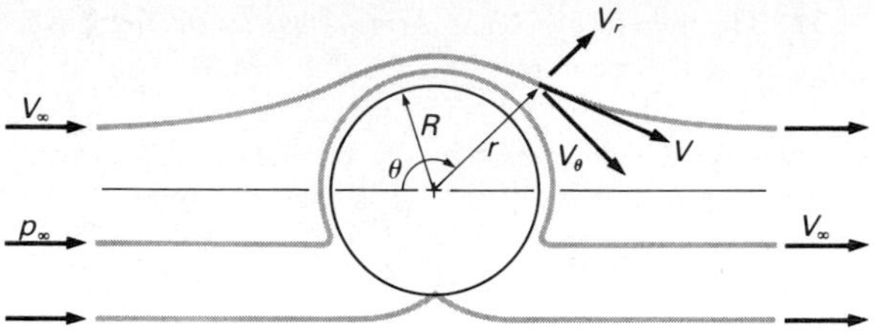

Figure P5.21

22. The fluid velocity for the air flow in Problem 20 is given by

$$V = V_\infty \sqrt{1 + \frac{R^4}{r^4} - \frac{2R^2}{r^2} \cos 2\theta}, \qquad \text{for } r \geq R.$$

Find the location on the cylinder ($r = R$) where the pressure, p, is a maximum.

23. The fluid velocity for the air flow in Problem 21 is given by

$$V = V_\infty \left(1 + \frac{R^2}{r^2} + \frac{2R}{r} \sin \theta\right), \qquad \text{for } r \geq R.$$

Find the location on the cylinder ($r = R$) where the pressure, p, is a minimum.

24. Water flows through an internal corner with the velocity given by $u = Cx$ and $v = -Cy$, with $C = 1.0/\text{s}$. Determine $p\{x, y\}$ for a constant-density fluid and inviscid and steady flow. Neglect gravity.

25. Consider the two-dimensional and two-directional air flow around the corner, as shown in Fig. P5.25. The x- and y-direction velocities are

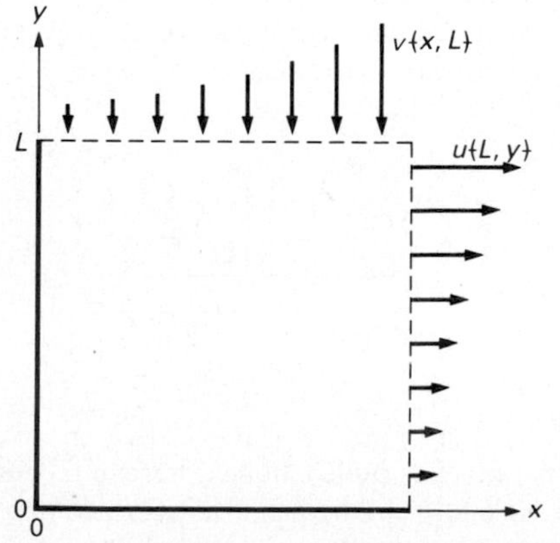

Figure P5.25

$$u = \frac{v_0}{L} \sinh\left(\frac{x}{L}\right) \cosh\left(\frac{y}{L}\right)$$

and

$$v = -\frac{v_0}{L} \cosh\left(\frac{x}{L}\right) \sinh\left(\frac{y}{L}\right),$$

respectively. Assume constant density, steady flow, negligible gravity, and inviscid flow. Find $p\{x, y\}$.

26. Sludge, which is formed from foreign impurities and oil, is denser than oil. A rotating cylinder is used as a centrifugal oil separator. Where on the bottom of the rotating cylinder should the drain pipes be located to remove the sludge? Where should the drain pipes be located to remove the oil?

27. A centrifugal water pump is advertised to have a shutoff head of 95 m. (The shutoff head occurs at zero flow rate.) The pump impeller has a radius of 0.46 m and rotates at a speed of 900 rpm. Can the advertisement be correct?

28. Find the forces F_x and F_y exerted on the stationary vane by the water jet shown in Fig. P5.28. Use the equation normal to flow to find the pressure distribution on the vane. Assume a uniform velocity over each area normal to flow; $V_0 = 12$ m/s and A_0 = flow area = 1.0 m^2. The radius of curvature R of the vane is given by

$$R = \frac{[1 + (dy/dx)^2]^{3/2}}{d^2y/dx^2}.$$

Assume inviscid flow and neglect gravity effects. The equation for the vane is $y = x^2/L$, where x and y are in meters and $L = 10.0$ m. The vane width $b = 4.0$ m.

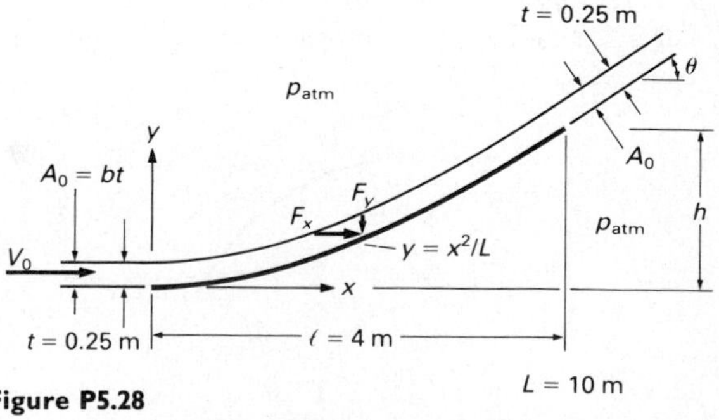

Figure P5.28

29. Find the forces F_x and F_y exerted on the vane moving with constant velocity U by the water jet, as shown in Fig. P5.29. Use the equation normal to flow to find the pressure distribution on the vane. Assume a uniform

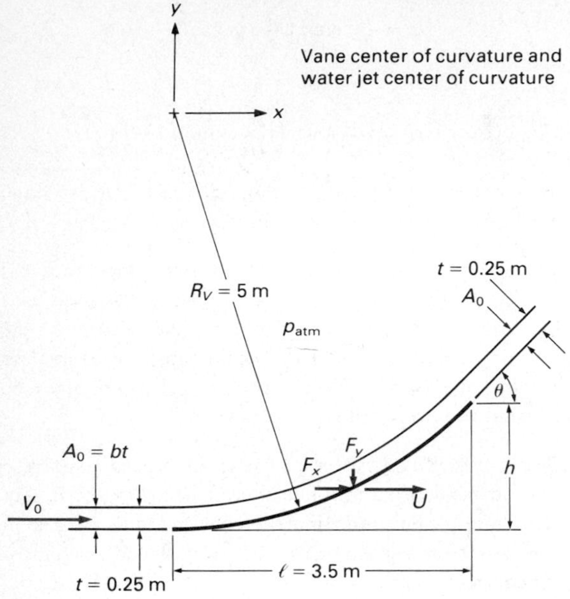

Figure P5.29

velocity over each area normal to flow; $V_0 = 12$ m/s, A_0 = flow area = 1.0 m^2, and $U = 4$ m/s. Assume inviscid flow and steady state and neglect gravity effects. The equation for the vane is $x^2 + y^2 = R_v^2$ where $R_v = 5$ m.

30. Water enters the two-dimensional elbow shown in Fig. P5.30. The elbow has a width $b = 0.5$ m, an inner radius $R_1 = 0.10$ m, and an outer radius $R_2 = 0.25$ m. The volume flow rate is 1.0 m^3/s. The flow is one-dimensional (radius r) and one-directional (V_θ), with $V_\theta = Cr$ and C a constant. The pressure at radius R_1 is $p_1 = 125$ kPa absolute. What is the pressure p_2 at radius R_2? Assume constant water density and inviscid, steady flow.

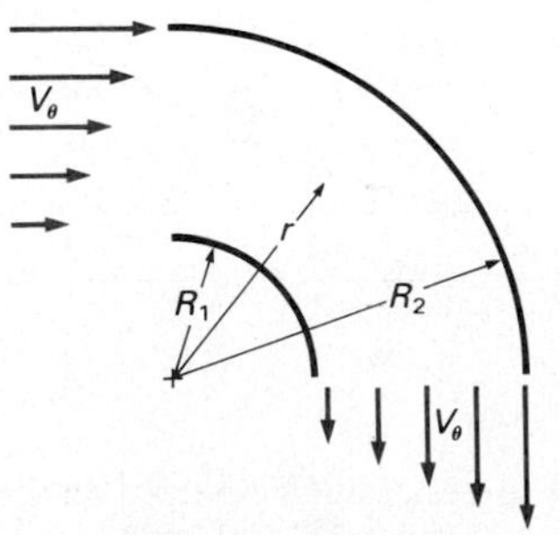

Figure P5.30

31. The fluid motion that forms above an open drain can be modeled as a free vortex (Fig. P5.31). For V_A = 0.50 m/s, $V_B = 0.10$ m/s, $r_A = 0.20$ m, and $r_B = 1.0$ m, calculate the difference in the elevation of the surface between A and B. The flow is incompressible and inviscid.

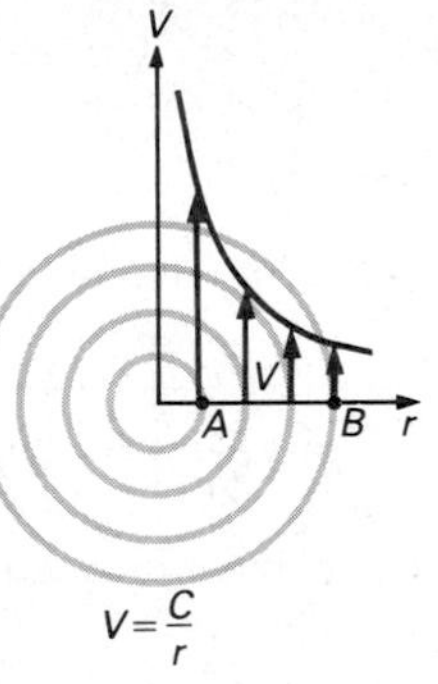

Figure P5.31

32. Model a cyclone as a free vortex. The velocity is 7.5 km/hr at a radius of 30 km from the center, where the atmospheric pressure is 105 kPa. Find the atmospheric pressure at a radius of 3 km from the center of the cyclone for the same elevation. Assume constant air density and inviscid flow.

33. Consider the velocity field

$$u = A(x^2 - y^2), \qquad v = -2Axy, \quad \text{and} \quad w = 0,$$

with $g_z = -g$. Determine the pressure distribution $p(x, y)$ for a Newtonian fluid with constant viscosity and density.

34. The disc shown in Fig. P5.34 has a diameter D of 1 m and a rotational speed of 1800 rpm. It is positioned 4 mm from a solid boundary. The gap is filled with 15°C SAE 10W oil. Find the torque required to overcome the frictional resistance of the oil on the disc. Assume that the circumferential velocity (V_θ) is linear across the gap at each radius r.

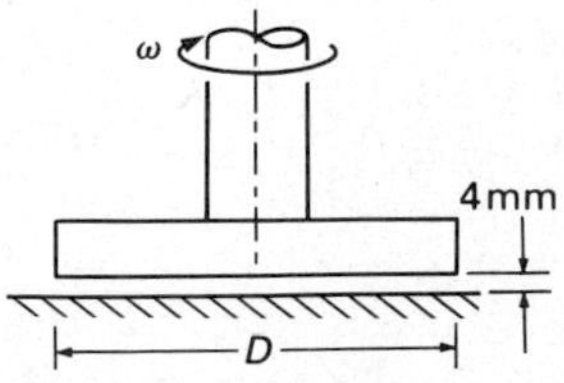

Figure P5.34

35. A flat block is pulled along a horizontal flat surface by a horizontal rope perpendicular to one of the sides. The block measures 1.0 m × 1.0 m, has a mass of 100 kg and a constant velocity of 1.0 m/s, and is separated from the flat surface by a 0.10-cm-thick oil layer of 15°C SAE 20 crankcase oil. Find the coefficient of friction for the block. Does this coefficient of friction change with the velocity?

36. Start with Newton's second law for a fluid element and derive an equation for the velocity profile of a viscous fluid entrained between two wide, long, horizontal, parallel flat plates. The plates are separated by a distance $2Y$. The upper plate moves with a constant horizontal velocity U_x, and the lower plate is fixed. [*Hint:* Assume fully developed flow with an axial pressure gradient of zero.] Assume steady flow.

• **37.** Derive an equation for the velocity profile of a viscous fluid entrained between two long concentric cylinders. The inner radius is R_i and the outer radius is R_o. The inner cylinder is stationary and the outer cylinder rotates with a constant peripheral speed of U_o. See Fig. P5.37.

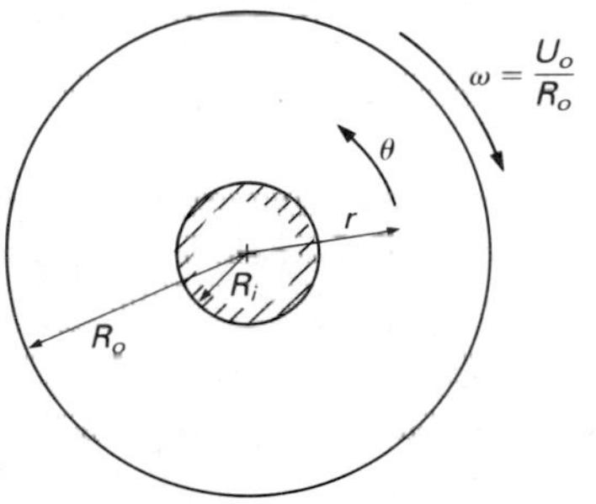

Figure P5.37

• **38.** In Example 1.3 in Chapter 1, a working formula for a concentric cylinder viscometer was developed:

$$\mu = \frac{\mathcal{T} h}{2\pi H \omega R_1^3}.$$

Torque $\mathcal{T}$ causes the inner cylinder to rotate as a constant angular velocity ω. The viscometer has a height H, a gap width h, and radii R_1 and R_2, where $h \ll R_1$ or R_2. Develop an improved working formula relating the fluid viscosity μ and the torque $\mathcal{T}$ by considering the viscous friction between the inner cylinder and the fluid along both the side and bottom of the cylinder. The thickness of the viscous fluid below the inner cylinder is h_1. See Fig. P5.38.

• **39.** Show that the velocity $u(y)$ for the simple model of the slipper bearing shown in Fig. P5.39 is given by

$$u = \frac{1}{2\mu}\frac{dp}{dx}(y^2 - hy) + U\left(\frac{y}{h}\right).$$

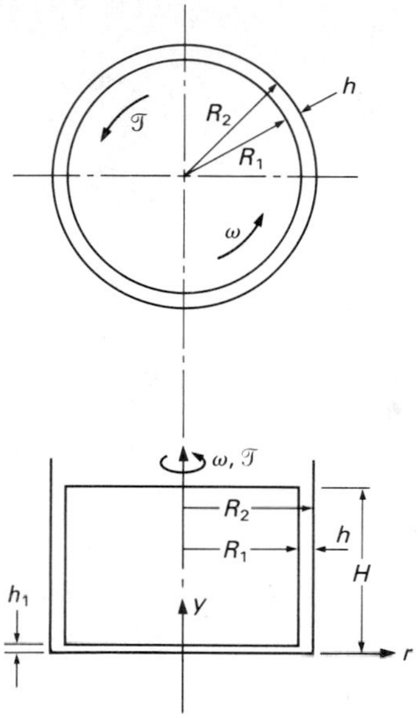

Figure P5.38

Assume that the bearing is infinitely wide. Also assume that $h_0 \ll \ell$, $h_\ell \ll \ell$, $\partial u/\partial x \approx 0$, and $\partial^2 u/\partial x^2 \approx 0$ and that the flow is steady.

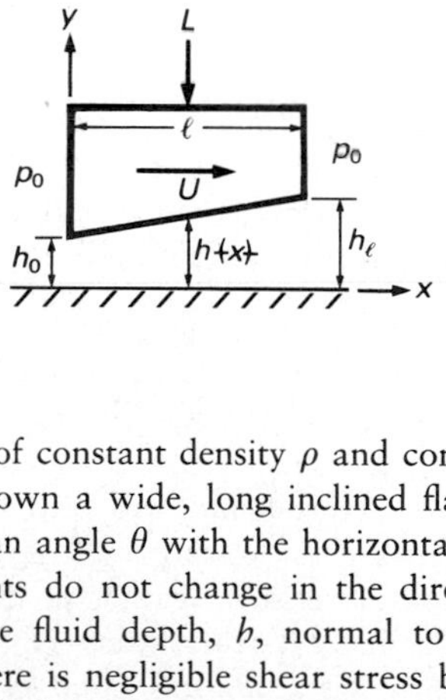

Figure P5.39

• **40.** A liquid of constant density ρ and constant viscosity μ flows down a wide, long inclined flat plate. The plate makes an angle θ with the horizontal. The velocity components do not change in the direction of the plate, and the fluid depth, h, normal to the plate is constant. There is negligible shear stress by the air on the fluid. Find the velocity profile $u(y)$, where u is the velocity parallel to the plate and y is measured perpendicular to the plate. Write an expression for the volume flow rate per unit width of the plate.

• **41.** The Navier–Stokes equations for steady flow, constant density, and constant viscosity are given in Appendix H. Simplify these equations to determine the differential equation for both the rotational velocity $V_\theta(r)$ and the axial velocity $V_x(r)$ for the annulus shown in

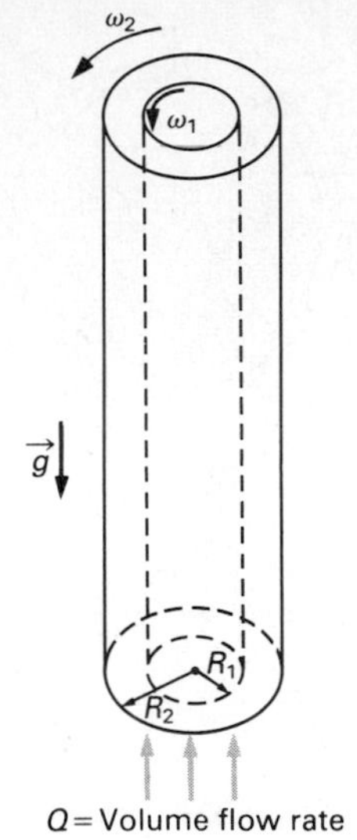

Figure P5.41

Fig. P5.41. Assume a fully developed, axial flow and angular symmetry ($\partial/\partial\theta = 0$).

• **42.** A Bingham plastic is a fluid in which the stress τ is related to the rate of strain du/dy by

$$\tau = \tau_0 + \mu\left(\frac{du}{dy}\right),$$

where τ_0 and μ are constants. Consider the flow of a Bingham plastic between two fixed, horizontal, infinitely wide, flat plates. For fully developed flow with $du/dx = 0$ and dp/dx constant, find $u\{y\}$.

• **43.** The bearing shown in Fig. P5.43 consists of two parallel discs of radius R separated from each other by a small distance $h (h \ll R)$. The lower disc is made of a porous material, and an incompressible, viscous fluid is pumped through it and into the gap, filling the space between the discs completely. The pores of the lower disc are closely spaced, so that the velocity of the fluid as it leaves the surface of the porous disc may be regarded as a uniform value w_0, that is, small compared to the mean radial velocity. Find the radial velocity u and the load L that the bearing can support as a function of w_0, μ, R, and h. The inertia terms are negligible, the circumferential velocity is zero ($V_\theta = 0$), and angular symmetry exists ($\partial/\partial\theta = 0$).

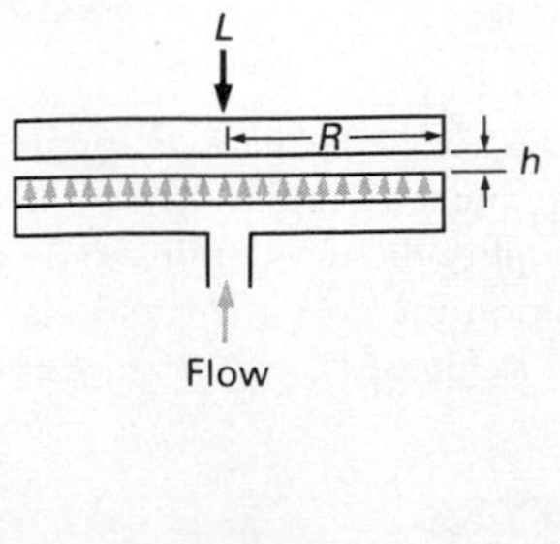

Figure P5.43

• **44.** A viscosity motor/pump is shown in Fig. P5.44. The rotor is concentric within a stationary housing. The clearance h between the housing and the rotor is small compared to the width w and radius R of the rotor. A seal divides the clearance space as shown. When the device is operated as a motor, the viscous liquid is introduced at the high-pressure side of the seal and leaves at the low-pressure side of the seal. This flow causes the rotor to turn and produce power. For $R = 1.0$ ft, $h = 0.50$ in., $w = 1.0$ ft, $\omega = 10$ rpm, and a flow rate of 1.5 ft^3/min, find the power output of the rotor. The fluid is 60°F, SAE 10 crankcase oil. Neglect gravity.

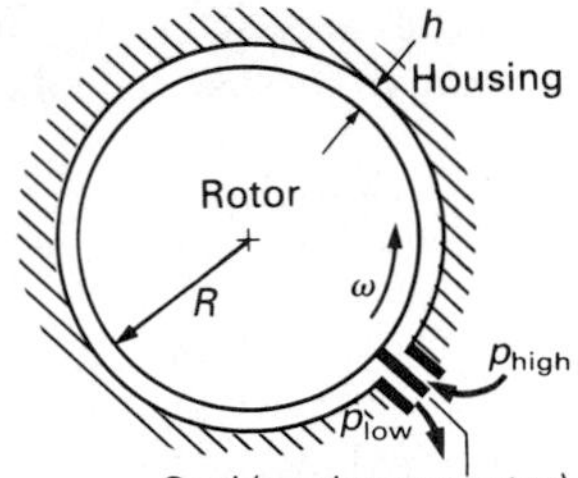

Figure P5.44

• **45.** The basic equations for the flow of a constant-density, constant-viscosity fluid in the entrance region of a circular pipe are

$$V_x \frac{\partial V_x}{\partial x} + V_r \frac{\partial V_x}{\partial r} = -\frac{1}{\rho}\frac{\partial p}{\partial x} + \nu\left[\frac{\partial^2 V_x}{\partial r^2} + \frac{1}{r}\frac{\partial V_x}{\partial r}\right]$$

and

$$\frac{\partial V_x}{\partial x} + \frac{1}{r}\frac{\partial (rV_r)}{\partial r} = 0.$$

Refer to Fig. P5.45 and write the appropriate boundary conditions.

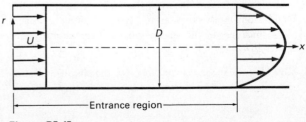

Figure P5.45

46. Figure P5.46 shows a jet of diameter d and velocity V_p mixed with the slower velocity stream of outer diameter D and velocity V_s. An incompressible Navier–Stokes equation and the continuity equation are used to find the jet velocity as it mixes with the slower stream. These equations are

$$V_x \frac{\partial V_x}{\partial x} + V_r \frac{\partial V_x}{\partial r} = \frac{\nu}{r} \frac{\partial}{\partial r}\left(\frac{r\,\partial V_x}{\partial r}\right)$$

and

$$\frac{\partial V_x}{\partial x} + \frac{1}{r}\left(\frac{\partial (rV_r)}{\partial r}\right) = 0.$$

Find the appropriate boundary conditions.

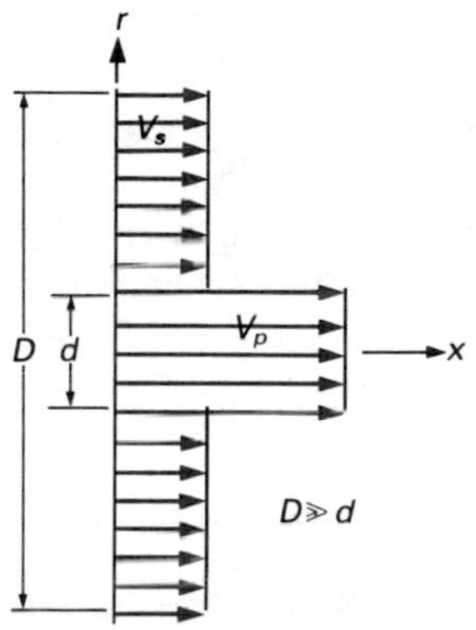

Figure P5.46

47. Natural convection occurs in a fluid when gravitational forces are present and density variations arise because of temperature variations. For many problems, pressure variations do not significantly change the density, and the density variations can be accounted for by using the coefficient of thermal expansion α_T. The corresponding equations for the condition described in Fig. P5.47 are

$$u\frac{\partial u}{\partial x} + v\frac{\partial u}{\partial y} = \nu\left(\frac{\partial^2 u}{\partial y^2}\right) + \alpha_T g_x (T - T_\infty),$$

$$\frac{\partial u}{\partial x} + \frac{\partial v}{\partial y} = 0, \quad \text{and} \quad u\frac{\partial T}{\partial x} + v\frac{\partial T}{\partial y} = \frac{k}{\rho c_p}\frac{\partial^2 T}{\partial y^2},$$

where k is the thermal conductivity and T is the fluid temperature. Determine the appropriate boundary conditions.

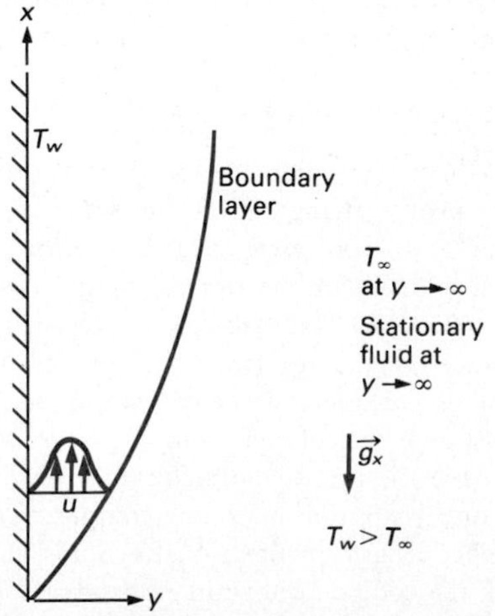

Figure P5.47

48. A constant-density fluid flows in laminar flow through a circular pipe of inside radius R and length L. The axial pressure gradient, dp/dx, is a constant C, and the flow is fully developed. The governing equation for the axial velocity V_x is

$$\frac{1}{\rho}\frac{dp}{dx} = \nu\left(\frac{d^2 V_x}{dr^2} + \frac{1}{r}\frac{dV_x}{dr}\right),$$

where x is the axial coordinate and r is the radial coordinate. Find the appropriate boundary conditions for V_x as a function of r and dp/dx.

49. A constant-density fluid flows in laminar flow through a rectangular duct $2A$ units wide and $2B$ units high; the duct length is L with axial pressure drop Δp and the flow is fully developed. The governing equation for the axial velocity u is

$$\frac{1}{\rho}\frac{dp}{dx} = \nu\left(\frac{\partial^2 u}{\partial y^2} + \frac{\partial^2 u}{\partial z^2}\right),$$

where x is the axial coordinate and y and z are the transverse coordinates. The duct walls are at $y = \pm A$ and $z = \pm B$. Find the appropriate boundary conditions for u as a function of y, z, Δp, and L.

50. "Stokes's first problem" involves the instantaneous acceleration at time $t = 0$ of a flat plate to a constant velocity U_0 while in contact with a "semi-infinite,"

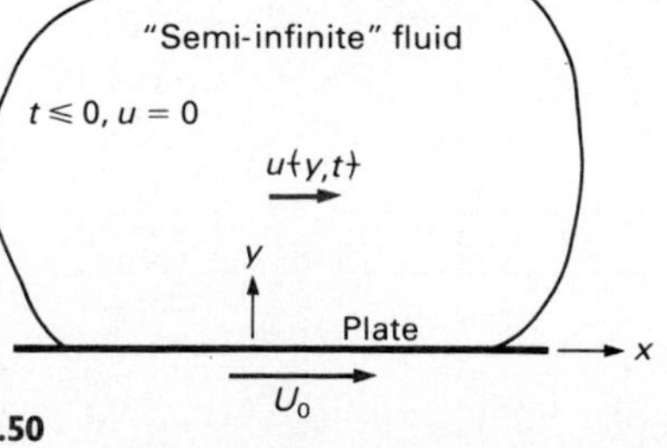

Figure P5.50

static fluid as shown in Fig. P5.50. For a constant fluid density and viscosity, the simplified Navier–Stokes equation is

$$\frac{\partial u}{\partial t} = \nu \frac{\partial^2 u}{\partial y^2},$$

where u is the fluid velocity in the x or velocity U_0 direction and y is a coordinate normal to the plate. Find the appropriate boundary conditions and initial conditions for this problem.

51. "Stokes's second problem" involves the oscillatory motion of a flat plate with velocity $U(0, t) = U_0 \cos \omega t$ while in contact with a "semi-infinite" fluid, as shown in Fig. P5.51, where U_0 and ω are constants. For a constant fluid density and viscosity, the simplified Navier–Stokes equation is

$$\frac{\partial u}{\partial t} = \nu \frac{\partial^2 u}{\partial y^2},$$

where u is the fluid velocity in the x direction, and y is a coordinate normal to the plate. Determine the appropriate boundary conditions for this problem.

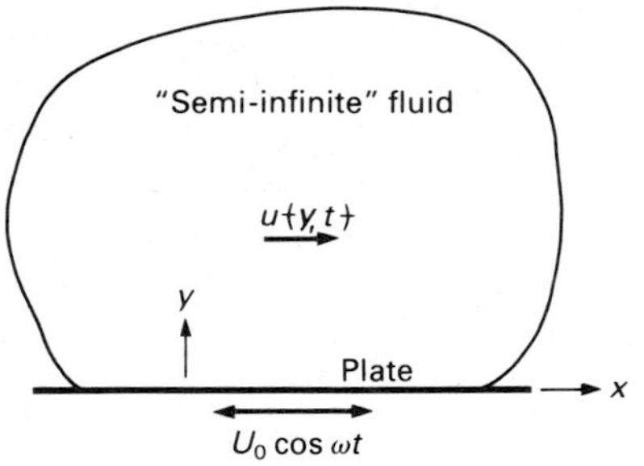

Figure P5.51

• **52.** Consider the fully developed flow in a wide rectangular channel of width $2w$ and height $2h$, where $2w \gg 2h$. The channel length is L, with axial pressure drop Δp. The Cauchy equation in the flow direction (x direction) for steady state is

$$0 = -\frac{dp}{dx} + \frac{\partial \tau_{yx}}{\partial y} + \frac{\partial \tau_{zx}}{\partial z},$$

where τ_{yx} and τ_{zx} are the shear stresses in the x direction. Find the boundary conditions for finding $\tau_{yx}(y, z)$, where y is the transverse coordinate in the height direction, as shown in Fig. P5.52. Neglect gravity.

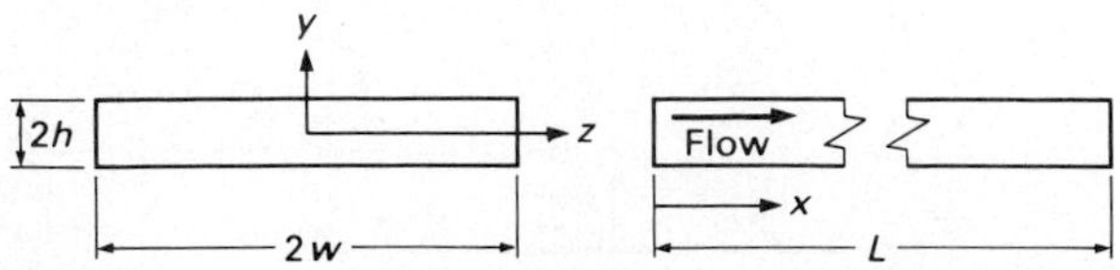

Figure P5.52

53. Compute the flow in a driven cavity with a cavity Reynolds number of 500, using the code described in Appendix I with a uniform grid spacing. After performing a grid convergence study to support your choice of a suitable mesh, compare your results to those presented in the text for a Reynolds number of 100.

54. Compute the flow in a driven cavity with a cavity Reynolds number of 1000 using the code described in Appendix I with a uniform grid spacing. After performing a grid convergence study to support your choice of a suitable mesh, compare your results to those presented in the text for a Reynolds number of 100.

55. Compute the flow in a driven cavity with a cavity Reynolds number of 100 using the code described in Appendix I with a nonuniform grid spacing. Although you are urged to use the minimum number of cells to obtain a converged solution, you should remember that large changes in cell size between adjacent cells introduce significant error into your solution. Compare the computational demands of your converged grid to that of the 40×40 uniform grid presented in the text.

56. When flow enters a pipe from a reservoir, the velocity profile changes from an approximately uniform profile at the entrance to a "fully developed" profile, which is fixed for the rest of the pipe run. The length over which this transition in profile occurs is called the entrance length. For laminar flow, the fully developed profile is parabolic, and the entrance length can be approximately calculated from: $L_e = 0.06\mathbf{R}$, where $\mathbf{R} = \rho V d/\mu$ is the Reynolds number. Evaluate the accuracy of this approximation by computing the flow in the entrance length of pipe at a Reynolds number of 1000, using the code described in Appendix I.

57. When flow enters a pipe from a reservoir, the velocity profile changes from an approximately uniform profile at the entrance to a "fully developed" profile, which is fixed for the rest of the pipe run. The length over which this transition in profile occurs is called the entrance length. For laminar flow, the fully developed profile is parabolic, and the entrance length can be approximately calculated from: $L_e = 0.06\mathbf{R}$, where $\mathbf{R} = 4\rho Q/(\mu \pi d)$ is the Reynolds number. Evaluate the accuracy of this approximation by computing the flow in the entrance length of pipe at a Reynolds number of 2000, using the code described in Appendix I.

58. Compute the flow in a right-angle bend of a high-aspect-ratio duct by modeling as laminar plane-two-dimensional flow using the code described in Appendix I. The flow is at a Reynolds number of 500 based on duct width ($\mathbf{R} = \rho V w/\mu$; V = average velocity). You will need to evaluate the length of duct to be modeled upstream and downstream from the bend, as well as select appropriate upstream and downstream velocity profiles (i.e., uniform or parabolic). What type of downstream boundary condition would be more suitable for this analysis than a specified velocity condition?

REFERENCES

1. White, F. M., *Viscous Fluid Flow,* McGraw-Hill, New York, 1974.
2. Daily, J. W., and D. R. F. Harleman, *Fluid Dynamics,* Addison-Wesley, Reading, Mass., 1966.

6 Organizing Information About Flow: Dimensional Analysis

The control volume approach described in Chapter 4 and the differential approach described in Chapter 5 are the analytic tools that engineers use to solve flow problems. If an engineer has information about every term in the equations and if the mathematical or computing tools are available to solve them, the information generated by a flow analysis can be used with confidence. Unfortunately, purely analytic methods are limited by lack of complete information about certain flow phenomena (such as turbulence) and the sheer difficulty of the mathematics and computations required.

One common practical problem is that of relating flow rate and mechanical energy loss in fully developed pipe flow. Suppose that steam is flowing in the pipe. We know that the flow may be either laminar or turbulent, but how do we decide which flow state is likely to exist in a particular case? Although steam is a compressible fluid, we know that sometimes we can assume that it is incompressible. When is such an assumption acceptable?

After answering these questions, we still face difficulties. If we decide that the flow is incompressible and laminar, we can obtain an excellent theoretical result. However, if the flow is turbulent, we cannot find a universally valid theoretical result. More than one hundred years of research has yet to yield a complete theory for turbulent flow in a pipe! Not surprisingly, more complicated geometries are even less likely to yield to purely theoretical analysis.

How then do engineers deal with complicated flow problems? How can airplanes, ships, harbors, and piping systems be successfully designed and operated? The answer is *experimentation*. Engineering fluid mechanics is a combination of theory and experiment, much more so than, say, structural mechanics or electric and electronic circuit analysis. For turbulent pipe flow, the "theoretical" energy equation must be supplemented by information gained from measurements of energy loss. Theoretical analysis suggests tentative designs for an airplane or a harbor. The candidate designs are evaluated and refined by extensive testing. An introduction to fluid mechanics would be incomplete without considering the contributions of experimentation.

In this chapter, we deal with the place of experimentation in fluid mechanics. It does not deal with the how-to of experimental techniques; in fact, we do not even present many experimental results. We do discuss a method by which experimental information can be packaged and transmitted efficiently from an experimenter to other engineers. Although almost all engineering fluid mechanics is, to some degree, based on experimental information, only a small percentage of fluids engineers are actually involved in performing experiments. The work of these engineers is frequently used by others. Our objective is to help you understand the forms in which experimental data are usually packaged.

The specific topic of this chapter is *dimensional analysis*. Dimensional analysis is a packaging or compacting technique used to reduce the complexity of experimental programs and at the same time increase the generality of experimental information. The information packaging produced by dimensional analysis also has other uses. It is equally applicable to theoretical (analytic) investigations and results. Also, dimensional analysis gives us the tools to decide a priori whether a flow is likely to be laminar or turbulent or whether simplifying assumptions, such as neglecting compressibility, are appropriate.

Before we discuss the subject in detail, we need to point out the following about dimensional analysis:

- Dimensional analysis is sometimes very confusing to beginners. The basic principle involved seems obvious, almost trivial. The execution of the technique, however, often seems almost magical.
- Confusion arises because we usually expect theories to be complete, that is, to begin at the beginning and carry through to the end. Dimensional analysis stops far short of providing a final answer to a particular engineering problem; it simply leads to a more efficient organization of the variables. The ultimate answer comes only in the laboratory or from analytic methods.
- Dimensional analysis tells us that we *can* organize variables efficiently and suggests possible ways that the organization can be carried out; however, it does not allow us to find unique results.

Two analogies help illustrate the utility of dimensional analysis. Dimensional analysis is similar to a trash compactor. Trash thrown into the compactor is crushed into a smaller space so it is easier to handle. The compactor does not solve the trash problem nor does it ultimately dispose of any of the trash put into it; it only reduces it to a more convenient package. In the same way, dimensional analysis reduces the number of variables that we must consider in an analytic or experimental investigation by "compacting" several of the raw variables into convenient packages.

Dimensional analysis also is similar to packing a suitcase for a trip. The traveler must select the items to take. Any item that is not selected will not be in the suitcase. The traveler can organize the items in the suitcase in one of several equally convenient ways. Different travelers prefer different arrangements, but that does not make any arrangement superior. After the suitcase is packed, it is up to the traveler to carry it to the ultimate destination. The suitcase does not go on its own, and just packing it does not accomplish the objective of the trip.

6.1 THE NEED FOR DIMENSIONAL ANALYSIS

Each experiment carried out in a laboratory is unique. The apparatus being tested has a unique geometry and a specific size. Only one fluid at a specific flow rate can be used in a particular experimental run. The unknown parameters whose evalua-

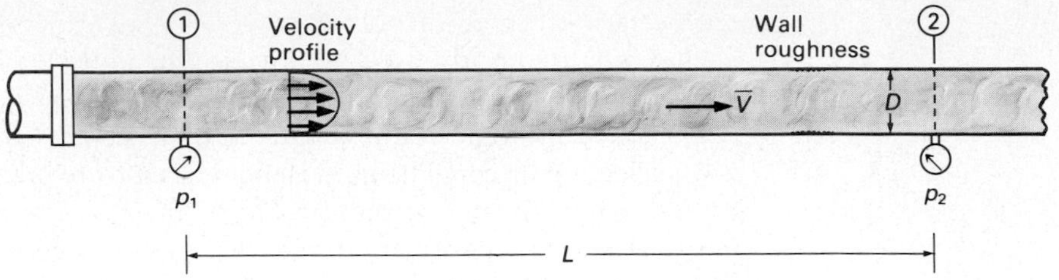

Figure 6.1 Incompressible turbulent flow in a circular pipe.

tion is the object of the experiment (e.g., energy loss, pressure distribution, or resistance force) have unique numerical values. Information about the effect of varying any "controllable" parameters, such as flow rate or fluid viscosity, can be obtained only by repeated experiments. Trends can be established only by systematic variation of the controllable parameters. Because experimentation is costly and time-consuming, experimental results should be documented so that a new experiment is not required the next time a similar problem arises. The investigator also would like to extend and generalize any information obtained from a particular experiment.

We examine two different types of flow problem that must be solved, at least in part, by experiment. The first is evaluation of mechanical energy (head) loss for fully developed incompressible turbulent flow in a pipe, illustrated in Fig. 6.1. The second problem is evaluation of the resistance force (drag) on a two-person submarine being designed for oceanographic research, illustrated in Fig. 6.2.

The first problem is quite general, because pipes appear in countless engineering applications. Thus generalization of experimental data that allows evaluation of energy loss in an arbitrary pipe flow would be extremely useful. Suppose that we are to carry out a systematic evaluation of the effects of various parameters on pipe flow energy loss. Figure 6.1 shows a possible experimental setup, with fully developed flow in a horizontal pipe. Applying the mechanical energy equation to a streamline between planes 1 and 2 gives

$$gh_L = \frac{p_1 - p_2}{\rho}. \tag{6.1}$$

We can evaluate the energy loss by measuring the pressure difference between stations 1 and 2, together with knowledge of the fluid density.

Satisfied that energy loss can be measured, we consider which parameters

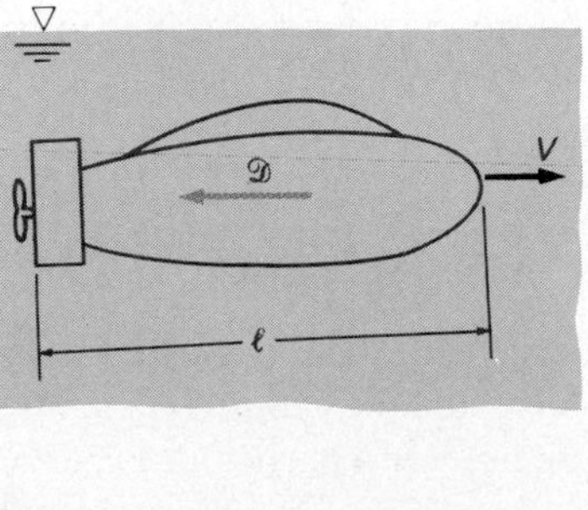

Figure 6.2 Drag force on small two-person submarine.

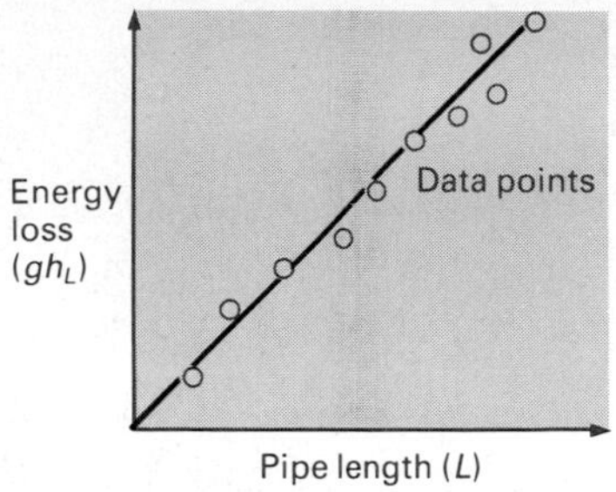

Figure 6.3 Typical results: pipe flow energy loss versus pipe length for fixed values of D, V, ρ, μ, and ε.

affect it. A little thought leads to the following list:

1. Length of pipe between 1 and 2 (L)
2. Pipe diameter (D)
3. Fluid velocity (V)
4. Fluid density (ρ)
5. Fluid viscosity (μ).

The items in this list should have been reasonably obvious. However, a final, somewhat subtle, parameter that might be missed is

6. Pipe wall roughness height (ε).

The shape of the pipe is also important, but we limit this discussion to straight pipes with circular cross section.

We might indicate that we expect *all* of these parameters to have an effect on the loss by writing the following *functional relation:*

$$gh_L = f\{L, D, V, \rho, \mu, \varepsilon\}. \tag{6.2}$$

Understanding what this relation means is extremely important. If we change *any one* of the arguments of f, we expect the energy loss to change, even if all other arguments remain fixed. The functional relation indicates only that there is interdependence between gh_L and the arguments of f. The form of the function f is not known; in fact, discovering the function is the objective of the experimental program. It need not be an analytic function; in fact, experiments simply produce tables of data (which can be presented as graphs and charts). Tables and charts are perfectly acceptable forms of f. So are curve fits of numerical data.

Now we ask: What do we need to do to discover the relation implied by Eq. (6.2), and how can we present the results? To discover the relation, we must systematically vary each parameter on the right-hand side. First we fix D, V, ρ, μ, and ε and systematically vary L. This would require, say, 10 runs (for 10 different L's) and produce results like those indicated in Fig. 6.3. These runs establish the dependence of gh_L on L only for the specific values of D, V, ρ, μ, and ε that we used in the 10 runs. Next we change to a new value of D, keep the same values of V, ρ, μ, and ε, and repeat the runs for all 10 values of L. We repeat this for, say, 10 values of D. Next we run through all combinations of L and D for, say, 10 different values of V. Figure 6.4 shows a few of the results we might expect. We proceed through the list of variables, each time repeating all combinations of the

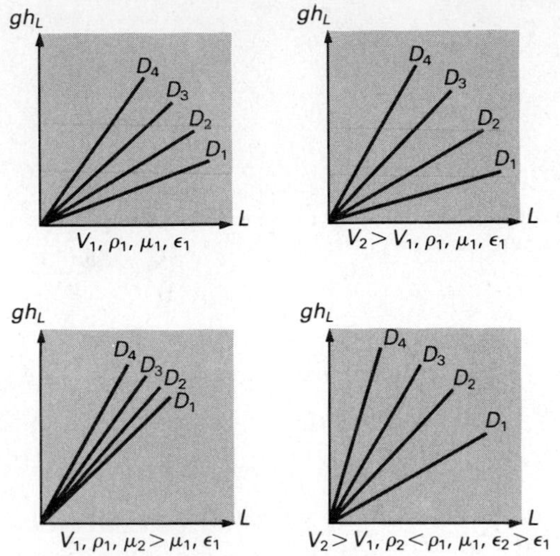

Figure 6.4 A few of the 10^4 plots necessary to present the results of a complete investigation of turbulent pipe flow.

variables preceding the current one. A complete investigation would require about 10^6 different experimental runs. The cost of such a program would be astronomical, not to mention that we would have to find 100 different fluids because we need 10 different ρ's and 10 different μ's!

To present the results of all our work, we would need about 10^5 plots like those of Fig. 6.3. We could save some paper by preparing parametric plots like those of Fig. 6.4, which would reduce the required number of plots to "only" 10^4!* The compilation of results would be so unwieldy that it would be useless.†

Dimensional analysis to the rescue! According to the fundamental theorem of dimensional analysis (see Section 6.3), we can replace the functional relation of Eq. (6.2) by a *completely equivalent* relation involving *groups* of variables, as follows:

$$\frac{gh_L}{\frac{1}{2}V^2} = F\left(\frac{L}{D}, \frac{\rho VD}{\mu}, \frac{\varepsilon}{D}\right). \tag{6.3}$$

Equation (6.3) is still a general functional relation, with the form of the *new* function F unspecified; however, it contains *all* the information implicit in Eq. (6.2). Instead of six controllable parameters, we now have only three. Even if no other simplifications could be found (and they can), we would need "only" 10^3 experiments. Only one fluid would be needed, because we can vary the second parameter on the right ($\rho VD/\mu$) by changing V instead of ρ and μ. Only 10 pages would be required to present the results. *Note, however, that someone still must run the 10^3 experiments and fill the 10 pages.*‡ The problem is much simpler but still not solved. We describe how to do this reduction in a later section.

---

* According to Kline [1], one independent parameter requires one line; two, a page; three, a book; and four, a library! We have six.

† According to "Murphy's law," any particular case you need does not exactly correspond to one of the 10^6 specifically documented, so you would have to do a six-parameter interpolation between cases.

‡ In the case of pipe flow losses, the experiments have been run and the information is available for general use. This information is discussed in Chapter 7.

Now we consider the second fluid mechanics problem, finding the drag force ($\mathscr{D}$) on a small submarine (Fig. 6.2). Because the shape of the submarine is rather complex and we would naturally expect the flow to be at least partially turbulent, accurate calculation of the drag force is very difficult and we must usually revert to experiment. Similar to the pipe problem, we first list the parameters that we expect to affect the drag force.

1. The size of the submarine. We can choose any single length dimension to characterize the size; we use the length (ℓ).
2. The speed of the submarine (V).
3. The fluid density (ρ). (Even though the submarine will operate only in water, saltwater and freshwater have different densities.)
4. The fluid viscosity (μ).

The force will, of course, depend on the specific shape of the submarine, but we suppose that we are investigating a specific design, so the shape is fixed. In functional notation, we write

$$\mathscr{D} = f(\ell, V, \rho, \mu). \tag{6.4}$$

To determine f, we must systematically vary ℓ, V, ρ, and μ.

If we compare this problem to the pipe energy loss investigation, we seem to be in better shape because we have fewer variables. A complete investigation of submarine drag would require "only" about 10^4 experiments. This case has problems of its own, however. Supposedly, this submarine is only in the design phase, and we may want to compare several different shapes (designs). Building and testing several different designs would be very costly. But we have an even greater problem; if we build the submarines full size, there is no laboratory where we can test them.* As a full-scale test program is out of the question, the only possible approach is to test a small replica (model) of the actual submarine. It seems reasonable that if the model is built *exactly* like the full-size (prototype) submarine, we could extrapolate measurements of forces on the model to the prototype—but how?

Once again, dimensional analysis comes to the rescue. If we apply dimensional analysis to the functional relation of Eq. (6.4), it reduces to the equivalent form

$$\frac{\mathscr{D}}{\frac{1}{2}\rho V^2 \ell^2} = F\left(\frac{\rho V \ell}{\mu}\right). \tag{6.5}$$

In Eq. (6.5), the size of the submarine has been combined with the other parameters in such a way that the relation is explicitly independent of size. If the model and prototype are exactly similar geometrically, the function F for the prototype will be identical to the function F for the model. Determining the latter experimentally gives us the former. Note also that the parameters have once again been "compacted," so only a few experimental runs and a single curve are required.

---

* This limitation is especially true for large aircraft, supertankers, skyscrapers, and the like.

These two examples should serve to whet your appetite for dimensional analysis. Although these examples may have overstated the case, they are representative of what can be done with dimensional analysis, from the viewpoints of both experimenters and users of experimental results.

6.2 THE FOUNDATIONS OF DIMENSIONAL ANALYSIS

Engineers are concerned with physical phenomena. The parameters of interest (such as force, velocity, energy) can be expressed in terms of a numerical magnitude and associated units of measurement. All physical entities have certain dimensions, and a given physical entity always has the same dimensions, even if the units employed to express the dimensions change.*

To deal with physical entities, engineers use the language and methods of mathematics. They write equations that allow them to calculate pressure, velocity, flow rate, and so on. Some typical equations in fluid mechanics are Bernoulli's equation, the linear momentum equation, the hydrostatic pressure formula, and the Navier–Stokes equations. These equations are not abstract mathematical expressions; they represent the physical laws of nature. All of these equations must obey the principle of dimensional homogeneity.

An equation is *dimensionally homogeneous* if it is valid regardless of the units of measurement assigned to the terms. An example of a dimensionally homogeneous equation is Bernoulli's equation:

$$\frac{p_1}{\rho} + \frac{V_1^2}{2} + gz_1 = \frac{p_2}{\rho} + \frac{V_2^2}{2} + gz_2.$$

The numerical values of the terms change if the units change, but the validity of the equation is not affected. The following two things are required for an equation to be dimensionally homogeneous:

1. The dimensions of the right-hand side must be the same as the dimensions of the left-hand side.
2. The dimensions of all additive terms must be the same.

If we write only the dimensions of Bernoulli's equation, using mass $[M]$, length $[L]$, and time $[T]$ as fundamental dimensions, we get

$$\frac{[M/LT^2]}{[M/L^3]} + \left[\frac{L}{T}\right]^2 + \left[\frac{L}{T^2}\right][L] = \frac{[M/LT^2]}{[M/L^3]} + \left[\frac{L}{T}\right]^2 + \left[\frac{L}{T^2}\right][L].$$

Simplifying, we get

$$\left[\frac{L^2}{T^2}\right] = \left[\frac{L^2}{T^2}\right].$$

As we claimed, Bernoulli's equation is dimensionally homogeneous.

---

* Reread Section 1.4 if you need to brush up on the concepts of units and dimensions.

The *principle of dimensional homogeneity* (PDH) can be stated as follows:

> Any equation that completely describes a physical phenomenon must be dimensionally homogeneous.

The PDH provides the entire foundation of dimensional analysis. All further developments of dimensional analysis are essentially mathematical.

The PDH is a helpful check of mathematical operations. If, after a lengthy derivation, you find the dimensions on the left-hand side of the equation are different from the dimensions on the right-hand side, you made an error somewhere. A comparison of the dimensions often helps you find the source of the error.

The PDH applies to derivatives and integrals, as well as to algebraic terms. The d or ∂ of a derivative and the $\int$ and d of an integral do not have dimensions, but the independent and dependent variables do. The following dimensional equations illustrate how this works:

$$\left[\frac{dV}{dt}\right] = \frac{[V]}{[t]} = \frac{[L/T]}{[T]} = \left[\frac{L}{T^2}\right];$$

$$\left[\frac{d^2p}{dx^2}\right] = \frac{[p]}{[x^2]} = \frac{[M/LT^2]}{[L^2]} = \left[\frac{M}{L^3T^2}\right];$$

$$\left[\int p\,dA\right] = [p][A] = \left[\frac{M}{LT^2}\right][L^2] = \left[\frac{ML}{T^2}\right].^*$$

You should be aware that some equations do not satisfy the PDH, such as Manning's formula for open channel flow (English units):

$$V = \frac{1.49}{n} R_h^{2/3} S^{1/2},$$

where V is velocity, R_h is radius, S is slope (dimensionless), and n is a number taken from tables. This equation is true only for a specific set of units; R_h must be in feet and V must be in feet per second. Such equations might be called *dimensionally inhomogeneous equations* or *unit-specific equations*. They are usually curve fits of experimental data.

If an equation satisfies the PDH, only the following three types of quantities should appear in it:

1. *Dimensional variables,* which include those parameters that we are trying to evaluate and those that represent known material properties or flow conditions. In Bernoulli's equation, p, ρ, V, and z are dimensional variables. In an experiment, dimensional variables are either measured or controlled.

2. *Dimensional constants,* which are associated with fundamental phenomena of nature, such as the gravitational constant G in Newton's law of universal gravitation,

$$F = G\frac{m_1 m_2}{r^2},$$

---

* If we use force $[F]$ instead of mass $[M]$ in our fundamental set of dimensions, the dimensions of this term are just $[F]$.

R_0, the universal gas constant, and c_0, the speed of light in a vacuum.

3. *Pure numbers,* which have no dimensions. They usually result from mathematical operations. The $\frac{1}{2}$ in Bernoulli's equation is a pure number. The exponent 2 on the velocity is also a pure number. Other pure numbers include e and π.

The distinction between dimensional variables and dimensional constants is a little fuzzy at times. Consider the acceleration of gravity, g, in Bernoulli's equation. We know that g is indeed variable, especially the farther we go from the Earth. However, most engineers who use Bernoulli's equation are interested only in happenings on the Earth's surface. Certainly, few experimenters plan to change g during the course of an experiment. In most cases, g would be considered a dimensional constant. We use the name *dimensional parameters* to refer to both dimensional variables and dimensional constants.

A fourth type of quantity that may sometimes appear in a dimensionally homogeneous equation is a *conversion factor*. Conversion factors are used to convert units and, if they appear in an equation, must be treated as pure numbers (equal to 1). In our opinion, conversion factors (including g_c) should not be written in equations but should be introduced as necessary in numerical calculations.

EXAMPLE 6.1 Illustrates Dimensional Homogeneity

The frictional pressure loss, Δp_L in the pipe flow shown in Fig E6.1, can be calculated from

$$\Delta p_L = 0.03\left(\frac{L}{D}\right)\left(\frac{\rho V^2}{2}\right),$$

where L and D are in meters, ρ in kilograms per cubic meter, and V in meters per second. Is this equation dimensionally homogeneous?

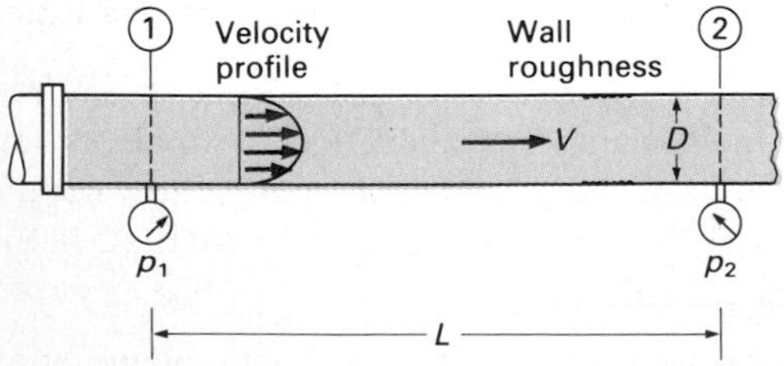

Figure E6.1 Incompressible turbulent flow in a circular pipe.

SOLUTION

Given

Equation:

$$\Delta p_L = 0.03\left(\frac{L}{D}\right)\left(\frac{\rho V^2}{2}\right)$$

Units:

L in meters
D in meters
ρ in kilograms per cubic meter
V in meters per second

Find

Whether the equation is dimensionally homogeneous

Solution

L, D, ρ, and V are expressed in SI units. Therefore we anticipate that Δp_L will be expressed in SI units of pressure, N/m^2. We try to determine whether the units of

$$\frac{L}{D}\left(\frac{\rho V^2}{2}\right)$$

are N/m^2. The units of this term are*

$$\left[\frac{L}{D}\left(\frac{\rho V^2}{2}\right)\right] = \left[\frac{\text{m}}{\text{m}}\left(\frac{\text{kg}}{\text{m}^3}\right)\left(\frac{\text{m}}{\text{s}}\right)^2\right]$$
$$= \left[\frac{\text{kg}\cdot\text{m}}{\text{s}^2}\left(\frac{1}{\text{m}^2}\right)\right] = \left[\frac{\text{N}}{\text{m}^2}\right],$$

which are the units of pressure in the SI system. Therefore:

The equation is dimensionally homogeneous. **ANSWER**

Discussion

We could also find the answer by using dimensions

$$\left[\frac{L}{D}\left(\frac{\rho V^2}{2}\right)\right] = \left[\frac{L}{L}\left(\frac{M}{L^3}\right)\left(\frac{L}{T}\right)^2\right]$$
$$= \left[\frac{ML}{T^2}\left(\frac{1}{L^2}\right)\right] = \left[\frac{F}{L^2}\right],$$

which are the dimensions of pressure.

* Note: These brackets mean that we are writing the units of the enclosed terms.

6.3 DIMENSIONLESS PARAMETERS AND THE PI THEOREM

We now consider how to use dimensional analysis to organize (experimental) information and reduce the complexity of an investigation. Recall the two flow problems discussed in Section 6.1. The various groups in Eqs. (6.3) and (6.5) are combinations of dimensional parameters. The dimensions of the groups in Eq. (6.3) are†

$$\left[\frac{gh_L}{\frac{1}{2}V^2}\right] = \frac{[L^2/T^2]}{[L/T]^2} = \frac{[L/T]^2}{[L/T]^2} = [1];$$

$$\left[\frac{L}{D}\right] = \frac{[L]}{[L]} = [1];$$

† See Table 1.1 in Chapter 1 for dimensions of the various quantities.

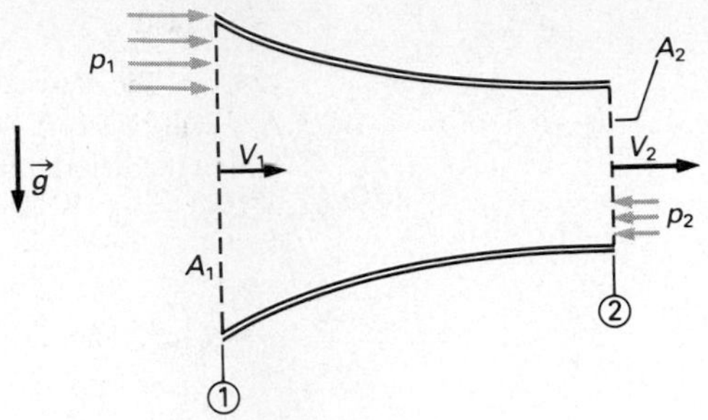

Figure 6.5 Flow in horizontal converging nozzle.

$$\left[\frac{\rho VD}{\mu}\right] = \frac{[M/L^3][L/T][L]}{[M/LT]} = \frac{[M/LT]}{[M/LT]} = [1];$$

$$\left[\frac{\varepsilon}{D}\right] = \frac{[L]}{[L]} = [1].$$

In these equations, [1] means dimensionless.

The dimensions of the groups in Eq. (6.5) are

$$\left[\frac{\mathscr{D}}{\frac{1}{2}\rho V^2 \ell^2}\right] = \frac{[F]}{[M/L^3][L/T]^2[L^2]} = \frac{[ML/T^2]}{[ML/T^2]} = [1]$$

and

$$\left[\frac{\rho V\ell}{\mu}\right] = \frac{[M/L^3][L/T][L]}{[M/LT]} = [1].$$

All the parameters in the reduced equations are dimensionless.* When two or more dimensional parameters are combined into a dimensionless group, the group is called a *dimensionless parameter*. Combining the dimensional parameters in Eqs. (6.2) and (6.4) into the dimensionless parameters in Eqs. (6.3) and (6.5) is the key to the resulting simplification.

To demonstrate (not derive) how this might occur, we consider a flow for which we have a theory. Figure 6.5 shows a horizontal converging nozzle. An incompressible fluid flows steadily through the nozzle. Suppose that we want to investigate the relation between the pressure drop $\Delta p_d (\Delta p_d \equiv p_1 - p_2)$, the fluid velocity V_1, and the nozzle areas A_1 and A_2. Applying the mechanical energy equation gives

$$\frac{p_1}{\rho} + \frac{V_1^2}{2} = \frac{p_2}{\rho} + \frac{V_2^2}{2} + gh_L.$$

Solving for the pressure drop, we obtain

$$\Delta p_d = \frac{1}{2}\rho(V_2^2 - V_1^2) + \rho g h_L. \tag{6.6}$$

* You might try calculating the dimensions of these parameters using a force-length-time-temperature $[F, L, T, \theta]$ system to verify that they are still dimensionless.

Because the flow is steady, the fluid velocities are related by the continuity equation

$$V_1 A_1 = V_2 A_2,$$

so the pressure drop can be computed by

$$\Delta p_d = \frac{1}{2}\rho V_1^2\left(\frac{A_1^2}{A_2^2} - 1\right) + \rho g h_L. \tag{6.7}$$

Equation (6.7) has the functional form

$$\Delta p_d = f\{V_1, \rho, A_2, A_1, gh_L\}. \tag{6.8}$$

Equation (6.7) and its generalized functional form, Eq. (6.8), are dimensionally homogeneous equations involving six dimensional parameters. We can find a corresponding dimensionless form by dividing Eq. (6.7) by $\frac{1}{2}\rho V_1^2$ to get

$$\frac{\Delta p_d}{\frac{1}{2}\rho V_1^2} = \left(\frac{A_1}{A_2}\right)^2 - 1 + \frac{gh_L}{\frac{1}{2}V_1^2}. \tag{6.9}$$

Now we check the dimensions of the terms of this equation:

$$\left[\frac{\Delta p_d}{\frac{1}{2}\rho V_1^2}\right] = \frac{[M/LT^2]}{[M/L^3][L/T]^2} = [1];$$

$$\left[\frac{A_1}{A_2}\right] = \frac{[L^2]}{[L^2]} = [1];$$

$$\left[\frac{gh_L}{\frac{1}{2}V_1^2}\right] = \frac{[L^2/T^2]}{[L/T]^2} = [1].$$

Equation (6.9) has the general functional form

$$\frac{\Delta p_d}{\frac{1}{2}\rho V_1^2} = F\left\{\frac{A_1}{A_2}, \frac{gh_L}{\frac{1}{2}V_1^2}\right\}. \tag{6.10}$$

Clearly, Eq. (6.9) and its generalization, Eq. (6.10), are in every way equivalent to Eq. (6.7) and its generalization, Eq. (6.8); however, Eq. (6.10) contains only three dimensionless parameters, whereas Eq. (6.7) contains six dimensional parameters. Note also that the reduction in the number of parameters by three equals the number of fundamental dimensions (M, L, and T) involved in the list of dimensional parameters.

The value of this reduction in the number of parameters is not apparent here, because we knew about the energy and continuity equations and could develop a complete theory for the nozzle. Equation (6.7) is perfectly adequate, and the dimensionless form, Eq. (6.9), is only a curiosity. Note, however, that if we did not have a complete theory, we could discover much more easily the three-parameter relation implied by Eq. (6.10) than the six-parameter relation implied by Eq. (6.8).

We are guaranteed that we can in every case reduce the number of independent parameters in a problem by compacting the dimensional parameters into a set of dimensionless parameters. The basis of this guarantee is the following theorem of dimensional analysis, called the *pi theorem*.

If it is known that a physical process is governed by a dimensionally homogeneous relation involving *n dimensional* parameters, such as

$$x_1 = f\{x_2, x_3, \ldots, x_n\},$$

where the x's are dimensional variables, there exists an equivalent relation involving a smaller number, $(n - k)$, of *dimensionless* parameters, such as

$$\Pi_1 = F\{\Pi_2, \Pi_3, \ldots, \Pi_{n-k}\},$$

where the Π's are dimensionless groups constructed from the x's. The reduction, k, is usually equal to, but never more than, the number of fundamental dimensions involved in the x's.

The name of Buckingham [2] is usually associated with this theorem, although it was apparently discovered independently by others. The name "pi theorem" is from assignment of the symbol Π (capital Greek pi) to the dimensionless parameters.

The pi theorem is provable by strictly mathematical manipulations. The only "physics" involved is the principle of dimensional homogeneity, which is part of the "if" statement of the theorem. The proof of the theorem is too involved to give here, but the idea is not too complicated. The general function involving the dimensional parameters is *constrained* by the principle of dimensional homogeneity—not every relation among the x's will do; only those relations that satisfy the principle of dimensional homogeneity are allowed. This constraint leads to a relation among a smaller number of Π's. A proof of the theorem is given in [3, 4, 5].

The pi theorem is the theoretical basis for dimensional analysis, but it only tells us that we *can* do the reduction. It does not tell us how to do it. In fact, there is no unique reduction of the n x's to the $(n - k)$ Π's. The following three questions come to mind immediately:

- How do we combine the x's into a set of Π's, making sure to get just the right number?
- If k is only usually equal to the number of fundamental dimensions involved, how do we know when it is not? If it is not, what is it?
- If a complicated physical phenomenon is under investigation, how do we even know what dimensional variables are involved?

Unfortunately, there is no single answer to the first and last questions. Dimensional analysis depends on experience, preference, judgment, and custom. In the next section, we present a way to do dimensional analysis that provides the answer to the second question. But our way is not the only way.

6.4 PUTTING THE PI THEOREM TO WORK

The pi theorem tells us that we can construct a set of dimensionless parameters, thereby reducing the complexity of a problem by reducing the number of independent parameters. Construction of dimensionless parameters from dimensional parameters is up to the analyst. The only help the pi theorem offers is to tell the analyst how many dimensionless parameters to construct.

6.4.1 Constructing a Set of Dimensionless Parameters

The following stepwise method can be used to construct a set of dimensionless parameters. We illustrate each step by carrying out a complete dimensional analysis for the pipe flow mechanical energy loss problem discussed in Section 6.1. We summarize the steps at the end of this section.

Step 1. Write a functional expression for the dimensional relation under investigation, being sure to include all relevant dimensional parameters.

Writing a functional relation is easy enough, but how do you know which dimensional parameters are relevant and which are not? The best guide is experience. A simple theory describing the phenomenon in question, no matter how crude, is helpful. The actual dimensional analysis does not begin until after step 1, so it is not much help. If you overlook a relevant parameter, experimental data will be confusing because the parameter will not be under control. However, if you include an irrelevant parameter, it will appear in at least one dimensionless group. Several experiments may be required to demonstrate that the parameter (and the group that it is in) is irrelevant to the particular phenomenon.

There is one very helpful check. Any dimension must appear in at least two parameters, or obtaining dimensionless groups would not be possible. Whenever a dimension is represented in only one parameter, either that parameter is irrelevant or you must include another parameter that also includes the unique dimension. Note that you may use dimensional constants (G, g, R_0) for this purpose.

An excellent illustration of these points is the problem of determining the period (Ω) of a simple pendulum (Fig. 6.6). With no knowledge of the physics of pendulums, we would probably propose a dimensional relation of the form

$$\Omega = f\{m, \ell\}.$$

This relation has a serious drawback: Each fundamental dimension—mass, length, and time—appears only once. You must add something and possibly remove something as well. You might add the acceleration of gravity g (a dimensional constant) or the weight, W. As weight and mass are always related by $W = mg$, including all three would be redundant. The new trial functional relation is

$$\Omega = f\{m, g, \ell\}.$$

You would still have a problem, because the dimension of mass appears in only one parameter. Including the weight does not help, because the only dimensionless

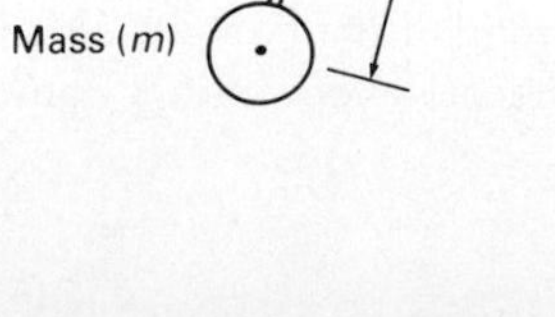

Figure 6.6 Simple pendulum.

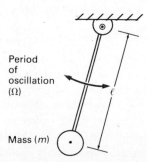

parameter that could be formed from W, m, and g is mg/W, which is always unity. Thus the pendulum's mass is irrelevant and m must be dropped from the functional relation, so the final form is

$$\Omega = f\{g, \ell\}.$$

Returning to the field of fluid mechanics, recall that we concluded that the functional relation, Eq. (6.2), for our pipe flow example is

$$gh_L = f\{L, D, V, \rho, \mu, \varepsilon\}.$$

Note that g and h_L are not considered separate parameters; gh_L is just our way of writing a single parameter, mechanical energy loss.

You may find it helpful to write a *dimension equation* that exhibits only the dimensions of the parameters involved in the postulated dimensional relation. The dimension equation corresponding to Eq. (6.2) is

$$\left[\frac{L^2}{T^2}\right] = f\left\{[L], [L], \left[\frac{L}{T}\right], \left[\frac{M}{L^3}\right], \left[\frac{M}{LT}\right], [L]\right\}. \tag{6.11}$$

Step 2. Determine the number of dimensionless parameters you need to construct.

This number is equal to the number of dimensional parameters (n) in the functional relation minus a number (k) that is equal to *the maximum number of dimensional parameters that cannot form a dimensionless group (pi) among themselves*. The number k is usually equal to the number of fundamental dimensions (m) involved in the dimensional parameters; it is never greater than the number of fundamental dimensions.

To find k, initially assume that it is equal to m and try to find a set of m of your dimensional parameters that cannot be formed into a pi group. If you succeed in finding any set, $k = m$, stop. If you do not find such a set, reduce k by 1 and try again. *Hint:* Try to find parameters that have only one dimension or are fundamentally of a given dimensional type.

In Eq. (6.2), there are seven dimensional parameters (don't forget to count the parameter on the left-hand side). The number of fundamental dimensions is three ($[M]$, $[L]$, $[T]$). Hoping that $k = 3$, you should look for three parameters that cannot be formed into a dimensionless group. If you pick L, ρ, and V, mass appears only in ρ and time only in V, so no combination of L, ρ, and V could be dimensionless, and k is equal to 3. Thus the number of dimensionless parameters that you need to construct is four,

$$n - k = 7 - 3 = 4.$$

The rare cases where k is less than the number of fundamental dimensions involved in the dimensional parameters occur when certain dimensions (usually mass and time) occur only in fixed combinations in the dimensional parameters. In stress analysis problems, for example, mass and time occur only in the combination $[M/T^2]$. If you were to analyze a stress problem in an $[F, L, T, \theta]$-dimension system instead of an $[M, L, T, \theta]$-dimension system, you would find only two dimensions occurring, namely, $[F]$ and $[L]$. A helpful check in determining k is to express the parameters of the problem in terms of both $[M, L, T, \theta]$ and $[F, L, T, \theta]$ systems and see whether the number of fundamental dimensions involved is the same in both systems. If not, then k will be the smaller number of fundamental dimensions.

Step 3. Select as *repeating variables* k of the dimensional parameters that contain among them all of the fundamental dimensions. Combine these parameters with the remaining $(n - k)$ dimensional parameters, one at a time, to form the required number, $(n - k)$, of dimensionless parameters.

Although it sounds difficult, this part is easy. Anyone can do it by following a few simple steps. You construct the dimensionless parameters as products of powers of the dimensional parameters. For the pipe flow example, you must select three of the parameters as repeating variables and use them with the remaining four parameters to form four dimensionless groups. Say that you select ρ, V, D as the repeating variables. You then take the remaining four parameters $(gh_L, L, \mu, \varepsilon)$ and combine them with ρ, V, D to form four dimensionless products (Π's).

Begin with (gh_L). A dimensionless product involving gh_L, ρ, V, and D has the general form

$$\Pi_1 = (gh_L)\rho^a V^b D^c, \tag{6.12}$$

where a, b, and c are exponents to be selected so that Π_1 is dimensionless. The dimensional equation equivalent to Eq. (6.12) is

$$[\Pi_1] = [gh_L][\rho]^a[V]^b[D]^c.$$

Substituting the appropriate dimensions yields

$$[\Pi_1] = \left[\frac{L^2}{T^2}\right]\left[\frac{M}{L^3}\right]^a\left[\frac{L}{T}\right]^b[L]^c.$$

Combining exponents gives

$$[\Pi_1] = [M]^a[L]^{2-3a+b+c}[T]^{-2-b}. \tag{6.13}$$

Now Π_1 is *dimensionless*, that is,

$$[\Pi_1] = [M]^0[L]^0[T]^0.$$

You must therefore equate the *net* exponent of any dimension on the right-hand side of Eq. (6.13) to zero. This gives the three equations

$$\begin{aligned} a &= 0 \qquad \text{(from } M\text{);} \\ 2 + b &= 0 \qquad \text{(from } T\text{);} \\ 2 - 3a + b + c &= 0 \qquad \text{(from } L\text{).} \end{aligned}$$

You can easily verify that the solution to these equations is

$$a = 0, \qquad b = -2, \quad \text{and} \quad c = 0.$$

Substituting back into Eq. (6.12) gives

$$\Pi_1 = \frac{gh_L}{V^2}. \tag{6.14}$$

The next dimensional parameter to be formed into a Π is L. Using the formal procedure just described, you would get

$$\Pi_2 = (L)\rho^a V^b D^c. \tag{6.15}$$

Note that a, b, and c are "new" values for this equation. The dimensional equation is

$$[\Pi_2] = [L]\left[\frac{M}{L^3}\right]^a \left[\frac{L}{T}\right]^b [L]^c,$$

or

$$[M]^0[L]^0[T]^0 = [M]^a[L]^{1-3a+b+c}[T]^{-b},$$

which implies that

$$a = 0;$$
$$b = 0;$$
$$1 - 3a + b + c = 0.$$

The solutions to these equations are

$$a = 0, \quad b = 0, \quad \text{and} \quad c = -1,$$

so

$$\Pi_2 = \frac{L}{D}. \tag{6.16}$$

Possibly, Π_2 was obvious to you at the start. If a dimensionless group is obvious, you do not need to go through all the mathematical manipulation; just write Π = (whatever) and go on to the next parameter. Note that Π_2 is a ratio of two lengths. The most obvious dimensionless parameters are ratios of like quantities.

To obtain the third Π, combine powers of ρ, V, D with μ:

$$\Pi_3 = (\mu)\rho^a V^b D^c. \tag{6.17}$$

The dimensional equation is

$$[\Pi_3] = \left[\frac{M}{LT}\right]\left[\frac{M}{L^3}\right]^a \left[\frac{L}{T}\right]^b [L]^c,$$

which implies that

$$1 + a = 0;$$
$$-1 - 3a + b + c = 0;$$
$$-1 - b = 0.$$

The solutions to these three equations are

$$a = -1, \quad b = -1, \quad \text{and} \quad c = -1,$$

so

$$\Pi_3 = \frac{\mu}{\rho V D}. \tag{6.18}$$

You obtain the fourth and final Π from ρ, V, D, and ε. The obvious parameter is

$$\Pi_4 = \frac{\varepsilon}{D}. \tag{6.19}$$

Finally, collect all your results into a dimensionless functional relation:

$$\frac{gh_L}{V^2} = F\left\{\frac{L}{D}, \frac{\mu}{\rho V D}, \frac{\varepsilon}{D}\right\}. \tag{6.20}$$

The mathematical manipulations required to form the pi's are rather simple. The only tricky element is selecting the dimensional parameters to be used as repeating variables. This selection is rather important because the parameters may appear in every one of the dimensionless groups. There is some room for choice in this selection, and the following rules should help:

- *Do not* select the primary *dependent* parameter (in our example, gh_L) as a repeating variable.
- *Do not* select any parameter whose influence on the problem is questionable or important only part of the time. (In the pipe flow example, for instance, viscosity and roughness are only sometimes relevant, as we show in Chapter 7.)
- Select parameters whose dimensions are as close to "pure" as possible. In the pipe flow case, pipe diameter has the pure dimension of length, density is as close as you can get to pure mass, and velocity is as close as you can get to pure time. If you are able to choose a set of parameters with pure dimensions, you can usually determine the various Π groups by inspection.

In case of conflict between these rules, they are listed in order of importance. Application of these rules to the pipe flow problem leaves only the choice between D and L. For fully developed pipe flow, D is more relevant than L, which is the reason for its selection.

Step 4. Rearrange the Π groups to please yourself or to correspond with customary usage.

This step is optional, although you will be unpopular with your professional colleagues if you insist on using nontraditional types of dimensionless parameters. You may have noted that the dimensionless pipe energy loss functional form derived as Eq. (6.20) does not correspond to the earlier stated form, Eq. (6.3). The differences are slight: a factor of 2 in Π_1 and a reciprocal in Π_3. The fact is that Eqs. (6.4) and (6.20) are completely equivalent theoretically. The parameters of Eq. (6.3) are traditional in fluid mechanics, but those of Eq. (6.20) are not.*

---

* The factors $\frac{1}{2}$ or 2 are irrelevant to the dimensions. They are included because we know that $\frac{1}{2}V^2$ is kinetic energy per unit mass.

Any set of $(n - k)$ independent dimensionless parameters formed from the n dimensional parameters is acceptable from the point of view of the pi theorem. A set of dimensionless parameters is independent if none of the parameters can be obtained from the others by multiplying parameters together or by combining powers of the parameters. The set of parameters

$$\left(\frac{L}{D}, \frac{\rho VD}{\mu}, \frac{\varepsilon}{D}, \frac{\rho VL}{\mu}\right)$$

is not an independent set, because the last parameter can be obtained by multiplying the first two.

According to the pi theorem, any set of independent dimensionless parameters is as good as any other set. The first three steps of the procedure always produce a set of independent dimensionless parameters (independence is guaranteed, because each of the $(n - k)$ nonrepeating dimensional parameters appears in only one Π). After you find this set, you can replace any member of the set by a different dimensionless parameter by

- multiplying the parameter by a constant,
- raising the parameter to *any* positive or negative power,
- multiplying any power of the parameter with any power of any of the other parameters, or
- any combination of the above.

To reconcile Eqs. (6.3) and (6.20), note that

$$\frac{gh_L}{\frac{1}{2}V^2} = 2\left(\frac{gh_L}{V^2}\right) \quad \text{and} \quad \frac{\rho VD}{\mu} = \left(\frac{\mu}{\rho VD}\right)^{-1}.$$

Similarly, the functional form, Eq. (6.3),

$$\frac{gh_L}{\frac{1}{2}V^2} = F\left(\frac{L}{D}, \frac{\rho VD}{\mu}, \frac{\varepsilon}{D}\right)$$

is equivalent* to

$$\frac{D\rho gh_L}{V\mu} = G\left(\frac{\rho VL}{\mu}, \frac{\rho VD}{\mu}, \frac{\rho V\varepsilon}{\mu}\right). \tag{6.21}$$

You obtain the dimensionless parameters in Eq. (6.21) by the following multiplications:

$$\frac{D\rho gh_L}{V\mu} = \frac{1}{2}\left(\frac{gh_L}{\frac{1}{2}V^2}\right)\left(\frac{\rho VD}{\mu}\right);$$

$$\frac{\rho VL}{\mu} = \left(\frac{L}{D}\right)\left(\frac{\rho VD}{\mu}\right);$$

* Note the word *equivalent,* which means "carries the same information," as opposed to the word *equal* or *identical.*

$$\frac{\rho V \varepsilon}{\mu} = \left(\frac{\varepsilon}{D}\right)\left(\frac{\rho V D}{\mu}\right).$$

Whichever form you choose for the dimensionless parameters, dimensional analysis ends with general functional statements such as Eqs. (6.3), (6.20), and (6.21). Finding the specific functional relations requires experiments or theories.

We can summarize the four-step method for applying the pi theorem to a particular problem as follows:

Step 1 Write the functional dependence between the dimensional parameters.

Step 2 Determine the number of dimensionless parameters to be constructed from the dimensional parameters.

Step 3 Select a set of *repeating variables* from the dimensional parameters that will be used to form each dimensionless group. Use the remaining dimensional parameters one at a time to combine with the repeating variables and form dimensionless groups. You may do so by analysis of the dimensional equation or, in simple cases, by inspection.

Step 4 Rearrange the dimensionless groups found in step 3 to correspond to customary usage, to suit yourself, or to facilitate an experimental program.

EXAMPLE 6.2 **Illustrates the Procedure for Determining Dimensionless Parameters**

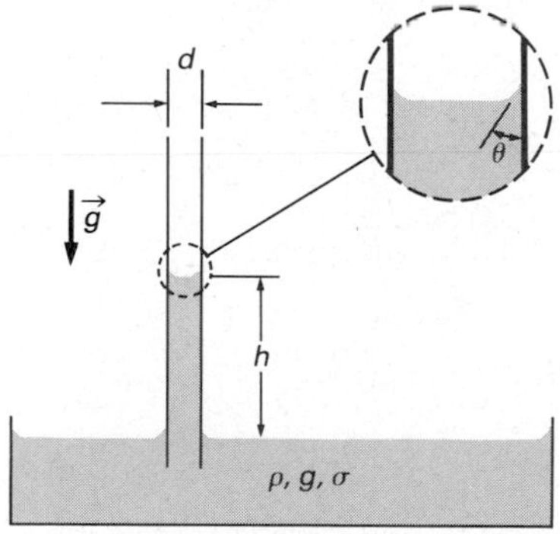

Figure E6.2 Straw inserted in liquid.

A soda straw of inside diameter d is placed in a pan of liquid of density ρ, as shown in Fig. E6.2. The water rises to height h in the straw because of the surface tension σ between the water and the straw and makes a contact angle θ at the water–straw interface. Find the appropriate dimensionless parameters describing this phenomenon.

SOLUTION

Given

Figure E6.2

Dimensional parameters d, ρ, g, h, σ, θ

Find

Appropriate dimensionless parameters

Solution

Step 1: Write the dimensional relation. Considering h as the dependent parameter gives

$$h = f\{d, \rho, g, \sigma, \theta\}.$$

Step 2: Determine the number of dimensionless parameters. There are six parameters, but θ is dimensionless because it can be expressed in radians. Therefore $n = 5$:

$$[d] = [L];$$
$$[\rho] = [M/L^3];$$
$$[g] = [L/T^2];$$
$$[h] = [L];$$
$$[\sigma] = [F/L] = [M/T^2].$$

There are three fundamental dimensions, so we expect $k = 3$; ρ, d, and σ cannot form a dimensionless group, so $k = 3$ and the number of dimensionless parameters is $(5 - 3) = 2$.

Step 3: Select k dimensional parameters that contain all the fundamental dimensions. We select d, ρ, and σ. Combining d, ρ, and σ with h gives

$$\Pi_1 = hd^a\rho^b\sigma^c.$$

The dimensional equation is

$$[h][d]^a[\rho]^b[\sigma]^c = [M]^0[L]^0[T]^0,$$

or

$$[L][L]^a\left[\frac{M}{L^3}\right]^b\left[\frac{M}{T^2}\right]^c = [M]^0[L]^0[T]^0.$$

Equating powers of each dimension gives

$$M: \quad b + c = 0;$$
$$L: \quad 1 + a - 3b = 0;$$
$$T: \quad -2c = 0.$$

Solving, we obtain

$$c = 0, \qquad b = 0, \quad \text{and} \quad a = -1.$$

The first dimensionless parameter is

$$\Pi_1 = \frac{h}{d}.$$ **ANSWER**

Combining d, ρ, and σ with g gives

$$[g][d]^a[\rho]^b[\sigma]^c = [M]^0[L]^0[T]^0,$$

or

$$\left[\frac{L}{T^2}\right][L]^a\left[\frac{M}{L^3}\right]^b\left[\frac{M}{T^2}\right]^c = [M]^0[L]^0[T]^0.$$

Equating powers of each dimension, we have

$$M: \quad b + c = 0;$$
$$L: \quad 1 + a - 3b = 0;$$
$$T: \quad -2 - 2c = 0.$$

Solving, we obtain

$$c = -1, \qquad b = -c = 1, \quad \text{and} \quad a = -1 + 3b = -1 + 3(1) = 2.$$

The second dimensionless parameter is

$$\Pi_2 = \frac{\rho g d^2}{\sigma}.$$ **ANSWER**

Step 4: Rearrange the Π groups. The Π groups are acceptable.

Recall that θ also influences the phenomenon. Thus the complete dimensionless equation is

$$\frac{h}{d} = F\left(\frac{\rho g d^2}{\sigma}, \theta\right).$$ **ANSWER**

Discussion

As a check on the number of dimensionless parameters, we list the fundamental dimensions of each dimensional parameter in the $[F, L, T, \theta]$ system:

$$[d] = [L];$$
$$[\rho] = [M/L^3] = [FT^2/L^4];$$
$$[g] = [L/T^2];$$
$$[h] = [L];$$
$$[\sigma] = [F/L].$$

We have three fundamental dimensions and five dimensional parameters, so $(n - k) = (5 - 3) = 2$, or two dimensionless parameters.

An interesting result is that the terms ρ and g appear only as the product ρg, which is the specific weight γ. This suggests that we could have started with the single dimensional parameter γ rather than the two separate parameters ρ and g. It would be instructive for you to start with the four dimensional parameters d, γ, h, and σ and develop the dimensionless parameters using both the $[M, L, T, \theta]$ and the $[F, L, T, \theta]$ system.

EXAMPLE 6.3 Illustrates the Procedure to Determine Dimensionless Parameters for a Case Where $k \neq m$

A cantilever beam of length ℓ and second moment of area I has a transverse force P applied at its free end. The bending stress σ at a distance c from its neutral axis is given by*

$$\sigma = f\{\ell, c, x, I, P\},$$

where x is defined in Fig. E6.3. Determine the dimensionless parameters describing the beam stress using

(a) fundamental dimensions $[F, L, T, \theta]$ and
(b) fundamental dimensions $[M, L, T, \theta]$.

---

* From strength of materials,

$$\sigma = \frac{P(\ell - x)c}{I}.$$

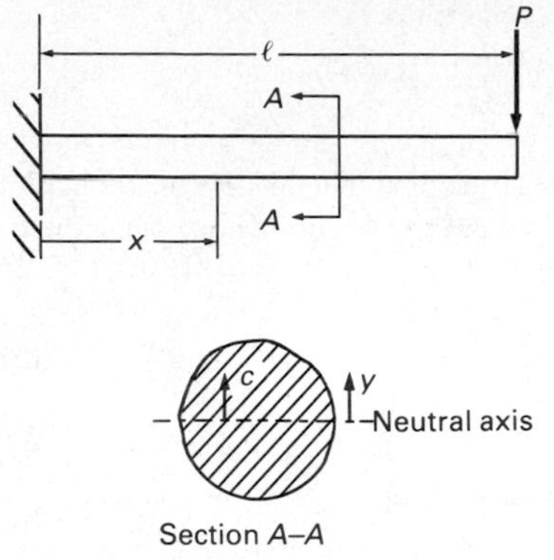

Figure E6.3 Cantilever beam.

SOLUTION

Given

Cantilever beam in Fig. E6.3

Length ℓ

Second moment of area I

Transverse force P

Bending stress σ at distance c from neutral axis at distance x from fixed end

Find

The dimensionless parameters using
fundamental dimensions $[F, L, T, \theta]$ and
fundamental dimensions $[M, L, T, \theta]$

Solution Using $[F, L, T, \theta]$ System

Step 1: Write the dimensional relation. Considering σ as the dependent variable gives

$$\sigma = f\{\ell, c, x, I, P\}.$$

Step 2: Determine the number of dimensionless parameters. There are six dimensional parameters, so $n = 6$. Their fundamental dimensions are

$$[\sigma] = [F/L^2]; \qquad [x] = [L];$$
$$[\ell] = [L]; \qquad [I] = [L^4];$$
$$[c] = [L]; \qquad [P] = [F].$$

There are two fundamental dimensions in the $[F, L, T, \theta]$ system, so we expect that $k = 2$. The number of dimensionless parameters is then $(n - k)$, or $(6 - 2) = 4$.

Step 3: Select k dimensional parameters that contain all the fundamental dimensions. As $k = 2$, we select ℓ and P. Combining ℓ and P with σ gives

$$[\sigma][\ell]^a[P]^b = [F]^0[L]^0,$$

or

$$\left[\frac{F}{L^2}\right][L]^a[F]^b = [F]^0[L]^0.$$

Equating powers of each dimension gives

$$F: \quad 1 + b = 0 \quad \text{and} \quad L: \quad -2 + a = 0.$$

Solving, we obtain

$$b = -1 \quad \text{and} \quad a = 2.$$

The first dimensionless parameter is

$$\Pi_1 = \frac{\sigma\ell^2}{P}.$$

ANSWER

Combining ℓ and P with c gives the obvious parameter

$$\Pi_2 = \frac{c}{\ell}.$$

ANSWER

We next combine ℓ and P with x. By inspection we get

$$\Pi_3 = \frac{x}{\ell}.$$

ANSWER

Combining ℓ and P with I gives

$$[I][\ell]^a[P]^b = [F]^0[L]^0 \quad \text{or} \quad [L^4][L]^a[F]^b = [F]^0[L]^0.$$

Equating powers of each dimension gives

$$F\colon \quad b = 0 \quad \text{and} \quad L\colon \quad 4 + a = 0.$$

Solving, we obtain

$$b = 0 \quad \text{and} \quad a = -4.$$

The fourth dimensionless parameter is

$$\Pi_4 = \frac{I}{\ell^4}.$$

ANSWER

Step 4: Rearrange the Π groups. The Π groups are acceptable.

Solution Using $[M, L, T, \theta]$ System

Step 1: Write the dimensional relation. Considering σ as the dependent variable gives

$$\sigma = f\{\ell, c, x, I, P\}.$$

Step 2: Determine the number of dimensionless parameters. There are six dimensional parameters, so $n = 6$. Their fundamental dimensions are

$$[\sigma] = [F/L^2] = [M/LT^2]; \qquad [x] = [L];$$
$$[\ell] = [L]; \qquad [I] = [L^4];$$
$$[c] = [L]; \qquad [P] = [F] = [ML/T^2].$$

There are three fundamental dimensions in the $[M, L, T, \theta]$ system, so we expect that $k = 3$. If this is true, the number of dimensionless parameters is $(n - k)$, or $(6 - 3) = 3$.

This value is not equal to that predicted by the $[F, L, T, \theta]$ system. We ignore this result for now and proceed with the $[M, L, T, \theta]$ dimensional analysis.

Step 3: Select k dimensional parameters that contain all the fundamental dimensions. As $k = 3$, we select ℓ, I, and P. Combining ℓ, I, and P with σ gives

$$[\sigma][\ell]^a[I]^b[P]^c = [M]^0[L]^0[T]^0,$$

or

$$\left[\frac{M}{LT^2}\right][L]^a[L^4]^b\left[\frac{ML}{T^2}\right]^c = [M]^0[L]^0[T]^0.$$

Equating powers of each dimension gives

$$M: \quad 1 + c = 0;$$
$$L: \quad -1 + a + 4b + c = 0;$$
$$T: \quad -2 - 2c = 0.$$

Note that the first and last equations both give $c = -1$; a and b both must be calculated from the second equation. Unfortunately, we have two unknowns and only one equation; solving for both a and b is impossible. Apparently, we cannot complete the dimensional analysis; we must have made an error.

Locating our error is not difficult. At step 2, we did not verify that $k = 3$. Apparently, we uncovered a case where k is less than the number of fundamental dimensions. Recall that k is equal to the maximum number of dimensional parameters that cannot form a dimensionless group. Some three-parameter combinations of our six dimensional parameters are

σ, ℓ, c	ℓ, c, x	c, x, I
σ, ℓ, x	ℓ, c, I	c, x, P
σ, ℓ, P		

We can form a dimensionless group from *any* of these combinations; for example,

σ, ℓ, c	form ℓ/c,	ℓ, c, x	form c/ℓ, ℓ/x,
σ, ℓ, x	form ℓ/x,	c, x, P	form c/x,

and so on.

All possible three-parameter combinations could form a dimensionless group, so k cannot equal three. We now assume that $k = 2$ and examine possible two-parameter combinations. We immediately see that σ and ℓ *cannot* form a dimensionless group, so $k = 2$. Then $(n - k) = (6 - 2) = 4$, and we must find four dimensionless parameters. This conclusion agrees with that from the $[F, L, T, \theta]$ system. We must now repeat step 3. We select ℓ and P as repeating variables. Combining ℓ and P with σ gives

$$[\sigma][\ell]^a[P]^b = [M]^0[L]^0[T]^0,$$

or

$$\left[\frac{M}{LT^2}\right][L]^a\left[\frac{ML}{T^2}\right]^b = [M]^0[L]^0[T]^0.$$

Equating powers of each dimension gives

$$M: \quad 1 + b = 0,$$
$$L: \quad -1 + a + b = 0,$$
$$T: \quad -2 - 2b = 0.$$

Solving,* we have $b = -1$ and $a = 2$. The first dimensionless parameter is

$$\Pi_1' = \frac{\sigma\ell^2}{P}.$$ **ANSWER**

* Note that we have two *independent* equations and two unknowns.

Combining ℓ and P with c gives the obvious parameter

$$\Pi_2' = \frac{c}{\ell}.$$ **ANSWER**

Combining ℓ and P with x gives our third dimensionless parameter by inspection:

$$\Pi_3' = \frac{x}{\ell}.$$ **ANSWER**

Combining ℓ and P with I gives

$$[I][\ell]^a[P]^b = [M]^0[L]^0[T]^0,$$

or

$$[L^4][L]^a\left[\frac{ML}{T^2}\right]^b = [M]^0[L]^0[T]^0.$$

Equating powers of each dimension gives

$$\begin{aligned} M: &\quad b = 0, \\ L: &\quad 4 + a + b = 0, \\ T: &\quad 2b = 0. \end{aligned}$$

Solving, we have

$$b = 0 \quad \text{and} \quad a = -4.$$

The fourth dimensionless parameter is

$$\Pi_4' = \frac{I}{\ell^4}.$$ **ANSWER**

Discussion

Properly applied, both methods give the same dimensionless parameters. This example illustrates that we must be careful in determining k. Because k is usually equal to m (the number of fundamental dimensions), we are tempted to set $k = m$ and bypass the lengthier procedure to check k. If you do this and k is actually less than m for your problem, you will invariably come up with an indeterminant set of equations in step 3. In mechanics problems, a quick check for the value of k is to list the fundamental dimensions in both the $[M, L, T, \theta]$ and the $[F, L, T, \theta]$ systems. If the number of fundamental dimensions involved is different for the two systems, k can be no larger than the smaller number of fundamental dimensions.

6.5 CUSTOMARY DIMENSIONLESS PARAMETERS IN FLUID MECHANICS

Generally, beginning each new problem in fluid mechanics with a formal dimensional analysis is unnecessary. Only a limited number of dimensional parameters can be relevant (length, velocity, density, viscosity, etc.). Although forming infinitely many dimensionless groups from a finite number of dimensional parameters is possible, only a fixed number of *independent* groups can be found. Certain dimensionless groups have become standard in fluid mechanics, and all engineers are more or less expected to use these parameters when conducting and reporting experimental investigations or proposing theories. It is important for beginning engineers to know and understand the standard dimensionless parameters.

6.5.1 Standard Dimensionless Parameters

In this section, we carry out a complete dimensional analysis of fluid mechanics problems and introduce you to the customary dimensionless parameters. Figures 6.7 and 6.8 illustrate two generic problems in fluid mechanics. For now, we assume that the flow in each case is steady. A general functional relation for these flows is

$$\begin{matrix}\text{Dimensional parameter}\\ \text{of primary interest}\end{matrix} = f\left\{\begin{matrix}\text{Size, shape, fluid velocity, fluid}\\ \text{properties, dimensional constants}\end{matrix}\right\}. \tag{6.22}$$

In general, the *type* of parameters on the right-hand side changes very little from problem to problem; however, the parameter on the left-hand side can be of several types. For external flow, it is probably the force exerted on the object by the fluid. Instead of the total force, we may be interested in components in the streamwise direction (called *drag force*) and the direction normal to the stream (called *lift force*). Rather than the vector components of the force, we may be interested in separating the force into pressure and shear stress contributions. For the internal flow, the parameter of interest may be the wall force or it may be the mechanical energy loss. For our discussion, we assume that we are interested in drag force ($\mathscr{D}$) for the external flow and mechanical energy loss (gh_L) for the internal flow.

Note that the size and shape of the object have been separated on the right-hand side of Eq. (6.22). The shape of an object is determined by a set of *dimensionless* parameters, essentially length ratios and angles. For a given shape, the size of an object is specified by a *single* length dimension. Because shape is defined by a (typically very large) set of already dimensionless parameters, it is usually dropped from the general functional equation. If shape is not shown explicitly, the relation

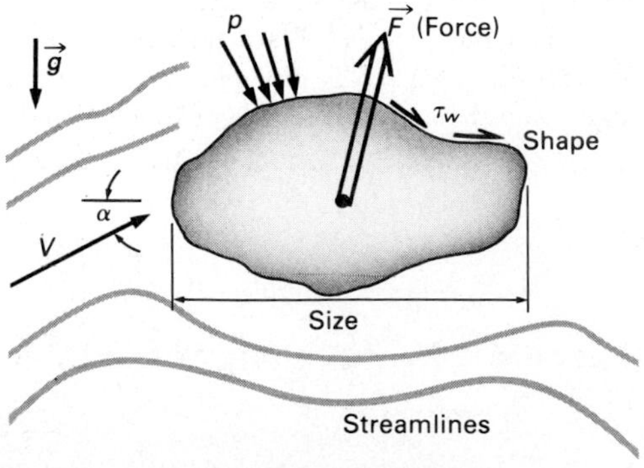

Figure 6.7 External flow problem.

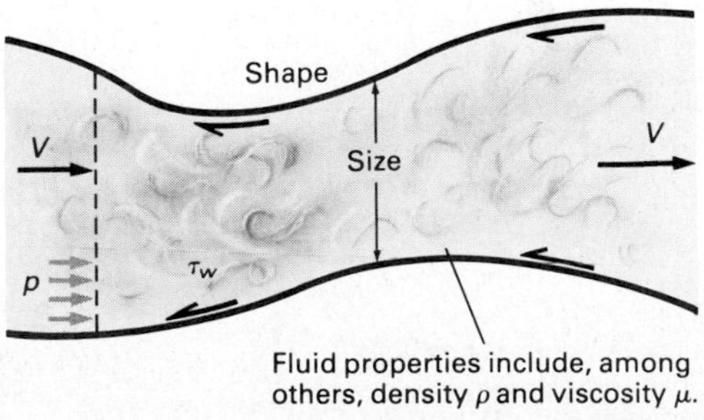

Figure 6.8 Internal flow problem.

is restricted to a specific shape, such as cubes or conical nozzles. Sometimes simple shape parameters such as the length-to-diameter ratio for pipes are retained in the equation.

Because velocity is a vector, specifying its direction as well as its speed may be necessary. In internal flow problems, velocity parallel to the axis of the channel is usually understood, or flow rate is used instead of velocity; in external flow problems, the angle between the approaching fluid velocity and an axis of the object must be specified. This is angle α in Fig. 6.7. Angles are already dimensionless, so they do not enter into the dimensional analysis but propagate directly to the dimensionless equation.

The fluid properties that are important in fluid mechanics problems include density, viscosity, modulus of elasticity, and surface tension. If the flow is highly compressible so that mechanical–thermal energy exchanges are important, then specific heats are also important. Note that pressure and temperature are not included in this list. The reason is that pressure is usually a parameter of primary interest and so is included on the left-hand side rather than on the right-hand side. For compressible fluids, pressure is also included on the right side in the modulus of elasticity because

$$E_v \propto p,$$

at least for an ideal gas. In incompressible flow, temperature itself is not relevant unless we are considering a heat transfer problem. In compressible flow, specification of density, pressure, and specific heats (and so, indirectly, gas constant) is equivalent to specification of temperature. In other words, for a given gas, we need to list any two of pressure, density, or temperature, but not all three.

If there are no electric, magnetic, or relativistic effects in the flow, the only relevant dimensional constant is the acceleration of gravity g. The dimensional functional relations for the flows of Figs. 6.7 and 6.8 become*

$$\mathscr{D} = f\{\ell,\ V,\ \rho,\ \mu,\ E_v,\ \sigma,\ c_p,\ c_v,\ g\} \tag{6.23}$$

and

$$gh_L = f\{\ell,\ V,\ \rho,\ \mu,\ E_v,\ \sigma,\ c_p,\ c_v,\ g\}. \tag{6.24}$$

As only the parameter on the left-hand side is different in Eqs. (6.23) and (6.24), we continue with only Eq. (6.23). The dimensional equation is

$$\left[\frac{ML}{T^2}\right] = f\left\{[L], \left[\frac{L}{T}\right], \left[\frac{M}{L^3}\right], \left[\frac{M}{LT}\right], \left[\frac{M}{LT^2}\right], \left[\frac{M}{T^2}\right], \left[\frac{L^2}{T^2\theta}\right], \left[\frac{L^2}{T^2\theta}\right], \left[\frac{L}{T^2}\right]\right\}.$$

It contains ten dimensional parameters and four fundamental dimensions ($[M]$, $[L]$, $[T]$, $[\theta]$). As ρ, V, ℓ, and c_p cannot form a dimensionless product, $k = 4$, and we can form six dimensionless parameters. We select ρ, V, ℓ and c_p as repeating

---

* The angle α has been omitted because it is already dimensionless. It must be reinserted in the final dimensionless relation.

variables. Performing the dimensional analysis and putting the resulting dimensionless groups in standard form gives

$$\frac{\mathscr{D}}{\frac{1}{2}\rho V^2 \ell^2} = F\left(\frac{\rho V \ell}{\mu}, \frac{V}{\sqrt{E_v/\rho}}, \frac{\rho V^2 \ell}{\sigma}, \frac{V}{\sqrt{g\ell}}, \frac{c_p}{c_v}, \alpha\right). \tag{6.25}$$

Each of the dimensionless parameters on the right-hand side of Eq. (6.25) is a "standard" dimensionless parameter in fluid mechanics. All but two of these parameters are named after persons famous for their contributions to fluid mechanics. The dimensionless character of the named parameters is emphasized by calling them "numbers."

In the introductory comments about dimensional analysis, we pointed out that dimensional analysis would develop the tools to help you classify flow, that is, to decide whether a given flow is laminar or turbulent, compressible or incompressible, and so on. The dimensionless "numbers" are these tools. The definitions, names, and relevance of these important parameters are as follows:

$\mathbf{R} \equiv \rho V\ell/\mu$ is the *Reynolds number,* probably the most important dimensionless parameter in fluid mechanics. **R** represents the effects of viscosity on the flow. Flows with large values of **R** ($\mathbf{R} \to \infty$) are turbulent.

$\mathbf{M} \equiv V/\sqrt{E_v/\rho}$ is the *Mach number,* which is important in compressible flows. The speed of sound (c) in a fluid (see Chapter 10) is given by

$$c^2 = \frac{E_v}{\rho},$$

so the Mach number is the ratio of the fluid speed to the speed of sound in the fluid. Flows with low Mach numbers ($\mathbf{M} \approx 0$) do not exhibit compressibility effects.*

$\mathbf{W} \equiv \rho V^2\ell/\sigma$ is the *Weber number,* which represents the effects of surface tension. Because surface tension is important only if there is an interface, a flow depends on **W** only if there is an interface between two fluids. If the passage in Fig. 6.8 is filled with a single fluid or if the object in Fig. 6.7 is far away from an interface, the Weber number can be dropped from the parameter list. Even if a two-fluid interface is present, surface tension effects are not relevant unless the length scale (size) of the problem is very small.

$\mathbf{F} \equiv V/\sqrt{g\ell}$ is the *Froude number,*† which represents free-surface effects. If there is no free surface, the Froude number can be dropped from the parameter list. Note that the Froude number does not deal with hydrostatic pressure, even though it is the only parameter that contains g. The reason is that the dimensionless parameters deal with dynamic effects, whereas the hydrostatic pressure is a static effect.

---

* For an ideal gas, $E_v = kp$ and $p = \rho RT$, so $\mathbf{M}^2 = \rho V^2/kp = V^2/k\mathbf{R}T$. The criteria for incompressibility suggested in Sections 1.3.6 and 4.3.5 are both equivalent to $\mathbf{M}^2 \ll 1$.

† Some authors define the Froude number as $V^2/g\ell$.

$k \equiv c_p/c_v$ is the *specific heat ratio,* which is relevant only in compressible flow problems. For an incompressible fluid, k may be dropped from the list.

The Reynolds, Mach, Weber, and Froude numbers and the specific heat ratio are the independent dimensionless parameters. They appear in all problems unless the effect specifically represented by the parameter is absent.

Next, we consider the parameters that could appear on the left-hand side of the functional relation. The dimensional parameters that usually appear are

- drag force ($\mathscr{D}$) or
- lift force ($\mathscr{L}$) or
- pressure difference ($\Delta p \equiv p - p_{\text{ref}}$)* or
- wall shear stress (τ_w) or
- mechanical energy loss (gh_L).

If we combine these dimensional parameters with ρ, V, and L, we get the following dimensionless coefficients.

$$\mathbf{C_D} \equiv \frac{\mathscr{D}}{\frac{1}{2}\rho V^2 \ell^2} \quad \text{Drag coefficient;}$$

$$\mathbf{C_L} \equiv \frac{\mathscr{L}}{\frac{1}{2}\rho V^2 \ell^2} \quad \text{Lift coefficient;}$$

$$\mathbf{C_p} \equiv \frac{\Delta p}{\frac{1}{2}\rho V^2} \quad \text{Pressure coefficient (sometimes called the Euler number);}$$

$$\mathbf{C_f} \equiv \frac{\tau_w}{\frac{1}{2}\rho V^2} \quad \text{Skin friction coefficient;}$$

$$\mathbf{K} \equiv \frac{gh_L}{\frac{1}{2}\rho V^2} \quad \text{Loss coefficient.}$$

These coefficients represent the dependent parameters, so we may express any one of them as a function of the five independent-variable dimensionless parameters. For example, an appropriate dimensionless relation for the internal flow of Fig. 6.8 might be

$$\mathbf{K} = F\{\mathbf{R}, \mathbf{M}, \mathbf{W}, \mathbf{F}, k\}.$$

Our list of relevant dimensional parameters and corresponding dimensionless parameters is not quite complete. Certain special cases involve a few more dimensional variables and so require a few more dimensionless parameters.

In the pipe flow discussed in Sections 6.1 and 6.4, we included the roughness of the pipe wall, ε, in the list of dimensional parameters. Therefore the parameter ε/D appeared in the dimensionless parameter list. As ε/D (more generally ε/ℓ) is a ratio of two lengths, we might be tempted to include it with the shape parameters and drop it from the list. This is not a good idea because roughness is a character-

* p_{ref} is a reference pressure.

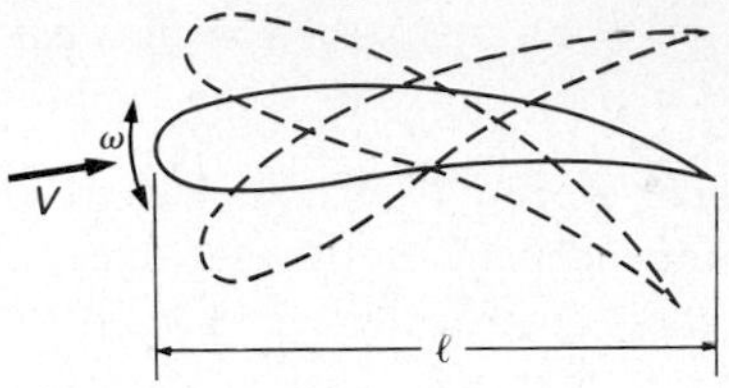

Figure 6.9 Airfoil oscillating with frequency ω.

istic of the type of surface rather than the shape of a passage or object. Laminar flow does not depend on surface roughness, but turbulent flow does. If we are considering a turbulent flow, then ε/ℓ, the *relative roughness,* should be included with the independent dimensionless parameters.

Many flow problems are unsteady. Suppose that the airfoil shown in Fig. 6.9 is oscillating with frequency ω. The frequency of oscillation definitely affects the flow. The reciprocal of the frequency is a characteristic time of the problem. If we add ω to the list of independent dimensional variables, the dimensionless parameter

$$\mathrm{S} \equiv \frac{\omega\ell}{V}$$

will appear on the list of independent dimensionless parameters. The S is called the *Strouhal number*. Sometimes an oscillatory phenomenon is the problem under investigation, that is, the frequency is the dependent parameter rather than an independent parameter. An important engineering problem of this type is flow-induced vibration of heat exchanger tubes or suspension bridge cables. In this instance, the Strouhal number would appear as a dependent parameter rather than an independent parameter, typically in an equation of the form

$$\mathrm{S} = f\{\mathrm{R}, \mathrm{M}\}.$$

A similar situation occurs when a problem contains two reference velocities. With flow in a centrifugal pump rotor, both the fluid velocity and the rotor velocity would be important dimensional parameters. We would have to include the rotor linear velocity U (or, twice its value, ND, where N is rotational speed and D is rotor diameter) on the list of independent dimensional parameters. The corresponding dimensionless parameter would be

$$\frac{V}{U} \quad \text{or} \quad \frac{V}{ND}.$$

The second of these parameters is actually a reciprocal Strouhal number. A similar parameter would be required to investigate a flow using a coordinate system that moves with velocity U (see Examples 4.22 and 4.23).

We could extend this list of special cases, but the important thing to remember is that each extra *dimensional* parameter that is important introduces a *dimensionless* parameter to represent it. We can usually express valid dimensionless parameters that represent these extra parameters in terms of length or velocity ratios.

Table 6.1 is a comprehensive list of the dimensionless parameters discussed in this section. Using this list, you do not need to do a dimensional analysis for each

Table 6.1 Customary dimensionless parameters of fluid mechanics.

Parameter	Symbol	Name	Relevance
Independent Variables			
$\dfrac{\rho V\ell}{\mu}$	**R**	Reynolds number	Laminar/turbulent flow
$\dfrac{V}{\sqrt{E_v/\rho}}$	**M**	Mach number	Compressible flow
$\dfrac{\rho V^2\ell}{\sigma}$	**W**	Weber number	Flow with two-fluid interface
$\dfrac{V}{\sqrt{g\ell}}$	**F**	Froude number	Flow with a free surface
c_p/c_v	k	Specific heat ratio	Compressible flow
$\dfrac{\omega\ell}{V}$	**S**	Strouhal number	Oscillating boundary
$\dfrac{V}{U}$	—	Velocity ratio	Two relevant velocities
ℓ'/ℓ	—	Length ratio	Two relevant lengths
$\dfrac{\varepsilon}{\ell}$	—	Relative roughness	Turbulent flow
α	—	Angle of attack	External flow over objects
Dependent Variables			
$\dfrac{\mathscr{D}}{\frac{1}{2}\rho V^2\ell^2}$	$\mathbf{C_D}$	Drag coefficient	External flow
$\dfrac{\mathscr{L}}{\frac{1}{2}\rho V^2\ell^2}$	$\mathbf{C_L}$	Lift coefficient	External flow
$\dfrac{\Delta p}{\frac{1}{2}\rho V^2}$	$\mathbf{C_p}$	Pressure coefficient	All flows
$\dfrac{\tau_w}{\frac{1}{2}\rho V^2}$	$\mathbf{C_f}$	Skin friction coefficient	All flows with shear stress
$\dfrac{gh_L}{\frac{1}{2}V^2}$	**K**	Loss coefficient	Internal flows
$\dfrac{\omega\ell}{V}$	**S**	Strouhal number	Flows with oscillatory output

situation you encounter. All you need to do is select the relevant parameters from the table and substitute them into the general functional relation

$$\begin{array}{c}\text{Dimensionless parameter}\\ \text{containing relevant}\\ \text{dependent variables}\end{array} = f\left\{\begin{array}{c}\text{Relevant dimensionless parameters}\\ \text{from independent parameters list}\end{array}\right\}. \tag{6.26}$$

A parameter is irrelevant if the effect it represents (e.g., surface tension for **W**, compressibility for **M** and k) is not important in the particular problem. If there is a compelling reason to use a different set of dimensionless parameters to describe a particular flow, you can construct the different parameters from the standard parameters by multiplying powers of the standard parameters.

EXAMPLE 6.4 Illustrates an Alternative Procedure for Identifying Dimensionless Parameters

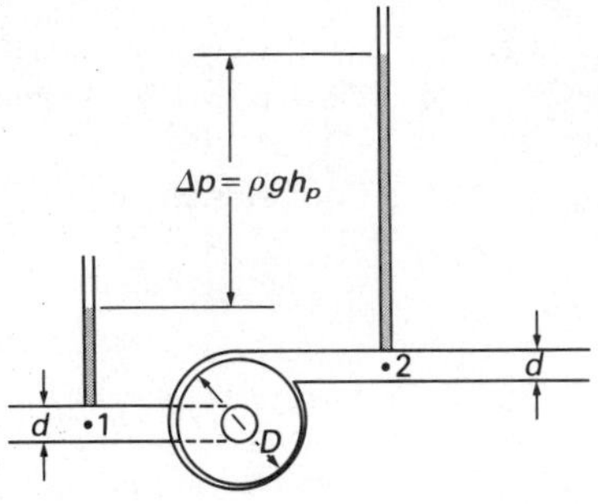

Figure E6.4 Centrifugal pump and its head rise h_p.

Figure E6.4 shows a centrifugal pump with a rotor of diameter D that rotates at speed N (rpm). Liquid flows through the pump with volume flow rate Q. The pressure of the liquid increases by Δp as it passes through the pump. Suggest a useful dimensionless functional relation to express the pump's performance.

SOLUTION

Given

Centrifugal pump in Fig. E6.4.

Find

A useful dimensionless functional relation to express pump performance

Solution

We can deduce a useful set of dimensionless parameters with the help of Table 6.1. We select the pressure rise as the primary dependent parameter. The primary *dimensionless* dependent parameter would then be a pressure coefficient:

$$\mathrm{C_p} = \frac{\text{Pressure rise}}{(\text{Density})(\text{Velocity})^2}.$$

We expect the pressure rise to depend on the liquid's viscosity. This relation implies that we should include the Reynolds number as an important dimensionless parameter:

$$\mathrm{R} = \frac{(\text{Density})(\text{Velocity})(\text{Length})}{\text{Dynamic Viscosity}} = \frac{(\text{Velocity})(\text{Length})}{\text{Kinematic viscosity}}.$$

There are two reference velocities in a centrifugal pump: a typical fluid velocity V and the pump rotor tip velocity U. The *velocity ratio*

$$\frac{V}{U}$$

is surely an important dimensionless parameter.

Effects of compressibility, surface tension, and surface gravity waves are irrelevant in a liquid pump, so we expect a dimensionless performance law of the form

$$\mathrm{C_p} = f\left(\frac{V}{U}, \mathrm{R}\right).$$

We now want to identify the most useful expressions for each dimensionless parameter. The pressure coefficient is

$$\mathbf{C_p} = \frac{\Delta p}{\rho(\text{Velocity})^2}.$$

The most obvious choice for the velocity is a typical fluid velocity. We can find a characteristic fluid velocity from

$$V = \frac{\text{Volume flow rate}}{\text{Area}};$$

thus

$$V \propto \frac{Q}{D^2}.$$

If we use V in the pressure coefficient, $\mathbf{C_p}$ will contain two parameters that vary while the pump operates, namely, Δp and Q. Centrifugal pumps are often driven by constant-speed AC electric motors, so the pump rotor velocity U is a more convenient velocity for the pressure coefficient. As

$$U \propto ND,$$

the most convenient pressure coefficient is

$$\mathbf{C_p} = \frac{\Delta p}{\rho U^2},$$

or, ignoring constants of proportionality,

$$\mathbf{C_p} = \frac{\Delta p}{\rho N^2 D^2}. \qquad \textbf{ANSWER}$$

We can proceed a step further by recognizing that we can express the pressure rise Δp as

$$\Delta p = \rho g h_p,$$

where h_p is the pump head. We can express the pressure coefficient in terms of pump head as

$$\mathbf{C_p} = \frac{\rho g h_p}{\rho N^2 D^2},$$

or

$$\mathbf{C_p} = \frac{g h_p}{N^2 D^2}. \qquad \textbf{ALTERNATIVE ANSWER}$$

This special form of pressure coefficient is called the *head coefficient* and is often denoted by $\mathbf{\Psi}$.

Although the Reynolds number is usually based on a fluid velocity, the rotor velocity (ND) is more convenient for expressing the Reynolds number in a pump. Using the rotor diameter as the relevant length in the Reynolds number gives

$$\mathrm{R} = \frac{ND^2}{\nu}. \qquad \textbf{ANSWER}$$

The final dimensionless parameter, velocity ratio, is given by

$$\frac{V}{U} = \frac{\text{Fluid velocity}}{\text{Rotor velocity}} \propto \frac{Q}{D^2(ND)}.$$

Ignoring constants of proportionality, we find the dimensionless parameter

$$C_Q = \frac{Q}{ND^3}. \qquad \textbf{ANSWER}$$

This parameter is called the *flow coefficient* and is often denoted by **Φ**.

Collecting all our results, we can write a useful dimensionless performance law for a centrifugal pump as

$$\frac{\Delta p}{\rho N^2 D^2} = \frac{gh_p}{N^2 D^2} = f\left(\frac{Q}{ND^3}, \frac{ND^2}{\nu}\right). \qquad \textbf{ANSWER}$$

Discussion

Using V instead of U in the pressure coefficient and the Reynolds number gives

$$\frac{gh_p D^4}{Q^2} = f\left(\frac{Q}{ND^3}, \frac{Q}{\nu D}\right),$$

an alternative, but not as convenient, performance law. We derive the recommended parameters from this set by multiplying the pressure coefficient by $(Q/ND^3)^2$ and dividing the Reynolds number by (Q/ND^3).

We also could have developed a set of dimensionless parameters for expressing pump performance using the pi theorem method. You should try this approach for practice and enlightenment. If you do, be sure to select ρ, N, and D as repeating variables.

6.5.2 Physical Significance of Dimensionless Parameters

You may be curious about why the particular dimensionless parameters of Table 6.1 are preferred. You may also be wondering whether any physical meaning can be attached to these parameters or whether they are just mathematical curiosities. The key to answering both of these questions is that each dimensionless parameter can be interpreted as a ratio of *typical* values of two physical entities. These entities are usually velocities or forces or energies. Sometimes interpreting a particular dimensionless parameter in two or more ways, usually as either a ratio of typical forces or a ratio of typical energies, is possible.

We can demonstrate the physical relevance of the various dimensionless parameters by identifying and estimating the magnitude of various forces on the fluid. If we accept the viewpoint advocated by d'Alembert, we can imagine a moving fluid particle as being in a state of dynamic equilibrium, influenced by an inertia "force" ($\vec{F}_I = -m\vec{a}$) as well as the actual forces due to pressure, shear stress, surface tension, and so on. We can estimate typical magnitudes of some of these forces in a flow.

We estimate the inertia "force" by

$$\text{Inertia force} \sim \text{"Dynamic Pressure"} \times \text{Area}.$$

A typical value of dynamic pressure is the value corresponding to the reference velocity, namely, $\frac{1}{2}\rho V^2$. A typical value of area is the square of the reference length dimension, ℓ^2. Using these estimates, we have

$$\text{Inertia force} \sim \rho V^2 \ell^2.$$

The symbol $\sim$ means that we are estimating the typical value of a quantity or that we are estimating its order of magnitude. The relationship is not quite as strong as a proportionality, indicated by the symbol $\propto$. Constant multipliers that do not change the order of magnitude, such as 2 and $\frac{1}{2}$, are usually irrelevant to the discussion and are dropped from the estimate.

We obtain an estimate of the pressure force from a typical pressure difference and a typical area:

$$\text{Pressure force} \sim \text{Pressure difference} \times \text{Area} \sim (\Delta p)\ell^2.$$

An estimate of the viscous shear force is

$$\text{Shear force} \sim \text{Shear stress} \times \text{Area}.$$

The shear stress is given by

$$\tau = \mu\left(\frac{\partial u}{\partial y}\right).$$

We estimate the velocity derivative by assuming that the velocity changes from zero to the typical value V over an appropriate length ℓ:

$$\frac{\partial u}{\partial y} \approx \frac{\Delta u}{\Delta y} \sim \frac{V-0}{\ell} = \frac{V}{\ell}.$$

Using this derivative estimate, we obtain the viscous force estimate:

$$\text{Viscous force} \sim \mu\left(\frac{V}{\ell}\right)\ell^2 = \mu V \ell.$$

We now form some ratios from our force estimates:

$$\frac{\text{Inertia force}}{\text{Viscous force}} \sim \frac{\rho V^2 \ell^2}{\mu V \ell} = \frac{\rho V \ell}{\mu} = \mathbf{R},$$

$$\frac{\text{Pressure force}}{\text{Inertia force}} \sim \frac{(\Delta p)\ell^2}{\frac{1}{2}\rho V^2 \ell^2} = \frac{\Delta p}{\frac{1}{2}\rho V^2} = \mathbf{C_p}.$$

We may interpret the dimensionless parameters $\mathbf{R}$ and $\mathbf{C_p}$ as ratios of typical forces in the flow field. We also may extend this procedure to other dimensionless groups.

Not all the standard dimensionless parameters are force ratios. The Strouhal number and velocity ratio are more readily interpreted as

$$\mathbf{S} = \frac{\omega \ell}{V} = \frac{(\ell/V)}{(1/\omega)} \sim \frac{\text{Characteristic flow time}}{\text{Characteristic oscillation time}}$$

and

$$\frac{V}{U} \sim \frac{\text{Fluid velocity}}{\text{Velocity of coordinate system}}.$$

Many of the dimensionless parameters may be expressed in terms of energy ratios. For example,

$$\frac{\text{Work done by pressure}}{\text{Fluid kinetic energy}} \sim \frac{\Delta p/\rho}{\frac{1}{2}V^2} = \frac{\Delta p}{\frac{1}{2}\rho V^2} = \mathrm{C_p}.$$

Table 6.2 gives interpretations of various dimensionless parameters.

The interpretation of dimensionless parameters as ratios of relevant physical entities that occur in a flow field greatly aids our understanding of why dimensionless parameters are so useful. When we choose ρ, V, and ℓ as repeating variables, we choose to compare all forces to the inertia "force" and all energies to the kinetic energy. Therefore, when the typical magnitude of another type of force, such as elastic, is either large or small relative to inertia, we expect one of the forces to be negligible. You can easily diagnose this condition by noting that the corresponding dimensionless parameter (Mach number in this case) is either very large or very small. If the Mach number is small, the elastic force resisting volumetric compression is so large relative to the inertia "force" available that you can neglect compression.

Table 6.2 Physical interpretation of common dimensionless parameters.

Parameter	Interpretation	Alternative Interpretation
R	$\frac{\text{Inertia force}}{\text{Viscous force}}$	$\frac{\text{Flow of kinetic energy}}{\text{Rate of mechanical energy loss}}$
M^2	$\frac{\text{Inertia force}}{\text{Elastic force resisting compression}}$	$\frac{\text{Kinetic energy}}{\text{Thermal energy}}$
W	$\frac{\text{Inertia force}}{\text{Surface tension force}}$	$\frac{\text{Kinetic energy}}{\text{Surface energy}}$
F^2	$\frac{\text{Inertia force}}{\text{Gravity force}}$	$\frac{\text{Kinetic energy}}{\text{Potential energy}}$
S	$\frac{\text{Time scale of flow}}{\text{Time scale of boundary}}$	$\frac{\text{Speed of boundary}}{\text{Speed of flow}}$
$\frac{V}{U}$	$\frac{\text{Speed of flow}}{\text{Speed of coordinate system}}$	$\frac{\text{Time scale of coordinate system}}{\text{Time scale of flow}}$
C_D, C_L	$\frac{\text{Resultant force on object}}{\text{Inertia force}}$	—
C_p	$\frac{\text{Pressure force}}{\text{Inertia force}}$	$\frac{\text{Work done by pressure}}{\text{Kinetic energy}}$
K	$\frac{\text{Mechanical energy loss}}{\text{Kinetic energy}}$	—

Ratios of pairs of forces that do not involve the inertia "force" can be obtained by multiplying or dividing the standard dimensionless parameters. For example,

$$\frac{\text{Pressure force}}{\text{Viscous force}} = \frac{\text{Inertia force}}{\text{Viscous force}} \times \frac{\text{Pressure force}}{\text{Inertia force}} \sim \mathbf{R} \times C_p$$

$$\sim \left(\frac{\rho V \ell}{\mu}\right) \times \left(\frac{\Delta p}{\rho V^2}\right) = \frac{(\Delta p)\ell}{\mu V}.$$

Sometimes using dimensional reasoning alone can generate most of the solution to a complex flow problem. One example of this technique is presented here, another in Section 7.2.8

EXAMPLE 6.5 **Illustrates Determining the Form of an Equation by Dimensional Analysis and That Standard Dimensionless Parameters are Not Always Best**

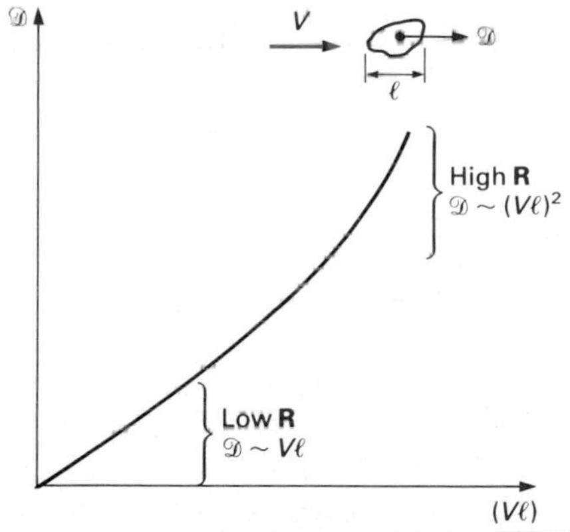

Figure E6.5 Drag variation at low and high Reynolds numbers.

Consider the external flow diagram in Fig. E6.5. At very low Reynolds numbers, the drag force ($\mathscr{D}$) is expected to depend on viscous effects only. At very high Reynolds numbers, the drag force is expected to depend on inertia effects only. Determine the form of the equations for the drag on the object for both cases. What conclusions can be drawn?

SOLUTION

Given

Flow over an object

At very low Reynolds numbers, drag depends on viscous effects only

At very high Reynolds numbers, drag depends on inertia effects only

Find

Form of:

$\mathscr{D} = f\{\text{Inertia effects}\}$ at very high $\mathbf{R}$

$\mathscr{D} = f\{\text{Viscous effects}\}$ at very low $\mathbf{R}$

Solution

Recall from the text that

$$\text{Inertia force} \sim \rho V^2 \ell^2 \quad \text{and} \quad \text{Viscous force} \sim \mu V \ell.$$

At very high Reynolds numbers, the dimensional relation involves ρ but not μ and is

$$\mathscr{D} = f_H\{\rho, V, \ell\}.$$

The dimensions of these parameters are

$$[\mathscr{D}] = \left[\frac{ML}{T^2}\right], \quad [\rho] = \left[\frac{M}{L^3}\right], \quad [V] = \left[\frac{L}{T}\right], \quad \text{and} \quad [\ell] = [L].$$

There are four dimensional parameters and three fundamental dimensions; therefore we expect one dimensionless parameter. This single parameter would have the form

$$\frac{\text{Drag force}}{\text{Inertia force}} \sim \frac{\mathscr{D}}{\frac{1}{2}\rho V^2 \ell^2}.$$

The "$\frac{1}{2}$" is retained to respect tradition. This expression represents the drag coefficient $\mathbf{C_D}$. Because there is only one dimensionless parameter,

$$\mathbf{C_D} = \text{Constant},$$

and the drag at high Reynolds numbers is easily calculated from

$$\mathscr{D} = \mathbf{C_D}(\tfrac{1}{2}\rho V^2 \ell^2) = \text{Constant}(\tfrac{1}{2}\rho V^2 \ell^2). \qquad \textbf{ANSWER}$$

At very low Reynolds numbers, the dimensional relation involves μ but not ρ and is

$$\mathscr{D} = f_L\{\mu, V, \ell\}.$$

The dimensions of these parameters are

$$[\mathscr{D}] = \left[\frac{ML}{T^2}\right], \quad [\mu] = \left[\frac{M}{LT}\right], \quad [V] = \left[\frac{L}{T}\right], \quad \text{and} \quad [\ell] = [L].$$

There are four dimensional parameters and three fundamental dimensions; therefore we expect no more than one dimensionless parameter. The form of this parameter is

$$\frac{\text{Drag force}}{\text{Viscous force}} \sim \frac{\mathscr{D}}{\mu V \ell}.$$

We call this expression the *viscous drag coefficient* $\mathbf{C'_D}$. Because there is only one dimensionless parameter,

$$\mathbf{C'_D} = \text{Constant},$$

and the drag at very low Reynolds number is calculated from

$$\mathscr{D} = \mathbf{C'_D}(\mu V \ell) = \text{Constant } (\mu V \ell). \qquad \textbf{ANSWER}$$

$\mathbf{C_D}$ and $\mathbf{C'_D}$ are related by

$$\mathbf{C_D} = \frac{\mathscr{D}}{\frac{1}{2}\rho V^2 \ell^2} = \frac{\mathscr{D}}{\mu V \ell}\left(\frac{\mu V \ell}{\frac{1}{2}\rho V^2 \ell^2}\right).$$

Simplifying, we obtain

$$\mathbf{C_D} = \left(\frac{\mathscr{D}}{\mu V \ell}\right)\left(\frac{2\mu}{\rho V \ell}\right) \quad \text{or} \quad \mathbf{C_D} = \frac{2\mathbf{C'_D}}{\mathbf{R}}.$$

Use of the standard drag coefficient, $\mathbf{C_D}$, is convenient for flow with high Reynolds numbers because it is constant. But $\mathbf{C_D}$ is not as convenient as $\mathbf{C'_D}$ for flow with very low Reynolds numbers because $\mathbf{C'_D}$ is constant and $\mathbf{C_D}$ depends on the exact value of the Reynolds number. We conclude:

The ordinary drag coefficient

$$\mathbf{C_D} = \frac{\mathscr{D}}{\frac{1}{2}\rho V^2 \ell}$$

is convenient for flow with high Reynolds numbers, but an alternative "drag coefficient"

$$C_D' \equiv \frac{\mathscr{D}}{\mu V \ell}$$

is more convenient for flow with very low Reynolds numbers.

Discussion

Two important questions are (1) How low is a "very low Reynolds number"? and (2) How high is a "very high Reynolds number"? In general, a very low Reynolds number is less than 1 (say, 10^{-1}) and a very high Reynolds number is greater than about 10^4. Between Reynolds numbers of 10^{-1} and 10^4, the drag is proportional to some power(s) of velocity between 1 and 2. At these intermediate Reynolds numbers, both C_D and C_D' depend on the exact value of the Reynolds number, and neither is especially convenient. Custom dictates the use of C_D instead of C_D' at intermediate Reynolds numbers.

Using dimensionless parameters to describe a flow is not simply an option that someone discovered by chance. Because the parameters represent ratios of the magnitudes of relevant physical entities, the parameters must be present in any mathematical model of a flow, whether we like it or not. Thus we can sometimes use this fact to find the dimensionless parameters themselves, as Example 6.6 illustrates.

EXAMPLE 6.6 Illustrates That Dimensionless Parameters are Buried in Mathematical Models of Flow and an Alternative Method for Finding Them

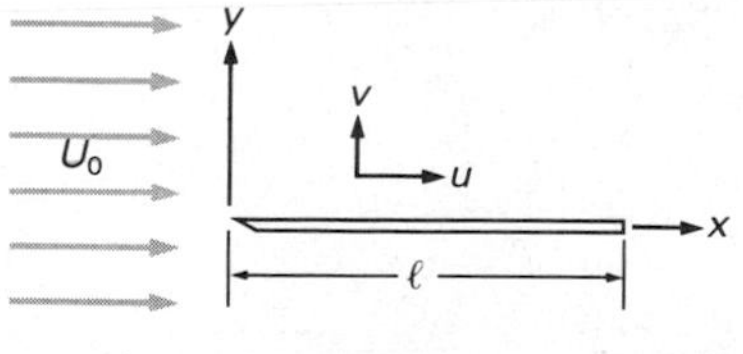

Figure E6.6 Flow of a fluid over a thin, flat plate.

We discussed the boundary layer model for flow near a surface in Section 5.6.2. For incompressible laminar flow over a thin, flat plate (Fig. E6.6), the boundary layer approximation to the Navier–Stokes equation and the continuity equation are

$$\rho\left[u\frac{\partial u}{\partial x} + v\frac{\partial u}{\partial y}\right] = \mu\frac{\partial^2 u}{\partial y^2} \quad \text{and} \quad \frac{\partial u}{\partial x} + \frac{\partial v}{\partial y} = 0,$$

where x and y are coordinates and u and v are the corresponding velocity components, ρ is the density, and μ the fluid viscosity. Convert the equations to dimensionless form and identify any familiar parameters.

SOLUTION

Given

Boundary layer equations for a flat plate

Find

Dimensionless form of the equations

Any familiar parameters

Solution

In this approach, we select a reference velocity U_0 and a reference length ℓ and define dimensionless velocities and lengths as follows:

$$U = \frac{u}{U_0}, \qquad V = \frac{v}{U_0}, \qquad X = \frac{x}{\ell}, \quad \text{and} \quad Y = \frac{y}{\ell}.$$

Substituting for u, v, x, and y in terms of U, V, X, and Y gives

$$\rho\left[U_0 U\left(\frac{\partial(U_0 U)}{\partial(\ell X)}\right) + U_0 V\left(\frac{\partial(U_0 U)}{\partial(\ell Y)}\right)\right] = \mu\left(\frac{\partial}{\partial(\ell Y)}\right)\left(\frac{\partial(U_0 U)}{\partial(\ell Y)}\right),$$

or, because U_0 and ℓ are constant,

$$\frac{\rho U_0^2}{\ell}\left[U\left(\frac{\partial U}{\partial X}\right) + V\left(\frac{\partial U}{\partial Y}\right)\right] = \frac{\mu U_0}{\ell^2}\left(\frac{\partial^2 U}{\partial Y^2}\right).$$

Similarly,

$$\frac{U_0}{\ell}\frac{\partial U}{\partial X} + \frac{U_0}{\ell}\frac{\partial V}{\partial Y} = 0.$$

Rearranging and simplifying give

$$U\left(\frac{\partial U}{\partial X}\right) + V\left(\frac{\partial U}{\partial Y}\right) = \frac{\mu}{\rho U_0 \ell}\frac{\partial^2 U}{\partial Y^2} \quad \text{and} \quad \frac{\partial U}{\partial X} + \frac{\partial V}{\partial Y} = 0.$$

The quantity $(\rho U_0 \ell/\mu)$ is a Reynolds number:

$$\frac{\rho U_0 \ell}{\mu} = \mathbf{R}.$$

The boundary layer equations in dimensionless form are

$$U\left(\frac{\partial U}{\partial X}\right) + V\left(\frac{\partial U}{\partial Y}\right) = \frac{1}{\mathbf{R}}\left(\frac{\partial^2 U}{\partial Y^2}\right)$$ **ANSWER**

and

$$\frac{\partial U}{\partial X} + \frac{\partial V}{\partial Y} = 0,$$ **ANSWER**

where

$$\mathbf{R} = \frac{\rho U_0 \ell}{\mu}.$$ **ANSWER**

Discussion

Solution of the dimensionless equations would exhibit a dependence on Reynolds numbers. Obviously, Reynolds number is a relevant dimensionless parameter for this type of flow.

This example demonstrates that relevant dimensionless parameters appear *naturally* when the equations describing a physical phenomenon are put in dimensionless form. We can use this method instead of the pi theorem to obtain dimensionless parameters; however, the method is limited to those phenomena for which we have appropriate equations.

Moreover, the boundary conditions may also introduce relevant dimensionless parameters.

6.6 AN APPLICATION OF DIMENSIONAL ANALYSIS: MODEL TESTING AND SIMILITUDE

Many engineering fluid mechanics problems must be solved by experimentation. A partial list of fluid mechanics problems for which there is no complete analytical solution includes aerodynamic drag on aircraft and road vehicles, surface wave resistance on ship hulls, and flow and erosion patterns in rivers and harbors. Engineering designs of airborne, land-based, and sea-going vehicles, river and harbor improvements, and large dams are based on extensive experimental testing. Obviously construction and testing of several alternative designs, all built to full scale and operated under all expected conditions of wind speed, water flow, and so on, is completely out of the question. The only possible approach is to test scale models that are smaller than the actual object.* Dimensional analysis provides the key to relating the data obtained from a model experiment to the expected values for the full-scale object (called the *prototype*). Dimensional analysis also allows engineers to determine the conditions that must be maintained in the model test if the conditions under which the prototype must operate are known.

6.6.1 Model Testing and Extrapolation of Results

Suppose that a group of engineers designing a two-person submarine for oceanographic research has proposed a design for the submarine's hull. They need to find the drag on the hull at various cruising speeds in order to determine the size of the submarine's propulsion system. The engineers plan to test models to study the drag. Figure 6.10 shows the proposed (prototype) hull shape and a model of the hull.† The first, almost obvious, requirement of model testing is that the model look exactly like the prototype, that is, have *geometric similarity*. Every model length dimension is related to the corresponding prototype length dimension by a constant scale factor (S):

$$\ell_{1,m} = S\ell_{1,p}, \qquad \ell_{2,m} = S\ell_{2,p}, \qquad \text{and so on.}$$

Model areas are S^2 times corresponding prototype areas, and model volume is S^3 times the prototype volume.

Applying what we know about drag and dimensional analysis, we may state the drag coefficient for the prototype submarine as

$$\mathbf{C}_{\mathbf{D},p} = F_p\{\mathbf{R}_p, \mathbf{M}_p, \mathbf{W}_p, \mathbf{F}_p, k_p, \alpha_p\},$$

where α, called the "angle of attack," is the angle between a reference axis on the submarine and the approach velocity vector. Because the prototype will operate in water, compressibility effects will no doubt be negligible, so a simpler relation is

$$\mathbf{C}_{\mathbf{D},p} = F_p\{\mathbf{R}_p, \mathbf{W}_p, \mathbf{F}_p, \alpha_p\}. \tag{6.27}$$

* Scale models larger than the actual object may be used when the actual device is too small to test conveniently.

† In this example, the prototype exists only on paper; it has not actually been built. Nevertheless, thinking of the prototype as actually having been built will prove helpful.

Figure 6.10 Prototype submarine and a scale model.

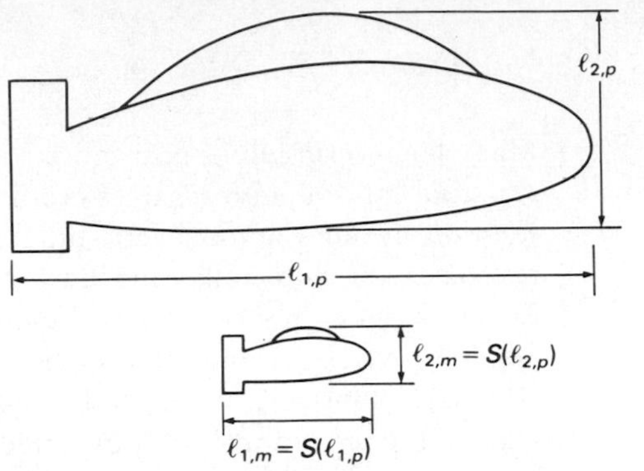

When the model is tested, its drag will obey a relation of the same form:

$$\mathbf{C}_{\mathbf{D},m} = F_m\{\mathbf{R}_m, \mathbf{W}_m, \mathbf{F}_m, \alpha_m\}. \tag{6.28}$$

Explicit effects of size, fluid properties, and velocity were removed by the use of dimensionless parameters, so the functions F_p and F_m are determined solely by the shape of the prototype and model. As the prototype and model have the same shape, F_p and F_m are the same function. Thus the relation

$$\mathbf{C}_{\mathbf{D}} = F\{\mathbf{R}, \mathbf{W}, \mathbf{F}, \alpha) \tag{6.29}$$

applies to both model and prototype. A function F determined from model tests can be used to calculate the prototype drag.

This principle applies point by point as well as to the entire function; if the model and prototype have the same Reynolds number, Weber number, Froude number, and orientation relative to the flow, they will have the same drag coefficient (and the same lift coefficient). That is, any pair of corresponding geometric points on the model and prototype (called *homologous points;* see Fig. 6.11) have the same pressure coefficient and skin friction coefficient.

We summarize our observations about the relationship between the model flow and the prototype flow as follows:

> If a model and its prototype have equal values of all of the relevant independent-variable dimensionless parameters, the model and prototype have equal values of all of the dependent-variable dimensionless parameters.

This principle can be used to scale up model test results and to determine the conditions (density, velocity, etc.) necessary in a model test to "duplicate" prototype operating conditions. The only difficulty in applying this principle seems to be in determining which parameters are relevant. Usually, the non-relevance of certain parameters is obvious, because the effects that they represent are not present in either model or prototype flow. For example, if there is no free surface, the Froude number is irrelevant. In other cases, design specifications for the model and prototype allow us to estimate a parameter, such as a Mach number, and decide whether compressibility—and hence the Mach number itself—is relevant. Consult Table 6.1 for relevance of the various parameters.

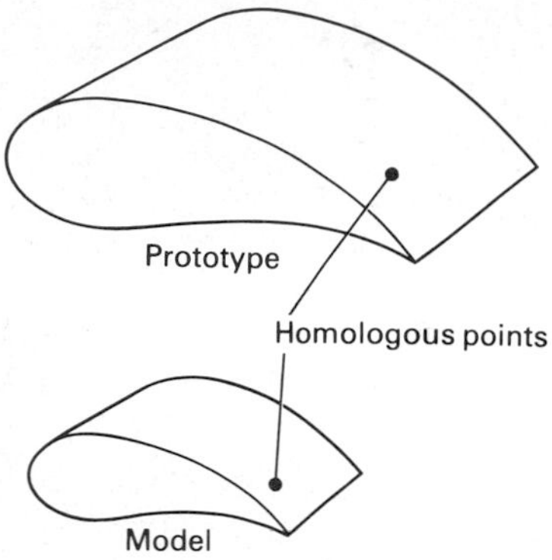

Figure 6.11
Corresponding points on model and prototype are called homologous points.

EXAMPLE 6.7 Illustrates Extrapolation of Model Test Results

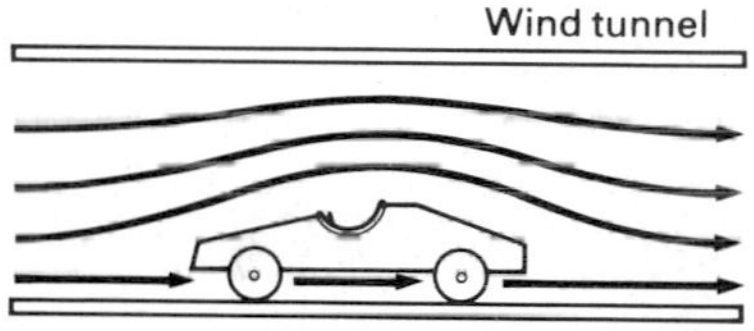

Figure E6.7 Soap box derby car in a wind tunnel.

A one-tenth-scale model of a soap box derby car, shown in Fig. E6.7, is tested in a wind tunnel. The air speed in the wind tunnel is 70 m/s, the air drag on the model car is 240 N, and the air temperature and pressure are identical to those expected when the prototype car is racing. Find the corresponding racing speed in still air and the drag on the car.

SOLUTION

Given

One-tenth-scale model test of soap box derby car

Air speed 70 m/s

Air temperature and pressure identical to actual racing conditions

Find

Corresponding racing speed in still air

Drag on the car

Solution

The relevant dependent-variable parameter is obviously the drag coefficient. In terms of the independent-variable parameters, the Froude and Weber numbers are irrelevant because there is no free surface. The Strouhal number is irrelevant because there is no frequency or second velocity. The Mach number (assuming atmospheric air) is

$$\mathbf{M} = \sqrt{\frac{\rho V^2}{E_v}} \approx \sqrt{\frac{\rho V^2}{kp}} = \sqrt{\frac{1.2\ \text{kg/m}^3\ (70\ \text{m/s})^2}{(1.4)(100{,}000)\ \text{N/m}^2}} = 0.203.$$

As $\mathbf{M}^2$ (≈ 0.04) $\ll 1$, compressibility is irrelevant, k and $\mathbf{M}$ are not relevant. Only the Reynolds number remains as a relevant independent-variable dimensionless parameter. The functional relation is

$$\mathbf{C_D} = f(\mathbf{R}),$$

where

$$\mathrm{R} = \frac{\rho V \ell}{\mu} \quad \text{and} \quad \mathrm{C_D} = \frac{\mathscr{D}}{\frac{1}{2}\rho V^2 \ell^2}.$$

For the wind tunnel test to simulate the actual (prototype) car, the Reynolds number of the model car must equal that of the actual car,

$$\mathrm{R}_{\text{model}} = \mathrm{R}_{\text{actual}} \quad \text{and} \quad \frac{\rho_m V_m \ell_m}{\mu_m} = \frac{\rho_p V_p \ell_p}{\mu_p}.$$

The problem statement shows that

$$\rho_m = \rho_p \quad \text{and} \quad \mu_m = \mu_p.$$

Then

$$V_p = V_m \frac{\ell_m}{\ell_p} = (70\ \text{m/s})\left(\frac{1}{10}\right),$$

or

$$V_p = 7.0\ \text{m/s}.$$ **ANSWER**

With this prototype velocity, the independent-variable dimensionless parameter has the same value for both the model and the actual car, and the dependent-variable dimensionless parameter has the same value. Therefore

$$\mathrm{C}_{\mathbf{D},\text{model}} = \mathrm{C}_{\mathbf{D},\text{prototype}}, \quad \text{or} \quad \frac{\mathscr{D}_m}{\frac{1}{2}\rho_m V_m^2 \ell_m^2} = \frac{\mathscr{D}_p}{\frac{1}{2}\rho_p V_p^2 \ell_p^2}.$$

The problem statement shows that

$$\rho_m = \rho_p, \quad \text{so} \quad \mathscr{D}_p = \mathscr{D}_m \left(\frac{\ell_p V_p}{\ell_m V_m}\right)^2.$$

The numerical values give

$$\mathscr{D}_p = 240\ \text{N}\left[\left(\frac{10}{1}\right)\left(\frac{7\ \text{m/s}}{70\ \text{m/s}}\right)\right]^2 = 240\ \text{N}\left(\frac{70\ \text{m/s}}{70\ \text{m/s}}\right)^2,$$

or

$$\mathscr{D}_p = 240\ \text{N}.$$ **ANSWER**

Discussion

Even though this problem contained one dependent-variable and one independent-variable dimensionless parameter, other problems may contain more than one dependent-variable and/or independent-variable parameter. The same principle holds; each dependent-variable dimensionless parameter has the same value for the model test and the prototype if each independent-variable dimensionless parameter has the same value for the model test and the prototype.

The equality of model and prototype drag in this example is an interesting but nongeneral result.

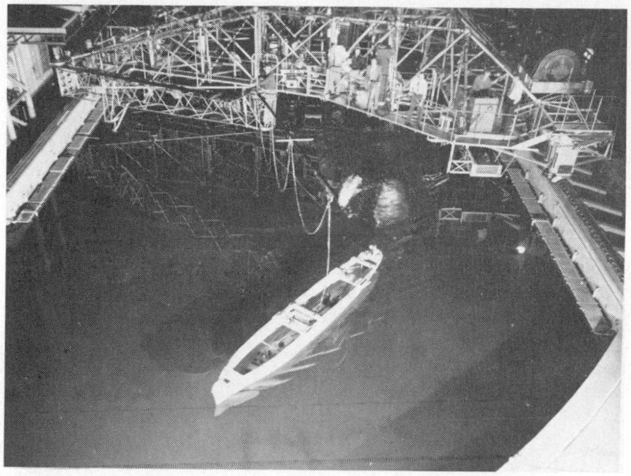

Figure 6.12 Ship model in towing tank facility (courtesy of U.S. Navy David Taylor Ship Research and Development Center).

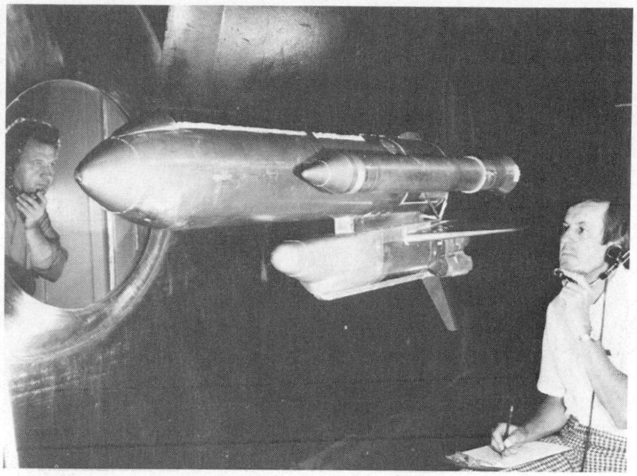

Figure 6.13 Model of the U.S. Space Shuttle in a supersonic wind tunnel (courtesy of NASA Lewis Research Center).

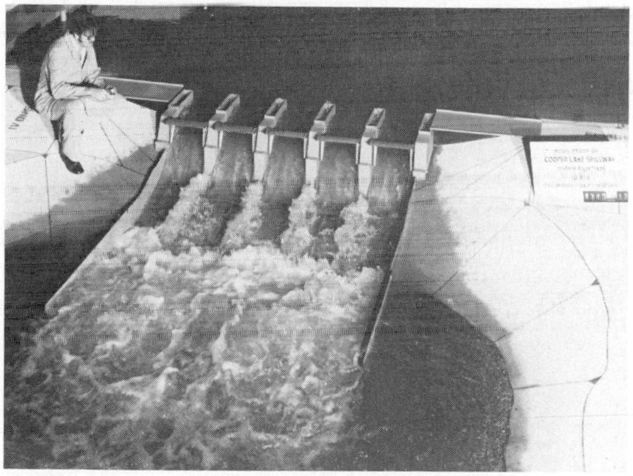

Figure 6.14 Hydraulic model of a dam and spillway (U.S. Army Corps of Engineers photograph).

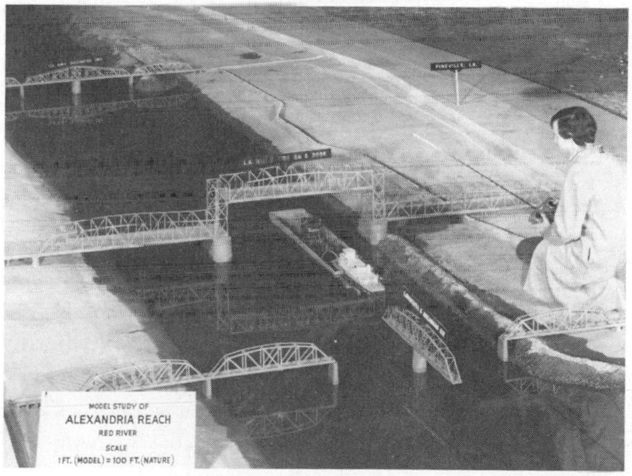

Figure 6.15 Hydraulic model used to determine effects of improvements to a river channel (U.S. Army Corps of Engineers photograph).

Extensive and costly facilities have been constructed for regular testing of models of aircraft, spacecraft, ships, dams, harbors, and rivers. Figures 6.12–6.15 illustrate some models and model-testing facilities. Sometimes, special one-of-a-kind model-testing facilities are constructed to obtain information about a specific piece of equipment. Figure 6.16 illustrates a model test of this type.

6.6.2 The Principle of Similitude

The principles underlying model testing and the equality of the various dimensionless parameters between model and prototype are illustrated by the concept of *similitude*. To explain this concept, we must introduce three types of similarity: *geometric similarity*, *kinematic similarity*, and *dynamic similarity*. Let's consider

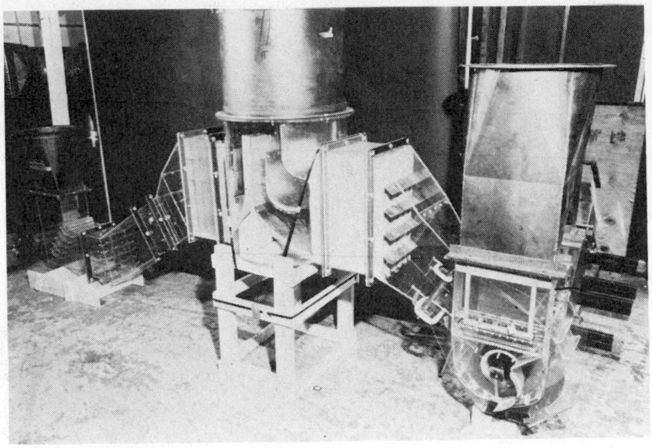

Figure 6.16 Scale-model test of a power plant induced draft fan, electrostatic precipitator, and stack (reproduced from Dynatech R/D Company Report No. 1928, Figure 2.2, page 22 of "An Experimental Gas Flow Model Study on the Precipitator, Fan and Silencer for the Michigan City Plant Boiler 12 of the Northern Indiana Public Service Company," completed under the direction of the Stone & Webster Engineering Corporation and NIPSCO).

the deflection of a jet of liquid by a turning vane attached to a moving cart. Figure 6.17 shows both a prototype and a model of this flow. Geometric similarity, which we have already discussed, requires that the shape of the model and the shape of the prototype be exactly the same. Formally, the requirement of *geometric similarity* is as follows:

> All lengths of the model and prototype must be in the same ratio. All corresponding angles must be equal.

Kinematic similarity requires that the ratio of all relevant velocities be the same for the model and prototype. In the moving vane example, the ratio of cart speed to fluid jet speed must be the same for the model and the prototype. Kinematic similarity also requires that the orientation of the flow with respect to the object be the same. If there are unsteady effects, the ratio of flow time scale to

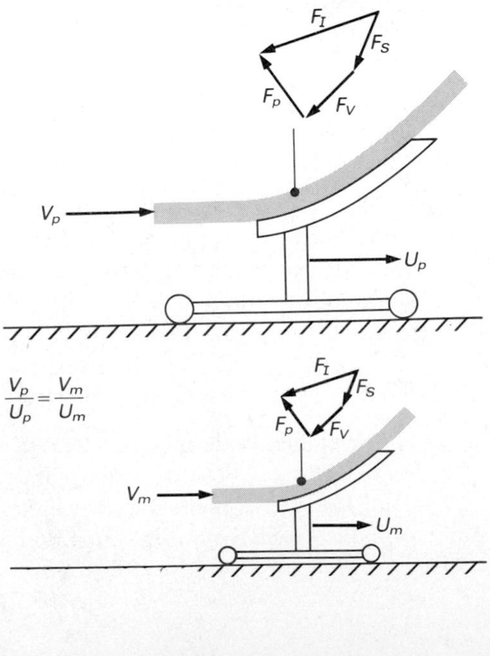

Figure 6.17 Prototype and model flow: fluid jet deflection by a moving cart; force polygons for typical point are shown.

boundary time scale (Strouhal number) must be the same. Formally, the requirement of *kinematic similarity* is as follows:

> Ratios of fluid velocity and other relevant velocities must be the same for the model and the prototype. The orientation of the flow with respect to the object must be the same. Ratios of flow time scale and boundary time scale must be the same.

Two flows that are geometrically and kinematically similar have geometrically similar streamline patterns.

Dynamic similarity involves similarity of the force polygons for the model and prototype flows. Figure 6.17 shows the polygons of the forces acting on fluid particles at homologous points in the model and prototype flows. In the specific case of the cart, we assume that inertia "forces (F_I)," pressure forces (F_p), viscous forces (F_V), and surface tension forces (F_S) are present. Similarity of force polygons requires that the ratios of all forces in one polygon be equal to the ratios of all forces in the other polygon:

$$\frac{\text{Inertia "force" in model flow}}{\text{Viscous force in model flow}} = \frac{\text{Inertia "force" in prototype flow}}{\text{Viscous force in prototype flow}},$$

$$\frac{\text{Pressure force in model flow}}{\text{Inertia "force" in model flow}} = \frac{\text{Pressure force in prototype flow}}{\text{Inertia "force" in prototype flow}},$$

and

$$\frac{\text{Surface tension force in model flow}}{\text{Inertia "force" in model flow}} = \frac{\text{Surface tension force in prototype flow}}{\text{Inertia "force" in prototype flow}}.$$

The requirements are equivalent to

$$\mathbf{R}_m = \mathbf{R}_p, \qquad \mathbf{C}_{\mathbf{p},m} = \mathbf{C}_{\mathbf{p},p}, \quad \text{and} \quad \mathbf{W}_m = \mathbf{W}_p.$$

If we have geometric and kinematic similarity, the angles in the force polygons are equal for the model and the prototype. Hence one requirement for dynamic similarity can be relaxed. If

$$\mathbf{R}_m = \mathbf{R}_p \quad \text{and} \quad \mathbf{W}_m = \mathbf{W}_p,$$

then

$$\mathbf{C}_{\mathbf{p},m} = \mathbf{C}_{\mathbf{p},p}$$

must be true, as stated by equations such as Eqs. (6.27) and (6.28). Formally, the complete *principle of similitude* is as follows:

> If model and prototype flows have geometric similarity and kinematic similarity in all relevant parameters and if there is dynamic similarity in all relevant independent-variable parameters, then there will be complete dynamic similarity, that is, all dependent-variable parameters will be equal.

There is some debate concerning which parameters are geometric, kinematic, and dynamic. The angle of orientation of flow with respect to the object could be considered geometric rather than kinematic. Froude and Mach numbers may be considered either kinematic or dynamic parameters (see Table 6.2). It really makes little difference in the end.

6.6.3 Difficulties in Model Testing

The rather well-developed theory of dimensional analysis and its interpretation in terms of similitude may lead you to believe that model testing is quite simple and straightforward. Unfortunately, this is not true, as the following situation demonstrates. Suppose that we want to determine the drag on a large ship hull by performing tests on a smaller scale-model ship (see Fig. 6.12). The viscous shear stress and surface wave resistance cause the drag, so the drag coefficient depends on Reynolds number and Froude number:

$$\mathbf{C_D} = F\{\mathbf{R}, \mathbf{F}\}.$$

Suppose that we decide to use a 1/25-scale model for the test. For similarity, we must have

$$\mathbf{F}_m = \mathbf{F}_p \quad \text{and} \quad \mathbf{R}_m = \mathbf{R}_p.$$

From the Froude number requirement,

$$\frac{V_m}{\sqrt{g_m \ell_m}} = \frac{V_p}{\sqrt{g_p \ell_p}}.$$

Undoubtedly $g_m = g_p$, so this equation requires that

$$\frac{V_m}{V_p} = \sqrt{\frac{\ell_m}{\ell_p}} = \sqrt{S}.$$

As $S = 1/25$,

$$\frac{V_m}{V_p} = \frac{1}{5}.$$

Next, we consider the Reynolds number requirement:

$$\frac{\rho_m V_m \ell_m}{\mu_m} = \frac{\rho_p V_p \ell_p}{\mu_p}.$$

The necessary kinematic viscosity ratio is

$$\frac{\nu_m}{\nu_p} = \frac{V_m}{V_p}\frac{\ell_m}{\ell_p} = \sqrt{\frac{\ell_m}{\ell_p}}\left(\frac{\ell_m}{\ell_p}\right) = \left(\frac{\ell_m}{\ell_p}\right)^{3/2} = (S)^{3/2}.$$

In this case, with $S = 1/25$, $\nu_m/\nu_p = 1/125$. Because the prototype ship will operate in water, we must perform the model test in a fluid with 1/125 the kinematic viscosity of water. No such fluid exists. In fact, the only practical test fluid is also water. Thus we cannot simultaneously match Reynolds number and Froude number in the scale-model test.

The same type of problem arises in tests of high-speed aerodynamic vehicles. In this case, Reynolds and Mach numbers should be matched. If we assume that the fluid is an ideal gas, the Mach number requirement is

$$\frac{\sqrt{\rho_m}\, V_m}{\sqrt{k_m}\sqrt{p_m}} = \frac{\sqrt{\rho_p}\, V_p}{\sqrt{k_p}\sqrt{p_p}}.$$

Using the ideal gas law, we have

$$\frac{p}{\rho} = RT,$$

and so

$$\frac{V_m}{\sqrt{k_m R_m T_m}} = \frac{V_p}{\sqrt{k_p R_p T_p}}, \quad \text{or} \quad \frac{V_m}{V_p} = \sqrt{\frac{k_m R_m T_m}{k_p R_p T_p}}.$$

The Reynolds number requirement is

$$\frac{\rho_m V_m \ell_m}{\mu_m} = \frac{\rho_p V_p \ell_p}{\mu_p}.$$

The necessary viscosity ratio is

$$\frac{\mu_m}{\mu_p} = \frac{\rho_m V_m}{\rho_p V_p}\left(\frac{\ell_m}{\ell_p}\right) = \frac{p_m R_p T_p}{p_p R_m T_m}\sqrt{\frac{k_m R_m T_m}{k_p R_p T_p}}\left(\frac{\ell_m}{\ell_p}\right) = \frac{p_m}{p_p}\sqrt{\frac{k_m R_p T_p}{k_p R_m T_m}}\left(\frac{\ell_m}{\ell_p}\right).$$

The prototype gas is air. The model test gas most likely is air also. The viscosity of a gas can be varied by changing its temperature. For air, an approximate equation for the viscosity ratio is

$$\frac{\mu_m}{\mu_p} \approx \left(\frac{T_m}{T_p}\right)^{0.75}, \quad \text{so} \quad \frac{\ell_m}{\ell_p} = S \approx \frac{p_p}{p_m}\left(\frac{T_m}{T_p}\right)^{1.25}.$$

For a small-scale-model test (S small), the pressure in the test gas must be high and the temperature must be low. For example, suppose that S again is 1/25. Also suppose that p_p is 70 kPa and T_p is 270 K (corresponding to about 3 km altitude in the standard atmosphere). Let's also assume that the temperature for the model test is 270 K. The required pressure in the wind tunnel would then be 1750 kPa. This situation would require prohibitively large compressor power. Similar to the ship model test, a completely similar model test is impractical. In most cases when two independent-variable dimensionless parameters must be matched, the only way to do so is to have the model be the same size as the prototype, which is not a model test at all.

If matching two (or more) independent similarity parameters is impractical, information must be obtained from a series of model tests or a combination of model test results with theoretical calculations. For example, a series of tests may be made at low Mach numbers (maybe even using water as the test fluid) to determine aircraft drag. These tests can establish the influence of Reynolds number on drag. This information can then be used as a basis for extrapolation of high Mach number (but low Reynolds number) test results to higher values of Reynolds number.

Alternatively, the low Mach number test results may be used to construct a theory that permits calculation of the drag from viscous forces. If we assume that viscous and compressibility effects on drag are independent, we can use the theory to "remove" viscous drag from model test data by calculating the viscous drag for the model and subtracting it. We convert the remaining "compressibility drag" to compressibility drag for the prototype by similarity considerations. We then use the theory to calculate the viscous drag for the prototype. We may also use this type of procedure to separate viscous and wave drag for ship model tests. Unfortunately, these procedures are approximate.

There are other, less obvious problems in model testing. One has to do with surface roughness. Surface roughness is a geometric parameter, so complete similarity requires that the relative roughness (ε/ℓ) be the same for a model and a prototype. In many cases, model roughness cannot be made small enough for complete similarity, and a "roughness correction" must be made. These corrections are essentially extrapolations based on a few studies of the effects of roughness.

The problems that we have discussed so far involve cases where it is impossible to properly model some effect that is actually present in the prototype flow. Another type of problem occurs when certain influences that are not relevant in the prototype flow occur in the model flow. For modeling harbors or rivers (see Fig. 6.15), a very small scale factor must be used. The water depth in the scale model becomes so small that surface tension forces, entirely irrelevant in the prototype, become important in the model flow. To eliminate this effect, the model must use deeper water, violating complete similarity. The data must be dealt with using a theory of dissimilar models, consideration of which is beyond the scope of this book.

In any multiparameter model test, a hierarchy of the dimensionless parameters should be followed if matching all of them is not possible. Table 6.3 lists such a hierarchy for most types of tests. The parameters indicated as "primary" must be matched for the specific type of test being conducted. The "secondary" parameters are also important but often cannot be matched simultaneously with the primaries in a scale-model test. The secondary parameters must be addressed by extrapola-

Table 6.3 Similarity parameters and their typical importance in various types of model tests.

Type of Test	Primary Similarity Parameter (Must be matched)	Secondary Similarity Parameter(s)	Normally Irrelevant Parameter(s)
Low-speed flow, no free surface (e.g., pipes, valves, and fittings, low-speed aerodynamics)	**R**, α	ε/ℓ	**W**, **F**, **M**, k
Surface vessels, ships	**F**	**R**, ε/ℓ	**W**, **M**, k
Dam spillways, canals, harbors, rivers	**F**	**R**, ε/ℓ	**W**, **M**, k
High-speed flows	**M**, k, α	**R**, ε/ℓ	**F**, **W**
Fluid machinery	V/U	**R**, ε/ℓ, **M**	**W**, **F**, k
Unsteady flows	**S**	Various	Various

tion of data, theoretical calculation, or shown to be irrelevant.* The parameters indicated as "normally irrelevant" are irrelevant only for the particular type of test indicated.

* For example, at high **R**, drag forces are independent of the exact **R** (Example 6.5).

EXAMPLE 6.8 Illustrates the Use of Dimensionless Parameters in Extrapolating Test Data and Some Consequences of a Multiparameter Similarity Law

Table E6.8a Test data for a centrifugal pump.

2000 rpm	
Q (gal/min)	h_p (ft)
0	115
400	122
800	122
1200	118
1600	112
2000	105
2400	96
2800	87
3200	78

3000 rpm	
Q (gal/min)	h_p (ft)
0	259
600	280
1200	280
1800	266
2400	250
3000	232
3600	215
4200	194
4800	170

Performance test data for a centrifugal pump is tabulated in Table E6.8a. The pump rotor diameter is 2.772 ft. The data were obtained at pump speeds of 2000 rpm and 3000 rpm. Find the head developed by the pump at a speed of 3500 rpm and a volume flow rate of 5000 gal/min. The fluid is water.

SOLUTION

Given

Test data in Table E6.8a for a centrifugal pump

Pump rotor diameter 2.772 ft

Rotor rotational speeds 2000 rpm and 3000 rpm

Fluid is water

See also Fig. E6.4

Find

Head developed by pump at 3500 rpm and volume flow rate of 5000 gal/min

Solution

Example 6.4 shows that the appropriate dimensionless performance function for a centrifugal pump is

$$\boldsymbol{\Psi} = f\{\boldsymbol{\Phi}, \mathbf{R}\},$$

where

$$\boldsymbol{\Psi} = \text{Head coefficient} = \frac{gh_p}{N^2D^2};$$

$$\boldsymbol{\Phi} = \text{Flow coefficient} = \frac{Q}{ND^3};$$

$$\mathbf{R} = \text{Reynolds number} = \frac{ND^2}{\nu}.$$

We want to find a value of the pump head at a certain speed and flow. The head can be calculated by

$$h_p = \frac{\boldsymbol{\Psi}(N^2D^2)}{g}.$$

We know N, D, and g but must find $\mathbf{\Psi}$, which depends on $\mathbf{\Phi}$ and $\mathbf{R}$; we calculate $\mathbf{\Phi}$ and $\mathbf{R}$ from the specified speed (3500 rpm), the specified flow (5000 gal/min), the pump rotor diameter, and the water viscosity. To establish the relation between $\mathbf{\Psi}$, $\mathbf{\Phi}$, and $\mathbf{R}$, we must use the test data of Table E6.8a. We calculate $\mathbf{\Phi}$ and $\mathbf{\Psi}$ from the test data by

$$\mathbf{\Phi} = \frac{Q}{ND^3} = \frac{Q(\text{gal/min})(1\ \text{ft}^3/7.48\ \text{gal})}{N(\text{rev/min})(2.772\ \text{ft})^3(2\pi\ \text{rad/rev})} = \frac{Q}{1000N},$$

where Q is in gal/min and N in rpm, and

$$\mathbf{\Psi} = \frac{gh_p}{N^2D^2} = \frac{(32.2\ \text{ft/sec}^2)h_p(60\ \text{sec/min})^2}{N^2(\text{rev/min})^2(2.772\ \text{ft})^2(2\pi\ \text{rad/rev})^2} = 382\frac{h_p}{N^2},$$

where h_p is in feet and N in rpm.

The Reynolds number is calculated by

$$\mathbf{R} = \frac{ND^2}{\nu} = \frac{N(\text{rev/min})(2.772\ \text{ft})^2(2\pi\ \text{rad/rev})}{(1.22 \times 10^{-5}\ \text{ft}^2\text{sec})(60\ \text{sec/min})} = (6.60 \times 10^4)N,$$

where N is in rpm and 60°F water is assumed.

Of the parameters appearing in the Reynolds number, only pump speed is different in the two tests; therefore the data contain only two different Reynolds numbers.

Table E6.8b shows the magnitudes of the dimensionless parameters calculated from the test data. Figure E6.8 is a plot of the dimensionless parameters. The table and the plot represent the functional relation

$$\mathbf{\Psi} = f\{\mathbf{\Phi}, \mathbf{R}\}$$

for this particular pump.

Table E6.8b Dimensionless parameters for Example 6.8.

2000 rpm ($\mathbf{R} \approx 1.31 \times 10^8$)	
$\mathbf{\Phi}$	$\mathbf{\Psi}$
0	0.0110
0.0002	0.0117
0.0004	0.0117
0.0006	0.0113
0.0008	0.0107
0.0010	0.0100
0.0012	0.0092
0.0014	0.0083
0.0016	0.0075
3000 rpm ($\mathbf{R} \approx 1.96 \times 10^8$)	
$\mathbf{\Phi}$	$\mathbf{\Psi}$
0	0.0110
0.0002	0.0119
0.0004	0.0119
0.0006	0.0113
0.0008	0.0106
0.0010	0.0098
0.0012	0.0091
0.0014	0.0082
0.0016	0.0072

From the plot and Table E6.8b, $\mathbf{\Psi}$ clearly depends on $\mathbf{\Phi}$, however, $\mathbf{\Psi}$ seemingly is independent of Reynolds number. Experience shows that if the Reynolds number is high enough, flow phenomena become independent of the exact value of Reynolds number. The data for this particular pump indicate that the range of Reynolds number independence has been reached, and we can assume that

$$\mathbf{\Psi} = f\{\mathbf{\Phi}\ \text{only}\},$$

at least at Reynolds numbers equal to or greater than the lowest value achieved in the test ($\mathbf{R} = 1.31 \times 10^8$).

At the new pump speed of 3500 rpm and flow of 5000 gal/min, the independent-variable dimensionless parameters are

$$\mathbf{\Phi} = \frac{Q}{1000N} = \frac{5000\ \text{gal/min}}{1000(3500\ \text{rpm})} = 0.00143,$$

and

$$\mathbf{R} = 6.60 \times 10^4(N) = (6.60)(3500) \times 10^4 = 2.31 \times 10^8.$$

Because $\mathbf{R} > 1.31 \times 10^8$, $\mathbf{\Psi}$ is independent of Reynolds number. We read the magnitude of $\mathbf{\Psi}$ at $\mathbf{\Phi} = 0.00143$ from Fig. E6.8

$$\mathbf{\Psi} = 0.0082.$$

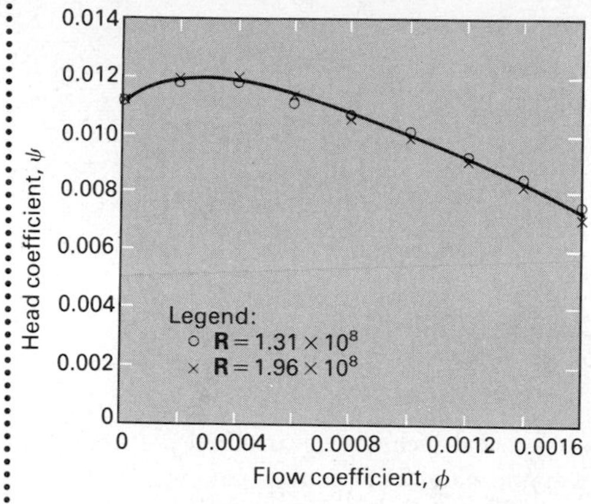

Figure E6.8 Head coefficient versus flow coefficient for two different Reynolds numbers.

The head at this operating point is

$$h_p = \frac{\Psi N^2}{382} = \frac{0.0082(3500 \text{ rpm})^2}{382};$$

$$h_p = 263 \text{ ft.}$$ **ANSWER**

Discussion

Note that the volume flow rates in the data were such that the flow coefficients in Table E6.8b were identical; this is normally not the case. If it is not, then a plot of Ψ versus Φ such as Fig. E6.8 is useful.

Strictly speaking, this example is not a model test, because we considered the same pump throughout. We might think of the low-speed pump as a model of the high-speed pump. At any rate, this example demonstrates the principles of similarity.

If the speed of the pump is reduced below 2000 rpm, the effects of Reynolds number may become important. For slower speeds, the simple dependence of Ψ on Φ only does not hold. This result typically occurs if **R** drops below about 10^5.

Approaches similar to that used in this example would be used to find

- the rotor diameter of a geometrically similar pump to deliver a given flow rate for a given pump speed;
- a substitute liquid to test a pump when Reynolds number effects must be considered; and
- a pump speed for a geometrically similar pump to deliver a specific flow rate at a specific head, and many others.

You might want to try solving these three cases.

PROBLEMS

1. A mixing basin in a sewage filtration plant is stirred by mechanical agitation (paddlewheel) with a power input $\dot{W}$ (ft·lb/sec). The degree of mixing of fluid particles is measured by a "velocity gradient" G given by

$$G = \sqrt{\frac{\dot{W}}{\mu \mathcal{V}}},$$

where μ is the fluid viscosity in lb·sec/ft^2 and $\mathcal{V}$ is the basin volume in ft^3. Find the units of the velocity gradient.

2. The Hazen–Williams equation for the head loss h_L in ft for pipe flow is

$$h_L = \frac{k_1 \ell}{C^{1.852} D^{4.8704}} Q^{1.852},$$

where C is a dimensionless constant, D the pipe diameter in ft, ℓ the pipe length in ft, and Q the volume flow rate in ft^3/sec. Determine the units of k_1.

3. An equation used to evaluate vacuum filtration is

$$Q = \frac{\Delta p A^2}{\alpha(\mathcal{V} R w + A R_f)},$$

where Q is the filtrate volume flow rate in $[L^3/T]$, Δp the vacuum pressure differential in $[F/L^2]$, A the filter area in $[L^2]$, α the filtrate "viscosity," $\mathcal{V}$ the filtrate volume in $[L^3]$, R the sludge specific resistance in $[L/F]$, w the weight of dry sludge per unit volume of filtrate in $[F/L^3]$, and R_f the specific resistance of the filter medium. What are the dimensions of R_f and α?

4. A large, hot plate hangs vertically in a room. Heat transfer from the plate to the air near it causes the air density to decrease. This lighter air rises upward with a velocity proportional to the group

$$[\alpha_T g L(T_p - T_f)]^{1/2},$$

where α_T is the coefficient of thermal expansion [units = $(°R)^{-1}$ or K^{-1}], g is the acceleration of gravity, L is the plate length, and T_p and T_f are the temperatures of the plate and fluid, respectively, in °R or K. The resulting heat transfer process depends on the Grashof Number, defined by

$$G = \frac{\rho^2 \alpha_T g(T_p - T_f)L^3}{\mu^2},$$

where ρ is density and μ is dynamic viscosity. Show that G is dimensionless. How is G related to the standard dimensionless parameters of fluid mechanics? Is it equivalent to a combination, power, or product of one or more of the standard dimensionless parameters listed in Table 6.1?

5. A mixing basin in a sewage filtration plant is stirred by a mechanical agitator with a power input $\dot{W}$ $[F \cdot L/T]$. Other parameters describing the performance of the mixing process are the fluid absolute viscosity μ $[F \cdot T/L^2]$, the basin volume $\forall$ $[L^3]$, and the velocity gradient G $[1/T]$. Determine the form of the dimensionless relationship.

6. In a sedimentation basin, the steady velocity at which suspended particles in a fluid settle to the bottom is determined by the particle weight, the buoyant force, and the drag force. The drag force $\mathscr{D}$ is given by

$$\mathscr{D} = \frac{\mathbf{C_D} A_p \rho V_p^2}{2},$$

where $\mathbf{C_D}$ is a drag coefficient, A_p the particle cross-sectional area perpendicular to the particle velocity V_p, and ρ is the fluid mass density. Denote the particle density by ρ_p and the particle volume by $\forall_p$ and identify the dimensionless groups.

7. The relevant quantities involved in vacuum filtration are the filtrate volume flow rate Q $[L^3/T]$, the vacuum pressure differential Δp $[F/L^2]$, the filter area A $[L^2]$, the filtrate absolute viscosity μ $[F \cdot T/L^2]$, the filtrate volume $\forall$ $[L^3]$, the sludge-specific resistance R $[T^2/F]$, the sludge weight per unit volume of the filtrate w $[F/L^3]$, and the filter medium specific resistance R_f $[T^2/L^2]$. Find the dimensionless groups.

8. The weir shown in Fig. P6.8 is used to measure the volume flow rate Q. The height H is a measure of this flow rate. The weir has length L (perpendicular to the

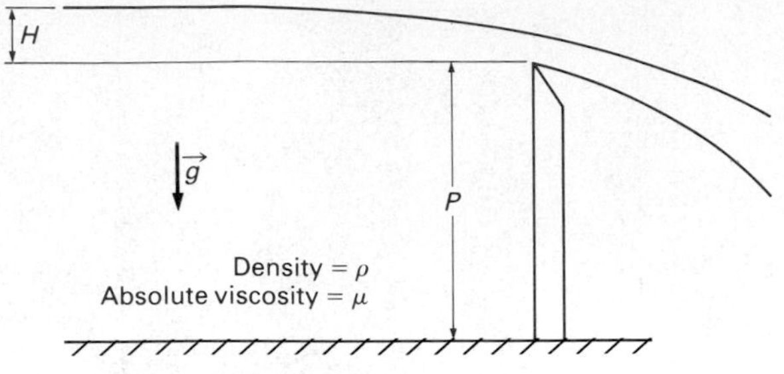

Figure P6.8

paper). Select and include relevant fluid properties and find the appropriate dimensionless parameters.

9. Experiments are conducted on a washing machine agitator. The relevant dimensional parameters are the driving torque, $\mathscr{T}$, the oscillation frequency, f, the angular velocity, ω, the number of paddles, N, the paddle height, H, and the paddle width, w. Specify the relevant fluid properties and find the appropriate dimensionless parameters.

10. The input power, $\dot{W}$, to a large industrial fan depends on the fan impeller diameter D, fluid viscosity μ, fluid density ρ, volumetric flow Q, and blade rotational speed ω. What are the appropriate dimensionless parameters?

11. The magnitude of a pressure pulse that travels through a liquid because of the quick closure of a valve depends on the isothermal bulk modulus of elasticity, the density, and the fluid velocity prior to closure of the valve. Find the dimensionless group.

12. Determine the appropriate dimensionless parameters for the jet pump shown in Fig. P6.12. The relevant dimensional parameters are the diameters D_o and D_p, the velocities V_p and V_s, and the pressure difference $p_2 - p_1$. Do not forget to include relevant fluid properties.

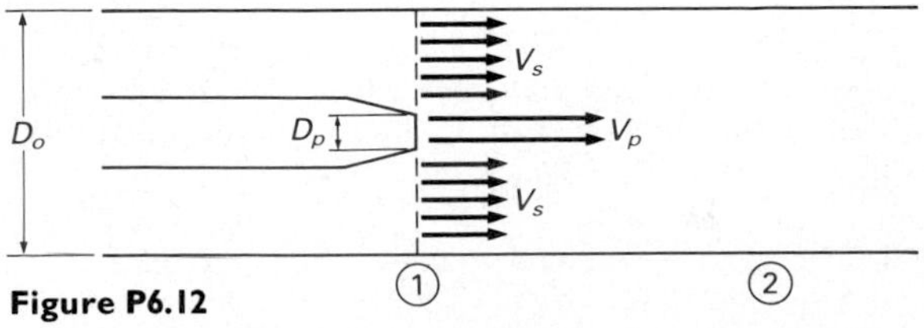

Figure P6.12

13. Figure P6.13 shows a soap bubble with internal pressure σ, external pressure p_o, radius R, and surface tension σ. Find the form of the dimensionless relationship.

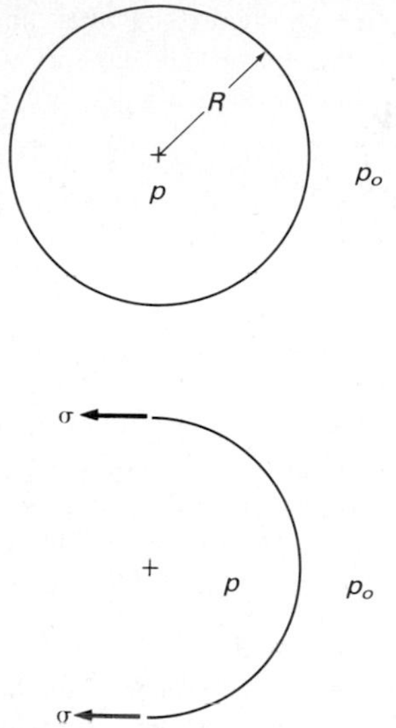

Figure P6.13

14. A student drops two spherical balls of different diameters and different densities. She has a stroboscopic photograph showing the positions of each ball as a function of time. However, she wants to express the velocity of each as a function of time in dimensionless form. Develop the dimensionless group. The equation of motion for each ball is

$$mg - \frac{\mathbf{C_D}}{2}\rho A V^2 = m\frac{dV}{dt},$$

where m is ball mass, g is acceleration of gravity, $\mathbf{C_D}$ is a dimensionless and constant drag coefficient, ρ is air mass density, A is ball cross-sectional area $(= \pi D^2/4)$ with D ball diameter, V is ball velocity, and t is time.

15. Liquid is to be drained from the two geometrically similar tanks with circular cross sections shown in Fig. P6.15. The liquid height h and velocity V are measured as a function of time if time $t = 0$ when the plug is pulled. Both the height h and the velocity V are to be plotted in dimensionless form versus time. Determine these dimensionless parameters for frictionless flow $(\mu = 0)$.

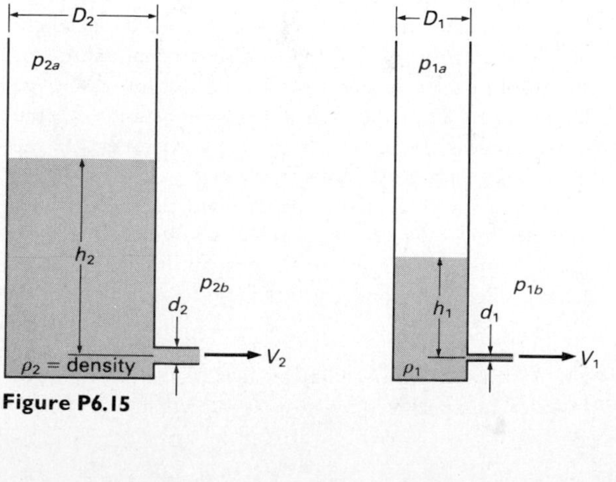

Figure P6.15

16. Find the appropriate dimensionless parameters for the maximum static deflection δ of a simply supported beam under load P. The other relevant dimensional parameters are the beam dimensions of length ℓ, width w, and thickness h, and the modulus of elasticity E.

17. Develop the appropriate dimensionless parameters for the period τ of transverse vibration of a turbine rotor of mass m connected to a shaft of stiffness k $[F/L]$ and length ℓ $[L]$. Other relevant dimensional parameters are the eccentricity ε $[L]$ of the center of mass of the rotor, the speed of rotation N $[1/T]$ of the shaft, and the amplitude A $[L]$ of the vibration.

18. Force F [N] acts on an electrical conductor carrying a current I (amps). A length ℓ [m] of the conductor experiences a magnetic field of flux density B [webers/m^2 or N/amp·m]. Find the appropriate dimensionless parameter(s).

19. The flow between the two fixed, parallel, circular discs shown in Fig. P6.19 is radial. The volume flow rate is Q, the fluid density ρ is constant, and the disc spacing $h \ll$ radius R. Therefore the fluid viscosity μ is important. What are the appropriate dimensionless parameters?

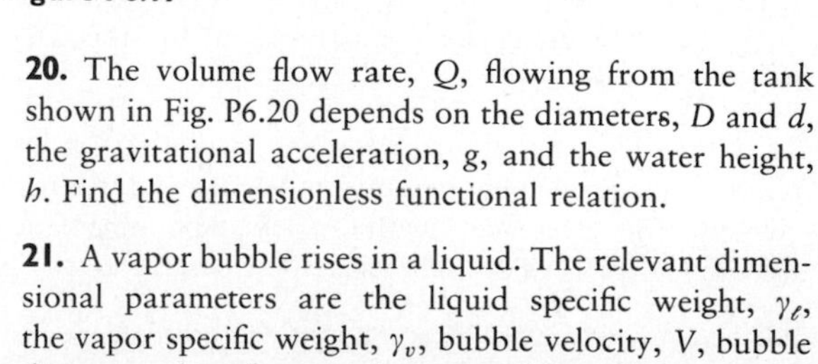

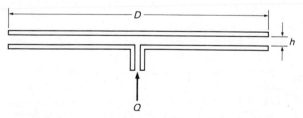

Figure P6.19

20. The volume flow rate, Q, flowing from the tank shown in Fig. P6.20 depends on the diameters, D and d, the gravitational acceleration, g, and the water height, h. Find the dimensionless functional relation.

21. A vapor bubble rises in a liquid. The relevant dimensional parameters are the liquid specific weight, γ_ℓ, the vapor specific weight, γ_v, bubble velocity, V, bubble diameter, d, surface tension, σ, and liquid viscosity, μ. Find appropriate dimensionless parameters.

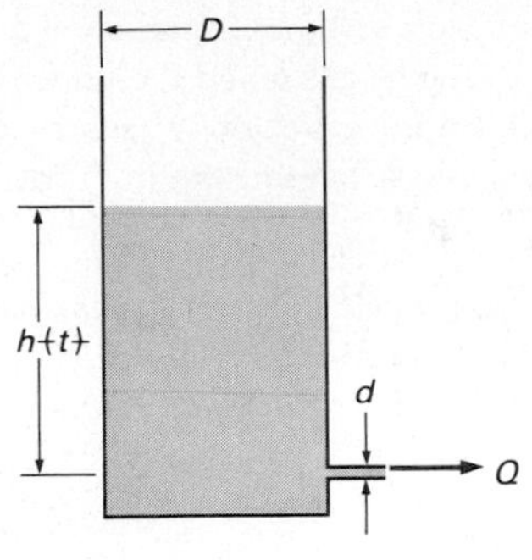

Figure P6.20

22. Determine the appropriate dimensionless parameters for the oscillatory motion of a mass m suspended by a spring with damping. The other relevant dimensional parameters are spring stiffness k in N/m, the damping coefficient c in N·s/m, and the period of oscillation τ.

23. The speed of deep ocean waves depends on the wave length and gravitational acceleration. What are the appropriate dimensionless parameters?

24. The frequency of a vertically oscillating buoy depends on the gravitational acceleration, g, the buoy density, ρ_b, the water density, ρ_w, and the volume, $\forall$, of the buoy. Find the appropriate functional relation in dimensionless form.

25. At very low velocities, the pressure drop Δp_d in a horizontal pipe of length L depends on the fluid viscosity, μ, the volume flow rate, Q, and the pipe inside diameter, d. Find the appropriate dimensionless parameter(s).

26. A concentric cylinder viscometer to measure fluid absolute viscosity was analyzed in Example 1.3. The viscometer had a height H, and the gap width h was quite small compared to R_1 and R_2. The fluid in the viscometer was Newtonian with constant viscosity μ. A torque $\mathcal{T}$ applied to the inner cylinder caused it to rotate with angular velocity ω. The outer cylinder was fixed. Determine the dimensionless parameters for this viscometer.

27. Using the four-step procedure outlined in this chapter, use the dimensional parameters density, ρ, reference velocity, U_0, reference length, ℓ, and fluid kinematic viscosity, ν, to develop a dimensionless parameter.

28. The power output $\dot{W}$ of a hydraulic turbine depends on the fluid density, ρ, the height, h, of the water surface above the turbine, the gravitational acceleration, g, the volume flow rate, Q, the turbine wheel angular velocity, ω, the turbine efficiency, η, and the turbine wheel diameter, D. Develop the dimensionless functional relation.

29. Find the dimensionless groups formed from the dimensional parameters: volume flow rate Q, fluid depth H, approach velocity V, and the gravitational acceleration g.

30. The shaft power ($\dot{W}$) input to a pump depends on the flow rate, Q, the pressure rise, Δp, the rotational speed, N, the fluid density, ρ, the impeller diameter, D, and the fluid viscosity, μ. How many dimensionless groups can be formed from these parameters? If Q, Δp, and ρ are selected as repeating variables, find a dimensionless group involving $\dot{W}$, Q, Δp, and ρ. Then, find a dimensionless group involving N, Q, Δp, and ρ.

31. A coach has been trying to evaluate the accuracy of a baseball pitcher. After two years of studying, he proposes a function that can be presented as the accuracy of any pitcher:

$$\text{Acc} = f(V, a, m, \rho, p, z),$$

where Acc is the dimensionless accuracy, V is the velocity of the ball, a is the age of the pitcher, m is the mass of the pitcher, ρ is the density of air where the game is played, p is the pressure where the game is played (which varies with elevation), and z is the elevation above sea level. Find the dimensionless groups for this function.

32. Arrange the following groups into dimensionless parameters:

(a) $Q\ [L^3/T]$, $N\ [1/T]$, $D\ [L]$, $\rho\ [M/L^3]$.
(b) $V\ [L/T]$, $\ell\ [L]$, $\rho\ [M/L^3]$, $\sigma\ [F/L]$.
(c) $\tau\ [F/L^2]$, $D\ [L]$, $N\ [1/T]$, $\rho\ [M/L^3]$.

• **33.** The relevant dimensional parameters for a centrifugal compressor delivering an ideal gas are the absolute inlet pressure, p_i, the absolute discharge pressure, p_d, the absolute inlet temperature, T_i, the absolute discharge temperature, T_d, the specific gas constant, R, the dimensionless specific heat ratio, k (see Chapter 10), the inlet gas density, ρ_i, the inlet gas velocity, V_i, the impeller rotational speed, N, the impeller diameter, D, the inlet volume flow rate, Q_i, the discharge volume flow rate, Q_d, the input power, $\dot{W}$, to the gas, and the gas specific heat, c_p. Perform a dimensional analysis on the compressor.

• **34.** A straw of inside diameter d is placed in a pan of liquid with specific gravity γ, as shown in Fig. P6.34.

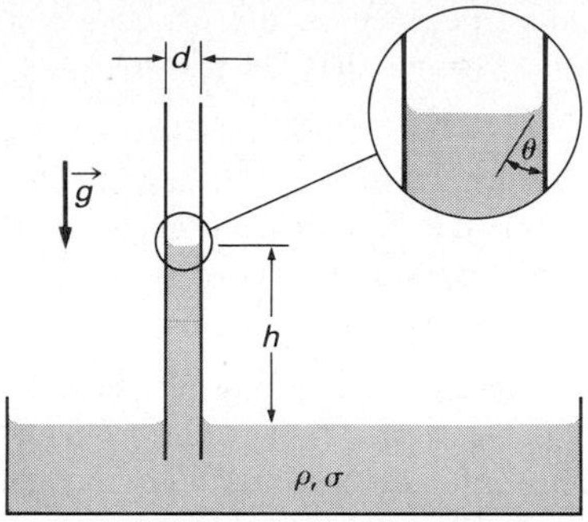

Figure P6.34

The water rises a height h in the straw because of surface tension σ between the water and the straw and makes a contact angle θ at the water–straw interface. Start with the dimensional parameters h, d, γ, and σ and develop the appropriate dimensionless parameters describing the phenomena. Use both the $[F, L, T]$ system and the $[M, L, T]$ system. Note and explain the differences that occur in the two methods.

•**35.** For the cantilever beam in Example 6.3, the relevant dimensional parameters and their fundamental dimensions are:

σ = bending stress	$ML^{-1}T^{-2}$
l = beam length	L
c = cross-sectional coordinate	L
x = axial position	L
I = second moment of area	L^4
P = transverse force	MLT^{-2}.

The powers of the fundamental dimensions of each relevant parameter are used as the elements of a matrix as follows:

$$\begin{array}{c} \\ M \\ L \\ T \end{array}\begin{array}{c} \begin{array}{cccccc} \sigma & \ell & c & x & I & P \end{array} \\ \begin{bmatrix} 1 & 0 & 0 & 0 & 0 & 1 \\ -1 & 1 & 1 & 1 & 4 & 1 \\ -2 & 0 & 0 & 0 & 0 & -2 \end{bmatrix} \end{array}$$

Show that the matrix has rank 2, which is the value of k in Example 6.3 by using the M, L, T system. The rank is the size of the largest nonzero determinant of the matrix.

36. Develop the Froude number by starting with expressions for the inertia and gravitational forces.

37. Develop the Weber number by starting with expressions for the inertia and surface tension forces.

38. Develop the Mach number **M** for an ideal gas in terms of the static pressure, p, the mass density, ρ, the fluid velocity, V, and the specific heat ratio, k.

39. By inspection, arrange the following dimensional parameters into dimensionless parameters: (a) kinematic viscosity, ν, length, ℓ, and time, t; and (b) volume flow rate, Q, pump diameter, D, and pump impeller rotation speed, N.

40. Seawater tests to determine drag force are run on a $\frac{1}{50}$-scale-model submarine traveling on the surface and also when deeply submerged. Specify the appropriate dimensionless parameter(s) for each situation.

•**41** Determine the relevant dimensional parameters for the liquid viscosity motor/pump shown in Fig. P6.41 and find the appropriate dimensionless parameters by either inspection or similarity. The output parameter is the power, $\dot{W}$.

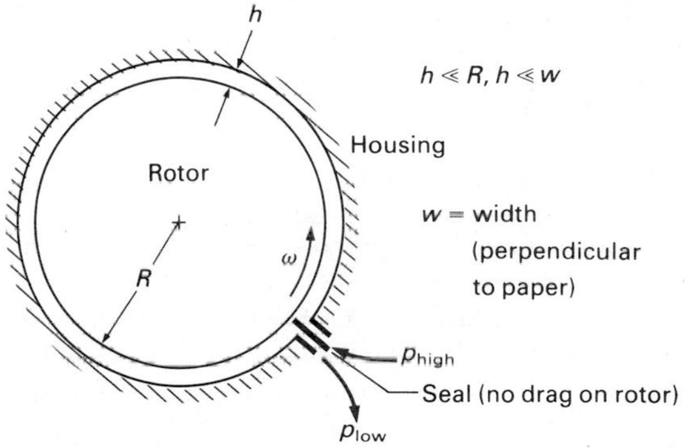

Figure P6.41

42. A screw propeller has the following relevant dimensional parameters: axial thrust, F, propeller diameter, D, fluid kinematic viscosity, ν, fluid density, ρ, gravitational acceleration, g, advance velocity, V, and rotational speed, N. Find appropriate dimensionless parameters to present the test data.

43. The basic equation that describes the motion of the fluid above a large oscillating flat plate is

$$\frac{\partial u}{\partial t} = \nu \frac{\partial^2 u}{\partial y^2},$$

where u is the fluid velocity component parallel to the plate, t is time, y is the spatial coordinate perpendicular to the plate, and ν is the fluid kinematic viscosity. The plate oscillating velocity is given by $U = U_0 \sin \omega t$. Find appropriate dimensionless parameters and the dimensionless differential equation.

44. The appropriate equation for fully developed flow in a circular pipe of radius R is

$$\frac{d^2u}{dr^2} + \frac{1}{r}\frac{du}{dr} = \frac{1}{\mu}\frac{dp}{dx}.$$

Find the appropriate dimensionless parameters and the dimensionless equation. The reference value for u is the centerline velocity U_0. Also find the dimensionless boundary conditions.

45. An incompressible liquid flows between two parallel plates, as shown in Fig. P6.45. The upper plate has a horizontal velocity U and is at temperature T_0, which is less than the fluid temperature T. The lower plate is fixed and also is at temperature T_0. The liquid has a velocity profile given by

$$u = \frac{Uy}{h}.$$

The temperature of the liquid is given by the differential equation

$$\rho u c_p \frac{\partial T}{\partial x} = k\left(\frac{\partial^2 T}{\partial x^2} + \frac{\partial^2 T}{\partial y^2}\right) + \mu\left(\frac{U}{h}\right)^2,$$

where ρ is the fluid density, c_p is a fluid specific heat (dimensions of $[FL/M\theta]$, $[\theta]$ a temperature dimension), T is temperature, k is a fluid property called the thermal conductivity with dimensions of $[FL]/[TL\theta] = [F/T\theta]$, h is the plate spacing, μ is the fluid viscosity, and x and y are spatial coordinates. Find the dimensionless parameters and the dimensionless differential equation.

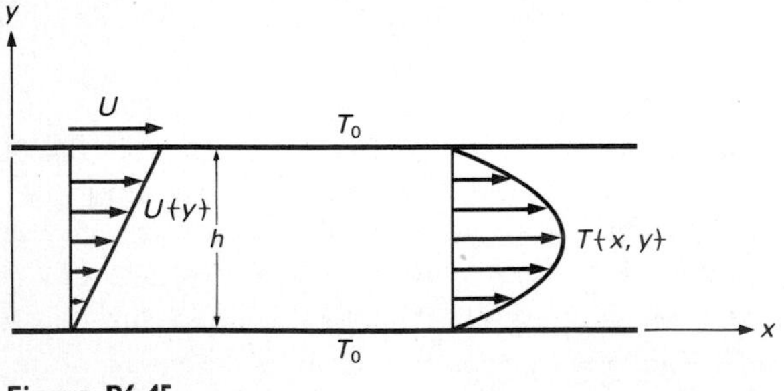

Figure P6.45

46. Hydrogen at 60°F is to be pumped through a 2-ft-I.D., 6000-ft-long pipeline. To determine information about the flow in this pipeline, tests are to be conducted in a geometrically similar pipe with an inside diameter of 0.02 ft and water as the fluid. The physical quantities are mass density, ρ, pressure drop, Δp, average fluid velocity, V, pipe inside diameter, D, pipe length, L, and kinematic viscosity, ν. Find the dimensionless groups pertinent to this problem and the water temperature. Assume that the hydrogen is at atmospheric pressure.

47. A model hydrofoil is to be tested. Is it practical to satisfy both the Reynolds number and the Froude number for the hydrofoil when it is operating near the water surface? Support your decision.

48. A $\frac{1}{5}$-scale model of a boat's propeller being tested at 1000 rpm in a towing tank requires a power input of 0.31 hp. For negligible Reynolds number effects, find the corresponding prototype values. The relevant dimensionless parameters are power coefficient ($\dot{W}/\rho N^3 D^5$), speed ratio (V/ND), and the Froude number ($V/\sqrt{gD}$).

49. A $\frac{1}{40}$-scale model of a dam spillway is to be tested using Froude number scaling. The prototype flow and velocity are 180 m³/s and 4.0 m/s, respectively. Find the corresponding model flow and velocity.

50. A dam spillway is 40 ft long and has a fluid velocity of 10 ft/sec. Considering Weber number effects as minor, calculate the corresponding model fluid velocity for a model length of 5 ft.

51. A $\frac{1}{4}$-scale model of an automobile is tested in a wind tunnel under dynamically similar conditions. The prototype speed is 75 km/hr, and the model drag is 300 N. What corresponding prototype drag and prototype horsepower are required to overcome this drag.

52. A torpedo 12 ft long has a velocity of 10 ft/sec. Find the corresponding velocity for a $\frac{1}{4}$-scale model tested in water. Assume that the torpedo is deeply submerged.

53. Assume that the pump performance characteristics shown in Fig. P6.53 are for a pump running at 1200 rpm. Assuming that the pump head, h_p, the pump impeller diameter, D, and the flow rate, Q, through

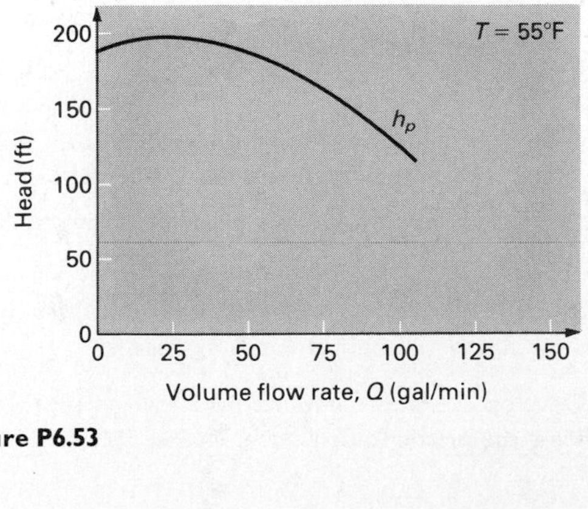

Figure P6.53

the pump are the other relevant parameters, calculate the pump performance characteristics at speeds of 1800 rpm and 2400 rpm. Neglect viscous effects.

54. Air at 70°F and 14.7 psia flows at 38 ft^3/min through a $2\frac{1}{2}$-in. schedule 40 steel pipe. A model test is to be run in 60°F water with a maximum flow rate of 3.0 gal/min. What is the standard size steel pipe that should be used for dynamic similarity? Discuss the effect of roughness.

55. A $\frac{1}{5}$-scale model of an automobile is to be tested in a water tunnel. What should be the velocity in the water tunnel to ensure dynamic similarity between the model and the prototype traveling in air at 55 mph? The water and air are at 60°F.

56. A $\frac{1}{20}$-scale model of a spillway measures 2 ft high and 1 ft wide. The scale-model flow rate is 1.0 ft^3/sec. What is the prototype flow rate?

57. The aluminum ball falling in water in Problem 94 is to be modeled with a magnesium ball (ρ_b = 1750 kg/m^3) falling in another fluid. Which fluid in Table A.5 would be the best fluid to use?

58. A stream of atmospheric air is used to keep a ping-pong ball aloft by blowing the air upward over the ball. The ping-pong ball has a mass of 2.5 g and a diameter D_1 = 3.8 cm, and the air stream has an upward velocity of V_1 = 0.942 m/s. This system is to be modeled by pumping water upward with a velocity V_2 over a solid ball of diameter D_2 and density ρ_{b_2} = 2710 kg/m^3. In both cases, the net weight of the ball W_b is equal to the air drag,

$$W_b = \frac{\mathbf{C_D}\rho A V^2}{2},$$

where $\mathbf{C_D}$ = 0.60, ρ is the fluid density, A the ball's projected area, and V the velocity of the fluid upstream from the ball. Determine all possible combinations of V_2 and D_2. [*Hint:* A force balance involving the drag on the ball, the buoyant force on the ball, and the weight of the ball is needed.]

59. A centrifugal fan operating in a duct has the dimensionless parameters

$$\mathbf{C_Q} = \mathbf{\Phi} = \frac{Q}{ND^3} \quad \text{and} \quad \mathbf{C_H} = \mathbf{\Psi} = \frac{\Delta p}{\rho N^2 D^2},$$

where $\mathbf{C_Q} = \mathbf{\Phi}$ is a flow coefficient, $\mathbf{C_H} = \mathbf{\Psi}$ is a head coefficient, Q is the volume flow rate, N is the fan speed, D is the fan diameter, ρ is the fluid density,

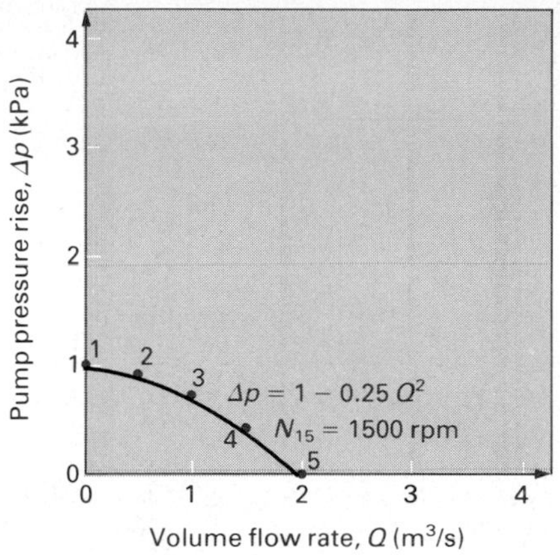

Figure P6.59

and Δp is the fan pressure rise. Figure P6.59 shows this fan's performance curve in dimensional form for a fan speed of N_{15} = 1500 rpm. Find the fan operating points (Q and Δp) for N_{30} = 3000 rpm and corresponding to points 1, 3, and 5 at N_{15} = 1500 rpm. Assume similarity between 1500 rpm and 3000 rpm.

60. A flagpole has a diameter of 1.0 ft and is 45 ft. long. The manufacturer wants to find the bending moment at the base of the pole for a wind speed of 60 mph. To do so, a scale model of the flagpole is built with a diameter of 1.0 in. and a length of 45 in. What must be the air speed at which the model is tested?

61. If the wind velocity over the flagpole in Problem 60 is reduced by 50%, how much would the bending moment at the base be reduced?

62. A large channel has a flow rate of 600 ft^3/min, a rectangular cross section, a flow area of 120 ft^2, and a width of 40 ft. Calculate a flow area and flow rate of a small model of the channel for laboratory studies if the model width is not to exceed 1.0 ft.

63. A ship's propeller operates close to the water surface. The ship is 150 m long and is to have a propeller of diameter 2.0 m that rotates at 60 rpm. A force of 100 N is required to tow a $\frac{1}{150}$-scale model of the ship through a water tank. What should be the model's propeller speed? What is the drag force on the prototype ship? The propeller must be matched in both Froude number and in the ratio of ship speed to propeller tip speed.

64. A $\frac{1}{200}$-scale model is made of a spillway. Assume that both the model and prototype flows are weakly dependent on the Reynolds number. The fluid velocity and flow rate at a particular point in the model spillway were measured to be 0.25 m/s and 0.1 m^3/s. Find the prototype values at the corresponding location on the spillway.

65. A designer wants to model a 150-ft-long airship traveling at 120 mph in the Standard Atmosphere at a 5000-ft altitude. The model is to be towed through 60°F water at 2 ft/sec. Find the necessary length of the model. What precautions should the designer take in the water test?

66. A $\frac{1}{10}$-scale model of an airplane is tested in a wind tunnel at 70°F and 14.40 psia. The model test results are:

Velocity (mph)	0	50	100	150	200
Drag (lb)	0	5	21	46	85.

Find the corresponding airplane velocities and drags if only fluid compressibility is important and the airplane is flying in the U.S. Standard Atmosphere at 30,000 ft. Assume that the air is an ideal gas.

67. Can the corresponding velocities and drags be found in Problem 66 if both compressibility and viscous effects were considered equally important? Support your decision.

68. A $\frac{1}{400}$-scale model of a beach is made to study tides. What length of time in the model would correspond to one hour in the prototype?

69. A company manufactures geometrically similar airplane propellers up to 4.0 m in diameter. Wind tunnel tests are run on geometrically similar propellers up to 0.33 m in diameter. In the test of a 0.33-m model of a 4.0-m propeller, the air velocity was 50 m/s. The model rotational speed was 2000 rpm, the thrust 100 N, and the propeller input torque 20 N·m. Calculate the corresponding prototype rotational speed, thrust, and input torque for an airplane speed of 100 m/s. The Reynolds numbers and Mach numbers are of minor importance.

70. A wind tunnel model test of a rocket capable of speeds up to 2000 mph at 30,000 ft in the Standard Atmosphere is to be run. Is it more important to match Mach number or Reynolds number? What should be the wind tunnel air speed if the wind tunnel is at sea level and 70°F? Is such a wind tunnel air speed practical?

71. Lubricating oil at 100°C flows through a 0.25-m-I.D.-diameter pipe with a velocity of 0.74 m/s. What must be the flow rate of 15°C water through the same pipe for dynamic similarity of the two flows?

72. An engineer wants to model the flight of an airplane flying at 30,000 ft, where the air is at −58°F and 4.4 psia. The engineer tests a $\frac{1}{20}$-scale model in a wind tunnel where the air is at 70°F and 14.40 psia. What wind tunnel air velocity is needed to simulate an airplane speed of 500 mph?

73. A breakwater is a wall built around a harbor so the incoming waves dissipate their energy against it. The significant dimensionless parameters are the Froude number **F** and the Reynolds number **R**. A particular breakwater measuring 450 m long and 20 m deep is hit by waves 5 m high and velocities up to 30 m/s. In a $\frac{1}{100}$-scale model of the breakwater, the wave height and velocity can be controlled. Can complete similarity be obtained using water for the model test?

74. An oil tanker is 700 ft long and travels at 12 knots. Determine the speed at which a 7-ft model of the tanker must be towed in 60°F freshwater to study the wave resistance on the tanker.

75. A torpedo 6 m long travels at 35 knots well below the ocean surface. What must be the length of a model torpedo being tested in 15°C air?

76. The flow of 500 gal/min of 68°F benzene through a 14-in. schedule 80, nominal diameter steel pipe is modeled by the flow of 60°F water through a 1-in. schedule 40 steel pipe. Find the flow rate through the 1-in. steel pipe.

77. An airplane travels at 160 mph in air at 14.7 psia and 0°F. A 1/3-scale model of the airplane is tested in a wind tunnel at 70°F. What must be the air pressure and velocity in the wind tunnel so that both the Reynolds number and the Mach number are satisfied?

78. An engineer wants to know the drag, $\mathscr{D}$, of 20°C kerosene flowing at 10.5 m/s over an object. For safety reasons, tests are to be run using air or water. The maximum available air and water velocities are 25 m/s and 2.0 m/s, respectively, and both can be heated to 90°C or cooled to 0°C. Which fluid, if either, can be used?

79. A very small needle valve is used to control the flow of air in a $\frac{1}{8}$-in. air line. The valve has a pressure drop of 4.0 psi at a flow rate of 0.005 ft^3/sec of 60°F air. Tests are performed on a large, geometrically similar valve and the results are used to predict the performance of the smaller valve. How many times larger can the model valve be if 60°F water is used in the test and the water flow rate is limited to 7.0 gal/min?

80. A designer wants to know the drag on an airship traveling at 15 mph in air at 60°F and 14.7 psia. She performs a test on a $\frac{1}{10}$-scale model in 60°F water at 14.7 psia. What is the necessary water velocity to simulate the 60 mph speed in air?

81. A centrifugal pump operates at 300 rpm to deliver 20°C lubricating oil. A $\frac{1}{5}$-size, geometrically similar pump delivering 15°C water is used to model the larger pump. How fast should the smaller pump run? Assume that Reynolds number effects are minor.

82. A $\frac{1}{10}$-scale model of a rocket is tested in 250 K, 100 kPa air at a Mach number of 2.0. Find the drag of the rocket at the same Mach number in the outer atmosphere at 250 K and 50 kPa. The rocket diameter is 4.0 m.

83. A 250-m-long ship has a wetted area of 8000 m². A $\frac{1}{100}$-scale model is tested in a towing tank with the prototype fluid, and the results are:

Model velocity (m/s)	0.57	1.02	1.40
Model drag (N)	0.50	1.02	1.65.

Calculate the prototype drag at 7.5 m/s and 12.0 m/s.

84. The appropriate dimensionless parameters for the drag on a ship are the Froude number and the Reynolds number. Experimental testing on the model at various Reynolds numbers produced the following data for the drag coefficient.

R	4.4×10^4	1×10^5	4×10^5	1×10^6
$\mathbf{C_D}$	0.44	0.55	0.61	0.67
R	4×10^6	1×10^7	2×10^7	6×10^7
$\mathbf{C_D}$	0.69	0.72	0.72	0.72

Estimate the drag coefficients $\mathbf{C_D}$ for the prototype at a Reynolds number of 5×10^8. The Froude number is the same for the model and the prototype.

85. Tests are run to determine the torque required on the inner of the two cylinders shown in Fig. P6.85. The results for $R_1 = 10$ cm and $R_2 = 15$ cm are:

Fluid	Viscosity (N·s/m²)	Rotational Speed (Hz)	Torque/length (N·m/m)
Lubricating oil			
20°C	1.31×10^{-2}	10.0	1.29
40°C	6.81×10^{-3}	10.0	6.72×10^{-1}
60°C	4.18×10^{-3}	10.0	4.12×10^{-1}
Water			
10°C	1.31×10^{-3}	5.0	6.46×10^{-2}
20°C	1.00×10^{-3}	5.0	4.93×10^{-2}.

Determine the torque per unit length required to rotate the inner cylinder at 10 Hz in 20°C kerosene. Are all the data in this table necessary? Support your answer.

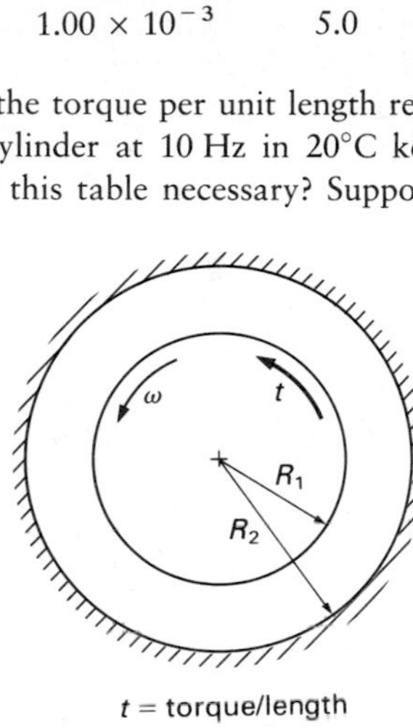

Figure P6.85

86. A pump testing setup is shown in Fig. P6.86. Water at 60°F is drawn from a sump and pumped though a pipe containing a valve. The water is discharged into a catch tank sitting on a scale. A test is being run on this system to study the flow through a valve and pump twice as large (the prototype). The fluid to be pumped by the larger pump also is 60°F water. The flow rate in the larger system is to be 3.5 ft³/sec and the loss coefficient for the valve is expected to be 6.3. The pump in the larger system rotates at 670 rpm. Find the test water flow rate to test the valve properly, the pump speed, and the test valve loss coefficient. The loss coefficient K is related to the valve energy loss gh_L by

$$gh_L = \mathrm{K}\frac{V^2}{2},$$

where V is the pipe fluid velocity.

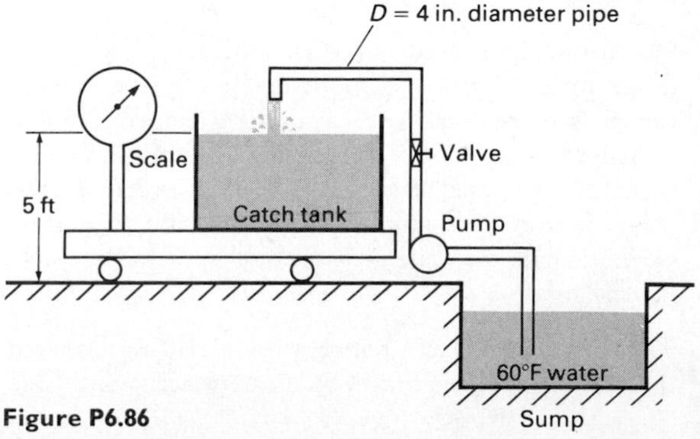

Figure P6.86

87. In low-speed external flow over a bluff object, vortices are shed from the object (as shown in Fig. P6.87). The frequency of vortex shedding, f [1/T], depends on ρ, μ, V, and d. Make a dimensional analysis of this phenomenon. Transform the dimensionless parameters that you developed into "standard" parameters. What are their names? Two identically shaped objects with a size ratio 2:1 are tested in air flow. What velocity ratio is necessary for dynamic similarity? What is the ratio of vortex-shedding frequencies?

Figure P6.87

88. In a nutritional products plant, a plate heat exchanger cools infant formula by approximately 25°C from an inlet temperature of 68°C as it flows through the plates shown in Fig. P6.88. The process engineer would like to be able to cool the infant formula by as much as 35°C as it passes through the plates. The plate heat exchanger is currently in use, so he would like to build a $\frac{1}{2}$-scale model for testing purposes. In the model, water will be used instead of infant formula. What flow rate is needed in the model?

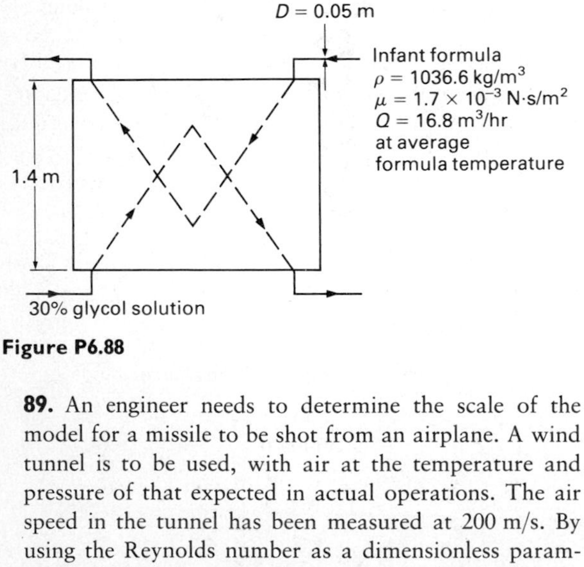

Figure P6.88

89. An engineer needs to determine the scale of the model for a missile to be shot from an airplane. A wind tunnel is to be used, with air at the temperature and pressure of that expected in actual operations. The air speed in the tunnel has been measured at 200 m/s. By using the Reynolds number as a dimensionless parameter, calculate the scale of the model if the missile normally flies at 38 m/s.

90. Molten plastic at a temperature of 510°F is augered through an extruder barrel by a screw occupying $\frac{3}{5}$ the

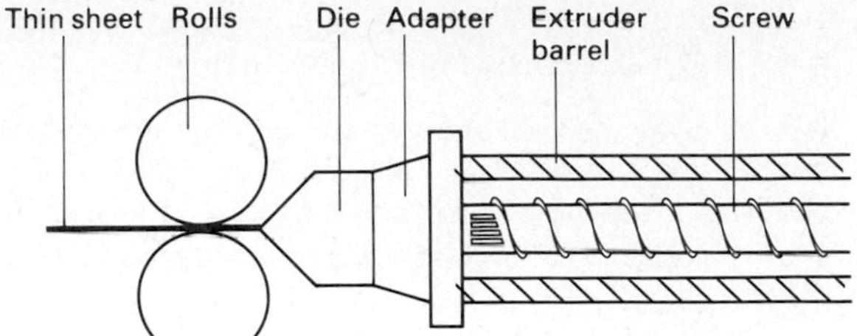

Figure P6.90

barrel volume, as shown in Fig. P6.90. The extruder is 16 ft long and has an inner diameter of 8 in. A scale model extruder screw ($S = \frac{1}{4}$) is tested to determine its output rate before an investment is made in a full-scale product. What dimensionless parameter should be approximately the same for both prototype and model? The model has a feed rate of 275 lbm/hr and a barrel inner diameter of 2.0 in. What would be the required feed rate on an 8 in. extruder with a screw occupying $\frac{3}{5}$ of its volume?

91. The black-liquor pump shown in Fig. P6.91 supplies a line of evaporators in a paper mill. The pump cannot be removed from service, yet performance data is needed. Given the information for a prototype pump (Fig. P6.91) and a smaller, identical pump that has a 5-cm inlet (for use with 50°C water), what velocity would the small pump need to perform the testing? Assume constant fluid densities.

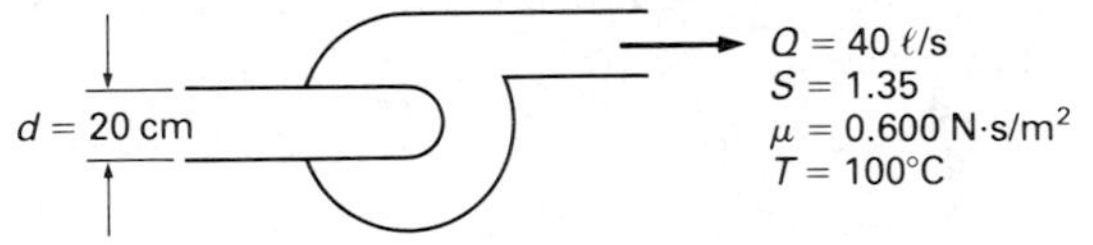

Figure P6.91

• **92.** Liquid is to be drained from the two geometrically similar tanks shown in Fig. P6.92. The dimensionless parameters for this problem are

$$\frac{h}{D}, \quad \frac{d}{D}, \quad \frac{Vt}{D}, \quad \text{and} \quad \frac{gt^2}{D}.$$

What must be the initial liquid height h_{20} for complete similarity if $h_{10} = 1.0$ m? Will both tanks empty (i.e., $h_2 = h_1 = 0$) in the same amount of time? Compare the times with those found by applying the continuity equation and assuming quasi-static conditions to calculate the velocity V to get

$$-\frac{\pi D^2}{4}\frac{dh}{dt} = \frac{\pi d^2}{4} V = \frac{\pi d^2}{4}\sqrt{2gh}.$$

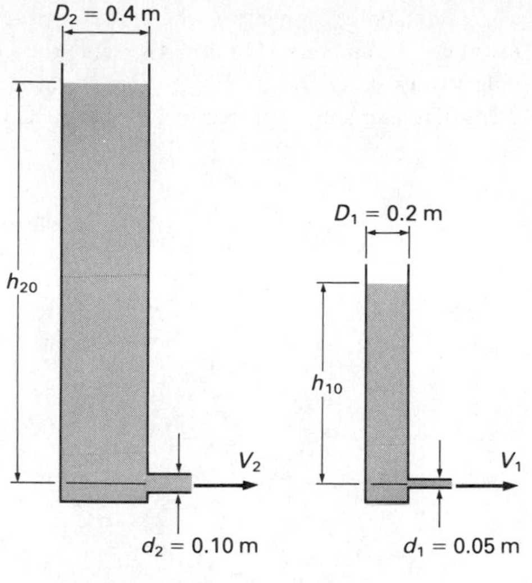

Figure P6.92

• **93.** A stream of atmospheric air is used to keep a ping-pong ball aloft by blowing the air upward over the ball. The ping-pong ball has a mass of 2.5 g and diameter $D_1 = 3.8$ cm, and the air stream has an upward velocity of $V = 1.90$ m/s. This system is to be modeled by pumping water upward over the ball with velocity V_2 over a ball with diameter $D_2 = 12.0$ cm and density ρ_{b_2}. For air blowing upward over a ball, the net weight of the ball W_b is equal to the air drag, or

$$W_b = \frac{\mathbf{C_D}\rho A V^2}{2},$$

where $\mathbf{C_D} = 0.60$, ρ is the fluid mass density, A is the ball's projected area, and V is the velocity of the fluid upstream from the ball. Find D_2 and ρ_{b_2}. What might be the material from which the ball is made? Assume that the model ball is solid. [*Note:* $W_b = (\rho_b - \rho)g\forall_b$.]

• **94.** The dimensionless parameters for a ball released and falling from rest in a fluid are

$$\mathbf{C_D}, \quad \frac{gt^2}{D}, \quad \frac{\rho}{\rho_b}, \quad \text{and} \quad \frac{Vt}{D},$$

where $\mathbf{C_D}$ is a drag coefficient (assumed to be constant at 0.4), g is the acceleration of gravity, D is the ball diameter, t is the time after it is released, ρ is the density of the fluid in which it is dropped, and ρ_b is the density of the ball. Ball 1, an aluminum ball ($\rho_{b_1} = 2710$ kg/m^3) having a diameter $D_1 = 1.0$ cm, is dropped in water ($\rho = 1000$ kg/m^3). The ball velocity V_1 is 0.733 m/s at $t_1 = 0.10$ s. Find the corresponding velocity V_2 and time t_2 for ball 2 having $D_2 = 2.0$ cm and $\rho_{b_2} = \rho_{b_1}$. Next, use the computed value of V_2 and the equation of motion,

$$\rho_b \forall g - \frac{\mathbf{C_D}}{2}\rho A V^2 = \rho_b \forall \frac{dV}{dt},$$

where $\forall$ is the volume of the ball and A is its cross-sectional area, to verify the value of t_2. Should the two values of t_2 agree?

• **95.** A prototype fan has a 20-ft diameter, an inlet pressure of 14.40 psia, an inlet temperature of 70°F, and a speed of 90 rpm. A $\frac{1}{10}$-scale model of the fan has the same inlet pressure and temperature, an inlet power of 1.24 hp, a flow rate of 220 ft^3/min, and a speed of 1800 rpm. Find the corresponding input power and flow rate of the prototype fan. Neglect Reynolds number effects.

• **96.** Two geometrically similar reciprocating pumps operate side by side. One is half the size of the other (S = 1/2). The smaller pump is operating at twice the speed of the larger one. Find the ratios of their volume flow rates and absolute discharge pressures. Neglect Reynolds number effects.

• **97.** The following data were taken for the flow of 60°C benzene in a horizontal steel pipe.

Pressure drop (kPa)	0.525	2.10	8.40	33.50
Flow rate (m^3/s)	0.10	0.20	0.40	0.80

What would be the pressure drop in the same pipe for a flow of 3.38 m^3/s of 40°C lubricating oil?

• **98.** Two canals are to have the same Reynolds number and the same Froude number. One canal is two thirds the size of the other. Both canals experience atmospheric pressure. The smaller canal is to carry water at 10°C. What should be the water temperature in the larger canal for dynamic similarity?

• **99.** A student at a famous research university is interested in the aerodynamic drag on spheres. She conducts a series of wind tunnel tests on a 10-cm-diameter sphere. The air in the wind tunnel is at 50°C and 101.3 kPa. She presents the results of her tests in the form of a correlation,

$$\mathcal{D} = 0.0015V^{1.92},$$

where $\mathcal{D}$ is the drag force in Newtons and V is the velocity in m/s. Another student realizes that the results would have been more effectively presented in terms of dimensionless parameters, say,

$$\Pi_1 = f(\Pi_2) = C\Pi_2^a,$$

where Π_1 and Π_2 are dimensionless parameters and C and a are constants. What are the most appropriate dimensionless parameters (Π_1 and Π_2) and the corresponding values of C and a? What would be the drag on a 2-cm-diameter sphere placed in a 1-m/s, 20°C water stream?

7 Steady Incompressible Flow in Pipes and Ducts

Now that you are familiar with the basic tools of flow analysis, we present some specific applications. In this chapter, we deal with flow in closed conduits. If a conduit has a circular cross section of constant area, it is called a pipe. Constant-area conduits with other cross sections (the most common being rectangular and annular) are called ducts. We consider only cases where the conduit is running full.*

Fluid transport systems composed of pipes or ducts have many practical applications. Chemical plants and oil refineries seem to be a maze of pipes. Power plants contain many pipes and ducts for transporting fluids involved in the energy-conversion process. City water supply and wastewater removal systems consist of many kilometers of pipe. Many machines are controlled by hydraulic systems with control fluid transported in hoses or tubes. Chemical, civil, and mechanical engineers are often involved in designing, maintaining, and operating these and other types of flow systems.

Piping systems are constructed by piecing together components such as those shown in Fig. 7.1. These components transport the fluid, change its direction, control its flow rate, divide it up, speed it up, or slow it down. In addition to conveying fluid, most of the components shown are responsible for a mechanical energy loss. Only the pump does not contribute a net loss† but instead increases the fluid's mechanical energy. Analysis of pipe and duct systems consists of relating flow variables, such as energy loss and flow rate, to pipe or duct system parameters, such as size, shape, length, and number and location of fittings. It also includes evaluating reaction forces that result from fluid momentum changes. This evaluation can be accomplished by applying the linear or angular momentum equation after fluid velocities and pressures have been established.

In this chapter, we discuss flow in straight pipes and ducts, in fittings and valves, and in the complete systems that are obtained by piecing these components together. We also describe some devices used to measure flow rate in pipe and duct systems and the effects that these devices have on the systems. We begin with consideration of flow in straight pipes and ducts.

---

* Partially full conduits are open channels and must be dealt with by the methods described in Chapter 11.

† There are losses even in a pump. The net increase of the fluid's mechanical energy is less the work done on the fluid; however, the *net* change of the fluid's mechanical energy is positive.

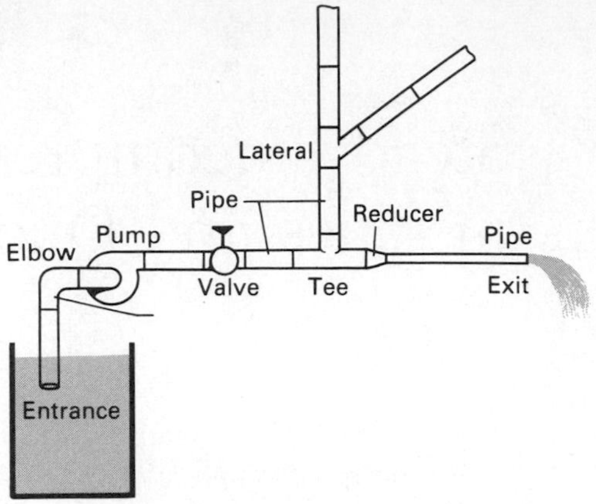

Figure 7.1 Piping system constructed by piecing together various components.

7.1 CLASSIFYING THE FLOW IN A PIPE OR DUCT

Before we can discuss or analyze flow in a particular pipe, we must decide whether the flow is *steady* or *unsteady, compressible* or *incompressible, laminar* or *turbulent,* and *developing* or *fully developed.*

We limit consideration in this chapter to steady flows. Steady flow results when the boundary conditions are constant over time. Circumstances that cause unsteady flow include the startup of a pump, closure of a valve, or pulsating flow or pressure at a pump.

We also limit consideration in this chapter to incompressible flow. This limitation means that the Mach number of the flow must be low ($\mathbf{M} \ll 1$), which usually applies to liquid flows (a notable exception is the unsteady flow that results from rapid closure of a valve in a liquid flow system) and to gas flows in which the maximum velocity of the gas is less than about 100 m/s. Analysis and discussion of compressible flow in a pipe is presented in Chapter 10.

7.1.1 Laminar Flow and Turbulent Flow

We introduced the nature of laminar and turbulent flows in Section 3.2.4 and discussed it further in Section 5.4.3. If the flow in a pipe is laminar, the fluid moves along smooth streamlines. The fluid velocity at any point in the pipe is constant in time. If the flow is turbulent, a rather violent mixing of the fluid occurs, and the fluid velocity at a point varies randomly with time. (See Figs. 3.15, 3.18, and 5.17.)

The differences between laminar and turbulent pipe flows were first clarified by Osborne Reynolds in 1883. Reynolds performed a series of experiments in which he injected dye into water flowing in a glass pipe. Figure 7.2 illustrates Reynolds's observations. At low flow rates, the dye streak remained smooth and regular as the dye flowed downstream. At higher flow rates, the dye seemed to explode, mixing rapidly throughout the entire pipe. A modern spark photograph of the mixing dye would reveal a very complex flow pattern, not discernible in Reynolds's experiments.

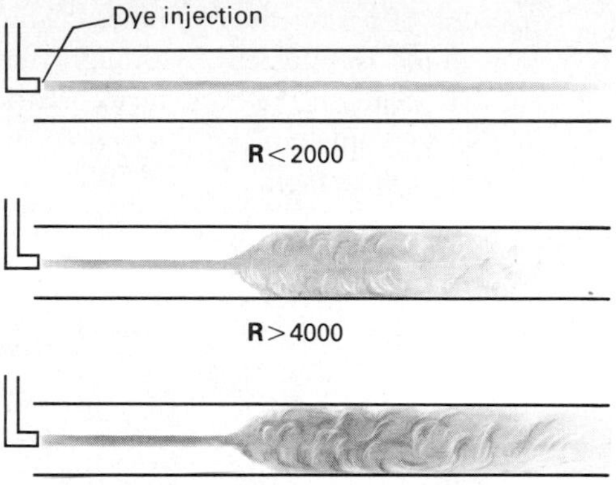

Figure 7.2 Schematic representation of Reynolds's observations of laminar and turbulent pipe flow.

Reynolds's experiments showed that the nature of pipe flow depends on the Reynolds number,

$$\mathbf{R} = \frac{\rho \bar{V} D}{\mu}, \tag{7.1}$$

in which $\bar{V}$ is the average velocity and D is the pipe inside diameter.* Reynolds found that if the value of **R** was below about 2000, the flow was always laminar, whereas at higher Reynolds numbers, the flow was turbulent. The exact value of Reynolds number that defined the boundary between laminar and turbulent flow depended on the experimental conditions. If the water in the inlet reservoir was very still and there was no vibration in the equipment, Reynolds found that laminar flow could be maintained at Reynolds numbers much higher than 2000. Reynolds also found that if he began at a high value of **R** with turbulent flow in the pipe and then decreased **R**, the flow became laminar at value of **R** of about 2000.

Although it is possible to obtain laminar flow at higher Reynolds numbers in the laboratory, most engineering situations do not qualify as "undisturbed." In engineering practice, the upper limit of Reynolds number for laminar flow in a pipe is taken to be

$$\mathbf{R} \approx 2300 = \text{Maximum for laminar flow in a pipe.}$$

For Reynolds numbers between 2300 and about 4000, the flow is unpredictable, sometimes pulsating or switching back and forth between laminar and turbulent. This type of flow is called *transitional flow*. If the Reynolds number is above 4000, the flow is usually turbulent:

$$\mathbf{R} \approx 4000 = \text{Minimum for stable turbulent flow in a pipe.}$$

The randomly fluctuating velocities and violent mixing action of turbulent flow make it significantly different from laminar flow. In most engineering appli-

---

* Note that for steady flow of an incompressible, constant-viscosity fluid, **R** is constant all along the pipe. In his experiments, Reynolds changed **R** by changing $\bar{V}$, that is, by changing the flow rate.

cations of turbulent pipe flow, dealing with the time-average velocity, defined by Eq. (5.66) is sufficient. Although the detailed flow is unsteady, the time-average flow is steady if the boundary conditions do not vary with time. The time-average velocity profile can be integrated over the pipe cross-sectional area to calculate the fluid flow rate:

$$Q = \int \langle V \rangle \, dA, \tag{7.2}$$

and an area-and-time-averaged velocity can be calculated by

$$V = \overline{\langle V \rangle} = \frac{Q}{A}. \tag{7.3}$$

The double average notation $\overline{\langle \ \rangle}$ is very cumbersome, so we do not use it further in this chapter; instead, we use the symbol V to refer to the area-and-time-averaged velocity.

Unfortunately, we cannot "average away" all the effects of turbulence. The random mixing action of turbulence creates a significant cross-stream momentum flux as small "lumps" of fluid are thrown about. Although this effect actually involves fluid "inertial forces," it is modeled as if it were additional shear stress in the fluid. We call the additional stresses *apparent stresses*. The apparent stress model is discussed in Section 5.4.3 and Example 5.6.

In turbulent flow, apparent stresses may be many orders of magnitude larger than laminar (viscous) stresses. Interpreting this phenomenon in terms of what we know about the Reynolds number is instructive. In Section 6.5, we pointed out that the Reynolds number represents the ratio of inertia and viscous forces for a typical fluid particle. If there are no velocity gradients (i.e., no fluid deformation), there are no stresses and the Reynolds number is irrelevant. If there are velocity gradients, experiments reveal that the flow is turbulent at high Reynolds numbers. In this case, the inertia forces in the Reynolds number represent the turbulent stresses. The high value of **R** then means

$$\langle \tau_{\text{app}} \rangle \gg \langle \tau_{\text{viscous}} \rangle.$$

The high stress levels in turbulent flow reflect the intense mixing of turbulence. These high stress levels cause much higher mechanical energy loss in turbulent pipe flow than in laminar pipe flow. Figure 7.3 shows typical trends of experimental measurements of energy loss in pipe flow. For a particular pipe of given

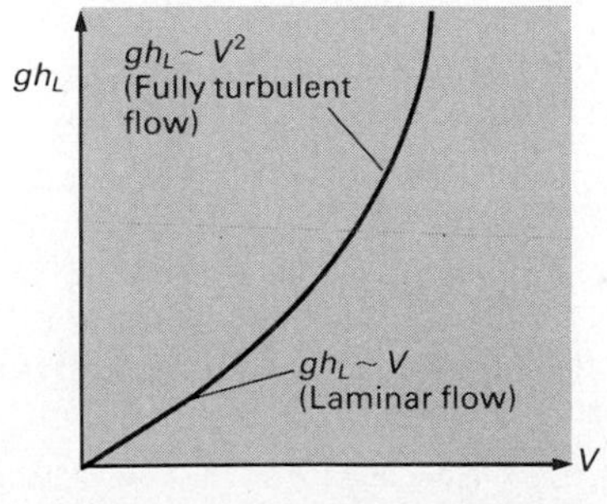

Figure 7.3 Variation of mechanical energy loss with velocity for flow of a specific fluid (ρ, μ fixed) in a specific pipe (D, L fixed).

diameter and for a particular fluid, increasing velocity corresponds to increasing Reynolds number. At low velocities (low **R**), energy loss varies as

$$gh_L \sim V,$$

which indicates that energy loss is due to viscous forces (see Section 6.5). At high velocities (high **R**), energy loss varies as

$$gh_L \sim V^2,$$

which indicates that energy loss is due to inertia effects (that is, turbulence).

7.1.2 Developing and Fully Developed Flow

The flow in a constant-area pipe or duct is said to be *fully developed* if the shape of the velocity profile is the same at all cross sections.* We discussed fully developed flow in Section 5.6.1. Fully developed flow is relatively simple to analyze because it is one-dimensional (the velocity varies with distance across the pipe but not along it) and one-directional (there is no velocity perpendicular to the pipe axis).

Fully developed flow seemingly would occur only under unique circumstances, but, fortunately, many pipe flows are fully developed. We can show this by considering the flow from a large reservoir into a pipe (Fig. 7.4). If the fluid in the reservoir is tranquil and the junction between the pipe and the reservoir is well rounded, the velocity of the fluid entering the pipe at cross section 1 will be nearly uniform; however, fluid in contact with the pipe wall has zero velocity (recall the no-slip condition). The velocity gradient near the pipe wall is associated with a retarding shear stress on the fluid. A *boundary layer* of slower-moving fluid is established at the wall (see Section 5.6.2). The boundary layer grows in thickness as the fluid proceeds downstream. The fluid outside the boundary layer, in the central region of the pipe, is called the *core flow*. The fluid velocity in the boundary layer is reduced, which means that the core fluid must accelerate as it moves downstream to maintain constant mass flow at all cross sections. Figure 7.4 also shows a velocity profile for the entrance (or developing) region. Clearly, the velocity profile changes as the fluid moves downstream.

Note that the distinction between laminar and turbulent flow in a pipe applies only to the fully developed flow condition. In the development zone, the core flow is irrotational and thus neither laminar nor turbulent. The flow in the boundary layer is laminar or turbulent. If the fully developed pipe flow is laminar ($\mathbf{R} < 2300$), the boundary layer is laminar, but if the fully developed pipe flow is turbulent ($\mathbf{R} \geq 4000$), the boundary layer is laminar near the pipe entrance, undergoes transition, and is turbulent as it approaches the fully developed condition.

Obviously, the boundary layer cannot continue to grow indefinitely; eventually it fills the entire pipe. This occurs at plane 2 in Fig. 7.4. Shortly downstream at plane 3, the velocity profile ceases to change and the flow is fully developed.

---

* For turbulent pipe flow, we interpret this statement as applying to the time-averaged velocity profile.

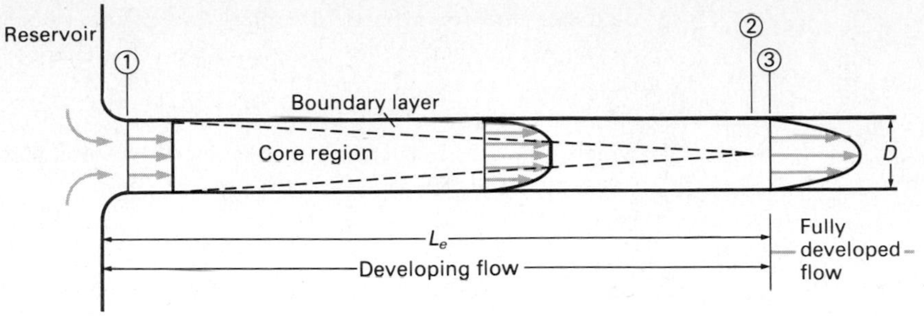

Figure 7.4 Developing flow.

The flow between planes 1 and 3 is said to be *developing*. The region between 1 and 3 is called the *entrance region*. The length L_e is called the *entrance length* or the *development length*. By dimensional analysis,

$$\frac{L_e}{D} = F\{\mathbf{R}\}.$$

The function F is different for laminar and turbulent flow. Analytic and experimental investigations have demonstrated the validity of the following correlations:

$$\frac{L_e}{D} \approx 0.06\mathbf{R}, \qquad \text{Laminar flow} \tag{7.4}$$

and

$$\frac{L_e}{D} \approx 4.4(\mathbf{R})^{1/6}. \qquad \text{Turbulent flow} \tag{7.5}$$

The longest practical development length corresponds to laminar flow with $\mathbf{R} \approx 2300$:

$$\left(\frac{L_e}{D}\right)_{\text{max}} \approx 140.$$

In many engineering pipe flows, $\mathbf{R}$ is between 10^4 and 10^5. For this case, the flow is turbulent and

$$\frac{L_e}{D} \approx 25. \qquad \text{Typical engineering flow}$$

With many pipes in engineering applications being hundreds or thousands of diameters long, the flow is fully developed over most of their lengths.

This discussion has concentrated on flow entering a pipe from a large reservoir. Any disturbance such as a valve, elbow, or change in pipe diameter causes the flow to depart from fully developed conditions; however, the flow returns to fully developed conditions several diameters downstream. Generally speaking, development lengths for other disturbances are shorter than for a reservoir entrance.

Aside from the analytic advantage of one-dimensionality and one-directionality, fully developed flow has several other interesting features.

- The streamlines are straight and parallel to the pipe axis.
- The shear stress at the pipe wall is the same at all locations.
- The momentum and kinetic energy correction factors β and α are constant.

Can you explain these features based on the condition that the shape of the velocity profile is constant?

EXAMPLE 7.1 Illustrates Developing Flow

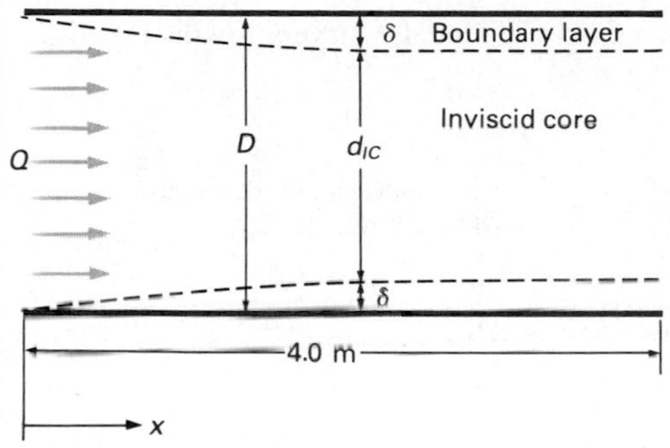

Figure E7.1 Expected boundary layer development in a wind tunnel.

A wind tunnel has a 3.0-m diameter, a 4.0-m-long test section, and a flow rate of 300 m^3/s, as shown in Fig. E7.1. Estimate the size of the inviscid core at the end of the test section.

SOLUTION

Given

Wind tunnel having 3.0-m diameter, 4.0-m length, and 300-m^3/s flow rate

Find

Size of the inviscid core at end of test section

Solution

The diameter d_{IC} of the inviscid core is related to the wind tunnel diameter D by

$$d_{IC} = D - 2\delta,$$

where δ is the boundary layer thickness. This thickness increases along the direction of flow, so the inviscid core diameter decreases along the direction of flow. The limit of this process is reached when the boundary layer fills the entire wind tunnel, that is, when the flow becomes fully developed.

We first calculate the Reynolds number and determine whether to use Eq. (7.4) or Eq. (7.5) for the development length. The average velocity is

$$V = \frac{Q}{A} = \frac{(300 \text{ m}^3/\text{s})}{(\pi/4)(3.0 \text{ m})^2} = 42.4 \text{ m/s}.$$

If we assume that the air is at 15°C, Table A.3 gives

$$\nu = 1.49 \times 10^{-5} \text{ m}^2/\text{s},$$

and so

$$\mathbf{R} = \frac{VD}{\nu} = \frac{(42.4 \text{ m/s})(3.0 \text{ m})}{(1.49 \times 10^{-5} \text{ m}^2/\text{s})} = 8.5 \times 10^6.$$

With $\mathbf{R} > 2300$, we use Eq. (7.5), which gives

$$L_e \approx 4.4(8.5 \times 10^6)^{1/6} D \approx 63(3.0 \text{ m}) \approx 190 \text{ m}.$$

The flow would require a length of 190 m to become fully developed.

Assuming that the boundary layer thickness increases linearly from zero at $x = 0$ to $\delta = D/2$ at $x = 190$ m, we estimate the thickness after 4 m as

$$\delta \approx \frac{4.0\text{ m}}{190\text{ m}}\left(\frac{3.0\text{ m}}{2}\right) = 0.032\text{ m}.$$

The inviscid core diameter is

$$d_{IC} \approx 3.0\text{ m} - 2(0.032\text{ m});$$

$$d_{IC} \approx 2.94\text{ m}.$$ **ANSWER**

Discussion

In Chapter 9, we show that the boundary layer growth is not linear but instead is approximated by

$$\delta \propto x^{4/5},$$

where x is measured from the wind tunnel inlet and along the direction of flow. For our wind tunnel with a 1.5-m radius,

$$\delta \approx 1.5\text{ m}\left(\frac{4.0}{190}\right)^{4/5} \approx 0.068\text{ m}.$$

A more accurate estimate of d_{IC} is

$$d_{IC} \approx 3.0\text{ m} - 2(0.068\text{ m}) \approx 2.86\text{ m}.$$

Wind tunnel test sections are kept short, so that a large region of uniform flow (the inviscid core) is available to simulate "free-flight" conditions in the atmosphere.

7.2 ANALYSIS OF FULLY DEVELOPED FLOW IN PIPES AND DUCTS

We now consider fully developed flow in a straight pipe or duct of constant cross section. Consideration is limited to steady flow of incompressible fluids with constant viscosity. Items of interest include relations between energy loss and flow rate, pressure and shear stress distributions, and characteristics of velocity profiles. By using all the tools that we have available (control volume approach, differential approach, and experimental results expressed in the language of dimensional analysis), we can obtain a good working knowledge of pipe flow. Although there are significant differences between fully developed laminar and turbulent flows, we carry the discussion as far as possible before introducing these differences.

7.2.1 Control Volume Analysis

Figure 7.5 illustrates fully developed flow occurring between two cross-sectional planes in a pipe or duct of constant area. For now, the cross-sectional shape is arbitrary, and we assume steady, incompressible flow. There are no pumps or turbines located between the two planes. Under these circumstances, the continuity

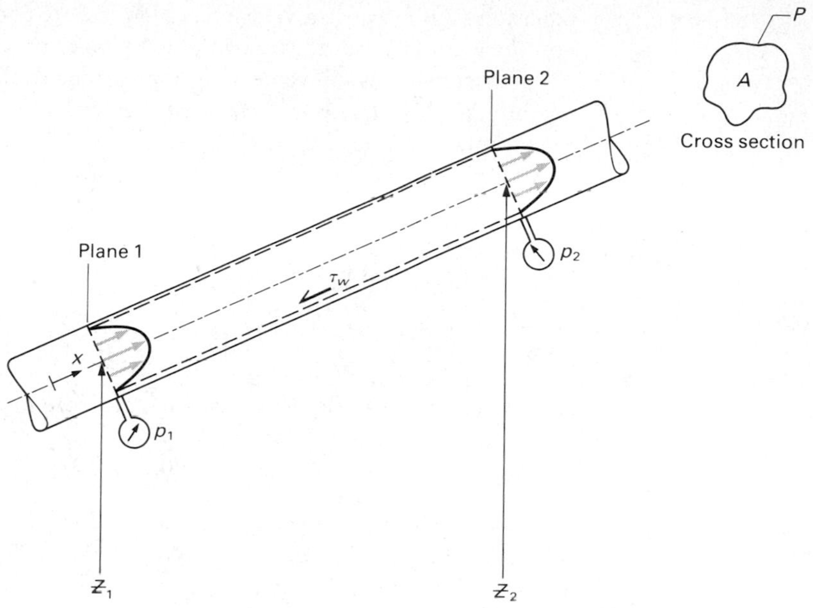

Figure 7.5 Fully developed flow (with control volume) in duct of arbitrary cross-sectional shape.

and energy equations for the control volume shown are

$$V_1 A_1 = V_2 A_2 \tag{7.6}$$

and

$$\frac{\bar{p}_1}{\rho} + \alpha_1 \frac{V_1^2}{2} + g\bar{z}_1 = \frac{\bar{p}_2}{\rho} + \alpha_2 \frac{V_2^2}{2} + g\bar{z}_2 + \overline{gh_L}. \tag{7.7}$$

The area is constant, so

$$V_1 = V_2 = V = \text{Constant}. \tag{7.8}$$

Using this result and knowing that, for fully developed flow, $\alpha_1 = \alpha_2$, the energy equation is

$$\frac{\bar{p}_1}{\rho} + g\bar{z}_1 = \frac{\bar{p}_2}{\rho} + g\bar{z}_2 + \overline{gh_L}. \tag{7.9}$$

The overbars representing averages are somewhat cumbersome, so we omit them unless they are necessary to clarify the difference between a local quantity and its area average. Because the streamlines are straight and parallel for fully developed flow, the variation of pressure over the cross section is hydrostatic. The sum of averages at any cross section $(\bar{p}/\rho + g\bar{z})$ is equal to the sum $(p/\rho + gz)$ at any point in the cross section. To avoid confusion, we assume that p and z are evaluated at the pipe centerline.

Dropping the overbar and solving for the energy loss, we get

$$gh_L = \frac{p_1 - p_2}{\rho} + g(z_1 - z_2). \tag{7.10}$$

Equation (7.10) reveals a very interesting feature about fully developed flow: Friction *does not* cause a decrease in fluid velocity or kinetic energy along the pipe.

It is sometimes convenient to express energy loss in terms of an equivalent potential energy loss or an equivalent pressure difference. The equivalent potential energy loss is the head loss, given by

$$h_L = \frac{(gh_L)}{g}. \tag{7.11}$$

For fully developed pipe flow,

$$h_L = \frac{p_1 - p_2}{\gamma} + z_1 - z_2. \tag{7.12}$$

Energy loss can also be expressed by a *pressure loss,* defined by

$$\Delta p_L \equiv \rho(gh_L). \tag{7.13}$$

Pressure loss is actually energy loss per unit volume. A more exact name for pressure loss is *stagnation pressure loss* (see Eq. 4.53 and the discussion following it). For fully developed flow,

$$\Delta p_L = (p_1 - p_2) + \gamma(z_1 - z_2). \tag{7.14}$$

Recognizing the difference between pressure change, pressure drop, and pressure loss is important. *Pressure change* between two points in a flow is

$$\Delta p = p_2 - p_1.$$

Pressure drop is the negative of pressure change,

$$\Delta p_d = p_1 - p_2.$$

In fully developed flow, pressure change is usually negative, and pressure drop is usually positive. Pressure loss is not necessarily equal to a difference between two pressures but is the energy loss expressed in units of pressure. For *fully developed* flow in a *horizontal pipe,* pressure drop and pressure loss are equal. In a nonhorizontal pipe or in a flow with acceleration, pressure drop and pressure loss are not equal. Unfortunately, engineers sometimes use the term "pressure drop" when they mean "pressure loss." As the change in elevation between two points in a pipe is usually known, relating pressure drop and pressure loss is usually simple.

Next we apply the linear momentum equation to the fully developed flow shown in Fig. 7.5. The forces on the fluid in the control volume are the result of pressure at the ends and along the wall, shear stress between the wall and the fluid, and gravity. To simplify the analysis, we assume that the pipe axis is horizontal ($z_1 = z_2$). The steady flow linear momentum equation for the axial (x) direction is

$$\sum \dot{M}_{x,\text{out}} - \sum \dot{M}_{x,\text{in}} = \sum F_x.$$

If we use various results from Section 4.3.2, the momentum equation becomes

$$\rho\beta_2 V_2^2 A_2 - \rho\beta_1 V_1^2 A_1 = p_1 A_1 - p_2 A_2 - \int_{A_w} \tau_w \, dA.$$

For fully developed flow,

$$\beta_2 = \beta_1 \quad \text{and} \quad V_2 = V_1.$$

The pipe area is constant, so

$$A_1 = A_2 = A.$$

With these simplifications, the axial momentum equation becomes

$$p_1 - p_2 = \frac{1}{A}\int_{A_w} \tau_w \, dA. \tag{7.15}$$

The "wetted area" on which the shear stress acts is

$$dA_w = (dx)(P),$$

where P is the perimeter of the duct. Let $\bar{\tau}_w$ be the average shear stress around the perimeter at any x location. Then

$$\int_{A_w} \tau_w \, dA = P \int_{x_1}^{x_2} \bar{\tau}_w \, dx.$$

For fully developed flow, the wall shear stress does not vary with x, so

$$\int_{A_w} \bar{\tau}_w \, dA = P\bar{\tau}_w \int_{x_1}^{x_2} dx = P\bar{\tau}_w(x_2 - x_1).$$

If we define $\Delta x \equiv x_2 - x_1$, Eq. (7.15) becomes

$$p_1 - p_2 = \frac{P\,\Delta x}{A}\,\bar{\tau}_w. \tag{7.16}$$

We can rearrange and generalize this equation. Although we assumed a horizontal pipe, we can extend this result to a nonhorizontal pipe by replacing p by $(p + \gamma z)$. Making this change and dropping the overbars, we have

$$(p_1 + \gamma z_1) - (p_2 + \gamma z_2) = \frac{P\tau_w}{A}\,\Delta x. \tag{7.17}$$

Combining Eqs. (7.11), (7.13), and (7.12) with Eq. (7.17) gives

$$\Delta p_L = \tau_w\left(\frac{P}{A}\right)\Delta x, \tag{7.18}$$

$$gh_L = \frac{\tau_w}{\rho}\left(\frac{P}{A}\right)\Delta x, \tag{7.19}$$

and

$$h_L = \frac{\tau_w}{\gamma}\left(\frac{P}{A}\right)\Delta x \tag{7.20}$$

for fully developed flow.

The most common duct cross section is circular. For a circular cross section (pipe),

$$\frac{P}{A} = \frac{\pi D}{\pi D^2/4} = \frac{4}{D},$$

and Eqs. (7.17)–(7.20) become

$$(p_1 + \gamma z_1) - (p_2 + \gamma z_2) = \Delta p_L = 4\tau_w\left(\frac{\Delta x}{D}\right), \tag{7.21}$$

$$g h_L = 4\frac{\tau_w}{\rho}\left(\frac{\Delta x}{D}\right), \tag{7.22}$$

and

$$h_L = 4\frac{\tau_w}{\gamma}\left(\frac{\Delta x}{D}\right). \tag{7.23}$$

Equations (7.11), (7.17), and (7.19) provide relations between three of the four most important parameters in pipe and duct flow: energy loss, pressure (plus elevation) change, and wall shear stress. Given any one of these parameters, together with fluid density and duct geometry, we can determine the other two for fully developed flow. The remaining problem is to relate any one of these parameters to the fourth parameter, flow rate. Because we have "used up" all the applicable control volume equations* (momentum, continuity, energy), we must turn to a differential approach *or* experimental data to complete the analysis. All these equations apply to either laminar or turbulent flow. The differential approach can solve the remainder of the problem if the flow is laminar; however, the turbulent flow problem can be solved only by referring to experimental data. These detailed considerations also require that we be specific about the particular geometry involved. When necessary, we limit discussion to flow in circular pipes.

7.2.2 Application of Dimensional Analysis: Friction Factors and Loss Coefficients

Anticipating the need to refer to experimental data for further information about pipe flow, we need to consider dimensional analysis before we proceed. We discussed dimensional analysis of energy loss in pipe flow extensively in Chapter 6. The result of the dimensional analysis, Eq. (6.4), was†

$$\frac{g h_L}{\frac{1}{2}V^2} = F\left\{\frac{L}{D}, \frac{\rho V D}{\mu}, \frac{\varepsilon}{D}\right\},$$

or

$$\mathbf{K} = F\left\{\frac{L}{D}, \mathbf{R}, \frac{\varepsilon}{D}\right\}, \tag{7.24}$$

---

* This situation clearly points up the most common failure of the finite control volume approach. Although the approach can provide valuable relations between key flow parameters, at least one of these parameters must often be evaluated by the differential or experimental approach.

† $F\{\ \}$ is used for the functional notation rather than $f\{\ \}$ to avoid confusion with the friction factor, to be introduced shortly.

where **K** is the loss coefficient. If we know **K**, we can calculate the energy loss by

$$gh_L = \mathbf{K}\left(\frac{V^2}{2}\right), \tag{7.25a}$$

and the head loss by

$$h_L = \mathbf{K}\left(\frac{V^2}{2g}\right). \tag{7.25b}$$

Using Eq. (7.13), we find that the pressure loss is

$$\Delta p_L = \mathbf{K}\left(\frac{\rho V^2}{2}\right). \tag{7.26}$$

To use Eqs. (7.24)–(7.26), we must know the dependence of **K** on the length-to-diameter ratio, Reynolds number, and relative roughness. If **K** had to be determined completely by experiment, about 10^3 runs would be required, because **K** depends on three parameters.

We simplify the problem of determining **K** by noting that, according to Eq. (7.22), energy loss in a pipe is directly proportional to the length-to-diameter ratio. The reason is that the wall shear stress is constant for fully developed flow. If we replace Δx with the pipe length L, the general function F in Eq. (7.24) must have the special form

$$F\left\{\frac{L}{D}, \mathbf{R}, \frac{\varepsilon}{D}\right\} = \left(\frac{L}{D}\right) G\left\{\mathbf{R}, \frac{\varepsilon}{D}\right\}.$$

The dimensionless parameter defined by

$$f \equiv G\left\{\mathbf{R}, \frac{\varepsilon}{D}\right\} \tag{7.27}$$

is called the *Darcy friction factor* (f is sometimes called the *Moody friction factor* for reasons that become apparent soon). The Darcy friction factor is related to the loss coefficient by

$$\mathbf{K} = f\frac{L}{D}. \tag{7.28}$$

In terms of the Darcy friction factor, the energy loss, head loss, and pressure loss are calculated by

$$gh_L = f\left(\frac{L}{D}\right)\frac{V^2}{2}, \tag{7.29a}$$

$$h_L = f\left(\frac{L}{D}\right)\frac{V^2}{2g}, \tag{7.29b}$$

and

$$\Delta p_L = f\left(\frac{L}{D}\right)\left(\frac{\rho V^2}{2}\right). \tag{7.30}$$

Equation (7.29b) is called the *Darcy–Weisbach equation*. This name could also be extended to the other two equations.

To determine f experimentally, we would need only about 10^2 experiments. Thus even a simple, incomplete theory (in this case, Eq. 7.22) can reduce the complexity of an investigation.

The Darcy friction factor is related to the skin friction coefficient. According to Eq. (7.22) with $\Delta x = L$,

$$gh_L = 4\,\frac{\tau_w}{\rho}\left(\frac{L}{D}\right).$$

Evaluating gh_L in terms of the Darcy friction factor from Eq. (7.29a), we have

$$f\left(\frac{L}{D}\right)\frac{V^2}{2} = 4\,\frac{\tau_w}{\rho}\left(\frac{L}{D}\right).$$

Rearranging gives

$$\frac{\tau_w}{\frac{1}{2}\rho V^2} = \mathbf{C_f} = \frac{f}{4}. \tag{7.31}$$

The skin friction coefficient ($\mathbf{C_f}$) for fully developed pipe flow is sometimes called the *Fanning friction factor*.

We can reduce the problem of predicting losses or pressure drop to the problem of determining any one of the three dimensionless parameters $\mathbf{K}$, f, or $\mathbf{C_f}$, because they are related by Eqs. (7.28) and (7.31). Whichever we select, we expect it to depend on the Reynolds number and relative roughness. To proceed further, we must consider laminar and turbulent flow separately.

7.2.3 Laminar Flow in a Circular Pipe

For fully developed laminar flow in a circular pipe, analytic methods produce an exact solution. The key is finding an expression for the velocity profile (Fig. 7.6):

$$u = u\left\{r, R, \mu, \frac{dp}{dx}\right\}.$$

After we obtain the velocity profile, we find the flow rate and average velocity by integration. The fluid viscosity and pressure gradient remain as parameters, so a

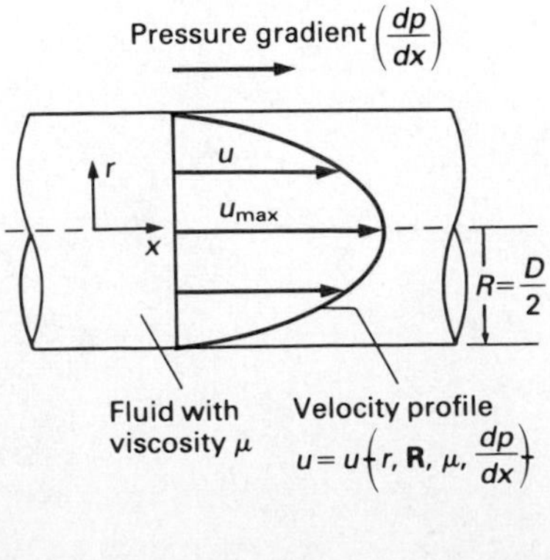

Figure 7.6 Schematic of velocity profile for fully developed pipe flow.

relation between pressure, pipe length and diameter, fluid viscosity, and flow rate can be found.

There are two ways to obtain a differential equation for the velocity distribution. One way is to start with the Navier–Stokes equations written in the appropriate coordinate system (cylindrical coordinates for a circular pipe) and simplify them for fully developed flow. We applied that method to fully developed laminar flow between two large, parallel plates in Example 5.8. You are invited to apply the same method to circular pipe flow. An alternative approach is to derive the differential equation from a force balance on a fluid element. We use this approach because it allows us to delay specialization to laminar flow.

Figure 7.7 shows a cylindrical fluid element of radius r and length δx. The fluid element is coaxial with the pipe, which we assume to be horizontal. Pressure and shear stress on the fluid element are shown. Since mechanical energy loss is always positive, Eq. (7.14) tells us that the pressure drops in the flow direction. Working with pressure drop δp as a positive quantity is convenient, so the pressure at $x + \delta x$ is $p - \delta p$.

Newton's second law for the fluid element is

$$(\delta m)a_x = \sum F_x.$$

With steady and fully developed flow, the fluid element has no acceleration, so

$$\sum F_x = 0.$$

Substituting appropriate expressions for shear and pressure forces, we have

$$(p)\pi r^2 - (p - \delta p)\pi r^2 - \tau(2\pi r\,\delta x) = 0,$$

which simplifies to

$$\tau = \frac{r}{2}\frac{\delta p}{\delta x}. \tag{7.32}$$

For fully developed flow, $\delta p/\delta x$ is a constant, given by

$$\frac{\delta p}{\delta x} = \left|\frac{dp}{dx}\right|,$$

so Eq. (7.32) shows that the shear stress increases linearly with radius. Putting

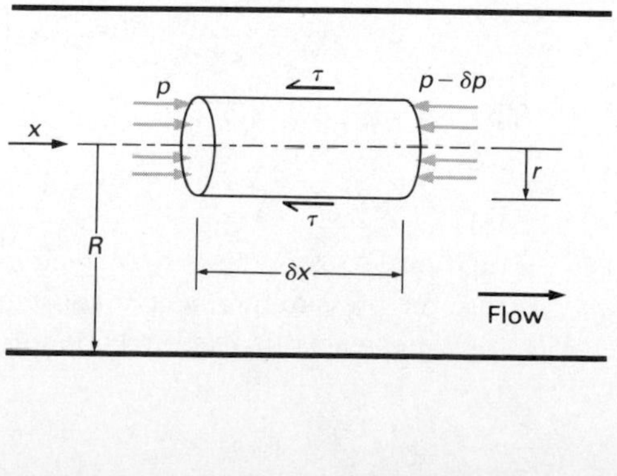

Figure 7.7 Cylindrical fluid element in pipe flow. Flow is from left to right.

$\tau = \tau_w$ at $r = R$ gives

$$\tau_w = \frac{R}{2}\left(\frac{\delta p}{\delta x}\right) = \frac{R}{2}\left(\frac{\Delta p_d}{\Delta x}\right),$$

where Δp_d is the pressure drop in length Δx. This expression is equivalent to Eq. (7.16). Equation (7.32) is valid for both laminar and turbulent flow.

We now specify that the flow is laminar and evaluate the shear stress from Newton's law of viscosity:

$$\tau = -\mu\left(\frac{du}{dr}\right).$$

The minus sign is necessary, because τ has been treated as a positive quantity, whereas du/dr is negative. Substituting into Eq. (7.32) gives

$$\mu\left(\frac{du}{dr}\right) = -\frac{r}{2}\left(\frac{\delta p}{\delta x}\right).$$

Separating variables and integrating yield

$$u = -\frac{1}{4\mu}\left(\frac{\delta p}{\delta x}\right)r^2 + \text{Constant.}$$

The constant of integration is evaluated from the condition

$$u = 0 \quad \text{at} \quad r = R.$$

The resulting velocity distribution is

$$u = \frac{1}{4\mu}\left(\frac{\delta p}{\delta x}\right)(R^2 - r^2). \tag{7.33}$$

The maximum velocity, $u_{\max}$, occurs at $r = 0$ and is

$$u_{\max} = u)_{r=0} = \frac{R^2}{4\mu}\left(\frac{\delta p}{\delta x}\right). \tag{7.34}$$

The average velocity is

$$V = \frac{\int V_n\,dA}{A} = \frac{2\pi \int_0^R ur\,dr}{\pi R^2} = \frac{2\pi \int_0^R (1/4\mu)(\delta p/\delta x)(R^2 - r^2)r\,dr}{\pi R^2}.$$

Evaluating the integral, we have

$$V = \frac{R^2}{8\mu}\left(\frac{\delta p}{\delta x}\right) = \frac{1}{2}u_{\max}. \tag{7.35}$$

The volume flow rate is

$$Q = VA = \frac{\pi R^4}{8\mu}\left(\frac{\delta p}{\delta x}\right). \tag{7.36}$$

Equation (7.36) is the desired relation between pressure drop and flow rate. Because the pressure gradient is constant, δp and δx need not be infinitesimal.

We extend this result to nonhorizontal pipes by replacing p with $p + \gamma z$. As the flow is fully developed, we use Eqs. (7.13) and (7.14) and write

$$(p_1 + \gamma z_1) - (p_2 + \gamma z_2) = \Delta p_L = \rho g h_L = \frac{8\mu L}{\pi R^4} Q, \tag{7.37}$$

Fully developed laminar flow in a circular pipe

where Δp_L and gh_L are the pressure loss and energy loss that occur in a pipe of length L. Equation (7.37) is called the *Hagen–Poiseuille formula* for pressure loss in laminar pipe flow. Note that the equation predicts that the loss is directly proportional to flow rate and hence directly proportional to the fluid's (average) velocity. The reason is that the loss is due to viscous stress only.

We find the friction factor for laminar pipe flow from the Hagen–Poiseuille formula, which can be written*

$$gh_L = \frac{8\nu L}{\pi R^4} Q.$$

But

$$Q = VA = V(\pi R^2),$$

so

$$gh_L = \frac{8\nu VL}{R^2}.$$

From Eq. (7.29a), we have

$$f = \frac{gh_L}{\frac{1}{2}V^2}\left(\frac{D}{L}\right).$$

Substituting $(R = D/2)$ and simplifying give

$$f = \frac{64\mu}{\rho VD} = \frac{64}{\mathbf{R}}. \tag{7.38}$$

Fully developed laminar flow in a circular pipe

The loss coefficient and skin friction coefficient are

$$\mathbf{K} = \left(\frac{L}{D}\right) f = \frac{64}{\mathbf{R}}\left(\frac{L}{D}\right) \tag{7.39}$$

and

$$\mathbf{C_f} = \frac{f}{4} = \frac{16}{\mathbf{R}}. \tag{7.40}$$

Equation (7.38) implies that the friction factor for laminar flow (and hence the loss) does not depend on pipe wall roughness. This surprising result is verified by experiment, as we show in the next section.

---

* Generally use of the kinematic viscosity (ν) is more convenient with energy loss, and use of the dynamic viscosity (μ) is more convenient with pressure loss.

Before closing this discussion of laminar pipe flow, we note that the momentum and kinetic energy correction factors can be calculated by substituting the parabolic velocity profile given by Eqs. (7.33) and (7.35) into the defining equations to obtain

$$\beta = \frac{\int u^2\, dA}{V^2 A} = \frac{4}{3}; \tag{7.41}$$

$$\alpha = \frac{\int u^3\, dA}{V^3 A} = 2. \tag{7.42}$$

Laminar fully developed flow in a circular pipe

Equations (7.37)–(7.42) apply only to fully developed laminar flow, so the Reynolds number must be less than 2300 for them to be used.

EXAMPLE 7.2 Illustrates Laminar Flow in a Circular Pipe

The suggestion is made that the standpipe system in Fig. E7.2 be used as a viscometer. Evaluate this proposal. Neglect energy loss in the entrance nozzle.

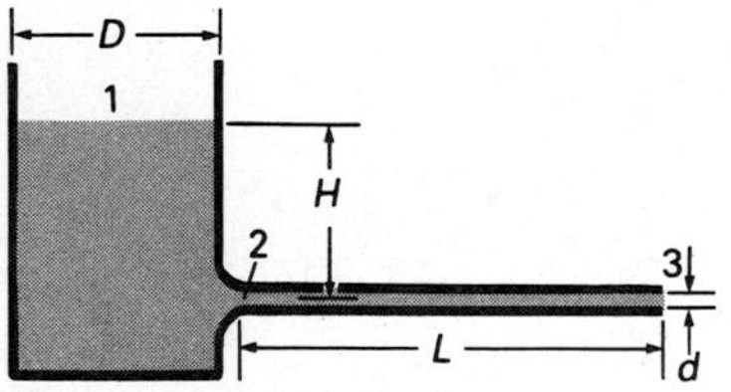

Figure E7.2 Standpipe system used as a viscometer.

SOLUTION

Given

Standpipe system in Fig. E7.2

Find

Whether the standpipe system can be used as a viscometer

Solution

The flow from point 1 to point 2 is assumed inviscid, but viscous effects are important from point 2 to point 3. We analyze each section differently and therefore break the pressure drop $(p_1 - p_3)$ into two parts:

$$p_1 - p_3 = (p_1 - p_2) + (p_2 - p_3).$$

We apply the mechanical energy equation, Eq. (4.48), to a streamline between points 1 and 2 to give

$$\frac{p_1}{\rho} + \alpha_1\left(\frac{V_1^2}{2}\right) + gz_1 - w_s = \frac{p_2}{\rho} + \alpha_2\left(\frac{V_2^2}{2}\right) + gz_2 + gh_L.$$

Now $V_1 \approx 0$, $w_s = 0$, and $z_1 - z_2 = H$. Neglecting losses between points 1 and 2, we have

$$gh_L = 0 \quad \text{and} \quad \alpha_2 = 1.0.$$

Then the mechanical energy equation gives

$$p_1 - p_2 = \rho\frac{V_2^2}{2} - \rho g H.$$

Next we apply the mechanical energy equation between points 2 and 3 to obtain

$$\frac{p_2}{\rho} + \alpha_2\left(\frac{V_2^2}{2}\right) + gz_2 - w_s = \frac{p_3}{\rho} + \alpha_3\left(\frac{V_3^2}{2}\right) + gz_3 + gh_L.$$

Now $w_s = 0$, $\alpha_2 = 1.0$, $z_2 - z_3 = 0$, and $V_2 = V_3$ from continuity. We assume that the flow between 2 and 3 is laminar. Equation (7.42) shows that $\alpha_3 = 2.0$ for fully developed laminar flow. The mechanical energy equation gives

$$p_2 - p_3 = \rho\left(\frac{\alpha_3 V_3^2}{2}\right) - \rho\left(\frac{\alpha_2 V_2^2}{2}\right) + \rho g h_L = \frac{\rho V_2^2}{2} + \rho g h_L.$$

If we assume that the development length is a small fraction of the total length of the pipe, Eq. (7.37) gives

$$\rho g h_L = \frac{8\mu L Q}{(\pi d^4)/16} = \frac{128\mu L Q}{\pi d^4} \quad \text{and} \quad p_2 - p_3 = \frac{\rho V_2^2}{2} + \frac{128\mu L Q}{\pi d^4}.$$

Substituting the expressions obtained for $(p_1 - p_2)$ and $(p_2 - p_3)$ into the equation for $(p_1 - p_3)$ gives

$$p_1 - p_3 = \left(\frac{\rho V_2^2}{2} - \rho g H\right) + \left(\frac{\rho V_2^2}{2} + \frac{128\mu L Q}{\pi d^4}\right).$$

With $p_1 = p_3$, we have

$$\rho V_2^2 - \rho g H + \frac{128\mu L Q}{\pi d^4} = 0.$$

The average velocity at point 2 is

$$V_2 = \frac{Q}{A_2} = \frac{4Q}{\pi d^2}.$$

Substituting this result into the previous equation yields

$$\rho\left(\frac{4Q}{\pi d^2}\right)^2 - \rho g H + \frac{128\mu L Q}{\pi d^4} = 0.$$

Solving for the viscosity μ gives

$$\mu = \frac{\pi d^4 \rho g H}{128 L Q} - \frac{\rho Q}{8\pi L}.$$

ANSWER

This result shows that viscosity can be calculated from known values of d and L and measured values of ρ, H, and Q. The density can be measured with a hydrometer, and Q can be obtained by collecting a volume of fluid in a measured time period. The ratio of tank cross-sectional area to pipe cross-sectional area should be large, so that the height H does not change appreciably during the time period required to collect the fluid.

Discussion

Note that the result is valid only for

- laminar flow ($\mathbf{R} \leq 2300$);
- velocity $V_1 \approx 0$;

- constant-density fluid;
- negligible energy loss in tank and in entrance nozzle;
- fully developed flow at pipe exit;
- development length short compared to overall pipe length; and
- horizontal pipe.

Make sure that you understand the reason(s) that each of these conditions must be satisfied.

Some numerical values would be instructive. An actual experiment was conducted using $H = 1.0$ ft, $D = 6.77$ in., $d = 0.125$ in., $L = 48$ in., and $\rho = 62.3\ \text{lbm/ft}^3$. The fluid was 70°F water. Q was measured as $1.74 \times 10^{-4}\ \text{ft}^3/\text{sec}$. The viscosity is then calculated to be

$$\mu = \frac{\pi(0.125\text{ in.} \times 1\text{ ft}/12\text{ in.})^4(62.3\text{ lbm/ft}^3)(32.2\text{ ft/sec}^2)(1.0\text{ ft})}{128(48\text{ in.} \times 1\text{ ft}/12\text{ in.})(1.74 \times 10^{-4}\text{ ft}^3/\text{sec})(32.2\text{ ft}\cdot\text{lbm/lb}\cdot\text{sec}^2)}$$
$$- \frac{(62.3\text{ lbm/ft}^3)(1.74 \times 10^{-4}\text{ ft}^3/\text{sec})}{8\pi(48\text{ in.} \times 1\text{ ft}/12\text{ in.})(32.2\text{ ft}\cdot\text{lbm/lb}\cdot\text{sec}^2)}$$
$$= 2.26 \times 10^{-5}\text{ lb}\cdot\text{sec/ft}^2,$$

which compares reasonably well with a value of $2.08 \times 10^{-5}\ \text{lb}\cdot\text{sec/ft}^2$ from Table A.6.

If we use the tabulated value of the kinematic viscosity from Table A.6, we get the Reynolds number for the experiment:

$$\mathbf{R} = \frac{\rho V d}{\mu} = \frac{Q}{(\pi d^2)/4}\left(\frac{\rho d}{\mu}\right) = \frac{4\rho Q}{\pi \mu d}$$
$$= \frac{4(62.3\text{ lbm/ft}^3)(1.74 \times 10^{-4}\text{ ft}^3/\text{sec})(12\text{ in./ft})}{\pi(2.08 \times 10^{-5}\text{ lb}\cdot\text{sec/ft}^2)(\frac{1}{8}\text{ in.})(32.2\text{ ft}\cdot\text{lbm/lb}\cdot\text{sec}^2)} = 1980.$$

Therefore the flow was laminar. Would you expect better agreement between the experimental value and the Table A.6 value if the pipe length were longer or shorter? Explain.

7.2.4 Turbulent Flow in a Circular Pipe: Experimental Information and the Moody Chart

Because our differential analysis of laminar pipe flow was so successful, we might try to analyze turbulent flow in the same way. However, we encounter a problem when we must substitute an expression for the shear stress into Eq. (7.32), because in turbulent flow the shear stress is composed of both viscous and turbulent parts. The turbulent stresses can be evaluated only from semi-empirical models that are based on experimental data. Much of the experimental data used to "calibrate" turbulent stress models was obtained from pipe flow experiments. As we must use experimental data for the turbulent stress, we might just as well rely on experimental data for the friction factor itself.

Pipe flow experiments have been conducted from as early as the mid-1800s. With hindsight, we can condense the available information and combine physical reasoning with experimental evidence to obtain working information about turbulent pipe flow.

Friction Factor Data. According to Eq. (7.27), we expect the friction factor for turbulent pipe flow to depend on Reynolds number and relative roughness. Early experiments, including those of Reynolds, verified the roughness effect. At a given Reynolds number, rough pipes have higher energy loss than smooth pipes. This condition is easily explained by the fact that roughness disturbs the flow, promoting more turbulence.

Experiments have shown that Eq. (7.27) has two limiting forms. For smooth pipe ($\varepsilon/D \approx 0$),

$$f = F_s\{\mathbf{R} \text{ only}\}. \tag{7.43}$$

For very rough pipe and high Reynolds number, we expect the flow to be highly turbulent. The turbulent stress would be much larger than the viscous stress and the loss would depend only on inertia forces. Because **R** is the ratio of inertia force to viscous force and because viscous force is irrelevant for highly turbulent flow, the exact value of the Reynolds number becomes irrelevant. For very rough pipe,

$$f = F_r\{\varepsilon/D \text{ only}\}. \tag{7.44}$$

A crucial question that we must answer in dealing with rough pipe is how to determine the surface roughness (ε). Most surfaces do not have uniform roughness. Pipes made of riveted steel or corrugated sheet metal have regular patterns of roughness. Pipes made of cast iron or concrete have random roughness distributions. How to quantify all types of roughness by a single parameter ε is far from clear.

Much of our detailed information about friction factor in rough pipes is based on experiments performed by J. Nikuradse in 1933 [1]. Nikuradse solved the problem of quantifying the roughness by using artificially roughened pipes. By a screening process, Nikuradse obtained sand grains of various sizes. He glued these sand grains to the inside of various pipes, producing a pipe wall roughness similar to sandpaper. Knowing the sand grain size, he could accurately determine the parameter

$$\frac{\varepsilon_s}{D}.$$

The subscript s reminds us that the roughness elements are sand grains of uniform size.

Some results of Nikuradse's experiments are shown in Fig. 7.8. The friction factor f is plotted versus the Reynolds number **R** with ε_s/D as a parameter. Logarithmic coordinates are used because f and **R** change by several orders of magnitude. The straight line labeled H–P represents the Hagen–Poiseuille equation

$$f = \frac{64}{\mathbf{R}}.$$

The straight line labeled B represents an early attempt by H. Blasius to correlate data on turbulent friction factor in smooth pipes. His equation,

$$f_{\text{smooth}} \approx 0.3164\mathbf{R}^{-1/4}, \tag{7.45}$$

is reasonably accurate for Reynolds numbers between 4000 and 10^5. The experimental data provide excellent verification of several points already discussed.

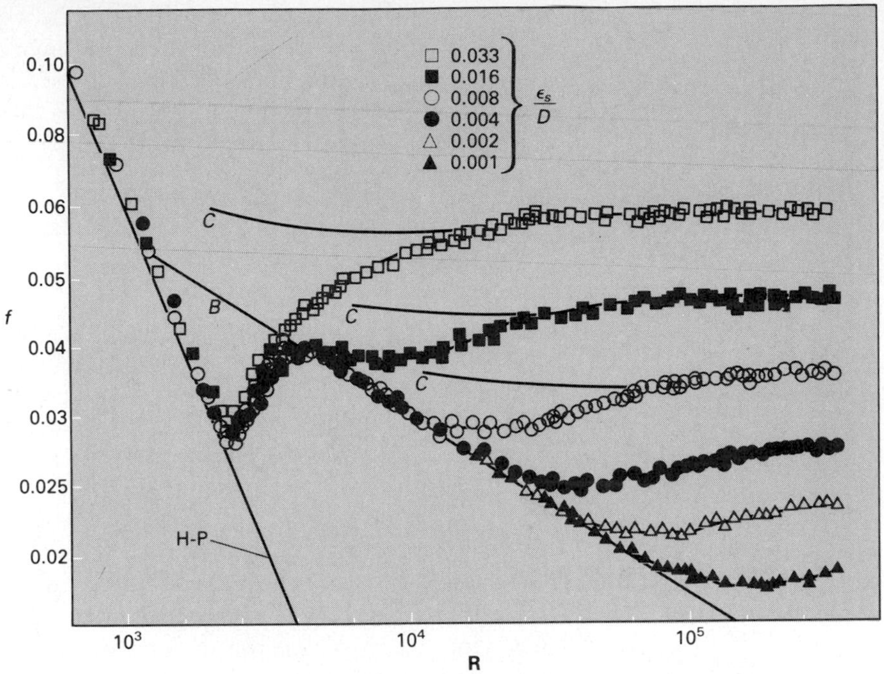

Figure 7.8 Friction factor for sand-roughened circular pipes as a function of Reynolds number and relative roughness. Based on the Nikuradse's tests [1].

- For low **R** (laminar flow), the Hagen–Poiseuille equation is accurate, and pipe roughness has no effect on f.
- For a given relative roughness, f becomes constant if **R** is large enough.
- For smooth pipe (ε_s/D small), f depends on **R** but not on ε_s/D.

The experimental data also show that pipe smoothness depends on Reynolds number. Up to a Reynolds number of about 4000, all pipes follow the same curve, indicating that they behave as if they were smooth. As the Reynolds number increases, pipes with different roughness break away from the group at various Reynolds numbers. Rougher pipes break away at lower Reynolds numbers.

One rather puzzling aspect of Nikuradse's data is the dip in f between Reynolds numbers of 2000 and 4000. This dip is characteristic of the uniform sand roughness used by Nikuradse and is not observed in commercial pipe. The curves labeled C in Fig. 7.8 are more typical of commercial pipe.

Knowledge gained from careful experiments, such as Nikuradse's, has aided development of semi-empirical theories and working charts for pipe friction factors. In 1935, L. Prandtl deduced the following formula for the friction factor in turbulent flow in smooth pipe:

$$\frac{1}{\sqrt{f}} = 2.0 \log(\mathbf{R}\sqrt{f}) - 0.8. \tag{7.46}$$

At about the same time, T. von Karman derived the following relation for friction factor for completely turbulent (very rough pipe) flow:

$$\frac{1}{\sqrt{f}} = -2.0 \log\left(\frac{\varepsilon/D}{3.7}\right). \tag{7.47}$$

Equations (7.46) and (7.47) are the limiting forms implied by Eqs. (7.43) and (7.44). The development of both equations is traced in Section 7.2.7.

In 1939, C. F. Colebrook [2] devised a formula that combines the smooth and rough limits, Eqs. (7.46) and (7.47), and also covers the transition range. The *Colebrook formula* is

$$\frac{1}{\sqrt{f}} = -2.0 \log\left(\frac{\varepsilon/D}{3.7} + \frac{2.51}{\mathbf{R}\sqrt{f}}\right), \tag{7.48a}$$

or, in slightly more convenient form,

$$f = \frac{0.25}{\{\log[(\varepsilon/D)/3.7 + 2.51/(\mathbf{R}\sqrt{f})]\}^2}. \tag{7.48b}$$

The difficulty with the Colebrook formula is that it is implicit in f. To overcome this difficulty, L. F. Moody [3], in 1944, published a chart of f versus $\mathbf{R}$ and ε/D based on the Colebrook formula. This chart, reproduced in Fig. 7.9, is called a *Moody chart*. Before the days of computers and calculators, the Moody chart was the primary source of information for friction factor, but the engineer of today and tomorrow might just as easily use the Colebrook formula.

There is one very big gap in the pipe flow information we have developed so far. The Nikuradse data and the von Karman and Colebrook formulas are based on a uniform sand grain roughness ε_s. But the engineer needs information on the roughness of commercial pipe. Colebrook and Moody have provided the necessary information. Figure 7.10 reproduces a second chart from Moody's paper. This chart gives the equivalent relative roughness (ε/D) of various types of commercial pipe and is based on information published by Colebrook.

Table 7.1 shows equivalent absolute roughness (ε) for various materials. The roughness values listed in Table 7.1 and implied in Fig. 7.10 have little to do with the actual *geometric* roughness of the surface. The given roughnesses are the values that make the Colebrook equation work; they were obtained by "backing out" a value of ε/D from pressure drop data on commercial pipe. The values of ε/D apply to clean, new pipe only. Pipe that has been in service for a long time experiences corrosion or scale buildup and may have an equivalent roughness that is orders of magnitude larger than the value implied by Table 7.1 and Fig. 7.10. Because of uncertainty in evaluating equivalent roughness, values of f from the Moody charts are accurate to only about 10 percent. Accordingly, the use of several significant figures in pipe flow calculations is *unjustified*.*

Now that we have discussed the friction factor for turbulent flow, we compare the effect of various pipe flow parameters on the energy loss for laminar and for fully turbulent flow. This comparison is presented in Table 7.2 and is based on the following formulas, adapted from Eqs. (7.29a), (7.38), and (7.47):

$$gh_L = \frac{64}{\mathbf{R}}\left(\frac{L}{D}\right)\left(\frac{V^2}{2}\right) = \frac{32\mu VL}{\rho D^2} \qquad \text{Laminar flow}$$

$$gh_L = \left\{\frac{0.5}{\log[3.7/(\varepsilon/D)]}\right\}^2\left(\frac{L}{D}\right)\left(\frac{V^2}{2}\right) \qquad \text{Completely turbulent flow}$$

---

* For this reason, more than two significant figures in the calculation of $\mathbf{R}$ usually is not justified.

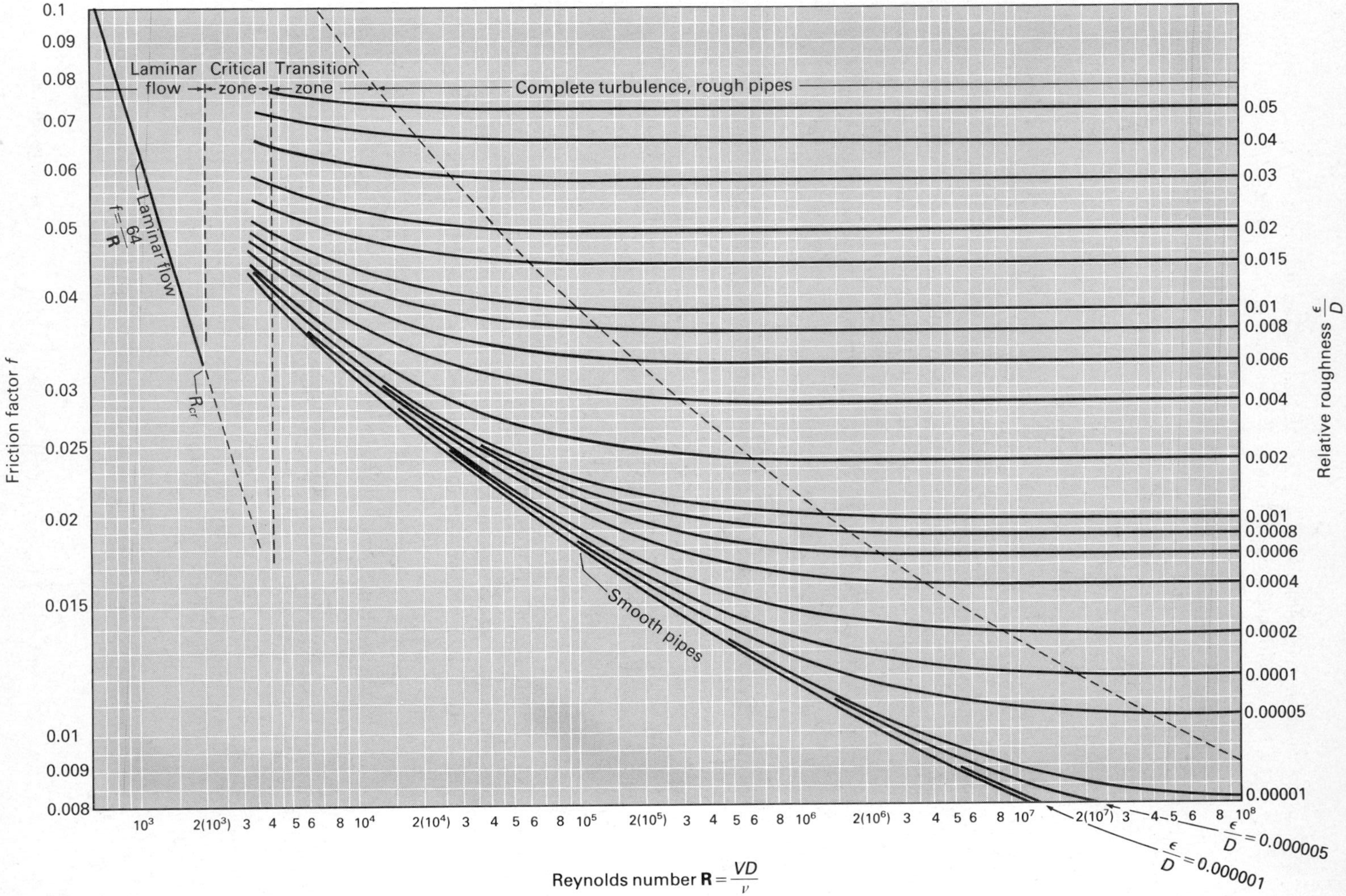

Figure 7.9 The Moody chart for friction factor (from [3]).

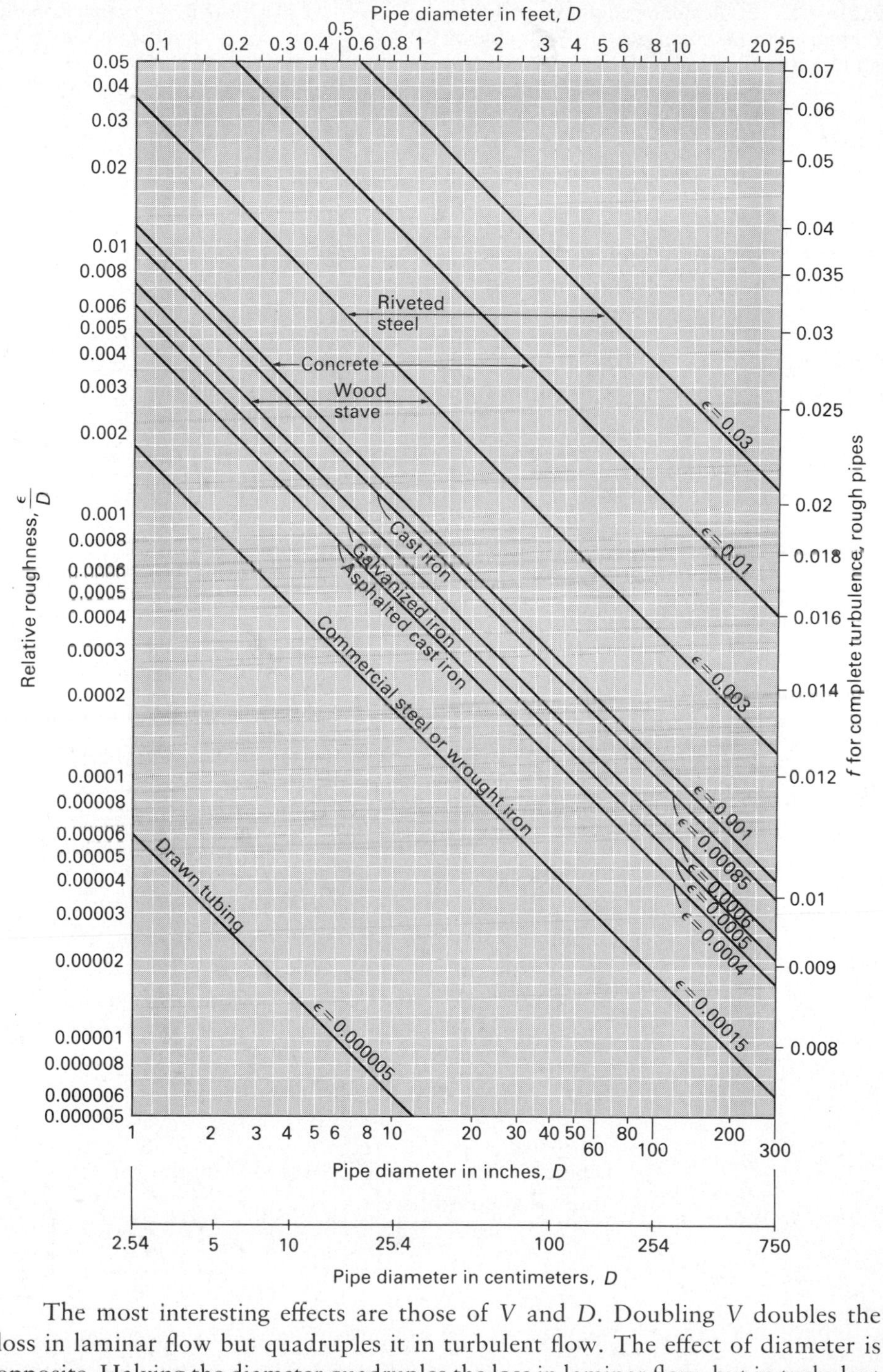

Figure 7.10 Moody's chart for relative roughness of commercial pipes (from [3]).

The most interesting effects are those of V and D. Doubling V doubles the loss in laminar flow but quadruples it in turbulent flow. The effect of diameter is opposite. Halving the diameter quadruples the loss in laminar flow, but in turbulent flow halving the diameter slightly more than doubles the loss. Note also that energy loss depends on both density and viscosity in laminar flow but is independent of

Table 7.1 Equivalent roughness of new commercial pipe materials (after Colebrook [2] and Moody [3]).

Material	ε (ft)	ε (mm)
Riveted steel	0.003–0.03	0.9–9.0
Concrete	0.001–0.01	0.3–3.0
Wood stave	0.0006–0.003	0.18–0.9
Cast iron	0.00085	0.26
Galvanized iron	0.0005	0.15
Asphalted cast iron	0.0004	0.12
Commercial steel or wrought iron	0.00015	0.046
Drawn tubing	0.000005	0.0015
Glass, plastic	Smooth	Smooth

Table 7.2 Comparison of energy loss dependence in laminar and completely turbulent pipe flow.

Parameter	Laminar Flow	Completely Turbulent Flow
Velocity (V)	V	V^2
Density (ρ)	$1/\rho$	No dependence
Viscosity (μ)	μ	No dependence
Diameter (D)	$1/D^2$	$\approx 1/D$
Length (L)	L	L
Roughness (ε)	No dependence	Depends on ε/D

both in fully turbulent flow. If we consider pressure loss instead of energy loss, we find that pressure loss depends on viscosity but not on density in laminar flow and on density but not on viscosity in fully turbulent flow.

Power Law Velocity Profiles. Our laminar pipe flow analysis gave an accurate equation for the velocity profile. An equation for the velocity profile in turbulent pipe flow is also useful.

Figure 7.11 shows velocity profiles for laminar and turbulent pipe flow. In Fig. 7.11(a), the profiles have the same *average* velocity, whereas in Fig. 7.11(b), the profiles have the same *maximum* velocity. The turbulent velocity profiles are much fuller than the laminar profiles and are relatively flat over most of the pipe, falling rapidly to zero velocity as the wall is approached.

Turbulent velocity profiles may be represented by a rather complicated logarithmic function (see Section 7.2.7) that has a firm physical basis. An alternative approximation for turbulent flow velocity profiles is the *power law formula:*

$$\frac{u}{u_{\max}} \approx \left(\frac{y}{R}\right)^{1/n}, \tag{7.49}$$

where y is the distance measured from the wall toward the centerline. In terms of the radial coordinate r,

$$\frac{u}{u_{\max}} \approx \left(\frac{R-r}{R}\right)^{1/n}. \tag{7.50}$$

The power law formula fits most of the profile but is unsatisfactory near the wall and the centerline. The value of n is often taken as 7; however, a single value of n cannot represent all conditions of Reynolds number and roughness. Table 7.3 gives values of n for various Reynolds numbers for smooth pipe. A correlation

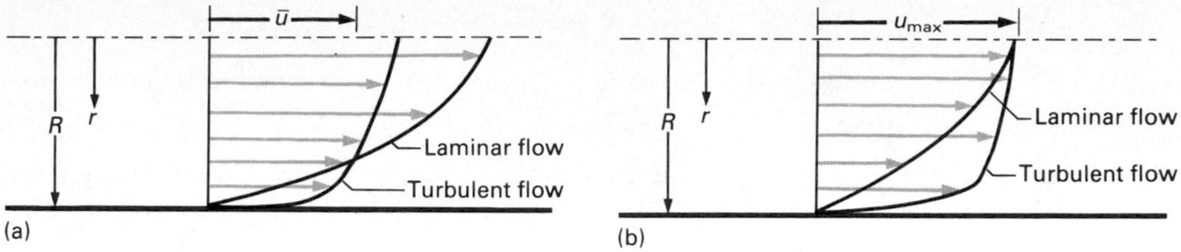

Figure 7.11 Comparison of laminar and turbulent velocity profiles for pipe flow: (a) same average velocity; (b) same maximum velocity.

that is reasonably accurate for smooth or rough pipes is

$$n \approx \frac{1}{\sqrt{f}}. \tag{7.51}$$

The power law profile may be used to evaluate momentum and kinetic energy correction factors for fully developed turbulent pipe flow. Substitution of Eq. (7.49) into the equations

$$V = \frac{\int u\,dA}{A}, \qquad \beta = \frac{\int u^2\,dA}{V^2 A}, \quad \text{and} \quad \alpha = \frac{\int u^3\,dA}{V^3 A}$$

gives

$$\frac{V}{u_{max}} = \frac{2n^2}{(n+1)(2n+1)}, \tag{7.52}$$

$$\beta = \frac{(n+1)(2n+1)^2}{4n^2(n+2)}, \tag{7.53}$$

and

$$\alpha = \frac{(n+1)^3(2n+1)^3}{4n^4(n+3)(2n+3)}. \tag{7.54}$$

Table 7.4 gives values of α and β for various values of n.

Table 7.3 Approximate values of power law profile parameter (n) for smooth pipe.

R	4×10^3	2.3×10^4	1.1×10^5	1.1×10^6	2×10^6	3.2×10^6
n	6.0	6.6	7.0	8.8	10	10

Table 7.4 Values of momentum and kinetic energy correction factors for various values of power law profile parameter (n).

n	5	6	7	8	9
β	1.037	1.027	1.020	1.016	1.013
α	1.106	1.077	1.058	1.046	1.037

7.2.5 Three Types of Pipe Flow Calculations

Engineering analysis of pipe flows and piping systems normally has one of three objectives:

- evaluating the energy loss or pressure drop for a given pipe or system and a known flow rate;
- calculating the flow rate or fluid velocity for a given pipe or system and a prescribed energy loss or pressure drop; or
- determining the appropriate pipe size (diameter) to convey a given flow rate with a prescribed energy loss or pressure drop.

In all three cases, the pipe and system configuration is known (i.e., pipe length, elevation change, and pipe material, as well as the number, type, and location of any fittings and valves) and the fluid is specified. With this information, the engineer bases the calculations on the relationships among energy loss, flow rate, and pipe size. In a complete system, the calculations must include the effects of fittings and valves and losses from pipe friction. In this section, we consider energy loss *from pipe friction only*. Such calculations are often sufficiently accurate for very long pipes with relatively few local losses or if the pipe friction energy loss can be isolated from other losses.

If pipe friction losses only are considered, all three types of calculation are based on the Darcy–Weisbach equation, Eq. (7.29a):

$$gh_L = f\left(\frac{L}{D}\right)\frac{V^2}{2},$$

with friction factors evaluated from Eq. (7.38) for laminar flow or the Colebrook formula, Eq. (7.48), for turbulent flow. The graphic equivalent of the friction factor formulas, the Moody chart, can also be used. For laminar flow, friction factor calculations can be bypassed with the Hagen–Poiseuille formula.

Flow rate and velocity are related by

$$Q = VA = \frac{\pi D^2}{4} V. \tag{7.55}$$

Laminar Flow. The three calculations are simple if the flow is laminar, because Eqs. (7.37) and (7.55) can be combined and solved for any single variable in terms of the others. The resulting equations are as follows:

- Energy/pressure loss:

$$gh_L = \frac{\Delta p_L}{\rho} = \frac{32\mu LV}{\rho D^2} = \frac{128\nu LQ}{\pi D^4}. \tag{7.56}$$

- Velocity/flow rate:

$$V = \frac{\Delta p_L D^2}{32\mu L} = \frac{(gh_L)D^2}{32\nu L} \tag{7.57}$$

or

$$Q = \frac{\pi \Delta p_L D^4}{128 \mu L} = \frac{\pi (g h_L) D^4}{128 \nu L}. \tag{7.58}$$

- Pipe sizing:

$$D = \left(\frac{32 \mu L V}{\Delta p_L} \right)^{1/2} = \left(\frac{32 \nu L V}{g h_L} \right)^{1/2} \tag{7.59}$$

or

$$D = \left(\frac{128 \mu L Q}{\pi \Delta p_L} \right)^{1/4} = \left(\frac{128 \nu L Q}{\pi g h_L} \right)^{1/4}. \tag{7.60}$$

Limitations:

These equations all apply if $\mathbf{R} < 2300$.

Turbulent Flow. If the flow is turbulent, the calculations are more difficult because V and D appear explicitly in the friction factor equations and the Moody chart. Both are in the Reynolds number. The term D also appears in the relative roughness. In addition to these complications, the Colebrook formula is implicit in f, which led to the popularity of the Moody chart that gives f directly if $\mathbf{R}$ and ε/D are known, whereas the Colebrook formula requires iteration. However, iteration is not as serious a drawback as it used to be, because calculators and computers now perform iterative calculations quickly.

Energy Loss/Pressure Drop in Turbulent Flow. Known information includes the fluid density and viscosity, pipe length, roughness, and diameter, and fluid velocity or flow rate. The Reynolds number and relative roughness can be calculated and the friction factor can be read from the Moody chart or calculated iteratively using the Colebrook formula, Eq. (7.48b). The energy loss, head loss, or pressure loss is then calculated from Eqs. (7.29a), (7.29b), or (7.30) as desired. The pressure drop is

$$\Delta p_d = \Delta p_L - \gamma (z_1 - z_2). \tag{7.61}$$

EXAMPLE 7.3 Illustrates the Energy/Pressure Loss Problem for Turbulent Flow

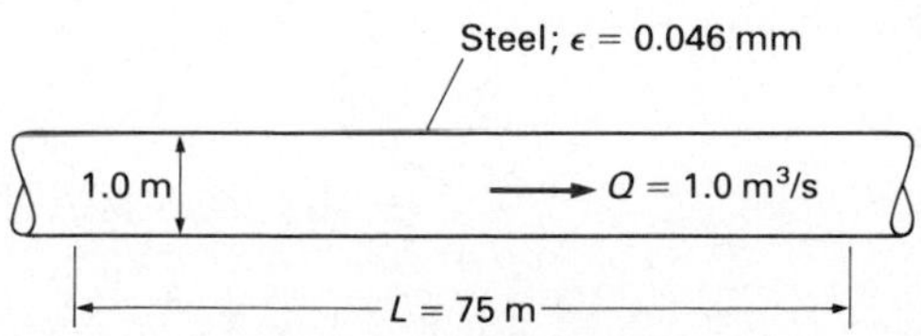

Figure E7.3 Horizontal steel pipe.

A horizontal commercial steel pipe, shown in Fig. E7.3, has an inside diameter of 1.0 m. Water at 20°C flows through the pipe at the rate of 1.0 m^3/s. Find the pressure drop and the energy loss in a 75-m length of the pipe. Assume fully developed flow.

SOLUTION

Given

Figure E7.3

Inside diameter 1.0 m

Length 75 m

Water flow rate 1.0 m^3/s

Water temperature 20°C

Find

Pressure drop

Energy loss

Solution

We solve this problem using two methods: the Moody chart and the Colebrook formula. In both methods, we find the energy loss from the Darcy–Weisbach formula:

$$gh_L = f\left(\frac{L}{D}\right)\left(\frac{V^2}{2}\right).$$

The relative roughness for the pipe from Fig. 7.10 is

$$\frac{\varepsilon}{D} = \frac{0.000046 \text{ m}}{1.0 \text{ m}} = 0.000046$$

Obtaining the kinematic viscosity from Table A.5, we calculate the Reynolds number:

$$\mathbf{R} = \frac{4Q}{\pi\nu D} = \frac{4(1.0 \text{ m}^3/\text{s})}{\pi(1.00 \times 10^{-6} \text{ m}^2/\text{s})(1.0 \text{ m})} = 1.27 \times 10^6.$$

The Moody chart gives $f = 0.0120$. Next we use Eq. (7.55) to obtain

$$V = \frac{Q}{A} = \frac{4Q}{\pi D^2} = \frac{4(1.0 \text{ m}^3/\text{s})}{\pi(1.0 \text{ m})^2} = 1.27 \text{ m/s}.$$

The Darcy–Weisbach formula gives

$$gh_L = 0.0120\left(\frac{75 \text{ m}}{1.0 \text{ m}}\right)\frac{(1.27 \text{ m/s})^2}{2},$$

or

$$gh_L = 0.73 \text{ m}^2/\text{s}^2.$$ **ANSWER**

The pressure drop is

$$\Delta p_d = \Delta p_L - \gamma(z_1 - z_2),$$

where

$$\Delta p_L = \rho g h_L.$$

For horizontal pipe, $z_1 = z_2$ and $\Delta p_d = \rho(gh_L)$. Table A.5 gives $\rho = 998 \text{ kg/m}^3$, so

$$\Delta p_d = (998 \text{ kg/m}^3)(0.73 \text{ N}\cdot\text{m/kg}) = 730 \text{ N/m}^2,$$

or

$$\Delta p_d = 730 \text{ Pa}.$$ **ANSWER**

To use the Colebrook formula, we first calculate the relative roughness and the Reynolds number, obtaining

$$\frac{\varepsilon}{D} = 0.000046 \quad \text{and} \quad \mathbf{R} = 1.27 \times 10^6.$$

Because the Colebrook formula is implicit in f, we use fixed-point iteration, which converges rapidly to an accurate value for f:

- Guess a value of f. The "fully turbulent" value for the given ε/D is usually a good guess if the pipe is rough. The Blasius formula, Eq. (7.45), gives a good first guess for smooth pipe ($f \approx 0.02$ is often a satisfactory first guess).
- Calculate an improved value of f from

$$f_{\text{new}} = \frac{0.25}{\{\log[(\varepsilon/D)/3.7 + 2.51/(\mathbf{R}\sqrt{f_{\text{old}}})]\}^2}.$$

- Repeat the iteration until you are satisfied that f_{new} is close enough to f_{old} (usually two iterations are sufficient).

As ε/D is small for our problem, we use the Blasius formula to make a first guess for f:

$$f \approx \frac{0.3164}{\mathbf{R}^{0.25}} = \frac{0.3164}{(1.27 \times 10^6)^{0.25}} = 0.0094.$$

Iterating gives

$$f_{\text{new}} = \frac{0.25}{\{\log[0.000046/3.7 + 2.51/(1.27 \times 10^6\sqrt{0.0094})]\}^2} = 0.0124.$$

Using this value of f on the right-hand side gives

$$f_{\text{new}} = \frac{0.25}{\{\log[0.000046/3.7 + 2.51/(1.27 \times 10^6\sqrt{0.0124})]\}^2} = 0.0122,$$

and, repeating the process,

$$f_{\text{new}} = \frac{0.25}{\{\log[0.000046/3.7 + 2.51/(1.27 \times 10^6\sqrt{0.0122})]\}^2} = 0.0122.$$

Convergence has occurred. Using this value of f (the same value we obtained from the Moody chart) with the Darcy–Weisbach and pressure drop equations, we have

$$gh_L = 0.73 \text{ m}^2/\text{s}^2 \qquad \textbf{ANSWER}$$

and

$$\Delta p_d = 730 \text{ Pa}. \qquad \textbf{ANSWER}$$

Discussion

Note that pressure *drop* and pressure *loss* are equal, because the pipe is horizontal and the flow is fully developed. How would the pressure drop and pressure loss change if the pipe were vertical ($z_1 - z_2 = L$)?

Flow Rate/Velocity Calculation for Turbulent Flow. The velocity is unknown and so the Reynolds number cannot be calculated until the entire problem is solved. Presumably, we have evidence that the flow is turbulent or a preliminary calculation has been made using Eq. (7.57) or Eq. (7.58) to verify that the flow is not laminar. We must know **R** in order to use the Moody chart (Fig. 7.9), so we are forced to use iterations if we want to use the chart. Iterations, however, are unnecessary, because we can cast the Colebrook formula in a form that permits direct calculation.

Recalling that gh_L, ε, and D are given, we define a new Reynolds number,

$$\mathbf{R_f} \equiv \sqrt{f}\,\mathbf{R},$$

and write the Colebrook formula:

$$f = \frac{0.25}{\{\log[(\varepsilon/D)/3.7 + 2.51/\mathbf{R_f}]\}^2}. \tag{7.62}$$

Evaluating $\mathbf{R_f}$ in terms of dimensional parameters, we have

$$\mathbf{R_f} = \sqrt{\frac{gh_L}{\frac{1}{2}V^2}\left(\frac{D}{L}\right)}\left(\frac{\rho VD}{\mu}\right) = \left[\frac{2(gh_L)D^3}{\nu^2 L}\right]^{1/2}. \tag{7.63}$$

$\mathbf{R_f}$ does not contain velocity and can be calculated from the given data. To solve the velocity/flow rate problem, we

- compute $\mathbf{R_f}$ by substituting given values of gh_L, D, ν, and L into Eq. (7.63);
- compute f from Eq. (7.62);
- compute V from the Darcy–Weisbach equation,*

$$V = \sqrt{\frac{2(gh_L)D}{fL}}, \tag{7.64}$$

or from

$$V = \frac{\nu \mathbf{R_f}}{D\sqrt{f}}; \tag{7.65}$$

- if desired, calculate Q from Eq. (7.55).

Moody's original chart had an auxiliary scale for $\mathbf{R_f}$ so that it could be used for direct calculation of velocity. We omitted this scale from our Moody chart in order to avoid clutter. We recommend the direct calculation using Eqs. (7.62)–(7.65) for flow rate/velocity calculation.

* Recall that gh_L in these equations represents energy loss from pipe friction only.

EXAMPLE 7.4 **Illustrates Velocity/Flow Rate Calculation for Turbulent Flow**

Water flows steadily down the inclined, 3/4-in.-inside-diameter copper pipe shown in Fig. E7.4. Two pressure gages 100 ft apart indicate

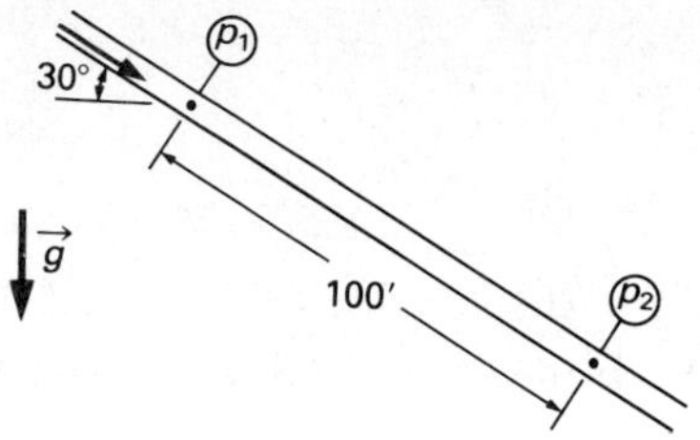

Figure E7.4 Flow through inclined pipe.

identical pressures. Find the volume flow rate. Assume that the flow is fully developed.

SOLUTION

Given

Figure E7.4

Fully developed flow in an inclined, 3/4-in.-inside-diameter copper pipe

$p_1 = p_2$

Find

Volume flow rate

Solution

We solve this problem first by using the Moody chart and then by using the Colebrook formula. Both methods start with the steady-state mechanical energy equation between 1 and 2:

$$\frac{p_1}{\rho} + \alpha_1\left(\frac{V_1^2}{2}\right) + gz_1 - w_s = \frac{p_2}{\rho} + \alpha_2\left(\frac{V_2^2}{2}\right) + gz_2 + gh_L.$$

In this equation, $p_1 = p_2$, $\alpha_1 = \alpha_2$, $V_1 = V_2$, and $w_s = 0$. Then we have

$$gh_L = g(z_1 - z_2).$$

Both methods also use the relative roughness for the copper pipe (drawn tubing). Using Table 7.1, we have

$$\frac{\varepsilon}{D} = \frac{0.000005\ \text{ft}}{0.75\ \text{in.}}\left(\frac{12\ \text{in.}}{\text{ft}}\right) = 8.0 \times 10^{-5}.$$

For the Moody chart method, we use Eq. (7.29a) for gh_L to get

$$f\left(\frac{L}{D}\right)\left(\frac{V^2}{2}\right) = g(z_1 - z_2).$$

Solving for V gives

$$V = \sqrt{\frac{2g(z_1 - z_2)D}{fL}}.$$

We do not know f, so iteration is necessary. Guessing that $f = 0.01$, we calculate V to be

$$V = \sqrt{\frac{2(32.2\ \text{ft/sec}^2)(100\ \text{ft})\sin 30°\ (0.75\ \text{in.})}{0.01(100\ \text{ft})(12\ \text{in./ft})}} = 14.2\ \text{ft/sec}.$$

For 60°F water, Table A.6 gives $\nu = 1.22 \times 10^{-5}\ \text{ft}^2/\text{sec}$. The Reynolds number is

$$\mathbf{R} = \frac{VD}{\nu} = \frac{(14.2\ \text{ft/sec})(0.75\ \text{in.})}{(1.22 \times 10^{-5}\ \text{ft}^2/\text{sec})(12\ \text{in./ft})} = 7.27 \times 10^4.$$

The Moody chart, Fig. 7.9 gives

$$f = 0.019.$$

Two more iterations finally give

$V = 9.84$ ft/sec, $\mathbf{R} = 5.1 \times 10^4$, and $f = 0.0208$.

The volume flow rate is

$$Q = VA = (9.84 \text{ ft/sec}) \frac{\pi(0.75 \text{ in.})^2}{4(144 \text{ in}^2/\text{ft}^2)};$$

$$Q = 0.0302 \text{ ft}^3/\text{sec}.$$ **ANSWER**

Using the Colebrook formula to solve the problem, we substitute

$$gh_L = g(z_1 - z_2)$$

into Eq. (7.63) and obtain

$$\mathbf{R_f} = \left[\frac{2g(z_1 - z_2)D^3}{\nu^2 L}\right]^{1/2}.$$

Substituting the known values gives

$$\mathbf{R_f} = \left[\frac{2(32.2 \text{ ft/sec}^2)(100 \text{ ft}) \sin 30^\circ \, (0.75 \text{ in.})^3}{(1.22 \times 10^{-5} \text{ ft}^2/\text{sec})^2(100 \text{ ft})(12 \text{ in./ft})^3}\right]^{1/2} = 7270.$$

Equation (7.48b) gives the friction factor:

$$f = \frac{0.25}{[\log(0.00008/3.7 + 2.51/7270)]^2} = 0.0212.$$

Equation (7.64) gives the velocity:

$$V = \sqrt{\frac{2g(z_1 - z_2)D}{fL}}$$

$$= \sqrt{\frac{2(32.2 \text{ ft/sec}^2)(100 \text{ ft}) \sin 30^\circ \, (0.75 \text{ in.})}{0.0212(100 \text{ ft})(12 \text{ in./ft})}} = 9.74 \text{ ft/sec}.$$

The volume flow rate is

$$Q = VA = (9.74 \text{ ft/sec}) \frac{\pi(0.75 \text{ in.})^2}{4(144 \text{ in}^2/\text{ft}^2)};$$

$$Q = 0.0300 \text{ ft}^3/\text{sec}.$$ **ANSWER**

Discussion

The answers are essentially equal, because the two methods are equivalent. However, the Moody chart method requires iteration. The Colebrook method does not and therefore simplifies solving this problem.

Note that there is no net pressure drop in the direction of flow. This intriguing phenomenon can be explained in terms of either energy or force.

In terms of energy, the work required to overcome the mechanical energy loss is provided by the decrease in potential energy. In terms of forces, the gravitational force exactly balances the resisting force from shear stress at the pipe walls.

In this example, the static pressures at planes 1 and 2 are equal. However, the pressure could increase or decrease from plane 1 to plane 2. If the pressure increases, the resisting force at the walls is less than the gravitational force. If the pressure decreases, both static pressure drop and the gravitational force are needed to overcome the wall resistance.

Pipe Sizing in Turbulent Flow. The diameter is unknown, so neither the Reynolds number nor the relative roughness can be calculated. Presumably, external evidence indicates that the flow is turbulent, or a preliminary calculation using Eq. (7.59) or Eq. (7.60) has verified that the flow is not laminar. The pipe diameter can be calculated from given values of V (or Q), L, and gh_L once the friction factor is known. Unfortunately, both **R** and ε/D are required in order to use the Moody chart, so iterations are necessary if we want to use the chart. As with the flow rate/velocity calculation, we formulate a direct calculation method that bypasses the Moody chart.

When working a pipe sizing problem, we may know either V or Q, but not both, because the diameter would then be determined solely by the continuity equation. We assume that Q is given here, because that is the more common occurrence in practice.

The key to direct solution of a pipe sizing problem is to rearrange the Colebrook formula by introducing new dimensionless parameters in such a way that the diameter appears in only one of them. Consider the following dimensionless parameters:

$$\mathbf{R} = \frac{VD}{\nu} = \frac{4Q}{\pi\nu D}; \tag{7.66}$$

$$f' \equiv f(\mathbf{R})^5 = \left(\frac{gh_L}{8Q^2}\right)\left(\frac{\pi^2 D^5}{L}\right)\left(\frac{4Q}{\pi\nu D}\right)^5 = \frac{128(gh_L)Q^3}{\pi^3 L\nu^5}; \tag{7.67}$$

$$e = \frac{4}{\pi}\left(\frac{\varepsilon}{D}\right)\left(\frac{1}{\mathbf{R}}\right) = \frac{4}{\pi}\left(\frac{\varepsilon}{D}\right)\left(\frac{\pi\nu D}{4Q}\right) = \frac{\varepsilon\nu}{Q}. \tag{7.68}$$

Note that f' and e can be calculated from given parameters and neither contains D. Substituting

$$f = f'\mathbf{R}^{-5} \quad \text{and} \quad \frac{\varepsilon}{D} = \frac{\pi \mathbf{R} e}{4}$$

into the Colebrook formula and rearranging give

$$\mathbf{R} = \left[-2.0\sqrt{f'}\,\log\left(\frac{\pi e \mathbf{R}}{14.8} + \frac{2.51}{\sqrt{f'}}\mathbf{R}^{1.5}\right)\right]^{0.4}. \tag{7.69}$$

Equation (7.69) is implicit in **R**, the only parameter that contains the unknown diameter, so it seems that we have to solve the problem by iteration, in which case all the advantages that we hoped to gain by reformulating the Colebrook formula would be lost. Fortunately, White [4] has shown that the relation implied by Eq. (7.69) is only weakly dependent on pipe roughness and can be approximated by

$$\mathbf{R} \approx 1.43(f')^{0.208}. \tag{7.70}$$

We may use Eq. (7.70) directly or substitute it into the *right-hand side* of Eq. (7.69) to produce the complicated but explicit equation

$$\mathbf{R} \approx \left[-2.0\sqrt{f'}\,\log\left(\frac{\pi e}{10.35}[f']^{0.208} + 4.29[f']^{-0.188}\right)\right]^{0.4}. \tag{7.71}$$

Then we solve a pipe sizing problem as follows:

- substitute given values* of gh_L, Q, ν, and L into Eqs. (7.67) and (7.68) to calculate f' and e;
- compute $\mathbf{R}$, using Eq. (7.70) for a "good" value or Eq. (7.71) for a "better" value; and
- calculate the required pipe size from

$$D = \frac{4Q}{\pi \nu \mathbf{R}}. \tag{7.72}$$

Commercial pipe and tubing come in standard sizes (see Appendix D). Pipe diameter calculated according to the procedure described seldom corresponds exactly to a standard size. Specifying a standard pipe size is usually more economical than ordering a custom fabrication just to match your carefully computed diameter. Specifying the standard pipe size that has the next largest diameter is customary. Note that the practice of rounding upward implies that there is little need to be extremely accurate in pipe sizing calculations (the "good" calculation using Eq. 7.70 is usually adequate). Note also that the energy loss and/or flow rate does not equal the originally specified values when the larger pipe is used, so you might have to do an energy loss or flow rate calculation for the pipe size you choose.

* Again, gh_L includes *only* pipe friction loss.

EXAMPLE 7.5 Illustrates Pipe Sizing for Turbulent Flow

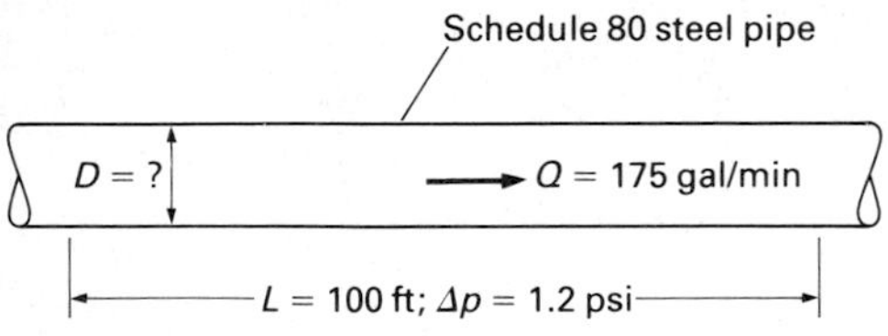

Figure E7.5 Steel pipe.

Find the minimum acceptable pipe size for a schedule 80, commercial steel pipe to permit the flow of 175 gal/min of 60°F water with a maximum pressure drop of 1.2 psi per 100 ft of pipe. Assume that the pipe is horizontal.

SOLUTION

Given

Figure E7.5

Flow of 175 gal/min

60°F water

Maximum pressure drop 1.2 psi per 100 ft of pipe

Find

Minimum acceptable pipe size

Solution

We could solve this problem by using either the Colebrook formula or the Moody chart. The Colebrook method is much simpler and direct, so we use it here.

We calculate the energy loss for a 100-ft length of pipe using

Eq. (7.10):

$$gh_L = \frac{p_1 - p_2}{\rho}$$

$$= \frac{(1.2 \text{ lb/in}^2)(144 \text{ in}^2/\text{ft}^2)}{(62.4 \text{ lbm/ft}^3)(1 \text{ lb}\cdot\text{sec}^2/32.2 \text{ ft}\cdot\text{lbm})} = 89.2 \text{ ft}^2/\text{sec}^2.$$

We use Eq. (7.67) to calculate f':

$$f' = \frac{128(gh_L)Q^3}{\pi^3 L v^5}$$

$$= \frac{128(89.2 \text{ ft}^2/\text{sec}^2)(175 \text{ gal/min})^3(1 \text{ ft}^3/7.48 \text{ gal})^3}{\pi^3(100 \text{ ft})(1.22 \times 10^{-5} \text{ ft}^2/\text{sec})^5(60 \text{ sec/min})^3} = 8.08 \times 10^{23}.$$

For commercial steel pipe, Table 7.1 gives $\varepsilon = 0.00015$ ft. Then, from Eq. (7.68),

$$e = \frac{\varepsilon v}{Q} = \frac{(0.00015 \text{ ft})(1.22 \times 10^{-5} \text{ ft}^2/\text{sec})}{(175 \text{ gal/min})(1 \text{ ft}^3/7.48 \text{ gal})(1 \text{ min}/60 \text{ sec})} = 4.69 \times 10^{-9}.$$

The approximate formula, Eq. (7.70), gives

$$\mathbf{R} = 1.43(f')^{0.208} = 1.43(8.08 \times 10^{23})^{0.208} = 1.34 \times 10^5.$$

Using Eq. (7.72), we find the diameter D:

$$D = \frac{4Q}{\pi v \mathbf{R}}$$

$$= \frac{4(175 \text{ gal/min})(1 \text{ ft}^3/7.48 \text{ gal})(1 \text{ min}/60 \text{ sec})}{\pi(1.22 \times 10^{-5} \text{ ft}^2/\text{sec})(1.34 \times 10^5)} = 0.304 \text{ ft} = 3.64 \text{ in.}$$

Table D.1 shows that the closest equal or larger diameter is 3.826 in., for a 4.5-in. schedule 80 pipe.

Use 4.5 in. (nominal) pipe. **ANSWER**

Discussion

Using the more exact Eq. (7.71) for the Reynolds number gives

$$\mathbf{R} = \left[-2.0\sqrt{8.08 \times 10^{23}} \log\left(\frac{\pi(4.69 \times 10^{-9})}{10.35} (8.08 \times 10^{23})^{0.208} + 4.29(8.08 \times 10^{23})^{-0.188} \right) \right]^{0.4} = 1.33 \times 10^5,$$

compared to the approximate value of 1.34×10^5.

If you want to solve this pipe sizing problem using the Moody chart, use Example 7.12 as a guide.

7.2.6 Fully Developed Flow in Noncircular Ducts

The Hagen–Poiseuille, Darcy–Weisbach, and Colebrook equations and the Moody chart provide adequate solutions for flow in *circular* pipes. Unfortunately, not all internal flows occur in circular pipes. Air- and gas-handling systems, such as home heating systems or power plant air and flue gas ducts, commonly have

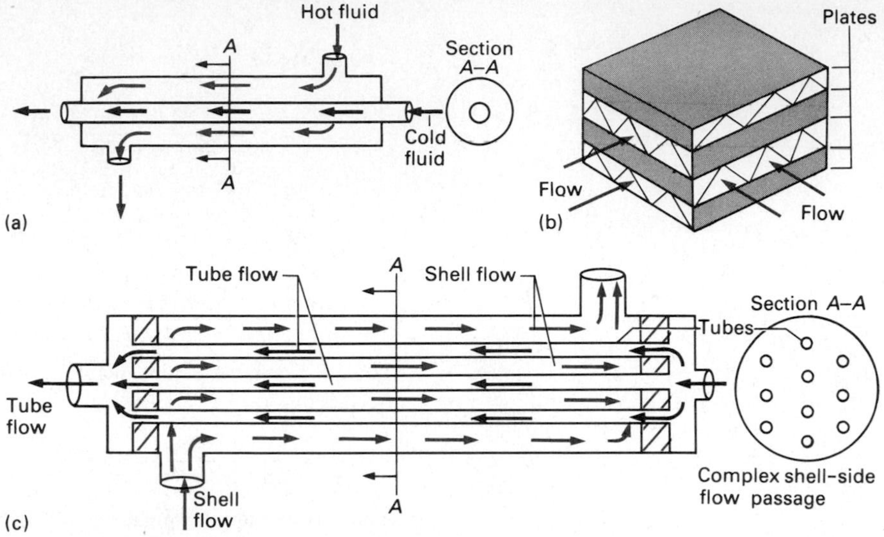

Figure 7.12 Some internal flow passages of complex shape: (a) double-pipe heat exchanger; (b) compact "plate fin" heat exchanger; (c) shell and tube heat exchanger.

rectangular ducts. A double-pipe heat exchanger (Fig. 7.12a) has flow in a concentric annulus, and a "compact" plate-fin heat exchanger (Fig. 7.12b) has flow in triangular passages. Flow in shell and tube heat exchangers (Fig. 7.12c) or in nuclear fuel-rod bundles occurs in passages of very complex shape. If the flow is laminar, obtaining an exact solution similar to the Hagen–Poiseuille equation is possible sometimes. If the flow is turbulent, no purely analytic solutions exist. Available experimental data are not nearly so abundant as for circular pipes, so general correlations are not readily available.

Practical analysis of flow in noncircular ducts is based on the idea of finding an "equivalent" circular pipe flow. The main question is how to determine the equivalence. The primary geometric characteristic of a circle is that it can be completely specified by a single characteristic dimension, the diameter. Maybe there is some way to define an equivalent single characteristic dimension for an arbitrary duct that can replace the circular pipe diameter in flow equations. The first such dimension that comes to mind is called the *equivalent diameter*. The equivalent diameter (D_{eq}) is the diameter of a circle that would have the same cross-sectional area (A_{NCD}) as the actual noncircular duct. Thus

$$\frac{\pi}{4}(D_{eq})^2 = A_{NCD}, \tag{7.73a}$$

or

$$D_{eq} \equiv \sqrt{\frac{4A_{NCD}}{\pi}}. \tag{7.73b}$$

The equivalent diameter is primarily a geometric idea. For a given flow rate, the average fluid velocity in a circular pipe with diameter equal to D_{eq} would be the same as the average velocity in the noncircular duct.

Experiments show that the equivalent diameter is not the best choice for evaluating friction factor and energy loss in noncircular ducts. Although the equivalent diameter sometimes appears in the literature and in handbooks, it is normally reserved for purely geometric purposes.*

A second possible characteristic dimension can be defined from the linear momentum equation, Eq. (7.18), for fully developed flow:

$$(p_1 + \gamma z_1) - (p_2 + \gamma z_2) = \tau_w\left(\frac{P}{A}\right)\Delta x.$$

For a circular pipe, the equation becomes Eq. (7.21):

$$(p_1 + \gamma z_1) - (p_2 + \gamma z_2) = \left(\frac{4}{D}\right)\tau_w(\Delta x).$$

Comparing Eqs. (7.18) and (7.21), we find that the *hydraulic diameter* (D_h), defined by

$$D_h \equiv \frac{4A}{P} = \frac{4(\text{Duct cross-sectional area})}{\text{Wetted perimeter}}, \tag{7.74}$$

is a characteristic dimension whose definition depends directly on the dynamics of the flow. Evaluating the wall shear stress from Eq. (7.31),

$$\tau_w = \mathbf{C_f}\left(\frac{1}{2}\rho V^2\right) = \frac{f}{4}\left(\frac{1}{2}\rho V^2\right),$$

using Eqs. (7.13), (7.14), and (7.21), and substituting L for Δx, we find

$$gh_L = f\frac{L}{D_h}\left(\frac{V^2}{2}\right). \tag{7.75}$$

The Darcy–Weisbach equation can be generalized by using the hydraulic diameter. For any duct shape except circular, the hydraulic diameter and the equivalent diameter are not equal. Expressions for calculating the hydraulic diameter of a few common shapes are given in Table 7.5. Because the hydraulic diameter is required for the Darcy–Weisbach equation, choosing the hydraulic diameter as the characteristic dimension of an equivalent circular pipe flow seems logical.

Changing from D to D_h in the Darcy–Weisbach equation is simple enough. The real question is "How do we evaluate the friction factor f?" Consider calculation of energy loss in a duct of arbitrary cross section. If we choose the hydraulic diameter as the relevant length dimension for the cross section, the energy loss would obey an equation of the form

$$gh_L = F\{V, L, D_h, \rho, \mu, \varepsilon\}.$$

---

* A typical use of the concept would be the statement "The Pitot tube must be located a distance equal to at least 10 equivalent diameters downstream from any flow disturbances."

In terms of dimensionless parameters,

$$\frac{gh_L}{\frac{1}{2}V^2} = F\left(\frac{L}{D_h}, \frac{\rho V D_h}{\mu}, \frac{\varepsilon}{D_h}\right).$$

With the aid of Eq. (7.75), this can be reduced to

$$\frac{gh_L}{\frac{1}{2}V^2} = \left(\frac{L}{D_h}\right) G\left(\mathbf{R_h}, \frac{\varepsilon}{D_h}\right),$$

where

$$\mathbf{R_h} = \frac{\rho V D_h}{\mu}. \tag{7.76}$$

The friction factor for the arbitrary duct is defined by

$$f \equiv G\left(\mathbf{R_h}, \frac{\varepsilon}{D_h}\right),$$

which is a generalization of Eq. (7.27).

Table 7.5 Hydraulic diameter of common geometric shapes.

Shape		Hydraulic Diameter $D_h = \frac{4A}{P}$
Circle	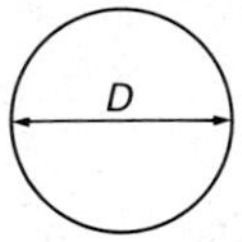	D
Rectangle	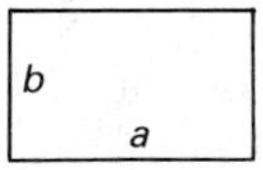	$\frac{2ab}{(a+b)}$
Isosceles triangle	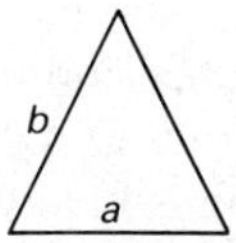	$\frac{\sqrt{4a^2b^2 - a^4}}{a + 2b}$
Ellipse	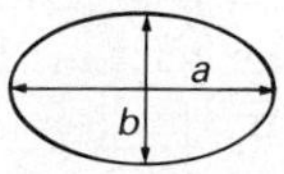	$\approx \frac{2\sqrt{2}ab}{\sqrt{a^2 + b^2}}$
Concentric annulus	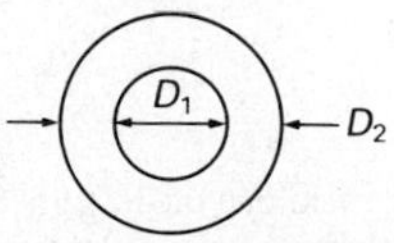	$D_2 - D_1$

The function G might be different for different duct shapes. We have rather complete information about G for one particular duct shape (circular). This information consists of the Hagen–Poiseuille equation, the Colebrook equation, and the Moody chart. We assume that we can use the function G for circular pipes as the general function for any duct simply by replacing the circular pipe diameter wherever it occurs in the circular pipe G by the hydraulic diameter. Based on this assumption, we approximate the friction factor for a noncircular duct by

$$f \approx \frac{64}{\mathbf{R_h}} \qquad \mathbf{R_h} < 2300, \tag{7.77}$$

$$f \approx \frac{0.25}{\{\log[(\varepsilon/D_h)/3.7 + 2.51/(\mathbf{R_h}\sqrt{f})]\}^2} \qquad \mathbf{R_h} > 4000, \tag{7.78}$$

or, reading f from the Moody chart with

$$\mathbf{R} = \mathbf{R_h}, \qquad \frac{\varepsilon}{D} = \frac{\varepsilon}{D_h}.$$

This assumption has small theoretical basis, but it seems reasonable. The proof of an assumption is comparison with more exact theories or experimental data. It turns out that the hydraulic diameter approach works remarkably well for turbulent flow, being accurate to ± 10 percent. As the Colebrook equation and Moody chart are accurate to only about ± 10 percent for circular pipe flow, the hydraulic diameter approach is quite acceptable for turbulent flow calculations.

The following items summarize the hydraulic diameter method:

- The duct shape is assumed to be known. Calculate the hydraulic diameter from Eq. (7.74) or use Table 7.5.
- Calculate $\mathbf{R_h}$ and ε/D_h.
- Using $\mathbf{R_h}$ and ε/D_h, calculate f from Eq. (7.77) *or* Eq. (7.78) or read f from the Moody chart.
- Calculate energy losses from Eq. (7.75).
- Be careful to use the true cross-sectional area and *not* $(\pi/4)(D_h^2)$ when relating average velocity and flow rate. Equation (7.55) must be replaced by

$$Q = VA_{\text{NCD}} = \frac{\pi}{4} V(D_h)^2 \left(\frac{4A_{\text{NCD}}}{\pi D_h^2}\right). \tag{7.79}$$

The last term in Eq. (7.79) is a numerical constant for a given shape of noncircular duct.

The hydraulic diameter model is not very accurate for laminar flow. Fortunately, using it for laminar flow is not necessary, because finding an exact solution of the Navier–Stokes equations for fully developed laminar flow in a duct with any cross-sectional shape that can be represented by an analytic function is possible. We have already found one such solution in Examples 5.8 and 4.11. The flow considered there—fully developed flow between infinite parallel plates—may

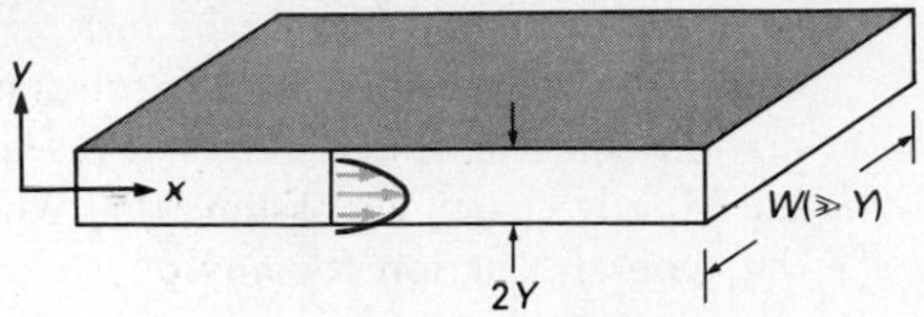

Figure 7.13 Flow in a very wide rectangular duct.

be thought of as flow in a rectangular duct whose dimension perpendicular to the xy plane is very large (Fig. 7.13). The hydraulic diameter of this duct is

$$D_h = \lim_{W \to \infty} \frac{4(2Y)(W)}{2(2Y + W)} = 4Y.$$

We find the friction factor for this flow with Eq. (7.75) and a result of Example 4.11:

$$f = \frac{24\mu}{\rho V Y} = \frac{96\mu}{\rho V D_h} = \frac{96}{\mathbf{R_h}}. \qquad (7.80)$$

Equation (7.80) is an *exact* expression for the friction factor in terms of the hydraulic diameter. If we use the hydraulic diameter in the Hagen–Poiseuille equation,

Table 7.6 Laminar friction constant for various duct shapes.

Duct Shape		Laminar Friction Constant $f\mathbf{R_h}$
Ellipse	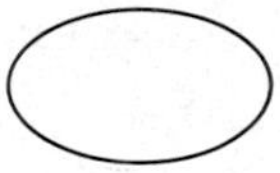	64
Equilateral triangle	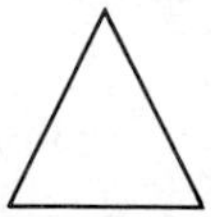	53.33
Concentric annulus	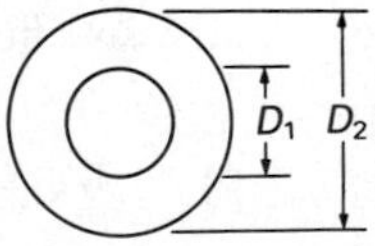 $\left(d = \frac{D_1}{D_a}\right)$	$64\left[\frac{(1-d)^2}{1 + d^2 + (1 - d^2)/\ln d}\right]$
Rectangle	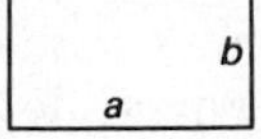	$\frac{b}{a}$ = 0.0: 96.00
		0.05: 86.91
		0.1: 84.68
		0.125: 82.34
		0.167: 78.81
		0.25: 72.93
		0.4: 65.47
		0.5: 62.19
		0.75: 57.89
		1.0: 56.91 $= f\mathbf{R_h}$

we get

$$f = \frac{64}{\mathbf{R_h}}.$$

This value of f is 33 percent too low.

Exact solutions for the laminar friction factor have been obtained for many geometries [5]. All these solutions are of the general form

$$f\mathbf{R_h} = \text{Constant.} \tag{7.81}$$

For circular pipe, the constant is 64, and for very wide rectangular duct, it is 96. The form of these equations is deducible from dimensional reasoning. The product

$$f\mathbf{R_h} \sim \left(\frac{\text{Force from pressure drop}}{\text{Inertia ``force''}}\right)\left(\frac{\text{``Inertia'' force}}{\text{Viscous force}}\right)$$

$$\sim \frac{\text{Force from pressure drop}}{\text{Viscous force}}.$$

In fully developed laminar flow, pressure drop is the result of viscous effects *only*, so the product $f\mathbf{R_h}$ should be constant. The constant is different for different geometries. Table 7.6 gives values of $f\mathbf{R_h}$ for several common geometries.

EXAMPLE 7.6 Illustrates Calculations for Noncircular Ducts

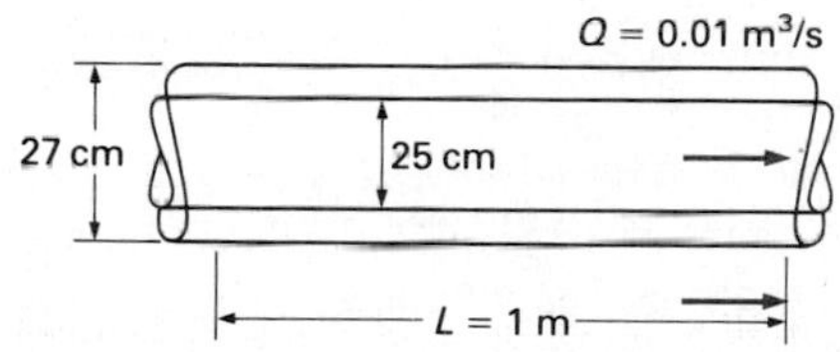

Figure E7.6 Annular channel.

A fluid flows through a horizontal, annular channel between concentric tubes at a rate of $0.010\ m^3/s$. The channel, shown in Fig. E7.6, has an inner diameter of 25 cm, an outer diameter of 27 cm, and a length of 1.0 m. Find the pressure loss if the fluid is machine oil, with $S = 0.85$ and $\nu = 1.3 \times 10^{-4}\ m^2/s$, and water, with $\nu = 1.0 \times 10^{-6}\ m^2/s$.

SOLUTION

Given

Figure E7.6

Fluid is machine oil ($S = 0.85$, $\nu = 1.3 \times 10^{-4}\ m^2/s$) or water ($\nu = 1.0 \times 10^{-6}\ m^2/s$)

Inner diameter 25 cm

Outer diameter 27 cm

Length 1.0 m

Find

Pressure loss for both fluids

Solution

We assume that the flow is fully developed. The pressure loss is

$$\Delta p_L = \rho g h_L = \rho f \frac{L}{D_h}\frac{V^2}{2}.$$

The velocity V is

$$V = \frac{Q}{A} = \frac{4Q}{\pi(D_2^2 - D_1^2)},$$

where D_1 is the inner diameter and D_2 the outer diameter. Substituting given values, we have

$$V = \frac{4(0.01\ \text{m}^3/\text{s})(100\ \text{cm/m})^2}{\pi[(27\ \text{cm})^2 - (25\ \text{cm})^2]} = 1.22\ \text{m/s}.$$

Table 7.5 gives

$$D_h = D_2 - D_1 = 27\ \text{cm} - 25\ \text{cm} = 2.0\ \text{cm}.$$

These data are common to both flows.

For the machine oil,

$$\mathbf{R_h} = \frac{VD_h}{\nu} = \frac{1.22\ \text{m/s}\ (0.02\ \text{m})}{1.3 \times 10^{-4}\ \text{m}^2/\text{s}} = 188.$$

The flow is laminar. Table 7.6 gives

$$d = \frac{D_1}{D_2} = \frac{25\ \text{cm}}{27\ \text{cm}} = 0.926,$$

and

$$f\mathbf{R_h} = 64\left[\frac{(1 - 0.926)^2}{1 + 0.926^2 + (1 - 0.926^2)/\ln 0.926}\right] = 64(1.500) = 96.0.$$

The friction factor is

$$f = \frac{96.0}{\mathbf{R_h}} = \frac{96.0}{188} = 0.51.$$

The mechanical energy loss (using D_h in L/D) is

$$gh_L = 0.51\left(\frac{1.0\ \text{m}}{2.0\ \text{cm}}\right)\frac{(1.22\ \text{m/s})^2}{2}\left(\frac{100\ \text{cm}}{\text{m}}\right) = 18.98\ \text{N}\cdot\text{m/kg}.$$

The pressure loss is

$$\Delta p_L = (850\ \text{kg/m}^3)(18.98\ \text{N}\cdot\text{m/kg});$$

$$\Delta p_L = 16.13\ \text{kPa}.$$ **ANSWER**

For the water,

$$\mathbf{R_h} = \frac{VD_h}{\nu} = \frac{1.22\ \text{m/s}\ (0.02\ \text{m})}{1.0 \times 10^{-6}\ \text{m}^2/\text{s}} = 24{,}400.$$

The flow is turbulent. For drawn tubing

$$\frac{\varepsilon}{D_h} = \frac{0.0000015}{0.02} = 0.000075.$$

At these values of $\mathbf{R_h}$ and ε/D_h, the channel effectively is smooth, so the friction factor is $f \approx 0.025$. The pressure loss is

$$\Delta p_L = \rho g h_L = f\frac{L}{D_h}\left(\tfrac{1}{2}\rho V^2\right)$$

$$= 0.025\left(\frac{100\text{ cm}}{2\text{ cm}}\right)(\tfrac{1}{2})(1000\text{ kg/m}^3)(1.22\text{ m/s})^2;$$

$$\Delta p_L = 0.93\text{ kPa.}$$ **ANSWER**

Discussion

We used the hydraulic diameter model for turbulent flow and the exact equation for laminar flow. If we had used the hydraulic diameter model for laminar flow, the friction factor would have been

$$f \approx \frac{64}{\mathbf{R_h}} \approx \frac{64}{188} \approx 0.34,$$

and the pressure loss would have been 10.75 kPa instead of 16.3 kPa. We used the exact approach, because it is more accurate. Note that the pressure loss in the (turbulent) water flow is an order of magnitude less than the pressure loss in the (laminar) oil flow, although the geometry and flow rate were the same.

7.2.7 Detailed Consideration of Fully Developed Turbulent Flow in a Circular Pipe

In this section, we fill in some of the background for the Colebrook formula and the Moody chart. We investigate the nature of turbulent flow near a surface and develop a semi-empirical theory for the velocity profile. As with laminar flow, integration of the velocity profile yields a useful relation between friction factor, average velocity, and other key parameters. The information about turbulent flow that we present here is a condensation of many years of research by many people. Using the benefit of hindsight, we present only the correct assumptions and deductions, disregarding all the blind alleys that were pursued in developing the information.

First we consider turbulent flow near a smooth surface, such as a pipe wall. Figure 7.14(a) shows the (time-average) velocity profile. If this profile represents a pipe flow, Eq. (7.32) tells us that the shear stress distribution is linear. The contributions of viscous stress and turbulent stress to the total stress are shown in Fig. 7.14(b). We can divide the flow into three regions:

- an "inner" *wall layer,* where turbulent stress is very small and viscous stress is large;
- an "outer" *fully turbulent layer,* where viscous stress is negligible and turbulent stress is large; and
- an *"overlap" layer,* where both stresses are significant.

This multilayer structure is the basis of most mathematical modeling and analysis of turbulent flow near a surface. An accurate theory describing the velocity profile can be constructed by examining each layer in detail and then patching the layers together. Although there are several possible approaches, we present the

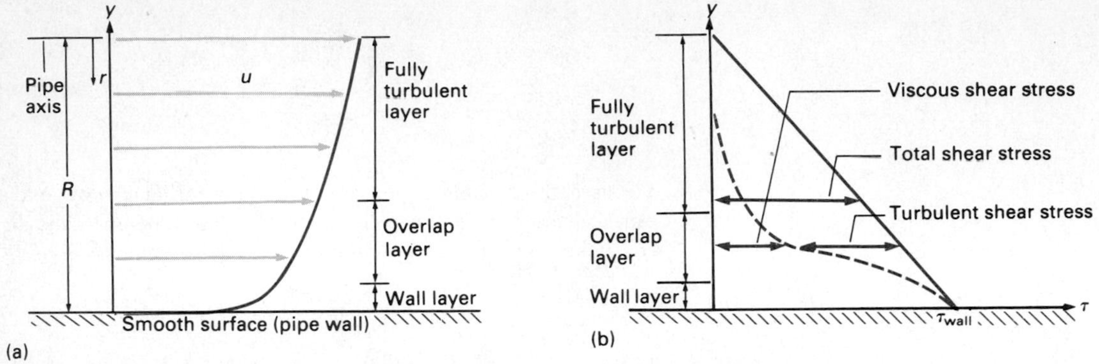

Figure 7.14 (a) Velocity profile and (b) shear stress distributions in turbulent flow near a smooth surface.

approach based on dimensional reasoning pioneered by L. Prandtl, T. von Karman, and C. Millikan.*

The presence of the inner layer can be understood by recalling that turbulent stress results from velocity fluctuations, which must vanish near the wall. The thickness of this layer is magnified in Fig. 7.14; it typically occupies about 2 percent of the entire layer. As turbulence effects are small in the inner layer, it is sometimes called the *viscous sublayer* or the *laminar sublayer*. The inner layer is very thin, so Prandtl assumed that the flow behaves as if it knew nothing about the outer flow; that is, it does not depend on what occurs near the pipe axis ($y = R$) or on the maximum (or average) velocity. Accordingly, the velocity in the inner layer is given by a function of the form

$$u = f\{\mu, \tau_w, \rho, y\}. \tag{7.82}$$

Applying dimensional analysis, we obtain

$$\frac{u\rho^{1/2}}{\tau_w^{1/2}} = F\left\{\frac{y\rho^{1/2}\tau_w^{1/2}}{\mu}\right\}. \tag{7.83}$$

Introducing the definitions

$$u_\tau \equiv \left(\frac{\tau_w}{\rho}\right)^{1/2} \tag{7.84}$$

$$u^+ \equiv u/u_\tau, \tag{7.85}$$

and

$$y^+ \equiv \frac{\rho y u_\tau}{\mu} = \frac{y u_\tau}{\nu}, \tag{7.86}$$

we write Eq. (7.83) as

$$u^+ = F\{y^+\}. \tag{7.87}$$

The parameter u_τ is called the *friction velocity* and has the dimensions $[L/T]$. The dimensionless velocity u^+ is related to the skin friction coefficient ($u^+ =$

* The most common alternative approach uses a "mixing length" model for turbulent shear stress. See Example 5.6 and Problem 92 at the end of this chapter.

$(u/V)\sqrt{2/C_f})$. The dimensionless distance y^+ is a type of Reynolds number. Equation (7.87) is called the *law of the wall*. Determination of the function F requires a specific theory or experimental data. One simple theory assumes that the stress in the inner layer is entirely the result of viscosity and is constant. According to these assumptions,

$$\tau = \mu \frac{du}{dy} = \tau_w.$$

Integrating and using the no-slip condition at $y = 0$ give

$$u = \frac{\tau_w}{\mu} y,$$

or

$$u^+ = y^+. \tag{7.88}$$

Equation (7.88) is valid only *very* near the wall.

Next we consider the outer layer. There, viscous stress is negligible, so viscosity is not important. Similarly, the details of the flow near the wall are not important; the only thing about the wall that affects the outer layer is the shear stress. The velocity distribution in the outer flow also depends on the total thickness of the layer (the radius R in pipe flow) and on the maximum velocity, u_{max}. We assume that the velocity in the outer layer obeys a relation of the form

$$u = g\{u_{max}, R, \tau_w, \rho, y\}. \tag{7.89}$$

Von Karman deduced that Eq. (7.89) can be written in the special form

$$(u_{max} - u) = \tilde{g}\{R, \tau_w, \rho, y\}, \tag{7.90}$$

where $(u_{max} - u)$ is called the *velocity defect* and is treated as a *single* parameter. Equation (7.90) can be reduced to the dimensionless form

$$\frac{u - u_{max}}{u_\tau} = u^+ - \frac{u_{max}}{u_\tau} = G\left\{\left(\frac{y}{R}\right)\right\}. \tag{7.91}$$

Equation (7.91) is called the *outer law* or the *velocity defect law*. Evaluating the function G requires a specific theory or experimental data.

Finally, we consider the overlap layer. In this layer, both velocity distributions, Eqs. (7.87) and (7.91), must apply. Equating u^+ from both equations, we have

$$\frac{u_{max}}{u_\tau} + G\left\{\left(\frac{y}{R}\right)\right\} = F\{y^+\}.$$

We introduce y^+ into the argument of G:

$$\frac{u_{max}}{u_\tau} + G\left\{\left(\frac{\nu}{u_\tau R} y^+\right)\right\} = F\{y^+\}. \tag{7.92}$$

This equation is rather cryptic, but we can deduce some valuable information from it. The parameters u_{max}, u_τ, R, and ν do not vary with y, and so the equation has the form

$$G\{\text{Constant } y^+\} = F\{y^+\} + \text{Constant}.$$

In the overlap layer, G and F have the property that a function of a constant times the argument equals a function of the argument plus another constant. Only one type of function, the logarithm, has this property:

$$\ln(Cy^+) = \ln y^+ + \ln C.$$

In the overlap layer, both F and G must be logarithmic. A general form of Eq. (7.87), with F a logarithmic function, is

$$\frac{u}{u_\tau} = \frac{1}{\kappa}\ln\left(\frac{yu_\tau}{\nu}\right) + B, \tag{7.93}$$

where $1/\kappa$ and B are constants. The logarithmic velocity profile in the overlap layer is probably the most important item of information available about turbulent flow. Note that this law does not apply all the way to the wall (ln 0 is undefined); it must be patched to an inner law, such as Eq. (7.88).

You are probably thinking that it took some very fancy "footwork" to derive the logarithmic velocity profile. Is it valid? We answer this question by comparing the logarithmic velocity profile with experimental data, which also is necessary to establish the constants κ and B.

Figure 7.15 shows typical pipe flow velocity profile data in semilogarithmic coordinates. Both the logarithmic law and the linear (near wall) law are verified by the data. Note that the difference between the outer layer velocities and the logarithmic law is very small for pipe flow, even though the logarithmic law is not supposed to apply in the outer region. The logarithmic scale for y^+ can be deceiving. In Fig. 7.15, the wall layer appears to take up about a third of the pipe,

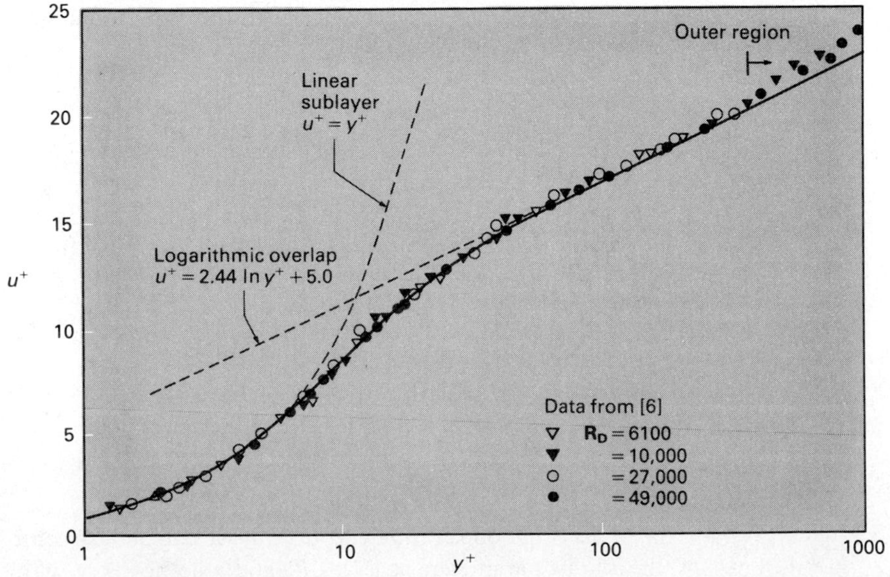

Figure 7.15 Comparison of log law and linear wall law velocity profiles with experimental data.

but in fact the fraction is only

$$\frac{y^+ \text{ (Edge of wall layer)}}{R^+} \approx \frac{20}{1000} = 0.02.$$

By fitting a straight line to the overlap region data, we determine the magnitudes of the constants κ and B:

$$\kappa \approx \frac{1}{2.44} = 0.41 \quad \text{and} \quad B \approx 5.0.$$

The need to determine these constants from experiments emphasizes that any theory of turbulent flow requires experimental data to be complete.

At this point, we compare the information we have about turbulent flow velocity profiles and our information about laminar profiles. In laminar flow, the velocity profile is related to viscosity, pipe radius, and pressure drop (Eq. 7.33). By integrating the velocity distribution across the pipe, we were able to relate pressure drop, viscosity, pipe radius, and average velocity. The ultimate result was the Hagen–Poiseuille equation for friction factor in terms of Reynolds number.

Let's now consider the velocity profile for turbulent flow. The logarithmic law is accurate over most of the cross section, failing badly only in the region very near the wall ($y^+ < 5$). Using the logarithmic profile for the whole cross section and noting that

$$y = R - r,$$

we have

$$u \approx 2.44 u_\tau \ln\left[\frac{u_\tau(R-r)}{\nu}\right] + 5.0 u_\tau. \tag{7.94}$$

Integrating this velocity distribution across the pipe and dividing by the pipe cross-sectional area gives an expression for the average velocity:

$$V = \frac{2}{R^2}\int_0^R \left[2.44 u_\tau \ln\left(\frac{u_\tau(R-r)}{\nu}\right) + 5.0 u_\tau\right] r\,dr.$$

Performing the integration, we get

$$\frac{V}{u_\tau} = 2.44 \ln\left(\frac{u_\tau R}{\nu}\right) + 1.34. \tag{7.95}$$

Now

$$\frac{V}{u_\tau} = \sqrt{\frac{\rho V^2}{\tau_w}} = \sqrt{\frac{2}{C_f}} = \sqrt{\frac{8}{f}} \quad \text{and} \quad \frac{u_\tau R}{\nu} = \left(\frac{u_\tau}{V}\right)\left(\frac{VD}{2\nu}\right) = \frac{1}{2}\sqrt{\left(\frac{f}{8}\right)}(\mathbf{R}).$$

Substituting into Eq. (7.95) and changing to common logarithms, we get

$$\frac{1}{\sqrt{f}} = 1.99 \log(\mathbf{R}\sqrt{f}) - 1.02.$$

This equation was first derived by Prandtl in 1935. Upon comparing it with Nikuradse's data for smooth pipes, Prandtl found it to be slightly inaccurate because of the approximation of the wall and outer layers by the logarithmic law. Prandtl obtained a better fit by adjusting the constants slightly to obtain

$$\frac{1}{\sqrt{f}} = 2.0 \log(\mathbf{R}\sqrt{f}\,) - 0.8,$$

which is the smooth pipe friction factor formula, Eq. (7.46), presented earlier.

The theory above applies only to smooth walls. A theory for rough walls can be developed by appropriate modifications. Roughness disturbs the flow in the wall region. If roughness elements are large enough, they completely break up the nearly laminar flow in the wall region. We can obtain a clue about the effects of roughness from Fig. 7.15. If the surface roughness is completely buried in the sublayer so that

$$\varepsilon^+ = \frac{\varepsilon u_\tau}{\nu} < 5,$$

then the pipe wall appears smooth. If the roughness is large enough to protrude into the overlap layer so that

$$\varepsilon^+ > 70,$$

then the sublayer is completely broken up and the flow is completely turbulent. For values of roughness given by

$$5 < \varepsilon^+ < 70,$$

the flow is *transitionally rough* (see the Moody chart, Fig. 7.9).

Von Karman developed a theory for the velocity profile in fully turbulent flow. If the flow is fully turbulent, the wall region velocity depends on roughness height but not on viscosity:

$$\frac{u}{u_\tau} = F\left(\frac{y}{\varepsilon}\right). \tag{7.96}$$

For fully turbulent flow, Eq. (7.96) replaces Eqs. (7.83) and (7.87). Because the flow in the outer layer does not depend on flow details near the wall, Eqs. (7.89)–(7.92) still apply. By the same arguments that led to Eq. (7.93), we find that the velocity profile in the overlap layer for fully rough flow is given by

$$\frac{u}{u_\tau} = C_1 \ln\left(\frac{y}{\varepsilon}\right) + C_2,$$

where C_1 and C_2 are constants.

Comparing this equation with experimental data for fully rough pipe flow shows that the logarithmic form is valid and gives values for the constants:

$$C_1 = \frac{1}{\kappa} = \frac{1}{0.41} \qquad \text{(same as that for smooth wall);}$$

$$C_2 = 8.5 \qquad \text{(different from that for smooth wall).}$$

Using these constants, we find the velocity profile for fully turbulent flow:

$$u \approx 2.44u_\tau \ln\left(\frac{y}{\varepsilon}\right) + 8.5u_\tau. \tag{7.97}$$

A comparison of smooth wall and fully rough wall velocity profiles is shown in Fig. 7.16. The rough wall velocity profile can be integrated to find the average velocity. The resulting equation is rearranged to give a friction factor equation. The procedure is exactly the same as that which led to Eq. (7.46). The result, Eq. (7.47), is

$$\frac{1}{\sqrt{f}} = -2.0 \log\left(\frac{\varepsilon/D}{3.7}\right),$$

which is valid for fully rough flow.

The Colebrook formula is constructed from the smooth and fully rough friction factor expressions. We write the smooth wall formula as

$$\begin{aligned}\frac{1}{\sqrt{f}} &= 2.0 \log(\mathbf{R}\sqrt{f}) - \log(10^{0.8})\\ &= -2.0[\log(10^{0.4}) - \log(\mathbf{R}\sqrt{f})]\\ &= -2.0 \log\left(\frac{2.51}{\mathbf{R}\sqrt{f}}\right).\end{aligned} \tag{7.98}$$

We rearrange the rough wall formula to get

$$\frac{1}{\sqrt{f}} = -2.0 \log\left[\frac{(\varepsilon u_\tau/\nu)[\nu/(u_\tau D)]}{3.7}\right] = -2.0 \log\left[\frac{\varepsilon^+}{3.7(u_\tau/V)\mathbf{R}}\right].$$

As

$$\frac{u_\tau}{V} = \sqrt{\frac{f}{8}},$$

then

$$\frac{1}{\sqrt{f}} = -2.0 \log\left(\frac{\varepsilon^+}{1.31\mathbf{R}\sqrt{f}}\right).$$

Colebrook constructed his formula by combining the smooth wall and rough wall

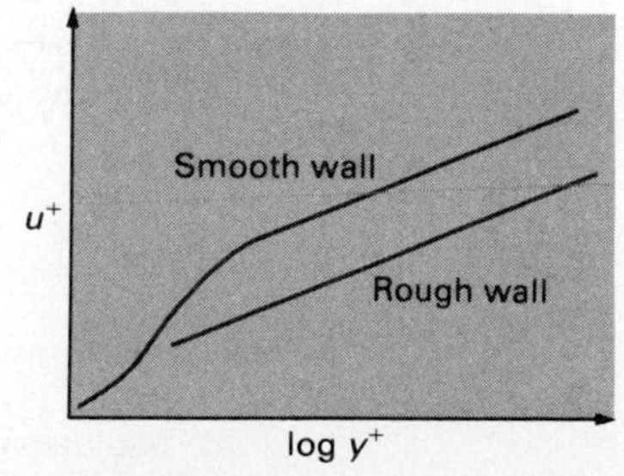

Figure 7.16 Comparison of smooth wall and rough wall velocity distribution in semilogarithmic coordinates.

parameters in the same logarithm to get:

$$\frac{1}{\sqrt{f}} = -2.0 \log\left[\frac{\varepsilon^+}{1.31\mathrm{R}\sqrt{f}} + \frac{2.51}{\mathrm{R}\sqrt{f}}\right].$$

This equation covers both limits, for if ε^+ is small ($\varepsilon^+ \ll 3.3$), the smooth wall equation results and if ε^+ is large ($\varepsilon^+ \gg 3.3$), the rough wall equation results.*

* Note that $(1.31)(2.51) \approx 3.3$.

EXAMPLE 7.7 Illustrates Calculations with Turbulent Velocity Profiles

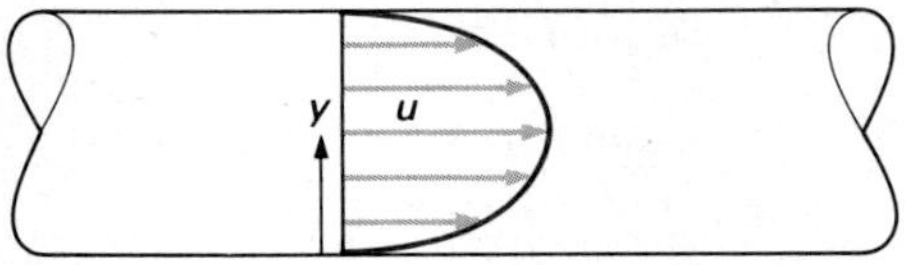

Figure E7.7 Horizontal smooth pipe.

Water at 20°C flows through a horizontal, smooth pipe, shown in Fig. E7.7, with an average velocity of 0.807 m/s and a Reynolds number of 2.3×10^4. Compare the velocity (u) calculated from the following:

- the linear law of the wall, the power law equation, and Fig. 7.15 for $y^+ = 5$,
- the logarithmic equation, the power law equation, and Fig. 7.15 for $y^+ = 100$ and $y^+ = 500$.

SOLUTION

Given

20°C water flowing through the pipe in Fig. E7.7

Average velocity 0.807 m/s

Reynolds number 2.3×10^4

Find

Comparison of dimensional velocities from linear law of the wall, power law equation, logarithmic velocity equation, and Fig. 7.17 for $y^+ = 5$, $y^+ = 100$, and $y^+ = 500$

Solution

The primary equations are

$$u = u^+ u_\tau, \qquad u^+ = y^+,$$

$$u^+ = \frac{1}{0.41} \ln y^+ + 5.0, \quad \text{and} \quad u = u_{\max}\left(\frac{y}{R}\right)^{1/n},$$

where R is the pipe radius and y the distance measured from the pipe wall. Equations (7.85) and (7.86) give the power law formula as

$$u = u_{\max}\left(\frac{y^+}{R^+}\right)^{1/n},$$

where

$$R^+ = \frac{R u_\tau}{\nu}.$$

To find the velocities, we need values for n, u_τ, ν, R^+, and $u_{\max}$. For

$\mathbf{R} = 2.3 \times 10^4$, Table 7.3 gives $n = 6.6$. For water at 20°C, Table A.5 gives

$$\nu = 1.00 \times 10^{-6}\ \text{m}^2/\text{s} \qquad \rho = 998\ \text{kg/m}^3.$$

For smooth pipe, Eq. (7.45) gives*

$$f \approx 0.3164(\mathbf{R})^{-0.25} = 0.3164(2.3 \times 10^4)^{-0.25} = 0.0257.$$

Equation (7.31) then gives

$$\frac{\tau_w}{\rho} = \frac{f}{8}V^2 = \frac{0.0257(0.807\ \text{m/s})^2}{8} = 0.00209\ \text{m}^2/\text{s}^2.$$

Next we have

$$u_\tau = \sqrt{\frac{\tau_w}{\rho}} = 0.0457\ \text{m/s}.$$

The definition of the Reynolds number gives

$$R = \frac{D}{2} = \frac{\nu\mathbf{R}}{2V} = \frac{(1.00 \times 10^{-6}\ \text{m}^2/\text{s})(2.3 \times 10^4)}{2(0.807\ \text{m/s})} = 0.0143\ \text{m}.$$

R^+ is calculated as†

$$R^+ = \frac{Ru_\tau}{\nu} = \frac{(0.0143\ \text{m})(0.0457\ \text{m/s})}{(1.00 \times 10^{-6}\ \text{m}^2/\text{s})} = 654.$$

We use Eq. (7.52) to calculate the maximum velocity u_{max}:

$$u_{max} = \frac{(n+1)(2n+1)}{2n^2}(V) = \frac{7.6(14.2)}{2(6.6)^2}(0.807\ \text{m/s}) = 1.0\ \text{m/s}.$$

We are now able to calculate the dimensionless velocities. For $y^+ = 5.0$, both Fig. 7.15 and the law of the wall give

$$u^+ = 5.0.$$

The fluid velocity is

$$u = u^+u_\tau = (5.0)(0.0457\ \text{m/s}) = 0.229\ \text{m/s}.$$

The power law at $y^+ = 5$ gives

$$u = u_{max}\left(\frac{y^+}{R^+}\right)^{1/n} = 1.0\ \text{m/s}\left(\frac{5}{654}\right)^{1/6.6} = 0.478\ \text{m/s}.$$

For $y^+ = 100$, the logarithmic velocity equation gives

$$u = u_\tau u^+ = (0.0457)\left(\frac{1}{0.41}(\ln 100) + 5.0\right) = 0.741\ \text{m/s},$$

the power law equation gives

$$u = 1.0\left(\frac{100}{654}\right)^{1/6.6} = 0.752\ \text{m/s},$$

---

* We could also use the Moody chart or Colebrook formula.

† We could also calculate R^+ from

$$R^+ = \frac{Ru_\tau}{\nu} = \frac{\mathbf{R}}{2}\sqrt{\frac{f}{8}}$$

and Fig. 7.15 gives $u^+ \approx 16.5$. Then

$$u = 16.5(0.0457) = 0.754 \text{ m/s}.$$

For $y^+ = 500$, the logarithmic velocity equation gives

$$u = u_\tau u^+ = (0.0457)\left[\frac{1}{0.41}(\ln 500) + 5.0\right] = 0.921 \text{ m/s},$$

the power law equation gives

$$u = 1.0\left(\frac{500}{654}\right)^{1/6.6} = 0.960 \text{ m/s},$$

and Fig. 7.15 gives $u \approx 21.5(0.0457) = 0.983$ m/s.

The following table shows the magnitudes of u determined by the various methods for the three y^+ values.

y^+	Linear Law of Wall	Logarithmic Equation	Power Law	Figure 7.15
5	0.229	Not applicable	0.478	0.229
100	Not applicable	0.741	0.752	0.754
500	Not applicable	0.921	0.960	0.983

ANSWER

Discussion

Figure 7.15 can be taken as the "exact" answer, because it is based on experimental data. The given values of y^+ are 0.8%, 15.3%, and 76.4% of the thickness of the velocity profile. As we noted previously, the power law profile is inaccurate near the wall (here giving about 100% error at $y^+ = 5$), and the logarithmic law is inaccurate near the outer edge.

7.3 FITTINGS, VALVES, AND LOCAL LOSSES

Figure 7.17 shows some auxiliary components that are used in a pipe or duct system, including

- *transitions* for changing pipe size;
- *elbows* and *bends* for changing flow (pipe) direction;
- *tees* and *laterals* for dividing or mixing streams;
- *valves* for controlling flow; and
- *entrances* and *exits,* special cases of transitions where the upstream (entrance) or downstream (exit) regions are considered infinite in extent.

All these components introduce disturbances that cause turbulence and mechanical energy loss in addition to that which occurs in the basic pipe flow. Figure 7.18(a) illustrates flow through a sudden enlargement, a rather crude transition. The flow

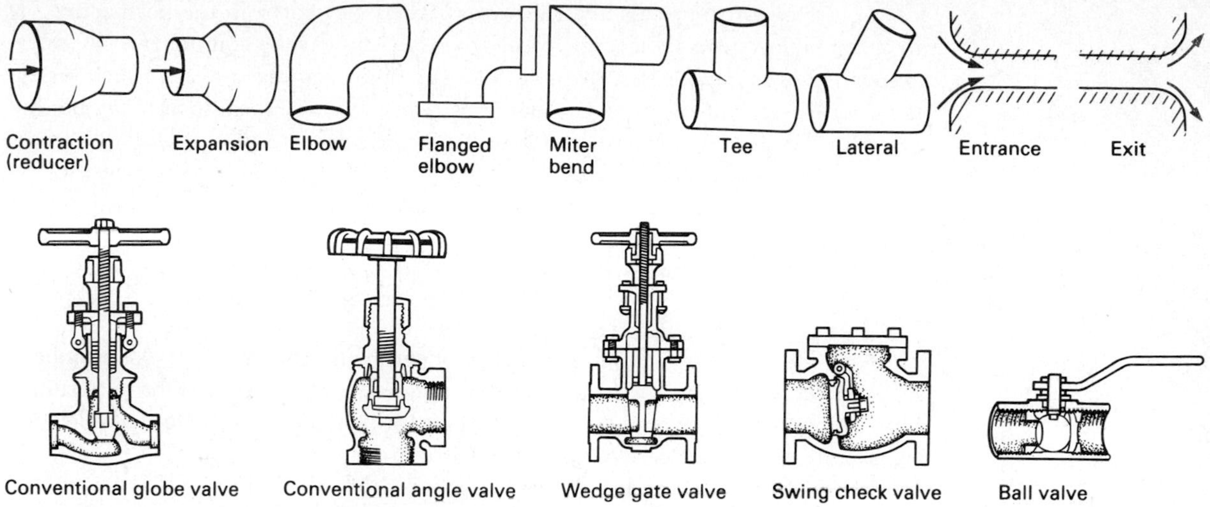

Figure 7.17 Some piping system components (valve diagrams courtesy of The Crane Co.).

is not able to turn the sharp corner, so it separates, generating a great deal of disorderly motion. The disorderly motion is often loosely called "turbulence"; however, the motion near the sudden enlargement is more orderly than true turbulence. The motion degenerates into true turbulence as the fluid proceeds downstream. The disturbance caused by the expansion persists for some distance downstream as the kinetic energy of the disorderly motion is dissipated and the flow gradually returns to a fully developed condition. This flow behavior is not unique to the sudden expansion; other components generate similar disturbances. These disturbances, loosely called "turbulence," are responsible for energy loss in the region immediately downstream from the component. A plot (Fig. 7.18b) of the mechanical energy downstream from the component shows that the energy

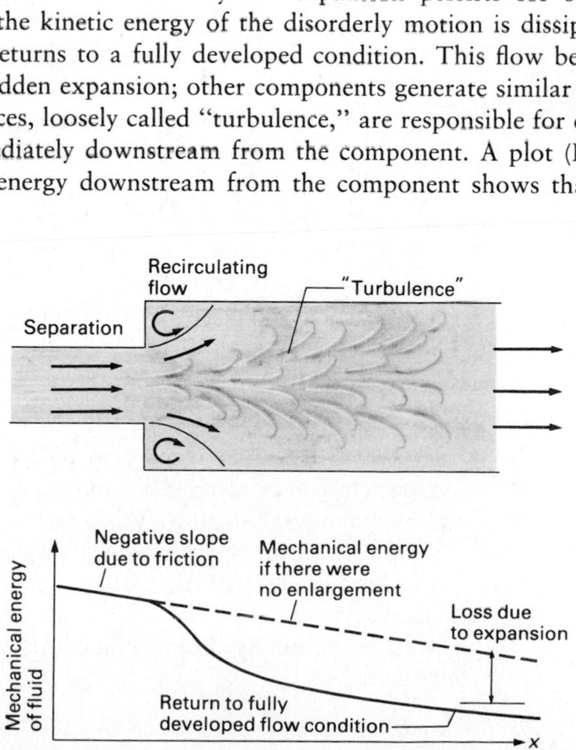

Figure 7.18 Flow detail and energy loss for flow through a sudden enlargement.

loss occurs over a finite distance; however, when viewed from the perspective of an entire pipe system, the energy losses are localized near the component. We refer to such losses as *local losses*. Most authors and many engineers refer to such losses as *minor losses*. This is an unfortunate misnomer, because local losses may be the dominant losses in some systems. Local losses are added to pipe friction losses to calculate the total energy loss in the system.

We calculate local losses by using a loss coefficient $\mathbf{K}$:

$$gh_L = \mathbf{K}\left(\frac{V^2}{2}\right). \tag{7.99}$$

The loss coefficient is a function of component geometry and Reynolds number. A size-effect parameter similar to relative roughness in pipes may also be important. Most components generate energy loss by promoting turbulence, so loss coefficients are usually nearly independent of Reynolds number.

An alternative to the loss coefficient method of calculating local losses is the "equivalent length" method. In this method, the component is "replaced" by a straight run of pipe (duct) that would have the same loss. Equating the equivalent pipe friction loss to the actual local loss, we have

$$f\frac{L_{eq}}{D_h}\frac{V^2}{2} = \mathbf{K}\frac{V^2}{2},$$

so we calculate the equivalent length by

$$L_{eq} = \frac{\mathbf{K}}{f}D_h, \tag{7.100}$$

where f is the friction factor in the pipe (duct) to which the component is attached. The equivalent length concept is reasonably straightforward for all components except transitions. For transitions, you must be careful to specify whether the equivalent length is added to the smaller or larger pipe (duct).

In practice, local loss coefficients or equivalent lengths are obtained from handbooks or, for best accuracy, from component manufacturers' specifications. These data have usually been obtained by experiment. References [7, 8, 9, 10] contain extensive information about local loss coefficients and equivalent lengths. We present a brief discussion of loss generation in various types of components and indicate typical values of $\mathbf{K}$.

Transitions. The sudden expansion is the only configuration for which the loss coefficient can be derived by simple analytic methods. Figure 7.19 shows a control volume for this flow. We assume that uniform flow enters the expansion at plane 1. Plane 2 is placed downstream a distance sufficient to ensure that the flow is uniform there also. We also assume that the pressure is uniform across planes 1 and 2.

The continuity, linear momentum, and energy equations for the control volume are

$$V_1A_1 = V_2A_2,$$

$$(p_1 - p_2)A_2 - \tau_{avg}A_{wall} = \rho(V_2^2A_2 - V_1^2A_1)$$

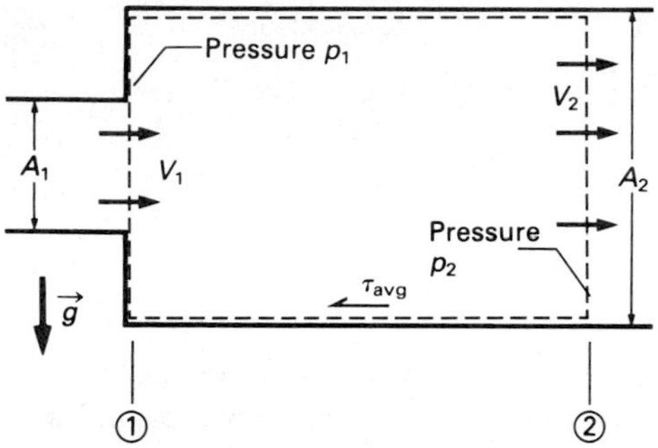

Figure 7.19 Control volume for analysis of flow in a sudden enlargement.

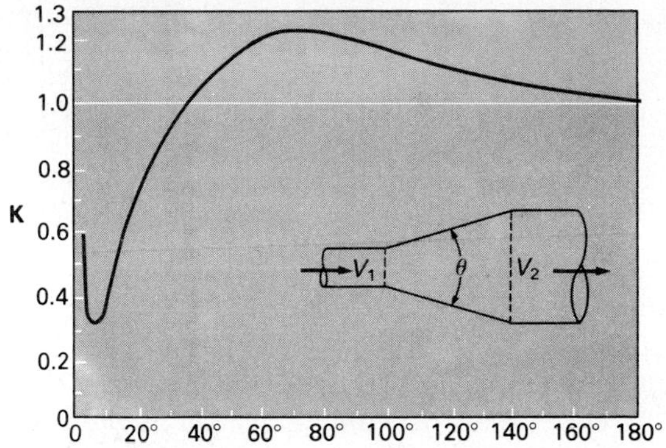

Figure 7.20 Typical values of loss coefficient for a conical gradual expansion.

and

$$\frac{p_1}{\rho} + \frac{V_1^2}{2} = \frac{p_2}{\rho} + \frac{V_2^2}{2} + gh_L,$$

respectively. For this particular flow, we neglect the average shear stress on the wall (τ_{avg}), because fluid motion near the wall is sluggish and random. We combine the continuity and momentum equations to obtain the pressure difference:

$$p_1 - p_2 = \rho V_1^2\left[\left(\frac{A_1}{A_2}\right)^2 - \frac{A_1}{A_2}\right].$$

Substituting this result into the energy equation and eliminating V_2 by the continuity equation give

$$gh_L = \left(1 - \frac{A_1}{A_2}\right)^2\left(\frac{V_1^2}{2}\right).$$

Comparing this equation with Eq. (7.99), we find

$$\mathbf{K} = \left(1 - \frac{A_1}{A_2}\right)^2. \tag{7.101}$$

Sudden expansion

Note that **K** is based on the smaller area (larger velocity). Moreover, note that local losses are not always interchangeable with pressure drop. The pressure actually *rises* in a sudden expansion because of the decrease in velocity.

Next, we consider a gradual expansion. Our ability to calculate **K** for a sudden expansion was based on the assumption of zero wall shear stress. This is not true for a gradual expansion; in fact, if the expansion is very gradual, the friction loss may be significant. Figure 7.20 shows typical loss coefficients for a gradual conical expansion as a function of included angle. An angle of 0° is a straight pipe, and an angle of 180° corresponds to a sudden enlargement. Note that for angles greater than about 40°, a sudden expansion has a lower loss coefficient. For $\theta > 40°$, separation losses are about the same as in a sudden expansion. Adding friction loss makes the net loss greater in the gradual expansion.

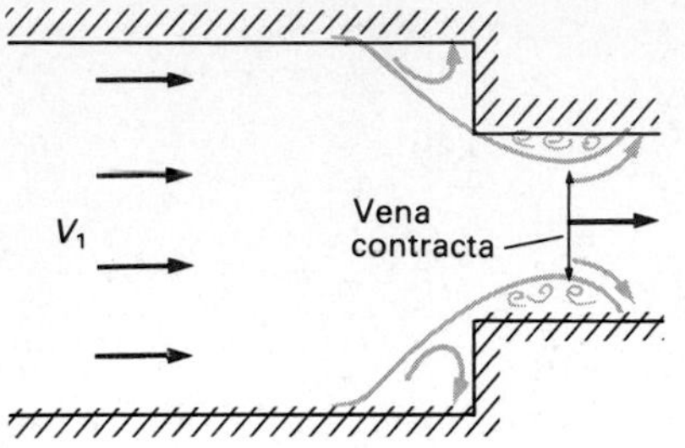

Figure 7.21 Sketch of the flow in a sudden contraction.

Table 7.7 Typical loss coefficients for gradual contractions.

Contraction Cone Angle (degrees)	K
30	0.02
45	0.04
60	0.07

Contractions generally have lower loss coefficients than do expansions. Figure 7.21 shows the flow in a sudden contraction. The flow separates at the corner, and a region of flow disturbance exists downstream. The main flow streamlines converge to a minimum area and then diverge to fill the pipe. The minimum flow area is called the *vena contracta*. An empirical equation for loss coefficient in a sudden contraction in a circular pipe is

$$\mathrm{K} \approx 0.42\left[1-\left(\frac{D_2}{D_1}\right)^2\right]. \tag{7.102}$$

Sudden contraction

This loss coefficient is based on the larger velocity (V_2). Loss coefficients for *gradual* contractions are given in Table 7.7.

Bends. Figure 7.22 shows the flow in a smooth bend of a circular pipe. Losses in bends result from

- ordinary friction losses, which are dependent on the length–diameter ratio of the bend and the relative roughness;
- flow separation on the downstream side of the bend; and
- secondary flow in the cross-sectional plane associated with centrifugal forces.

Figure 7.23 gives coefficients for calculating losses caused by separation and secondary flow in bends. Loss from friction in the bend must be added. The total loss coefficient in a 90° smooth bend is

$$\mathrm{K}_{\text{total}} = \mathrm{K}_{\text{Fig. 7.23}} + f\frac{\pi R}{2D}. \tag{7.103}$$

Smooth bends

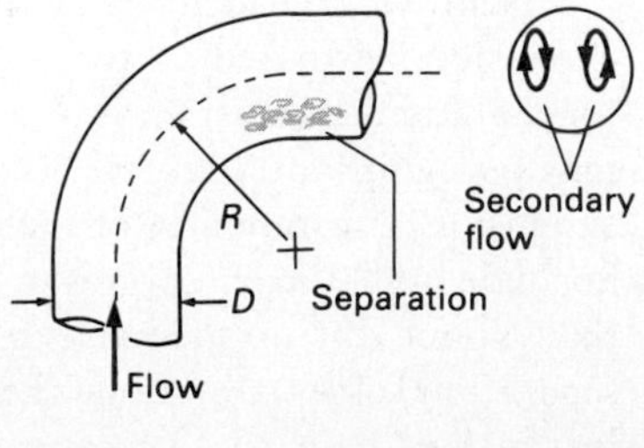

Figure 7.22 Flow in a bend in a circular pipe.

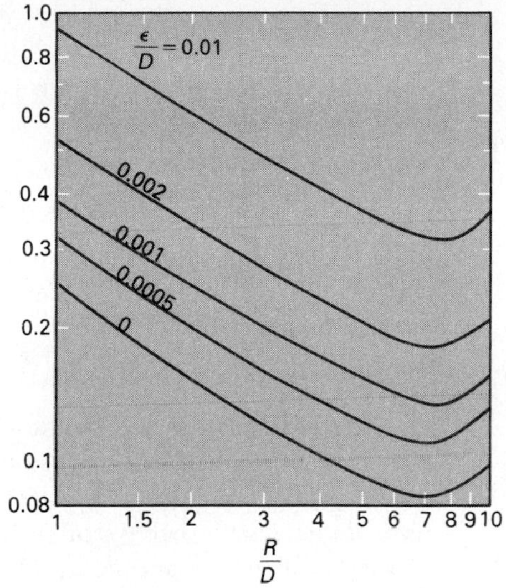

Figure 7.23 Loss coefficients for separation and secondary flow losses in circular pipe bends.

Miter bends are often used where space or construction costs must be minimized. Figure 7.24 gives equivalent lengths for miter bends. The addition of turning vanes can reduce loss by about 80 percent.

Entrances and Exits. Figure 7.25 shows three types of entrances with corresponding loss coefficients. Note that sharp corner entrances exhibit flow separa-

Figure 7.24 Equivalent length for miter bends without turning vanes.

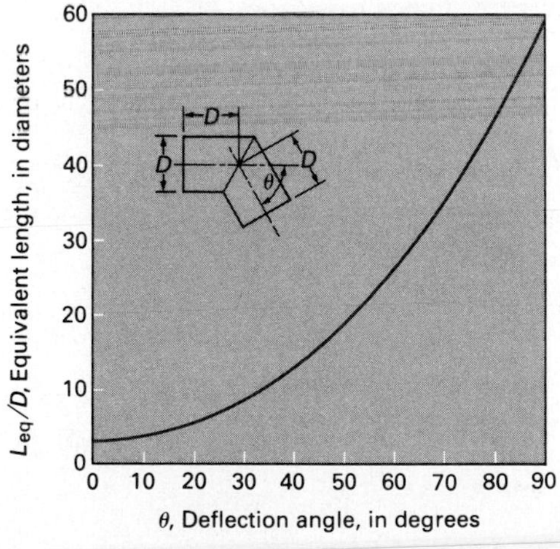

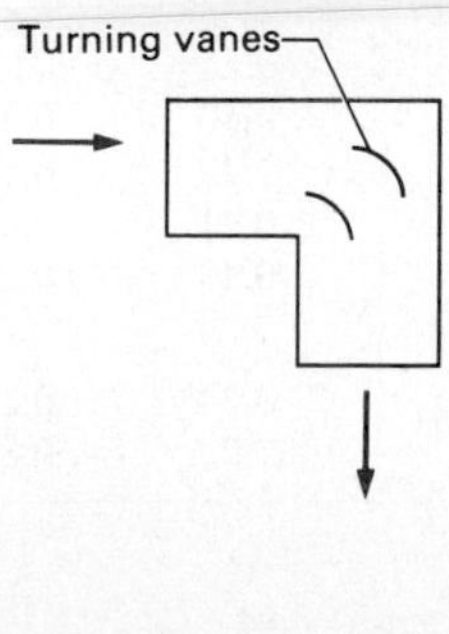

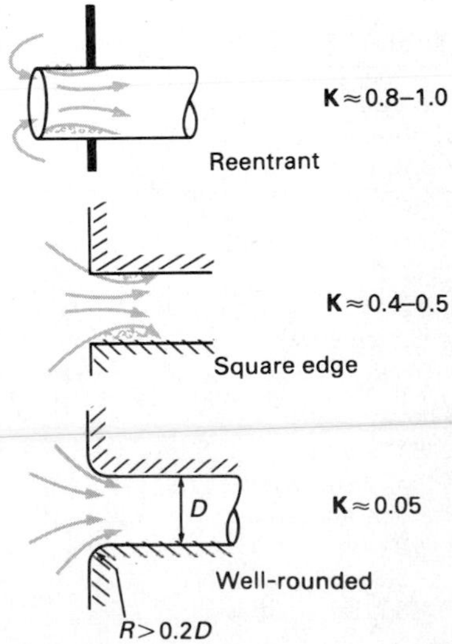

Figure 7.25 Loss coefficients for pipe entrances.

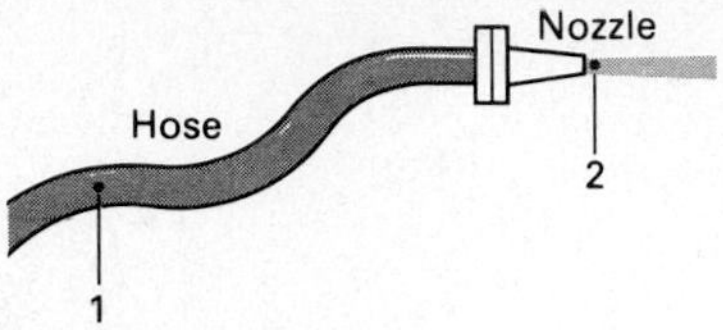

Figure 7.26 Flow through a fire hose and nozzle.

tion and a vena contracta. Loss coefficients for entrances apply only to local energy loss; pipe frictional energy loss is not included. The total energy loss between the inlet and the fully developed location is

$$gh_L = K\frac{V^2}{2} + f\left(\frac{L_e}{D}\right)\left(\frac{V^2}{2}\right),$$

where f is evaluated as if the flow were fully developed.

An exit is a sudden expansion with a very large downstream area. No matter what its shape, the loss coefficient for an exit is 1.0. A word of caution: Exit loss is just the kinetic energy of the flow at the exit. Consider calculation of flow through a fire hose and nozzle (Fig. 7.26). The energy equation for flow between points 1 and 2 is

$$\frac{p_1}{\rho} + \frac{V_1^2}{2} + gz_1 = \frac{p_2}{\rho} + \frac{V_2^2}{2} + gz_2 + gh_L.$$

Because the exit kinetic energy has already been included as $V_2^2/2$, it must not be included in the losses. Exit loss is considered only when the downstream point in the energy equation lies beyond the exit a distance sufficient to ensure that the exit kinetic energy has been dissipated.

Commercial Pipe Fittings and Valves. Smaller pipes and tubing are assembled with commercial fittings screwed, cemented, or soldered onto the pipe or

Table 7.8 Typical loss coefficients for commercial pipe fittings. Data from [10].

Nominal Diameter (in.)	Screwed or Soldered				Flanged				
	$\frac{1}{2}$	1	2	4	1	2	4	8	20
Valves (fully open)									
Globe	14	8.2	6.9	5.7	13	8.5	6.0	5.8	5.5
Gate	0.30	0.24	0.16	0.11	0.80	0.35	0.16	0.07	0.03
Swing check	5.1	2.9	2.1	2.0	2.0	2.0	2.0	2.0	2.0
Angle	9.0	4.7	2.0	1.0	4.5	2.4	2.0	2.0	2.0
Elbows									
45° standard	0.39	0.32	0.30	0.29					
45° long radius					0.21	0.20	0.19	0.16	0.14
90° standard	2.0	1.5	0.95	0.64	0.50	0.39	0.30	0.26	0.21
90° long radius	1.0	0.72	0.41	0.23	0.40	0.30	0.19	0.15	0.10
180° standard	2.0	1.5	0.95	0.64	0.41	0.35	0.30	0.25	0.20
180° long radius					0.40	0.30	0.21	0.15	0.10
Tees									
Line flow	0.90	0.90	0.90	0.90	0.24	0.19	0.14	0.10	0.07
Branch flow	2.4	1.8	1.4	1.1	1.0	0.80	0.64	0.58	0.41

Table 7.9 Loss coefficient multipliers for partially closed valves.

Condition	Ratio K/K_{open}	
	Gate Valve	Globe Valve
Open	1.0	1.0
Closed		
25%	3.0–5.0	1.5–2.0
50%	12–22	2.0–3.0
75%	70–120	6.0–8.0

bolted by flanges. Table 7.8 gives typical loss coefficients for commercial fittings. Note that a size effect is present, with larger fittings having lower loss coefficients.

Table 7.8 also contains loss coefficients for fully open valves. Valves control the flow rate by introducing large losses, especially when partly closed. Table 7.9 gives loss coefficient multipliers for partially closed valves.

The data in this section have two important limitations.

- With the exception of sudden expansions and exits, the values of **K** are based on experimental measurements and have an uncertainty of about 10 percent for elbows to as much as 50 percent for valves [10].
- The value of **K** for any particular component applies only to the component in isolation with either uniform or fully developed flow upstream. If two components are placed close together, the resulting losses usually are greater than the sum of the losses of the components in isolation.

EXAMPLE 7.8 Illustrates Calculation of Local Losses and the Difference between Pressure Loss and Pressure Drop

Find the pressure loss and pressure drop between cross sections 1 and 2 for each of the two conical expansions in Fig. E7.8a and b. Water flows at 2.0 m^3/s, $D_1 = 1.0$ m, $D_2 = 2.0$ m, $L = 2.75$ m, and $\theta = 20°$.

SOLUTION

Given

Two conical expansions in Fig. E7.8

Water flow rate 2.0 m^3/s, $D_1 = 1.0$ m, $D_2 = 2.0$ m, $L = 2.75$ m, $\theta = 20°$

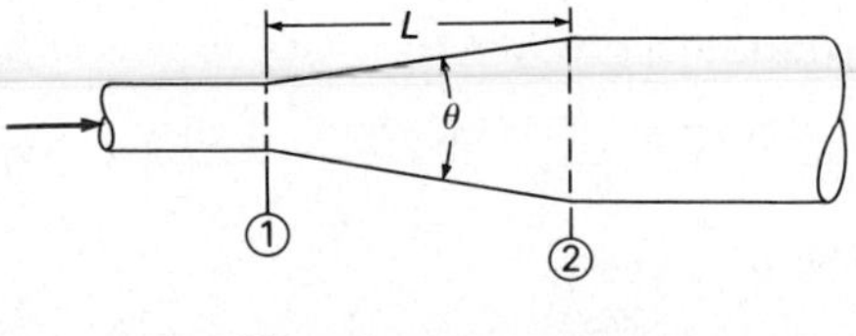

Figure E7.8a Horizontal conical gradual expansion.

Find

Pressure loss and pressure drop between sections 1 and 2 for both expansions

Solution

In either expansion, the pressure loss is

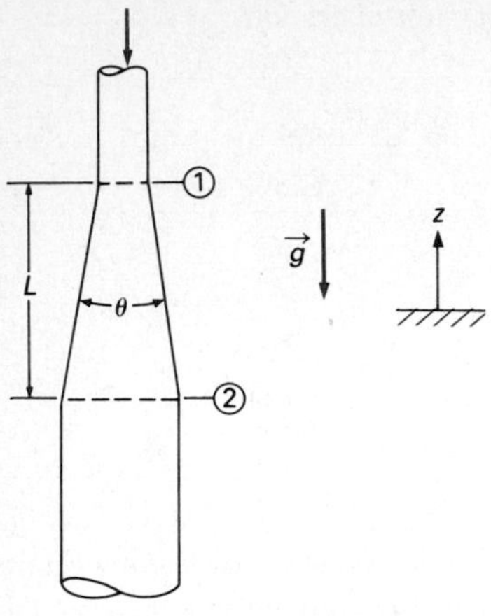

Figure E7.8b Vertical conical gradual expansion.

$$\Delta p_L = \rho g h_L = \mathbf{K}\frac{\rho V^2}{2},$$

where V is the inlet velocity. For 20°C water, Table A.5 gives $\rho = 998\ \text{kg/m}^3$. For an expansion angle of 20°, Fig. 7.20 gives $\mathbf{K} = 0.75$ based on the velocity V_1, which is calculated by

$$V_1 = \frac{Q}{A} = \frac{4Q}{\pi D_1^2} = \frac{4(2.0\ \text{m}^3/\text{s})}{\pi(1.0\ \text{m})^2} = 2.55\ \text{m/s}.$$

The pressure loss is then

$$\Delta p_L = \frac{0.75(998\ \text{kg/m}^3)(2.55\ \text{m/s})^2}{2} = 2430\ \text{kg}\cdot\text{m/s}^2\cdot\text{m}^2,$$

or

$$\Delta p_L = 2.43\ \text{kPa},$$ **ANSWER**

for both the horizontal and vertical conical expansion.

Assuming constant density, we find the pressure drop by applying the mechanical energy equation between sections 1 and 2 to obtain

$$\frac{p_1}{\rho} + \alpha_1\left(\frac{V_1^2}{2}\right) + gz_1 = \frac{p_2}{\rho} + \alpha_2\left(\frac{V_2^2}{2}\right) + gz_2 + gh_L.$$

The continuity equation gives

$$V_2 = V_1\left(\frac{A_1}{A_2}\right) = V_1\left(\frac{D_1}{D_2}\right)^2.$$

Solving the mechanical energy equation for the pressure drop ($\Delta p_a = p_1 - p_2$) gives

$$p_1 - p_2 = \rho g(z_2 - z_1) + \rho\left(\frac{V_1^2}{2}\right)\left[\alpha_2\left(\frac{D_1}{D_2}\right)^2 - \alpha_1\right] + \rho g h_L.$$

Using Eq. (7.13), we have

$$p_1 - p_2 = \rho g(z_2 - z_1) + \rho\left(\frac{V_1^2}{2}\right)\left[\alpha_2\left(\frac{D_1}{D_2}\right)^2 - \alpha_1\right] + \Delta p_L,$$

which shows that the pressure drop $(p_1 - p_2)$ also depends on the elevation change and the velocity profiles. Assuming uniform velocity at both sections 1 and 2, $\alpha_1 = \alpha_2 = 1.0$. Then

$$p_1 - p_2 = \rho g(z_2 - z_1) + \rho\left(\frac{V_1^2}{2}\right)\left[\left(\frac{D_1}{D_2}\right)^2 - 1.0\right] + \Delta p_L.$$

Substituting the numerical values for the horizontal expansion gives

$$p_1 - p_2 = \rho g(0) + \frac{(998\ \text{kg/m}^3)(2.55\ \text{m/s})^2}{2}\left[\left(\frac{1.0\ \text{m}}{2.0\ \text{m}}\right)^2 - 1.0\right] + 2430\ \text{N/m}^2 = (-2430 + 2430)\ \text{N/m}^2,$$

or

$$p_1 - p_2 = 0\ \text{Pa} \quad \text{(for horizontal expansion).}$$ **ANSWER**

This result illustrates a situation where the pressure drop caused by viscous effects is exactly offset by the pressure rise resulting from the decrease in velocity.

Substituting the numerical values for the vertical expansion gives

$$\begin{aligned} p_1 - p_2 &= (998\ \text{kg/m}^3)(9.806\ \text{m/s}^2)(-2.75\ \text{m}) \\ &\quad + \frac{(998\ \text{kg/m}^3)(2.55\ \text{m/s})^2}{2}\left[\left(\frac{1.0\ \text{m}}{2.0\ \text{m}}\right)^2 - 1.0\right] \\ &\quad + 2430\ \text{Pa} \\ &= (-26{,}900 - 2430 + 2430)\ \text{N/m}^2, \end{aligned}$$

or

$$p_1 - p_2 = -27\ \text{kPa} \qquad \text{(for vertical expansion).}$$ **ANSWER**

This negative pressure drop (i.e., a pressure rise) indicates that the pressure increase from the decrease in elevation and velocity exceeds the pressure drop from viscous effects.

Discussion

The pressure loss is the same in both cases; however, the pressure drop is not the same because it reflects the difference in the average static pressure as would be measured by two pressure gages at sections 1 and 2.

Note that the pressure loss is always positive for a real fluid flowing through a fitting, valve, or pipe. This is true regardless of the orientation of the fitting, valve, or pipe.

EXAMPLE 7.9 Illustrates Energy/Pressure Loss Calculation for a Piping System

Heavy oil having a specific gravity of 0.85 and an absolute viscosity of $4 \times 10^{-2}\ \text{N}\cdot\text{s/m}^2$ is pumped through 20 m of 0.052-m-inside-diameter PVC pipe. The pipeline is shown in Fig. E7.9 and contains one swing check valve, two gate valves, four 45° standard elbows, and a nozzle with a throat diameter of 0.026 m. A manometer connecting the inlet and throat of the nozzle reads 2.0 m of mercury. All fittings and valves have cemented connections. Find the pressure loss between points 2 and 3. Assume frictionless flow in the nozzle.

SOLUTION

Given

Sketch of piping system in Fig. E7.9

Find

Pressure loss Δp_L from point 2 to point 3

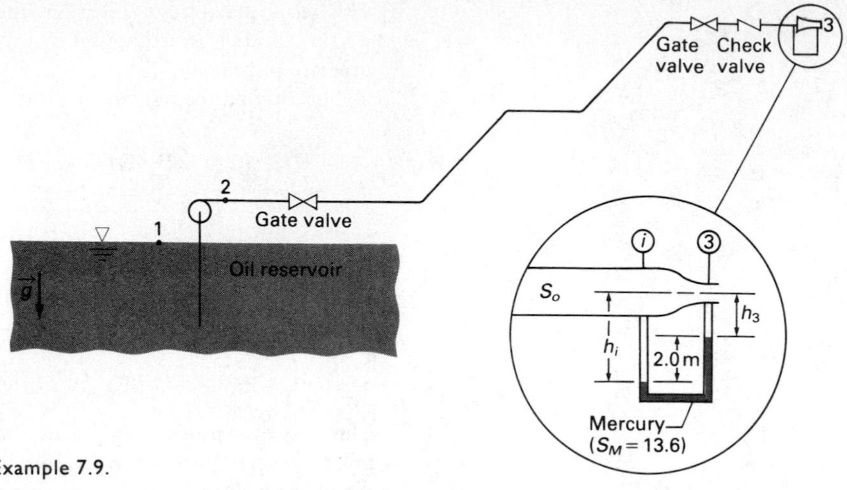

Figure E7.9 Pipe system for Example 7.9.

Solution

The pressure loss is

$$\Delta p_L = \rho g h_L.$$

We express the total energy loss as the sum of the individual energy losses:

$$g h_L = \left(\mathrm{K}\frac{V^2}{2}\right)_{\substack{\text{check}\\ \text{valve}}} + 2\left(\mathrm{K}\frac{V^2}{2}\right)_{\substack{\text{gate}\\ \text{valve}}} + 4\left(\mathrm{K}\frac{V^2}{2}\right)_{\substack{45\\ \text{standard}\\ \text{elbow}}}$$

$$+ \left[f\left(\frac{L}{D}\right)\left(\frac{V^2}{2}\right)\right]_{\text{pipe}}.$$

We do not yet know V, the average velocity. For constant density, the mechanical energy equation between the inlet and outlet of the nozzle gives

$$\frac{p_i}{\rho_0} + \alpha_i\left(\frac{V_i^2}{2}\right) + g z_i = \frac{p_3}{\rho_0} + \alpha_3\left(\frac{V_3^2}{2}\right) + g z_3 + g h_L,$$

where ρ_0 is the oil density. The nozzle is horizontal, so $z_i = z_3$. The continuity equation gives

$$V_3 = V_i\left(\frac{A_i}{A_3}\right) = V_i\left(\frac{D_i}{D_3}\right)^2.$$

Table 7.7 shows that energy loss in nozzles is small. We neglect loss in the nozzle. Also, we assume that $\alpha_i = \alpha_3 = 1.0$. The mechanical energy equation for the nozzle gives

$$\frac{V_i^2}{2}\left[\left(\frac{D_i}{D_3}\right)^4 - 1\right] = \frac{p_i - p_3}{\rho_0}.$$

We next apply the manometer rule to the nozzle manometer to obtain

$$p_3 = p_i + S_O \gamma_w h_i - S_M \gamma_W (h_i - h_3) - S_O \gamma_W h_3,$$

where S_O and S_M are the specific gravities of the oil and the mercury and γ_W is the specific weight of water at 4°C and 101,330 Pa. Simplifying, we obtain

$$p_i - p_3 = (h_i - h_3)\gamma_W(S_M - S_O),$$

or, since $V_i = V$, the mechanical energy equation for the nozzle gives

$$V = \sqrt{\frac{2g(h_i - h_3)(S_M - S_O)}{S_O[(D_i/D_3)^4 - 1]}}.$$

Using Table A.5, the numerical values give

$$V = \sqrt{\frac{2(9.807\ \text{m/s}^2)(2.0\ \text{m})(13.6 - 0.85)}{(0.85)[(0.052/0.026)^4 - 1]}} = 6.26\ \text{m/s}.$$

Noting that 0.052 m ≈ 2.0 in. and assuming that cemented connections are equivalent to screwed connections, we find values from Table 7.8:

$$\underset{\text{check valve}}{K} = 2.1, \qquad \underset{\text{gate valve}}{K} = 0.16, \qquad \underset{\text{standard elbow}}{K_{45}} = 0.30.$$

The oil kinematic viscosity is

$$\nu_o = \frac{\mu_o}{\rho_o} = \frac{(4 \times 10^{-2}\ \text{N}\cdot\text{s/m}^2)}{(0.85)(1000\ \text{kg/m}^3)} = 4.71 \times 10^{-5}\ \text{m}^2/\text{s}.$$

The pipe Reynolds number is

$$\mathsf{R} = \frac{VD}{\nu} = \frac{(6.26\ \text{m/s})(0.052\ \text{m})}{(4.71 \times 10^{-5}\ \text{m}^2/\text{s})} = 6900.$$

Table 7.1 shows that plastic pipe has zero relative roughness. Therefore the Moody chart gives $f = 0.033$. The total energy loss then is

$$gh_L = \left(\underset{\text{check valve}}{K} + 2\underset{\text{gate valve}}{K} + 4\underset{\text{standard elbow}}{K_{45}} + f\frac{L}{D}\right)\frac{V^2}{2}$$

$$= \left[2.1 + 2(0.16) + 4(0.30) + 0.033\left(\frac{20\ \text{m}}{0.052\ \text{m}}\right)\right]\frac{(6.26\ \text{m/s})^2}{2}$$

$$= 320\ \text{m}^2/\text{s}^2\ (\text{N}\cdot\text{s}^2/\text{kg}\cdot\text{m}) = 320\ \text{N}\cdot\text{m/kg}.$$

The pressure loss is

$$\Delta p_L = 0.85(1000\ \text{kg/m}^3)(320\ \text{N}\cdot\text{m/kg}),$$

or

$$\Delta p_L = 272\ \text{kPa}.$$ **ANSWER**

Discussion

Note that no energy loss term was included for the exit, because the kinetic energy at the exit was included explicitly in the energy equation.

7.4 FLOW MEASUREMENT

Up to this point, our discussion of internal flow has concentrated on calculation of energy loss or flow rate. An equally important practical problem is measurement of these quantities. Although this book is not intended to teach measurement methods, a brief discussion of internal flow measurements serves to illustrate an important application of the concepts and methods discussed in this chapter. Also, devices used for flow measurement are often permanently installed in piping systems and their energy loss/flow characteristics must be accounted for in system analysis or design.

As we have previously pointed out, measurement of losses is accomplished by measuring pressure, elevation, and, sometimes, velocity. We discussed pressure measurement in Section 2.3. Elevation measurement is quite straightforward. We discussed at least one device for measuring velocity, the Pitot-static tube, in Section 4.3.4. Measurement of losses need not be discussed further.

Flow measurement is important in laboratory experiments and is also necessary to monitor plant operation or to provide information for process control. Many different devices and techniques are available for flow measurement. Generally, a device for flow measurement is called a *flow meter* or a *fluid meter*. We discuss only the most common type of fluid meter, called the *obstruction type* or *differential pressure type*.

Figure 7.27 shows a pipe with a flow obstruction. Holes are tapped into the pipe wall upstream of the obstruction and downstream at the vena contracta, and instruments are attached to measure the pressures p_1 and p_2. Initially, we assume that losses between the upstream point 1 and downstream point 2 in the vena contracta are negligible and apply Bernoulli's equation:

$$p_1 + \frac{\rho V_1^2}{2} + \gamma z_1 = p_2 + \frac{\rho V_2^2}{2} + \gamma z_2.$$

We also use the continuity equation and assume uniform flow at points 1 and 2:

$$V_1 A_1 = V_2 A_2 = Q.$$

Combining these two equations gives

$$V_2 = \sqrt{\frac{2[(p_1 + \gamma z_1) - (p_2 + \gamma z_2)]}{\rho[1 - (A_2/A_1)^2]}}. \tag{7.104}$$

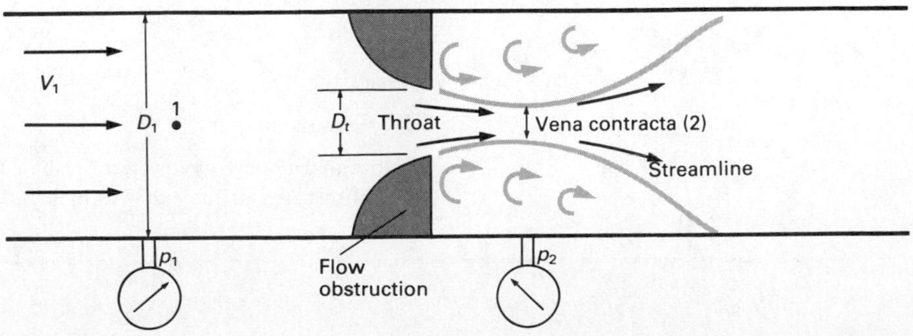

Figure 7.27 Generalized flow obstruction in a pipe.

so

$$Q = V_2 A_2 = A_2 \sqrt{\frac{2[(p_1 + \gamma z_1) - (p_2 + \gamma z_2)]}{\rho[1 - (A_2/A_1)^2]}}. \tag{7.105}$$

According to Eqs. (7.104) and (7.105), all that is necessary for flow determination (and velocity determination as well) is the measurement of pressure differential and some geometric quantities. There are three difficulties with these equations. First, the area of the vena contracta is difficult to determine. Use of the throat area of the obstruction (A_t) would be much better. Second, as there are losses in the flow, the velocity will be lower than Eq. (7.104) indicates. Third, the exact location of the vena contracta (for the pressure taps) is difficult to determine.

We handle the first two problems by defining a *contraction coefficient,*

$$\mathbf{C_c} \equiv \frac{A_2}{A_t},$$

and a *velocity coefficient,*

$$\mathbf{C_v} \equiv \frac{V_{2,\text{real}}}{V_{2,\text{ideal}}}.$$

Substituting these definitions into Eqs. (7.104) and (7.105), we have

$$V_2 = \mathbf{C_v} \sqrt{\frac{2\,\Delta p_z}{\rho[1 - (\mathbf{C_c} A_t/A_1)^2]}} \quad \text{and} \quad Q = \mathbf{C_c}\mathbf{C_v} A_t \sqrt{\frac{2\,\Delta p_z}{\rho[1 - (\mathbf{C_c} A_t/A_1)^2]}},$$

where

$$\Delta p_z \equiv (p_1 + \gamma z_1) - (p_2 + \gamma z_2).$$

The contraction coefficient and velocity coefficient are rather difficult to obtain experimentally and actually represent more detail than we need. They are both absorbed into a single *discharge coefficient* ($\mathbf{C_d}$) so that the flow equation becomes

$$Q = \mathbf{C_d} A_t \sqrt{\frac{2\,\Delta p_z}{\rho[1 - (A_t/A_1)^2]}}. \tag{7.106}$$

We also define a *flow coefficient** ($\mathbf{C_Q}$) according to

$$\mathbf{C_Q} \equiv \frac{\mathbf{C_d}}{\sqrt{1 - (A_t/A_1)^2}} = \frac{\mathbf{C_d}}{\sqrt{1 - \beta^4}}, \tag{7.107}$$

where

$$\beta \equiv \sqrt{\frac{A_t}{A_1}} \qquad (= D_t/D_1 \text{ for circular cross sections}).$$

The term $\sqrt{1 - \beta^4}$ is called the *velocity of approach factor.* Using $\mathbf{C_Q}$, we have

* The most common symbol for flow coefficient is **K**. We have used $\mathbf{C_Q}$ to avoid confusion with the loss coefficient.

for Eq. (7.106)

$$Q = \mathbf{C_Q} A_t \sqrt{\frac{2\,\Delta p_z}{\rho}}. \tag{7.108}$$

Both $\mathbf{C_d}$ and $\mathbf{C_Q}$ are functions of Reynolds number, diameter ratio, and the specific geometry of the obstruction and must be determined by experiment. The process of determining $\mathbf{C_d}$ or $\mathbf{C_Q}$ is called *calibration.* Calibration also allows us to avoid the problem of locating the vena contracta. The downstream pressure tap may be placed at any convenient location near the obstruction. Calibration gives the value of $\mathbf{C_Q}$ for that particular pressure tap location.

Any type of obstruction can be used as a fluid meter, but each requires calibration. Certain types of obstructions have been standardized, and calibration data for them are available in the literature. Professional organizations, primarily the American Society of Mechanical Engineers (ASME) and the International Organization for Standardization (ISO), have established standard designs for obstruction-type flow meters. Their publications [11, 12] provide values of the various calibration coefficients that can be used for flow meters of standard design. If you choose a standard obstruction meter, specific calibration is not necessary unless a high degree of accuracy is required.

The desirable characteristics of a fluid meter are that it

- has reliable, repeatable calibration (that is, the published values of $\mathbf{C_Q}$ and $\mathbf{C_d}$ can be used with a high degree of certainty);
- introduces small energy loss into the system;
- is inexpensive; and
- requires minimum space.

No single type of obstruction fluid meter satisfies all these requirements, so there are three standard types of obstruction flow meters (Fig. 7.28). The *Venturi* meter meets the first two requirements and the *thin-plate orifice* meets the last two requirements. The *flow nozzle* represents a compromise between them. The pressure drop Δp_d indicates the energy loss due to the device.

The Venturi meter actually is a combined nozzle and diffuser. The problem of locating the vena contracta is not present in a Venturi meter, because the maximum velocity and minimum pressure occur at the throat. The design of Venturi meters is not completely standardized because the "ideal" flow equations are reasonably accurate for a Venturi. Published calibration information for Venturi meters is usually given in terms of the discharge coefficient $\mathbf{C_d}$, which is a weak function of Reynolds number and β. Typically,

$$0.94 < \mathbf{C_d} < 0.98.$$

Consult references [11] and [12] for further details.

There are three standard designs for flow nozzles. The ASME has published both a short-radius and a long-radius design. The ISO has published a design called the "type 1932" flow nozzle. Both ASME and ISO standards specify the

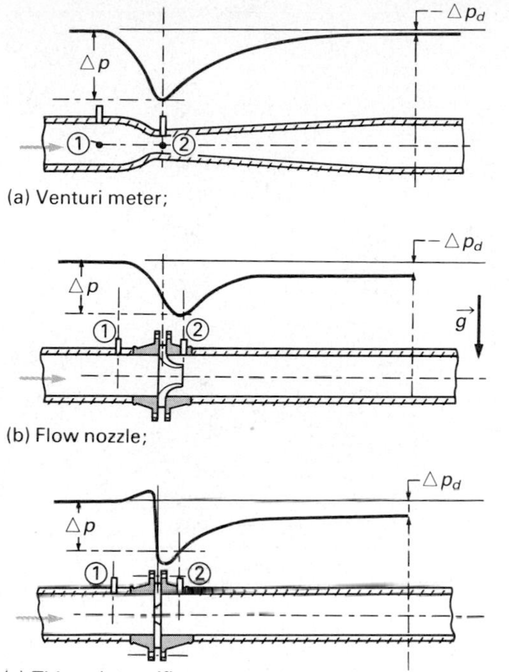

Figure 7.28 Three common types of fluid meters installed in a pipe: (a) Venturi meter; (b) flow nozzle; (c) thin-plate orifice.

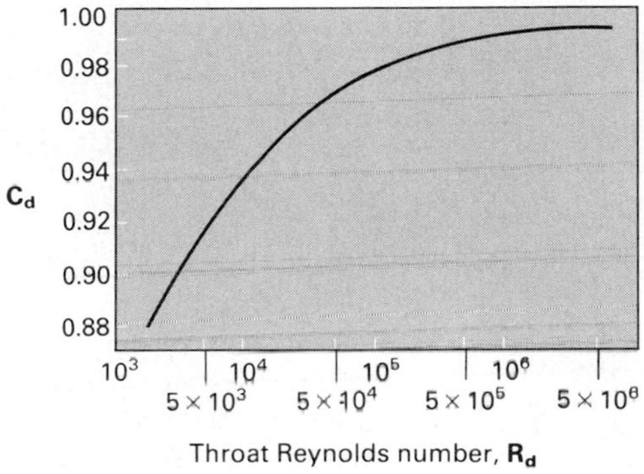

Figure 7.29 Discharge coefficient for ASME flow nozzles. Note that Reynolds number is based on throat diameter.

location of the upstream pressure tap. The downstream tap is located at the nozzle exit. Calibration data for flow nozzles are usually presented in terms of flow coefficient $\mathbf{C_Q}$, which is a function of β and Reynolds number. Figure 7.29 shows a typical plot of $\mathbf{C_d}$ versus $\mathbf{R_d}$. Curve fit equations for calculating $\mathbf{C_Q}$ and $\mathbf{C_d}$ are also available. Consult references [11] and [12] for details.

Thin-plate orifices are the cheapest type of flow meter. They are also very easy to install, because the orifice simply clamps between pipe flanges. Orifices may be of sharp edge or square edge design. There are three standard locations for upstream and downstream pressure taps for orifices.

- Vena Contracta taps. The upstream tap is one pipe diameter upstream from the orifice, and the downstream tap is located at the plane of the vena contracta. Vena contracta location is determined from an experimental curve [11].
- D and $D/2$ taps. The upstream tap is one pipe diameter upstream from the orifice, and the downstream tap is one-half diameter downstream.
- Flange taps. The pressure taps are located one inch upstream and one inch downstream from the orifice.

Each pressure tap location requires a different calibration curve, as do sharp edge and square edge orifices. Figure 7.30 shows the variation of $\mathbf{C_Q}$ with β and $\mathbf{R_d}$ for a sharp edge orifice with D and $D/2$ taps. Further information, including

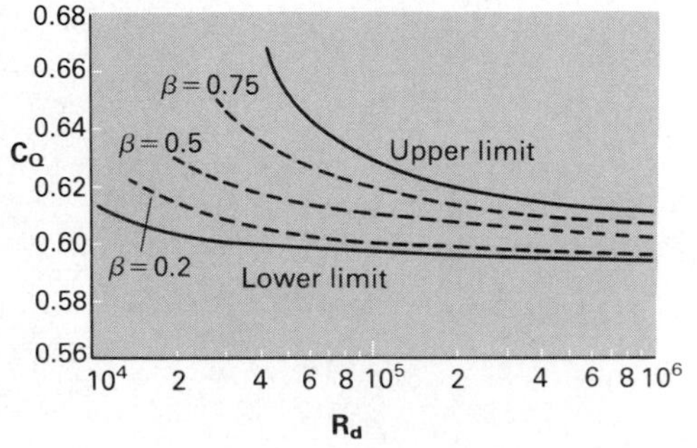

Figure 7.30 Flow coefficients for sharp-edged orifice plate with *D* and *D*/2 taps.

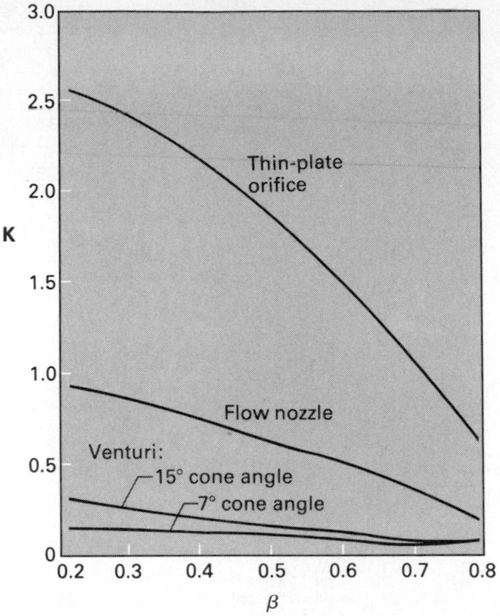

Figure 7.31 Typical loss coefficients for fluid meters. **K** is based on throat velocity.

tables and curve fit equations for flow coefficient and discharge coefficient, is available in references [11, 12, 13]. They also give information on eccentric orifices in which the hole is not concentric with the pipe.

If a fluid meter is part of a complete flow system, energy loss because of the meter must be included in the system resistance. Figure 7.31 shows typical loss coefficients (based on the throat velocity $V_t = Q/A_t$) for fluid meters.

If you use fluid meters of standard design, you can use the published information on calibration coefficients, but you must exercise some care. The published coefficients are valid only if the flow entering the meter is fully developed; accordingly, the fluid meter must be located a sufficient distance downstream from flow disturbances such as elbows, valves, and so on. Information on acceptable locations is given in reference [11]. Even if all restrictions are met, the calibration information is accurate only to about ± 2 percent (Venturi meters have better accuracy and orifices may have worse). For extreme accuracy, a flow meter should be calibrated in place.

Many other flow measurement methods are available. Consult references [11, 14, 15] for information. Two basic methods are the "catch-and-weigh" technique and the velocity traverse method. In the catch-and-weigh method, the flow from the piping system is collected in a tank. The quantity of fluid collected in a given time period (measured with a stopwatch) is determined by weighing. The flow rate is the quantity collected divided by the elapsed time. Obviously, this method works only for liquid flows.

In the velocity traverse method, a velocity-measuring instrument, typically a Pitot-static tube, is positioned at various locations in a cross-sectional plane. The measured velocity distribution can be integrated (graphically or numerically) over the cross section to compute the flow. Details of the velocity traverse method can be found in references [15, 16, 17]. Both the catch-and-weigh and velocity traverse methods can be used to calibrate other types of fluid meters.

EXAMPLE 7.10 Illustrates Considerations in Selecting Flow Meters

You are an engineer for a small company and are to select an appropriate fluid meter from your warehouse stock to measure the water flow rate in a 6-in. schedule 40, horizontal commercial steel pipe. The fluid meter is needed immediately, so no time is available for machining or modification. The flow rate is estimated to be between 100 and 400 gal/min. A mercury manometer is to be used to measure the appropriate pressure difference to determine the flow rate.

Your instructions are to choose a fluid meter to determine the flow rate with a maximum uncertainty of 10 percent because of errors in reading the manometer. You estimate that the manometer can be read with uncertainty of 0.05 in. The net pressure drop across the meter must not exceed 5.0 psi. The following meters are available:

Type of Fluid Meter	Throat Diameter
Venturi	4.5 in.
ASME long-radius flow nozzle	2.5 in.
Thin-plate orifice	2.5 in.

SOLUTION

Given

Three fluid meters:

ASME long-radius flow nozzle with 2.5 in. throat diameter
Thin-plate orifice with 2.5 in. throat diameter
Venturi with 4.5 in. throat diameter

Water flow rate estimated between 100 and 400 gal/min

Mercury manometer that can be read reliably to within 0.05 in.

6-in. schedule 40, horizontal commercial steel pipe

Net pressure drop (pressure loss) not to exceed 5.0 psi

Figure E7.10

Find

Which meter(s) measure the flow rate with a maximum uncertainty of 10 percent in the manometer reading and maximum pressure loss of 5.0 psi

Solution

Each fluid meter and manometer combination must satisfy two requirements.

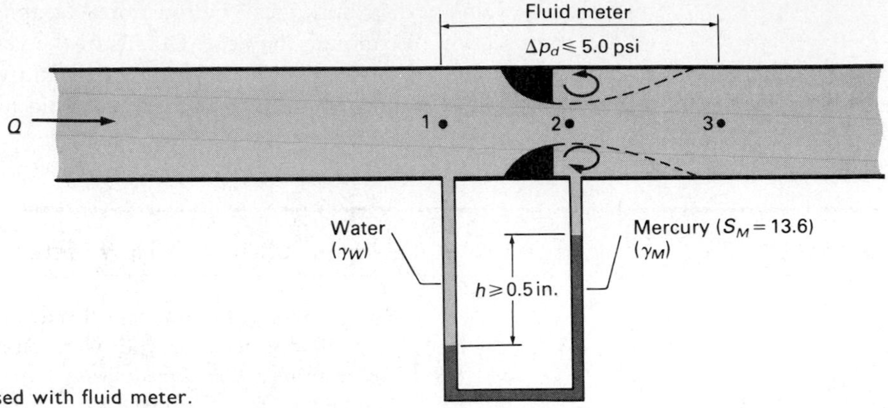

Figure E7.10 Manometer used with fluid meter.

- The minimum manometer reading must be at least 0.5 in., so that a 0.05-in. reading error represents no more than 10 percent of the reading. This reading must be for 100 gal/min since a 400-gal/min flow rate would give a higher reading.
- Net pressure drop cannot exceed 5.0 psi for the maximum flow rate of 400 gal/min.

The relation between the flow rate and the maximum pressure drop in a fluid meter is given by Eq. (7.106)

$$Q = \mathbf{C_d} A_t \sqrt{\frac{2\,\Delta p_z}{\rho[1-(A_t/A_1)^2]}}.$$

For a horizontal pipe, $\Delta p_z = p_1 - p_2 = \Delta p$. Solving for Δp gives

$$\Delta p = \frac{\rho}{2}\left[1-\left(\frac{A_t}{A_1}\right)^2\right]\left(\frac{Q}{\mathbf{C_d} A_t}\right)^2.$$

Applying the manometer rule (Fig. E7.10) gives

$$p_2 = p_1 + (\gamma_W - \gamma_M)h \quad \text{or} \quad \Delta p = p_1 - p_2 = (\gamma_M - \gamma_W)h.$$

Eliminating Δp gives the manometer reading h in terms of the flow rate Q. Introducing the diameters results in

$$h = \frac{\rho[1-(D_t/D_1)^4]}{2(\gamma_M - \gamma_W)}\left[\frac{4Q}{\mathbf{C_d}\pi D_t^2}\right]^2.$$

For 60°F water, Table A.6 gives

$$\rho = 62.4 \text{ lbm/ft}^3, \qquad \gamma_W = 62.4 \text{ lb/ft}^3,$$

and

$$\gamma_M = 13.6(62.4 \text{ lb/ft}^3) = 849 \text{ lb/ft}^3.$$

For a 6-in. schedule 40 commercial steel pipe, Table D.1 gives $D_1 =$ 6.065 in. We now calculate the manometer reading for each fluid meter with a 100-gal/min flow rate. For a Venturi meter, the text shows $0.94 < \mathbf{C_d} < 0.98$. The minimum manometer reading for a Venturi–manometer system and 100 gal/min is obtained using

$\mathbf{C_d} = 0.94$:

$$h = \frac{(62.4 \text{ lbm/ft}^3)[1 - (4.5 \text{ in.}/6.065 \text{ in.})^4]}{2(849 - 62.4) \text{ lb/ft}^3}$$
$$\times \left[\frac{4(100 \text{ gal/min})(1 \text{ ft}^3/7.48 \text{ gal})(1 \text{ min}/60 \text{ sec})}{(0.94)\pi(4.5 \text{ in.})^2(1 \text{ ft}/12 \text{ in.})^2}\right]^2$$
$$= 0.254 \text{ lbm}\cdot\text{ft}^2/\text{lb}\cdot\text{sec}^2 \text{ (12 in./ft)}\left(\frac{1 \text{ lb}\cdot\text{sec}^2}{32.2 \text{ ft}\cdot\text{lbm}}\right) = 0.095 \text{ in.}$$

The Venturi meter is not acceptable, because its minimum manometer reading is less than 0.5 in.

Next, we consider the ASME flow nozzle. We first calculate the Reynolds number based on the throat diameter for 60°F water; Table A.6 gives $\nu = 1.22 \times 10^{-5}$ ft²/sec.

Then

$$\mathbf{R_d} = \frac{V_t D_t}{\nu} = \frac{4Q}{\pi \nu D_t}$$
$$= \frac{4(100 \text{ gal/min})(1 \text{ ft}^3/7.48 \text{ gal})(1 \text{ min}/60 \text{ sec})}{\pi(1.22 \times 10^{-5} \text{ ft}^2/\text{sec})(2.5 \text{ in.})(1 \text{ ft}/12 \text{ in.})} = 1.12 \times 10^5.$$

Figure 7.29 gives $\mathbf{C_d} = 0.98$. The minimum manometer reading for the flow nozzle-manometer system then is

$$h = \frac{(62.4 \text{ lbm/ft}^3)[1 - (2.5 \text{ in.}/6.065 \text{ in.})^4]}{2(849 - 62.4) \text{ lb/ft}^3}$$
$$\times \left[\frac{4(100 \text{ gal/min})(1 \text{ ft}^3/7.48 \text{ gal})(1 \text{ min}/60 \text{ sec})}{(0.98)\pi(2.5 \text{ in.})^2(1 \text{ ft}/12 \text{ in.})^2}\right]^2$$
$$= 1.72 \text{ lbm}\cdot\text{ft}^2/\text{lb}\cdot\text{sec}^2 \text{ (12 in./ft)}\left(\frac{1 \text{ lb}\cdot\text{sec}^2}{32.2 \text{ ft}\cdot\text{lbm}}\right) = 0.64 \text{ in.}$$

The flow nozzle provides a high-enough reading. Now we consider the orifice. The Reynolds number based on the throat diameter is the same as for the flow nozzle, so $\mathbf{R_d} = 1.12 \times 10^5$. The β ratio is

$$\beta = \frac{D_t}{D_1} = \frac{2.5 \text{ in.}}{6.065 \text{ in.}} = 0.412.$$

Figure 7.30 gives $\mathbf{C_d} = 0.61$. The minimum manometer reading for the orifice-manometer system is

$$h = \frac{(62.4 \text{ lbm/ft}^3)[1 - (2.5 \text{ in.}/6.065 \text{ in.})^4]}{2(849 - 62.4) \text{ lb/ft}^3}$$
$$\times \left[\frac{4(100 \text{ gal/min})(1 \text{ ft}^3/7.48 \text{ gal})(1 \text{ min}/60 \text{ sec})}{(0.61)\pi(2.5 \text{ in.})^2(1 \text{ ft}/12 \text{ in.})^2}\right]^2$$
$$= 4.42 \text{ lbm}\cdot\text{ft}^2/\text{lb}\cdot\text{sec}^2 \text{ (12 in./ft)}\left(\frac{1 \text{ lb}\cdot\text{sec}^2}{32.2 \text{ ft}\cdot\text{lbm}}\right) = 1.65 \text{ in.}$$

The orifice also provides a high-enough manometer reading. At this point, only the nozzle and the orifice satisfy the first requirement. We now check to determine whether these two satisfy the second requirement that the flow meter must have a net pressure drop less than 5.0 psi. This net pressure drop is found by applying the mechanical energy equation between inlet 1 and outlet 3 (see Fig. E7.6):

$$\frac{p_1}{\rho} + \frac{V_1^2}{2} + gz_1 = \frac{p_3}{\rho} + \frac{V_3^2}{2} + gz_3 + gh_L,$$

where p, V, and z represent average values. The fluid meter is horizontal, so $z_1 = z_3$. The pipe flow Reynolds number for 400 gal/min is

$$\mathbf{R} = \frac{V_1 D_1}{\nu} = \frac{4Q}{\pi \nu D_1}$$

$$= \frac{4(400 \text{ gal/min})(1 \text{ ft}^3/7.48 \text{ gal})(1 \text{ min}/60 \text{ sec})}{\pi(1.22 \times 10^{-5} \text{ ft}^2/\text{sec})(6.065 \text{ in.})(1 \text{ ft}/12 \text{ in.})} = 1.84 \times 10^5.$$

The pipe flow is turbulent. Noting that $D_1 = D_3$, we get, from the continuity equation, $V_1 = V_3$. The mechanical energy equation reduces to $p_1 - p_3 = \rho g h_L$. Using Eq. (7.99) to express the energy loss, gh_L, in terms of a loss coefficient, $\mathbf{K}$, we obtain

$$p_1 - p_3 = \rho \mathbf{K} \frac{V_t^2}{2},$$

where $p_1 - p_3$ is the net pressure drop across the fluid meter. We first consider the flow nozzle; its β ratio is the same as that for the orifice:

$$\beta = \frac{D_t}{D_1} = 0.412.$$

Figure 7.31 gives $\mathbf{K} = 0.75$. The throat velocity is

$$V_t = \frac{Q}{A_t} = \frac{4Q}{\pi D_t^2} = \frac{4(400 \text{ gal/min})(1 \text{ min}/60 \text{ sec})(1 \text{ ft}^3/7.48 \text{ gal})}{\pi(2.5 \text{ in.})^2(1 \text{ ft}/12 \text{ in.})^2}$$

$$= 26.13 \text{ ft/sec}.$$

Then

$$p_1 - p_3 = \frac{(62.4 \text{ lbm/ft}^3)(0.75)(26.13 \text{ ft/sec})^2}{2(32.2 \text{ ft}\cdot\text{lbm/lb}\cdot\text{sec}^2)(144 \text{ in}^2/\text{ft}^2)} = 3.44 \text{ psi}.$$

The flow nozzle-manometer system satisfies both requirements, so it is acceptable. Next we consider the orifice. It has the same β ratio and pipe velocity as the flow nozzle. Figure 7.31 gives $\mathbf{K} = 2.1$ so

$$p_1 - p_3 = \frac{(62.4 \text{ lbm/ft}^3)(2.1)(26.13 \text{ ft/sec})^2}{2(32.2 \text{ ft}\cdot\text{lbm/lb}\cdot\text{sec}^2)(144 \text{ in}^2/\text{ft}^2)} = 9.63 \text{ psi}.$$

The orifice is not acceptable, because its net pressure drop is too high. Our conclusion is:

Use the ASME flow nozzle. **ANSWER**

Discussion

The actual uncertainty of the measured flow rate is larger than the uncertainty owing to the manometer reading alone. One source of additional uncertainty is the discharge coefficient. See references [14] and [15] for discussion of uncertainty analysis.

We presented the minimum manometer readings for the Venturi meter, the flow nozzle, and orifice meters. Only those calculations that were essential to a decision were presented in this solution. We presented a so-called worst-case analysis. In practice, you may not know beforehand which of several calculations are the crucial ones and may make more than the minimum necessary.

The net pressure drop would be smaller if the Venturi meter could be used. It might be worth the trouble to search for a more sensitive pressure measuring device. Could you drain the mercury from the manometer and use water? Explain.

7.5 ANALYSIS AND DESIGN OF PIPE AND DUCT SYSTEMS

The analysis or design of a pipe or duct system is accomplished by patching together the information that we presented about pipes, ducts, and components. In an analysis problem, the system configuration is given and the engineer wants to calculate flow rate, energy loss, and pressure drop or required pumping power to maintain flow in the system. In a design problem, the engineer needs to select pipe sizes, specify number and location of pumps or fans, and select other system geometric and operating parameters to obtain a given flow rate (or flow rates if a multiple branch system is required) at an economical cost.

7.5.1 Working Equations and Grade Lines

The mathematical tools available for pipe system analysis are the mechanical energy equation, the continuity equation, and the friction factor and loss coefficient information that we developed. We also consider a graphic tool that helps engineers visualize the energy and pressure distributions throughout the system.

Let's consider the pipe system shown in Fig. 7.32. The energy equation for flow between planes 1 and 2 is

$$\dot{m}\left[\frac{p_1}{\rho} + \alpha_1\left(\frac{V_1^2}{2}\right) + gz_1\right] - \dot{W}_s = \dot{m}\left[\frac{p_2}{\rho} + \alpha_2\left(\frac{V_2^2}{2}\right) + gz_2\right] + \dot{m}(gh_L),$$

where $\dot{W}_s$ is the sum of all shaft work done on the fluid; in the system shown in Fig. 7.32, $\dot{W}_s$ includes both pump work and turbine work. The term $\dot{m}(gh_L)$ represents the sum of all mechanical energy losses in the system. It includes friction loss

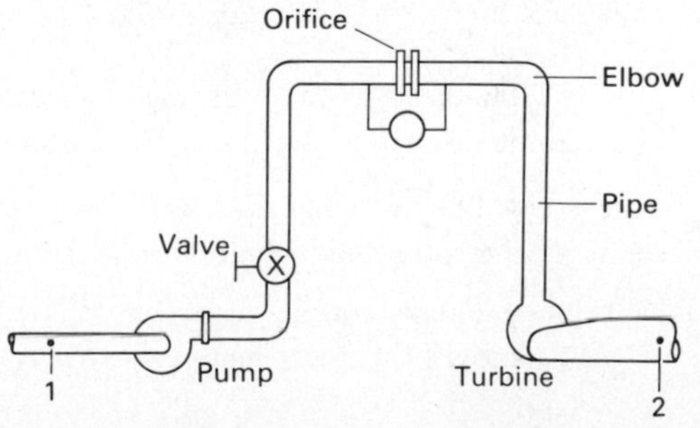

Figure 7.32 Pipe system.

in the pipes, local losses at the fittings, and losses within the pump and turbine. We rewrite the mechanical energy equation as

$$\dot{m}\left[\frac{p_1}{\rho} + \alpha_1 \frac{V_1^2}{2} + gz_1\right] + \dot{W}_p - \dot{m}(gh_{L,p})$$
$$= \dot{m}\left[\frac{p_2}{\rho} + \alpha_2 \frac{V_2^2}{2} + gz_2\right] + \dot{m}(gh_{L,\text{sys}}) + \dot{W}_t + \dot{m}(gh_{L,t}).$$

In this equation, $\dot{W}_p$ is the power input to the fluid by the pump, $\dot{W}_t$ is the power delivered to the turbine by the fluid, $gh_{L,p}$ is the mechanical energy loss within the pump, $gh_{L,t}$ is the mechanical energy loss within the turbine, and $gh_{L,\text{sys}}$ is the mechanical energy loss in the passive components of the system (pipes and fittings).

The net effect of a pump is to increase the fluid's mechanical energy and the net effect of a turbine is to decrease the fluid's mechanical energy. We define *pump head* (h_p) and *turbine head* (h_t) as follows:

$$h_p \equiv \frac{\dot{W}_p/\dot{m} - gh_{L,p}}{g} \tag{7.109}$$

and

$$h_t \equiv \frac{\dot{W}_t/\dot{m} + gh_{L,t}}{g}. \tag{7.110}$$

If we divide the mechanical energy equation by $\dot{m}g$, we get the "head form" of the mechanical energy equation:

$$\frac{p_1}{\gamma} + \alpha_1\left(\frac{V_1^2}{2g}\right) + z_1 + h_p = \frac{p_2}{\gamma} + \alpha_2\left(\frac{V_2^2}{2g}\right) + z_2 + h_t + h_{L,\text{sys}}. \tag{7.111}$$

Customarily, we ignore the kinetic energy correction factors in Eq. (7.111). Some justifications of this simplification are that 1 and 2 often lie in reservoirs where the velocity is zero; that loss coefficients are often obtained from experiments that do not consider the effect of α, so it would be inconsistent to include such effects in the energy equation; and that the slight increase in accuracy that might be gained by using α's is completely offset by the uncertainty in typical values of loss coefficients and friction factors. Also, we drop the subscript, sys, from $h_{L,\text{sys}}$, because we no longer need to distinguish between system and pump losses. The working equation for pipe system analysis is thus

$$\frac{p_1}{\gamma} + \frac{V_1^2}{2g} + z_1 + h_p = \frac{p_2}{\gamma} + \frac{V_2^2}{2g} + z_2 + h_t + h_L. \tag{7.112}$$

The head form lends itself to a graphic interpretation of the mechanical energy balance of a flowing fluid. We define two grade lines.

- The *energy grade line* (EGL) is a line drawn above the $z = 0$ datum that shows the total head of the fluid. The total head is the sum $p/\gamma + V^2/2g + z$.
- The *hydraulic grade line* (HGL) is a line drawn above the $z = 0$ datum that shows the piezometric head of the fluid. The piezometric head is the sum $p/\gamma + z$.

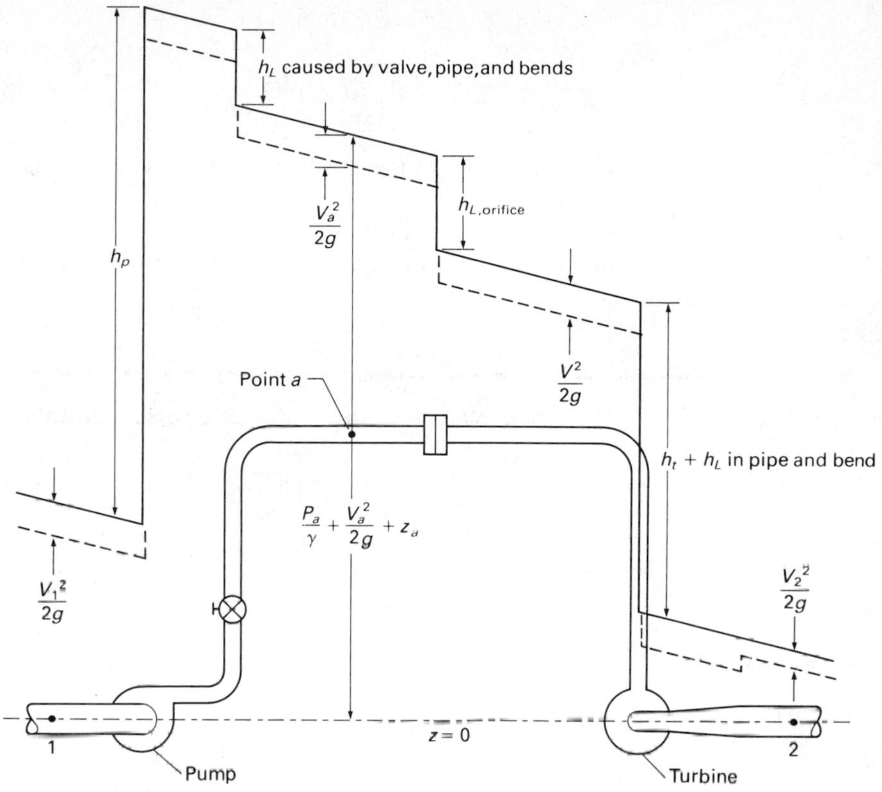

Figure 7.33 EGL(——) and HGL(– – –) for piping system. Note gradual downslopes resulting from pipe friction.

Figure 7.33 shows the EGL and HGL for the piping system of Fig. 7.32. These lines rise abruptly at pumps because of the pump head, fall abruptly at fittings because of local losses, fall abruptly at turbines because of the turbine head, and fall gradually along the length of a pipe because of pipe friction losses. Sketching the EGL and/or the HGL often helps piping engineers visualize the energy relationships in a flow.

The system head losses are calculated from the pipe friction and local loss models that we developed in Sections 7.2–7.4:

$$h_L = \sum h_{L,\text{friction}} + \sum h_{L,\text{local}} = \sum f\left(\frac{L}{D_h}\right)\left(\frac{V^2}{2g}\right) + \sum \mathrm{K}\left(\frac{V^2}{2g}\right). \tag{7.113}$$

If any local losses are evaluated by the equivalent length method, the equivalent lengths are included in L.

If all pipes have the same cross-sectional area, they all have the same average velocity, so Eq. (7.113) becomes

$$h_L = \left(f\frac{L}{D_h} + \sum \mathrm{K}\right)\frac{V^2}{2g}. \tag{7.114}$$

If the pipe/duct has elements of different cross-sectional area, the flow rate may be a more convenient parameter than the velocity:

$$h_L = \left(\sum \frac{f}{A^2}\frac{L}{D_h} + \sum \frac{K}{A^2}\right)\frac{Q^2}{2g}, \tag{7.115}$$

where A is the inside cross-sectional area of the segment of pipe or duct under consideration.

Example 7.9 illustrated calculation of energy loss/pressure drop in a single-branch piping system. The next two examples illustrate flow rate calculation and pipe sizing.

EXAMPLE 7.11 Illustrates Velocity/Flow Rate Calculation in a Piping System

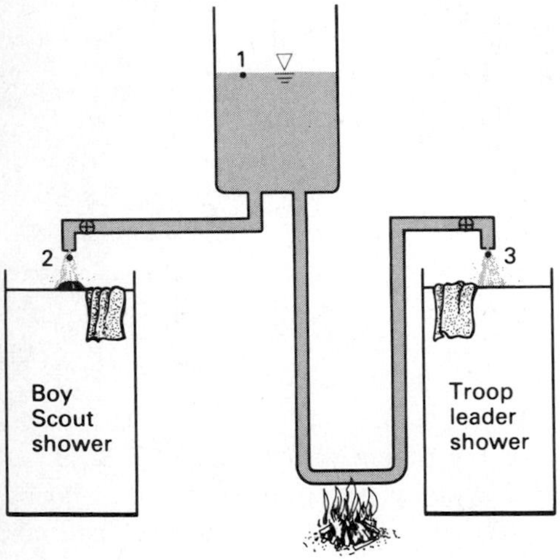

Figure E7.11 Boy Scout shower.

This example takes a closer look at the Boy Scout camp shower in Example 4.9. The Scouts used 55°F water, while the leaders heated their water to 100°F. Both showers used about 25 ft of $\frac{1}{2}$-in. (inside diameter = 0.569 in.) copper pipe and 90° standard elbows. Estimate the volume flow rate to each shower when the water level in the drum is 3 ft above the pipe discharge.

SOLUTION

Given

Camp showers shown in Fig. E7.11

25 ft of $\frac{1}{2}$-in. copper pipe to each shower with 90° standard elbows

Scouts' water 55°F

Leaders' water 100°F

Find

Volume flow rate to each shower when drum water level is 3 ft above pipe discharge

Solution

We start with the steady-state, mechanical energy equation in head form:

$$\frac{p_1}{\gamma} + \frac{V_1^2}{2g} + z_1 = \frac{p_2}{\gamma} + \frac{V_2^2}{2g} + z_2 + h_L$$

and

$$\frac{p_1}{\gamma} + \frac{V_1^2}{2g} + z_1 = \frac{p_3}{\gamma} + \frac{V_3^2}{2g} + z_3 + h_L,$$

where p, V, and z represent average values at each location. The h_p and h_t terms were omitted because there are no pumps or turbines, and $V_1 \approx 0$ for each case. Also $p_1 = p_2 = p_3$.

With the subscript B denoting the boys' shower and L denoting the leaders' shower, the mechanical energy equation gives

$$V_B^2 = 2g(z_1 - z_2 - h_L)_B \quad \text{and} \quad V_L^2 = 2g(z_1 - z_3 - h_L)_L.$$

In each case, the total head loss is the sum of the individual head losses:*

$$h_L = \left(K\frac{V^2}{2g}\right)_{\text{entrance}} + 2\left(K\frac{V^2}{2g}\right)_{\substack{90\\\text{standard}\\\text{elbow}}} + \left[f\left(\frac{L}{D}\right)\left(\frac{V^2}{2g}\right)\right]_{\text{pipe}}$$

$$= \left(K_{\text{entrance}} + nK_{\substack{90\\\text{standard}\\\text{elbow}}} + f\frac{L}{D}\right)\frac{V^2}{2g} \quad (n = 2 \text{ for boys, 4 for leaders}).$$

Substituting into the mechanical energy equation, we obtain

$$V_B = \sqrt{\frac{2g(z_1 - z_2)}{1 + K_{\text{entrance}} + 2K_{\substack{90\\\text{standard}\\\text{elbow}}} + f(L/D)}};$$

$$V_L = \sqrt{\frac{2g(z_1 - z_3)}{1 + K_{\text{entrance}} + 4K_{\substack{90\\\text{standard}\\\text{elbow}}} + f(L/D)}}.$$

Assuming a reentrant pipe entrance for the $\frac{1}{2}$-in. copper pipe, we find from Fig. 7.25 and Table 7.8 that

$$K_{\text{entrance}} \approx 1.0 \quad \text{and} \quad K_{\substack{90\\\text{standard}\\\text{elbow}}} = 2.0,$$

assuming crude field installation of the entrance and soldered or screwed connections at the elbow. We find the friction factor by using the relative roughness and the Reynolds number. The (unknown) velocity is involved in the Reynolds number, so iteration is necessary. For copper pipe (drawn tubing), Table 7.1 gives

$$\frac{\varepsilon}{D} = \frac{0.000005 \text{ ft}}{0.569 \text{ in.}}(12 \text{ in./ft}) = 0.00011.$$

Guessing that $f = 0.013$ (the complete turbulence value from Fig. 7.9), we calculate each velocity as

$$V_B = 3.88 \text{ ft/sec}$$
$$V_L = 3.38 \text{ ft/sec}.$$

We next check our guess of the friction factor by calculating the Reynolds number. From Table A.6,

$$\nu(55°\text{F}) = 1.33 \times 10^{-5} \text{ ft}^2/\text{sec}$$

and

$$\nu(100°\text{F}) = 0.739 \times 10^{-5} \text{ ft}^2/\text{sec}.$$

The corresponding Reynolds numbers are

$$R_B = \frac{VD}{\nu} = \frac{(3.88 \text{ ft/sec})(0.569 \text{ in.})}{(1.33 \times 10^{-5} \text{ ft}^2/\text{sec})}\left(\frac{1 \text{ ft}}{12 \text{ in.}}\right) = 1.4 \times 10^4$$

*Note that an exit energy loss term is not included, because the exit kinetic energy is already included.

and

$$\mathbf{R}_L = \frac{VD}{\nu} = \frac{(3.38 \text{ ft/sec})(0.569 \text{ in.})}{(0.739 \times 10^{-5} \text{ ft}^2/\text{sec})}\left(\frac{1 \text{ ft}}{12 \text{ in.}}\right) = 2.2 \times 10^4.$$

Figure 7.9 gives $f_B = 0.028$ and $f_L = 0.025$.

Repeating the calculations with these new values of friction factor and iterating until friction factors and velocities do not change, we obtain

$$\mathbf{R}_B = 1.1 \times 10^4, \qquad \mathbf{R}_L = 1.8 \times 10^4, \qquad f_B = 0.030, \qquad f_L = 0.026,$$
$$V_B = 2.98 \text{ ft/sec}, \quad \text{and} \quad V_L = 2.85 \text{ ft/sec}.$$

The flow rates are

$$Q_B = V_B A = V_B\left(\frac{\pi D^2}{4}\right)$$
$$= (2.98 \text{ ft/sec})\,\frac{\pi}{4}\left(\frac{0.569 \text{ in.}}{12 \text{ in./ft}}\right)^2 (60 \text{ sec/min}),$$

or

$$Q_B = 0.32 \text{ ft}^3/\text{min},$$ **ANSWER**

and

$$Q_L = V_L A = V_L\left(\frac{\pi D^2}{4}\right)$$
$$= (2.85 \text{ ft/sec})\,\frac{\pi}{4}\left(\frac{0.569 \text{ in.}}{12 \text{ in./ft}}\right)^2 (60 \text{ sec/min}),$$

or

$$Q_L = 0.30 \text{ ft}^3/\text{min}.$$ **ANSWER**

Discussion

Note the drastic change in the calculated exit velocity (and the flow rate) caused by including the resistance of the copper pipe, the entrance, and the elbows.

Lossless Flow (Example 4.9)	Flow with Losses (Example 7.11)
$V_B = 13.9$ ft/sec	$V_B = 2.98$ ft/sec
$V_L = 13.9$ ft/sec	$V_L = 2.85$ ft/sec

This comparison also shows that heating the fluid has a slight effect on the fluid velocity, as well as providing a more enjoyable shower.

We had to solve this problem by iteration. This type of problem cannot be solved directly with Eqs. (7.63)–(7.65), because local losses, as well as pipe friction losses, are present.

EXAMPLE 7.12 Illustrates the Pipe Sizing Problem in a Pipe System

Find the schedule 40 pipe size required to deliver at least 700 gal/min from the higher reservoir to the lower reservoir in Fig. E7.12 for the water levels shown. The line contains 250 ft of straight galvanized iron pipe, three fully open globe valves, and six 90° standard elbows. All connections are flanged.

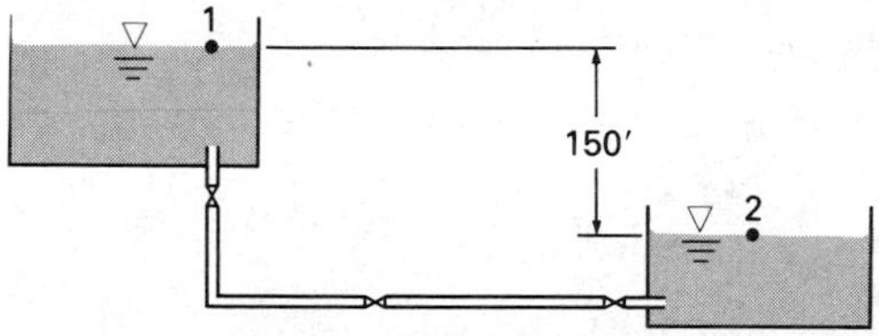

Figure E7.12 Flow between reservoirs.

SOLUTION

Given

System shown in Figure E7.12

Water flow rate 700 gal/min

Flanged connections

Schedule 40 pipe

Find

Required pipe size

Solution

We apply the mechanical energy equation,

$$\frac{p_1}{\gamma} + \frac{V_1^2}{2g} + z_1 = \frac{p_2}{\gamma} + \frac{V_2^2}{2g} + z_2 + h_L.$$

Now $p_1 = p_2$ and $V_1 \approx V_2 \approx 0$. The mechanical energy equation becomes

$$h_L = (z_1 - z_2),$$

where $z_1 - z_2$ is 150 ft. Equation (7.113) gives*

$$h_L = \left[\mathrm{K}_{\text{entrance}} + 3\mathrm{K}_{\substack{\text{globe}\\\text{valve}}} + 6\mathrm{K}_{\substack{90\\\text{standard}\\\text{elbow}}} + \mathrm{K}_{\text{exit}} + \left(f\frac{L}{D}\right)_{\text{pipe}} \right] \frac{V^2}{2g},$$

where V is the average fluid velocity in the pipe,

$$V = \frac{Q}{A} = \frac{4Q}{\pi D^2}.$$

Substituting,

$$h_L = \left[\mathrm{K}_{\text{entrance}} + 3\mathrm{K}_{\substack{\text{globe}\\\text{valve}}} + 6\mathrm{K}_{\substack{90\\\text{standard}\\\text{elbow}}} + \mathrm{K}_{\text{exit}} + \left(f\frac{L}{D}\right)_{\text{pipe}} \right] \frac{8Q^2}{\pi^2 D^4 g}.$$

Equating the expressions for h_L and rearranging give

$$\frac{\pi^2 D^4 g(z_1 - z_2)}{8Q^2} - \left(f\frac{L}{D}\right)_{\text{pipe}} = \mathrm{K}_{\text{entrance}} + 3\mathrm{K}_{\substack{\text{globe}\\\text{valve}}} + 6\mathrm{K}_{\substack{90\\\text{standard}\\\text{elbow}}} + \mathrm{K}_{\text{exit}}.$$

For a reentrant pipe, Fig. 7.25 gives $\mathrm{K}_{\text{entrance}} \approx 0.9$; also, $\mathrm{K}_{\text{exit}} = 1.0$. Substituting numerical values gives

$$\frac{\pi^2 D^4 g(z_1 - z_2)}{8Q^2} = \frac{\pi^2 D^4 (32.2\ \text{ft/sec}^2)(150\ \text{ft})}{8(700\ \text{gal/min})^2(1\ \text{ft}^3/7.48\ \text{gal})^2(1\ \text{min}/60\ \text{sec})^2}$$

$$= 2.45 \times 10^3 \left(\frac{D^4}{\text{ft}^4}\right)$$

* Why do we include the exit loss coefficient here?

and

$$f\left(\frac{L}{D}\right) = \frac{f(250\text{ ft})}{D}.$$

Table 7.8 shows that loss coefficients for globe valves and standard elbows depend on the pipe diameter. Therefore

$$\mathrm{K}_{\text{entrance}} + 3\mathrm{K}_{\substack{\text{globe}\\ \text{valve}}} + 6\mathrm{K}_{\substack{90\\ \text{standard}\\ \text{elbow}}} + \mathrm{K}_{\text{exit}} = 1.9 + 3\mathrm{K}_{\substack{\text{globe}\\ \text{valve}}} + 6\mathrm{K}_{\substack{90\\ \text{standard}\\ \text{elbow}}}.$$

The mechanical energy equation is then

$$2.45 \times 10^3\left(\frac{D}{\text{ft}}\right)^4 - 250f\left(\frac{\text{ft}}{D}\right) - \left[1.9 + 3\mathrm{K}_{\substack{\text{globe}\\ \text{valve}}} + 6\mathrm{K}_{\substack{90\\ \text{standard}\\ \text{elbow}}}\right] = 0,$$

where D is measured in feet. We must now find D by trial and error. For each value of D chosen, we find a corresponding value of f from the Moody chart after computing the Reynolds number and relative roughness. The problem statement requires that we use schedule 40 steel pipe, so we need consider only values of D for that type of pipe. We first assume a 6-in. pipe with $D = 6.065$ in. $= 0.505$ ft. Noting that $\nu(55°\text{F}) = 1.33 \times 10^{-5}\ \text{ft}^2/\text{sec}$, we calculate

$$\begin{aligned}\mathrm{R} &= \frac{VD}{\nu} = \frac{4Q}{\pi\nu D}\\ &= \frac{4(700\text{ gal/min})(1\text{ ft}^3/7.48\text{ gal})(1\text{ min}/60\text{ sec})}{\pi(1.33 \times 10^{-5}\text{ ft}^2/\text{sec})(0.505\text{ ft})} = 2.96 \times 10^5.\end{aligned}$$

Using Table 7.1 for the relative roughness for galvanized pipe, we have

$$\frac{\varepsilon}{D} = \frac{0.00015\text{ ft}}{0.505\text{ ft}} = 0.0003.$$

The Moody chart gives $f = 0.017$, and Table 7.8 implies

$$\mathrm{K}_{\substack{\text{glove}\\ \text{valve}}} \approx 5.9 \quad \text{and} \quad \mathrm{K}_{\substack{90\\ \text{standard}\\ \text{elbow}}} = 0.28.$$

The mechanical energy equation gives

$$\begin{aligned}&2.45 \times 10^3\left(\frac{0.505\text{ ft}}{1\text{ ft}}\right)^4 - 250(0.017)\left(\frac{1\text{ ft}}{0.505\text{ ft}}\right)\\ &\qquad - [1.9 + 3(5.9) + 6(0.28)] = 129.6 \neq 0.\end{aligned}$$

As a second guess, we choose 5-in. schedule 40 pipe with $D = 5.047$ in. Repeating the calculations gives

$$\mathrm{R} = 3.6 \times 10^5, \qquad \frac{\varepsilon}{D} = 0.00036, \qquad f = 0.017,$$

$$\mathrm{K}_{\substack{\text{glove}\\ \text{valve}}} \approx 5.9, \quad \text{and} \quad \mathrm{K}_{\substack{90\\ \text{standard}\\ \text{elbow}}} = 0.28.$$

The mechanical energy equation gives

$$\begin{aligned}&2.45 \times 10^3\left(\frac{5.047/12\text{ ft}}{1\text{ ft}}\right)^4 - 250(0.017)\left(\frac{1\text{ ft}}{5.047/12\text{ ft}}\right)\\ &\qquad - [1.9 + 3(5.9) + 6(0.28)] = 55.4 \neq 0.\end{aligned}$$

As a third guess, we choose a 4-in. nominal diameter schedule 40 pipe, with a 4.026-in. inside diameter. Repeating the calculations gives

$$\mathrm{R} = 4.5 \times 10^5, \qquad \frac{\varepsilon}{D} = 0.00045, \qquad f = 0.017,$$

$$\mathrm{K}_{\substack{\text{glove}\\\text{valve}}} = 6.0, \quad \text{and} \quad \mathrm{K}_{\substack{90\\\text{standard}\\\text{elbow}}} = 0.30.$$

The mechanical energy equation gives

$$2.45 \times 10^3\left(\frac{4.026/12\ \text{ft}}{1\ \text{ft}}\right)^4 - 250(0.017)\left(\frac{1\ \text{ft}}{4.026/12\ \text{ft}}\right) - [1.9 + 3(6.0) + 6(0.30)] = -3.12 \approx 0.$$

We use

$$D_{\text{nominal}} = \text{4-in. schedule 40 galvanized pipe.} \quad \textbf{ANSWER}$$

Discussion

Because the mechanical energy equation evaluates to a negative value (−3.12), the 4-in. pipe is slightly smaller than necessary; thus the flow is slightly smaller than 700 gal/min.

Note that, once again, we cannot use the direct calculation method of Eqs. (7.70)–(7.72), because local losses are a significant part of gh_L.

Application to Gas Flow Systems. In our discussions and examples so far, we have concentrated on liquid flow systems. If the Mach number is low enough that the flow can be considered incompressible, we may apply similar methods to the analysis and design of gas flow systems. A few changes in viewpoint and technique, however, are necessary.

When analyzing gas or air flow systems, engineers usually prefer the pressure form of the mechanical energy equation:

$$p_1 + \tfrac{1}{2}\rho V_1^2 + \gamma z_1 + p_{FT} = p_2 + \tfrac{1}{2}\rho V_2^2 + \gamma z_2 + \Delta p_{L,\text{sys}}. \tag{7.116}$$

Equation (7.116) is based on the assumption that no turbines or motors are present in the flow stream. Useful energy is supplied by fans (or blowers or "air pumps"). It is measured by the *fan total pressure rise,* defined by

$$p_{FT} \equiv \rho w_{\text{FAN}} - \Delta p_{L,\text{FAN}}. \tag{7.117}$$

The mechanical energy losses are expressed in terms of pressure loss:

$$\Delta p_L = \rho g h_L = \sum f\left(\frac{L}{D_h}\right)(\tfrac{1}{2}\rho V^2) + \sum \mathrm{K}(\tfrac{1}{2}\rho V^2), \tag{7.118}$$

where the friction factors and loss coefficients are evaluated in exactly the same fashion as in Eqs. (7.113)–(7.115). Finally, energy grade lines and hydraulic grade lines are not widely used by engineers who deal with air/gas handling systems.

7.5.2 Systems with Pumps (or Fans)

The mechanical energy required to move fluid through a piping system is usually supplied by a pump (or a fan, if the fluid is a gas). Here, we briefly consider the integration of a pump or fan with a pipe or duct system but focus on pumps. References [18, 19] provide a wealth of practical information for engineers who use and design pumps and fans.

There are hundreds of different pump designs, and millions of pumps are in use all over the world. Most pumps can be classified as either positive displacement type or dynamic (turbomachine) type. *Positive displacement pumps* operate by "trapping" the fluid in a definite volume and then "squeezing" or "pushing" it. The pumps illustrated in Figure 7.34 are common examples of positive displacement pumps. Positive displacement pumps are preferred for large heads, low flow rates, and highly viscous fluids.

Dynamic pumps operate by increasing the fluid momentum in a nearly steady flow process. All pumps of this type feature a spinning rotor through which the fluid passes. Work is done on the fluid by force interaction with the spinning rotor. In Section 4.4.6, we presented a rudimentary analysis of turbomachine processes, and Fig. 4.27 illustrated some typical dynamic pumps. Dynamic pumps are further classified as *radial flow,* in which the fluid flows through the rotor primarily perpendicular to the axis of rotation; *axial flow,* in which the throughflow is primarily parallel to the axis of rotation; and *mixed flow,* in which the flow is between axial and radial. Dynamic pumps are preferred for lower heads and high flow rates and are suitable for less viscous fluids. Here, we concentrate on the performance of dynamic pumps.

Performance Curves. The working equation for pipe system analysis, Eq. (7.112), indicates that the purpose of a pump is to supply head h_p to the fluid. The pump head is the difference between the work (per unit mass) done on the fluid by the pump and the mechanical energy losses within the pump. The analysis in Section 4.4.6 showed that work depends on the rotor diameter and rotational speed and the fluid velocity. The hydraulic losses within the pump depend on fluid velocity, fluid viscosity, and rotor geometry and roughness. For a given pump

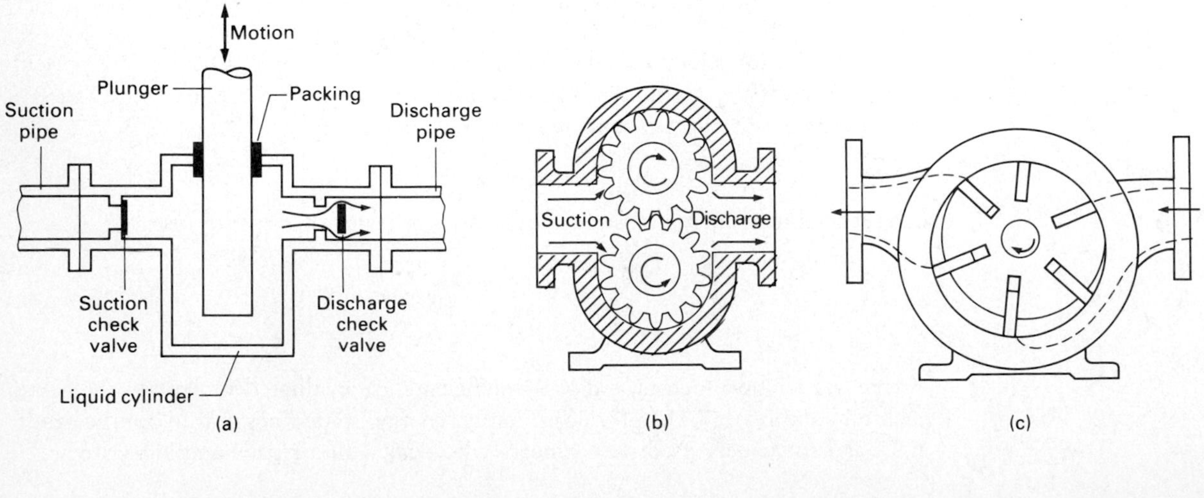

Figure 7.34 Typical positive displacement pumps: (a) plunger pump; (b) gear pump; (c) sliding vane pump.

handling a given fluid (so that size, roughness, density, and viscosity are fixed), the pump head still depends on rotational speed (N, usually measured in revolutions per minute) and fluid flow rate (a measure of fluid velocity):

$$h_p = h_p\{Q, N\}.$$

Given pump handling given fluid

An additional item of interest is the shaft power input required to drive the pump. Shaft power for a given pump and a given fluid also depends on pump rotational speed and flow rate:

$$\dot{W}_s = \dot{W}_s\{Q, N\}.$$

Given pump handling given fluid

An important parameter is the pump's overall efficiency, which is the ratio of the useful power supplied to the fluid to the shaft power input to the pump. Useful energy is measured by the pump head, so the overall efficiency is

$$\eta \equiv \frac{\dot{m}gh_p}{\dot{W}_s} = \frac{\rho Q g h_p}{\dot{W}_s} = \frac{\gamma Q h_p}{\dot{W}_s}. \tag{7.119}$$

The useful energy added to the fluid is less than the shaft work for two reasons: hydraulic losses in the flow path and mechanical losses from friction in bearings, seals, and the like. Pump efficiency also depends on the flow rate and rotational speed:

$$\eta = \eta\{Q, N\}.$$

Given pump handling given fluid

Pump performance is normally determined by experimental measurements of head, flow, and shaft power at various speeds and is available from the pump manufacturer. Curves are usually plotted for operation at constant speed with flow rate as the independent variable. Head, power, and efficiency curves are often plotted simultaneously on the same graph. Figure 7.35 shows a typical format. *Such curves are limited to a particular pump handling a particular fluid.*

Important points on these curves are the *shut off* (no flow), *maximum flow* (zero head), and *best efficiency point* (BEP). Obviously, the optimum application

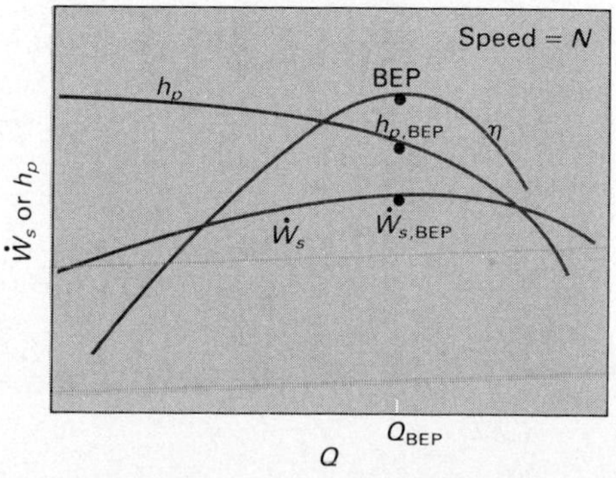

Figure 7.35 Typical pump performance curve format.

for this pump would be at the best efficiency point; in fact, the values of head and flow at this point may be quoted as the point of rating for this pump.

If pump speed or size or the working fluid is changed, a particular type of pump will generate performance curves that are similar in shape but with different magnitudes for the parameters. Different types (designs) of pumps generate performance curves of different shapes. Figure 7.36 compares typical performance curve shapes for axial flow and radial flow pumps. Note that the curves are normalized by dividing each parameter by its value at the BEP.

Pump–System Matching. When a pump is installed in a system, the flow rate through the pump and the flow rate through the system must be equal. Moreover, the head supplied by the pump must equal the head required by the system. The existing flow and head must be calculated by matching pump performance with system requirements; neither the pump alone nor the system alone dictates the flow and head.

The basis for matching is the energy equation, Eq. (7.112). The *system head* is defined as the algebraic sum of all heads except the pump head (turbine head would be included if a turbine were present; however, we neglect it in this discussion):

$$h_{sys} \equiv \frac{p_2 - p_1}{\gamma} + \frac{V_2^2 - V_1^2}{2g} + (z_2 - z_1) + h_L, \qquad (7.120)$$

where h_L is calculated by Eqs. (7.113)–(7.115). System head is a function of system configuration and flow rate, because head losses and kinetic energy change, if present, depend on fluid velocity. A plot of system head versus flow rate is called a *system head curve* or a *system resistance curve* and has a generally parabolic shape, as shown in Figure 7.37.

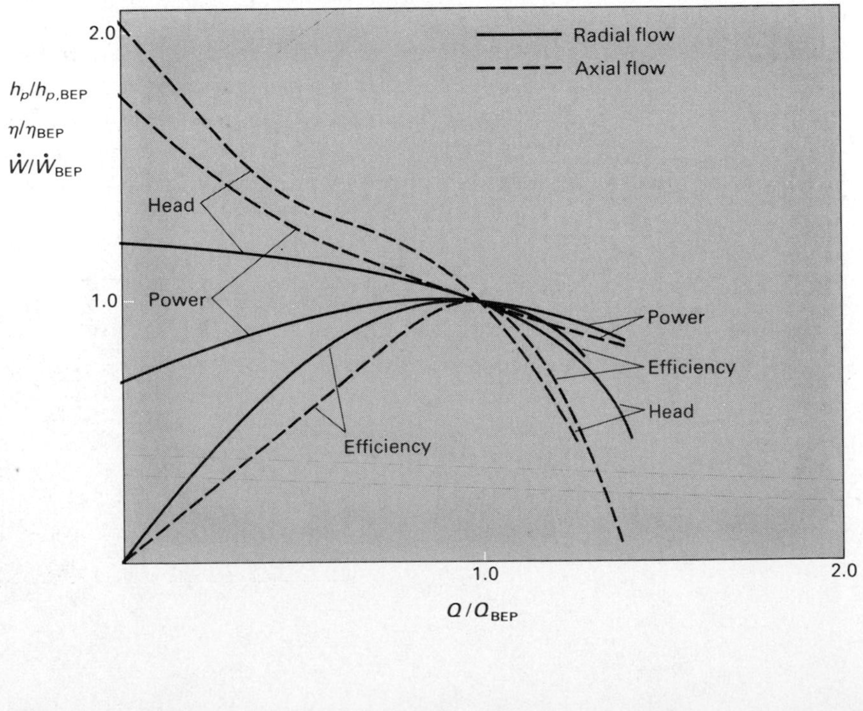

Figure 7.36 Comparison of performance curve shapes.

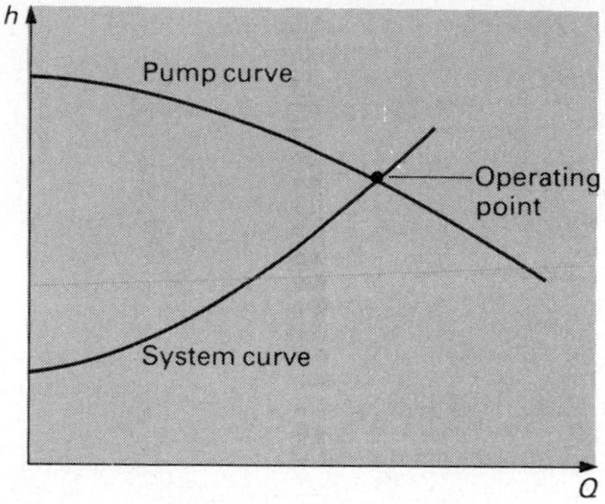

Figure 7.37 Pump curve, system curve, and operating point.

The matching requirements for a system with a pump are

$$h_p = h_{sys}$$

from the energy equation and

$$Q_p = Q_{sys}$$

from continuity. Figure 7.37 shows a system resistance curve and a pump head curve (for the pump operating speed) plotted on the same axes. The point of intersection of these two curves is the only point that satisfies both the energy and continuity equations and is therefore the unique operating point. The operating point can be changed only by changing the system's resistance (e.g., by changing a valve setting) or by changing the pump curve (e.g., by changing the speed).

We determine the operating point for a system that includes a pump by using one of three procedures:

- by iterative calculation, as illustrated in Example 7.13;
- graphically, by plotting the pump curve and the system resistance curve on the same axes, as illustrated in Figure 7.37; or
- algebraically, if expressions for $h_{sys}(Q)$ and $h_p(Q)$ are available, which usually requires a polynomial curve fit to the pump performance data.

EXAMPLE 7.13 Illustrates Velocity/Flow Rate Calculation with a Pump in the Pipe System

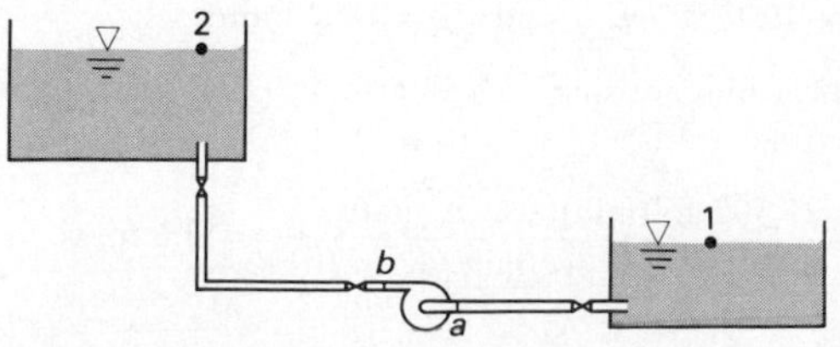

Figure E7.13a Modification of Example 7.12 piping system.

A centrifugal pump having the characteristics shown in Fig. E7.13b is used in the pipe system of Example 7.12 to pump water from the low reservoir to the high reservoir. Find the flow rate and the required motor power. The pump speed is 1700 rpm.

SOLUTION

Given

Pipe system in Example 7.12, modified as shown in Fig. E7.13a
Pump characteristics as in Fig. E7.13b
Pump speed of 1700 rpm

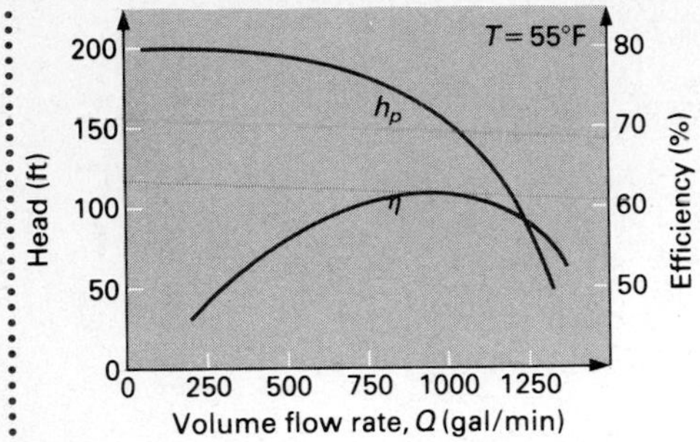

Figure E7.13b Pump characteristics.

Find

Volume flow rate and required motor power with pump installed

Solution

The system head is given by Eq. (7.120):

$$h_{\text{sys}} = \frac{p_2 - p_1}{\gamma} + \frac{V_2^2 - V_1^2}{2g} + (z_2 - z_1) + h_L.$$

As in Example 7.12, $p_1 - p_2 = 0$ and $V_1 \approx V_2 \approx 0$. The system head is then

$$h_{\text{sys}} = (z_2 - z_1) + h_L,$$

where $(z_2 - z_1)$ is 150 ft and h_L is found from Example 7.12.

$$h_L = \left(\mathbf{K}_{\text{entrance}} + 3\mathbf{K}_{\substack{\text{globe}\\\text{valve}}} + 6\mathbf{K}_{\substack{90\\\text{standard}\\\text{elbow}}} + \mathbf{K}_{\text{exit}} + f\frac{L}{D}\right)\frac{V^2}{2g}.$$

A pump is installed in the system, so

$$h_{\text{sys}} = h_p.$$

Substituting for h_{sys} and h_L and solving for the velocity V, we obtain

$$V = \sqrt{\frac{2g[h_p - (z_2 - z_1)]}{\mathbf{K}_{\text{entrance}} + 3\mathbf{K}_{\substack{\text{globe}\\\text{valve}}} + 6\mathbf{K}_{\substack{90\\\text{standard}\\\text{elbow}}} + \mathbf{K}_{\text{exit}} + [f(L/D)]_{\text{pipe}}}}.$$

The unknowns in this equation are the velocity V, the pump head h_p, and the friction factor f; h_p and f depend on the volume flow rate and V. Using the numerical values from Example 7.12 and for 55°F water (Table A.6) gives

$$V = \sqrt{\frac{2(32.2\ \text{ft/sec}^2)(h_p - 150\ \text{ft})}{0.9 + 3(6.0) + 6(0.30) + f(250\ \text{ft}/4.026\ \text{in.})(12\ \text{in./ft})}}.$$

Simplifying, we have

$$V = \sqrt{\frac{2(32.2\ \text{ft/sec}^2)(h_p - 150\ \text{ft})}{20.7 + 745f}},$$

where h_p is in ft. We must now solve for V. The velocity/flow rate calculation in Example 7.4 was most easily done using the Colebrook formula, because $\mathbf{R}_f(= \sqrt{f}\mathbf{R})$ could be conveniently grouped in the mechanical energy equation. We cannot use that method in this problem, because a pump is present and there are other losses besides pipe friction. We use an iterative solution method. For 55°F water,

$$\nu = 1.33 \times 10^{-5}\ \text{ft}^2/\text{sec} \quad \text{and} \quad \rho = 62.4\ \text{lbm/ft}^3.$$

We begin the iteration by choosing* a value for Q. Choosing $Q =$ 500 gal/min, we obtain

$$V = \frac{4Q}{\pi D^2} = \frac{4(500\ \text{gal/min})(231\ \text{in}^3/\text{gal})}{\pi(4.026\ \text{in.})^2(60\ \text{sec/min})(12\ \text{in./ft})} = 12.6\ \text{ft/sec}$$

---

* In a velocity/flow rate problem, the iteration often converges faster if you select f rather than Q or V. Do you see why selecting f is not the better approach in this case?

and

$$\mathbf{R} = \frac{VD}{\nu} = \frac{(12.6 \text{ ft/sec})(4.026 \text{ in.})}{(1.33 \times 10^{-5} \text{ ft}^2\text{/sec})(12 \text{ in./ft})} = 3.18 \times 10^5.$$

With $\varepsilon/D = 0.00045$ from Example 7.12, the Moody chart gives $f = 0.017$. For $Q = 500$ gal/min, Fig. E7.13b gives $h_p \approx 190$ ft. The mechanical energy equation gives

$$V = \sqrt{\frac{(64.4 \text{ ft/sec}^2)(190 \text{ ft} - 150 \text{ ft})}{20.7 + 745(0.017)}} = 8.8 \text{ ft/sec}.$$

This velocity is not equal to the velocity found from the assumed flow rate, so our guess is not correct. Repeating the calculations for $V = 8.8$ ft/sec gives

$$Q = \frac{\pi D^2}{4}(V) = \frac{\pi}{4}\frac{(4.026 \text{ in.})^2(8.8 \text{ ft/sec})(7.48 \text{ gal/ft}^3)}{(144 \text{ in}^2)(1 \text{ min/60 sec})}$$
$$= 349 \text{ gal/min}$$

and

$$\mathbf{R} = \frac{VD}{\nu} = \frac{(8.8 \text{ ft/sec})(4.026 \text{ in.})}{(1.33 \times 10^{-5} \text{ ft}^2\text{/sec})(12 \text{ in./ft})} = 2.22 \times 10^5.$$

This Reynolds number and flow rate give $f = 0.018$ and $h_p = 198$ ft, and the mechanical energy equation gives

$$V = \sqrt{\frac{(64.4 \text{ ft/sec}^2)(198 \text{ ft} - 150 \text{ ft})}{20.7 + 745(0.018)}} = 9.52 \text{ ft/sec}.$$

Repeating the calculations until the velocity converges, we have

$$V = 9.30 \text{ ft/sec}, \quad \mathbf{R} = 2.35 \times 10^5, \quad f = 0.018, \quad h_p = 196 \text{ ft},$$
$$Q = 369 \text{ gal/min}.$$ **ANSWER**

Using Eq. (7.119), we write the motor power as

$$\dot{W}_{\text{shaft}} = \frac{\gamma Q h_p}{\eta}.$$

From Fig. E7.13b, the efficiency at the calculated flow rate of 369 gal/min is

$$\eta \approx 65\% = 0.65.$$

Then

$$\dot{W}_{\text{shaft}} = \frac{(62.4 \text{ lb/ft}^3)(369 \text{ gal/min})(196 \text{ ft})}{0.65[449 \text{ (gal/min)/(ft}^3\text{/sec)}][550 \text{ (ft}\cdot\text{lb)/(hp}\cdot\text{sec)}]};$$
$$\dot{W}_{\text{shaft}} = 28.1 \text{ hp}.$$ **ANSWER**

Discussion

An iteration of this type often converges faster if the new guess is taken as the average of the past trial value and the newly calculated value; that is,

$$Q_{\text{guess}} = \tfrac{1}{2}(Q_{\text{old}} + Q_{\text{new}}).$$

You might repeat the calculations using this idea to see how rapidly they converge.

Alternative methods of solving this problem involve plotting the system curve on Fig. E7.13b to determine its intersection with the pump curve and fitting a polynomial curve to the pump head ($h_p = a + bQ + cQ^2$) and using this expression to obtain an analytical solution. We invite you to try either or both alternatives to verify the result.

Similarity Laws for Pumps. Pump performance curves are usually obtained from tests run at a particular speed and with a particular fluid. Often, running a pump at a different speed (say, to obtain more head and a higher flow rate) or pumping a different fluid than that for which the pump was rated is necessary. Sometimes, performance tests run on a smaller pump must be scaled up to predict the curves for a geometrically similar larger pump. Prediction of the effects of changes in speed, size, and working fluid on pump performance curves can be made using *similarity laws,* which are derived from dimensional analysis. Examples 6.4 and 6.8 illustrated the development and use of such laws.

For dynamic pumps, the most important dependent variable measures of pump performance are the head, h_p, the shaft power, $\dot{W}_s$, and the efficiency, η. If we limit our consideration to a family of geometrically similar pumps, the independent variable parameters are flow rate, Q, pump speed, N, pump size, measured by the impeller diameter, D, fluid density, ρ, fluid viscosity, μ, and surface roughness, ε. Dimensional analysis yields the following parameters:

Dependent Parameters		**Independent Parameters**	
Head coefficient	$\Psi = gh_p/N^2D^2$	Flow coefficient	$\Phi = Q/ND^3$
Power coefficient	$\mathbf{P} = \dot{W}_s/\rho N^3D^5$	Reynolds number	$\mathbf{R} = \rho ND^2/\mu$
Efficiency	$\eta = \rho Qgh_p/\dot{W}_s$	Relative roughness	ε/D

Any of the dependent parameters may be considered a function of all three independent parameters; for example,

$$\Psi = f\{\Phi, \mathbf{R}, \varepsilon/D\}.$$

Such three-parameter similarity laws are not especially useful, because of the difficulty in changing any operating or performance parameters in such a way that all three independent dimensionless parameters can be matched. Suppose, however, that we restrict consideration to a specific pump or a pair of identical pumps. Then, the relative roughness is the same, and we can neglect its effect. If we change only speed, size, or fluid, then only flow coefficient and Reynolds number need be matched for similarity.

In many practical cases, the similarity law can be simplified even further. Our considerations of flow in pipes, fittings, and valves revealed that for very "rough" surfaces and/or high Reynolds numbers, the performance parameters (that is, friction factor or loss coefficient) became independent of the exact value of Reynolds number. Experiments have shown that above a Reynolds number of about 3×10^5, the exact value of Reynolds number has no effect on pump performance, and so Reynolds number need not be matched for similarity. We then have the simplified similarity law:

$$\Psi \text{ (or } \eta\text{, or } \mathbf{P}) = f\{\Phi, \text{ only}\}.$$

Specific machine, Reynolds number > 3 × 10^5

This law also applies to pumps of different sizes if the relative roughnesses are equal (that is, if they scale with D).

Using this similarity law, we transform a point ("1") on a pump curve for a particular speed and a particular fluid into a *corresponding point* ("2") at another speed or for another fluid by the following equations:

From equating Φ:	$Q_2 = Q_1(N_2D_2^3/N_1D_1^3)$	(7.121)
From equating Ψ:	$h_{p2} = h_{p1}(N_2^2D_2^2/N_1^2D_1^2)$	(7.122)
From equating $\mathbf{P}$:	$\dot{W}_{s2} = \dot{W}_{s1}(\rho_2N_2^3D_2^5/\rho_1N_1^3D_1^5)$	(7.123)
From equating η:	$\eta_2 = \eta_1$	(7.124)

Scale effects and Reynolds number effects negligible ($\mathbf{R} > 3 \times 10^5$)

The relationship between corresponding points is illustrated in Fig. 7.38.

The following points about the similarity laws are important:

- The similarity laws relate points on two different performance curves; they cannot be used to relate points lying on the same curve.
- Suppose that a pump is installed in a system and that the speed of the pump is changed. The new operating point does not necessarily correspond to the old operating point; the two points correspond *only* if the system resistance curve is purely parabolic (i.e., only if $h_{sys} = KQ^2$).

Several modifications of the simple similarity laws, Eqs. (7.121)–(7.124), to account for Reynolds number and scale effects are available, but none is based on a well-developed theory and all are highly empirical. A popular equation for estimating the combined Reynolds number and scale effects on efficiency is Moody's formula:

$$\frac{1-\eta_2}{1-\eta_1} \approx \left(\frac{D_1}{D_2}\right)^{0.25}. \tag{7.125}$$

Specific Speed. Often, a pump user knows the desired magnitude of the head and flow to be produced by the pump and also can estimate or select the speed

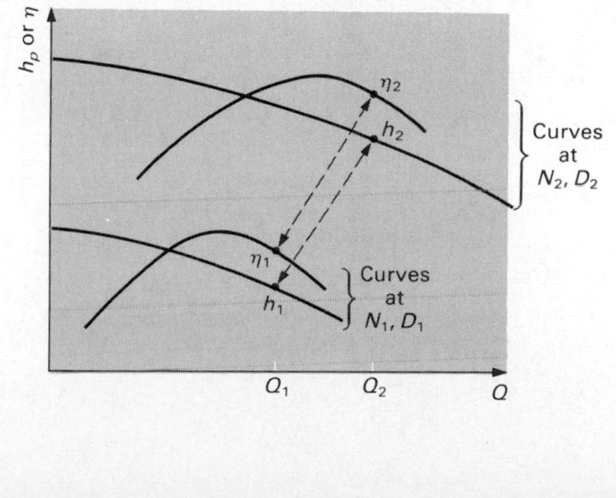

Figure 7.38 Corresponding points related by Eqs. (7.121), (7.122), and (7.124).

of the pump. Using these data, the user can determine the most efficient type of pump, estimate the (BEP) efficiency of the pump, and also estimate the impeller size. Pump designers use such information to begin the design process.

The key to all this valuable information is a dimensionless parameter called the *specific speed,* as defined by

$$\boldsymbol{\Omega} \equiv \frac{N\sqrt{Q_{\mathrm{BEP}}}}{(gh_{p,\mathrm{BEP}})^{3/4}}. \tag{7.126}$$

Note that the specific speed is not an independent dimensionless parameter; it can be formed from $\boldsymbol{\Psi}$ and $\boldsymbol{\Phi}$ by

$$\boldsymbol{\Omega} = \frac{\boldsymbol{\Phi}^{1/2}}{\boldsymbol{\Psi}^{3/4}}.$$

To be truly dimensionless, the parameters in $\boldsymbol{\Omega}$ must be in consistent units (N in radians/sec, Q in ft^3/sec or m^3/s, and gh_p in ft^2/sec^2 or m^2/s^2). In the United States, engineers usually use the commercial units of pump performance and drop the g altogether. The resulting parameter is denoted N_s:

$$N_s = \frac{N(\mathrm{rpm})\sqrt{Q\ (\mathrm{gal/min})}}{[h_p\ (\mathrm{ft})]^{3/4}}. \tag{7.127}$$

Note that we dropped the BEP subscript—that BEP values of Q and h_p are to be used is understood. Substituting the appropriate conversion factors, we find that

$$N_s = 2734\boldsymbol{\Omega}. \tag{7.128}$$

Experience has shown that different types of pumps yield maximum efficiency at different specific speeds. This result is quite logical; we would expect wide-open axial flow designs to perform better at high flow and low head (i.e., high specific speed) but radial flow designs to be better suited for lower flow and higher head (i.e., lower specific speed). Figure 7.39 shows peak (BEP) efficiency and corresponding impeller type and geometry as a function of specific speed.

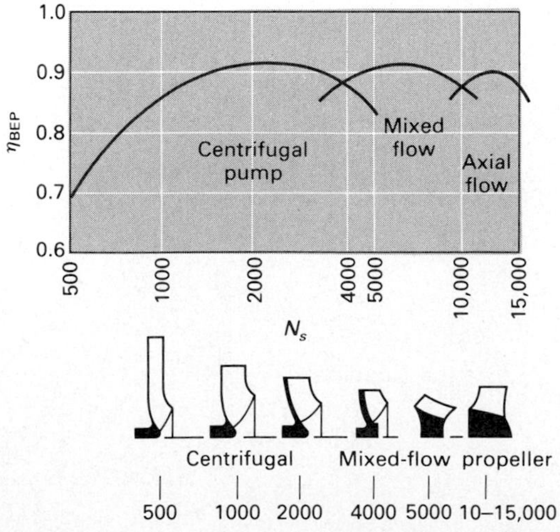

Figure 7.39 Typical (BEP) efficiency and pump impeller cross section as a function of specific speed.

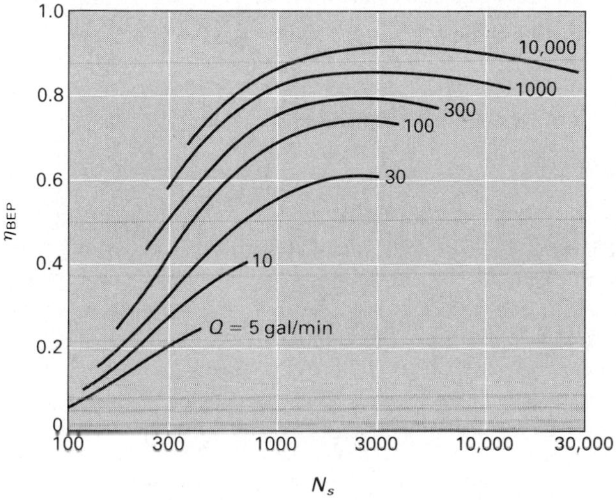

Figure 7.40 Typical (BEP) efficiency of dynamic pumps as a function of capacity and specific speed.

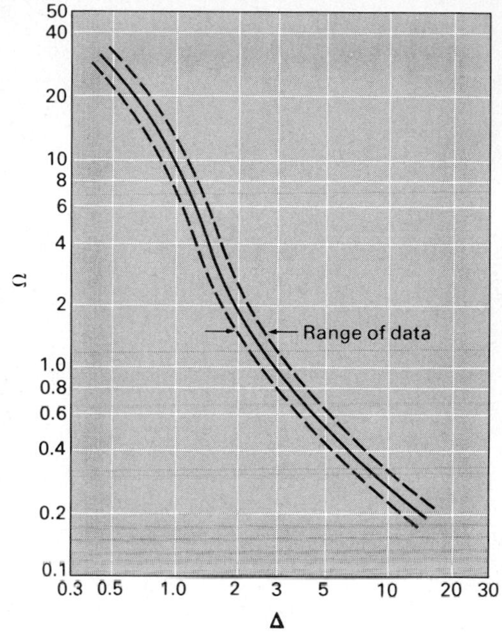

Figure 7.41 Correlation between specific diameter and specific speed for dynamic pumps.

Many studies have demonstrated the importance of specific speed as a correlating parameter for pump performance. If there were no scale effects or Reynolds number effects, efficiency would correlate perfectly with N_s. Figure 7.40 shows expected (BEP) efficiency of dynamic pumps as a function of N_s and (BEP) flow rate. As flow is a measure of size, it represents both Reynolds number and scale effects. You should use information from this figure cautiously, first because the assumption is that the optimum design for that specific speed is being used and second because the curves give the BEP efficiency *only*, not the entire η versus Q curve.

Specific speed is a reliable indicator of pump size as well as shape. Another dimensionless parameter, the *specific diameter*, defined by

$$\mathbf{\Delta} \equiv \frac{D(gh_{p,\mathrm{BEP}})^{1/4}}{Q_{\mathrm{BEP}}^{1/2}}, \tag{7.129}$$

D, gh, Q in consistent units

can be roughly correlated with $\mathbf{\Omega}$, as shown in Fig. 7.41.

EXAMPLE 7.14 **Illustrates Pump Similarity Laws and Performance Correlations**

The task is to deliver 1000 gal/min of water from the lower reservoir to the upper reservoir in the system considered in Example 7.13. There, the pump speed was 1700 rpm. Estimate the pump speed required for 1000 gal/min and the power required. If possible, recommend a better pump for the new application.

SOLUTION

Given

System defined in Examples 7.12 and 7.13

Pump with performance as shown in Figure E7.13b

Pump speed 1700 rpm

Find

Pump speed required for 1000 gal/min

Better pump, if possible

Solution

As we increase the speed, the pump performance curve shifts upward and to the right. The operating point is the intersection of the pump curve and the system resistance curve. *The system resistance curve remains unchanged.* From Examples 7.12 and 7.13, the system resistance curve is

$$h_{sys} = 150 \text{ ft} + (20.7 + 745f)\left(\frac{V^2}{2g}\right).$$

The velocity is

$$V = \frac{4Q}{\pi D^2} = \frac{4(1000/449 \text{ ft}^3/\text{sec})}{\pi(4.026/12 \text{ ft})^2} = 25.2 \text{ ft/sec},$$

which, in turn, gives

$$\mathbf{R} = 6.36 \times 10^5 \quad \text{and} \quad f = 0.018.$$

(See Examples 7.12 and 7.13.) The system head at 1000 gal/min is

$$h_{sys} = 150 \text{ ft} + [20.7 + 745(0.018)]\left(\frac{25.2^2}{2(32.2)} \text{ ft}\right) = 486.4 \text{ ft}.$$

As $h_p = h_{sys}$ and $Q_p = Q_{sys}$, we must now find the speed at which the point (1000 gal/min, 486.4 ft) lies on the pump curve.

Because we are not changing the pump, there are no scale effects. We also assume that the (pump) Reynolds number is sufficiently high to avoid Reynolds number effects. Because we are changing only the speed, not the diameter, Eqs. (7.121) and (7.122) imply that corresponding points are related by

$$\frac{Q_2}{N_2} = \frac{Q_1}{N_1} \quad \text{and} \quad \frac{h_{p,2}}{N_2^2} = \frac{h_{p,1}}{N_1^2}.$$

We know the desired values of Q_2 and $h_{p,2}$ and the speed N_1. We do not know Q_1 and $h_{p,1}$, because we do not know what point on the original pump curve corresponds to our desired final point. Point "2" does not correspond to the old operating point found in Example 7.13, because the system curve is not purely parabolic. Substituting known values gives

$$Q_1 = (1000 \text{ gal/min})\left(\frac{1700 \text{ rpm}}{N_2 \text{ rpm}}\right) \tag{E7.1}$$

and

$$h_{p,1} = (486.4 \text{ ft})\left(\frac{1700 \text{ rpm}}{N_2 \text{ rpm}}\right)^2. \quad \text{(E7.2)}$$

Equations (E7.1) and (E7.2) relate the three unknowns Q_1, $h_{p,1}$, and N_2. A third relationship, involving Q_1 and $h_{p,1}$ is the pump curve at 1700 rpm, given in Figure E7.13b. If we eliminate N_2 from Eqs. (E7.1) and (E7.2), we get

$$h_{p,1} = 486.4\left(\frac{Q_1}{1000}\right)^2. \quad \text{(E7.3)}$$

We solve Eq. (E7.3) together with the pump curve to find the point at 1700 rpm that corresponds to the desired operating point. We can do so in one of three ways: (1) iteration, as in Example 7.13; (2) graphically, by plotting Eq. (E7.3) together with the pump curve; or (3) analytically, using a fitted polynomial for the pump curve. We choose the graphic method. Figure E7.14 shows the pump curve (repeated from Figure E7.13b) and a plot of Eq. (E7.3). The curves intersect at

$$Q_1 = 595 \text{ gal/min} \quad \text{and} \quad h_{p,1} = 172 \text{ ft}.$$

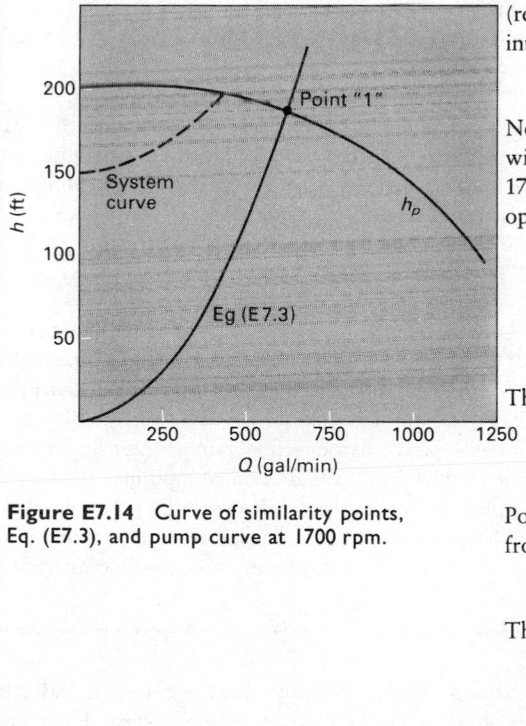

Figure E7.14 Curve of similarity points, Eq. (E7.3), and pump curve at 1700 rpm.

Note that Fig. E7.14 also shows the system curve and its intersection with the 1700 rpm pump curve. This point is the operating point at 1700 rpm, as determined in Example 7.13. Point "1" *is not* the operating point at 1700 rpm.

We now calculate the desired speed from

$$N_2 = \left(\frac{1000}{Q_1} \text{ gal/min}\right)(1700 \text{ rpm});$$

$$N_2 = 2860 \text{ rpm}.$$ **ANSWER**

The shaft power is

$$\dot{W}_s = \frac{\gamma Q_2 h_{p,2}}{\eta}.$$

Point "2" corresponds to point "1," so we can read the efficiency from Figure E7.13b for $Q = 595$ gal/min, or

$$\eta_2 = \eta_1 = 62\%.$$

Then

$$\dot{W}_s = \frac{(62.4 \text{ lb/ft}^3)(1000/449 \text{ ft}^3/\text{sec})(486.4 \text{ ft})}{0.62(550 \text{ ft}\cdot\text{lb/hp}\cdot\text{sec})};$$

$$\dot{W}_s = 198.3 \text{ hp}.$$ **ANSWER**

The specific speed of the existing pump is

$$N_s = \frac{N\sqrt{Q_{\text{BEP}}}}{(h_{p,\text{BEP}})^{3/4}}.$$

Reading (BEP) values from Fig. E7.13b, we have

$$N_s = \frac{1700\sqrt{900}}{(160)^{3/4}} = 1130.$$

Figure 7.40 indicates that a pump of this specific speed and size should yield a peak efficiency of about 79 percent. Figure E7.13b shows that the peak efficiency of the existing pump falls below this prediction. In addition, we are not operating at the BEP at 1000 gal/min and 2860 rpm. From Fig. 7.40, a 1000 gal/min pump with $N_s \approx 2000$ would yield a (BEP) efficiency of about 85 percent. If we change the pump to a more typical design with $N_s = 2000$ and operate at the BEP, we could improve the efficiency from 62 percent to about 85 percent. The speed for the new pump would be

$$N = \frac{N_s(h_p)^{3/4}}{\sqrt{Q}} = \frac{2000(486.4)^{3/4}}{\sqrt{1000}} \approx 6550 \text{ rpm}.$$

Unfortunately, this speed is somewhat fast for an ordinary water pump. The most practical recommendation is:

Use a pump with $N \approx 3500$ rpm, with its BEP near the desired operating point (1000 gal/min, 490 ft head). Such a pump would have $N_s \approx 1100$ and $\eta \approx 82$ percent. **ANSWER**

Discussion

The specific speed of a pump is defined by its BEP values of head and flow at any speed that the pump can be operated. Changing the speed and/or the fluid or making a geometrically similar pump of different size does not change N_s for that design. As a result, we could evaluate the specific speed of the installed pump from its BEP performance at 1700 rpm and use that specific speed to discuss pump performance at 2860 rpm.

The replacement pump would be very similar in design to the existing pump. The primary gain in efficiency would come from changing the speed and size to move the BEP head and flow to the desired magnitudes (1000 gal/min, 486.2 ft).

Note the difference between the actual pump/system operating point at 1700 rpm (369 gal/min, 196 ft) and our point "1" on the 1700 rpm curve. Point "1" is not a possible operating point with this pump and system; however, point "1" corresponds (in the dimensionless sense) to point "2" at 2860 rpm, which is a possible operating point.

Fans. Most of the information and analytic methods that we developed for pumps may also be applied to fans, with a slight change in viewpoint. Fans are usually rated in terms of fan (total) pressure, as defined in Eq. (7.117), and volumetric flow. In the United States, fan pressure is usually expressed in inches of water and volume flow in cubic feet per minute. Fan efficiency is

$$\eta \equiv \frac{Qp_{FT}}{\dot{W}_s}. \tag{7.130}$$

Fan performance curves are plots of p_{FT} and η versus Q, with rotational speed as a parameter.

The operating point for a fan installed in a system is obtained by matching the fan curve and the system curve, given by

$$p_{T,\text{sys}} = (p_2 - p_1) + \gamma(z_2 - z_1) + \frac{\rho(V_2^2 - V_1^2)}{2} + \Delta p_L,$$

with the pressure loss, Δp_L, given by Eq. (7.118). The matching requirement is

$$Q_{\text{fan}} = Q_{\text{sys}} \quad \text{and} \quad p_{FT} = p_{T,\text{sys}}.$$

Similarity laws for fans follow from the laws for pumps by substituting

$$h_p = \frac{p_{FT}}{\rho g} = \frac{p_{FT}}{\gamma}. \tag{7.131}$$

Note that although fluid density has no effect on head, it does on pressure.

Specific speed for a fan is formally defined by Eq. (7.126). In practice, we calculate a "dimensional" specific speed based on customary units:

$$N_{s,F} = \frac{N\ (\text{rpm})\sqrt{Q\ (\text{ft}^3/\text{min})}}{[p_{FT}(\text{in. H}_2\text{O})/\rho(\text{lbm/ft}^3)]^{3/4}}.$$

Expected (BEP) efficiency of fans can be correlated with specific speed; see reference [19] for further information.

7.5.3 Series and Parallel Lines

Pipe systems often contain lines or elements in series or parallel. Figure 7.42 shows both series and parallel arrangements. In a *series arrangement,* all elements have the same flow, and the total loss is the sum of all the element losses:

Series arrangement

$$Q_A = Q_B = \cdots = Q_N; \tag{7.132a}$$

$$h_L = h_{L,A} + h_{L,B} + \cdots + h_{L,N}. \tag{7.132b}$$

In a *parallel arrangement,* the total flow is the sum of the flows through all elements, and the loss is the same for each element. The reason is that there can be only one pressure and velocity at point 1 and only one pressure and velocity at point 2:*

Parallel arrangement

$$Q = Q_A + Q_B + Q_C + \cdots + Q_N; \tag{7.133a}$$

$$h_{L,A} = h_{L,B} = h_{L,C} = \cdots = h_{L,N}. \tag{7.133b}$$

Equations (7.113)–(7.115) are directly applicable to series systems. In general, series calculations are quite simple if flow rate and geometry are given but are more complicated if losses are given and either flow rate or pipe sizes are to be

* A graphic interpretation is that the hydraulic grade line for each pipe must pass through a common point at 1 and 2.

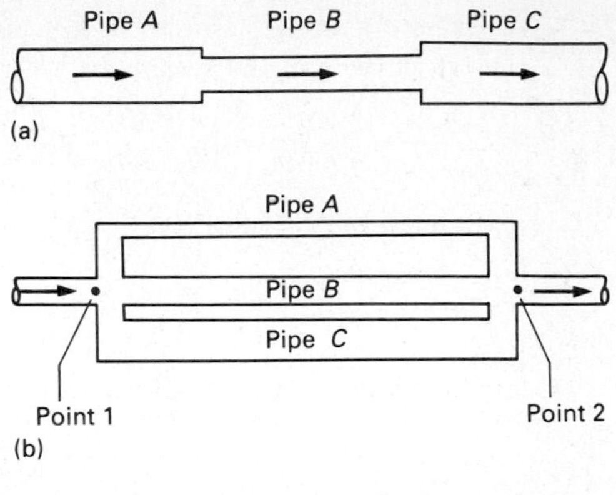

Figure 7.42 Pipes in series and parallel.

found. The equivalent length concept is quite helpful in these latter cases because the direct calculation methods of Section 7.2.5 can be used if friction is the only loss.

Parallel system calculations are easiest if the loss is given. With loss known, the flow rate (or pipe size, depending on the problem) can be separately calculated for each element in the system. The total flow is obtained by summing the element flows.

A common problem for a parallel system involves determining the head loss and division of flow between parallel branches for given system geometry and total flow rate. Solution of this problem typically requires iteration. The procedure is as follows:

- Guess a flow rate for one branch of the system. A good rule of thumb is

$$Q_A \approx \left(\frac{[A/L]_A}{[A/L]_A + [A/L]_B + \cdots + [A/L]_N} \right) Q_{\text{total}}.$$

- Calculate the head loss for the selected branch using the guessed flow and the appropriate loss equations (energy/pressure loss problem).

- As the loss is the same for all branches, calculate the flow in each of the remaining branches using the tentative loss value (flow rate/velocity problem).

- Sum the flow rates in all the branches. If the sum equals the known total flow rate through the system, the calculations are complete.

- If the sum of the branch flows does not equal the total flow, adjust the trial flow rate according to

$$Q_{A,\text{new}} = \left[\frac{Q_{\text{total, given}}}{(Q_A + Q_B + \cdots + Q_N)_{\substack{\text{last}\\ \text{try}}}} \right] Q_{\substack{A,\text{last}\\ \text{try}}}.$$

Return to the second step and repeat the calculations until you obtain convergence.

This procedure usually results in convergence after a few cycles. Note that internal iterations may be built into each step, such as calculating f for known Q and geometry using the Colebrook formula.

EXAMPLE 7.15 Illustrates How to Solve a Parallel Piping System Problem

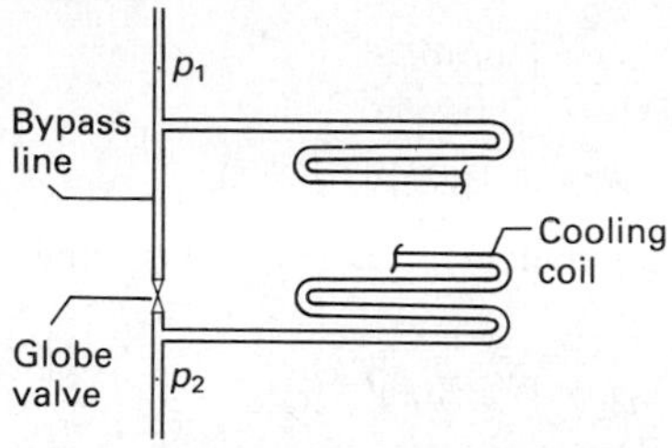

Figure E7.15 Parallel piping system.

Figure E7.15 shows a bypass line containing a globe valve in parallel with a cooling coil. Both the cooling coil and the bypass piping are 1.34-cm-inside-diameter copper pipe. The cooling coil contains 25 m of straight pipe and thirty-five 180° standard elbows. The bypass piping contains 5 m of straight pipe and a 50% closed globe valve. All connections are soldered. Find the flow rate through the cooling coil and the bypass for a total flow rate of 0.00040 m^3/s. The fluid is water at 10°C.

SOLUTION

Given

Cooling coil in parallel with bypass line

Cooling coil 25 m of 1.34-cm-inside-diameter copper pipe and thirty-five 180° standard elbows

Bypass piping 5 m of 1.34-cm-inside-diameter copper pipe and 50% closed globe valve

Soldered connections

Figure E7.15

Total flow rate 0.00040 m^3/s

Find

Flow rate through cooling coil and bypass

Solution

Assume constant density. For two lines in parallel,

$$Q_{\text{total}} = Q_{\text{bypass}} + Q_{\text{coil}} \quad \text{and} \quad h_{L,\text{bypass}} = h_{L,\text{coil}}.$$

The head loss in each line is

$$h_{L,\text{bypass}} = \left[\left(f\frac{L}{D}\right)_{\text{pipe}} + 2\text{K}_{\substack{\text{tee}\\\text{line}\\\text{flow}}} + \text{K}_{\substack{\text{globe}\\\text{valve}}}\right]\frac{8Q^2_{\text{bypass}}}{\pi^2 g D^4}$$

and

$$h_{L,\text{coil}} = \left[\left(f\frac{L}{D}\right)_{\text{pipe}} + 2\text{K}_{\substack{\text{tee}\\\text{branch}\\\text{flow}}} + 35\text{K}_{\substack{180^\circ\\\text{standard}\\\text{elbow}}}\right]\frac{8Q^2_{\text{coil}}}{\pi^2 g D^4}.$$

Noting that 1.34 cm ≈ 1/2 in. and assuming that the losses in soldered fittings are equal to those in screwed fittings, we find from Tables 7.8 and 7.9 that

$$h_{L,\text{bypass}} = \left[f\left(\frac{5\text{ m}}{1.34\text{ cm}}\right)(100\text{ cm/m}) + 2(0.90) + (2.5\times 14)\right]$$
$$\times \frac{8Q^2_{\text{bypass}}(100\text{ cm/m})^4}{\pi^2(9.807\text{ m/s}^2)(1.34\text{ cm})^4}$$
$$= 9.42\times 10^7(10.1f + 1)(\text{s}^2/\text{m}^5)(Q^2_{\text{bypass}})$$

and

$$h_{L,\text{coil}} = \left[f\left(\frac{25 \text{ m}}{1.34 \text{ cm}}\right)(100 \text{ cm/m}) + 2(2.4) + 35(2.0) \right]$$
$$\times \frac{8Q_{\text{coil}}^2 (100 \text{ cm/m})^4}{\pi^2 (9.807 \text{ m/s}^2)(1.34 \text{ cm})^4}$$
$$= 1.91 \times 10^8 (24.9f + 1)(\text{s}^2/\text{m}^5)(Q_{\text{coil}}^2).$$

As each parallel path has the same head loss, we equate the two expressions for h_L to get

$$9.42 \times 10^7 (10.1f + 1)(\text{s}^2/\text{m}^5)(Q_{\text{bypass}}^2)$$
$$= 1.91 \times 10^8 (24.9f + 1)(\text{s}^2/\text{m}^5)(Q_{\text{coil}}^2),$$

or

$$Q_{\text{coil}} = \sqrt{\frac{0.493(10.1f + 1)_{\text{bypass}}}{(24.9f + 1)_{\text{coil}}}} \, (Q_{\text{bypass}}).$$

The problem statement and the continuity equation give

$$Q_{\text{bypass}} = 0.00040 \text{ m}^3/\text{s} - Q_{\text{coil}}.$$

Substituting into the energy equation, we have

$$Q_{\text{coil}} = \sqrt{\frac{0.493(10.1f + 1)_{\text{bypass}}}{(29.4f + 1)_{\text{coil}}}} \, (0.0004 \text{ m}^3/\text{s} - Q_{\text{coil}}).$$

Solving for Q_{coil}, we have

$$Q_{\text{coil}} = \frac{0.0004 \text{ m}^3/\text{s} \, F}{1 + F},$$

where

$$F = \sqrt{\frac{0.493(10.1f + 1)_{\text{bypass}}}{(29.4f + 1)_{\text{coil}}}}.$$

The factor F depends on Q_{coil} and Q_{bypass} through the friction factors, but this dependence is fairly weak. We can rapidly generate a solution by guessing values for the two f's and so on until convergence. Preparing for the friction factor calculations, we obtain the Reynolds number for either path:

$$\mathbf{R} = \frac{4Q}{\pi \nu D} = \frac{4(100 \text{ cm/m})Q}{\pi(1.31 \times 10^{-6} \text{ m}^2/\text{s})(1.34 \text{ cm})} = 7.25 \times 10^7 \, Q.$$

The relative roughness is

$$\frac{\varepsilon}{D} = \frac{0.00015 \text{ cm}}{1.34 \text{ cm}} = 0.00011.$$

We now guess that

$$f_{\text{coil}} = f_{\text{bypass}} = 0.03.$$

Then

$$F = \sqrt{\frac{(0.493)[10.1(0.03) + 1]}{(29.4)(0.03) + 1}} = 0.584,$$

$$Q_{coil} = \left(\frac{0.584}{1.584}\right)(0.0004\ m^3/s) = 0.000148\ m^3/s,$$

and

$$Q_{bypass} = 0.0004\ m^3/s - 0.000148\ m^3/s = 0.000252\ m^3/s.$$

Then

$$\mathbf{R}_{coil} = 10{,}730 \quad \text{and} \quad \mathbf{R}_{bypass} = 18{,}270.$$

From the Moody chart,

$$f_{coil} \approx 0.031 \quad \text{and} \quad f_{bypass} \approx 0.027.$$

These values give

$$F = 0.573,$$

$$Q_{coil} = \frac{0.573}{1.573}(0.0004\ m^3/s) = 0.000146\ m^3/s,$$

and

$$Q_{bypass} = 0.0004\ m^3/s - 0.000146\ m^3/s = 0.000254\ m^3/s.$$

The Reynolds numbers are

$$\mathbf{R}_{coil} = 10{,}600 \quad \text{and} \quad \mathbf{R}_{bypass} = 18{,}400,$$

giving

$$f_{coil} \approx 0.031 \quad \text{and} \quad f_{bypass} \approx 0.027.$$

The calculations have converged, and

$$Q_{bypass} = 0.000254\ m^3/s.$$ **ANSWER**

Discussion

Note how we constructed our iteration to isolate that portion of the equation with the weakest dependence on the unknown Q's. The trick was to substitute the continuity equation for $Q_{bypasss}$ into the energy equation. Had we not done this but instead solved the continuity and energy equations by straightforward iteration, at least twice as many trials would have been required to generate the same answers. You should be able to show that the pressure drop $p_1 - p_2$ is equal to 70.7 kPa.

7.5.4 Complicated Pipe Networks

The ultimate problem in pipe and duct system analysis involves a complex network with many branches. Such a network is illustrated schematically in Fig. 7.43. Two examples of such networks are municipal water supply systems and air-handling ducts for large air-conditioning systems. Methods for calculating flow and losses in such systems are similar to methods for analyzing electrical networks. The following rules are the basis of any calculation procedure:

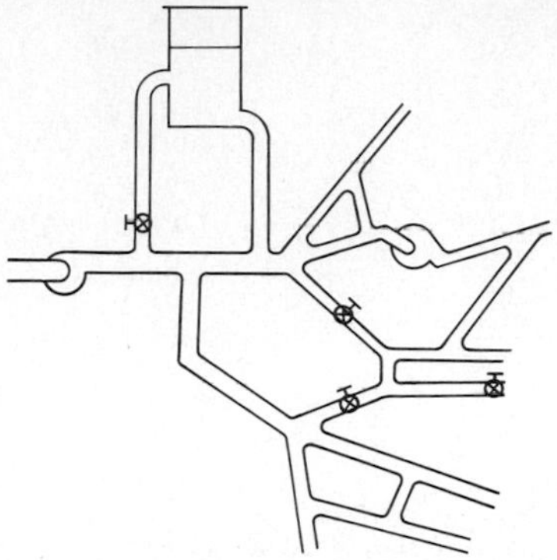

Figure 7.43 Complex multibranch pipe or duct network.

- The net flow into any junction must equal the net flow out of the junction. This requirement is most conveniently handled by letting each flow have an algebraic sign so that the net flow into any junction can be set to zero.
- The sum of head (or pressure) increases and losses around any loop must be zero. Alternative ways of stating this requirement are that the static pressure and velocity at any junction must be the same regardless of which path is followed to arrive at the junction and that the hydraulic grade lines for all pipes that come into a junction must intersect at the junction.
- All losses must satisfy the pipe friction equations or the local loss equations. All pumps must operate at a point on their pump curve.

The primary difference between electrical networks and hydraulic networks is that hydraulic networks are governed by nonlinear equations. Application of the preceding rules to a hydraulic network produces a set of nonlinear algebraic equations (some of these "equations" may even be graphic, such as pump curves) for the flows and pressure at each junction. Hardy Cross [20] developed a clever calculation technique for making the iterations. In the past, his equations had to be solved by tedious hand calculations, but now the digital computer handles this type of problem in nothing flat [21, 22].

PROBLEMS

1. An experiment is to be conducted to demonstrate the flow development in the laminar entrance region of a pipe. The experimental setup is shown in Fig. P7.1. Should cold water (60°F), warm water (120°F), or kerosene (77°F) be used to ensure the greatest entrance length?

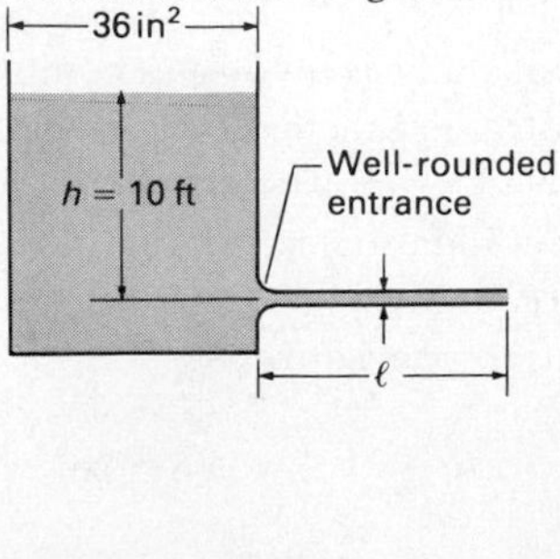

Figure P7.1

2. The experimental setup shown in Fig. P7.1 has 50°F water in the tank. Estimate the water heights h for which the flow in the pipe is laminar and fully developed over 90% of the length ℓ. The pipe is copper with a 0.315-in. I.D.

3. The experimental setup shown in Fig. P7.3 has 50°F water in the two tanks. Estimate the water height difference h for which the flow in the pipe is laminar and fully developed over 90% of the length ℓ. The pipe is copper with a 0.315-in. I.D.

4. Verify that the kinetic energy correction factor α for fully developed, laminar flow in a circular pipe is 2.0. See Eq. (7.42).

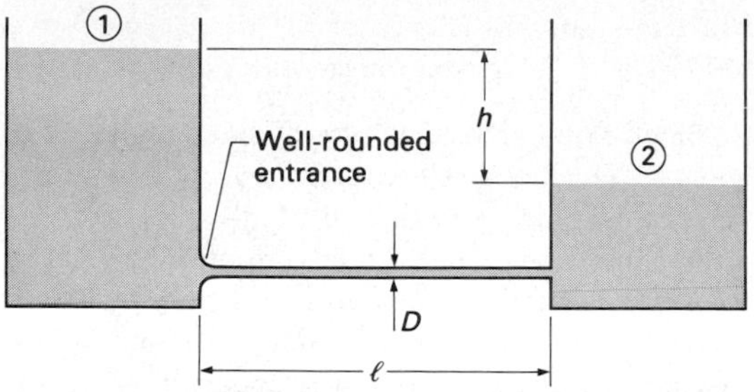

Figure P7.3

5. Verify that the momentum correction factor β for fully developed, laminar flow in a circular pipe is 4/3. See Eq. (7.41).

6. Assume that laminar flow of 60°F water has been maintained in a 1-in. schedule 160, horizontal, 100-ft-long commercial steel pipe at a Reynolds number of 25,000. Compare the mechanical energy loss, pressure loss, and pressure drop for this flow and that of turbulent flow for the same Reynolds number.

7. Lubricating oil at 20°C flows in a steady, laminar manner through an annulus of inside radius $R_i = 0.04$ m and outer radius $R_o = 0.06$ m. The annulus is 1.0 m long. Find the energy loss and pressure loss for a fully developed flow of 0.00005 m^3/s.

8. Water at 60°F flows at a rate of 4.0 gal/min through a 6-in. I.D. plastic pipe. The pipe is 500 ft long and rises a vertical height of 40 ft over the 500 ft. Find the energy loss, pressure loss, and pressure drop.

9. Fluid with kinematic viscosity ν flows down an inclined circular pipe of length ℓ and diameter D with flow rate Q. Find the vertical drop per unit length of the pipe so that the pressure drop $(p_1 - p_2)$ is zero for laminar flow.

10. The piston in the syringe shown in Fig. P7.10 is pushed at the rate of 0.10 cm/s. Find the pressure drop $(p_1 - p_2)$ in the 1.0-m-long tubing having an inside diameter of 0.003 m. The fluid has the properties of 20°C water.

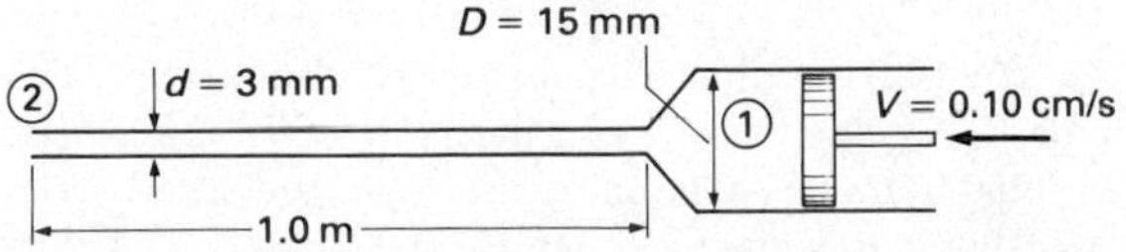

Figure P7.10

11. Consider laminar flow in a horizontal circular pipe. For a given pressure difference over a length L, how will the flow rate change if the pipe inside diameter is doubled? Assume fully developed flow.

12. Air at 20°C flows in a duct with inside dimensions of 8.0 cm × 8.0 cm. A flow straightener measuring 8.0 cm × 8.0 cm × 12.0 cm long has one hundred 0.75-cm-diameter holes drilled in the 12.0-cm direction. The pressure drop across the holes is 0.25 Pa. Find the flow rate.

13. Lubricating oil at 20°C flows from tank A to tank B, as shown in Fig. P7.13. Find the flow rate for a pipe of 1.0-cm I.D. All elbows are standard and all connections are screwed.

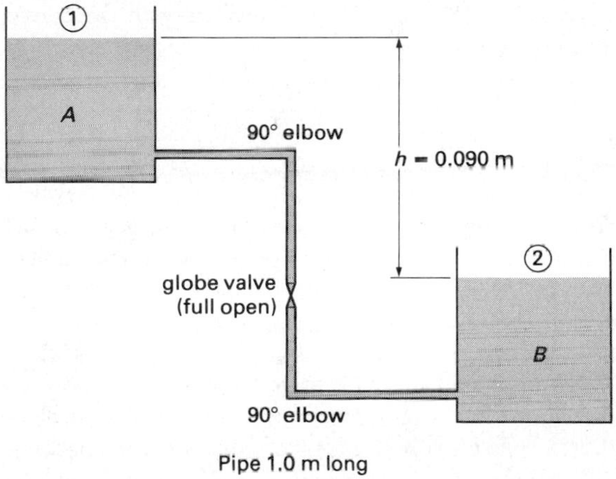

Figure P7.13

14. A 3-in. schedule 40 commercial steel pipe carries 210°F SAE 40 crankcase oil at the rate of 6.0 gal/min. The oil specific gravity is 0.89. For the same pressure drop, what must the new pipe size be to carry the oil at 10.7 gal/min?

15. A 3-in. schedule 40 commercial steel pipe carries 210°F SAE 40 crankcase oil at the rate of 6.0 gal/min. The oil specific gravity is 0.89. Calculate the pipe size required to carry the same flow rate at approximately one-half the pressure drop of the 3-in. pipe. Both pipes are horizontal.

16. Water at 60°F flows from a large reservoir having a water surface 50 ft above the point of usage, where the discharge pressure is also atmospheric. Find the pipe size(s) for which the flow in the pipe is definitely laminar if the pipe length is also 50 ft.

17. Water at 20°C flows down a vertical pipe with no pressure drop. Find the range of pipe diameters D (if any) for which the flow is definitely laminar.

18. Kerosene at 20°C flows down an inclined pipe making an angle of 10° with the horizontal. The pressure drop is 20 Pa over 4.0 m and the flow rate is 1.0 l/min. Find the pipe inside diameter.

19. Consider horizontal, fully developed laminar flow of 20°C lubricating oil through an annulus of inner radius $R_1 = 0.5$ cm and outer radius $R_2 = 0.75$ cm. The Reynolds number is 1000. Find the pressure drop per unit length of the channel by using the hydraulic diameter approach and the equivalent diameter approach. The equivalent diameter is

$$D_{eq} = \sqrt{D_2^2 - D_1^2},$$

where $D = 2R$. Compare the answers with the correct answer of

$$\Delta p = \frac{8LQ\mu}{\pi[R_2^4 - R_1^4 - (R_2^2 - R_1^2)^2/\ln(R_2/R_1)]}.$$

Which answer is more accurate?

• **20.** A person is donating blood. The pint bag in which the blood is collected is initially flat and is at atmospheric pressure. Neglect the initial mass of air in the 1/8-in. I.D., 4 ft-long plastic tube carrying blood to the bag. The average blood pressure in the vein is 40 mm Hg above atmospheric pressure. Estimate the time required for the person to donate one pint of blood. Assume that blood has a specific gravity of 1.06 and a viscosity of 1.0×10^{-4} lb·sec/ft². The needle's I.D. is 1/16 inch and the needle length is 2.0 in. The bag is 1.0 ft below the needle inlet and the vein's I.D. is 1/8 in. *Optional:* Donate a pint of blood and check your answer.

• **21.** A force of $F = 1.0$ N is applied to the hypodermic needle shown in Fig. P7.21. Find the flow rate Q. Assume that the piston moves with a constant velocity. The fluid has a specific gravity of 0.93 and a viscosity of 0.60 N·s/m². The piston moves with no friction.

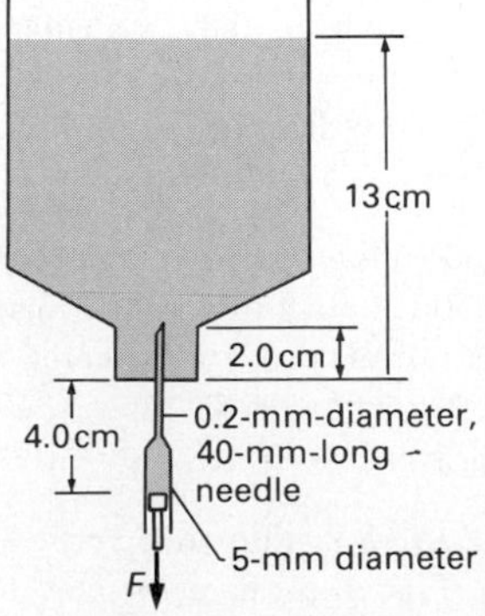

Figure P7.21

• **22.** Blood flows at volume rate Q in a circular tube of radius R. The blood cells concentrate and flow near the center of the tube, while the cell-free fluid (plasma) flows in the outer region. The center core of radius R_c has a viscosity μ_c, and the plasma has a viscosity μ_p. Assume laminar, fully developed flow for both the core and plasma flows and show that an "apparent" viscosity is defined by

$$\mu_{app} \equiv \frac{\pi R^4 \Delta p}{8LQ}$$

is given by

$$\mu_{app} = \frac{\mu_p}{1 - (R_c/R)^4(1 - \mu_p/\mu_c)}.$$

23. The following equation shows the axial velocity $u(x, r)$ for the laminar flow of a constant density, Newtonian fluid flowing in the entrance region of length ℓ_e of a horizontal pipe of radius R. The fluid entering this entrance region has a uniform velocity U.

$$u = U\left[2 - \left(1 + \frac{x}{L_e}\right)\left(\frac{r^2}{R^2}\right)^{x/L_e}\right].$$

$$0 \le x \le L_e$$

$$0 \le r \le R$$

Show this equation satisfies:

(a) the inlet velocity profile $u = U$.

(b) the fully developed velocity profile at the end of the entrance length,

$$u = 2U\left(1 - \frac{r^2}{R^2}\right).$$

(c) a finite, nonzero shear stress at the wall ($r = R$).

(d) conservation of mass where the volume flow rate Q is

$$Q = \pi R^2 U$$

for all x. Even though this equation satisfies all the conditions, what is wrong with it?

24. A 24-in. schedule 40, horizontal, commercial steel pipe (ID = 23.624-in.) has a flow rate of 77°F kerosene of 1200 ft³/min. At point A the static pressure is 50.6 psig. Point B is 2000 ft downstream from point A. The pipeline has no fittings or valves. Find the mechanical energy loss, the pressure loss, and the pressure drop between points A and B.

25. The 40°F water velocity in the pipe in Fig. E7.2 is 10.0 ft/sec for steady-state conditions. Find the energy loss gh_L, the pressure loss Δp_L, and the pressure change between points 1 and 3. Assume that the entrance nozzle is lossless and that the entrance length in the pipe is negligible. The copper tubing's inside diameter is 0.128 in., and its length is 6.0 ft.

26. The large blood vessel leading from the human heart, the aorta, has an inside diameter of 1.0 cm and a length of 40 cm. The flowing blood ($S = 1.06$) has an absolute viscosity of 0.0035 N·s/m² and an average velocity of 50 cm/s. Find the pressure drop if the aorta is (a) horizontal and (b) vertical with downward flow. Express your answers in mm Hg. Assume that blood is a Newtonian fluid.

27. Methanol at 20°C flows through a 25-cm I.D., clear plastic tube at the rate of 1.2 m³/min. The tube is inclined upward in the direction of flow at an angle of 30° with the horizontal. Find the pressure drop, pressure loss, and energy loss for a 50-m-long section of the tube.

28. Find the pressure drop, pressure loss, and energy loss for a 70°F water flow rate of 500 gal/min flowing upward in a vertical 6-in. schedule 80 pipe. The pipe is 400 ft long.

29. A 2-in. schedule 40 commercial steel pipe has a flow rate of 40 gal/min. Find the energy loss and pressure loss in 100 ft for 60°F water.

30. A pipeline has an inside diameter of 5.25 cm, is 30 m long, has a rise of 1.0 m in the direction of flow, and has a pressure loss of 47.3 kN/m². Find the pressure drop in the direction of flow. Consider 20°C water in turbulent flow.

31. Methanol at 32°F flows through a 3-in. schedule 40 pipe having a length of 700 ft. Find the flow rate through the pipe if the pipe slopes downward at an angle of 20° and the static pressure is constant along the pipe.

32. Lubricating oil at 20°F flows through a horizontal, $2\frac{1}{2}$-in. schedule 80 commercial steel pipe with a head loss (h_L) of 0.75 ft per 100 ft of pipe. Find the flow rate.

33. Kerosene at 0°F flows through a 2-in. schedule 40, horizontal, 100-ft-long commercial steel pipe at a rate of 200 gal/min. Find the increase in flow rate for the same pressure drop if the kerosene is heated to 80°F.

34. For the standpipe system shown in Fig. E7.2, calculate the flow rate for $H = 4.0$ ft, $D = 6.77$ in., $d =$ 0.125 in., and $L = 48$ in. The fluid is 70°F water. Assume steady flow and neglect the energy loss in the entrance nozzle. The pipe is commercial steel.

35. Consider fully turbulent flow in a horizontal, commercial steel, circular pipe. For a given pressure difference over a length L, how will the flow rate change if the pipe's inside diameter is doubled from 2.0 cm to 4.0 cm? Consider fully developed flow.

36. Water at 20°C flows from a tank, through a frictionless nozzle, and then through a 1.0-cm I.D., smooth plastic pipe as shown in Fig. P7.36. Find the average fluid velocity and pressure at cross-section AA when $\ell = 4.0$ cm.

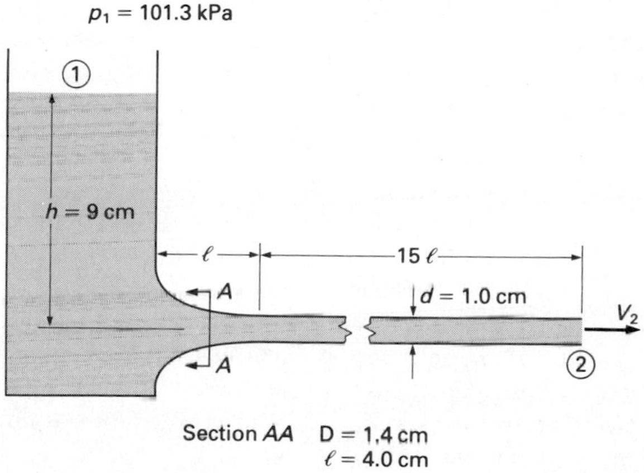

Figure P7.36

37. Water at 20°C flows through a 2.0-cm I.D. cast iron pipe with a pressure drop of 100 kPa in a length of 100 m and an elevation drop of 5.0 m. Use the Colebrook formula to find the water flow rate.

38. Kerosene at 77°F flows through a schedule 40, 3-in., commercial steel pipe with a pressure rise of 4.0 psi and an elevation drop of 200 ft over a pipe length of 250 ft. Find the kerosene flow rate. Use the Colebrook formula.

39. A $1\frac{1}{2}$-in. I.D. cast iron line has a natural gas flow rate of 100 ft³/min at 70°F and 14.95 psia. How much will the flow rate increase if the pipe inside diameter is increased to $2\frac{1}{2}$ in. with the same energy loss? The gas density may be assumed constant and equal to that at 70°F and 14.95 psia, and the gas absolute viscosity is 2.30×10^{-7} lb·sec/ft². The pipe has no valves or fittings and the pipe length is unchanged. The natural gas specific gas constant is 3090 ft·lb/slug·°R.

40. Water at 20°C is to flow through a 3-cm I.D. plastic pipe at the rate of 0.001 m^3/s. Find the incline angle of the pipe needed to make the static pressure constant along the pipe.

41. A type L copper pipe is horizontal and has a 20°C air flow rate of 0.005 m^3/s. Find the minimum pipe diameter for a pressure drop of no more than 1.0 kPa per 1.0 m of pipe length. The average air pressure is approximately 104 kPa.

42. A galvanized iron circular duct has an air flow rate of 5000 ft^3/min and a pressure loss of 0.20 in. of water at 68°F per 100 ft of duct. The average air pressure is approximately 14.9 psia and the air temperature is also 68°F. Find the duct size.

43. Water at 70°F enters a schedule 80, 4-in.-diameter pipeline at 100 gal/min and a pressure of 50 psig, with fully developed flow. Find the length L so that the pressure change is 40 psi if the flow is vertically upward.

44. A company markets ethylene glycol antifreeze in half-gallon bottles. A machine fills and caps the bottles at a rate of 60 per minute. The 68°F ethylene glycol is pumped to the machine from a tank 36 ft away. The pressure at the discharge of the pump is 35 psig, and the pressure at the inlet to the filling machine must be at least 15 psig. Select a suitable diameter for a feed pipe made of drawn, type K, copper pipe. The pipe has no elevation change.

45. Liquid mercury at 20°C flows through the horizontal channel shown in Fig. P7.45. Assume fully developed flow and find the mechanical energy loss and pressure drop per unit length. The channel walls are made of copper plate (copper plate is considered drawn tubing for purposes of determining the plate absolute roughness).

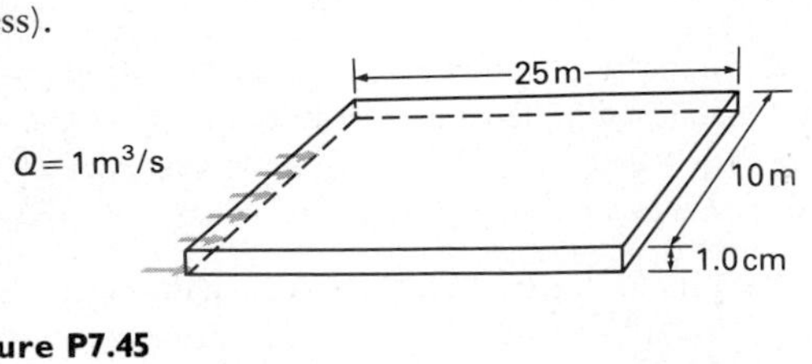

Figure P7.45

46. Air at 101 kPa and 20°C is to flow through a duct at 1.0 m^3/s. The duct designer is considering either a square duct or a circular duct with the same flow area. Which duct will have the smaller frictional pressure loss for equal lengths of smooth duct with turbulent flow?

47. Air at 101 kPa and 20°C is to flow through a duct at 1.0 m^3/s. The duct designer is considering both a square duct, measuring 20 cm × 20 cm, and a rectangular duct, measuring 10 cm × 40 cm. Even though both ducts have the same flow area, which duct will have the smaller frictional pressure loss for equal lengths of smooth duct with turbulent flow?

48. Water at 20°C flows through a concentric annulus of inner diameter $D_1 = 2.0$ cm and outer diameter $D_2 = 4.0$ cm. The surface roughness is 0.002 cm. Calculate the frictional pressure loss per unit length for a flow rate of 0.01 m^3/s.

49. Normal octane at 68°F flows through a channel with a triangular cross section measuring 3.0 in. on a side. The flow rate is 0.15 ft^3/min. Find the pressure loss per unit length.

50. A concentric annulus has an inside diameter $D_1 =$ 1.5 in. and an outside diameter $D_2 = 3.0$ in. Ethanol at 68°F flows vertically upward at a rate of 1.0 lb/min. Find the pressure drop $p_A - p_B$ over a length of 40 in.

51. A commercial steel flow channel has an equilateral triangle cross section with each side measuring 5.0 in. and a length measuring 96.0 in. Water at 60°F flowing through the channel has a pressure loss of 0.10 psi. Find the water flow rate.

52. A vertical channel with a rectangular flow area measuring 6.0 cm × 12.0 cm is made of commercial steel, and has an upward flow of 20°C water. The pressure drop per unit length is 20 kPa/m. Calculate the water flow rate using both the Moody chart and the Colebrook formula.

53. A horizontal duct has an annular cross section with an inside diameter of 0.5 m and an outside diameter of 1.0 m. The pressure loss for 20°C air flow is 1.0 cm H_2O/m. The duct is made of galvanized iron. Calculate the air flow rate using the Colebrook formula.

54. Helium at 200°F and approximately atmospheric pressure flows upward through a commercial steel duct having a rectangular cross section measuring 1.0 ft × 2.0 ft. The pressure drop is 0.05 in. 60°F water over a height of 100 ft. Find the helium flow rate using the Colebrook formula.

55. Nitrogen at 200°F and approximately atmospheric pressure flows downward through a duct having a square cross section measuring 0.025 ft × 0.025 ft. Calculate the flow rate if the pressure is uniform along the duct and the flow is laminar.

56. A 10-m-long, 5.042-cm I.D. copper pipe has two fully open gate valves, a swing check valve, and a sudden enlargement to a 9.919-cm I.D. copper pipe. The 9.919 cm copper pipe is 5.0 m long and then has a sudden contraction to another 5.042-cm copper pipe. Find the energy loss for soldered fittings and a 20°C water flow rate of 0.05 m^3/s.

57. Calculate the energy loss, pressure loss, and pressure drop for a sudden pipe contraction. The volume flow rate is 2.0 m^3/s, and elevation changes are negligible. Assume that the velocity is uniform over each flow area. The fluid is 20°C ethanol, and $D_1 = 1.0$ m and $D_2 = 0.50$ m.

58. Figure P7.58 shows the 60°F water flow rates from the branches of a main supply line. Find the total pressure drop ($p_A - p_E$) for soldered copper pipe. Assume that the loss coefficient for each tee in Table 7.8 is based on the upstream velocity before the flow divides.

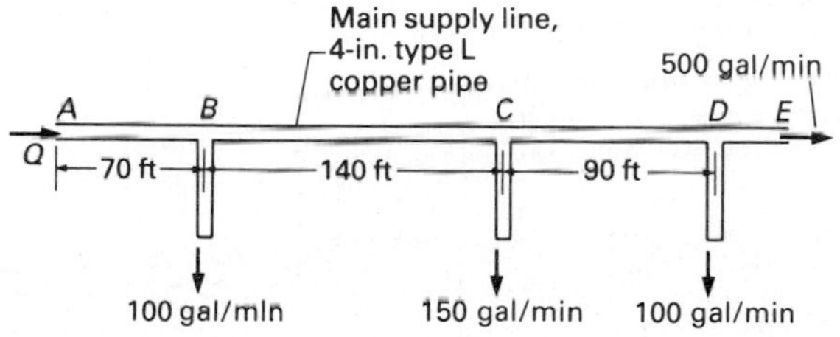

Figure P7.58

59. Normal octane at 68°F flows through the tee shown in Fig. P7.59. Calculate the energy losses and pressure losses from A to B and A to C for type L 1-in. copper pipe. Assume that the loss coefficients for the tee in Table 7.8 are based on the upstream velocity before the flow divides and that the connections are soldered.

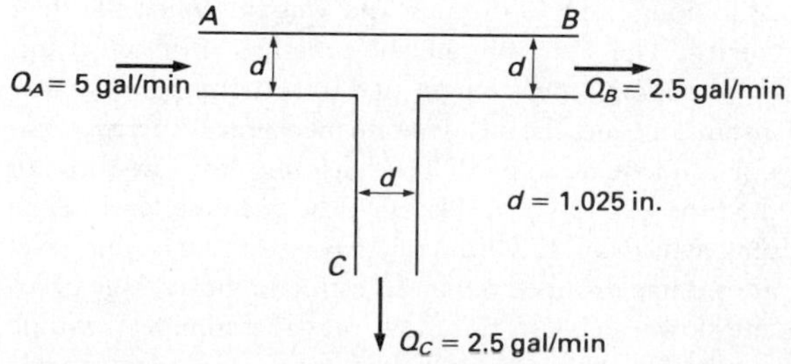

Figure P7.59

60. What gage pressure p_1 is necessary in the tank shown in Fig. P7.60 to produce a jet flow of 1 m^3/min? The connections are screwed.

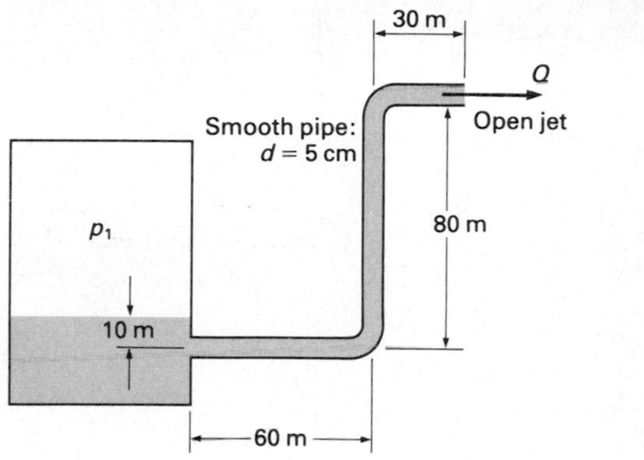

Figure P7.60

61. Water (20°C) flows at the rate of 0.008 m^3/s in the pipeline shown in Fig. P7.61. Calculate the pressure at point 2. The connections are screwed.

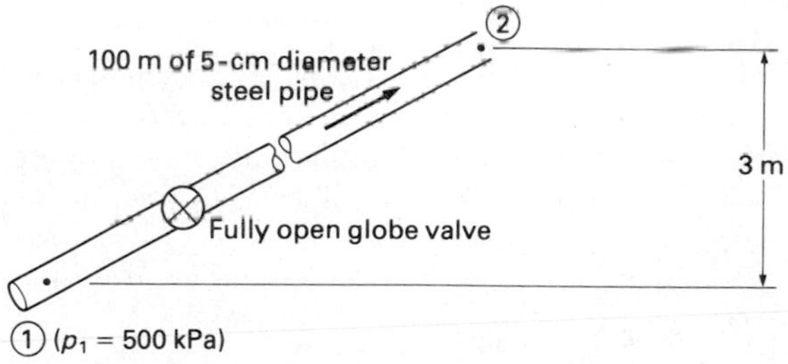

Figure P7.61

62. A 30% glycol solution ($\rho = 67.2$ lbm/ft^3, $\nu = 6.28 \times 10^{-5}$ ft^2/sec) flows out of a centrifugal chiller at 285 gal/min. The glycol solution flows through 50 ft of 2-in. I.D. steel pipe before reaching a plate heat exchanger. The pipeline contains a swing check valve, three fully opened globe valves, and four 90° standard elbows. All fittings are soldered. Find the pressure loss between the chiller and the plate heat exchanger.

63. The piping system shown in Fig. P7.63 is used to connect a large reservoir to a pump where it is then pumped to a hostel. The pipe is made of cast iron and 2-in. I.D. and must deliver 80 gal/min of 60°F water. The pipeline contains four swing check valves and five 90° standard elbows and is 300 ft long. All connections are screwed. Find the head loss between points 1 and 2.

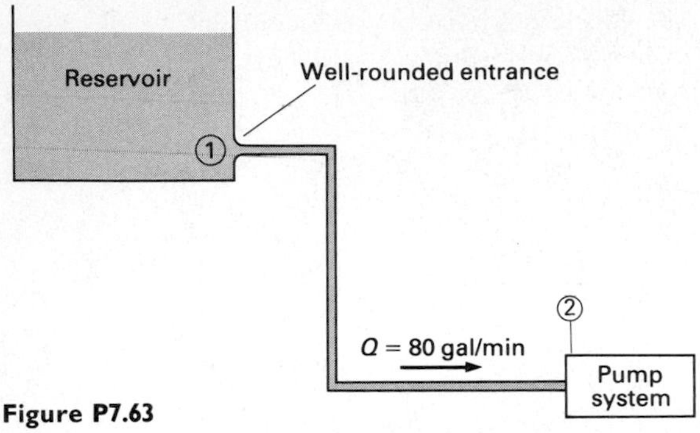

Figure P7.63

64. Kerosene at 77°F is flowing through a 3-in. schedule 80 commercial steel pipe at 50 gal/min. The pipe is 100 ft long and contains one fully open globe valve. Find the pressure loss for screwed connections.

65. Ethylene glycol at 20°C flows through a 7.36-cm I.D., smooth plastic pipe with two fully open gate valves. The flow rate is 0.02 m^3/s, and the pipe falls 10.0 m in 100 m. Find the pressure drop in this length of pipe. The plastic pipe is equivalent to soldered pipe.

66. Air flows in a horizontal 100-ft-long, 24-in. × 24-in. duct at the rate of 5000 ft^3/min. The air then flows through an expansion into 200 ft of 36-in. × 36-in. duct. The expansion has a loss coefficient of 0.80, based on the higher inlet velocity. Calculate the static pressure change across the expansion and the static pressure loss across the entire duct system. The duct is made of galvanized sheet metal. Use a constant air density of 0.0024 $slug/ft^3$.

67. Find the flow rate of 20°C kerosene from tank *A* to tank *B* for the conditions shown in Fig. P7.67. The orifice loss coefficient K = 2.5 is based on the orifice velocity. The pipe entrance is square edged.

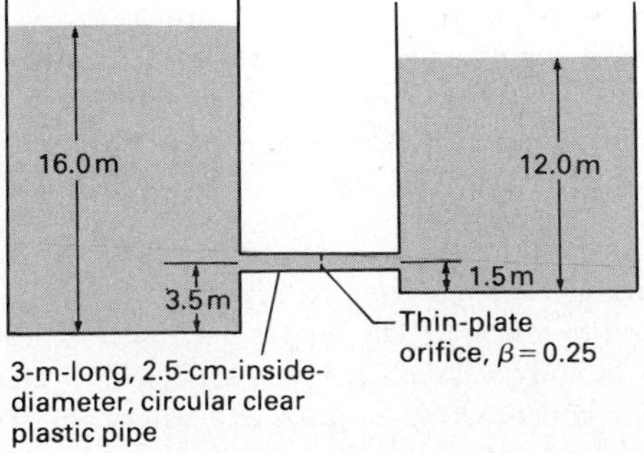

Figure P7.67

68. Calculate the water flow rate in the system shown in Fig. P7.68. The piping system includes four gate valves, two half-open globe valves, fourteen 90° standard elbows, and 250 ft of 2-in. schedule 40 commercial steel pipe. Assume screwed connections and a square-edged pipe entrance.

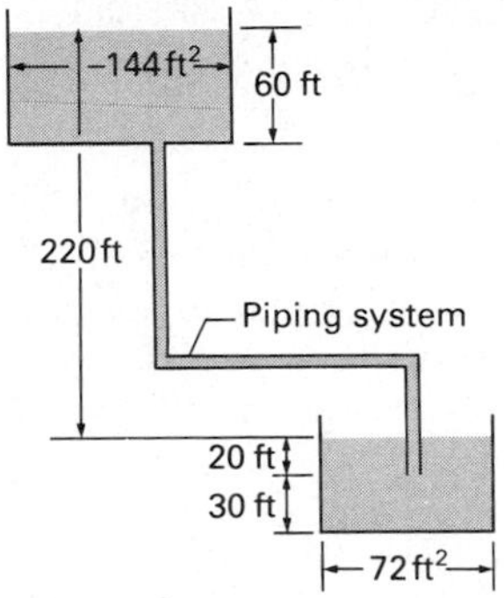

Figure P7.68

69. Find the flow rate of 68°F wine from the wine vat shown in Fig. P7.69. The wine vat is full. The wine has the properties of ethyl alcohol.

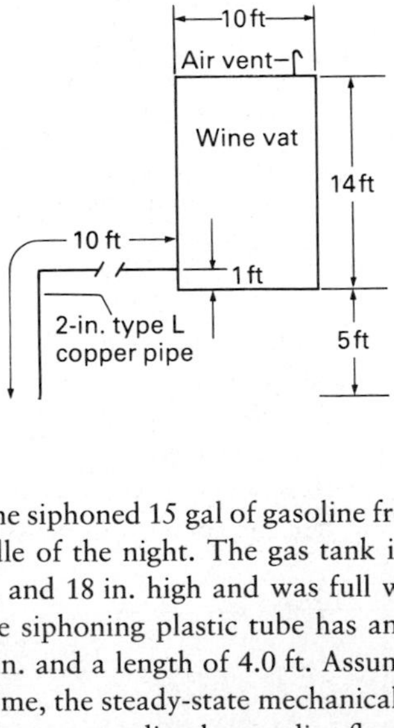

Figure P7.69

70. Someone siphoned 15 gal of gasoline from a gas tank in the middle of the night. The gas tank is 12 in. wide, 24 in. long, and 18 in. high and was full when the thief started. The siphoning plastic tube has an inside diameter of 0.5 in. and a length of 4.0 ft. Assume that at any instant of time, the steady-state mechanical energy equation is adequate to predict the gasoline flow rate through the tube. As 15 gal is 3465 in^3, the gasoline level in the tank will drop 12.0 in. You may use the gasoline level after it has dropped 6.0 in. to estimate the average gasoline flow rate. Use this flow rate to estimate the time needed to siphon the 15 gal of gasoline. Compare your answer with the answer of 190 sec found in Chapter 4 using Bernoulli's equation. The siphon discharges at the level of the bottom of the gasoline tank.

71. Ethanol at 20°C flows through 40 m of 2.0-cm I.D. copper pipe, through a 40° conical expansion, and then through 125 m of 3.0-cm I.D. copper pipe. The pressure loss is 150 kPa when the pipe is horizontal. Obtain the volume flow rate.

72. Find the flow rate between tanks *A* and *B* shown in Fig. P7.72. The piping consists of 100 ft of 2-in. schedule 80 galvanized pipe, a 30° conical expansion, and 200 ft of 3-in. schedule 80 galvanized pipe.

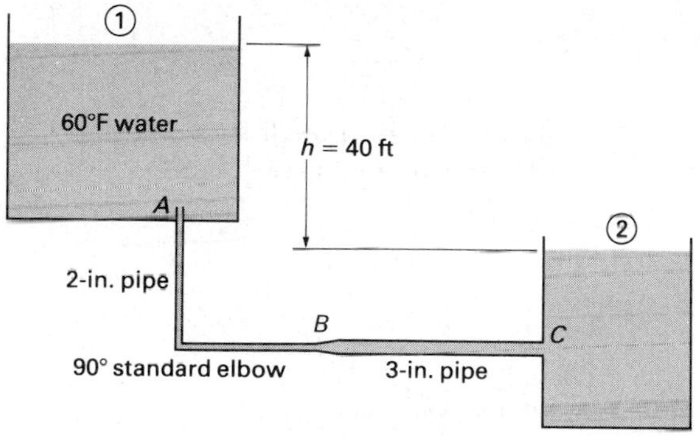

Figure P7.72

73. Figure P7.73 shows a penstock for transporting water to a hydraulic turbine. Calculate the flow rate in the penstock. The discharge nozzle loss coefficient based on the inlet velocity *V* is 0.75. Connections are flanged.

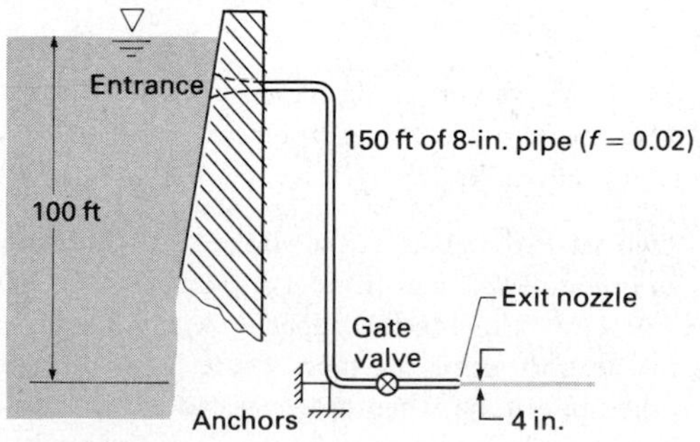

Figure P7.73

74. For the conditions shown in Fig. P7.74, determine the magnitude and direction of the volumetric flow rate. The fluid is 60°F water, the connections are screwed, and the pipe is schedule 40.

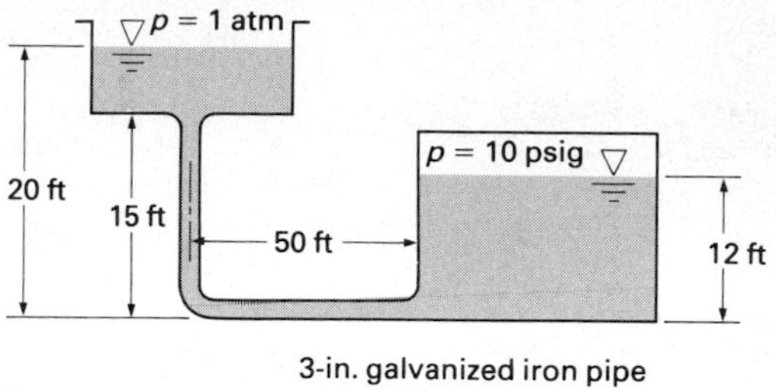

Figure P7.74

75. The vented storage tank shown in Fig. P7.75 is used to refuel race cars at a race track. A total of 40 ft of steel pipe (I.D. = 0.957 in.), two 90° standard elbows, and a globe valve make up the system. Calculate the time needed to put 20 gal of fuel in a car tank. The pressures, p_2 and p_1, are equal, the connections are screwed, and the fuel has the properties of 68°F normal octane.

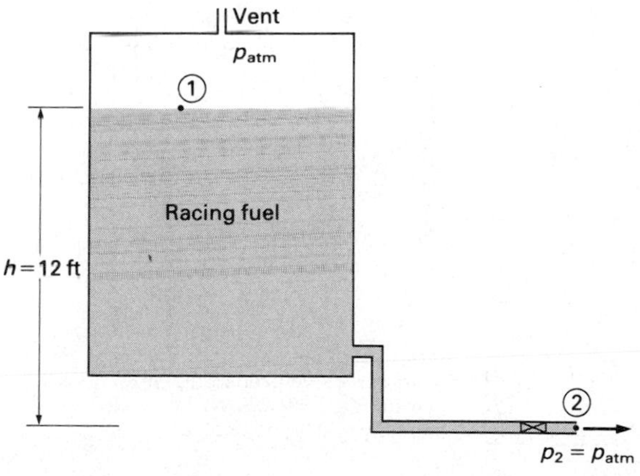

Figure P7.75

76. A galvanized iron pipe with an inside diameter of 5 in. and total length of 50 ft is used to transfer water at 60°F. The pipeline contains two swing check valves, two fully open globe valves and a one 90° standard elbow. The pressure at point 1 is measured as 50 psig, and the pressure at point 2 is measured as 40 psig. Find the average velocity of water in the pipe. Consider flanged connections. Points 1 and 2 are at the same elevation.

77. For the piping system shown in Fig. P7.77, determine the volume flow rate of oil ($\nu = 1 \times 10^{-5}$ ft²/sec) in the 2-in. I.D., 1200-ft-long cast iron pipe. Consider screwed connections and $p_1 \cong p_2 = p_{atm}$.

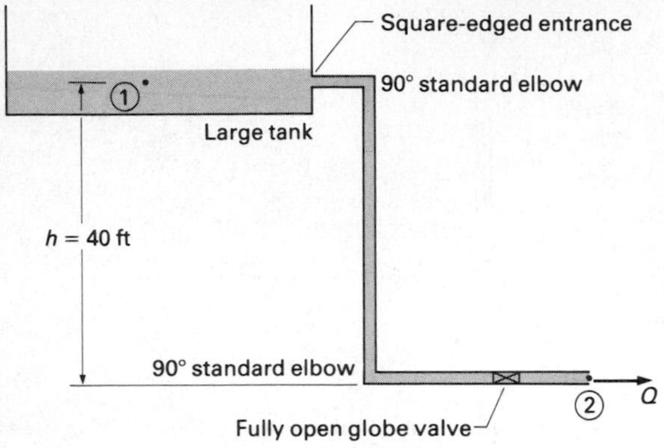

Figure P7.77

• **78.** Estimate the time required for the water depth in the reservoir shown in Fig. P7.78 to drop from a height of 25 m to 5 m. The connections are screwed.

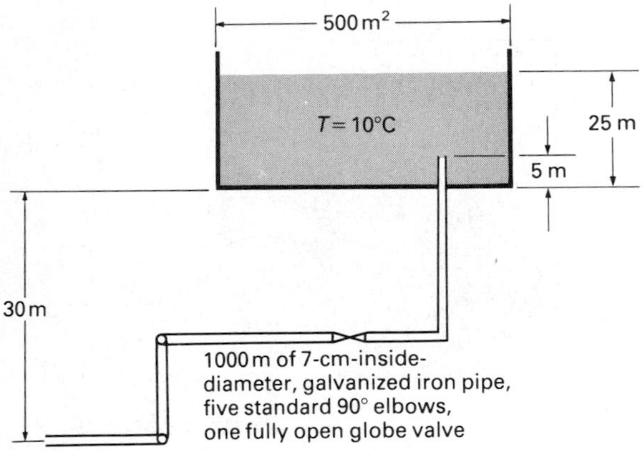

Figure P7.78

• **79.** For large tanks and steady-flow conditions, calculate the net rate of 60°F water flow from each tank shown in Fig. P7.79. Assume screwed connections.

80. Ethylene glycol is to be dispensed from a 4 ft × 4 ft × 4 ft storage tank through a 15-ft-long, 1-in. I.D. hose ($\varepsilon = 0.002$ in.) hanging from the bottom of the tank. How fast can a full tank be drained in an emergency?

81. A tank of water with a cross section given by $y = x^2$, as shown in Fig. P7.81 is connected to a 50-ft-long, 0.5-in. I.D. steel pipe that contains four 90° standard elbows and a gate valve. If the 10-ft-long tank is initially filled to a depth of 3 ft and the pipe outlet is 20 ft below the

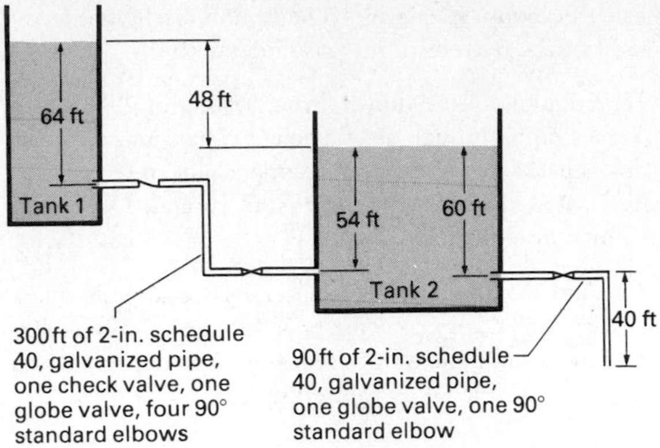

Figure P7.79

bottom of the tank, how long will it take to empty the tank with a fully open gate valve?

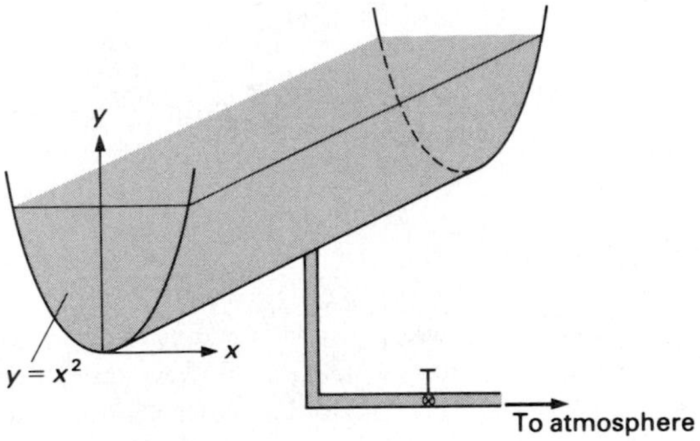

Figure P7.81

82. The pump shown in Fig. P7.82 provides a constant 12 kW to the fluid. How long will it take to fill an initially empty storage tank?

83. Two 25-ft-diameter tanks, linked by a siphon as shown in Fig. P7.83, constitute one part of a water purification plant. While in the upper tank, solid debris settles to the bottom of the tank. Water begins to drain from the upper tank when the level exceeds the crest of the siphon and stops draining when the siphon mouth is uncovered. The siphon is made of commercial steel and has a total length of 75 ft. If there is no inflow into the upper tank and the lower tank level is initially at the level of the siphon exit, how long will it take for the upper pond level to fall from the siphon crest to the siphon mouth?

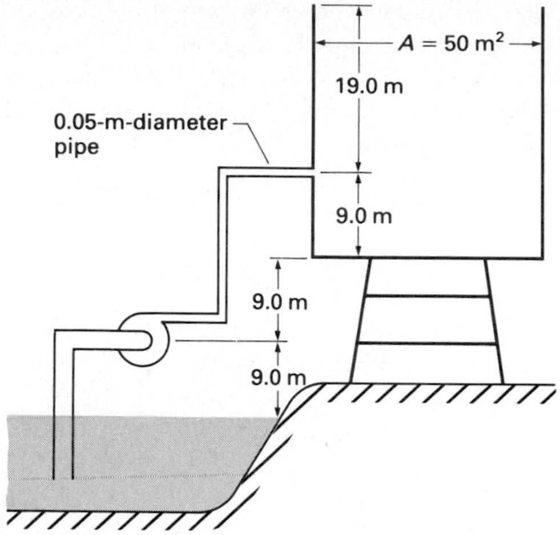

Figure P7.82

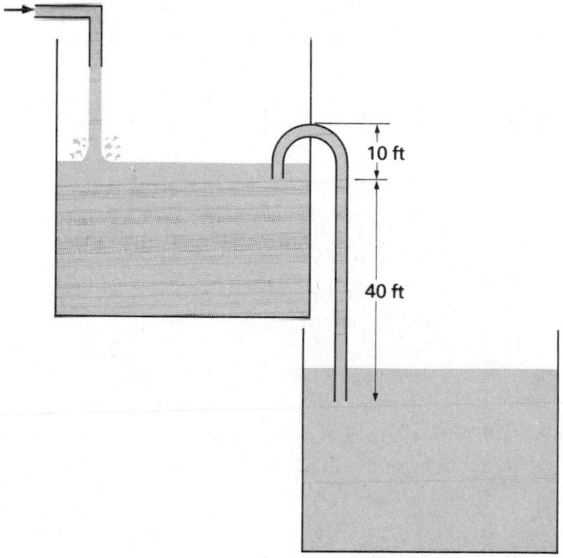

Figure P7.83

84. How long will it take for the level in the upper tank in Problem 83 to drop from siphon crest to siphon mouth if there is a constant inflow to the upper tank of 50 gal/min?

85. Calculate the diameter of a wrought-iron pipe required to deliver 20°C kerosene from one tank to another at the rate of 2.0 m^3/s. The pipe is 1500 m long and has two fully open globe valves and six 45° standard elbows. The two free liquid surfaces in the tanks have an elevation difference of 120 m. The entrance is square edged, and the connections are flanged.

86. It is necessary to deliver 270 ft^3/min of water from reservoir A to reservoir B, as shown in Fig. P7.86. The connecting piping consists of four fully open gate valves, twelve standard 90° elbows, one swing check valve, two fully open globe valves, and 3000 ft of commercial steel pipe. Find the minimum diameter for steel pipe needed to obtain this flow rate. All fittings and valves have flanged connections. Assume a square-edged entrance.

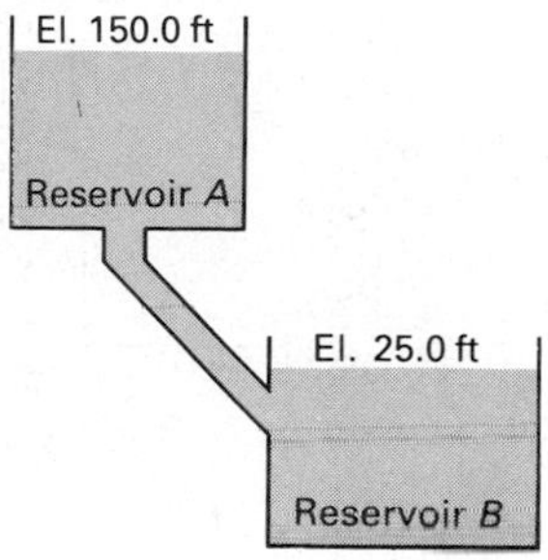

Figure P7.86

87. A 5000-ft cast iron pipeline originates in a reservoir and terminates in a 2-ft outlet diameter nozzle (K = 0.40, based on the pipeline velocity). The nozzle discharges to the atmosphere. The nozzle is 200 ft below the reservoir water level, and the pipeline has seventeen 90° standard elbows (K = 0.21) and two fully open gate valves (each at K = 0.03). Determine the pipe size required to deliver 10,000 ft^3/min. The pipeline entrance is square edged.

88. Normal octane at 68°F is to be delivered at a flow rate of 3.0 gal/min through a 2.0-in. schedule 40 commercial steel pipe. Energy losses are important, so consideration is being given to expanding the pipe to a 3.0-in. schedule 40 commercial steel pipe, using a 10° conical expansion. After a length ℓ of the 3.0-in. pipe, it is decreased to a 2.0-in. schedule 40 commercial steel pipe with a gradual contraction (K = 0.10, based on the smaller diameter pipe flow). Find the minimum length ℓ so that the energy loss will be the same as in a 2.0-in. pipe of length $\ell + 2$ (5.7 in.). See Fig. P7.88.

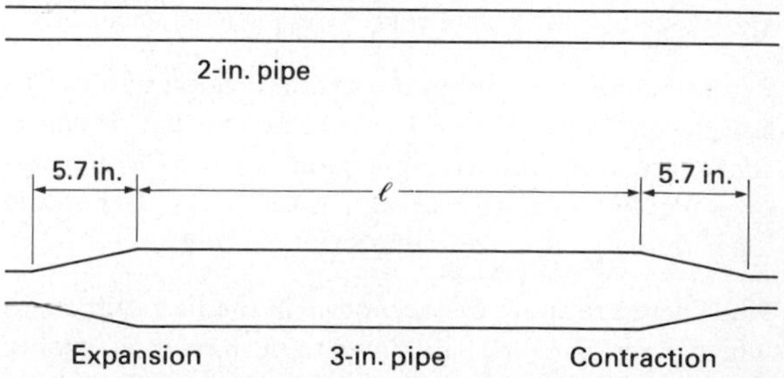

Figure P7.88

89. Water must be delivered from the lower tank *A* to the upper tank *B* shown in Fig. P7.89 at the rate of 10 gal/min. The pipeline is to be 1000 ft long and screw connected and is to have fourteen 45° standard elbows. The lower tank is to be pressurized to 50 psig. Find the minimum nominal pipe size of schedule 40 galvanized pipe required.

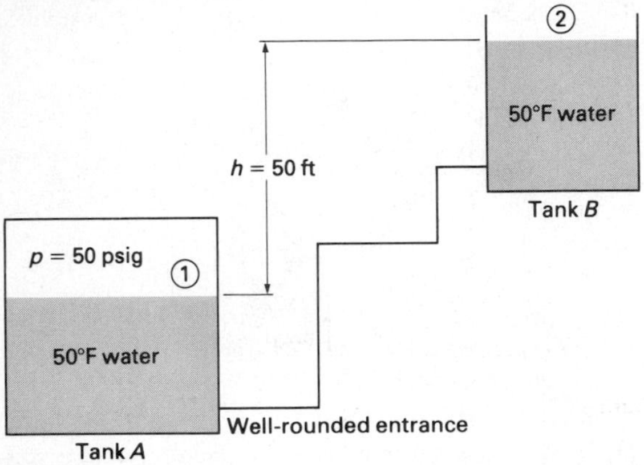

Figure P7.89

90. At various stages in plastic formulation, large quantities of nonvolatile chemicals are stored in vented tanks. One such setup has a 1000-gallon tank that empties into a larger holding bin. Before the chemicals actually arrive at the tank they drop 95 ft in elevation and pass through 100 ft of straight commercial steel pipe. There are two fully open globe valves and one 180° standard elbow (all have flanged connections). If a 43-gal/min flow rate is desired, what size of schedule 80 pipe would be necessary? The chemical in storage has properties similar to those of ethylene glycol.

91. Use the high-level computer language or spreadsheet of your choice and develop a tool for pipe sizing for a specified flow rate, an allowable pressure drop, a total length, and a total local loss coefficient. Use this tool to compute the minimum I.D. copper tube that can carry 20 gal/min of water with a total pressure drop of 25 psi a distance of 50 ft with a total local loss coefficient of 15.

92. The mixing length in the overlap region of a turbulent flow is given by $\ell = \kappa y$. Assume that the turbulent shear stress in the overlap region is constant and approximately equal to the wall shear stress. Derive the logarithmic velocity distribution.

93. There are some discrepancies in the literature as to whether Eq. (7.45) is valid down to the Reynolds number of 30,000 or 4000. To determine the proper lower limit, calculate the friction factor for smooth pipes from Eq. (7.45) for Reynolds numbers of 4000, 10,000, 20,000, 30,000, 50,000, 75,000, and 100,000. Compare these values with the values from the Moody chart and draw your own conclusions.

94. Compare the values of n in Table 7.3 for Reynolds numbers of 4×10^3 and 2.3×10^4 with those values of n obtained from Eqs. (7.45) and (7.51).

95. Von Karman suggested that the fully rough friction factor be expressed by the equation

$$f = \frac{1}{4[0.57 - \log(\varepsilon/D)]^2},$$

where ε is the absolute roughness of the pipe. Compare the values predicted by this equation and those indicated on the Moody chart.

96. The following equation for the Darcy friction factor f is sometimes used for computer calculations:

$$f = a + b/\mathbf{R}^c,$$

where

$$a = 0.104\left(\frac{\varepsilon}{D}\right)^{0.225} + 0.532\left(\frac{\varepsilon}{D}\right);$$

$$b = 88\left(\frac{\varepsilon}{D}\right)^{0.44};$$

$$c = 1.62\left(\frac{\varepsilon}{D}\right)^{0.134}.$$

Compare this equation for f for $\varepsilon/D = 0.00001$, 0.0001, 0.001, and 0.01 and Reynolds numbers of 10^4, 10^5, 10^6, and 10^7 with the Moody chart and decide whether it is an acceptable replacement for the Colebrook formula.

97. The Swamee and Jain formula for the friction factor is

$$f = \frac{0.25}{[\log(\varepsilon/3.7D + 5.74/\mathbf{R}^{0.9})]^2}.$$

Compare this equation for $\varepsilon/D = 0.00001$, 0.0001, 0.001, and 0.01 and Reynolds numbers of 10^4, 10^5, 10^6, and 10^7 with the Moody chart and decide whether it is an acceptable replacement for the Colebrook formula.

98. The Haaland formula for the friction factor is

$$f = \frac{0.3086}{\{\log[6.9/\mathbf{R} + (\varepsilon/3.7D)^{1.11}]\}^2}.$$

Compare this equation for f for $\varepsilon/D = 0.00001$, 0.0001, 0.001, and 0.01 and Reynolds numbers of 10^4, 10^5, 10^6, and 10^7 with the Moody chart and decide whether it is an acceptable replacement for the Colebrook formula.

99. The Churchill formula for the friction factor is

$$f = 8\left[\left(\frac{8}{\mathbf{R}}\right)^{12} + \frac{1}{(A+B)^{1.5}}\right]^{1/12},$$

where

$$A = \left\{-2.457 \ln\left[\left(\frac{7}{\mathbf{R}}\right)^{0.9} + \frac{\varepsilon}{3.7D}\right]\right\}^{16};$$

$$B = \left(\frac{37{,}530}{\mathbf{R}}\right)^{16}.$$

Compare this equation for f for both the laminar and turbulent regimes for $\varepsilon/D = 0.00001$, 0.0001, 0.001, and 0.01 and Reynolds numbers of 10, 10^2, 10^3, 10^4, 10^5, 10^6, and 10^7 with the Moody chart and decide whether it is an acceptable replacement for the Colebrook formula.

100. An approximate formula for the static pressure drop Δp (in. H_2O) for air flowing in a horizontal circular duct of diameter D (in.) and length L (ft) with average velocity V (ft/min) is

$$\Delta p = 0.027\left(\frac{L}{D^{1.22}}\right)\left(\frac{V}{1000}\right)^{1.82}.$$

Tabulate Δp versus the average velocity V for a duct 100 ft long with a diameter of 24 in. Consider turbulent flow, a galvanized sheet-metal duct, and 65°F air. Compare your result with that obtained from the Moody chart.

• **101.** Approximate the velocity distribution in turbulent flow in a smooth annulus by a pair of logarithmic velocity profiles joined in the center of the annulus and develop an expression for the friction factor.

102. Use the high-level computer language or spreadsheet of your choice and develop a tool for solving the Colebrook formula for friction factor given values for relative roughness and Reynolds number. Test this tool by comparing your results to the Moody chart for the following cases: (a) $\varepsilon/D = 0.004$, $\mathbf{R} = 6 \times 10^3$; (b) $\varepsilon/D = 0.03$, $\mathbf{R} = 4 \times 10^7$; (c) $\varepsilon/D = 0.00001$, $\mathbf{R} = 8 \times 10^6$; and (d) $\varepsilon/D = 0.001$, $\mathbf{R} = 5 \times 10^5$.

103. For a pipe with a relative roughness of 0.001, use the Colebrook formula to compute a curve of friction factor versus Reynolds number for $5 \times 10^3 < \mathbf{R} < 5 \times 10^7$.

104. A thin-plate, sharp-edged orifice is to measure flow rates of 60°F water between 50 gal/min and 250 gal/min. The pressure drop is to be measured by a vertical mercury manometer 1 yd long. What orifice diameter should be used for a β ratio of 0.5?

105. An ASME flow nozzle has a β ratio of 0.5 and is to measure water flow rates between 50 gal/min and 250 gal/min. Recommend a suitable fluid for a vertical manometer that can be no more than 1 yd long. The temperature is 50°F.

106. A thin-plate, sharp-edged orifice with a throat diameter of 2.1 in. is used in a 3-in. type K copper pipe to measure the flow of 68°F lubricating oil. Determine the pressure drop for a flow rate of 600 gal/min for D and $(1/2)D$ taps and for both upward and downward flow.

107. A thin-plate, sharp-edged orifice with D and $(1/2)D$ taps and a β ratio of 0.25 is used in a piping system. The pressure drop across the orifice is to be measured by a mercury manometer 1 m high. What is the maximum flow rate that can be indicated for $D = 20$ cm and 10°C water?

108. An ASME flow nozzle with a β ratio of 0.25 is used in a piping system. The pressure drop across the nozzle is to be measured by a mercury manometer 1 m high. What is the maximum flow rate of 10°C water that can be indicated? The throat diameter is 5.0 cm.

109. Find the instantaneous mass flow rate of 60°F kerosene from tank A to tank B for the situation shown in Fig. P7.109. Treat the hole as a thin-plate orifice.

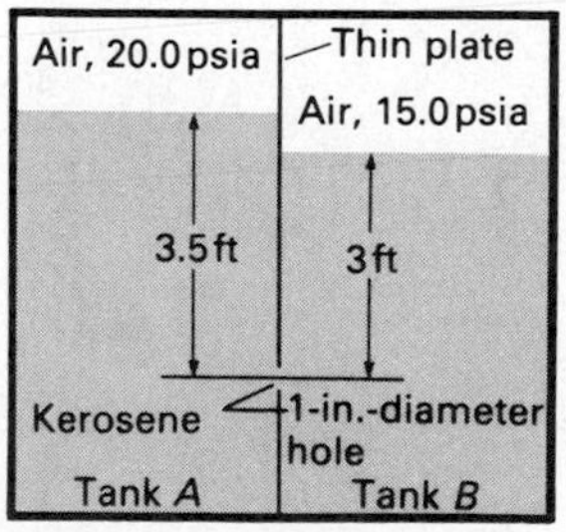

Figure P7.109

110. Air at 70°F and 20 psia flows through a 4-in. schedule 40 commercial steel pipe. A horizontal ASME long-radius flow nozzle having a β ratio of 0.70 is located in the pipeline. A manometer connected between the inlet and throat for the nozzle indicates a level difference of 0.40 in. of the two legs of the manometer. The manometer contains 68°F lubricating oil. Calculate the air flow rate.

111. The minimum flow area A, or vena contracta, for the fluid leaving an orifice is

$$A = [0.62 + 0.38(A_2/A_1)^3]A_2,$$

where A_2 is the orifice throat area and A_1 is the pipe area. Find the maximum β ratio for an area ratio A/A_2 of at least 95%.

112. Figure P7.112 shows a portion of a piping system in which a differential pressure gage is used to measure the pressure drop across a flow nozzle. Valves C and D are normally open and the flow rate in the pipeline containing the flow nozzle is obtained by opening valves A and B, recording the pressure drop across the nozzle, and then using a calibration curve that gives the flow rate through the nozzle as a function of the pressure drop. Valves A and B are closed after the reading is taken. On a particular occasion, the pressure difference indicated by the gage continued to increase after valves A and B were closed. Valve C also was observed to have a small leak from the valve stem. Is the leakage from valve C a possible explanation for the increasing pressure difference observed after valves A and B were closed? Give reasons for your answer. Valves A, B, C, and D are lever-handle cock valves.

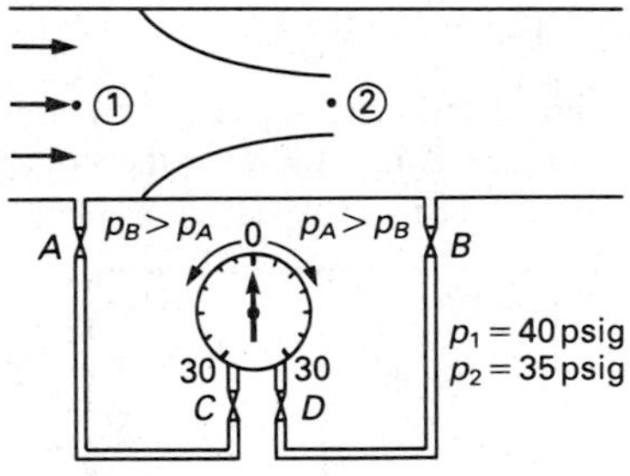

Figure P7.112

113. Plot the hydraulic grade line for the piping system in Fig. P7.113. The pipe is galvanized steel with a reentrant entrance loss coefficient of 1.0, a friction factor of 0.0126, a fully open gate value at the 1000-ft. point, and a 90° bend (K = 0.21) at the 200-ft. point. The pipe's inside diameter is 45 in. and the pipe's length is 4000 ft.

114. Plot the hydraulic grade line for the piping system in Fig. P7.114. The pressure is 200 kPa gage at point 1 and the flow rate is 0.010 m^3/s. The friction factor for the pipe is 0.015. The fluid density is 1000 kg/m^3. The connections are soldered.

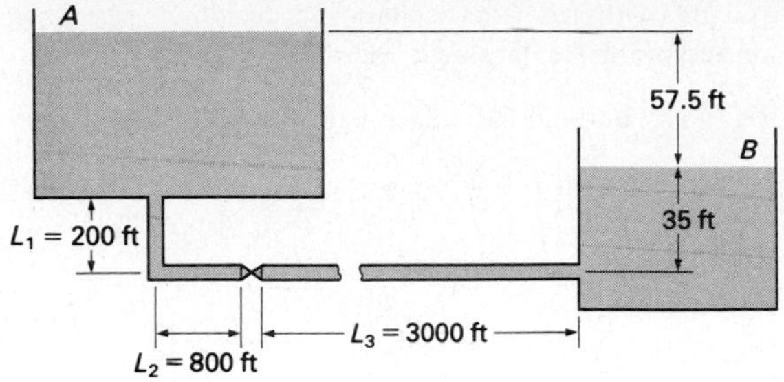

Figure P7.113

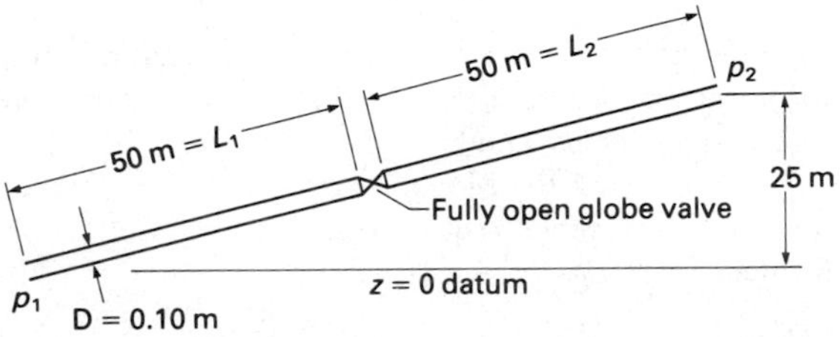

Figure P7.114

115. The piping system in Fig. P7.115 has a well-rounded entrance, 500 ft of suction line, and 500 ft of discharge piping. The pipeline from B to C has four fully open gate valves and three standard 90° elbows. The pipeline from D to E has four fully open gate valves, six 90° standard elbows, and two 45° standard elbows. The pump input work is 8330 ft·lb/slug of fluid. The average fluid velocity is 12 ft/sec in the 2-in. schedule 40 pipe. The connections are screwed, and the water is at 60°F. The pipe friction factor is 0.027. The pump's overall efficiency is 85%. Plot the hydraulic and energy grade lines.

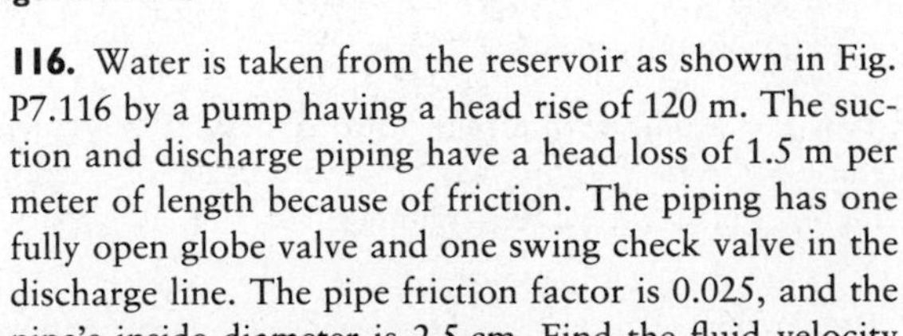

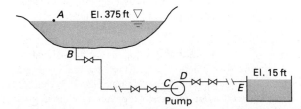

Figure P7.115

116. Water is taken from the reservoir as shown in Fig. P7.116 by a pump having a head rise of 120 m. The suction and discharge piping have a head loss of 1.5 m per meter of length because of friction. The piping has one fully open globe valve and one swing check valve in the discharge line. The pipe friction factor is 0.025, and the pipe's inside diameter is 2.5 cm. Find the fluid velocity and plot the hydraulic grade line.

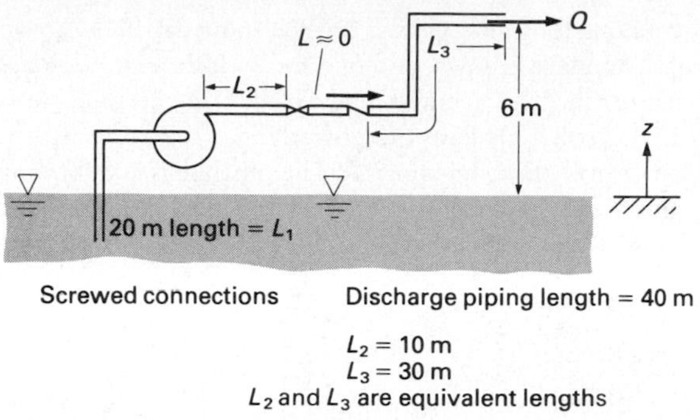

Figure P7.116

117. Find the fluid velocity and plot the energy grade line for the system described in Problem 116.

118. Plot the hydraulic grade line for Fig. P7.118. The friction loss is 2.0 m per meter of pipe length. The pipe friction factor is 0.020 and the pipe's inside diameter is 2.0 cm. Also find the fluid velocity in the pipe.

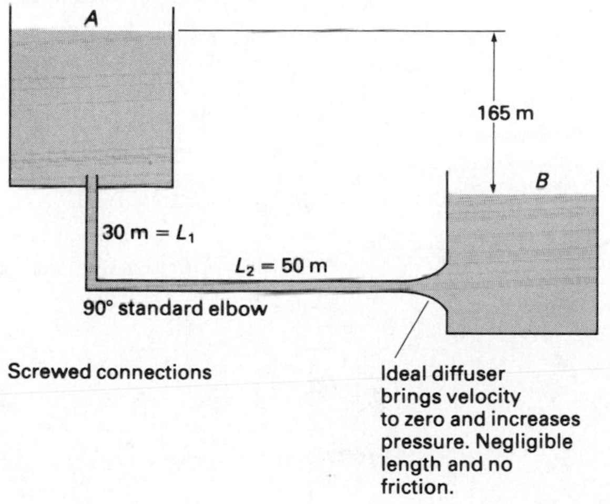

Figure P7.118

119. A 250-ft-high building has a 6-in.-diameter steel standpipe and a 100-ft-long $2\frac{9}{16}$-in. diameter fire hose on each floor. The nearest fireplug is 100 ft from the standpipe's ground-level connection. Assume that firefighters connect a 6-in.-diameter, 50-ft-long fire hose from the fireplug to the fire truck and a 4-in.-diameter, 50-ft-long fire hose from the fire truck to the standpipe's ground-level connection. The National Fire Protection Association (NFPA) requires that a minimum pressure of 65 psig be maintained at the connection of the $2\frac{9}{16}$-in.-diameter hose and the standpipe while maintaining a flow rate of 500 gal/min through the fire hose. What pressure rise must the pump on the fire engine supply to satisfy the NFPA requirement for this building? The fire hydrant water pressure is 80 psig and the water temperature is 60°F. The connections are screwed.

120. Crude oil having a specific gravity of 0.80 and a viscosity of 6.0×10^{-5} ft^2/sec flows through a pumping station at a rate of 10,000 barrels/hr. The oil then flows through 120,000 ft of 24-in. ID, horizontal, commercial steel pipe, enters another pumping station at 20 psig, and leaves at a discharge pressure equal to the discharge pressure of the previous pumping station. Calculate the discharge pressures and the power supplied to the crude oil at the second pumping station. See Fig. P7.120.

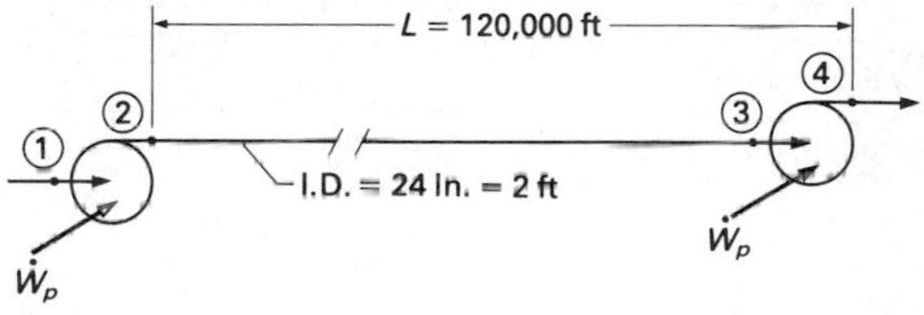

Figure P7.120

121. A pump has the head and efficiency curves shown in Fig. P7.121. Can this pump deliver any flow through the piping system shown? Why?

122. Consider the forced draft fan and air duct shown in Fig. P7.122. The air flow rate through the fan at normal operating conditions is 4×10^5 ft^3/min. The fan total pressure rise in inches of water is given by $p_{F,T} = 17.0 - 0.125(Q \times 10^{-5})^2$, with Q in ft^3/min. All the losses in the inlet box, ducts, and heaters are represented by

$$gh_L = \mathrm{K}\left(\frac{V_{\text{duct}}^2}{2}\right).$$

Find the value of K. Assume constant air density and negligible kinetic and potential energy changes.

123. An experienced engineer suggests that the flow in the system in Fig. P7.122 can be increased and the mechanical energy losses decreased by rounding the inlet box entrance (giving K = 0.05) and adding turning vanes in the miter bends. The total pressure rise developed by the fan is

$$p_{F,T} = 17.0 - 0.125(Q \times 10^{-5})^2,$$

where $p_{F,T}$ is in in. H_2O and Q in ft^3/min. Without the modifications, the flow rate is 4×10^5 ft^3/min. Determine the flow rate with the suggested modifications.

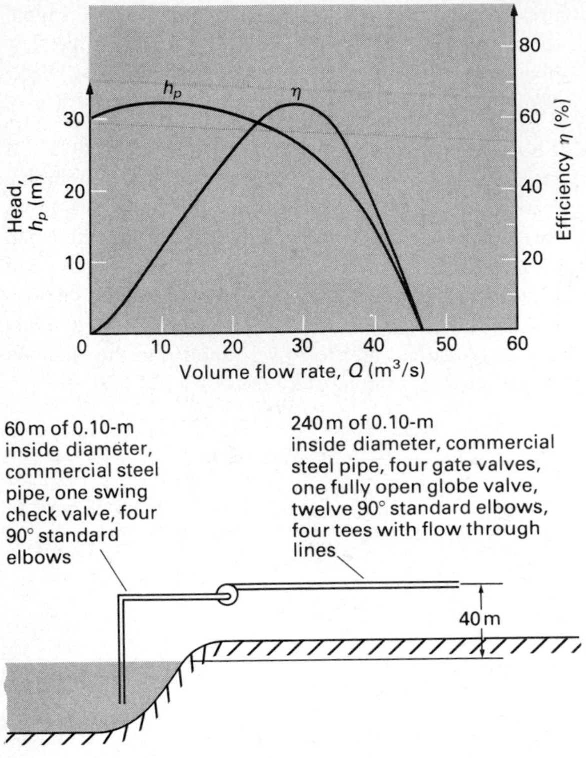

Figure P7.121

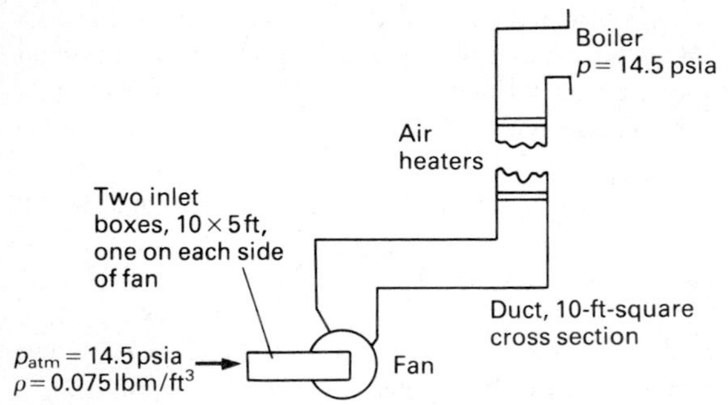

Figure P7.122

Assume constant air density and negligible kinetic and potential energy changes. Turning vanes in a miter bend reduce the loss coefficient by 80%.

124. Water flows from a large reservoir to a storage tank, as shown in Fig. P7.124. The pipe entrance at *B* is well rounded. The pipeline from *B* to *C* contains four gate valves, three standard 90° elbows, and one tee with flow through the main run. The pipeline from *D* to *E* contains four gate valves, six 90° standard elbows, two 45° standard elbows, and one tee with flow through the main run. The average fluid velocity in the pipeline is 12.0 ft/sec. Calculate the power input to the pump for an overall efficiency of 85%. The pipeline is 1000 ft long and is a 2-in schedule 40 galvanized steel pipe. Assume screwed connections and 60°F water.

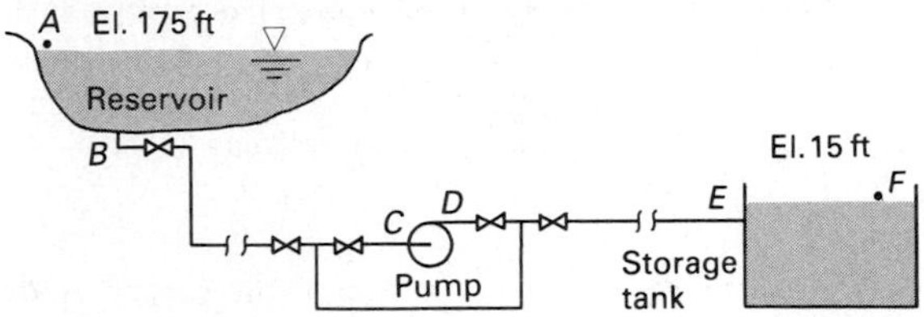

Figure P7.124

125. Wine having a specific gravity of 1.02 and a viscosity of 6.72×10^{-4} lbm/ft·sec is pumped from the large wine vat shown in Fig. P7.125 by a positive displacement pump at the rate of 100 gal/min. Find the power required to pump the wine through a 2-in. copper pipe that is 250 ft long, has fourteen standard 90° elbows, four fully open gate valves, two half-open globe valves, and one swing check valve. The pump's overall efficiency is 89%. Type *L* copper pipe and soldered connections were used.

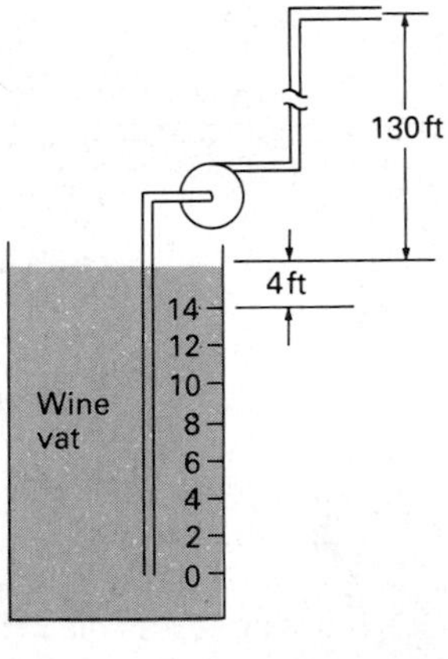

Figure P7.125

126. A pump delivers water from a lake at a maximum rate of 4.0 m^3/min through a 15-cm I.D. cast iron pipeline to atmospheric pressure at an elevation of 150 m above the lake's surface. The pipeline is 2000 m long and has three flanged elbows and two flanged gate valves. Determine the minimum power output of the motor driving the pump. Consider all flanged connections and a reentrant pipe entrance.

127. Kerosene at 15°C is pumped through 200 m of smooth 50-cm I.D. pipe at a rate of 10 m^3/min from zero velocity, atmospheric pressure, and an elevation of 150 m to zero velocity, atmospheric pressure, and an elevation of 155 m. Find the power input to the fluid by the pump. The pipe entrance is square edged.

128. A motor-driven centrifugal pump delivers 15°C water at the rate of 10 m^3/min from a reservoir, through a 2500-m-long, 30-cm I.D. plastic pipe, to a second reservoir. The water level in the second reservoir is 40 m above the water level in the first reservoir. The pump efficiency is 75%. Find the motor output power. The pipe entrance is square edged.

129. An emergency flooding system for a nuclear reactor core is shown in Fig. P7.129. Find the power input required to flood the core at the rate of 5000 gal/min. Assume a square-edged entrance, 60°F water, and screwed connections.

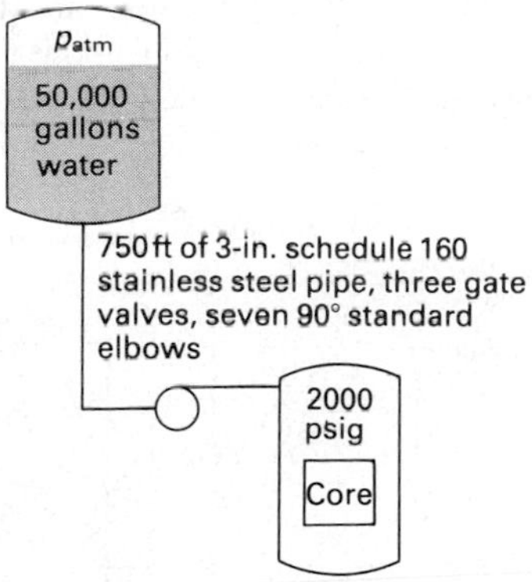

Figure P7.129

130. What horsepower is needed to pump 120 gal/min of 140°F benzene through 500 ft of 3-in., type K, horizontal copper pipe?

131. The system resistance for a pipeline is given by $\Delta p_{sys} = 4.0 + 2.0Q^2$, where Δp_{sys} is the pressure rise required of a pump to deliver the flow rate Q through the piping system. Δp_{sys} is in psi and Q is in gal/min. A pump has the pressure–rise–flow curve given by $\Delta p_p = 30.0 - 3.0Q^2$ where Δp_p is in psi and Q is in gal/min. Find the flow rate Q and the corresponding values of Δp_{sys} and Δp_p.

132. The system resistance for a pipeline is given by $\Delta p_{sys} = 2.0Q^2$, where Δp_{sys} is the pressure rise required of a pump to deliver the flow rate Q through the piping system. A pump has the pressure–rise–flow characteristic given by $\Delta p_p = 30.0 - 3.0Q^2$. In both curves, Δp is in kPa and Q is in m^3/s. Find the pump input power if this pump is placed in this piping system and the pump overall efficiency is 90%.

133. Figure P7.133 shows a piping system with frictional losses of $h_{L1-2} = 1.5Q^2$, with h_{L1-2} in m and Q in m^3/s. The pump head curve is given by $h_p = 50 - 2.0Q^2$, with h_p in m and Q in m^3/s. Calculate the flow rate Q.

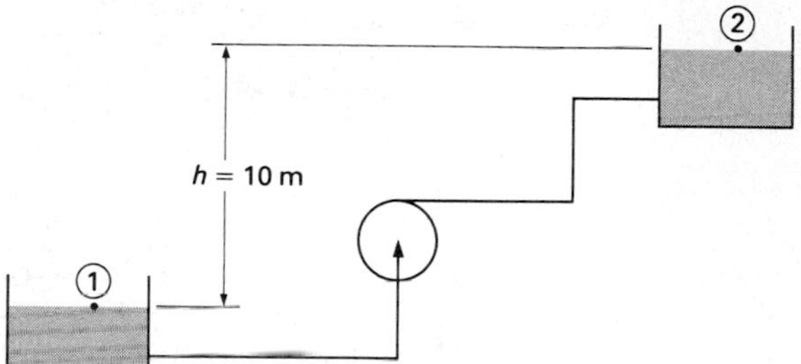

Figure P7.133

134. Figure P7.134 shows a piping system with frictional losses of $h_{L1-2} = 4.0Q^2$, with h_{L1-2} in ft and Q in gal/min. The turbine performance characteristics are given by $h_t = 20 + 12.0Q$, where h_t is the turbine head in ft and Q is in gal/min. Find the flow rate Q.

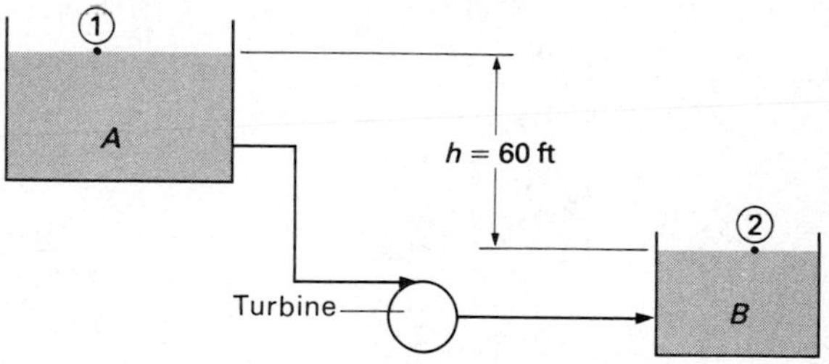

Figure P7.134

135. A factory draws its 60°F water supply from a lake having a water surface 200 ft above the supply pipe discharge. The 5-in. cast iron supply pipe is 1000 ft long and has four gate valves, one globe valve, and twenty 90° standard elbows. Calculate the pump head required for a flow rate of 1000 gal/min if the pipe outlet is open to the atmosphere. Consider a square-edged inlet, flanged connections, and schedule 40 pipe.

136. A hydraulic turbine takes water from a lake with the piping system shown in Figure P7.136. Find the head

h_t available to the turbine for flow rates of 0, 5000, 10,000, 15,000, and 20,000 ft³/min.

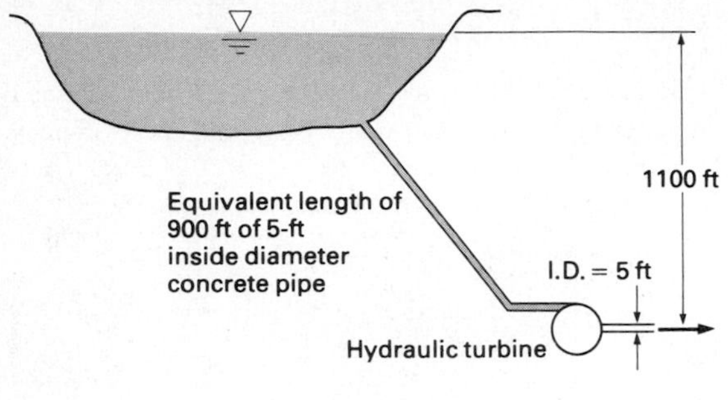

Figure P7.136

137. A pump is used to supply agitation water to a wet pit, as shown in Fig. P7.137. A pipe with 15-$\frac{1}{4}$-in., diameter holes runs along the bottom of the pit. The pump is supplied with 1 hp from an electric motor. If the pump delivers 10 gal/min, find the pressure p_2 available to agitate the pit. The pump's overall efficiency is 0.64. Assume smooth pipe and soldered connections.

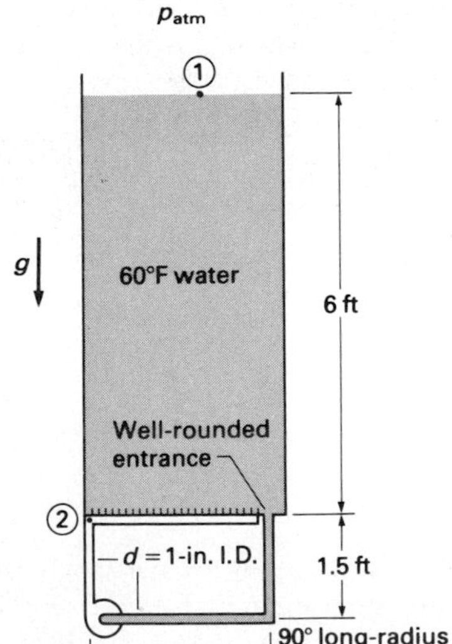

Figure P7.137

138. A car wash uses two high-pressure water pumps to spray cars going through the wash, as shown in Fig. P7.138. The water pumps supply water at 120°F and 30 gal/min total. If the pumps require a total of 5 hp, what is the nozzle inlet pressure? Assume that there are no energy losses in the pumps, that the pipes are cast iron, and that the connections are screwed.

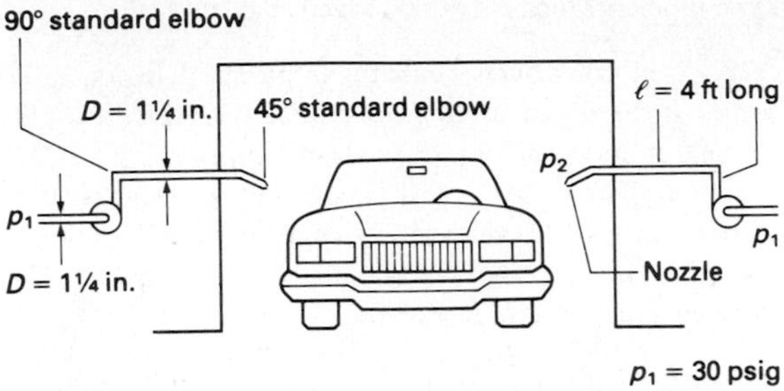

Figure P7.138

139. A farmer in a remote location uses a solar-powered submersible pump to supply water for irrigation, as shown in Fig. P7.139. If the flow rate into the reservoir is 7 gal/min, how much pump input power is required from the solar power source? The pump is 55% energy efficient and $z_2 - z_1 = 109$ ft.

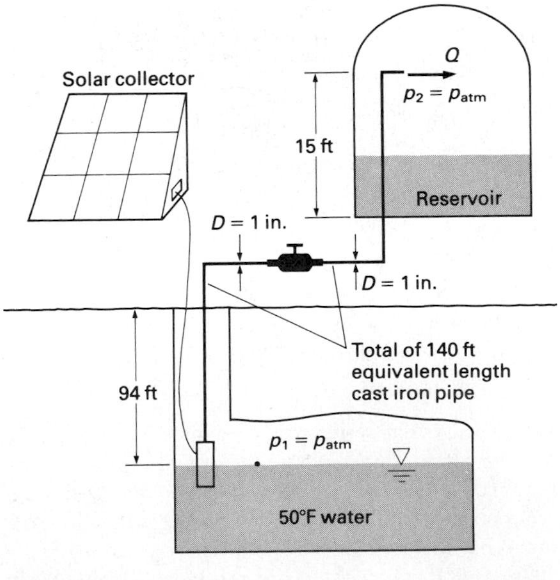

Figure P7.139

• **140.** A pump has the head and efficiency characteristics shown in Fig. P7.140 and delivers 20°C water from the lake to the reservoir. Determine the average fluid velocity in the piping, which has screwed connections.

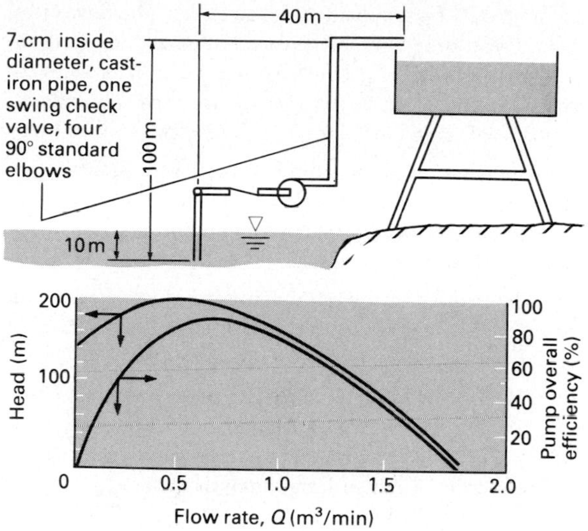

Figure P7.140

141. Both the suction and discharge piping for the pump shown in Fig. P7.141 consist of 4-in. I.D. 40-ft-long plastic pipe. Find the volume flow rate of 60°F water through the pump. The connections are glued (equivalent to soldered connections).

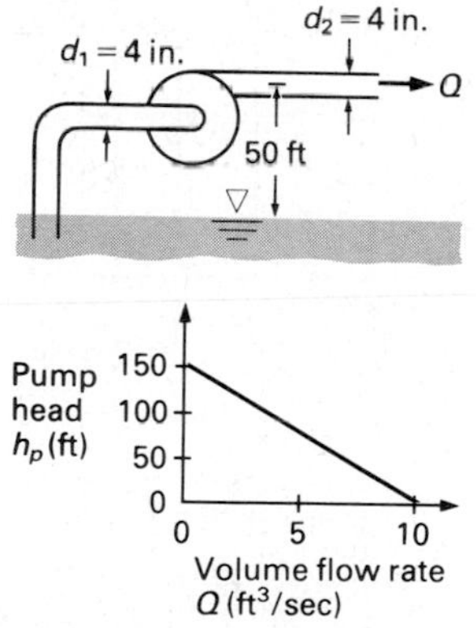

Figure P7.141

142. The pump in the piping system shown in Fig. P7.142 has the head curve shown in Fig. E7.13(b). Find the pump operating point. All fittings have screwed connections. Assume the loss coefficient for a 45° long-radius screwed elbow is one-half the loss coefficient for a 90° long-radius screwed elbow.

143. Figure P7.143 shows a pumped storage system. During a 12-hr period at night when the electrical demand is low, electrical motor-driven pumps transfer

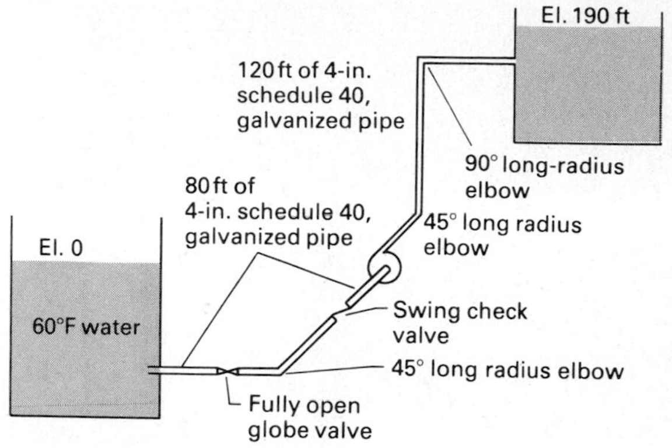

Figure P7.142

water from the lower reservoir to the higher reservoir. During a 12-hour period in the daytime when the electrical demand is higher than available generation capacity, water flows from the higher reservoir, through the turbine–generator set to generate electricity, and into the lower reservoir. Assume that the electrical motors, pump, turbine, and generator all have no energy losses. Determine the mechanical energy loss in transferring 1×10^6 m^3 of 20°C water from the lower reservoir to the higher reservoir and back again.

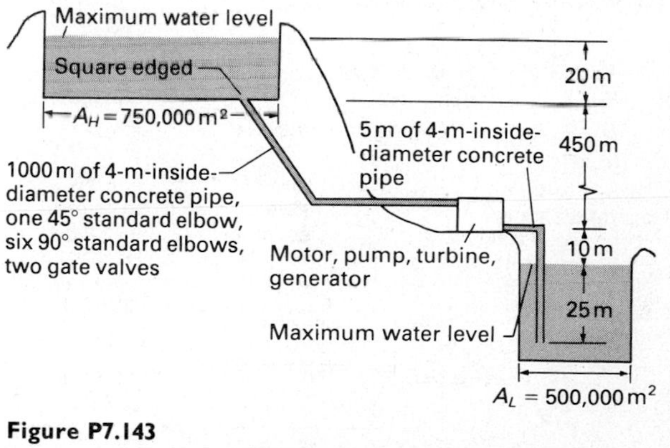

Figure P7.143

•**144.** An axial fan operating at 1000 rpm has the characteristics shown in Fig. P7.144. It delivers 15°C air through a 50-cm I.D., galvanized, sheet-metal, horizontal duct having a length of 175 m and seven 90° long-radius elbows. For constant air density what is the flow rate if the duct discharges to the atmosphere?

•**145.** Two of the pumps in Fig. P7.145 are operated in series to supply water through the piping system. Determine the flow rate through the piping system for 10°C water and screwed connections. Then find the total power input to the two pumps.

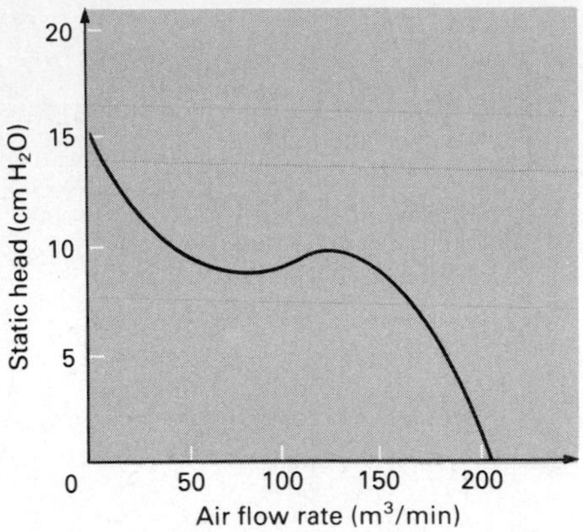

Figure P7.144

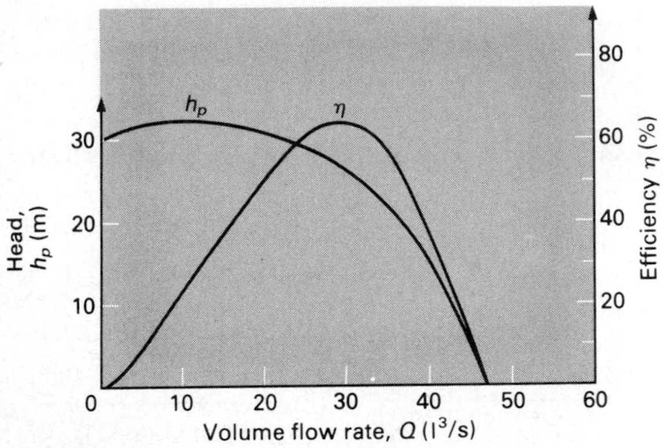

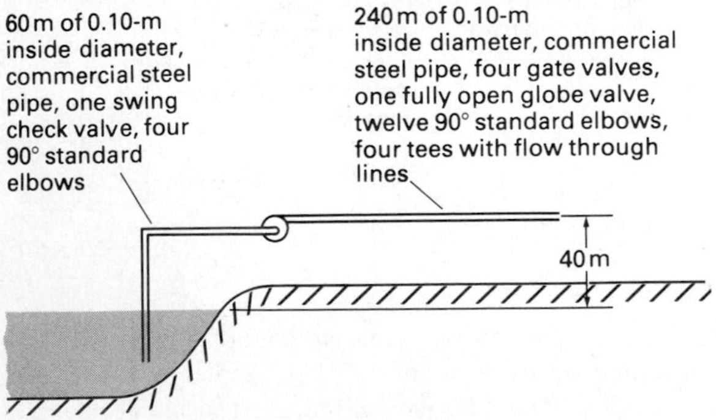

Figure P7.145

• 146. The Air Force is designing a new air-to-air missile. The full-scale missile is shown in Fig. P7.146. A $\frac{1}{5}$-scale model of the missile is tested in a wind tunnel. To power the model's rocket engine, kerosene (specific gravity = 0.75, $\mu = 5 \times 10^{-4}$ lb·sec/ft^2) and liquid oxygen (specific gravity = 1.1, $\mu = 5 \times 10^{-5}$ lb·sec/ft^2) are pumped to the model from outside storage tanks. The flow rates of kerosene and liquid oxygen are 10 lbm/sec and 30 lbm/sec, respectively. The fuel/oxidizer supply system is shown in Fig. P7.146. Assume square-edged tubing entrances, 90° standard elbows, and screwed connections.

(a) The pressure in the supply tanks is atmospheric and the pressure in the combustion chamber is 350 psia. Calculate the power required to pump the kerosene and liquid oxygen to the model.

(b) The density of the combustion products at the exit of the rocket nozzle in the model is 0.03 lbm/ft^3. Calculate the exit velocity and the thrust.

(c) The thrust (T) of a rocket is a function of the combustion chamber pressure (p_c), ambient air pressure (p_e), nozzle exit diameter (D_e), and the geometric properties of the nozzle. A dimensional analysis yields

$$\frac{T}{p_c D_e^2} = f\left(\frac{p_e}{p_c}, \text{geometry}\right).$$

Calculate the thrust developed by the full-scale missile if the combustion chamber pressure is 500 psia and p_e/p_c is the same as in the model test.

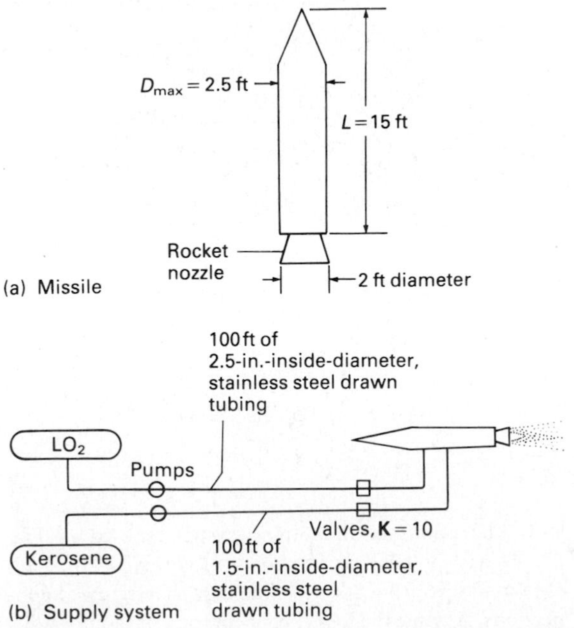

Figure P7.146

147. Marco and Tom come home from a long day of studying to discover that their basement is flooded with water. They have a submersible pump for such an emergency and connect the pump discharge to a 12-ft-long garden hose, as shown in Fig. P7.147. The garden hose may be considered a smooth plastic pipe with a 1.0-in. inside diameter. If the initial water depth is 6.0 in. and the pump has an input power of 0.25 hp and an overall efficiency of 0.60, how long will it take to pump out the basement? The pump inlet piping length is negligible, and the garden hose outlet is 7.0 ft above the basement floor.

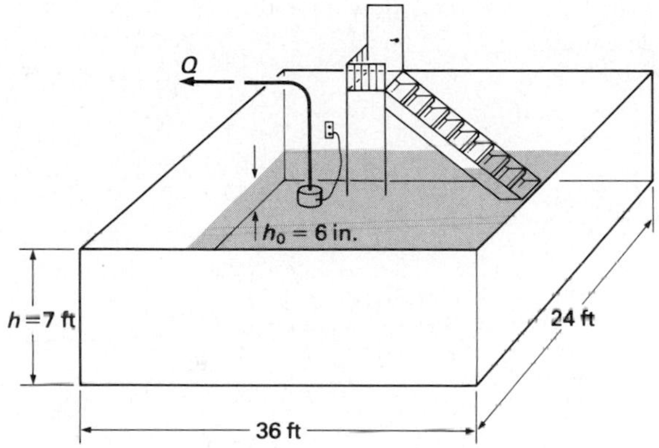

Figure P7.147

148. Each pipeline in Fig. P7.148 is cast iron and $p_1 - p_2 = 20$ kPa. Neglecting minor losses, find the flow rate in each pipeline for 20°C kerosene. Also neglect elevation changes. Compare the result with those of Problem 149.

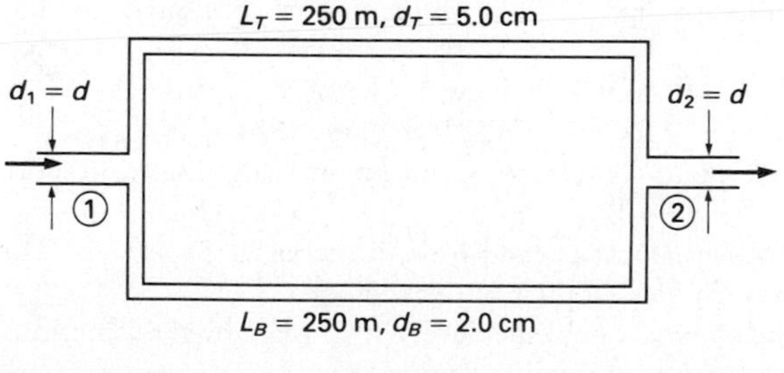

Figure P7.148

149. Each pipeline in Fig. P7.148 is smooth and $p_1 - p_2 = 20$ kPa. Neglecting minor losses, find the flow rate in each pipeline for 20°C kerosene. Neglect elevation changes. Compare the result with those of Problem 148.

150. Each pipeline in Fig. P7.150 is smooth and $p_1 - p_2$ is zero. Neglecting minor losses, find the flow rate in each pipeline for 68°F benzene.

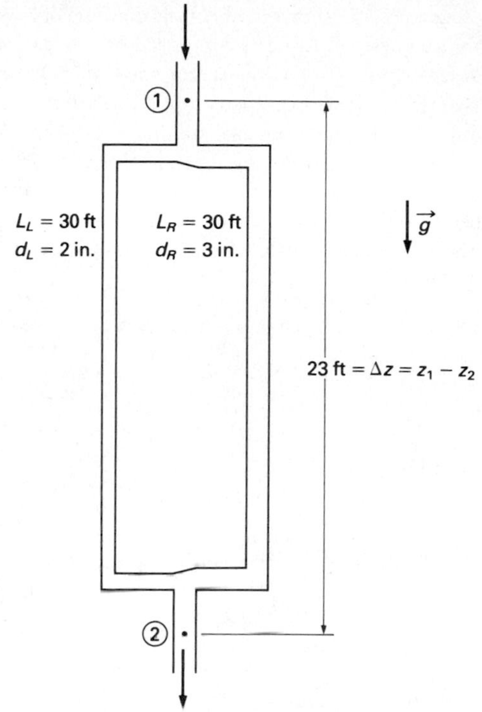

Figure P7.150

151. Water at 20°C flows at a rate of 0.10 m^3/s through the three parallel lines in Fig. P7.151. Find the flow rate in each line and the pressure drop $p_A - p_B$. All pipes are smooth plastic with an inside diameter of 3.0 cm. The pipeline is in a horizontal plane, and $D_1 = D_2 = D_3 = 3.0$ cm.

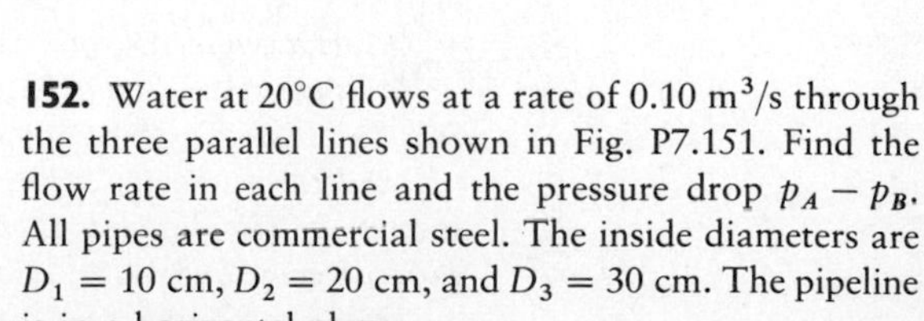

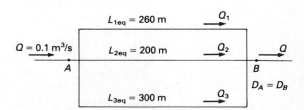

Figure P7.151

152. Water at 20°C flows at a rate of 0.10 m^3/s through the three parallel lines shown in Fig. P7.151. Find the flow rate in each line and the pressure drop $p_A - p_B$. All pipes are commercial steel. The inside diameters are $D_1 = 10$ cm, $D_2 = 20$ cm, and $D_3 = 30$ cm. The pipeline is in a horizontal plane.

153. Figure P7.153 shows a schematic diagram of a water distribution system. Leg 1 is composed of 200 ft of 2-in. schedule 40 steel pipe with twelve standard 90° bends and two gate valves. Leg 2 is composed of 50 ft of 1-in. schedule 40 steel pipe with two standard 180° bends and two swing check valves. Leg 3 is composed of 75 ft of 0.75-in. I.D. smooth plastic pipe with three standard 45° bends and one swing check valve. What is the flow rate through each branch when the pump is delivering 500 gal/min?

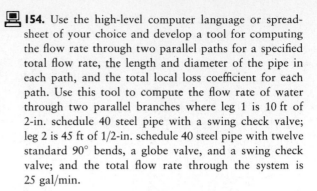

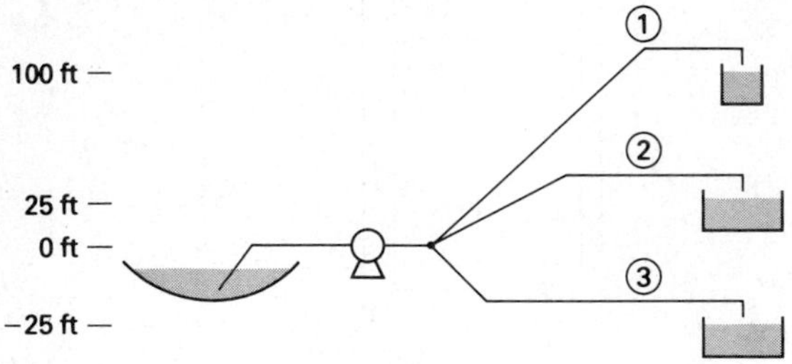

Figure P7.153

154. Use the high-level computer language or spreadsheet of your choice and develop a tool for computing the flow rate through two parallel paths for a specified total flow rate, the length and diameter of the pipe in each path, and the total local loss coefficient for each path. Use this tool to compute the flow rate of water through two parallel branches where leg 1 is 10 ft of 2-in. schedule 40 steel pipe with a swing check valve; leg 2 is 45 ft of 1/2-in. schedule 40 steel pipe with twelve standard 90° bends, a globe valve, and a swing check valve; and the total flow rate through the system is 25 gal/min.

REFERENCES

1. Nikuradse, J., "Stomungsgesetze in Rauhen Rohren," *VDI-Forschungsch,* no. 361 (1933). Also available as NACA Tech Memo 1292.

2. Colebrook, C. F., "Turbulent Flow in Pipes with Particular Reference to the Transition Between the Smooth and Rough Pipe Laws," *J. Inst. Civ. Eng. Lond.*, vol. 11, 1939.

3. Moody, L. F., "Friction Factors for Pipe Flow," *Transactions of the ASME,* vol. 66, 1944.

4. White, F. M., *Fluid Mechanics,* McGraw-Hill, New York, 1979.

5. White, F. M., *Viscous Fluid Flow,* McGraw-Hill, New York, 1974.

6. Lindgren, E. R., Oklahoma State University Civil Engineering Dept., Report 1AD621071, 1965.

7. "Flow of Fluids Through Valves, Fittings, and Pipe," Technical Paper No. 410 (1978), The Crane Co. available through The Crane Co., New York.

8. *ASHRAE Handbook of Fundamentals,* published every four years by the American Society of Heating, Refrigeration and Air-Conditioning Engineers.

9. Idel'chek, I. E., "Handbook of Hydraulic Resistance" (2nd ed.), Hemisphere, Washington, 1986.

10. *Hydraulic Institute Engineering Data Book,* Hydraulic Institute, Cleveland, 1988.

11. Bean, H. S. (ed.), *Fluid Meters: Their Theory and Application* (6th ed.), American Society of Mechanical Engineers, New York, 1971.

12. "Measurement of Fluid Flow by Means of Orifice Plates, Nozzles, and Venturi Tubes Inserted in Circular Cross Section Conduits Running Full," International Standards Organization Report DIS-5167, 1976.

13. "The ISO-ASME Orifice Coefficient Equation," *Mechanical Engineering,* July, 1981.

14. Beckwith, T., N. Buck, and R. Marangoni, *Mechanical Engineering Measurements* (3rd ed.), Addison-Wesley, Reading, Mass, 1981.

15. Benedict, R. P., *Fundamentals of Temperature, Pressure and Flow Measurements* (3rd ed.), John Wiley & Sons, New York, 1984.

16. "Laboratory Methods of Testing Fans for Rating," AMCA Bulletin 210-74, ASHRAE Standard 210-75, a Joint Bulletin of the Air Moving and Conditioning Association and the American Society of Heating, Refrigeration and Air Conditioning Engineers, 1974.

17. "ASME Performance Test Code No. 11—Fans," American Society of Mechanical Engineers, New York, 1984.

18. I. J. Karassick et al., *Pump Handbook* (2d ed.), McGraw-Hill, New York, 1985.

19. Jorgensen, R. (ed.), *Fan Engineering,* (8th ed.), Buffalo Forge Co., Buffalo, NY, 1983.

20. Cross, H., "Analysis of Flow in Networks of Conduits or Conductors," University of Illinois Bulletin No. 286, November, 1936.

21. Jepson, R. W., *Analysis of Flow in Pipe Networks,* Ann Arbor Science Publishers, Ann Arbor, MI, 1976.

22. Wood, D. J., and A. G. Ruyes, "Reliability of Algorithms for Pipe Network Analysis," *J. Hydraulics Div., ASCE,* Vol. 107, NO. HY 10, 1981.

8 Steady, Incompressible External Flow

In this chapter, we consider flow over bodies immersed in a stream of fluid. Such flows are called *external flows,* because the fluid is external to a surface. Figure 8.1 shows a few external flows that are of interest to engineers. The fluid stream in which the body is immersed is usually considered to be infinite in extent. The items of interest are the forces and moments that the fluid exerts on the body and, to a lesser extent, the details of the flow pattern near the body. The study and engineering application of external flow is often called *aerodynamics,* although this title is inappropriate if the fluid is not air.

Complete information on aerodynamic forces and flow patterns can be obtained from a combination of theory, experiment, and computer calculation. Current theories for flow over immersed bodies provide excellent qualitative explanations for nearly all external flow situations. Purely analytic predictions are available for a few simple flows. A considerable amount of experimental data is available because, until the 1970s, almost all practical engineering problems involving external flow relied on experimental data for solution. The prospect of using computational fluid dynamics (CFD) for solving external flow problems has been a major driving force in the development of powerful computing hardware and software. Today, the computer and the wind tunnel are of equal importance in aerodynamic engineering.

Both qualitative and quantitative theories of external flow rely heavily on the concept of dividing the flow field into potential flow and boundary layer regions (see Section 5.6.2). In this chapter, we consider the qualitative aspects of this

Figure 8.1 Some external flows.

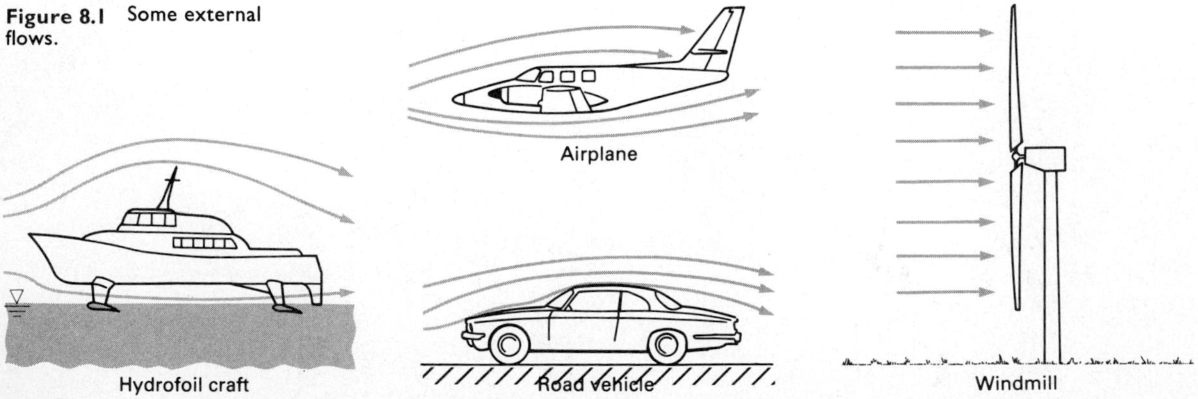

model. Quantitatively, we concentrate on experimental information for standard simple shapes. In addition to our emphasis on aircraft, we consider road vehicle aerodynamics and, on the lighter side, applications of aerodynamics in sports.

8.1 FORCE ON BODIES: DRAG AND LIFT

A body immersed in a moving stream (or alternatively, a body moving through a still fluid) experiences a force. In this chapter, we are concerned only with those forces that are the result of relative motion between the fluid and the body; we do not include buoyancy forces that are the result of hydrostatic pressure in this discussion.

8.1.1 Body Geometry and Force Components

In this chapter, we refer to two-dimensional bodies, three-dimensional bodies, and axisymmetric bodies. A *two-dimensional body* has the same shape in all planes perpendicular to an infinitely long axis. Figure 8.2(a) shows some two-dimensional bodies. The infinite axis is perpendicular to the paper. A *three-dimensional body* is finite in all directions. Figure 8.2(b) illustrates some three-dimensional bodies. An *axisymmetric body* is a body of revolution with circular cross sections. An axisymmetric body has the same shape in all meridional (x, r) planes. Figure 8.2(c) illustrates some axisymmetric bodies.

When a body is immersed in a stream, the dimensionality and directionality of the flow are determined by the dimensionality of the body and the alignment between the approach flow and the body. Two-dimensional bodies produce two-dimensional, two-directional flow if there is no flow along the body's long axis. Three-dimensional bodies produce three-dimensional, three-directional flow. Even if the body shape is the same in all planes perpendicular to the width axis and the approaching flow is perpendicular to this axis, there will still be flow around the "ends" of the body. If a three-dimensional body is very wide, we may approximate the flow in the central portion as two-dimensional. For example, the flow over the central portion of the cylinder of Fig. 8.2(b) might be approximated by the two-dimensional flow over the infinitely long cylinder of Fig. 8.2(a). The flow over an

Figure 8.2 (a) Some two-dimensional bodies; (b) some three-dimensional bodies; (c) some axisymmetric bodies.

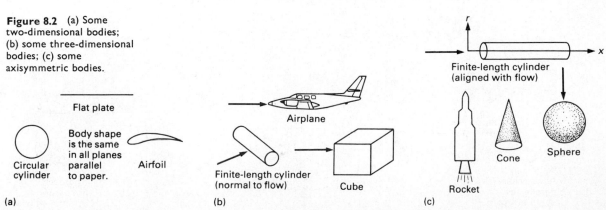

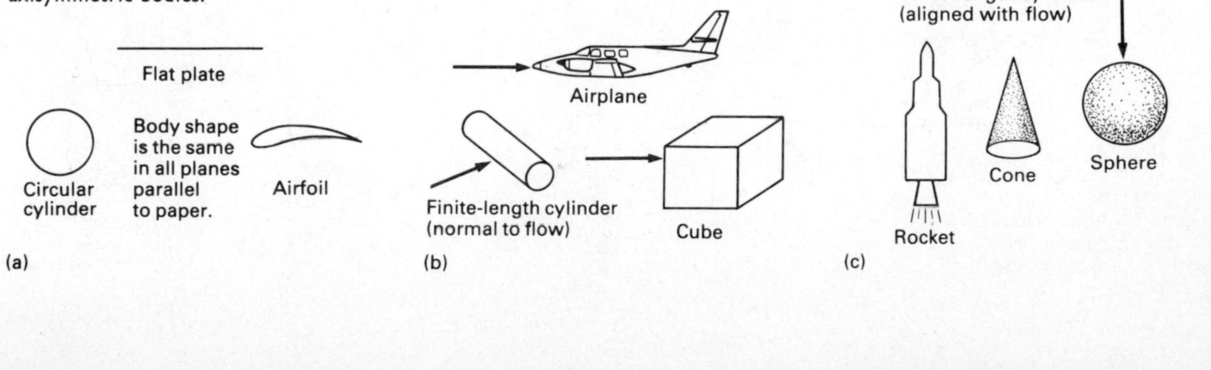

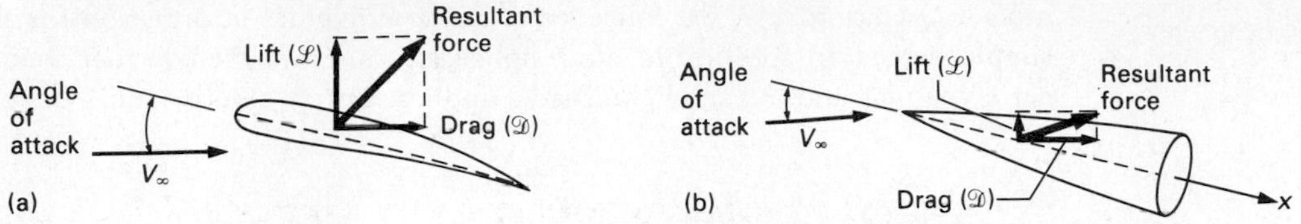

Figure 8.3 Aerodynamic forces on two-dimensional and axisymmetric bodies.

axisymmetric body with its axis parallel to the approaching flow is two-dimensional and two-directional in a meridional (x, r) plane. If the approach flow is not parallel to the axis of symmetry, the flow is three-dimensional and three-directional.

We may break the aerodynamic force on a body into components along any arbitrary set of axes. Let's first consider a two-dimensional or axisymmetric body (Fig. 8.3). For the two-dimensional body, the resultant force is in the plane of the body. For the axisymmetric body, the resultant force lies in the plane defined by the body axis and the approaching flow velocity vector. In either case, the force component in the direction of the approaching flow is called *drag* and the force component perpendicular to the approaching flow is called *lift*.

Experiments show that any body placed in a moving stream experiences drag. If a body moves relative to a still fluid, drag force resists the motion. The drag force vector always points downstream.

Lift forces are not necessarily present in all flows; they occur only if there is asymmetry. This asymmetry may be caused by asymmetry of the body or it may be caused by a misalignment between the body and the approaching flow. The angle of misalignment is called the *angle of attack*, as shown in Fig. 8.3.

Let's now consider a three-dimensional body. Many three-dimensional bodies have at least one plane of symmetry (Fig. 8.4a). If the approach flow velocity vector is parallel to this symmetry plane, the resultant force on the body lies in the symmetry plane and therefore has only two components, still identified as drag and lift. If the flow is completely asymmetric, either because the three-dimensional body has no plane of symmetry (Fig. 8.4b) or because the approach flow is not parallel to the symmetry plane (Fig. 8.4c), the resultant force has three components, called drag, lift, and *side force*. We concentrate on those cases in which there is at least one plane of symmetry, so the only forces present are drag and lift.

Figure 8.4 Flow over three-dimensional bodies.

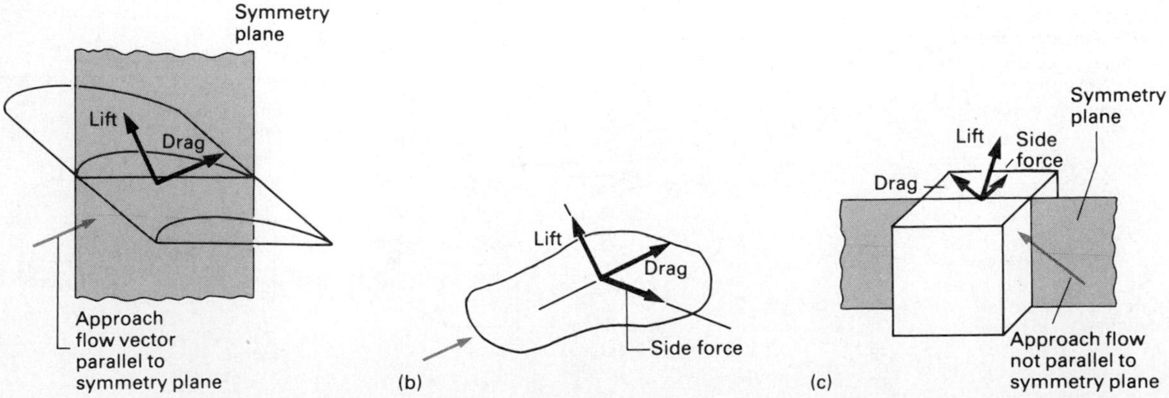

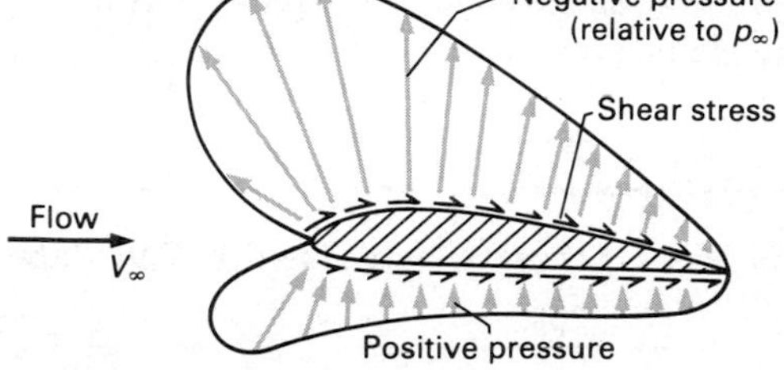

Figure 8.5 Pressure and shear stress distributions on an airfoil.

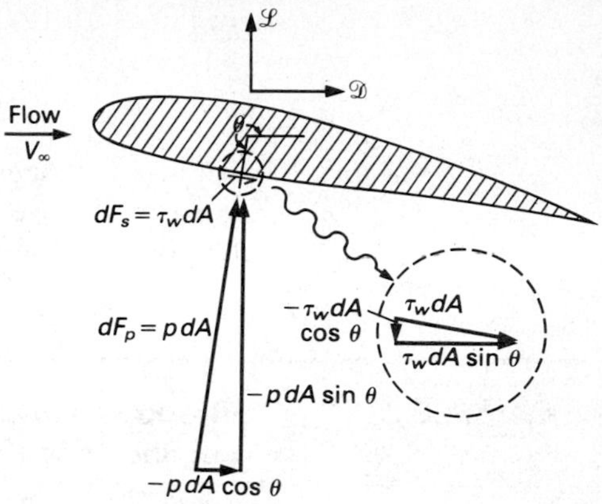

Figure 8.6 Geometry of elemental forces on an airfoil.

8.1.2 Calculation of Drag and Lift from Pressure and Shear Stress

Drag and lift are the resultants of forces from pressure and shear stress on the body surface. If the pressure and shear stress distributions over the surface can be determined, either by calculation or by experiment, the aerodynamic forces can be obtained by integration. We illustrate the basic approach by reference to an airfoil. An *airfoil* is a long, thin, two-dimensional body designed to produce large lift and low drag. Figure 8.5 shows a typical airfoil and the pressure and shear stress distributions on it. The magnitude of the pressure at a point on the airfoil surface is indicated by the length of an arrow perpendicular to the surface. Pressure is plotted relative to the pressure in the approaching (undisturbed) stream. Note that pressures on the top of the airfoil are negative (lower than free-stream pressure), whereas pressures on the bottom are positive. The shear stresses at the airfoil surface are indicated by arrows parallel to the surface.

The net force on the airfoil can be calculated by integrating the pressure and shear stress over the surface:

$$\vec{F} = -\oint p\hat{n}\,dA + \oint \tau_w \hat{t}\,dA, \tag{8.1}$$

where $\hat{n}$ and $\hat{t}$ are unit vectors perpendicular and tangent to the airfoil surface. In this expression, τ_w is positive if the shear force points in the same direction as $\hat{t}$.

Drag and lift are the force components in the flow direction and perpendicular to it. With the aid of Fig. 8.6, we find that*

$$\mathscr{D} = \oint (-p\cos\theta + \tau_w \sin\theta)\,dA \tag{8.2}$$

* τ_w is positive if it is clockwise in these expressions.

and

$$\mathcal{L} = \oint (-p \sin\theta + \tau_w \cos\theta)\, dA. \qquad (8.3)$$

Because θ is approximately 90° over most of the top of the airfoil and approximately 270° over most of the bottom of the airfoil, the lift force is mainly caused by pressure, and the drag force is mainly caused by shear stress for this airfoil.

EXAMPLE 8.1 Illustrates How Pressure and Shear Stresses Are Used to Calculate Lift and Drag Forces

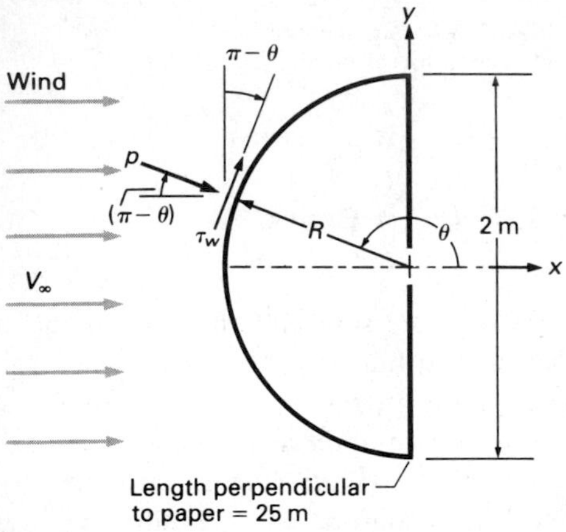

Figure E8.1 Rearward wind on sign.

A highway sign has the shape of a half-cylinder with radius 1.0 m and length 25.0 m. A rearward 10-km/hr wind is blowing over the sign as shown in Fig. E8.1. The pressure and shear stress distributions on the circular surface are given by

$$p - p_\infty = 4.63(1 - 4\sin^2\theta)\ \text{N/m}^2$$

and

$$\tau_w = [1.74 \times 10^{-3}(\pi - \theta) - 1.59 \times 10^{-5}(\pi - \theta)^2]\ \text{N/m}^2,$$

where θ is in radians and $\pi/2 \le \theta \le \pi$. The pressure on the flat face is constant and equal to the pressure at the corner. Find the lift and drag on the sign. Neglect end effects.

SOLUTION

Given

Sign having shape of half-cylinder, shown in Fig. E8.1

Sign radius 1.0 m, length 25.0 m

Rearward, 10-km/hr wind

Pressure and shear stress distribution on circular surface given by

$$p - p_\infty = 4.63(1 - 4\sin^2\theta)\ \text{N/m}^2$$

and

$$\tau_w = [1.74 \times 10^{-3}(\pi - \theta) - 1.59 \times 10^{-5}(\pi - \theta)^2]\ \text{N/m}^2,$$

where θ is in radians

Neglect end effects.

Find

Lift and drag on the sign

Solution

The directions of positive lift and drag are the positive y and x directions, respectively. We use Eq. (8.2) to find the drag. As

$$dA = bR\, d\theta,$$

where b is the sign length and R the radius, the drag is

$$\mathscr{D} = \int_0^{2\pi} (-p \cos\theta + \tau_w \sin\theta) bR \, d\theta.$$

Recognizing that the drag on the top half equals the drag on the bottom half gives

$$\mathscr{D} = 2\int_0^{\pi} (-p \cos\theta + \tau_w \sin\theta) bR \, d\theta = \mathscr{D}_p + \mathscr{D}_\tau.$$

Evaluating the pressure and shear stress contributions to the drag separately is informative. We evaluate the pressure integral in two "pieces," one for the flat face and one for the cylindrical portion. On the flat face, the pressure is given by

$$p - p_\infty = 4.63\left[1 - 4\sin^2\left(\frac{\pi}{2}\right)\right] = 4.63(-3).$$

Thus

$$\mathscr{D}_p = 2\int_0^{\pi/2} - [p_\infty + 4.63(-3)] \cos\theta \, bR \, d\theta$$
$$+ 2\int_{\pi/2}^{\pi} - [p_\infty + 4.63(1 - 4\sin^2\theta)] \cos\theta \, bR \, d\theta.$$

Integrating, we obtain

$$\mathscr{D}_p = -2bR(p_\infty - 13.9)[\sin\theta]_0^{\pi/2}$$
$$- 2bR\left[p_\infty \sin\theta + 4.63 \sin\theta - 18.5\left(\frac{\sin^3\theta}{3}\right)\right]_{\pi/2}^{\pi}.$$

Substituting limits gives

$$\mathscr{D}_p = 2bR(12.36 \text{ N/m}^2) = 2(25 \text{ m})(1 \text{ m})(12.36 \text{ N/m}^2)$$
$$= 618 \text{ N}.$$

Now we evaluate the shear stress drag. The shear stress force on the flat face has no component in the streamwise direction. The shear drag is

$$\mathscr{D}_\tau = 2\int_{\pi/2}^{\pi} [1.74 \times 10^{-3}(\pi - \theta) - 1.59 \times 10^{-5}(\pi - \theta)^2] \sin\theta \, bR \, d\theta$$
$$= 3.48 \times 10^{-3} \, bR \int_{\pi/2}^{\pi} (\pi - \theta) \sin(\pi - \theta) \, d\theta$$
$$- 3.18 \times 10^{-5} \, bR \int_{\pi/2}^{\pi} (\pi - \theta)^2 \sin(\pi - \theta) \, d\theta.$$

Letting $\phi = \pi - \theta$, so that $d\theta = -d\phi$, gives

$$\mathscr{D}_\tau = -3.48 \times 10^{-3} \, bR \int_{\pi/2}^{0} \phi \sin\phi \, d\phi$$
$$+ 3.18 \times 10^{-5} \, bR \int_{\pi/2}^{0} \phi^2 \sin\phi \, d\phi$$
$$= -3.48 \times 10^{-3} \, bR[\sin\phi - \phi\cos\phi]_{\pi/2}^{0}$$
$$+ 3.18 \times 10^{-5} \, bR[2\phi \sin\phi - (\phi^2 - 2)\cos\phi]_{\pi/2}^{0}.$$

Substituting limits and values of b and R gives

$$\mathscr{D}_\tau = 0.086 \text{ N}.$$

The total drag is

$$\mathscr{D} = \mathscr{D}_p + \mathscr{D}_\tau = 618 \text{ N} + 0.086 \text{ N} = 618 \text{ N}.$$ **ANSWER**

We use Eq. (8.3) to find the lift. Recognizing that $dA = bR\,d\theta$ and that there is no lift component on the flat face, we have

$$\mathscr{L} = \int_{\pi/2}^{3\pi/2} (-p \sin\theta - \tau_w \cos\theta) bR\, d\theta.$$

Substituting for p and τ_w gives

$$\mathscr{L} = \int_{\pi/2}^{3\pi/2} -[p_\infty + 4.63(1 - 4\sin^2\theta)] \sin\theta\, bR\, d\theta$$
$$+ \int_{\pi/2}^{3\pi/2} [1.74 \times 10^{-3}(\pi - \theta)$$
$$- 1.59 \times 10^{-5}(\pi - \theta)^2 \cos\theta\, bR\, d\theta.$$

We now use the trigonometric identity

$$\sin^2\theta = 1 - \cos^2\theta$$

in the first integral and $\cos\theta = -\cos(\pi - \theta)$, $\pi - \theta = \phi$, and $-d\theta = d\phi$ in the second integral. The lift equation then is

$$\mathscr{L} = bR \int_{\pi/2}^{3\pi/2} -[p_\infty + 4.63(-3 + 4\cos^2\theta)] \sin\theta\, d\theta$$
$$+ 1.74 \times 10^{-3}\, bR \int_{\pi/2}^{-(\pi/2)} \phi \cos\phi\, d\phi$$
$$- 1.58 \times 10^{-5}\, bR \int_{\pi/2}^{-(\pi/2)} \phi^2 \cos\phi\, d\phi.$$

Integrating yields

$$\mathscr{L} = bR\left[(p_\infty - 13.9)\cos\theta + \frac{18.52\cos^3\theta}{3}\right]_{\pi/2}^{3\pi/2}$$
$$+ 1.74 \times 10^{-3}\, bR[\cos\phi + \phi\sin\phi]_{\pi/2}^{-(\pi/2)}$$
$$- 1.59 \times 10^{-5}\, bR[2\phi\cos\phi + (\phi^2 - 2)\sin\phi]_{\pi/2}^{-(\pi/2)},$$

which reduces to

$$\mathscr{L} = (0)bR,$$

so

$$\mathscr{L} = 0.$$ **ANSWER**

Discussion

The zero lift should not be surprising, because the vertical component of the pressure or shear stress at any point on the top half of the sign is exactly balanced by an equal but oppositely directed vertical component of the pressure or shear stress on the bottom half. Also note that, unlike that of the airfoil, this object's drag is dominated by pressure force rather than by shear stress force.

8.1.3 Drag and Lift Coefficients

Equations (8.2) and (8.3) are useful for calculating drag and lift only if we know the pressure and shear stress distributions. Frequently, such information is determined by experiment. The engineer's ultimate interest is often the drag and lift forces rather than the pressure and stress distributions, so experimental and ana-

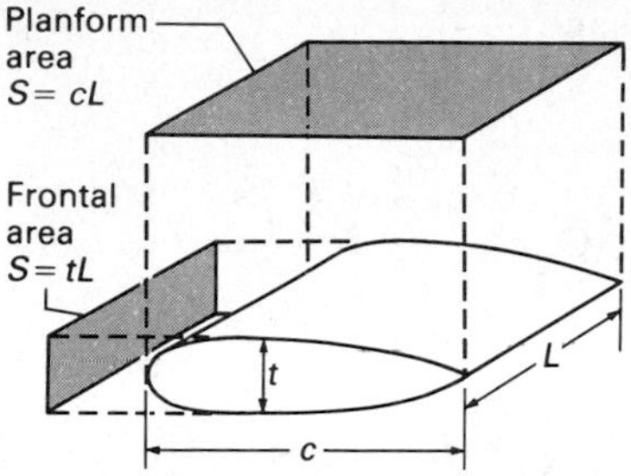

Figure 8.7 Definition of frontal area and planform area.

lytic results are often presented directly as drag and lift. The data are usually presented in terms of dimensionless *drag coefficient* and *lift coefficient*, defined by

$$\mathbf{C_D} \equiv \frac{\mathscr{D}}{\frac{1}{2}\rho V_\infty^2 S} \tag{8.4}$$

and

$$\mathbf{C_L} \equiv \frac{\mathscr{L}}{\frac{1}{2}\rho V_\infty^2 S}, \tag{8.5}$$

where V_∞ is the velocity of the fluid relative to the object and S is a reference area. According to pure dimensional analysis (discussed in Chapter 6), the reference area should be taken as the square of a single reference length ($S = \ell^2$). In practice, we use various definitions for the area. The two most common are the *frontal area* (the area that you would see if you looked at the body from the direction of the approach flow) and the *planform area* (the area you would see if you looked at the body from above), as shown in Fig. 8.7. For two-dimensional bodies, the area is based on a unit width ($L = 1$). From dimensional analysis, drag and lift coefficients for a given shape in steady flow are functions of the primary independent-variable dimensionless parameters (see Table 6.1). That is,

$$\mathbf{C_D} = \mathbf{C_D}\{\alpha, \mathbf{R}, \mathbf{M}, \mathbf{W}, \mathbf{F}\} \quad \text{and} \quad \mathbf{C_L} = \mathbf{C_L}\{\alpha, \mathbf{R}, \mathbf{M}, \mathbf{W}, \mathbf{F}\}.$$

In most flows of practical interest, surface tension (Weber number) effects are irrelevant. In almost all cases (except for hydrofoils operating near a free surface), gravity wave (Froude number) effects also are irrelevant. With these simplifications, we write

$$\mathbf{C_D} = \mathbf{C_D}\{\alpha, \mathbf{R}, \mathbf{M}\} \tag{8.6}$$

and

$$\mathbf{C_L} = \mathbf{C_L}\{\alpha, \mathbf{R}, \mathbf{M}\}. \tag{8.7}$$

In this chapter, we limit our attention to incompressible flow ($\mathbf{M} < 0.3$),* for which Mach number effects are irrelevant. Thus

$$\mathbf{C_D} = \mathbf{C_D}\{\alpha, \mathbf{R}\} \tag{8.8}$$

* The effect of compressibility on drag and lift is discussed briefly in Chapter 10.

and

$$C_L = C_L(\alpha, R). \tag{8.9}$$

We relate drag and lift coefficients to pressure and skin friction coefficients (C_p and C_f) by equations similar to Eqs. (8.2) and (8.3):

$$C_D = \oint (-C_p \cos\theta + C_f \sin\theta)\, d\left(\frac{A}{S}\right) \tag{8.10}$$

and

$$C_L = \oint (-C_p \sin\theta - C_f \cos\theta)\, d\left(\frac{A}{S}\right). \tag{8.11}$$

We devote the remainder of this chapter to presentation of typical values of C_D and C_L and development of a qualitative understanding of drag and lift.

EXAMPLE 8.2 Illustrates Lift and Drag Coefficients

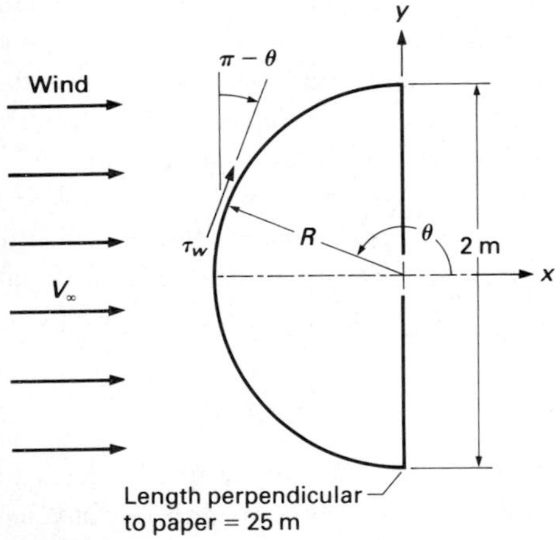

Figure E8.2 Rearward wind on sign.

Calculate the lift and drag coefficients for the sign in Example 8.1. What would be the lift and drag if the sign were reduced in size by one half and the wind velocity doubled?

SOLUTION

Given

Highway sign in Fig. E8.2

Drag $\mathscr{D} = 618$ N

Lift $\mathscr{L} = 0$

Find

Drag coefficient C_D and lift coefficient C_L

Values of lift and drag on a similar sign of half size in a wind with double the original velocity

Solution

The drag coefficient is defined by Eq. (8.4):

$$C_D = \frac{\mathscr{D}}{\frac{1}{2}\rho V_\infty^2 S}.$$

In our case, V_∞ is 10 km/hr and S is the frontal area,

$$S = 2Rb = 2(1.0 \text{ m})(25.0 \text{ m}) = 50 \text{ m}^2.$$

Assuming 20°C air at standard atmospheric pressure, we get from Table A.3

$$\rho = 1.20 \text{ kg/m}^3.$$

The numerical values give

$$C_D = \frac{618 \text{ N}}{\frac{1}{2}(1.20 \text{ kg/m}^3)(10 \text{ km/hr})^2(50 \text{ m}^2)(1000 \text{ m/km})^2(1 \text{ hr}/3600 \text{ s})^2};$$

$$C_D = 2.67. \qquad \textbf{ANSWER}$$

The lift coefficient is defined by Eq. (8.5):

$$C_L = \frac{\mathscr{L}}{\frac{1}{2}\rho V_\infty^2 S}.$$

As $\mathscr{L} = 0$,

$$C_L = 0. \qquad \textbf{ANSWER}$$

With incompressible flow, Eqs. (8.8) and (8.9) show that the lift and drag coefficients are functions of the angle of attack α and the Reynolds number $\mathbf{R}$. Assuming that α is the same for both the full-size and half-size signs, we have

$$C_D = C_D\{\mathbf{R}\} \quad \text{and} \quad C_L = C_L\{\mathbf{R}\}.$$

An appropriate Reynolds number for the sign is

$$\mathbf{R} = \frac{V_\infty R}{\nu}.$$

Because

$$R_{\text{full-size}} = 2R_{\text{half-size}} \quad \text{and} \quad (V_\infty)_{\text{full-size}} = \tfrac{1}{2}(V_\infty)_{\text{half-size}},$$

both signs have the same $\mathbf{R}$. Therefore both signs have the same values of C_D and of C_L. Solving for $\mathscr{D}$ and $\mathscr{L}$ for the half-size sign gives

$$\mathscr{D}_{\text{half-size}} = (\tfrac{1}{2}C_D\rho V_\infty^2 S)_{\text{half-size}}$$

and

$$\mathscr{L}_{\text{half-size}} = (\tfrac{1}{2}C_L\rho V_\infty^2 S)_{\text{half-size}}.$$

Substituting for the velocity and frontal area S gives

$$\mathscr{D}_{\text{half-size}} = \tfrac{1}{2}C_D\rho[(2V_\infty)^2(\tfrac{1}{2})(2Rb)] = 2[\tfrac{1}{2}C_D\rho V_\infty^2(2Rb)]$$

and

$$\mathscr{L}_{\text{half-size}} = \tfrac{1}{2}C_L\rho[(2V_\infty)^2(\tfrac{1}{2})(2Rb)] = 2[\tfrac{1}{2}C_L\rho V_\infty^2(2Rb)].$$

Then

$$\mathscr{D}_{\text{half-size}} = 2\mathscr{D}_{\text{full-size}},$$

or

$$\mathscr{D}_{\text{half-size}} = 1240 \text{ N} \qquad \textbf{ANSWER}$$

and

$$\mathscr{L}_{\text{half-size}} = 2\mathscr{L}_{\text{full-size}},$$

or

$$\mathscr{L}_{\text{half-size}} = 0. \qquad \textbf{ANSWER}$$

Discussion

We were able to calculate the drag and lift for the half-size sign only because both signs had the same Reynolds numbers and the same angles of attack. Unless the drag and lift coefficients are independent of Reynolds number, we would not be able to do so if the Reynolds numbers are not equal.

The calculated drag coefficient (2.67) is not the true value for flow over a semicircular cylinder; this number is about 70 percent too large. The reason is that we chose to give an easy-to-integrate expression for the pressure distribution, rather than the real distribution.

8.2 DRAG

The phenomenon of aerodynamic drag is familiar to anyone who has ever tried walking or bicycling into the wind or watched leaves flutter to the ground. As children, we learned from experience that "heavier" objects like rocks fall faster than "lighter" objects like paper wads. We were surprised when we learned in high school or college physics that an object's rate of fall in a vacuum is independent of its weight. Aerodynamic drag is the culprit in our childhood experience; heavier objects experience a larger acceleration because the *net* force-to-mass ratio (weight minus drag divided by mass) is larger than on a lighter object. In the same way, drag obscured the true laws of mechanics for centuries. The ancient Greeks believed that a force must be exerted on a body to move it at *constant* velocity. Isaac Newton in about 1670 realized that moving bodies experience a resisting force and deduced that force was necessary to *change* a body's velocity.

After the existence of aerodynamic drag was recognized, it remained a confusing phenomenon. The fluid mechanics of Euler and Bernoulli was incapable of predicting drag forces because the equations did not contain viscosity.* Not only were theories incapable of predicting drag, but no one was able to explain the fundamental mechanisms by which significant drag forces could arise in fluids of very small viscosity, such as air and water. Not until the twentieth century was even a qualitative explanation of drag developed. By about 1970, a few limited theoretical predictions and a large mass of experimental data on drag accumulated. During the past 20 years, computational fluid dynamics has made a priori prediction of drag a realistic possibility.

In this book, we mainly review currently available information (mostly based on experiments) about drag on simple shapes and on quantitative explanations of this complicated phenomenon.†

8.2.1 Drag of Simple Shapes: Types of Drag

Equations (8.2) and (8.10) show that drag arises from two sources: pressure and shear stress. Drag from pressure is called *form drag* because it depends on the

* This is called d'Alembert's paradox.

† An excellent presentation of the principles of drag is the film *The Fluid Dynamics of Drag* [1].

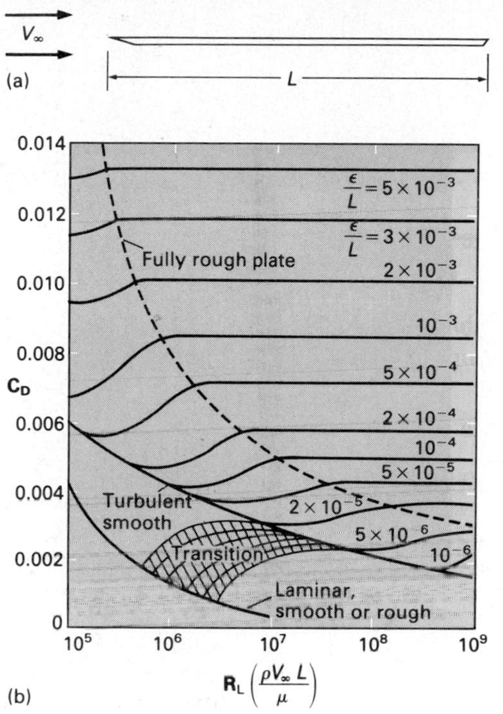

Figure 8.8 Flow over a thin, flat plate aligned with the flow: (a) geometry; (b) drag coefficient for smooth and rough plates (negligible free-stream turbulence).

shape or form of the body. Drag from shear stress is called *skin friction drag* or *friction drag*. Friction drag depends primarily on the amount of surface in contact with the fluid.

The drag on most bodies is a combination of form and friction drag. We begin our discussion by considering bodies that have only one type of drag. First, we look at a thin flat plate that is aligned with the approach flow (Fig. 8.8a). For the present, we assume that the plate extends infinitely perpendicular to the paper. As long as the plate is parallel to the flow, the situation would not change appreciably if the plate had finite width. Because the plate is thin, it has negligible area perpendicular to the flow and experiences no form drag. The fluid velocity at the plate surface is zero and increases to the free-stream velocity a short distance from the plate. The velocity gradient is not zero at the plate surface, so there is a shear stress there and the plate experiences friction drag.

Figure 8.8(b) shows the drag coefficient of a flat plate aligned with the flow as a function of Reynolds number, plate roughness, and flow state (laminar or turbulent). The reference area is the planform area. Note the similarity between the flat plate drag coefficient curves and those for the friction factor for pipe flow shown previously in Fig. 7.9.

The flow near the plate surface may be laminar or turbulent. Which flow state exists depends on the plate length (Reynolds number), plate roughness, and turbulence level in the free-stream flow. Surface roughness and free-stream turbulence cause transition to turbulence to occur at lower Reynolds numbers than on smooth plates in a quiet free stream. The drag coefficient for laminar flow is quite small; the drag coefficient for turbulent flow is much larger because of the increased shear stress in turbulent flow.

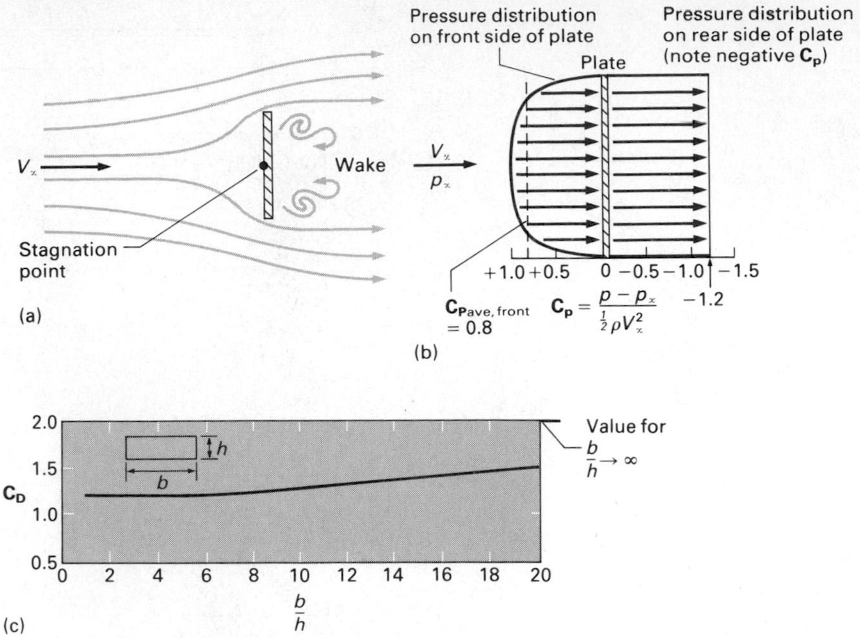

Figure 8.9 Flow over a flat plate normal to the flow: (a) flow schematic; (b) pressure distribution on the plate; (c) effect of width on $\mathbf{C_D}$ for finite plate ($\mathbf{R}_h > 10^3$).

Now suppose that we rotate the plate so that it is perpendicular to the approach flow. We obtain a completely different flow pattern (Fig. 8.9). The fluid must turn as shown to flow over the plate. At the edges of the plate, the fluid is incapable of turning to flow over the back, so the streamlines *separate* at the corner. The region behind the plate, called the *wake,* contains low-pressure, low-velocity fluid.

The forces caused by shear stresses at the plate surface are perpendicular to the approaching flow and do not contribute to drag. The drag on a normal plate is caused by pressure only; it is all form drag. Figure 8.9(b) shows a typical pressure distribution for flow over a two-dimensional normal flat plate. At the center of the plate (the stagnation point), the pressure is equal to the free-stream *total* pressure because the velocity is zero. Near the edges of the plate, the pressure falls rapidly, dropping to a value well below the free-stream pressure. The pressure over the back of the plate is constant and equal to the pressure at the edge of the plate. We calculate the drag coefficient by

$$\mathbf{C_D} = \int_{A_{\text{front}}} -\,\mathbf{C_p}\cos(180°)\,d\left(\frac{A}{S}\right) + \int_{A_{\text{back}}} -\,\mathbf{C_p}\cos(0°)\,d\left(\frac{A}{S}\right)$$
$$= \bar{\mathbf{C}}_{\mathbf{p,front}} - \bar{\mathbf{C}}_{\mathbf{p,back}}.$$

Obtaining the average pressure coefficients from Fig. 8.9(b), we have*

$$\mathbf{C_D} = 0.80 - (-1.20) = 2.0.$$

* Simple "logic" might imply that $\mathbf{C_D}$ should be 1.0, assuming that the plate experiences a force equal to the free-stream dynamic pressure times the area. Because the pressure on the back side of the plate is lower than the free-stream pressure, this "logic" is in error by as much as 100 percent!

Table 8.1 Comparison of flat plate drag.

	Plate Aligned with Flow	Plate Normal to Flow
Drag type	Friction drag only	Form drag only
Magnitude of C_D	Small	Large
Reynolds number dependence	Strong	Weak or none
Effect of plate roughness	Strong	None
Difference between two-dimensional and three-dimensional plate	Minor	Significant

This drag coefficient magnitude is valid for all Reynolds numbers above about 10^3.

If the plate has finite width (a three-dimensional plate), the drag coefficient depends on the plate width-to-height ratio (Fig. 8.9c). Note that these drag coefficients are based on frontal area. Table 8.1 summarizes the differences in drag coefficients for flat plates aligned with and normal to the flow.

EXAMPLE 8.3 Illustrates Flat Plate Drag

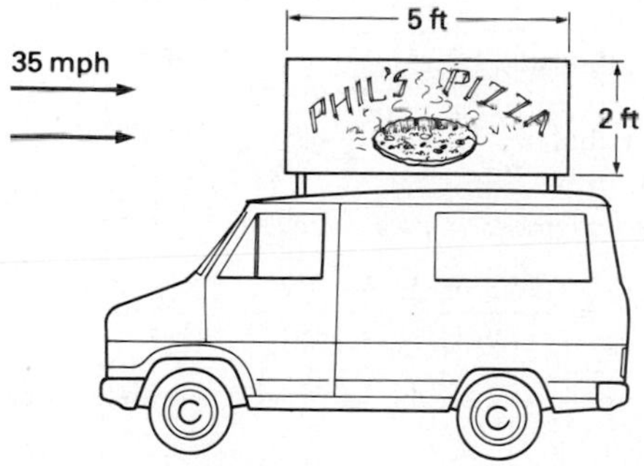

Figure E8.3 Phil's pizza van.

Phil's Pizza Parlor decides to place a thin, rectangular, plastic sign on top of its delivery van as shown in Figure E8.3. The sign measures 2 ft by 5 ft. Estimate the extra power required to drive the van in still air at 35 mph if the sign faces forward rather than sideways.

SOLUTION

Given

Sign on top of van traveling at 35 mph in still air

Sign dimensions 2 ft by 5 ft

Find

Extra power required to drive van if sign faces forward rather than sideways

Solution

We assume that the sign is far enough away from the van that the air flow over the sign is not influenced by the van, so we may use Figs. 8.8 and 8.9. We first consider the sign placed sideways. Assuming 60°F air at atmospheric pressure, we have from Table A.4 $\nu = 1.59 \times 10^{-4}$ ft^2/sec. The Reynolds number is $\mathbf{R}_L = V_\infty L/\nu$, where $L = 5$ ft and $V_\infty = 35$ mph. Then

$$V_\infty = \frac{(35 \text{ mi/hr})(5280 \text{ ft/mi})}{(3600 \text{ sec/hr})} = 51.3 \text{ ft/sec}$$

and

$$\mathbf{R}_L = \frac{(51.3 \text{ ft/sec})(5 \text{ ft})}{1.59 \times 10^{-4} \text{ ft}^2/\text{sec}} = 1.61 \times 10^6.$$

Table 7.1 shows that plastic has $\varepsilon = 0$ (smooth surface). The flow is likely to be turbulent, so from Fig. 8.8, we have $\mathbf{C_D} = 0.004$. We use Eq. (8.4) to calculate the drag:

$$\mathscr{D} = \tfrac{1}{2}\rho \mathbf{C_D} V_\infty^2 S,$$

where S is the planform area. Using Table A.4, we have $\rho = 0.077$ lbm/ft^3. Then

$$\mathscr{D} = \tfrac{1}{2}(0.077 \text{ lbm/ft}^3)(0.004)(51.3 \text{ ft/sec})^2(2 \text{ ft} \times 5 \text{ ft})$$
$$= (4.05 \text{ ft}\cdot\text{lbm/sec}^2)\left(\frac{1 \text{ lb}\cdot\text{sec}^2}{32.2 \text{ ft}\cdot\text{lbm}}\right) = 0.125 \text{ lb}.$$

The power required to overcome this air drag is the product of the drag force and the van speed:

$$\dot{W}_1 = \mathscr{D} \cdot V_\infty = \frac{(0.11 \text{ lb})(51.3 \text{ ft/sec})}{(550 \text{ ft}\cdot\text{lb/sec}\cdot\text{hp})} = 0.012 \text{ hp}.$$

We now consider the forward-facing sign. For 60°F air,

$$\mathbf{R} = \frac{V_\infty h}{\nu} = \frac{(51.3 \text{ ft/sec})(2 \text{ ft})}{1.59 \times 10^{-4} \text{ ft}^2/\text{sec}} = 6.5 \times 10^5.$$

We then use Fig. 8.9 with $b/h = 2.5$ to obtain $\mathbf{C_D} \approx 1.2$. Equation (8.4) gives

$$\mathscr{D} = \tfrac{1}{2}\rho \mathbf{C_D} V_\infty^2 S$$
$$= \tfrac{1}{2}(0.077 \text{ lbm/ft}^3)(1.2)(51.3 \text{ ft/sec})^2(2 \text{ ft} \times 5 \text{ ft})$$
$$= (1215 \text{ ft}\cdot\text{lbm/sec}^2)\left(\frac{1 \text{ lb}\cdot\text{sec}^2}{32.2 \text{ ft}\cdot\text{lbm}}\right) = 37.8 \text{ lb}.$$

The power required to overcome this air drag is

$$\dot{W}_2 = \mathscr{D} \cdot V_\infty = \frac{(37.8 \text{ lb})(51.3 \text{ ft/sec})}{(550 \text{ ft}\cdot\text{lb/sec}\cdot\text{hp})} = 3.53 \text{ hp}.$$

The extra power required to drive the van if the sign faces forward rather than sideways is

$$\dot{W}_2 - \dot{W}_1 = 3.53 \text{ hp} - 0.01 \text{ hp};$$
$$\dot{W}_2 - \dot{W}_1 = 3.52 \text{ hp}.$$ **ANSWER**

Discussion

The magnitudes of 0.01 hp and 3.53 hp represent the power required to overcome air drag at 35 mph. Considering that the van's drive-train efficiency might be as low as 85 percent, the corresponding engine output powers would be

$$\dot{W}'_1 = \frac{0.012 \text{ hp}}{0.85} = 0.016 \text{ hp} \quad \text{and} \quad \dot{W}'_2 = \frac{3.52 \text{ hp}}{0.85} = 4.15 \text{ hp}.$$

The extra *engine* power could then be as large as 4.14 hp. Fuel economy might be lowered by about 1 mile per gallon.

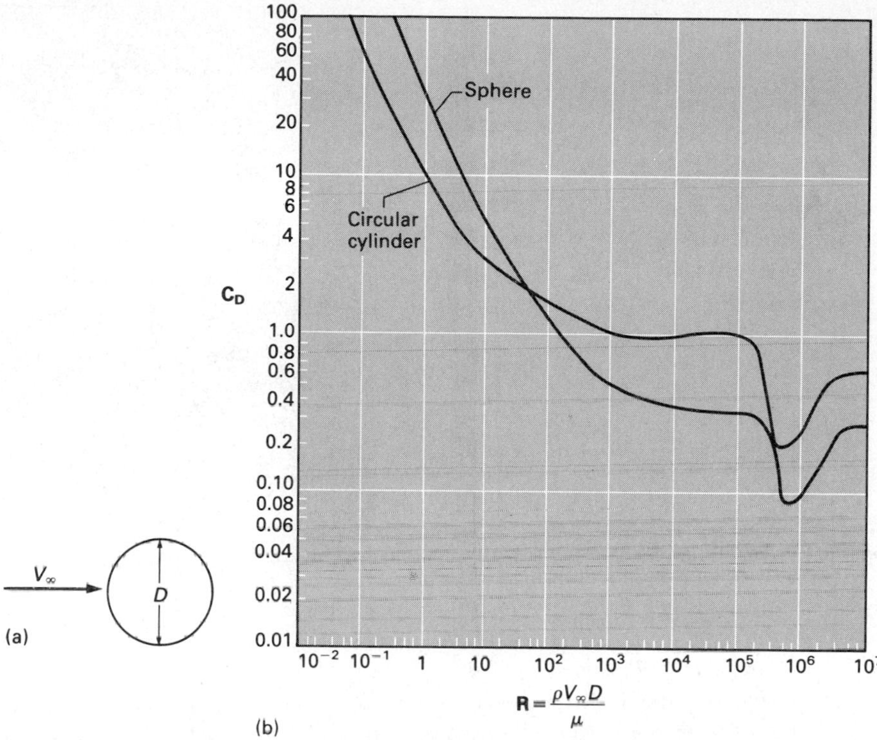

Figure 8.10 Drag of smooth body of circular cross section (cylinder or sphere): (a) schematic; (b) drag coefficients as a function of Reynolds number.

The nature of drag on flat plates aligned with or normal to the flow is relatively easy to understand, because only one type of drag is present in each case. Let's now turn to simple shapes for which both types of drag are present. Figure 8.10(a) illustrates flow over a smooth body of circular cross section. If the body is two-dimensional, it is a cylinder; if it is axisymmetric, it is a sphere. Both a cylinder and a sphere have equal amounts of area parallel and perpendicular to the flow and experience both form drag and friction drag. Drag coefficients for smooth cylinders and spheres are plotted in Fig. 8.10(b). For both bodies, the Reynolds number is based on diameter and the drag coefficient is based on frontal area. Therefore

$$S = \begin{cases} D(1), & \text{Cylinder} \\ \dfrac{\pi}{4} D^2. & \text{Sphere} \end{cases}$$

These curves are based almost entirely on experimental data. For hand calculations, drag coefficients can be read from the curves. White [2] suggested the following curve fits for numerical work:

$$C_{\mathbf{D},\text{cylinder}} \approx 1.0 + 10.0(\mathbf{R})^{-0.67}, \qquad \mathbf{R} < 2 \times 10^5, \tag{8.12a}$$

$$C_{\mathbf{D},\text{sphere}} \approx \frac{24}{\mathbf{R}} + \frac{6}{1 + \sqrt{\mathbf{R}}} + 0.4, \qquad \mathbf{R} < 2 \times 10^5. \tag{8.12b}$$

Comparing the cylinder and sphere drag coefficients with those of the smooth parallel flat plate and the normal flat plate is instructive. At low Reynolds num-

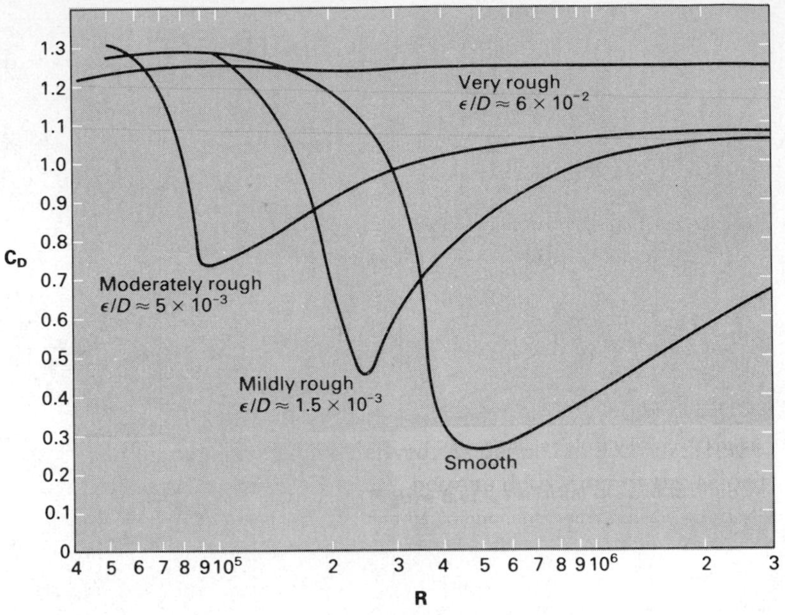

Figure 8.11 Effects of roughness on $\mathbf{C_D}$ for a cylinder.

bers, the cylinder and sphere drag coefficients vary with Reynolds number, decreasing as Reynolds number increases. This behavior is similar to that of the flat plate aligned with the flow. At a Reynolds number of about 10^3, the drag coefficient becomes essentially constant and the curve is flat. This behavior is similar to that of the normal flat plate. At a Reynolds number of about 5×10^5, the cylinder and sphere drag coefficient dips and then returns to a lower constant value. Note that Eqs. (8.12a) and (8.12b) do not apply in this region. The sudden change in drag coefficient is similar to that which occurs for the aligned flat plate, except that the cylinder and sphere coefficient *drops sharply,* whereas the flat plate coefficient *rises gradually.*

Figure 8.11 illustrates the effects of surface roughness on the drag of a cylinder. The figure concentrates on that part of the range of Reynolds number in which the "dip" in drag coefficient occurs. Note that the mildly and moderately rough surfaces cause the dip to occur at lower values of **R**, with the resulting high **R** drag coefficient being about 50 percent higher than for the smooth surface. The very rough surface does not cause a dip and the resulting high **R** coefficient is about twice as large as that for the smooth surface.

If we attempted to enter the drag characteristics of the cylinder and sphere in a table such as Table 8.1, we would not learn very much. Each space would contain several entries, based on the value of the Reynolds number. A complete explanation of the cylinder and sphere drag coefficient requires separate consideration of high Reynolds number and low Reynolds number regimes. We develop this explanation in the next two sections.

EXAMPLE 8.4 **Illustrates Sphere Drag and a Simple Method to Measure a Drag Coefficient**

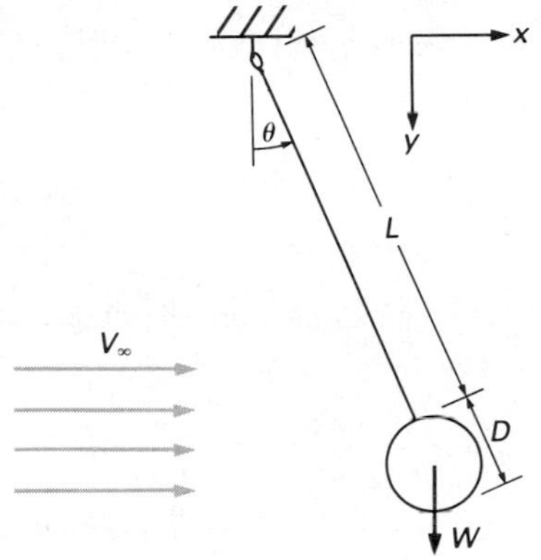

Figure E8.4a Simple experiment to determine drag coefficients for a sphere.

The smooth sphere shown in Fig. E8.4a has weight W and hangs by a weightless wire of length L. A wind of velocity V_∞ blows over the sphere. Develop an expression relating the drag coefficient C_D of the sphere, the velocity V_∞, and the angle θ. The air density is ρ and the sphere diameter is D. Use the resulting expression to calculate the angle θ for a 15-cm diameter sphere weighing 3.5 N, a wind velocity of 17 m/s, and a 60-cm wire. Assume that the sphere diameter is very large compared to the diameter of the wire so that the air drag on the wire may be neglected.

SOLUTION

Given

Figure E8.4a

$W = 3.5$ N, $D = 15$ cm, $V_\infty = 17$ m/s, $L = 60$ cm

Find

Relation between angle θ, drag coefficient, and properties of the sphere and the wind

Calculate θ for a specific case.

Solution

We first apply Newton's first law in the horizontal (x) direction. Using the free-body diagram in Fig. E8.4b, we get

$$\mathscr{D} = T \sin\theta,$$

where T is unknown but can be found by applying Newton's first law in the vertical (y) direction. Referring to Fig. E8.4b, we see that

$$W = T\cos\theta, \quad \text{so} \quad T = \frac{W}{\cos\theta}.$$

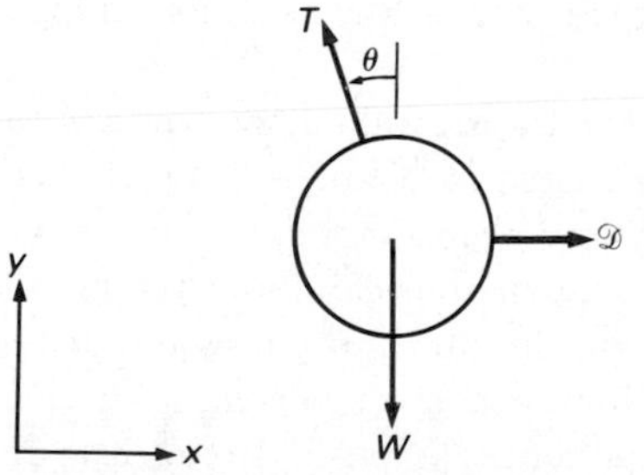

Figure E8.4b Free-body diagram of sphere.

Substituting this expression for T into the equation for $\mathscr{D}$ gives

$$\mathscr{D} = W\tan\theta.$$

Equation (8.4) gives

$$\mathscr{D} = \tfrac{1}{2}\rho C_D V_\infty^2 S.$$

Eliminating $\mathscr{D}$ from the last two equations gives

$$W\tan\theta = \tfrac{1}{2}\rho C_D V_\infty^2 S,$$

where S is the frontal area, $S = \pi D^2/4$.

Solving for θ gives

$$\theta = \tan^{-1}\left(\frac{\rho C_D V_\infty^2 \pi D^2}{8W}\right).$$ **ANSWER**

If we assume 20°C air, Table A.3 gives $\rho = 1.2\ \text{kg/m}^3$ and $\nu = 1.51 \times 10^{-5}\ \text{m}^2/\text{s}$. The Reynolds number is

$$\mathbf{R} = \frac{V_\infty D}{\nu} = \frac{(17\ \text{m/s})(0.15\ \text{m})}{1.5 \times 10^{-5}\ \text{m}^2/\text{s}} = 1.69 \times 10^5.$$

Figure 8.10 or Eq. (8.12b) gives $\mathbf{C_D} = 0.42$. The numerical values give

$$\theta = \tan^{-1}\left[\frac{(1.2\ \text{kg/m}^3)(0.42)(17\ \text{m/s})^2\pi(0.15\ \text{m})^2}{8(3.5\ \text{N})}\right];$$

$$\theta = 20.2°.$$ **ANSWER**

Discussion

We could use this simple pendulum to measure the drag coefficient for an object by solving the equation for $\mathbf{C_D}$:

$$\mathbf{C_D} = \frac{8W \tan\theta}{\rho V_\infty^2 \pi D^2}.$$

For known values of W, ρ, V_∞, and D, measuring the angle θ gives the value of $\mathbf{C_D}$. Such a pendulum could also serve as a wind-speed indicator. In this case, measurement of θ would give a value for V_∞.

8.2.2 The Boundary Layer and Its Effect on Drag at High Reynolds Number

Most flows of interest to engineers have rather large Reynolds numbers, because the size of the object of interest is significant and the (kinematic) viscosity of the fluid involved is not large. As Reynolds number is the ratio of inertia "force" to viscous force in the fluid (see Chapter 6), we could reasonably assume that high Reynolds number flow is approximately inviscid. If we discard viscosity, we would have no mechanism for friction drag, but the inviscid flow assumption might give us a good understanding of form drag.

If we neglect viscous forces, we may calculate streamline patterns and pressure distribution for flow over simple bodies (discussed in Chapter 9). Figure 8.12 illustrates results of an inviscid calculation for flow over a circular cylinder. Both the flow pattern and the pressure distribution are symmetrical about the midplane of the cylinder. Consider the calculation of form drag from the pressure distribution shown in Fig. 8.12(b). The pressure distribution is symmetrical, so each surface element on the front of the cylinder that experiences a "push" in the flow direction is balanced by a corresponding element on the back that experiences an equal push opposite to the flow direction. All such elements occur in pairs, so the net drag force would be zero! The prediction of zero drag for an inviscid fluid even extends to the *normal* flat plate. Figure 8.13 shows streamlines for flow over a normal flat plate, which we calculated using the inviscid fluid assumption. The predicted pressure distribution on the plate is the same on the front and back, so no drag would result. Apparently, the inviscid fluid assumption does not allow us to investigate form drag, even approximately (d'Alembert's paradox again).

In 1904, L. Prandtl introduced the boundary layer hypothesis, which explains how both form drag and friction drag arise in a high Reynolds number flow. For

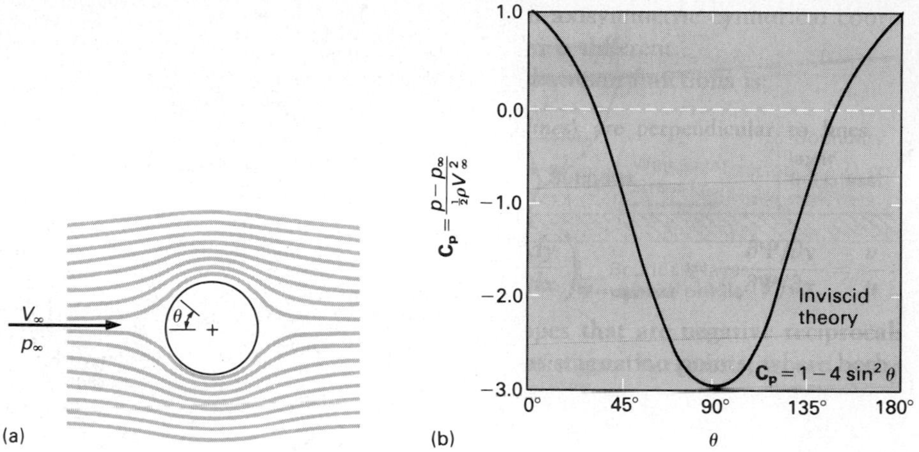

Figure 8.12 Calculated streamline pattern and pressure distribution for inviscid flow over a circular cylinder.

flow over a long, slender body, such as a parallel flat plate or airfoil (Fig. 8.14), Prandtl realized that no matter how small the viscosity, the fluid must still satisfy the no-slip condition at the surface. The fluid velocity must then rise from zero at the surface to a value that is the order of magnitude of the free-stream velocity a short distance above the surface. This region where the velocity increases from zero to the free-stream value is called the *boundary layer*. Prandtl noted that for large Reynolds numbers, the boundary layer is very thin; in fact, he showed that

$$\frac{\delta}{L} \sim \sqrt{\frac{\nu}{V_\infty L}} = \frac{1}{\sqrt{R_L}}. \tag{8.13}$$

Of course, the thinner the boundary layer is, the larger the value of $\partial u/\partial y$ in the boundary layer becomes. Significant shear stresses are present in the boundary layer fluid, even if the viscosity is small.

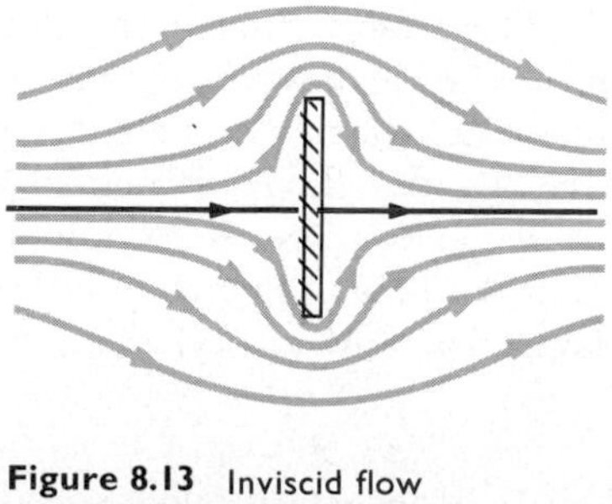

Figure 8.13 Inviscid flow streamlines for flow over a normal flat plate.

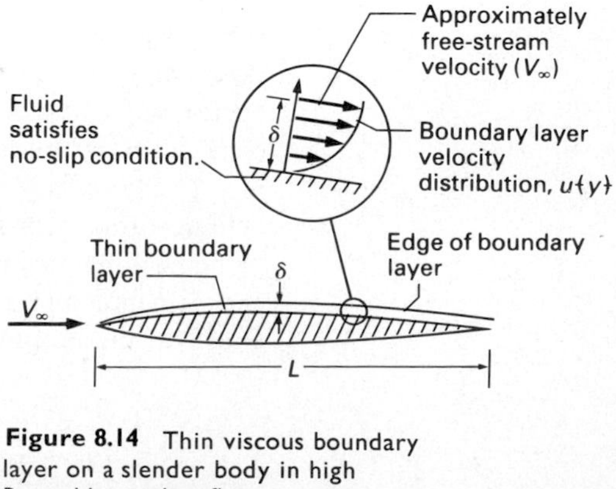

Figure 8.14 Thin viscous boundary layer on a slender body in high Reynolds number flow.

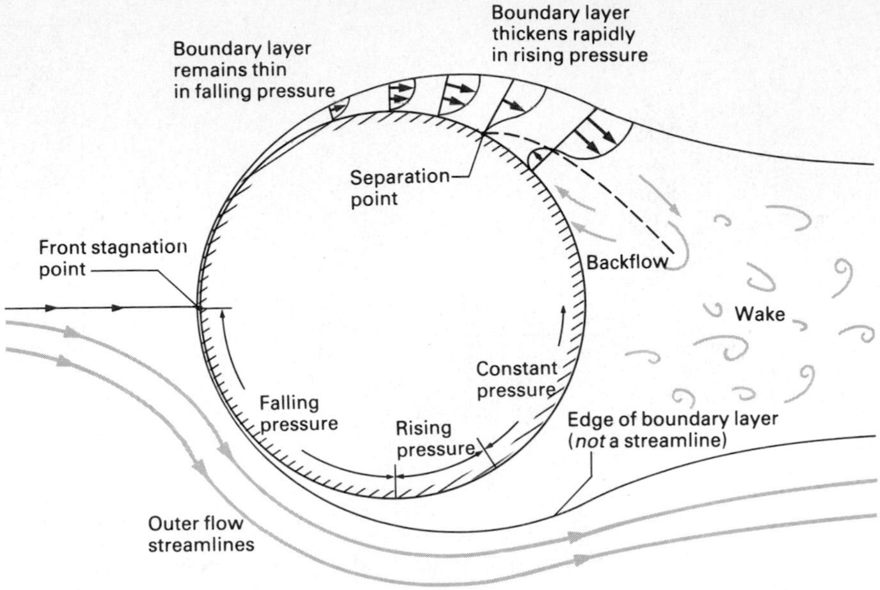

Figure 8.15 Boundary layer on a circular cylinder.

As long as the boundary layer is thin, its presence has little effect on the pressure distribution or the flow pattern near the body. For the flat plate or thin airfoil, the streamline shapes are essentially unchanged by the boundary layer. By analyzing the boundary layer, we may calculate the shear stress and thereby calculate the drag on long, slender shapes such as the parallel flat plate (we consider this type of calculation in Chapter 9). As the boundary layer is a region where the fluid experiences shear deformation, the flow may be laminar or turbulent, depending on the magnitude of the Reynolds number. Flow in the boundary layer near the leading edge of a plate is always laminar, because the (local) Reynolds number is small; if the plate is long enough, transition to turbulence occurs and the flow is turbulent over the downstream portion of the plate.* If the flow is disturbed by surface roughness or turbulence in the stream outside the boundary layer, transition occurs nearer the leading edge. Because of enhanced mixing in turbulent flow, the turbulent boundary layer has larger shear stresses than the laminar boundary layer. Similarly, rough surfaces experience larger stress than do smooth ones (see Sections 7.1.1 and 7.2.7). These facts explain the behavior of the drag coefficient for a flat plate aligned with the flow (as shown in Fig. 8.8b).

The boundary layer model adequately explains flat plate drag, but what about cylinder drag? Consider the boundary layer on a circular cylinder (Fig. 8.15). The primary difference between flat plate and cylinder flows is that the cylinder's shape causes a pressure change in the flow. The pressure distribution has a profound effect on the boundary layer. The flow outside the boundary layer is determined

---

* Note that the same plate can have laminar flow over part of it and turbulent flow over another part. This is not true of *fully developed* pipe flow, which must be either laminar or turbulent.

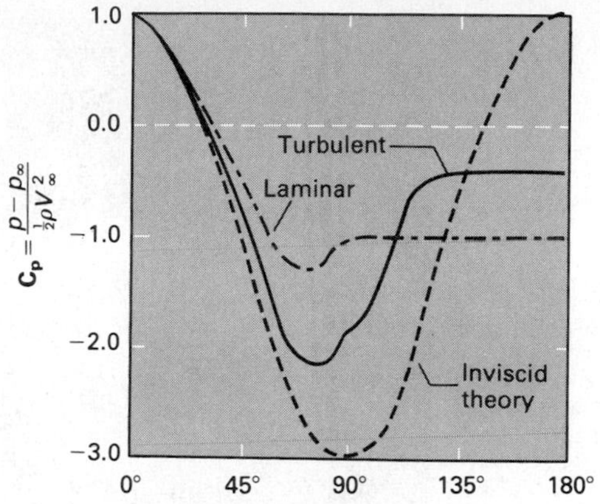

Figure 8.16 Pressure distribution on cylinder surface for laminar and turbulent boundary layers.

by a balance between pressure force and fluid momentum. On the front of the cylinder, the fluid momentum increases and the pressure drops, while on the back side the fluid exchanges momentum for increasing pressure. For the boundary layer, the flow is determined by a balance between momentum and pressure *plus* viscous forces. Because the velocity of the boundary layer fluid is less, it has less momentum than the fluid outside the boundary layer. This situation is not too serious on the front side of the cylinder, where the falling pressure helps "push" the fluid in the boundary layer along. On the back side of the cylinder, the situation is different. As the pressure begins to rise, the fluid must exchange momentum for pressure. Because the boundary layer is deficient in momentum, it is unable to penetrate very far into the rising pressure. The pressure causes the fluid in the boundary layer to stop and, ultimately, reverse its direction. The boundary layer separates from the cylinder surface and a broad wake forms behind the cylinder. The separating boundary layer pushes the flow streamlines outward and so alters the entire flow pattern and pressure distribution. The pressure on the rear of the cylinder, in the wake, is low and approximately constant. The pressure on the front of the cylinder is high, so a net pressure force (form drag) acts on the cylinder. The shear stress on the cylinder surface produces friction drag. At Reynolds numbers greater than about 10^3, form drag accounts for about 95 percent of the total drag, so the drag coefficient becomes independent of Reynolds number,* except near the dip at a Reynolds number of about 5×10^5.

The dip in drag coefficient—and the resulting lower value of $\mathbf{C_D}$ after the dip—is the result of transition from laminar to turbulent flow in the boundary layer; however, the effect is opposite that in the flat plate flow, where turbulence causes a drag increase. To understand the effects of turbulence in the cylinder flow, you need to recall that a turbulent boundary layer contains more momentum than a laminar boundary layer, because the velocity profile is fuller (see Fig. 7.11). Because the turbulent boundary layer contains more momentum, it can penetrate farther into the rising pressure on the rear of the cylinder, and separation is delayed. The wake is smaller, and the pressure in the wake is higher. Figure 8.16 shows

* Reynolds number independence means that the drag depends on inertia "force" but not on viscous force.

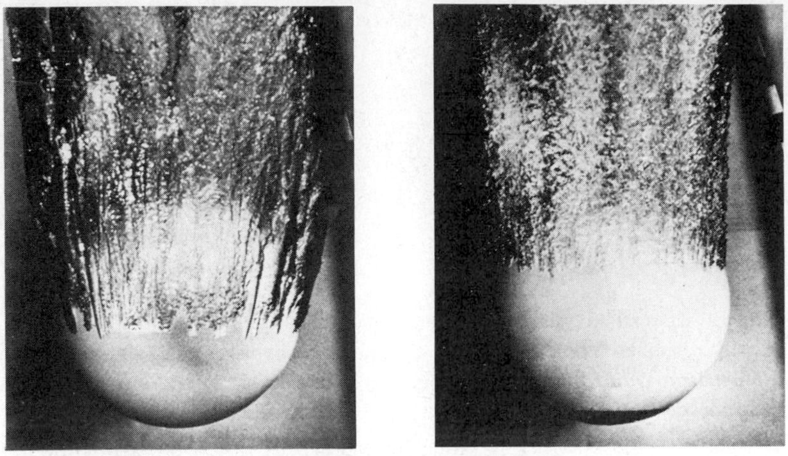

Figure 8.17 Effect of turbulence in boundary layer on delaying separation. An 8.5-in. bowling ball is dropped into water at a speed of 25 ft/sec. (a) The ball is smooth; (b) the ball has a patch of sandpaper on the nose that causes transition. (U.S. Navy photographs.)

pressure distributions on cylinders with laminar and turbulent boundary layers. The higher average pressure in the back of the cylinder with a turbulent boundary layer results in lower form drag and a lower drag coefficient, even though the friction drag of the turbulent boundary layer is higher than that of the laminar layer.

The effects of turbulence in delaying separation are graphically illustrated in Fig. 8.17. The two spheres are identical, except that one has a patch of sandpaper on its nose. This sandpaper causes the boundary layer to become turbulent, which delays separation. Note that the wake is turbulent in both cases; there is no visible difference between the laminar and turbulent boundary layers on the spheres.

You are now in a position to understand better the drag on a normal flat plate. The flow is unable to undergo the acceleration necessary to turn the sharp corner at the edge of the plate. The separation point is fixed at the plate edge for all Reynolds numbers, so the drag coefficient is independent of Reynolds number.

These considerations help explain the *mechanisms* of drag at high Reynolds number, but, except for the parallel plate, analytically predicting drag coefficients with confidence is difficult. We can design low-drag shapes based on our qualitative understanding of drag, keeping in mind two important principles.

- If the body is long and thin, such as a parallel flat plate or an airfoil, the drag is primarily caused by friction. This drag can be reduced by keeping the flow laminar as much as possible. This condition implies smooth surfaces and, if the body has thickness (like an airfoil), means selecting the shape that delays transition as long as possible.

- If the body is blunt, the (high Reynolds number) drag is primarily form drag. This drag can be reduced by delaying separation as long as possible.

One way to delay boundary layer separation on a blunt (thick) body is to promote transition to a turbulent boundary layer (Fig. 8.17). A better method is *streamlining*, that is, elongating the rear portion of the body. It turns out that boundary layer separation depends on the rate of pressure rise, that is, on the pressure gradient dp/dx, where x is the distance along the surface.

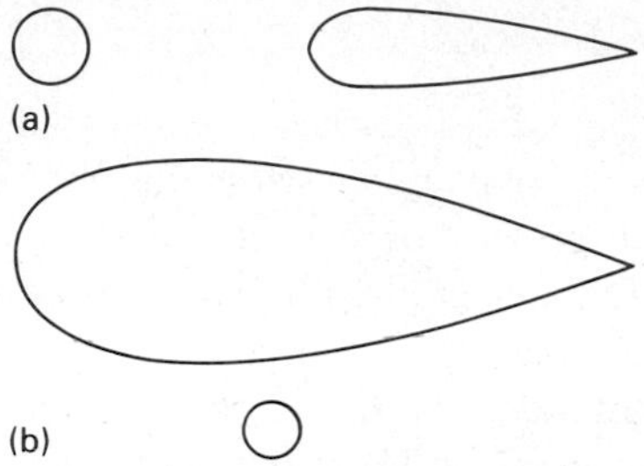

Figure 8.18 Illustration of the effects of streamlining on drag at high Reynolds number: (a) cylinder and strut with the same frontal area; $C_{D,\text{cylinder}} \approx 5C_{D,\text{strut}}$; (b) cylinder and strut with equal drag force at a given flow speed.

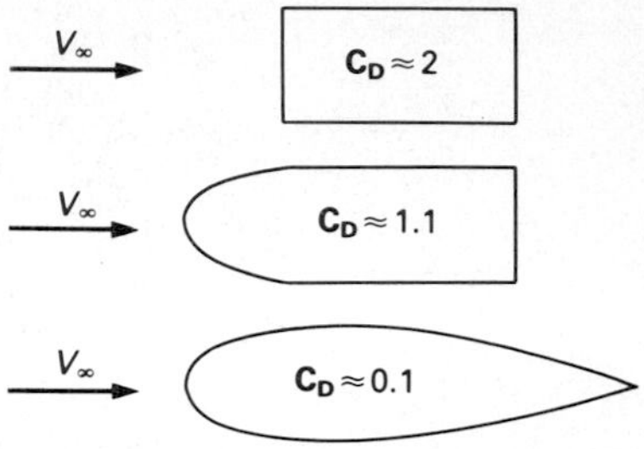

Figure 8.19 Effect of streamlining on drag reduction. Drag coefficients are based on frontal area. All bodies are two-dimensional.

Figure 8.18(a) shows a circular cylinder and a streamlined strut with the same frontal area. The streamlined strut will have a drag coefficient about one fifth that of the cylinder. Alternatively, Fig. 8.18(b) shows a cylinder and strut that have equal drag for a given flow speed.

If the nose of a body is flat, nose rounding can also reduce the drag. Figure 8.19 illustrates the benefits of nose rounding and rearward taper. The round nose-rear tapered body would have higher drag if it were reversed, with the tapered section pointing forward. The forward portion of a body can be quite blunt with a large (negative) pressure gradient, because separation does not occur in falling pressure. It is the rearward (rising) pressure change that must be accomplished gradually.

There is a limit to the amount of drag reduction that can be accomplished by streamlining. As the body becomes more elongated, friction drag increases because there is more "wetted" surface (Fig. 8.20). The minimum drag body provides the proper balance between friction drag and form drag.

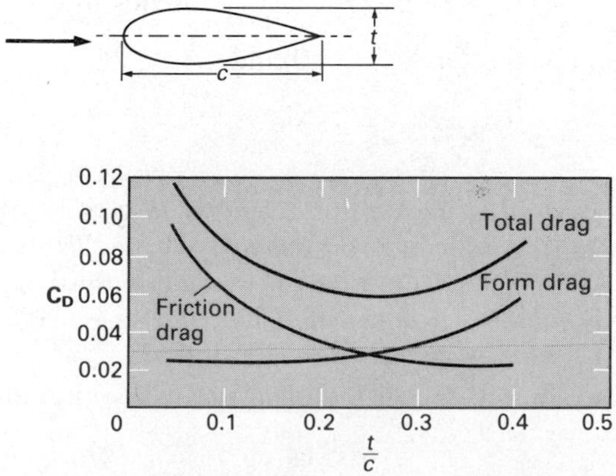

Figure 8.20 Total drag is the sum of form drag, which increases as a body becomes more blunt, and friction drag, which increases as a body becomes longer.

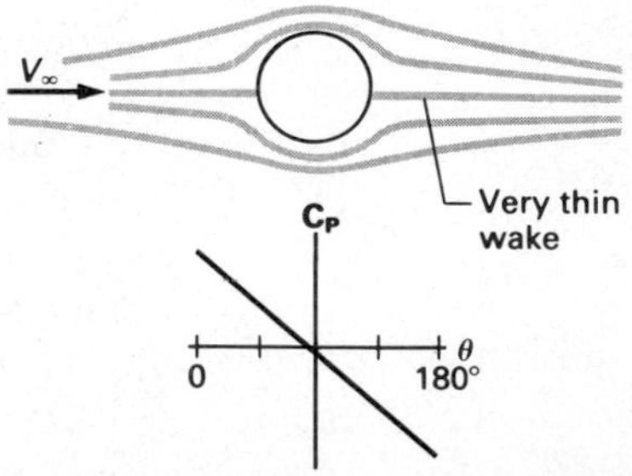

Figure 8.21 Streamline pattern and pressure distribution for very low Reynolds number flow over a sphere.

8.2.3 Drag at Low Reynolds Number

Occasionally an engineer must deal with a low Reynolds number flow. Low Reynolds numbers result from small body sizes (dust or pulverized coal particles), large fluid viscosities (oil, grease), or very low density (high-altitude flight). The mechanisms of drag are fundamentally different in low and high Reynolds number flow. The boundary layer concept, so valuable in high Reynolds number flow, is invalid at low Reynolds number. At low Reynolds number, viscous forces are not confined to a thin region near the body but are significant everywhere. That is, the boundary layer is so thick that it distorts the streamline pattern everywhere. At very low Reynolds number (less than 1), inertia "forces" are negligible and only pressure and shear forces are present. Of course, low Reynolds number flow is always laminar.

Consider the flow over a sphere at very low Reynolds number. This flow was successfully analyzed by G. G. Stokes in 1851. The streamline pattern and pressure distribution are shown in Fig. 8.21. The streamlines are nearly symmetric and have only a thin wake (no separation); however, the pressure distribution is asymmetric. Stokes found that the drag on a sphere at very low Reynolds number is given by

$$\mathscr{D} = 3\pi\mu D V_\infty. \tag{8.14}$$

Equation (8.14) is called Stokes's law. The drag is not from friction only; one third of the drag is from pressure and two thirds is from friction. The drag coefficient for a sphere at very low Reynolds number is*

$$\mathbf{C_D} = \frac{3\pi\mu D V_\infty}{\frac{1}{2}\rho V_\infty^2(\pi D^2/4)} = \frac{24\mu}{\rho V D} = \frac{24}{\mathbf{R}}. \tag{8.15}$$

Actually, the "usual" drag coefficient defined by Eq. (8.4) is not the best definition for low Reynolds number flow, because the drag does not depend on fluid inertia (see Example 6.5). A *viscous drag coefficient*, defined by

$$\mathbf{C'_D} \equiv \frac{\mathscr{D}}{\mu V_\infty \ell} \qquad \left(\sim \frac{\text{Drag force}}{\text{Viscous force}}\right), \tag{8.16}$$

is more useful. For the sphere, $\ell = D$ and

$$\mathbf{C'_D} = 3\pi,$$

a constant.

The viscous drag coefficient for any body in very low Reynolds number flow is a constant and depends only on the body shape. Table 8.2 gives values of $\mathbf{C'_D}$ for various body shapes. Note that $\mathbf{C'_D}$ is roughly the same for all shapes except long, slender ones. As a first approximation, $\mathbf{C'_D}$ of a sphere can be used for all blunt bodies in very low $\mathbf{R}$ flow.

Table 8.3 compares drag in high and low Reynolds number flows. These com-

---

* Note the similarity with the Hagen–Poiseuille equation for friction factor in laminar pipe flow. Also note the leading term of Eq. (8.12b).

Table 8.2 Viscous drag coefficients for low Reynolds number flow (valid for $\mathbf{R} < 1$).

Shape		Reference Length	C'_D
Sphere →		D	3π
Hemisphere →	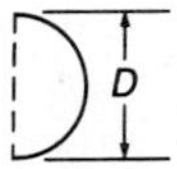	D	8.7
Normal disc →	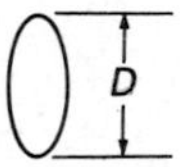	D	8
Parallel disc →	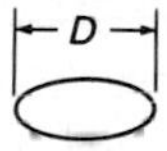	D	$\frac{16}{3}$
Normal rod →		L	$\frac{4\pi}{\ln(2L/D) + 0.5}$
Parallel rod →	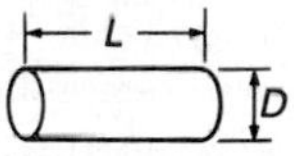	L	$\frac{2\pi}{\ln(2L/D) - 0.72}$

parisons are based on the approximate equations:

$$\mathscr{D}(\text{High } \mathbf{R}) \approx (\text{Constant})(\tfrac{1}{2}\rho V_\infty^2)(\text{Area}); \tag{8.17a}$$

$$\mathscr{D}(\text{Low } \mathbf{R}) \approx (\text{Constant})\mu V_\infty(\text{Length}). \tag{8.17b}$$

Steps taken to reduce drag in high Reynolds number flow are ineffective or even counterproductive in low Reynolds number flow. In particular, streamlining usually results in a drag *increase* in low **R** flow because body length and wetted area are greater.

Drag in the intermediate Reynolds number range ($10 < \mathbf{R} < 10^3$) is most difficult to analyze, even qualitatively, because the boundary layer concept is not valid but inertia "forces" are significant. Data are usually presented in terms of $\mathbf{C_D}$. Both $\mathbf{C_D}$ and $\mathbf{C'_D}$ are functions of Reynolds number.*

* Note that $\mathbf{C'_D} = \mathbf{C_D} \times \mathbf{R}$; $\mathbf{C'_D}$ is an alternative dimensionless group.

Table 8.3 Dependence of drag on various parameters in high **R** and low **R** flow (compare to Table 7.2).

Parameter	High R Flow	Low R Flow
Velocity (V)	V^2	V
Density (ρ)	ρ	None
Viscosity (μ)	Slight to none	μ
Size (ℓ)	ℓ^2	ℓ

EXAMPLE 8.5 Illustrates Low Reynolds Number Drag

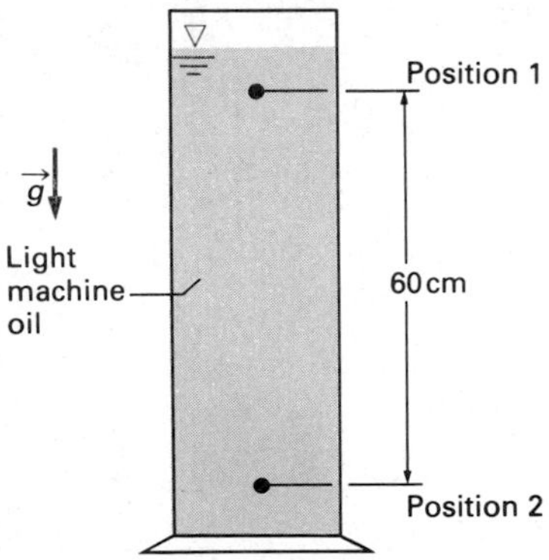

Figure E8.5 Steel ball falling in light machine oil.

The 1.0-mm-diameter steel ball in Fig. E8.5 is released from rest at the oil surface and reaches a constant velocity at point 1. Find the velocity of the steel ball and estimate the time for the ball to fall from position 1 to position 2. The fluid is light machine oil with an absolute viscosity of 0.10 N·s/m^2 and a specific weight of 8340 N/m^3. Steel has a specific gravity of 7.8.

SOLUTION

Given

1.0-mm-diameter steel ball falling in light machine oil

Ball falling with constant velocity for a distance of 60 cm

Oil viscosity 0.10 N·s/m^2

Oil specific weight 8340 N/m^3

Figure E8.5

Steel specific gravity 7.8

Find

Time for steel ball to fall 60 cm

Constant velocity during time of fall

Solution

We first apply Newton's second law in the vertical direction, with the positive direction taken as upward. The velocity is constant, there is no acceleration, and Newton's second law gives

$$\sum F = \mathscr{D} - W + F_B = 0,$$

where $\mathscr{D}$ is the viscous drag on the ball, W its weight, and F_B the buoyant force on the ball. If we assume that the ball falls very slowly, Stokes's law (Eq. 8.14) gives

$$\mathscr{D} = 3\pi\mu D V_\infty .$$

The ball's weight is $W = \gamma_s \forall$, where γ_s is the specific weight of steel and $\forall$ is the ball's volume. Now

$$\gamma_s = S\gamma_w = 7.8(9810\ \text{N/m}^3) = 76{,}500\ \text{N/m}^3.$$

The buoyant force is $F_B = \gamma_0 \forall$, where γ_0 is the oil specific weight.
Substituting for $\mathscr{D}$, W, and F_B gives

$$3\pi\mu D V_\infty - \gamma_s \forall + \gamma_0 \forall = 0.$$

Using $\forall = \pi D^3/6$ gives

$$V_\infty = \frac{(\gamma_s - \gamma_0)D^2}{18\mu}.$$

The numerical values give

$$V_\infty = \frac{(76{,}500\ \text{N/m}^3 - 8340\ \text{N/m}^3)(1.0\ \text{mm})^2}{18(0.10\ \text{N}\cdot\text{s/m}^2)(1000\ \text{mm/m})^2};$$

$$V_\infty = 0.0379\ \text{m/s}.$$ **ANSWER**

Before proceeding, we check our assumption of very low Reynolds number. The oil kinematic viscosity is

$$\nu_0 = \frac{\mu_0}{\rho_0} = \frac{g\mu_0}{\gamma_0}.$$

Assuming sea-level elevation, the numerical values give

$$\nu_0 = \frac{(9.81\ \text{m/s}^2)(0.10\ \text{N}\cdot\text{s/m}^2)}{(8340\ \text{N/m}^3)} = 1.18 \times 10^{-4}\ \text{m}^2/\text{s}$$

and

$$\mathsf{R} = \frac{V_\infty D}{\nu_0} = \frac{(0.0379\ \text{m/s})(1.0\ \text{mm})}{(1.18 \times 10^{-4}\ \text{m}^2/\text{s})(1000\ \text{mm/m})} = 0.321.$$

We are in the Reynolds number range where Stokes's law is valid, so our velocity estimate is also valid. The time required for the sphere to fall 60 cm is

$$t = \frac{\text{Distance}}{\text{Velocity}} = \frac{L}{V_\infty}.$$

The numerical values give

$$t = \frac{(60\ \text{cm})(1\ \text{m}/100\ \text{cm})}{(0.0379\ \text{m/s})},$$

or

$$t = 15.8\ \text{s}.$$ **ANSWER**

Discussion

If we combine the equations for V_∞ and t, we obtain

$$\mu = \frac{(\gamma_s - \gamma_0)D^2 t}{18\ L}.$$

Could we use this experiment and equation to find the viscosity of fluids? Must the fluid be a particular type? Do the walls of the container have any effect?

8.2.4 Working Data on Drag Coefficients

A large amount of data on drag coefficients has been compiled over the years, much of it the result of experimental investigations. Reference [3] presents an extensive compilation of such information. In this textbook, Figs. 8.8(b), 8.9(c), 8.10(b), and 8.11 and Table 8.2 are examples of such data. Tables 8.4, 8.5, and 8.6 present more data on a few simple shapes. Most of the bodies included in these tables are rather angular, with separation points fixed at sharp edges or corners. As a result, the listed drag coefficients are insensitive to Reynolds number beyond about 10^4 and may be taken as constant above that magnitude. The experimental uncertainty of this information is about 5 percent.

Sometimes we can estimate drag on shapes other than those listed by using the listed data. For example, we might assume that the ratio of drag for a cylinder of finite length to drag for an infinite cylinder is the same as the ratio of drag for a finite normal plate to drag for an infinite plate. However, we must use such estimates with extreme caution; for example, the preceding assumptions might be valid for form drag, but they are not valid for friction drag.

Another common pitfall in extrapolating this information is *interference effects*. The data presented are for an object in isolation in an otherwise infinite

Table 8.4 Drag coefficients for various two-dimensional bodies.

Shape		Reynolds Number	C_D
Parallel flat plate	→ ▭	$> 10^5$	See Fig. 8.8(b).
Normal flat plate	→ ▯	$> 10^3$	2.0
Circular cylinder	→ ◯	All	See Fig. 8.10.
Square rod	→ ▢	$> 10^4$	2.0
Square rod	→ ◇	$> 10^4$	1.50
Equilateral triangular rod	→ ◁	$> 10^4$	Sharp edge forward: 1.40 Flat face forward: 2.0
C-section	→)	$> 10^4$	2.30
C-section	→ (	$> 10^4$	1.20
Airfoils	→ ⌒	Various	(See reference [5].)

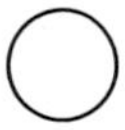

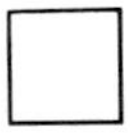

Table 8.5 Drag coefficients for various three-dimensional objects.

Shape		Reynolds Number	C_D
Parallel flat plate → (finite width)	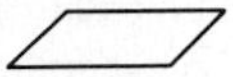	$>10^5$	Use Fig. 8.8(b).
Normal plate →	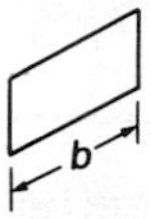	$>10^3$	See Fig 8.9(c).
Cube →	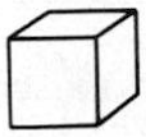	$>10^4$	1.10
Cube →	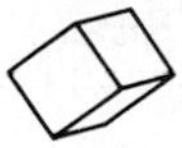	$>10^4$	0.81

Table 8.6 Drag coefficients for various axisymmetric objects.

Shape		Reynolds Number	C_D	
Sphere →		All	See Fig. 8.10.	
Cylinder/disc ($L = 0$) →	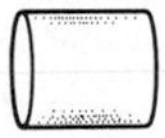	$>10^4$	$L/D = 0$	1.17
			$L/D = 0.5$	1.15
			$L/D = 1$	0.90
			$L/D = 2$	0.85
			$L/D = 4$	0.87
			$L/D = 8$	0.99
Hemispherical cup →		$>10^4$	1.40	
Hemispherical cup →		$>10^4$	0.40	
60° cone →		$>10^4$	0.50	
Parachute		$>10^5$	1.2	

stream. If we made an automobile rearview mirror in the shape of a hemispherical cup, the drag of the mirror installed on an auto would most likely be greater than the hemisphere's drag in isolation. Our calculations in Example 8.3 are probably somewhat nonconservative because of interference effects. Whenever possible, you should use data for the exact configuration that you are considering.

EXAMPLE 8.6 Illustrates Use of Drag Coefficient Data

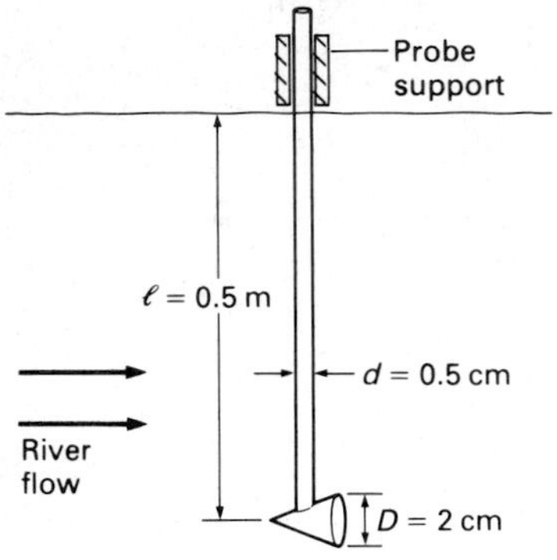

Figure E8.6a Probe in a river.

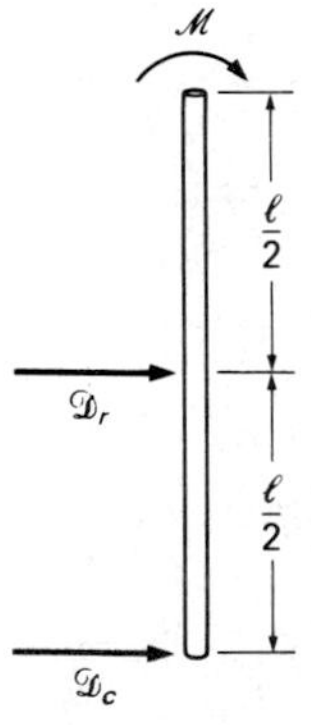

Figure E8.6b Free-body diagram of probe.

Figure E8.6a shows a probe used to measure contaminants in a river. The probe's body is a 60° cone mounted on a long rod. A particular probe has a 2-cm diameter cone base and is mounted on a 0.5-cm diameter rod. The probe is inserted to a depth of 0.5 m in a river flowing at 1.5 m/s. Calculate the bending moment at the base of the probe (a) if the point of the cone is facing the flow and (b) if the point is facing away from the flow.

SOLUTION

Given

Cone mounted atop cylindrical rod and immersed in water

Cone base 2 cm in diameter

Rod 0.5 m in length and 0.5 cm in diameter

Water velocity 1.5 m/s

Cone faces either into or away from the flow

Find

Bending moment for both cone orientations

Solution

First, we draw a free-body diagram of the rod, as shown in Fig. E8.6b. The forces on the rod are the drag on the cone and the drag on the rod itself. The cone drag acts at the tip of the rod and the rod drag acts at the midpoint of the rod. We neglect interference effects. The bending moment is

$$\mathcal{M} = \mathcal{D}_c \ell + \mathcal{D}_r \frac{\ell}{2}.$$

The drag forces are

$$\mathcal{D}_c = \mathbf{C}_{\mathbf{D},c}\left(\frac{1}{2}\right)\rho V^2 \frac{\pi}{4} D^2 \quad \text{and} \quad \mathcal{D}_r = \mathbf{C}_{\mathbf{D},r}\left(\frac{1}{2}\right)\rho V^2 \ell d.$$

The drag coefficient is different for the forward- and backward-facing cones, but the rod drag coefficient is the same for each case. We find the rod drag coefficient from Fig. 8.10 [or, possibly, Eq. (8.12a)]. The Reynolds number for the rod is

$$\mathbf{R} = \frac{Vd}{\nu} = \frac{(1.5\text{ m/s})(0.005\text{ m})}{1.0 \times 10^{-6}\text{ m}^2/\text{s}} = 7500.$$

Using Eq. (8.12a), we get

$$\mathbf{C}_{\mathbf{D},r} = 1.0 + 10.0(7500)^{-0.67} = 1.03.$$

The moment from rod drag is

$$\mathbf{C}_{\mathbf{D},r}\left(\frac{1}{2}\right)\rho V^2 \ell d\left(\frac{\ell}{2}\right) = (1.03)\left(\frac{1}{2}\right)(1000 \text{ kg/m}^3)(1.5 \text{ m/s})^2(0.5 \text{ m}) \times (.005 \text{ m})\left(\frac{0.5 \text{ m}}{2}\right) = 0.724 \text{ N}\cdot\text{m}.$$

We now consider the cone. For the point facing into the flow, Table 8.6 gives $\mathbf{C}_{\mathbf{D},c} = 0.5$ (note that $\mathbf{R} = (1.5)(0.02)/(1 \times 10^{-6}) = 3 \times 10^4$). Then

$$\mathcal{M} = 0.724 \text{ N}\cdot\text{m} + (0.5)\left(\frac{1}{2}\right)(1000 \text{ kg/m}^3)(1.5 \text{ m/s})^2\left(\frac{\pi}{4}\right) \times (0.02 \text{ m})^2(0.5 \text{ m});$$

$$\mathcal{M} = 0.812 \text{ N}\cdot\text{m}.$$ **ANSWER**

Finally, we consider the cone with the base facing the flow. A check of Tables 8.4–8.6 shows that we do not have any specific data for this situation. Our understanding of the nature of high Reynolds number blunt body flows suggests that we may probably neglect the effects of the rearward taper, because the separation point is at the sharp edge of the base. Therefore we use the drag coefficient of a circular disc to model the "backwards" cone. From Table 8.6, we find $\mathbf{C}_{\mathbf{D},\text{disc}} = 1.17$. Then

$$\mathcal{M} = 0.724 \text{ N}\cdot\text{m} + (1.17)\left(\frac{1}{2}\right)(1000 \text{ kg/m}^3)(1.5 \text{ m/s})^2\left(\frac{\pi}{4}\right) \times (0.02 \text{ m})^2(0.5 \text{ m});$$

$$\mathcal{M} = 0.931 \text{ N}\cdot\text{m}.$$ **ANSWER**

Discussion

We made two assumptions in solving this problem: We neglected interference effects and modeled the base-forward cone as a disc. Both assumptions seem reasonable and, in this case, they are fairly accurate. Note that we assumed the point of application of the drag force on an object to be at the center of the object. This condition is true in the absence of interference effects.

8.3 LIFT

Drag is present in all external flows. Lift is present only if there is asymmetry. Usually (but not always), drag is undesirable, leading to increased fuel consumption in vehicles, wind loading on structures, and the like. Lift is usually beneficial; for example, aerodynamic lift on the wings of an airplane allows it to fly. Lift forces also do most of the useful work in flow over propeller, compressor, and turbine blades.

8.3.1 Mechanism of Lift Generation: The Simple Airfoil

To explain how lift is generated, we consider a shape designed to produce high lift with low drag—a two-dimensional airfoil. Figure 8.22 shows a typical airfoil shape, with a rounded nose and a sharp trailing edge. Airfoils are typically thin, with a thickness-to-length ratio of less than 0.2. A straight line that connects the nose to the trailing edge is called the *chordline*. The length of this line, called the *chord,* is the basic measure of airfoil size. The chordline serves as the reference for other airfoil dimensions. A line midway between the upper and lower surfaces is called the *midline* or the *camberline*. For a symmetric airfoil, the camberline and chordline are identical. An airfoil with a curved camberline is asymmetric and is said to be *cambered*. The airfoil camber (h) at any point is the distance between the chordline and the camberline. Airfoil camber and thickness (t) are measured perpendicular to the chordline.

When an airfoil is immersed in a flow, the angle between the approaching flow velocity vector and the chordline is called the *angle of attack*. Figure 8.23 shows both symmetric and cambered airfoils in a flow with angle of attack.

Airfoil lift is primarily the result of surface pressure, so fluid viscosity would seem to have little effect on lift. This conclusion is almost true; measured lift coefficients show very little dependence on Reynolds number, and calculations of lift based on an assumed inviscid fluid are reasonably accurate, provided that we take into account one extremely important "viscosity" effect, without which there would be no lift. If we assume an absolutely inviscid fluid and calculate the streamline pattern for flow over an airfoil at an angle of attack (we discuss how this can be done in Chapter 9), the results are as shown in Fig. 8.24(a). The flow over the front half of the airfoil is reasonable, but the flow over the back half is not. Note that the flow on the lower side is expected to turn the sharp corner at the trailing edge and flow "backward" over the top surface to the rear stagnation point. If we calculate the pressure distribution on the airfoil surface (using Bernoulli's equation) and then calculate the lift using Eq. (8.3) (with $\tau_w = 0$), we would predict zero lift.

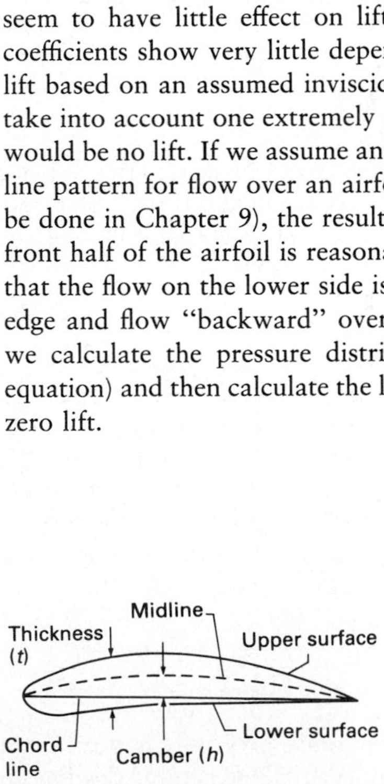

Figure 8.22 Typical airfoil shape.

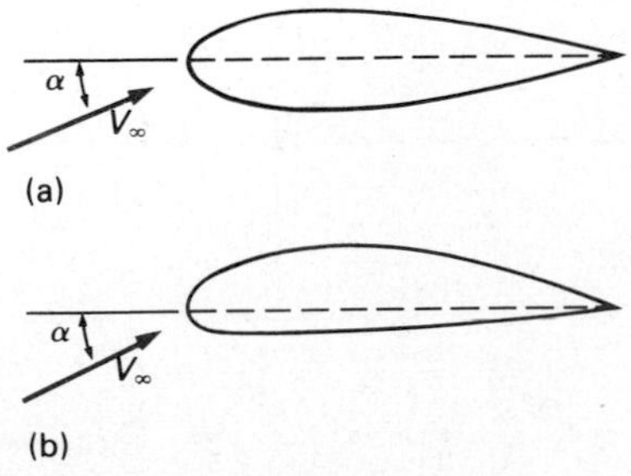

Figure 8.23 Two-dimensional airfoil profiles: (a) symmetric; (b) cambered.

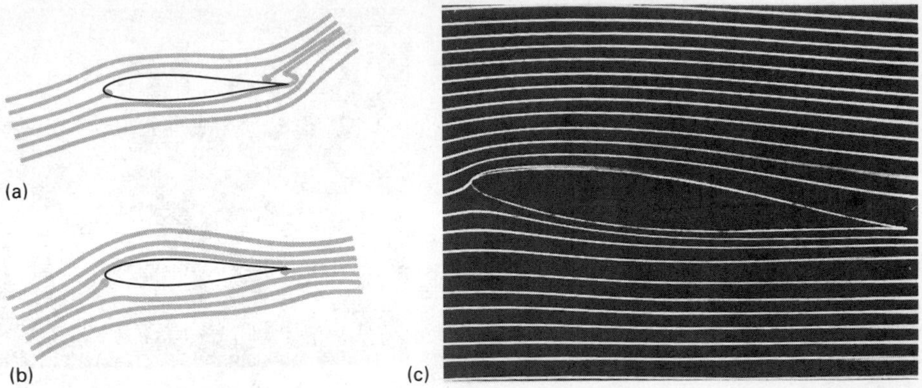

Figure 8.24 Streamline pattern for airfoil at angle of attack: (a) theoretical pattern for inviscid flow; (b) more realistic trailing edge flow; (c) streamlines for real airfoil flow. (Part c from "NCFMF Book of Film Notes," 1974, The MIT Press with Education Development Center, Inc., Newton, MA.)

The obvious problem is the trailing edge flow. A real fluid could not turn the sharp corner but would separate at the trailing edge, and the upper and lower streams would rejoin smoothly. The real streamline pattern would be as sketched in Fig. 8.24(b). Figure 8.24(c) shows a photograph of streamlines (actually streaklines marked by smoke) of flow over an airfoil at a small angle of attack. Note the streamline pattern near the sharp trailing edge in this photograph.

We can generate a realistic flow pattern if we force the rear separation point to lie at the trailing edge of the airfoil. This is called the *Kutta condition*. As we discuss in the next section, we can do so while still modeling the fluid as inviscid. With the separation point at the trailing edge, the fluid passing over the "top" of the airfoil must travel farther than the fluid passing over the "bottom." In a steady flow, this condition means that the "topside" fluid must have, on average, a higher velocity than the "bottomside" fluid. Bernoulli's equation implies, again on average, that the "topside" pressure is lower than the "bottomside" pressure, generating a net upward (lift) force. (See Fig. 8.5 for detailed pressure distribution.)

The effect just described occurs only if the flow is asymmetric. A symmetric airfoil at zero angle of attack does not generate lift. Airfoil camber and/or angle of attack are required for lift generation. A cambered airfoil produces lift at zero angle of attack. Increasing the angle of attack increases the flow asymmetry and increases the lift. For thin airfoils, inviscid flow theory predicts that

$$C_L \approx 2\pi\left(\alpha + \frac{2h_{\text{max}}}{c}\right) \qquad \left[\left(\alpha + \frac{2h_{\text{max}}}{c}\right) < 0.3\right], \tag{8.18}$$

where α is in radians and h_{max} is the maximum camber. In practice, the increase of C_L with α predicted by Eq. (8.18) persists to an angle of about 15°, when the airfoil stalls and the lift drops rapidly. *Stall* occurs when the upper separation point moves from the trailing edge toward the nose. Figure 8.25 shows an airfoil in stall at a high angle of attack.

8.3.2 Vortexes, Circulation, and Flow over Lifting Bodies

A *vortex* is a line or tube in a fluid that induces a circular motion in the surrounding fluid. You might have seen the vortex that forms as a bathtub drains or a tornado spins, which are examples of naturally occurring vortex flows. For purposes of analysis, we introduce two idealized vortexes.

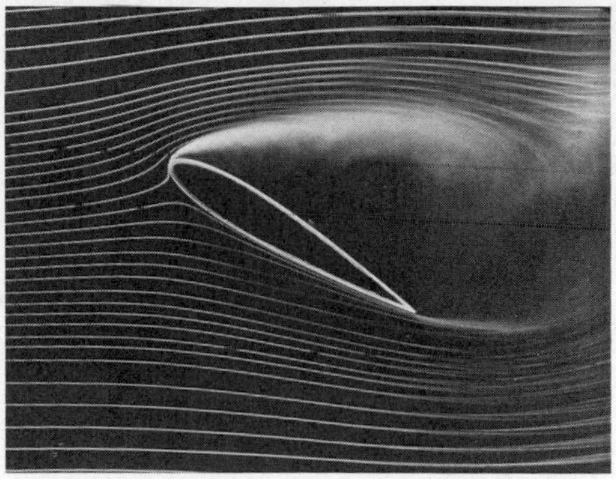

Figure 8.25 Airfoil in stall at high angle of attack; upper surface boundary layer separates near leading edge. (From "NCFMF Book of Film Notes," 1974, The MIT Press with Education Development Center, Inc., Newton, MA.)

A *forced vortex* is a line that induces a flow in the surrounding fluid, which moves in concentric circles about the vortex core and in which the fluid velocity increases linearly with radius. Figure 8.26(a) illustrates forced vortex flow. The vortex core is perpendicular to the paper, and the fluid velocity is

$$V_\theta = \omega r, \tag{8.19}$$

Forced vortex

where ω is a constant.

A *free vortex* is a line that induces a flow in the surrounding fluid, which moves in concentric circles about the vortex core and in which the fluid velocity decreases inversely with radius. Figure 8.26(b) illustrates free vortex flow. The vortex core is perpendicular to the paper, and the fluid velocity is

$$V_\theta = \frac{\Gamma}{2\pi r}, \tag{8.20}$$

Free vortex

where Γ is a constant.

Both the forced and free vortex are simplified models. Actual vortexes have a slender, forced-vortexlike core, but the flow field farther away from the core resembles a free vortex.

An important quantity associated with many flow fields, but especially with vortex flows, is *circulation*. For any closed curve superimposed on a flow field, as

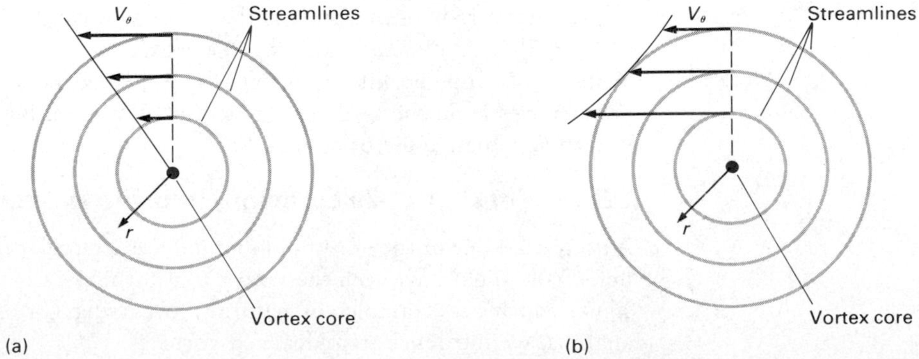

Figure 8.26 (a) Forced vortex flow; (b) free vortex flow.

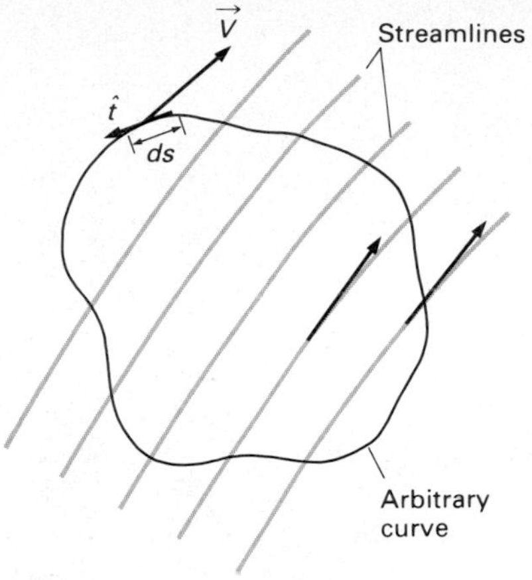

Figure 8.27 Arbitrary curve superimposed on a flow field for definition of circulation.

shown in Fig. 8.27, the circulation, Γ, for the curve is defined as

$$\Gamma \equiv \oint \vec{V} \cdot d\vec{s} = \oint \vec{V} \cdot \hat{t}\, ds, \tag{8.21}$$

where ds is an increment of path length and $\hat{t}$ is a unit vector tangent to the path. Circulation is usually a function of the path chosen, as well as the velocity field.

By choosing a circular path with the vortex core at its center, we may determine circulation for both vortex flows. For forced vortex flow, the circulation is

$$\Gamma = \int_0^{2\pi} \omega r(r\, d\theta) = 2\pi\omega r^2.$$

Forced vortex

Note that the circulation depends on the path chosen, through r. For free vortex flow, the circulation is

$$\Gamma = \int_0^{2\pi} \frac{\Gamma}{2\pi r}(r\, d\theta) = 2\pi \frac{\Gamma}{2\pi} = \Gamma.$$

Free vortex

The circulation is a constant for any path enclosing a free vortex [hence the choice of the symbol Γ in Eq. (8.20)].

We use a free vortex flow field to adjust the unrealistic streamline pattern predicted for inviscid flow over an airfoil into a realistic streamline pattern. Figure 8.28 illustrates this idea. Figure 8.28(a) shows a streamline pattern predicted by inviscid flow theory for a symmetrical airfoil at angle of attack. Figure 8.28(b)

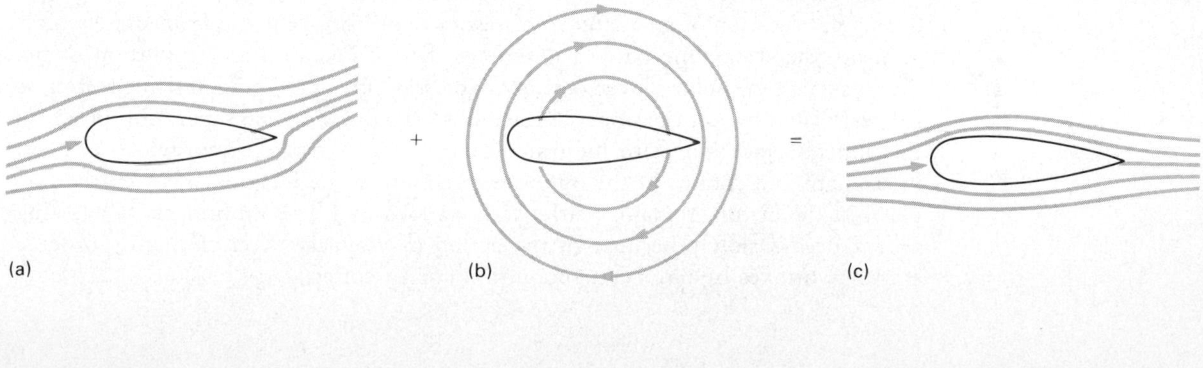

Figure 8.28 Realistic streamline pattern obtained by adding circulatory flow to theoretical inviscid flow over airfoil.

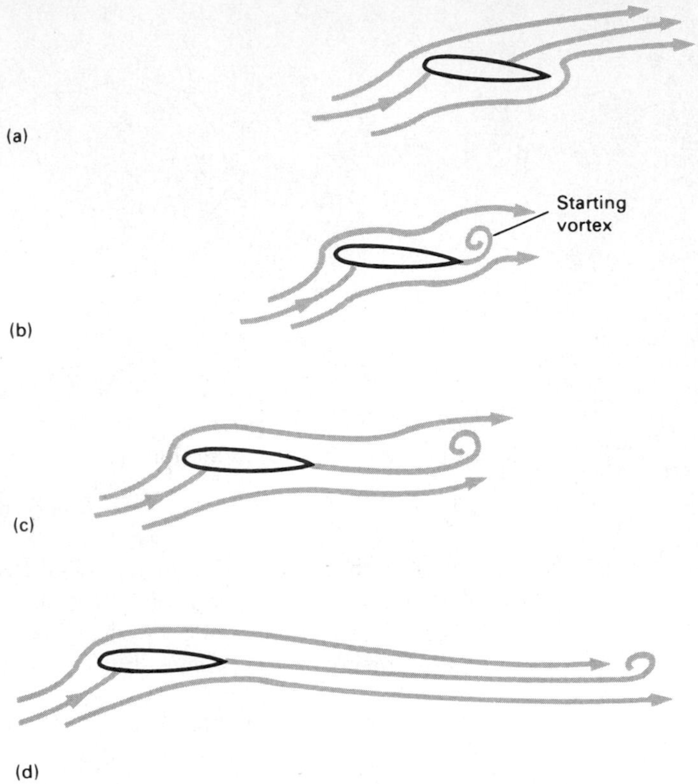

Figure 8.29 Sketches of the establishment of flow during acceleration of an airfoil from rest: (a) instant of start-up; (b) lower surface boundary layer separates; (c) nearly steady flow; (d) steady flow pattern (starting vortex decays with time).

shows a free vortex with a clockwise (negative) circulation. We use a free vortex rather than a forced vortex, because the free vortex velocity vanishes far from the airfoil and does not disturb the approach flow. We "bury" the vortex core inside the airfoil. Figure 8.28(c) shows that adding the free vortex flow in just the right amount brings the rear separation point to the trailing edge and produces a realistic streamline pattern.

Although this model for flow over a lifting body may seem unrealistic, it has a firm physical basis. The vortex core replaces the vorticity in the thin boundary layers on the airfoil surfaces. When the flow over an airfoil starts from rest, an actual vortex is formed in the fluid. Figure 8.29 shows the processes that occur during acceleration of an airfoil from rest to a uniform velocity. Immediately after start-up, the streamline pattern resembles that of Fig. 8.24(a). Separation of the lower surface boundary layer occurs at the trailing edge, and a starting vortex is formed. The flow on the upper surface fills the void, and the separation point on the upper surface moves to the trailing edge. The starting vortex, which is equal in strength and rotates in the opposite direction to the imaginary vortex that we place in the airfoil, remains at the starting location. The motion of the starting vortex decays slowly because of the action of viscosity. You can easily observe starting vortexes by moving a spoon in a cup of coffee.

8.3.3 The Kutta–Joukowsky Theorem

A theorem derived independently by W. M. Kutta in 1902 and N. E. Joukowsky in 1906 shows the strong connection between a circulating flow pattern and lift. Figure 8.30 shows an airfoil in an infinite stream. The fluid velocity and pressure far from the airfoil are V_∞ and p_∞. We assume that the fluid is inviscid and the flow is irrotational (actually a very good assumption outside the thin boundary layer near the airfoil surface and the downstream wake). Figure 8.30 also shows a coordinate system and control volume for analysis of the airfoil forces. Note that the airfoil is excluded from the control volume. The y-direction force that the airfoil exerts on the fluid at the inner portion of the control surface is the reaction to the lift force; that is,

$$\mathscr{L} = -F_y. \tag{8.22}$$

No fluid flows across the inner boundary of the control volume, because it is adjacent to the airfoil.

We write the fluid velocity components at the outer boundary of the control volume as

$$u = V_\infty + u' \tag{8.23a}$$

and

$$v = 0 + v', \tag{8.23b}$$

where u' and v' are *perturbations* from the values of u and v far ahead of the airfoil. By selecting the outer boundary of the control surface far from the airfoil, we may make u' and v' very small compared to V_∞; that is, far from the airfoil,

$$u'/V_\infty \ll 1 \tag{8.24a}$$

and

$$v'/V_\infty \ll 1. \tag{8.24b}$$

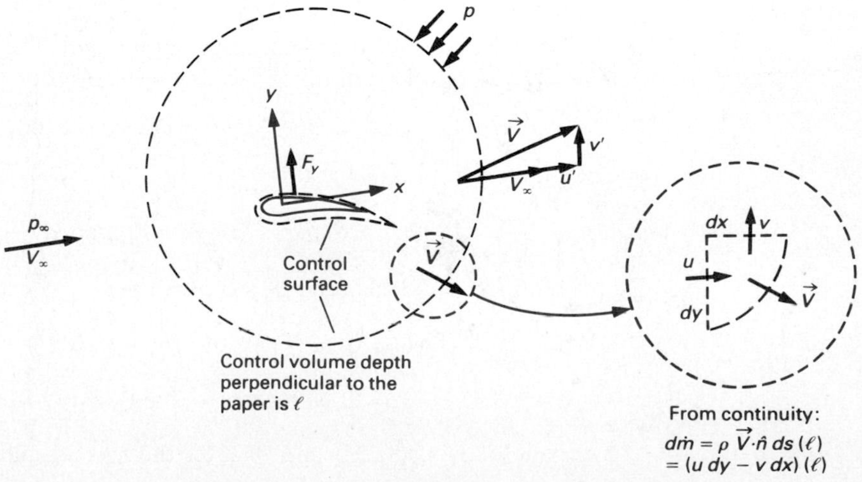

Figure 8.30 Airfoil and control volume for derivation of Kutta–Joukowsky theorem.

We now apply the y-direction linear momentum equation to the control volume. The flow is two-dimensional and two-directional. The control volume is ℓ units deep perpendicular to the page. The forces are F_y at the inner boundary and the pressure force at the outer boundary. Momentum flows across the outer boundary only. The momentum equation is

$$F_y + \oint_{\substack{\text{outer}\\\text{boundary}}} p\,dx(\ell) = \oint_{\substack{\text{outer}\\\text{boundary}}} \rho v(\vec{V}\cdot\hat{n})\,ds(\ell).$$

We evaluate the integrals by moving counterclockwise around the boundary. In the pressure integral, $dx(\ell)$ is the projected area that results in a y-direction pressure force. Solving for F_y, we have

$$F_y = -\oint p\ell\,dx + \oint \rho v(\vec{V}\cdot\hat{n})\ell\,ds. \tag{8.25}$$

The integrals are to be evaluated at the outer boundary only.

We may evaluate the pressure integral because the flow is assumed ideal and irrotational. Bernoulli's equation gives

$$p + \tfrac{1}{2}\rho V^2 + \gamma z = p_\infty + \tfrac{1}{2}\rho V_\infty^2 + \gamma z_\infty,$$

and so

$$p = p_\infty + \gamma(z_\infty - z) + \tfrac{1}{2}\rho(V_\infty^2 - V^2). \tag{8.26}$$

Now

$$\begin{aligned} V^2 &= u^2 + v^2 = (V_\infty + u')^2 + v'^2 = V_\infty^2 + 2u'V_\infty + u'^2 + v'^2 \\ &= V_\infty^2\left[1 + 2\left(\frac{u'}{V_\infty}\right) + \left(\frac{u'}{V_\infty}\right)^2 + \left(\frac{v'}{V_\infty}\right)^2\right]. \end{aligned} \tag{8.27}$$

By virtue of Eqs. (8.24a) and (8.24b), the last two terms are very small compared to the first two, so

$$\oint p\,dx \approx 0 + 0 - \rho \oint u'V_\infty\,dx(\ell). \tag{8.28}$$

Next, we consider the momentum flux integral. The term $\rho(\vec{V}\cdot\hat{n})\,ds$ represents mass flow across the control surface, so we can write it as (see Fig. 8.30):

$$\rho(\vec{V}\cdot\hat{n})\,ds = \rho(u\,dy - v\,dx). \tag{8.29}$$

The momentum flux integral is

$$\dot{M}_y = \oint \rho v(\vec{V}\cdot\hat{n})\,\ell\,ds = \oint \rho v(u\,dy - v\,dx)(\ell)$$

Substituting Eqs. (8.23a) and (8.23b), we have

$$\begin{aligned} \dot{M}_y &= \oint \rho v'[(V_\infty + u')\,dy - v'\,dx](\ell) \\ &= \oint \rho v'V_\infty\ell\,dy + \oint \rho v'u'\ell\,dy - \oint \rho v'^2\ell\,dx. \end{aligned}$$

By virtue of Eqs. (8.23a) and (8.23b), the second and third integrals are much smaller than the first, so

$$\dot{M}_y \approx \oint \rho v' V_\infty \, dy(\ell). \tag{8.30}$$

We may make this approximation as exact as we want by selecting the outer control surface farther from the airfoil.

We now substitute the momentum flux and pressure integrals into Eq. (8.25) to get

$$F_y = \rho V_\infty \oint (u' \, dx + v' \, dy)(\ell). \tag{8.31}$$

As V_∞ is a constant,

$$\oint V_\infty \, dx = 0. \tag{8.32}$$

Adding Eqs. (8.31) and (8.32), we have

$$\begin{aligned} F_y &= \rho V_\infty \oint [(V_\infty + u') \, dx + v' \, dy](\ell) \\ &= \rho V_\infty \oint (u \, dx + v \, dy)(\ell). \end{aligned} \tag{8.33}$$

The path length vector is

$$d\vec{s} = dx \, \hat{i} + dy \hat{j},$$

so

$$\vec{V} \cdot d\vec{s} = (u\hat{i} + v\hat{j}) \cdot (dx\hat{i} + dy\hat{j}) = u \, dx + v \, dy.$$

Using Eq. (8.22), we obtain the following equation for the lift force:

$$\mathscr{L} = -\rho V_\infty \oint \vec{V} \cdot d\vec{s}(\ell). \tag{8.34}$$

The integral is the circulation Γ, so

$$\mathscr{L} = -\rho V_\infty \ell \Gamma. \tag{8.35}$$

Equation (8.35) is the Kutta–Joukowsky theorem.

The Kutta–Joukowsky theorem shows that adding a free vortex to the otherwise unrealistic airfoil flow pattern was not only a convenient way to approximate the real flow, but that it also was actually necessary to introduce circulation in order to produce any lift at all. The theorem also demonstrates that control volume analysis alone cannot solve an external flow problem completely. In order to calculate the lift, we must know the circulation, and in order to calculate the circulation, we must know something about the velocity field "near" the airfoil. Only a differential analysis can yield the necessary level of detail about the velocity field.

8.3.4 Lift on Spinning Cylinders and Spheres: The Magnus Effect

The Kutta–Joukowsky theorem shows that any body that causes circulation in an otherwise uniform stream experiences lift. An airfoil generates the required

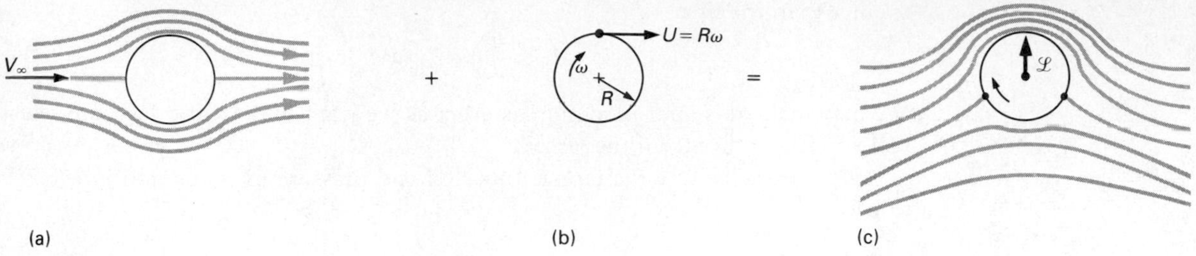

Figure 8.31 Streamline pattern produced by spinning cylinder in a uniform stream.

circulation by causing a larger velocity over its "top" side than over its "bottom" side. However, airfoils are not the only body shapes that produce lift. For a circular cylinder immersed in a uniform inviscid flow, the streamline pattern appears as shown in Fig. 8.31(a). The flow is symmetric, the velocity over the "top" is identical to that over the "bottom," and there is no lift.

Now suppose that we spin the cylinder clockwise with angular velocity ω, as shown in Fig. 8.31(b). We allow *one* viscosity effect; the fluid at the cylinder surface sticks to it and rotates with it at speed U ($U = R\omega$). The fluid at and near the top of the cylinder speeds up, because the velocity is additive, whereas the fluid at and near the bottom slows down. These effects diminish as distance from the surface increases; Fig. 8.31(c) shows the resulting flow. This flow has circulation and hence lift.

We model the flow over the spinning cylinder by superimposing a *clockwise* free vortex onto the flow for the stationary cylinder. The circulation for the required vortex immediately adjacent to the cylinder surface is given by

$$\Gamma = \oint \vec{V} \cdot d\hat{s} = -\int_0^{2\pi} UR\, d\theta = -2\pi UR.$$

The negative sign indicates that U is clockwise. Applying the Kutta–Joukowsky theorem, we have

$$\mathscr{L} = -\rho V_\infty \Gamma \ell = 2\pi\rho U V_\infty \ell R$$

and

$$\mathbf{C_L} = 2\pi\left(\frac{U}{V_\infty}\right), \tag{8.36}$$

which is the theoretical value for a spinning cylinder. The ratio of surface speed to approach flow velocity, U/V_∞, is analogous to the angle of attack for airfoils. The flow model that leads to Eq. (8.36) is incapable of predicting drag.

This is fine theory, but does it work? Figure 8.32 shows measured lift and drag coefficients for a spinning cylinder. Although the high-lift, zero-drag prediction is quantitatively incorrect, considerable lift is generated on the spinning cylinder. Roughening the surface of the cylinder produces *more* lift, because the surface more effectively causes circulation.

The phenomenon of lift generation by spinning bodies is called the *Magnus effect*. Spinning spheres are also capable of generating significant lift, as shown in Fig. 8.33. In engineering, the Magnus effect is mainly a curiosity, although a few devices that rely on the Magnus effect have been proposed or actually built. Magnus

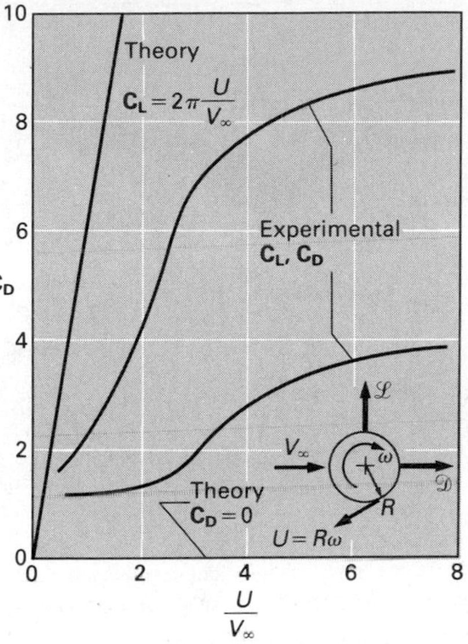

Figure 8.32 Lift and drag coefficients for spinning cylinders in uniform flow.

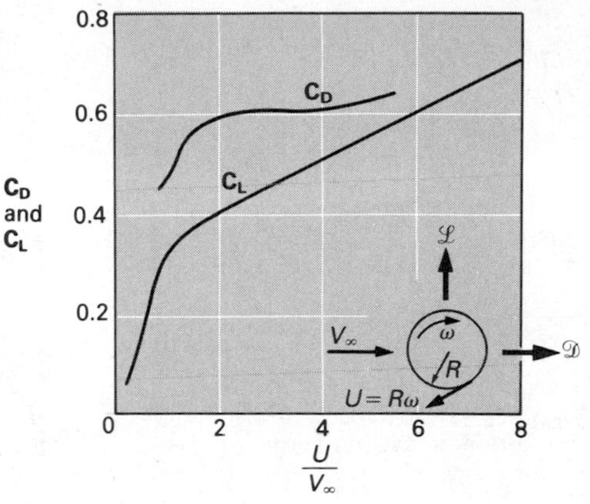

Figure 8.33 Lift and drag coefficients for spinning spheres in uniform flow.

effect devices are attractive in applications where the approach flow speed is low, because they have large lift and drag coefficients. Two of the more interesting applications are the rotor ship and the Madaras wind energy plant.

A German, A. Flettner, built a rotor ship that used rotating vertical cylinders to generate propulsive thrust from the wind. The ship had two rotors 50 ft (15 m) high and 9 ft (3 m) in diameter that rotated at 750 rpm. Flettner's rotor ship never became popular because the era of cheap energy and efficiency improvements made engine/propeller propulsion systems more attractive. Recently, the rotor ship has made a comeback. Several demonstration models have been built, the French have an oceanographic research rotor vessel, and the Japanese have built a large tanker rotor ship.

A large plant for generating electric power from the wind was proposed by J. P. Madaras in 1927. The plant was to feature eighteen 90-ft (27-m) high by 23-ft (7-m) diameter vertical rotating cylinders, each mounted on a railroad car. The cars were to move around a closed 1500-ft (460-m) diameter track, propelled by the lift force generated by the wind (Fig. 8.34). Electric generators attached to the car wheels were to generate power. Several electric utility companies funded development of the Madaras rotor plant concept, but it was never built. More recent studies [4] have indicated that in a high-energy-cost economy, commercial utility-sized Madaras plants, with a power-generating capacity of over 230 megawatts are economically and technologically feasible.

8.3.5 Lift (and Drag) Information for Standard Airfoils

Theories based on the inviscid flow assumption and the Kutta–Joukowsky theorem are capable of reasonably accurate predictions of lift for thin airfoils at low angles of attack but are incapable of predicting airfoil drag or stall. In practice, lift and drag on airfoils are evaluated from experimental data. Several agencies, most notably the U.S. National Advisory Committee for Aeronautics (NACA) (the

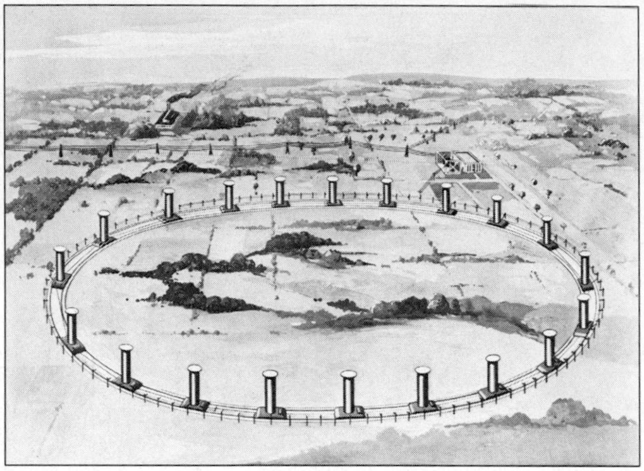

Figure 8.34 Artist's conception of Madaras rotor power plant.

forerunner of the National Aeronautics and Space Administration, NASA), have produced a large number of standardized shapes of airfoils and have published information on their lift and drag coefficients established from wind tunnel tests. The amount of information is so extensive (see reference [5]) that we present only typical examples here.

A (NACA) family of airfoil shapes is determined by specifying a thickness distribution and a midline shape. A well-known family of airfoil shapes is the NACA four-digit series. The midline shape and thickness distribution are polynomials with the maximum camber, maximum camber location, and maximum thickness specified as parameters. The profile number conveys information about these parameters. For example, a 2415 profile has the following characteristics:

2	4	15
Maximum camber is 2% of chord.	Maximum camber occurs at 40% of chord.	Maximum thickness is 15% of chord.

In addition to the four-digit series, NACA developed other families of airfoil shapes. A family of particular interest is the 6-series, which was designed to produce laminar boundary layers over a large portion of the surface, resulting in low friction drag. This series also shows good performance at high subsonic Mach numbers. The designations of 6-series airfoil shapes are rather complicated, a typical designation being 63_2-615. We do not interpret these numbers here; see reference [5] if you are interested.

Performance information for standard airfoils is given as plots of lift and drag coefficients versus angle of attack. Figure 8.35 illustrates such plots for NACA 2415 and 63_2-615 airfoils, as well as the shapes of the profiles themselves. Lift and drag coefficients are based on planform area (chord times length), not frontal area. Note that these nonsymmetric airfoils produce lift at zero angle of attack. Note also that the slope of the lift curves ($dC_L/d\alpha$) is nearly 2π, except near and after stall.

Airfoil lift and drag coefficients are nearly independent of Reynolds number so long as the Reynolds number (based on airfoil chord) is high. The drag coefficient is sensitive to surface roughness and can increase up to 300 percent of its

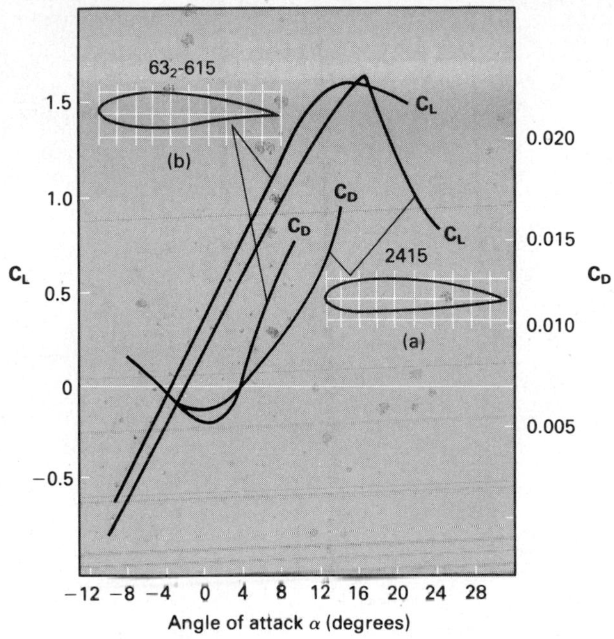

Figure 8.35 Lift and drag coefficients for two NACA airfoil shapes: (a) 2415 profile; (b) 63_2-615 profile. (Adapted from [5].)

smooth surface value (the value indicated in Fig. 8.35) if the surface is rough. This behavior is similar to that of the flat plate parallel to the flow stream, which should come as no surprise since an airfoil is a thin, elongated body.

EXAMPLE 8.7 Illustrates Airfoil Performance

A NACA 2415 profile is placed in a flow at a 4° angle of attack. Compare the lift coefficient to the value estimated from thin airfoil theory and the drag coefficient to that of a smooth flat plate. Assume that the Reynolds number is 1×10^6.

SOLUTION

Given

NACA 2415 airfoil, angle of attack 4°, and $\mathbf{R} = 1 \times 10^6$
Figure E8.7

Find

Lift and drag coefficients

Compare with simplified theoretical models

Solution

From Figure E8.7, the values are

$$C_L = 0.60 \quad \text{and} \quad C_D = 0.0087. \qquad \textbf{ANSWER}$$

We estimate the lift coefficient from Eq. (8.18):

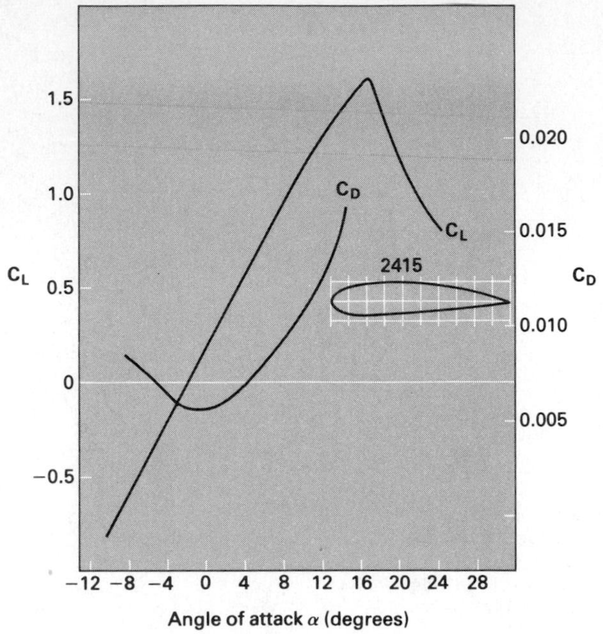

Figure E8.7 Lift and drag coefficients.

$$C_L = 2\pi\left(\alpha + \frac{2h_{max}}{c}\right).$$

The maximum camber, as a fraction of chord, for a 2415 profile is 0.02 (2%). The lift coefficient is

$$C_L = 2\pi\left[4^\circ \frac{\pi \text{ rad}}{180^\circ} + 2(0.02)\right] = 0.69. \qquad \textbf{ANSWER}$$

From Fig. 8.8—and assuming turbulent flow in the airfoil boundary layer and a smooth surface—we have

$$C_D = 0.0044. \qquad \textbf{ANSWER}$$

Comparing theoretical estimates with actual values,

the lift coefficient estimate is about 15 percent too high, **ANSWER**

and

the drag coefficient estimate is about 50 percent too low. **ANSWER**

Discussion

The theoretical lift estimate is reasonable, given the approximate nature of the theory and the fact that this is a fairly thick airfoil. It would be less accurate for larger angles of attack.

The drag estimate is not accurate. This fairly thick airfoil at angle of attack would have some form drag. Also, the wetted surface would be larger than the planform area on which the drag is based.

EXAMPLE 8.8 **Illustrates Airfoil Lift**

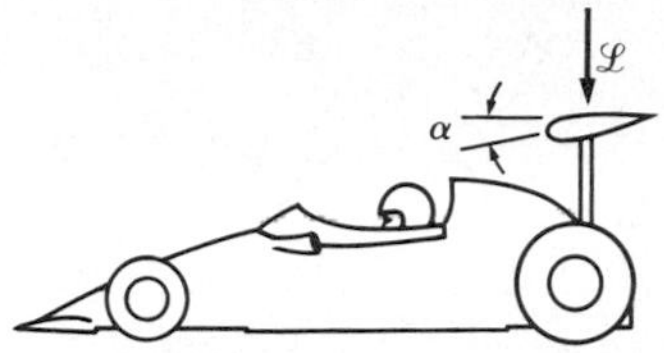

Figure E8.8 Inverted airfoil on race car to provide downthrust.

A race car uses an inverted NACA 63_2-615 airfoil to provide downthrust on its rear wheels (Fig. E8.8). The airfoil has two end plates, so the flow is assumed to be two-dimensional. The airfoil is 4 ft long and has a chord of 1 ft. Find the angle of attack to produce a downward thrust of 300 lb at a car speed of 200 mph.

SOLUTION

Given

Inverted NACA 63_2-615 airfoil 4 ft long with 1-ft chord

Two-dimensional flow

Find

Angle of attack to produce 300-lb downward thrust at 200 mph speed

Solution

Equation (8.4) relates the downward thrust (or lift) $\mathscr{L}$ and the lift coefficient C_L; that is,

$$\mathscr{L} = \tfrac{1}{2} C_L \rho V_\infty^2 S,$$

when S is the planform area, or

$$S = (1 \text{ ft} \times 4 \text{ ft}) = 4 \text{ ft}^2.$$

Solving for the lift coefficient gives

$$C_L = \frac{2\mathscr{L}}{\rho V_\infty^2 S}.$$

Assuming 60°F air, Table A.4 gives $\rho = 0.0024$ slug/ft^3. The numerical values give

$$C_L = \frac{2(300 \text{ lb})(1.0 \text{ ft}\cdot\text{slug/lb}\cdot\text{sec}^2)}{\left(\dfrac{0.0024 \text{ slug}}{\text{ft}^3}\right)\left(\dfrac{200 \text{ mi}}{\text{hr}}\right)^2\left(\dfrac{5280 \text{ ft}}{\text{mi}}\cdot\dfrac{1 \text{ hr}}{3600 \text{ sec}}\right)^2(4 \text{ ft}^2)} = 0.73.$$

Figure 8.35 shows that the angle of attack must be

$$\alpha = 3°.$$

ANSWER

8.3.6 Lift Augmentation Devices and Three-Dimensional Effects

Let's consider what happens when an airplane takes off and lands. As in level flight, aircraft weight must be counterbalanced by wing lift. The lift *force* varies with the square of the velocity, which typically is low during takeoff and landing. To provide the required lift at low speeds, the lift coefficient must increase. One way to increase the lift coefficient is to raise the angle of attack. During landing, the pilot does so by raising the plane's nose and lowering its tail. This works only up

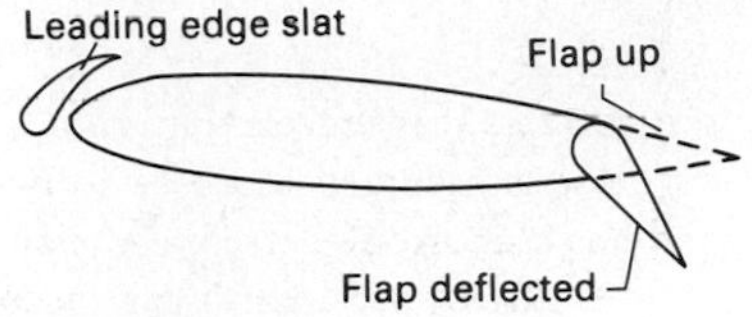

Figure 8.36 Airfoil with leading edge slat and trailing edge flap.

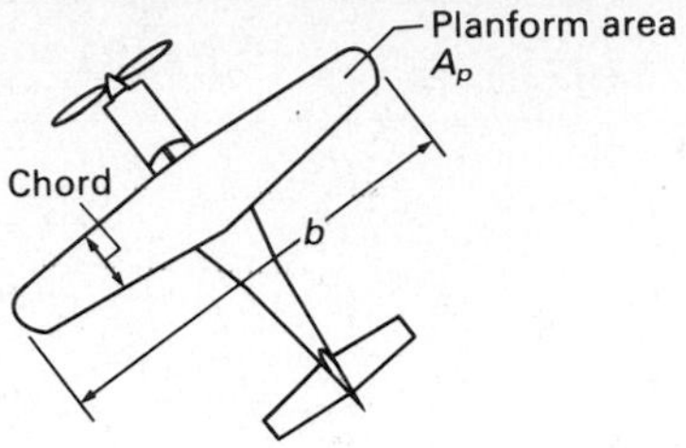

Figure 8.37 Wing of finite span, viewed from above.

to a point; if the angle of attack gets too high, the wing will stall with sudden loss of lift—usually resulting in a crash. To keep takeoff and landing speeds reasonable and avoid catastrophic stall, airplane wings are fitted with lift augmentation devices to increase the lift coefficient at low speeds. The most common lift augmentation devices are trailing edge flaps and leading edge slats (Fig. 8.36). Flaps increase the effective angle of attack, increasing the lift coefficient. Leading edge slats produce a fast-moving jet of air over the upper surface, giving a low pressure and at the same time energizing the boundary layer and delaying stall. These lift augmentation devices also produce high drag and so are used only during takeoff and landing.

Our previous discussion of lift assumed two-dimensional, two-directional flow over a two-dimensional airfoil. Real wings have finite span and are three-dimensional bodies. Figure 8.37 shows a wing with span b. An important parameter of such a wing is the *aspect ratio, AR,* defined by

$$AR = \frac{b^2}{A_p},$$

where A_p is the planform area. For a nontapered wing (constant chord along the span) without sweepback, the aspect ratio reduces to the ratio of wing span to chord length.

Finite-span wings have a larger drag coefficient and a smaller lift coefficient than infinite-span wings (e.g., airfoils). The reason is that *trailing vortexes* are shed from finite wings. The primary component of the trailing vortex is a strong vortex generated at the wing tip as fluid "leaks" from the higher pressure lower surface around the wing tip. This tip vortex is joined by vorticity shed all along the wing span to form the trailing vortex, which extends far behind the aircraft (see Fig. 8.38).* The trailing vortex causes the fluid behind the wing to have an induced downward velocity, called the *downwash velocity,*† w. The downwash causes the effective angle of attack of the wing to decrease. Figure 8.39 shows that

* Light planes must avoid the trailing vortexes of large commercial airliners, which may extend 10 mi downstream from the aircraft. Theoretically, the vortexes extend all the way back to the takeoff point, where they join up with the starting vortex that was shed at takeoff (see Fig. 8.29). The imaginary bound vortex in the wing, the trailing vortexes, and the starting vortex then form one giant closed loop.

† Note that there is an induced *upwash* velocity outboard of the wingtips. When migrating birds fly in a V formation, the downstream birds can take advantage of the upwash of the upstream birds [6].

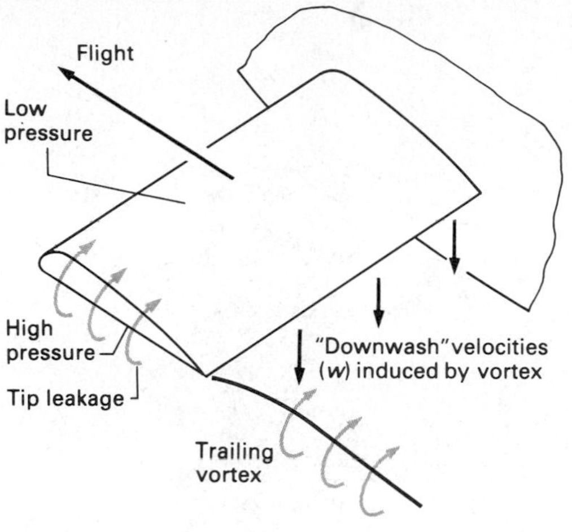

Figure 8.38 Detail of flow near the tip of a finite wing.

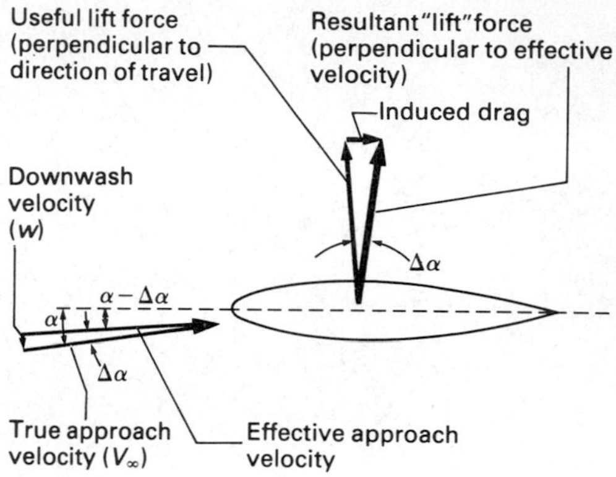

Figure 8.39 Downwash velocity reduces the effective angle of attack and rotates the lift force backward from true direction of travel.

this reduction has two effects. First, because lift is perpendicular to the effective approach velocity, the lift vector is rotated backward. The lift vector now has a component in the (true) downstream direction that adds to the drag of the wing. This is called *induced drag*. The second effect is lower lift (lower lift is also explained by the smaller average pressure on the lower wing surface, which is caused by tip leakage). A theory based on Prandtl's work shows that if the shape of the spanwise lift distribution curve is elliptical (a reasonably good approximation for nonswept wings), the downwash velocity is

$$w = \frac{2\mathscr{L}}{\pi \rho V_\infty b^2}, \tag{8.37}$$

the reduction in effective angle of attack is

$$\Delta\alpha \approx -\frac{w}{V_\infty} = -\frac{2\mathscr{L}}{\pi \rho V_\infty^2 b^2} = -\frac{\mathbf{C_L}}{\pi AR}, \tag{8.38}$$

and the induced drag is

$$\mathscr{D}_{\text{ind}} = \frac{2\mathscr{L}^2}{\pi \rho V_\infty^2 b^2}. \tag{8.39}$$

The induced drag coefficient is

$$\mathbf{C_{D,ind}} = \frac{\mathscr{D}_{\text{ind}}}{\frac{1}{2}\rho V_\infty^2 A_p} = \frac{\mathbf{C_L^2}}{\pi AR}. \tag{8.40}$$

Problems of induced drag and lower effective angle of attack are especially critical during takeoff and landing, because the lift coefficient is especially large. Increasing the aspect ratio of the wing reduces these effects. Figure 8.40 shows the effect of aspect ratio on the lift and drag of a finite wing. Wings that produce high lift and low drag must have a large aspect ratio. The wing of a modern sailplane

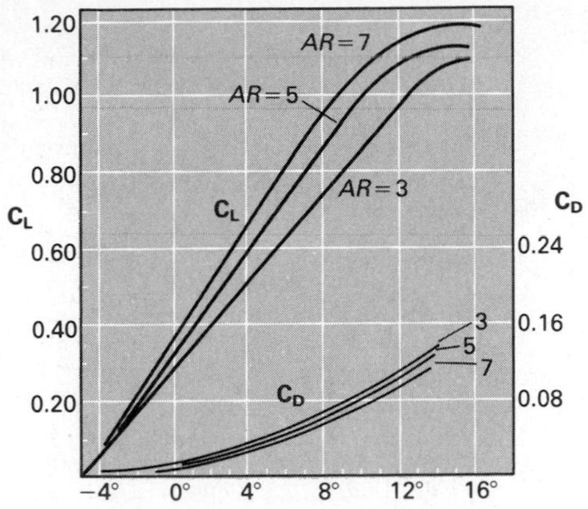

Figure 8.40 Effect of aspect ratio on lift and drag coefficient of a typical finite wing. All wings have the same (airfoil) profile shape.

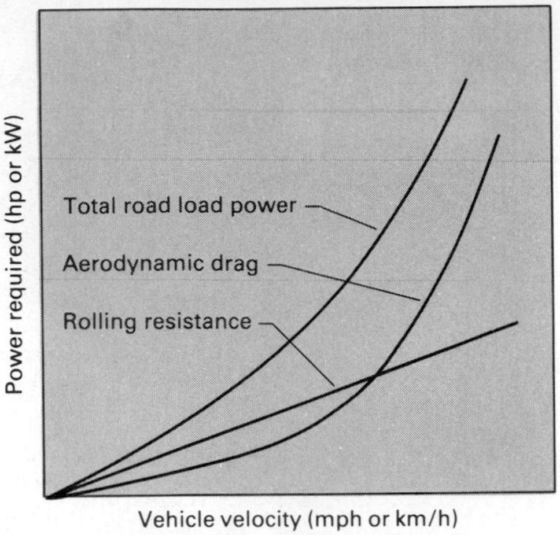

Figure 8.41 Variation of road load power and its components with vehicle velocity.

(glider) may have $AR \approx 30$. Birds that fly long distances, such as seagulls, have wings of large aspect ratio. Modern sport sailboats, which use "lift" sails rather than "drag" sails, normally have tall, narrow sails.

In some applications, it is practical to fit endplates on a lifting surface to suppress formation of the tip vortex. Note the endplates on the cylinders of the Madaras wind plant shown in Fig. 8.34.

8.4 NONAVIATION APPLICATIONS OF AERODYNAMICS

Practical applications of aerodynamics are not limited to aircraft. Any object moving through air or water—or wind or water flowing over a stationary object—generates aerodynamic forces. These aerodynamic forces are often of considerable importance in the operation, stability, and durability of engineering devices and systems. We next consider three nonaviation applications of aerodynamics.

8.4.1 Road Vehicle Aerodynamics

Aerodynamic considerations are important in the design of road vehicles such as trucks and automobiles. The most important aerodynamic force on a road vehicle is drag. When a vehicle travels at constant speed on a level road, the vehicle experiences two forces that resist its motion: rolling resistance and aerodynamic drag. *Rolling resistance* results from continuous deformation of the vehicle's tires. The sum of the aerodynamic drag and rolling resistance is called the *road load*. The power required to overcome the road load is equal to the product of the road load force and the vehicle velocity. Figure 8.41 shows the variation of road load power with vehicle velocity for a typical automobile. Rolling resistance power varies almost linearly with velocity, whereas aerodynamic drag power varies with the cube of the velocity (the drag coefficient is approximately constant). The rolling resistance and aerodynamic drag curves typically intersect (each contributes equally to road load) at a speed between about 50 mph (80 km/hr) and 60 mph (96 km/hr). Above this speed, the power required to overcome aerodynamic drag increases rapidly and becomes the controlling factor in the vehicle's speed. Fuel economy depends strongly on the road load power curve; reducing the magnitude of road load improves fuel economy. As most road vehicles normally operate at speeds ranging from 35 mph to 70 mph, significant reductions in road load power can be realized by reducing either rolling resistance or aerodynamic drag. The optimum design must address both road load components [7].

Road vehicles operate at high Reynolds numbers (a typical five-passenger sedan traveling at 55 mph has a Reynolds number of about 10^7, based on length) and are rather blunt bodies, so most of the drag on them is form drag. The primary method of reducing form drag is well known: streamlining. A second method, obvious but often overlooked by the beginner, is to reduce the size (frontal area) of the vehicle. Early (1920 vintage) automobiles had high profiles (large frontal area per unit car volume) and boxy shapes that produced drag coefficients of about 0.9–1.0. Modern automobiles have much more streamlined shapes with lower profiles. Current production cars have drag coefficients in the range of 0.3–0.6. The ultimate streamlined car would have a nearly circular cross section, a rounded blunt nose, and a long, pointed tail (like a teardrop shape) and might be expected to yield a value of $\mathbf{C_D}$ as low as 0.05–0.10; however, such a car would probably be unacceptable in terms of cost, load-carrying volume, and aesthetics.

Drag reduction for automobiles is complicated by problems introduced by underbody drag and interference drag. The underbody of a road vehicle is aerodynamically complicated by protruding surfaces, such as oil pan, muffler, and suspension parts. Each of these surfaces experiences a relatively large drag force. Interference drag is caused by protrusions from the vehicle surface, such as rearview mirrors, door handles, and luggage racks. These objects not only experience drag on themselves but, equally important, they disturb the flow over the basic body shape, influencing its drag as well. The total drag caused by a surface protrusion is usually larger than the sum of the drags of the basic body in isolation and the protrusion in isolation.

Aerodynamic drag on trucks and buses presents its own set of problems. Legal regulations set limits on the length, width, and height of these commercial vehicles. Truck and bus operators normally want to maximize the quantity of goods or number of people that can be carried per trip, so these vehicles are designed with a basically rectangular shape. The shape limits the amount of drag reduction that can be realized; however, attention to a few details, such as rounding corners on the front of a semitrailer or bus, can substantially reduce drag with little sacrifice in interior volume. In recent years, cab rooftop air deflectors have been installed on many semitractors to improve the flow pattern over the front of the trailer, as shown in Fig. 8.42.

Although drag reduction is probably the most important aerodynamic problem in road vehicle design, it is by no means the only area of concern. Other aerody-

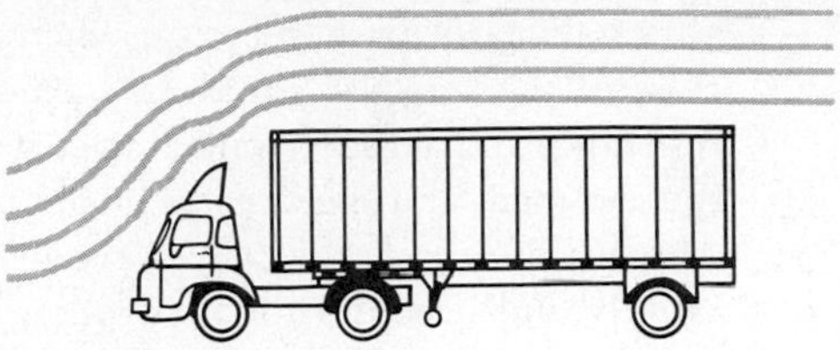

Figure 8.42 Semitraller with an air deflector to reduce drag.

namic considerations in road vehicles include

- secondary aerodynamic forces such as lift and side force;
- provision of air flow to the interior of the vehicle for engine cooling and passenger compartment ventilation;
- provision of an acceptable flow pattern near the car surface to minimize dirt deposits on the vehicle and transport exhaust gases away from the vehicle; and
- aerodynamically generated noise.

Figures 8.43 and 8.44 show the pressure distribution and streamline pattern for a typical automobile. Note that the lower surface pressure is considerably larger (smaller negative value) than the upper surface pressure, resulting in a lift force. Upward lift is an undesirable phenomenon, because it reduces traction and can adversely affect handling characteristics. Problems caused by lift forces are seldom significant for ordinary passenger and commercial vehicles but can be important for high-speed race cars. Many race cars are fitted with "negative lift" devices (basically upside-down wings) to improve traction (see Fig. E8.8).

Air flow into the vehicle usually occurs at three locations. Engine cooling and combustion air usually enter the front and underside of the vehicle. The pressure drop of the air as it flows through the radiator contributes to the vehicle's drag. Passenger ventilation air is normally taken in at the higher pressure zone at the base of the windshield.

Dirt and other particles deposited by air flow on vehicle windows can reduce visibility. This problem is especially severe in vans and station wagons, where the recirculating flow behind the vehicle picks up dirt or water from the road and

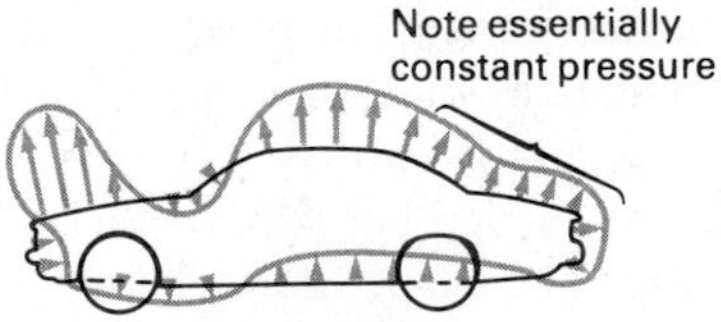

Figure 8.43 Typical pressure distribution on symmetry plane of a sedan: Pressure curve lying outside the contour indicates negative (suction) pressure; pressure curve lying inside the contour indicates positive pressure. Pressures are gage pressure.

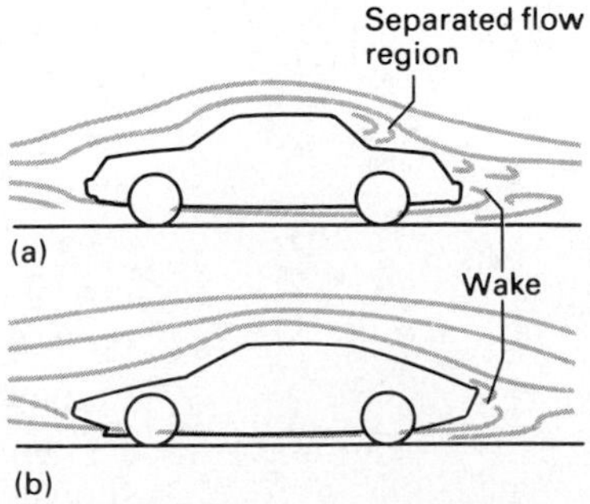

Figure 8.44 Streamline patterns for (a) sedan and (b) "fastback" automobiles.

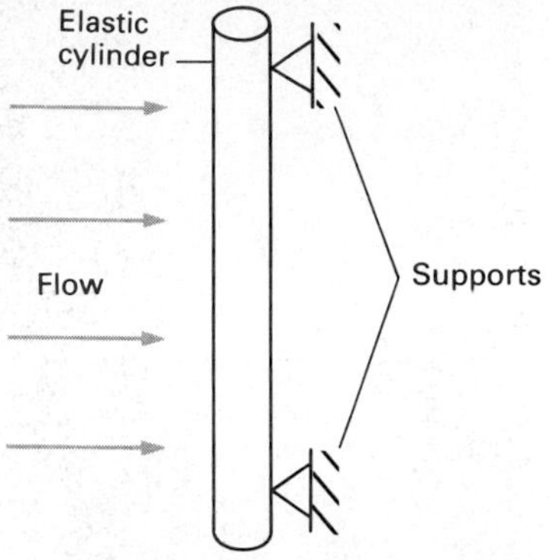

Figure 8.45 Flow perpendicular to axis of long, flexible elastic cylinder.

deposits it on the rear windows. These vehicles are sometimes equipped with deflecting vanes to jet air from the top downward over the back window to keep it clean.

Aerodynamic considerations are an important part of road vehicle design, although, unlike in the case of aircraft, aerodynamics is not the primary consideration. References [8, 9, 10] provide excellent insight into this field.

8.4.2 Flow-Induced Vibration

We now consider a situation in which the flow over an object and the object itself interact dynamically. Many practical applications involve flow perpendicular to the axis of a long flexible cylinder (Fig. 8.45). Important examples include wind blowing over power poles, power lines, suspension bridge cables, and smokestacks, water flowing over bridge pilings, and liquid or gas flowing over banks of tubes in a heat exchanger. Because of the cylinder's flexibility, the aerodynamic load causes the cylinder to deflect (bend), which is balanced by an elastic restoring force. If the aerodynamic load is removed, the cylinder springs back to its original shape. If the aerodynamic load fluctuates, the cylinder vibrates under the influence of the fluctuating load. The frequency and amplitude of the vibration depends on the frequency and magnitude of the fluctuating load and the elastic and support characteristics of the cylinder. All elastic structures possess one or more *natural frequencies,* which are the frequencies at which the structure would vibrate under the influence of its mass and elastic force conditions alone. If the structure is subjected to a fluctuating load whose frequency is equal to (or near) one of the natural frequencies, the amplitude of vibration becomes very large, often resulting in destruction of the structure.* Catastrophic failures have resulted from this type of phenomenon, the most dramatic being the failure of the Tacoma Narrows suspension bridge just a few weeks after its opening in 1940 [11]. Less drastic problems associated with flow-induced vibration include gradual fatigue of heat exchanger tubes and wear of tubes as they rub against their supports.

---

* The action of pushing a child on a swing illustrates this concept. If the swing is pushed each time it instantaneously comes to a stop at the high point of its travel, it will keep going higher and higher. This once-per-cycle push is exactly in time with the "natural frequency" of the swing.

Figure 8.46 Vortex street shed from a circular cylinder. Photo courtesy of Sadatoshi Taneda. Reprinted with permission of Milton Van Dyke, from *An Album of Fluid Motion*, by Milton Van Dyke, The Parabolic Press, 1982.

We can explain some of the causes of flow-induced vibration by considering the details of flow over an isolated circular cylinder. Experiments show that at Reynolds numbers greater than about 50, vortexes are shed on the downstream side of the cylinder. These vortexes are shed alternately from the top and the bottom of the cylinder with a definite frequency. The vortexes trail behind the cylinder in two rows (Fig. 8.46), called a *Karman vortex street.* The oscillating streamline pattern caused by the alternate vortex shedding causes a fluctuating pressure force on the cylinder and hence a time-variant load. The frequency of the fluctuating force is equal to the frequency of the vortex shedding. Dimensional analysis shows that the frequency of vortex shedding is governed by a relation of the form

$$\mathrm{S} = f(\mathrm{R}), \tag{8.41}$$

where

$$\mathrm{S} = \frac{\omega_v D}{V_\infty},$$

is the Strouhal number of the vortex pattern,

$$\mathrm{R} = \frac{V_\infty D}{\nu},$$

and

$$\omega_v = \text{Frequency of vortex shedding in Hz.}$$

The functional relation implied by Eq. (8.41) has been the subject of many experimental investigations. Figure 8.47 represents its typical behavior. A struc-

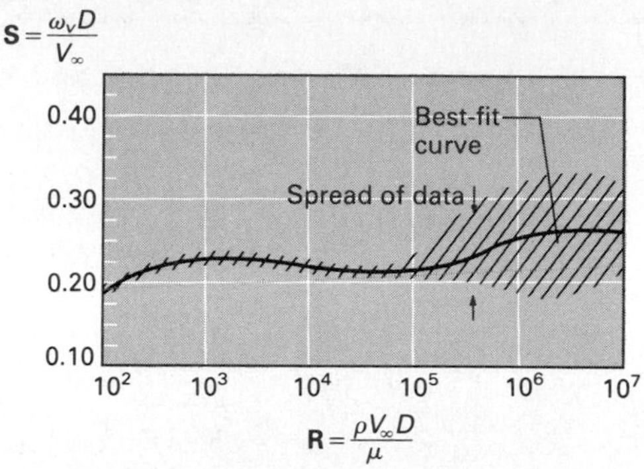

Figure 8.47 Strouhal number versus Reynolds number for vortex shedding from a single circular cylinder.

tural engineer must be sure that the natural frequency of the structure is not too close to the vortex-shedding frequency. This type of vortex shedding is responsible for wires "singing" in the wind, a phenomenon extensively investigated by Strouhal.

If more than one cylinder is present in the flow, other mechanisms may induce vibration. Obviously, if a second circular cylinder were placed in the wake of the cylinder in Fig. 8.46, the shed vortexes would generate a fluctuating flow over the second cylinder. In a bank of tubes, the Karman vortex pattern rapidly breaks up into turbulence. Large amounts of turbulence in the flow cause fluctuation and may lead to flow-induced vibration.

A third mechanism for generating flow-induced vibration is *hydroelastic instability*. Let's consider flow between two circular cylinders, as shown in Fig. 8.48. Suppose that some unknown perturbation deflects the two cylinders in such a way that the gap between them decreases. Narrowing the gap causes a higher fluid velocity in the gap by virtue of conservation of mass. According to Bernoulli's equation, the higher velocity in the gap is accompanied by a lower pressure. The lower pressure on the gap side causes the cylinders to deflect inward, narrowing the gap and further decreasing the pressure, which causes more deflection. Eventually, the elastic forces in the deflected cylinders act to restore the cylinders to their initial position; however, if the cylinders overshoot their equilibrium position, the process of gap narrowing will start on the other side, repeating the process. At some flow velocities, the frequency of vibration induced by the process may match the natural frequency of the cylinders, and the resulting vibration can become quite large.

A great deal of research is currently being conducted in the field of flow-induced vibrations. Consult reference [12] for further details.

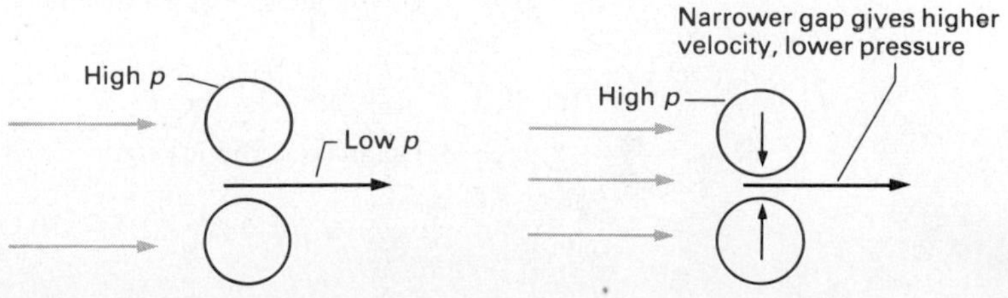

Figure 8.48 Flow in the gap between two circular cylinders. Narrowing the gap causes increased velocity and lower pressure and tends to narrow the gap further.

EXAMPLE 8.9 **Illustrates Vortex Shedding and Its Relation to a Structural Natural Frequency**

A 150-ft-high sign pole is a steel pipe with an inside diameter $d =$ 11.94 in. and an outside diameter $D = 12.75$ in. Its natural frequency is

$$\omega = \frac{\pi^2}{4}\sqrt{\frac{EI}{m'\ell^4}},$$

where E is the pipe's modulus of elasticity, m' its mass per unit length, I its cross-sectional moment of inertia, and ℓ its length; $E = 30 \times 10^6$ psi for steel and $I = \pi(D^4 - d^4)/64$. Find the critical wind velocity for vortex shedding if the wind velocity is uniform along the pole. Steel density is 490 lbm/ft^3.

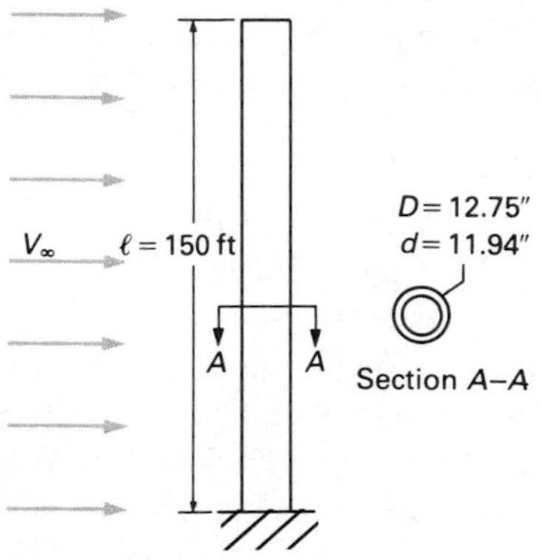

Figure E8.9 Given information for Example 8.9.

SOLUTION

Given

Pole in Fig. E8.9

Pole inside diameter $d = 11.94$ in. and outside diameter $D =$ 12.75 in.

Natural frequency

$$\omega = \frac{\pi^2}{4}\sqrt{\frac{EI}{m'\ell^4}}$$

with

$$I = \frac{\pi}{64}(D^4 - d^4)$$

Find

Wind velocity at which the pole vibrates at its natural frequency due to vortex shedding

Solution

Figure 8.47 relates the Strouhal number for vortex shedding to Reynolds number. Assuming 60°F air, Table A.4 gives $\nu = 1.7 \times 10^{-4}$ ft^2/sec, and the Reynolds number is

$$\mathbf{R} = \frac{V_\infty D}{\nu} = \frac{V_\infty(12.75 \text{ in.})(1 \text{ ft}/12 \text{ in.})}{(1.7 \times 10^{-4} \text{ ft}^2/\text{sec})} = (6250 \text{ sec/ft})V_\infty.$$

The moment of inertia of the pipe cross section is

$$I = \frac{\pi}{64}(D^4 - d^4) = \frac{\pi}{4}[(12.75 \text{ in.})^4 - (11.94 \text{ in.})^4] = 4793 \text{ in}^4.$$

The mass per unit length of the pipe is

$$m' = \rho A = \rho\frac{\pi}{4}(D^2 - d^2)$$

$$= \frac{(490 \text{ lbm/ft}^3)(\pi/4)[(12.75 \text{ in})^2 - (11.94 \text{ in})^2]}{(144 \text{ in}^2/\text{ft}^2)} = 53.4 \text{ lbm/ft}.$$

The pipe's natural frequency is

$$\omega = \frac{\pi^2}{4}\sqrt{\frac{(30 \times 10^6 \text{ lb/in}^2)(4793 \text{ in}^4)(32.2 \text{ ft}\cdot\text{lbm/lb}\cdot\text{sec}^2)}{(53.4 \text{ lbm/ft})(150 \text{ ft})^4(144 \text{ in}^2/\text{ft}^2)}}$$
$$= 2.69 \text{ rad/sec}.$$

For resonance to occur,

$$\omega_v = \omega$$

and

$$\mathbf{S} = \frac{\omega D}{V_\infty} = \frac{(2.69 \text{ rad/sec})(12.75 \text{ in.})(1 \text{ ft/12 in.})}{V_\infty} = \frac{(2.86 \text{ ft/sec})}{V_\infty}.$$

The wind velocity V_∞ is an unknown in both the Reynolds number and the Strouhal number, so we must find it by trial and error. Assuming a Strouhal number of 0.20, we get

$$V_\infty = \frac{(2.86 \text{ ft/sec})}{0.20} = 14.3 \text{ ft/sec}.$$

The Reynolds number for this wind velocity is

$$\mathbf{R} = (6250 \text{ sec/ft})(14.3 \text{ ft/sec}) = 8.94 \times 10^4.$$

Figure 8.47 shows that the Strouhal number equals 0.20 for a Reynolds number of 8.94×10^4, so our initial guess of the Strouhal number was correct. Therefore

$$V_\infty = 14.3 \text{ ft/sec} = (14.3 \text{ ft/sec})(1 \text{ mi/5280 ft})(3600 \text{ sec/hr});$$
$$V_\infty = 9.75 \text{ mph}.$$ **ANSWER**

Discussion

In a calculation of this nature, you can usually obtain convergence faster by assuming a value for the Strouhal number rather than a value for the Reynolds number. The reason is that the Strouhal number is relatively constant over a wide range of Reynolds numbers.

8.4.3 Aerodynamics in Sports

The science of aerodynamics is not all work and no play. Various flow phenomena play an important part in many sports. The application of aerodynamics to hang gliding, kite flying, and sailing is rather obvious. Aerodynamics has always played an important role in auto racing, and race cars have always been more aerodynamically styled than ordinary cars.

Golf and baseball provide several interesting applications of aerodynamic principles. Did you ever wonder why a golf ball has dimples? A golf ball in flight has a Reynolds number of about 10^5. Check the drag coefficient of a smooth sphere at this Reynolds number (Fig. 8.10b). The dimples on a golf ball promote transi-

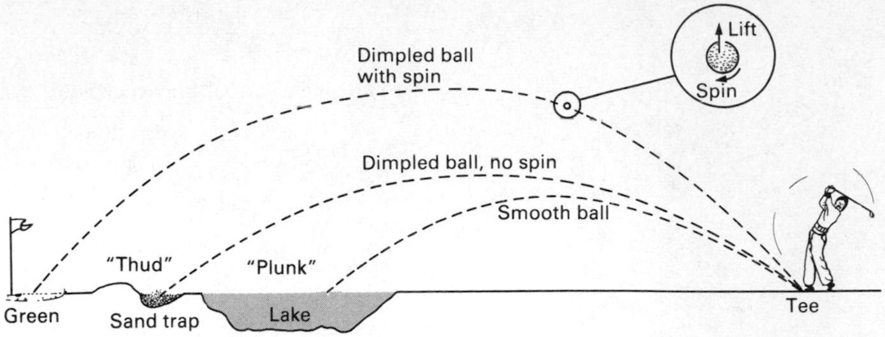

Figure 8.49 Various golf ball trajectories.

tion to a turbulent boundary layer on the ball, delaying separation and lowering the drag coefficient. "Carefully controlled experiments" (i.e., people hitting several drives with dimpled and smooth golf balls) indicate that a smooth ball can be driven only about 60 percent as far as a dimpled ball [13].

A well-hit golf ball has underspin in addition to its linear motion. This spin, together with the dimpled surface, is responsible for lift on the ball (Fig. 8.33), which helps it carry farther. Figure 8.49 illustrates various effects on golf ball trajectories. Spin-generated side force is responsible for hooking and slicing of balls that are not squarely hit. Ping-Pong and tennis players also use spin to control the trajectory of the ball.

Baseball pitchers are among the most clever practical aerodynamicists. A baseball is a sphere about 3 in. (8 cm) in diameter; however, the stitches on the ball give it both roughness and a certain asymmetry. By varying the spin on the ball, a pitcher can throw a wide variety of pitches. A good fastball has a spin that generates vertical lift, allowing the ball to follow a flatter trajectory or even rise. A curveball is the result of spin that causes a side force. A knuckleball pitcher throws the ball with very little spin and at a relatively slow speed. The irregular pattern of the stitches causes irregular transition back and forth between laminar and turbulent boundary layers. The fluctuating forces thus generated cause the ball to wobble on its path to the plate, confusing the batter. Some unscrupulous pitchers have been known to cut or scratch the cover of the ball in order to cause unpredictable aerodynamic forces and ball motions.

EXAMPLE 8.10 Illustrates Applications of Aerodynamics in Sports and the Use of Lift and Drag Data in a Complex Calculation

A pitcher releases a baseball at a velocity of 100 mph parallel to the ground. The ball has an underspin of 40 rev/sec in a vertical plane, as shown in Fig. E8.10a. Estimate the y coordinate of the ball when it reaches home plate (54 ft away).

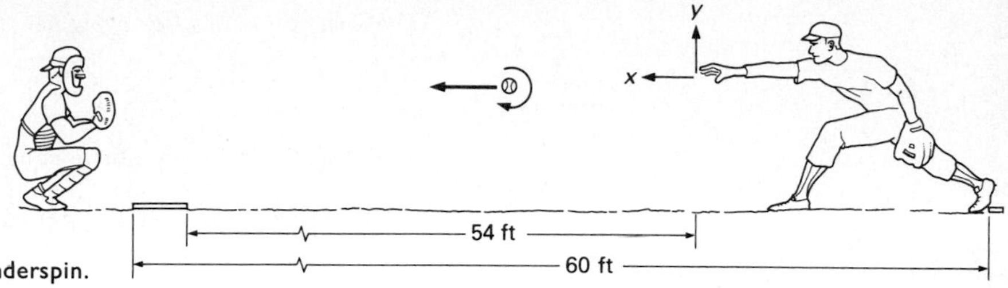

Figure E8.10a Fastball with underspin.

SOLUTION

Given

Baseball thrown at 100 mph with underspin of 40 rev/sec in vertical plane
Figure E8.10a

Find

The y coordinate of the ball after it travels 54 ft horizontally

Solution

Figure E8.10a shows the coordinate system; the origin is at the point where the ball is released by the pitcher at time $t = 0$, the x axis is horizontal and directed toward the catcher, and the y axis is directed vertically upward. The gravitational force on the ball is downward while the lift due to the underspin is upward. The ball may rise or drop, but it will remain in the xy plane. The ball's trajectory may not be parallel to the x axis. The angle θ (Fig. E8.10b) measures the difference between the ball's trajectory and the x axis.

We start with a free-body diagram of the ball (Fig. E8.10b). Applying Newton's second law in both the x and y directions, we obtain

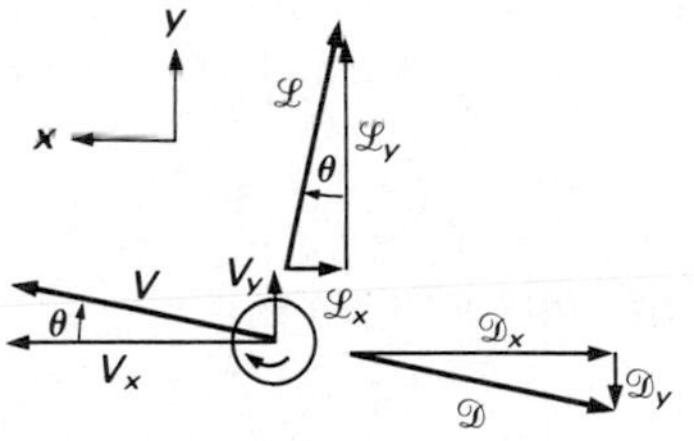

Figure E8.10b Free-body diagram of pitched baseball with underspin.

$$\sum F_x = m\left(\frac{dV_x}{dt}\right) \quad \text{and} \quad \sum F_y = m\left(\frac{dV_y}{dt}\right).$$

Substituting for the forces gives

$$-\mathscr{L}_x - \mathscr{D}_x = m\left(\frac{dV_x}{dt}\right) \quad \text{and} \quad -mg + \mathscr{L}_y - \mathscr{D}_y = m\left(\frac{dV_y}{dt}\right).$$

To simplify, we assume that the trajectory of the ball is relatively flat* ($\theta \approx 0$), so $\mathscr{L}_x \approx 0$ and $\mathscr{D}_y \approx 0$ and thus $\mathscr{L}_y \approx \mathscr{L}$ and $\mathscr{D}_x \approx \mathscr{D}$. The lift and drag are given by Eqs. (8.4) and (8.5):

$$\mathscr{L} = \tfrac{1}{2}\rho C_L V^2 S \quad \text{and} \quad \mathscr{D} = \tfrac{1}{2}\rho C_D V^2 S.$$

The square of the velocity is

$$V^2 = V_x^2 + V_y^2 = V_x^2\left[1 + \left(\frac{V_y}{V_x}\right)^2\right].$$

* We check this assumption later.

The assumption of a relatively flat trajectory implies that

$$\left(\frac{V_y}{V_x}\right)^2 \ll 1,$$

so $V^2 \approx V_x^2$. The approximate equations of motion for the ball are

$$-\tfrac{1}{2}\rho \mathbf{C_D} V_x^2 S = m\left(\frac{dV_x}{dt}\right) \quad \text{and} \quad -mg + \tfrac{1}{2}\rho \mathbf{C_L} V_x^2 S = m\left(\frac{dV_y}{dt}\right).$$

We now assume that the lift and drag coefficients of the baseball are approximated by the values for a spinning sphere given in Fig. 8.33. Both $\mathbf{C_D}$ and $\mathbf{C_L}$ vary with the velocity ratio U/V_∞, whose value at the time of release, $t = 0$, is

$$\left(\frac{U}{V_\infty}\right)_0 = \left(\frac{R\omega}{V_x}\right)_0.$$

A telephone call to a local sporting goods store gave us the information that the circumference of a baseball is 9 in. and the mass is 5 oz. The radius is

$$R = \frac{9 \text{ in.}}{2\pi} (1 \text{ ft}/12 \text{ in.}) = 0.119 \text{ ft}$$

and

$$\left(\frac{U}{V_\infty}\right)_0 = \frac{(0.119 \text{ ft})(40 \text{ rev/sec})(2\pi \text{ rad/rev})}{(100 \text{ mi/hr})(5280 \text{ ft/mi})(1 \text{ hr}/3600 \text{ sec})} = 0.204.$$

Figure 8.33 shows the ball to be in a range where $\mathbf{C_D}$ and $\mathbf{C_L}$ change significantly with U/V_∞. We assume that U/V_∞ is constant (to be checked later), so $\mathbf{C_D}$ and $\mathbf{C_L}$ are constant. With this assumption, we can integrate both equations of motion. Separating the variables in the x direction equation gives

$$\int_{V_{x0}}^{V_x} \frac{dV_x}{V_x^2} = -\frac{\rho \mathbf{C_D} S}{2m} \int_0^t dt,$$

where V_{x0} is the x component of velocity at time $t = 0$. Integrating and solving for V_x, we obtain

$$V_x = \frac{1}{1/V_{x0} + \rho \mathbf{C_D} St/2m} = \frac{V_{x0}}{1 + \beta t},$$

where

$$\beta = \frac{\rho \mathbf{C_D} S V_{x0}}{2m}.$$

Substituting into the y-direction equation gives

$$-mg + \frac{\rho \mathbf{C_L} S}{2}\left(\frac{V_{x0}}{1 + \beta t}\right)^2 = m\left(\frac{dV_y}{dt}\right).$$

Rearranging and integrating, we have

$$\int_0^{V_y} dV_y = \int_0^t \left[-g + \frac{(\rho \mathbf{C_L} S V_{x0}^2/2m)}{(1 + \beta t)^2}\right] dt,$$

where $V_y = 0$ at $t = 0$, because the ball is released parallel to the ground. Integrating, we get

$$V_y = -gt - \frac{\mathbf{C_L}}{\mathbf{C_D}}\left(\frac{V_{xo}}{1 + \beta t}\right) = -gt + \frac{\mathbf{C_L}}{\mathbf{C_D}}(V_{xo} - V_x).$$

After substituting this equation into

$$\frac{dy}{dt} = V_y,$$

multiplying by dt and integrating, we obtain

$$\int_0^y dy = \int_0^t \left[-gt + \frac{\mathbf{C_L}}{\mathbf{C_D}} V_{xo}\left(1 - \frac{1}{1 + \beta t}\right)\right] dt,$$

where $y = 0$ at $t = 0$. Performing the integration, we have

$$y = \frac{-gt^2}{2} + \frac{\mathbf{C_L}}{\mathbf{C_D}}(V_{xo}t) - \frac{2m\mathbf{C_L}}{\rho \mathbf{C_D^2} S}\ln(1 + \beta t)$$

$$= \frac{-gt^2}{2} + \frac{\mathbf{C_L}}{\mathbf{C_D}}(V_{xo}t) - \frac{2m\mathbf{C_L}}{\rho \mathbf{C_D^2} S}\ln\left(\frac{V_{xo}}{V_x}\right).$$

For 60°F air, Table A.4 gives $\rho = 0.077$ lbm/ft³. For $U/V_\infty = 0.204$, Fig. 8.33 gives* $\mathbf{C_D} \approx 0.35$ and $\mathbf{C_L} \approx 0.070$. The only remaining unknown quantity needed to calculate the y coordinate is the time t. We find it by substituting the equation for V_x into

$$\frac{dx}{dt} = V_x$$

and integrating. Thus

$$\int_0^x dx = \int_0^t \frac{V_{xo}\,dt}{1 + \beta t},$$

where $x = 0$ at $t = 0$. Carrying out the indicated operations gives

$$x = \frac{2m}{\rho \mathbf{C_D} S}\ln(1 + \beta t).$$

Solving for t gives

$$t = \frac{1}{\beta}\left[\exp\left(\frac{\rho \mathbf{C_D} S x}{2m} - 1\right)\right].$$

The numerical values of the various parameters are

$$\frac{\rho \mathbf{C_D} S}{2m} = \frac{(0.077\text{ lbm/ft}^3)(0.35)\pi(0.119\text{ ft})^2}{2(5\text{ oz})(1\text{ lbm/16 oz})} = 0.0019/\text{ft},$$

$$V_{xo} = (100\text{ mi/hr})(5280\text{ ft/mi})(1\text{ hr/3600 sec}) = 147\text{ ft/sec},$$

$$\frac{1}{\beta} = \frac{2m}{\rho \mathbf{C_D} S V_{xo}} = \frac{1}{(0.0019/\text{ft})(147\text{ft/sec})} = 3.58\text{ sec},$$

---

* We find these values by extrapolating the curves using $\mathbf{C_L} = 0$ and $\mathbf{C_D} = 0.30$ at $U/V_\infty = 0$ (see Fig. 8.10 and note that $\nu = 1.7 \times 10^{-4}$ ft²/sec and $\mathbf{R} = 2.0 \times 10^5$).

and

$$\frac{2m\mathbf{C_L}}{\rho \mathbf{C_D^2} S} = \left(\frac{2m}{\rho \mathbf{C_D} S}\right)\left(\frac{\mathbf{C_L}}{\mathbf{C_D}}\right) = \frac{1}{(0.0019/\text{ft})}\left(\frac{0.070}{0.35}\right) = 105.3 \text{ ft}.$$

The time required for the ball to travel 54 ft is

$$t = 3.58 \text{ sec } [\exp[(0.0019/\text{ft})(54 \text{ ft})] - 1] = 0.387 \text{ sec}.$$

Finally, the y coordinate is

$$y = \frac{-(32.2 \text{ ft/sec}^2)(0.387 \text{ sec})^2}{2} + \left(\frac{0.070}{0.35}\right)(147 \text{ ft/sec})(0.387 \text{ sec})$$
$$- (105.3 \text{ ft}) \ln\left(1 + \frac{0.387 \text{ sec}}{3.58 \text{ sec}}\right);$$
$$y = -1.84 \text{ ft}.$$

ANSWER

Discussion

We must now check our assumptions. The first assumption is that the trajectory is flat. If we assume a straight-line path for the ball, the actual path is at an angle of

$$\theta \approx \tan^{-1}\left(\frac{y}{x}\right) = \tan^{-1}\left(\frac{-1.84 \text{ ft}}{54 \text{ ft}}\right) \approx -2° \text{ (below horizontal)}.$$

Clearly, $\theta \approx 0$.

The second assumption is that

$$\left(\frac{V_y}{V_x}\right)^2 \ll 1.$$

The *largest* value of V_y and the *smallest* value of V_x occur when $x = x_p = 54$ ft. The value of V_y is

$$V_y = (-32.2 \text{ ft/sec}^2)(0.387 \text{ sec})$$
$$+ \left(\frac{0.070}{0.35}\right)(147 \text{ ft/sec})\left(1 - \frac{1}{1 + 0.387 \text{ sec}/3.58 \text{ sec}}\right)$$
$$= -9.6 \text{ ft/sec}.$$

Also

$$V_x = \frac{(147 \text{ ft/sec})}{1 + 0.387 \text{ sec}/3.58 \text{ sec}} = 133 \text{ ft/sec}.$$

The maximum value of $(V_y/V_x)^2$ is

$$\left(\frac{V_y}{V_x}\right)^2_{\max} = \left(\frac{9.6}{133}\right)^2 = 0.0052.$$

The third assumption is that U/V_∞ is constant. A complicated analysis involving surface shear stresses is necessary to determine how much the ball's angular velocity decreases as the ball travels 54 ft. Both U and V_x decrease, so we can evaluate a limiting case by assuming that U remains constant and using the maximum and minimum values of V_x. These assumptions give

$$U = R\omega = (0.119 \text{ ft})(40 \text{ rev/sec})(2\pi \text{ rad/rev}) = 29.9 \text{ ft/sec},$$

$$\left(\frac{U}{V_{x_{\max}}}\right) = \left(\frac{29.9 \text{ ft/sec}}{147 \text{ ft/sec}}\right) = 0.204,$$

and

$$\left(\frac{U}{V_{x_{\min}}}\right) = \left(\frac{29.9 \text{ ft/sec}}{133 \text{ ft/sec}}\right) = 0.225.$$

The difference is roughly 10 percent. The values of C_D and C_L may differ from the values used by about 10 percent. A more accurate analysis might use the values of C_D and C_L for the average value of U/V_x; however, the accuracy with which values from C_D and C_L can be read from Fig. 8.33 does not justify the more refined solution.

The final assumption to consider is use of Fig. 8.33 to obtain values for C_L and C_D. The coefficients given are for a smooth sphere; however, a baseball is rather rough because of the stitches. The lift coefficient for a baseball may be as much as three times larger than that for a smooth sphere. The drag coefficient for a baseball also is somewhat larger. You might repeat the calculations using a three-times-larger value of C_L and two-times-larger value of C_D and compare your answers to ours.

For the interest of baseball fans, the fastball released by our pitcher at 100 mph arrives at the plate at 90.7 mph (133 ft/sec).

The quantity $\rho C_D S/2m$ is called the *ballistic drag parameter.* "Heavier" objects fall faster than "lighter" objects because heavier objects have lower values of this parameter.

This example provides a fitting summary of the material presented in this chapter. It shows how readily available data on lift and drag coefficients can provide models of these aerodynamic forces for use in analyses of motion of objects in a fluid medium. If you were to redo this problem, neglecting both lift and drag (the standard model of high school and college first-year physics), and compare the answers with those we obtained here, you would find the results very informative.

PROBLEMS

1. The pressure distribution over a normal flat plate is shown in Fig. P8.1. Find the drag force on the plate if it measures 2.0 m high and 10.0 m wide (perpendicular to the paper).

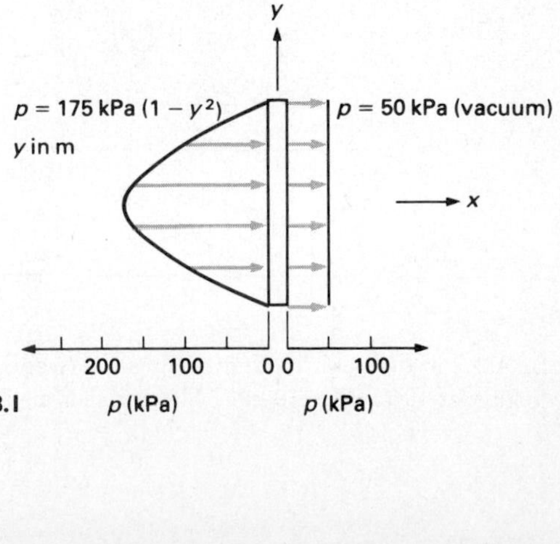

Figure P8.1

2. When warm water is running in a shower, the shower curtain is pulled inward. Explain why this happens. Will it happen if cold water is used? Explain.

3. Why do airplanes try to take off and land while headed into the wind?

4. A ball has a diameter of 0.10 m and a density of 500 kg/m^3. It is anchored to the bottom of a flowing river by a straight, very thin, massless string that makes an angle of 60° with the river bed. Determine the drag on the ball. Does the string length make a difference?

5. A balloon having a diameter of 1.0 m is filled with helium at 2.0 kPa gage and the ambient temperature of 20°C. The balloon ascends through the atmosphere at a steady rate of 0.5 m/s while carrying a payload of 1.0 N. The balloon itself weighs 0.25 N. Calculate the drag force on the balloon and its payload for $p_{\text{atm}} =$ 101 kPa.

6. A 40-mph wind is blowing across the quonset hut shown in Fig. P8.6. The pressure distribution is

$$p = p_\infty + \frac{\rho_\infty V_\infty^2}{2}(1 - 4\sin^2\theta),$$

where θ is the angle shown and p_∞, ρ_∞, and T_∞ are the ambient pressure, density, and temperature, respectively, far upstream from the quonset hut. The pressure inside the hut is p_∞. Find the lift force on the quonset hut for this 40-mph wind. The hut's length is 40 ft.

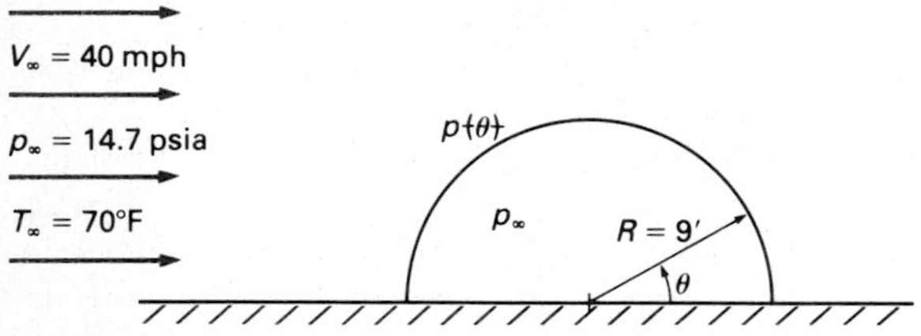

Figure P8.6

7. Figure P8.7 shows the pressure distributions on the top and bottom surfaces of an airfoil. Calculate the lift force per unit length (perpendicular to the paper) for $V_\infty =$ 120 mph and $\rho_\infty = 0.075$ lbm/ft^3.

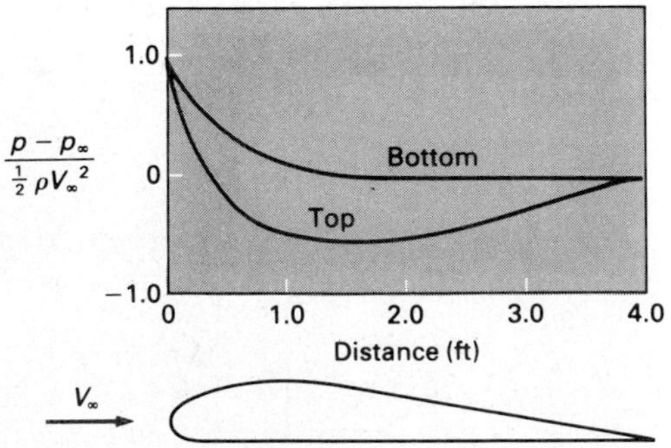

Figure P8.7

8. A 1.2 m diameter balloon weighs 1.0 N and contains helium at 10°C and 110 kPa. The ambient air is at 10°C and 101 kPa and has a horizontal velocity of 20 km/hr. The balloon is fastened to the ground by a 10-m long, weightless string. Determine the air drag on the balloon if the string from the ground makes an angle of 57.3° with the vertical. Neglect the air drag on the string.

9. A 1.0-m diameter balloon weighs 1.0 N and contains helium at 20°C and 110 kPa. The ambient air is at 20°C and 101 kPa, and the balloon is rising through the still air with a constant upward velocity of 0.25 m/s. What is the air drag on the balloon?

10. A steel ($\gamma = 500$ lb/ft^3) ball having a diameter of $\frac{1}{8}$ in. is falling through 60°F water with a terminal (i.e., constant) velocity of 3.13 ft/sec. Find the drag force of the water on the ball.

11. A nonspinning ball having a mass of 3 oz. is thrown vertically upward with a velocity of 100 mph and has zero velocity at a height 250 ft above the release point. Assume that the air drag on the ball is constant and find this constant "average" air drag. Neglect the buoyant force of air on the ball.

12. The power supplied to the wheels to move a large tractor-trailer along a horizontally level road at 55 mph is estimated at 120 hp. Determine the total drag force on the tractor-trailer. There is no wind.

13. An airplane flying at 10,000 m in the Standard Atmosphere has a weight of 400 kN and a drag at 32 kN while traveling at 450 km/hr. Calculate the lift and drag coefficents for a wing area of 120 m^2 if the wing provides 90% of the lift and 60% of the drag.

14. An automobile engine has a maximum power output of 70 hp, which occurs at an engine speed of 2200 rpm. A 10% power loss occurs through the transmission and differential. The rear wheels have a radius of 15.0 in. and the automobile is to have a maximum speed of 75 mph along a level road. At this speed, the power absorbed by the tires because of their continuous deformation is 27 hp. Find the maximum permissible drag coefficient for the automobile at this speed. The car's frontal area is 24.0 ft^2.

15. A man riding his bicycle is traveling along a horizontal road at 10 mph. The air is still and is at 85°F and 14.67 psia. Estimate the drag coefficient of this bicyclist if his frontal area is 4.5 ft^2 and his body burns up 80 ft·lb per second with 1/4 of it going to work in pedaling the bicycle.

16. A woman riding her bicycle is sitting upright and coasting down a hill making an angle of 10° with the horizontal. Her velocity is 25 km/hr into an oncoming 25-km/hr wind. The air is at 15°C and 101 kPa. Estimate the drag coefficient of this bicyclist if her frontal area is 0.40 m^2. The woman and her bicycle weigh 520 N.

17. A glider that weighs 4.0 lb has a wing area of 7.0 ft^2 and is "diving" to pick up velocity. During the dive, it comes to a terminal velocity of 25 mph while moving at an angle of 20° below the horizontal. Find the lift and drag coefficient for the dive. The air temperature is 60°F.

18. The ideal pressure coefficient distribution for the flow over a circular cylinder in cross-flow is given by $C_p = 1 - 4\sin^2\theta$. Determine $\mathbf{C_D}$ and $\mathbf{C_L}$ by integrating pressure over the surface.

19. A circular cylinder of infinite span is immersed in air in cross-flow. The air speed is V_∞ and its pressure is p_∞, as shown in Fig. P8.19. Find the coefficient of drag using the control volume shown, together with the incoming and outgoing profiles indicated. Take into account that flow is leaving the control volume through the top and bottom, where its horizontal velocity component is constant and has value of V_∞. Comment about the usefulness of this procedure. Could it be implemented in testing? If so, how?

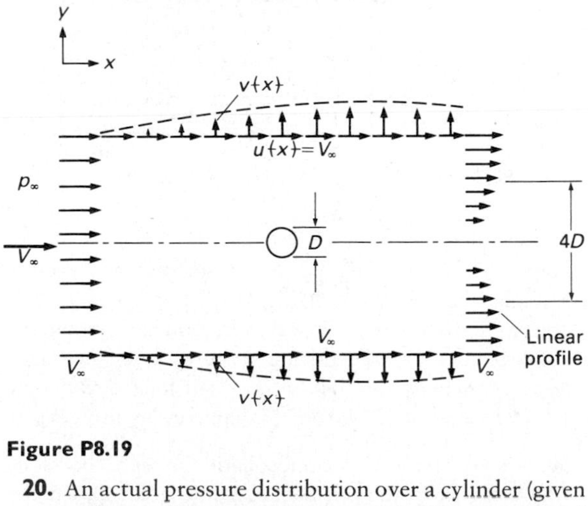

Figure P8.19

20. An actual pressure distribution over a cylinder (given as coefficient of pressure) is plotted in Fig. P8.20. Find $\mathbf{C_D}$ by numerical integration (using a computer might speed things up, but it is not necessary).

21. A large automobile manufacturer is conducting aerodynamic tests of a new automobile model. The drag force versus speed data are shown in Fig. P8.21. The model car ($\frac{1}{10}$ scale) has a 0.03 m^2 frontal area. Find the $\mathbf{C_D}$ versus Reynolds number graph for this car (using the square root of the frontal area at the characteristic length). Assume that the model and prototype operate at standard air conditions. Determine whether the data obtained are applicable for a full-scale car traveling at speeds between 35 and 55 mph. Why or why not?

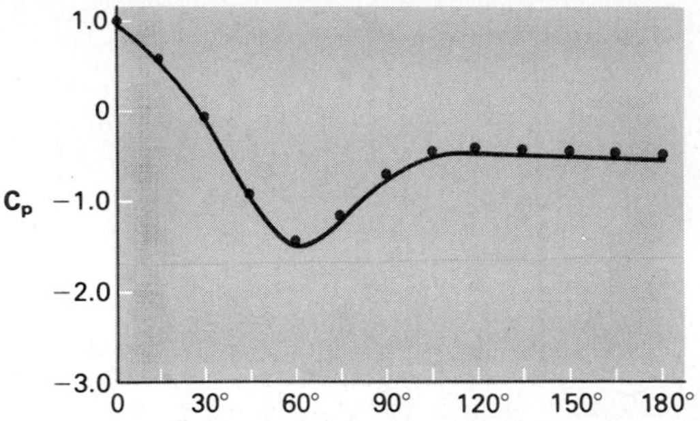

Figure P8.20

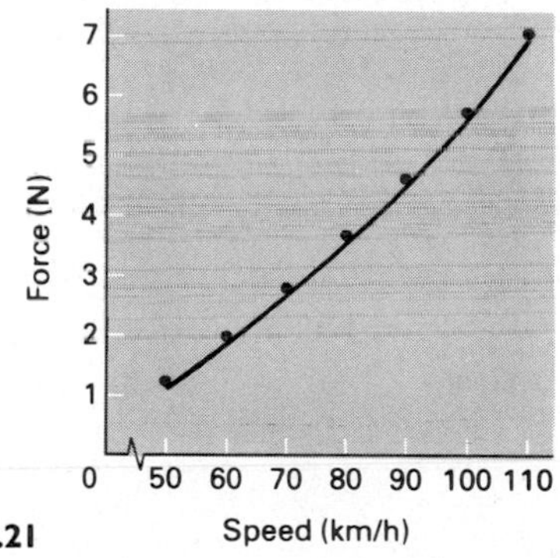

Figure P8.21

22. A $\frac{1}{8}$-scale model for an automobile is tested in a wind tunnel with air having a temperature of 20°C at atmospheric pressure. The drag force of the model is measured at air speeds of $V = 50$, 80, and 128 km/hr yielding drag forces of 1.56, 3.83, and 9.70 N, respectively. The prototype has a frontal area of 1.8 m^2. Determine the drag coefficient for each model test, and estimate the drag of the prototype and the power needed to overcome air resistance at 50 mph.

23. A truck model is to be tested in a wind tunnel to obtain an approximate aerodynamic coefficient of drag. The truck model is made to span the width of the wind tunnel channel in order to obtain a 2-dimensional coefficient. The surface C_p data are shown in Fig. P8.23, which has the scale shown. Find the model $\mathbf{C_D}$ per unit depth for the truck model. Neglect skin friction.

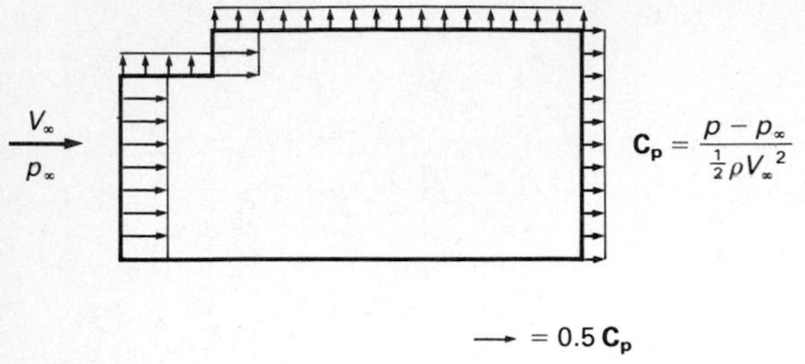

Figure P8.23

24. In wind tunnel testing, model blockage can affect surface pressure distribution and therefore (in the case of negligible skin friction effects) the drag coefficient. As a rule of thumb, model cross-sectional blockage should be limited to 5 to 10% of the wind tunnel cross section. You are presented with an oversized circular cylinder for which you have to estimate whether the C_D measured in the wind tunnel will be higher or lower than the C_D found without blockage for similar Reynolds numbers. The wind tunnel, as shown in Fig. P8.24, is 10 cm high and the cylinder measures 5 cm in diameter. Explain your answer.

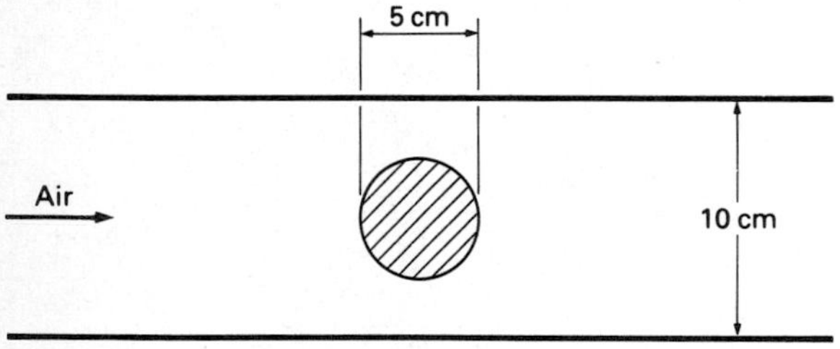

Figure P8.24

• **25.** The addition of an air deflector on top of the tractor has increased the gasoline mileage of a tractor-trailer from 4 mi/gal to 4.25 mi/gal at 55 mph. The engine provides 11,000,000 ft·lb of energy to the drive wheels from every gallon of fuel. Half of this energy is used to overcome the drag resistance of the air on the tractor-trailer. Find the drag coefficient for the tractor-trailer with and without the air deflector. The tractor-trailer has a frontal area of 82 ft^2.

• **26.** Students are doing a coast-down study to determine the drag coefficient on a small car. The car's frontal area is 24 ft^2, and the car weighs 2600 lb. The air density is 0.075 lbm/ft^3. The coast-down test was conducted on a level road, in neutral gear, and with no wind. The test results are

Distance Traveled (ft)	Time Elapsed (sec)	Car Velocity (mph)
0	0	56.7
500	6.0	55.3
1000	12.3	47.6
1500	20.3	44.8
2000	27.5	46.3
2500	35.0	45.4
3000	42.5	42.5
3500	51.0	40.0
4000	59.5	38.4
4500	68.7	36.6
5000	78.1	36.2

Find the drag coefficient (which is assumed constant for this velocity range).

27. A 180-lb man parachutes from a plane using a hemispherical parachute in air at 20°F and 14.60 psia. Calculate the parachute diameter required for the man's terminal velocity not to exceed 20 ft/sec. Neglect the weight of the parachute.

28. The railroad boxcar shown in Fig. P8.28 weighs 200 kN and rides on a track whose rails are 1.44 m apart. What wind velocity normal to the side of the boxcar will topple it? The air is at 15°C.

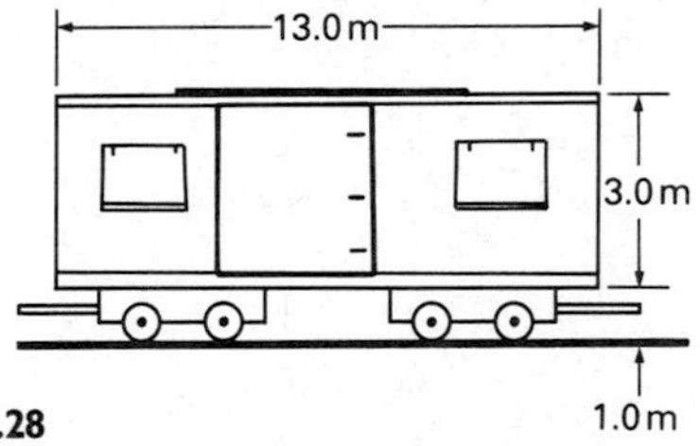

Figure P8.28

29. The dirigible *Akron* had a length of 785 ft, a maximum diameter of 132 ft, and a maximum speed of 84 mph. The lifting force was 182,000 lb. Estimate the power required to overcome skin friction in the Standard Atmosphere at 5000 ft. Assume turbulent boundary layer flow. Model the dirigible as a 653-ft-long cylinder of 132 ft diameter and two 132-ft-diameter hemispheres.

30. Air at 100°F and atmospheric pressure flows at 10,000 cfm through a horizontal, 24-in. square duct made of sheet metal. A 0.125-in.-diameter rod spans the duct. Determine the drag on the rod if the velocity is uniform over the flow area.

31. A particle of sand is assumed to be a sphere of 1-mm diameter. It is dropped into a river flowing at 5 km/hr.

The sphere is carried horizontally with the water velocity and soon drops through the water at a terminal velocity. Find the terminal velocity V_y if sand has a specific gravity of 1.6.

32. A smooth steel ball ($\gamma = 500$ lb/ft^3) is dropped into a lake ($\gamma = 62.4$ lb/ft^3). Calculate its terminal (i.e., constant) velocity as it falls toward the bottom of the lake if the ball diameter is $\frac{1}{8}$ in. and the lake temperature is 60°F.

33. A car-top carrier measuring 1 ft high, 4 ft wide, and 4 ft long is used for a trip. How much extra power (in hp) does the addition of the carrier require at 55 mph on a level road? Assume a sizeable air space between the car and the carrier. The air is at 65°F.

34. A 24-in.-diameter smooth ball rests on top of a thin, 48-in.-long rod. Air at 70°F blows over the ball and deflects the rod back. Find the bending moment at the base of the rod for a 35-mph wind on the ball. Assume a small transverse deflection of the rod.

35. Find the terminal velocity of a 2.0-mm-diameter, smooth, spherical air bubble as it rises in 10°C water.

36. The supports for Phil's Pizza Parlor sign in Example 8.3 consists of two steel pipes 25 cm long and of 3.34 cm outside diameter. Estimate the power required to overcome the air drag on these two supports for an advertising sign facing forward. The air temperature is 10°C, and the van speed is 15.6 m/s (35 mph).

37. A water tower is a spherical shell with a diameter of 10 m. The shell is supported by a single, 30-m-long, vertical pipe. A wind with a uniform velocity of 20 km/hr blows over the water tower. Calculate the drag force on the spherical shell and the bending moment on the base of the tower from this drag force. The air temperature is 20°C.

38. Determine the drag force $\mathcal{D}$ on the 30-m-long vertical pipe in Problem 37. The pipe diameter is 1.0 m. Then find the bending moment at the base of the tower from this drag force.

39. A vertical radio antenna on an automobile is 3.0 ft long and has a diameter of 0.125 in. Find the drag on the antenna at a car speed of 55 mph in still air. Consider U.S. Standard Atmospheric air at zero elevation.

40. A speed limit sign along a highway is shown in Fig. P8.40. Calculate the bending moment at the base of the 2-in.-outside-diameter support poles for a 60-mph wind. The air temperature is 60°F.

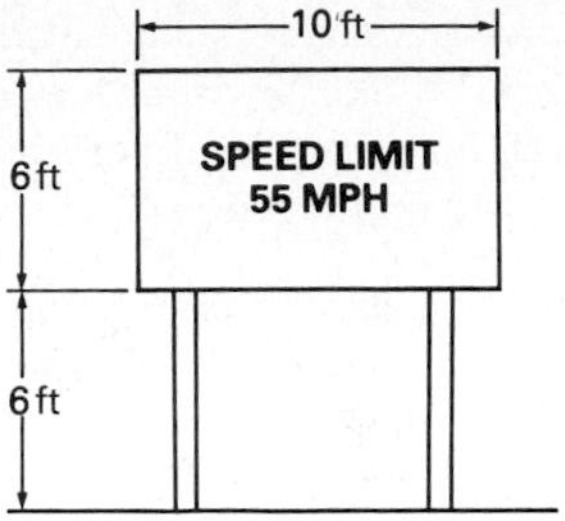

Figure P8.40

41. A 1.2-m-diameter spherical balloon weighs 1.0 N and contains helium at 10°C and 150 kPa. The ambient air is at 10°C and 101 kPa. Air blows horizontally over the balloon at 20 km/hr. Determine the angle that a 10-m weightless string restraining the balloon makes with the ground; also find the tension in the string. Assume that the balloon surface is smooth and that the string remains straight.

42. A flat movie screen at an outdoor theater is 15 m square. Find the bending moment at the base of the structure that supports the bottom of the screen 15 m above the ground. The wind velocity is 60 km/hr, and the atmospheric conditions are 15°C and 100 kPa. Neglect the drag force of the wind on the support structure.

43. A smokestack has an outside diameter of 10 m and a height of 120 m. Find the overturning moment caused by a uniform wind velocity of 70 km/hr. The atmospheric conditions are 10°C and 101 kPa.

44. Show that the terminal velocity V of a sphere of diameter D_s and density ρ_s falling in a liquid of density ρ_ℓ is given by

$$V = \sqrt{\frac{4gD_s}{3C_D}\left(\frac{\rho_s}{\rho_\ell} - 1\right)}.$$

45. A smooth orange ball weighs $\frac{1}{64}$ lb (at sea level) and has a diameter of 1.5 in. The discharge of a vacuum cleaner is directed upward and supports the ball $\frac{1}{2}$ in. above the hose outlet. If the hose inside diameter is 4.0 in., estimate the volume flow rate through the vacuum cleaner. The air temperature is 100°F.

46. As an engineer working for Amtrak, you want to estimate the aerodynamic drag force on a train 10 cars long and traveling at 55 mph. You know that each car is 15 m long and is roughly square with a cross section measuring 9 m^2. Skin friction drag is estimated at 3 N/m^2. Neglect undercarriage drag and estimate the total drag using the upper three sides.

47. Two pilings of square cross section are placed in 20 ft of water, as shown in Fig. P8.47. The water velocity is 5 ft/sec. Which piling will have the larger bending moment at its base? Assume 60°F water and $w = 1.0$ft.

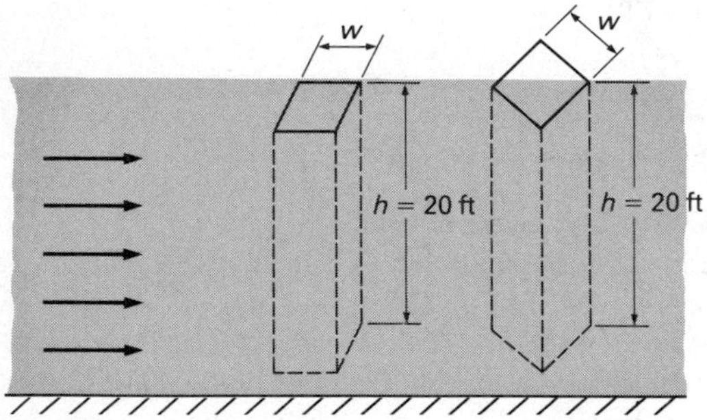

Figure P8.47

48. A 70-kg soldier on a secret mission has to parachute from an airplane over the desert. It takes five seconds for him to release his parachute after he jumps. Taking the mass of the parachute as 5 kg, estimate the force acting on the soldier as he releases the parachute. The diameter of the parachute is 2.0 m. Neglect drag on the soldier's body prior to the opening of the parachute. The air is at 20°C.

49. On a processing line, glass nursette bottles travel at a speed of 1 m/s. The bottles are 5.08 cm in diameter and stand 15.24 cm tall. What is the drag force on one bottle as it travels down the line? (Assume that the air temperature is approximately 25°C.)

50. Calculate the drag that the plastic rearview mirror of a motorcycle experiences at 60 mph. The coating used on the plastic tends to crack when drag exceeds 17.3 pounds. Is the mirror's casing in danger of cracking? Model as a 3″ × 3″ plate with flow of 60°F air normal to the plate surface.

51. A crane sits outside at a construction site. Today, the wind velocity is 60 km/hr in the same direction as the view in Fig. P8.51 (i.e., perpendicular to the paper). The crane boom is made up of five identical sections for a total length of 60 m. The cross section *A–A* is a square-shaped construction measuring 2.0 m on a side. Each element of each section is made up of 12-cm-outside-diameter pipe. Neglecting the gib and the cables, calculate the moment at the base of the boom caused by the wind. Assume that the drag force on any element "hidden" by another element in Fig. P8.51 is zero. Assume 20°C air.

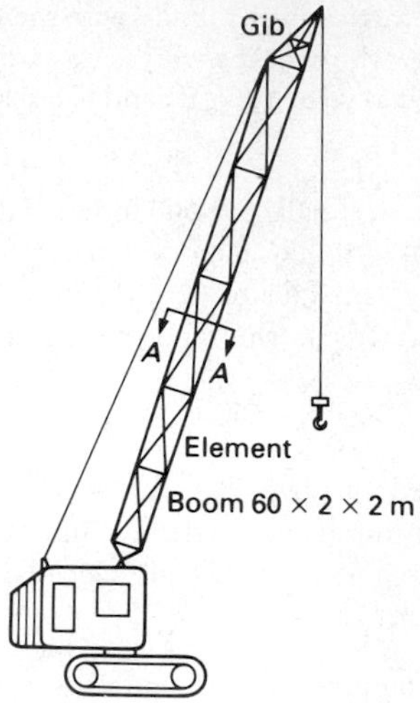

Figure P8.51

52. A marine location marker is a smoke-producing device usually dropped from an airplane and used to mark a reference point in the ocean. One is being tested in a wind tunnel to determine the drag force when it is carried by an airplane at a velocity of 200 mph. The marker is a cylinder with flat ends, has a diameter of 12 in., and is 48 in. long. Calculate the drag force on the marker in 10°F air if the air flow is parallel to the cylinder's axis.

53. A sky diver having mass of 60 kg jumps from an airplane at 3000 m above the ground. Estimate how far she has fallen in 45 s and her velocity at that instant if she has not yet opened her parachute and has provided the largest possible drag coefficient. Assume that there is no horizontal velocity, that the arms, legs, and body are cylinders, and that the arms and legs are fully extended perpendicular to the direction of travel.

54. Two carbon dioxide gas bubbles are rising in a glass of soda pop. The diameter of one is larger than the diameter of the other. Which bubble has the larger terminal velocity?

• **55.** Two smooth wood spheres ($\rho = 500\ kg/m^3$) are dropped in air. The diameter, D, of one is larger than the diameter, d, of the other. Which ball has the larger terminal velocity? Does the ball with the larger terminal velocity falling in air have the larger terminal velocity when rising in water?

• **56.** Repeat Problem 43 using the velocity distribution $V = Ay^2$, where $A = 0.00139\ \text{km/m}^2\cdot\text{hr}$ with y measured vertically above the ground.

• **57.** Boy Scouts are taught to "feather" oars on the backstroke when rowing a boat. Estimate the energy saved in 100 oar strokes when they do so under Standard Atmosphere conditions. The boat is traveling at 15 km/hr. The oars are moved from $-45°$ to $+45°$ with respect to an axis perpendicular to the boat's axis and at a speed of 3.76 rad/s (see Fig. P8.57).

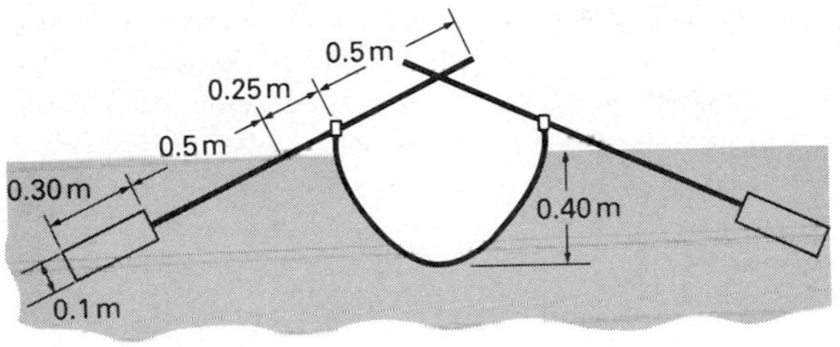

Figure P8.57

58. A domestic engineer dusts off a small table 24 in. high. The dust particles from the table are spherical with a diameter of 0.001 in. and a density of 90 lbm/ft³. The air temperature is 60°F. If the dust particle is moving with its terminal velocity the entire fall, how long does it take for the dust particle to settle to the floor?

59. Two identical, small-diameter aluminum rods that have a diameter of 0.05 mm and a length of 5 mm are placed in 20°C kerosene. One rod falls with the length L vertical, and the second falls with the length L horizontal. Find the terminal velocity of each. The mass density of aluminum is 2690 kg/m³.

60. A $\frac{1}{16}$-in.-diameter steel ball bearing is dropped into 60°F SAE 10 oil. The density of steel is 490 lbm/ft³, and the density of 60°F SAE 10 oil is 54.3 lbm/ft³. Find the terminal velocity of the ball bearing.

61. Dust particles fall in 10°C air with a downward terminal velocity of 0.5 cm/s. The dust particles have a density of 1500 kg/m³. Assume that the dust particles are spheres and estimate their diameters.

62. A 0.250-mm-diameter brass sphere is falling through a 20°C liquid listed in Table A.5. The liquid density is 875 kg/m³ and the brass sphere's density is 8490 kg/m³. The sphere has a terminal velocity of 0.020 m/s. Identify the liquid if the Reynolds number is less than unity.

63. A 0.025-in.-diameter wood sphere is rising with its terminal velocity in a 68°F liquid listed in Table A.6. The liquid's density is 54.7 lbm/ft³, and the wood sphere's density is 32.0 lbm/ft³. The sphere has an upward terminal velocity of 0.020 ft/sec. Identify the liquid if the Reynolds number is less than unity.

64. A simple experiment is to be devised whereby a wood ball ($S = 0.50$) rises in 20°C kerosene ($S = 0.814$) with a Reynolds number of 0.50. Determine the ball's diameter and terminal velocity.

65. Apply the linear momentum equation to a control volume enclosing an airfoil (similar to the derivation of the Kutta–Joukowsky theorem) and obtain an expression for the drag in terms of measured upstream and downstream velocity profiles. Assume a constant-density fluid.

66. Show that the Kutta–Joukowsky theorem, Eq. (8.35), is valid for a free vortex superimposed on a uniform flow.

67. The two hydrofoils on a watercraft have a platform area of 50 ft² and a cross-sectional profile of an NACA 2415 airfoil. How great a load can the hydrofoil support at a speed of 20 mph in still, 60°F water if the angle of attack is 12°?

68. The cross section of an airplane wing is an NACA 2415 airfoil. The wing has a constant chord length of 3.5 m and a length of 15 m. Calculate the lift and drag on this wing at an airplane speed of 200 km/hr in air at 98 kPa and 0°C and angle of attack of 8°.

69. An airplane weighs 320,000 lb, has a wing area of 2800 ft², and has a wing length of 140 ft. The atmospheric pressure and temperature are 14.67 psia and 60°F, respectively. The airplane is moving along a runway at 200 mph, and the wing has the lift and drag coefficients given in Fig. 8.39. Find the lift and drag on the airplane wings for an angle of attack of 10°.

70. An airfoil is mounted above the rear wheels of a race car for downthrust. The airfoil makes an angle of 10° with the track and has a chord length of 1.0 ft and a length of 3.0 ft. The race car is traveling at 180 mph in air at 90°F and 14.72 psia. Find the downthrust and drag on the airfoil if it has the lift and drag coefficient characteristics of Fig. 8.39.

71. A design group has two possible wing designs (A and B) for an airplane wing. The planform area of either wing is 130 m² and each must provide a lift of 1,550,000 N.

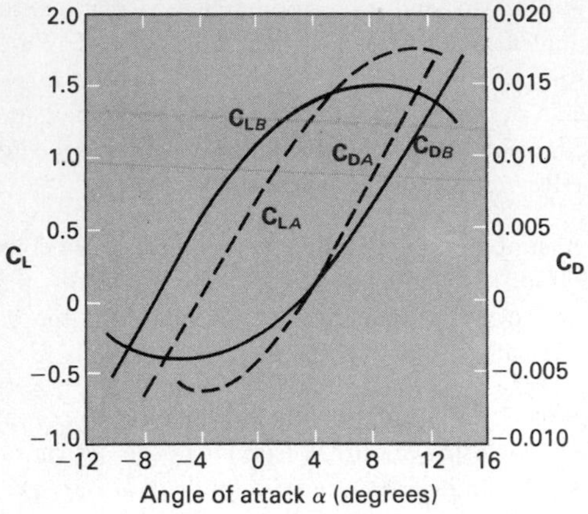

Figure P8.71

The airplane is to fly at 700 km/hr at an altitude of 10,000 m in the Standard Atmosphere. The lift and drag coefficients are shown in Fig. P8.71. Both sets of lift and drag coefficients give the total lift $\mathcal{L}$ and the total drag $\mathcal{D}$ on the airplane per unit area of the wing so that

$$\mathcal{L}_{\text{airplane}} = \tfrac{1}{2}C_L \rho A_{\text{wing}} V^2$$

and

$$\mathcal{D}_{\text{airplane}} = \tfrac{1}{2}C_D \rho A_{\text{wing}} V^2,$$

where V is the airplane velocity. Which wing design would you recommend? Support your recommendation.

72. The lift and drag coefficients for a flat plate inclined at angle of attack α are shown in Fig. P8.72. A ceiling fan with a vertical shaft rotates at 60 Hz with four flat blades that make an angle of 12° with the horizontal. Each blade is 50 cm long and 20 cm wide. The hub radius is 10 cm. Find the vertical force on the bearings in the fan.

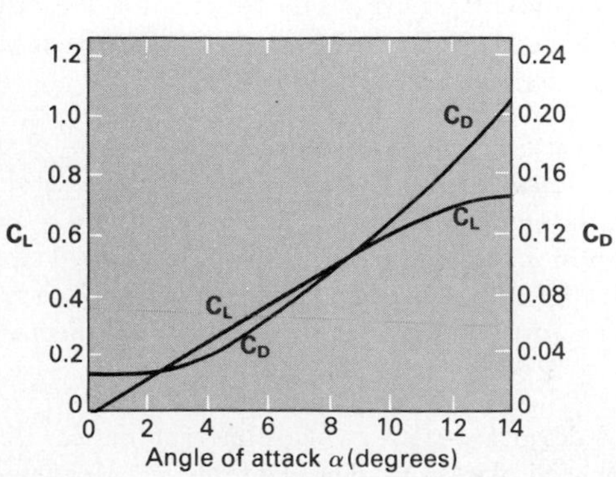

Figure P8.72

73. A gasoline sign sits atop a 50-m-high cylindrical pole with a diameter of 0.5 m. What is the reduction in the bending moment at the base of the pole if the pole is covered with a symmetric airfoil that is always aligned with the wind, measures 0.55 m thick by 1.65 m chord length in cross section, and has a drag coefficient of 0.06. The wind velocity is 20 km/hr, and the air temperature is 20°C.

74. A cross-wind propels a small boat with an NACA 2415 airfoil mounted on it along a river, as shown in Fig. P8.74. For a cross-wind of $V = 20$ km/hr and a boat velocity of $U = 4$ km/hr, calculate the forward thrust T of the wind on the boat. The airfoil is 1 m high and has a 4 m chord length. Recall that the angle of attack is the angle that the velocity of the air relative to the airfoil (W) makes with the chord line. The air temperature is 25°C.

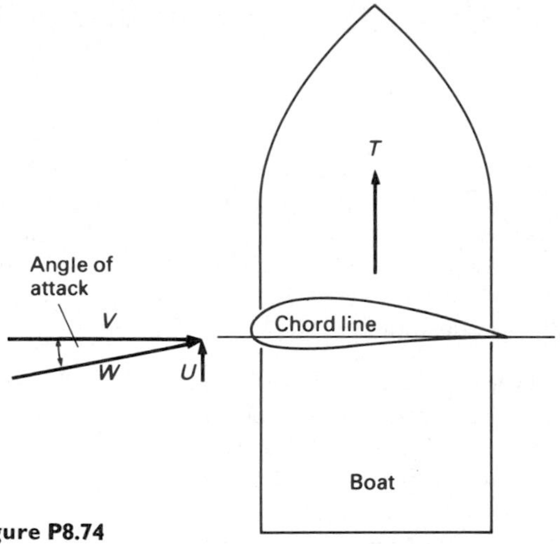

Figure P8.74

75. A cross-wind propels a small boat with an NACA 2415 airfoil mounted on it along a river, as shown in Fig. P8.74. For a cross-wind of $V = 20$ km/hr and a boat velocity of $U = 4$ km/hr, find the side thrust, F, of the wind on the boat. The airfoil is 1 m high and has a 4 m chord length. Recall that the angle of attack is the angle that the velocity of the air relative to the airfoil (W) makes with the chord line. The air temperature is 25°C.

76. A model airplane has a wing with an NACA 2415 airfoil cross section, a constant chord length of 12 in. and a wing span of 8 ft. The airplane is traveling with a velocity of 60 mph in still air with a wing angle of attack of 8°. The airplane encounters a headwind of 15 mph. How much does the wing lift increase if the airplane remains parallel to its original line of travel and maintains its 60-mph velocity into the 15-mph wind. The air temperature is 60°F. Does this calculation suggest why it is advantageous for an airplane to take off while headed into the wind? If so, why?

77. Figure P8.77 shows the lift and drag coefficients for an airfoil that has a chord length of 2 ft and a wing span of 10 ft for Reynolds number of 4×10^5 and 1×10^6. For 60°F air, find the maximum lift for each Reynolds number and the velocity at which it can be obtained.

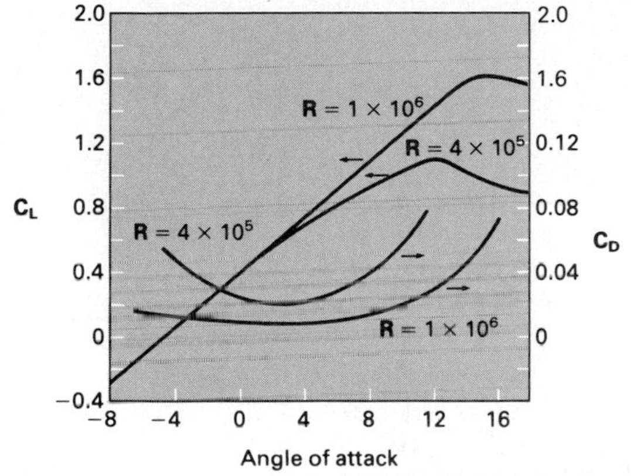

Figure P8.77

78. Calculate the engine thrust required to overcome the wing drag on the airfoil in Problem 77 for the maximum lift conditions for Reynolds numbers 4×10^5 and 1×10^6.

• **79.** A windmill consists of four flat plates measuring 15.0 ft long and 1.0 m wide and connected to a 3.0-ft-diameter hub. The blades rotate at 0.1 Hz. The blades have a twist so that each segment of each blade makes an angle of 14° with the relative velocity of the wind and the blade segment. Find the torque developed by the four blades for a wind velocity of 15 mph parallel to the windmill's axis of rotation. Refer to Fig. P8.72 for the lift and drag coefficients of a flat plate at angles of attack. Assume standard atmospheric conditions at sea level.

• **80.** An SAE Supermileage vehicle is racing along a straight track at 25 mph with a 3 mph cross-wind. The vehicle is covered with a two-dimensional, symmetric airfoil having the lift and drag coefficients given in Fig. P8.80. The airfoil is 4 ft high. Does the airfoil vehicle experience a forward or retarding thrust due to the cross-wind? Neglect end effects on the airfoil. The air temperature is 60°F.

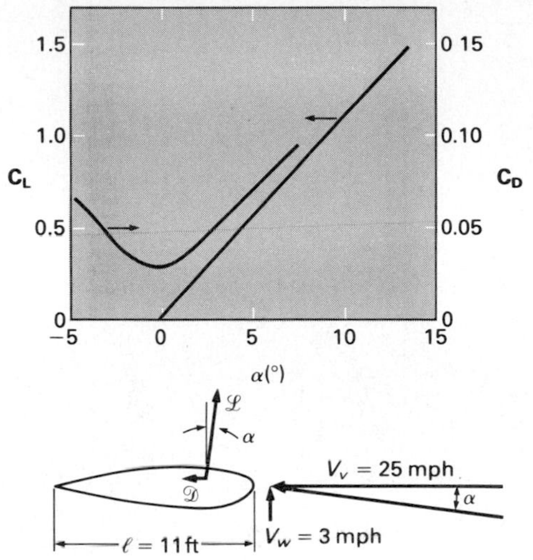

Figure P8.80

• **81.** A windmill has a wind approach velocity $V =$ 24 mph (35.3 ft/sec) and a blade diameter of 100 ft. The blade rotates at 15 rpm. Figure P8.81 shows the blade cross-sectional profile and approach velocity diagram for a 1-ft-long section of the blade at a radius of $r = 25.0$ ft (i.e., the section extends from $r = 24.5$ ft to $r = 25.5$ ft). The blade cross-sectional profile is that of an NACA 2415 airfoil with a 2-ft chord length. Find the tangential force F exerted on this 1-ft-long section by the air. Assume 60°F and 14.4 psia air.

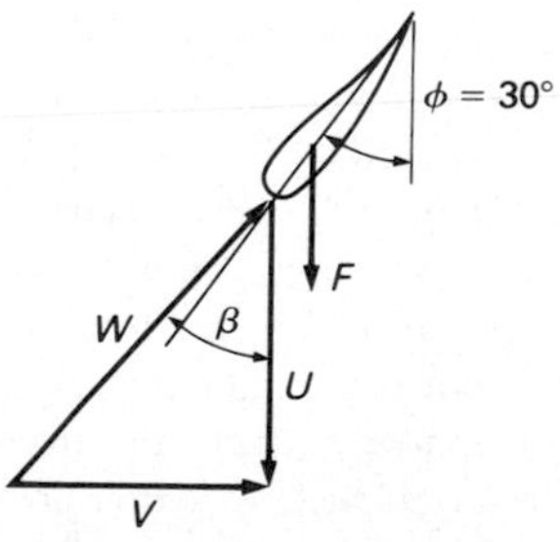

Figure P8.81

• **82.** A propeller-driven airplane is traveling with a velocity $V = 200$ mph (294 ft/sec). The propeller diameter is 10 ft and rotates at 3000 rpm. Fig. P8.82 shows the propeller cross-sectional profile and upstream velocity diagram for a 6-in. section of the propeller centered at a radius $r = 3.0$ ft. The blade cross-sectional profile is that of an NACA 2415 airfoil with a 1-ft chord length. Determine the tangential force F required to rotate this short section of the propeller in 60°F and 14.4 psia air.

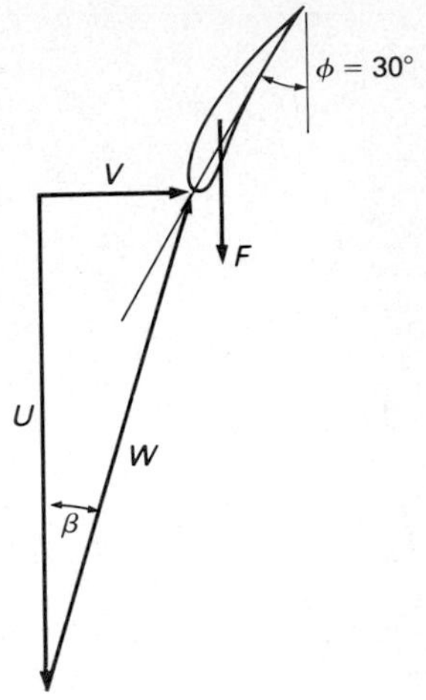

Figure P8.82

• **83.** Find the forward thrust, T, produced by the 6-in. section of the propeller in Problem 82. This thrust is in the opposite direction of the velocity vector V in Fig. P8.82.

• **84.** A 9000-N vehicle traveling at 11.4 m/s has a constant drag coefficient of 0.7 and a frontal area of 1.2 m^2. All four wheels skid on the ice when the brakes are applied. The coefficient of sliding friction between the tires and the ice is 0.10. Calculate the time required for the vehicle to slow to 8 m/s. The air is at 0°C.

• **85.** A full-sized automobile has a frontal area of 24 ft^2, and a compact car has a frontal area of 13 ft^2. Both have a drag coefficient of 0.5 based on the frontal area. Find the horsepower required to move each automobile along a level road in still air at 55 mph. Assume that the power required to deform the tires continuously at this speed (called rolling resistance) is equal to the power to overcome the air resistance. Estimate the gas mileage of both automobiles if the energy supplied to the drive wheels is $\frac{1}{4}$ that available in the fuel. A gallon of fuel has 1.0×10^8 ft·lb available energy.

• **86.** An automobile initially traveling at 35 mph experiences a constant force of 400 lb. The automobile has a frontal area of 25 ft^2, a constant drag coefficient of 0.30, and a weight of 3000 lb. Find the time required for the automobile to reach a speed of 55 mph. The drag coefficient accounts for both air drag and rolling resistance. The air density is 0.075 lbm/ft^3.

87. The solid steel circular rod shown in Fig. P8.87 is supported by two springs, each having a spring constant k of 50 lb/in. (i.e., a force of 50 lb will compress or stretch each spring a distance of 1.0 in.). The natural frequency of vibration of the rod–spring system is

$$\omega = \frac{1}{2\pi}\sqrt{\frac{k}{m\ell}},$$

where m is the mass per unit length and ℓ is the length of the rod. Find the wind velocity V_∞ that causes vortex-shedding resonance at the natural frequency of the rod. Use 60°F air and a steel density of 490 lbm/ft^3.

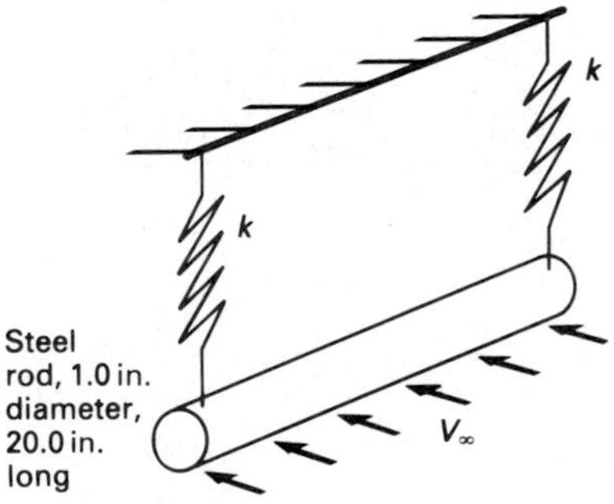

Figure P8.87

88. A solid aluminum rod has a length of 4.0 ft and a diameter of 2.0 in. It is pinned at its ends and has a spring constant k for a center deflection of

$$k = \frac{384EI}{5\ell^3},$$

where E is the modulus of elasticity of the rod, I the second moment of area about the diameter of its circular cross section, and ℓ its length. Find the wind velocity V_x that causes vortex-shedding resonance at the natural frequency of vibration of the rod. Refer to Problem 87 for a description of how to calculate the natural frequency of the rod. Assume 60°F air. For aluminum, $E = 10.6 \times 10^6$ lb/in. and $\rho = 0.100$ lbm/in^3.

89. During a winter storm, a variety of sounds can be heard as the wind strikes the bare tree branches and utility poles. If the wind velocity ranges from 50 to 100 mph and the tree branches measure 0.5 to 1.5 in. in diameter, estimate the range of frequencies of the sound generated during the storm.

90. An engineering–architectural firm has been awarded a project to build the tallest skyscraper in the world. The owner wants it to be 150 stories high (each story is 2.5 m high) and cylindrical in shape (15 m in diameter). The project engineer has decided that there are two structural criteria of aerodynamic origin: overturning moment and resonant frequency. Find both if winds (neglecting the Earth's boundary layer) are expected to reach 100 mph. Neglect the effect of surrounding buildings and assume that the building faces Lake Michigan.

91. Estimate the energy required for a 6-ft-tall person to run a mile in 4 min in still, 60°F air. Let the person be a cylinder 6 ft tall and 2 ft in diameter.

92. A vertical wind tunnel is to be built with an upward flow to keep people in suspension to learn sky-diving techniques. What would be the velocity of the 60°F air past the sky divers? Model a person as a cylinder 6 ft long, 2 ft in diameter, and weighing 150 lb.

93. A tennis ball traveling in a straight line and hit 10 ft above the ground at the baseline must be served at an angle θ below the horizontal to land in the opponent's forecourt (see Fig. P8.93). Assume that the ball is served at a speed of $V_0 = 70$ mph with a top spin of 40 rev/sec. Find the differential equations for $V_x(t, \theta)$ and $V_y(t_0, \theta)$ where V_x is the horizontal component of the total velocity V, V_y is the vertical component of V, θ is the angle below the horizontal that the velocity V makes with the horizontal, and t is time. The numerical values are given for reference only.

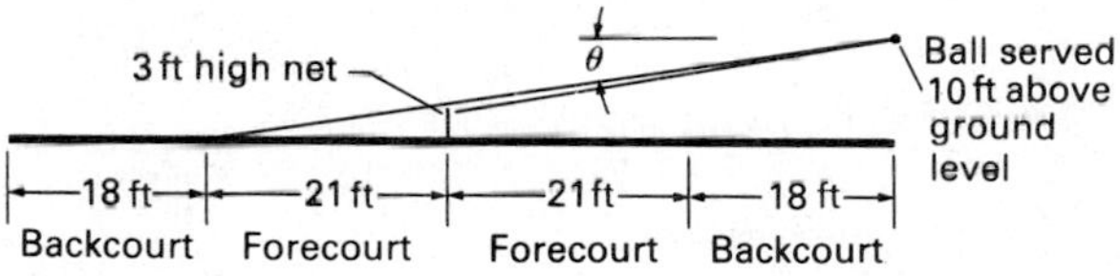

Figure P8.93

94. A baseball leaves a bat with a speed of $V_0 = 120$ mph and no spin, and at an angle of 30° above the horizontal. Write the differential equation of motion. Is this differential equation of motion any different if the ball has an overspin or an underspin ω? The numerical values are given for reference only. The ball's circumference is 9.0 in.

• **95.** A baseball leaves the pitcher's hand with a horizontal velocity of 90 mph and travels a distance of 45 ft. Neglect air drag and gravity, so the ball moves in a horizontal plane. The ball has a mass of 5 oz, a circumference of 9 in., a rotational speed of 1600 rev/min, and a baseball lift coefficient of 0.75. How far does the baseball "break" in the horizontal plane in 60°F, still air?

• **96.** A baseball leaves a bat with a horizontal velocity of 150 mph. The ball has no spin and has the same drag coefficient as the smooth sphere in Fig. 8.10. How far does the baseball travel before it falls five feet. The baseball weighs 5 oz, the air temperature is 60°F with no wind, and the ball's circumference is 9 in.

• **97.** A golf ball has a mass of 1.62 oz and a diameter of 1.68 in. The ball is hit at an angle of 20° above the horizontal in Standard Atmosphere air at sea level with an initial speed of 220 ft/sec. Find the distance the ball travels in still air, assuming that the drag coefficient is given by Eq. (8.12b).

REFERENCES

1. Shapiro, A. H. (Principal), *The Fluid Dynamics of Drag* (4 parts), NCFMF Film, available from Encyclopaedia Britannica Films, Chicago, IL.
2. White, F. M., *Viscous Fluid Flow,* McGraw-Hill, New York, 1974.
3. Hoerner, S. F., *Fluid Dynamic Drag,* published by Author, Midland Park, NJ, 1965.
4. Whitford, D. H., and J. E. Minardi, "Utility-Sized Madaras Wind Plants," *The International Journal of Ambient Energy,* vol. 2, no. 1, 1981.
5. Abbott, I. H., and A. E. von Doenhoff, *Theory of Wing Sections,* Dover Books, New York, 1959.
6. Lissaman and Shollenberger, "Formation Flight of Birds," *Science,* vol. 168, no. 3934, May 1970.

7. *Design for Fuel Economy: The General Motors X Cars,* SAE/SP-80/452, Society of Automotive Engineers, Warrendale, PA, 1980.

8. Scibor-Rylski, A. J., *Road Vehicle Aerodynamics,* Halsted Press, New York, 1975.

9. *Automotive Aerodynamics,* SAE/PT-78/16, Society of Automotive Engineers, Warrendale, PA, 1978.

10. Morel, T., and C. Dalton (Eds.), *Aerodynamics of Transportation,* proceedings of a symposium held at the Joint ASME-CSME Applied Mechanics, Fluids Engineering and Bioengineering Conference, Niagara Falls, NY, June 1979.

11. "Tacoma Narrows Bridge Collapse" (a film), available from Ohio State University, Motion Picture Division, Columbus, OH.

12. Blevins, R. D., *Flow Induced Vibration,* Van Nostrand-Reinhold, New York, 1977.

13. Fox, R. W., and A. T. MacDonald, *Introduction to Fluid Mechanics* (2nd ed.), Wiley, New York, 1978.

9 Potential Flow and Boundary Layer Theory

In Chapter 7 on internal flow and Chapter 8 on external flow, we showed how to apply information based on the control volume model and experimental data to analyze many practically important flows. We also demonstrated the use of data on friction factor and loss coefficients in designing and analyzing piping systems and data on drag and lift coefficients in evaluating aerodynamic forces. Unfortunately, there are two shortcomings in the information presented in those two chapters. First, many details of the flow pattern are not available. Second, and most importantly, information is available *only* for flow systems that have actually been built and tested. There is no information on how flattening the nose of an NACA 2415 airfoil would affect the drag, for example.

In Chapter 5, we pointed out that the differential approach reveals all the details of any flow. In the differential approach, we formulate the fundamental physical laws as a set of differential equations and describe the specific flow geometry through boundary conditions. The difficulties with the differential approach are primarily mathematical; solutions to the general differential equations are hard to obtain. Practical flow calculations are usually based on simplified models and simplified equations.

One of the most valuable models is the potential flow and boundary layer model introduced in Section 5.6.2. The essential feature of the model is the division of a *high Reynolds number* flow into two regions: a thin layer, in which the flow is viscous and rotational, and an outer flow, in which the effects of viscosity and rotation can be neglected. We used the boundary layer concept extensively in Chapter 8 to explain drag and lift in high Reynolds number flows. Figure 9.1 shows the boundary layer and potential flow regions for typical internal and external flows.

The mathematical advantage of the potential flow-boundary layer model is that each region is described by a set of differential equations that are easier to solve than the full Navier–Stokes equations. To a first approximation, we obtain the entire flow field by calculating the two regions independently and then combining the results. Figure 9.2 illustrates the procedure schematically. We obtain the pressure distribution and streamline pattern from the potential flow calculation. We obtain surface shear stresses from the boundary layer calculation. The boundary layer calculation also yields the thickness of the boundary layer and any indication of flow separation. The boundary layer thickness allows us to account for the effect of the boundary layer on the potential flow. Separation signals the breakdown of the model.

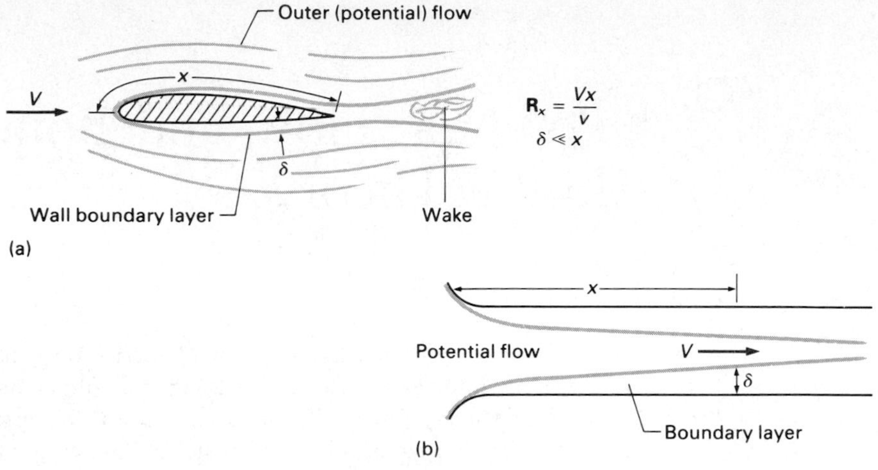

Figure 9.1 Division of high Reynolds number flow into potential flow and boundary layer regions: (a) boundary layer and wake in external flow; (b) boundary layer and ideal flow regions in developing flow.

We need to note how the physical boundary conditions are handled in the potential flow-boundary layer model. The potential flow calculation handles the uniform flow at "infinity" condition for external flow or the specified mass flow condition for internal flows. The boundary layer calculation handles the no-slip condition at a solid surface. Both calculations use the zero normal (relative) velocity condition at a solid surface. The potential flow calculation produces a finite tangential (slip) velocity at a solid surface (it does not satisfy the no-slip condition). This slip velocity is taken to be the velocity at the *outer edge* of the boundary layer in the boundary layer calculation.

We devote the remainder of this chapter to separate consideration of potential flow and boundary layer calculation. Our treatment of both subjects allows

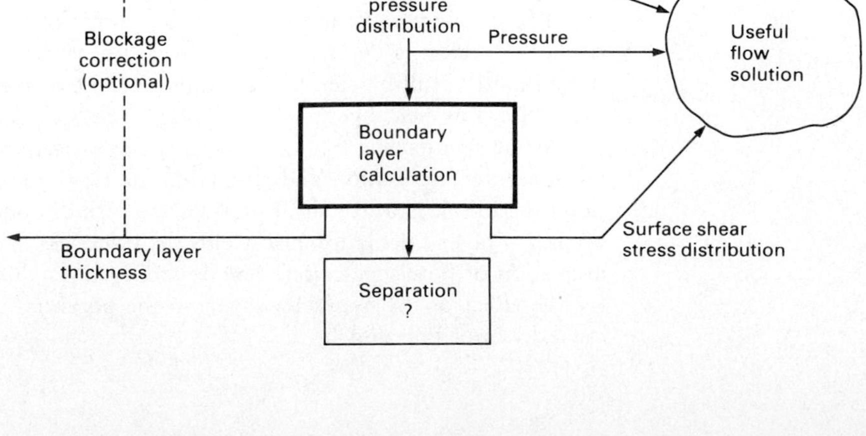

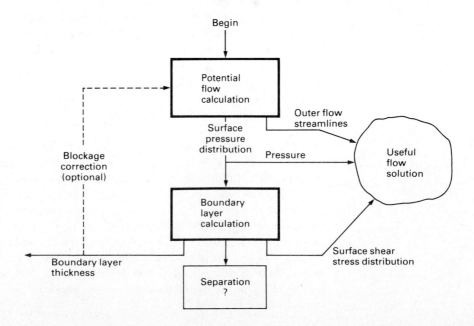

Figure 9.2 Schematic of calculations in the high **R** potential flow-boundary layer model.

you to study them in any order, or to study only one of them. Entire books have been written on each subject ([1, 2] and [3, 4, 5]), so our consideration here is only a brief introduction. If you are interested in further study, also see references [6, 7, 8, 9].

9.1 POTENTIAL FLOW THEORY

9.1.1 Dynamics of Inviscid, Irrotational Flow

Figure 9.3 illustrates the type of flows that we want to calculate. We assume that the fluid is inviscid and incompressible. The governing equations, in Cartesian coordinates, are the continuity equation, Eq. (5.24),

$$\frac{\partial u}{\partial x} + \frac{\partial v}{\partial y} + \frac{\partial w}{\partial z} = 0,$$

and Euler's equations, Eqs. (5.41)–(5.43),

$$\frac{\partial u}{\partial t} + u\frac{\partial u}{\partial x} + v\frac{\partial u}{\partial y} + w\frac{\partial u}{\partial z} = -\frac{1}{\rho}\frac{\partial p}{\partial x} + g_x;$$

$$\frac{\partial v}{\partial t} + u\frac{\partial v}{\partial x} + v\frac{\partial v}{\partial y} + w\frac{\partial v}{\partial z} = -\frac{1}{\rho}\frac{\partial p}{\partial y} + g_y;$$

$$\frac{\partial w}{\partial t} + u\frac{\partial w}{\partial x} + v\frac{\partial w}{\partial y} + w\frac{\partial w}{\partial z} = -\frac{1}{\rho}\frac{\partial p}{\partial z} + g_z.$$

We simplify these equations slightly by defining

$$p' \equiv p - \rho g h = p - \gamma h, \tag{9.1}$$

where h is the depth below a specified horizontal datum. Assuming that ρg is constant, we get

$$\frac{\partial p'}{\partial x} = \frac{\partial p}{\partial x} - \rho g\frac{\partial h}{\partial x} = \frac{\partial p}{\partial x} - \rho g_x.$$

(See Fig. 5.15.) Similar relations hold for $\partial p'/\partial y$ and $\partial p'/\partial z$, so we may write the

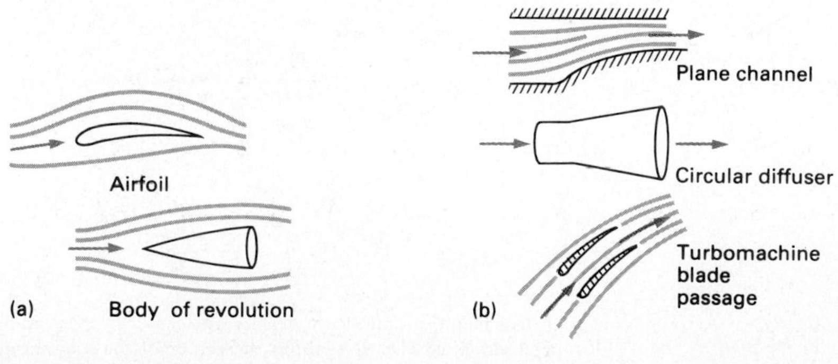

Figure 9.3 Typical flow geometries for potential flow problems: (a) external flows; (b) internal flows.

Euler equations (using the Eulerian derivative operator D/Dt) as

$$\frac{Du}{Dt} = -\frac{1}{\rho}\frac{\partial p'}{\partial x}; \tag{9.2}$$

$$\frac{Dv}{Dt} = -\frac{1}{\rho}\frac{\partial p'}{\partial y}; \tag{9.3}$$

$$\frac{Dw}{Dt} = -\frac{1}{\rho}\frac{\partial p'}{\partial z}. \tag{9.4}$$

In Eqs. (9.2)–(9.4), p' is the difference between the fluid static pressure (p) and the hydrostatic pressure (γh). For incompressible flow, we can solve for p' and then *add* the hydrostatic pressure to obtain p.

If the flow is steady and irrotational, Euler's equations, Eqs. (9.2)–(9.4), can be integrated once and for all as Bernoulli's equation. In the potential flow model, the outer flow is assumed to be inviscid *and* irrotational. The assumption that the flow is inviscid usually implies that the flow is also irrotational. This result can be derived from the Navier–Stokes equations and also justified by physical reasoning.

First, let's consider the physical argument. Recall that an irrotational flow is one in which the average angular velocity of all fluid particles is zero. For a fluid at rest or in uniform, parallel (one-directional, zero-dimensional) motion, all the fluid particles obviously have zero angular velocity. The only way that a fluid particle can obtain any angular velocity is from the action of a torque on it. If there are no shear stresses (inviscid fluid), the only forces on the particle are normal forces (pressure, which acts through the geometric center) and gravity (which acts through the center of mass), neither of which is capable of exerting a net torque on a fluid particle.* Even if the still fluid is disturbed by a solid moving through it or if an object is placed in the moving stream, there is no rotation if there are no shear stresses. We conclude that in the absence of viscosity, a fluid particle that initially has zero angular velocity maintains that condition.

We now describe development of the connection between viscosity and rotation from the Navier–Stokes equations. Fluid rotation is quantified by the vorticity (Section 5.1.3), Eq. (5.13),

$$\vec{\zeta} \equiv \nabla \times \vec{V}.$$

An irrotational flow has $\vec{\zeta} = 0$ everywhere.

We write the Navier–Stokes equations, Eqs. (5.63)–(5.65), in vector notation as

$$\frac{D\vec{V}}{Dt} = -\frac{1}{\rho}\nabla p + \vec{g} + \frac{\mu}{\rho}\nabla^2\vec{V}, \tag{9.5}$$

where

$$\nabla = \hat{i}\frac{\partial}{dx} + \hat{j}\frac{\partial}{\partial y} + \hat{k}\frac{\partial}{\partial z}$$

* In a fluid of nonuniform density, the center of mass and the geometric center of a fluid particle do not coincide, so a torque about the center of mass of the particle may be generated. We do not consider such cases here.

and

$$\nabla^2 = \frac{\partial^2}{\partial x^2} + \frac{\partial^2}{\partial y^2} + \frac{\partial^2}{\partial z^2}. \tag{9.6}$$

Taking the curl ($\nabla \times$) of this equation (details omitted) gives the following *vorticity transport equation:*

$$\frac{D\vec{\zeta}}{Dt} = (\vec{\zeta} \cdot \nabla)\vec{V} + \frac{\mu}{\rho}\nabla^2\vec{\zeta}. \tag{9.7}$$

Note that the vorticity transport equation does not contain pressure. If the fluid is inviscid ($\mu = 0$), the second term on the right vanishes, and we get

$$\frac{D\vec{\zeta}}{Dt} = (\vec{\zeta} \cdot \nabla)\vec{V}.$$

Recall that D/Dt of any quantity is the time rate of change of that quantity for a specific fluid particle. If a fluid particle initially has $\vec{\zeta} = 0$, then $D\vec{\zeta}/Dt$ is zero and the fluid particle never gains vorticity. Because the vorticity equation applies to all fluid particles, an initially irrotational flow field remains irrotational if viscosity is zero.

In summary, we can describe the dynamics of incompressible, inviscid flow as follows:

- If we neglect fluid viscosity, an initially irrotational flow remains irrotational.
- If the flow is irrotational and steady, we can replace Euler's (differential) equations by Bernoulli's (algebraic) equation,

$$\frac{p}{\rho} + \frac{V^2}{2} + gz = \frac{p'}{\rho} + \frac{V^2}{2} = \text{Constant} \tag{9.8}$$

for the entire flow field.

The problem reduces to that of finding the velocity.

9.1.2 Velocity Potential and Stream Function

Velocity has three components, so we assume that we need three equations to calculate them. Suppose that we could find the velocity from a single scalar function $\Phi(x, y, z)$ by

$$\vec{V} = \text{Gradient }(\Phi) = \nabla\Phi. \tag{9.9}$$

In component form,

$$u = \frac{\partial\Phi}{\partial x}; \tag{9.10}$$

$$v = \frac{\partial\Phi}{\partial y}; \tag{9.11}$$

$$w = \frac{\partial\Phi}{\partial z}. \tag{9.12}$$

Mathematically necessary and sufficient conditions for the existence of such a function are

$$\frac{\partial}{\partial x}\left(\frac{\partial \Phi}{\partial y}\right) - \frac{\partial}{\partial y}\left(\frac{\partial \Phi}{\partial x}\right) = 0;$$

$$\frac{\partial}{\partial z}\left(\frac{\partial \Phi}{\partial x}\right) - \frac{\partial}{\partial x}\left(\frac{\partial \Phi}{\partial z}\right) = 0;$$

$$\frac{\partial}{\partial y}\left(\frac{\partial \Phi}{\partial z}\right) - \frac{\partial}{\partial z}\left(\frac{\partial \Phi}{\partial y}\right) = 0.$$

Substituting Eqs. (9.10)–(9.12) gives

$$\frac{\partial v}{\partial x} - \frac{\partial u}{\partial y} = 0;$$

$$\frac{\partial u}{\partial z} - \frac{\partial w}{\partial x} = 0;$$

$$\frac{\partial w}{\partial y} - \frac{\partial v}{\partial z} = 0.$$

From Eqs. (5.10)–(5.12), the terms on the left are the three components of the vorticity; therefore the function Φ exists if the flow is irrotational.

The function Φ is called the *velocity potential.* A velocity potential can be defined only for irrotational flow, and any velocities derived from a potential function always constitute an irrotational flow. Solving for a single scalar potential function is usually simpler than solving for two or three velocity components, so inviscid-irrotational flow problems are almost always analyzed in terms of velocity potential. That is why the entire field of study is often called *potential flow theory.* A velocity potential can even be defined for a compressible fluid. Potential flow theory limited to incompressible fluids is sometimes called *ideal flow theory* (ideal = incompressible + inviscid) or *hydrodynamics.*

In order to determine the velocity potential for a particular flow geometry, we must solve a single equation. The appropriate equation can be derived from the continuity equation, Eq. (5.24). Substituting the velocity components from Eqs. (9.10)–(9.11) gives

$$\frac{\partial}{\partial x}\left(\frac{\partial \Phi}{\partial x}\right) + \frac{\partial}{\partial y}\left(\frac{\partial \Phi}{\partial y}\right) + \frac{\partial}{\partial z}\left(\frac{\partial \Phi}{\partial z}\right) = 0,$$

which simplifies to

$$\frac{\partial^2 \Phi}{\partial x^2} + \frac{\partial^2 \Phi}{\partial y^2} + \frac{\partial^2 \Phi}{\partial z^2} = \nabla^2 \Phi = 0, \tag{9.13}$$

where ∇^2 is defined in Eq. (9.6). Equation (9.13) is called *Laplace's equation* and is probably the best known partial differential equation in all of mathematics. In more compact and more general vector notation, the equation is

$$\nabla^2 \Phi = \nabla \cdot (\nabla \Phi) = 0.$$

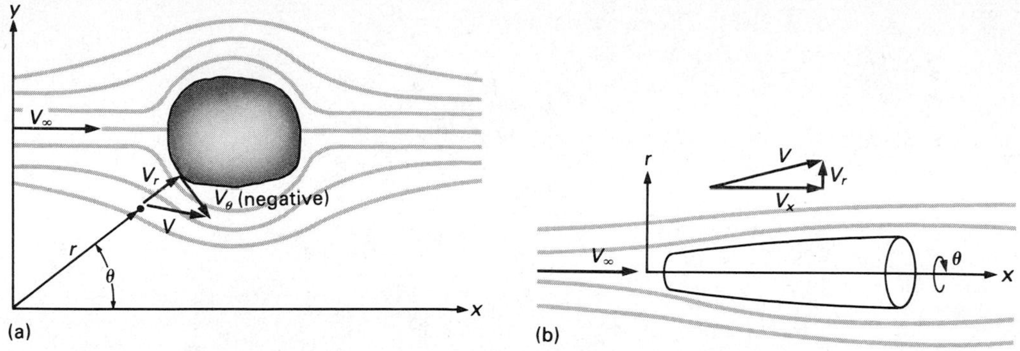

Figure 9.4 Alternative coordinate systems for potential flow analysis: (a) plane polar coordinates; (b) cylindrical coordinates.

The vector form is applicable in any coordinate system.

Calculation of three-dimensional, three-directional potential flow is difficult. Two-dimensional, two-directional flows exhibit most of the important characteristics of potential flow and are easier to solve. If we limit consideration to two-dimensional, two-directional flows and choose Cartesian coordinates (x, y), the velocities (u, v) are given by Eqs. (9.10) and (9.11) and Φ satisfies the two-dimensional Laplace equation

$$\frac{\partial^2 \Phi}{\partial x^2} + \frac{\partial^2 \Phi}{\partial y^2} = 0. \tag{9.14}$$

Cartesian coordinates are not always the most convenient for analyzing potential flow. For two-dimensional flow in a plane, using polar coordinates (Fig. 9.4a) is sometimes convenient. In plane polar coordinates, the velocity potential equation is

$$\nabla^2 \Phi = \frac{\partial}{\partial r}\left(r \frac{\partial \Phi}{\partial r}\right) + \frac{1}{r}\left(\frac{\partial^2 \Phi}{\partial \theta^2}\right) = 0, \tag{9.15}$$

and the velocity components are

$$V_r = \frac{\partial \Phi}{\partial r} \tag{9.16}$$

and

$$V_\theta = \frac{1}{r}\left(\frac{\partial \Phi}{\partial \theta}\right). \tag{9.17}$$

Flow over an axisymmetric body at zero angle of attack or through a passage of circular cross section is conveniently analyzed in cylindrical coordinates

(Fig. 9.4b). The equations are

$$\nabla^2\Phi = \frac{1}{r}\left(\frac{\partial}{\partial r}\right)\left(r\frac{\partial\Phi}{\partial r}\right) + \frac{\partial^2\Phi}{\partial x^2} = 0, \tag{9.18}$$

$$V_x = \frac{\partial\Phi}{\partial x}, \tag{9.19}$$

and

$$V_r = \frac{\partial\Phi}{\partial r}. \tag{9.20}$$

For two-dimensional flow, we can define a *second* scalar function from which the fluid velocity components may be derived. As defined in Section 5.1, the stream function (or streamline function), $\Psi\{x, y\}$, permits derivation of the velocity components from

$$u = \frac{\partial\Psi}{\partial y} \tag{9.21}$$

and

$$v = -\frac{\partial\Psi}{\partial x}. \tag{9.22}$$

In Section 5.1, we showed that

- the (two-dimensional) continuity equation is the necessary and sufficient condition for Ψ to exist;
- velocity fields derived from Ψ automatically satisfy the continuity equation; and
- lines with Ψ = Constant are tangent to the velocity vector and are streamlines.

Note that it is not necessary for the flow to be irrotational for Ψ to exist, although Ψ can exist for an irrotational flow.

The stream function definition can be put into plane polar coordinates. The continuity equation is

$$\frac{1}{r}\frac{\partial(rV_r)}{\partial r} + \frac{1}{r}\frac{\partial V_\theta}{\partial\theta} = 0. \tag{9.23}$$

Defining

$$V_r = \frac{1}{r}\frac{\partial\Psi}{\partial\theta} \tag{9.24}$$

and

$$V_\theta = -\frac{\partial \Psi}{\partial r} \tag{9.25}$$

gives

$$\frac{1}{r}\frac{\partial}{\partial r}\left(\frac{\partial \Psi}{\partial \theta}\right) - \frac{1}{r}\frac{\partial}{\partial \theta}\left(\frac{\partial \Psi}{\partial r}\right) \equiv 0.$$

We can also define a stream function for axisymmetric flow, but the stream function must have a slightly different form. For axisymmetric flow, the continuity equation is

$$\frac{1}{r}\frac{\partial}{\partial r}(rV_r) + \frac{\partial V_x}{\partial x} = 0. \tag{9.26}$$

Multiplying by r and rearranging give

$$-\frac{\partial}{\partial r}(-rV_r) + \frac{\partial}{\partial x}(rV_x) = 0.$$

This result allows us to define a function Ψ_s such that

$$rV_r = -\frac{\partial \Psi_s}{\partial x} \quad \text{and} \quad rV_x = +\frac{\partial \Psi_s}{\partial r},$$

or

$$V_r = -\frac{1}{r}\frac{\partial \Psi_s}{\partial x}$$

and

$$V_x = \frac{1}{r}\frac{\partial \Psi_s}{\partial r}. \tag{9.28}$$

The term Ψ_s is called *Stokes's stream function*. Stokes's stream function (Ψ_s) has all the physical characteristics of the plane two-dimensional stream function (Ψ). Proof is left to an exercise.

We derive another important property of the stream function as follows. Consider the change of Ψ between two streamlines (Fig. 9.5a):

$$\Psi_2 - \Psi_1 = \int_1^2 d\Psi = \int_1^2 \frac{\partial \Psi}{\partial x}\,dx + \int_1^2 \frac{\partial \Psi}{\partial y}\,dy.$$

Substituting from Eqs. (9.21) and (9.22) gives

$$\Psi_2 - \Psi_1 = \int_1^2 (-v\,dx + u\,dy) = \int_2^1 v\,dx + \int_1^2 u\,dy.$$

From Fig. 9.5(b) the integral represents the volume flow rate (per unit depth) between the two curves; thus

$$\Psi_2 - \Psi_1 = \int_2^1 v\,dx + \int_1^2 u\,dy = \int_1^2 dQ = Q_{1-2}.$$

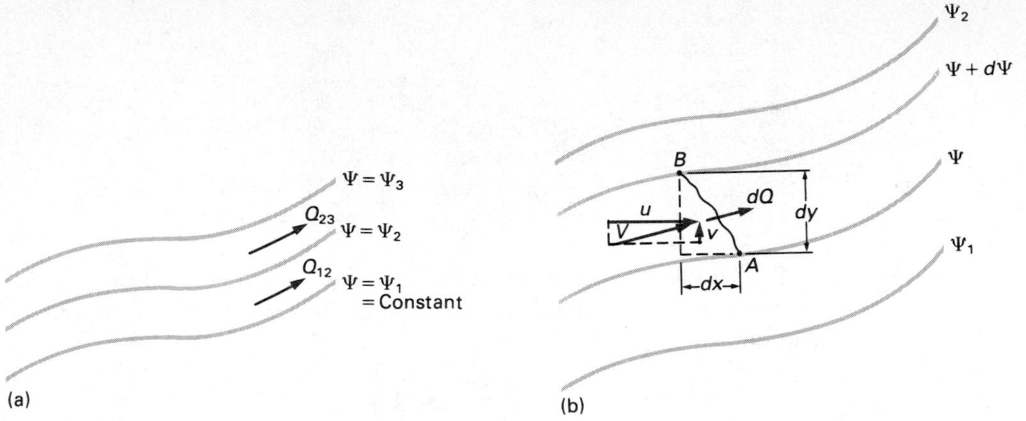

Figure 9.5 Geometry of lines of constant Ψ (streamlines): (a) lines with Ψ = Constant (streamlines); (b) geometry for calculating flow rate between two streamlines. Going from A to B, dy is positive and dx is negative.

Therefore:

> The difference in Ψ between two streamlines is the volume flow rate (per unit depth) between the two streamlines.

To determine the stream function for a particular flow, we must solve an appropriate governing differential equation. For the velocity potential, the irrotationality condition *defined* Φ, and the continuity equation provided the appropriate governing equation. For the stream function, the continuity equation defines Ψ, and we can find the appropriate governing equation from the irrotationality (zero vorticity) condition. For Cartesian coordinates, we express vorticity in terms of Ψ by

$$\zeta_z = \frac{\partial v}{\partial x} - \frac{\partial u}{\partial y} = \frac{\partial}{\partial x}\left(-\frac{\partial \Psi}{\partial x}\right) - \frac{\partial}{\partial y}\left(\frac{\partial \Psi}{\partial y}\right) = -\nabla^2\Psi.$$

For irrotational flow, ζ_z is zero and Ψ must satisfy

$$\frac{\partial^2 \Psi}{\partial x^2} + \frac{\partial^2 \Psi}{\partial y^2} = 0. \tag{9.29}$$

This expression is the same form of equation (Laplace's) that governs Φ, but of course this does not mean that Φ and Ψ are the same function because each depends on different boundary conditions. In plane polar coordinates, the differential equation for Ψ is

$$\frac{\partial}{\partial r}\left(r\frac{\partial \Psi}{\partial r}\right) + \frac{1}{r}\frac{\partial^2 \Psi}{\partial \theta^2} = 0. \tag{9.30}$$

The differential equation for Stokes's stream function is obtained from the irrotationality condition

$$-\zeta_\theta = \frac{\partial^2 \Psi_s}{\partial r^2} - \frac{1}{r}\frac{\partial \Psi_s}{\partial r} + \frac{\partial^2 \Psi_s}{\partial x^2} = 0. \tag{9.31}$$

This expression is *almost* Laplace's equation in axisymmetric cylindrical coordinates, but not quite; the sign of the second term is different.

A final important property of potential and stream functions is:

> Lines of Φ = Constant (called *equipotential lines*) are perpendicular to lines of Ψ = Constant (*streamlines*).

We demonstrate this property by noting that

$$\left.\frac{dy}{dx}\right)_{\Phi=\text{constant}} = -\frac{\partial\Phi/\partial x}{\partial\Phi/\partial y} = -\frac{u}{v} \quad \text{and} \quad \left.\frac{dy}{dx}\right)_{\Psi=\text{constant}} = -\frac{\partial\Psi/\partial y}{\partial\Psi/\partial x} = \frac{v}{u}.$$

As equipotential lines and streamlines have slopes that are negative reciprocals, they are perpendicular. This relation fails only at stagnation points, where both u and v are zero.

Figure 9.6 shows equipotential lines and streamlines for external and internal flows. Determining this network of lines for a particular boundary shape is tantamount to solving the corresponding potential flow problem. Solving potential flow problems is essentially a geometric exercise that might even be done (approximately) by graphic construction of a flow net like those shown in Fig. 9.6. Of particular interest is that the primary equations of potential flow, such as Eqs. (9.13) and (9.29), do not contain any fluid properties (not even density). There are no dynamic similarity parameters in (incompressible) potential flow; geometric similarity alone is sufficient to guarantee kinematic similarity. The only thing that distinguishes one potential flow from another is the shape of the boundaries.

To solve for Φ, Ψ, or Ψ_s, we have to specify boundary conditions. We derive them from boundary conditions on velocity by noting that the derivative of Φ in a given direction is equal to the velocity in that direction and that the derivative of Ψ (or Ψ_s) in a given direction is the velocity perpendicular to that direction. Various boundary conditions are listed in Table 9.1. Also see Fig. 9.7.

Note that the stream function is constant on a solid surface; that is, a solid surface is a streamline. We can put this knowledge to great use: We can replace any streamline in a potential flow with a solid surface. Many useful potential flow solutions are obtained by finding an interesting streamline pattern and then replacing one or more of the streamlines by a solid surface. The remaining streamlines represent the potential flow over that surface.

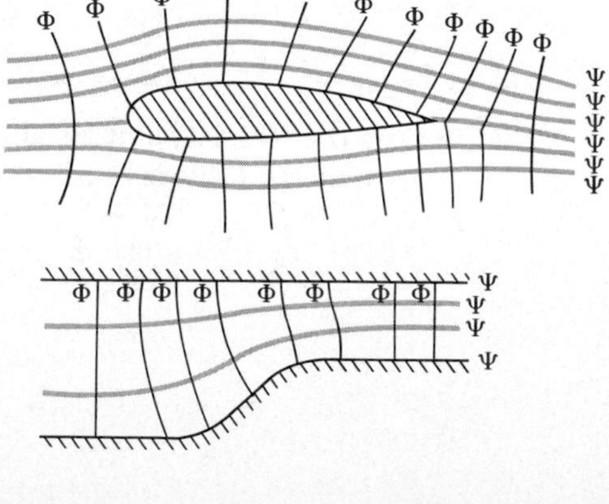

Figure 9.6 Equipotential (Φ) lines and streamlines (Ψ) for external and internal flow.

Table 9.1 Boundary conditions for potential flow.

Boundary	Physical Boundary Condition	Condition on Φ	Condition on Ψ	Condition on Ψ_s
Solid surface	Velocity normal to surface = 0	$\dfrac{\partial \Phi}{\partial n} = 0$	$\dfrac{\partial \Psi}{\partial s} = 0$, $\Psi = \text{Constant}$	$\dfrac{\partial \Psi_s}{\partial s} = 0$, $\Psi_s = \text{Constant}$
Inlet or outlet	Velocity a known function of x, y	$\dfrac{\partial \Phi}{\partial x}$ and $\dfrac{\partial \Phi}{\partial y}$ known	$\dfrac{\partial \Psi}{\partial x}$ and $\dfrac{\partial \Psi}{\partial y}$ known	$\dfrac{1}{r}\left(\dfrac{\partial \Psi_s}{\partial x}\right)$ and $\dfrac{1}{r}\left(\dfrac{\partial \Psi_s}{\partial r}\right)$ known
Inlet or outlet	Uniform parallel flow in x direction	$\Phi = V_\infty x + C$	$\Psi = V_\infty y + C$	$\Psi_s = \frac{1}{2}V_\infty r^2 + C$
Outer boundary far from surface	Uniform parallel flow in x direction	$\Phi = V_\infty x + C$	$\Psi = V_\infty y + C$	$\Psi_s = \frac{1}{2}V_\infty r^2 + C$

Notes: n = Coordinate normal to surface; s = Coordinate parallel to surface; C = Constant of integration.

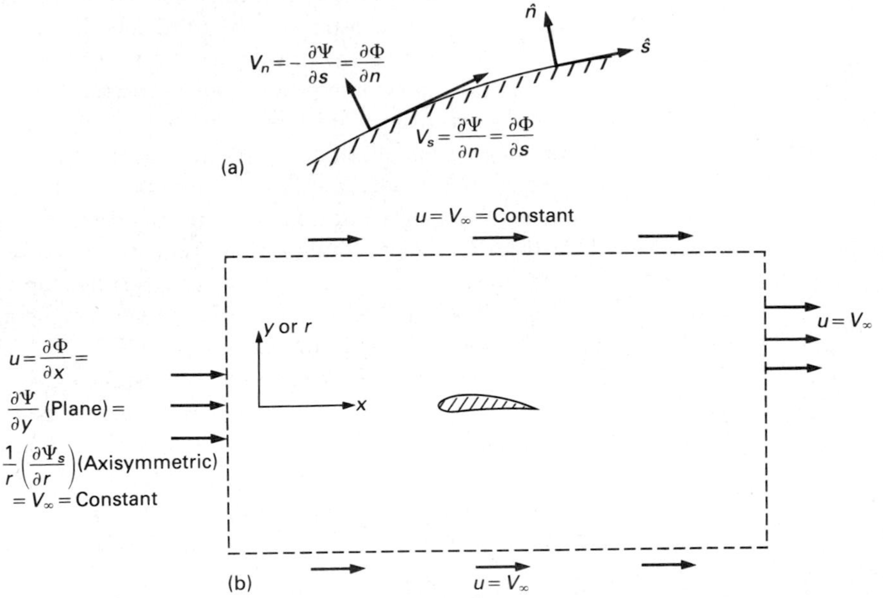

Figure 9.7 Boundary conditions for potential flow: (a) velocities near solid surface; (b) velocities at faraway boundaries.

EXAMPLE 9.1 **Illustrates the Velocity Potential and Stream Function for Flow over an Object in a Uniform Stream**

Show that the velocity potential Φ and the stream function Ψ given by the following:

$$\Phi = V_\infty\left(r + \frac{R^2}{r}\right)\cos\theta, \qquad r \geq R,$$

and

$$\Psi = V_\infty\left(r - \frac{R^2}{r}\right)\sin\theta, \qquad r \geq R,$$

satisfy the continuity equation and the irrotationality condition and yield a velocity field that satisfies the boundary conditions:

$$V_r = 0 \qquad \text{at } r = R,$$
$$V_r = V_\infty \cos\theta \qquad \text{at } r \to \infty,$$
$$V_\theta = -V_\infty \sin\theta \qquad \text{at } r \to \infty.$$

Therefore, demonstrate that the given potential and stream functions describe the two-dimensional flow over a cylinder of radius R.

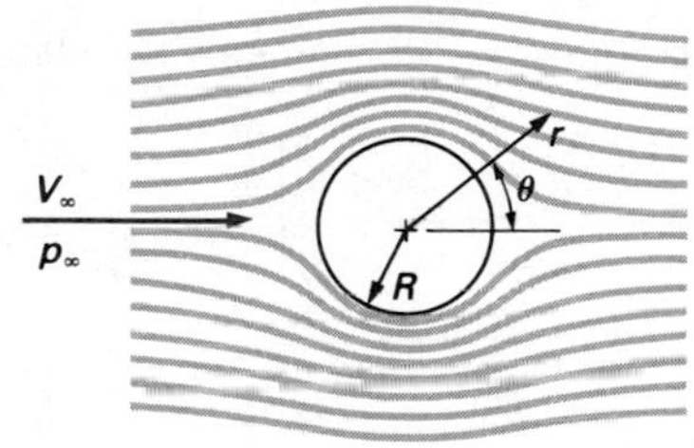

Figure E9.1 Flow over circular cylinder.

SOLUTION

Given

Velocity potential:

$$\Phi = V_\infty\left(r + \frac{R^2}{r}\right)\cos\theta, \qquad r \geq R$$

Stream function:

$$\Psi = V_\infty\left(r - \frac{R^2}{r}\right)\sin\theta, \qquad r \geq R$$

Figure E9.1

Show

That Φ and Ψ both satisfy the continuity equation and the irrotationality condition and yield a velocity field that satisfies the boundary conditions:

$$V_r = 0 \qquad \text{at } r = R,$$
$$V_r = V_\infty \cos\theta \qquad \text{at } r \to \infty,$$
$$V_\theta = -V_\infty \sin\theta \qquad \text{at } r \to \infty$$

Solution

We will use polar coordinates. In terms of Φ, the continuity equation is Eq. (9.15), or

$$\frac{\partial}{\partial r}\left(r\frac{\partial \Phi}{\partial r}\right) + \frac{1}{r}\left(\frac{\partial^2 \Phi}{\partial \theta^2}\right) = 0.$$

Substituting for Φ gives

$$\frac{\partial}{\partial r}\left[rV_\infty\left(1 - \frac{R^2}{r^2}\right)\cos\theta\right] + \frac{1}{r}\left[V_\infty\left(r + \frac{R^2}{r}\right)(-\cos\theta)\right]$$
$$= V_\infty\left(1 + \frac{R^2}{r^2}\right)\cos\theta - V_\infty\left(1 + \frac{R^2}{r^2}\right)\cos\theta = 0,$$

and

Φ satisfies the continuity equation. **ANSWER**

In terms of Ψ, the continuity equation is Eq. (9.30),

$$\frac{\partial}{\partial r}\left(r\frac{\partial \Psi}{\partial r}\right)+\frac{1}{r}\left(\frac{\partial^2 \Psi}{\partial \theta^2}\right)=0.$$

Substituting for Ψ gives

$$\begin{aligned}\frac{\partial}{\partial r}&\left[rV_\infty\left(1+\frac{R^2}{r^2}\right)\sin\theta\right]+\frac{1}{r}\left[V_\infty\left(r-\frac{R^2}{r}\right)(-\sin\theta)\right]\\ &=V_\infty\left(1-\frac{R^2}{r^2}\right)\sin\theta+V_\infty\left(1-\frac{R^2}{r^2}\right)(-\sin\theta)=0,\end{aligned}$$

and

Ψ satisfies the continuity equation. **ANSWER**

The irrotationality condition for plane flow in polar coordinates is

$$\zeta_z=\frac{1}{r}\left[\frac{\partial}{\partial r}(rV_\theta)-\frac{\partial}{\partial \theta}(V_r)\right]=0.$$

Now

$$V_r=\frac{\partial \Phi}{\partial r}=V_\infty\left(1-\frac{R^2}{r^2}\right)\cos\theta$$

and

$$V_\theta=\frac{1}{r}\left(\frac{\partial \Phi}{\partial \theta}\right)=-\frac{V_\infty}{r}\left(r+\frac{R^2}{r}\right)\sin\theta.$$

Substituting gives

$$\begin{aligned}\zeta_z&=\frac{1}{r}\frac{\partial}{\partial r}\left[-V_\infty\left(r+\frac{R^2}{r}\right)\sin\theta\right]\\ &\quad-\frac{1}{r}\frac{\partial}{\partial \theta}\left[V_\infty\left(1-\frac{R^2}{r^2}\right)\cos\theta\right]\\ &=\frac{1}{r}\left[-V_\infty\left(1-\frac{R^2}{r^2}\right)\sin\theta\right]+\frac{1}{r}\left[V_\infty\left(1-\frac{R^2}{r^2}\right)\sin\theta\right]\\ &=0\end{aligned}$$

Therefore

Φ satisfies the irrotationality condition. **ANSWER**

Noting that

$$\begin{aligned}V_r&=\frac{1}{r}\left(\frac{\partial \Psi}{\partial \theta}\right)=\frac{1}{r}\left[V_\infty\left(r-\frac{R^2}{r}\right)\cos\theta\right]\\ &=V_\infty\left(1-\frac{R^2}{r^2}\right)\cos\theta\end{aligned}$$

and

$$V_\theta=-\frac{\partial \Psi}{\partial r}=-V_\infty\left(1+\frac{R^2}{r^2}\right)\sin\theta$$

are the same expressions obtained for V_r and V_θ from Φ, we conclude that

$$\Psi \text{ satisfies the irrotationality condition.} \qquad \textbf{ANSWER}$$

Last we consider the boundary conditions. The velocity components are given by Eqs. (9.16), (9.17), (9.24), and (9.25):

$$V_r = \frac{\partial \Phi}{\partial r}, \qquad V_r = \frac{1}{r}\left(\frac{\partial \Psi}{\partial \theta}\right), \qquad V_\theta = \frac{1}{r}\left(\frac{\partial \Phi}{\partial \theta}\right), \quad \text{and}$$

$$V_\theta = -\frac{\partial \Psi}{\partial r}.$$

At $r = R$,

$$V_r = \left.\frac{\partial \Phi}{\partial r}\right|_{r=R} = \left[V_\infty\left(1 - \frac{R^2}{r^2}\right)\cos\theta\right]_{r=R};$$

$$V_r = 0 \qquad \text{for } \Phi \text{ at } r = R; \qquad \textbf{ANSWER}$$

$$V_r = \frac{1}{r}\left.\left(\frac{\partial \Psi}{\partial \theta}\right)\right|_{r=R} = \left[\frac{V_\infty}{r}\left(r - \frac{R^2}{r}\right)\cos\theta\right]_{r=R};$$

$$V_r = 0 \qquad \text{for } \Psi \text{ at } r = R. \qquad \textbf{ANSWER}$$

At $r \to \infty$,

$$V_r = \left.\frac{\partial \Phi}{\partial r}\right|_{r\to\infty} = \left[V_\infty\left(1 - \frac{R^2}{r^2}\right)\cos\theta\right]_{r\to\infty};$$

$$V_r = V_\infty \cos\theta \qquad \text{for } \Phi \text{ at } r \to \infty; \qquad \textbf{ANSWER}$$

$$V_r = \frac{1}{r}\left.\left(\frac{\partial \Psi}{\partial \theta}\right)\right|_{r\to\infty} = \left[\frac{V_\infty}{r}\left(r - \frac{R^2}{r}\right)\cos\theta\right]_{r\to\infty};$$

$$V_r = V_\infty \cos\theta \qquad \text{for } \Psi \text{ at } r \to \infty; \qquad \textbf{ANSWER}$$

$$V_\theta = \frac{1}{r}\left.\left(\frac{\partial \Phi}{\partial \theta}\right)\right|_{r\to\infty} = \left[\frac{-V_\infty}{r}\left(r + \frac{R^2}{r^2}\right)\sin\theta\right]_{r\to\infty};$$

$$V_\theta = -V_\infty \sin\theta \qquad \text{for } \Phi \text{ at } r \to \infty; \qquad \textbf{ANSWER}$$

and

$$V_\theta = -\frac{\partial \Psi}{\partial r} = -\left[V_\infty\left(1 + \frac{R^2}{r^2}\right)\sin\theta\right]_{r\to\infty};$$

$$V_\theta = -V_\infty \sin\theta \qquad \text{for } \Psi \text{ at } r \to \infty. \qquad \textbf{ANSWER}$$

Discussion

Because $V_r = 0$ at $r = R$, no fluid crosses the circle given by $r = R$. The velocity field thus represents flow over a circle in the $r\theta$ plane or flow over an infinitely long cylinder perpendicular to the plane.

9.1.3 Methods for Solving Potential Flow Problems

Solving a potential flow problem means determining the potential, stream function, and velocity distributions. As these quantities are related by Eqs. (9.10), (9.11), (9.21), (9.22), and so on, finding one of the three for a particular flow geometry is sufficient to determine all of them. Potential flow problems can be solved by mathematical means, without the need for any experimental data.

There are two general categories of potential flow calculation methods. In the *indirect method,* we construct potential and stream functions that satisfy the governing differential equations and then investigate the resulting streamline pattern to see whether it represents an interesting or useful flow geometry. Example 9.1 illustrated this idea. In the *direct method,* we attempt to determine the potential and stream functions for a given flow (boundary) geometry. The direct method is clearly more useful, but it is also more difficult. One way to obtain "direct" results is to investigate a large number of indirect ones. Another direct method is to "build" a solution by combining or transforming indirect solutions.

Indirect methods include the following:

- *The method of singularities,* which we consider in some detail shortly.
- *The method of complex variables,* in which a few basic flow solutions are geometrically and mathematically transformed into other flows by using the theory of complex variables. This method is most useful for thin airfoil theory and is the method by which the lift coefficient formula given in Eq. (8.18) is derived. We do not consider this method in this book because many undergraduate fluid mechanics students lack the necessary mathematical background.

Direct methods include the following:

- *The graphical flow net method,* in which an orthogonal grid of potential and streamlines, such as that shown in Fig. 9.6, is developed by hand sketching.
- *Analog methods,* which rely on the direct geometric equivalence between potential/streamlines and voltage/current lines in an electric conductor or temperature/heat flux lines in a heat conducting medium.
- *Finite difference and finite element methods,* in which the field is subdivided into a large number of small subregions and the differential equations are replaced by algebraic approximations. We consider the finite difference method shortly.
- *Distributed singularity,* or *panel, methods,* in which an arbitrary flow geometry is mathematically generated by distributing singularities over lines or planes (panels) in the field. We also briefly consider this method.

We limit consideration to plane two-dimensional potential flows. Solution methods for axisymmetric flows are similar in principle and are considered in advanced texts [2, 6].

9.1.4 Plane Potential Flows from Singularities

The elementary plane potential flows are those caused by *singularities.* A singularity is a point* where the governing equations are *violated.* At these singular points, mass is created or destroyed, or the fluid has infinite vorticity. It is paradoxical

* In two-dimensional flow in a plane, this point can be viewed as the piercing point of a line perpendicular to the plane.

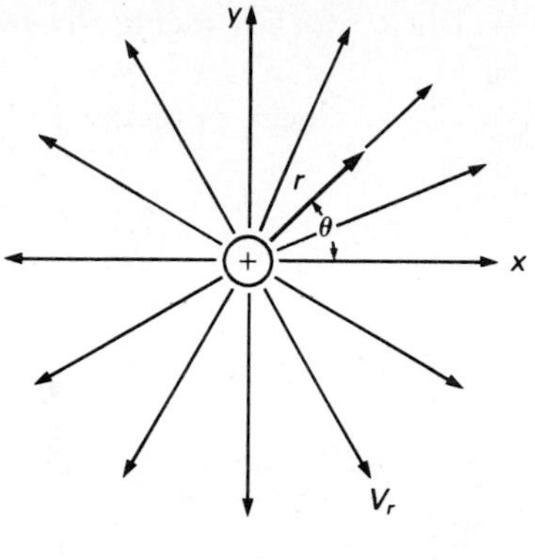

Figure 9.8 Source located at origin.

that mass-conserving irrotational flows can be built up from singularities. The reason is that the singularities that we consider *do* generate flows that obey the governing equations everywhere *except* at the singularity itself. Singularities are just mathematical "tricks" to generate realistic flows. The basic singularities are the *source* (and *sink*) and the *line vortex*.

Figure 9.8 shows a source located at the origin. The source introduces fluid into the field at a constant volume flow rate, λ, called the *strength* of the source. The fluid flows radially outward from the source, uniformly in all directions. The streamlines are straight radial lines. Because of symmetry, the velocity is purely radial and varies only with radius. This flow is most easily described in plane polar coordinates. To calculate the velocity, we consider a circular control volume of radius r enclosing the source. The integral continuity equation gives

$$Q = \int_0^{2\pi} V_r r\, d\theta = \lambda.$$

As V_r is independent of θ,

$$2\pi r V_r = \lambda,$$

or

$$V_r = \frac{\lambda}{2\pi r}; \tag{9.32}$$

by symmetry, (Plane source)

$$V_\theta = 0. \tag{9.33}$$

To demonstrate that this flow does indeed satisfy the governing equations, we first note that the continuity equation requires

$$\frac{1}{r}\frac{\partial}{\partial r}(rV_r) + \frac{1}{r}\frac{\partial}{\partial \theta}(V_\theta) = 0 \quad \text{or} \quad \frac{1}{r}\frac{\partial}{\partial r}\left(\frac{\lambda}{2\pi}\right) = 0,$$

which is obviously true, except possibly at $r = 0$. In addition, the irrotationality condition requires that

$$\zeta_z = \frac{1}{r}\left[\frac{\partial(rV_\theta)}{\partial r}\right] - \frac{1}{r}\left(\frac{\partial V_r}{\partial \theta}\right) = 0,$$

which is satisfied identically everywhere, because $V_\theta = 0$ and V_r is independent of θ.

The velocity potential and stream functions for a plane source are obtained from

$$V_r = \frac{\partial \Phi}{\partial r} = \frac{1}{r}\left(\frac{\partial \Psi}{\partial \theta}\right) = \frac{\lambda}{2\pi r}.$$

Integrating gives

$$\Phi = \frac{\lambda}{2\pi} \ln r + C \quad (9.34)$$

and

$$\Psi = \frac{\lambda}{2\pi} \theta + C. \quad (9.35)$$

Plane source

The constants of integration are customarily assigned values of zero. The streamlines are straight radial lines (θ = Constant) and the equipotential lines are circles (r = Constant).

A second singularity, easily related to the source, is the *sink*. A sink is simply a negative source, ingesting fluid at a constant rate. The velocity, velocity potential, and stream function for a sink are simply those for a source with a negative quantity substituted for the strength ($\lambda < 0$).

The velocity, velocity potential, and stream functions for a source or sink can be transformed into Cartesian coordinates:

$$u = V_r \cos\theta = \frac{\lambda}{2\pi}\left(\frac{x}{\sqrt{x^2 + y^2}}\right); \quad (9.36)$$

$$v = V_r \sin\theta = \frac{\lambda}{2\pi}\left(\frac{y}{\sqrt{x^2 + y^2}}\right); \quad (9.37)$$

$$\Phi = \frac{\lambda}{4\pi} \ln(x^2 + y^2); \quad (9.38)$$

and

$$\Psi = \frac{\lambda}{2\pi} \tan^{-1}\left(\frac{y}{x}\right). \quad (9.39)$$

These results are for a source/sink located at the origin. If the singularity is located at point (a, b), the quantity $(x - a)$ replaces x, and $(y - b)$ replaces y. For example, the x component of velocity generated by a source at point (a, b) is

$$u = \frac{\lambda}{2\pi}\left(\frac{(x - a)}{\sqrt{(x - a)^2 + (y - b)^2}}\right).$$

Another basic singularity is the *line vortex*. We can obtain equations describing this singularity by noting that, as Φ and Ψ are a set of orthogonal lines, a known potential flow yields a "new" one if we interchange Φ and Ψ. In this case, we

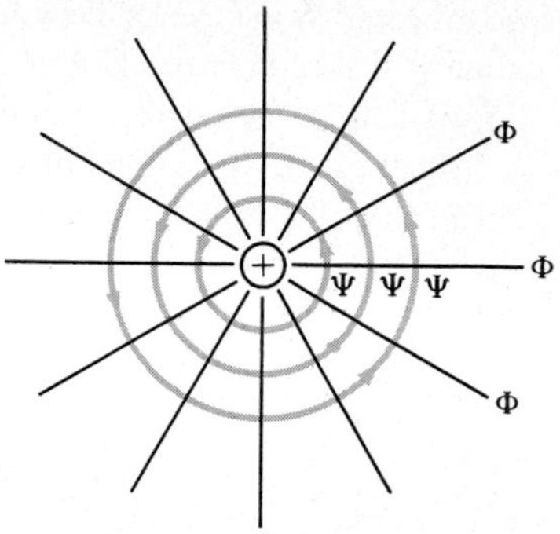

Figure 9.9 Equipotential and streamlines for line vortex.

simply switch the roles of Φ and Ψ that we obtained from the source, letting

$$\Phi = \frac{\Gamma}{2\pi}\theta \qquad (9.40)$$

and*

$$\Psi = -\frac{\Gamma}{2\pi}\ln r. \qquad (9.41)$$

(Line vortex)

Streamlines are circles, and equipotential lines are radial lines[†] (Fig. 9.9). The velocities are

$$V_r = \frac{\partial \Phi}{\partial r} = \frac{1}{r}\left(\frac{\partial \Psi}{\partial \theta}\right) = 0 \qquad (9.42)$$

and

$$V_\theta = \frac{1}{r}\left(\frac{\partial \Phi}{\partial \theta}\right) = -\frac{\partial \Psi}{\partial r} = \frac{\Gamma}{2\pi r}. \qquad (9.43)$$

(Line vortex)

Substituting these velocities into the continuity equation, we have

$$\frac{1}{r}\frac{\partial}{\partial r}(rV_r) + \frac{1}{r}\left(\frac{\partial V_\theta}{\partial \theta}\right) = 0 + 0 = 0.$$

The vorticity equation,

$$\zeta_z = \frac{1}{r}\left[\frac{\partial (rV_\theta)}{\partial r}\right] - \frac{1}{r}\left(\frac{\partial V_r}{\partial \theta}\right) = \frac{1}{r}\frac{\partial}{\partial r}\left(\frac{\Gamma}{2\pi}\right),$$

shows that the flow is irrotational everywhere except at the origin. The parameter Γ is the circulation, defined in Eq. (8.21). If we choose a circular path about the origin and calculate the circulation, we get

$$\Gamma = \oint \vec{V}\cdot d\hat{s} = \int_0^{2\pi} V_\theta r\, d\theta = \int_0^{2\pi} \frac{\Gamma}{2\pi}\, d\theta = \frac{\Gamma}{2\pi}(2\pi) = \Gamma.$$

We can show that the circulation about any path that encloses the origin is Γ and that the circulation about any path that does not enclose the origin is zero. We

---

* The minus sign is inserted to make later results come out right. The reason for writing the constant as $\Gamma/2\pi$ will also become apparent soon.

[†] Again, the origin can be viewed as the piercing point of a line perpendicular to the plane.

encountered this type of flow before, in Example 5.5 and Section 8.3.2, where we called it a *free vortex*. The vortex may be positive or negative, depending on the sign of Γ and the corresponding direction of V_θ.

Writing the equations in Cartesian coordinates for a free vortex located at an arbitrary point (a, b), we have

$$u = -\frac{\Gamma}{2\pi}\frac{y-b}{\sqrt{(x-a)^2+(y-b)^2}}; \tag{9.44}$$

$$v = \frac{\Gamma}{2\pi}\frac{x-a}{\sqrt{(x-a)^2+(y-b)^2}}; \tag{9.45}$$

$$\Phi = \frac{\Gamma}{2\pi}\tan^{-1}\left(\frac{y-b}{x-a}\right); \tag{9.46}$$

$$\Psi = -\frac{\Gamma}{4\pi}\ln[(x-a)^2+(y-b)^2]. \tag{9.47}$$

An important basic potential flow (but not a singularity) is *uniform, parallel flow*. For flow aligned at an arbitrary angle α with the x axis, the velocities are

$$u = V_\infty \cos\alpha \tag{9.48}$$

and

$$v = V_\infty \sin\alpha. \tag{9.49}$$

The velocity potential and stream function are

$$\Phi = V_\infty x\cos\alpha + V_\infty y\sin\alpha, \tag{9.50}$$

$$\Psi = V_\infty y\cos\alpha - V_\infty x\sin\alpha. \tag{9.51}$$

Uniform parallel flow

9.1.5 Body Flows from Superposition of Simple Singularities

We can generate some interesting flows by combining several singularities in the same plane. This method is called *superposition* and works only because the potential and stream function equations (Eqs. 9.14 and 9.29) are linear. A linear equation has the property that if Φ_1 and Φ_2 are both solutions to the equations, then $(\Phi_1 + \Phi_2)$ is also a solution. If, for example, we put a source and a vortex in the same plane, the potential and stream functions for the resulting flow are

$$\Phi = \Phi_{\text{source}} + \Phi_{\text{vortex}} \quad \text{and} \quad \Psi = \Psi_{\text{source}} + \Psi_{\text{vortex}}.$$

We plot the streamlines for the resulting flow by adding the two (algebraic) stream functions, setting the result equal to various constants, and plotting the resulting xy curves. An easier procedure is to draw the streamlines of the basic singularities and then combine them graphically, as in Fig. 9.10.

Figure 9.11 shows the flow resulting from a combined source and vortex, both located at the origin. The potential and stream functions for this flow are

$$\Phi = \Phi_{\text{source}} + \Phi_{\text{vortex}} = \frac{\lambda}{2\pi}\ln r + \frac{\Gamma}{2\pi}\theta \tag{9.52}$$

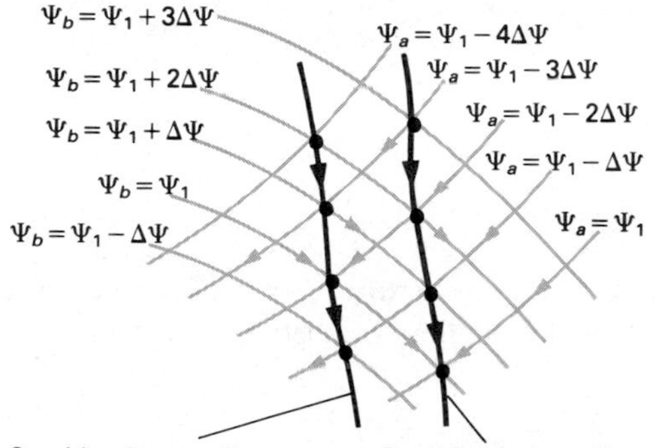

Figure 9.10 Graphic combination of streamlines.

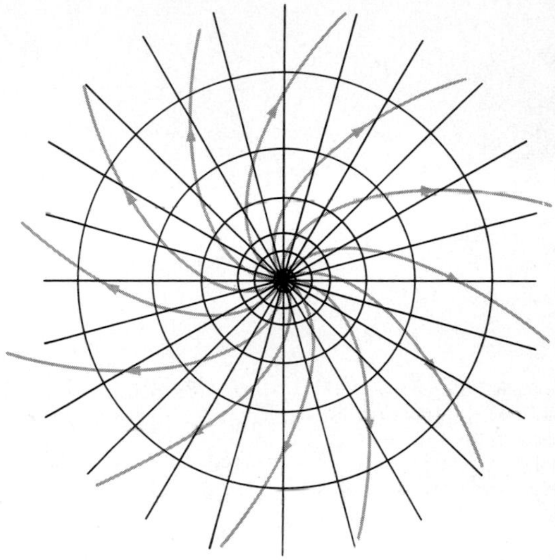

Figure 9.11 Streamline pattern for combined source and vortex.

and

$$\Psi = \Psi_{\text{source}} + \Psi_{\text{vortex}} = \frac{\lambda}{2\pi}\theta - \frac{\Gamma}{2\pi}\ln r. \tag{9.53}$$

The streamlines are logarithmic spirals, given by

$$r = \exp\left(-\frac{2\pi\Psi}{\Gamma}\right)\exp\left(\frac{\lambda\theta}{\Gamma}\right).$$

Recall that λ and Γ are constants and that Ψ is constant for a particular streamline.*

The most interesting flows of this type are obtained by combining a uniform flow with one or more singularities. Suppose we combine a source and a uniform flow parallel to the x axis ($\alpha = 0$). The stream function (in polar coordinates) is

$$\Psi = V_\infty r \sin\theta + \frac{\lambda}{2\pi}\theta. \tag{9.54}$$

A few streamlines are sketched in Fig. 9.12(a). This flow looks relatively uninteresting until we examine the streamline marked A. This streamline separates the fluid introduced by the source from the uniform stream. The point 0 is a stagnation point. If we use the fact that any streamline can be replaced by a solid surface, the flow outside of streamline A represents flow over a blunt half-body (Fig. 9.12b). The half-body is open in the downstream direction. We might use the resulting flow pattern to model the flow over the front of a blunt object. We are not bothered by the fact that the flow pattern inside the body looks very unrealistic and even contains a point where conservation of mass is violated. We have hidden this point inside the body.

* What are some practical examples of this flow pattern?

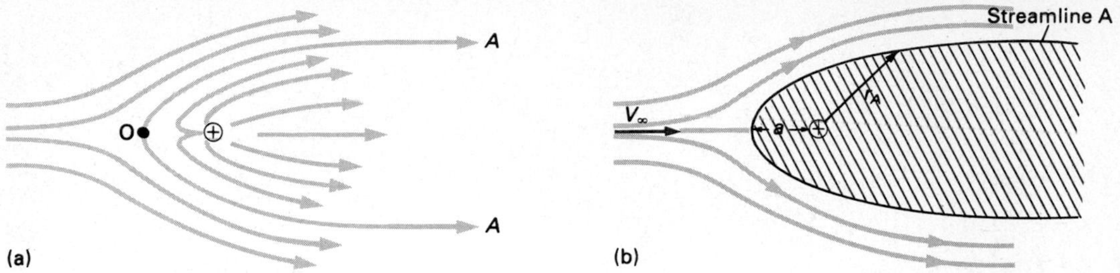

Figure 9.12 Flow over half-body: (a) streamline pattern for source in uniform stream; (b) flow over blunt half-body.

We obtain the contour of streamline A by setting $\Psi = 0$ along the centerline and noting that exactly one-half of the source flow lies between the centerline and streamline A; thus

$$\Psi_A = \frac{\lambda}{2},$$

and the streamline contour is given by

$$r_A = \frac{\lambda(\pi - \theta)}{2\pi V_\infty \sin\theta}. \tag{9.55}$$

We find the location of the front stagnation point, $x = -a$, by setting

$$u = \frac{\partial \Psi}{\partial x} = 0 \qquad \text{at } r = a, \quad \theta = \pi,$$

with the result that

$$a = \frac{\lambda}{2\pi V_\infty}. \tag{9.56}$$

The general expression

$$V^2 = V_r^2 + V_\theta^2 = \left(\frac{1}{r}\frac{\partial \Psi}{\partial \theta}\right)^2 + \left(\frac{\partial \Psi}{\partial r}\right)^2$$

yields the velocity on the body surface:

$$V^2 = V_\infty^2\left(1 + \frac{a^2}{r^2} + 2\frac{a}{r}\cos\theta\right). \tag{9.57}$$

We can obtain a closed body by introducing a sink with strength equal to that of the source downstream from the source. The sink ingests all the output from the source. The resulting streamline pattern is sketched in Fig. 9.13. The most convenient placement of the origin is midway between the source and sink, with the source located at $(-a, 0)$ and the sink located at $(+a, 0)$. The stream function becomes

$$\Psi = \Psi_{\substack{\text{uniform}\\ \text{flow}}} + \Psi_{\substack{\text{source at}\\ -a}} + \Psi_{\substack{\text{sink at}\\ +a}}.$$

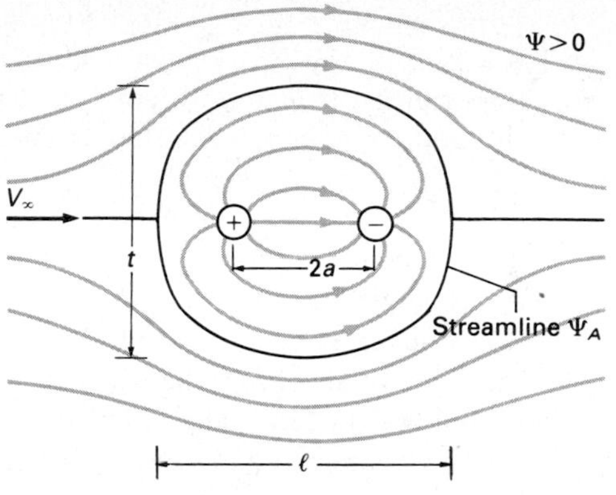

Figure 9.13 Flow over Rankine oval produced by source–sink pair in uniform stream.

Substituting the various stream functions, we have

$$\Psi = V_\infty y + \frac{\lambda}{2\pi}\tan^{-1}\left(\frac{y}{x+a}\right) - \frac{\lambda}{2\pi}\tan^{-1}\left(\frac{y}{x-a}\right).$$

Rearranging yields

$$\Psi = V_\infty y - \frac{\lambda}{2\pi}\tan^{-1}\left(\frac{2ay}{x^2+y^2-a^2}\right). \tag{9.58}$$

The streamline marked Ψ_A in Fig. 9.13 separates the fluid introduced and ingested by the source–sink pair from the main stream fluid. The net flow from the source–sink pair is zero, so

$$\Psi_A = 0.$$

Streamline Ψ_A is a closed curve and can be replaced by a solid surface. This surface is called a *Rankine oval* and is somewhat blunter than an ellipse. The streamlines lying outside of Ψ_A (that is, streamlines with $\Psi > 0$) form the flow pattern over the oval. The internal streamlines ($\Psi < 0$) are unrealistic but are hidden inside the body contour. The geometric parameters of the body, t and ℓ, are related to the source–sink spacing $2a$, which is in turn related to the strength of the source and sink and the free-stream velocity. The relations are

$$\frac{t}{a} = \cot\frac{t/a}{\lambda/(\pi V_\infty a)} \tag{9.59}$$

and

$$\frac{\ell}{a} = \left(1 + \frac{\lambda}{\pi V_\infty a}\right)^{1/2}. \tag{9.60}$$

The maximum velocity, which occurs at the point of maximum thickness on the body, is

$$\frac{V_{\max}}{V_\infty} = 1 + \frac{\lambda/(\pi V_\infty a)}{1 + (t/a)^2}. \tag{9.61}$$

Now let's find out what happens if we vary the source–sink spacing. If $a \to \infty$,

$t \to 0$, and $\ell \to \infty$, we obtain a uniform, parallel flow over a very thin, flat plate. If $a \to 0$ (λ and V_∞ remaining constant), both t and ℓ approach zero, but, more importantly, their ratio becomes unity. As we bring the source and sink together, the body becomes a vanishingly small circular cylinder. Increasing λ makes the body larger for a given source–sink spacing, so we generate the flow over a finite circular cylinder by letting λ increase without bound as a decreases to zero; in other words, we keep the *product* λa constant as $a \to 0$. What we are actually doing is generating a new type of singularity called a *doublet*. A doublet is the direct superposition of a source and sink, both of infinite strength. The fact that this type of singularity has almost no physical significance should not bother us. We carefully "hide" any doublets inside a body contour.

The potential and stream functions for a doublet are obtained by

$$\Phi_{\text{doublet}} = \lim_{\substack{a \to 0 \\ \lambda a = \text{constant}}} \left(\Phi_{\substack{\text{source} \\ \text{at } -a}} + \Phi_{\substack{\text{sink} \\ \text{at } +a}}\right)$$

and

$$\Psi_{\text{doublet}} = \lim_{\substack{a \to 0 \\ \lambda a = \text{constant}}} \left(\Psi_{\substack{\text{source} \\ \text{at } -a}} + \Psi_{\substack{\text{sink} \\ \text{at } +a}}\right).$$

Substituting the appropriate source and sink singularities, defining μ as the constant value of λa, and taking the limits, we obtain

$$\Phi_{\text{doublet}} = \frac{\mu}{2\pi}\left(\frac{x}{x^2 + y^2}\right) = \frac{\mu}{2\pi}\left(\frac{\cos\theta}{r}\right) \tag{9.62}$$

and

$$\Psi_{\text{doublet}} = \frac{\mu}{2\pi}\left(\frac{y}{x^2 + y^2}\right) = \frac{\mu}{2\pi}\left(\frac{\sin\theta}{r}\right). \tag{9.63}$$

If we combine a doublet with a uniform stream, we expect to obtain the flow over a circular cylinder. The stream function is

$$\Psi = \Psi_{\substack{\text{uniform} \\ \text{stream}}} + \Psi_{\text{doublet}},$$

or

$$\Psi = V_\infty y - \frac{\mu}{2\pi}\left(\frac{y}{r^2}\right). \tag{9.64}$$

Setting Ψ equal to zero to find a closed contour, we get

$$r)_{\Psi=0} = \sqrt{\frac{\mu}{2\pi V_\infty}} \equiv R, \qquad \text{a constant,}$$

which of course is a circle. Figure 9.14 illustrates some other streamlines of the flow. As usual, the streamlines inside the contour are quite unrealistic, but the flow outside is just as we would expect. The velocity components are given by*

* Note that θ in these equations is measured from the *rear* stagnation point.

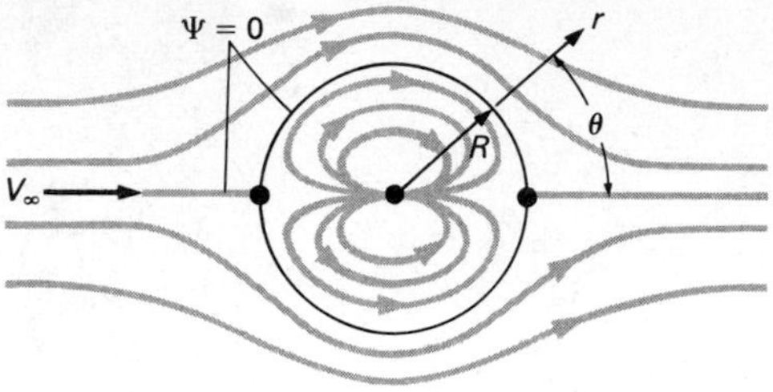

Figure 9.14 Flow resulting from combination of doublet and uniform stream.

(see Example 9.1):

$$V_r = V_\infty \cos\theta \left(1 - \frac{R^2}{r^2}\right) \tag{9.65}$$

and

$$V_\theta = -V_\infty \sin\theta \left(1 + \frac{R^2}{r^2}\right). \tag{9.66}$$

On the surface of the cylinder, $r = R$,

$$V_r)_{r=R} = 0; \tag{9.67a}$$

$$V_\theta)_{r=R} = -2V_\infty \sin\theta. \tag{9.67b}$$

The maximum velocity occurs at $\theta = \pi/2$ and is

$$V_{\max} = \left|V_\theta)_{\theta=\pi/2}\right| = 2V_\infty.$$

The pressure distribution on the surface of the cylinder is obtained from Bernoulli's equation,

$$p = p_\infty + \tfrac{1}{2}\rho(V_\infty^2 - V_\theta^2),$$

or, in terms of pressure coefficient,

$$\mathbf{C_p} \equiv \frac{p - p_\infty}{\frac{1}{2}\rho V_\infty^2} = 1 - \left(\frac{V_\theta}{V_\infty}\right)^2. \tag{9.68}$$

Substituting Eq. (9.67b), we have

$$\mathbf{C_p} = 1 - 4\sin^2\theta. \tag{9.69}$$

This pressure distribution was plotted in Fig. 8.12(b).*

An item of considerable interest is the force on the cylinder. The pressure distribution is symmetrical both front to back and top to bottom, so the net force is zero (d'Alembert's paradox).

In Section 8.3.4, we learned that a *spinning* cylinder generates lift. We simulate this type of flow by placing a line vortex at the origin, buried inside the circular cylinder. The lift is proportional to the circulation of this vortex. To obtain positive lift, we must introduce negative circulation (see Eq. 8.35). The resulting stream function is

$$\Psi = V_\infty r \sin\theta - R^2 V_\infty \left(\frac{\sin\theta}{r}\right) + \frac{\Gamma}{2\pi}\ln r.$$

* Refer to Section 8.2.2 for a discussion of the realism of these potential flow results.

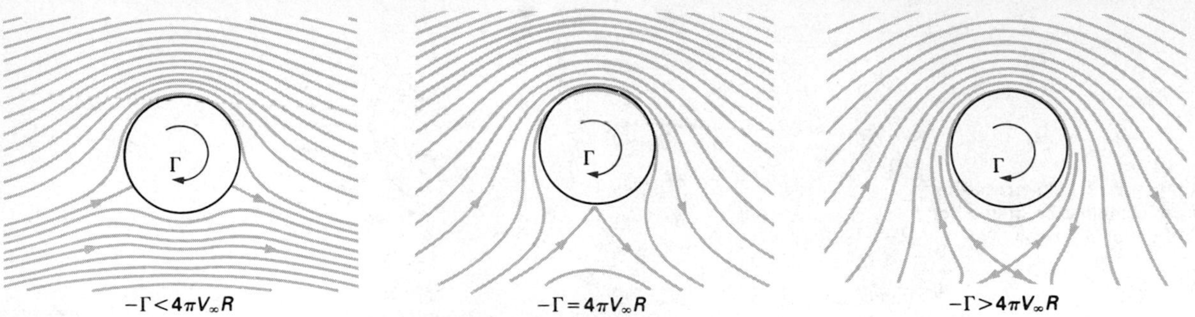

Figure 9.15 Streamlines for flow over circular cylinder with circulation. Circulation is negative so as to generate positive lift.

Some streamlines are sketched in Fig. 9.15. The surface velocity is*

$$V_\theta)_{r=R} = -\frac{\partial \Psi}{\partial r}\bigg)_{r=R} = -2V_\infty \sin\theta - \frac{\Gamma}{2\pi R}. \tag{9.70}$$

Stagnation points are located on the surface at points where $V_\theta = 0$. These points are located at

$$\theta = \sin^{-1}\left(-\frac{\Gamma}{4\pi V_\infty R}\right).$$

If $-\Gamma > 4\pi V_\infty R$, no stagnation points occur on the surface, but a single stagnation point occurs in the fluid below the cylinder.

The lift on the cylinder is obtained from the Kutta–Joukowsky theorem, Eq. (8.35). If we had a real flow in which the cylinder were spinning, rather than a fixed cylinder and a vortex, the surface velocity of the cylinder (U) would be given by the second term of Eq. (9.70),

$$U = \frac{\Gamma}{2\pi R}.$$

The circulation necessary to model the spin is

$$\Gamma = -2\pi R U, \quad \text{so} \quad \mathcal{L} = -\rho V_\infty \Gamma \ell = 2\pi\rho R V_\infty U \ell,$$

and

$$\mathbf{C_L} = \frac{\mathcal{L}}{\frac{1}{2}\rho V_\infty^2 (2R)\ell} = 2\pi\left(\frac{U}{V_\infty}\right). \tag{9.71}$$

This relation is plotted in Fig. 8.33.

---

* Recall that θ is positive *counterclockwise.*

EXAMPLE 9.2 Illustrates the Addition of Potential Flow Streamlines to Produce an Interesting Flow

Investigate the addition of the stream function Ψ_U for a uniform flow parallel to the x axis (flowing right to left) and the stream function $\Psi_\ominus$ for a sink. Show that the streamlines simulate the flow over the blunt tail of a body.

SOLUTION

Given

$$\Psi = \Psi_U + \Psi_\ominus = V_\infty r \sin\theta + \frac{\lambda}{2\pi}\theta,$$

where $\lambda < 0$ and is the strength of the sink per unit depth (perpendicular to the page) and V_∞ is also negative.

Show

That the superposition produces the flow over the blunt tail of a body

Solution

We plot lines of

$$\Psi_U = V_\infty r \sin\theta = \text{Constant}$$

and lines of

$$\Psi_\ominus = \frac{\lambda}{2\pi}\theta = \text{Constant}.$$

For illustrative purposes, we choose $\lambda = -32.0\ \text{m}^2/\text{s}$ and $V_\infty = -1\ \text{m/s}$ and measure θ in radians. Then

$$r\sin\theta = -\Psi_U \quad \text{and} \quad \theta = \frac{2\pi\Psi_S}{\lambda} = -\frac{\pi}{16}\Psi_\ominus.$$

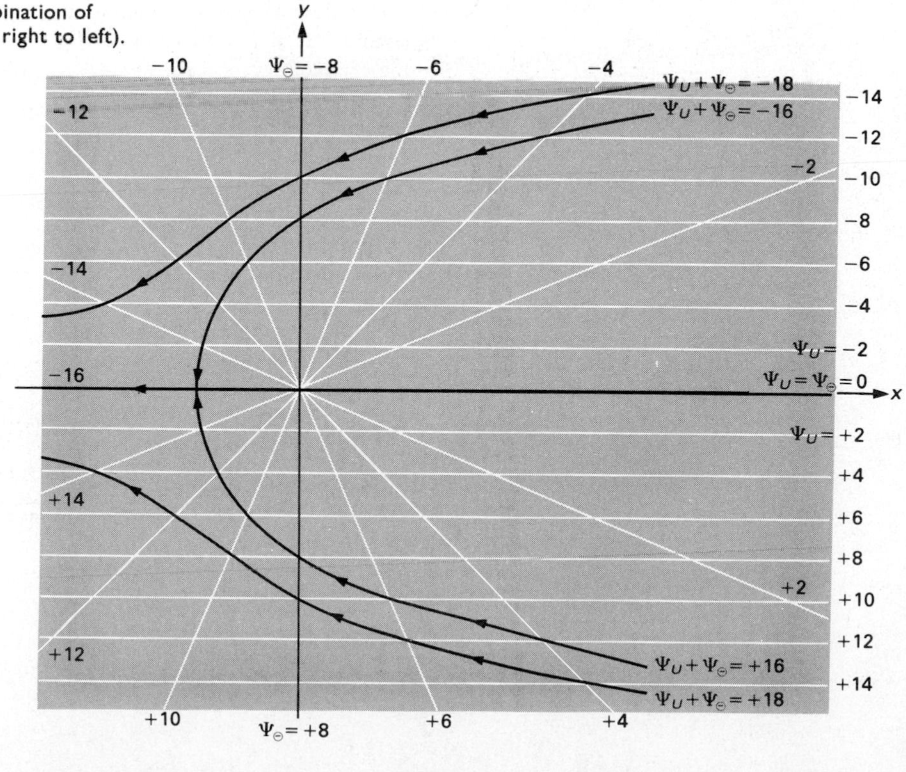

Figure E9.2 Graphic combination of streamlines (note flow from right to left).

We choose values of $\Psi_U = 0, \pm 2, \pm 4, \pm 6, \pm 8, \pm 10, \pm 12, \pm 14$, and ± 16 and identical values for Ψ_Θ. The corresponding streamlines are plotted in Fig. E9.2. Lines of constant Ψ_U are parallel to the x axis, and lines of constant Ψ_Θ are radial lines. Lines for which $\Psi_U + \Psi_\Theta = \pm 16$ and ± 18 are also plotted in the figure. Lines for values of ± 16 form the rearward surface of a blunt body. Lines for values of ± 18 represent two streamlines of the flow over the rearward surface or tail.

The answer is Fig. E9.2. **ANSWER**

Discussion

Note that the surface of the body is formed by two streamlines, one by $\Psi = +16$ and one by $\Psi = -16$.

We chose convenient values for λ and V_∞ to produce an easy numerical superposition. Other values of λ and V_∞ would be slightly more cumbersome; however, the procedure would be identical to that used here, but the size and shape of the body would be somewhat different.

EXAMPLE 9.3 Illustrates Calculation of Pressure and Force in a Potential Flow

The following stream function represents the flow of a uniform stream over a cylinder of radius R with a line vortex at the center of the cylinder:

$$\Psi = V_\infty r \sin\theta - \frac{\mu}{2\pi}\left(\frac{\sin\theta}{r}\right) - \frac{\Gamma}{2\pi}\ln\frac{r}{R},$$

with

$$\Gamma = +2\pi R V_\infty \quad \text{and} \quad \mu = 2\pi R^2 V_\infty.$$

Determine the pressure distribution over the cylinder and calculate the lift.

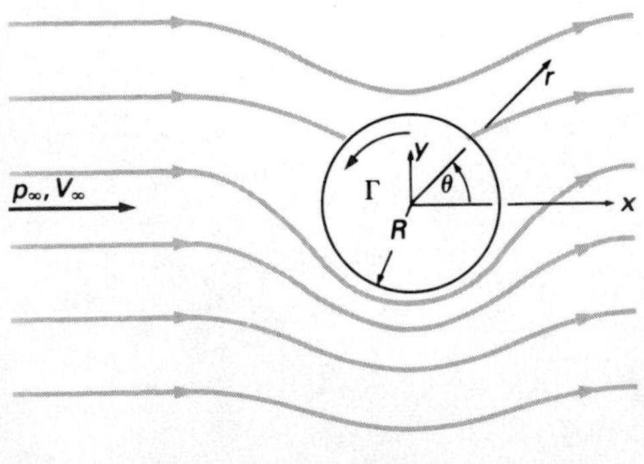

Figure E9.3a Uniform flow over a cylinder with imbedded vortex.

SOLUTION

Given

Stream function

$$\Psi = V_\infty r \sin\theta - R^2 V_\infty\left(\frac{\sin\theta}{r}\right) - R V_\infty \ln\frac{r}{R}$$

Find

Pressure distribution over the cylinder and the lift

Solution

A sketch of the flow field is shown in Fig. E9.3a. Equations (9.42) and

(9.43) give

$$V_r = \frac{1}{r}\left(\frac{\partial \Psi}{\partial \theta}\right) = \frac{V_\infty}{r}\frac{\partial}{\partial \theta}\left[\left(r - \frac{R^2}{r}\right)\sin\theta - R\ln\frac{r}{R}\right]$$
$$= V_\infty\left(1 - \frac{R^2}{r^2}\right)\cos\theta$$

and

$$V_\theta = -\frac{\partial \Psi}{\partial r} = -V_\infty\frac{\partial}{\partial r}\left[\left(r - \frac{R^2}{r}\right)\sin\theta - R\ln\frac{r}{R}\right]$$
$$= V_\infty\left[\frac{R}{r} - \left(1 + \frac{R^2}{r^2}\right)\sin\theta\right].$$

We now apply Bernoulli's equation to streamlines immediately adjacent to the surface of the cylinder. See Fig. E9.3a. With $V_r = 0$ at the cylinder surface, Bernoulli's equation gives

$$\frac{p_\infty}{\rho} + \frac{1}{2}V_\infty^2 = \frac{p_s}{\rho} + \frac{1}{2}V_\theta^2,$$

where p_s is the pressure on the cylinder surface. Substituting for V_θ, we have

$$\frac{p_\infty}{\rho} + \frac{1}{2}V_\infty^2 = \frac{p_s}{\rho} + \frac{V_\infty^2}{2}\left[\frac{R}{R} - \left(1 + \frac{R^2}{R^2}\right)\sin\theta\right]^2,$$

or

$$\frac{p_\infty}{\rho} + \frac{1}{2}V_\infty^2 = \frac{p_s}{\rho} + \frac{V_\infty^2}{2}[1 - 2\sin\theta]^2,$$

or

$$\frac{p_s - p_\infty}{\rho} = \frac{V_\infty^2}{2}[1 - (1 - 4\sin\theta + 4\sin^2\theta)].$$ **ANSWER**

As ρ, V_∞, and p_∞ are not known, we plot the pressure coefficient:

$$C_P = \frac{p_s - p_\infty}{\frac{1}{2}\rho V_\infty^2} = 4\sin\theta - 4\sin^2\theta.$$ **ANSWER**

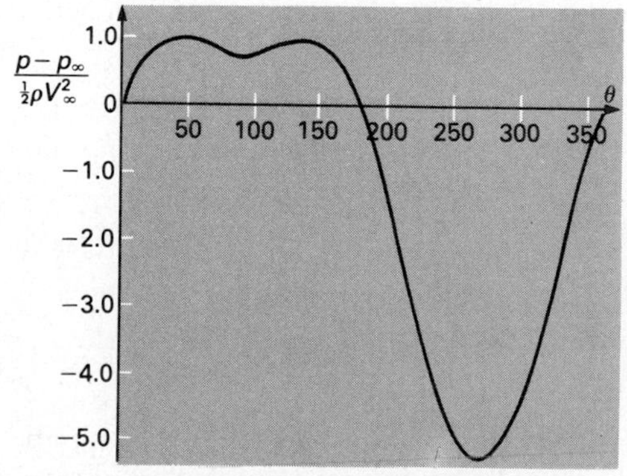

Figure E9.3b Pressure coefficient for a uniform stream flowing over a circular cylinder with imbedded vortex.

This result is plotted in Fig. E9.3b.

We find the lift $\mathscr{L}$ by using Eq. (8.3), with $p = p_s$ and $\tau_w = 0$:

$$\mathscr{L} = \oint (-p_s \sin\theta)\, dA = \int_0^{2\pi} (-p_s\sin\theta)(2R\ell)\, d\theta,$$

where ℓ is the cylinder length. Substituting for p_s and dA gives

$$\mathscr{L} = -2R\ell\int_0^{2\pi}[p_\infty + \tfrac{1}{2}\rho V_\infty^2(4\sin\theta - 4\sin^2\theta)]\sin\theta\, d\theta.$$

Integrating gives

$$\mathscr{L} = -2R\ell\left[-p_\infty\cos\theta + \frac{4}{2}\rho V_\infty^2\left(\frac{\theta}{2} - \frac{\sin 2\theta}{4}\right)\right.$$
$$\left.\left. + \frac{4}{2}\rho V_\infty^2\left(\frac{\cos\theta}{3}(\sin^2\theta + 2)\right)\right]_0^{2\pi}$$

$$= -2R\ell\left[-p_\infty(1-1) + \frac{4}{2}\rho V_\infty^2\left(\frac{2\pi - 0}{2} - \frac{0-0}{4}\right) + \frac{4}{2}\rho V_\infty^2\left(\frac{1(0+2) - 1(0+2)}{3}\right)\right];$$

$$\mathscr{L} = -2\pi\rho V_\infty^2 R\ell.$$ **ANSWER**

Discussion

This result is equal to that of the Kutta–Joukowsky theorem, Eq. (8.35),

$$\mathscr{L} = -\rho V_\infty \Gamma \ell,$$

since $\Gamma = 2\pi R V_\infty$.

The negative (downward) lift results from the counterclockwise (positive) circulation. A negative circulation produces a positive (upward) lift. Compare with Fig. 9.15.

9.1.6 The Finite-Difference Method for Direct Calculation of Potential Flows

The finite-difference method is a useful technique for generating approximate solutions to differential equations of all types. When applied to calculation of fluid flow, the finite-difference method is one approach to computational fluid dynamics (CFD). The differences between the finite-difference approach and the finite-volume approach described in Section 5.6.3 are subtle. In the finite-difference approach, we formulate the computational equations by beginning with the differential equations and then approximating the derivatives in the equations by finite differences. We normally think of approximating a flow by obtaining values at grid points rather than for tiny volumes that enclose the points.

The following are the basic steps in a finite-difference calculation:

- A grid is superimposed on the flow field in order to define a large number of points. The flow variables (velocity, pressure, potential, stream function, etc.) are defined at these points.
- At each point, appropriate differential equations are approximated by replacing any derivatives with finite difference quotients involving flow variables at neighboring points.
- Similarly, boundary conditions are approximated for points on or near the flow boundaries.
- A set of algebraic equations results. The equations are solved for the flow variables at the grid points using any applicable and efficient method.

As with all CFD methods, accuracy usually requires a fine grid, a large number of equations, and extensive calculations. Usually, computer solutions are the only realistic choice; however, potential flow is one area where reasonable approximations can sometimes be obtained by hand calculation. The following development illustrates the major concepts and computational steps for the finite-difference method as applied to potential flow.

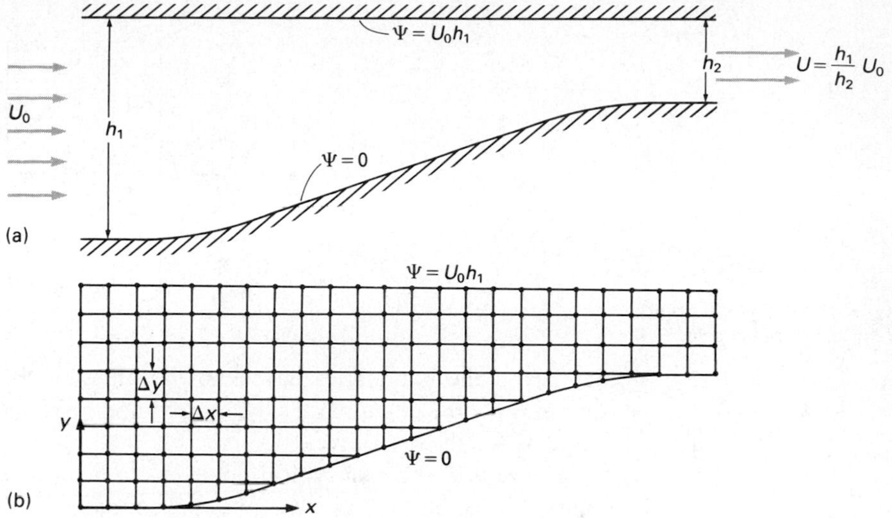

Figure 9.16 Finite-difference setup for flow in a converging channel.

Let's consider the problem of calculating the stream function for potential flow in a converging passage (Fig. 9.16). If we find the stream function, we can then obtain the velocity by differentiation and plot the streamlines by connecting points having the same value of Ψ. To calculate the stream function in the passage, we must find Ψ so that

$$\frac{\partial^2 \Psi}{\partial x^2} + \frac{\partial^2 \Psi}{\partial y^2} = 0 \tag{9.29}$$

everywhere in the passage. We must also satisfy the boundary conditions

$$\Psi = 0 \quad \text{(on the lower wall);*} \tag{9.72}$$

$$\Psi = U_0 h_1 \quad \text{(on the upper wall);*} \tag{9.73}$$

$$\Psi = U_0 y \quad \text{(at the inlet);} \tag{9.74}$$

$$\Psi = U_0\left(\frac{h_1}{h_2}\right)[y - (h_1 - h_2)] \quad \text{(at the outlet).} \tag{9.75}$$

The outlet condition is obtained from the continuity equation and the inlet condition. We formulated the problem in terms of Ψ rather than Φ, because the boundary conditions are easier. For a channel, boundary conditions for Ψ are simply known values on the boundaries, but boundary conditions for Φ involve the derivative normal to the boundary. Consideration of methods for calculating Φ are presented in advanced texts [2].

We overlay the field with a rectangular grid, as shown in Fig. 9.16(b). We seek values for the stream function at the points defined by this grid. We must have as many independent equations as there are unknown values of the stream function; therefore we need an equation for each *interior* point in the field. Points

* Remember that the change in Ψ between two streamlines is equal to the volume flow between the streamlines and that solid surfaces are streamlines.

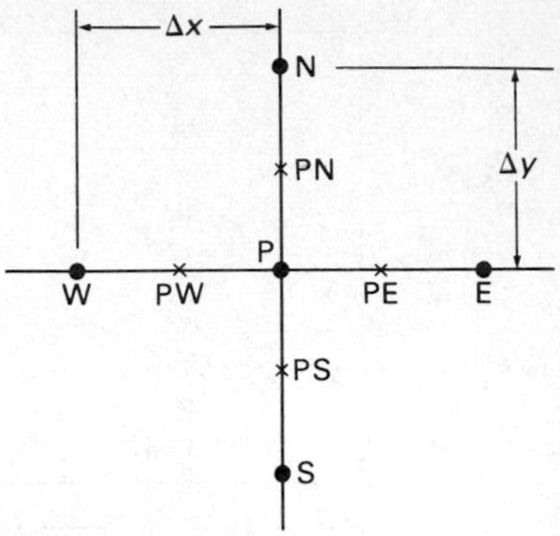

Figure 9.17 Typical points in finite-difference grid.

that lie on the boundaries have known values of Ψ. We obtain the necessary equations for the unknown Ψs by approximating Eq. (9.29), using finite-difference quotients for the derivatives.

Figure 9.17 shows a typical point in the field, labeled P, and its neighboring points, labeled N, E, S, and W.* We approximate the first derivative of Ψ by either

$$\left.\frac{\partial \Psi}{\partial x}\right)_{\mathrm{P}} \approx \frac{\Psi_{\mathrm{E}} - \Psi_{\mathrm{P}}}{\Delta x} \tag{9.76}$$

or

$$\left.\frac{\partial \Psi}{\partial x}\right)_{\mathrm{P}} \approx \frac{\Psi_{\mathrm{P}} - \Psi_{\mathrm{W}}}{\Delta x}. \tag{9.77}$$

These two approximations, called a *forward difference* and a *backward difference,* become more accurate as Δx becomes smaller. The error in these approximations is proportional to $(\Delta x)^2$. Although intended as approximations of the derivative at P, they are better approximations for the derivative at point PE midway between P and E and point PW, midway between P and W. At PE and PW the error in the approximation is smaller, being proportional to $(\Delta x)^3$. We obtain an approximation to the second derivative by

$$\frac{\partial^2 \Psi}{\partial x^2} = \frac{\partial}{\partial x}\left(\frac{\partial \Psi}{\partial x}\right) \approx \frac{\partial \Psi/\partial x)_{\mathrm{PE}} - \partial \Psi/\partial x)_{\mathrm{PW}}}{\Delta x}.$$

Equations (9.76) and (9.77) are good approximations for the derivative at points PE and PW, so

$$\frac{\partial^2 \Psi}{\partial x^2} \approx \frac{(\Psi_{\mathrm{E}} - \Psi_{\mathrm{P}})/\Delta x - (\Psi_{\mathrm{P}} - \Psi_{\mathrm{W}})/\Delta x}{\Delta x} = \frac{\Psi_{\mathrm{E}} + \Psi_{\mathrm{W}} - 2\Psi_{\mathrm{P}}}{(\Delta x)^2}. \tag{9.78}$$

This approximation for the second derivative is most accurate at P, because it is "centered" there and has error proportional to $(\Delta x)^3$.

By similar reasoning, the y derivative for Eq. (9.29) is approximated by

$$\frac{\partial^2 \Psi}{\partial y^2} \approx \frac{\Psi_{\mathrm{N}} + \Psi_{\mathrm{S}} - 2\Psi_{\mathrm{P}}}{(\Delta y)^2}, \tag{9.79}$$

* The labeling scheme is adapted from the points of a compass.

and Eq. (9.29) is approximated by

$$\left(\frac{1}{\Delta x}\right)^2 (\Psi_E + \Psi_W - 2\Psi_P) + \left(\frac{1}{\Delta y}\right)^2 (\Psi_N + \Psi_S - 2\Psi_P) = 0.$$

If $\Delta x = \Delta y$ (a square grid), we get

$$\left(\frac{1}{\Delta x}\right)^2 (\Psi_E + \Psi_W + \Psi_N + \Psi_S - 4\Psi_P) = 0,$$

or, as Δx is finite,

$$\Psi_P = \tfrac{1}{4}(\Psi_E + \Psi_W + \Psi_N + \Psi_S). \tag{9.80}$$

Equation (9.80) may be written for every point in the grid (except points on the boundaries). Of course, we have to change the PNESW notation to a global point-numbering system. For points adjacent to the boundaries, at least one of the surrounding points is on a boundary and the corresponding value of Ψ is known. Sometimes the point located a distance Δx away may lie *outside* the boundary (Fig. 9.18), in which case we find a value for Ψ at the point by extrapolation:

$$\Psi_S = \Psi_P + \frac{\Delta x}{\Delta x_B}(\Psi_B - \Psi_P). \tag{9.81}$$

Substituting this expression into Eq. (9.80) and solving for Ψ_P give the modified equation

$$\Psi_P = \frac{1}{3 + a}(\Psi_E + \Psi_W + \Psi_N + a\Psi_B), \tag{9.82}$$

where $a = \Delta x/\Delta x_B$. After writing equations for each internal point, we find that

- there are as many equations as there are unknowns;
- the known boundary values of Ψ appear in the equations; and
- the equation for Ψ at a point contains the values for Ψ at the *four* surrounding points.

There are many ways to solve the resulting set of equations for the unknown values of Ψ. One of the simplest is the *Gauss–Seidel iteration* method, which proceeds as follows:

- Guess values for all unknown Ψs.

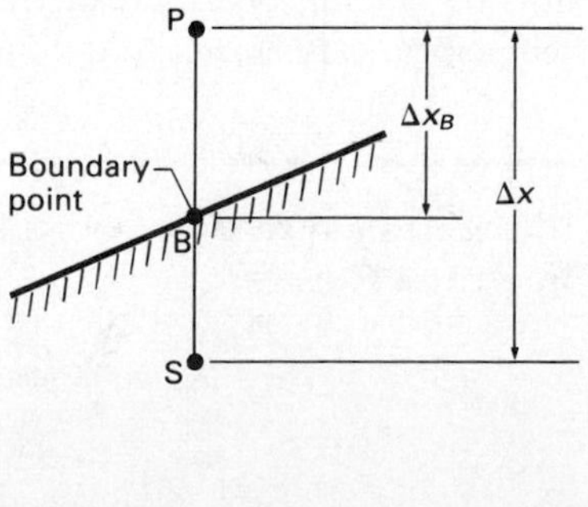

Figure 9.18 Regular point (S) lies outside a boundary.

- "Visit" each point, calculating a new value of Ψ_P from Eq. (9.80) for fully interior points and equations similar to Eq. (9.82) for points near boundaries. The most recently calculated values for Ψ are used on the right side. If we move through the grid from top to bottom and left to right, the values for Ψ_N and Ψ_E will be "new" values, and the values for Ψ_W and Ψ_S will be "old" values.
- Continue the process until the largest change in Ψ_P is as small as you want it to be.

This process can be easily programmed for a computer or programmable calculator.

After calculating the Ψs, we plot streamlines by connecting points having the same value of Ψ. Interpolation is usually required, because Ψ is known only at discrete points. Velocities in the field can be calculated by

$$u = \frac{\partial \Psi}{\partial y} \approx \frac{\Delta \Psi}{\Delta y} \tag{9.83}$$

and

$$v = -\frac{\partial \Psi}{\partial x} \approx -\frac{\Delta \Psi}{\Delta x}. \tag{9.84}$$

These values are most accurate at the midpoints between grid points; that is, Eq. (9.83) is most accurate at PN or PS, and Eq. (9.84) is most accurate at PE or PW. To find velocities at the boundaries, the following *three-point formulas* should be used:

$$u \approx \frac{3\Psi_P - 4\Psi_S + \Psi_{S+1}}{2\Delta y} \quad \text{(for } \textit{upper} \text{ boundaries);} \tag{9.85}$$

$$u \approx \frac{-3\Psi_P + 4\Psi_N - \Psi_{N+1}}{2\Delta y} \quad \text{(for } \textit{lower} \text{ boundaries);} \tag{9.86}$$

$$v \approx \frac{-3\Psi_P + 4\Psi_E - \Psi_{E+1}}{2\Delta x} \quad \text{(for } \textit{left} \text{ boundaries);} \tag{9.87}$$

$$v \approx \frac{3\Psi_P - 4\Psi_W + \Psi_{W+1}}{2\Delta x} \quad \text{(for } \textit{right} \text{ boundaries).} \tag{9.88}$$

In these formulas, N + 1 is the next point beyond N and so on.

This introduction is rather brief, but we covered the important ideas. For further discussion, consult reference [2]. An undergraduate heat transfer text would also be helpful, because exactly the same numerical method is used to calculate temperatures in steady heat conduction in solids.

EXAMPLE 9.4 Illustrates Finite-Difference Calculation of a Simple Potential Flow

A two-dimensional potential flow in a converging channel is shown in Fig. E9.4a. Using a uniform grid spacing of 0.5 cm, calculate the

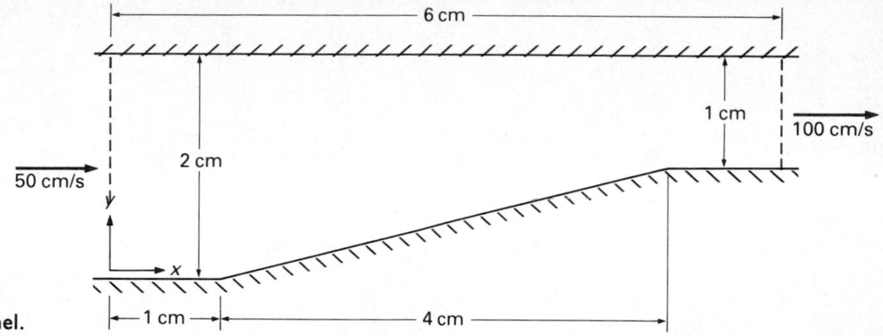

Figure E9.4a Converging channel.

stream function in the channel and the velocity distribution along the upper (flat) wall.

SOLUTION

Given

Potential flow in converging channel with flat upper wall and straight sloped lower wall

Flow velocity at inlet 50 cm/s and at outlet 100 cm/s

Grid spacing 0.5 cm

Find

Stream function at grid points

Velocity along upper wall

Solution

Figure E9.4b shows the grid and the boundary values. The stream function is taken as zero along the lower wall. By Eq. (9.73), Ψ is 100 along the upper wall. Equations (9.74) and (9.75) give values for Ψ at the inlet and outlet. The interior points are numbered from 1 to 25.

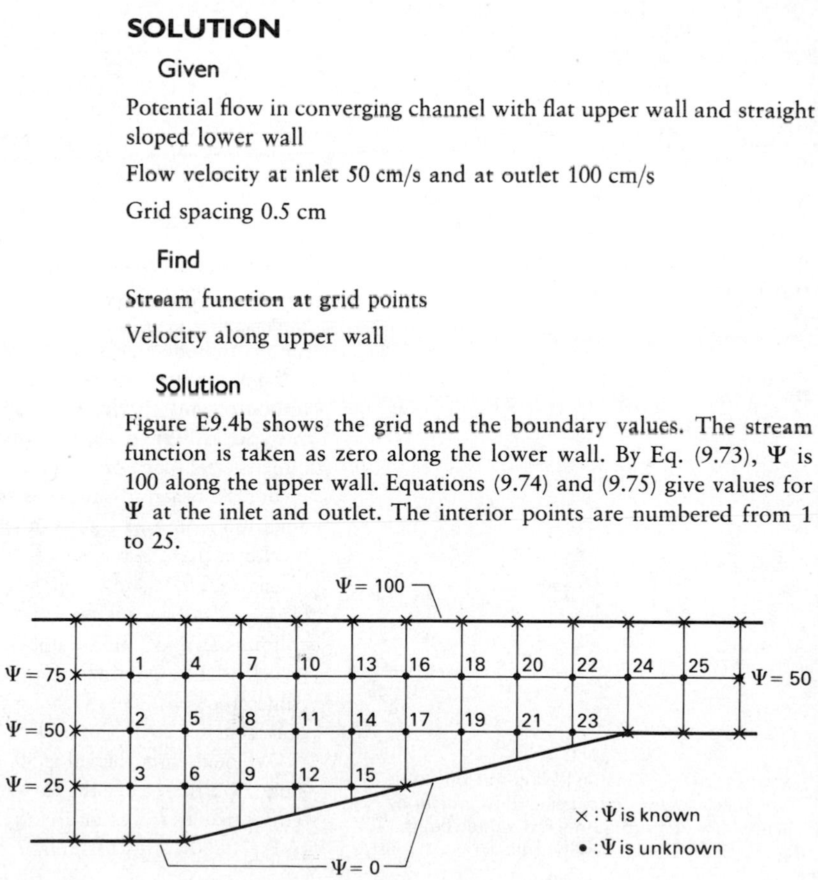

Figure E9.4b Grid with boundary values of Ψ.

Equation (9.80) describes all interior points except 9, 12, 15, 19, 21, and 23, which are near boundaries. At the fully interior points,

$$\Psi_1 = \frac{75 + 100 + \Psi_2 + \Psi_4}{4};$$

$$\Psi_2 = \frac{50 + \Psi_1 + \Psi_5 + \Psi_3}{4};$$

$$\vdots$$

$$\Psi_{10} = \frac{100 + \Psi_7 + \Psi_{11} + \Psi_{13}}{4};$$

$$\Psi_{11} = \frac{\Psi_8 + \Psi_{10} + \Psi_{12} + \Psi_{14}}{4};$$

$$\vdots$$

From Eq. (9.82), equations for points 9, 12, 15, 19, 21, and 23 are

$$\Psi_9 = \frac{\Psi_6 + \Psi_8 + \Psi_{12}}{4.3333}; \tag{E9.1}$$

$$\Psi_{12} = \frac{\Psi_9 + \Psi_{11} + \Psi_{15}}{5}; \tag{E9.2}$$

$$\Psi_{15} = \frac{\Psi_{12} + \Psi_{14}}{7}; \tag{E9.3}$$

$$\Psi_{19} = \frac{\Psi_{17} + \Psi_{18} + \Psi_{21}}{4.3333}; \tag{E9.4}$$

$$\Psi_{21} = \frac{\Psi_{19} + \Psi_{20} + \Psi_{23}}{5}; \tag{E9.5}$$

$$\Psi_{23} = \frac{\Psi_{21} + \Psi_{22}}{7}. \tag{E9.6}$$

We solve these equations by Gauss–Seidel iteration, after guessing values for Ψ_{1-25} to start the calculation. A uniform value of 50 is a simple and effective starting value.

This problem can be set up and solved effectively using a personal computer and an electronic spreadsheet. A particular advantage of the spreadsheet is that the equations can be laid out in a two-dimensional plane on it in a format that coincides with the grid-point layout. Another advantage is the ability to COPY the computational equation from one cell to other cells. The following calculation uses the Lotus 123® spreadsheet.

Figure E9.4c shows the boundary conditions entered in computational cells. Row 3 is the upper boundary. The grid points lie between columns C and O. The upper wall stream function (100) is entered in cells C3 . . O3. Cells C3 . . C7 contain the inlet stream function values and cells O3 . . O5 contain the outlet values. Along the bottom, only cells C7 . . E7, I6, and M5 . . O5 lie on the boundary.

We next enter the value 50 in all interior cells to provide a starting value. We then enter the formula +(C4+D3+E4+D5)/4 in cell D4. This calculates a value for the stream function at point 1 according to Eq. (9.80). This formula is now COPYed to all the other fully in-

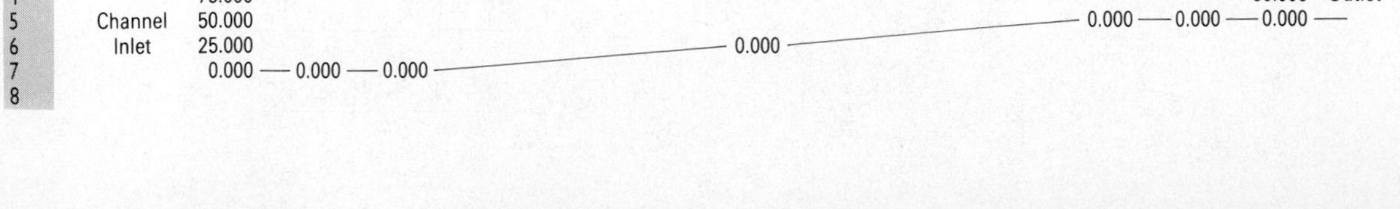

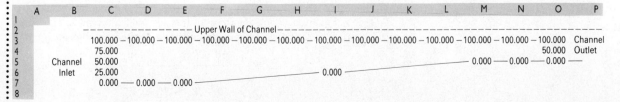

Figure E9.4c "Channel" laid out on Lotus 123® spreadsheet, with heavy line indicating channel "wall" and numerical values being values of Ψ at boundary points.

terior cells (E4 . . N4, D5 . . I5, and D6 . . E6). Cell addresses are automatically adjusted to compute the value in any cell in terms of the values in the four surrounding cells. Finally, we enter modified formulas based on Eqs. (E9.1)–(E9.6) into cells adjacent to boundaries; for example, +(F6+G5+H6)/5 is entered at cell G6 and +(K5+L4)/7 is entered at L5.

	A	B	C	D	E	F	G	H	I	J	K	L	M	N	O	P
1																
2						Upper Wall of Channel										
3			100.000	100.000	100.000	100.000	100.000	100.000	100.000	100.000	100.000	100.000	100.000	100.000	100.000	Channel
4			75.000	74.475	73.715	72.476	70.658	68.766	66.231	63.809	60.085	55.537	51.493	50.373	50.000	Outlet
5		Channel	50.000	49.185	47.806	45.401	41.789	37.210	32.910	27.238	19.656	10.843	0.000	0.000	0.000	
6		Inlet	25.000	24.298	22.922	19.179	13.754	6.497	0.000							
7			0.000	0.000	0.000											
8					CIRC											

Figure E9.4d Spreadsheet display immediately after initial entry of computational equations.

The monitor now displays a first pass calculation of the stream functions (Fig. E9.4d). The CIRC message reminds us that the calculation is "circular," that is, there are interconnections between cell formulas. (We knew that all along; that is the nature of the problem that we are solving!) To complete the calculation of the stream function, we repeatedly force recalculation of the entire spreadsheet (by hitting the F9 key) until we observe no change in any of the displayed values. We used about 20 recalculations. As a final step, we format all the cells to display only three decimal places.

The solution for Ψ is shown in Fig. E9.4e. **ANSWER**

	A	B	C	D	E	F	G	H	I	J	K	L	M	N	O	P
1																
2						Upper Wall of Channel										
3			100.000	100.000	100.000	100.000	100.000	100.000	100.000	100.000	100.000	100.000	100.000	100.000	100.000	Channel
4			75.000	74.404	73.579	72.327	70.641	68.538	66.004	62.979	59.419	55.381	51.435	50.359	50.000	Outlet
5		Channel	50.000	49.038	47.583	45.090	41.697	37.507	32.501	26.492	19.316	10.671	0.000	0.000	0.000	
6		Inlet	25.000	24.166	22.626	18.754	13.549	7.294	0.000							
7			0.000	0.000	0.000											
8																

Figure E9.4e Converged solution for stream function.

In order to calculate the velocity on the upper wall, we use the three-point formula, Eq. (9.85). Note that the velocity is parallel to the upper wall; it has only a "u" component. We calculate velocity values on the spreadsheet by entering the formula +(3*C3−4*C4+C5)/(2*0.5) in cell C10 and then COPYing to D10 . . O10. The velocity distribution is now displayed in C10 . . O10. By entering the x coordinates in cells C14 . . O14, we can generate a computer plot of the velocity distribution.

The velocity along the upper wall is shown in Figures E9.4f and E9.4g. **ANSWER**

Discussion

This problem is a fairly simple example of its type. A finer grid gives more accuracy but also generates many more equations. An electronic spreadsheet calculation would probably not converge for more than about 100 points. In larger problems, a BASIC or FORTRAN program would be more effective.

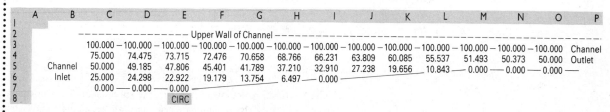

	A	B	C	D	E	F	G	H	I	J	K	L	M	N	O	P
1																
2			---------------- Upper Wall of Channel ----------------													
3			100.000	100.000	100.000	100.000	100.000	100.000	100.000	100.000	100.000	100.000	100.000	100.000	100.000	Channel
4			75.000	74.404	73.579	72.327	70.641	68.538	66.004	62.979	59.419	55.381	51.435	50.359	50.000	Outlet
5		Channel	50.000	49.038	47.583	45.090	41.697	37.507	32.501	26.492	19.316	10.671	0.000	0.000	0.000	
6		Inlet	25.000	24.166	22.626	18.754	13.549	7.294	0.000							
7			0.000	0.000	0.000											
8																
9		Velocity														
10		on	50.000	51.421	53.268	55.781	59.135	63.355	68.483	74.576	81.640	89.146	94.260	98.565	100.000	
11		Upper														
12		Wall														
13																
14		x	0	0.5	1	1.5	2	2.5	3	3.5	4	4.5	5	5.5	6	
15		Values														
16																

Figure E9.4f Stream function solution with velocity distribution on upper wall.

The "natural" calculation order in the spreadsheet is from top to bottom and left to right; therefore our manual "recalculations" were effectively Gauss–Seidel-type iterations.

Calculation of the fluid velocity along the lower wall would be more difficult than along the upper wall for two reasons. Most critically, the points near the lower wall are not at a uniform distance above the wall, so Eqs. (9.85)–(9.88) cannot be used. Modified formulas that permit nonuniform spacing must be used. Also, the velocity at the lower wall is parallel to the wall and has both u and v components. This complication is most easily handled by calculating u and using the known wall slope to find V from $V = u/\cos\theta$.

Figure E9.4g Velocity distribution.

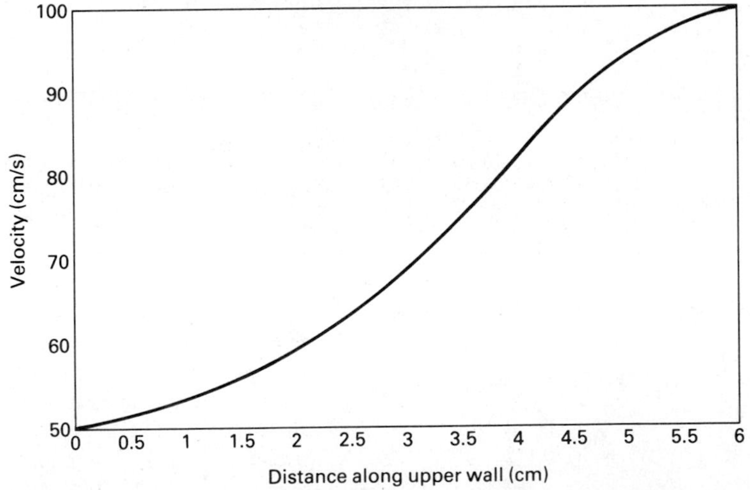

9.1.7 The Distributed Singularity (Panel) Method for Direct Calculation of Potential Flows

The finite difference method is not used just for potential flow; it also is a very common and widely used tool in engineering analysis. The distributed singularity method, however, is restricted mainly to potential flow problems.* It is most

* The so-called boundary element method for stress analysis and heat conduction problems is essentially identical to the distributed singularity method.

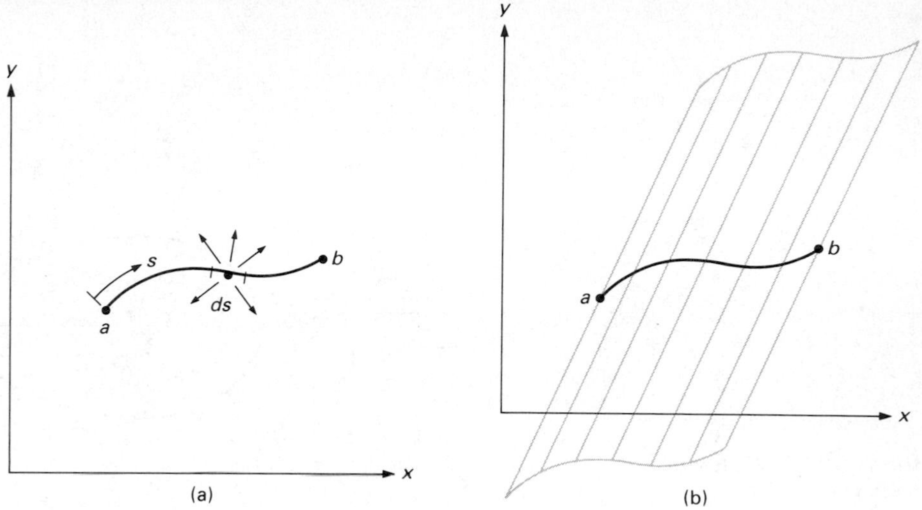

Figure 9.19 (a) Curve in *xy* plane with source–sink singularity distributed along it; (b) curve as intersection of infinite sheet with the *xy* plane.

effective for external flows, especially when compared to finite-difference methods that require a very large number of grid points to represent the "infinite" flow field.

Earlier, we discovered that we could simulate flow over a few simple body shapes by combining one or more singularities with a uniform, parallel flow. We had slight control over body shape by positioning of point singularities. We could change a half-body into an oval by adding a sink, and we could change the shape of the oval by changing the distance between the singularities or altering their strength. This suggests the possibility of simulating flow over an arbitrary body by the proper combination of a uniform stream with sources and sinks (and vortexes, if a lifting body is to be simulated). This can be done, although it requires continuous distribution of singularities.

Let's consider an arbitrary curve in the xy plane (Fig. 9.19a). Suppose that every point on this curve is a tiny source or sink. In three dimensions (Fig. 9.19b), the curve is the intersection of an infinitely long source–sink sheet with the xy plane. The strength of the source* in the vicinity of a particular point is defined as $\lambda'\{s\}\,ds$, where λ' is the strength per unit length and is a function of s.

Figure 9.20 shows a point P in the field near the curve. The velocity potential at point P *owing to the source at s* is obtained from Eqs. (9.34) and (9.36):

$$d\Phi_P = \frac{\lambda'\{s\}\,ds}{2\pi}\ln(r_{P,s}) = \frac{\lambda'\{s\}\,ds}{4\pi}\ln[(x_P - x_s)^2 + (y_P - y_s)^2].$$

We obtain the total potential at P by integrating over the entire line of distributed source:

$$\Phi_P = \int_a^b \frac{\lambda'\{s\}}{4\pi}\ln[(x_P - x_s)^2 + (y_P - y_s)^2]\,ds. \tag{9.89}$$

If we add a uniform stream, parallel to the x axis, the total potential becomes

$$\Phi = V_\infty x + \int_a^b \frac{\lambda'\{s\}}{4\pi}\ln[(x - x_s)^2 + (y - y_s)^2]\,ds. \tag{9.90}$$

* From here on, we refer only to source and let the sign of the strength differentiate between sources and sinks.

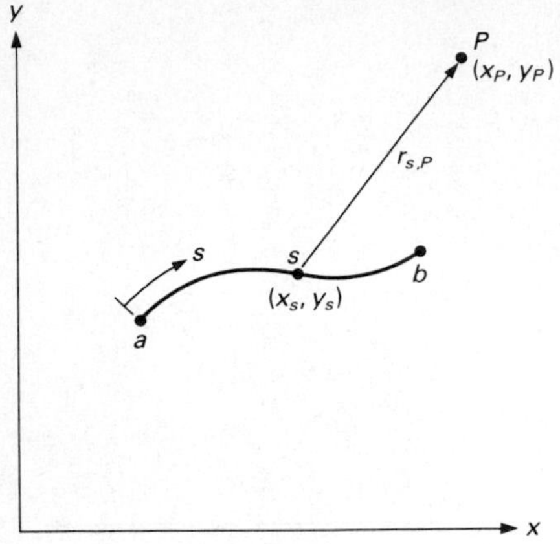

Figure 9.20 Point *P* near point *s* on the curve.

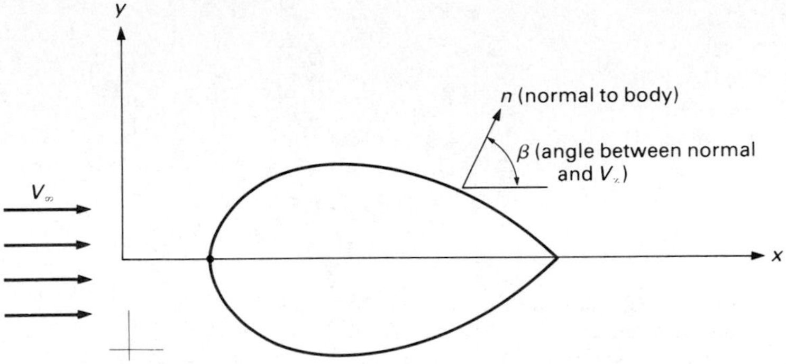

Figure 9.21 Arbitrary body in uniform stream.

We dropped the subscript *P* from Eq. (9.90) to emphasize its applicability to any point (x, y).

Obviously, we can generate an infinite variety of flows from Eq. (9.90). We can arbitrarily choose the location, size, and shape of the source sheet, and we can arbitrarily choose its strength distribution, $\lambda'(s)$. Any choices with the net source strength equal to zero, that is,

$$\int \lambda'(s)\, ds = 0, \tag{9.91}$$

simulate flow over a closed body. Of course, choosing these parameters arbitrarily and then examining the resulting flow is an indirect method.

Suppose that we want to calculate the potential flow over a specific shape, such as that shown in Fig. 9.21. We limit consideration to symmetric shapes to avoid the complication of lift generation and the Kutta condition. Two particular locations and shapes for a source sheet suggest themselves:

- the axis of symmetry (x axis); and
- the surface of the body itself.

Experience has shown that simulating flow over a shape by sources distributed on the axis of symmetry is not always possible, but simulating any shape by distributing sources on the surface itself is possible. The problem then reduces to determining the specific source distribution $\lambda'(s)$ that, when combined with a uniform stream, will generate a streamline that coincides with the surface itself. Another way of stating this requirement is that the fluid velocity component normal to the surface, evaluated at the surface, must be zero. Using Eq. (9.90) for the velocity potential, this requirement yields the following equation, which we must solve for $\lambda'(s)$:

$$\frac{\partial}{\partial n} \oint_{\text{body}} \frac{\lambda'(s)}{4\pi} \ln[(x - x_s)^2 + (y - y_s)^2]\, ds = -V_\infty \cos\beta, \tag{9.92}$$

where n is normal to the body and β is defined in Fig. 9.21.

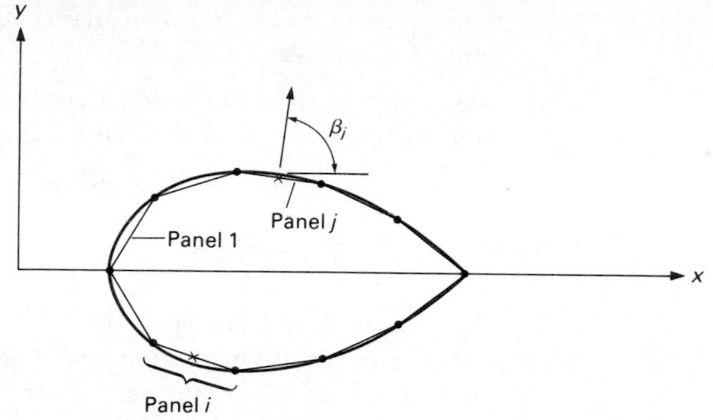

Figure 9.22 Surface approximated by line segments (panels). Source strength on panel i is λ_i', etc.

In practice, we take a simplified approach and use a numerical approximation to Eq. (9.92). The surface is approximated by a number of straight-line segments, as shown in Fig. 9.22. In three dimensions, the line segments represent flat sheets, called *panels* (hence the alternative name, *panel method*). The source strength on each panel (i) is constant but is different from panel to panel. Thus we need find only as many values for the source strength as there are panels. Usually, 20–50 panels provide a highly accurate solution; sometimes as few as 5–10 panels may be used. Equations for the panel source strength are obtained by enforcing the flow tangency ($V_n = 0$) condition at the midpoint of each panel. This method produces as many equations as unknowns λ_i's.

Mathematical development of the equations for the λ_i's is tedious and is not given here. The result is a set of *linear* algebraic equations for the λ_i's:

$$A_{ji}\lambda_i' = -V_\infty \cos \beta_j. \tag{9.93}$$

The coefficients A_{ji} depend only on the geometry of the surface, as represented by the approximating panels. Physically, A_{ji} is the normal velocity induced at the midpoint of panel j by a unit strength source distribution on panel i.

We solve Eq. (9.93) for the λ_i's by any convenient method for solving a set of linear algebraic equations. We then calculate the fluid velocity (tangent to the surface) at each panel midpoint from a relationship of the form:

$$V_j = B_{ji}\lambda_i' + V_\infty \sin \beta_j,$$

where the B_{ji} are similar in form to the A_{ji} and depend only on the geometry of the panels. The pressure distribution on the body surface then follows from Bernoulli's equation.

The source panel method successfully calculates (potential flow) velocity and pressure distributions; in fact, the calculated pressure distributions are usually far better approximations of the exact value than the "paneled" body is of the true body shape. Several extensions of the basic panel method presented here have been developed, including:

- extension to axisymmetric flow over axisymmetric bodies;
- use of vortex panels rather than source panels to represent nonsymmetric and lifting body flows;

- use of linear or quadratic singularity distribution on each panel, instead of uniform distribution; and
- extension to three-dimensional geometries, where the finite planes of distribute singularity truly merit the name "panels."

References [1, 10, 11] provide more details.

9.2 BOUNDARY LAYERS

In a high Reynolds number flow, the effects of fluid viscosity and rotation are confined to a relatively thin region near solid surfaces or lines of discontinuity, such as wakes. Because the boundary layer is thin, certain simplifications can be introduced into the equations of motion; however, retaining both inertia (acceleration) terms and stress (viscous) terms is necessary. Pressure terms may or may not be present, depending on the nature of the flow outside the boundary layer. Because the vorticity of the boundary layer fluid is not zero, there is no velocity potential function for boundary layer flow. The equation of motion must be attacked directly. This equation, even with the boundary layer simplifications included, is much more difficult to solve than the equation of potential flow. Further complications are introduced by the fact that the boundary layer flow may be either laminar or turbulent.

Only a few exact solutions of the equations of motion for boundary layers are known, and these are all for laminar flow. The mathematics involved in obtaining these solutions is too complicated and lengthy to consider in this text. There are excellent approximate methods for calculating boundary layer flow, which we do consider, and digital computer-based numerical methods for calculating boundary layer flow, which we only mention. Our aim is to help you establish a feel for boundary layer flow and to examine a few simple cases in detail. To keep things as simple as possible, we limit the discussion to steady, two-dimensional, two-directional, incompressible flow.

9.2.1 Dimensional Analysis and Qualitative Estimates of Boundary Layer Parameters

Figure 9.23 shows a boundary layer on a flat plate. The following boundary layer characteristics are of primary importance:

- The boundary layer is thin ($\delta \ll x$).
- The thickness of the boundary layer increases in the downstream direction, but δ/x is always small.
- The boundary layer velocity profile satisfies the no-slip condition at the wall and merges smoothly into the free-stream velocity at the edge of the layer.
- There is a shear stress at the wall.
- The streamlines of the boundary layer flow are *approximately* parallel to the surface; that is, the velocity parallel to the surface is considerably larger than the velocity normal to the surface. Note that the locus of points that defines the outer edge of the boundary layer is not a streamline.

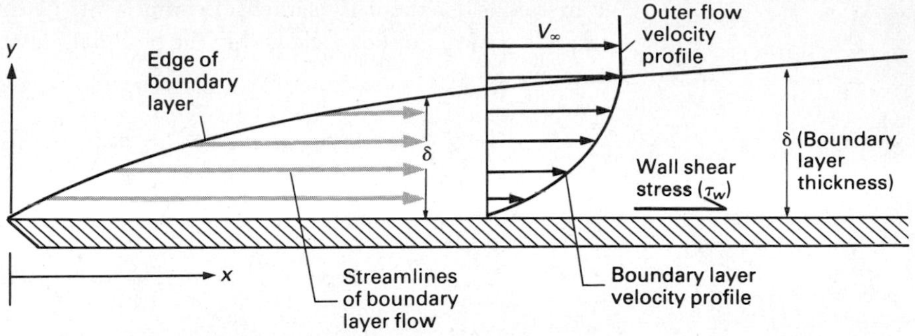

Figure 9.23 Details of boundary layer near a thin flat plate.

In this section, we demonstrate the validity of these characteristics and obtain *estimates* of key boundary layer parameters. We show that most of the listed characteristics depend on large Reynolds number. Most boundary layer flows are at least qualitatively similar to flow over a flat plate. Flat plate flow has a uniform free-stream velocity (outside the boundary layer) and hence constant pressure. Most of our qualitative results for flat plate flow can be extended to flows with variable free-stream velocity by replacing V_∞, the constant free-stream velocity, with $U(x)$, the local free-stream velocity.*

As usual in complicated flow problems, we begin with a dimensional analysis. The primary independent dimensionless parameter is the Reynolds number. If we adopt the local coordinate x as the reference length, we find that

$$\frac{\delta}{x} = f_1(\mathrm{R}_x), \tag{9.94}$$

$$\mathrm{C_f} = \frac{\tau_w}{\frac{1}{2}\rho V_\infty^2} = f_2(\mathrm{R}_x), \tag{9.95}$$

and

$$\frac{v_e}{V_\infty} = f_3(\mathrm{R}_x), \tag{9.96}$$

where

$$\mathrm{R}_x = \frac{\rho V_\infty x}{\mu} = \frac{V_\infty x}{\nu} \tag{9.97}$$

and v_e is the y velocity at the outer edge of the boundary layer.

We estimate the form of the functions f_1, f_2, and f_3 by some clever, if not exactly rigorous, reasoning. The key is to recognize that, by definition, the boundary layer is the region of flow where viscous forces and inertia forces are the same order of magnitude. We relate the flow parameters by examining this order of magnitude requirement. We use the symbol $\sim$ to represent "of the same order of magnitude." Thus, in the boundary layer,

$$\text{Inertia "force" } (\delta F_I) \sim \text{Viscous force } (\delta F_v).$$

---

* Recall that this free-stream velocity can be obtained from a potential flow analysis that neglects the boundary layer.

We first consider the x (streamwise) components of the forces.

For a typical fluid particle within the boundary layer* (Fig. 9.24), the respective forces are

$$\delta F_{I,x} = \rho(\delta \forall)a_x = \rho u\left(\frac{\partial u}{\partial x}\right)(\delta \forall)$$

and

$$\delta F_{v,x} = \frac{\partial \tau}{\partial y}\,\delta y\,\delta A_y = \mu\left(\frac{\partial^2 u}{\partial y^2}\right)(\delta \forall).$$

Thus

$$\rho u\left(\frac{\partial u}{\partial x}\right) \sim \mu\left(\frac{\partial^2 u}{\partial y^2}\right). \tag{9.98}$$

We now proceed to estimate the quantities in this "equation." If the particle is near the center of the layer,

$$u \sim \tfrac{1}{2}(u_{\text{wall}} + u_{\text{free stream}}) = \tfrac{1}{2}(0 + V_\infty) = \tfrac{1}{2}V_\infty.$$

To estimate $\partial u/\partial x$, we note that, at the leading edge of the plate, $x = 0$; the velocity at the center of the layer is V_∞; and at x, the velocity at the center of the layer is approximately $V_\infty/2$. Thus

$$\frac{\partial u}{\partial x} \approx \frac{\Delta u}{\Delta x} \sim \frac{\tfrac{1}{2}V_\infty - V_\infty}{x - 0} = -\frac{V_\infty}{2x},$$

and thus

$$\rho u\left(\frac{\partial u}{\partial x}\right) \sim -\tfrac{1}{4}\rho\,\frac{V_\infty^2}{x}.$$

To estimate $\partial^2 u/\partial y^2$, we first estimate the derivative $\partial u/\partial y$. The velocity changes from $u = 0$ to $u = V_\infty$ in a distance δ, so

$$\frac{\partial u}{\partial y} \sim \frac{\Delta u}{\Delta y} = \frac{V_\infty}{\delta}.$$

The velocity gradient is greater than this magnitude at the wall and less than this magnitude (namely zero) at the boundary layer edge. The second derivative is estimated by

$$\frac{\partial^2 u}{\partial y^2} = \frac{\partial}{\partial y}\left(\frac{\partial u}{\partial y}\right) \sim \frac{\Delta(\partial u/\partial y)}{\Delta y} = \frac{\partial u/\partial y)_{\text{edge of layer}} - \partial u/\partial y)_{\text{wall}}}{\delta}.$$

We estimate the velocity gradient at the wall is twice as large as the "typical" value (V_∞/δ), we find that

$$\mu\,\frac{\partial^2 u}{\partial y^2} \sim \mu\,\frac{0 - 2(V_\infty/\delta)}{\delta} = -\frac{2\mu V_\infty}{\delta^2}.$$

* The particle is located somewhere between the wall and the edge of the layer. At the wall, the inertia force is zero, and at the edge, the viscous force is zero.

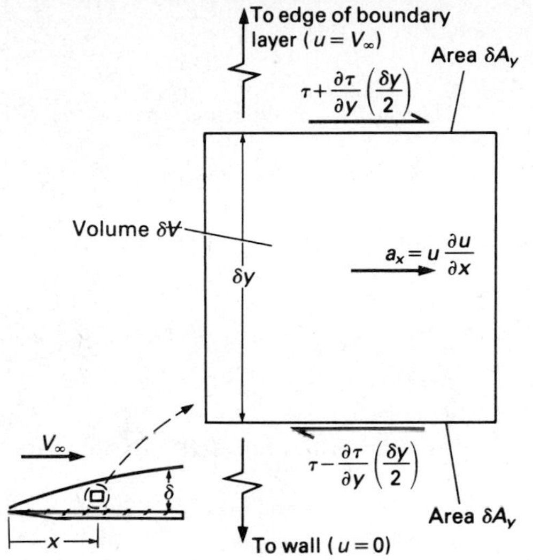

Figure 9.24 Stress and acceleration on fluid particle from "center" of boundary layer.

Substituting our estimates into "Eq." (9.98), we get

$$\tfrac{1}{4}\rho\left(\frac{V_\infty^2}{x}\right) \sim 2\mu\left(\frac{V_\infty}{\delta^2}\right),$$

or, solving for δ,

$$\delta \sim \sqrt{8}\sqrt{\frac{\mu x}{\rho V_\infty}}.$$

Of course we cannot put very much faith in the $\sqrt{8}$ in this expression. Recognizing that we are dealing with an order of magnitude estimate, we drop it. Upon dividing by x and replacing μ/ρ with ν, we get

$$\frac{\delta}{x} \sim \sqrt{\frac{\nu}{V_\infty x}} = \frac{1}{\sqrt{\mathbf{R}_x}}, \tag{9.99}$$

which not only verifies the relation implied by Eq. (9.94) but also verifies that the boundary layer is thin for large Reynolds number. Equation (9.99) is limited to laminar boundary layers because only viscous stress was included in the estimate. Attempts at making a corresponding estimate for turbulent flow fail because a universal relation for turbulent stress is not available.

We continue our investigation by estimating the wall shear stress and skin friction coefficient. The wall shear stress is

$$\tau_w = \mu\left(\frac{\partial u}{\partial y}\right)_{y=0}$$

An order of magnitude estimate gives*

* No numerical "constants" are included in our estimates from here on.

$$\tau_w \sim \mu\left(\frac{V_\infty}{\delta}\right).$$

Using Eq. (9.99) for δ gives

$$\tau_w \sim \sqrt{\frac{\rho\mu V_\infty^3}{x}}.$$

Thus

$$\mathbf{C_f} = \frac{2\tau_w}{\rho V_\infty^2} \sim \frac{1}{\sqrt{\mathbf{R}_x}}, \tag{9.100}$$

which verifies Eq. (9.95). Note that, with Eq. (9.99),

$$\frac{\mathbf{R}_\delta}{\mathbf{R}_x} = \frac{(V_\infty \delta)/\nu}{(V_\infty x)/\nu} = \frac{\delta}{x} \sim \frac{1}{\sqrt{\mathbf{R}_x}}, \quad \text{so} \quad \mathbf{R}_\delta \sim \sqrt{\mathbf{R}_x}.$$

Combining this expression with Eq. (9.100), we obtain

$$\mathbf{C_f} \sim \frac{1}{\mathbf{R}_\delta}; \tag{9.101}$$

that is, the friction coefficient is inversely proportional to Reynolds number based on a *transverse* length dimension. We relate this result to the Hagen–Poiseuille formula for laminar pipe flow (Eq. 7.40) by noting that δ corresponds to $D/2$.

We can estimate the transverse velocity at the edge of the boundary layer from the differential continuity equation,

$$\frac{\partial u}{\partial x} + \frac{\partial v}{\partial y} = 0.$$

Estimating orders of magnitude, we have

$$\frac{\partial v}{\partial y} \approx \frac{\Delta v}{\Delta y} = \frac{v_e - v_w}{\delta} = \frac{v_e - 0}{\delta} = \frac{v_e}{\delta} \quad \text{and} \quad -\frac{\partial u}{\partial x} \approx -\frac{\Delta u}{\Delta x} \sim \frac{V_\infty}{x},$$

so

$$\frac{v_e}{\delta} \sim \frac{V_\infty}{x}$$

and

$$\frac{v_e}{V_\infty} \sim \frac{\delta}{x} \sim \frac{1}{\sqrt{\mathbf{R}_x}}. \tag{9.102}$$

Equation (9.102) not only verifies Eq. (9.96) but, more importantly, shows that the streamlines in the boundary layer are approximately parallel to the surface for large Reynolds number.

We now estimate the variation of pressure across a typical boundary layer. The complete force balance considers pressure, inertia, and stress:

$$\delta F_{\text{pressure}} + \delta F_{\text{inertia}} + \delta F_{\text{stress}} = 0.$$

In a boundary layer, both inertia and stress are important. We estimate the largest possible pressure force in a boundary layer by assuming that both inertia and stress contribute to the pressure variation:

$$\delta F_{\text{pressure}} = -(\delta F_{\text{inertia}} + \delta F_{\text{stress}}).$$

We are interested in the pressure variation across the boundary layer, so we apply this equation in a direction normal to the surface (the y direction). For a typical fluid particle, we get

$$\frac{\partial p}{\partial y}(\delta \mathcal{V}) = -\rho a_y \delta \mathcal{V} + \delta F_{\text{stress}}.$$

In a boundary layer, the inertia force and stress force are the same order of magnitude, so

$$\frac{\partial p}{\partial y} \sim 2\rho a_y.$$

As y is essentially normal to the streamlines, from Eq. (5.20) we have

$$a_y \approx \frac{V^2}{R} = \frac{u^2 + v^2}{R},$$

where R is the radius of curvature of the streamlines. In the boundary layer, $v \ll u$, and the streamlines are approximately parallel to the body, so

$$a_y \approx \frac{u^2}{R_B},$$

where R_B is the radius of curvature of the body profile (Fig. 9.25). With these results, our estimate of the transverse pressure gradient becomes

$$\frac{\partial p}{\partial y} \sim 2\rho\left(\frac{u^2}{R_B}\right).$$

For a typical point in the boundary layer,

$$u^2 \sim \tfrac{1}{2}V_\infty^2,$$

so

$$\frac{\partial p}{\partial y} \sim \rho\left(\frac{V_\infty^2}{R_B}\right). \tag{9.103}$$

The change of pressure across the boundary layer is approximated by

$$\Delta p = \int_0^\delta \frac{\partial p}{\partial y}\,dy \approx \left(\frac{\partial p}{\partial y}\right)\delta, \quad \text{so} \quad \Delta p \sim \rho V_\infty^2\left(\frac{\delta}{R_B}\right),$$

or, in terms of a pressure coefficient,

$$C_{\Delta p} = \frac{2\Delta p}{\rho V_\infty^2} \sim \frac{\delta}{R_B}. \tag{9.104}$$

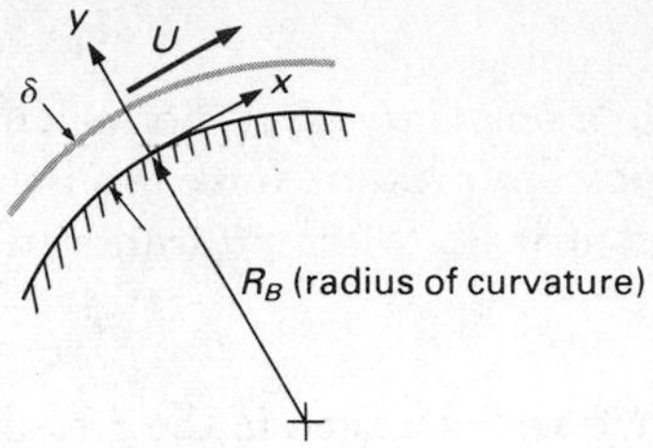

Figure 9.25 Boundary layer on curved surface.

Equation (9.104) shows that:

> If the boundary layer thickness is small compared to the radius of curvature of the body surface, the pressure can be considered constant across the boundary layer.

In this section, we used mathematics a bit loosely. You should realize that we were interested only in estimates and general conclusions, not in producing equations for boundary layer calculation. We verified all the important assumptions of boundary layer theory. In the next section, we use these assumptions to derive a simplified equation of motion for the boundary layer region. Our later solutions of this equation verify our order of magnitude estimates as well as fill in the gaps. Although they *are not* computational equations, you should remember the principles of boundary layer theory that are embodied in Eqs. (9.99), (9.100), (9.102), and (9.103).

9.2.2 The Differential Equations of Boundary Layer Analysis

The important parameters of a boundary layer are the thickness, which is a function of distance along the surface, and the shear stress at the surface. The blockage that the boundary layer presents to the outer inviscid flow and information concerning the possibility of boundary layer separation also are important. All these quantities can be determined from the boundary layer velocity field $u\{x, y\}$ and $v\{x, y\}$. The velocities are the solution to differential equations expressing the law of conservation of mass and the linear momentum equation.

Figure 9.26 shows a control volume located in a boundary layer. Our analysis can be extended to curved surfaces (with $R_B \gg \delta$) by choosing coordinates parallel and perpendicular to the surface. To formulate the equations for the control volume, we use the following results from the previous sections.

- The pressure is constant across the boundary layer.
- The pressure is a known function of x and is related to the velocity immediately outside the boundary layer by Bernoulli's equation.
- The velocity changes much more rapidly across the layer than along it $[(\partial u/\partial x \sim V_\infty/x) \ll (V_\infty/\delta \sim \partial u/\partial y)]$.

The final point allows us to neglect the normal viscous stresses in the boundary layer.

We make short work of the principle of conservation of mass by noting that, as far as mass flow is concerned, Fig. 9.26 is identical to Fig. 5.10 (except that ρ

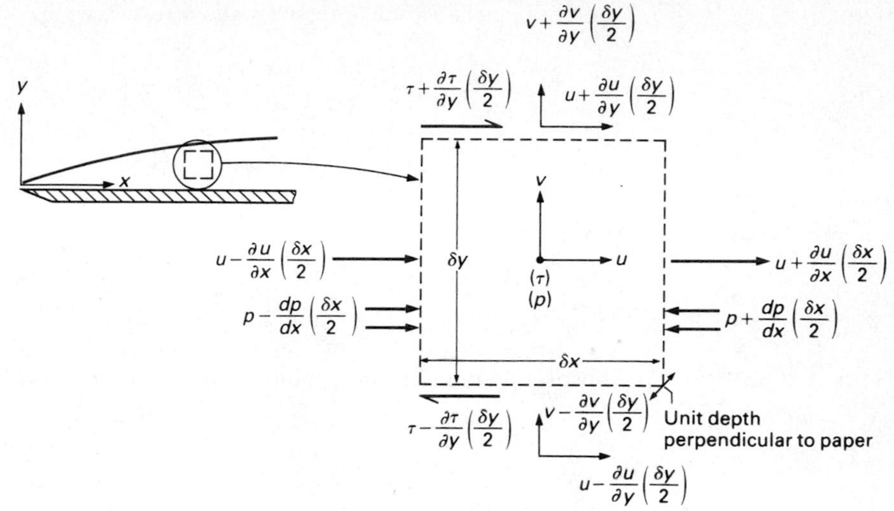

Figure 9.26 Control volume for boundary layer analysis. Only velocities and stresses relevant to continuity and x-momentum equations are shown.

is constant), and so the continuity equation is

$$\frac{\partial u}{\partial x} + \frac{\partial v}{\partial y} = 0 \tag{9.105}$$

as we have assumed two-dimensional, two-directional, incompressible flow.

Next, we apply the linear momentum equation in the x direction. The x-momentum flux terms are

$$\begin{aligned}\delta\dot{M}_{x,\text{out}} - \delta\dot{M}_{x,\text{in}} &= \delta\dot{M}_{x,\text{right}} + \delta\dot{M}_{x,\text{top}} - \delta\dot{M}_{x,\text{left}} - \delta\dot{M}_{x,\text{bottom}}\\ &= \left[u + \frac{\partial u}{\partial x}\left(\frac{\delta x}{2}\right)\right]\rho\left[u + \frac{\partial u}{\partial x}\left(\frac{\delta x}{2}\right)\right]\delta y(1)\\ &\quad + \left[u + \frac{\partial u}{\partial y}\left(\frac{\delta y}{2}\right)\right]\rho\left[v + \frac{\partial v}{\partial y}\left(\frac{\delta y}{2}\right)\right]\delta x(1)\\ &\quad - \left[u - \frac{\partial u}{\partial x}\left(\frac{\delta x}{2}\right)\right]\rho\left[u - \frac{\partial u}{\partial x}\left(\frac{\delta x}{2}\right)\right]\delta y(1)\\ &\quad - \left[u - \frac{\partial u}{\partial y}\left(\frac{\delta y}{2}\right)\right]\rho\left[v - \frac{\partial v}{\partial y}\left(\frac{\delta y}{2}\right)\right]\delta x(1).\end{aligned}$$

Carrying out the multiplications, canceling terms where possible, and dropping higher-order terms, we obtain

$$\delta\dot{M}_{x,\text{out}} - \delta\dot{M}_{x,\text{in}} = \rho\left[u\left(\frac{\partial u}{\partial x}\right) + v\left(\frac{\partial u}{\partial y}\right) + u\left(\frac{\partial u}{\partial x} + \frac{\partial v}{\partial y}\right)\right]\delta x\,\delta y(1).$$

The term $(\partial u/\partial x + \partial v/\partial y)$ is zero by Eq. (9.105), so

$$\delta\dot{M}_{x,\text{out}} - \delta\dot{M}_{x,\text{in}} = \rho\left[u\frac{\partial u}{\partial x} + v\frac{\partial u}{\partial y}\right]\delta x\,\delta y(1).$$

The net x force is*

$$\delta F_x = \delta F_{p,\text{left}} - \delta F_{p,\text{right}} + \delta F_{\tau,\text{top}} - \delta F_{\tau,\text{bottom}}.$$

---

* Gravity forces can be combined with pressure in the same way as in Eqs. (9.1)–(9.4).

Substituting the various pressures and stresses, we have

$$\delta F_x = \left[p - \frac{dp}{dx}\left(\frac{\delta x}{2}\right)\right]\delta y(1) - \left[p + \frac{dp}{dx}\left(\frac{\delta x}{2}\right)\right]\delta y(1)$$
$$+ \left[\tau + \frac{\partial \tau}{\partial y}\left(\frac{\delta y}{2}\right)\right]\delta x(1) - \left[\tau - \frac{\partial \tau}{\partial y}\left(\frac{\delta y}{2}\right)\right]\delta x(1).$$

Simplifying gives

$$\delta F_x = \left(-\frac{dp}{dx} + \frac{\partial \tau}{\partial y}\right)\delta x\, \delta y(1).$$

Substituting into the linear momentum equation and canceling $\delta x\, \delta y(1)$, we obtain

$$\rho\left(u\frac{\partial u}{\partial x} + v\frac{\partial u}{\partial y}\right) = -\frac{dp}{dx} + \frac{\partial \tau}{\partial y}. \tag{9.106}$$

We can apply this equation to either laminar or turbulent flow, because we have not yet specified how τ is obtained. If we limit consideration to laminar flow of a Newtonian fluid with constant viscosity, the boundary layer momentum equation becomes (on dividing by ρ):

$$u\left(\frac{\partial u}{\partial x}\right) + v\left(\frac{\partial u}{\partial y}\right) = -\frac{1}{\rho}\left(\frac{dp}{dx}\right) + \nu\left(\frac{\partial^2 u}{\partial y^2}\right). \tag{9.107}$$

Equations (9.105) and (9.107) are two equations for the two unknowns u and v. These are the *Prandtl boundary layer equations*. We assume that the pressure is known, so that the condition

$$\frac{\partial p}{\partial y} \approx 0$$

replaces the entire y-momentum equation. We can write Bernoulli's equation for the flow just outside the boundary layer* as

$$p + \frac{\rho U^2}{2} = \text{Constant},$$

which, on differentiating, gives

$$-\frac{1}{\rho}\left(\frac{dp}{dx}\right) = U\left(\frac{dU}{dx}\right). \tag{9.108}$$

We can use Eq. (9.108) to eliminate the pressure from Eq. (9.107).

To solve the equations, we must specify boundary conditions. At $y = 0$ (the wall),

$$u = 0 \qquad \text{at } y = 0; \tag{9.109}$$

$$v = 0 \qquad \text{at } y = 0. \tag{9.110}$$

---

* We write U for the velocity outside the boundary layer rather than V_∞ to emphasize that the equations are not restricted to a flat plate.

A third condition seems obvious when we recognize that the boundary layer must join with the outer inviscid flow at its edge. We would like to say

$$u = U \qquad \text{at } y = \delta;$$

however, we do not know where the edge of the boundary layer is (δ is an unknown function of x). Therefore we specify that

$$u \rightarrow U \quad \text{as} \quad y \rightarrow \infty. \tag{9.111}$$

This condition is quite realistic, because the boundary layer merges smoothly into the outer flow. We do have a problem calculating a value for the boundary layer thickness δ when condition (9.111) is used to obtain a velocity profile. Conventionally, δ is defined as *the value of y at which u is* $0.99U$.

A fourth condition is that the velocity profile must be known at some upstream or inlet position,

$$u = u_0(y) \qquad \text{at } x = x_0. \tag{9.112}$$

Equations (9.109)–(9.112) are all the boundary conditions that are needed. The boundary layer equations do not need and *cannot use* conditions on the velocity at an outlet (downstream) station or conditions on v at the outer edge of the layer.

9.2.3 Evaluation of Key Boundary Layer Parameters

A complete boundary layer calculation requires solution of Eqs. (9.105) and (9.107) to obtain the fluid velocities $u(x, y)$ and $v(x, y)$. This method is difficult because the momentum equation is nonlinear. For now, we imagine that we have such a solution and consider how to evaluate some key boundary layer parameters.

The parameters of primary interest are the boundary layer thickness and the wall shear stress or skin friction coefficient. The boundary layer thickness is obtained from the velocity profile by the definition

$$\delta \equiv y)_{u/U=0.99}. \tag{9.113}$$

The wall shear stress is calculated by

$$\tau_w = \mu\left(\frac{\partial u}{\partial y}\right)_{\text{wall}} = \mu\left(\frac{\partial u}{\partial y}\right)_{y=0}, \tag{9.114}$$

or, in terms of the skin friction coefficient,

$$C_f = \frac{2\tau_w}{\rho U^2} = \frac{2\nu}{U^2}\left(\frac{\partial u}{\partial y}\right)_{y=0}. \tag{9.115}$$

We may also be interested in the two boundary layer effects on the outer flow. The first is streamline deflection. If there were no boundary layer, the streamlines near the surface would be parallel to it. The boundary layer produces a transverse velocity at its edge that deflects the outer flow streamlines outward slightly (Fig. 9.27a). The second effect is blockage. The mass flow in the boundary layer is reduced because the velocity is lowered. The streamline deflection and blockage effects are related, because a thicker body (more blockage) would also cause more

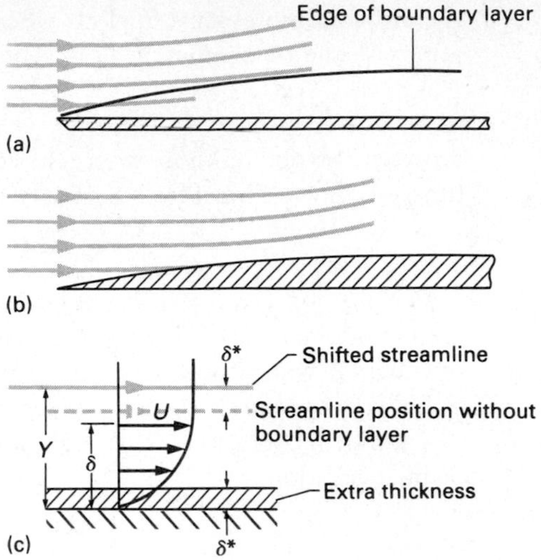

Figure 9.27 Streamline shift and blockage effects of a boundary layer: (a) deflection of outer flow streamlines; (b) flow over thicker body; (c) equivalence of extra thickness and streamline shift.

deflection of the outer flow (Fig. 9.27b). We can quantify both effects by introducing the *displacement thickness* (δ^*), defined as either the distance that the outer flow streamlines are shifted (displaced) by the boundary layer or, equivalently, as the extra thickness that would be added to a body or surface to account for the extra blockage. Both concepts are illustrated in Fig. 9.27(c).

The displacement thickness is calculated from a mass balance:

$$\int_0^Y \rho u \, dy = \int_{\delta^*}^Y \rho U \, dy.$$

The integral on the left is the actual mass flow between the wall and a point in the free stream. The integral on the right is the mass that would be carried by a uniform velocity $U(x)$ over an imaginary surface that is thicker by amount δ^* than the real surface. Because ρ and U are constants, we can cancel ρ and evaluate the right-hand integral:

$$\int_{\delta^*}^Y u \, dy = U(Y - \delta^*).$$

Solving for δ^*, we have

$$\delta^* = Y - \int_0^Y \frac{u}{U} dy = \int_0^Y dy - \int_0^Y \frac{u}{U} dy = \int_0^Y \left(1 - \frac{u}{U}\right) dy.$$

The actual value of Y is arbitrary so long as $Y > \delta$; however, note that

$$u = U \qquad \text{if } Y \geq \delta,$$

so the upper limit can be changed to δ (it could just as easily be changed to ∞):

$$\delta^* = \int_0^\delta \left(1 - \frac{u}{U}\right) dy. \tag{9.116}$$

The boundary layer blockage and streamline displacement effects are obtained by calculating δ^* from the velocity profile. The effect of the boundary layer on the outer flow is obtained by adding the displacement thickness to the surface and recalculating the inviscid flow over the resulting body. This effect is especially important in internal flows, such as the development region of a pipe (Fig. 7.4), where the displacement effect causes an acceleration of the inviscid core flow.

Besides accounting for blockage, δ^* provides a more rigorous and repeatable measurement of boundary layer thickness than δ does. This is especially true when we consider experimentally measured boundary layer velocity profiles. The point at which $u/U = 0.99$ is a matter of judgment in fitting a curve to the data; however, the integral in Eq. (9.116) gives almost the same value of δ^* no matter where the outer limit is located, so long as $Y \geq \delta$.

9.2.4 Momentum Integral Equation for Boundary Layers

Solutions to the Prandtl boundary layer equations have been found for only a few special (but nevertheless important) cases. If we consider the topics discussed in the preceding section, we note that a complete solution (that is, u and v as functions of x and y) is not very important anyway. All that we really need for most practical applications are values for the boundary layer thickness (both δ and δ^*) and the skin friction coefficient. For most engineering purposes, reliable estimates of these parameters are sufficient. This situation is tailor-made for application of the control volume approach. We already know that δ and C_f are functions of x (Eqs. 9.99 and 9.100), so we use a "hybrid" control volume that is finite in the y direction but differentially small in the x direction. This approach gives us an ordinary differential equation involving the key boundary layer parameters.

Figure 9.28 shows a control volume that spans the entire boundary layer, extending from the wall to an arbitrary height Y, outside the boundary layer. The length of the control volume is δx. The control volume has unit depth perpendicular to the paper. The velocity distributions at the inlet (left) and outlet (right) faces of the control volume are not uniform. The pressure is constant over each x face. We first write the continuity equation for the control volume. Noting that

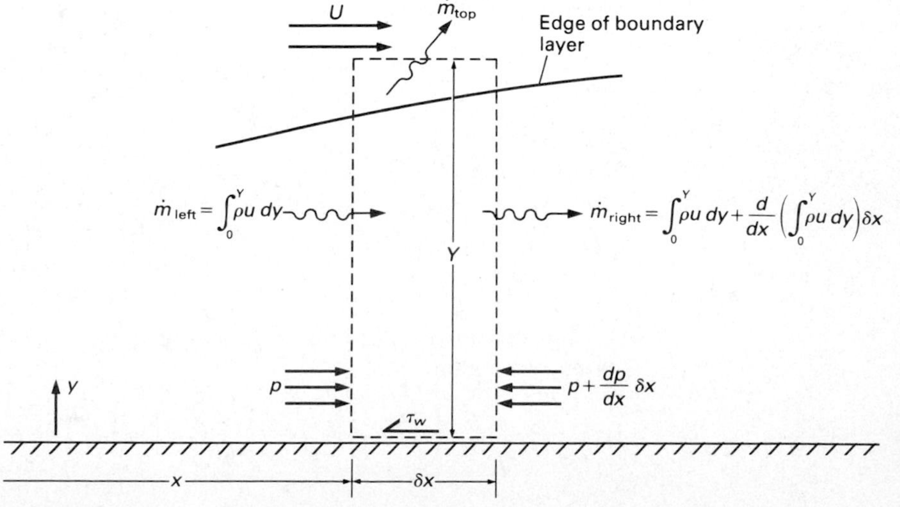

Figure 9.28 "Hybrid" control volume for integral analysis of boundary layer.

no fluid crosses the lower surface, we get

$$\dot{m}_{\text{left}} = \dot{m}_{\text{right}} + \dot{m}_{\text{top}}. \tag{9.117}$$

The mass flow across the left (inlet) face is

$$\dot{m}_{\text{left}} = \left.\int_0^Y \rho u \, dy\right)_x. \tag{9.118}$$

The mass flow rate through the right face is

$$\dot{m}_{\text{right}} = \dot{m}_{x+\delta x} = \left.\int_0^Y \rho u \, dy\right)_x + \frac{d}{dx}\left(\int_0^Y \rho u \, dy\right)\delta x. \tag{9.119}$$

Substituting into Eq. (9.117) and solving for the mass flow across the top of the control volume give

$$\dot{m}_{\text{top}} = -\frac{d}{dx}\left(\int_0^Y \rho u \, dy\right)\delta x. \tag{9.120}$$

Now we consider the momentum equation. Momentum flows across the control surface at the right and left faces and at the top of the control volume. The forces on the fluid are caused by pressure and shear stress. The x-direction momentum equation is

$$\dot{M}_{x,\text{right}} + \dot{M}_{x,\text{top}} - \dot{M}_{x,\text{left}} = F_{x,\text{pressure}} + F_{x,\text{stress}}. \tag{9.121}$$

The x-momentum flux at the left (inlet) face is

$$\dot{M}_{x,\text{left}} = \int_0^Y \rho u V_n \, dA = \left.\int_0^Y \rho u^2 \, dy\right)_x. \tag{9.122}$$

The x-momentum flux at the right face is

$$\dot{M}_{x,\text{right}} = \dot{M}_x)_{x+\delta x} = \left.\int_0^Y \rho u^2 \, dy\right)_x + \frac{d}{dx}\left(\int_0^Y \rho u^2 \, dy\right)\delta x, \tag{9.123}$$

and the x-momentum flux at the top is

$$\dot{M}_{x,\text{top}} = \dot{m}_{\text{top}} u_{\text{top}} = \dot{m}_{\text{top}} U.$$

Using Eq. (9.120), we have

$$\dot{M}_{x,\text{top}} = -U \frac{d}{dx}\left(\int_0^Y \rho u \, dy\right)\delta x. \tag{9.124}$$

The forces are

$$F_{x,\text{pressure}} = pY - \left(p + \frac{dp}{dx}\,\delta x\right)Y = -\frac{dp}{dx}\,Y\,\delta x \tag{9.125}$$

and

$$F_{x,\text{stress}} = -\tau_w \, \delta x. \tag{9.126}$$

Substituting Eqs. (9.122)–(9.126) into Eq. (9.121), canceling terms, and dividing by δx, we obtain

$$\frac{d}{dx}\left(\int_0^Y \rho u^2 \, dy\right) - U \frac{d}{dx}\left(\int_0^Y \rho u \, dy\right) = -Y\left(\frac{dp}{dx}\right) - \tau_w. \tag{9.127}$$

At the beginning of this section, we implied that we would obtain an equation for the boundary layer parameters that is easier to solve than the Prandtl boundary layer equations. Equation (9.127) certainly does not seem to be it! To get Eq. (9.127) into useful form, we have to work with it a bit. Our first priority is elimination of the arbitrary height Y. To accomplish this, we first eliminate the pressure gradient, using Eq. (9.108). Noting that

$$Y = \int_0^Y dy,$$

multiplying by -1, dividing by ρ, and rearranging, we get

$$U \frac{d}{dx} \int_0^Y u\, dy + U\left(\frac{dU}{dx}\right) \int_0^Y dy - \frac{d}{dx} \int_0^Y u^2\, dy = \frac{\tau_w}{\rho}. \tag{9.128}$$

Next consider the quantity*

$$\frac{d}{dx}\left(U \int_0^Y u\, dy\right) = \frac{dU}{dx}\left(\int_0^Y u\, dy\right) + U \frac{d}{dx} \int_0^Y u\, dy.$$

The second term on the right appears in Eq. (9.128). We solve for it in terms of the other two and substitute the result into Eq. (9.128) to get

$$\frac{d}{dx}\left(U \int_0^Y u\, dy\right) - \frac{dU}{dx}\left(\int_0^Y u\, dy\right) + U\left(\frac{dU}{dx}\right) \int_0^Y dy - \frac{d}{dx} \int_0^Y u^2\, dy = \frac{\tau_w}{\rho}.$$

As U is not a function of y, we can move it inside the first and third integrals. Performing this operation and grouping terms, we have

$$\frac{d}{dx} \int_0^Y u(U-u)\, dy + \frac{dU}{dx} \int_0^Y (U-u)\, dy = \frac{\tau_w}{\rho}.$$

Note that if $y > \delta$, the integrands are zero, so we can change the upper limit of the integrals to δ to get

$$\frac{d}{dx} \int_0^\delta u(U-u)\, dy + \frac{dU}{dx} \int_0^\delta (U-u)\, dy = \frac{\tau_w}{\rho}. \tag{9.129}$$

Equation (9.129) is *von Karman's momentum integral equation.* It probably does not seem any simpler to you than the Prandtl boundary layer equations, but note that it *is* an ordinary differential equation and that two key boundary layer parameters (δ and τ_w) appear explicitly. Also, the second integral reminds us strongly of the displacement thickness. Equation (9.129) has the additional advantage that it can be applied to either laminar or turbulent flow, because no specific expression for τ_w has been used.

To discover the real advantages of Eq. (9.129), we must manipulate it a bit more. We multiply the first integral by U^2/U^2 and the second integral by U/U to

---

* This quantity does not appear in Eq. (9.128). It is not equal to the first integral in Eq. (9.128), because U is a function of x.

get

$$\frac{d}{dx}\left[U^2 \int_0^\delta \frac{u}{U}\left(1-\frac{u}{U}\right)dy\right] + U\left(\frac{dU}{dx}\right)\int_0^\delta \left(1-\frac{u}{U}\right)dy = \frac{\tau_w}{\rho}.$$

Next we multiply each integral by δ/δ (noting that δ is independent of y) to get

$$\frac{d}{dx}\left\{\delta U^2\left[\int_0^1 \frac{u}{U}\left(1-\frac{u}{U}\right)d\left(\frac{y}{\delta}\right)\right]\right\} + \delta U\left(\frac{dU}{dx}\right)\int_0^1 \left(1-\frac{u}{U}\right)d\left(\frac{y}{\delta}\right) = \frac{\tau_w}{\rho}. \tag{9.130}$$

The integrals are complicated, but in this form they involve only the dimensionless shape of the velocity profile. Because the variable of integration has been changed from y to y/δ, boundary layer thickness does not matter so long as we know the shape of the velocity profile. We *assume* a reasonable shape for the boundary layer velocity profile:

$$\frac{u}{U} \approx f\left(\frac{y}{\delta}\right).$$

Then, the integrals can be evaluated, and Eq. (9.130) becomes a differential equation involving U (which is known from potential flow analysis), δ, and τ_w. We obtain some guidance in assuming a velocity profile function from the boundary conditions that it must satisfy (Eqs. 9.109–9.111). We do not have to know the exact shape; integration "smooths" our assumption.

We can write Eq. (9.130) in a more abbreviated form. Note that the second integral is the displacement thickness. The first integral also has the dimension of length and may be considered as a type of boundary layer thickness. This integral is called the *momentum thickness* and is assigned the symbol Θ:

$$\Theta \equiv \int_0^\delta \frac{u}{U}\left(1-\frac{u}{U}\right)dy. \tag{9.131}$$

In terms of δ^* and Θ, we write von Karman's momentum integral equation as

$$\frac{d}{dx}[U^2\Theta] + \delta^* U\left(\frac{dU}{dx}\right) = \frac{\tau_w}{\rho}. \tag{9.132}$$

We may also write the equation as

$$\frac{d\Theta}{dx} + (2 + H)\frac{\Theta}{U}\left(\frac{dU}{dx}\right) = \frac{C_f}{2}, \tag{9.133}$$

where

$$H \equiv \frac{\delta^*}{\Theta} \tag{9.134}$$

is called the *shape factor*.

EXAMPLE 9.5 Illustrates Boundary Layer Parameters

A 4-in.-nominal-diameter schedule 40 steel pipe has a water flow rate of 50 gal/min. Determine the values of the displacement thickness δ^*, momentum thickness Θ, and the shape factor H, for the fully developed flow velocity profile. The water temperature is 60°F.

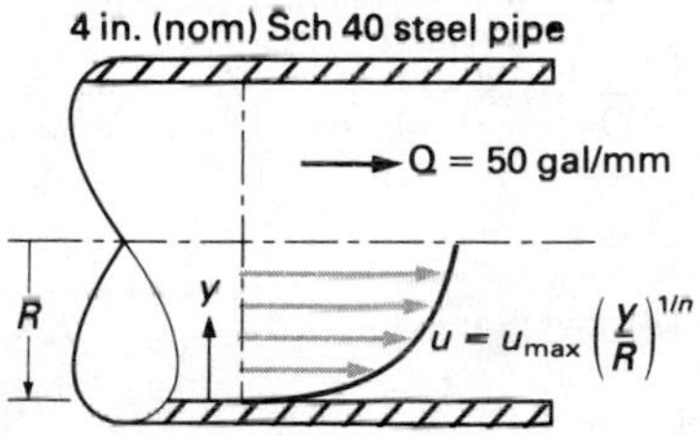

Figure E9.5 Velocity profile in pipe flow.

SOLUTION

Given

4-in.-nominal-diameter schedule 40 steel pipe

Flow rate 50 gal/min of 60°F water

Fully developed flow

Figure E9.5

Find

Displacement thickness δ^*

Momentum thickness Θ

Shape factor H

Solution

Appendix *D* shows that the pipe inside diameter is 4.026 in. For 60°F water, Table A.6 gives $\nu = 1.22 \times 10^{-5}$ ft²/sec. The average velocity is

$$V = \frac{Q}{A} = \frac{4Q}{\pi D^2}$$

$$= \frac{4(50 \text{ gal/min})(1 \text{ ft}^3/7.48 \text{ gal})(1 \text{ min}/60 \text{ sec})}{\pi(4.026 \text{ in.})^2(1 \text{ ft}/12 \text{ in.})^2} = 1.26 \text{ ft/sec},$$

and the Reynolds number is

$$\mathbf{R} = \frac{VD}{\nu} = \frac{(1.26 \text{ ft/sec})(4.026 \text{ in.})}{(1.22 \times 10^{-5} \text{ ft}^2/\text{sec})(12 \text{ in./ft})} = 3.5 \times 10^4.$$

For 4-in. commercial steel pipe, Fig. 7.10 gives $\varepsilon/D = 0.00043$, the Moody chart gives $f = 0.025$, and Eq. (7.49) and (7.51) give

$$\frac{u}{u_{\text{max}}} = \left(\frac{y}{R}\right)^{1/n},$$

with

$$n \approx \frac{1}{\sqrt{f}} = \frac{1}{\sqrt{0.025}} = 6.3.$$

We are now able to calculate the boundary layer parameters. The displacement thickness δ^* is

$$\delta^* = \int_0^R \left(1 - \frac{u}{u_{\text{max}}}\right) dy,$$

where the boundary layer thickness δ is equivalent to the pipe radius R. Substituting the velocity profile, we have

$$\delta^* = \int_0^R \left[1 - \left(\frac{y}{R}\right)^{1/n}\right] dy = \left[y - \frac{n}{n+1}\left(\frac{y^{(n+1)/n}}{R^{1/n}}\right)\right]_0^R$$

$$= \left[R - \frac{n}{n+1} R\right] = \frac{R}{n+1} = \frac{2.013 \text{ in.}}{7.3};$$

$$\delta^* = 0.276 \text{ in.}$$ **ANSWER**

The momentum thickness Θ is

$$\Theta = \int_0^R \frac{u}{u_{\max}}\left(1 - \frac{u}{u_{\max}}\right) dy = \int_0^R \left[\frac{u}{u_{\max}} - \left(\frac{u}{u_{\max}}\right)^2\right] dy,$$

or

$$\Theta = \int_0^R \left[\left(\frac{y}{R}\right)^{1/n} - \left(\frac{y}{R}\right)^{2/n}\right] dy$$

$$= \left[\frac{n}{n+1}\left(\frac{y^{(n+1)/n}}{R^{1/n}}\right) - \frac{n}{n+2}\left(\frac{y^{(n+2)/n}}{R^{2/n}}\right)\right]_0^R,$$

$$= \left[\left(\frac{n}{n+1}\right)R - \left(\frac{n}{n+2}\right)R\right]$$

$$= \left[\frac{n(n+2) - n(n+1)}{(n+1)(n+2)}\right]R = \left(\frac{n}{(n+1)(n+2)}\right)R,$$

$$= \frac{6.3}{(7.3)(8.3)}(2.013 \text{ in.}),$$

$$\Theta = 0.209 \text{ in.}$$ **ANSWER**

The shape factor, from Eq. (9.134), is

$$H = \frac{\delta^*}{\Theta} = \frac{0.276 \text{ in.}}{0.209 \text{ in.}};$$

$$H = 1.32.$$ **ANSWER**

Discussion

The problem statement specifies fully developed flow, so the "boundary layer thickness" is taken as the pipe radius, 2.013 in. The traditional boundary layer thickness, δ, is given by

$$\delta = y_{u=0.99u_{\max}},$$

or

$$\delta = R\left(\frac{u}{u_{\max}}\right)^n\Bigg]_{u=0.99u_{\max}} = 2.013(0.99)^{6.3} \text{ in.} = 1.89 \text{ in.}$$

9.2.5 Evaluation of Boundary Layer Parameters for Flow over a Flat Plate

A flat plate is aligned parallel to a uniform flow with velocity V_∞, (as shown in Fig. 9.29). If the plate is very thin and we neglect the boundary layer displacement

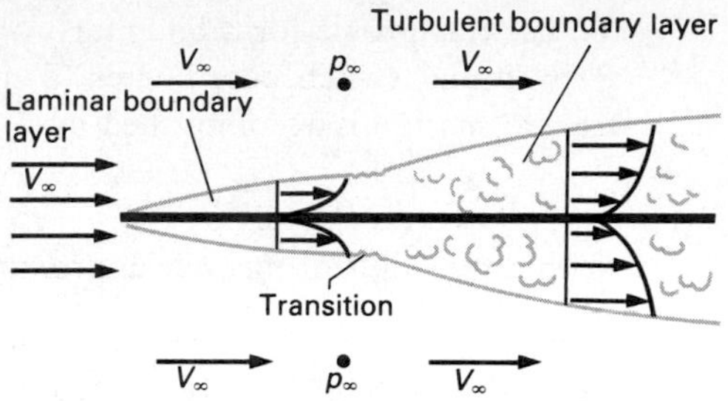

Figure 9.29 Boundary layers on a thin, flat plate parallel to the main flow.

effect, the inviscid outer flow solution is

$$U = V_\infty = \text{Constant} \quad \text{and} \quad p = p_\infty = \text{Constant}. \tag{9.135}$$

The boundary layer on a flat plate is the simplest boundary layer to calculate because there is no free-stream pressure or velocity gradient.

Near the leading edge of the plate, the Reynolds number is small and the flow is laminar. If the plate is long enough, eventually the boundary layer becomes thick enough and the Reynolds number becomes large enough that transition occurs and the flow becomes turbulent. The laminar and turbulent portions of the boundary layer must be considered separately.

Laminar Boundary Layer. First, we use the momentum integral equation to obtain an estimate of the laminar boundary layer parameters. With constant free-stream velocity, Eq. (9.130) reduces to

$$U^2 \frac{d}{dx}\left[\delta \int_0^1 \frac{u}{U}\left(1 - \frac{u}{U}\right) d\left(\frac{y}{\delta}\right)\right] = \frac{\tau_w}{\rho}. \tag{9.136}$$

For laminar flow, the wall shear stress is

$$\tau_w = \mu\left(\frac{\partial u}{\partial y}\right)_{y=0}.$$

Noting that U and δ are not functions of y, we rewrite this expression as

$$\tau_w = \frac{\mu U}{\delta}\left[\frac{\partial(u/U)}{\partial(y/\delta)}\right]_{y/\delta=0}. \tag{9.137}$$

Substituting into Eq. (9.136) and dividing by U^2, we get

$$\frac{d}{dx}\left[\delta \int_0^1 \frac{u}{U}\left(1 - \frac{u}{U}\right) d\left(\frac{y}{\delta}\right)\right] = \frac{\nu}{U\delta}\left[\frac{\partial(u/U)}{\partial(y/\delta)}\right]_{y/\delta=0}. \tag{9.138}$$

If we knew how u/U depends on y/δ, we could evaluate the integral on the left and the derivative on the right. All we need is a velocity profile that looks like a boundary layer. What is available? We might try various assumptions, but we already know at least one velocity profile for laminar flow—that for fully developed flow in a pipe. Combining Eqs. (7.33) and (7.34), we obtain the velocity profile for fully developed laminar *pipe* flow:

$$\frac{u}{u_{\max}} = 1 - \left(\frac{r}{R}\right)^2.$$

To use this profile for a boundary layer analysis, we must replace u_{max} by U and R by δ and switch coordinates to measure outward from the wall. This latter transformation is accomplished by

$$y = R - r.$$

With these adjustments, our approximate boundary layer velocity profile becomes

$$\frac{u}{U} \approx 2\left(\frac{y}{\delta}\right) - \left(\frac{y}{\delta}\right)^2 = 2\eta - \eta^2,$$

where

$$\eta = \frac{y}{\delta}.$$

Note that this profile satisfies the conditions

$$u = 0 \qquad \text{at } y = 0 \quad \text{and} \quad u = U \qquad \text{at } y = \delta.$$

Note that we are not assuming that the boundary layer is in a pipe and that we definitely are not assuming that U (which corresponds to u_{max}) is in any way related to the parameter $(R^2/4\mu)(dp/dx)$. We use the dimensionless pipe flow velocity profile only because it looks reasonably like a laminar boundary layer velocity profile.

The dimensionless slope of the parabolic profile, evaluated at the wall, is

$$\left.\frac{\partial(u/U)}{\partial(y/\delta)}\right)_0 = \left.\frac{d(u/U)}{d\eta}\right)_{\eta=0} = 2.$$

Substituting this result and the equation for u/U into the momentum integral equation, Eq. (9.138), gives

$$\frac{d}{dx}\left[\delta \int_0^1 (2\eta - \eta^2)(1 - 2\eta + \eta^2)\, d\eta\right] = 2\frac{\nu}{U\delta}.$$

Integrating, we obtain

$$\frac{2}{15}\left(\frac{d\delta}{dx}\right) = \frac{2\nu}{U\delta}.$$

Noting that ν is constant and that the velocity at the boundary layer edge (U) is constant and equal to V_∞, we separate the variables,

$$\delta\, d\delta = 15\left(\frac{\nu}{V_\infty}\right) dx,$$

and integrate from $x = 0$, where $\delta = 0$, to $x = x$ to get

$$\frac{\delta^2}{2} = 15\left(\frac{\nu x}{V_\infty}\right).$$

Solving for δ, we obtain

$$\delta = 5.48\sqrt{\frac{\nu x}{V_\infty}}. \tag{9.139a}$$

Dividing both sides by x, we get

$$\frac{\delta}{x} = 5.48\sqrt{\frac{\nu}{V_\infty x}} = \frac{5.48}{\sqrt{\mathbf{R}_x}}. \tag{9.139b}$$

Equation (9.139b) not only verifies Eq. (9.99), but it also provides a value for the proportionality constant.

The wall shear stress is calculated from Eq. (9.137):

$$\tau_w = \frac{2\mu V_\infty}{5.48\sqrt{(\nu x)/V_\infty}} = 0.365\sqrt{\frac{\rho\mu V_\infty^3}{x}}. \tag{9.140}$$

The skin friction coefficient is

$$\mathbf{C_f} = \frac{2\tau_w}{\rho V_\infty^2} = 0.730\sqrt{\frac{\nu}{V_\infty x}} = \frac{0.730}{\sqrt{\mathbf{R}_x}}, \tag{9.141}$$

verifying Eq. (9.100). We calculate the displacement thickness and momentum thickness from

$$\delta^* = \delta\int_0^1\left(1-\frac{u}{U}\right)d\left(\frac{y}{\delta}\right) \approx \delta\int_0^1 (1-2\eta+\eta^2)\,d\eta$$

and

$$\Theta = \delta\int_0^1 \frac{u}{U}\left(1-\frac{u}{U}\right)d\left(\frac{y}{\delta}\right) \approx \delta\int_0^1 (2\eta-\eta^2)(1-2\eta+\eta^2)\,d\eta.$$

The results are

$$\frac{\delta^*}{x} = \frac{1.83}{\sqrt{\mathbf{R}_x}} \tag{9.142}$$

and

$$\frac{\Theta}{x} \approx \frac{0.73}{\sqrt{\mathbf{R}_x}}. \tag{9.143}$$

These results are nice, but how accurate are they? Also, how much do they depend on the form of the function selected to approximate the velocity profile? To answer the second question, we note that changing the assumed velocity profile changes the numerical values of the definite integral and the dimensionless wall slope in Eq. (9.138).* These changes in turn cause slight changes in the numerical constants in Eqs. (9.139)–(9.143). Table 9.2 lists various approximations for the velocity profile and their effect on the results.

The accuracy of our momentum integral estimate is evaluated by comparing the results to an exact solution of the Prandtl boundary layer equations. The first boundary layer calculation ever made was the solution of the Prandtl boundary

* Note that the fundamental nature of δ versus x dependence indicated by $d\delta/dx \propto \nu/V_\infty\delta$ *is not* changed.

Table 9.2 Effects of various velocity profile assumptions on momentum integral calculation of flat plate laminar boundary layer parameters.

	$\frac{\delta}{x}\sqrt{\mathbf{R}_x}$	$\mathbf{C_f}\sqrt{\mathbf{R}_x}$	$\frac{\delta^*}{x}\sqrt{\mathbf{R}_x}$
Linear profile $\frac{u}{U} \approx \frac{y}{\delta}$	3.46	0.578	1.73
Parabolic profile $\frac{u}{U} \approx 2\left(\frac{y}{\delta}\right) - \left(\frac{y}{\delta}\right)^2$	5.48	0.730	1.83
Cubic profile $\frac{u}{U} \approx \frac{3}{2}\left(\frac{y}{\delta}\right) - \frac{1}{2}\left(\frac{y}{\delta}\right)^3$	4.64	0.646	1.74
Sine wave profile $\frac{u}{U} \approx \sin\left[\frac{\pi}{2}\left(\frac{y}{\delta}\right)\right]$	4.79	0.656	1.74
Exact solution	5.0	0.664	1.721

layer equations for laminar flow over a flat plate, carried out by H. Blasius, one of Prandtl's students [12]. Blasius assumed that the velocity profile in the boundary layer at any location along the plate is similar to the velocity profile at any other location; that is, the velocity function is of the special form

$$\frac{u(x, y)}{V_\infty} = f\left(\frac{y}{\delta(x)}\right) = f\left(\frac{y}{\sqrt{(\nu x)/V_\infty}}\right).$$

Using this assumption, Blasius was able to reduce the Prandtl boundary layer equations to a single ordinary differential equation. The details of the reduction and the solution of the differential equation are available elsewhere [3–9]. Figure 9.30 is a plot of the resulting velocity profile in Blasius's coordinates. The boundary layer parameters can be evaluated from this velocity profile. They are

$$\delta = 5.0\sqrt{\frac{\nu x}{V_\infty}}; \tag{9.144}$$

$$\mathbf{C_f} = \frac{0.664}{\sqrt{\mathbf{R}_x}}; \tag{9.145}$$

$$\delta^* = 1.721\sqrt{\frac{\nu x}{V_\infty}}; \tag{9.146}$$

$$\Theta = 0.664\sqrt{\frac{\nu x}{V_\infty}}. \tag{9.147}$$

Comparing Blasius's "exact" results with our momentum integral estimates, we find that we are about 10 percent too high—not bad in boundary layer theory. Blasius's results are also included in Table 9.2, so that you can evaluate the accuracy of other velocity profile approximations in the momentum integral equation.

With a boundary layer solution available, we can calculate the drag coefficient for a plate of length L and unit width if the flow is laminar over the entire length.

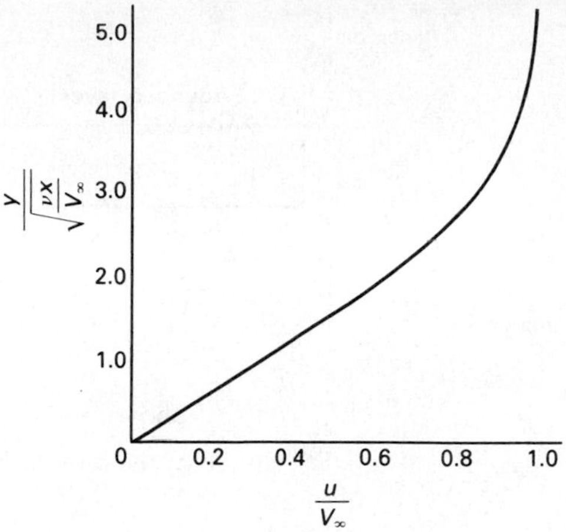

Figure 9.30 Velocity profile in laminar boundary layer on a flat plate from exact calculation.

The drag on a flat plate is entirely the result of shear stress, so

$$\mathscr{D} = \int_0^L \tau_w \, dx.$$

This drag is for only one side of the plate. The drag coefficient is

$$\mathbf{C_D} = \frac{\mathscr{D}}{\frac{1}{2}\rho V_\infty^2 L} = \frac{1}{L}\int_0^L \frac{\tau_w}{\frac{1}{2}\rho V_\infty^2}\, dx = \frac{1}{L}\int_0^L \mathbf{C_f}\, dx.$$

Choosing Blasius's equation for $\mathbf{C_f}$, Eq. (9.145), as the most accurate gives

$$\mathbf{C_D} = \frac{1.328\sqrt{\nu}}{\sqrt{V_\infty L}} = \frac{1.328}{\sqrt{\mathbf{R}_L}}. \tag{9.148}$$

Interestingly,

$$\mathbf{C_D} = 2\mathbf{C_f}(L) = 2\frac{\Theta(L)}{L}. \tag{9.149}$$

EXAMPLE 9.6 Illustrates Estimation of Boundary Layer Parameters

Equation (7.4) gives the entrance length for laminar flow in a circular pipe. See Fig. E9.6. Use Eq. (9.139b) to verify the form of Eq. (7.4).

SOLUTION

Given

Equation (9.139b):

$$\frac{\delta}{x} = 5.48\sqrt{\frac{\nu}{V_\infty x}},$$

where δ is the boundary layer thickness

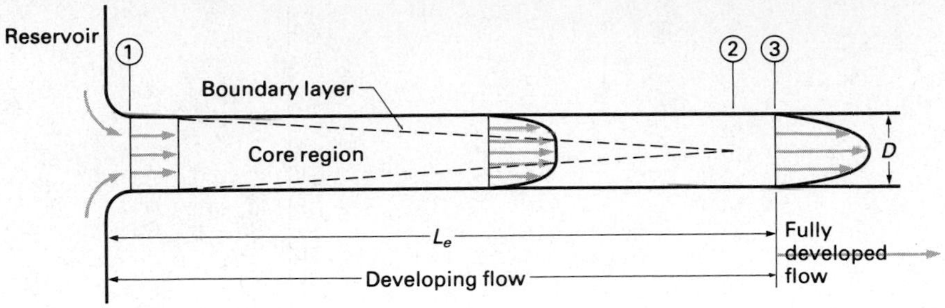

Figure E9.6 Developing flow.

Show

That the form of Eq. (7.4) is

$$\frac{L_e}{D} = (\text{Constant})\left(\frac{V_\infty D}{\nu}\right),$$

where D is the pipe diameter and V is the average pipe velocity

Solution

Since $\delta = D/2$ at the end of the entrance length L_e, Eq. (9.139b) gives

$$\frac{D}{2L_e} \approx 5.48\sqrt{\frac{\nu}{V_\infty L_e}}.$$

Rearranging and squaring give

$$\frac{1}{L_e} = \frac{(10.96)^2}{D^2}\left(\frac{\nu}{V_\infty}\right).$$

Then

$$\frac{D}{L_e} = (10.96)^2\left(\frac{\nu}{V_\infty D}\right)$$

or

$$\frac{L_e}{D} = \frac{1}{(10.96)^2}\left(\frac{V_\infty D}{\nu}\right) \approx 0.01\left(\frac{V_\infty D}{\nu}\right).$$

The suggested form is then

$$\frac{L_e}{D} = (\text{Constant})\left(\frac{V_\infty D}{\nu}\right).$$ **ANSWER**

Discussion

Equation (9.139b) was developed for a laminar flow over a flat plate, and Eq. (7.4) describes laminar flow in a circular pipe. Therefore, Eq. (9.139b) should not be expected to provide an exact result. In actual pipe flow, the acceleration of the inviscid core is accompanied by a pressure drop. This pressure drop is superimposed on the boundary layer to accelerate the boundary layer flow and reduce the growth rate of the boundary layer. The result is that the actual development length is about six times larger than our simple estimate predicts.

Turbulent Boundary Layer. If the laminar boundary layer is allowed to develop along the plate, it eventually becomes unstable and the process of transition to turbulent flow begins. Although the transition process is not completely understood, we do know that at some location the boundary layer becomes fully turbulent. Transition occurs over some finite distance, but assuming that it occurs instantaneously at a point is often sufficiently accurate. The location of the transition point depends on the Reynolds number and other parameters, such as surface roughness and the amount of turbulence in the flow adjacent to the boundary layer. For flow over a smooth plate, the transition Reynolds number is often taken to be

$$\mathbf{R}_{x,\mathrm{tr}} = \frac{V_\infty x_{\mathrm{tr}}}{\nu} \approx 5 \times 10^5. \tag{9.150}$$

The boundary layer downstream from the transition point is turbulent.

There is no "exact" solution of the Prandtl boundary layer equations for turbulent flow, even for a flat plate. The reason is that there is no rigorous expression for shear stress in turbulent flow. However, there are several numerical (computer) solutions of turbulent boundary layer flow. These solutions are based on semi-empirical models of the turbulent shear stress, of which the mixing length model (Example 5.6) is typical. These stress models are usually constructed from data on flat plate boundary layers, so any subsequent flat plate boundary layer calculation actually has the answer built in.

We estimate the key parameters for a turbulent boundary layer on a flat plate by using the momentum integral approach; however, we also have to use some empirical information. The turbulent boundary layer on a flat plate is described by the momentum integral equation with $dU/dx = 0$:

$$\frac{d}{dx}\left[\delta \int_0^1 \frac{u}{U}\left(1 - \frac{u}{U}\right) d\left(\frac{y}{\delta}\right)\right] = \frac{\tau_w}{\rho U^2}. \tag{9.151}$$

Equation (9.151) is identical to that for the laminar boundary layer, Eq. (9.136), except that the equation has been divided by U^2. All that we needed to do for the laminar boundary layer was to select an appropriate velocity profile. Turbulent boundary layer velocity profiles do not look like laminar ones, so we cannot use the same profiles; however, borrowing the idea that pipe flow velocity profiles are reasonably accurate for flat plate boundary layers, we can use a power law velocity profile,

$$\frac{u}{U} \approx \left(\frac{y}{\delta}\right)^{1/n} = \eta^{1/n}. \tag{9.152}$$

The most common choice for n is 7.

For laminar flow, the surface shear stress could be calculated from Newton's law of viscosity by differentiating the assumed velocity profiles. This does not work for turbulent boundary layers because the slope of a power law profile is infinite at $y = 0$. To complete the problem, we must introduce independent information about τ_w. We do so by again using the idea that the flat plate boundary layer is similar to fully developed pipe flow and by using a "friction law" for pipe flow to get τ_w. We might be tempted to use the Colebrook formula (Eq. 7.48) as a starting point, but that equation is implicit and would be difficult to use. If we limit consideration to smooth surfaces and relatively low Reynolds numbers, we can use the Blasius correlation, Eq. (7.45),

$$f \approx 0.3164\mathbf{R_D}^{-1/4} \qquad (\mathbf{R_D} < 10^5),$$

where

$$\mathbf{R_D} = \frac{\bar{V}D}{\nu}.$$

This expression can be converted to an equation for surface shear stress in a boundary layer by

- writing $\tau_w = (f/8)(\rho\bar{V}^2)$ (Eq. 7.31);
- substituting 2δ for D; and
- substituting

$$\frac{2n^2V_\infty}{(n+1)(2n+1)}$$

for $\bar{V}$ (Eq. 7.52).

The resulting shear stress equation, for $n = 7$, is

$$\tau_w = 0.0225\rho V_\infty^{7/4}\left(\frac{\nu}{\delta}\right)^{1/4} \tag{9.153}$$

We now substitute Eqs. (9.152) and (9.153) into Eq. (9.151) to get

$$\frac{d}{dx}\left[\delta\int_0^1 \eta^{1/7}(1-\eta^{1/7})\,d\eta\right] = 0.0225\left(\frac{\nu}{V_\infty}\right)^{1/4}\delta^{-1/4}.$$

The magnitude of the definite integral is 7/72. The differential equation for δ becomes

$$\delta^{1/4}\left(\frac{d\delta}{dx}\right) = 0.231\left(\frac{\nu}{V_\infty}\right)^{1/4}.$$

Integrating gives

$$\frac{4}{5}\delta^{5/4} = 0.231\left(\frac{\nu}{V_\infty}\right)^{1/4}x + C.$$

To evaluate the constant of integration, we specify that $\delta = 0$ at $x = 0$, neglecting the laminar portion of the boundary layer. This gives $C = 0$ and

$$\delta = 0.370\left(\frac{\nu}{V_\infty}\right)^{1/5}x^{4/5}, \tag{9.154}$$

or, on dividing by x,

$$\frac{\delta}{x} \approx 0.370\mathbf{R}_x^{-1/5}. \tag{9.155}$$

Note that Eq. (9.154) shows that the turbulent boundary layer grows at a faster rate ($\delta \sim x^{4/5}$) than a laminar boundary layer ($\delta \sim x^{1/2}$). Substituting Eq. (9.154) into Eq. (9.153) gives a skin friction equation

$$\mathbf{C_f} \approx 0.0577\mathbf{R}_x^{-1/5}. \tag{9.156}$$

The displacement thickness is evaluated from

$$\delta^* = \delta \int_0^1 \left(1 - \frac{u}{U}\right) d\left(\frac{y}{\delta}\right) = \delta \int_0^1 (1 - \eta^{1/7})\, d\eta,$$

which gives

$$\frac{\delta^*}{x} \approx \frac{1}{8}\left(\frac{\delta}{x}\right) = 0.046\mathbf{R}_x^{-1/5}. \tag{9.157}$$

As with the integral momentum solution for laminar boundary layers, we must ask two questions: How accurate are these results? Can they be improved by different choices for the velocity profile and/or the skin friction equation? Unlike laminar flow, there are no exact boundary layer calculations for turbulent flow, so the question of accuracy can be answered only by comparing predictions with experimental data. This comparison reveals that Eq. (9.156) is accurate to ± 3 percent in the Reynolds number range $5 \times 10^5 < \mathbf{R}_x < 10^7$ (not bad for such a relatively crude analysis) but becomes progressively inaccurate at higher values of $\mathbf{R}_x$.

An improved momentum integral analysis involves using the logarithmic velocity profile, Eq. (7.93):

$$\frac{u}{u_\tau} = \frac{1}{0.41} \ln\left(\frac{yu_\tau}{\nu}\right) + 5.0,$$

which we discussed in connection with pipe flow. A review of Section 7.2.7 reveals that this velocity profile was not specifically restricted to pipe flow. The logarithmic velocity profile equation contains its own friction law if we assume that it is valid all the way to the edge of the boundary layer. Recalling that

$$u_\tau = \sqrt{\frac{\tau_w}{\rho}} = \sqrt{\frac{\mathbf{C_f} V_\infty^2}{2}}$$

and putting $u = V_\infty$ at $y = \delta$ in Eq. (7.94), we have

$$\left(\frac{\mathbf{C_f}}{2}\right)^{-1/2} = 2.44 \ln\left[\mathbf{R}_\delta\left(\frac{\mathbf{C_f}}{2}\right)^{1/2}\right] + 5.0, \tag{9.158}$$

an accurate but complicated implicit equation for $\mathbf{C_f}$ as a function of $\mathbf{R}_\delta$. White [8] used Eqs. (7.93) and (9.158) in a momentum integral analysis of flat plate boundary layers and obtained the following results:

$$\frac{\delta}{x} = 0.14\mathbf{R}_x^{-1/7} \tag{9.159}$$

and

$$C_f = 0.027 R_x^{-1/7}. \tag{9.160}$$

These equations are accurate to ± 5 percent over the range $10^5 < R_x < 10^9$.

The drag coefficient of a flat plate with turbulent flow over its entire length can be obtained from

$$C_D = \frac{1}{L}\int_0^L C_f\,dx.$$

Using Eq. (9.156) for C_f gives

$$C_D = 0.072 R_L^{-1/5}. \tag{9.161}$$

If we account for the laminar boundary layer on the front part of the plate, the drag coefficient is

$$C_D = \frac{1}{L}\left(\int_0^{x_{tr}} C_{f,lam}\,dx + \int_{x_{tr}}^L C_{f,turb}\,dx\right).$$

The transition location is calculated from

$$\frac{V_\infty x_{tr}}{\nu} \approx 5 \times 10^5.$$

Substituting Eqs. (9.145) and (9.156) and integrating, we obtain

$$C_D = 0.072 R_L^{-1/5} - \frac{1700}{R_L}. \tag{9.162}$$

This equation provides a reasonable fit to the "smooth plate" curves of Fig. 8.8.

As lengthy as this development was, we still considered only smooth plates. The analysis based on the logarithmic velocity profile can be extended to rough plates as well. Consult reference [8] for details.

9.2.6 Calculation of Boundary Layers with Pressure Gradient

The flat plate boundary layer solutions—a collection of algebraic equations for the key boundary layer parameters—are easy to use, but many practical flows occur with geometries other than a flat plate. In reality, the flatness of the surface is not of primary importance, but the absence of a streamwise pressure (or free-stream velocity) gradient is. Many practical applications involve flow with variable pressure: flow over airfoils, flow in nozzles and diffusers, and even flow in the entrance region of a pipe or channel. If the streamwise pressure gradient is small (and especially if it is negative), flat plate boundary layer results can be used to approximate the boundary layer parameters; however, such approximations are unsatisfactory if the pressure gradient is significant (Example 9.6). In this section, we briefly consider calculation of boundary layers in a flow with nonuniform pressure. We concentrate on integral analysis of laminar boundary layers.

Figure 9.31 shows boundary layers developing on surfaces with a pressure gradient. The pressure and free-stream velocity are related by Eq. (9.112). The free-stream pressure and velocity are known functions of x. Although we talk about

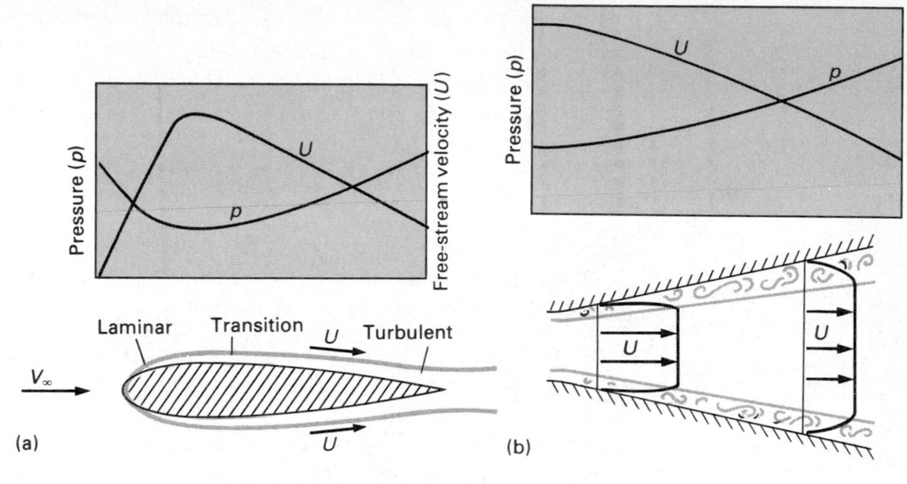

Figure 9.31 Examples of boundary layers in flow with pressure gradient: (a) airfoil boundary layer; (b) diffuser wall boundary layer.

boundary layers with *pressure* gradient, the analysis deals with the free-stream *velocity* gradient. The boundary layer parameters are related by the momentum integral equation, which, for laminar flow, is

$$\frac{d}{dx}\left[\delta U^2 \int_0^1 \frac{u}{U}\left(1-\frac{u}{U}\right) d\eta\right] + \delta U \frac{dU}{dx}\int_0^1 \left(1-\frac{u}{U}\right) d\eta = \frac{\nu U}{\delta}\left(\frac{\partial(u/U)}{\partial\eta}\right)_{\eta=0}, \quad (9.163)$$

where

$$\eta = \frac{y}{\delta}.$$

Unlike flat plate flow, U is a variable, which causes two immediate difficulties:

- U^2 cannot be factored out of the first term; and
- dU/dx is nonzero, so the second term on the left does not vanish, as it did for flat plate flow.

We expect these difficulties to lead to increased mathematical complexity, but, as $U\{x\}$ and dU/dx are known, we should be able to handle it. *If* we assume that

$$\frac{u}{U} = f\{\eta\},$$

as we did in flat plate flow, and select a "reasonable" profile (maybe the pipe flow profile again), we evaluate the two definite integrals on the left and the derivative on the right and obtain an equation of the form

$$C_1 \frac{d}{dx}(\delta U^2) + C_2\,\delta U\left(\frac{dU}{dx}\right) = C_3\left(\frac{\nu U}{\delta}\right),$$

where C_1, C_2, and C_3 are constants. Presumably this equation could be solved for $\delta\{x\}$ once $U\{x\}$ is specified. This approach seems straightforward enough; however, it does not work. The problem lies in the assumption that the velocity profile still depends on y/δ only. This assumption means that the shape of the velocity profile is the same at all x stations, which is simply not true for boundary layers in a pressure gradient. The shape of the velocity profile depends on the pressure gradient, so the velocity profile function must be of the form

$$\frac{u}{U} = f\left(\frac{y}{\delta}, \Lambda\right),$$

where Λ is a dimensionless pressure gradient parameter.

We can determine at least one form for Λ by dimensional analysis. We first assume

$$u = f\left(U, y, \delta, \frac{dU}{dx}, \nu\right). \tag{9.164}$$

Equation (9.164) contains six dimensional parameters and two fundamental dimensions ([L] and [T]), so we expect four dimensionless parameters. Three obvious parameters are u/U, y/δ, and $\mathbf{R}_\delta = U\delta/\nu$. A fourth is

$$\Lambda = \frac{\delta^2}{\nu}\left(\frac{dU}{dx}\right), \tag{9.165}$$

so the dimensionless velocity profile function is

$$\frac{u}{U} = F\{\eta, \Lambda, \mathbf{R}_\delta\}.$$

Our experience with the flat plate boundary layer indicates that the dependence of velocity profile on Reynolds number is contained in the dependence of δ on $\mathbf{R}_\delta$ so, for boundary layers in a pressure gradient, we assume that

$$\frac{u}{U} = f\{\eta, \Lambda\}. \tag{9.166}$$

Similarly, we expect that

$$\mathbf{C_f} = \mathbf{C_f}\{\Lambda, \mathbf{R}_\delta\}. \tag{9.167}$$

To calculate boundary layers with pressure gradient, we must find a velocity profile function of the form required by Eq. (9.166). When this profile is substituted into Eq. (9.163), a differential equation will result. Because we had such good luck with a second-degree polynomial profile for a flat plate, we assume that the velocity profile can be represented by a polynomial,

$$\frac{u}{U} \approx a + b\eta + c\eta^2 + d\eta^3 + e\eta^4. \tag{9.168}$$

We might use more terms, but a fourth-order polynomial proves adequate. We determine the coefficients of the polynomial by making the velocity distribution satisfy certain conditions at the wall ($\eta = 0$) and at the edge of the boundary layer ($\eta = 1$). Two of these conditions are

$$\frac{u}{U} = 0 \qquad \text{at } \eta = 0 \tag{9.169a}$$

and

$$\frac{u}{U} = 1 \qquad \text{at } \eta = 1. \tag{9.169b}$$

We obtain a third condition by requiring that the boundary layer velocity profile join smoothly with the outer flow:

$$\frac{d}{d\eta}\left(\frac{u}{U}\right) = 0 \qquad \text{at } \eta = 1. \tag{9.169c}$$

If we try to obtain a fourth condition by specifying a value for the first derivative of the velocity at the wall, we fail because the wall derivative is related to the wall shear stress ($\partial u/\partial y = \tau_w/\mu$), which is unknown. We can obtain a fourth condition by evaluating Prandtl's boundary layer momentum equation, Eq. (9.107), at the wall, where u and v are both zero:

$$-\frac{1}{\rho}\left(\frac{dp}{dx}\right) + \nu\left(\frac{\partial^2 u}{\partial y^2}\right) = 0 \qquad \text{at } y = 0.$$

Substituting Eq. (9.108) and multiplying by δ^2/U, we have

$$\frac{d^2}{d\eta^2}\left(\frac{u}{U}\right) = -\frac{\delta^2}{\nu}\left(\frac{dU}{dx}\right) = -\Lambda \qquad \text{at } \eta = 0. \tag{9.169d}$$

Note that this fourth condition automatically introduces Λ into the velocity profile. The fifth condition is obtained by returning to the outer edge of the boundary layer and requiring a "very smooth" patching with the outer flow by specifying

$$\frac{d^2(u/U)}{d\eta^2} = 0 \qquad \text{at } \eta = 1. \tag{9.169e}$$

Using Eqs. (9.169a)–(9.169e), we evaluate the coefficients in Eq. (9.168) and obtain,

$$\frac{u}{U} = 1 - (1 + \eta)(1 - \eta)^3 + \frac{\Lambda}{6}\eta(1 - \eta)^3. \tag{9.170}$$

Velocity profiles are plotted for various values of Λ in Fig. 9.32. In an actual boundary layer, Λ is a function of x that must be determined from the momentum equation. Once we determine Λ, we know the velocity profile shape.

Substituting Eq. (9.170) into Eq. (9.163) and carrying out the integration and differentiation yield an ordinary differential equation for δ (or Λ, because δ can be evaluated from Eq. 9.165 once Λ is known). The resulting differential equation is difficult to solve. In practice, it is rarely used to calculate boundary layers, so we do not include it here. Consult references [3] and [7] if you are interested.

In what follows, we outline a much simpler procedure for calculating laminar boundary layers with pressure gradient. This simple method, developed by Thwaites [13], is based on the ideas involved in the more complicated approach just developed. Suppose that we use the polynomial velocity profile (Eq. 9.170) to evaluate the boundary layer shape factor (H) and the skin friction coefficient. The results are

$$H = \frac{\delta^*}{\Theta} = \frac{\frac{3}{10} - (\Lambda/120)}{\frac{37}{315} - (\Lambda/945) - (\Lambda^2/9072)} \tag{9.171}$$

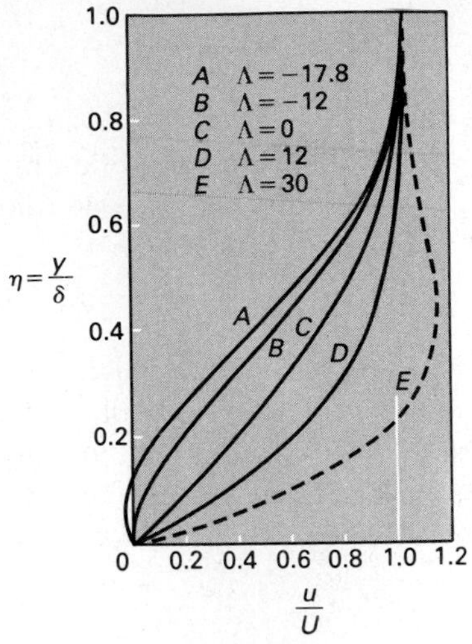

Figure 9.32 Fourth-order polynomial velocity profiles for various values of pressure gradient parameter Λ.

and

$$\frac{\mathbf{C_f}}{2} = \mathbf{R}_\delta^{-1}\left(2 + \frac{\Lambda}{6}\right). \tag{9.172}$$

The details of these equations are relatively unimportant; what is important is that they have the form

$$H = H\{\Lambda\} \tag{9.173}$$

and

$$\frac{\mathbf{C_f}}{2}\mathbf{R}_\delta = T_\delta\{\Lambda\}. \tag{9.174}$$

Thwaites, as well as others before him, realized that the whole problem with the "polynomial profile" method of calculating boundary layers was that it was based on the wrong type of boundary layer thickness. Considering the integral momentum equation, Eq. (9.133), in the form

$$\frac{d\Theta}{dx} + (2 + H)\frac{\Theta}{U}\left(\frac{dU}{dx}\right) = \frac{\mathbf{C_f}}{2},$$

we see that Θ, not δ, is the most "natural" thickness parameter. Thus a dimensionless pressure gradient parameter

$$\lambda \equiv \frac{\Theta^2}{\nu}\left(\frac{dU}{dx}\right) \tag{9.175}$$

should be used rather than Λ. Equation (9.133) can be converted into an equation for λ by multiplying by $\mathbf{R_\Theta}$ $(= U\Theta/\nu)$ to give

$$\frac{U}{2}\left(\frac{d}{dx}\right)\left(\frac{\lambda}{U'}\right) + (2 + H)\lambda = T, \tag{9.176}$$

where

$$T \equiv \frac{\tau_w \Theta}{\mu U} \quad \left(= \frac{\mathbf{C_f R_\Theta}}{2} \right) \quad \text{and} \quad U' = \frac{dU}{dx}.$$

Equations (9.173) and (9.174) imply that H and T are functions of the pressure gradient parameter, so Eq. (9.176) is equivalent to

$$U\left(\frac{d}{dx}\right)\left(\frac{\lambda}{U'}\right) = 2[T\{\lambda\} - \lambda(2 + H\{\lambda\})] \equiv F\{\lambda\}. \tag{9.177}$$

Thwaites realized that there is no real need to relate λ to the velocity profiles, deciding instead to seek the H and T functions more directly. Examining a large amount of boundary layer data, both from experiment and the few available exact solutions, Thwaites determined a relation between T and λ, between H and λ, and even between F [the entire right-hand side of Eq. (9.177)] and λ. These relations are represented by the following curve fits [4, 13]:

$$T\{\lambda\} \approx (\lambda + 0.09)^{0.62}; \tag{9.178}$$

$$H\{\lambda\} \approx \begin{cases} 2.61 - 3.75\lambda + 5.24\lambda^2 & (\lambda > 0); \\ \dfrac{0.0731}{0.15 + \lambda} + 2.088 & (\lambda < 0); \end{cases} \tag{9.179}$$

$$F\{\lambda\} \approx 0.45 - 6.0\lambda. \tag{9.180}$$

If we use Eq. (9.180), Eq. (9.177) becomes

$$U\left(\frac{d}{dx}\right)\left(\frac{\Theta^2}{\nu}\right) \approx 0.45 - 6.0\left(\frac{\Theta^2}{\nu}\right)\left(\frac{dU}{dx}\right),$$

which has the solution

$$\Theta^2 - \Theta_0^2 = \frac{0.45\nu}{U^6} \int_{x_0}^{x} U^5\, dx, \tag{9.181}$$

where Θ_0 is the value of Θ at $x = 0$. Using this equation, we have a simple calculation of the laminar boundary layer with a pressure gradient. The steps are as follows:

- $U\{x\}$, x_0, and Θ_0 must be known.
- At any x, calculate Θ from Eq. (9.181). This step is the most difficult, because evaluating the integral graphically or numerically is often necessary.
- Calculate λ [$= (\Theta^2/\nu)(dU/dx)$] and $\mathbf{R_\Theta}$ ($= U\Theta/\nu$).
- Calculate T from Eq. (9.178) and H from Eq. (9.179).
- Calculate $\mathbf{C_f}$ from

$$\mathbf{C_f} = \frac{2T}{\mathbf{R_\Theta}} \tag{9.182}$$

or calculate τ_w from

$$\tau_w = \frac{\mu U}{\Theta} T. \tag{9.183}$$

- Calculate δ^* from

$$\delta^* = H\Theta. \tag{9.184}$$

As a check, we can use Thwaites's method to calculate a flat plate boundary layer. Putting $U = V_\infty = \text{Constant}$ and $\Theta_0 = 0$, we obtain the following results:

$$\frac{\Theta}{x} = 0.671\mathrm{R}_x^{-1/2} \quad \text{and} \quad \mathrm{C_f} = 0.671\mathrm{R}_x^{-1/2}.$$

Compared to Blasius's values, these are quite satisfactory.

Unfortunately, no accurate, simple method like that of Thwaites is available for calculating turbulent boundary layers with pressure gradient. Many methods have been proposed over the years. In 1968, a milestone "showdown" between the various methods was held [14]. The best methods available were found to be equally distributed between momentum integral methods that used a logarithmic velocity profile, appropriately modified to account for pressure gradient, together with a consistent skin friction law and a third auxiliary equation based on a further integration of the boundary layer equations across the layer; and numerical finite-difference methods for solving the Prandtl boundary layer equations, together with an appropriate model of the turbulent shear stresses. All currently available accurate methods of calculating turbulent boundary layers with pressure gradient require computer solution, even those that involve only ordinary differential equations.

EXAMPLE 9.7 Illustrates Thwaites's Method and a Boundary Layer Application

Use Thwaites's method to verify the shear stress distribution

$$\tau_w = [1.74 \times 10^{-3}\,(\pi - \theta) - 1.59 \times 10^{-5}\,(\pi - \theta)^2]\ \mathrm{N/m^2}$$

used for the air flowing over the highway sign in Example 8.1. The air temperature is 20°C.

SOLUTION

Given

20°C air flowing over the highway sign of Example 8.1 at 10 km/hr

The sign has the shape of a half-cylinder with diameter 2.0 m and length 25.0 m.

Show

That the shear stress can be approximated by

$$\tau_w = [1.74 \times 10^{-3}\,(\pi - \theta) - 1.59 \times 10^{-5}\,(\pi - \theta)^2]\ \mathrm{N/m^2}$$

Solution

We first calculate the Reynolds number to verify that the boundary layer is laminar. Table A.3 gives $\nu = 1.51 \times 10^{-5}$ m²/s. Then

$$\mathbf{R} = \frac{V_\infty D}{\nu} = \frac{(10 \text{ km/hr})(1000 \text{ m/km})(1 \text{ hr}/3600 \text{ sec})(2.0 \text{ m})}{(1.51 \times 10^{-5} \text{ m}^2/\text{s})}$$
$$= 3.7 \times 10^5.$$

Our consideration of drag on a circular cylinder (Section 8.2.2) concluded that the boundary layer experiences transition somewhere on the cylinder if $\mathbf{R} > 5 \times 10^5$; therefore our boundary layer remains laminar,* and Thwaites's method may be used.

We calculate the shear stress using Eqs. (9.183), (9.178), (9.175), and (9.181):

$$\tau_w = \frac{\mu U}{\Theta} T,$$

where

$$T \approx (\lambda + 0.09)^{0.62}, \qquad \Theta^2 = \Theta_0^2 + \frac{0.45\nu}{U^6} \int_{x_0}^{x} U^5 \, dx,$$

and

$$\lambda = \frac{\Theta^2}{\nu} \left(\frac{dU}{dx} \right).$$

For a circular cylinder, Eq. (9.66) gives

$$V_\theta = -V_\infty \left(1 + \frac{R^2}{r^2} \right) \sin \theta.$$

On the cylinder surface, $r = R$ and†

$$U = -V_\theta = 2V_\infty \sin \theta.$$

Then using $\theta = \pi - \phi$ and $x = \phi R$, we have

$$U = 2V_\infty \sin(\pi - \phi) = 2V_\infty \sin \phi$$

and

$$\Theta^2 = \Theta_0^2 + \frac{0.45(32V_\infty^5)\nu}{64V_\infty^6 \sin^6 \phi} \int_0^\phi \sin^5 \phi' \, R \, d\phi'.$$

Noting that $\Theta_0 = 0$, simplifying, and integrating,‡ we obtain

$$\Theta^2 = \frac{0.45\nu R}{2V_\infty \sin^6 \phi} \int_0^\phi (1 - \cos^2 \phi')^2 \sin \phi' \, d\phi'$$
$$= \frac{0.45\nu R}{2V_\infty \sin^6 \phi} \left[-\cos \phi + \frac{2}{3} \cos^3 \phi - \frac{\cos^5 \phi}{5} + \frac{8}{15} \right].$$

Then

$$\Theta = \sqrt{\frac{0.45\nu R}{2V_\infty \sin^6 \phi} \left[-\cos \phi + \frac{2}{3} \cos^3 \phi - \frac{\cos^5 \phi}{5} + \frac{8}{15} \right]}.$$

* Example 9.8 gives an alternative procedure for estimating the state (laminar or turbulent) of the boundary layer.

† $U(x)$ is opposite V_θ for the direction of increasing x.

‡ Note that ϕ is a dummy variable of integration.

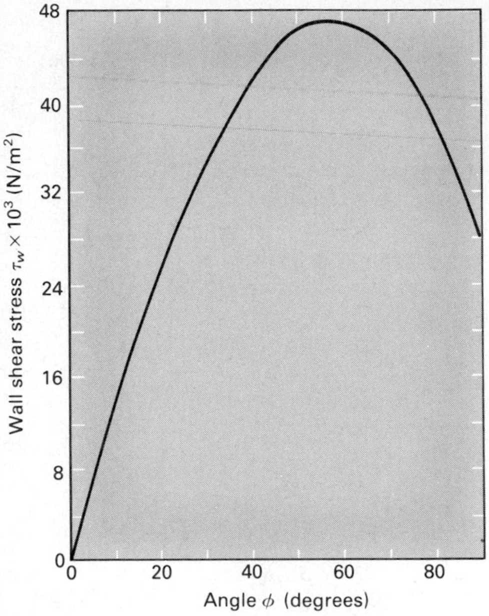

Figure E9.7 Wall shear stress over cylinder of Example 8.1.

To find the shear stress, we note that

$$T = (\lambda + 0.09)^{0.62} = \left[\frac{\Theta^2}{\nu}\left(\frac{dU}{dx}\right) + 0.09\right]^{0.62}$$

$$= \left[\frac{\Theta^2}{\nu R}\left(\frac{dU}{d\phi}\right) + 0.09\right]^{0.62}.$$

Substituting $U(\phi)$ gives

$$T = \left(\frac{2\Theta^2 V_\infty}{\nu R}\cos\phi + 0.09\right)^{0.62}.$$

The wall shear stress is

$$\tau_w = \frac{\mu U}{\Theta} T = \frac{2\mu V_\infty \sin\phi\,[(2\Theta^2 V_\infty/\nu R)\cos\phi + 0.09]^{0.62}}{\Theta},$$

where $\Theta(\phi)$ is given above. Now τ_w is plotted in Fig. E9.7 for the conditions of Example 8.1:

$$\mu = 1.81 \times 10^{-5}\ \text{N}\cdot\text{s/m}^2, \qquad \nu = 1.51 \times 10^{-5}\ \text{m}^2/\text{s},$$

$$R = 1.0\ \text{m}, \qquad V_\infty = 10\ \text{km/hr} = 2.78\ \text{m/s}.$$

A second-order polynomial approximation of the curve is

$$\tau_w \approx (1.74 \times 10^{-3}\,\phi - 1.59 \times 10^{-5}\,\phi^2)\ \text{N/m}^2,$$

or, because $\phi = (\pi - \theta)$ on the upper half of the cylinder,

$$\tau_w = [1.74 \times 10^{-3}\,(\pi - \theta) - 1.59 \times 10^{-5}\,(\pi - \theta)^2]\ \text{N/m}^2.$$

ANSWER

Discussion

We made the simplifying assumption that the flow over half a circular cylinder can be approximated by the flow over the front half of a complete cylinder. Our calculations are inaccurate near $\phi = 90°$. See Example 9.8.

9.2.7 Boundary Layer Separation and the Breakdown of the Potential Flow and Boundary Layer Model

The potential flow and boundary layer model of a high Reynolds number flow is based on the assumption that the boundary layer has only a minor effect on the external flow. This assumption is reasonably accurate, and the model produces acceptable results so long as the boundary layer is thin and remains attached to the surface. Under certain conditions, the boundary layer thickens rapidly and separates from the surface, forming a broad region of recirculating flow and grossly disturbing the outer flow streamlines (see Figs. 8.26, 8.47, and especially 8.15). In this case, although the concept of patching together a viscous-rotational flow and an inviscid-irrotational flow is still valid, simple boundary layer theory is no longer capable of calculating the viscous-rotational region, and the potential flow and boundary layer model fails.

Although the potential flow and boundary layer model is not capable of predicting the complete flow field, it is capable of yielding some very important information. Generally, the model is able to

- explain the conditions necessary for separation;
- indicate whether separation will occur; and
- indicate the approximate location where separation will occur.

Figure 9.33 shows a boundary layer developing toward separation. As the point of separation is approached, the boundary layer thickens rapidly. Downstream from separation is a region of backflow near the surface. At the separation point, the velocity profile has a vertical slope at the wall and $\tau_w = 0$ (at separation). Physical reasoning (see Section 8.2.2) indicates that separation occurs in a rising pressure gradient, usually called an *adverse pressure gradient.* This fact can be verified with the aid of Prandtl's boundary layer momentum equation, Eq. (9.107), together with a consideration of the shape of the boundary layer velocity profiles. The velocity profiles in Fig. 9.33 indicate that for backflow to develop, a boundary layer must have a point of inflection where the second derivative $(\partial^2 u/\partial y^2)$ changes sign to obtain the S-shape characteristic of a separating profile. The second derivative of the velocity is always negative near the edge of the layer, because the boundary layer must join smoothly with the free stream. The second derivative at the wall can be found from Eq. (9.107) evaluated at $y = 0$ where $u = v = 0$:

$$\left.\frac{\partial^2 u}{\partial y^2}\right)_{y=0} = \frac{1}{\mu}\left(\frac{dp}{dx}\right). \tag{9.185}$$

This relation is valid for laminar or turbulent boundary flow, because the flow very near the wall is always laminar. According to this equation, the sign of the second derivative at the wall is the same as the sign of the pressure gradient. If the pressure gradient is positive, the second derivative must change sign between the wall and the edge of the layer, and the flow is susceptible to separation. If the pressure gradient is negative (falling pressure), there is no point of inflection, and the flow will not be susceptible to separation.

The location of the separation point (and hence whether separation will occur) can, in principle, be determined by calculating the boundary layer up to the point where

$$C_f = \tau_w = 0,$$

which is the separation point. In practice, boundary layer calculations often be-

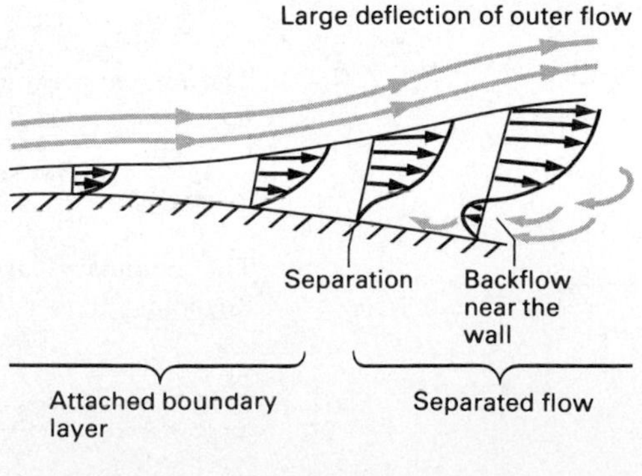

Figure 9.33 Separating boundary layer.

have poorly near separation, and location of the zero stress point is difficult. Monitoring the value of the shape factor H provides a reliable indication of separation. As a boundary layer approaches separation, the value of H increases rapidly. Separation is generally taken to occur at the point where

$$\underset{\text{Laminar flow}}{H > 3.5} \quad \text{and} \quad \underset{\text{Turbulent flow}}{H > 2.4.} \tag{9.186}$$

The lower value of H for turbulent separation (lower even than the laminar flat plate value of H) reflects the fact that the characteristic S-shape of separation occurs in the viscous sublayer near the wall, whereas the outer turbulent portion of the profile remains somewhat fuller.

EXAMPLE 9.8 Illustrates How to Determine the Location of Boundary Layer Separation

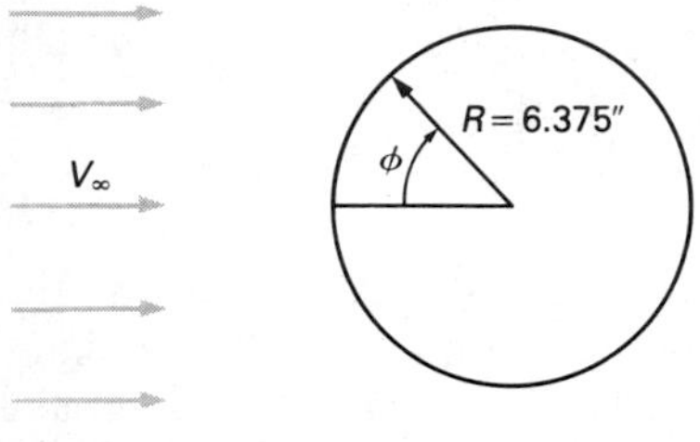

Figure E9.8 Cross section of sign pole in Example 8.9.

The critical wind velocity for vortex shedding for the 12.75-in.-outside-diameter sign pole in Example 8.9 was found to be 9.75 mph (14.3 ft/sec). Find the angle ϕ in Fig. E9.8 where the boundary layer separates from the pole for this flow condition. The free-stream velocity is given by $U(\phi) = 2V_\infty \sin\phi$ (see Example 9.7).

SOLUTION

Given

Wind at 14.3 ft/sec flowing over 12.75-in.-outside-diameter steel pipe
Air 60°F

Find

Angle ϕ where boundary layer separates from pole surface

Solution

The location of boundary layer separation is the point where $\tau_w = 0$. We must develop an expression for τ_w as a function of the angle ϕ for $U(\phi) = 2V_\infty \sin\phi$. An expression was developed in Example 9.7 for a laminar boundary layer. We therefore check that the boundary layer around the sign pole is laminar. The Reynolds number is

$$\mathbf{R}_x = \frac{Vx}{\nu} = \frac{V\phi R}{\nu}.$$

The air temperature in Example 8.9 is 60°F; Table A.4 gives $\nu = 1.7 \times 10^{-4}$ ft²/sec. Then

$$\mathbf{R}_x = \frac{(14.3 \text{ ft/sec})(\frac{1}{2})(12.75 \text{ in.})(1 \text{ ft}/12 \text{ in.})\phi}{(1.7 \times 10^{-4} \text{ ft}^2/\text{sec})} = 4.78 \times 10^4\,\phi.$$

The boundary layer becomes turbulent (transition occurs) at a Reynolds number (based on x) of about* 5×10^5. Therefore we ex-

---

* This transition value is for flow over a flat plate. We assume that it also applies for flow over a cylinder.

pect a laminar boundary layer for

$$\phi < \frac{5 \times 10^5}{4.78 \times 10^4} = 10.5 \text{ rad.}$$

As the entire cylinder is covered by $0 \le \phi \le \pi$ rad, the boundary layer will be laminar. Example 9.7 gives

$$\tau_w = \frac{2\mu V_\infty \sin\phi\,[(2\Theta^2 V_\infty/\nu R)\cos\phi + 0.09]^{0.62}}{\Theta},$$

where

$$\Theta = \sqrt{\frac{0.45\nu R}{2V_\infty \sin^6\phi}\left[-\cos\phi + \frac{2}{3}\cos^3\phi - \frac{\cos^5\phi}{5} + \frac{8}{15}\right.}.$$

The numerical values are

$$\mu = 3.77 \times 10^{-7}\ \text{lb}\cdot\text{sec/ft}^3, \qquad \nu = 1.70 \times 10^{-4}\ \text{ft}^2/\text{sec},$$
$$R = 6.375\ \text{in.} = 0.531\ \text{ft}, \qquad V_\infty = 14.3\ \text{ft/sec.}$$

Then

$$2\mu V_\infty = 1.08 \times 10^{-5}\ \text{lb/ft},$$
$$\frac{2V_\infty}{\nu R} = 3.17 \times 10^5/\text{ft}^2,$$

and

$$\frac{0.45\nu R}{2V_\infty} = 1.42 \times 10^{-6}\ \text{ft}^2.$$

Defining

$$\Omega \equiv \frac{1}{\sin^3\phi}\left[-\cos\phi + \frac{2}{3}\cos^3\phi - \frac{\cos^5\phi}{5} + \frac{8}{15}\right]^{1/2}$$

gives

$$\tau_w = \frac{(1.08 \times 10^{-5}\ \text{lb/ft})\sin\phi}{(1.42 \times 10^{-6}\ \text{ft}^2)^{1/2}\,\Omega}$$
$$\times [(3.17 \times 10^5/\text{ft}^2)(1.42 \times 10^{-6}\ \text{ft}^2)\Omega^2\cos\phi + 0.09]^{0.62}.$$

Simplifying, we obtain

$$\tau_w = 0.00204(\sin\phi)\left[\frac{(1 + 5\Omega^2\cos\phi)^{0.62}}{\Omega}\right]\ \text{lb/ft}^2.$$

We must solve for the angle ϕ where $\tau_w = 0$. We may do so graphically or by trial and error. The result is

$$\phi_{sep} = \phi_{\tau_w = 0} = 104°.$$ **ANSWER**

Discussion

Actually, the separated boundary layer affects the pressure and free-stream velocity distribution, and the velocity distribution is approximately

$$\frac{U}{V_\infty} \approx 1.814\phi - 0.271\phi^3 - 0.0471\phi^5,$$

rather than $U/V_\infty = 2\sin\phi$. Using this more realistic velocity distribution, we find that the calculated separation point is at about 80° (on the front of the cylinder). The 80° separation point is verified by experiments (see Figs. 8.16 and 8.17). The discrepancy between our calculated separation point and the experimental value is not caused by inaccuracies in Thwaites's method but by the breakdown of the potential flow and boundary layer model near separation. Given the correct free-stream velocity distribution, Thwaites's method is reasonably accurate.

PROBLEMS

1. Show that the slope for a constant-value Stokes stream function Ψ_s for axisymmetric flow is parallel to the velocity vector and is therefore a streamline.

2. Show that the difference $\Delta\Psi_s$ in the value of a Stokes stream function between two streamlines for axisymmetric flow is the volume flow rate between the streamlines.

3. Use the irrotationality condition for axisymmetric flow to develop Eq. (9.31).

4. Show that lines of constant Φ for steady, axisymmetric, irrotational flow are normal to streamlines.

5. Show that if Φ_1 and Φ_2 are both solutions of Laplace's equation, the sum $(\Phi_1 + \Phi_2)$ is also a solution.

6. Find the corresponding stream function for which

$$\Phi = \frac{AV_\infty \cos\theta}{r}$$

if both A and V_∞ are positive constants.

7. Consider the flow of a liquid of viscosity μ and density ρ down an inclined plate making an angle θ with the horizontal. The film thickness is t and is constant. The fluid velocity parallel to the plate is given by

$$V_x = \frac{\rho t^2 g \cos\theta}{2\mu}\left[1 - \left(\frac{y}{t}\right)^2\right],$$

where y is the coordinate normal to the plate. Calculate Φ and Ψ for this flow and show that neither satisfies Laplace's equation. Why not?

8. The dispersion of salt in a still liquid by diffusion is given by

$$j_x = -\mathscr{D}\left(\frac{\partial c}{\partial x}\right), \qquad j_y = -\mathscr{D}\left(\frac{\partial c}{\partial y}\right),$$

and $$j_z = -\mathscr{D}\left(\frac{\partial c}{\partial z}\right),$$

where j_x, j_y, and j_z are the mass fluxes (kg/m$^2\cdot$s), $\mathscr{D}$ is the diffusivity (m^2/s), and c is the salt concentration (kg/m^3) in the liquid. Consider $\mathscr{D}$ constant and show that c satisfies Laplace's equation.

9. A horizontal oil-bearing stratum is 3 m high and provides a volume flow rate of $Q = 1000$ m^3/hr. The flow moves radially inward and is collected by a vertical porous pipe having an outer radius of 1.0 m. The Laplace equation for the potential flow model of the oil is

$$\frac{1}{r}\left(\frac{\partial\Phi}{\partial r}\right) + \frac{\partial^2\Phi}{\partial r^2} + \frac{1}{r^2}\left(\frac{\partial^2\Phi}{\partial\theta^2}\right) = 0.$$

Find $\Phi\{r, \theta\}$.

10. Air at 25°C flows normal to the axis of an infinitely long cylinder of 1.0-m radius. The cylinder is rotating at 10 rad/s, and the approach velocity is 100 km/hr. Find the maximum and minimum pressures on the cylinder surface. Assume potential flow.

11. Flow stagnating against a flat wall, with the stream function $\Psi = Cxy$ (C a constant), is used sometimes to approximate flow near the stagnation point of a blunt body. Consider modeling flow near the front stagnation point of a circular cylinder in this fashion. Assume that the fluid is water, the cylinder has a radius of 1 ft, and the velocity far ahead of the cylinder is 20 ft/sec. Further, assume that the stagnation pressure is the same for both flows. Calculate and plot the pressure distribution, $(p_2 - p)$ vs x, where x is the distance measured *along the body surface*, for both flows. Also assume that the static pressure in both flows is equal at $x = \pi/6$, which is the point on the cylinder where the pressure equals its free-stream value, p_∞.

12. Show that

$$\Phi = Ax, \qquad \Phi = Ax + By,$$

and $$\Phi = x^2y - xy^2 + \frac{x^3}{3} - \frac{y^3}{3}$$

satisfy the two-dimensional Laplace equation in Cartesian coordinates if A and B are constants.

13. Show that

$$\Phi = C \ln[(x^2 + y^2)^{1/2}]$$

satisfies the two-dimensional Laplace equation in Cartesian coordinates for constant C.

14. The velocity potential for a horizontal flow field is

$$\Phi = K \ln r.$$

Find the pressure $p(r, \theta)$ in terms of the stagnation pressure p_0.

15. The stream function for a particular incompressible flow is

$$\Psi = Cxy,$$

where C is a constant. Is this flow irrotational? Sketch a few streamlines for this flow. What practical situation might it model? Calculate the pressure coefficient distribution.

16. A 15-mph wind flows over a Quonset hut having a radius of 12 ft and a length of 60 ft, as shown in Fig. P9.16. The upstream pressure and temperature are equal to those inside the Quonset hut: 14.696 psia and 70°F. Estimate the upward force on the Quonset hut. Find the location θ on the roof of the Quonset hut where the pressure is p_∞.

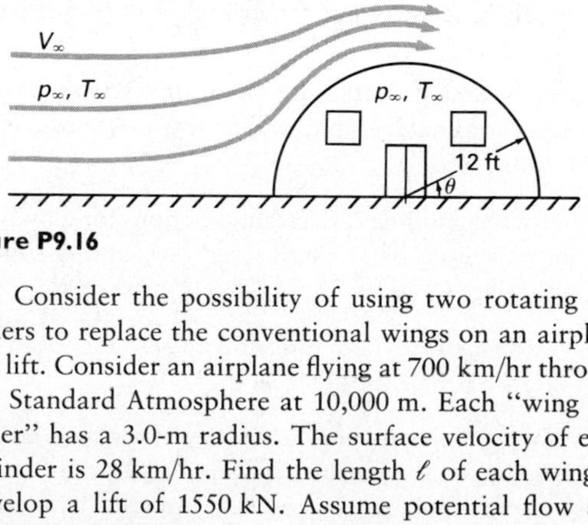

Figure P9.16

17. Consider the possibility of using two rotating cylinders to replace the conventional wings on an airplane for lift. Consider an airplane flying at 700 km/hr through the Standard Atmosphere at 10,000 m. Each "wing cylinder" has a 3.0-m radius. The surface velocity of each cylinder is 28 km/hr. Find the length ℓ of each wing to develop a lift of 1550 kN. Assume potential flow and neglect end effects.

18. A cylinder on a stationary rotor ship has a diameter of 9 ft and a height of 50 ft and rotates at 125 rpm. Calculate the lift on the cylinder in a 20-mph crosswind of 60°F air. Compare your result with that obtained using Fig. 8.32.

19. A stream function is given by

$$\Psi = \sin\frac{x}{X}\sinh\frac{y}{Y},$$

where X and Y are constants, $0 \le x \le \pi X$, and $y \ge 0$. Can Ψ represent a potential flow? If so, locate any stagnation points and sketch several streamlines.

•**20.** In plane polar coordinates, Laplace's equation can be written in the form

$$\nabla^2\Phi = \frac{1}{r}\frac{\partial}{\partial r}\left(r\frac{\partial\Phi}{\partial r}\right) + \frac{1}{r^2}\frac{\partial^2\Phi}{\partial\theta^2} = 0.$$

Here, let $\Phi(r, \theta) = R(r)\Theta(\theta)$, where R and Θ are functions of the coordinates r and θ, respectively.

(a) By direct substitution, show that $\Phi(r, \theta)$ satisfies Laplace's equation provided that $R(r) = Ar^n$ and $\Theta(\theta) = B\sin(n\theta)$, where A, B, and n are constants.

(b) What special values must n take?

(c) Repeat (a) and (b) if $R(r) = Ar^n$ and $\Theta(\theta) = B\cos(n\theta)$.

(d) If $\Phi(r, \theta) = (Ar^2)(B\sin 2\theta)$ express this potential in terms of Cartesian coordinates to find $\Phi(x, y)$.

(e) Does your answer for $\Phi(x, y)$ obtained in (d) satisfy Laplace's equation in two dimensions in Cartesian coordinates?

•**21.** Consider potential flow around a 90° corner.

(a) Given $\Phi(x, y) = A(x^2 - y^2)$, where A is a constant, compute the two-dimensional velocity components u and v.

(b) Determine the value of A if $u = -1$ m/s and $v = +1$ m/s at the point $x = 1$ m, $y = 1$ m.

(c) On a Cartesian coordinate system draw the velocity vector for this flow at the points (0, 0), (1, 0), (0, 1), (2, 1), (3, 1), (1, 1), and (1, 2).

(d) On a separate graph plot the lines of constant potential for the cases of $\Phi = 0$, $\Phi = 4$, and $\Phi = -4$ m²/s.

(e) Graphically compute $u(1, 2)$ and $v(1, 2)$ from your potential lines drawn in part (d). *Note:* $u \approx \Delta\Phi/\Delta x$ and $v \approx \Delta\Phi/\Delta y$. Compare your graphic result with your analytic result obtained in part (c).

(f) Show analytically that the equation for the streamlines is of the form $\Psi = Axy + C$.

(g) Plot the streamlines of constant value 1, 2, and 4 m^2/s on your graph with the existing potential lines.

(h) What is the volume flow rate through a $\sqrt{2}$-m × 1-m rectangular screen (that does not disturb the flow) that is placed between the points (1, 1) and (2, 2) so that the screen extends 1 m in depth in the z direction?

(i) Find both Φ and Ψ as functions of r and θ by using two-dimensional plane polar coordinates.

•**22.** Consider ideal potential flow inside a corner, as shown in Fig. P9.22.

(a) Starting with the potential function $\Phi\{r, \theta\} = Ar^n \cos(n\theta)$, where A and n are constants, find the radial and tangential velocity components denoted by $V_r = \partial\Phi/\partial r$ and $V_\theta = (1/r)\partial\Phi/\partial\theta$.

(b) Determine the constants A and n from the boundary conditions that $V\{r = 1 \text{ m}, \theta = 30^0\} = -10$ m/s and $V_\theta\{r = 1 \text{ m}, \theta = 30^0\} = 0$ m/s.

(c) Find the equation for the streamline function $\Psi\{r, \theta\}$.

(d) Sketch curves for constant potential lines and streamlines between $0 < \theta < 30^0$ for $\Phi = 0$, 1, and -1 m^2/s; and for $\Psi = 0$, 1/10, 1, and 100 m^2/s.

(e) Let the fluid density be 1000 kg/m^3. If the pressure at $r = 1$ m, $\theta = 30^0$ is $\frac{1}{2}$ atm gage, follow the $\Psi = 100$ m^2/s streamline and compute the gauge pressure at the point $r = 1.97$ m, $\theta = 15^0$. Is your answer reasonable?

(f) Find the locations of the stagnation points. What is the stagnation pressure at these points?

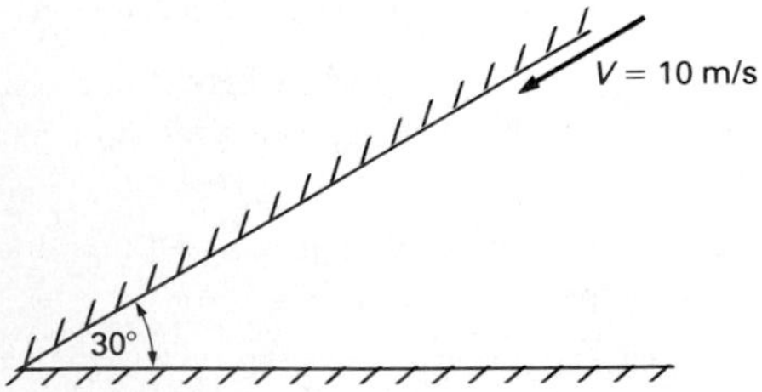

Figure P9.22

•**23.** A uniform flow in the positive x direction is expressed in spherical polar coordinates by the independent coordinates r', θ', and ϕ so that

$$V_{r'} = U_0 \cos\theta' \quad \text{and} \quad V_\theta = -U_0 \sin\theta',$$

where U_0 is a constant. See Fig. P9.23.

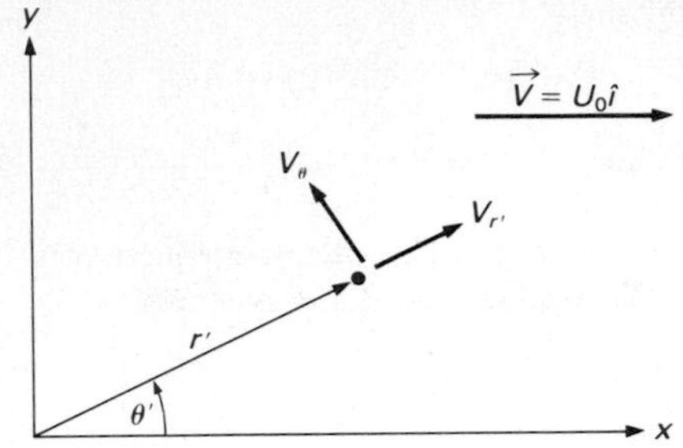

Figure P9.23

(a) Find the equation for the velocity potential $\Phi\{r', \theta'\}$, starting with $\vec{V} = \nabla\Phi$ and $\vec{V} = U_0\ \hat{x}_j$. Neglect the azimuth dependence of Φ.

(b) Find the equation of the streamlines $\Psi\{r', \theta'\}$.

24. Show that the circulation of a free vortex for any closed path that does not enclose the origin is zero.

25. Show that the circulation of a free vortex for any closed path that encloses the origin is Γ.

•**26.** A point source (placed at the origin) for three-dimensional potential flow is a point where an incompressible ideal flow is created and sent out evenly in all directions. In spherical polar coordinates let $V_{r'} = Q/4\pi r'^2$, $V_{\theta'} = 0$, $V_\phi = 0$, where Q is a constant.

(a) Show that the velocity potential (in spherical polar coordinates) is given by $\Phi = -(Q/4\pi r') + \text{constant}$.

(b) What are the dimensions of Q?

(c) Find the equation of the streamlines $\Psi\{r', \theta', \phi\}$.

(d) Find the volume flow per unit time through a 1-m radius sphere (centered at the origin) if $V_{r'}\{r' = 1 \text{ m}, \theta', \phi\} = +5$ m/s.

(e) Sketch some potential lines and some streamlines on a polar coordinate graph, where the ϕ coordinate is taken to be zero.

27. Show that adding the stream functions for a uniform stream, a source of strength λ at $-a$, and a sink of strength λ at $+a$ gives Eq. (9.58).

28. Verify that the length ratios t/a and ℓ/a for a Rankine oval are given by Eqs. (9.59) and (9.60).

29. Show that the maximum velocity of the fluid around the Rankine oval described by Eq. (9.58) occurs at the point of maximum thickness of the Rankine oval and is given by Eq. (9.61).

30. The flow in the impeller of a centrifugal pump is modeled by the superposition of a source and a free vortex. The impeller has an outer diameter of 0.5 m and an inner diameter of 0.3 m. At the outlet from the impeller, the flowing water has the following velocity components, relative to the impeller: radial component 2 m/s and tangential component 7 m/s.

(a) Find the strength of the source and the vortex required to model this flow.

(b) Assume that the impeller blades are shaped like the streamlines and plot an impeller blade shape.

(c) Find the radial and tangential components of velocity at the inlet to the impeller.

31. The rotating infinite cylinder (of radius R) in a free-stream flow of speed U_0 has the velocity potential given by $\Phi = U_0(r + R^2/r) - \Gamma\theta/2\pi$, where Γ is the circulation. Here, let $U_0 = 1$ m/s and $R = 1$ m, as shown in Fig. P9.31.

(a) The averaged tangential velocity, V_θ, of the cylinder is $U_0/3$. Compute the location of the stagnation points if they exist.

(b) Find the velocity components V_r and V_θ at the points $(r = R, \theta = 90°)$ and $(r = R, \theta = 270°)$.

(c) Compute the force per unit length on the cylinder.

(d) If the free-stream pressure at $r = \infty$ is denoted by p_∞, find $\{p(r = R, \theta) - p_\infty\}$ at $\theta = 0°, 90°, 180°$, and $270°$.

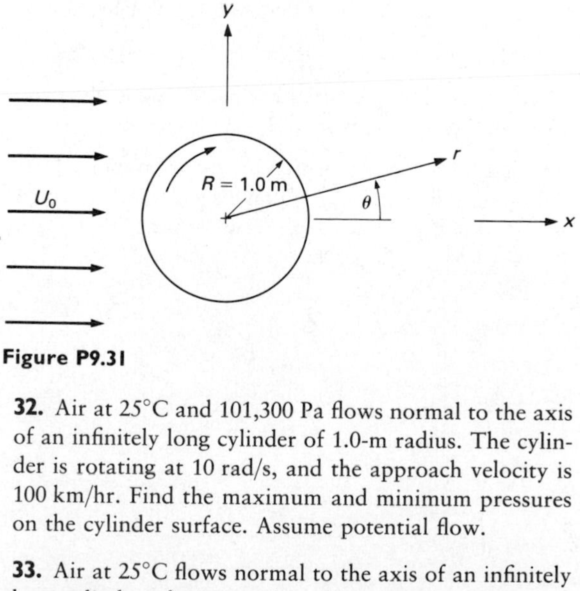

Figure P9.31

32. Air at 25°C and 101,300 Pa flows normal to the axis of an infinitely long cylinder of 1.0-m radius. The cylinder is rotating at 10 rad/s, and the approach velocity is 100 km/hr. Find the maximum and minimum pressures on the cylinder surface. Assume potential flow.

33. Air at 25°C flows normal to the axis of an infinitely long cylinder of 1.0-m radius. The approach velocity is 100 km/hr. Find the rotational velocity of the cylinder so that a single stagnation point occurs on the lowermost part of the cylinder. Assume potential flow.

34. The velocity potential for the combination of a source and a vortex is given by

$$\Phi = (3.0 \text{ m}^3/\text{s})(\ln r - \theta).$$

Calculate the pressure at $(4, \pi/4)$ if the pressure is 120 kPa at $(1, \pi/2)$. The fluid density is 1000 kg/m³.

35. Air at 25°C flows normal to the axis of an infinitely long cylinder of 1.0-m radius. The approach velocity is 100 km/hr. Find the maximum and minimum pressures on the cylinder surface. Assume potential flow.

36. A velocity potential is given by

$$\Phi = \frac{\cos\theta}{r} - V_\infty r\cos\theta.$$

Find $\Psi(r, \theta)$ and locate any stagnation points. Sketch several streamlines and determine a body shape about which Φ could represent the potential flow.

37. Plot lines of constant Ψ_1 for a source and lines of constant Ψ_2 for a free vortex. Graphically add Ψ_1 and Ψ_2 and show that the sum produces the streamlines in Fig. 9.11.

38. For the two-dimensional velocity potential (from the superposition of a uniform flow with a line source)

$$\Phi = -U_0x + (Q/2\pi L)\ln(x^2 + y^2),$$

where U_0, Q, and L are constants:

(a) Find the velocity components u and v.

(b) Locate any stagnation points.

(c) Show that the equation for the streamlines is $\Psi(x, y) = -U_0y + (Q/2\pi L)\tan^{-1}(y/x)$.

(d) Plot the streamline that goes through the lone stagnation point.

(e) What is the value of Ψ that goes through the stagnation point?

39. The superposition of a three-dimensional point source and a uniform stream results in a flow pattern that describes ideal flow past a solid semi-infinite body of revolution, as shown in Fig. P9.39.

(a) For $\Phi_1(r', \theta') = U_0r'\cos\theta'$ and $\Phi_2 = -Q/4\pi r'$, find the velocity components $V_{r'}$, $V_{\theta'}$, and V_ϕ in spherical coordinates.

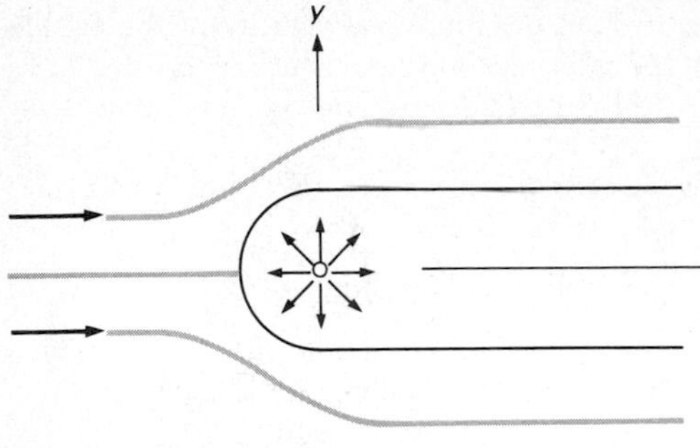

Figure P9.39

(b) At what position along the x axis ($\theta' = 0°$) is the velocity zero?

(c) What is the stagnation pressure (in terms of the pressure, p_∞, at $r' = \infty$, $\theta' = 0°$) at the point found in (b)?

(d) Plot the streamline, $\Psi = 0$, that corresponds to the shape of the body.

(e) Show that the streamline in (d) is a horizontal straight line until it reaches the stagnation point and then traces a blunt semi-infinite body of revolution.

•**40.** (a) Find the velocity potential (as a function of r', θ') from the superposition of a three-dimensional point source, $\Phi_1 = (Q/4\pi r'_1)$, placed at $r' = d$, $\theta' = 0°$, and a three-dimensional point sink, $\Phi_2 = (Q/4\pi r'_2)$, placed at $r' = d$, $\theta' = 180°$. Refer to Fig. P9.40 and determine $\Phi_{tot} = \Phi_1 + \Phi_2$ by expressing the magnitudes for r'_1 and r'_2 in terms of r', d, and θ'.

(b) Sketch lines of equipotential for $\Phi = -1$, 0, and $+1\ m^2/s$. Here, let $d = 1$ m and $Q = 1\ m^3/s$.

(c) Find the equation for the streamlines $\Psi(r', \theta')$. Sketch the three individual streamlines that go through the points ($r' = 1$ m, $\theta' = 90°$), ($r' = 1$ m, $\theta' = -90°$), and ($r' = 0$ m).

(d) If the three-dimensional point source and sink are placed close together, so that $d \ll r'$, show that $\Phi_{tot} = \Phi_1 + \Phi_2 \approx [Qd(\cos\theta')]/2\pi r'^2$. This condition is sometimes called a three-dimensional doublet. (*Hint:* Write $(1/r'_1) - (1/r'_2) = (r'_2 - r'_1)/r'_1 r'_2$.)

(e) From part (d) find the equation for the streamlines valid in the limit that $d \ll r'$.

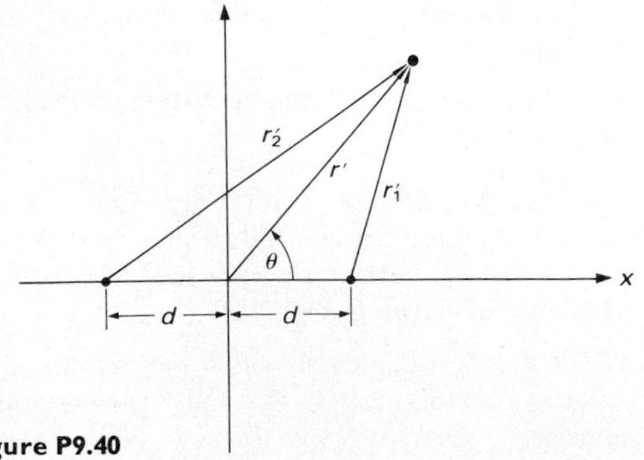

Figure P9.40

•**41.** The superposition of the ideal flow for a three-dimensional doublet (see Problem 40) with a uniform flow in the x direction results in a potential that describes ideal flow past a sphere.

(a) Let $\Phi = Ar'\cos\theta' + B\cos\theta'/r'^2$ represent this flow in spherical polar coordinates. Use the boundary conditions $V_{r'}(r' = \infty, \theta') = U_0\cos\theta'$, $V_{\theta'}(r' = \infty, \theta') = -U_0\sin\theta'$, and $V_{r'}(r' = R, \theta') = 0$ and solve for the constants A and B. The sphere's radius is R.

(b) What are the velocity components $V_{r'}(r', \theta')$ and $V_{\theta'}(r', \theta')$?

(c) Let $U_0 = 1$ m/s and $R = 1$ m. Plot potential lines and streamlines on the same polar graph paper. In particular, what value of Ψ corresponds to the streamline that goes through the stagnation points?

(d) If $p(r' = \infty, \theta' = 0°) = p_\infty$, find the pressure coefficient $C_p = [p(r' = R, \theta') - p_\infty]/(\rho U_0^2/2)$. How does your answer compare with C_p obtained from the two-dimensional ideal flow past a cylinder of radius R?

•**42.** The two-dimensional velocity potential for a continuous distribution of (infinitely long) line sources of constant strength per unit length λ' is given by

$$\Phi(x, y) = (\lambda'/4\pi)\int_{curve} \ln[(x - x')^2 + (y - y')^2]\,ds',$$

where ds' is the differential arc length along the curve representing the continuous source distribution.

(a) Find the velocity potential for the continuous distribution of sources placed continuously along the segment $x' = 0$, $0 < y' < L$. See Fig. P9.42. [*Hint:* The integral

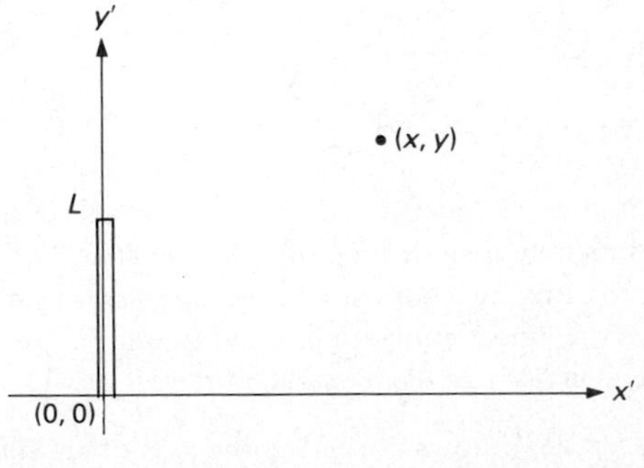

Figure P9.42

$\int \ln(z^2 + a^2)\,dz = z\ln(z^2 + a^2) - 2z + 2a\tan^{-1}(z/a)$ will be useful.]

(b) Evaluate the velocity potential in the limit that $L \ll x$, $L \ll y$. This condition is called the farfield limit.

(c) From part (b) find the velocity components u and v. Next find the equation for the streamlines $\Psi\{x, y\}$.

(d) Evaluate the velocity potential in the limit that $L \gg x$ and $L \gg y$ and let the segment length approach infinity. This condition is called the nearfield case with a large source dimension.

(e) From part (d) find the equation for the streamlines $\Psi\{x, y\}$ in this limit.

(f) Try to explain the results in both limiting cases in terms of potentials and streamlines that you have already encountered.

• **43.** The two-dimensional velocity potential [in polar coordinates (r, θ)] for an infinitely long line source of strength per unit length λ' is

$$\Phi\{r, \theta\} = (\lambda'/4\pi)\ln(r^2),$$

where the source is located at the origin.

(a) Show that the corresponding two-dimensional stream function $\Psi\{r, \theta\}$ is given by

$$\Psi\{r, \theta\} = (\lambda'/2\pi)\theta.$$

(b) The two-dimensional stream function for a continuous distribution of (infinitely long) line sources of constant strength per unit length λ' is given by

$$\Psi\{r, \theta\} = (\lambda'/2\pi)\int_{\text{curve}} \theta\{s'\}\,ds'$$

where ds' is the differential arc length along the curve representing the continuous source distribution. Here, $\theta\{s'\}$ represents how the angle changes from a point on the arc length to the observation point. Find an exact integral expression for $\Psi\{r, \theta\}$ for the continuous distribution of infinitely long line sources of constant strength per unit length λ' along the segment ds', where $y' = 0$ and $0 < x' < L$. Refer to Fig. P9.43.

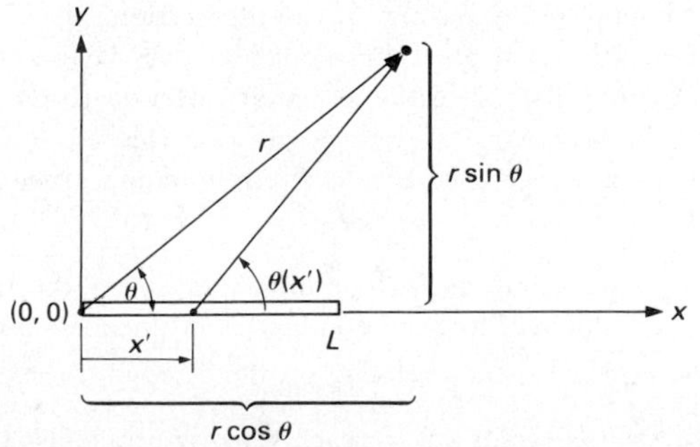

Figure P9.43

44. Consider a two-dimensional diverging passage that has straight sides, a constant width, and rectangular cross sections. Fluid enters with a uniform axial velocity U_0. The boundary layer is thin enough that it may be neglected. The inlet measures 0.10 m on a side, the outlet measures 0.10 m × 0.90 m, and the length is 4.0 m. Obtain a set of finite difference equations for the stream function. Use a 1.0-cm × 1.0-cm grid.

45. In Fig. P9.45, the velocity potential to the left and right of the channel are given. Write the finite difference equations for the velocity potential for the eight points indicated by the lowercase letters a through h.

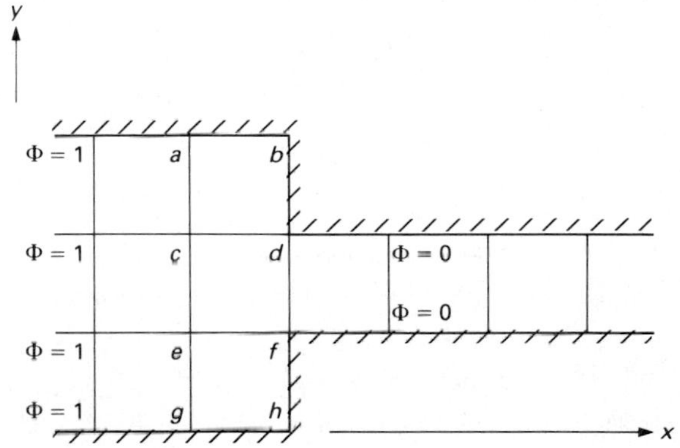

Figure P9.45

46. In Fig. P9.46, the potential flow to the left and right of the thin obstruction is a uniform flow in a rectangular channel of velocity $u = 1$ m/s, $v = 0$ m/s far to the left of the obstruction. Write the finite difference equations for the velocity potential for the 12 points indicated by the lowercase letters a through l.

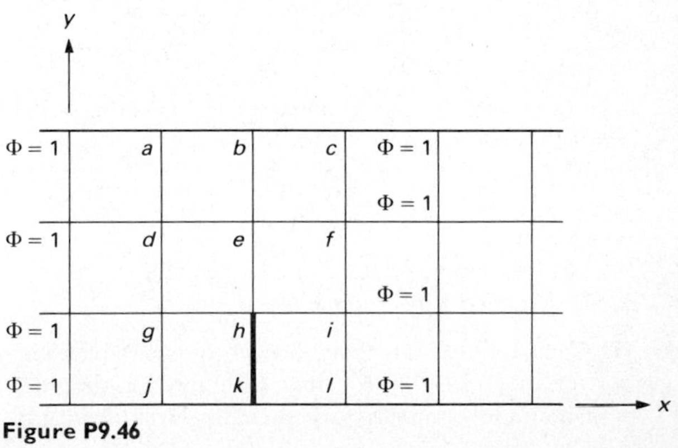

Figure P9.46

47. Figure P9.47 shows a circular cylinder placed in a wind tunnel. Air ($p = 105$ kPa, $\rho = 1.25$ kg/m^3) approaches the cylinder at 10 m/s. Use the finite difference method with a 1-cm × 1-cm grid to solve for the stream function. Use your solution to:

(a) Plot streamlines for the flow.

(b) Calculate and plot the pressure distribution on the upper and lower walls.

(c) Calculate and plot the pressure distribution on the cylinder. Also plot the pressure distribution for the cylinder in an unbounded stream. (*Hint:* Use symmetry to reduce the number of points in your finite difference model.)

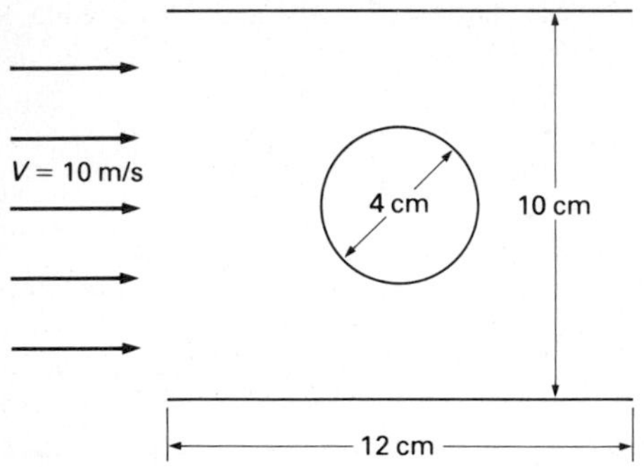

Figure P9.47

48. Derive the momentum integral equation for a boundary layer by integrating the Prandtl boundary layer equations between the limits of zero and δ.

49. Use the Blasius transformation

$$\Psi = \sqrt{\nu V_\infty x} f(\eta) \quad \text{and} \quad \eta = \frac{y}{\sqrt{\nu x / V_x}}$$

along with

$$u = \frac{\partial \Psi}{\partial y} \quad \text{and} \quad v = -\frac{\partial \Psi}{\partial x}$$

to show that the Prandtl boundary layer equations (9.105 and 9.107) reduce to

$$f''' + \frac{1}{2} ff'' = 0.$$

50. Find the appropriate boundary conditions for the Blasius function f in Problem 49.

51. Consider the Reynolds–Navier–Stokes equations, Eqs. (5.71) and (5.72), together with appropriate order of magnitude estimates for boundary layer flow, and develop a set of partial differential equations for plane, two-dimensional, steady, incompressible, turbulent boundary layer flow.

52. A coal barge 1000 ft long and 100 ft wide is submerged a depth of 12 ft in 60°F water. It is being towed at a speed of 12 mph. Estimate the friction drag on the barge.

53. The open wind tunnel shown in Fig. P9.53 has a test section that measures 2 ft × 2 ft × 10 ft. Consider the case of an inlet velocity of 100 mph to the test section. Assume that the laminar boundary layer starts at the test section inlet and develops as that defined for a flat plate with a free-stream velocity of 100 mph. Show that the length of this laminar boundary is 5.4% of the test section's total length.

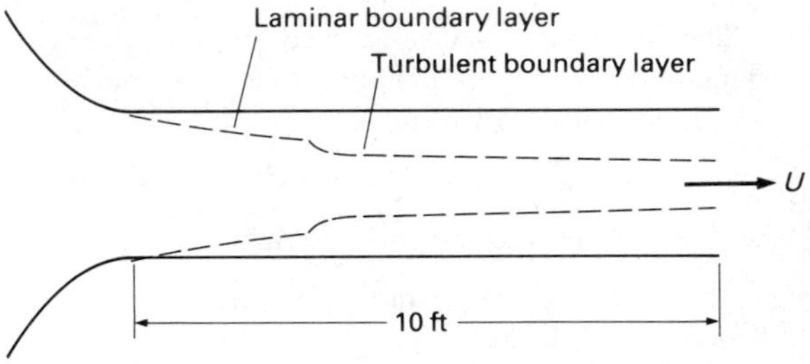

Figure P9.53

54. The analysis of a turbulent flat plate boundary layer given in Eqs. (9.152)–(9.157) neglects the laminar boundary layer near the leading edge of the plate and assumes that the turbulent layer starts at $x = 0$. Improve the model by assuming that a laminar layer exists up to a transition point given by $\mathbf{R}_{x,\mathrm{tr}} = 5 \times 10^5$ and the turbulent layer begins at that point, with $\delta_{\mathrm{turb}} = \delta_{\mathrm{lam}}$ and redo the analysis.

55. Assume that the length of the laminar boundary layer in Problem 53 may be neglected and that the turbulent boundary layer starts at the test section inlet. Also assume that a flat plate turbulent boundary layer analysis describes the test section boundary layer development. Find the boundary layer thickness δ at the test section exit for a constant core velocity of 100 mph along the entire test section's length.

56. Consider the laminar boundary layer for the flow over a flat plate. Plot the boundary layer thickness δ versus $\sqrt{\nu x / V_\infty}$ for each velocity profile listed in Table 9.2.

57. Plot u/U versus y/δ for each velocity profile listed in Table 9.2. Make any significant observations.

58. Use the control volume shown in Fig. 9.28 and the exact results for a laminar boundary layer on a flat plate to find an expression for the y velocity at the edge of the boundary layer, v_e, as a function of V_∞, ν, and x.

59. Use the sine wave velocity profile and the corresponding boundary layer parameters given in Table 9.2 to develop an equation for the y component of velocity, $v(x, y)$, in the laminar boundary layer on a flat plate.

60. Repeat the analysis for a turbulent boundary layer on a flat plate given in Eq. (9.152)–(9.157), except use $\frac{1}{5}$ as the power law exponent instead of $\frac{1}{7}$.

61. Figure P9.61 shows a thin triangular wing, viewed from above. Assume that the flow over the wing can be modeled as an infinite number of two-dimensional, side-by-side flat plate boundary layers. Estimate the skin friction drag on the wing.

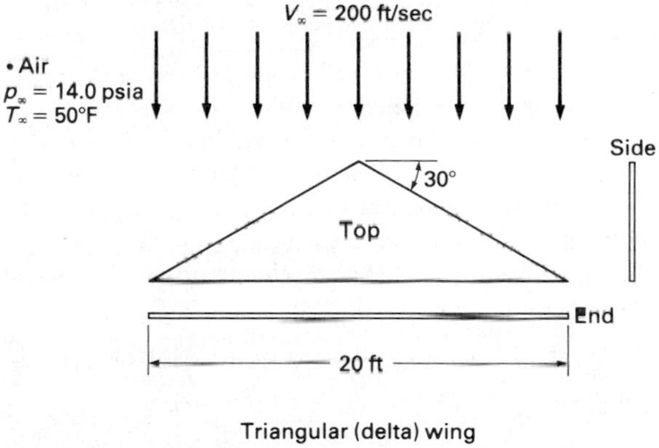

Figure P9.61

62. An experiment is set up in the wind tunnel to obtain air velocity profiles for flow over a flat plate such as the one shown in Fig. P9.62. A Pitot tube is used to obtain velocity pressure measurements, with a water manometer, as a function of distance from the plate. For air traveling at standard atmospheric conditions, calculate the velocity profiles at stations 1 and 2, using the results in Table P9.62. If the profile follows $u/u_\infty = (y/\delta)^{1/n}$, find the coefficient n. (*Hint:* $\log(x^a) = a \log x$.)

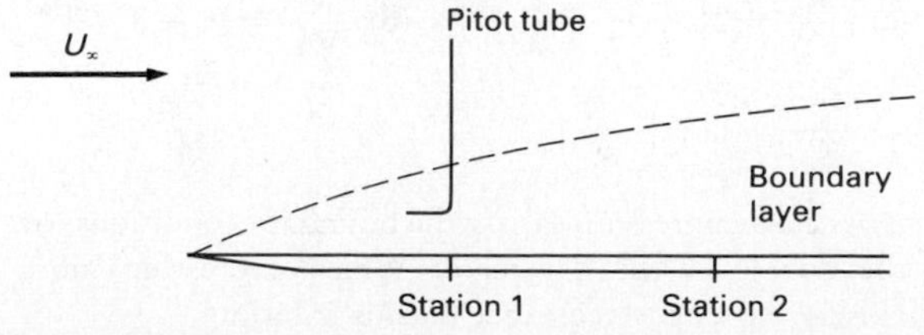

Figure P9.62

Table P9.62

Pitot Tube Position	Distance from Plate (mm)	Velocity Pressures, 1 (mm H_2O)	Velocity Pressures, 2 (mm H_2O)
1	0	0.00	0.00
2	2	30.59	31.61
3	3	35.90	38.75
4	4	39.06	43.85
5	5	42.93	45.89
6	6	45.07	47.11
7	7	48.95	49.87
8	8	50.28	51.09
9	9	53.03	53.95
10	10	54.86	55.27
11	11	55.27	55.58
12	12	55.58	54.86
13	15	54.97	55.17

63. An engineer wants to calculate the change in skin drag (and its share of the propulsive power) for an underwater missile as the missile goes from the water to the air. For simplicity, let's assume that the steady-state underwater speed of the missile is 30 m/s and when airborne (near the surface) is steady at 150 m/s. Find the reduction in drag if the missile surface is smooth and its dimensions are as shown in Fig. P9.63. Assume that the water temperature is 20°C and that the air temperature is 10°C. (*Hint:* For simplicity, calculate the skin drag from the point at which the diameter is constant.)

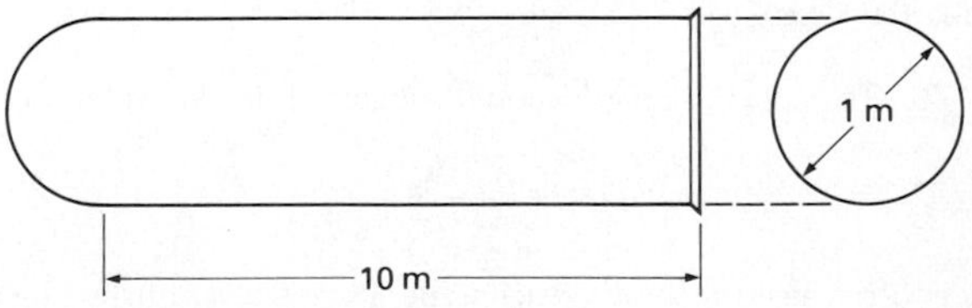

Figure P9.63

64. The expression $\tau_w = \mu\, \partial u/\partial y)_{y=0}$ implies a no-slip condition exists at the wall for most fluids ($\partial u/\partial y \neq 0$). Show that viscous flow is rotational (use the incompressible flow assumption).

65. Estimate the total skin drag for a passenger train traveling at 50 mph. The train measures 600 ft long and has a square cross-sectional area of 100 ft^2. (*Hint:* Consider that the train has smooth surfaces with no interruptions, and neglect the bottom side.)

66. Refer to Table 9.2 for laminar flow on a flat plate and Eq. (9.160). Plot C_f as a function of $\mathbf{R}_x$ for $5000 < \mathbf{R}_x < 10^7$. What parallels can be drawn with the drag over a circular cylinder?

67. Choose two of the velocity profiles and corresponding boundary layer parameter formulas presented in Table 9.2. Plot the thickness of the boundary layer and the wall shear stress versus x for values of x from zero to the laminar–turbulent transition point. The fluid is 20°C water and the free-stream velocity is 10 m/s. Ask your instructor to specify which velocity profile models you should use.

68. Air at 20 psia and 90°F flows over a flat plate at 100 ft/sec. Find the value of x at which the Reynolds number is 10^4. Choose three velocity profile models from Table 9.2 and plot the velocity profiles (u versus y) on the same axes. Ask your instructor to specify which velocity profiles to choose.

69. For the boundary layer profile $u/U = y/\delta$, $v/U = 0$, where $\delta = \sqrt{Lx}$ and L is a constant, find the circulation over the path shown in Fig. P9.69.

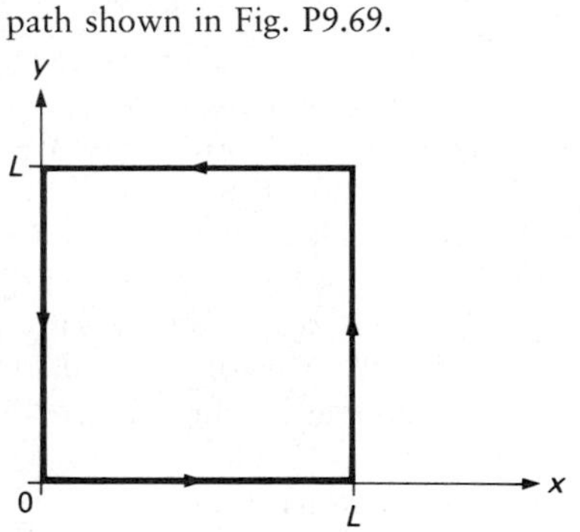

Figure P9.69

70. The mixing length model suggested in Example 5.6,

$$\tau \approx \rho l^2 \left(\frac{du}{dy}\right)^2,$$

is often used to model turbulent shear stress. For a turbulent boundary layer near a smooth wall, l is given by

$$l = 0.4y.$$

Assume that the shear stress in the fluid near the wall is constant and equal to the value at the wall, τ_w, and derive the velocity profile

$$u/u_\tau = 2.5 \ln\left(\frac{u_\tau y}{\nu}\right) + \text{constant}.$$

•**71.** Calculate the numerical values for the quantities

$$\frac{\delta}{x}\sqrt{\mathbf{R}_x}, \quad C_f\sqrt{\mathbf{R}_x}, \quad \text{and} \quad \frac{\delta^*}{x}\sqrt{\mathbf{R}_x}$$

in the laminar boundary layer over a flat plate by assuming a sinusoidal profile,

$$\frac{u}{V_\infty} = \sin\left[\frac{\pi}{2}\left(\frac{y}{\delta}\right)\right].$$

Do your values agree with those in Table 9.2?

•**72.** Repeat Problem 9.71 but use

$$\frac{u}{V_\infty} = \frac{3}{2}\left(\frac{y}{\delta}\right) - \frac{1}{2}\left(\frac{y}{\delta}\right)^3.$$

•**73.** Calculate the numerical values for the quantities

$$\frac{\delta}{x}\sqrt{\mathbf{R}_x}, \quad C_f\sqrt{\mathbf{R}_x}, \quad \text{and} \quad \frac{\delta^*}{x}\sqrt{\mathbf{R}_x}$$

in the laminar boundary layer over a flat plate by assuming a linear profile, $u/V_\infty = y/\delta$. Do your values agree with those in Table 9.2?

74. Consider 20°C water flowing over a thin, wide, smooth flat plate aligned with the flow. The approach velocity is 60 km/hr and the Reynolds number based on the plate length is 2×10^5. Calculate the drag per unit width of the plate for each of the velocity profiles listed in Table 9.2. Next calculate the drag coefficient C_D as used in Fig. 8.8. Which velocity profile(s) provide the best estimation of C_D? Why?

75. Turbulent boundary layer velocity profiles are sometimes approximated by the equation

$$\frac{u}{u_\tau} = 2.44 \ln\left(\frac{yu_\tau}{\nu}\right) + 5.0 + 2.44\Pi\left[1 - \cos\left(\frac{\pi y}{\delta}\right)\right].$$

Determine an expression for u/U. Flat plate turbulent boundary layers have $\Pi \approx 0.5$. Plot turbulent velocity profiles for this value of Π and for a power law profile with $n = 7$. If necessary, use $\mathbf{R}_x = 10^6$.

•**76.** Develop expressions for the wall shear stress and the displacement thickness by solving the momentum integral equation (9.151) for laminar flow over a flat plate. Use the following expressions for the velocity u/V_∞:

(a) $\dfrac{u}{V_\infty} = 2\left(\dfrac{y}{\delta}\right) - \left(\dfrac{y}{\delta}\right)^3$ (b) $\dfrac{u}{V_\infty} = (1 - e^{-10y/\delta})$

(c) $\dfrac{u}{V_\infty} = \tanh\left(\dfrac{5y}{\delta}\right)$

Does each expression satisfy the boundary conditions for the flow over the flat plate? Which expression more closely represents the exact Blasius solution?

77. Consider the flow of 15°C air over the thin and wide flat plate shown in Fig. P9.77. The flow on the lower side is turbulent over the entire plate and the flow on the upper side is laminar on the front portion and then becomes turbulent. Compare the drag per unit width on the upper half with that on the lower half. The velocity $V_\infty = 10$ m/s; consider lengths x_{trans} and $2x_{trans}$.

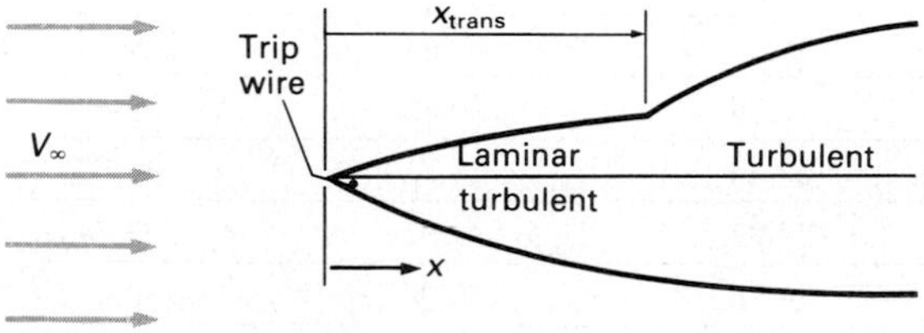

Figure P9.77

78. Solve the Blasius equation for f by a convenient numerical method. (*Hint:* Guess values of $f''(0)$ until you find the correct value. See Problems 49 and 50 for the Blasius equation and its boundary conditions.)

79. Verify the exact values of

$$\frac{\delta}{x}\sqrt{R_x}, \quad C_f\sqrt{R_x}, \quad \text{and} \quad \frac{\delta^*}{x}\sqrt{R_x}$$

from Table 9.2. You must use information from Problems 49, 50, and 78.

80. The experimental results shown in Table P9.80 were obtained by laser Doppler anemometry (LDA) of 15°C air flowing over a flat plate with no pressure gradient. If the velocity profile follows $u/U = (y/\delta)^{1/n}$, find coefficient n. Determine whether the boundary layer is laminar or turbulent and calculate (by numerical integration of profiles) δ^*, Θ, and C_f (if possible).

Table P9.80

Position	Distance from Plate (mm)	Velocity, u (m/s)
1	0	0
2	0.5	31
3	1	35.1
4	1.5	37.4
5	2	39
6	2.5	40.4
7	3	41.4
8	3.5	42.2
9	4	43.42
10	4.5	44.2
11	5	45
12	5.5	45.7
13	6	46.1
14	6.5	46.3
15	7	47.5
16	7.5	47.6
17	8	48.5
18	8.5	49
19	9	49.2
20	9.5	49.65
21	10	50.1
22	10.5	50.4
23	11	50.1
24	11.5	49.98

81. Use the Pohlhausen polynomial velocity profile, Eq. (9.170) to derive a relationship between the von Karman–Pohlhausen parameter Λ, Eq. (9.165), and Thwaites parameter λ, Eq. (9.175).

82. Turbulent boundary layer velocity profiles can be approximated by

$$\frac{u}{u_\tau} = 2.44\ln\left(\frac{yu_\tau}{\nu}\right) + 5.0 + 2.44\Pi\left[1 - \cos\left(\frac{\pi y}{\delta}\right)\right],$$

where Π is called the *wake parameter*. Find an (implicit) expression for the skin friction coefficient C_f as a function of $R_\delta(\delta U/\nu)$ and Π.

83. Use the velocity profile in Problem 82 and find expressions for δ^*/δ, Θ/δ, and H as functions of u_τ/U and Π.

84. Using the (implicit) expression for the skin friction coefficient C_f as a function of R_δ $(\delta U/\nu)$ and Π found in Problem 82 as the basis, fit an explicit expression $C_f = f(R_\delta, \Pi)$.

85. Consider a body whose free-stream velocity $U(x)$ is given by

$$U(x) = U_\infty\left(1 - \frac{x}{\ell}\right),$$

where x is the coordinate along the surface of the body in the direction of flow. Use Thwaites's method to cal-

Table P9.90

Pos. No.	Distance from Surface y (mm)	Velocity, u (m/s) $x_1 = 0.15$ m	$x_2 = 0.17$ m	$x_3 = 0.19$ m	$x_4 = 0.21$ m	$x_5 = 0.23$ m
1	0	0	0	0	0	0
2	1	9	10.24	11.22	12.24	13.68
3	2	12.9	14.08	15.3	17.64	19
4	3	16.5	18.24	19.38	20.88	22.8
5	4	18.9	20.48	22.44	23.76	25.84
6	5	21	22.4	24.82	25.92	27.74
7	6	23.1	24.96	26.86	28.08	30.4
8	7	24.9	26.56	28.9	31.32	33.82
9	8	26.4	29.12	30.6	31.68	34.96
10	9	28.5	30.4	32.3	34.2	36.1
11	10	30	32	34	36	38
12	11	30.3	32.64	33.66	36.36	37.62
13	12	30	31.68	34.34	36	38

culate the value of x/ℓ where the laminar boundary layer separates from the wall.

86. Repeat Problem 85 but use

$$U(x) = U_x\left(1 - \frac{x^2}{\ell^2}\right).$$

87. A horizontal 100-km/hr wind blows over a 200-m-high, 10-m-diameter circular building. Separation occurs at an angle $\beta = 80°$, as shown in Fig. P9.87. Assume potential flow up to the separation point and a constant pressure from $\beta = 80°$ to $\beta = 180°$. Find the force tending to topple the building. Compare your result with that obtained by using the drag coefficient from Fig. 8.33. Assume Standard Atmosphere conditions at sea level.

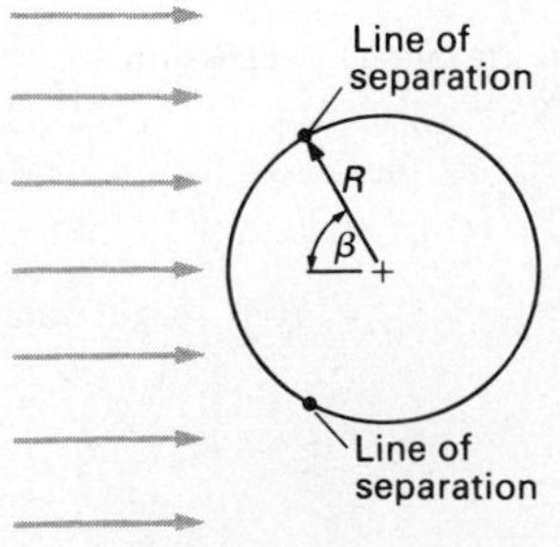

Figure P9.87

88. Consider the plane wall diffuser in Fig. P9.88. Use Thwaites's method to estimate the angle β at which separation of a laminar boundary layer occurs at the exit plane. The diffuser is wide, with $d = 0.25$ m, $\ell = 1.0$ m, and $U_\infty = 2.0$ m/s.

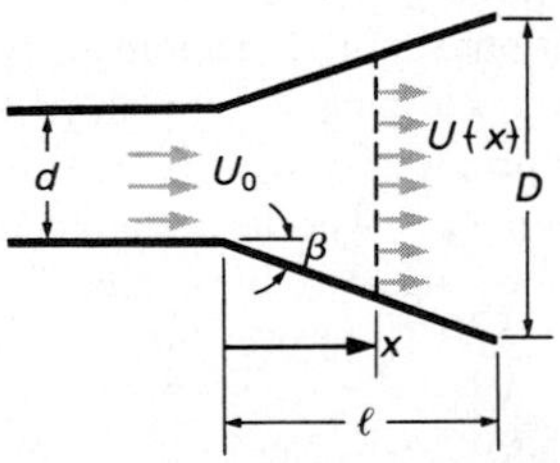

Figure P9.88

• **89.** Use Thwaites' model for the laminar boundary layer and a one-dimensional model for the inviscid core, $U(\delta - \delta^*) = \dot{m} = \rho U_0(2Y)$, and calculate the development length for the two-dimensional channel shown in Fig. P9.89. $\mathbf{R} = 2U_0Y/\nu = 1000$.

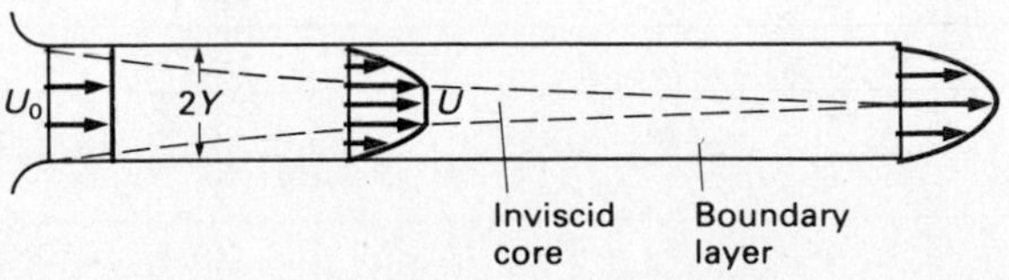

Figure P9.89

Table P9.91

Pos. No.	Distance from Surface, y(mm)	Velocity, u_1(m/s)	Distance from Surface, y(mm)	Velocity, u_2(m/s)	Distance from Surface, y(mm)	Velocity, u_3(m/s)
1	0	0	0	0.00	0	0
2	0.5	31	2	30.59	0.25	5.45
3	1	35.1	3	35.90	0.5	7.6
4	1.5	37.4	4	39.06	0.75	9.6
5	2	39	5	42.93	1	10.95
6	2.5	40.4	6	45.07	1.25	12.25
7	3	41.4	7	48.95	1.5	13.38
8	3.5	42.2	8	50.28	1.75	14.53
9	4	43.42	9	53.03	2	15.3
10	4.5	44.2	10	54.86	2.25	16.5
11	5	45	11	55.27	2.5	17.3
12	5.5	45.7	12	55.58	2.75	18.2
13	6	46.1	15	54.97	3	19
14	6.5	46.3			3.25	19.75
15	7	47.5			3.5	20.4
16	7.5	47.6			3.75	21.1
17	8	48.5			4	21.92
18	8.5	49			4.25	22.4
19	9	49.2			4.5	23.3
20	9.5	49.65			4.75	24
21	10	50.1			5	24.5
22	10.5	50.4			5.25	24.6
23	11	50.1			5.5	24.4
24	11.5	49.98			5.75	24.5

90. Consider Eq. (9.133),

$$\frac{d\Theta}{dx} + (2 + H)\frac{\Theta}{U}\frac{dU}{dx} = \frac{C_f}{2},$$

and the five laminar flow velocity distributions shown in Table P9.90. Find C_f as a function of distance x.

91. Table P9.91 shows velocity profiles: At least one is for a turbulent boundary layer and at least one is for a laminar boundary layer. For these profiles find boundary layer thickness, δ, displacement thickness, δ^*, momentum thickness, Θ, and shape factor, H, by numerical integration. (Do not attempt to develop an expression approximating the profile.) Draw conclusions based on the values calculated. Which profiles are laminar, and which are turbulent?

• **92.** Consider an infinitely long cylinder of radius R in an air stream of approach velocity V_∞. Assume that separation occurs at angle β (defined in Fig. P9.92). Assume potential flow up to the separation line. Estimate the form (or pressure) drag component of a unit length of the cylinder for $R = 1.0$ in. and $V_\infty = 100$ mph. Calculate a drag coefficient based on the pressure profile. Do calculations for $\beta = 70°$ and $\beta = 85°$ and compare with published values. Draw any worthwhile conclusions.

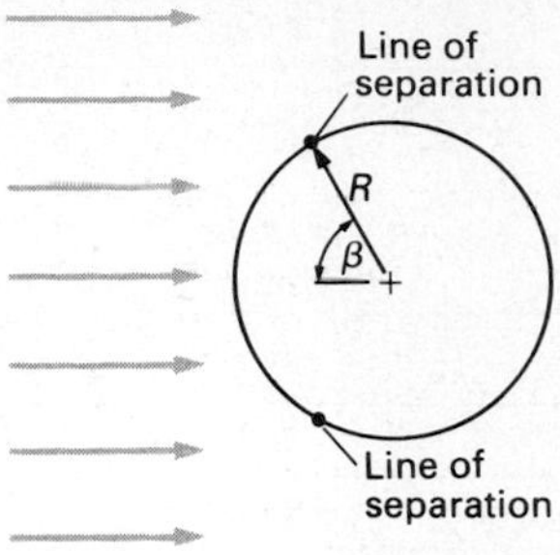

Figure P9.92

93. A telephone wire has a 1.5-cm diameter and is in a 10-m/s cross wind. Estimate the location on the wire where flow separation occurs. Assume Standard Atmosphere conditions at sea level.

• **94.** The one-dimensional plane wall diffuser follows the exponential profile, as shown in Fig. P9.94.

(a) Use Thwaites' method to estimate the distance at which separation of a laminar boundary layer occurs from the exit plane. Here, $U_0 = 2$ m/s, $d = 0.1$ m, $L = 1.0$ m, and

$$\ell(x) = \frac{d}{2} e^{px},$$

where $p = 3.0$/m and x is measured in meters. Then

$$U(x) = \frac{U_0 d}{2\ell(x)} = \frac{U_0}{2} e^{-3x/m}.$$

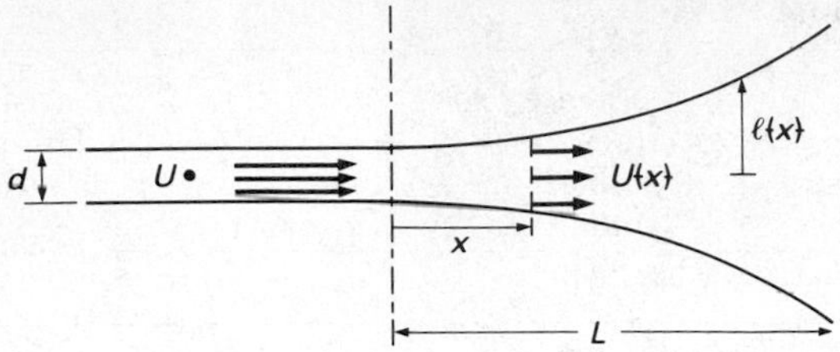

Figure P9.94

(b) What value for p is needed in the diffuser so that laminar separation will occur in the exit plane?

• **95.** Consider the flow normal to a right circular cylinder. Use the realistic velocity distribution

$$\frac{U}{V_\infty} = 1.814\phi - 0.271\phi^3 - 0.0471\phi^5$$

and develop an expression for the angle ϕ where the boundary layer separates from the right circular cylinder. See Fig. E9.8.

96. Write a computer program or develop a spreadsheet template that calculates laminar boundary layers by Thwaites' method. The free-stream velocity distribution $U(x)$ is input as a table of values. Output should be a table of values of C_f, Θ, and H versus x, as well as any indication of boundary layer separation.

REFERENCES

1. Milne-Thompson, L. M., *Theoretical Hydrodynamics* (4th ed.), Macmillan, New York, 1960.
2. Robertson, J. M., *Hydrodynamics in Theory and Application,* Prentice-Hall, Englewood Cliffs, N.J., 1965.
3. Schlichting, H., *Boundary Layer Theory* (7th ed.), McGraw-Hill, New York, 1979.
4. Cebeci, T., and P. Bradshaw, *Momentum Transfer in Boundary Layers,* McGraw-Hill-Hemisphere, New York, 1977.
5. Schetz, J. A., *Foundations of Boundary Layer Theory for Momentum, Heat, and Mass Transfer,* Prentice-Hill, Englewood Cliffs, N.J., 1984.
6. Panton, R. L., *Incompressible Flow,* Wiley-Interscience, New York, 1984.
7. Curie, I. G., *Fundamental Mechanics of Fluids,* McGraw-Hill, New York, 1974.
8. White, F. M., *Viscous Fluid Flow* (2nd ed.), McGraw-Hill, New York, 1990.
9. Pao, R. H. F., *Fluid Dynamics,* Merrill, Columbus, Oh., 1967.
10. Hess, J. L., "Review of Integral-Equation Techniques for Solving Potential-Flow Problems with Emphasis on the Surface-Source Method," *Computer Methods in Applied Mechanics and Engineering,* 5, 1975, pp. 145–196.

11. Anderson, J. D. Jr., *Fundamentals of Aerodynamics,* McGraw-Hill, New York, 1984.

12. Blasius, H., "Grenzschechten in Flussigkerten mit kleiner Reibung," *Z. Mat. Physik,* 56, 1. (Available in English as NACA Tech Memo 1256).

13. Thwaites, B., "Approximate Calculation of the Laminar Boundary Layer," *Aeronautical Quarterly,* 1, 1949.

14. Kline, S. J., M. Morkovin, G. Sovran, and D. Cockrell, *Computation of Turbulent Boundary Layers—1968 AFOSR–IFP–Stanford Conference,* Department of Mechanical Engineering, Stanford University, Palo Alto, Calif., 1968.

10 Compressible Flow

Most of our discussions in the previous chapters have concentrated on incompressible flow. The incompressibility condition permits two important simplifications.

- The density is known and can be treated as a constant in the continuity, momentum, and energy equations.
- The coupling between mechanical and thermal forms of energy is weak, permitting use of a special form of the energy equation.

Assuming that fluids are incompressible is not always realistic, however, so we cannot always rely on results based on these two simplifications. All fluids, even liquids, are compressible. If the pressure changes that occur in a flow are sufficiently large to cause significant density changes, we have to abandon the incompressible flow assumption. This situation is much more likely to occur in a gas or vapor than in a liquid. A pressure change of 500 kPa (72.6 lb/in^2) causes a density change of about 0.024 percent in water but a density change of about 250 percent in standard air! Because compressibility is more likely to be significant in gas flow than in liquid flow, the study of compressible flow is sometimes called *gas dynamics*.

In this chapter, we consider flows in which the effects of fluid compressibility are significant. Thus we must treat density as a variable and introduce a relation between density and pressure. The simplest available relation is the ideal gas law, Eq. (1.1):

$$p = \rho R T.$$

This equation also involves temperature, which becomes an additional variable. The general energy equation and a relation between internal energy and temperature must also be considered. Analysis of compressible flow requires simultaneous application of the continuity, momentum, and energy equations and a property relation to calculate fluid velocity, pressure, density, and temperature. This requirement holds even if the fluid is not an ideal gas.

There are two important elements in a study of compressible flow: consideration of the physical effects of fluid compressibility and development of analysis techniques for engineering calculations. The latter task is greatly simplified if we assume that the fluid is an ideal gas. Many of the equations that we develop in this chapter are based on the ideal gas model. Generally, the information that we present about the physical behavior of compressible fluids is not limited to ideal gases. If you need to analyze flow of a fluid that is not adequately described by the ideal gas model, be sure to recognize that results based on the equations and methods that we present in this chapter will be qualitatively correct, but numerical values should be treated with caution.

Both internal and external flows can be affected by fluid compressibility. In this chapter, we consider both types, but the major emphasis is on internal flow. We can present only a brief introduction to compressible flow in a single chapter. Several books devoted to compressible flow (gas dynamics) are available [1–6].

10.1 SOME CONCEPTS FROM THERMODYNAMICS

Concepts from thermodynamics and concepts from fluid mechanics are equally important in compressible flow analysis. Before we begin the study of compressible flow, we should consider a few basic concepts from thermodynamics. If you have already studied thermodynamics, this section will serve as a review. In this section, we present all the concepts and equations needed for compressible flow analysis, but the developments and explanations are brief. Consult references [7–9] for more detail.

10.1.1 Properties, State, and Process

The following properties are important in thermodynamics:

- Density (ρ)
- Pressure (p)
- Temperature (T)
- Specific internal energy ($\tilde{u}$)
- Specific enthalpy ($\tilde{h} \equiv \tilde{u} + p/\rho$)
- Specific entropy ($\tilde{s}$)

We introduced all these properties except entropy in Chapter 1. We define entropy later in this section.

Properties are used to describe the state of a system. Recall that a system is a collection of mass of fixed identity; a system can be as small as a single fluid particle. An important principle of thermodynamics is the *state postulate:*

> The state of a system is fixed if a certain minimum number of its properties are specified.

For a pure, compressible substance (gas or vapor), the minimum number of properties is two. For example, oxygen with $p = 14.70$ lb/in^2 and $T = 0°$F *always* has $\rho = 0.0954$ lbm/ft^3; no other density is possible at that pressure and temperature.

The state postulate tells us that the properties of any substance are related to each other. Using *equations of state* such as the ideal gas law, Eq. (1.1), we can quantify this relation. Unfortunately, the ideal gas law is not complete because internal energy cannot be determined from it. For an ideal gas, internal energy is related to temperature by the specific heat at constant volume (c_v):

$$\tilde{u}_2 - \tilde{u}_1 = \int_{T_1}^{T_2} c_v\{T\}\, dT.$$

In this chapter, we assume that c_v is constant, so

$$\tilde{u}_2 - \tilde{u}_1 = c_v(T_2 - T_1). \tag{10.1}$$

The enthalpy of an ideal gas is obtained from the temperature and specific heat at constant pressure (c_p):†

$$\tilde{h}_2 - \tilde{h}_1 = \int_{T_1}^{T_2} c_p\{T\}\, dT.$$

For an ideal gas with constant c_p,

$$\tilde{h}_2 - \tilde{h}_1 = c_p(T_2 - T_1). \tag{10.2}$$

Because $\tilde{h} = \tilde{u} + p/\rho$,

$$\tilde{h}_2 - \tilde{h}_1 = \tilde{u}_2 - \tilde{u}_1 + \left(\frac{p}{\rho}\right)_2 - \left(\frac{p}{\rho}\right)_1 = c_v(T_2 - T_1) + R(T_2 - T_1).$$

Comparing this result with Eq. (10.2), we note that

$$c_p = c_v + R. \tag{10.3}$$

Tables A.3 and A.4 give values of the gas constants R, c_v, c_p, and $k(\equiv c_p/c_v)$ for several gases. These tables show that c_p, c_v, and k vary slightly with temperature; but we assume that they are constant. Although we normally use values at standard sea-level temperature, we can improve accuracy slightly by using values appropriate for the actual temperatures in a particular problem.

Thermodynamics and fluid mechanics deal with changes of state. When a gas flows through a nozzle, the pressure and temperature of the gas at the nozzle outlet are, in general, different from the pressure and temperature at the nozzle inlet, so the state of the gas will be different at the inlet and the outlet. If we follow a typical gas particle from the inlet to the outlet, we observe that its state changes continuously. This sequence of different states is called a *process:*

A process describes a sequence of states of a system.

There are many different types of processes. A process that occurs in such a way that the temperature remains constant is called an *isothermal* process. In an isothermal process, the pressure and density may change but the temperature does not. A particularly important type of process is the *adiabatic* process, in which no heat is transferred.

Processes can be represented as lines on a *property diagram*. Figure 10.1 illustrates a pressure–density (p–ρ) diagram. Because only two properties are required to specify the state of a substance, any point on a property diagram represents a specific state. For an ideal gas, $p = (RT)\rho$, so an isothermal process plots as a

† The terms "constant volume" and "constant pressure" do not mean that these equations apply only if volume (density) or pressure is constant [7].

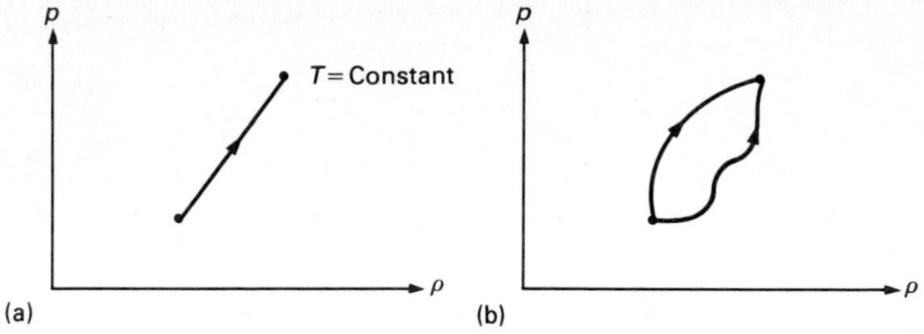

Figure 10.1 Processes on a property diagram: (a) isothermal process on a p–ρ diagram; (b) two processes that have the same end states.

straight line on a p–ρ diagram. An adiabatic process is represented by a different line.

An important characteristic of properties and processes is that:

> Changes of properties in a process depend only on the end points of the process and not on the path of states between the end points.

Figure 10.1(b) illustrates two different processes that have the same end points. Obviously, the changes of p and ρ for the two processes are identical, as the beginning and end points are identical. Moreover the changes of all other properties (such as $T_2 - T_1$, and $\tilde{u}_2 - \tilde{u}_1$) are identical for the two processes.

10.1.2 The Laws of Thermodynamics

The laws of thermodynamics deal with energy transformations and relate energy transfers to or from a system to changes of the system's properties. We introduced the first and second laws of thermodynamics in Section 3.3.1.

We considered the first law of thermodynamics extensively in Section 4.3 and we developed several equations representing the first law for a control volume. Let's consider, for example, Eq. (4.39), the general energy equation for steady flow along a streamline:

$$q - w_s = (\tilde{u}_2 - \tilde{u}_1) + \left(\frac{p_2}{\rho_2} - \frac{p_1}{\rho_1}\right) + \left(\frac{V_2^2}{2} - \frac{V_1^2}{2}\right) + (gz_2 - gz_1).$$

This equation relates energy transfers to or from a fluid particle (the terms on the left) to changes of the particle's properties (the terms on the right-hand side). The energy transfers are associated with the process that occurs between points 1 and 2. Because both heat and work appear on the left-hand side, we cannot determine both if we know only the change of fluid properties. Heat and work depend on the exact nature of the process that occurs between the end states. The two processes illustrated in Fig. 10.1(b) would each have different amounts of heat transfer and work, although the sum $q - w_s$ would be the same for both. This means that when we use the general energy equation to analyze a flow process, we must make an explicit statement about either the heat transfer or the work.†

† To illustrate, in Example 4.8 it was necessary to assume that no heat was transferred to the air during compression.

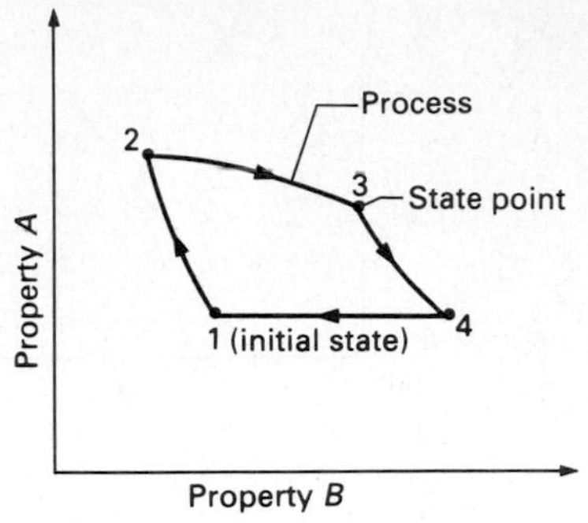

Figure 10.2 A cycle is a sequence of processes that returns the fluid to its initial state.

The second law of thermodynamics deals with conversions between heat and work. Let's consider a sequence of processes that eventually returns the fluid to its initial state (Fig. 10.2). Such a sequence of processes is called a *cycle*. Heat and/or work is transferred to the fluid in some of the processes and is transferred from the fluid in others. Applying the first law to the entire sequence of processes gives

$$\sum q - \sum w_s = 0.$$

The right-hand side is zero, because the sequence returns the fluid to its initial state. Recall the sign conventions for heat and work:

- heat addition is positive; and
- work production is positive.

Using subscripts "in" and "out" rather than signs, we get

$$\sum q_{\text{in}} - \sum q_{\text{out}} = \sum w_{\text{out}} - \sum w_{\text{in}} = w_{\text{out,net}}.$$

The second law of thermodynamics states that it is impossible to devise a sequence of processes that has no effect other than the conversion of a given amount of heat input into an equal amount of work output. In a *single process,* it is possible that work can be produced that is equal to or even greater than the amount of heat input during the process, but this effect can occur only with a change of properties, which is an additional effect. A concise statement of the second law of thermodynamics is as follows:

> It is impossible to convert a given amount of heat completely into useful work (or other form of mechanical energy) with no additional effect on the system or surroundings.

Of course, a given amount of work (or other form of mechanical energy) can be converted completely into heat; this process is the mechanical energy "loss" that is so important in viscous flow.

Closely associated with the second law of thermodynamics is the concept of the *reversible process.* A reversible process is one that can be reversed or "run backward" with no net effect on either the system or the surroundings. Suppose that a process changed the fluid from state 1 to state 2 while absorbing heat q and producing work w. The reversed process must not only return the fluid from state 2 to state 1 but must also reject an amount of heat *exactly* equal to q and absorb an amount of work *exactly* equal to w. Any process that converts work (or other form of mechanical energy) into heat (by friction, for example) cannot be reversible because, according to the second law, converting the reverse heat completely into work is impossible. Other occurrences that render a process irreversible are mixing of dissimilar substances and heat transfer across a finite temperature difference. Because all real-world processes involve friction, mixing, or heat transfer, they are irreversible.

10.1.3 Entropy and the Isentropic Process

The second law of thermodynamics implies the existence of a property called *entropy*. Consider two states of a substance, state 1 and state 2. The difference in specific entropy between the two states is defined by

$$\tilde{s}_2 - \tilde{s}_1 \equiv \int_1^2 \frac{dq_{\text{rev}}}{T}. \tag{10.4}$$

In this definition, q_{rev} is the heat transfer per unit mass that would occur in a *reversible* process between 1 and 2. We can show [7, 8] that

- $\int (dq_{\text{rev}}/T)$ is the same for any reversible path between states 1 and 2; and
- $\tilde{s}_2 - \tilde{s}_1$ can be related to only the values of the other properties at states 1 and 2.

These facts tell us that entropy is truly a thermodynamic property, whose change between two state points depends only on the state points themselves and not on the process between them.

The second law of thermodynamics can be used to show that entropy changes in a *real* (irreversible) process can be related to actual heat transfer and temperature by

$$\tilde{s}_2 - \tilde{s}_1 \geq \int_1^2 \frac{dq}{T}. \tag{10.5}$$

The equality holds if the process is reversible and the inequality holds if the process is irreversible. The entropy gives us a way to evaluate the irreversibility of a process by comparing actual entropy change to the integral in Eq. (10.5).

As entropy is a property, we can relate it to other properties. The appropriate equations are

$$T\,d\tilde{s} = d\tilde{h} - \frac{dp}{\rho} \tag{10.6}$$

and

$$T\,d\tilde{s} = d\tilde{u} - \frac{p}{\rho}\left(\frac{d\rho}{\rho}\right). \tag{10.7}$$

Equations (10.6) and (10.7) are called the *Gibbs equations* or, sometimes, the *T-d$\tilde{s}$ equations*.

If we specify an ideal gas with constant specific heat, the Gibbs equations can be integrated to give

$$\tilde{s}_2 - \tilde{s}_1 = c_p \ln\left(\frac{T_2}{T_1}\right) - R \ln\left(\frac{p_2}{p_1}\right) \tag{10.8}$$

and

$$\tilde{s}_2 - \tilde{s}_1 = c_v \ln\left(\frac{T_2}{T_1}\right) - R \ln\left(\frac{\rho_2}{\rho_1}\right). \tag{10.9}$$

Note that Eqs. (10.8) and (10.9) allow calculation of entropy from other fluid properties. You do not have to know the heat transfer to calculate entropy change.

In analyzing compressible flow, we often assume that there is no heat transfer to or from the fluid; that is, we assume an adiabatic process. If we also assume that the process is reversible, the entropy remains constant. Such a process (reversible, adiabatic) is called an *isentropic process.* If the fluid is an ideal gas, Eqs. (10.8) and (10.9) yield simple and convenient process laws for an isentropic process. Considering Eq. (10.8) applied to an isentropic process, we have

$$\tilde{s}_2 - \tilde{s}_1 = 0 = c_p \ln\left(\frac{T_2}{T_1}\right) - R \ln\left(\frac{p_2}{p_1}\right).$$

Dividing by R and taking the antilogarithm give

$$\frac{p_2}{p_1} = \left(\frac{T_2}{T_1}\right)^{c_p/R}.$$

Using Eq. (10.3), we have

$$\frac{c_p}{R} = \frac{k}{k-1}, \tag{10.10}$$

and we can write

$$\frac{p_2}{p_1} = \left(\frac{T_2}{T_1}\right)^{k/(k-1)} \tag{10.11}$$

Similarly,

$$\frac{\rho_2}{\rho_1} = \left(\frac{T_2}{T_1}\right)^{1/(k-1)} \tag{10.12}$$

and

$$\frac{p_2}{p_1} = \left(\frac{\rho_2}{\rho_1}\right)^{k}. \tag{10.13}$$

Isentropic process in an ideal gas

Note that Eqs. (10.11)–(10.13) are process laws relating changes of properties that occur in a specific type of process. They *are not* equations of state.

10.1.4 Stagnation Properties

The thermodynamic state of a fluid particle is defined by its properties (p, ρ, T, $\tilde{u}$, $\tilde{h}$, $\tilde{s}$); but from the viewpoint of mechanics, we must also know the velocity of the particle and, possibly, its position in a gravitational field.

The thermodynamic properties are called *static properties;* they are the values that would be measured by instruments that are static with respect to the fluid. The static properties represent the molecular structure of the fluid and obey all

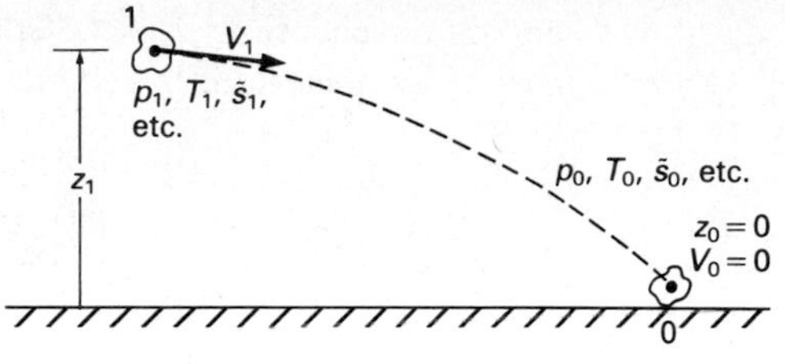

Figure 10.3 Process for defining stagnation properties.

equations of state and other property-related laws and thermodynamic equations. The particle's velocity and elevation are specified separately.

A useful approach is to combine the (static) thermodynamic properties with the velocity and elevation to obtain equivalent thermodynamic properties that represent the total (thermodynamic *and* mechanical) state. This is done by using *stagnation properties:*

> Stagnation properties are the properties that the fluid would obtain if it were brought to a condition of zero velocity and zero elevation in a reversible process with no heat transfer and no work.

We can obtain equations for stagnation properties with the aid of Fig. 10.3, which shows a (possibly imaginary) process that brings a fluid particle to rest and zero elevation. Applying the energy equation to the streamline between 1 and 0 gives

$$q - w_s = \tilde{h}_0 - \tilde{h}_1 + \frac{V_0^2}{2} - \frac{V_1^2}{2} + gz_0 - gz_1.$$

By definition, q, w_s, V_0, and z_0 are all zero, so we get

$$\tilde{h}_0 = \tilde{h}_1 + \frac{V_1^2}{2} + gz_1,$$

where $\tilde{h}_0$ is called the *stagnation enthalpy.*† In compressible flow analysis, potential energy and gravity force are almost always negligible, and the definition of stagnation enthalpy is usually simplified to

$$\tilde{h}_0 \equiv \tilde{h} + \frac{V^2}{2}. \tag{10.14}$$

The subscript 1 has been dropped to obtain a general definition of stagnation enthalpy. Note that we did not need to use the specification of reversibility to obtain the stagnation enthalpy.

If the fluid is an ideal gas with constant specific heat, Eq. (10.14) becomes

$$c_p T_0 = c_p T + \frac{V^2}{2},$$

and the *stagnation temperature* is given by

$$T_0 = T + \frac{V^2}{2c_p}. \tag{10.15}$$

† The subscript zero on stagnation properties means "zero velocity (and elevation)."

We can obtain equations for the *stagnation pressure* and *stagnation density* by noting that, by definition, the stagnation process is isentropic. For an ideal gas,

$$\frac{p_0}{p} = \left(\frac{T_0}{T}\right)^{k(k-1)} \tag{10.16}$$

and

$$\frac{\rho_0}{\rho} = \left(\frac{T_0}{T}\right)^{1/(k-1)}. \tag{10.17}$$

Dividing Eq. (10.15) by T, we obtain

$$\frac{T_0}{T} = 1 + \frac{V^2}{2c_pT}. \tag{10.18}$$

Substituting into Eqs. (10.16) and (10.17), we have

$$p_0 = p\left(1 + \frac{V^2}{2c_pT}\right)^{k/(k-1)} \tag{10.19}$$

and

$$\rho_0 = \rho\left(1 + \frac{V^2}{2c_pT}\right)^{1/(k-1)}. \tag{10.20}$$

Equations (10.14)–(10.20) show that the stagnation properties are determined by the static properties *and* the fluid velocity. Because any fluid particle always has a pressure, a temperature, and a velocity (the velocity might be zero), any fluid particle can be described by both its static properties and its stagnation properties. We may say that "the fluid has a pressure of 100 kPa and a stagnation pressure of 125 kPa." The static properties actually are the properties of the fluid, and the stagnation properties are the properties that the fluid would have if it were brought to zero velocity in an isentropic process with no work. If (and only if) the fluid particle has zero velocity, its static and stagnation properties are equal.

Stagnation properties represent the maximum possible values of the fluid properties unless energy is added to the fluid by heat transfer or work. The postulated stagnation process represents the ideal conversion of the fluid's kinetic and potential energies into pressure and internal energy.

Stagnation properties are sometimes called *total properties*. Because the process of stagnation (that is, bringing the fluid to rest) can be accomplished by an irreversible process† as well as by a reversible process, there is some possible ambiguity in using the term *stagnation* in cases where viscous flow processes are considered. This ambiguity can also be remedied by use of the term "isentropic stagnation properties."

---

† An example is the deceleration that occurs in a viscous boundary layer (see Chapter 9).

EXAMPLE 10.1 Illustrates Basic Concepts and Calculations in Thermodynamics

A turbine expands 10 lbm/sec of air from an inlet pressure and temperature of 40 psia† and 500°F to an outlet pressure of 14.9 psia. The turbine delivers 760 hp to an electric generator. The air velocity is 500 ft/sec at the turbine inlet and 100 ft/sec at the turbine outlet. The turbine casing is well insulated. The flow through the turbine is steady. Calculate the static temperature at the turbine outlet, the stagnation temperature and pressure at the turbine inlet and outlet, and the change of specific entropy between the inlet and outlet. Is the expansion a reversible process?

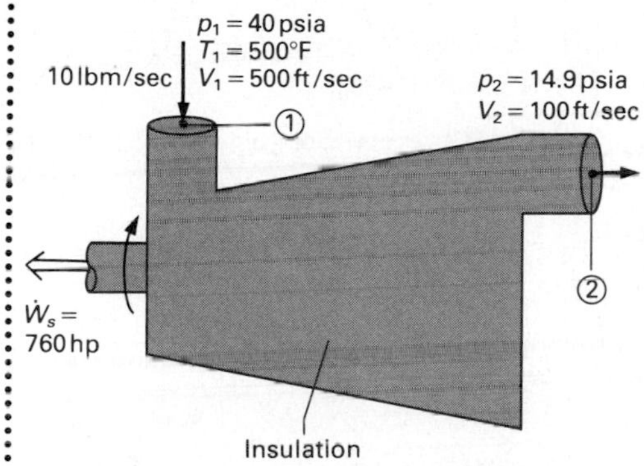

Figure E10.1 Schematic drawing of hot air turbine.

SOLUTION

Given

A turbine, which expands hot air and produces power

Figure E10.1, a schematic of a turbine

Data on flow rate, air properties, and velocity given in Fig. E10.1

Steady flow

Find

Static temperature at turbine outlet (T_2)

Stagnation pressure and temperature at turbine inlet (p_{0_1}, T_{0_1})

Stagnation pressure and temperature at turbine outlet (p_{0_2}, T_{0_2})

Entropy change ($\tilde{s}_2 - \tilde{s}_1$)

Whether expansion is reversible

Solution

The energy equation for a control volume enclosing the turbine is obtained from Eq. (4.38):

$$\dot{Q} - \dot{W}_s = \dot{m}_2\left[\bar{\tilde{u}}_2 + \overline{\left(\frac{p_2}{\rho_2}\right)} + \alpha_2\left(\frac{\bar{V}_2^2}{2}\right) + g\bar{z}_2\right] - \dot{m}_1\left[\bar{\tilde{u}}_1 + \overline{\left(\frac{p_1}{\rho_1}\right)} + \alpha_1\left(\frac{\bar{V}_1^2}{2}\right) + g\bar{z}_1\right].$$

To simplify this equation, we assume that α_1 and α_2 are unity, that elevation changes are negligible, and that the given data represent proper averages. The continuity equation for the control volume is

$$\dot{m}_2 = \dot{m}_1 = \dot{m},$$

so the energy equation becomes

$$\dot{Q} - \dot{W}_s = \dot{m}\left[\left(\tilde{h}_2 + \frac{V_2^2}{2}\right) - \left(\tilde{h}_1 + \frac{V_1^2}{2}\right)\right].$$

The turbine is well insulated, so $\dot{Q}$ is small (we assume that it is zero).

---

† In this chapter, we follow the customary procedure in thermodynamics and write "psia" for lb/in^2, abs.

Solving for the enthalpy change of the air gives

$$\tilde{h}_2 - \tilde{h}_1 = -\frac{\dot{W}_s}{\dot{m}} + \frac{V_1^2 - V_2^2}{2}.$$

If we assume that the air is an ideal gas, Eq. (10.2) relates enthalpy to temperature. Substituting this relation and solving for T_2, we have

$$T_2 = T_1 - \frac{\dot{W}_s}{\dot{m}c_p} + \frac{V_1^2 - V_2^2}{2c_p}.$$

Substituting numbers gives (from Table A.4, $c_p = 186.7$ ft·lb/lbm·°R)

$$T_2 = (500 + 459.7)°\text{R} - \frac{760 \text{ hp}(550 \text{ ft}\cdot\text{lb/sec}\cdot\text{hp})}{10 \text{ lbm/sec}(186.7 \text{ ft}\cdot\text{lb/lbm}\cdot°\text{R})}$$
$$+ \frac{(500)^2 \text{ ft}^2/\text{sec}^2 - (100)^2 \text{ ft}^2/\text{sec}^2}{2(186.7 \text{ ft}\cdot\text{lb/lbm}\cdot°\text{R})(32.2 \text{ ft}\cdot\text{lbm/lb}\cdot\text{sec}^2)},$$

or

$$T_2 = 755.8°\text{R} = 296.1°\text{F}.$$ **ANSWER**

The stagnation temperature and pressure are calculated from Eqs. (10.15) and (10.16). At the turbine inlet,

$$T_{0_1} = T_1 + \frac{V_1^2}{2c_p}$$
$$= 959.7°\text{R} + \frac{(500)^2 \text{ ft}^2/\text{sec}^2}{2(186.7 \text{ ft}\cdot\text{lb/lbm}\cdot°\text{R})(32.2 \text{ ft}\cdot\text{lbm/lb}\cdot\text{sec}^2)},$$

or

$$T_{0_1} = 980.5°\text{R} = 520.8°\text{F},$$ **ANSWER**

and

$$p_{0_1} = p_1\left(\frac{T_{0_1}}{T_1}\right)^{k/(k-1)} = 40 \text{ psia}\left(\frac{980.5}{959.7}\right)^{1.4/0.4},$$

or

$$p_{0_1} = 43.1 \text{ psia}.$$ **ANSWER**

At the turbine outlet,

$$T_{0_2} = T_2 + \frac{V_2^2}{2c_p}$$
$$= 755.8°\text{R} + \frac{(100)^2 \text{ ft}^2/\text{sec}^2}{2(186.7 \text{ lb}\cdot\text{sec/lbm}\cdot°\text{R})(32.2 \text{ ft}\cdot\text{lbm/lb}\cdot\text{sec}^2)},$$

or

$$T_{0_2} = 756.6°\text{R} = 296.9°\text{F},$$ **ANSWER**

and

$$p_{0_2} = p_2\left(\frac{T_{0_2}}{T_2}\right)^{k/(k-1)} = 14.9 \text{ psia}\left(\frac{756.6}{755.8}\right)^{3.5},$$

or

$$p_{0_2} = 14.96 \text{ psia}.$$ **ANSWER**

We calculate the entropy change across the turbine by

$$\tilde{s}_2 - \tilde{s}_1 = c_p \ln\left(\frac{T_2}{T_1}\right) - R \ln\left(\frac{p_2}{p_1}\right).$$

Substituting numbers, we have

$$\tilde{s}_2 - \tilde{s}_1 = 186.7 \frac{\text{ft}\cdot\text{lb}}{\text{lbm}\cdot{}^\circ\text{R}} \ln\left(\frac{755.8}{959.7}\right) - 53.35 \frac{\text{ft}\cdot\text{lb}}{\text{lbm}\cdot{}^\circ\text{R}} \ln\left(\frac{14.9}{40}\right),$$

or

$$\tilde{s}_2 - \tilde{s}_1 = 8.1 \text{ ft}\cdot\text{lb/lbm}\cdot{}^\circ\text{R}.$$ **ANSWER**

The expansion is adiabatic, so

$$\int \frac{dq}{T} = 0 \quad \text{and} \quad \tilde{s}_2 - \tilde{s}_1 > \int \frac{dq}{T}.$$

Thus

The expansion is irreversible. **ANSWER**

Discussion

Several important points require emphasis. First note that the turbine power is treated as positive, because the turbine delivers work. Second, note that temperatures can be expressed in either Fahrenheit or Rankine (absolute), but temperature ratios must be in terms of absolute temperature. Third, note that the difference between the static and stagnation properties is significant at the turbine inlet where the velocity is 500 ft/sec but nearly insignificant at the turbine outlet where the velocity is 100 ft/sec. Fourth, because the entropy at a stagnation state is equal to the entropy at the corresponding static state, note that

$$\tilde{s}_{0_2} - \tilde{s}_{0_1} = c_p \ln \frac{T_{0_2}}{T_{0_1}} - R \ln \frac{p_{0_2}}{p_{0_2}} = \tilde{s}_2 - \tilde{s}_1.$$

You might substitute the appropriate numbers to verify this equality. Fifth, note that we used the room temperature values of c_p and k in the calculation. The accuracy could be improved slightly by using values appropriate to the "average" temperature, about 400°F.

Finally, note that Eq. (10.11) is not valid for relating the pressure and temperature between the turbine inlet and outlet:

$$\frac{p_2}{p_1} = \frac{14.9}{40} = 0.3725 \neq \left(\frac{T_2}{T_1}\right)^{k/(k-1)} = \left(\frac{755.9}{959.7}\right)^{3.5} = 0.4334.$$

The reason is that the expansion is not isentropic, even though it is adiabatic.

10.2 WAVES IN COMPRESSIBLE FLUIDS

Disturbances introduced in one part of a fluid are communicated to other parts of the fluid. If a valve in a pipeline is opened or closed, the flow in the entire pipeline is changed. Examination of Figs. 3.4 and 3.5 shows that fluid responds to the presence and motion of a surface well upstream of the surface itself.[†] The influence of a disturbance at a point is propagated to other points by waves that travel at a large, but finite, speed with respect to the fluid.

Consider the effect of suddenly closing a valve at the end of a long pipe. Before the valve is closed, fluid flows through the pipe at a steady rate. At the instant the valve is closed, the fluid at the valve is brought to rest. If the fluid continued to flow toward the valve, it would "pile up" at the valve. A wave propagates from the closed valve back into the fluid, informing the fluid about the closed valve. The wave also causes a pressure increase as the kinetic energy of the fluid is rapidly changed. This pressure change can lead to a significant density change. The wave is called a compression wave. Experimental and theoretical investigations of such waves show them to be very thin (in a gas, wave thickness is a few molecular mean free paths). The speed at which such a wave travels in a particular fluid is a measure of the fluid's compressibility. The more compressible the fluid, the slower is the wave. Because the existence and speed of such disturbance waves are intimately connected with fluid compressibility, analysis of compressible flow often involves consideration of wave speed and motion.

10.2.1 Wave Speed, Speed of Sound, and Mach Number

Figure 10.4(a) depicts a wave propagating into a still fluid. Ahead of the wave, fluid pressure and density are p and ρ. The wave causes the pressure and density to change by Δp and $\Delta\rho$. The velocity of the wave is ω. The fluid is set in motion by the wave, and the velocity of the fluid behind the wave is ΔV; the changes Δp, $\Delta\rho$, and ΔV are not necessarily small, and ω is finite. We choose a control volume that moves with the wave and so transform the flow into the situation shown in Fig. 10.4(b). The wave is very thin, so we pretend that it has no thickness at all, making the areas of the front and back faces of the control volume equal and the volume zero. Applying the continuity equation, we get

$$\dot{m} = \rho A\omega = (\rho + \Delta\rho)A(\omega - \Delta V).$$

Canceling A and solving for ΔV, we have

$$\Delta V = \omega\left(\frac{\Delta\rho}{\rho + \Delta\rho}\right). \tag{10.21a}$$

Note that if $\Delta\rho$ is small,

$$\Delta V \approx \omega\left(\frac{\Delta\rho}{\rho}\right), \tag{10.21b}$$

and ΔV will be much smaller than ω.

Next, we apply the linear momentum equation. Because the "sides" of the control volume are vanishingly small, the only force is from pressure on the inlet

[†] Note the curvature of the streaklines (streamlines) upstream of the airfoil in Fig. 3.4.

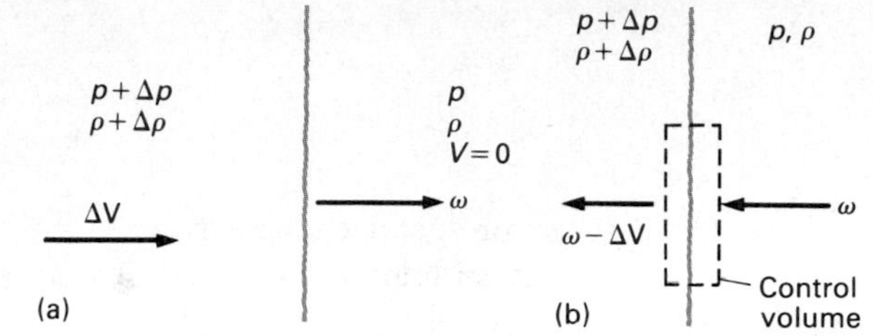

Figure 10.4 Compression wave in a fluid: (a) wave moving relative to fluid; (b) wave motion removed by choosing a control volume fixed to the wave.

and outlet faces:

$$\sum F_x = pA - (p + \Delta p)A = \dot{m}[(\omega - \Delta V) - \omega].$$

Substituting for $\dot{m}$ and solving for Δp give

$$\Delta p = \rho\omega\,\Delta V. \tag{10.22}$$

Equation (10.22) shows that waves can generate very large pressure changes if ω is large.

An expression for wave speed is obtained by substituting Eq. (10.21a) into Eq. (10.22):

$$\omega^2 = \frac{\Delta p}{\Delta \rho}\left(1 + \frac{\Delta \rho}{\rho}\right). \tag{10.23}$$

The *speed of sound* (c) is defined as the speed of a vanishingly weak wave:†

$$c^2 \equiv \lim_{\substack{\Delta p \to 0 \\ \Delta \rho \to 0}} \left[\frac{\Delta p}{\Delta \rho}\left(1 + \frac{\Delta \rho}{\rho}\right)\right] = \frac{\partial p}{\partial \rho}.$$

The limiting process produces a derivative that can be evaluated only if the thermodynamic process caused by the wave is specified. Two (relatively) obvious choices are an isothermal process and an adiabatic (isentropic) process. The correct choice is the isentropic process, because the opportunity for heat transfer to or from the fluid is minimal as it passes rapidly through a very thin wave. The speed of sound in any fluid is therefore defined by

$$c^2 \equiv \left.\frac{\partial p}{\partial \rho}\right)_s. \tag{10.24}$$

For an ideal gas, an isentropic process obeys the equation

$$\frac{p}{p_{\text{ref}}} = \left(\frac{\rho}{\rho_{\text{ref}}}\right)^k, \quad \text{so} \quad \left.\frac{\partial p}{\partial \rho}\right)_s = k\left(\frac{p}{\rho}\right) = kRT.$$

The speed of sound in an ideal gas is calculated by

$$c = \sqrt{k\frac{p}{\rho}}, \tag{10.25}$$

Ideal gas

† Disturbances that the human ear perceives as sound are very weak, typically having a pressure change between 10^{-9} and 10^{-4} atm.

or

$$c = \sqrt{kRT}. \qquad (10.26)$$

Ideal gas

The speed of sound changes from point to point in a flow if the temperature changes. Substituting gas constants for air gives the useful equations

$$c_{\text{air}} \approx 20.05\sqrt{T(\text{K})}\ \text{m/s} \quad \text{and} \quad c_{\text{air}} \approx 49.02\sqrt{T(°\text{R})}\ \text{ft/sec}.$$

Speed of sound in liquids and solids is related to the isentropic bulk modulus of elasticity by

$$c = \sqrt{\frac{E_{v,s}}{\rho}}, \qquad (10.27)$$

where

$$E_{v,s} \equiv \rho\left(\frac{\partial p}{\partial \rho}\right)_s.$$

For most liquids and solids, the modulus of elasticity is nearly the same for any type of process, so the subscript s is dropped and c^2 depends on the general bulk modulus of elasticity E_v.

In Chapter 6, we stated that the primary dimensionless parameter of compressible flow is the Mach number, defined by

$$\mathbf{M} \equiv \frac{V}{\sqrt{E_v/\rho}}.$$

We now recognize that the term in the denominator of the Mach number is actually the speed of sound, so the Mach number is

$$\mathbf{M} = \frac{V}{c}. \qquad (10.28)$$

If the Mach number is large, the effects of fluid compressibility are significant, but if it is small, compressibility effects are negligible. If we square both sides of Eq. (10.28) and assume an ideal gas, we get

$$\mathbf{M}^2 = \frac{V^2}{kRT} = \frac{\rho V^2}{kp}. \qquad (10.29)$$

The criteria for neglecting fluid compressibility suggested in Sections 1.3.6 and 4.3.5 are equivalent to specifying that the Mach number is small.

The importance of Mach number as a parameter in compressible flow can be illustrated by considering the (ideal gas) equations for stagnation properties. The stagnation-to-static-temperature ratio, Eq. (10.18), is

$$\frac{T_0}{T} = 1 + \frac{V^2}{2c_pT}.$$

Multiplying and dividing the second term on the right by kR and using Eqs. (10.3), (10.26), and (10.29), we obtain

$$\frac{V^2}{2c_pT} = \frac{k-1}{2}\left(\frac{V^2}{c^2}\right) = \left(\frac{k-1}{2}\right)\mathbf{M}^2,$$

so

$$\frac{T_0}{T} = 1 + \frac{k-1}{2}\mathbf{M}^2. \tag{10.30}$$

Using Eqs. (10.19) and (10.20), we get

$$\frac{p_0}{p} = \left(1 + \frac{k-1}{2}\mathbf{M}^2\right)^{k/(k-1)} \tag{10.31}$$

and

$$\frac{\rho_0}{\rho} = \left(1 + \frac{k-1}{2}\mathbf{M}^2\right)^{1/(k-1)}. \tag{10.32}$$

Because of the exponential form of Eqs. (10.31) and (10.32), hand or slide-rule calculations involving these equations were quite tedious and extensive tables of property ratios versus Mach number were prepared and used. Even with electronic calculators, use of tables for such calculations is still quite common, so we have included a brief table of values for $k = 1.4$ (Table E.1 in Appendix E). Note that T/T_0 and p/p_0 (numbers less than 1.0) are tabulated with $\mathbf{M}$ as an independent parameter. There is no column for density ratio because

$$\frac{\rho}{\rho_0} = \left(\frac{p}{p_0}\right)\Big/\left(\frac{T}{T_0}\right).$$

The remaining column of the table (A/A^*) will be explained later.

Even use of the tables can be quite cumbersome. The reason is that a new set of tables must be prepared for each different value of k. Writing a computer program to evaluate Eqs. (10.30)–(10.32) and even to solve for $\mathbf{M}$ when, say, p/p_0 is known is a relatively easy task. A program for evaluation of these and other compressible flow functions is included on the personal computer disk that is a companion to this textbook.

Various flow regimes are defined by reference to Mach number:

Incompressible flow	has	$\mathbf{M} \ll 1$.
Subsonic flow	has	$\mathbf{M} < 1$.
Transonic flow	has	$\mathbf{M} \approx 1$.
Supersonic flow	has	$\mathbf{M} > 1$.
Hypersonic flow	has	$\mathbf{M} \gg 1$.

These are *local* definitions. A flow field may contain regions of subsonic flow, regions of transonic flow, and regions of supersonic flow.

EXAMPLE 10.2 Illustrates Mach Number, Stagnation Properties, and the Use of Mach Number Tables

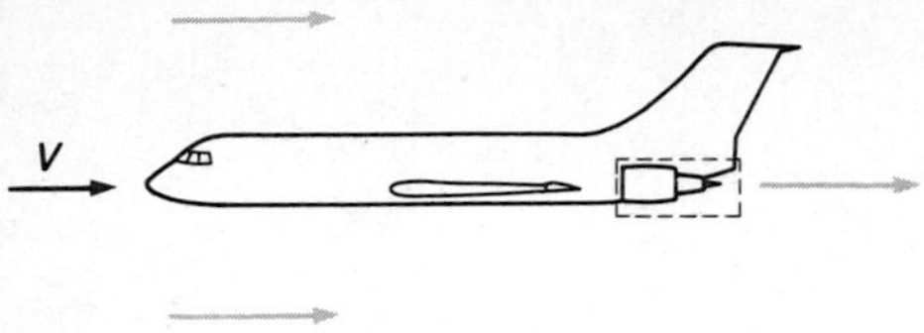

Figure E10.2 An airliner.

An airliner cruises at 250 m/s at an altitude of 10 km above sea level. Calculate the airliner's Mach number and the stagnation pressure and temperature relative to the airliner. Then compute the same quantities for an airliner speed of 500 m/s. Finally compute the same quantities for a speed of 250 m/s but change the altitude to 1 km above sea level.

SOLUTION

Given

Airliner cruising at

250 m/s at 10 km

500 m/s at 10 km

250 m/s at 1 km

Figure E10.2

Find

Airliner Mach number

Stagnation pressure and temperature relative to the airliner

Solution

We first find the atmospheric pressure and temperature at the two altitudes. We assume that the atmospheric conditions are those of Standard Atmosphere. The atmospheric properties can be calculated from Eqs. (2.7)–(2.9) or read from Fig. 2.6 or Table A.1:

$$p_{10\text{ km}} = 26.50\text{ kPa} \quad \text{and} \quad T_{10\text{ km}} = 223.3\text{K};$$

$$p_{1\text{ km}} = 89.88\text{ kPa} \quad \text{and} \quad T_{1\text{ km}} = 281.7\text{K}.$$

The Mach number is

$$\mathbf{M} = \frac{V}{c}.$$

From Eq. (10.26),

$$c = \sqrt{kRT} = 20.05\sqrt{T(\text{K})}\text{ m/s}.$$

At 10 km,

$$c = 20.05\sqrt{223.3}\text{ m/s} = 299.6\text{ m/s}.$$

Thus when $V = 250$ m/s,

$$\mathbf{M} = \frac{250\text{ m/s}}{299.6\text{ m/s}};$$

$$\mathbf{M} = 0.834 \text{ at 10 km and 250 m/s.}$$ **ANSWER**

The airliner is cruising at subsonic speed.

From Table E.1 (or the supplied software), the pressure and temperature ratios at this Mach number are

$$\frac{p}{p_0} = 0.633 \quad \text{and} \quad \frac{T}{T_0} = 0.878.$$

The static properties are those of the atmosphere. The stagnation properties relative to the airliner are

$$p_0 = \frac{p_{10\text{ km}}}{p/p_0} \quad \text{and} \quad T_0 = \frac{T_{10\text{ km}}}{T/T_0}.$$

Substituting, we have

$$p_0 = \frac{26.50 \text{ kPa}}{0.633};$$

$$p_0 = 41.86 \text{ kPa at 10 km and 250 m/s.}$$ **ANSWER**

Then

$$T_0 = \frac{223.3\text{K}}{0.878};$$

$$T_0 = 254.3\text{K at 10 km and 250 m/s.}$$ **ANSWER**

Next, consider 500 m/s at 10 km. The speed of sound and the static pressure and temperature are still the values appropriate to the 10-km altitude. The Mach number is

$$\mathbf{M} = \frac{V}{c} = \frac{500 \text{ m/s}}{299.6 \text{ m/s}};$$

$$\mathbf{M} = 1.669 \text{ at 10 km and 500 m/s.}$$ **ANSWER**

This is a supersonic cruise condition. At this Mach number, the pressure and temperature ratios are

$$\frac{p}{p_0} = 0.212 \quad \text{and} \quad \frac{T}{T_0} = 0.642.$$

Calculating p_0 and T_0, we have

$$p_0 = \frac{26.50 \text{ kPa}}{0.212},$$

or

$$p_0 = 125.0 \text{ kPa at 10 km and 500 m/s;}$$ **ANSWER**

and

$$T_0 = \frac{223.3\text{K}}{0.642\text{K}},$$

or

$$T_0 = 347.8\text{K at 10 km and 500 m/s.}$$ **ANSWER**

Finally, consider 250 m/s at 1 km. The speed of sound is now

$$c = \sqrt{kRT} = 20.05\sqrt{281.7\text{K}} \text{ m/s} = 336.5 \text{ m/s.}$$

The Mach number is

$$\mathbf{M} = \frac{V}{c} = \frac{250 \text{ m/s}}{336.5 \text{ m/s}};$$

$$\mathbf{M} = 0.743 \text{ at 1 km and 250 m/s.}$$ **ANSWER**

Again, this is a subsonic cruise condition. To calculate the stagnation pressure and temperature, we must use the atmospheric properties at 1 km. It is getting boring to use Table E.1, so we use Eqs. (10.30) and (10.31) instead. They give

$$p_0 = p\left(1 + \frac{k-1}{2}\mathrm{M}^2\right)^{k/(k-1)} = 89.88 \text{ kPa } [1 + 0.2(0.743)^2]^{3.5},$$

or

$$p_0 = 129.7 \text{ kPa at 1 km and 250 m/s;} \qquad \textbf{ANSWER}$$

and

$$T_0 = T\left(1 + \frac{k-1}{2}\mathrm{M}^2\right) = 281.7\mathrm{K}[1 + 0.2(0.743)^2],$$

or

$$T_0 = 312.8\mathrm{K} \text{ at 1 km and 250 m/s.} \qquad \textbf{ANSWER}$$

Discussion

Note that the Mach number changed with altitude, even for the same velocity. The reason is that the temperature and, hence, the speed of sound changed. The stagnation properties changed with speed and altitude. Either Table E.1 (or a table from some other source) or equations can be used to calculate static-stagnation-property ratios. A pocket calculator or computer is often easier to use than to look up and interpolate values in a table. Calculating the temperature ratio first is usually best, because this ratio is then raised to various powers to obtain pressure and density ratios.

10.2.2 The Radically Different Nature of Subsonic and Supersonic Flow

The response of a flowing fluid to a disturbance differs, depending on whether the Mach number is less or greater than unity. Imagine a large body of fluid that is influenced by a single point disturbance (Fig. 10.5). We assume that the point is fixed and that the fluid flows steadily past the point.† The point influences the flow by changing its direction or speed or pressure; however, because we are considering only a single point, the influence is infinitesimal. The disturbance is propagated into the fluid by waves that are generated at the point. For an infinitesimal disturbance, the waves are infinitesimal and travel at the speed of sound relative to the fluid.

Suppose that there is essentially no fluid motion relative to the disturbance. Disturbance waves propagate uniformly from the point. Figure 10.5(a) shows the positions of a few waves. These waves were all generated at times earlier than the current time, t. If we select a uniform time interval Δt, the wave generated Δt sec-

† This development can also be carried out by considering a moving point in a still fluid [1, 2]. Identical results are obtained; it is the relative motion between point and fluid that is important.

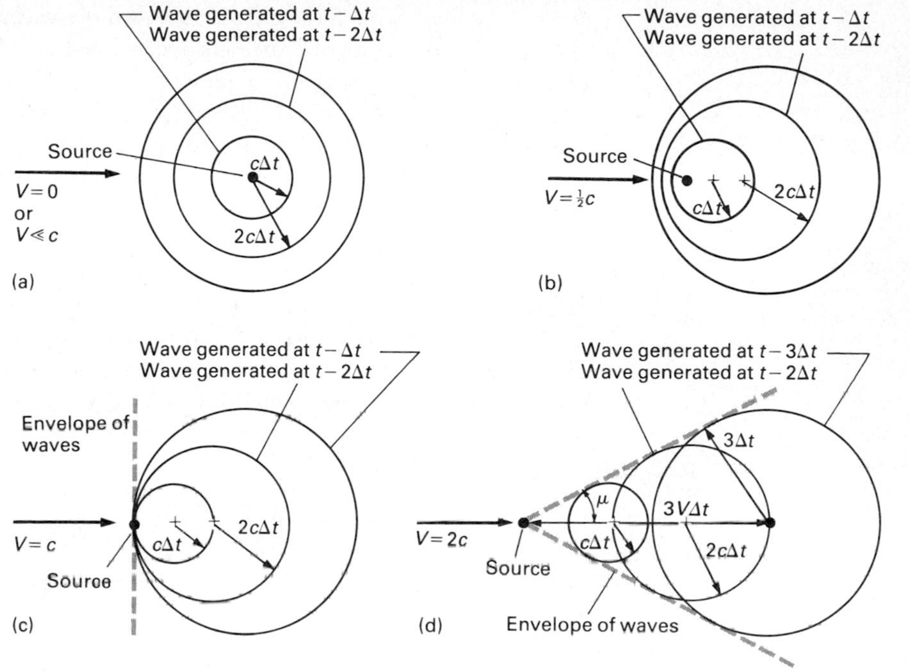

Figure 10.5 Wave pattern generated by flow over an infinitesimal (point) disturbance: (a) incompressible flow ($\mathbf{M} \approx 0$); (b) $\mathbf{M} = 0.5$; (c) $\mathbf{M} = 1.0$; (d) $\mathbf{M} = 2.0$.

onds earlier (at time $t - \Delta t$) has spread into a circle of radius $c\,\Delta t$, the wave generated at $t - 2\,\Delta t$ has spread to a circle of radius $2c\,\Delta t$, and so on. The wave pattern is symmetric about the point, and all points at a given distance from the disturbance are influenced equally.

Now suppose that the fluid is moving past the disturbance with a velocity very much smaller than the speed of sound ($\mathbf{M} \ll 1$). Figure 10.5(a) closely approximates the wave pattern because, in a time lapse (Δt) small enough to capture the wave motion, the fluid motion is barely perceptible.

Next, suppose that the fluid is flowing past the disturbance at a speed that is the order of, but less than, the speed of sound, say, $V = 0.5c$ (i.e., $\mathbf{M} = 0.5$). This situation is illustrated in Fig. 10.5(b). In addition to spreading into the fluid, the waves are swept downstream with the flow. As a result, the waves bunch up on the upstream side of the disturbance and spread out on the downstream side. The waves still propagate to all parts of the fluid; however, a fluid particle approaching the disturbance from upstream must get closer to the point to experience the same perturbation that it would experience farther away in a very small Mach number flow. If the waves actually represent sound, an upstream listener would hear a different frequency (pitch) than a downstream listener. This is the well-known *Doppler effect*.

Now suppose that the fluid flows past the disturbance with a speed that is exactly equal to the sonic speed ($\mathbf{M} = 1$). In this case, the disturbance waves are swept downstream exactly as rapidly as they spread into the fluid, and the wave pattern appears as shown in Fig. 10.5(c). In this case, the waves do not propagate upstream and the fluid is not affected by the disturbance until the instant it arrives! Note also that the waves form a straight envelope, perpendicular to the fluid velocity and passing through the disturbance. All the waves are part of this envelope, so the influence of the disturbance is concentrated and a fluid particle experiences it all at once upon crossing the envelope.

Finally, suppose that the fluid flows past the point with a speed greater than the sonic speed ($\mathbf{M} > 1$). Figure 10.5(d) illustrates the resulting wave pattern for the specific case of $\mathbf{M} = 2$. At supersonic speeds, the waves are swept downstream faster than they spread. All the waves are confined to a triangular (in two-dimensional flow in a plane) or conical (in three-dimensional or axisymmetric flow) region extending downstream from the point. The waves form an envelope of half-angle μ. Only the fluid inside the envelope is affected by the disturbance. The effect of the disturbance is concentrated at the envelope, and the fluid adjusts suddenly upon crossing the envelope. The envelope of infinitesimal waves is called a *Mach wave* in two-dimensional flow and a *Mach cone* in three-dimensional or axisymmetric flow. The half-angle of the envelope can be calculated from the geometry of Fig. 10.5(d):

$$\sin \mu = \frac{c\,\Delta t}{V\,\Delta t} = \frac{2c\,\Delta t}{2V\,\Delta t} = \cdots = \frac{c}{V} = \frac{1}{\mathbf{M}},$$

so the *Mach angle* (μ) is given by

$$\mu = \sin^{-1}\frac{1}{\mathbf{M}}. \tag{10.33}$$

We can summarize the results of this discussion in the following manner:

- In incompressible flow ($\mathbf{M} \ll 1$), the influence of any point is felt at all other points in the field. The influence is equally strong upstream and downstream. Fluid property and velocity variations must be continuous.
- In subsonic flow, the influence of any point is felt at all points. The effect is concentrated nearer the point in the upstream region; however, fluid property and velocity variations must be continuous.
- In sonic flow, the influence of any point is not felt upstream. The response of the flow to the point's presence occurs abruptly upon passing the point.
- In supersonic flow, the influence of any point is confined to a limited region downstream. There is no upstream influence. The higher the Mach number, the smaller the downstream region of influence is. The flow responds to the point's presence abruptly upon entering the region of the influence. Fluid property and velocity variations may be discontinuous.

The first two items tell us that compressible subsonic flow is qualitatively similar to incompressible flow, so we should expect subsonic flow in ducts or around bodies to behave similarly to the incompressible flows examined in Chapters 7–9. Transonic and supersonic flow is radically different from incompressible flow because of the removal of upstream influence. In almost all cases, supersonic flow responds differently to a given perturbation (such as area increase in a duct) than does incompressible or subsonic flow.

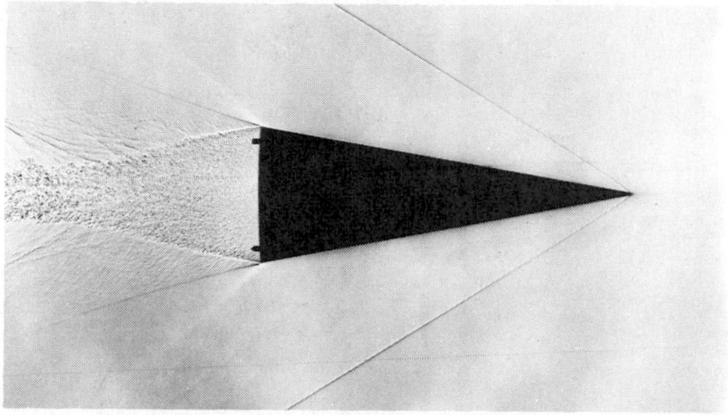

Figure 10.6 Oblique shock in supersonic flow over a right circular cone. (Courtesy of Launch and Flight Division, Ballistic Research Laboratory/ARRADCOM, Aberdeen Proving Ground, Md.)

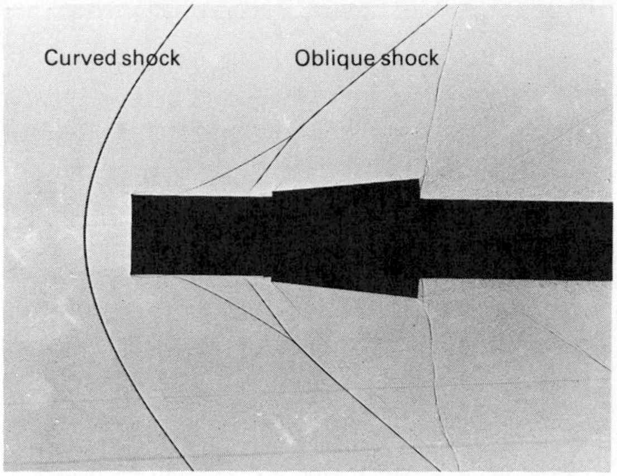

Figure 10.7 Curved shock standing ahead of a blunt body in supersonic flow. (Courtesy of Launch and Flight Division, Ballistic Research Laboratory/ARRADCOM, Aberdeen Proving Ground, Md.)

10.2.3 Shock Waves

Since we introduced the definition of the speed of sound, we have concentrated on infinitesimal disturbances and waves of infinitesimal strength. Let's now consider waves associated with finite disturbances. In the preceding section, we saw that introducing an infinitesimal disturbance into a supersonic flow generates a stationary (with respect to the disturbance) Mach wave. Placing a finite disturbance (say, a sharp wedge or cone) in a steady supersonic flow, generates a finite-strength wave (Fig. 10.6). This wave turns the fluid as it passes over the wedge or cone. Placing a blunt object in a supersonic flow generates a finite-strength wave some distance ahead of the object (Fig. 10.7).†

Finite-strength disturbance waves are called *shock waves*. Shock waves are extremely thin, and fluid properties and velocity change by large amounts (tens or even hundreds of percent) across a shock wave. Shock waves are actually rather common occurrences; examples of shock waves include explosion waves from dynamite (or maybe an extra-powerful Fourth of July firecracker), sonic "booms," which are actually the shock waves generated by supersonic aircraft, and water hammer, which can occur when a water faucet or valve is opened or closed rapidly.

Shock waves may be moving, like the wave that results from rapidly closing a valve in a pipe, or stationary, as in flow over a stationary wedge or blunt body.‡ In either case, a shock wave adjusts the flow to some *downstream* condition. A

---

† The waves shown in the photographs in Figs. 10.6 and 10.7 are made visible by *schlieren* or *shadograph* photography. These flow visualization methods rely on the effects of density changes on light passing through the flow field.

‡ In reality, it's all relative. A wedge or blunt body moving in a still fluid would generate an identical shock to that generated by fluid flowing over a stationary body.

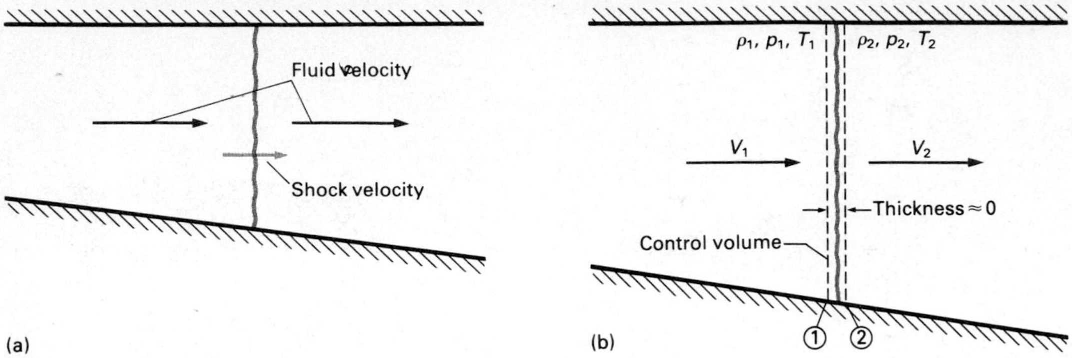

Figure 10.8 Normal shock in a duct: (a) flow schematic; (b) flow view from control volume fixed to shock.

normal shock is perpendicular to the upstream flow. An *oblique shock* is inclined at a constant angle to the upstream flow (Fig. 10.6). A *curved shock* has a variable angle between the wave and the upstream flow (Fig. 10.7). Flows that contain oblique or curved shocks are at least two-dimensional and two-directional, because the shock changes the direction of the flow. One-dimensional, one-directional flow can contain only normal shocks. For the sake of simplicity, our consideration is limited to normal shocks. For the same reason, we consider shocks only in an ideal gas.

Figure 10.8(a) shows a normal shock in a duct. The shock may be moving or stationary with respect to the duct. We can develop an analysis that applies to either case by using a control volume and coordinate system fixed to the shock. Velocities (and, therefore, stagnation properties) are defined relative to the shock. Figure 10.8(b) shows the flow into and out of the control volume. We assume that the velocity is uniform at the inlet and at the outlet. We assume that the shock has zero thickness, so

$$A_1 = A_2 = A,$$

and the volume of the control volume is zero.

The continuity equation is

$$\rho_1 A V_1 = \rho_2 A V_2. \tag{10.34}$$

From the ideal gas law,

$$\rho = \frac{p}{RT}. \tag{10.35}$$

We can write the velocity as

$$V = \mathbf{M}c = \mathbf{M}\sqrt{kRT}. \tag{10.36}$$

Substituting into Eq. (10.34) and canceling A, k, and R, we obtain

$$\frac{p_1 \mathbf{M}_1}{\sqrt{T_1}} = \frac{p_2 \mathbf{M}_2}{\sqrt{T_2}}, \tag{10.37a}$$

or

$$\frac{\mathbf{M}_2}{\mathbf{M}_1} = \frac{p_1}{p_2}\sqrt{\frac{T_2}{T_1}}. \tag{10.37b}$$

The linear momentum equation for the flow direction (x) is

$$\sum F_x = \rho_2 V_2^2 A - \rho_1 V_1^2 A.$$

Because the sides of the control volume in contact with the duct wall have zero thickness, there is no force on them. The only forces are from pressure on the inlet and outlet planes, so the equation becomes

$$p_1 A - p_2 A = \rho_2 V_2^2 A - \rho_1 V_1^2 A.$$

From Eq. (10.29),

$$\rho V^2 = kp\mathbf{M}^2,$$

and the momentum equation reduces to

$$p_1(1 + k\mathbf{M}_1^2) = p_2(1 + k\mathbf{M}_2^2), \tag{10.38a}$$

or

$$\frac{p_2}{p_1} = \frac{1 + k\mathbf{M}_1^2}{1 + k\mathbf{M}_2^2}. \tag{10.38b}$$

Next, we apply the general energy equation,

$$\dot{Q} - \dot{W}_s = \dot{m}\left(\tilde{h}_2 - \tilde{h}_1 + \frac{V_2^2}{2} - \frac{V_1^2}{2}\right).$$

Because the control volume has zero thickness, there can be neither heat transfer nor work, so the energy equation reduces to

$$\tilde{h}_2 + \frac{V_2^2}{2} = \tilde{h}_1 + \frac{V_1^2}{2},$$

or, alternatively, in terms of stagnation enthalpy,

$$\tilde{h}_{0_2} = \tilde{h}_{0_1}. \tag{10.39}$$

For an ideal gas, the energy equation gives

$$T_{0_2} = T_{0_1}. \tag{10.40}$$

Substituting Eq. (10.30) into Eq. (10.40), we have

$$T_2\left(1 + \frac{k-1}{2}\mathbf{M}_2^2\right) = T_1\left(1 + \frac{k-1}{2}\mathbf{M}_1^2\right), \tag{10.41a}$$

or

$$\frac{T_2}{T_1} = \frac{1 + [(k-1)/2]\mathbf{M}_1^2}{1 + [(k-1)/2]\mathbf{M}_2^2}. \tag{10.41b}$$

Equations (10.37b), (10.38b), and (10.41b) relate the ratio of temperature and pressure across a shock to the Mach numbers upstream and downstream from the shock. If we select a single parameter, say, upstream Mach number $\mathbf{M}_1$, we can express the other three parameters in terms of the selected parameter. To do this, we square Eq. (10.37b):

$$\frac{\mathbf{M}_2^2}{\mathbf{M}_1^2} = \left(\frac{p_1}{p_2}\right)^2\left(\frac{T_2}{T_1}\right).$$

Then we substitute Eqs. (10.38b) and (10.41b) into the right side. The result is a quadratic equation relating $\mathbf{M}_2^2$ to $\mathbf{M}_1^2$; k appears in the equation as a parameter. Because the equation is quadratic, there are two solutions for $\mathbf{M}_2^2$:

$$\mathbf{M}_2^2 = \mathbf{M}_1^2 \tag{10.42a}$$

and

$$\mathbf{M}_2^2 = \frac{\mathbf{M}_1^2 + 2/(k-1)}{[2k/(k-1)]\mathbf{M}_1^2 - 1}. \tag{10.42b}$$

The first solution seems trivial, but it shows that the Mach number need not change discontinuously at a plane. This solution is valid if there is no shock in the control volume. If there is a shock in the control volume, the second solution, Eq. (10.42b), gives the downstream Mach number in terms of the upstream Mach number. For any realistic value of k (that is, $k > 1$), calculations with Eq. (10.42b) show that $\mathbf{M}_1$ and $\mathbf{M}_2$ always lie on opposite sides of $\mathbf{M} = 1.0$; that is, if $\mathbf{M}_1 > 1$, then $\mathbf{M}_2 < 1$ and vice versa.

We can now substitute Eqs. (10.42a) and (10.42b) into the pressure ratio and temperature ratio equations, Eqs. (10.38b) and (10.41b). The constant Mach number solution gives pressure and temperature ratios equal to unity; that is, there can be no discontinuous change of pressure and temperature at a plane if there is no shock at the plane. If there is a shock, the pressure and temperature ratios are

$$\frac{p_2}{p_1} = \frac{2k}{k+1}\mathbf{M}_1^2 - \frac{k-1}{k+1} \tag{10.43}$$

and

$$\frac{T_2}{T_1} = \left(1 + \frac{k-1}{2}\mathbf{M}_1^2\right)\left(\frac{2k}{k-1}\mathbf{M}_1^2 - 1\right)\left[\frac{2(k-1)}{(k+1)^2\mathbf{M}_1^2}\right]. \tag{10.44}$$

A shock may change stagnation properties as well as static properties. According to Eq. (10.40), the stagnation temperature ratio across a shock is[†]

$$\frac{T_{0_2}}{T_{0_1}} = 1. \tag{10.45}$$

[†] Recall in both Eq. (10.45) and Eq. (10.46) that these are stagnation properties relative to the shock.

Table 10.1 Relative changes of properties as predicted by the normal shock equations.

	$M_1 > 1$ (Supersonic Upstream)	$M_1 < 1$ (Subsonic Upstream)
M_2	<1 (subsonic)	>1 (supersonic)
$\frac{p_2}{p_1}$	>1	<1
$\frac{T_2}{T_1}$	>1	<1
$\frac{T_{0_2}}{T_{0_1}}$	$=1$	$=1$
$\frac{p_{0_2}}{p_{0_1}}$	<1	>1
$\tilde{s}_2 - \tilde{s}_1$	>0	<0 (Not physically possible)

We obtain the stagnation pressure ratio across a shock by first expressing stagnation pressure in terms of static pressure and Mach number. Using Eq. (10.31), we have

$$\frac{p_{0_2}}{p_{0_1}} = \frac{p_2\{1 + [(k-1)/2]M_2^2\}^{k/(k-1)}}{p_1\{1 + [(k-1)/2]M_1^2\}^{k/(k-1)}}.$$

We reduce this expression to an equation involving only M_1^2 and k by substituting Eq. (10.43) for the static pressure ratio and Eq. (10.42b) for M_2^2. After some heavy algebra, we find that†

$$\frac{p_{0_2}}{p_{0_1}} = \left(\frac{k+1}{2} M_1^2\right)^{\frac{k}{(k-1)}} \left(1 + \frac{k-1}{2} M_1^2\right)^{\frac{-k}{(k-1)}} \left(\frac{2k}{k+1} M_1^2 - \frac{k-1}{k+1}\right)^{\frac{-1}{(k-1)}}. \tag{10.46}$$

Equations (10.42b)–(10.46) enable us to calculate properties on the downstream side of a shock if we know the properties on the upstream side and the upstream Mach number M_1. As the equations are algebraically complicated, determining the effects of a shock on the fluid by examining them is difficult. Table 10.1 indicates typical results of calculations made with the equations. Note the second column, which tabulates results for a (hypothetical) shock that occurs in a subsonic flow. Our discussion of the different nature of subsonic and supersonic flow indicates that discontinuous changes of fluid properties (such as shock waves) do not occur in subsonic flow. Is there any way to decide, mathematically, whether the results in the right-hand column are realistic? Yes, you can use the second law of thermodynamics, in the mathematical form of Eq. (10.5):

$$\tilde{s}_2 - \tilde{s}_1 \geq \int_1^2 \frac{dq}{T}.$$

† Recall in both Eq. (10.45) and Eq. (10.46) that these are stagnation properties relative to the shock.

With no heat transfer to or from the gas as it passes through a shock wave,

$$\tilde{s}_2 - \tilde{s}_1 \geq 0.$$

Entropy change across a shock wave can be calculated by using Eq. (10.8). Working with a dimensionless entropy change is convenient. Dividing by R and using Eq. (10.10) give

$$\frac{\tilde{s}_2 - \tilde{s}_1}{R} = \frac{k}{k-1} \ln\left(\frac{T_2}{T_1}\right) - \ln\left(\frac{p_2}{p_1}\right).$$

Substituting Eqs. (10.43) and (10.44) for the pressure and temperature ratios yields an expression for entropy change in terms of Mach number. By selecting various values of $\mathbf{M}_1$, we can determine $(\tilde{s}_2 - \tilde{s}_1)$. The entropy change is negative for subsonic inlet Mach numbers; therefore no such "shock wave" can exist.[†] Shock waves can exist only if the upstream flow (relative to the shock) is supersonic. Real shock waves always increase the static pressure, density, and temperature and decrease the stagnation pressure (relative to the shock).

If you are analyzing a supersonic flow and need to calculate property changes across a shock wave, you can use Eqs. (10.42b)–(10.46); however, even with a calculator, these equations are quite tedious. This is especially true if you know, say, T_2/T_1 and are trying to calculate Mach number. Tables are helpful in such cases. Table E.2 presents numerical values of the normal shock functions for $k = 1.4$. Note that the table is prepared with $\mathbf{M}_1$ as the independent parameter. A computer program that evaluates Eqs. (10.42b)–(10.46) is a useful alternative to a set of tables. Such a program is provided on the personal computer disk that is a companion to this text. If you want either density or velocity change across a shock, you may easily calculate it from the ideal gas law and the continuity equation.

† It might be a good idea to cross out the right-hand column of Table 10.1.

EXAMPLE 10.3 Illustrates Use of Normal Shock Tables

A stream of nitrogen flowing in a duct passes through a normal shock. A Pitot tube and thermometer are placed downstream from the shock, and a Bourdon tube pressure gage is attached to a wall tap upstream from the shock, as shown in Fig. E10.3. The pressure and temperature readings are $p_A = 2$ lb/in^2, $p_B = 40$ lb/in^2, and $T = 100°$F. Calculate the Mach number and velocity of the nitrogen upstream of the shock.

SOLUTION

Given

Nitrogen flowing through a shock

Measurements made as indicated in Fig. E10.3

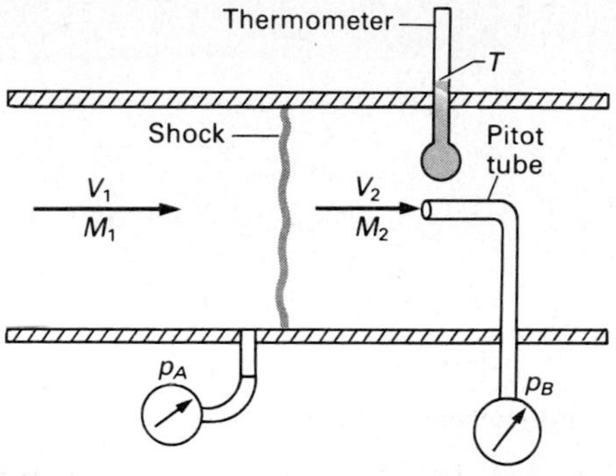

Figure E10.3 Shock in a duct with instruments.

Find

Mach number and velocity upstream of shock

Solution

The first step is to convert all given data to absolute units. We assume that the atmospheric pressure in the vicinity of the pressure gages is sea-level standard, 14.7 psia. Then

$$p_A = 2\ \text{lb/in}^2 + 14.7\ \text{lb/in}^2 = 16.7\ \text{psia};$$

$$p_B = 40\ \text{lb/in}^2 + 14.7\ \text{lb/in}^2 = 54.7\ \text{psia};$$

$$T = 100°\text{F} + 459.7°\text{F} = 559.7°\text{R}.$$

For nitrogen, Table A.4 gives

$$R = 55.2\ \text{ft·lb/lbm·°R} \quad \text{and} \quad k = 1.4.$$

Next, we interpret the data. Because the flow is parallel to the duct and the streamlines are straight, p_A is the *static* pressure upstream from the shock; p_B is the *stagnation* pressure downstream from the shock. The gas comes to rest at the thermometer bulb, so T approximates the stagnation temperature. From the given data, for the shock

$$\frac{p_1}{p_{0_2}} = \frac{16.7\ \text{psia}}{54.7\ \text{psia}} = 0.3053.$$

We want to obtain the corresponding Mach number $\mathbf{M}_1$, but a check of Eqs. (10.42)–(10.46) and Table E.2 reveals that we do not have a function or table of p_1/p_{0_2}. Note that we *cannot* use the isentropic table (Table E.1) or Eq. (10.31), because the static and stagnation pressures lie on different sides of the shock. We can, however, write

$$\frac{p_1}{p_{0_2}} = \left(\frac{p_1}{p_{0_1}}\right)\left(\frac{p_{0_1}}{p_{0_2}}\right) = \frac{p_1/p_{0_1}}{p_{0_2}/p_{0_1}}.$$

We do have functions and tables for p_1/p_{0_1} and p_{0_2}/p_{0_1}. To find $\mathbf{M}_1$, we must iterate. A simple procedure is as follows:

- Guess a trial value of $\mathbf{M}_1$.
- Obtain a value of p_{0_2}/p_{0_1} from Table E.2.
- Calculate p_1/p_{0_1} from

$$\frac{p_1}{p_{0_1}} = \left(\frac{p_1}{p_{0_2}}\right)\left(\frac{p_{0_2}}{p_{0_1}}\right) = 0.3053\left(\frac{p_{0_2}}{p_{0_1}}\right).$$

- Obtain a new value for $\mathbf{M}_1$ from Table E.1.
- Continue the process until $\mathbf{M}_1$ doesn't change.

We first guess $\mathbf{M}_1 = 1.5$, giving

$$\frac{p_{0_2}}{p_{0_1}} = 0.9298.$$

The trial value of p_1/p_{0_1} is

$$\frac{p_1}{p_{0_1}} = (0.9298)(0.3053) = 0.2837.$$

This implies a value

$$\mathbf{M}_1 = 1.473.$$

This Mach number gives

$$\frac{p_{0_2}}{p_{0_1}} = 0.9381 \quad \text{and} \quad \frac{p_1}{p_{0_1}} = (0.3053)(0.9381) = 0.2864.$$

From Table E.1, the corresponding Mach number is

$$\mathbf{M}_1 = 1.466.$$

Repeating the process until convergence, we have

$$\mathbf{M}_1 = 1.463. \qquad \textbf{ANSWER}$$

The nitrogen velocity is calculated by

$$V_1 = \mathbf{M}_1 c_1 = \mathbf{M}_1 \sqrt{kRT_1}.$$

T_1 is computed by

$$T_1 = T_{0_1}\left(\frac{T_1}{T_{0_1}}\right),$$

where T_1/T_{0_1} is found for $\mathbf{M}_1 = 1.463$ and is

$$\frac{T_1}{T_{0_1}} = 0.7002.$$

The stagnation temperature is 559.7°R, because a stationary shock does not change T_0. Thus

$$T_1 = 559.7°\text{R}(0.7002) = 391.9°\text{R}.$$

The nitrogen velocity is

$$V_1 = 1.463\sqrt{(1.4)(55.2\ \text{ft}\cdot\text{lb/lbm}\cdot°\text{R})(391.9°\text{R})(32.2\ \text{ft}\cdot\text{lbm/lb}\cdot\text{sec}^2)};$$

$$V_1 = 1440\ \text{ft/sec}. \qquad \textbf{ANSWER}$$

Discussion

Note particularly the comments made during the solution. Do you understand why the isentropic table cannot be used with p_1/p_{0_2}? We do use the isentropic table to relate $\mathbf{M}_1$ and p_1/p_{0_1}, even though the flow across the shock itself is not isentropic. You know, of course, that we could not use Bernoulli's equation to calculate the nitrogen velocity, but you might try it to see how close you come to the correct answer.

Substituting the appropriate Mach number functions allows us to find a new function

$$\frac{p_1}{p_{0_2}} = \left(\frac{k+1}{2}\mathbf{M}_1^2\right)^{-k/(k-1)}\left(\frac{2k\mathbf{M}_1^2}{k+1} - \frac{k-1}{k+1}\right)^{-1/(k-1)}.$$

This function is called the supersonic Pitot-tube formula, because a Pitot tube will always generate a shock upstream from itself in supersonic flow. If you had a lot of supersonic Pitot-tube data to reduce, you might make a table of values of this function.

The temperature of the thermometer bulb is actually slightly less than the stagnation temperature of the gas. See the discussion following Example 10.10 and Chapter 7 of reference [1].

10.3 ISENTROPIC FLOW AND BERNOULLI'S EQUATION

Although real flows are often influenced by viscosity and heat transfer, we sometimes ignore these effects. The resulting flow model, in addition to its (relative) simplicity, is reasonably accurate for external flow outside of viscous boundary layers and for flow in ducts if the boundary layers on the duct walls are thin and rates of heat transfer to or from the fluid are small. The usefulness of an inviscid flow model is immediately apparent when we recall that Bernoulli's equation, Eq. (4.51) or Eq. (4.52), describes steady flow of an *incompressible* inviscid fluid. In this section, we consider the compressible flow equivalent(s) of Bernoulli's equation and demonstrate the inapplicability of Bernoulli's equation for high Mach number flow.

10.3.1 Steady Isentropic Flow of an Ideal Gas

For flow of a compressible fluid in a small streamtube, we make the following assumptions:

- The flow is steady.
- There are no body forces (gravity, electromagnetic, or others)
- There are no shear stresses.
- There is no heat transfer.
- The streamtube does not pass through a shock wave.†

The last three assumptions taken together imply that the flow is adiabatic and reversible; therefore, the entropy of the fluid is constant:

$$\tilde{s} = \text{Constant};$$

or, in terms of two locations along the streamtube,

$$\tilde{s}_1 = \tilde{s}_2.$$

Flow with constant entropy is called *isentropic flow*.

For steady flow, the energy equation is

$$q - w_{\text{shaft}} - w_{\text{shear}} = \left(\tilde{h}_2 + \frac{V_2^2}{2}\right) - \left(\tilde{h}_1 + \frac{V_1^2}{2}\right).$$

† Assuming that there are no shocks if the flow is subsonic everywhere along the streamtube is not necessary, because shocks are possible only if the inlet flow is supersonic.

According to our assumptions, q and w_{shear} are zero. But w_{shaft} is not necessarily zero, and the energy equation becomes

$$-w_{\text{shaft}} = \left(\tilde{h}_2 + \frac{V_2^2}{2}\right) - \left(\tilde{h}_1 + \frac{V_1^2}{2}\right).$$

When we use the definition of stagnation enthalpy,

$$-w_{\text{shaft}} = \tilde{h}_{0_2} - \tilde{h}_{0_1}.$$

If the fluid is an ideal gas,

$$-w_{\text{shaft}} = c_p(T_{0_2} - T_{0_1}). \tag{10.47}$$

Doing work on the fluid increases its *stagnation temperature,* even if there is no heat transfer.

Now we consider changes of stagnation pressure in isentropic flow. By definition of the stagnation state,

$$\tilde{s}_{0_1} = \tilde{s}_1 \quad \text{and} \quad \tilde{s}_{0_2} = \tilde{s}_2.$$

For isentropic flow,

$$\tilde{s}_{0_2} - \tilde{s}_{0_1} = \tilde{s}_2 - \tilde{s}_1 = 0;$$

thus

$$\frac{p_{0_2}}{p_{0_1}} = \left(\frac{T_{0_2}}{T_{0_1}}\right)^{c_p/R} = \left(\frac{T_{0_2}}{T_{0_1}}\right)^{k/(k-1)}. \tag{10.48}$$

Using Eq. (10.47), we can also write

$$\frac{p_{0_2}}{p_{0_1}} = \left(1 - \frac{w_{\text{shaft}}}{c_p T_{0_1}}\right)^{k/(k-1)}. \tag{10.49}$$

A situation of considerable interest in compressible flow is the case with no shaft work. The following statement applies to this case:

In isentropic flow with no shaft work, all stagnation properties are constant.

Note that stagnation properties are constant in isentropic flow only if there is no shaft work. This special case ($w_{\text{shaft}} = 0$) is so important in compressible flow that some engineers and authors use the phrase "isentropic flow" to mean "isentropic flow with no work." We use the phrase "isoenergetic-isentropic flow" to refer to isentropic flow with no work.†

Equations relating property and velocity changes in isoenergetic-isentropic flow of an ideal gas can easily be developed from the stagnation property definitions and the constancy of stagnation properties. Take, for example, the calculation of the change in temperature between two points in an isoenergetic-isentropic

† Isoenergetic means that there is no energy transfer (neither heat nor work) to the fluid from outside.

flow. Because T_0 is constant, we can write

$$T_{0_1} = T_1 + \frac{V_1^2}{2c_p} = T_2 + \frac{V_2^2}{2c_p} = T_{0_2}. \tag{10.50a}$$

Using Mach number instead of velocity is sometimes convenient. From Eq. (10.30),

$$\frac{T_2}{T_1} = \frac{T_{0_2}\{1 + [(k-1)/2]\mathbf{M}_1^2\}}{T_{0_1}\{1 + [(k-1)/2]\mathbf{M}_2^2\}},$$

but, as T_0 is constant,

$$\frac{T_2}{T_1} = \frac{1 + [(k-1)/2]\mathbf{M}_1^2}{1 + [(k-1)/2]\mathbf{M}_2^2}. \tag{10.50b}$$

A third approach involves use of the tables relating stagnation property ratios and Mach number. We calculate the temperature ratio by

$$\frac{T_2}{T_1} = \frac{T}{T_0}\{\mathbf{M}_2\} \Big/ \frac{T}{T_0}\{\mathbf{M}_1\}, \tag{10.50c}$$

where

$$\frac{T}{T_0}\{\mathbf{M}\}$$

is a numerical value read from Table E.1 at Mach number $\mathbf{M}$. In a fourth method, we make the calculation in two steps:

$$T_{0_2} = T_{0_1} = \frac{T_1}{\dfrac{T}{T_0}\{\mathbf{M}_1\}}$$

and

$$T_2 = T_{0_2}\left[\frac{T}{T_0}\{\mathbf{M}_2\}\right]. \tag{10.50d}$$

Any of Eqs. (10.50) could be used to relate temperature and velocities (or Mach numbers) at points 1 and 2. Which equation is most convenient depends on the exact information available. Do not fall into the trap of always using Mach number tables; direct calculations are sometimes easier. Similarly, pressure and velocity (or Mach number) in isoenergetic-isentropic flow can be related by any of the following equations:

$$p_1\left(1 + \frac{V_1^2}{2c_pT_1}\right)^{k/(k-1)} = p_2\left(1 + \frac{V_2^2}{2c_pT_2}\right)^{k/(k-1)}; \tag{10.51a}$$

$$\frac{p_2}{p_1} = \frac{\{1 + [(k-1)/2]\mathbf{M}_1^2\}^{k/(k-1)}}{\{1 + [(k-1)/2]\mathbf{M}_2^2\}^{k/(k-1)}}; \tag{10.51b}$$

$$\frac{p_2}{p_1} = \frac{p}{p_0}\{\mathbf{M}_2\} \Big/ \frac{p}{p_0}\{\mathbf{M}_1\}; \tag{10.51c}$$

$$p_2 = p_{0_2}\left[\frac{p}{p_0}\{\mathbf{M}_2\}\right]; \tag{10.51d}$$

where

$$p_{0_2} = p_{0_1} = \frac{p_1}{\dfrac{p}{p_0}\{\mathbf{M}_1\}}.$$

Don't forget you can also use Eqs. (10.11)–(10.13), for the flow is isentropic.

EXAMPLE 10.4 Illustrates "Routine" Use of Isentropic Flow Functions and Tables for Computation of Isoenergetic-Isentropic Flow

Air flows in a duct. At one point in the flow, the pressure, temperature, and velocity are 50 psia, 100°F, and 500 ft/sec. At a downstream point, the Mach number is 1.80. The flow is steady, isoenergetic, and isentropic. Calculate the air pressure, temperature, and velocity at the downstream point. Do the calculation twice, once using isentropic flow equations and once using Table E.1.

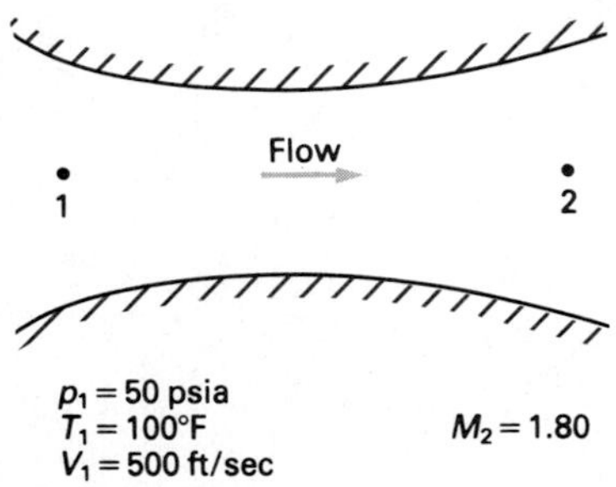

Figure E10.4 Flow between two points in a duct.

SOLUTION

Given

Air flow in a duct

At point 1, $p_1 = 50$ psia, $T_1 = 100°\text{F}$, and $V_1 = 500$ ft/sec

At a downstream point (point 2), $\mathbf{M}_2 = 1.80$

Figure E10.4

Find

T_2, p_2, and V_2 using both isentropic flow equations and Table E.1

Solution

The first step is to convert all given data into absolute units. The only conversion necessary is temperature:

$$T_1 = 100°\text{F} + 459.7°\text{F} = 559.7°\text{R}.$$

We first do the calculations with isentropic flow equations. The energy equation for adiabatic flow with no work between 1 and 2 is

$$c_p T_1 + \frac{V_1^2}{2} = c_p T_2 + \frac{V_2^2}{2}.$$

The velocity at 2 is

$$V_2 = \mathbf{M}_2 c_2 = \mathbf{M}_2\sqrt{kRT_2}, \quad \text{so} \quad c_p T_1 + \frac{V_1^2}{2} = \left(c_p + \frac{kR\mathbf{M}_2^2}{2}\right)T_2,$$

or

$$T_2 = \frac{c_p T_1 + V_1^2/2}{c_p + kR\mathbf{M}_2^2/2}.$$

Substituting numbers, we have

$$T_2 = \frac{(186.7\ \text{ft}\cdot\text{lb/lbm}\cdot°\text{R})(559.7°\text{R}) + \dfrac{(500)^2\ \text{ft}^2/\text{sec}^2}{2(32.2\ \text{ft}\cdot\text{lbm/lb}\cdot\text{sec}^2)}}{186.7\ \text{ft}\cdot\text{lb/lbm}\cdot°\text{R} + \dfrac{1.4(53.35\ \text{ft}\cdot\text{lb/lbm}\cdot\text{R})(1.80)^2}{2}};$$

$$T_2 = 352.2°\text{R}. \qquad \textbf{ANSWER}$$

The velocity at 2 is

$$V_2 = \mathbf{M}_2 c_2 = [(1.80)(49.02)\sqrt{352.2}]\ \text{ft/sec};$$

$$V_2 = 1660\ \text{ft/sec}. \qquad \textbf{ANSWER}$$

We compute the pressure at 2 from the isentropic process equation, Eq. (10.11):

$$p_2 = p_1\left(\frac{T_2}{T_1}\right)^{k/(k-1)} = 50\ \text{psia}\left(\frac{352.2}{559.7}\right)^{3.5};$$

$$p_2 = 9.88\ \text{psia}. \qquad \textbf{ANSWER}$$

Now we repeat the calculation using Table E.1. The Mach number at 1 is

$$\mathbf{M}_1 = \frac{V_1}{c_1} = \frac{V_1}{\sqrt{kRT_1}} = \frac{500\ \text{ft/sec}}{(49.02\sqrt{559.7})\ \text{ft/sec}} = 0.43.$$

Next, we use Table E.1 to obtain the pressure and temperature ratio functions at 1 and at 2. At $\mathbf{M} = \mathbf{M}_1 = 0.43$,

$$\frac{p}{p_0} = 0.8806 \quad \text{and} \quad \frac{T}{T_0} = 0.9643.$$

At $\mathbf{M} = \mathbf{M}_2 = 1.80$,

$$\frac{p}{p_0} = 0.1740 \quad \text{and} \quad \frac{T}{T_0} = 0.6068.$$

Because the flow is isoenergetic and isentropic,

$$p_2 = p_1 \frac{p}{p_0}\{\mathbf{M}_2\} \Big/ \frac{p}{p_0}\{\mathbf{M}_1\} = 50\ \text{psia}\left(\frac{0.1740}{0.8806}\right);$$

$$p_2 = 9.88\ \text{psia}. \qquad \textbf{ANSWER}$$

Now

$$T_2 = T_1 \frac{T}{T_0}\{\mathbf{M}_2\} \Big/ \frac{T}{T_0}\{\mathbf{M}_1\} = 559.7°\text{R}\left(\frac{0.6068}{0.9643}\right);$$

$$T_2 = 352.2°\text{R}. \qquad \textbf{ANSWER}$$

The velocity at 2 is

$$V_2 = \mathbf{M}_2\sqrt{kRT_2} = (1.80)(49.02)\sqrt{352.2}\ \text{ft/sec};$$

$$V_2 = 1660\ \text{ft/sec}. \qquad \textbf{ANSWER}$$

Discussion

Of course the answers determined by using Table E.1 are the same as those determined by using equations. We leave the decision to you as to which method is easier.

Always convert pressures and temperatures to absolute units as the first step in solving a compressible flow problem.

10.3.2 Relation to Bernoulli's Equation and a Criterion for Incompressible Flow

One of the most widely used (and misused) equations in fluid mechanics is Bernoulli's equation:

$$\frac{p}{\rho} + \frac{V^2}{2} + gz = \text{Constant.}$$

The assumptions used in deriving Bernoulli's equation are similar to those we just used in deriving the isoenergetic-isentropic flow equations, except that whereas Bernoulli's equation assumes incompressible flow, the isoenergetic-isentropic equations assume adiabatic flow of an ideal gas with no shocks. In this section, we show that, in the limit of very small Mach number, the isoenergetic-isentropic equation for pressure becomes identical to Bernoulli's equation. In addition, we develop a criterion for deciding whether a gas flow can be treated as incompressible.

Consider a steady flow with no shear stress, shaft work, or heat transfer. Under these conditions, the stagnation pressure is constant. We also assume that elevation changes are negligible. If the fluid is incompressible, we can calculate the pressure at any location from the "pressure form" of Bernoulli's equation:

$$p = p_0 - \tfrac{1}{2}\rho V^2. \qquad \text{Incompressible flow} \qquad (10.52)$$

If the fluid is compressible and an ideal gas, the static pressure and stagnation pressure are related by

$$p_0 = p\left(1 + \frac{k-1}{2}\mathbf{M}^2\right)^{k/(k-1)}. \qquad \text{Compressible flow} \qquad (10.53)$$

If we restrict consideration to Mach numbers less than 1, we can expand the Mach number term into an infinite series using the binomial theorem:

$$(1+x)^n = 1 + nx + \frac{n(n-1)}{2!}x^2 + \frac{n(n-1)(n-2)}{3!}x^3 + \cdots. \qquad (10.54)$$

Letting

$$x = \frac{k-1}{2}\mathbf{M}^2 \quad \text{and} \quad n = \frac{k}{k-1},$$

we get

$$\left(1 + \frac{k-1}{2}\mathbf{M}^2\right)^{k/(k-1)} = 1 + \frac{k}{2}\mathbf{M}^2 + \frac{k}{8}\mathbf{M}^4 + \text{H.O.T.},$$

where H.O.T. involves higher *even* powers of $\mathbf{M}$. Substituting into Eq. (10.53) gives

$$p_0 = p\left(1 + \frac{k\mathbf{M}^2}{2} + \frac{k\mathbf{M}^4}{8} + \text{H.O.T.}\right).$$

We are interested in small values of $\mathbf{M}$. Thus the higher-order terms are negligible, and we can write the equation

$$p_0 \approx p + \frac{kp\mathbf{M}^2}{2}\left(1 + \frac{\mathbf{M}^2}{4}\right). \tag{10.55}$$

From Eq. (10.29), we have

$$kp\mathbf{M}^2 = \rho V^2.$$

Substituting and rearranging, we get

$$p \approx p_0 - \frac{1}{2}\rho V^2\left(1 + \frac{\mathbf{M}^2}{4}\right). \tag{10.56}$$

If the Mach number is small, then $\mathbf{M}^2/4$ is small compared to 1, and we can write

$$p \approx p_0 - \tfrac{1}{2}\rho V^2; \qquad \text{Compressible flow, small } \mathbf{M} \tag{10.57}$$

therefore Bernoulli's equation is an approximation to the isoenergetic-isentropic flow pressure relation for small Mach number. The accuracy of this approximation depends on the "smallness" of the Mach number. Equation (10.56) shows that the error is proportional to $\mathbf{M}^2/4$ at low Mach numbers. If we want to limit the error in using Bernoulli's equation for calculation of pressure to no more than, say, 2 percent, then

$$\mathbf{M} \leq \sqrt{4(0.02)} = 0.283.$$

There is nothing unique about 2 percent error. For rough estimates, 5 percent error may be acceptable, in which case the Mach number must be less than about 0.45. The most widely used criterion for the boundary between compressible and incompressible flow puts the threshold Mach number at 0.3:

Generally, a flow with $\mathbf{M} < 0.3$ everywhere can be assumed to be incompressible.

10.3.3 The Critical State

We have shown that the nature of a compressible flow changes drastically when the fluid velocity exceeds the speed of sound. The flow state that corresponds to exactly sonic flow is called the *critical state*. The fluid properties at the critical state are called the *critical properties*. Consider acceleration of a gas from zero velocity to the local sonic speed in an isoenergetic-isentropic process. As the fluid's velocity increases, its temperature and speed of sound decrease.† The stagnation temperature is constant for this process, so we can write

$$T_0 = T\left(1 + \frac{k-1}{2}\mathbf{M}^2\right).$$

At the critical state, the Mach number is unity. Using the superscript * to denote conditions at the critical state, we have

$$T_0 = T^*\left(1 + \frac{k-1}{2}(1)^2\right),$$

† Note that $c/c_0 = (T/T_0)^{1/2}$ for an ideal gas.

or

$$\frac{T^*}{T_0} = \frac{2}{k+1}. \tag{10.58}$$

For the specified isoenergetic-isentropic process, the pressure and density at the critical state are given by

$$\frac{p^*}{p_0} = \left(\frac{T^*}{T_0}\right)^{k/(k-1)} = \left(\frac{2}{k+1}\right)^{k/(k-1)} \tag{10.59}$$

and

$$\frac{\rho^*}{\rho_0} = \left(\frac{T^*}{T_0}\right)^{1/(k-1)} = \left(\frac{2}{k+1}\right)^{1/(k-1)}. \tag{10.60}$$

For an isoenergetic-isentropic flow, properties at the critical state are uniquely determined by the stagnation properties and k. The critical properties T^*, p^*, and ρ^* are constant if the stagnation properties are constant.

10.4 INTERNAL FLOW: ONE-DIMENSIONAL FLOW IN A VARIABLE AREA PASSAGE

When a fluid flows through a passage, the fluid motion is affected by changes in the passage cross-sectional area, friction at the walls, and any heat transfer or work. Simultaneous consideration of all these effects is difficult. In this section, we consider internal flow of a compressible fluid, and we assume that there is no friction, no work, and no heat transfer; we consider only the effects of area change.† We also assume that the flow is one-dimensional and one-directional.

The primary effect of area change is fluid acceleration or deceleration. A device for accomplishing fluid acceleration is called a *nozzle,* and a device for accomplishing fluid deceleration is called a *diffuser*. The assumptions of one-dimensional flow and negligible friction are much more realistic for nozzles than for diffusers, because the adverse-pressure gradient in diffusers causes rapid boundary layer growth and separation (see Section 9.2.7). We concentrate on nozzle flow and assume steady or quasi-steady flow.

10.4.1 Isoenergetic-Isentropic Flow with Area Change: Preliminary Considerations

If there is no heat transfer or work, flow in a variable-area passage is isoenergetic. If there are no shocks or friction, the flow is also isentropic. Without limiting consideration to a specific type of fluid (such as an ideal gas), the fluid properties

† The film *Channel Flow of a Compressible Fluid* [10] provides an excellent introduction to this topic and is highly recommended.

and velocity must satisfy the following equations:

- Continuity equation:

$$\rho V A = \dot{m} = \text{Constant.}$$

- Energy equation:

$$\tilde{h} + \frac{V^2}{2} = \tilde{h}_0 = \text{Constant.}$$

- Property relation [Gibbs equation, Eq. (10.6), for isentropic flow]:

$$d\tilde{h} - \frac{dp}{\rho} = T\, d\tilde{s} = 0.$$

Because the Gibbs equation is a differential equation, working with differential forms of the other equations is convenient. Taking the logarithm of the continuity equation and differentiating, we obtain

$$\frac{d\rho}{\rho} + \frac{dV}{V} + \frac{dA}{A} = 0. \tag{10.61}$$

Differentiating the energy equation gives

$$d\tilde{h} + V\, dV = 0. \tag{10.62}$$

With three equations, we can eliminate two variables. From the Gibbs equation,

$$d\tilde{h} = \frac{dp}{\rho}.$$

Substituting into Eq. (10.62), we have†

$$\frac{dp}{\rho} + V\, dV = 0. \tag{10.63}$$

The process is isentropic, so we write, using Eq. (10.24),

$$dp = \left.\frac{\partial p}{\partial \rho}\right)_s d\rho = c^2\, d\rho,$$

and Eq. (10.63) becomes

$$c^2\left(\frac{d\rho}{\rho}\right) + V\, dV = 0. \tag{10.64}$$

We now solve the differential continuity equation for $d\rho/\rho$ and substitute into Eq. (10.64) to obtain

$$\left(\frac{V^2}{c^2} - 1\right)\frac{dV}{V} = \frac{dA}{A},$$

† Note that Eq. (10.63) is equivalent to the momentum equation (Euler's equation) for one-dimensional flow (Section 5.3.1).

or, using the definition of Mach number,

$$\frac{dV}{V} = \left(\frac{1}{\mathbf{M}^2 - 1}\right)\frac{dA}{A}. \tag{10.65}$$

We can learn a great deal about the effect of area changes on the flow by examining this algebraic relation between dV and dA. We can determine the effect on fluid properties by noting that, according to Eqs. (10.62)–(10.64), changes of enthalpy, pressure, and density must have the opposite algebraic sign of velocity changes; when velocity increases, enthalpy, pressure, and density decrease.

Equation (10.65) reveals that the relation between velocity change and area change depends on the Mach number and in fact *changes sign* at $\mathbf{M} = 1$. Let's consider flow in a converging passage ($dA < 0$). If the flow is subsonic ($\mathbf{M} < 1$), the velocity increases, and hence the enthalpy, pressure, and density decrease. This behavior is familiar from consideration of incompressible flow. If the flow is supersonic ($\mathbf{M} > 1$), a converging passage causes the velocity to *decrease* (with corresponding increases in enthalpy, pressure, and density). A divergent passage ($dA > 0$) has the opposite effect, decelerating a subsonic flow and accelerating a supersonic flow. These observations are summarized in Table 10.2.

In isoenergetic-isentropic flow, an increase in velocity always corresponds to a Mach number increase and vice versa; therefore a *converging* passage always drives the Mach number *toward* unity and a *diverging* passage always drives the Mach number *away* from unity. These conditions imply that the Mach number cannot pass through unity (either accelerating or decelerating) in a passage that is only convergent or only divergent. If we consider the process of accelerating a gas from rest or a low subsonic Mach number in a simply converging passage, the *maximum* possible Mach number that can be achieved is unity. This maximum value can occur *only* at the end of the converging passage.

Suppose that the Mach number is unity at some point in a flow. Since infinite acceleration ($dV \to \infty$) is not possible unless there is a shock wave, which is specifically excluded by the assumptions, Eq. (10.65) implies that

$$dA = 0 \qquad \text{when } \mathbf{M} = 1.$$

Table 10.2 Effects of area change in isoenergetic-isentropic compressible flow.

		$\mathbf{M} < 1$	$\mathbf{M} > 1$
Converging passage	Velocity (V)	Increases	Decreases
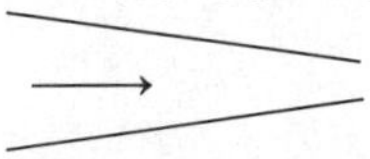	Enthalpy ($\tilde{h}$)	Decreases	Increases
	Pressure (p)	Decreases	Increases
	Density (ρ)	Decreases	Increases
$dA < 0$	Passage acts as a	Nozzle	Diffuser
Diverging passage	Velocity (V)	Decreases	Increases
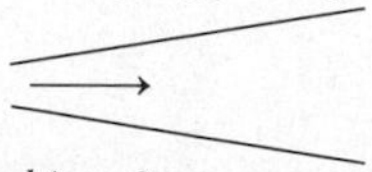	Enthalpy ($\tilde{h}$)	Increases	Decreases
	Pressure (p)	Increases	Decreases
	Density (ρ)	Increases	Decreases
$dA > 0$	Passage acts as a	Diffuser	Nozzle

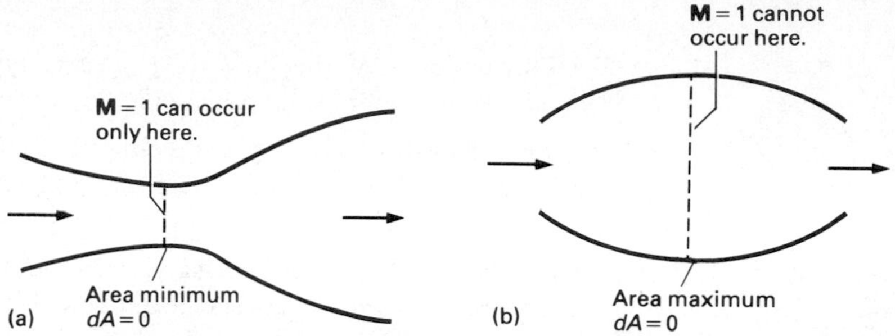

Figure 10.9 Passages with $dA = 0$.

In isoenergetic-isentropic flow, sonic flow ($\mathbf{M} = 1$) can occur only at an area minimum (Fig. 10.9). The occurrence of sonic flow at an area maximum is excluded by the considerations identified in the preceding paragraph. To accelerate a fluid from rest (or a subsonic Mach number) to a supersonic Mach number, a *converging-diverging nozzle* is required.

Finally, suppose that a compressible fluid is flowing in a passage that actually has an area minimum (called a *throat*). Is it always possible to conclude that the Mach number is unity at the throat? To investigate this question, we rearrange Eq. (10.65) as follows:

$$\frac{dA}{A} = (\mathbf{M}^2 - 1)\frac{dV}{V}.$$

If dA is zero, then two options are possible:

$$\mathbf{M} = 1 \quad \text{or} \quad dV = 0.$$

Simply knowing that $dA = 0$ does not allow us to determine which condition occurs. Figure 10.10 illustrates possible velocity (or Mach number) distributions in a passage with a mininum area. If the throat Mach number is not unity, the velocity must have a local maximum or minimum at the throat ($dV_t = 0$). This behavior is observed in a Venturi tube in incompressible flow (Section 7.4 and Example 4.13). If the throat Mach number is 1, dV_t can be either positive or negative, and the velocity can either increase or decrease downstream from the throat, depending on downstream conditions.

The following points summarize our findings about isoenergetic-isentropic

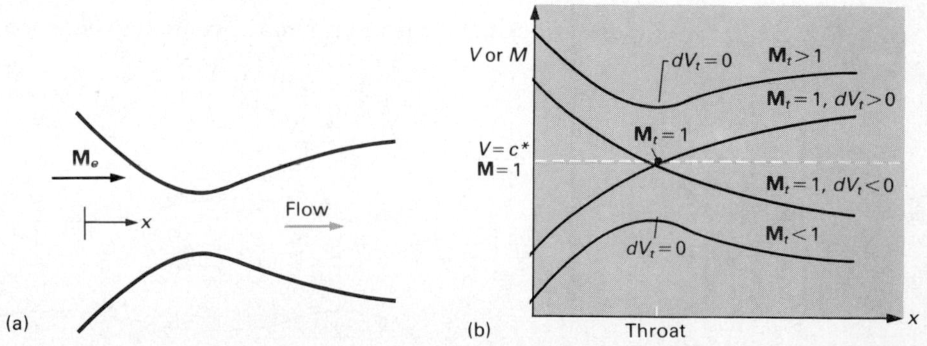

Figure 10.10 Possible velocity (or Mach number) distributions in passage with an area minimum.

flow:

- The response of the flow to a specific type of area change is exactly opposite for subsonic and supersonic flow. Details are given in Table 10.2.
- Sonic flow ($\mathbf{M} = 1$) can occur only at a minimum area. Minimum areas occur at the inlet of a simply diverging passage, the outlet of a simply converging passage, and the throat of a converging–diverging passage.
- It is possible, but not necessary, that the Mach number at the throat of a converging-diverging passage be equal to 1. If the Mach number at a throat is not 1, the velocity must pass through a maximum or minimum. If the throat Mach number is 1, the fluid may either accelerate or decelerate downstream from the throat.

We used a completely general property relation (Gibbs equation), so these conclusions are valid for one-dimensional isoenergetic-isentropic flow of any compressible fluid, ideal gas or not.

10.4.2 Isoenergetic-Isentropic Flow of an Ideal Gas

By assuming that the fluid is an ideal gas, we can obtain working equations for flow in variable-area ducts. Figure 10.11 illustrates the general flow geometry. The typical problem involves relating changes of fluid properties and velocity to the passage area. Calculation of mass flow rate or force on the duct walls is also important. For isoenergetic-isentropic flow, we can use Eqs. (10.50a)–(10.51d) and Table E.1 to relate pressure, temperature, velocity, and Mach number for flow between any two planes in the duct. All we need for a complete analysis are equations relating Mach number to passage area and mass flow rate to Mach number, flow properties, and passage area.

Area–Mach Number Relation. To obtain an equation relating area and Mach number, we use the continuity equation for a control volume with inflow at plane 1 and outflow at plane 2 (see Fig. 10.11):

$$\rho_1 V_1 A_1 = \rho_2 V_2 A_2.$$

For an ideal gas,

$$\rho = \frac{p}{RT} \quad \text{and} \quad V = \mathbf{M}c = \mathbf{M}\sqrt{kRT}.$$

Substituting these equations into the continuity equation gives

$$\sqrt{\frac{k}{R}}\left(\frac{p_1 A_1 \mathbf{M}_1}{\sqrt{T_1}}\right) = \sqrt{\frac{k}{R}}\left(\frac{p_2 A_2 \mathbf{M}_2}{\sqrt{T_2}}\right).$$

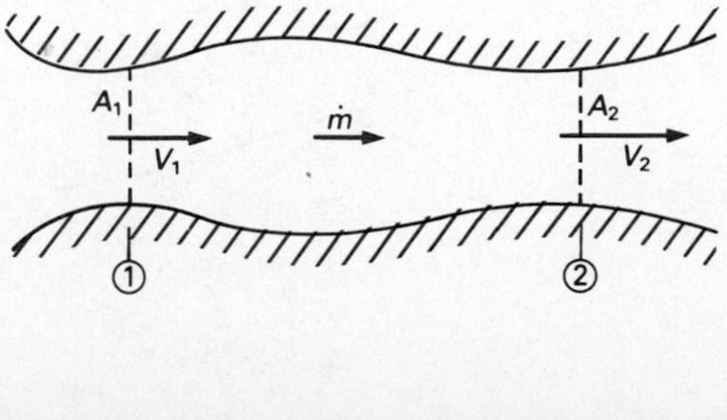

Figure 10.11 General flow between planes 1 and 2 in a variable-area duct.

Canceling k and R and solving for the area ratio, we have

$$\frac{A_2}{A_1} = \frac{\mathbf{M}_1}{\mathbf{M}_2}\left(\frac{p_2}{p_1}\right)^{-1}\left(\frac{T_2}{T_1}\right)^{1/2}.$$

For isoenergetic-isentropic flow, we can substitute Eqs. (10.50b) and (10.51b) for the temperature and pressure ratios, giving

$$\frac{A_2}{A_1} = \frac{\mathbf{M}_1}{\mathbf{M}_2}\left\{\frac{1 + [(k-1)/2]\mathbf{M}_2^2}{1 + [(k-1)/2]\mathbf{M}_1^2}\right\}^{(k+1)/2(k-1)}. \qquad (10.66)$$

If any three of the quantities A_1, $\mathbf{M}_1$, A_2, and $\mathbf{M}_2$ are specified, the fourth can be calculated. Usually A_1 and $\mathbf{M}_1$ are known and either $\mathbf{M}_2$ or A_2 is to be found. If $\mathbf{M}_2$ is given and A_2 is to be calculated, the calculation is simple; however, if A_2 is given and $\mathbf{M}_2$ must be calculated, that is not so easy. We can simplify the calculations by making a table. To tabulate area ratio, we select a reference area with known Mach number. The obvious[†] choice is the area corresponding to a Mach number of 1. We set $\mathbf{M}_1 = 1$, $A_1 = A^*$ and drop the subscript 2 to get

$$\frac{A}{A^*} = \frac{1}{\mathbf{M}}\left[\frac{2}{k+1}\left(1 + \frac{k-1}{2}\mathbf{M}^2\right)\right]^{(k+1)/2(k-1)}. \qquad (10.67)$$

In Eq. (10.67), A is the area at the location where the Mach number is $\mathbf{M}$, and A^* (the critical area) is the area that would correspond to a Mach number of unity. Note that A^* is a reference area; the actual area need not be A^* anywhere in the flow. Values of A/A^* for $k = 1.4$ are included in the isentropic flow table (Table E.1). The relationship between A/A^* and $\mathbf{M}$ for any value of k can be obtained from the personal computer disk that is a companion to this textbook. The area ratio function can be used to relate areas and Mach numbers at two planes in a duct by

$$\frac{A_2}{A_1} = \frac{A}{A^*}\{\mathbf{M}_2\}\bigg/\frac{A}{A^*}\{\mathbf{M}_1\}. \qquad (10.68)$$

A plot of the area ratio function versus Mach number (Fig. 10.12) reveals some interesting information and confirms some of the observations made in the preceding section. First, we note that A/A^* is always greater than unity except at $\mathbf{M} = 1$, where A/A^* is 1. This confirms that in isoenergetic-isentropic flow, the Mach number can be 1 only at the minimum area. Next, we note that if Mach number is given, there is a single corresponding area ratio, so the problem of calculating the area required to obtain a given Mach number seems easy. We must be careful, however. If the initial Mach number ($\mathbf{M}_1$) is less than 1 and the desired final Mach number ($\mathbf{M}_2$) is greater than 1, a minimum area exactly equal to A^* must occur between planes 1 and 2; otherwise the flow cannot pass through $\mathbf{M} = 1$.

Finally, we note that if A/A^* is known, there are *two* corresponding Mach numbers. Choice of the proper Mach number for a known area ratio requires more information than just the area ratio itself. Generally, we must know the pressure

[†] You might think that we would choose a "stagnation area" to correspond to the stagnation temperature and pressure. Why won't this work?

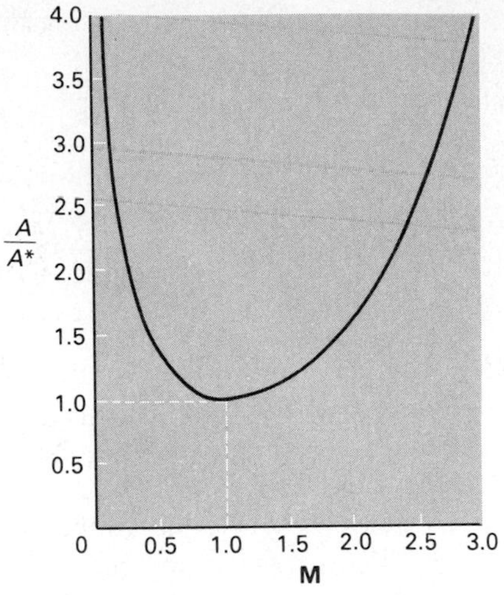

Figure 10.12 Area ratio function for $k = 1.4$.

or some other item of information besides the area ratio to obtain the Mach number. We can sometimes logically eliminate one of the possibilities. If we know that the flow is subsonic at plane 1 and that a minimum area exactly equal to A^* does not occur between planes 1 and 2, the flow at 2 cannot be supersonic.

Before concluding this section on area ratio, we offer a few comments and warnings. The critical area A^* is the area necessary to accelerate or decelerate a flow to sonic conditions. We do not necessarily identify A^* with any *real* area in a flow passage. If we know that the flow is sonic somewhere in a passage, the passage area equals the critical area at the sonic location. We also know that it must be the minimum area if the flow is isoenergetic-isentropic. However, the converse is not always true. If there is a minimum area in a flow passage, we *cannot* assume that the minimum area is equal to A^* without further evidence that the flow actually is sonic at the minimum area. Stated simply, in a passage with a throat, the throat area (A_t) does not necessarily equal the critical area (A^*) unless other evidence indicates it.

EXAMPLE 10.5 Illustrates Use of the Area Ratio Function and the Logic Necessary for Analysis of Flow in a Converging-Diverging Passage

The small laboratory wind tunnel shown in Fig. E10.5 draws air from the atmosphere. The pressure and temperature of the laboratory air are 100 kPa and 20°C. Schlieren photographs reveal a shock wave in the nozzle, as shown. Calculate the air temperature and pressure at the throat and at planes 1, 2, and 3. Assume isoenergetic-isentropic flow except for the shock.

SOLUTION

Given

Air flow in laboratory wind tunnel

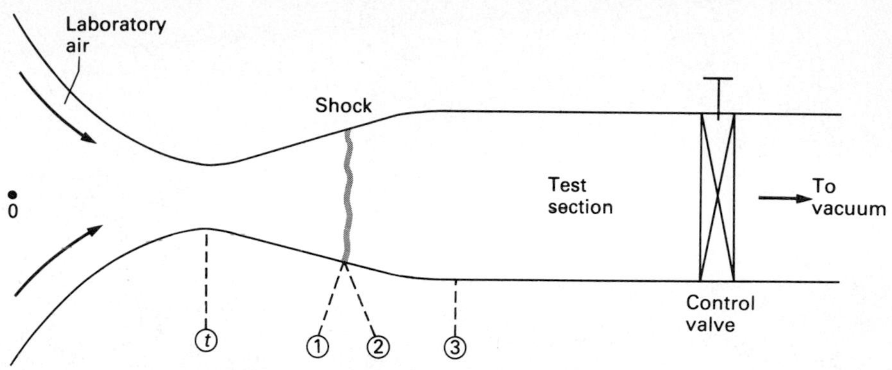

Figure E10.5 Small laboratory wind tunnel.

Shock wave at specific location in the nozzle

Laboratory air at 100 kPa and 20°C

Areas given in Fig. E10.5

Isoenergetic-Isentropic flow

Find

Air temperature and pressure at all areas shown

Solution

First convert the given air temperature to absolute units:

$$T_{\text{lab}} = 20°\text{C} + 273.2°\text{C} = 293.2\text{K}.$$

The passage is converging–diverging, so we must be alert to the possibilities of sonic flow at the throat and supersonic flow in the diverging section. We note that there is a shock in the diverging section of the nozzle and conclude that the flow is supersonic at 1. Because the flow is subsonic ($\mathbf{M} \approx 0$) in the laboratory, the air must pass through $\mathbf{M} = 1$ somewhere upstream from plane 1. The *only* place this can occur is the throat, so

$$\mathbf{M}_t = 1.0.$$

The process is isentropic ahead of the shock, so the stagnation properties are constant up to the shock and equal to the laboratory pressure and temperature (since $\mathbf{M} \approx 0$ in the ambient laboratory air). The critical area for the flow upstream from the shock is

$$A^*_{0 \to 1} = A_t.$$

We calculate the throat pressure and temperature with the aid of Table E.1:

$$p_t = p_0\left(\frac{p}{p_0}\{\mathbf{M}_t\}\right) = 100 \text{ kPa } (0.5283), \qquad \textbf{ANSWER}$$

or

$$p_t = 52.8 \text{ kPa};$$

and

$$T_t = T_0\left(\frac{T}{T_0}\{\mathbf{M}_t\}\right) = 293.2\text{K}(0.8333),$$

or

$$T_t = 244.3\text{K}. \qquad \textbf{ANSWER}$$

The flow at plane 1, just ahead of the shock, has

$$A_1^* = A_t = 1.0 \times 10^{-3}\ \text{m}^2.$$

The area ratio for plane 1 is

$$\left.\frac{A}{A^*}\right)_1 = \frac{2 \times 10^{-3}\ \text{m}^2}{1 \times 10^{-3}\ \text{m}^2} = 2.000.$$

From Table E.1, we find the *supersonic* Mach number corresponding to this area ratio:

$$\mathbf{M}_1 = 2.20.$$

The pressure and temperature ratios (from Table E.1) are

$$\frac{p_1}{p_{0_1}} = 0.0935 \quad \text{and} \quad \frac{T_1}{T_{0_1}} = 0.5081.$$

We now calculate

$$p_1 = 0.0935 p_{0_1} = 0.0935(100\ \text{kPa}),$$

or

$$p_1 = 9.35\ \text{kPa}; \qquad \textbf{ANSWER}$$

and

$$T_1 = 0.5081 T_{0_1} = 0.5081(293.2\text{K}),$$

or

$$T_1 = 149.0\text{K}. \qquad \textbf{ANSWER}$$

Plane 2 lies downstream from the shock. We must use the shock functions in Table E.2 to relate properties at plane 2 to those at plane 1. For $\mathbf{M}_1 = 2.20$, Table E.2 gives

$$\frac{p_2}{p_1} = 5.480, \qquad \frac{T_2}{T_1} = 1.857, \quad \text{and} \quad \mathbf{M}_2 = 0.547.$$

Then

$$p_2 = \left(\frac{p_2}{p_1}\right)p_1 = (5.480)(9.35\ \text{kPa}),$$

or

$$p_2 = 51.2\ \text{kPa}; \qquad \textbf{ANSWER}$$

and

$$T_2 = \left(\frac{T_2}{T_1}\right)T_1 = (1.857)(149.0\text{K}),$$

or

$$T_2 = 276.7\text{K}. \qquad \textbf{ANSWER}$$

To calculate the properties at plane 3, we note that the flow from plane 2 to plane 3 is isentropic, although with different values of p_0 and A^* than the flow upstream from the shock. In particular, A_3^*, which is equal to A_2^*, *is not* equal to the throat area. We calculate the area ratio function at plane 3 from

$$\left.\frac{A}{A^*}\right)_3 = \left(\frac{A_3}{A_2}\right)\left(\frac{A}{A^*}\{\mathbf{M}_2\}\right).$$

From Table E.1,

$$\frac{A}{A^*}\{\mathbf{M}_2\} = \frac{A}{A^*}\{0.547\} = 1.259,$$

so

$$\left.\frac{A}{A^*}\right)_3 = \left(\frac{3 \times 10^{-3}\ \text{m}^2}{2 \times 10^{-3}\ \text{m}^2}\right)(1.259) = 1.888.$$

This value of A/A^* corresponds to a supersonic Mach number and a subsonic Mach number. As there is no throat between planes 2 and 3, the flow at plane 3 must be subsonic. From Table E.1,

$$\mathbf{M}_2 = 0.547, \qquad \frac{p_2}{p_{0_2}} = 0.8159,$$

$$\mathbf{M}_3 = 0.326, \qquad \frac{p_3}{p_{0_3}} = 0.9290, \quad \text{and} \quad \frac{T_3}{T_{0_3}} = 0.9792.$$

We calculate p_3:

$$p_3 = p_2 \frac{p}{p_0}\{\mathbf{M}_3\} \bigg/ \frac{p}{p_0}\{\mathbf{M}_2\} = 51.2\ \text{kPa}\left(\frac{0.9290}{0.8159}\right),$$

or

$$p_3 = 58.3\ \text{kPa}. \qquad \textbf{ANSWER}$$

The expression for T_3 is

$$T_3 = T_0\left(\frac{T_3}{T_0}\right),$$

but as the shock does not affect T_0,

$$T_3 = (293.2\text{K})(0.9792),$$

or

$$T_3 = 287.1\text{K}. \qquad \textbf{ANSWER}$$

Discussion

Note the logic used to decide that A^* upstream of the shock is equal to A_t, the geometric area of the throat. The value of A^* on the downstream side of the shock can be calculated as

$$A_2^* = \frac{A_2}{(A/A^*)\{\mathbf{M}_2\}} = 1.59 \times 10^{-3}\ \text{m}^2.$$

Note that $A_2^* > A_1^*$.

There are alternative ways to manipulate the functions and tables to obtain the answers. For example, we could use

$$T_3 = T_2 \frac{T}{T_0}\{\mathbf{M}_3\} \Big/ \frac{T}{T_0}\{\mathbf{M}_2\} \text{ and } p_3 = p_{0_1}\left(\frac{p_{0_2}}{p_{0_1}}\{\mathbf{M}_1\}\right)\left(\frac{p}{p_0}\{\mathbf{M}_3\}\right)$$

instead of the equations we used. You should be able to think of other possibilities. Suppose that only p_3 and T_3 were requested. Could they be figured without calculating p_1, T_1, p_2, T_2, and so on? Explain.

This example demonstrates how to use the shock functions and tables to patch together different isentropic flows at a shock. You could also use computer evaluation of isentropic and shock functions instead of looking up quantities in tables.

Mass Flow Relations and Choking. Several alternative equations can be used for calculation of mass flow rate in a duct. The "primitive variable" equation for mass flow rate is

$$\dot{m} = \rho V A. \tag{10.69}$$

If the fluid is an ideal gas,

$$\rho = \frac{p}{RT} \quad \text{and} \quad V = \mathbf{M}c = \mathbf{M}\sqrt{kRT},$$

and an alternative equation for mass flow rate is

$$\dot{m} = \sqrt{\frac{k}{R}}\left(\frac{pA\mathbf{M}}{\sqrt{T}}\right). \tag{10.70}$$

Both Eqs. (10.69) and (10.70) evaluate the mass flow rate in terms of static properties. By substituting

$$p = p_0\left[1 + \left(\frac{k-1}{2}\right)\mathbf{M}^2\right]^{-k/(k-1)} \quad \text{and} \quad T = T_0\left[1 + \left(\frac{k-1}{2}\right)\mathbf{M}^2\right]^{-1},$$

we obtain the mass flow rate in terms of stagnation properties:

$$\dot{m} = \sqrt{k}\left(\frac{p_0 A}{\sqrt{RT_0}}\right)\mathbf{M}\left[1 + \left(\frac{k-1}{2}\right)\mathbf{M}^2\right]^{-(k+1)/2(k-1)}. \tag{10.71}$$

Although this equation is complicated, it has certain advantages for isoenergetic-isentropic flow because p_0 and T_0 are constants. Any of Eqs. (10.69)–(10.71) can be used to calculate mass flow. The choice between them depends on what data are available.

By setting $\mathbf{M} = 1$ and $A = A^*$ in Eq. (10.71), we get

$$\dot{m} = \frac{p_0 A^*}{\sqrt{RT_0}}\sqrt{k}\left(\frac{2}{k+1}\right)^{(k+1)/2(k-1)}. \tag{10.72}$$

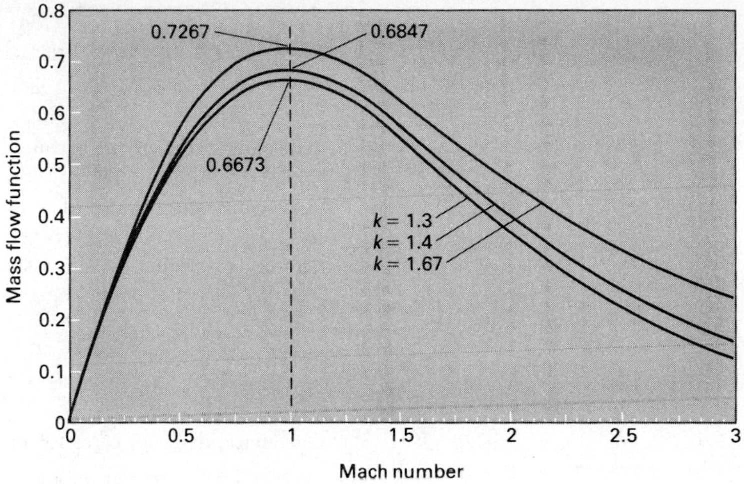

Figure 10.13
Dimensionless mass flow function $[(\dot{m}\sqrt{RT_0})/(p_0A)]$.

Equation (10.72) shows that A^* is constant in steady isoenergetic-isentropic flow with constant values of $\dot{m}$, p_0, T_0, R, and k.

We can obtain a dimensionless mass flow function from Eq. (10.71):

$$\frac{\dot{m}\sqrt{RT_0}}{p_0A} = \sqrt{k}\mathbf{M}\left(1 + \frac{k-1}{2}\mathbf{M}^2\right)^{-(k+1)/2(k-1)}. \tag{10.73}$$

A plot of this function is shown in Fig. 10.13. The key fact about Fig. 10.13 for this discussion is the occurrence of the maximum at $\mathbf{M} = 1$:

$$\left.\frac{\dot{m}\sqrt{RT_0}}{p_0A}\right)_{\max} = f(k \text{ only}).$$

We can draw several conclusions from this fact.

- Suppose that p_0 and T_0 are fixed (isoenergetic-isentropic flow). A specified mass flow rate can be forced through a limiting minimum area. No smaller area can pass that flow.
- Suppose that p_0 and T_0 are fixed (isoenergetic-isentropic flow). Any particular fixed-geometry duct, with fixed minimum area, can pass only a certain maximum mass flow. Passage of this maximum mass flow occurs when the Mach number at the minimum area is 1. Once the Mach number at the minimum area becomes 1, it is not possible to increase the mass flow rate.
- It is possible to increase the limiting mass flow in a fixed geometry duct or to force a specified mass flow through a smaller area by *increasing* $p_0/\sqrt{T_0}$. Doing work on a gas (compression) increases both p_0 and T_0; however, the ratio $p_0/\sqrt{T_0}$ is usually increased by compression.

The preceding statements describe the phenomenon called *choking*. Simply stated:

> The choking phenomenon occurs in compressible duct flow when the local Mach number reaches 1 at the minimum area in the duct. When this occurs, the mass flow rate through the duct cannot be increased unless the ratio of stagnation pressure to square root of stagnation temperature is increased.

EXAMPLE 10.6 Illustrates Use of Compressible Flow Equations and Tables in Design of a Flow Passage

The duct between the turbine and exhaust nozzle of a jet engine is 0.60 m in diameter. The gas flowing in the duct has stagnation pressure and temperature of 275 kPa and 700K, respectively. The gas flow rate is 82 kg/s. Specify the relevant areas of a nozzle for expanding the gas to standard atmospheric pressure. Calculate the force required to hold the nozzle to the duct. The gas has $k = 1.4$ and $R = 265$ N·m/kg·K.

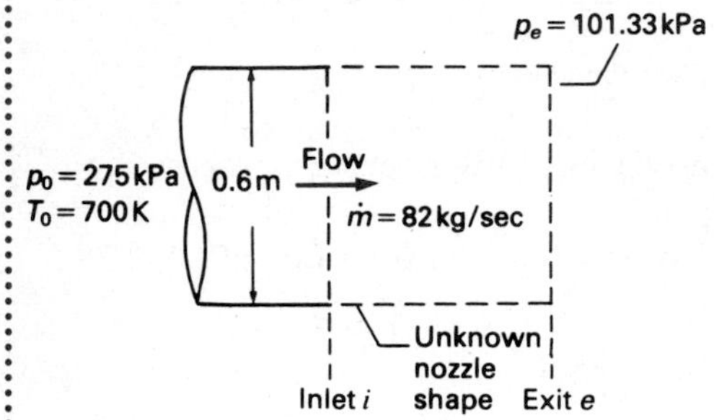

Figure E10.6a Schematic diagram of nozzle design problem.

SOLUTION

Given

Schematic diagram of problem in Fig. E10.6a

Nozzle inlet diameter 0.6 m

Gas with $k = 1.4$ and $R = 265$ N·m/kg·K, flowing at 82 kg/s

Gas with $p_0 = 275$ kPa and $T_0 = 700$K

Find

Relevant nozzle areas for expanding gas to standard atmospheric pressure (101.33 kPa)

Force required to hold nozzle to duct

Solution

First verify that all given pressures and temperatures are absolute. Next, establish what the nozzle geometry should be. The desired static-stagnation pressure ratio at the exit plane is

$$\frac{p_e}{p_{0_e}} = \frac{101.33 \text{ kPa}}{275.0 \text{ kPa}} = 0.3684.$$

From Table E.1, the desired exit Mach number is

$$\mathbf{M}_e = 1.285.$$

We assume that the inlet Mach number will be less than 1, so we need a minimum area (throat) between the entrance and the exit. We calculate the throat area from Eq. (10.72), with $k = 1.4$:

$$\dot{m} = 0.6847 \frac{p_0 A^*}{\sqrt{RT_0}}.$$

For $A_t = A^*$,

$$A_t = \frac{\dot{m}\sqrt{RT_0}}{0.6847 p_0} = \frac{82 \text{ kg/s} \sqrt{(265\text{N}\cdot\text{m/kg}\cdot\text{K})(700\text{K})}}{0.6847(275{,}000 \text{ N/m}^2)};$$

$$A_t = 0.1876 \text{ m}^2.$$ **ANSWER**

The throat diameter is

$$D_t = \sqrt{\frac{4A_t}{\pi}} = 0.4887 \text{ m}.$$

The inlet area of the nozzle is

$$A_i = \frac{\pi}{4} D_1^2 = \frac{\pi}{4} (0.6 \text{ m})^2;$$

$$A_i = 0.2874 \text{ m}^2. \qquad \textbf{ANSWER}$$

The nozzle exit area is that required to produce a Mach number of 1.285:

$$A_e = A_t \left[\frac{A}{A^*} \{\mathbf{M}_e\} \right].$$

From Table E.1,

$$\frac{A}{A^*} = 1.0601 \text{ at } \mathbf{M} = 1.285, \quad \text{so} \quad A_e = 0.1876 \text{ m}^2 \ (1.0601);$$

$$A_e = 0.1989 \text{ m}^2. \qquad \textbf{ANSWER}$$

The corresponding diameter is

$$D_e = \sqrt{\frac{4A_e}{\pi}} = 0.5032 \text{ m}.$$

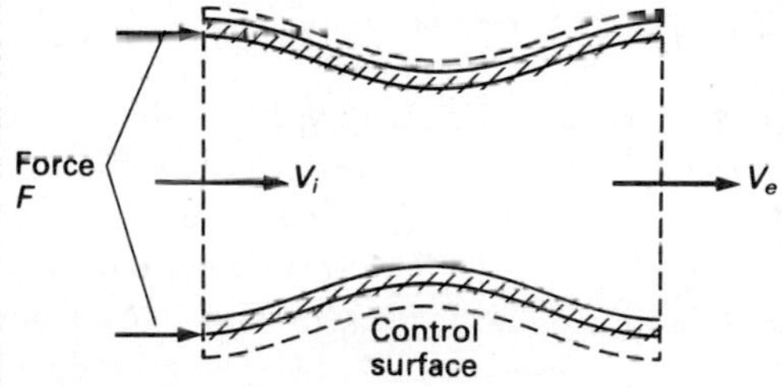

Figure E10.6b Control volume for nozzle force analysis.

To calculate the force, we apply the linear momentum theorem to a control volume enclosing the nozzle, as shown in Fig. E10.6b. We assume that the anchoring force, F, acts in the flow direction. The momentum equation gives

$$F = (p_e + \rho_e V_e^2)A_e - (p_i + \rho_i V_i^2)A_i + p_{\text{atm}}(A_i - A_e).$$

From Eq. (10.29),

$$\rho V^2 = kp\mathbf{M}^2,$$

so

$$F = p_e A_e (1 + k\mathbf{M}_e^2) - p_i A_i (1 + k\mathbf{M}_i^2) + p_{\text{atm}}(A_i - A_e).$$

We find $\mathbf{M}_i$ and p_i from

$$\frac{A}{A^*} \{\mathbf{M}_i\} = \frac{A_i}{A_t} = \frac{0.2874 \text{ m}^2}{0.1876 \text{ m}^2} = 1.5319.$$

From Table E.1 (using subsonic Mach number),

$$\mathbf{M}_i = 0.419 \quad \text{and} \quad \frac{p_i}{p_0} = 0.8862.$$

Then

$$p_i = (0.8862)(275 \text{ kPa}) = 243.7 \text{ kPa}.$$

The force F is

$$\begin{aligned} F = {} & (101{,}330 \text{ N/m}^2)(0.1989 \text{ m}^2)[1 + 1.4(1.285)^2] \\ & - (243{,}700 \text{ N/m}^2)(0.2874 \text{ m}^2)[1 + 1.4(0.419)^2] \\ & + 101{,}330 \text{ N/m}^2 \ (0.2874 \text{ m}^2 - 0.1989 \text{ m}^2) \\ = {} & -11{,}500 \text{ N}. \end{aligned}$$

$$F = 11{,}500 \text{ N} \quad \text{Opposite the flow direction.} \qquad \textbf{ANSWER}$$

Discussion

A "force function,"

$$\frac{F}{F^*} \equiv \frac{pA(1 + k\mathbf{M}^2)}{p^*A^*(1 + k)} = \frac{1 + k\mathbf{M}^2}{\mathbf{M}\sqrt{2(k + 1)\left(1 + \frac{k - 1}{2}\mathbf{M}^2\right)}},$$

is listed in some compressible flow tables [1].

We treated p_{atm} and p_e as separate entities in the momentum equation even though they are equal for this problem. In some cases, the pressure at the exit of a supersonic nozzle is not equal to the pressure of the surrounding atmosphere (Section 10.4.4).

The nozzle flow was calculated as isoenergetic-isentropic because the exit Mach number was to be supersonic; therefore no shocks occur in the nozzle.

The flow in the nozzle is choked. The given flow rate (82 kg/s) with the given stagnation properties (275 kPa, 700K) cannot pass through any area smaller than 0.1876 m^2.

10.4.3 Flow in a Converging Nozzle

We can show how the information about compressible flow in variable-area ducts fits together by examining the flow in a converging nozzle. Figure 10.14 shows a converging nozzle that exhausts gas from a large reservoir to a variable-pressure region. We assume that the pressure and temperature in the reservoir are constant. We can vary the pressure in the exhaust region (called the *back pressure*) by means of the control valve that connects the exhaust region to a downstream vacuum pump. By opening or closing the control valve, we decrease or increase the back pressure and therefore expect to increase or decrease the nozzle flow rate. We assume that there is no wall friction and no heat transfer or work.

We can make several preliminary observations about the flow in the nozzle.

- The nozzle is only convergent, so the flow cannot pass *through* $\mathbf{M} = 1$.
- The flow at the nozzle inlet (in the large reservoir) is obviously subsonic ($\mathbf{M} \approx 0$), so the flow in the entire nozzle is subsonic, with the possible exception of the nozzle exit.

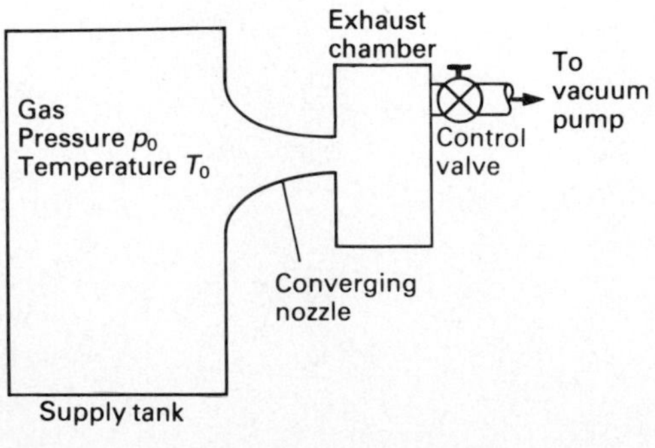

Figure 10.14 Apparatus for studying flow in a converging nozzle.

- The flow cannot be supersonic in the nozzle, so there can be no shocks and the flow is isoenergetic-isentropic everywhere in the nozzle. The stagnation properties are constant and equal to the gas properties in the reservoir.
- The maximum possible Mach number in the nozzle is 1.0. This value can occur only at the nozzle exit (minimum area).
- There is a maximum possible mass flow rate that can occur. This maximum is determined by the gas constants (k and R), the values of the reservoir (stagnation) properties, and the exit area. The maximum occurs only when the Mach number is 1 at the nozzle exit.

The flow through the nozzle is described in terms of the pressure and Mach number distribution in the nozzle and the (dimensionless) mass flow rate ($\dot{m}\sqrt{RT_0}/p_0A_e$), all as functions of the ratio of back pressure to supply pressure and k. We also plot the ratio of pressure in the nozzle exit plane (p_e) to the back pressure (p_e/p_B). The various curves are shown in Fig. 10.15.

The curves and points labeled *a* correspond to a closed control valve. There is no flow. The pressure equals the reservoir pressure everywhere, and the Mach number is zero everywhere.

Case *b* corresponds to a slight opening of the control valve. The back pressure is less than the supply pressure and there is flow. The gas accelerates from the reservoir to the nozzle exit, and the pressure drops. Minimum pressure and maximum Mach number are at the nozzle exit. The pressure at the nozzle exit is equal to the back pressure. If we know the ratio p_B/p_0, the equations

$$\frac{p_e}{p_0} = \frac{p_B}{p_0} \quad \text{and} \quad \mathbf{M}_e = \mathbf{M}\left(\frac{p_e}{p_0}\right)$$

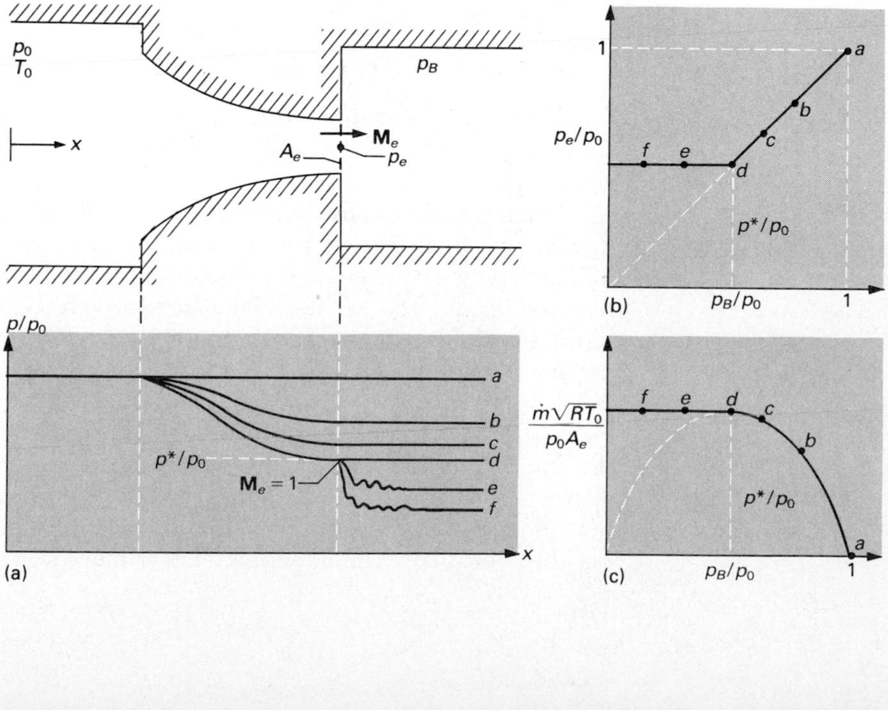

Figure 10.15 Performance of a converging nozzle at various pressure ratios.

enable us to determine the Mach number at the nozzle exit. We can then calculate the mass flow at the exit (and hence in the entire nozzle) from Eq. (10.73).

Case *c* is similar to case *b*, except that a larger control valve opening permits a lower back pressure with correspondingly higher mass flow and exit Mach number and a lower exit pressure. Note that the maximum Mach number still occurs at the nozzle exit and is less than 1.

In case *d*, the control valve has been opened a sufficient amount to bring the exit Mach number to a value of 1.0. The exit pressure now equals the critical pressure (p^*) and also equals the back pressure. Knowing $\mathbf{M}_e = 1.0$ allows calculation of the mass flow rate.

Case *e* corresponds to opening the control valve farther than in case *d*. When we try this, we find that no changes occur in the nozzle. In case *d*, we reached the limit of the nozzle's capability. The exit Mach number cannot exceed 1, the exit pressure cannot drop below the critical pressure, which is determined solely by the supply pressure and the specific heat ratio (Eq. 10.59), and the mass flow cannot exceed the choked value. The only difference between case *d* and case *e* is that the back pressure and exit pressure are no longer equal. The flow must adjust to the lower back pressure *after* leaving the nozzle. The downstream flow is multidimensional, so the pressure curve is shown as a wavy line downstream from the nozzle exit. Opening the control valve with further lowering of the back pressure does not change the nozzle flow.

At first, these results may seem strange; however, they have a simple physical explanation. Once the exit flow becomes sonic, pressure waves from the exhaust region cannot propagate upstream into the nozzle and supply reservoir to cause any flow adjustment.

There are two flow regimes in a simple converging nozzle: *unchoked flow* and *choked flow*. Which flow regime occurs in a given case depends on the relative values of the back pressure and critical pressure.

- If $p_B/p_0 > p^*/p_0$, the flow is unchoked.
- If $p_B/p_0 \leq p^*/p_0$, the flow is choked.

The critical pressure ratio (p^*/p_0) depends only on specific heat ratio. For a converging nozzle, the occurrence of choked flow does not depend on nozzle geometry.

You might be curious about the generality of these results. Although it appears that we have been discussing a very specific system, we did so only for the purpose of illustration. The large reservoir represents a supply of gas with specific stagnation conditions. The exhaust region/control valve/vacuum pump represents the resistance to flow downstream from the nozzle. In principle, considering a compressor and control valve upstream from the nozzle and a constant-pressure exhaust region would be equally valid. In this case, all of the *dimensionless* plots of Fig. 10.15 would still be valid; however, if we permit p_0 to increase, we can increase the *dimensional* mass flow rate ($\dot{m}$), even at choked conditions. Finally, we have shown a long, smoothly contoured nozzle. The principles apply to converging nozzles of any length, even orifices; however, the departure from one-dimensional, one-directional flow conditions becomes significant in short nozzles.

10.4.4 Flow in a Converging–Diverging Nozzle

We complete our consideration of compressible flow in variable-area ducts by discussing the flow in a converging–diverging nozzle. Figure 10.16 shows a converging–diverging nozzle that exhausts gas from a large reservoir at constant pressure and temperature to a variable-pressure exhaust region. The back pressure can be varied by opening or closing a control valve (similar to Fig. 10.14). Generalizations of this arrangement are similar to those discussed for the converging nozzle. We assume that there is no wall friction, heat transfer, or work.

Preliminary observations about the flow in the converging–diverging nozzle are as follows:

- The nozzle is convergent–divergent, so the flow can pass through $\mathbf{M} = 1$. The flow also can be subsonic everywhere.
- If $\mathbf{M} = 1$ anywhere, it must be at the throat.
- There may be supersonic flow in the divergent portion of the nozzle, so there may be shocks in the flow. If there are shocks, the flow is not completely isentropic, although it is isoenergetic.
- If there are no shocks, the flow is isentropic. If there are shocks, the flow from the reservoir up to the first shock is isentropic. Flow downstream from a shock also is isentropic but with different values of entropy, stagnation pressure, and critical area.
- The maximum possible Mach number that can occur anywhere in the passage corresponds to acceleration of the fluid in an isentropic process from the reservoir to the nozzle exit. The maximum possible Mach number can occur only at the exit and is determined by the ratio of exit area to throat area.
- The maximum possible mass flow rate in the nozzle is determined by the gas constants, the reservoir properties, and the minimum area, which occurs at the throat.

The flow in the converging–diverging nozzle can be described by plots of pressure and Mach number distributions in the nozzle, the dimensionless mass flow rate $(\dot{m}\sqrt{RT_0}/p_{0_1}A_t)$, and the ratio of nozzle exit pressure to supply pressure (p_e/p_{0_1}), all as functions of the ratio of back pressure to supply pressure (p_B/p_{0_1}).

The nozzle performance curves are shown in Fig. 10.17. The case labeled *a*

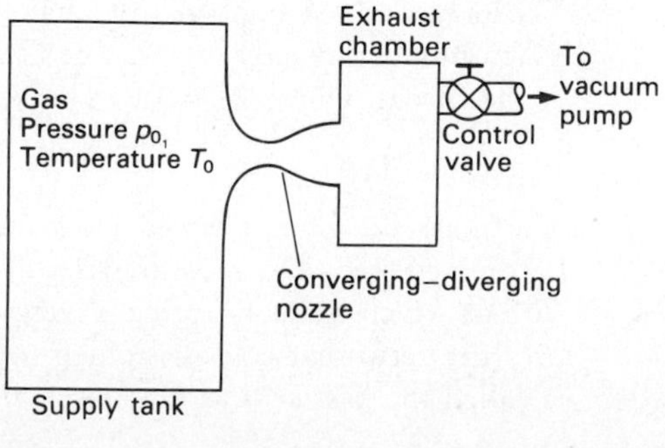

Figure 10.16 Apparatus for studying flow in a converging–diverging nozzle.

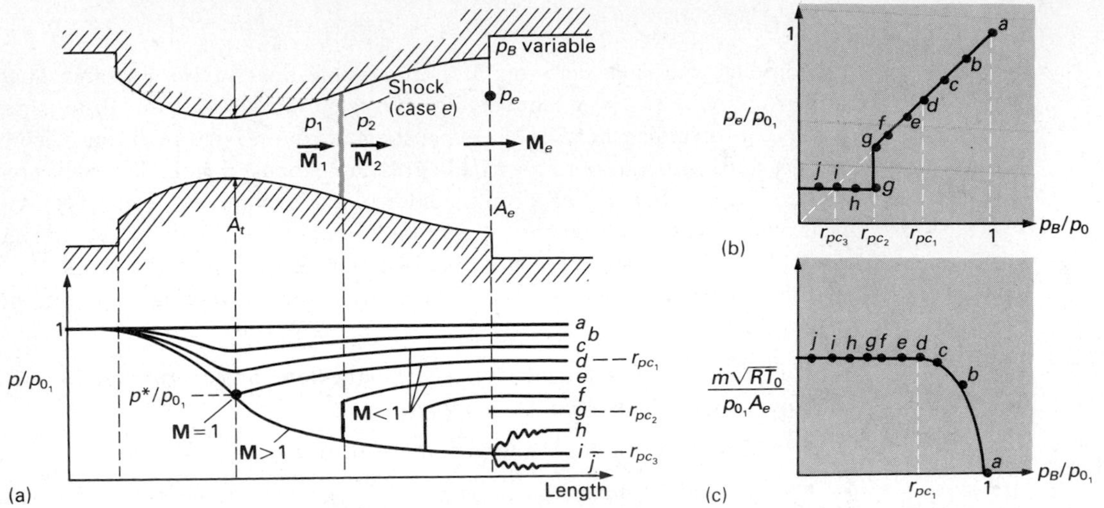

Figure 10.17 Performance of converging–diverging nozzle at various pressure ratios.

corresponds to complete closure of the control valve. There is no flow, the pressure is uniform, and the Mach number is zero everywhere.

Curves and points labeled b represent a slightly open control valve. The gas accelerates in the convergent portion of the nozzle, and the pressure drops. The throat Mach number is considerably less than 1 (subsonic flow). The flow decelerates in the divergent portion of the nozzle, and the pressure rises. The pressure and Mach number distributions are roughly symmetrical about the throat. There are no shocks, because the flow is subsonic everywhere. The flow at the nozzle exit is subsonic, and the exit and back pressures are equal. The exhaust Mach number can be found from

$$\frac{p}{p_0}(\mathbf{M}_e) = \frac{p_B}{p_{0_1}},$$

and the mass flow can be calculated from Eq. (10.73). A value of A^* can be calculated from Eq. (10.67) or Eq. (10.72) applied at the exhaust plane. A^* will be less than A_t.

Curve c represents a slightly larger control valve opening. The situation is qualitatively similar to case b, but note that the mass flow is larger. The pressure and Mach number distributions are still roughly symmetrical about the throat. The maximum Mach number occurs at the throat.

In case d, the control valve has been opened just enough to bring the throat Mach number exactly to 1.0. The flow is still subsonic everywhere except exactly at the throat, where $\mathbf{M} = 1.0$. The flow is isentropic, but now

$$p_t = p^* \quad \text{and} \quad A_t = A^*.$$

The mass flow has reached its maximum possible value and the flow has just become choked. The pressure rises downstream from the throat, so the back pressure at which the converging–diverging nozzle chokes is greater than p^*/p_0. The pressure ratio that causes choking to first occur in a converging–diverging nozzle is called the *first critical pressure ratio* (r_{pc_1}) and can be calculated by

$$r_{pc_1} = \frac{p}{p_0}\left(\frac{A}{A^*} = \frac{A_e}{A_t}, \mathbf{M} < 1\right); \tag{10.74}$$

that is, by finding the pressure ratio corresponding to the subsonic value of A/A^*.

What happens when the valve is opened beyond the case *d* point? Because the throat is choked, the mass flow cannot be increased, and conditions upstream of the throat cannot be affected. There is a response in the flow downstream from the throat, because, at case *d*, the downstream flow is subsonic. When the valve is opened and the back pressure is lowered, the fluid begins to accelerate as it enters the divergent portion of the nozzle; that is, the flow becomes supersonic. If the control valve is opened only slightly past the case *d* point, the downstream resistance is too large for complete acceleration in the divergent portion of the nozzle, so a shock occurs in the divergent portion.

Curves and points labeled *e* represent the flow for a slightly more open control valve. The flow accelerates in the convergent portion of the nozzle, reaches sonic speed at the throat, and accelerates to supersonic speed downstream from the throat. The supersonic acceleration terminates in a shock wave. Downstream from the shock, the flow experiences subsonic deceleration and exits the passage with $\mathbf{M}_e < 1$. The exit pressure and back pressure are equal. The exact position of the shock depends on the back pressure, and for a given back pressure, is fixed. The mass flow rate for case *e*, as well as for all lower values of back pressure, is the same as for case *d*.

Opening the control valve and lowering the back pressure causes the shock to move downstream (case *f*). Note that once the shock passes a plane, the flow *up to* that plane is no longer affected by lowering the back pressure. As the valve is opened farther, the shock is eventually drawn to the nozzle exit. The flow accelerates isentropically all the way from the reservoir to the exit shock. Exactly at the exit, the pressure jumps to the back pressure as the fluid exits through the shock. This situation is shown as case *g*. Note that the exit pressure is double-valued at case *g*.

Lowering the back pressure further causes the shock to move out of the nozzle and become multidimensional (case *h*). The flow accelerates isentropically from the reservoir to the exit. The gas exits the nozzle supersonically with $\mathbf{M}_e$ corresponding to the nozzle's exit-to-throat area ratio. The exit pressure is determined only by the stagnation pressure (p_{0_1}) and the exit Mach number and is not equal to the back pressure. The gas adjusts to the back pressure externally. If we continue to lower the back pressure, it eventually reaches equality with the exit pressure (case *i*). At this condition, there is no pressure adjustment in the exiting gas. Lowering the pressure further (case *j*) requires external expansion pressure adjustments but does not affect the nozzle flow.

We can summarize this information about converging–diverging nozzle flow as follows. There are four regimes of flow in a converging–diverging nozzle.

- *Venturi regime* (cases *a–d*). The flow is subsonic and isentropic everywhere. The flow accelerates in the convergent portion and decelerates in the divergent portion. Maximum Mach number and minimum pressure occur at the throat.

- *Shock regime* (cases *d–g*). The flow is subsonic in the convergent portion, sonic at the throat, and partly supersonic in the divergent portion. The acceleration terminates in a shock that stands in the divergent portion at a location determined by the exact value of the back pressure. The flow experiences subsonic deceleration from the shock to the exit. The flow is choked.
- *Overexpanded regime* (cases *g–i*). The flow accelerates throughout the nozzle. The throat flow is sonic, and the exit flow is supersonic. The pressure of the gas increases to the back pressure downsteam from the nozzle exit.
- *Underexpanded regime* (cases *i–j*). This case is similar to the overexpanded regime, except that the external pressure adjustments are expansive rather than compressive.

The boundaries between the four flow regimes are indicated by curves *d*, *g*, and *i*. The ratios of back pressure to supply pressure that correspond to these cases are called the *first critical pressure ratio* (r_{pc_1}), the *second critical pressure ratio* (r_{pc_2}), and the *third critical pressure ratio* (r_{pc_3}). These ratios are calculated in the following manner.

- First critical pressure ratio: See Eq. (10.74).
- Second critical pressure ratio: Assume that a shock stands at the nozzle exit and calculate the ratio of *downstream static* to *upstream stagnation* pressure:

$$r_{pc_2} = \frac{p_2}{p_1}\{\mathbf{M}_e\} \times \frac{p}{p_0}\{\mathbf{M}_e\}, \tag{10.75}$$

where ($p_2/p_1\{\mathbf{M}\}$) is the shock static pressure ratio function and $\mathbf{M}_e$ is given by

$$\mathbf{M}_e = \mathbf{M}\ \{A/A^* = A_e/A_t\}.$$

- Third critical pressure ratio:

$$r_{pc_3} = \frac{p}{p_0}\left\{\frac{A}{A^*} = \frac{A_e}{A_t}, \mathbf{M} > 1\right\}. \tag{10.76}$$

Calculation of these ratios is easier than it looks. Note that the three critical pressure ratios for a converging–diverging nozzle depend on the nozzle geometry as well as the specific heat ratio of the flowing gas. The first step in calculating flow through a converging–diverging nozzle is to classify the flow. You do so by calculating the three critical pressure ratios and comparing them to the actual ratio of back pressure to supply pressure. After you have identified the proper flow regime, the rest of the problem falls into place easily.

EXAMPLE 10.7 Illustrates Performance Analysis for Converging and Converging–Diverging Nozzles

A thick-walled pressure vessel has a hole in its side. Figure E10.7 shows two possible configurations for the hole. The pressure and

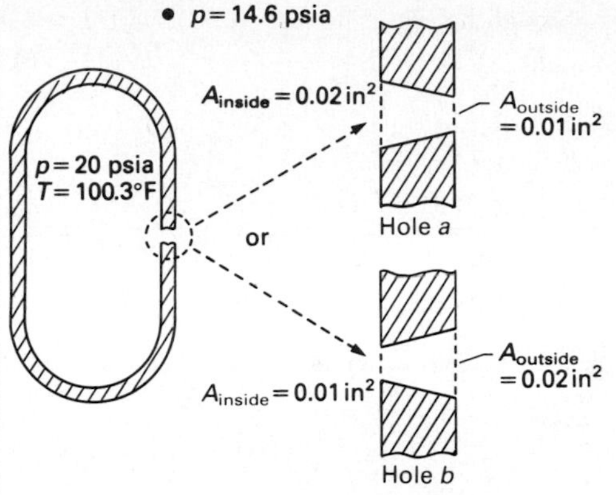

Figure E10.7 Thick-walled pressure vessel with two possible hole configurations.

temperature of the air inside the vessel are 20 psia and 100.3°F, respectively. The pressure outside the vessel is 14.6 psia. Calculate the leak rate of air for each hole configuration.

SOLUTION

Given

Pressure vessel containing air at 20 psia and 100.3°F

Pressure of surrounding atmosphere 14.6 psia

Two possible configurations of hole in vessel wall, both with the same maximum and minimum areas

Find

Leak rate of air for each hole configuration

Solution

First convert given data to absolute units. The air temperature in the vessel is

$$T = 100.3°\text{F} + 459.7°\text{F} = 560°\text{R}.$$

Next, recognize that hole a acts like a simply converging nozzle and hole b acts like a converging–diverging nozzle. The "converging portion" of hole b is formed by streamlines within the tank. Both holes have the same minimum area and the supply temperature and pressure are the same, so the maximum possible flow rate is the same for both holes. If both holes are choked, they have the same flow rate. We first check both holes for choking.

Hole a
Hole a will be choked if

$$\frac{p_B}{p_{0_1}} < \frac{p^*}{p_0} = 0.5283.$$

For hole a,

$$\frac{p_B}{p_{0_1}} = \frac{14.6 \text{ psia}}{20 \text{ psia}} = 0.730.$$

Hole a will not be choked.

Hole b
Hole b will be choked if

$$\frac{p_B}{p_{0_1}} < r_{pc_1},$$

where

$$r_{pc_1} = \frac{p}{p_0}\left(\frac{A}{A^*} = \frac{0.02 \text{ in}^2}{0.01 \text{ in}^2} = 2, \mathbf{M} < 1\right).$$

From Table E.1,

$$r_{pc_1} = 0.9371.$$

The back pressure ratio is still 0.730, so hole b is choked. Next, calculate the mass flow through hole b using Eq. (10.72), with $k = 1.4$ and $\mathbf{M} = 1$:

$$\dot{m}_b = 0.6847\left(\frac{p_0 A_{\min}}{\sqrt{RT_0}}\right)$$

$$= 0.6847\left[\frac{(20\ \text{lb/in}^2)(0.01\ \text{in}^2)\sqrt{32.2\ \text{ft}\cdot\text{lbm/lb}\cdot\text{sec}^2}}{\sqrt{(53.35\ \text{ft}\cdot\text{lb/lbm}\cdot{}^\circ\text{R})(560^\circ\text{R})}}\right];$$

$$\dot{m}_b = 0.00450\ \text{lbm/sec}. \qquad \textbf{ANSWER}$$

To calculate the leak rate in hole a, we need the Mach number at the exit. For

$$\frac{p_e}{p_0} = \frac{p_b}{p_0} = 0.730,$$

Table E.1 gives

$$\mathbf{M}_e = 0.686.$$

Using Eq. (10.73), we have

$$\dot{m}_a = \frac{(20\ \text{lb/in}^2)(0.01\ \text{in}^2)\sqrt{32.2\ \text{ft}\cdot\text{lbm/lb}\cdot\text{sec}^2}}{\sqrt{(53.35\ \text{ft}\cdot\text{lb/lbm}\cdot{}^\circ\text{R})(560^\circ\text{R})}} \times \sqrt{1.4}(0.686)[1 + (0.2)(0.686)^2]^{-3};$$

$$\dot{m}_a = 0.00407\ \text{lbm/sec}. \qquad \textbf{ANSWER}$$

Discussion

These answers are only approximations, because the flow through real holes in the wall of a pressure vessel would probably not be one-dimensional. The third and second critical pressure ratios for hole b are

$$r_{pc_3} = \frac{p}{p_0}\left\{\frac{A}{A^*} = 2, \mathbf{M}_e > 1\right\} = 0.0935 \qquad (\text{note: } \mathbf{M}_e = 2.20)$$

and

$$r_{pc_2} = \frac{p_2}{p_1}\{\mathbf{M} = 2.20\} \times \frac{p}{p_0}\{\mathbf{M} = 2.20\}.$$

Using Table E.2, we have

$$r_{pc_2} = (5.480)(0.0935) = 0.512.$$

Hole b is operating in the shock regime.

10.5 INTERNAL FLOW: CONSTANT-AREA DUCT FLOW WITH FRICTION

We now examine the effects of wall friction in compressible internal flow. To simplify our discussion, we limit consideration to flow in ducts† of constant cross-

† In this section, the word *ducts* includes circular pipes as well as other cross-sectional shapes.

sectional area. We assume that the flow is one-dimensional, with the velocity varying only with duct length (x) but *not* with distance across the duct. This situation is somewhat artificial, because a transverse velocity gradient is always associated with shear stress. The situation that we have assumed is most closely approximated by turbulent flow. The single value of fluid velocity associated with any duct cross section should be interpreted as an average velocity.

10.5.1 Preliminary Consideration: Comparison with Incompressible Duct Flow

Flow of an incompressible fluid in a constant-area duct is one of the most important practical problems in fluid mechanics. Incompressible flow in long ducts eventually becomes fully developed, and the velocity profile does not change shape. As the density is constant in incompressible flow, continuity requires that the average velocity remain constant, so neither velocity profile shape nor magnitude vary in fully developed incompressible flow. Therefore the fluid's kinetic energy and momentum are constant. The retarding effect of fluid friction is balanced by a pressure drop in the flow direction. This pressure drop is directly proportional to the irreversible loss of mechanical energy caused by fluid friction.

What changes does fluid compressibility introduce? A compressible flow can become fully developed in the sense that the velocity profile does not change *shape,* so we could write

$$\frac{u}{u_{\max}} = f\left(\frac{r}{R}\text{ only}\right)$$

for flow in a circular pipe. If the density changes, the average velocity must also change to satisfy the continuity equation. Changing the average velocity changes the magnitude of the entire velocity profile, even if its shape is fixed. More important, if the velocity changes, the fluid's kinetic energy and momentum flux also change. Even if the flow is fully developed, the retarding effect of fluid friction is balanced by *both* pressure *and* momentum changes. Although fluid friction causes an irreversible loss of mechanical energy, this loss cannot be interpreted in terms of a (static) pressure drop. Pressure changes in *compressible* flow *cannot* be calculated from the Darcy–Weisbach equation. In compressible flow, friction appears explicitly only in the momentum equation. The effects of friction on flow can be determined only by simultaneous solution of the equations of continuity, momentum, energy, and state.

We must use the general energy equation, so we must also make an explicit assumption about work and heat transfer or include equations for their calculation. Work transfer is usually "concentrated" in compressors or turbines and does not occur in long runs of duct. The question of heat transfer to or from the fluid is more complicated, because the mechanisms of fluid friction and heat transfer are closely related. Three types of problems may be considered:

- adiabatic flow, with no heat transfer;
- isothermal flow, in which sufficient heat is transferred to or from the fluid to maintain its static temperature constant; and
- flow with simultaneous friction and heat transfer.

If the duct is reasonably short, the adiabatic flow assumption is accurate. Because compressibility effects usually occur only if the fluid velocity is large and because, on dimensional grounds, mechanical energy loss increases as the square of the velocity, ducts in which friction and compressibility are both significant are usually kept short (duct length less than about 100 diameters) and heat transfer can be neglected.

Isothermal flow occurs in long pipelines that are exposed to a constant-temperature environment. A practical example is underground natural gas transmission lines. After traveling a few hundred diameters along the pipe, the gas attains the temperature of its surroundings. Sufficient heat is transferred to or from the gas to maintain constant temperature.

We limit our consideration in this textbook to adiabatic flow in a constant-area duct. This type of flow is sometimes called *Fanno flow*. Many of the qualitative features of isothermal flow are similar to those of Fanno flow; however, different working equations result from the $q = 0$ or $T =$ Constant assumptions. Analysis of flow with simultaneous friction and heat transfer is not terribly difficult but is beyond the scope of a single chapter's introduction to compressible flow. Consult references [1–4] for more information on isothermal flow and references [1, 3] for analysis of flow with simultaneous friction and heat transfer.

10.5.2 Workless Adiabatic Flow (Fanno Flow) of an Ideal Gas

To obtain qualitative and quantitative information about Fanno flow, we assume that the flowing fluid is an ideal gas. The general conclusions that we draw are valid for nonideal gas and vapor flows, but the equations are restricted to an ideal gas.

Figure 10.18(a) illustrates the fundamental problem. A gas flows from plane 1 to plane 2 in a constant-area duct. There is no work or heat transfer, and elevation changes are negligible. There is shear stress at the duct wall. We want to relate changes of gas properties, velocity, and Mach number between planes 1 and 2 to the length, Δx, and other duct characteristics.

For an ideal gas, treating the pressure, temperature, and Mach number as dependent variables is convenient. Because the shear stress at the wall varies with location along the duct, we have to apply the governing equations to a differential control volume lying somewhere between plane 1 and plane 2 (Fig. 10.18b). We then integrate the resulting ordinary differential equations between planes 1 and 2. Figure 10.18(c) shows the details of the flow through the control volume.† The continuity equation is

$$\rho V A = (\rho + d\rho)(V + dV)(A),$$

which simplifies to

$$\rho\, dV + V\, d\rho = 0.$$

† Formally, we should consider a small control volume of length δx and fluid property changes δp, δV, and so on. The exact differential equations then result from taking the limit at the end of the derivation. This more exact procedure was used extensively in Chapters 2 and 5.

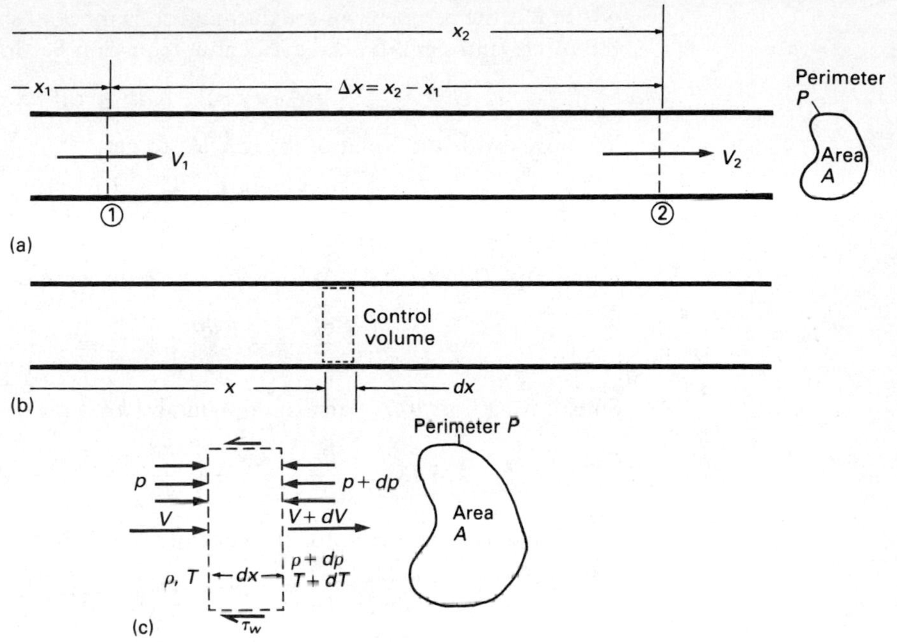

Figure 10.18 Analysis of flow in constant-area duct: (a) fundamental problem geometry; (b) differential control volume; (c) details of flow through control volume.

Dividing by ρV gives

$$\frac{dV}{V} + \frac{d\rho}{\rho} = 0. \tag{10.77}$$

From the equation of state,

$$\frac{d\rho}{\rho} = \frac{dp}{p} - \frac{dT}{T}. \tag{10.78}$$

The velocity equation is

$$V = \mathbf{M}c = \mathbf{M}\sqrt{kRT}.$$

Differentiating logarithmically gives

$$\frac{dV}{V} = \frac{d\mathbf{M}}{\mathbf{M}} + \frac{1}{2}\left(\frac{dT}{T}\right). \tag{10.79}$$

Substituting Eqs. (10.78) and (10.79) into (10.77) gives

$$\frac{d\mathbf{M}}{\mathbf{M}} + \frac{dp}{p} - \frac{1}{2}\left(\frac{dT}{T}\right) = 0. \tag{10.80}$$

The momentum equation for the control volume is

$$\sum dF_x = (\rho VA)[(V + dV) - V].$$

The forces are those from pressure and shear stress; that is,

$$\sum dF_x = pA - (p + dp)A - \tau_w P\,dx,$$

where P is the perimeter of the duct and τ_w is the wall shear stress. If we substitute the forces and simplify, the momentum equation becomes

$$\rho VA\,dV = -A\,dp - \tau_w P\,dx.$$

We now divide the equation by $(\rho V^2 A)$ to get

$$\frac{dV}{V} + \frac{dp}{\rho V^2} + \frac{\tau_w}{\rho V^2}\left(\frac{P}{A}\right) dx = 0. \tag{10.81}$$

Using Eqs. (10.29), (7.31), and (7.75), we reduce the momentum equation to

$$\frac{dV}{V} + \frac{1}{k\mathbf{M}^2}\left(\frac{dp}{p}\right) + \frac{1}{2}\,(4\mathbf{C_f})\,\frac{dx}{D_h} = 0.$$

Substituting Eq. (10.79) and multiplying by $k\mathbf{M}^2$ give

$$k\mathbf{M}\,d\mathbf{M} + \frac{dp}{p} + \frac{k\mathbf{M}^2}{2}\left(\frac{dT}{T}\right) + \frac{k\mathbf{M}^2}{2}\,(4\mathbf{C_f})\,\frac{dx}{D_h} = 0. \tag{10.82}$$

The energy equation for the control volume is

$$dq - dw_s = \left[(\tilde{h} + d\tilde{h}) + \frac{1}{2}\,(V + dV)^2\right] - \left[\tilde{h} + \frac{V^2}{2}\right].$$

For adiabatic flow and no work, this equation gives

$$d\tilde{h} + V\,dV = 0.$$

For an ideal gas,

$$c_p\,dT + V\,dV = 0. \tag{10.83}$$

Multiplying Eq. (10.83) by c_p gives

$$c_p\left[dT + d\left(\frac{V^2}{2c_p}\right)\right] = c_p\,dT_0 = 0;$$

thus the stagnation temperature of the gas is constant. Dividing Eq. (10.83) by $c_p T$ and using Eq. (10.79) and

$$\frac{V^2}{c_p T} = \frac{kR}{c_p}\left(\frac{V^2}{kRT}\right) = (k-1)\mathbf{M}^2,$$

we have

$$\left(1 + \frac{k-1}{2}\,\mathbf{M}^2\right)\frac{dT}{T} + (k-1)\mathbf{M}\,d\mathbf{M} = 0. \tag{10.84}$$

Equations (10.80), (10.82), and (10.84) contain four (dimensionless) differentials, dp/p, dT/T, $d\mathbf{M}$, and dx/D_h. As there are three equations, we can express three of the differentials in terms of one independent differential. The obvious choice for the independent variable is duct length dx/D_h. From Eq. (10.84),

$$\frac{dT}{T} = -\frac{(k-1)\mathbf{M}}{1 + [(k-1)/2]\mathbf{M}^2}\,d\mathbf{M}. \tag{10.85}$$

Substituting Eq. (10.85) into Eq. (10.80) gives

$$\frac{dp}{p} = -\left\{\frac{1}{\mathbf{M}} + \frac{[(k-1)/2]\mathbf{M}^2}{1+[(k-1)/2]\mathbf{M}^2}\right\} d\mathbf{M}. \tag{10.86}$$

Substituting Eqs. (10.85) and (10.86) into Eq. (10.82) gives, after simplification,

$$d\mathbf{M} = \frac{\mathbf{M}\{1+[(k-1)/2]\mathbf{M}^2\}}{(1-\mathbf{M}^2)}\left(\frac{k\mathbf{M}^2}{2}\right)\left(4\mathbf{C_f}\frac{dx}{D_h}\right). \tag{10.87}$$

Back substitution into Eqs. (10.85) and (10.86) gives

$$\frac{dT}{T} = -\frac{(k-1)\mathbf{M}^2}{(1-\mathbf{M}^2)}\left(\frac{k\mathbf{M}^2}{2}\right)\left(4\mathbf{C_f}\frac{dx}{D_h}\right) \tag{10.88}$$

and

$$\frac{dp}{p} = -\frac{1+(k-1)\mathbf{M}^2}{(1-\mathbf{M}^2)}\left(\frac{k\mathbf{M}^2}{2}\right)\left(4\mathbf{C_f}\frac{dx}{D_h}\right). \tag{10.89}$$

We can draw several conclusions about the nature of adiabatic frictional flow from an examination of Eqs. (10.87)–(10.89). We consider how the pressure, temperature, and Mach number change as the flow proceeds downstream (dx is positive). Note that changes of pressure, temperature, and Mach number can be either positive or negative, depending on the sign of $(1 - \mathbf{M}^2)$. If the flow is subsonic, friction causes the Mach number to increase and the pressure and temperature to decrease. The *pressure* change in subsonic flow is qualitatively similar to that in incompressible flow; surprisingly, however, friction causes the fluid to accelerate. If the flow is supersonic, exactly the opposite happens! In supersonic flow, friction causes the Mach number to decrease while the pressure and temperature increase.

Let's consider the special case of sonic flow ($\mathbf{M} = 1$). Since infinite values of $d\mathbf{M}$, dp, and dT are not physically possible, the Mach number can be unity *only* if $4\mathbf{C_f}\,(dx/D_h)$ is zero. This situation can occur in two ways: $\mathbf{C_f} = 0$ or $dx = 0$. If $\mathbf{C_f} = 0$, there is no friction and we are not considering a frictional flow but an isentropic flow. The condition $dx = 0$ means that the fluid cannot proceed down the duct any farther. This does not mean that the duct is closed but that the fluid is at the end of the duct. In frictional adiabatic flow, the Mach number can be 1 only at the end of the duct.

Frictional adiabatic flow in a constant-area duct is similar to isentropic flow in a converging nozzle in two respects.

- The Mach number is always driven toward a value of unity.
- The Mach number can reach unity only at the end of the passage.

Note that, although we can say that if $\mathbf{M} = 1$, then $dx = 0$, we cannot always conclude that if $dx = 0$, then $\mathbf{M} = 1$. Flow may be sonic at the end of a frictional duct, but it does not have to be.

Considering the effect of friction on the stagnation properties is instructive. For adiabatic flow with no work, the stagnation temperature is constant, and

$$dT_0 = 0.$$

The stagnation pressure is related to the static pressure and Mach number by

$$p_0 = p\left(1 + \frac{k-1}{2}\mathbf{M}^2\right)^{k/(k-1)}.$$

Differentiating logarithmically, we get

$$\frac{dp_0}{p_0} = \frac{dp}{p} + \left\{\frac{2k\mathbf{M}}{1 + [(k-1)/2]\mathbf{M}^2}\right\} d\mathbf{M}.$$

Substituting Eqs. (10.87) and (10.89) and simplifying, we get

$$\frac{dp_0}{p_0} = -\frac{k\mathbf{M}^2}{2}\left(4\mathbf{C_f}\frac{dx}{D_h}\right), \tag{10.90}$$

which tells us that the stagnation pressure always decreases. The minimum value of stagnation pressure occurs at the end of the duct. The dimensionless mass flow rate

$$\frac{\dot{m}\sqrt{RT_0}}{p_0 A}$$

is always largest at the end of the duct, because $\dot{m}$, T_0, and A are the same at all locations along the duct and p_0 is smallest at the end. Choking of the flow, if it occurs, can occur only at the *end* of the duct.

We can obtain equations for calculating pressure, temperature, and Mach number by integrating the differential equations from plane 1 to plane 2 in the duct. Equation (10.87) contains only Mach number and duct length as variables, so we can rearrange and integrate it:

$$\int_{x_1}^{x_2} 4\mathbf{C_f}\frac{dx}{D_h} = \int_{\mathbf{M}_1}^{\mathbf{M}_2} \frac{2(1-\mathbf{M}^2)}{k\mathbf{M}^3\{1 + [(k-1)/2]\mathbf{M}^2\}} d\mathbf{M}.$$

If we assume that $\mathbf{C_f}$ is constant, the integral on the left will be simple enough; however, the integral on the right will obviously produce a rather complicated function of k, $\mathbf{M}_1$, and $\mathbf{M}_2$. We can simplify calculations by introducing a reference state at which the value of $\mathbf{M}$ is known and then constructing tables of the resulting function. Because we know that the flow approaches a limiting value of $\mathbf{M} = 1$, we write

$$\int_{x}^{x+\ell^*} 4\mathbf{C_f}\frac{dx}{D_h} = \int_{\mathbf{M}}^{1} \frac{2(1-\mathbf{M}^2)}{k\mathbf{M}^3\{1 + [(k-1)/2]\mathbf{M}^2\}} d\mathbf{M},$$

where ℓ^* is the critical length, the length of duct necessary to change the Mach number from $\mathbf{M}$ at x to 1 at $x + \ell^*$. Integrating gives

$$4\mathbf{C_f}\frac{\ell^*}{D_h} = \frac{1-\mathbf{M}^2}{k\mathbf{M}^2} + \frac{k+1}{2k}\ln\left(\frac{(k+1)\mathbf{M}^2}{2\{1 + [(k-1)/2]\mathbf{M}^2\}}\right). \tag{10.91}$$

Equation (10.91) defines the *critical length function.* A table of values of $4\mathbf{C_f}\ell^*/D_h$ versus $\mathbf{M}$ is provided in Table E.3. The function is also evaluated by the compressible flow function program on the personal computer disk that is a companion to this textbook.

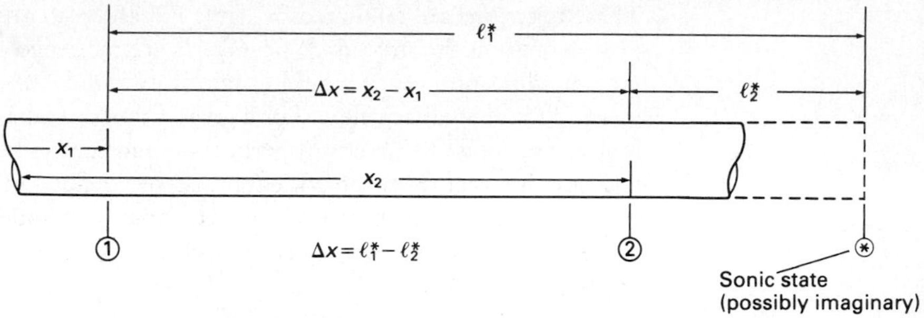

Figure 10.19 Relation between critical length and actual duct length.

Figure 10.19 illustrates use of the choking length to relate Mach number to duct length. The Mach numbers at 1 and 2 are related by

$$4C_f \frac{\Delta x}{D_h} = 4C_f \frac{\ell^*}{D_h} \{M_1\} - 4C_f \frac{\ell^*}{D_h} \{M_2\}. \tag{10.92}$$

The quantity on the left-hand side is calculated from the duct geometry and friction characteristics. The quantities on the right-hand side are calculated from Eq. (10.91) or, more frequently, obtained from Table E.3. Note that ℓ^* is a reference length; the actual duct length may be less than ℓ^*. Table E.3 shows that the value of $4C_f\ell^*/D_h$ becomes very large for small Mach number and that small changes of **M** cause large changes of the critical length function, making calculations difficult. If the largest Mach number in the duct is less than 0.2, modeling the flow as incompressible and using the methods of Chapter 7 is the best approach.

Static and stagnation pressure and static temperature are most easily calculated by relating them to Mach number and so indirectly to duct length via the critical length function. To accomplish this, we must integrate Eqs. (10.85), (10.86), and (10.90). To obtain an easily tabulated function, we take one limit of the integration to be the sonic condition, so, for example, the pressure integration is

$$\int_{p^*}^{p} \frac{dp}{p} = -\int_{1}^{M} \left\{ \frac{1}{M} + \frac{[(k-1)/2]M^2}{1 + [(k-1)/2]M^2} \right\} dM,$$

giving the pressure ratio function

$$\frac{p}{p^*} = \frac{1}{M} \left\{ \frac{(k+1)/2}{1 + [(k-1)/2]M^2} \right\}^{1/2}. \tag{10.93}$$

Similarly, we find

$$\frac{T}{T^*} = \frac{(k+1)/2}{1 + [(k-1)/2]M^2} \tag{10.94}$$

and

$$\frac{p_0}{p_0^*} = \frac{1}{M} \left\{ \frac{1 + [(k-1)/2]M^2}{(k+1)/2} \right\}^{(k+1)/2(k-1)}. \tag{10.95}$$

These functions are tabulated in Table E.3 and also are available on the computer disk. Note that the starred (*) properties are reference values, defined as the values that the fluid properties would obtain if the fluid were brought to the sonic state in a frictional adiabatic flow. For a given frictional adiabatic flow, there is a unique limiting state, and starred properties are constant. The starred properties defined here are different from the critical properties defined in Section 10.3.3, because the earlier definition was based on an isentropic (frictionless) process.

Static and stagnation pressures and static temperatures at two points in a duct are related by

$$\frac{p_2}{p_1} = \frac{p}{p^*}\{\mathbf{M}_2\} \bigg/ \frac{p}{p^*}\{\mathbf{M}_1\}; \tag{10.96}$$

$$\frac{p_{0_2}}{p_{0_1}} = \frac{p_0}{p_0^*}\{\mathbf{M}_2\} \bigg/ \frac{p_0}{p_0^*}\{\mathbf{M}_1\}; \tag{10.97}$$

$$\frac{T_2}{T_1} = \frac{T}{T^*}\{\mathbf{M}_2\} \bigg/ \frac{T}{T^*}\{\mathbf{M}_1\}. \tag{10.98}$$

The terms on the right are calculated by Eqs. (10.93)–(10.95) or, more frequently, obtained from tables.

EXAMPLE 10.8 Illustrates Computations in Fanno Flow

A converging nozzle is connected to a large tank of air by a long pipe (see Fig. E10.8). The flow rate through the nozzle is 10 lbm/sec and the nozzle is choked. Calculate the static and stagnation pressures at planes 1 and 2.

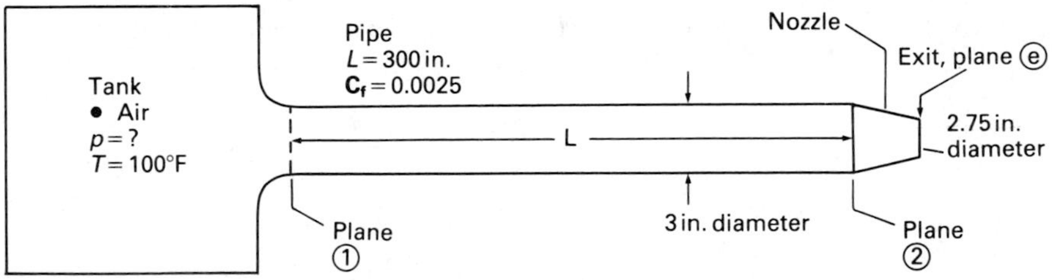

Figure E10.8 Tank–pipe–nozzle system.

SOLUTION

Given

Converging nozzle connected to air tank by pipe with $L = 300$ in., $D = 3.0$ in., and $\mathbf{C_f} = 0.0025$

Nozzle exit diameter 2.75 in.

Nozzle passes 10 lbm/sec of air and is choked.

Air temperature in tank 100°F

Find

Static and stagnation pressures at pipe exit/nozzle entrance and at tank exit/pipe entrance

Solution

First convert given temperatures to absolute units:

$$T_{\text{tank}} = 100°\text{F} + 459.7°\text{F} = 559.7°\text{R}.$$

To solve the problem, we break the system into parts. We assume isoenergetic-isentropic flow in the exit nozzle and isoenergetic frictional (Fanno) flow in the pipe. The exit nozzle is choked, so we start there and work backward.

The nozzle flow rate is related to the nozzle stagnation pressure by Eq. (10.72), which, for $k = 1.4$, is

$$\dot{m} = 0.6847\left(\frac{p_{0_e} A_e}{\sqrt{R T_{0_e}}}\right),$$

where

$$A_e = \frac{\pi}{4} D_e^2 = 5.94 \text{ in}^2 \quad \text{and} \quad T_{0_e} = T_0 = T_{\text{tank}} = 559.7°\text{R}.$$

Solving for p_{0_e}, we have

$$p_{0_e} = \frac{\dot{m}\sqrt{R T_0}}{0.6847 A_e}.$$

For isentropic nozzle flow,

$$p_{0_2} = p_{0_e} = \frac{10 \text{ lbm/sec} \sqrt{53.35 \text{ ft}\cdot\text{lb/lbm}\cdot°\text{R}\ (559.7°\text{R})}}{(0.6847)(5.94 \text{ in}^2)\sqrt{32.2 \text{ ft}\cdot\text{lbm/lb}\cdot\text{sec}^2}};$$

$$p_{0_2} = 74.9 \text{ psia}.$$ **ANSWER**

The static pressure at 2 is

$$p_2 = p_{0_2}\left(\frac{p_2}{p_{0_2}}\right).$$

To find p_2/p_{0_2}, we note that

$$\left.\frac{A}{A^*}\right)_2 = \frac{A_2}{A_e} = \left(\frac{D_2}{D_e}\right)^2 = \left(\frac{3.0 \text{ in.}}{2.75 \text{ in.}}\right)^2 = 1.190.$$

From Table E.1,

$$\mathbf{M}_2 = 0.599 \quad \text{and} \quad \frac{p_2}{p_{0_2}} = 0.7846.$$

Thus

$$p_2 = 74.9 \text{ psia } (0.7846);$$

$$p_2 = 58.8 \text{ psia}.$$ **ANSWER**

To calculate the static and stagnation pressures at plane 1, we apply Fanno flow analysis between planes 1 and 2. From Eq. (10.92),

$$\left.\frac{4C_f \ell^*}{D_h}\right)_1 = 4C_f \left.\frac{\ell^*}{D_h}\right)_2 + 4C_f \frac{\Delta x}{D_h}.$$

From Table E.3, with $\mathbf{M}_2 = 0.599$,

$$\left.\frac{4C_f \ell^*}{D_h}\right)_2 = 0.4948.$$

Also,

$$4C_f \frac{\Delta x}{D_h} = 4C_f \frac{L}{D} = 4(0.0025)\left(\frac{300 \text{ in.}}{3 \text{ in.}}\right) = 1.0,$$

so,

$$4C_f \left.\frac{\ell^*}{D_h}\right)_1 = 1.0 + 0.4948 = 1.4948.$$

From Table E.3,

$$\mathbf{M}_1 = 0.456.$$

The static pressure at plane 1 is given by Eq. (10.96):

$$p_1 = p_2\left(\frac{p}{p^*}\{\mathbf{M}_1\} \Big/ \frac{p}{p^*}\{\mathbf{M}_2\}\right).$$

With the aid of Table E.3, we have

$$p_1 = 58.8 \text{ psia}\left(\frac{2.354}{1.767}\right);$$

$$p_1 = 78.3 \text{ psia.}$$ **ANSWER**

The stagnation pressure at plane 1 is

$$p_{0_1} = \frac{p_1}{\frac{p_1}{p_0}\{\mathbf{M}_1\}} = \frac{78.3}{0.8671} = \text{psia};$$

$$p_{0_1} = 90.3 \text{ psia.}$$ **ANSWER**

Discussion

This complicated-looking problem became relatively simple when we broke it into pieces. Note that although we cannot use isentropic tables and functions to relate the flow at plane 1 to the flow at plane 2, we do use the isentropic tables to relate static and stagnation properties at the same location.

An important question for many practical applications is how to find C_f. From dimensional analysis, we expect that

$$C_f = C_f\left\{\mathbf{R}, \mathbf{M}, \frac{\varepsilon}{D}\right\}.$$

Based on a rather small amount of experimental evidence, dependence of C_f on $\mathbf{M}$ appears to be weak, at least for subsonic flow; accordingly, values of C_f for $\mathbf{M} = 0$ (incompressible flow) are usually sufficient. For fully developed flow, we obtain values for C_f from the Moody chart (Fig. 7.9), the Colebrook equation, Eq. (7.48), or the Hagen–Poiseuille equation, Eq. (7.38), by noting that the Darcy friction factor thus obtained is related to the skin friction coefficient by

$$C_f = \frac{f}{4}. \tag{10.99}$$

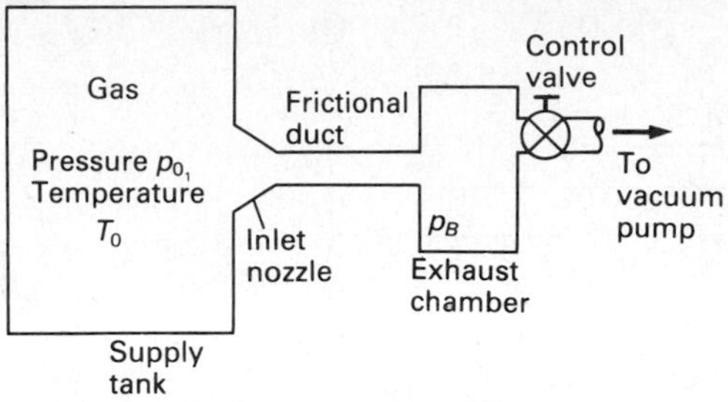

Figure 10.20 Apparatus for studying (subsonic) flow in a frictional duct.

The factor of 4 is rather convenient, as we can write

$$4C_f \frac{\Delta x}{D_h} = f \frac{\Delta x}{D_h}.$$

Because f depends on Reynolds number, our assumption of a constant value of C_f along the duct seems questionable. The Reynolds number can be written

$$\mathbf{R} = \frac{\rho V D_h}{\mu} = \frac{\dot{m}}{A}\frac{D_h}{\mu} = \frac{\dot{m}}{\mu}\frac{D_h}{A}. \tag{10.100}$$

The mass flow rate, hydraulic diameter, and cross-sectional area are all constant, so the variation of Reynolds number along the duct results only from viscosity. For a gas, viscosity varies rather slowly with temperature ($\mu \sim T^{0.7}$). In addition, the dependence of f on **R** is weak in turbulent flow, which is the usual condition at velocities high enough for compressibility to be significant.

10.5.3 Frictional Flow in a Duct

We now consider the various possibilities for frictional flow in a duct. Figure 10.20 shows a duct that exhausts fluid from a large reservoir to a region with constant back pressure. We assume that the connection between the duct and the inlet reservoir is a converging nozzle with minimum area equal to the duct area.† The flow entering the duct must then be subsonic. The maximum possible Mach number anywhere in the flow is 1, which can occur only at the *end* of the *duct*. The nozzle exit-duct entrance Mach number *cannot* be 1 unless the duct is frictionless. Figure 10.21 shows the pressure and Mach number distributions along the duct, the dimensionless mass flow rate, and the duct exit pressure, all as functions of the ratio of back pressure to supply pressure (p_B/p_{0_1}). In the case labeled a, the pressures are equal and there is no flow. In case b, the back pressure is lowered and there is flow. The gas accelerates in the inlet nozzle, enters the duct with a low subsonic Mach number, and accelerates in the duct. At the end of the duct, the flow is still subsonic, and the exit pressure and back pressure are equal. The inlet Mach number, $\mathbf{M}_1$, is small enough that

$$4C_f \frac{\ell^*}{D_h}\{\mathbf{M}_1\} > 4C_f \frac{L}{D_h}.$$

† The inlet nozzle need not be smoothly contoured. The duct may be directly attached to a hole in the tank.

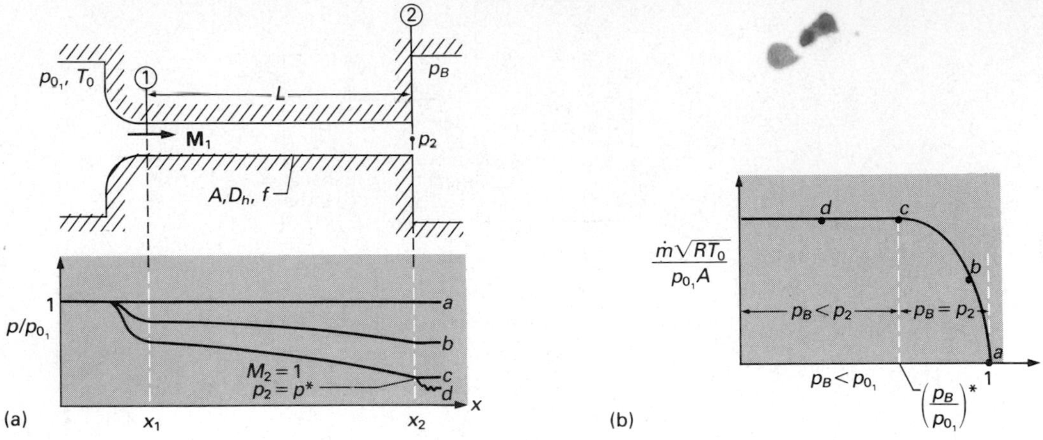

Figure 10.21 Performance of a frictional duct fed by a converging nozzle.

In case *c*, the back pressure has been lowered to the point where the flow has just become sonic at the exit. In this case, the entering flow is still subsonic, but the gas accelerates to $\mathbf{M} = 1$ at the duct exit. The exit pressure is equal to p^* and also equal to p_B. The mass flow rate is greater than in case *b*. The duct entrance Mach number satisfies the condition

$$4\mathbf{C_f}\frac{\ell^*}{D_h}(\mathbf{M}_1) = 4\mathbf{C_f}\frac{L}{D_h}. \tag{10.101}$$

Case *d* represents a lower back pressure than case *c* does. The flow was choked in case *c*, so lowering the back pressure has no effect on the flow in the nozzle and duct. The entrance Mach number is still subsonic, and Eq. (10.101) still applies. The mass flow rate is the same as in case *c*. The exit plane pressure is equal to p^* and is higher than the back pressure.

The procedure for analysis of flow in a frictional duct depends on whether the flow is choked or unchoked. We decide by comparing the actual pressure ratio, p_B/p_{0_1}, to the value that corresponds to choking $(p_B/p_{0_1})^*$. The latter is calculated in the following manner:

- Calculate $4\mathbf{C_f}\ell^*/D_h$ from Eq. (10.101).
- Find $\mathbf{M}_1$ from Eq. (10.91) or Table E.3.
- Find p_1/p_{0_1} from Eq. (10.31) or Table E.1.
- Find p_1/p^* from Eq. (10.93) or Table E.3.
- Then

$$\left(\frac{p_B}{p_{0_1}}\right)^* = \frac{p_1/p_{0_1}}{p_1/p^*}. \tag{10.102}$$

If p_B/p_{0_1} is less than $(p_B/p_{0_1})^*$, the flow is choked, and the fact that $\mathbf{M}_{\text{exit}} = 1$ is the key to solving the problem. If p_B/p_{0_1} is greater than $(p_B/p_{0_1})^*$, the flow is not choked, and the key to the problem is the equality of p_e and p_B.

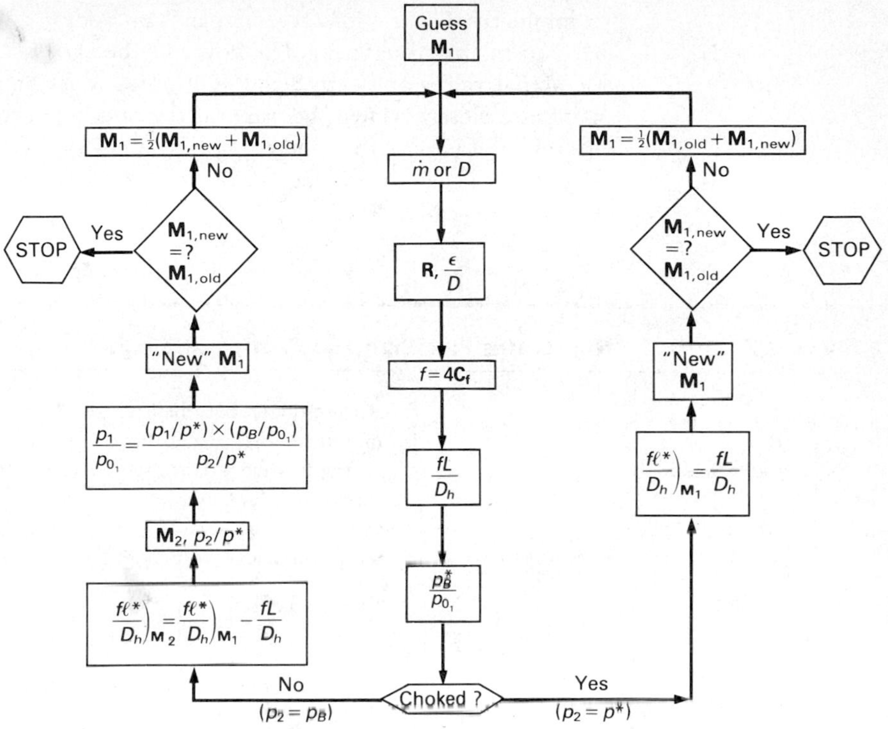

Figure 10.22 Flow chart for frictional duct flow calculation.

As in incompressible duct flow, there are three types of problems:

- pressure change calculation;
- mass flow rate calculation; and
- duct sizing.

The pressure change calculation involves fluid property changes between two locations in the duct. Usually, one of these locations is the duct exit. In this type of problem, the duct dimensions, mass flow rate, and fluid properties at one location are given either explicitly or implicitly.† The Reynolds number and relative roughness can be calculated from the given data, and the friction coefficient is readily obtained from the Moody diagram or Colebrook formula. Fluid properties and velocity or Mach number at other locations in the duct can be calculated by direct application of Eqs. (10.91)–(10.98) or by using tables. Example 10.8 illustrated this type of problem.

Both mass flow rate calculation and duct sizing require iterative solution. A flow chart that works equally well for either problem is shown in Fig. 10.22. In both cases, duct length, inlet (static or stagnation) pressure, and exhaust region pressure are given. The fluid stagnation temperature is also given, either explicitly

† If pressure, temperature, Mach number, and cross-sectional area are known, mass flow rate can be immediately calculated.

or implicitly. The friction factor depends on both $\dot{m}$ and D_h via the Reynolds number and relative roughness. The flow may be choked or unchoked, depending on the actual values of p_B, p_{0_1}, and $4\mathbf{C_f}(L/D_h)$. Mass flow rate calculation and duct sizing are closely related, because, at the duct inlet, the mass flow and duct area are related to each other and to p_{0_1}, T_0, and $\mathbf{M}_1$ by Eq. (10.71). Either the mass flow rate calculation or duct sizing is carried out by iterating on the duct inlet Mach number $\mathbf{M}_1$.

EXAMPLE 10.9 Illustrates Pipe Sizing in Compressible Flow

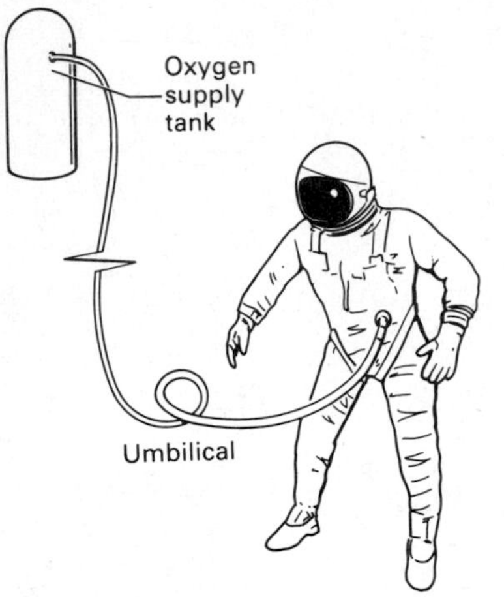

Figure E10.9a Astronaut being supplied with oxygen through an umbilical.

An astronaut breathes oxygen supplied to a spacesuit through an umbilical cord (Fig. E10.9a). The umbilical is 7 m long. The oxygen is supplied from a storage tank at 200 kPa and 9.8°C. The suit pressure is 20 kPa. Compute the required inside diameter of the umbilical to supply 0.05 kg/s of oxygen. The inside of the umbilical is lined with a material with roughness of 0.01 mm.

SOLUTION

Given

7-m-long umbilical (duct) with surface roughness 0.01 mm ($= 10^{-5}$ m)

Flow rate 5×10^{-2} kg/s to be delivered from reservoir at 200 kPa and 9.8°C to region (spacesuit) at 20 kPa

Find

Required inside diameter of umbilical

Solution

Figure E10.9b is a schematic diagram of the duct flow problem. First convert all given pressures and temperatures to absolute units:

$$T_{\text{tank}} = 9.8°\text{C} + 273.2°\text{C} = 283\text{K}.$$

The flow in the umbilical obviously is subsonic (no converging–diverging nozzle at the inlet), so the problem can be solved by following the flow chart in Fig. 10.22. We first obtain the properties of oxygen. They are $R = 260$ N·m/kg·K, $k = 1.4$, and $\mu = 1.97 \times 10^{-5}$ N·s/m² from Table A.3.

For a circular duct, from Eq. (10.100),

$$\mathbf{R} = \frac{\dot{m}}{\mu}\frac{D}{\pi D^2/4} = \frac{4\dot{m}}{\pi \mu D}.$$

Substituting numbers where possible, we have

$$\mathbf{R} = \frac{(4)(0.05\ \text{kg/s})}{(\pi)(1.97 \times 10^{-5}\ \text{N}\cdot\text{s/m}^2)D} = \frac{3230}{D}, \qquad \text{(E10.1)}$$

where D is in meters.

The equation for flow rate at plane 1 is Eq. (10.71). As we know, $\dot{m}$ but do not know the area A, we rearrange the equation and sub-

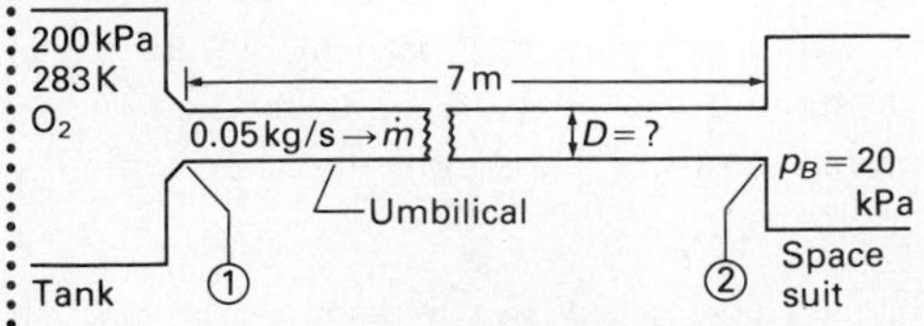

Figure E10.9b Schematic diagram of space suit umbilical.

stitute $A = \pi D^2/4$ to get the following:

$$D^2 = \frac{4}{\pi}\left(\frac{\dot{m}\sqrt{RT_0}}{p_{0_1}\sqrt{k}}\right)\left[\frac{1}{\mathbf{M}_1}\left(1 + \frac{k-1}{2}\mathbf{M}_1^2\right)^{(k+1)/2(k-1)}\right].$$

We can speed up the calculations considerably if we recognize that the quantity in brackets (see Eq. 10.67) is

$$\left(\frac{k+1}{2}\right)^{(k+1)/2(k-1)} \frac{A}{A^*}\{\mathbf{M}_1\} = 1.728\,\frac{A}{A^*}\{\mathbf{M}_1\},$$

for $k = 1.4$. Thus

$$D^2 = \frac{4}{\pi}\left[\frac{(0.05 \text{ kg/s})\sqrt{(260 \text{ N}\cdot\text{m/kg}\cdot\text{K})(283 \text{ K})}}{(200{,}000 \text{ N/m}^2)\sqrt{1.4}}\right](1.728)\,\frac{A}{A^*}\{\mathbf{M}_1\}$$

and

$$D = \sqrt{D^2} = 0.0112\left(\sqrt{\frac{A}{A^*}\{\mathbf{M}_1\}}\right). \qquad \text{(E10.2)}$$

The ratio of back pressure to supply pressure is

$$\frac{p_B}{p_{0_1}} = \frac{20 \text{ kPa}}{200 \text{ kPa}} = 0.1.$$

We now begin following the flow chart of Fig. 10.22. We guess that

$$\mathbf{M}_1^{(1)} = 0.5.$$

The superscript (1) indicates that this is our first guess. From Table E.1,

$$\frac{A_1}{A^*} = 1.340.$$

From Eq. (E10.2),

$$D = 0.0112\sqrt{1.340} = 0.013 \text{ m}.$$

From Eq. (E10.1),

$$\mathbf{R} = \frac{3230}{0.013} = 2.48 \times 10^5.$$

Also,

$$\frac{\varepsilon}{D} = \frac{1 \times 10^{-5} \text{ m}}{1.3 \times 10^{-2} \text{ m}} = 0.00077.$$

From Fig. 7.9 (the Moody chart),

$$f = 4\mathrm{C_f} = 0.020.$$

The duct length parameter is

$$4\mathbf{C_f}\frac{L}{D} = (0.020)\left(\frac{7 \text{ m}}{0.013 \text{ m}}\right) = 10.77.$$

Next, we calculate the choking pressure ratio for this duct length parameter. We let $\mathbf{M}_{1_c}$ be the inlet Mach number corresponding to choking. From Table E.3, for $4C_f\ell^*/D_h = 10.77$,

$$M_{1_c} = 0.227 \quad \text{and} \quad \frac{p}{p^*}\{\mathbf{M}_{1_c}\} = 4.801.$$

From Table E.1,

$$\frac{p}{p_0}\{\mathbf{M}_{1_c}\} = 0.9648.$$

From Eq. (10.102),

$$\frac{p_B^*}{p_{0_1}} = \frac{0.9648}{4.801} = 0.201.$$

Because

$$\frac{p_B}{p_{0_1}} = 0.10 < \frac{p_B^*}{p_{0_1}} = 0.201,$$

the duct seems to be choked, and we follow the right branch of Fig. 10.22. Setting $4C_f\ell^*/D_{h_1} = 4C_f(L/D) = 10.77$, we find that the "new" Mach number is

$$\mathbf{M}_{1,\text{new}} = \mathbf{M}_{1_c}.$$

As $\mathbf{M}_{1_c} = 0.227 \neq 0.5 = \mathbf{M}_1^{(1)}$, we now try

$$\mathbf{M}_1^{(2)} = \frac{0.227 + 0.5}{2} = 0.36.$$

From Table E.1,

$$\frac{A_1}{A^*} = 1.736.$$

From Eqs. (E10.2) and (E10.1), $D = 0.0148$ m and $\mathbf{R} = 2.18 \times 10^5$. Also,

$$\frac{\varepsilon}{D} = \frac{10^{-5}\ \text{m}}{1.48 \times 10^{-2}\ \text{m}} = 6.76 \times 10^{-4}$$

and, from Fig. 7.11,

$$f = 4C_f = 0.019.$$

This gives

$$4C_f\frac{L}{D} = 0.019\left(\frac{7\ \text{m}}{0.0148\ \text{m}}\right) = 8.99.$$

From Tables E.3 and E.1,

$$\mathbf{M}_{1_c} = 0.244, \qquad \frac{p}{p^*}\{\mathbf{M}_{1_c}\} = 4.463, \quad \text{and} \quad \frac{p}{p_0}\{\mathbf{M}_{1_c}\} = 0.9594,$$

so

$$\frac{p_B^*}{p_{0_1}} = \frac{0.9594}{4.463} = 0.215.$$

The duct seems to be choked and so the "new" Mach number is $\mathbf{M}_{1_c}$. We now repeat the procedure with

$$\mathbf{M}_1^{(3)} = \tfrac{1}{2}(0.36 + 0.244) = 0.30.$$

The results of this calculation are

$$D = 0.016 \text{ m}, \quad \mathbf{R} = 2.02 \times 10^5, \quad \frac{\varepsilon}{D} = 6.25 \times 10^{-4}, \quad f = 0.019,$$

$$4\mathbf{C_f}\frac{L}{D} = 8.31, \qquad \mathbf{M}_{1_c} = 0.252, \quad \text{and} \quad \frac{p_B^*}{p_{0_1}} = 0.221 \qquad \text{(Choked)}.$$

The implication is

$$\mathbf{M}_1^{(4)} = \tfrac{1}{2}(0.30 + 0.252) = 0.28.$$

The results for this trial are

$$D = 0.0165 \text{ m}, \quad \mathbf{R} = 1.96 \times 10^5, \quad \frac{\varepsilon}{D} = 6.07 \times 10^{-4}, \quad f = 0.019,$$

$$4\mathbf{C_f}\frac{L}{D} = 8.060, \quad \mathbf{M}_{1_c} = 0.255, \quad \text{and} \quad \frac{p_B^*}{p_{0_1}} = 0.224 \qquad \text{(Choked)}.$$

Examining the calculations, we note that $\mathbf{M}_{1_c}$ has been changing very little. We make a calculation with

$$\mathbf{M}_1^{(5)} = 0.26.$$

The parameters of this calculation are

$$D = 0.017 \text{ m}, \qquad \mathbf{R} = 1.90 \times 10^5,$$

$$\frac{\varepsilon}{D} = 5.88 \times 10^{-4}, \qquad f = 0.019, \qquad 4\mathbf{C_f}\frac{L}{D} = 7.82,$$

$$\mathbf{M}_{1_c} = 0.258, \quad \text{and} \quad \frac{p_B^*}{p_0} = 0.226 \qquad \text{(Choked)}.$$

Because $\mathbf{M}_{1c} = 0.258 \approx 0.260 = \mathbf{M}_1^{(5)}$, the iteration has converged, and we specify

$$D = 0.017 \text{ m} = 1.7 \text{ cm}$$ **ANSWER**

Discussion

The iteration procedure converged slowly. The reason, in part, is that we averaged the "new" and "old" Mach numbers before repeating calculations. A table of values of "old" and "new" Mach numbers might be useful:

Trial	$\mathbf{M}_{old}$	$\mathbf{M}_{new}$
1	0.5	0.227
2	0.36	0.244
3	0.30	0.252
4	0.28	0.255
5	0.26	0.258

The "new" Mach numbers are changing very slowly, so they likely are close to the correct value. We probably should "weigh" them more heavily by an equation such as

$$\mathbf{M}_1^{(i+1)} = a\mathbf{M}_{1,\text{old}}^{(i)} + (1 - a)\mathbf{M}_{1,\text{new}}^{(i)},$$

where a is selected between 0 and 1. Selecting $a = 0$ would produce more rapid convergence in the example, because the "new" Mach numbers are always smaller than the "old" Mach numbers. If the "new" Mach numbers are alternately larger and smaller than the "old" ones, the iteration can become unstable if $a = 0$. Unstable iteration can occur on the left branch of the flow chart of Fig. 10.22 (nonchoked flow), so sticking with $a = 0.5$ in that case is a good idea. Note that we used $a = 0.2$ in our last iteration.

A Note about Supersonic Flow. Our discussion has concentrated on subsonic frictional flow. A discussion of supersonic frictional flow would be interesting but of limited practical value compared to the effort expended for two reasons. First, Eq. (10.90) shows that loss of stagnation pressure is very large if the Mach number is greater than 1, so supersonic flow is usually avoided if friction is significant. Second, interaction between shocks, which can occur in supersonic flow, and boundary layers is a complex multidimensional process. The one-dimensional Fanno flow model is not sufficiently accurate to be of value. If you are interested in discussions of supersonic Fanno flow, see references [1–4].

10.6 EFFECTS OF COMPRESSIBILITY IN EXTERNAL FLOW

Before ending our consideration of compressible flow, let's briefly consider the effects of compressibility in external flow. This discussion is mainly qualitative. Suppose that an object such as an airfoil is moving at high speed through a compressible fluid (Fig. 10.23). The nature of the flow over the airfoil is different for subsonic, transonic, and supersonic motion. The key to understanding the differences is the discussion of Section 10.2.2. If the flow is subsonic everywhere, disturbances introduced by the airfoil motion are propagated to all parts of the fluid. The flow pattern (Fig. 10.23a) is similar to that which we would expect in a low-speed (incompressible) flow.

If the flow is supersonic (Fig. 10.23b), the upstream influence of the airfoil is limited, extending no farther than a leading shock wave. The disturbances caused by the airfoil will be concentrated in waves. In general, the fluid contains both

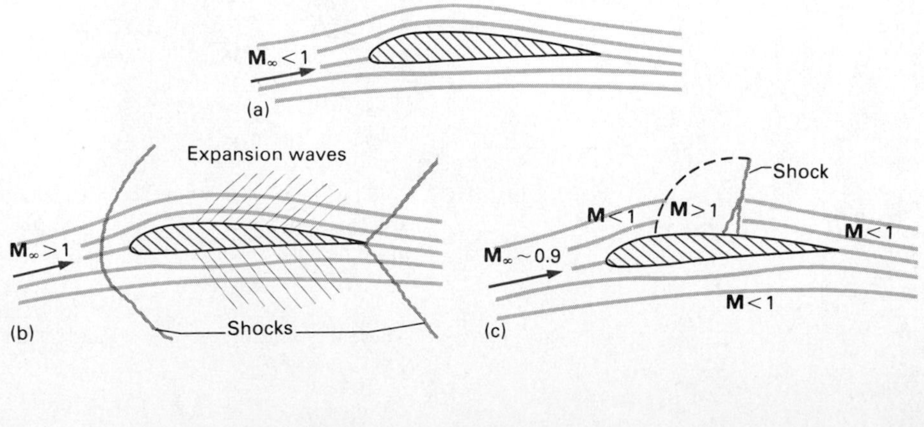

Figure 10.23 Airfoil in compressible flow: (a) subsonic flow; (b) supersonic flow; (c) transonic flow.

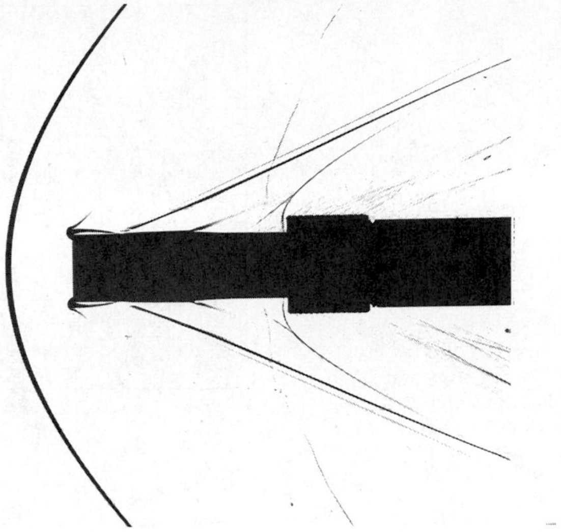

Figure 10.24 Supersonic flow over an object is influenced by a complex wave pattern. (Courtesy of Launch and Flight Division, Ballistic Research Laboratory/ARRADCOM, Aberdeen Proving Ground, MD.)

waves of finite strength (shocks) and waves of infinitesimal strength (Mach waves). Shock waves result from sudden compression of the fluid, and Mach waves result from expansion or gradual compression of the fluid. In external flow, these waves are multidimensional and cause multidimensional flow. Figure 10.24 illustrates these waves in an actual flow. Although the wave pattern and the resulting flow appear to be complex, two- and three-dimensional supersonic flow can be analyzed by first analyzing the basic waves (oblique shocks and Mach waves) and then "patching" the waves together to obtain a complete flow field [1–6].

Figure 10.23(c) illustrates transonic flow, in which some regions of the field are subsonic and some regions are supersonic. Shock waves exist in the supersonic portion of the flow. Analysis of transonic flow is very difficult because it is neither entirely like subsonic flow nor entirely like supersonic flow.

Dimensional analysis of compressible external flow indicates that drag and lift coefficients depend on Mach number as well as Reynolds number:

$$\mathbf{C_D} = \mathbf{C_D}\{\alpha, \mathbf{R}, \mathbf{M}_\infty\} \quad \text{and} \quad \mathbf{C_L} = \mathbf{C_L}\{\alpha, \mathbf{R}, \mathbf{M}_\infty\}.$$

Obtaining test data in which both Reynolds number and Mach number are controlled is very difficult (see Section 6.6). Most studies concentrate on the effects of Reynolds number alone (usually at low Mach number; see Chapter 8) or on the effects of Mach number alone. For high Reynolds number,

$$\mathbf{C_D} \approx \mathbf{C_D}\{\alpha, \mathbf{M}_\infty \text{ only}\} \quad \text{and} \quad \mathbf{C_L} \approx \mathbf{C_L}\{\alpha, \mathbf{M}_\infty \text{ only}\}.$$

Figure 10.25 illustrates the effect of Mach number on drag coefficient for three axisymmetric bodies. Note the following important effects:

- The drag coefficient is roughly constant for very low and very high Mach numbers.
- The drag coefficient at supersonic speeds is higher than the drag coefficient at low subsonic speeds.

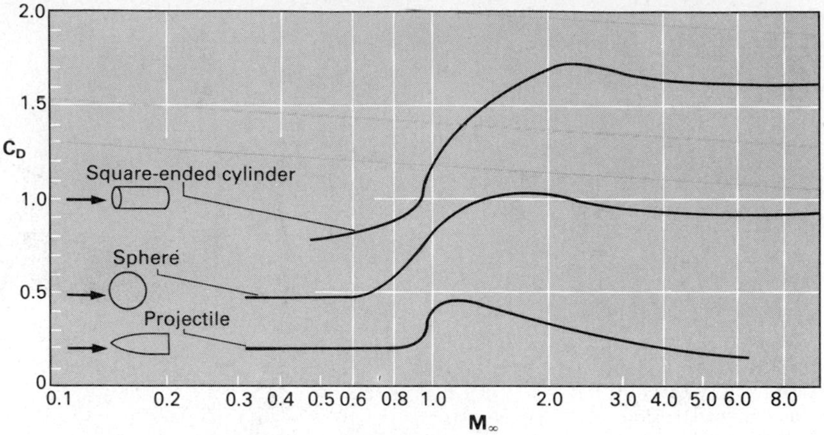

Figure 10.25 Effect of Mach number on drag of three different axisymmetric objects. Reynolds number effects are negligible.

- The drag coefficient rises sharply near $M_\infty = 1$.
- The value of the Mach number at which the "drag rise" occurs is pushed closer to 1 by making the body more "streamlined."

Some of these effects can be explained as follows. In supersonic flow, energy is required to maintain the wave pattern, so the drag is higher than in subsonic flow with no waves. The drag rise near $M_\infty = 1$ is associated with the sudden appearance of (nearly) normal shock waves in locally supersonic regions of the flow. The (slight) drag drop after $M_\infty = 1$ corresponds to a relative weakening of these shock waves as they become inclined to the flow. The drag rise delay for a more streamlined body is due to the fact that the first appearance of locally supersonic flow and shock waves is delayed by streamlining.

The value of M_∞ at which the highest local Mach number in the field first becomes 1 is called the *critical Mach number*. Note that, as the Mach number near the maximum body thickness is always greater than the free-stream Mach number, the critical Mach number is always less than 1. The "drag rise" occurs at a Mach number slightly higher than the critical Mach number. The drag rise that occurs near "Mach One" is the true sound barrier that had to be overcome before supersonic flight was possible.

The effects of Mach number on lift and drag at subsonic and supersonic (but *not* transonic) speed are illustrated by the following equations [1, 3, 6].

- Prandtl–Glauert rules for subsonic flow:

$$C_D\{M_\infty\} \approx \frac{C_D\{M_\infty = 0\}}{\sqrt{1 - M_\infty^2}}; \tag{10.103}$$

$$C_L\{M_\infty\} \approx \frac{C_L\{M_\infty = 0\}}{\sqrt{1 - M_\infty^2}}; \tag{10.104}$$

where $M_\infty = 0$ means incompressible flow.

- Ackeret rules for supersonic flow:

$$C_D\{M_\infty\} \approx \frac{\text{Constant}}{\sqrt{M_\infty^2 - 1}}; \tag{10.105}$$

$$C_L\{M_\infty\} \approx \frac{\text{Constant}}{\sqrt{M_\infty^2 - 1}}. \tag{10.106}$$

These rules apply only to *thin, two-dimensional* objects. The equations obviously cannot be valid when M_∞ is 1 and become less accurate as M_∞ approaches 1. The angle of attack must be constant and, strictly speaking, so must the Reynolds number. The "constant" in the Ackeret rules for supersonic flow is not the incompressible flow value of the corresponding coefficient, because supersonic flow is not qualitatively or quantitatively similar to incompressible flow. The "incompressible" flow coefficients and the "constants" in the Ackeret rules are functions of body shape and Reynolds number. You should consider Eqs. (10.103)–(10.106) as illustrative, rather than directly useful for calculations.

EXAMPLE 10.10 Illustrates Aerodynamic Force Calculations in Compressible Flow

The astronaut in Example 10.9 enters a spherical reentry capsule and returns to Earth. The reentry capsule is 2 m in diameter and has a mass of 1250 kg. At one point during the reentry, the capsule's altitude is 30 km, and it is traveling with a speed of 2300 m/s at an angle of 20° below the horizontal. Calculate the acceleration of the capsule in g's. Also estimate the temperature on the windward surface of the capsule.

SOLUTION

Given

1250-kg spherical reentry capsule traveling at 2300 m/s at altitude of 30 km

Capsule velocity vector 20° below horizontal

Capsule diameter 2 m

Figure E10.10a

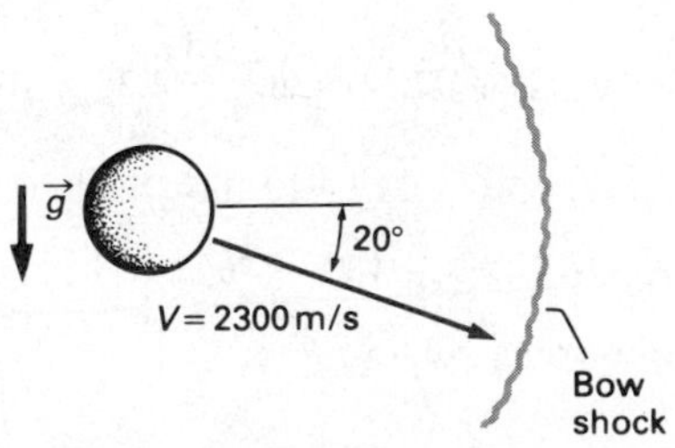

Figure E10.10a Spherical reentry capsule in flight.

Find

Capsule acceleration in g's

Temperature of windward surface

Solution

The capsule acceleration is computed from Newton's second law:

$$\vec{a} = \frac{\sum \vec{F}}{m} = \frac{\vec{F}_R}{m}.$$

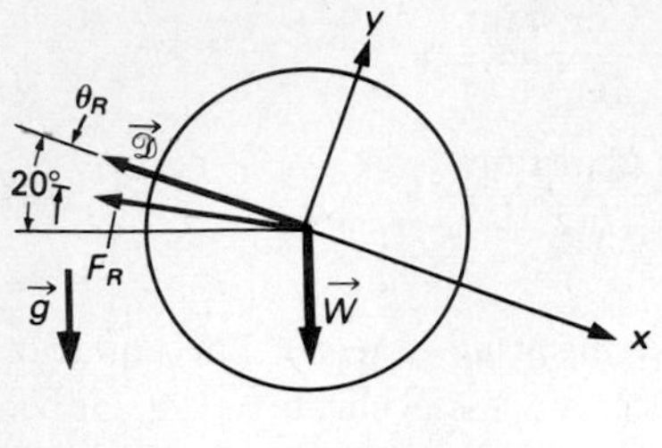

Figure E10.10b Free-body diagram of reentry capsule. $\vec{F}_R$ is the resultant force.

Figure E10.10b shows a free-body diagram of the capsule. The x axis is aligned with the direction of travel. The mass m includes the mass of the capsule and the mass of the astronaut inside. The astronaut's mass is not given, so we have to assume a reasonable value, say, 80 kg. Then

$$m = 1250 \text{ kg} + 80 \text{ kg} = 1330 \text{ kg}.$$

The forces on the capsule are the gravity force $\vec{W}$ and the aerodynamic drag, $\vec{\mathscr{D}}$. The magnitude of $\vec{W}$ is

$$W = mg.$$

At an altitude of 30 km, the value of g is found in Table A.1, or

$$g = 9.72 \text{ m/s}^2,$$

which gives

$$W = (1330 \text{ kg})(9.72 \text{ m/s}^2) = 1.29 \times 10^4 \text{ N}.$$

The magnitude of the drag force is

$$\mathscr{D} = \mathbf{C_D}\left(\frac{1}{2}\rho V^2\right)\left(\frac{\pi}{4}D^2\right),$$

where D is the capsule diameter. From Eq. (10.29), we have

$$\rho V^2 = kp\mathbf{M}^2, \quad \text{so} \quad \mathscr{D} = \frac{\pi}{8}\mathbf{C_D}kp\mathbf{M}^2D^2.$$

Figure 10.25 gives drag coefficients as a function of Mach number for spheres. To compute the Mach number, we need the temperature. From Eqs. (2.7)–(2.9), Fig. 2.6, or Table A.1, $T = 226.5\text{K}$ and $p = 1.20$ kPa at 30-km altitude. Then

$$c = \sqrt{kRT} = (20.05\sqrt{226.5}) \text{ m/s} = 302 \text{ m/s}$$

and

$$\mathbf{M} = \frac{V}{c} = \frac{2300 \text{ m/s}}{302 \text{ m/s}} = 7.62.$$

From Fig. 10.25,

$$\mathbf{C_D} \approx 0.9,$$

so

$$\mathscr{D} = \frac{\pi}{8}(0.9)(1.4)(1.20 \times 10^3 \text{ N/m}^2)(7.62)^2(2\text{m})^2 = 1.38 \times 10^5 \text{ N}.$$

Summing forces in the y direction (see Fig. E10.10b), we have

$$\sum F_y = W\cos 20° = -1.21 \times 10^4 \text{ N}.$$

Summing forces in the x direction, we have

$$\begin{aligned}\sum F_x &= W\sin 20° - \mathscr{D}\\ &= (1.29 \times 10^4 \text{ N})\sin 20° - 13.8 \times 10^4 \text{ N} = -13.4 \times 10^4 \text{ N}.\end{aligned}$$

The resultant force has magnitude

$$F_R = \sqrt{(1.21)^2 + (13.4)^2} \times 10^4\ \text{N} = 13.5 \times 10^4\ \text{N}.$$

The angle of the resultant force is

$$\theta_R = \tan^{-1}\left(\frac{F_y}{F_x}\right) = \tan^{-1}\left(\frac{-1.21}{-13.4}\right) = 5.2^\circ.$$

The acceleration has the same direction as the resultant force; that is, it is directed away from the direction of travel and $(20^\circ - 5.2^\circ) = 14.8^\circ$ *above* the horizontal. The magnitude of the acceleration is

$$a = \frac{F_R}{m} = \frac{135{,}000\ \text{N}}{1330\ \text{kg}} = 101.5\ \text{m/s}^2.$$

Expressing the answer in *g*'s ($1\ g = 9.807\ \text{m/s}^2$), we have

$a = 10.35$ *g*'s, opposite direction of travel
and 14.8° above the horizontal. **ANSWER**

The temperature on the windward surface of the reentry capsule is approximately equal to the stagnation temperature relative to the capsule. Note that the bow shock does not affect the stagnation temperature. From Eq. (10.30),

$$T_0 = T\left(1 + \frac{k-1}{2}\mathbf{M}^2\right) = 226.5\text{K}[1 + 0.2(7.62)^2];$$

$$T_{\substack{\text{windward}\\\text{surface}}} \approx T_0 = 2860\text{K}\ (4690^\circ\text{F}).$$ **ANSWER**

Discussion

Note that the strong bow shock ahead of the reentry capsule had no effect on our calculations. All the objects in Fig. 10.25 generate shocks when traveling at supersonic speed. The Mach number must be interpreted as the value far ahead of the object (i.e., upstream from the shock).

The deceleration of 10.35 *g*'s is a little severe. A peak value of about 7 *g*'s is generally taken as the safe upper limit for sustained exposure of human beings. Note that the standard *g*, $9.807\ \text{m/s}^2$, is used.

Our surface temperature estimate is too high for two reasons. First, nonideal gas effects are important at such high temperatures, so the true value of T_0 is about 10 percent lower than the calculated value. Second, the surface temperature is somewhat lower than the stagnation temperature of the gas. The true temperature of the surface is called the *recovery temperature* and is calculated by [1]

$$T_{\text{surf}} = T\left[1 + r\left(\frac{k-1}{2}\right)\mathbf{M}^2\right],$$

where r, the recovery factor, is about 0.9. Combining recovery factor and nonideal gas effects yields a surface temperature estimate of

$$T_{\substack{\text{windward}\\\text{surface}}} \approx 2350\text{K}\ (3800^\circ\text{F}).$$

Obviously some sort of heat shield is necessary to prevent an astronaut roast!

PROBLEMS

1. A jet is flying at a flight (or local) Mach number of 2.2 at 15,000 m in the Standard Atmosphere. Find the jet's velocity in km/hr.

2. An airplane is flying at a flight (or local) Mach number of 0.70 at 10,000 m in the Standard Atmosphere. Find the ground speed (a) if the air is not moving relative to the ground and (b) if the air is moving at 30 km/hr in the opposite direction from the airplane.

3. The stagnation pressure in a Mach 2 wind tunnel operating with air is 900 kPa. A 1.0-cm-diameter sphere positioned in the wind tunnel has a drag coefficient of 0.95. Calculate the drag force on the sphere.

4. An airplane is flying at Mach 1.8 at 10,000 m, where the temperature is −44°C and the pressure is 30.5 kPa.

(a) How fast is the airplane traveling in kilometers per hour?

(b) The stagnation temperature is an *estimate* of the surface temperature on the aircraft. What is the stagnation temperature under these conditions? Why is the stagnation temperature only an estimate of the surface temperature?

(c) What is the stagnation pressure?

(d) If the airplane slows down, at what speed (kilometers per hour) will it be traveling in the subsonic range?

5. Calculate the speed of sound in air, helium, and hydrogen. The temperature is 70°F.

6. A missile is moving parallel to the ground at Mach 4.0 at an altitude of 10,000 ft. How long after the missile is directly overhead will an observer on the ground hear any sound? Assume the Standard Atmosphere and still air. Also assume a constant air temperature of 463.6°R from the ground level to 10,000 ft.

7. Consider the flow process in Fig. P10.7. Does the fluid flow from left to right or from right to left? Justify your answer. Assume an ideal gas with constant specific heats.

8. An airplane flies at Mach 2.0 at an altitude where the static pressure and temperature are 5 psia and 0°F. A normal shock stands just ahead of the engine inlet. Calculate the stagnation pressure and air velocity (relative to the airplane) just downstream from the shock.

9. At some point for air flow in a duct, $p = 20$ psia, $T = 500°\text{R}$, and $V = 500$ ft/sec. Can a normal shock occur at this point?

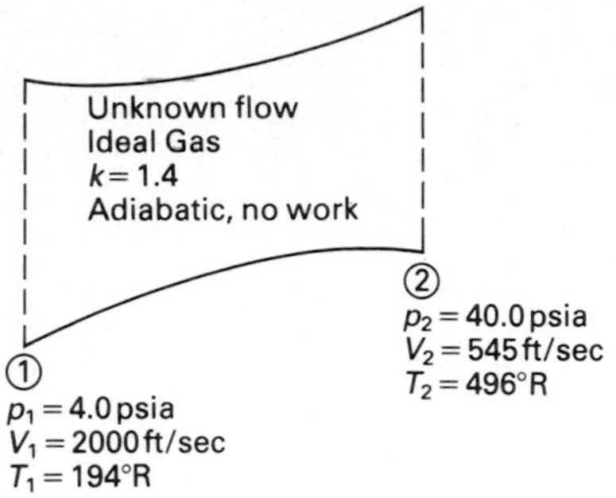

Figure P10.7

10. A normal shock exists in a 400-m/s stream of nitrogen. The upstream static pressure and temperature are −40°C and 70 kPa. Calculate the Mach number, pressure, and temperature downstream from the shock and the entropy change across the shock.

11. A Pitot tube is used to measure a supersonic Mach number. A detached normal shock appears just upstream from the tube, which measures a stagnation pressure of 20.0 psia. The static pressure upstream from the shock is 7.0 psia. Determine the Mach number just upstream from the normal shock. The fluid is air.

12. A normal shock propagates at 2000 ft/sec into the still air in a tube. The temperature and pressure of the air are 80°F and 14.7 psia before "hit" by the shock. Calculate the air temperature, pressure, and velocity after the shock, the stagnation temperature and pressure *relative to the shock* ahead of and behind the shock, and the stagnation temperature and pressure *relative to the tube* ahead of and behind the shock.

13. Air at $V_1 = 800$ m/s, $p_1 = 100$ kPa, and $T_1 = 300$ K passes through a normal shock. Calculate the velocity V_2, temperature T_2, and pressure p_2 after the shock. What would be the values of T_2 and p_2 if the same velocity change were accomplished isentropically?

14. The cylindrical rocket shown in Fig. P10.14 moves upward through the Standard Atmosphere at a constant velocity of 320 m/s. The drag coefficient for the rocket also is shown in the figure. Calculate the drag force at altitudes of 3500 m and 20,000 m, the altitude at which a shock wave first appears ahead of the rocket, and the pressure on the nose of the rocket at altitudes of 3500 m and 20,000 m.

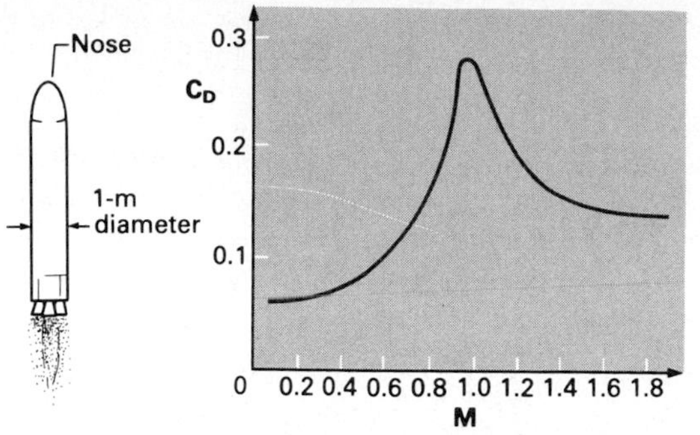

Figure P10.14

15. The air velocity in the duct in Fig. P10.15 is 750 ft/sec. The air static temperature is 100°F. Use the mercury manometer measurement to calculate the static and stagnation pressures of the flowing air.

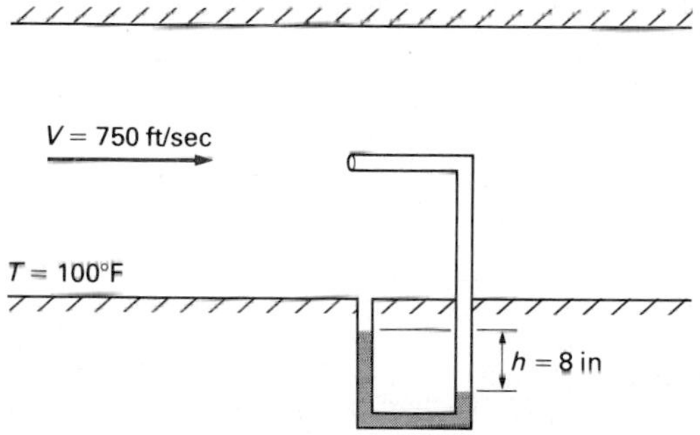

Figure P10.15

16. Air is flowing in a duct as shown in Fig. P10.16. A Pitot-static tube and a thermocouple are inserted into the flow stream as shown. Calculate the air velocity and the air-mass flow rate.

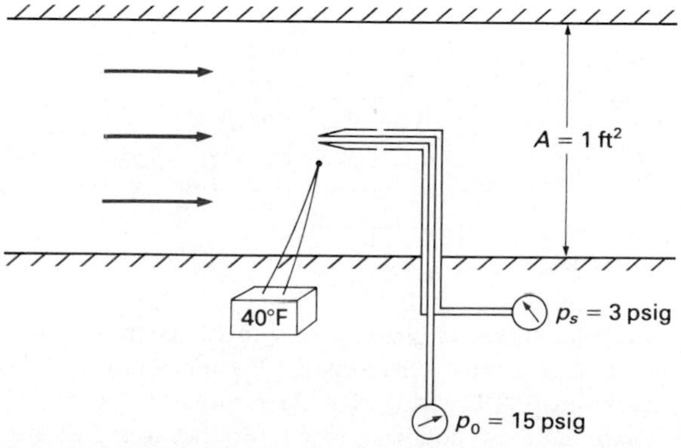

Figure P10.16

17. Air flows isentropically through a duct as shown in Fig. P10.17. For the conditions shown, find the Mach number at both stations 1 and 2 and the flow rate.

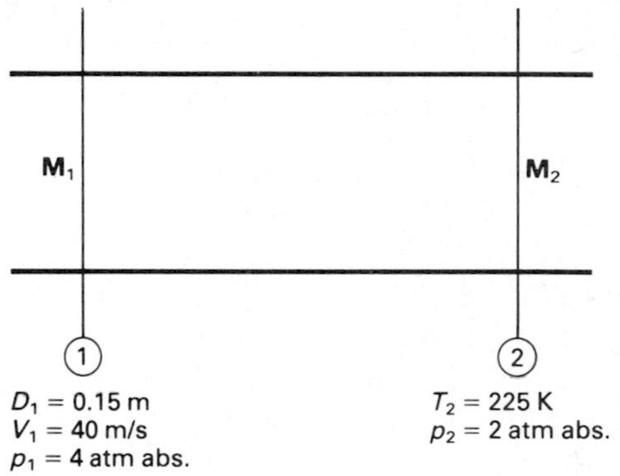

Figure P10.17

18. A jet-propelled airliner flies at an altitude of 15,000 m and a velocity of 265 m/s. What is the stagnation pressure and stagnation temperature of the air entering the engine? If the airliner is stationary on the ground, the engines are running with the same Mach number flow into the engines, and the air flow to each engine is 40 kg/s, what is the airflow to each engine at the above flight conditions?

• **19.** Develop an expression for the thrust developed by the air as it flows through the channel in Fig. P10.19. The expression should be a function of M_1, M_2, p_1, p_2, A_1, and A_2.

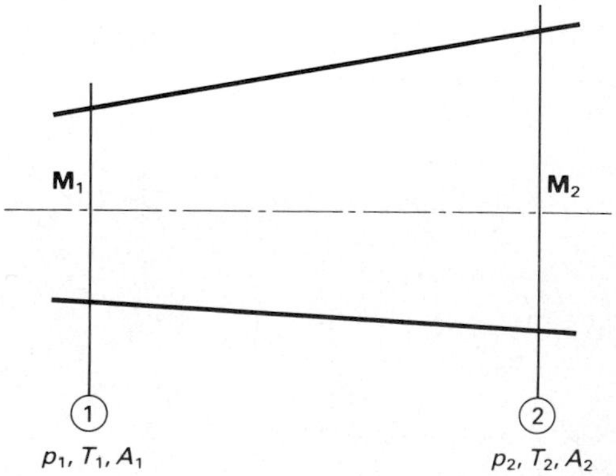

Figure P10.19

• **20.** A simplified schematic diagram of a carburetor of a gasoline engine is shown in Fig. P10.20. The throat area is 0.5 in^2. The engine draws air downward through the

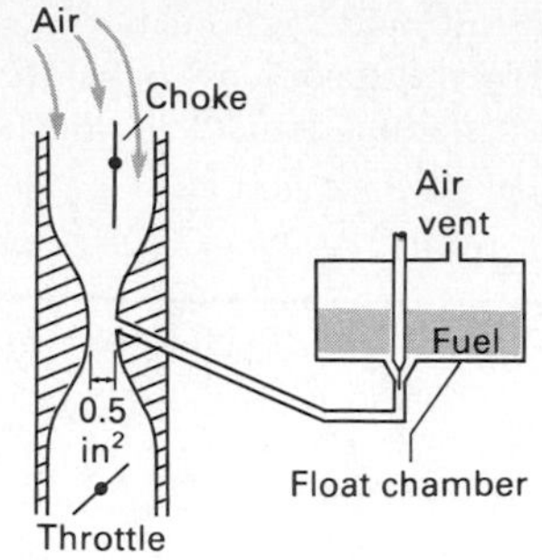

Figure P10.20

carburetor Venturi and maintains a throat pressure of 14.3 psia. This low throat pressure draws fuel from the float chamber and into the air stream. The energy losses in the 0.06-in. I.D. fuel line and valve are given by

$$h_L = \frac{KV^2}{2g},$$

with K = 6.0. The fuel specific gravity is 0.75. Assume an atmospheric pressure of 14.7 psia. Find the air–fuel ratio (that is, the ratio of the air mass flow rate to the fuel mass flow rate) based on an (a) ideal (constant density and inviscid) air flow and (b) an isentropic air flow.

•**21.** An engineering student wants to satisfy her curiosity about the compressibility of air in motion. She has set up a converging nozzle in which air discharges into the atmosphere. Figure P10.21 shows the nozzle with the necessary information. For these conditions, find the velocity at the exit by using the incompressible Bernoulli's equation, a compressible isothermal process, and a compressible isentropic process. Comment on your results.

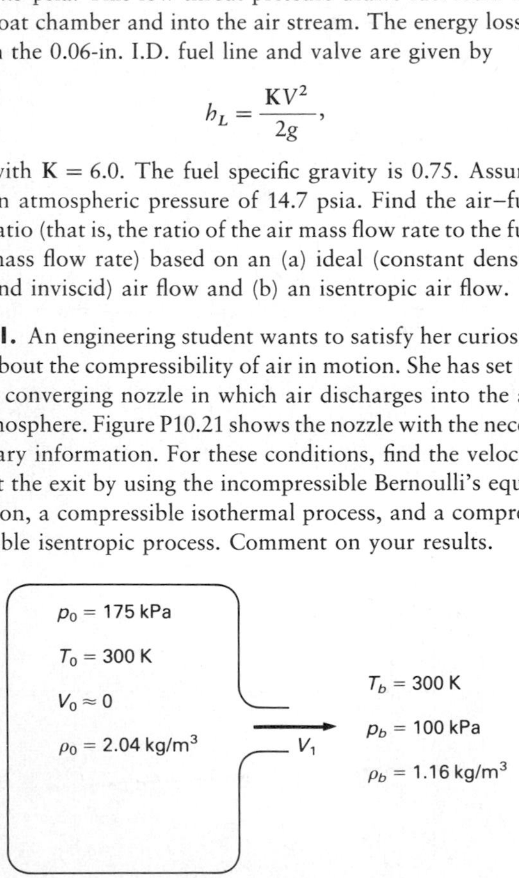

Figure P10.21

22. In a jet engine running on a test stand, the stagnation pressure and stagnation temperature just upstream from the convergent nozzle are measured to be 70 kPa gage and 700°C. The diameter of the nozzle exit is 0.5 m. Calculate the mass flow, jet exit velocity, and thrust of the nozzle. The nozzle exit pressure is 101 kPa.

23. An airliner is flying at 35,000 ft, where the ambient pressure and temperature are 3.468 psia and 395°R. The pressure and temperature in the aircraft cabin are 14.7 psia and 75°F. A terrorist fires a shot that makes a 1.0-in.-diameter hole in the aircraft. Assume that the hole behaves like a converging nozzle with an exit diameter of 1.0 in. and that the flow out of the hole is quasi-steady. Find the following:

(a) The initial Mach number of the flow from the hole;

(b) The initial mass flow rate from the hole;

(c) The cabin pressure at which the Mach number at the hole reaches half the value calculated in (a).

24. At a certain point in a pipe, air flows steadily with a velocity of 150 m/s and has a static pressure of 70 kPa and a static temperature of 4°C. The flow is isoenergetic and reversible.

(a) Calculate the maximum possible reduction in area and the following quantities for that minimum area: stagnation pressure, stagnation temperature, static pressure, static temperature, velocity, and Mach number.

(b) Calculate the quantities listed in (a) at a point where the area is 15% smaller than the *initial* area.

25. A tank of oxygen has a hole of area 0.5 cm^2 in its wall. The temperature of the oxygen in the tank is 25°C. Calculate the rate (kg/s) at which oxygen leaks to the atmosphere for tank pressures of 135 kPa and 375 kPa. Assume frictionless flow.

26. A simple converging nozzle is connected to the end of a 0.5-m-diameter pipe, as shown in Fig. P10.26. The nozzle exit diameter is 0.2 m. The air flowing in the pipe has stagnation properties p_0 = 200 kPa and T_0 = 600K. Compute the maximum possible air velocity in the pipe. Assume frictionless flow.

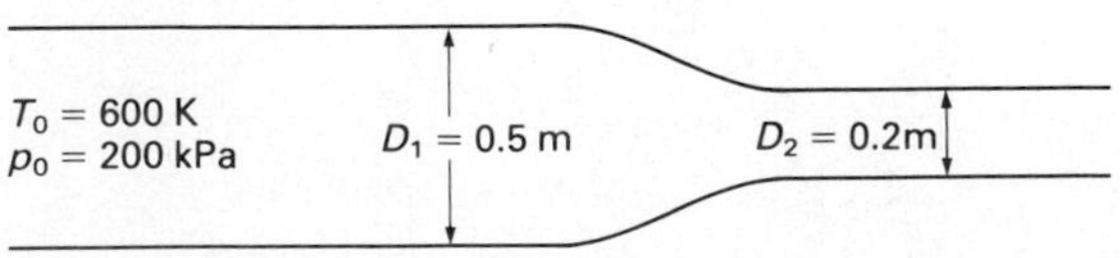

Figure P10.26

27. A simply converging nozzle with a 12.0-cm^2 exit area is supplied from an air reservoir. The temperature in the reservoir is 38°C and the back pressure is 100 kPa. Calculate the mass flow rate of air for reservoir pressures of 130 kPa and 350 kPa.

•**28.** A pressure vessel is tested for leakage in the following manner. The vessel is filled with Helium gas at 29 in. Hg, abs, and 70°F. The vessel is then placed inside a vacuum chamber in which the pressure is 50 μ Hg, abs. A Helium detector measures the rate at which Helium leaks from the test vessel into the vacuum chamber as 10^{-5} standard cubic centimeters per second. If the test vessel were filled with air at 22 psig and 70°F and surrounded by the atmosphere (p = 14.25 psia), what would be the rate of air leakage.

29. What is the increase of net thrust developed by a sonic converging nozzle if a divergent portion is added to the choked nozzle. The exit-to-throat area ratio is 1.75, and the working fluid is air. Assume that the nozzle has a stagnation pressure of 3 MPa, a stagnation temperature of 2000 K, and discharges to a back pressure of 100 kPa.

30. High pressure air is stored in a tank of 3 m³ at 2 MPa and 500 K, as shown in Fig. P10.30. Assume a quasi-steady, adiabatic process in the tank and isentropic flow through the well-rounded hole. Find the time needed for the tank pressure to drop to 300 kPa.

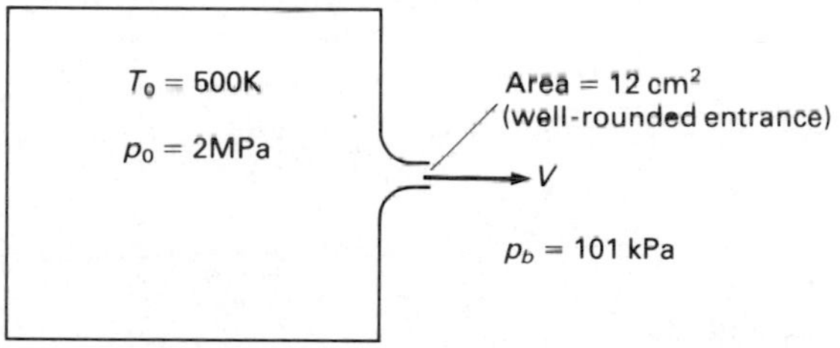

Figure P10.30

•**31.** Consider Problem 23, where a terrorist has put a 1.0-in.-diameter hole in an aircraft. Assume that flow is one-dimensional at the hole, that the pressure is uniform throughout the cabin at any instant, that the flow at the hole depends only on the instantaneous values of the cabin temperatures and pressure and the fixed atmospheric pressure, and that the expansion of the air in the cabin is reversible and adiabatic. Set up a differential equation for the cabin pressure as a function of time, the initial cabin temperature and pressure, the area of the hole, and appropriate gas constant(s). Assume isentropic air flow.

•**32.** Plot the isentropic operating conditions for a converging nozzle (i.e., the following graphs):

$$\frac{\dot{m}\sqrt{T_0}}{A_e p_0} \text{ versus } \frac{p_B}{p_0} \quad \text{and} \quad \frac{p_e}{p_0} \text{ versus } \frac{p_B}{p_0}$$

for $0 \leq p_B/p_0 \leq 1$.

33. Develop an analysis to predict the pressure–time history in "blowing down" a pressurized air tank of volume $\forall$ through an isentropic convergent nozzle of exit area A_e. The gas in the tank has an initial pressure p_i and an initial temperature T_i and exhausts to the atmosphere. Assume that the gas in the tank expands (a) isentropically and (b) isothermally. Illustrate your analysis by plotting the two pressure–time histories for a tank with $\forall = 1.0\ \text{m}^3$, $A_e = 5 \times 10^{-4}\ \text{m}^2$, $p_i = 500$ kPa, $p_{\text{final}} = 120$ kPa, and $T_i = 300$K. Comment on the applicability of the two cases to a real process.

34. Compute M_3, p_3, and p_{0_3} for the flow situation shown in Fig. P10.34.

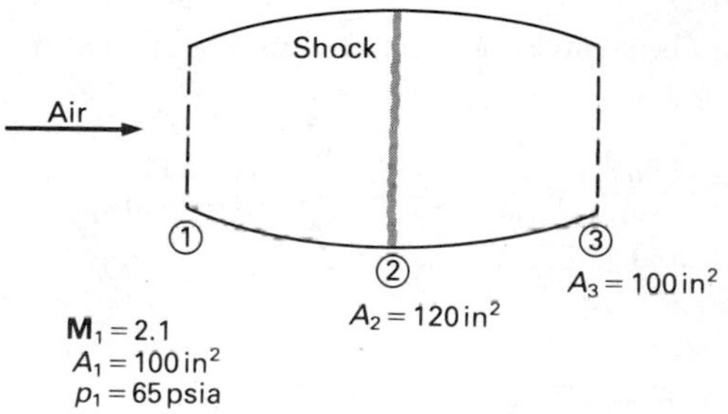

Figure P10.34

35. The nozzle of a rocket is to expand the gases from the combustion chamber to an exit pressure of 2.6×10^4 Pa. The combustion chamber temperature is 3000K, and the pressure is 1500 kPa. The exit area is 1.0 m². Find the flow rate for k = 1.30 and R = 190 N·m/kg·K. Assume that combustion chamber conditions are approximately stagnation conditions.

36. The gas entering a rocket nozzle has a stagnation pressure of 1500 kPa and a stagnation temperature of 3000°C. The rocket is traveling in the still Standard Atmosphere at 30,000 m. Find the throat and exit area for a flow rate of 10 kg/s. Assume k = 1.35, R = 287.0 N·m/kg·K. The gas is perfectly expanded to the ambient pressure.

37. Design a nozzle to isentropically expand steam from 150 psia and 1000°F to 15 psia at a flow rate of 1.0 lbm/sec. The steam may be considered an ideal gas with k = 1.3.

38. Figure P10.38 shows a rocket engine on a test stand. The ambient pressure is p_a and the exit plane pressure is p_e. The exhaust gases have specific heat ratio k and exit Mach number M_e through exit area A_e. Show that the thrust developed by the rocket engine (that is, the force on the restraint) is $F_T = kp_eA_eM_e^2 + (p_e - p_a)A_e$.

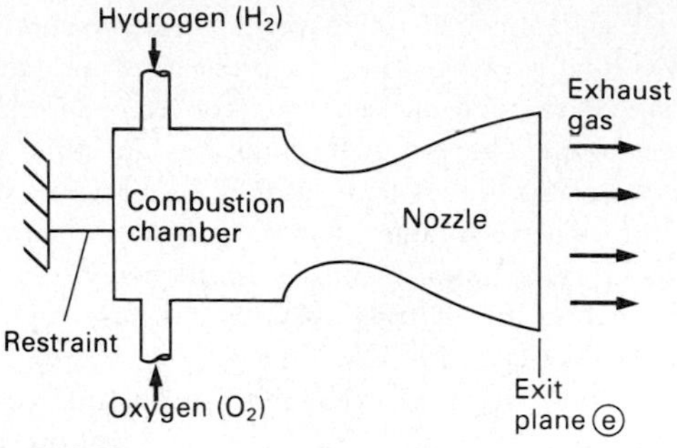

Figure P10.38

39. The combustion chamber reaction in Problem 38 is

$$H_2 + \tfrac{1}{2}O_2 \rightarrow H_2O.$$

The rocket engine is designed to produce 100 kN of thrust at an altitude of 10 km, where the ambient pressure is 26.5 kPa. The combustion chamber temperature and pressure are 6.9 MPa and 3300K. Calculate the design throat and exit areas and the hydrogen and oxygen mass flow rates ($k = 1.3$ for H_2O).

40. Air flows at 12.5 kg/s in a well-insulated nozzle. At one point in the nozzle, the cross-sectional area is 0.01 m^2, the air pressure is 120 kPa, and the air temperature is 0°C. At a point farther downstream, the air pressure is 90 kPa. The flow between the two points is reversible. Calculate the air velocity at both points and the temperature and area at the downstream point. Assuming that no work is done on the air, what is the *maximum* possible velocity that the air could achieve?

41. Steam enters the nozzles of a turbine stage with $T = 700°F$, $p = 100$ psia, and $V = 100$ ft/sec. At the nozzle outlet, $p = 30$ psia. Heat transfer is negligible and work is zero for the nozzle flow. Calculate the steam velocity and temperature at the nozzle outlet and the stagnation pressure and stagnation temperature at both the nozzle inlet and the nozzle outlet. Also calculate the ratio of the nozzle outlet area to the nozzle inlet area. Assume a reversible process and that the steam is an ideal gas with $k = 1.3$.

42. A jet engine is to be designed for an altitude of 12,000 m, where the atmospheric pressure is 19.3 kPa. The jet nozzle has a supersonic exit Mach number and is perfectly expanded. The stagnation pressure and temperature of the gas are 100 kPa and 600°C. The flow rate of gas is 45 kg/s. Calculate the throat area, exit area, and exit velocity. Use $k = 1.4$ and $R = 260$ J/kg·K for the gas.

43. A convergent–divergent nozzle with a throat area of half the exit area is supplied with air at a stagnation pressure of 140 kN/m^2. It discharges into an atmosphere of static pressure 100 kPa. Find the stagnation pressure of the gas at the nozzle exit.

44. A rocket nozzle with an exit-to-throat-area ratio of 5:1 is operating at a stagnation pressure of 1.3 MPa and exhausting to an atmospheric pressure of 30 kPa. Determine whether the nozzle is overexpanded, underexpanded, or ideally expanded if $k = 1.3$.

45. Figure P10.45 shows a nozzle for the first (high-pressure) stage of a steam turbine. Steam enters the nozzle with $p = 500$ psia, $T = 800°F$, and negligible velocity. The nozzle exit pressure is 100 psia and the mass flow rate is to be 5.0 lb/sec. Assume that the steam is an ideal gas with $k = 1.3$ and find V_e, A_e, and A_t.

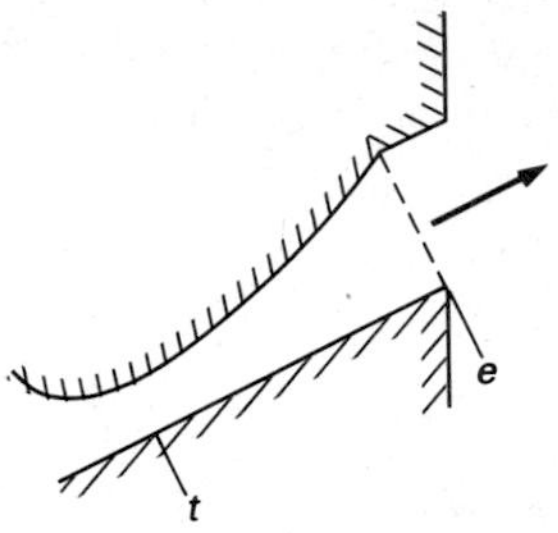

Figure P10.45

46. Air flows in the channel in Fig. P10.46. Determine the Mach number, static pressure, and stagnation pressures at station 3. Assume isentropic flow except for the normal shock wave.

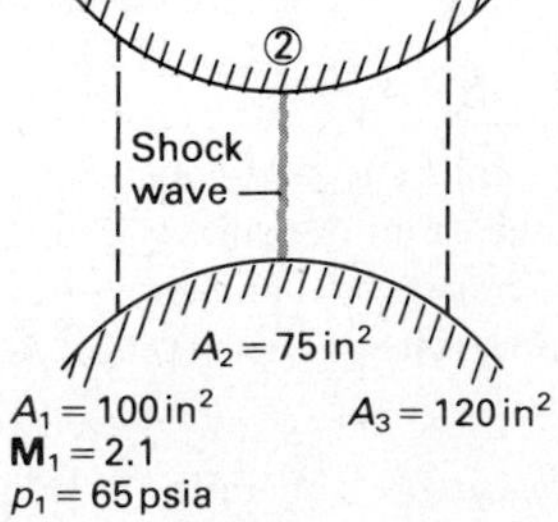

Figure P10.46

47. Assume that the engine described in Problem 38 and Fig. P10.38 has a throat area of 0.321 m^2 and an exit area of 0.495 m^2. The engine is started on the ground, where the atmospheric pressure is 101 kPa. During the starting process, the stagnation pressure increases as shown in Fig. P10.47. Calculate the stagnation pressure when the nozzle first chokes and when the exit Mach number first becomes supersonic.

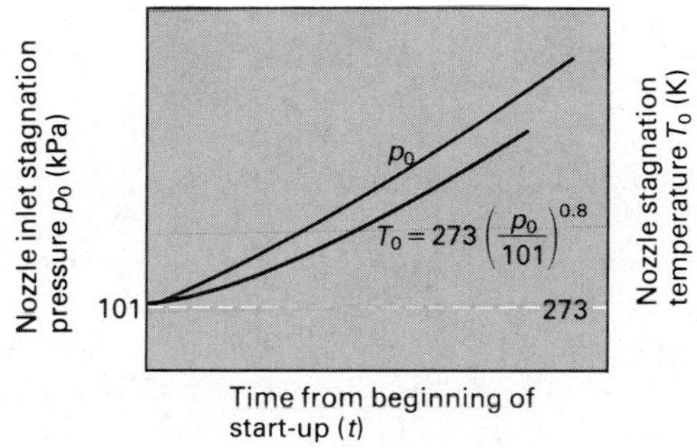

Figure P10.47

48. A converging–diverging nozzle exhausts air from a large tank to the atmosphere (see Fig. P10.48). The nozzle has a first critical pressure ratio (r_{pc_1}) of 0.90.

(a) Calculate the second and third critical pressure ratios for the nozzle.

(b) Calculate the tank pressure that will cause a normal shock to stand at area $A_s = \frac{1}{2}(A_t + A_e)$.

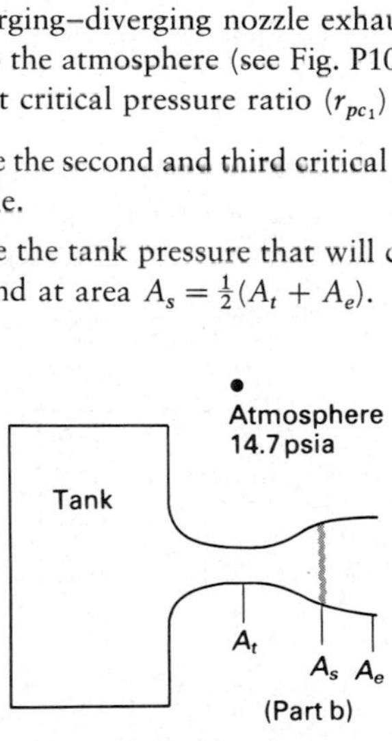

Figure P10.48

49. Hydrogen flows through a converging–diverging nozzle. The inlet stagnation conditions are 200 kPa and 310 K. The throat area is 12.5 cm^2, and the exit area is 25.0 cm^2. Find the exit pressures for normal shocks to occur at (a) the throat and (b) the exit.

50. Design a turbine nozzle to accelerate 1 lbm/sec of air from a stagnation pressure of 100 psia and temperature of 600°F to a pressure of 20 psia. Assume isentropic flow.

51. Air flows from the reservoir shown in Fig. P10.51 where $p_0 = 500$ kPa and $T_0 = 20$°C through a converging–diverging nozzle with throat area 10 cm^2. A normal shock is formed downstream from the throat at a plane where the area is 15 cm^2. Calculate the pressure just downstream from the shock and the mass flow rate in the nozzle.

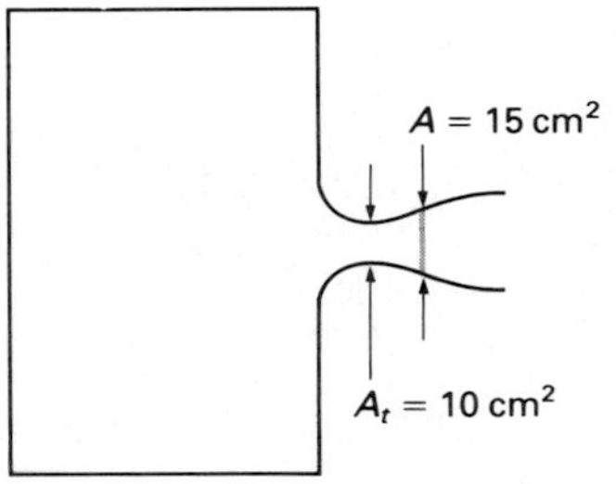

Figure P10.51

52. Air is flowing in the converging–diverging nozzle shown in Fig. P10.52. Determine the three critical pressure ratios and the Mach numbers immediately upstream and immediately downstream from the shock.

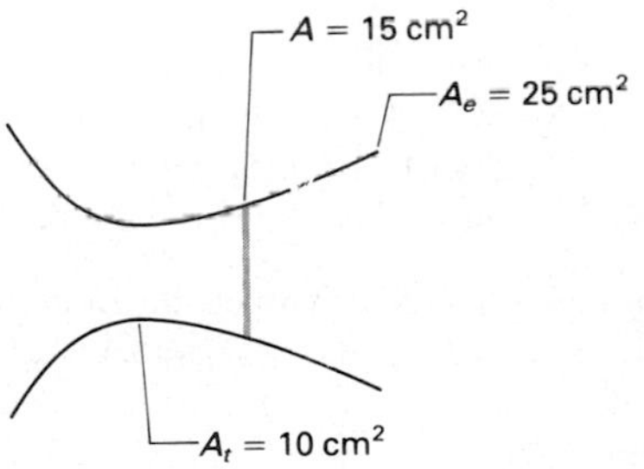

Figure P10.52

53. A supersonic nozzle (converging–diverging) has an exit area three times that of the throat. For steady isentropic air flow discharging to the atmosphere ($p_B = 1$ atm), find the Mach numbers at the throat and exit for

$$\frac{p_B}{p_0} = 0.04 \quad \text{and} \quad \frac{p_B}{p_0} = 0.98,$$

where p_0 is the nozzle stagnation pressure.

54. A nozzle for a supersonic wind tunnel is designed to achieve a Mach number of 3.0, with a velocity of 2000 m/s, and a density of 1.0 kg/m^3 in the test section. Find the temperature and pressure in the test section and the upstream stagnation conditions. The fluid is helium.

55. A converging–diverging nozzle is to be used to maintain a constant air flow rate of 3.0 kg/min from the high-pressure tank to the low-pressure tank shown in Fig. P10.55. The pressure in the downstream tank varies from 90 kPa to 10 kPa. The pressure in the upstream tank is constant at 100 kPa. Determine the nozzle throat and exit diameters. (*Hint:* To maintain a constant flow rate, the nozzle must always be choked.)

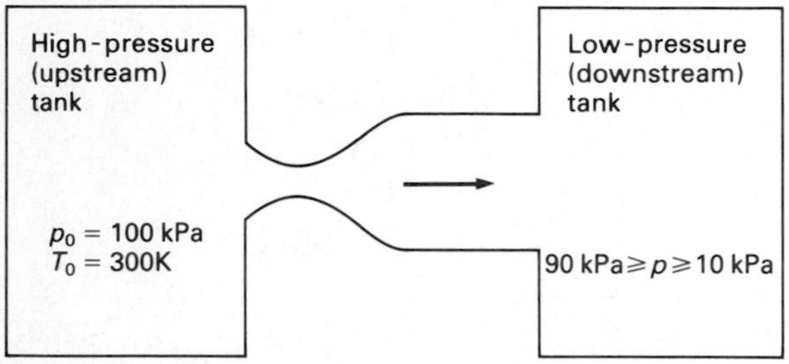

Figure P10.55

56. Plot the isentropic operating conditions for a converging–diverging nozzle (i.e., the following graphs):

$$\frac{\dot{m}\sqrt{T_0}}{A_e p_0} \text{ versus } \frac{p_B}{p_0} \quad \text{and} \quad \frac{p_e}{p_0} \text{ versus } \frac{p_B}{p_0},$$

for $0 \le p_B/p_0 \le 1.0$. The nozzle has a geometric area ratio (exit to throat) of 2.0. The nozzle is connected to a reservoir filled with air at $p_0 = 500$ kPa, abs, and $T_0 = 350$K. The nozzle exit area is $A_e = 0.10\ \text{m}^2$.

57. Air is flowing in the channel shown in Fig. P10.57. The air discharges to the atmosphere (at 101 kPa). Cross-sectional areas 3 and 1 are equal and measure 0.2 m², cross-sectional area 2 measures 0.15 m², and cross-sectional area 4 measures 0.1 m². Assume isentropic flow and find (a) the pressure at cross section 4, (b) flow rate in kg/s, and (c) Mach number at cross section 4.

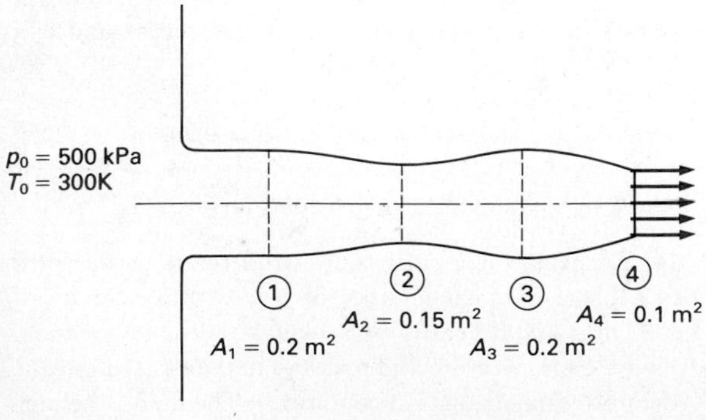

Figure P10.57

58. A convergent–divergent nozzle has an exit throat area ratio of 3.0. It is to be supplied with air. Find:

(a) The first, second, and third critical pressure ratios;

(b) The exit plane Mach number in each case;

(c) The throat Mach number in each case.

• **59.** Air is supplied to a convergent–divergent nozzle from a reservoir where the pressure is 100 kPa. The air is then discharged through a short pipe into another reservoir where the pressure can be varied. The cross-sectional area of the pipe is twice the area of the throat of the nozzle. Friction and heat transfer may be neglected throughout the flow. If the discharge pipe has constant cross-sectional area, determine the range of static pressure in the pipe for which a normal shock will stand in the divergent section of the nozzle. If the discharge pipe tapers so that its cross-sectional area is reduced by 25%, show that a normal shock cannot be drawn to the end of the divergent section of the nozzle. Find the maximum strength of shock (as expressed by the upstream Mach number) that can be formed.

• **60.** Air enters the primary nozzle of the jet pump in Fig. P10.60 at 300 kPa, 350K, and 20 m/s. The nozzle exit area and pressure are 0.5 m² and 50 kPa. The secondary flow rate is to be three times the primary flow rate. Find the total flow rate and the pressure and area at cross section 2.

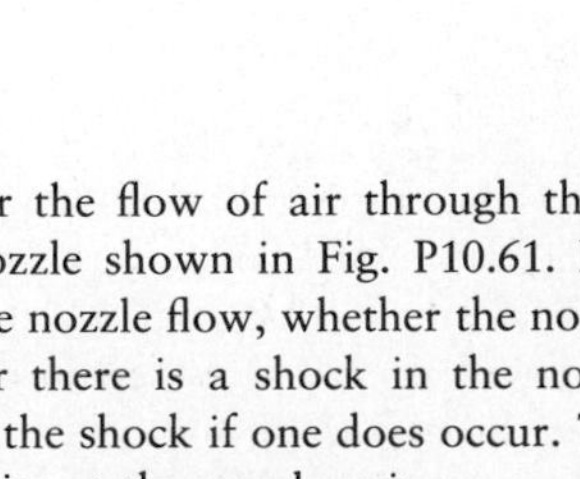

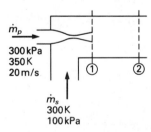

Figure P10.60

• **61.** Consider the flow of air through the converging–diverging nozzle shown in Fig. P10.61. Determine the regime of the nozzle flow, whether the nozzle is choked, and whether there is a shock in the nozzle. Find the flow area at the shock if one does occur. Then calculate the air velocity at the nozzle exit.

• **62.** Figure P10.62 shows the converging–diverging nozzle for a supersonic wind tunnel. The nozzle width (perpendicular to the paper) is 2.0 in. During operation, there is a viscous boundary layer on the nozzle walls and

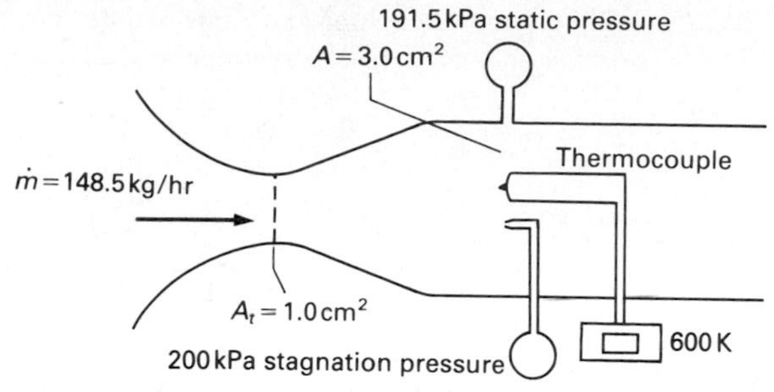

Figure P10.61

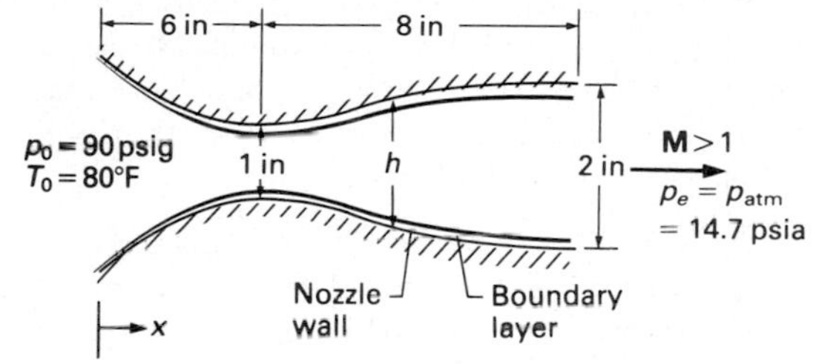

Figure P10.62

the side walls. These boundary layers get thicker in the downstream direction. The flow blockage owing to the boundary layers is given by the displacement thickness δ^*. The effective cross-sectional area is

$$A_{\text{eff}} = (h - 2\delta^*)(W - 2\delta^*).$$

Calculate the Mach number and velocity at both the nozzle throat and exit if the boundary layer is not present. Then calculate the exit Mach number and velocity by accounting for the "blockage" of the boundary layer. Use methods and equations from Chapter 9 to estimate δ^*. Is it true that $\mathbf{M} = 1$ at the geometric throat? Explain.

• **63.** Consider the turbojet engine shown in Fig. P10.63. Assume isentropic flow in the diffuser and nozzle. Calculate the Mach number $\mathbf{M}_1$ and the mass flow rate through the engine. Next calculate $p_2 \cdot p_{0_2}$, the critical area for the diffuser, A_4 and A_5. Compare A_4 with the critical area for the diffuser. If they are not the same, explain why.

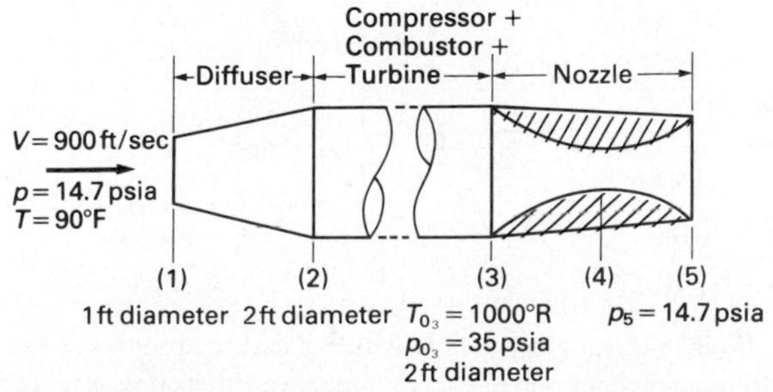

Figure P10.63

• **64.** Models of high-speed aircraft often are tested in wind tunnels similar to that shown in Fig. P10.64. To test a certain model, a test section cross-sectional area of 2.0 ft^2 is required. The test section static properties are to be 14.6 psia and 60°F.

(a) Sketch the nozzle shape necessary to obtain test section Mach numbers of (i) 0.60, (ii) 1.10, and (iii) 1.80. Indicate the necessary minimum nozzle area.

(b) Calculate the stagnation properties and air mass flow rate for $\mathbf{M} = 0.6$ and $\mathbf{M} = 1.80$.

(c) An important problem in wind tunnel testing is *blockage*. Because a model reduces the test section cross-sectional area, the flow properties at the model are different from those in the empty test section. Estimate the percentage error in air velocity at a model for $\mathbf{M} = 0.6$ and $\mathbf{M} = 1.10$. The model cross-sectional area is 1.5 in^2 (approximately 0.5% blockage).

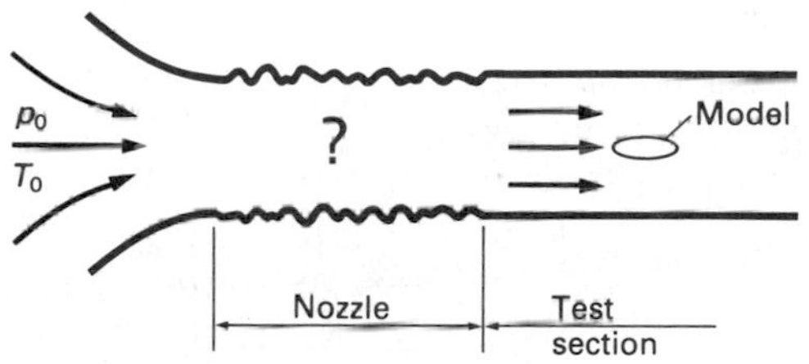

Figure P10.64

65. Air flows through a constant-area, insulated passage. The inlet and exit Mach numbers are 0.80 and 1.00, respectively. Find T_2 and p_2, the entropy change $(\tilde{s}_2 - \tilde{s}_1)$, and the wall frictional force. The duct has a 1.0-ft diameter, and the inlet temperature and pressure are 60°F and 20 psia.

66. An air blower and pipe system is to be designed to convey pulverized coal through an 8-in. schedule 40 steel pipe. The pipe is to be 600 ft long and the outlet velocity is to be 150 ft/sec. If the pipe discharges to atmospheric pressure, what must be the pressure, velocity, and density of the air at the inlet end of the pipe? Assume isoenergetic flow with $T_0 = 50$°F and neglect the effect of the coal particles.

67. Consider the flow of air through the piping system shown in Fig. P10.67. If the system is choked, determine the location where the Mach number is 1. Calculate the pressure ratio p_B/p_{0_1} that causes choking in this system.

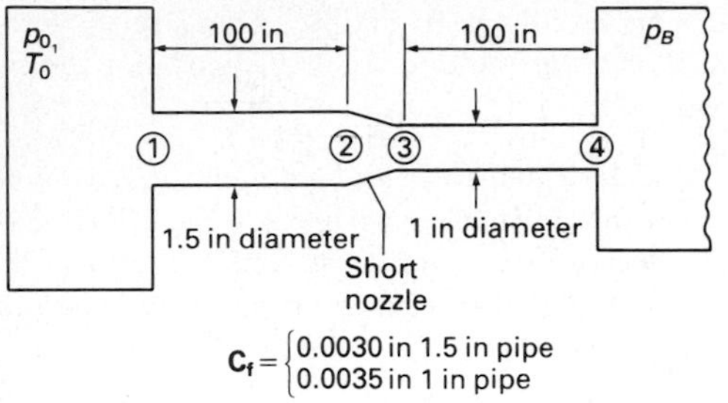

Figure P10.67

68. A thermally insulated pipe has inside diameter of 0.50 m. The pipe is 100 m long and has a relative roughness of 0.001. The air at the inlet is at 275 kPa and 300K, and the pipe discharges to an ambient pressure of 100 kPa. Find the flow rate for choked conditions.

69. A compressor supplies air at 100 psia and 100°F to a 12-in.-diameter schedule 40 steel pipe. The pipe is 100 ft long, is thermally insulated, and discharges to an ambient at 40°F and 14.7 psia. Find the mass flow rate through the pipe.

70. Estimate the maximum mass flow rate of air that can be passed by the duct shown in Fig. P10.70 for $\mu = 4 \times 10^{-7}$ lb·sec/ft^2.

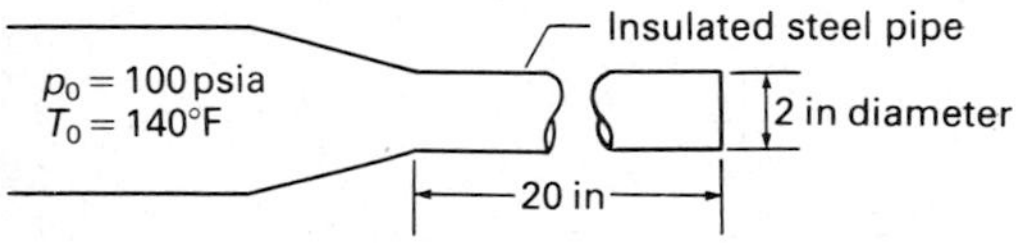

Figure P10.70

71. The stagnation temperature and Mach number of air flowing out of a 1.0-in., type L copper pipe are 550°R and 0.65, respectively. The atmospheric pressure is 14.7 psia. Find the distance upstream from the exit where the Mach number is 0.25. Assume adiabatic flow.

72. Air with $p_0 = 50$ psia, $T_0 = 150$°F is to be transported through a 100-ft pipe. What minimum pipe diameter is required to transport the air without choking? The flow rate of air is 10 lbm/sec. The pipe is smooth and the friction coefficient is given by Blasius's formula:

$$4C_f = \frac{0.3164}{(\mathrm{R})^{0.25}}.$$

73. Waste gas (CO_2) is vented to outer space from a spacecraft through a circular pipe 0.2 m long. The pressure and temperature in the spacecraft are 35 kPa and 25°C. The gas must be vented at the rate of 0.01 kg/s. The friction factor for the flow in the pipe is given by

$$C_f = \frac{16}{\mathrm{R}}, \qquad \mathrm{R} < 5000,$$

$$C_f = 0.0032, \qquad \mathrm{R} > 5000, \qquad \mathrm{R} = \frac{4\dot{m}}{\pi\mu D}.$$

The viscosity (μ) is 4×10^{-4} N·s/m^2. Determine the required pipe diameter.

74. A 10-ft-long hose is connected to the outlet of the regulator valve on a nitrogen bottle. The hose must deliver 0.065 lb/sec when the regulator pressure is 42 psia. The hose exhausts to a back pressure of 8.0 psia. The stagnation temperature is 120°F. The hose has a roughness equivalent to galvanized iron. Calculate the required hose diameter. Assume adiabatic flow.

75. The velocity of nitrogen in a 1-in.-diameter commercial steel pipe is 150 ft/sec. The static temperature and pressure are 67°F and 30 psia, respectively. Calculate the length of pipe needed to achieve sonic flow. Calculate also the pressure at the end of the pipe. Assume adiabatic flow.

76. Air at an absolute pressure of 500 kPa and temperature of 300K is contained in a pressure vessel. The vessel has a wall thickness of 6 mm and is found to be leaking to the atmosphere at a rate of 2×10^{-6} kg/s. Assume that there is a small circular hole of length equal to the wall thickness and determine its diameter. The friction factor is given by $f = (4\pi\mu/\dot{m})D$, where μ is the air viscosity, $\dot{m}$ the mass flow rate, and D the hole diameter. The viscosity is constant and equal to 3×10^{-5} kg/m·s. Assume that the flow is choked at the hole outlet and the fluid acceleration up to the hole entrance is isentropic. Assume adiabatic flow.

77. Air is to be delivered at the rate of 10 lbm/sec from tank *A* to tank *B*, as shown in Fig. P10.77. The two tanks are connected by 20 ft of insulated steel pipe. Find the required diameter to deliver the necessary flow rate.

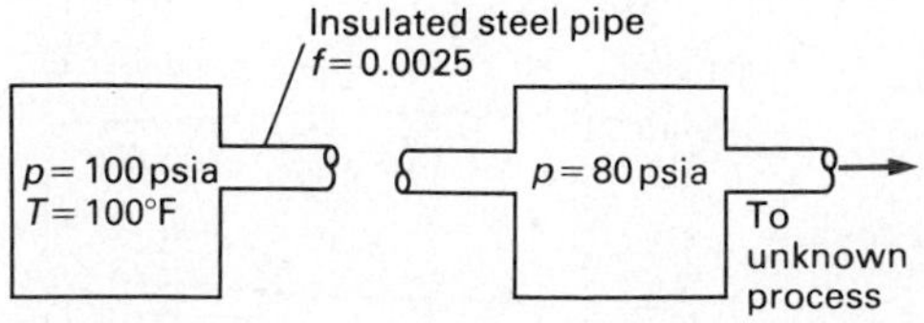

Figure P10.77

78. A natural-gas pipeline is being designed with 0.3-m-diameter commercial steel pipe, an inlet pressure of

1 MPa, $\mu = 1.1 \times 10^{-5}$ N·s/m^2, and an inlet temperature of 300K. Natural gas has $k = 1.3$ and $R =$ 520 J/kg K. For an inlet velocity of 27 m/s, find the maximum pipe length for the highest mass flow rate and the exit pressure and temperature for these conditions.

79. Air flows isentropically through a duct to a section where $p_1 = 25$ kPa, $T_1 = 300$K, and $V_1 = 900$ m/s. For these conditions:

(a) Determine the stagnation conditions for the flow.

(b) What is the Mach number at station 1? Show a T–s diagram displaying stagnation and static conditions.

(c) Is the flow choked? Is the throat behind or ahead of section 1? Label this state on the T–s diagram.

80. Air flows in an insulated pipe with the conditions shown in Fig. P10.80. Calculate the maximum possible length of the pipe downstream from the shock.

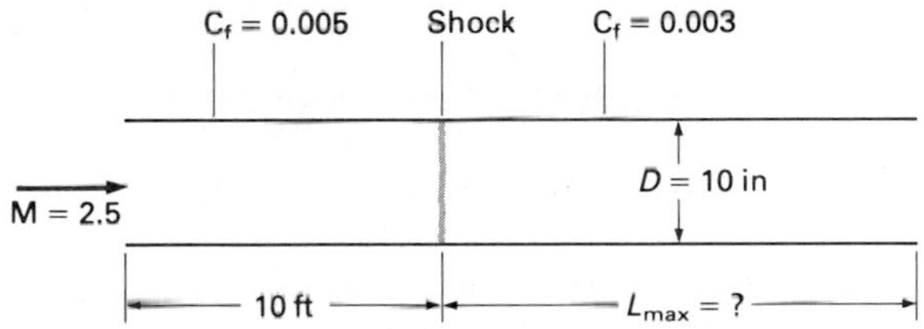

Figure P10.80

81. Figure P10.81 shows an insulated pipe attached to a tank of air. Estimate the maximum mass flow rate that the pipe could exhaust from the tank.

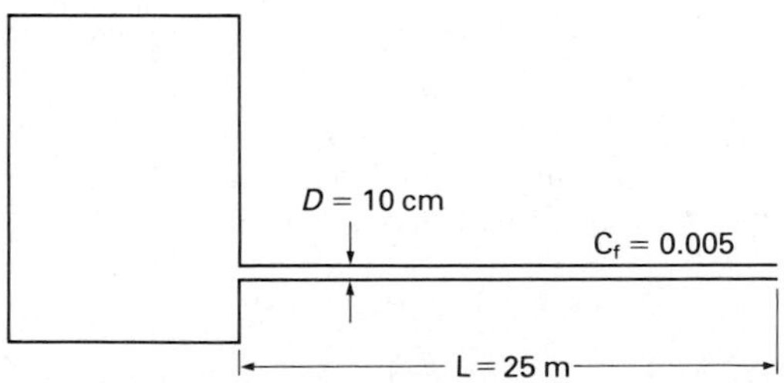

Figure P10.81

•**82.** Consider the frictional flow of an ideal gas through a long pipe in a constant-temperature environment, with heat transfer between the gas and the environment. The pipe is long enough that the fluid can be considered to have constant temperature throughout the entire length (i.e., the pipe entrance length along which the gas temperature changes is relatively short). Develop an expression for the pressure gradient (dp/dx) along the pipe.

•**83.** Air enters a 4-cm-square galvanized steel duct with $p_0 = 150$ kPa, $T_0 = 400$K and $V_1 = 120$ m/s. (*Note:* $\mu =$ 2.2×10^{-5} N·s/m^2).

(a) Compute the maximum possible duct length for these conditions.

(b) If the actual duct length is 0.75 times the maximum value, calculate the mass flow rate, the exit pressure, and the stagnation pressure.

(c) If the actual duct length is 1.3 times the maximum value for the stated conditions, compute the new mass flow rate and inlet velocity. Assume a low back pressure and use the same value of $\mathbf{C_f}$ as used in the maximum possible duct length case.

•**84.** A 9-m-long pipe is fabricated from brass and must carry an air flow of 0.18 kg/s. The stagnation temperature of the air is 373K, and the tolerable pressure drop is 130 kPa. The pipe discharges to a pressure of 100 kPa. Determine the diameter of thc pipe.

•**85.** Repeat the analysis related to Fig. 10.19 for a frictional duct fed by a converging–diverging nozzle. Note the various regimes that can occur, specifically noting when the flow is choked, when shocks appear in either the nozzle or the duct, and when and where the flow is supersonic.

•**86.** The jet pipe on a jet engine installation is 10 m long and 0.50 m in diameter and terminates in a short converging nozzle with an exit diameter of 0.46 m. The stagnation temperature and pressure at entry to the pipe are 800°C and 300 kPa, respectively. The engine is run with the aircraft stationary on the ground, where the atmospheric pressure is 101.3 kPa. Calculate the net thrust of the engine when (a) friction in the jet pipe is neglected and (b) the friction coefficient in the jet pipe is 0.005. The gas may be assumed to be ideal, with $k = 1.3$ and $c_p = 1.150$ kJ/kg·K.

•**87.** An air compressor has the performance curve shown in Fig. P10.87. It discharges 700°F air through a 12-in., type L copper pipe. The pipe is 100 ft long and discharges to the 14.7 psia atmosphere. Find the mass flow rate through the copper pipe.

•**88.** Develop differential equations for the temperature and pressure changes for the flow of a gas through a constant-area duct with friction and variable specific heat $c_p\{T\}$. The gas still obeys $p = \rho RT$. This result would be useful for gas flows with a large temperature change along the duct.

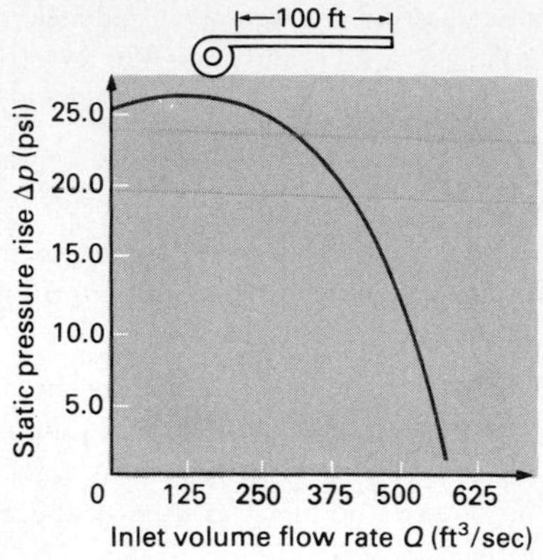

Figure P10.87

89. A streamlined, two-dimensional body has a length-to-maximum-thickness ratio of 5 : 1 and a drag coefficient of 0.032 at a Mach number of $\mathbf{M}_\infty = 0.132$. Find the drag coefficient at $\mathbf{M}_\infty = 0.661$.

REFERENCES

1. Shapiro, A. H., *Dynamics and Thermodynamics of Compressible Fluid Flow,* vols. 1 and 2, Ronald Press, New York, 1953.

2. Zucker, R. D., *Fundamentals of Gas Dynamics,* Matrix Publishers, Portland, Ore., 1977.

3. Zucrow, M., and J. D. Hoffman, *Gas Dynamics,* vols. 1 and 2, John Wiley & Sons, New York, 1976.

4. John, J. E. A., *Gas Dynamics* (2nd ed.), Allyn & Bacon, Boston, 1984.

5. Anderson, J. D., *Modern Compressible Flow* (*with Historical Perspective*) (2nd ed.), McGraw-Hill, New York, 1990.

6. Liepmann, H. W., and A. Roshko, *Elements of Gasdynamics,* John Wiley & Sons, New York, 1957.

7. Van Wylen, G. J., and R. E. Sonntag, *Fundamentals of Classical Thermodynamics* (2nd ed.), John Wiley & Sons, New York, 1973, 1976. Available in both English units and SI units versions.

8. Holman, J. P., *Thermodynamics* (3rd ed.), McGraw-Hill, New York, 1980.

9. Sears, F. W., M. W. Zemansky, and H. D. Young, *University Physics* (6th ed.), chaps. 14–19, Addison-Wesley, Reading, Mass., 1982.

10. Coles, D. (Principal), *Channel Flow of a Compressible Fluid,* National Committee for Fluid Mechanics Films. Distributed by Encyclopaedia Britannica Film Corporation.

11 Liquid Flow in Open Channels

In this chapter, we consider open channels in which liquid flows with a free surface. Probably the most familiar examples of open channels are natural streams and rivers. Man-made open channels include irrigation and navigation canals, drainage ditches, and sewer and culvert pipes running partially full. Water running along the surface of a road after a heavy rain is another example of open channel flow.

Although strong similarities between open channel flow and flow in closed pipes and ducts (discussed in Chapter 7) might be expected, there are profound differences between them. In closed conduit flow, the extent of the flow stream is always known, because the fluid fills the flow channel. A pressure change is usually associated with the flow. In an open channel flow, the free surface is always at atmospheric pressure, and the only pressure variation is hydrostatic. The driving force for open channel flows is *gravity:* there must be a net change in elevation of the channel to generate flow. The primary complicating factor in open channel flow analysis is that the location of the free surface is unknown a priori; in fact, the free surface rises and falls in response to perturbations to the flow (such as changes in channel slope or geometry). Thus the exact geometry of the flow stream itself is unknown and must be determined as part of the solution. This fact—and the great variety in types of open channels (ranging from natural streams to carefully designed flumes or spillways)—makes analysis of open channel flow more difficult than analysis of pipe flow. In practice, open channel flow analysis makes more use of empirical data and is more approximate than pipe flow analysis.

Most practical open channels contain water, and almost all the available experimental data were obtained with water. The general principles of open channel flow analysis are applicable to other liquids, but specific results may not be accurate. In this chapter, we discuss the fundamentals of open channel flow analysis, usually concentrating on simple channel geometries. More complete treatments of the subject are contained in references [1–3].

11.1 INTRODUCTORY CONCEPTS

In this section, we examine some introductory concepts that provide the basis for open channel flow analysis.

11.1.1 Velocity Distributions and the One-Dimensional Flow Approximation

An open channel cross section has two types of boundaries: solid surfaces and a free surface. The fluid in contact with the solid surfaces must satisfy the no-slip

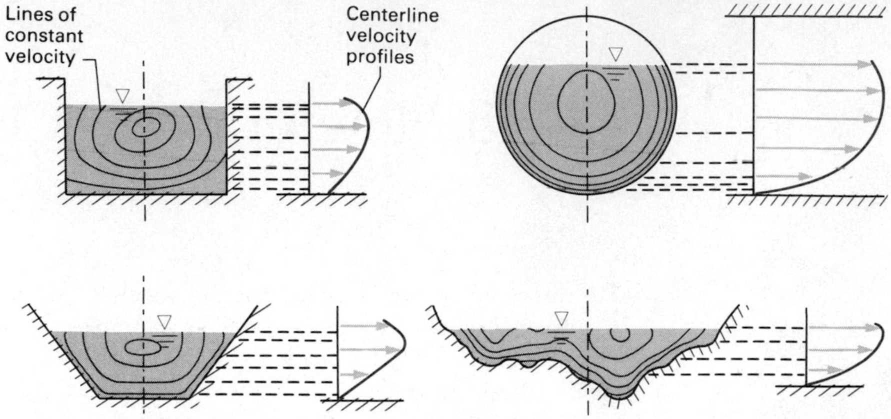

Figure 11.1 Typical velocity profiles in various open channels.

condition while the fluid at the free surface has nonzero velocity. The velocity distribution in an open channel is generally three-directional and three-dimensional. If the channel is straight and has a regular and constant cross section, nearly one-directional flow may occur with the main velocity component parallel to the channel axis. Figure 11.1 illustrates fully developed velocity distributions for prismatic* channels of various shapes. Note that the maximum velocity always occurs at a point below the free surface, usually at 20–30 percent of the channel depth. The point of maximum velocity is nearer the surface in shallow channels.

If the channel cross section varies in the flow direction and/or the channel axis is curved, the velocities are even more complicated. In these cases, significant velocity components in the cross-sectional plane exist and the flow is three-directional. These secondary motions are especially strong when a channel bends sharply because of centripetal acceleration of fluid particles, and can cause significant bottom erosion.

Few theories are available for predicting velocity distributions in open channel flow. Engineering analysis methods are based on the assumption of one-dimensional, one-directional flow. Secondary motions in the channel cross-sectional plane are neglected, and the velocity is assumed to be uniform and parallel to the channel axis. Of course, the uniform velocity should be taken as the average†

$$V = \frac{\int (\vec{V} \cdot \hat{n})\, dA}{A} = \frac{Q}{A}, \tag{11.1}$$

where Q is the volume flow rate in the channel and A is the cross-sectional area of the liquid in the channel (*not* including parts of the channel above the free surface).

When applying the mechanical energy and linear momentum equations to open channel flow, we can use the kinetic energy correction factor (α) and the momentum correction factor (β) to account for the nonuniformity of velocity. Measurements indicate that values of α range from 1.03 to 1.36 and values of β range from 1.01

---

* A prismatic channel has constant cross-sectional shape and a straight axis.

† The usual overbar ( ) is omitted for convenience.

to 1.12 for prismatic channels. These values are somewhat larger than typical values for fully developed turbulent pipe flow; nevertheless, most open channel flow analyses are based on the assumption that $\alpha = \beta = 1.0$. We follow this practice; however, if accurate estimates of α and β are available, flow calculations can be improved accordingly.

11.1.2 Dimensionless Parameters and Wave Speed

Water flowing in an open channel is influenced by forces resulting from

- gravity,
- friction at the channel walls, and
- surface tension at the free surface.

The effects of pressure and fluid compressibility are negligible. Table 6.1 gives the relevant dimensionless parameters:

- Froude number **F** ($\mathbf{F} = V/\sqrt{g\ell}$);
- Reynolds number **R** ($\mathbf{R} = V\ell/\nu$);
- Weber number **W** ($\mathbf{W} = \rho V^2\ell/\sigma$).

The relevant dimension, ℓ, is either the depth, hydraulic diameter, or hydraulic depth of the channel. In most open channel flows, ℓ is large and σ is small, so surface tension effects are negligible; therefore the important dimensionless parameters are the Froude and Reynolds numbers. By now the significance of Reynolds number should be familiar; however, the Froude number has not concerned us thus far. Because **F** is dimensionless, the denominator, $\sqrt{g\ell}$, must have dimensions of velocity. According to the principles of similarity (Section 6.5), $\sqrt{g\ell}$ must represent a velocity significant to the flow. With proper selection of ℓ, $\sqrt{g\ell}$ represents the speed of propagation of surface waves.

Consider a still liquid in an open channel (Fig. 11.2a), and suppose that a disturbance is introduced. Typical disturbances might be caused by a stone thrown into the liquid or a finger dipped into it. The depth of the liquid changes in the vicinity of the disturbance. Gravity forces act to level the surface, so the depth

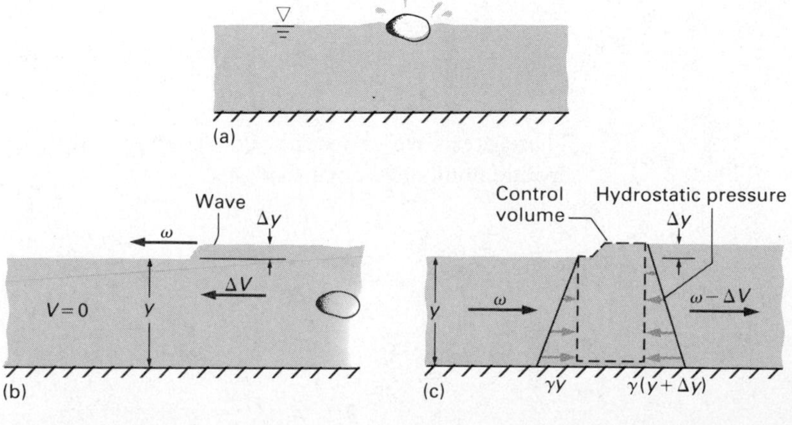

Figure 11.2 Free-surface gravity wave: (a) disturbance in open channel; (b) disturbance produces surface wave; (c) control volume moving with wave.

change propagates as a surface wave. Ripples on a pond are familiar examples of this type of wave. We can determine the speed of free-surface waves by applying the continuity and momentum equations. Figure 11.2(b) shows a wave propagating in a still liquid. We attach a control volume to the wave and obtain the flow situation shown in Fig. 11.2(c), in which the wave appears fixed. Fluid approaches the control volume with speed ω and leaves the control volume with speed $\omega - \Delta V$, where ΔV is the velocity change induced by the wave. The disturbance increases the fluid depth an amount Δy; ΔV and Δy are not necessarily small. We assume that the geometry is uniform in all planes parallel to the paper. As the control volume is very thin, there is no storage of mass or momentum. Applying the continuity equation gives

$$\dot{m} = \rho\omega y(1) = \rho(\omega - \Delta V)(y + \Delta y)(1), \qquad (11.2)$$

where (1) is the unit width perpendicular to the paper. Solving for ΔV, we have

$$\Delta V = \omega\left(\frac{\Delta y}{y + \Delta y}\right). \qquad (11.3)$$

Next, we apply the linear momentum equation. The control volume is thin, so the force from shear stress on the channel bottom is negligible and the only force is caused by hydrostatic pressure. The momentum equation is

$$\dot{m}(\omega - \Delta V) - \dot{m}\omega = \tfrac{1}{2}\gamma y(y)(1) - \tfrac{1}{2}\gamma(y + \Delta y)(y + \Delta y)(1).$$

Substituting the mass flow rate from Eq. (11.2), noting that $\gamma = \rho g$, and simplifying give

$$\omega\,\Delta V = g\left(1 + \frac{\Delta y}{2y}\right)\Delta y.$$

Substituting ΔV from Eq. (11.3) and rearranging give

$$\omega^2 = gy\left(1 + \frac{\Delta y}{2y}\right)\left(1 + \frac{\Delta y}{y}\right). \qquad (11.4)$$

Stronger waves, with larger Δy, travel faster. If we consider a weak wave with $\Delta y \to 0$, and use the symbol c to represent the speed of such a wave, we get

$$c^2 = \lim_{\Delta y \to 0} \omega^2 = gy. \qquad (11.5)$$

Therefore if we choose the local water depth y as the length parameter in the Froude number, we see that

$$\mathbf{F} = \frac{V}{\sqrt{gy}} \qquad (11.6)$$

represents the ratio of flow speed to the speed of an infinitesimal surface wave.

EXAMPLE 11.1 Illustrates Surface Wave Speed

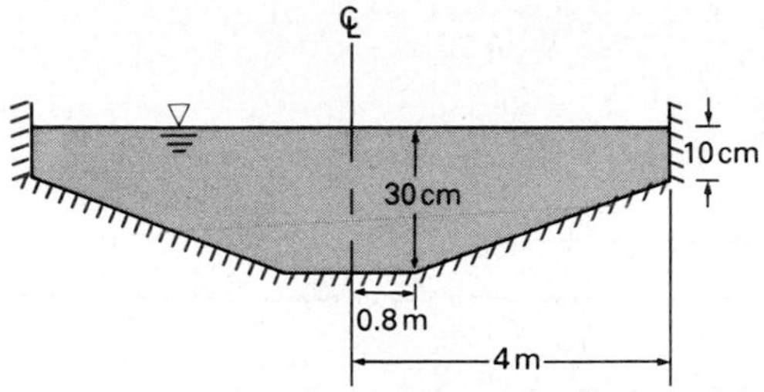

Figure E11.1 Cross section of a small circular pool.

A child throws a pebble into the center of a circular pool. The bottom of the pool is shaped as shown in Fig. E11.1. Compute the time for the first ripple to travel to the edge of the pool.

SOLUTION

Given

Pebble thrown into center of circular pool in Fig. E11.1

Find

Time for first ripple to travel to edge of pool

Solution

Assume that the wave is weak and may be approximated by a wave of infinitesimal strength. The wave speed, c, is then given by Eq. (11.5):

$$c = \sqrt{gy},$$

where y is the pool depth. Because the pool depth changes with the radial distance from the center, c varies with the radial coordinate r. The time, dt, for the wave to travel a short distance, dr, is

$$dt = \frac{dr}{c}.$$

The total time, t_0, for the wave to travel to the edge of the pool is

$$t_0 = \int_0^{t_0} dt = \int_0^{R_0} \frac{dr}{c},$$

where R_0 is the pool radius. We must evaluate the integral in two parts:

$$t_0 = \int_0^{R_i} \frac{dr}{c} + \int_{R_i}^{R_0} \frac{dr}{c},$$

where $R_i = 0.8$ m and $R_0 = 4$ m. The wave speeds are

$$c = \sqrt{gy_i}, \qquad 0 \leq r \leq R_i,$$

and

$$c = \sqrt{g\left[y_i - (y_i - y_0)\left(\frac{r - R_i}{R_0 - R_i}\right)\right]}, \qquad R_i \leq r \leq R_0,$$

where $y_i = 0.3$ m and $y_0 = 0.1$ m. Substituting and integrating give

$$t_0 = \frac{R_i}{\sqrt{gy_i}} + \frac{2}{\sqrt{g}}\left(\frac{R_0 - R_i}{\sqrt{y_i} + \sqrt{y_0}}\right).$$

For sea-level conditions, the numerical values give

$$t_0 = \frac{0.8 \text{ m}}{\sqrt{(9.81 \text{ m/s}^2)(0.30 \text{ m})}} + \frac{2}{\sqrt{9.81 \text{ m/s}^2}}\left(\frac{4.0 \text{ m} - 0.8 \text{ m}}{\sqrt{0.30 \text{ m}} + \sqrt{0.10 \text{ m}}}\right);$$

$$t_0 = 2.83 \text{ s}.$$ **ANSWER**

Discussion

Note that a surface wave travels more slowly in shallow water.

11.1.3 Classification of Open Channel Flow

An orderly classification system for open channel flow helps engineers visualize what is likely to occur in a specific situation and select the proper methods for analysis. Some of the terms used to classify open channel flows should already be familiar.

An open channel flow may be *steady* or *unsteady,* depending on whether the flow rate is constant or varies with time. Although many open channel flows are unsteady (such as rainwater runoff in drainage ditches), we limit our consideration to steady flows.

Open channel flow may be *laminar* or *turbulent.* We decide which flow state is likely to exist by examining the Reynolds number. Most open channel flows have rather large dimensions and involve water, which has a relatively low viscosity, so Reynolds numbers are quite large and the flow is usually turbulent. An exception to this generality is flow in very thin sheets, such as rainwater runoff from streets and airport runways. Sometimes an engineer may be interested in the flow in a very short section (short *reach*) of a channel, in which case the flow may be assumed frictionless (i.e., neither laminar nor turbulent).

An important problem in open channel flow analysis involves predicting the change in liquid depth,* and hence in the profile of the free surface, caused by disturbances in the channel. The most common disturbances are changes of bottom slope, changes of channel cross-sectional shape or size, and the placing of obstructions in the stream. Flow with constant depth in a prismatic channel is called *uniform flow* (UF). Flow with varying liquid depth is called *varied flow.* Varied flow is further subdivided into *gradually varying flow* (GVF), in which the rate of change of depth with distance along the channel is small, and *rapidly varying flow* (RVF), in which the rate of change of depth is not small. Figure 11.3 illustrates the various types of flow. Obviously, a single channel can contain regions of all types, although the flow at each location is of a specific type.

The flow state at a particular location in an open channel is characterized by the local Froude number. The classifications are

$\mathrm{F} < 1$ *subcritical* or *tranquil* flow,

$\mathrm{F} = 1$ *critical* flow, and

$\mathrm{F} > 1$ *supercritical* or *rapid* flow.

* At times, distinguishing between *depth,* which is measured parallel to the gravity vector, and flow *thickness,* which is measured perpendicular to the channel bottom, is important.

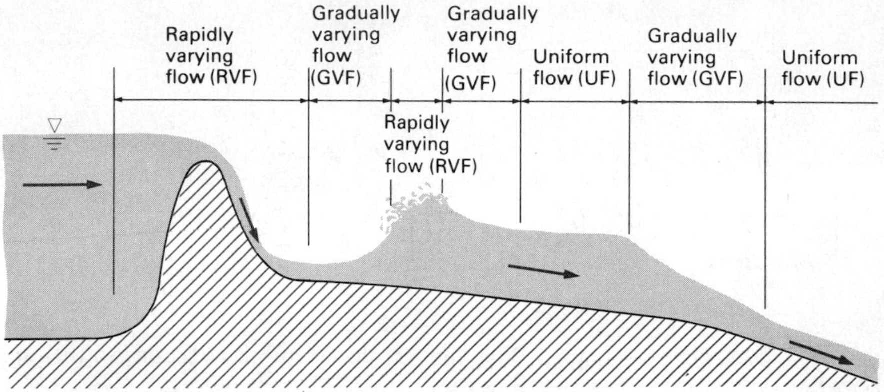

Figure 11.3 Illustration of varied flow and uniform flow in an open channel.

The response of the channel flow to disturbances is different for subcritical and supercritical conditions. If the flow past the disturbance is supercritical, the waves generated by the disturbance cannot travel upstream because the wave speed is less than the flow speed. The waves are swept downstream and only the downstream flow is affected by the disturbance. If the flow is subcritical, waves generated by a disturbance travel both upstream and downstream and the upstream flow is "warned" of the disturbance.

The physical differences between subcritical and supercritical flow are reflected in methods of problem solution. In supercritical flow, we can start the solution at a point and proceed downstream; however, in subcritical flow, we usually have to begin at a point and calculate upstream.

11.2 UNIFORM FLOW

Uniform flow is flow with a constant depth and constant velocity. It is the open channel equivalent of fully developed pipe flow. Uniform flow can occur only in a straight prismatic channel with a constant bottom slope. When liquid enters a reach of such a channel, there is a development region of gradually varying flow, called the *transitory zone.* The flow is determined by the balance between gravity and wall resistance forces and the inertia "force." In the transitory zone, the gravity force exceeds the wall force and the flow accelerates. The higher velocity increases wall shear stress. If the channel is long enough, a condition of equilibrium between gravity force and wall force ultimately results, and the flow becomes uniform.

Many channels are designed for the condition of uniform flow. The depth corresponding to uniform flow in a particular channel is called the *normal depth* (y_n). The normal depth is an important design parameter, even if the flow in the channel is nonuniform.

11.2.1 Chezy and Manning Formulas

The fundamental problem of uniform flow is determining a relation between (average) fluid velocity, normal depth, and channel slope and geometry. Figure 11.4(a) illustrates uniform flow. The channel has an arbitrary cross-sectional

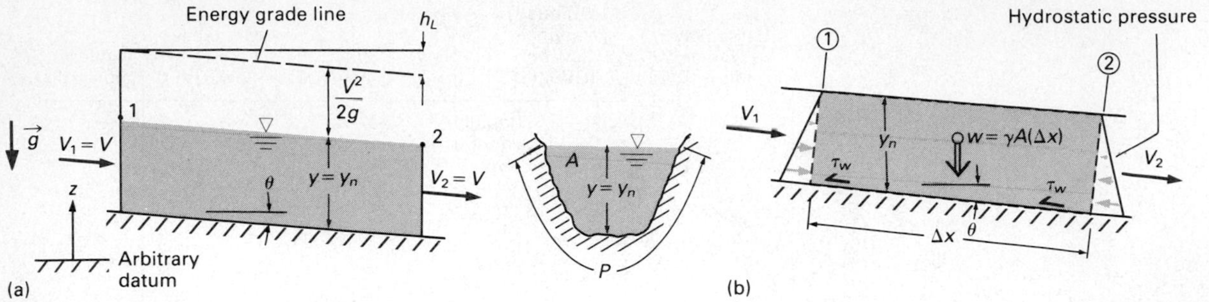

Figure 11.4 Details of uniform flow in an open channel: (a) geometry; (b) control volume for momentum analysis.

shape.* The uniform flow depth is constant, so the liquid surface is parallel to the channel bottom.

We now analyze the flow through the control volume, shown in Fig. 11.4(b). The continuity equation is

$$Q = \bar{V}_1 A_1 = \bar{V}_2 A_2,$$

where A is the area of the liquid cross section. For a given cross-sectional shape, the liquid area and depth are uniquely related, so if we know any two of velocity, flow rate, or depth, we can calculate the remaining quantity.

We now write the energy equation for the control volume. The most convenient form of the energy equation for this flow is the specific mechanical energy equation, Eq. (4.48). As there is no shaft or shear work, the equation becomes

$$\alpha_1 \frac{\bar{V}_1^2}{2} + \frac{\bar{p}_1}{\rho} + \overline{gz}_1 = \alpha_2 \frac{\bar{V}_2^2}{2} + \frac{\bar{p}_2}{\rho} + \overline{gz}_2 + \overline{gh}_L.$$

In uniform flow, the streamlines are straight and parallel, so the piezometric head $(p/\gamma + z)$ is constant over any cross section and equal to the value at the free surface. Taking the free surface pressure as the zero datum,

$$\frac{\bar{p}}{\rho} + \overline{gz} = gz,$$

where z is the elevation of the free surface. In uniform flow,

$$\bar{V}_1 = \bar{V}_2 \quad \text{and} \quad \alpha_1 = \alpha_2,$$

so

$$h_L = z_1 - z_2, \tag{11.7}$$

and we conclude that the energy grade line is also parallel to the bottom (and the free surface).

Next, we apply the momentum equation to the control volume. From here on, we omit the overbar denoting an average quantity. For uniform flow, the liquid's velocity entering the control volume and exiting the control volume are equal and

* In practice, only engineered channels with simple shapes have long enough runs of constant shape and slope to establish uniform flow.

there is no momentum change. Constant depth means that the forces from hydrostatic pressure at the control volume entrance and exit are equal and opposite. The momentum equation reduces to a balance between the gravity force and the resisting force due to shear stress at the walls:

$$\gamma(A\,\Delta x)\sin\theta = \bar{\tau}_w P\,\Delta x,$$

where P is the wetted perimeter of the channel cross section and $\bar{\tau}_w$ is the average shear stress. Channels with uniform flow generally have very small slopes, so

$$\sin\theta \approx \tan\theta \approx \theta \equiv S_0,$$

where S_0 is the slope of the channel bottom in radians. The momentum equation then gives

$$\bar{\tau}_w = \gamma S_0 R_h = \rho g S_0 R_h, \tag{11.8}$$

where R_h is the *hydraulic radius* defined by

$$R_h \equiv \frac{A}{P}.$$

The hydraulic radius is one fourth the hydraulic diameter D_h defined in Section 7.2.6. Note that the hydraulic radius depends on both the liquid depth in the channel and the channel geometry.

The shear stress, in terms of the skin friction coefficient, is

$$\bar{\tau}_w = C_f(\tfrac{1}{2}\rho V^2).$$

Because uniform open channel flow is somewhat similar to fully developed pipe flow, common practice is to use a Darcy friction factor,

$$f = 4C_f,$$

giving

$$\bar{\tau}_w = \frac{f}{8}\rho V^2. \tag{11.9}$$

Substituting Eq. (11.9) into Eq. (11.8) and solving for the velocity, we have

$$V = \sqrt{\frac{8g}{f}}\sqrt{R_h S_0}. \tag{11.10}$$

The dependence of liquid velocity (and hence flow rate) on channel slope and hydraulic radius implied by Eq. (11.10) was first discovered by the French engineer A. Chezy in 1769. The equation is sometimes written as

$$V = C\sqrt{R_h S_0}. \tag{11.11}$$

Equation 11.11 is called the *Chezy formula*. The coefficient C is called the *Chezy coefficient*. Obviously,

$$C = \sqrt{\frac{8g}{f}}. \tag{11.12}$$

A dimensional coefficient, C ranges in value from about 60 $ft^{1/2}/sec$ (33 $m^{1/2}/s$) for small, rough channels to about 160 $ft^{1/2}/sec$ (88 $m^{1/2}/s$) for large, smooth channels. Early research on open channel flow was involved with attempts to correlate C with channel geometry. Equation (11.12) indicates that C depends on the same parameters as f, namely, Reynolds number ($\mathbf{R} \equiv VD_h/\nu = 4\,VR_h/\nu$) and channel surface roughness (ε/D_h). Many experiments have established that C can be evaluated by using the hydraulic diameter concept (see Section 7.2.6) together with the Colebrook formula or Moody chart to evaluate the friction factor f. Equation (11.12) then gives a value for C. We call this the *friction-factor method.* By analogy with fully developed pipe flow, uniform open channel flow is laminar if $\mathbf{R} < 2300$ and turbulent if $\mathbf{R} > 4000$.

We can determine a more direct equation for C by noting that most open channel flows occur at very large Reynolds numbers and on moderately rough to very rough surfaces. In this case, the friction factor depends only on roughness and is given by (see Eq. 7.47)

$$\frac{1}{\sqrt{f}} \approx -2.0 \log\left(\frac{\varepsilon/D_h}{3.7}\right)$$

or, in terms of R_h ($= D_h/4$),

$$\frac{1}{\sqrt{f}} \approx -2.0 \log\left(\frac{\varepsilon/R_h}{14.8}\right).$$

Substituting into Eq. (11.12) gives

$$C \approx -\sqrt{32g}\, \log\left(\frac{\varepsilon/R_h}{14.8}\right). \qquad (11.13)$$

Equation (11.13) has one serious drawback. Values of surface roughness can be obtained for several materials from Fig. 7.12 or Table 7.1; however, these are manufactured materials. Open channels are often lined with stones, rough earth, or vegetation. Natural channels may have extremely irregular bottom surfaces. Direct information on the absolute roughness (ε) of these types of channel surfaces is not easy to obtain. Information about roughness of open channel surfaces is available in a form more suitable for use with *Manning's formula.* In 1891, Robert Manning, an Irish engineer, proposed the following form of Chezy's formula:*

$$V = \frac{1.0}{n} (R_h)^{2/3} (S_0)^{1/2} \qquad (11.14a)$$

for SI units (R_h in meters) or

$$V = \frac{1.49}{n} (R_h)^{2/3} (S_0)^{1/2} \qquad (11.14b)$$

* Note that 1.0 has units of $m^{1/3}/s$ and 1.49 has units of $ft^{1/3}/sec$.

Table 11.1 Values of Manning's *n* for various channel surfaces (uncertainty of values approximately 20%).

Channel Surface	n	Channel Surface	n
Artificial lined channels		Excavated earth channels	
Glass	0.010	Clean	0.022
Brass	0.011	Gravelly	0.025
Steel, smooth	0.012	Weedy	0.030
Painted	0.014	Stony, cobbles	0.035
Riveted	0.015	Natural channels	
Cast iron	0.013	Clean and straight	0.030
Concrete, finished	0.012	Sluggish, deep pools	0.040
Unfinished	0.014	Major rivers	0.035
Planed wood	0.012	Floodplains	
Clay tile	0.014	Pasture, farmland	0.035
Brickwork	0.015	Light brush	0.05
Asphalt	0.016	Heavy brush	0.075
Corrugated metal	0.022	Trees	0.15
Rubble masonry	0.025		

for BG or EE units (R_h in feet). In terms of the Chezy coefficient, Manning's proposal is equivalent to

$$C = \frac{1.0\ (\mathrm{m}^{1/3}/\mathrm{s})}{n}[R_h\ (\mathrm{m})]^{1/6} \qquad \text{(SI units)} \tag{11.15a}$$

or

$$C = \frac{1.49\ (\mathrm{ft}^{1/3}/\mathrm{sec})}{n}[R_h\ (\mathrm{ft})]^{1/6} \qquad \text{(BG and EE units).} \tag{11.15b}$$

The parameter n, called *Manning's n,* is a dimensionless measure of the roughness of the channel surface. Manning's research produced values of n for many surfaces; a partial list of values is presented in Table 11.1. As n is a measure of surface roughness, there is some equivalence between n and surface roughness ε. White [4] has shown that in the range $0.001 \leq \varepsilon/D_h \leq 0.1$, the relation can be represented by

$$n \approx 0.031[\varepsilon\ (\mathrm{ft})]^{1/6}. \tag{11.16}$$

Thus n varies much less rapidly than ε, with a thousandfold change in ε producing only a threefold change in n.

In practice, Eq. (11.12), Eq. (11.13), or Eq. (11.15) may be used with Eq. (11.11) to predict the flow velocity in an open channel, provided information on ε or n is available. You do not get identical answers, however, because of the approximations of the data correlations and because Manning's method is based on the assumption of fully rough flow.

EXAMPLE 11.2 Illustrates Calculations in Uniform Flow and Differences between Equations for Estimation of the Chezy Coefficient

A 24-in.-diameter cast-iron storm sewer pipe is laid on a slope of 0.1° as shown in Fig. E11.2. The pipe runs half-full of water. Calculate the flow rate in the pipe using

a) Eqs. (11.11) and (11.12), with f evaluated from the Moody chart;

b) Eqs. (11.11) and (11.13);

c) Eq. (11.14b).

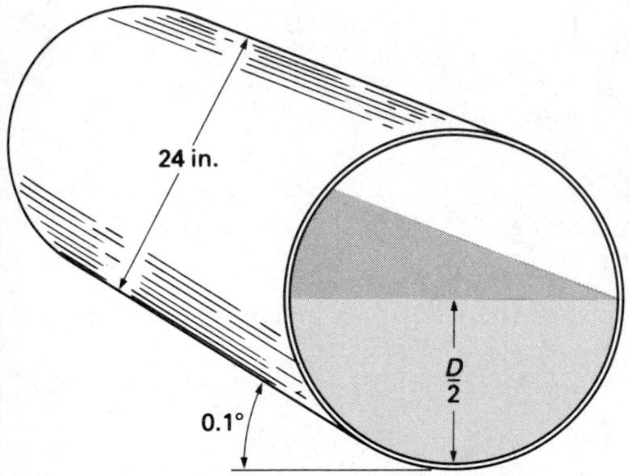

Figure E11.2 Storm sewer pipe.

SOLUTION

Given

24-in.-diameter cast-iron pipe

Slope 0.1°

Pipe half-full of water

Find

Flow rate using

a) Eqs. (11.11) and (11.12) and Moody chart;

b) Eqs. (11.11) and (11.13);

c) Eq. (11.14b).

Solution

a) The volume flow rate, Q, is given by

$$Q = VA = V\left(\frac{1}{2}\right)\left(\frac{\pi D^2}{4}\right) = \frac{\pi D^2 V}{8},$$

where D is the pipe inside diameter and V the average water velocity. Equations (11.11) and (11.12) give

$$V = C\sqrt{R_h S_0} = \sqrt{\frac{8g}{f}}\sqrt{R_h S_0}.$$

The hydraulic radius R_h is

$$R_h = \frac{A}{P} = \left(\frac{\pi D^2}{8}\right)\left(\frac{2}{\pi D}\right) = \frac{D}{4} = \frac{24 \text{ in.}}{4} = 6 \text{ in.}$$

The slope S_0 is

$$S_0 = 0.1°\left(\frac{\pi}{180°}\right) = 0.00175.$$

With data from Table 7.1, the relative roughness is

$$\frac{\varepsilon}{D_h} = \frac{\varepsilon}{4R_h} = \frac{8.5 \times 10^{-4} \text{ ft}}{4(6.0 \text{ in.})(1 \text{ ft}/12 \text{ in.})} = 4.25 \times 10^{-4}.$$

If we assume 60°F water, Table A.6 gives $\nu = 1.22 \times 10^{-5}$ ft²/sec. The Reynolds number is

$$\mathbf{R} = \frac{VD_h}{\nu} = \frac{4VR_h}{\nu} = \frac{4V(6 \text{ in.})(1 \text{ ft}/12 \text{ in.})}{(1.22 \times 10^{-5} \text{ ft}^2/\text{sec})}$$
$$= 1.64 \times 10^5 \; V \text{ sec/ft}.$$

We have to use an iterative procedure to find the velocity V; the fastest convergence is obtained by guessing a value for f, calculating V from

$$V = \sqrt{\frac{8(32.2 \text{ ft/sec}^2)}{f}}\sqrt{(6 \text{ in.})(1 \text{ ft}/12 \text{ in.})(0.00175)}$$
$$= \frac{0.475}{\sqrt{f}} \text{ ft/sec},$$

calculating the Reynolds number, finding f from the Moody chart, and comparing this value of f with the one chosen. Beginning with $f = 0.0162$ (the fully turbulent value), we have

$$V = \frac{0.475}{\sqrt{0.0162}} \text{ ft/sec} = 3.73 \text{ ft/sec},$$
$$\mathbf{R} = 1.64 \times 10^5 (3.73 \text{ ft/sec})(\text{sec/ft}) = 6.11 \times 10^5,$$

and

$$f = 0.0171.$$

The next iteration gives

$$V = \frac{0.475}{\sqrt{0.0171}} \text{ ft/sec} = 3.64 \text{ ft/sec},$$
$$\mathbf{R} = 1.64 \times 10^5 (3.64 \text{ ft/sec})(\text{sec/ft}) = 6.0 \times 10^5,$$

and

$$f = 0.0171.$$

Convergence has been obtained, and the flow rate is

$$Q = \frac{\pi D^2 V}{8} = \frac{\pi (24 \text{ in.})^2 (1 \text{ ft}/12 \text{ in.})^2 (3.64 \text{ ft/sec})}{8},$$

or

$$Q = 5.72 \text{ ft}^3/\text{sec}.$$ **ANSWER**

b) Combining Eqs. (11.11) and (11.13) gives

$$V = -\sqrt{32g}\,\log\left(\frac{\varepsilon/R_h}{14.8}\right)\sqrt{R_h S_0},$$

where

$$\frac{\varepsilon}{R_h} = \frac{8.5 \times 10^{-4} \text{ ft}}{6 \text{ in.}(1 \text{ ft}/12 \text{ in.})} = 17.0 \times 10^{-4}.$$

Then

$$V = -\sqrt{32(32.2 \text{ ft/sec}^2)} \log\left(\frac{17.0 \times 10^{-4}}{14.8}\right)$$

$$\times \sqrt{(6 \text{ in.})\left(\frac{1 \text{ ft}}{12 \text{ in.}}\right)(0.00175)}$$

$$= 3.74 \text{ ft/sec}.$$

The volume flow rate is

$$Q = \frac{\pi D^2 V}{8} = \frac{\pi(24 \text{ in.})^2(1 \text{ ft/12 in.})^2(3.74 \text{ ft/sec})}{8},$$

$$Q = 5.87 \text{ ft}^3\text{/sec}.$$ **ANSWER**

c) Equation (11.14b) gives

$$V = \frac{1.49}{n}(R_h)^{2/3}(S_0)^{1/2}.$$

For cast-iron pipe, Table 11.1 gives $n = 0.013$, so

$$V = \frac{1.49}{0.013}\left(6.0 \text{ in.} \times \frac{1 \text{ ft}}{12 \text{ in.}}\right)^{2/3}(0.00175)^{1/2} = 3.02 \text{ ft/sec}$$

and

$$Q = \frac{\pi D^2 V}{8} = \frac{\pi(24 \text{ in.})^2(1 \text{ ft/12 in.})^2(3.02 \text{ ft/sec})}{8},$$

or

$$Q = 4.74 \text{ ft}^3\text{/sec}.$$ **ANSWER**

Discussion

Note that Eq. (11.14b) gives the velocity in ft/sec if R_h is in ft. Although the values of Q calculated by the three methods differ by as much as 20 percent, selecting the most accurate value is not possible. The close agreement between the values calculated by the friction factor equations does not imply that they are more correct than the Manning's formula calculation.

11.2.2 Analysis and Design of Uniform Flow Channels

Uniform flow problems may involve either analysis or design of channels. In the analysis problem, channel cross section, surface type, and bottom slope are given and the objective is to determine liquid velocity, flow rate, or depth, given any one of them. The design problem involves selecting channel shape and/or slope to convey a given flow rate with a given depth. Often an engineer wants to maximize the flow rate for a given slope or, conversely, to minimize the necessary slope to convey a given flow rate. We first consider a few aspects of the analysis problem and then consider the design problem.

Analysis Problem. In the analysis problem, channel slope, shape, and surface condition are given. The value of n or ε can be estimated from tables. For any one of flow rate (Q), velocity (V), or normal depth (y_n), we can find the remaining two from the continuity equation, the Chezy formula or Manning's formula, and geometric relations (relating A, P, and y_n). The calculations are straightforward if depth is given but require iteration if velocity or flow rate is given. Example 11.2 illustrated the procedure if the depth is given. The following example illustrates the procedure for calculating the depth for a specified flow rate.

EXAMPLE 11.3 Illustrates Normal Depth Calculation in Uniform Open Channel Flow

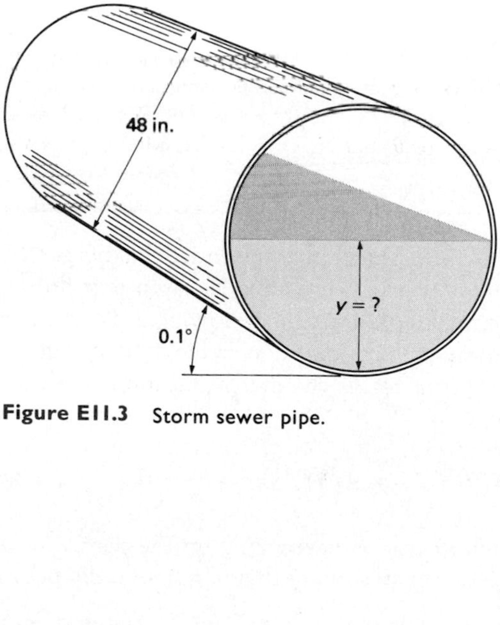

Figure E11.3 Storm sewer pipe.

Five of the storm sewer pipes described in Example 11.2 empty into a 48-in.-diameter, unfinished concrete pipe laid on a slope of 0.1° as shown in Fig. E11.3. Compute the depth of flow in the 48-in. pipe.

SOLUTION

Given

Five storm sewer pipes of Example 11.2

Pipes emptying into 48-in.-diameter, unfinished concrete pipe

Pipe slope 0.1°

Find

Liquid depth in 48-in. pipe

Solution

If we assume that Manning's formula provides the most reliable estimate of the flow in the five smaller storm sewer pipes, the flow rate in the 48-in.-diameter pipe is

$$Q = 5(4.74 \text{ ft}^3/\text{sec}) = 23.7 \text{ ft}^3/\text{sec}.$$

The flow rate is given by

$$Q = VA,$$

where V is obtained from Eq. (11.14b) and Appendix B gives

$$A = \frac{R^2}{2}(\theta - \sin\theta),$$

where R is the radius of the concrete pipe and θ is defined in Appendix B. Also $P = R\theta$. Substituting for Q, V, and A gives

$$23.7 \text{ ft}^3/\text{sec} = \frac{1.49}{n}(R_h)^{2/3}(S_0)^{1/2}\frac{R^2}{2}(\theta - \sin\theta).$$

For the concrete pipe,

$$R_h = \frac{A}{P} = \frac{(R^2/2)(\theta - \sin\theta)}{R\theta} = \frac{R(\theta - \sin\theta)}{2\theta}.$$

Substituting the hydraulic radius gives

$$23.7 \text{ ft}^3/\text{sec} = \frac{1.49}{n}\left[\frac{R(\theta - \sin\theta)}{2\theta}\right]^{2/3}(S_0)^{1/2}\frac{R^2}{2}(\theta - \sin\theta).$$

Using Table 11.1 and Example 11.2, we substitute the remaining known values

$$23.7\ \text{ft}^3/\text{sec} = \frac{1.49}{0.014}\left[\frac{(24\ \text{in.})(1\ \text{ft}/12\ \text{in.})}{2}\left(\frac{\theta - \sin\theta}{\theta}\right)\right]^{2/3}$$
$$\times\ (0.00175)^{1/2}\,\frac{(24\ \text{in.})^2(1\ \text{ft}/12\ \text{in.})^2}{2}\,(\theta - \sin\theta),$$

or

$$2.66 = \frac{(\theta - \sin\theta)^{5/3}}{\theta^{2/3}}.$$

Solving for θ by iteration gives $\theta = 2.96$ rad. The liquid depth is

$$y_n = R - R\cos\frac{\theta}{2} = R\left(1 - \cos\frac{\theta}{2}\right) = 24\ \text{in.}\left(1 - \cos\frac{2.96}{2}\right),$$

or

$$y_n = 21.8\ \text{in.}$$ **ANSWER**

Discussion

Note that we actually calculated the thickness of the flow, t_n, rather than the depth. The two quantities are related by

$$y = \frac{t}{\cos S_0}.$$

Most channels have small bottom slopes, so $\cos S_0$ is usually very nearly 1, and the difference between thickness and depth is insignificant. In this problem, $\cos S_0 = \cos 0.1° = 0.99999$. We used fixed-point iteration to solve for θ, but Newton-Raphson iteration is also a good choice.

In the problems considered thus far, we assumed that the entire channel surface can be characterized by a single value of ε or n. However, let's consider flow in the channel shown in Fig. 11.5, which may represent a river flowing at flood stage. The main channel does not have the same surface condition as the flow outside the riverbank. We can analyze this flow by breaking the channel into sections as shown and writing, say, Manning's equation as

$$Q_{\text{total}} = \frac{1.49}{n_1}A_1R_{h_1}^{2/3}S_0^{1/2} + \frac{1.49}{n_2}A_2R_{h_2}^{2/3}S_0^{1/2} + \cdots. \tag{11.17}$$

Note that the hydraulic radius of each section is based only on the perimeter of the channel surface and does not include the imaginary dividing line in the liquid.

Design Problem. When designing channels for uniform flow, an engineer must select the channel shape and slope to convey liquid at a certain rate and with a specific depth or velocity. Many criteria may be involved in the selection process. In one case, perhaps the channel must convey as large a flow rate as possible, with a large velocity desirable. In another case, perhaps the velocity must be kept small to avoid erosion. Selecting all the channel's geometric parameters based on

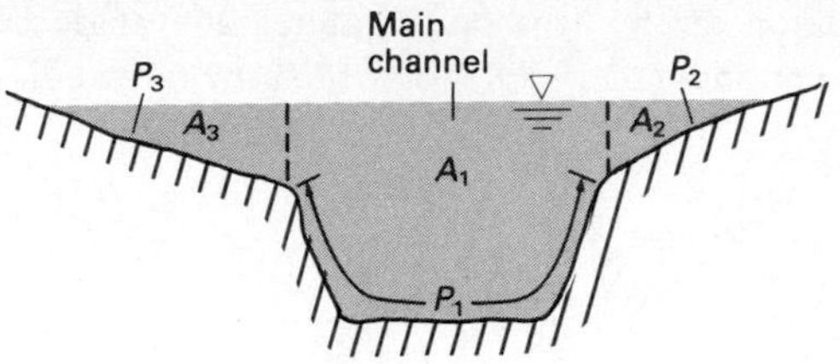

Figure 11.5 Flow in channel with sections with different characteristics.

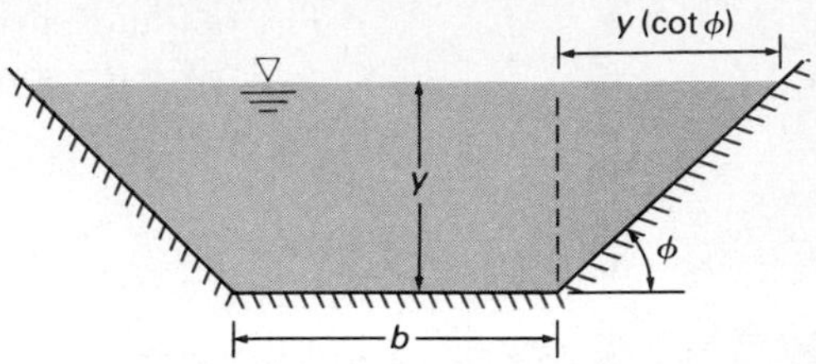

Figure 11.6 Trapezoidal channel.

specific criteria is usually not possible. For example, channel shape may be arbitrarily selected or may be dictated by the method of excavation.

If channel slope is to be determined, it is easily calculated from the Chezy or Manning formula, provided depth and flow rate are specified and channel shape and surface condition are known. An important practical problem involves selecting the channel shape and depth to most efficiently carry a particular flow rate, that is, the shape that conveys the flow with the smallest bottom slope. Combining Manning's formula with the continuity equation gives

$$Q = \frac{1.49}{n}(R_h)^{2/3} S_0^{1/2} A.$$

Substituting the definition of R_h, we have

$$Q = \frac{1.49}{n} S_0^{1/2} \frac{A^{5/3}}{P^{2/3}}.$$

For a given surface type (n), S_0 is minimized for a given Q by making $A^{5/3}/P^{2/3}$ as large as possible. This problem is purely geometric; the most efficient channel section has the smallest perimeter for a given area. A totally unspecified shape presents a difficult problem; however, if a class of shapes is chosen, optimizing that class is not too difficult. Consider a trapezoidal channel filled with liquid to a depth y (Fig. 11.6). The area is*

$$A = by + y^2 \cot \phi \tag{11.18}$$

and the wetted perimeter is

$$P = b + 2y(1 + \cot^2 \phi)^{1/2}. \tag{11.19}$$

Eliminating b between these two equations gives

$$P = \frac{A}{y} - (\cot \phi)y + 2y(1 + \cot^2 \phi)^{1/2}. \tag{11.20}$$

To find the optimum depth, we take $\partial P/\partial y$ and set the result equal to zero, giving

$$y = \left[\frac{A}{2(1 + \cot^2 \phi)^{1/2} - \cot \phi}\right]^{1/2}. \tag{11.21}$$

* Again, depth is used instead of the more exactly correct thickness.

The optimum depth is a function of the area and channel side angle ϕ. A particularly interesting shape is a rectangle ($\phi = 90°$) for which the most efficient flow has depth

$$y = \sqrt{\frac{A}{2}}$$

or, because $A = by$ for a rectangle,

$$y = \frac{b}{2}.$$

We can find the most efficient trapezoidal channel shape by taking $\partial P/\partial\phi$ and setting the result to zero, yielding

$$\phi = 60°. \tag{11.22}$$

If a trapezoidal channel is specified, with values of Q and S_0 given, we can combine Eqs. (11.21) and (11.22) with Manning's equation to select the optimum channel dimensions.

This type of calculation can be repeated for other geometries. Not surprisingly (if you think about it), the most efficient channel is one of semicircular shape.

11.3 CONCEPTS FOR ANALYSIS OF VARIED FLOW

We now present some of the phenomena associated with varied flow and introduce the parameters that are commonly used to describe them.

11.3.1 Specific Energy and Critical Depth

Figure 11.7 illustrates varied flow in an open channel. For simplicity, we assume that the channel is rectangular and prismatic. The specific mechanical energy equation, Eq. (4.48), may be written for the flow between cross sections 1 and 2. There is no work, so we get

$$\alpha_1 \frac{\bar{V}_1^2}{2} + \frac{\bar{p}_1}{\rho} + \overline{gz}_1 = \alpha_2 \frac{\bar{V}_2^2}{2} + \frac{\bar{p}_2}{\rho} + \overline{gz}_2 + \overline{gh}_L.$$

We assume that the streamlines are nearly straight and parallel so that, within any

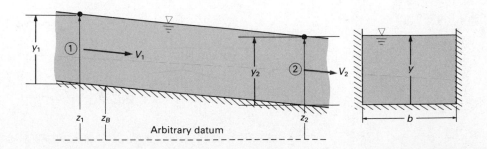

Figure 11.7 Varied flow.

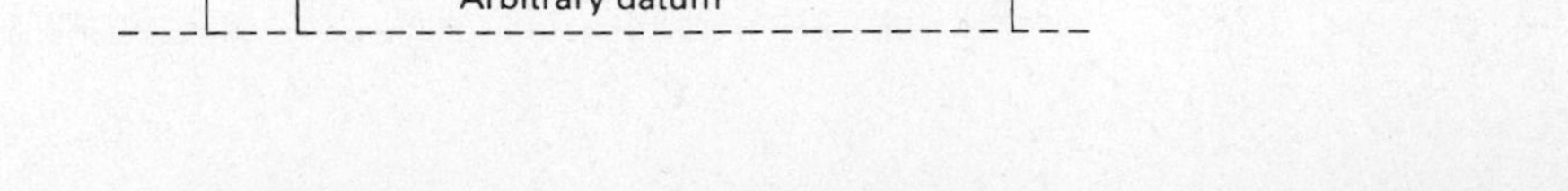

cross section,

$$\frac{\bar{p}}{\rho} + \overline{gz} = 0 + gz,$$

where z is the elevation of the free surface at that cross section. Also, we assume that $\alpha \approx 1$, drop the overbars denoting average quantities, and divide by g to convert the equation to head form. The result is

$$\frac{V_1^2}{2g} + z_1 = \frac{V_2^2}{2g} + z_2 + h_L.$$

We divide the elevation into two parts,

$$z = z_B + y,$$

where z_B is the elevation of the channel bottom and y is the depth of liquid in the channel. The energy equation can then be written

$$\left(\frac{V_1^2}{2g} + y_1\right) - \left(\frac{V_2^2}{2g} + y_2\right) = z_{B_2} - z_{B_1} + h_L. \tag{11.23}$$

The quantity E, defined by

$$E \equiv \frac{V^2}{2g} + y, \tag{11.24}$$

is called the *specific energy** and is the energy per unit weight of the liquid, measured with the channel bottom as datum. At any cross section, the specific energy has a unique value, although the distribution of specific energy between kinetic and potential forms is not necessarily unique. The specific energy changes along the channel because of changes of bottom elevation and mechanical energy loss. Specific energy is constant in a horizontal, frictionless channel, even if the flow is varied. In uniform flow, $\Delta z_B = -h_L$ and the specific energy is constant.

For a given value of specific energy and a given flow rate, there may be zero, one, or two possible flow states. For a rectangular channel, the flow rate is

$$Q = Vby = qb,$$

where b is the width of the channel and q ($\equiv Q/b$) is the flow rate per unit width. The velocity is

$$V = \frac{q}{y}. \tag{11.25}$$

Substituting Eq. (11.25) into Eq. (11.24) gives

$$E = y + \frac{q^2}{2gy^2}. \tag{11.26}$$

For a given value of q, we can plot a specific energy curve, shown in Fig. 11.8.

* A better name would be *specific head;* however, the term *specific energy* is widely used.

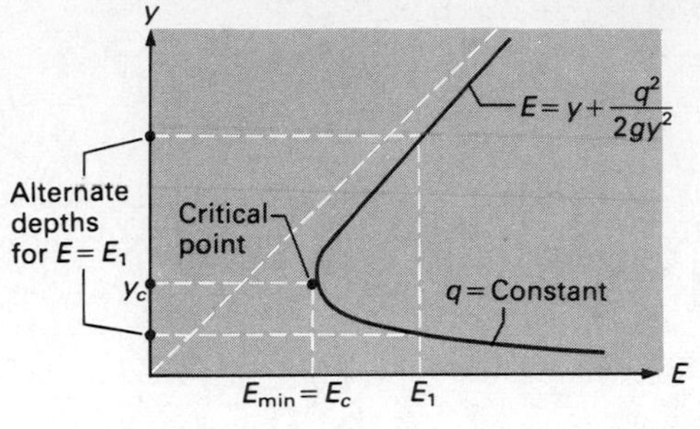

Figure 11.8 Specific energy curve for constant q.

This curve shows that for given values of q and E, two depths, and hence two velocities, are possible. The two depths are called *alternate depths*.

Note also that there is a minimum possible value of E for a given value of q. At this minimum energy point, the alternate depths are equal. We can evaluate the flow state at the minimum energy point by noting that

$$\left.\frac{\partial E}{\partial y}\right)_q = 0$$

there. Differentiating Eq. (11.26) and setting the result to zero, we obtain

$$y_c = \left(\frac{q^2}{g}\right)^{1/3} \tag{11.27}$$

and hence

$$E_c = \frac{3}{2}\left(\frac{q^2}{g}\right)^{1/3} = \frac{3}{2}y_c. \tag{11.28}$$

The velocity at the minimum energy point is

$$V_c = \frac{q}{y_c} = (gy_c)^{1/2}. \tag{11.29}$$

The quantity $(gy_c)^{1/2}$ is the small disturbance wave speed, so, at the minimum energy point, the Froude number is 1 and the flow is critical.* The quantity y_c is called the *critical depth*. The critical depth depends only on the flow rate and g.

We can now label the upper and lower branches of the specific energy curve. Along the upper branch, the depth is large and the velocity is small, with

$$V < \sqrt{gy_c}$$

and the flow is subcritical. Along the lower branch, the depth is small and the velocity is large, with

$$V > \sqrt{gy_c}$$

* Hence the subscript c to identify parameters at the minimum energy point.

and the flow is supercritical. For a given value of E, one of the alternate depths corresponds to subcritical flow (upper branch) and the other corresponds to supercritical flow (lower branch).

We can make several observations about varied flow by examining Fig. 11.8, including the following:

- When the flow depth is greater than the critical depth (subcritical flow), an increase in specific energy causes an increase in depth.
- When the flow depth is less than the critical depth (supercritical flow), an increase in specific energy causes a decrease in depth.
- If the flow is critical, small changes in specific energy cause large changes in depth. In practice, sustained critical flow (over a long reach of channel) is unstable.

The specific energy diagram of Fig. 11.8 is for a specific flow rate. By selecting various flow rates, we can generate a family of specific energy curves as shown in Fig. 11.9. Note that the larger the flow rate is, the larger the minimum value of specific energy is. For any value of E, there is a maximum flow rate, given by

$$q_{\max} = q_c = \sqrt{g\left(\frac{2}{3}E_c\right)^3}. \tag{11.30}$$

This discussion of specific energy and critical depth has been limited to rectangular channels. Extending the discussion to nonrectangular channels is not difficult. Figure 11.10 shows a channel of arbitrary cross section. The specific energy is

$$E = y + \frac{Q^2}{2gA^2}, \tag{11.31}$$

where Q is the total flow rate and A is the cross-sectional area of the liquid. Because A is a function of y for a given cross section, we can plot a specific energy curve similar to that in Fig. 11.8 for a given value of Q. We locate the critical (minimum energy) point by setting the derivative of E with respect to y equal to zero and noting that A is a function of y, which gives

$$\left.\frac{dA}{dy}\right)_c = \frac{gA_c^3}{Q^2}.$$

Figure 11.9 Specific energy curves for various flow rates.

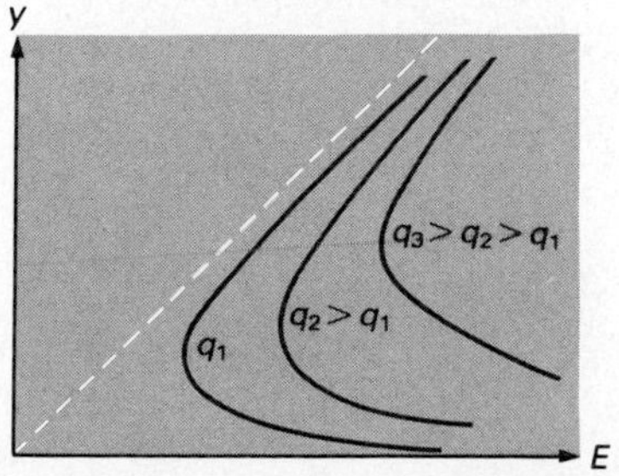

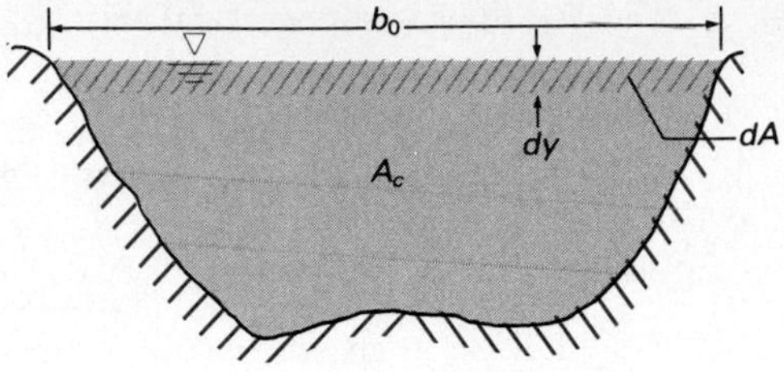

Figure 11.10 Open channel with arbitrary cross section for defining critical conditions.

From Fig. 11.10, we have

$$\frac{dA}{dy} = b_0.$$

The critical area and velocity are given by

$$A_c = \left(\frac{b_0 Q^2}{g}\right)^{1/3} \tag{11.32}$$

and

$$V_c = \frac{Q}{A_c} = \left(\frac{gA_c}{b_0}\right)^{1/2}. \tag{11.33}$$

The Froude number for a nonrectangular channel is

$$\mathbf{F} = \frac{V}{V_c} = \frac{V}{(gA_c/b_0)^{1/2}}. \tag{11.34}$$

The geometric quantity

$$y_h \equiv \frac{A}{b_0}$$

is called the *hydraulic depth* and is an important reference length in open channel flow.

Uniform flow in an open channel can be subcritical, critical, or supercritical, depending on whether the normal depth (y_n) is greater than, equal to, or less than the critical depth (y_c). The critical slope is the value of the bottom slope that yields uniform critical flow. The critical slope is evaluated by equating the critical velocity from Eq. (11.33) to the velocity from Manning's (or Chezy's) formula, Eq. (11.14):

$$\left(\frac{gA_c}{b_0}\right)^{1/2} = \frac{\xi}{n}\left(\frac{A_c}{P_c}\right)^{2/3}(S_c)^{1/2},$$

where $\xi = 1.0$ for SI units and $\xi = 1.49$ for EE or BG units. Solving for S_c, the critical slope, gives

$$S_c = \frac{gn^2}{b_0\xi^2}\frac{(P_c)^{4/3}}{(A_c)^{1/3}}. \tag{11.35}$$

For a wide rectangular channel ($b_0 \gg y$), Eq. (11.35) reduces to

$$S_c \approx \frac{gn^2}{\xi^2} (y_2)^{-1/3}. \tag{11.36}$$

Channel slopes less than the critical are called *mild* and channel slopes greater than the critical are called *steep*. Note that the same channel can be either steep or mild, depending on the flow rate.

EXAMPLE 11.4 Illustrates the Calculation of Critical Flow and Critical Slope

Figure E11.4 illustrates flow over a free overfall in a wide, rectangular channel. Measurements indicate that the depth at the brink is about 0.72 times the critical depth and that critical depth occurs about 5 times the critical depth upstream from the brink if the approaching flow is subcritical. In the channel shown, the depth at the brink is 0.5 m. Compute the flow rate per unit channel width. Calculate the critical slope if the channel is lined with brick and, alternatively, with rubble masonary. What is the Froude number at the overfall?

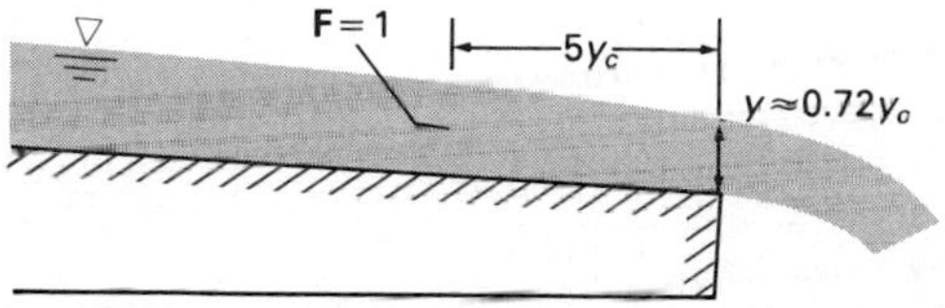

Figure E11.4 Water flow at free overfall.

SOLUTION

Given

Free overfall in Fig. E11.4

Brink depth of 0.5 m is 0.72 times critical depth

Critical depth occurs upstream from brink at distance 5 times critical depth

Subcritical flow

Find

Flow rate per unit channel width

Critical slopes for brick and rubble masonary channels

Froude number at overfall

Solution

The critical depth of the approaching flow is

$$y_c = \frac{y_{\text{brink}}}{0.72} = \frac{0.5 \text{ m}}{0.72} = 0.694 \text{ m}.$$

We use Eq. (11.27) to find the flow rate per unit width, q:

$$q = \sqrt{gy_c^3} = \sqrt{(9.81 \text{ m/s}^2)(0.694 \text{ m})^3};$$

$$q = 1.81 \text{ m}^2/\text{s}.$$ **ANSWER**

The critical slope, Eq. (11.36), is

$$S_c = \frac{gn^2}{1.0} (y_c)^{-1/3},$$

for SI units, with n from Table 11.1. For a brick-lined channel, $n =$

0.015 and

$$S_c = (9.81 \text{ m/s}^2)(0.015)^2(0.694 \text{ m})^{-1/3};$$

$$S_c = 0.0025.$$ **ANSWER**

For a rubble masonary-lined channel, $n = 0.025$ and

$$S_c = (9.81 \text{ m/s}^2)(0.025)^2(0.694 \text{ m})^{-1/3};$$

$$S_c = 0.0069.$$ **ANSWER**

We find the Froude number at the overfall by applying Eq. (11.23) between the critical depth (c) and the brink (b). Neglecting the slight drop of the channel bottom and assuming frictionless flow, we have

$$\frac{V_c^2}{2g} + y_c = \frac{V_b^2}{2g} + y_b.$$

At the critical depth, Eq. (11.29) gives

$$V_c^2 = gy_c, \quad \text{so} \quad \frac{y_c}{2} + y_c = \frac{V_b^2}{2g} + y_b.$$

The problem statement gives $y_b = 0.72y_c$, so

$$\frac{3}{2}y_c = \frac{V_b^2}{2g} + 0.72y_c \quad \text{and} \quad V_b = \sqrt{2g(0.78y_c)} = \sqrt{1.56gy_c}.$$

The Froude number at the brink is

$$\mathbf{F}_b = \frac{V_b}{\sqrt{gy_b}} = \sqrt{\frac{1.56gy_c}{gy_b}} = \sqrt{\frac{1.56y_c}{y_b}}.$$

The numerical values give

$$\mathbf{F}_b = \sqrt{\frac{1.56(0.694 \text{ m})}{(0.50 \text{ m})}};$$

$$\mathbf{F}_b = 1.47.$$ **ANSWER**

Discussion

The approach flow is subcritical, and the brink flow is supercritical ($\mathbf{F} > 1$). The flow must pass through the critical depth for the transition to occur from subcritical to supercritical flow. The given values of brink depth ($0.72y_c$) and upstream distance from brink to critical depth ($\approx 5y_c$) are generally valid for subcritical approach flow.

11.3.2 Frictionless Flow in a Rectangular Channel

We can illustrate the usefulness of the specific energy concept by considering frictionless flow in a rectangular channel. For flow between any two locations 1 and 2, Eq. (11.23) can be written

$$E_2 - E_1 = -\Delta z_B - h_L, \tag{11.37}$$

where Δz_B ($= z_{B_2} - z_{B_1}$) is the increase in bottom elevation. The specific energy is changed by changes in bottom elevation (which may be positive or negative) or by energy loss, which must be positive. Uniform flow has $\Delta z_B = -h_L$, while frictionless flow has $h_L = 0$.

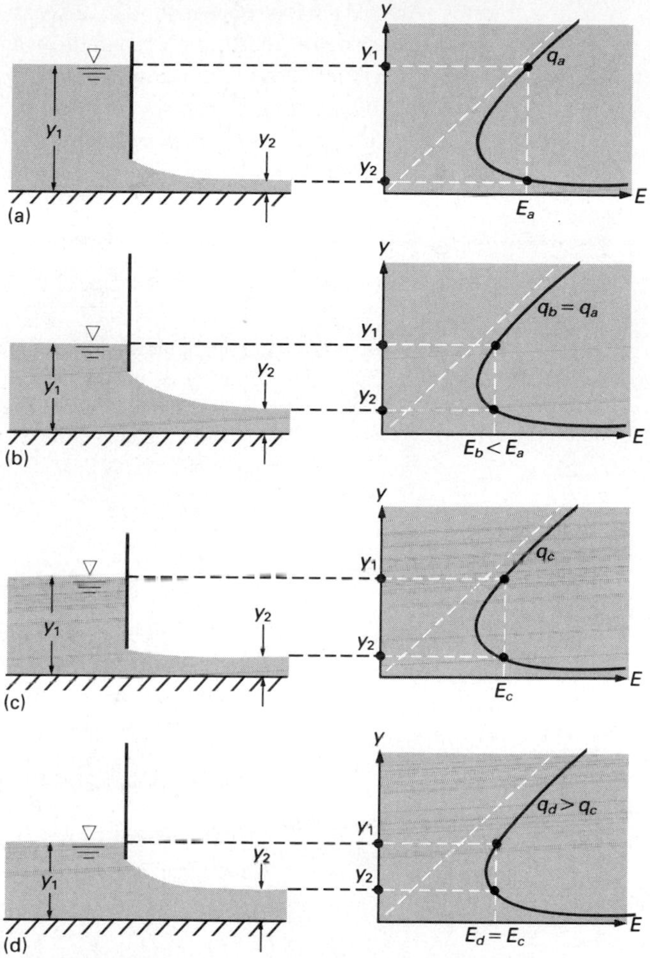

Figure 11.11 Specific energy diagrams for flow under sluice gate: (a) and (b) q is constant; (c) and (d) E is constant.

For a frictionless flow with constant specific energy, the channel bottom must be horizontal. These conditions are closely approximated by flow under a *sluice gate* (Fig. 11.11a), a device for regulating flow in open channels. For a given opening of the gate (y_G), a specific flow rate (q) exists. If we neglect friction, Eq. (11.37) indicates that the upstream and downstream values of E are equal; thus the upstream and downstream flow states are alternate states. The upstream state is subcritical and the downstream state is supercritical. Theoretically, the critical state occurs right at the opening, with $y_G = y_c$; however, the flow near the opening is not one-directional and one-dimensional.

If we specify that the flow rate remain constant and consider a larger gate opening (Fig. 11.11b), the upstream depth is less and the downstream depth is greater; the specific energy is smaller. For successively larger gate openings, the upstream and downstream states approach the critical state, which occurs just as the gate no longer touches the stream.

Next, imagine that the specific energy remains constant as the gate is opened (Fig. 11.11c and d). As the gate is opened, the flow rate increases, and the upstream and downstream state points are located on curves corresponding to larger flow rates. As the gate is opened and q increases, the upstream depth decreases, the downstream depth increases, and the flow again approaches the critical condition. When the critical flow rate is reached, further opening of the gate will not increase the flow rate, so long as the specific energy remains constant.

Considering the sequence of events as the gate is opened with the upstream depth fixed would be interesting. We leave that as an exercise for you.

EXAMPLE 11.5 Illustrates Flow with Constant Specific Energy

A rectangular, concrete-lined irrigation canal is 80 ft wide and carries 1200 ft^3/sec of water at a (normal) depth of 4 ft. A bridge is built over the canal. The bridge is supported by five concrete piers, each 3 ft wide. Calculate the water depth as it passes under the bridge. Assume frictionless flow in the vicinity of the bridge piers.

SOLUTION

Given

Canal water flow rate of 1200 ft^3/sec with depth of 4 ft and width of 80 ft

Canal width reduced by five 3-ft-wide piers

Frictionless flow

Find

Water depth in vicinity of piers

Solution

Combining Eqs. (11.23) and (11.25) gives

$$\left(y_1 + \frac{q^2}{2gy_1^2}\right) - \left(y_2 + \frac{q^2}{2gy_2^2}\right) = z_{B_2} - z_{B_1} + h_L,$$

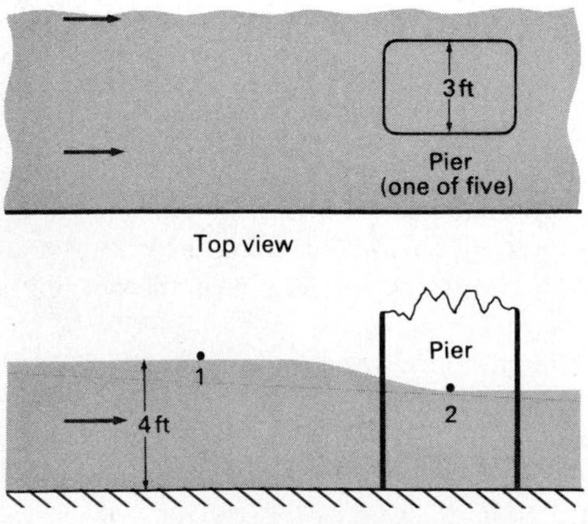

Figure E11.5a Flow around bridge piers in a canal.

where points 1 and 2 are shown in Fig. E11.5a. According to the problem statement, frictionless flow is assumed, so $h_L = 0$. The canal bed is assumed to be horizontal between points 1 and 2. Equation (11.26) gives

$$y_2 + \frac{q_2^2}{2gy_2^2} = y_1 + \frac{q_1^2}{2gy_1^2} \quad \text{or} \quad E = \text{Constant}.$$

The numerical values in the problem statement give

$$q_1 = \frac{(1200 \text{ ft}^3/\text{sec})}{(80 \text{ ft})} = 15 \text{ ft}^2/\text{sec},$$

$$E = E_1 = y_1 + \frac{q_1^2}{2gy_1^2} = 4 \text{ ft} + \frac{(15 \text{ ft}^2/\text{sec})^2}{2(32.2 \text{ ft/sec}^2)(4 \text{ ft})^2} = 4.22 \text{ ft},$$

and thus

$$E_2 = y_2 + \frac{q_2^2}{2gy_2^2} = 4.22 \text{ ft}.$$

The continuity equation gives

$$b_2 q_2 = b_1 q_1, \quad \text{so} \quad y_2 + \frac{1}{2g}\left(\frac{b_1 q_1}{b_2 y_2}\right)^2 = 4.22 \text{ ft}.$$

At the piers,

$$b_2 = 80 \text{ ft} - 5(3 \text{ ft}) = 65 \text{ ft}$$

and

$$y_2 + \frac{1}{2g}\left(\frac{b_1 q_1}{b_2 y_2}\right)^2 = y_2 + \frac{1}{2(32.2 \text{ ft/sec}^2)}\left(\frac{(80 \text{ ft})(15 \text{ ft}^2/\text{sec})}{(65 \text{ ft})y_2}\right)^2$$
$$= y_2 + \frac{5.29 \text{ ft}^3}{y_2^2}.$$

Simplifying, we have

$$y_2^3 - 4.22 \text{ ft } y_2^2 + 5.29 \text{ ft}^3 = 0.$$

This equation has three roots. Solving for the first root by trial and error* gives

$$y_2 = 1.36.$$

We find the remaining roots by factoring out the first root, leaving the quadratic equation

$$y_2^2 - 2.86 y_2 - 3.89 = 0.$$

Using the quadratic formula gives

$$y_2 = \frac{2.86}{2} \pm \sqrt{\left(\frac{2.86}{2}\right)^2 + 3.89} = 3.87, \ -1.01.$$

The two realistic roots are

$$y_2 = 1.36 \text{ ft} \quad \text{and} \quad y_2 = 3.87 \text{ ft}.$$

We must choose one of these as the correct answer. Noting that $q_1 =$ 15 ft^2/sec and

$$q_2 = \frac{1200 \text{ ft}^3/\text{sec}}{65 \text{ ft}} = 18.5 \text{ ft}^2/\text{sec},$$

we have as the corresponding critical depths

$$y_{c_1} = \left(\frac{q_1^2}{g}\right)^{1/3} = \left[\frac{(15 \text{ ft}^2/\text{sec})^2}{32.2 \text{ ft/sec}^2}\right]^{1/3} = 1.91 \text{ ft}$$

and

$$y_{c_2} = \left(\frac{q_2^2}{g}\right)^{1/3} = \left[\frac{(18.5 \text{ ft}^2/\text{sec})^2}{32.2 \text{ ft/sec}^2}\right]^{1/3} = 2.20 \text{ ft}.$$

Thus one of the two calculated depths at location 2 corresponds to subcritical flow and the other corresponds to supercritical flow. Perhaps these conditions are the key to the correct root. The specific

---

* We may use other root-finding methods, such as the Newton-Raphson method or the secant method.

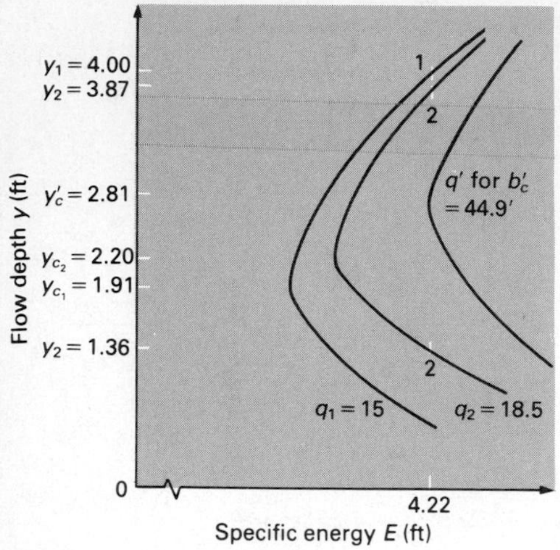

Figure E11.5b Specific energy curve for flow around pier.

energy curves for q_1 and q_2 are shown in Fig. E11.5b. With the specific energy constant from 1 to 2, the flow must pass through a critical depth y_c' corresponding to a flow rate q' to pass from subcritical at 1 to supercritical at 2. This critical depth y_c' corresponds to a critical width b_c' and critical velocity V_c'. At this critical condition,

$$E_c' = E_1 = 4.22 \text{ ft}.$$

Then Eqs. (11.24) and (11.29) give

$$E_c' = \frac{V_c'^2}{2g} + y_c' = \frac{y_c'}{2} + y_c' = 1.5y_c';$$

thus

$$y_c' = \frac{E_c'}{1.5} = \frac{4.22 \text{ ft}}{1.5} = 2.81 \text{ ft}$$

and

$$V_c' = (gy_c')^{1/2} = [(32.2 \text{ ft/sec}^2)(2.81 \text{ ft})^{1/2} = 9.51 \text{ ft/sec}.$$

We obtain the corresponding channel width, b_c', from the continuity equation:

$$b_c' = \frac{Q}{V_c' y_c'} = \frac{1200 \text{ ft}^3\text{/sec}}{(9.51 \text{ ft/sec})(2.81 \text{ ft})} = 44.9 \text{ ft}.$$

As the irrigation canal does not narrow to 44.9 ft between locations 1 and 2, the flow does not pass from subcritical to supercritical, and the supercritical depth at 2 is not possible; therefore

$$y_2 = 3.87 \text{ ft}$$ **ANSWER**

under the bridge.

Discussion

Considering all possible solutions is important when a physical phenomen is described by a higher-order equation. If we had stopped at the first realistic root ($y_c = 1.36$ ft), we would have had the wrong answer.

If you have studied Chapter 10 on compressible flow, you might have noted the similarity between this flow and isentropic flow in a converging–diverging nozzle. We discuss this similarity in detail in a later section.

Note that the flow, surprisingly, becomes shallower as the channel width is decreased.

Effects of Changes in Specific Energy. Let's now consider a frictionless flow in which the specific energy changes along the channel. According to Eq. (11.37), we can modify specific energy levels by changing the channel bottom elevation. Figure 11.12(a) shows a horizontal channel with a bump on the bottom. The bump is higher than the upstream channel bottom, so the specific energy is less over the bump. Because we know the upstream specific energy, the specific energy at any

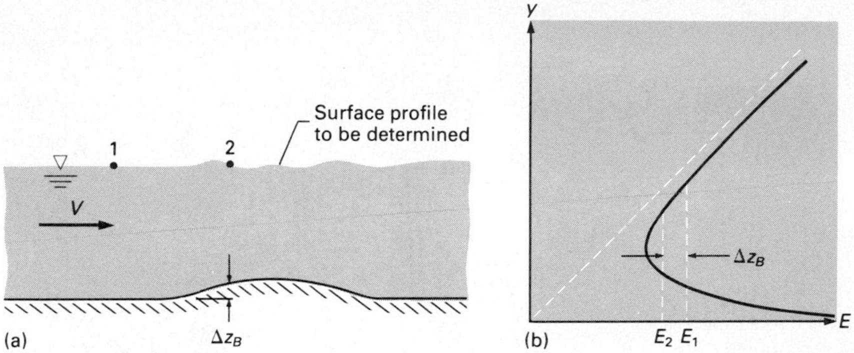

Figure 11.12 Frictionless flow over a bump in a horizontal channel: (a) flow schematic; (b) change of *E*.

downstream point is

$$E_2 = E_1 - \Delta E = E_1 - \Delta z_{B_2}. \tag{11.38}$$

Figure 11.12(b) illustrates the process on a specific energy diagram.

The change in the liquid surface profile over the bump depends on whether the flow approaching the bump is subcritical (Fig. 11.13) or supercritical (Fig. 11.14). For subcritical flow, the state point upstream of the bump is shown as point 1 in Fig. 11.13(b). When the specific energy is lowered to E_2, the flow state is either point 2 or point 2′ on the specific energy diagram (Fig. 11.13b). With a constant flow rate, the only way to get to point 2′ is to pass through the critical state, moving along the specific energy curve. This flow is impossible because the specific energy has not been lowered to E_c upstream of point 2 in the channel, so state point 2 describes the flow. The state is still subcritical and the depth is decreased. By choosing several points over the bump, we can work out the complete liquid surface profile.

When the flow approaching the bump is supercritical, the initial state is shown as point 1 in Fig. 11.14(b). Possible states with specific energy E_2 are shown as state points 2 and 2′ on the specific energy curve. The flow cannot possibly pass through the critical state, so the downstream state must correspond to state point 2. The flow is still supercritical, and the depth increases over the bump. By selecting several points, we can determine the complete surface profile.

Now, let's find out what would happen if we made the bump higher. The minimum specific energy always occurs at the crest of the bump. Suppose that the flow approaching the bump is subcritical. Raising the bump causes the depth at the crest to decrease and the velocity over the crest to increase. The state point at

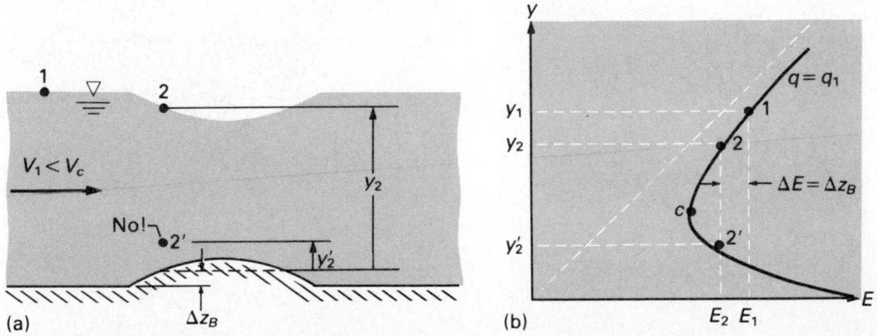

Figure 11.13 Subcritical flow over a bump.

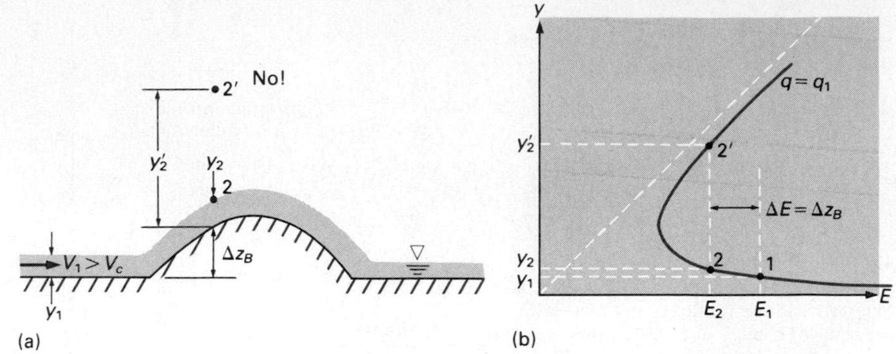

Figure 11.14 Supercritical flow over a bump.

the crest moves toward the critical state. If the flow state does not become critical over the crest, then on the downstream side of the bump the depth increases, the velocity decreases, and the flow returns to its initial state. If we make the bump high enough, the flow over the crest can become critical. If it does, the flow on the downstream side of the crest either may return to the initial subcritical state or may accelerate to the alternate supercritical state, depending on conditions far downstream from the bump.

We can calculate the height of the bump necessary to cause critical flow over the crest from Eq. (11.38) by setting $E_2 = E_c$:

$$\Delta z_{B,\max} = E_1 - E_c.$$

Using Eqs. (11.26) and (11.28), we have

$$\Delta z_{B,\max} = y_1 + \frac{q^2}{2gy_1^2} - \frac{3}{2}\left(\frac{q^2}{g}\right)^{1/3}. \tag{11.39}$$

Now we look at what happens if the bump is made higher than $\Delta z_{B,\max}$. This configuration would imply $E_2 < E_c$; however, it would put the downstream point off the specific energy curve for the given value of q. Physically, maintaining the given flow with the given upstream state 1 would no longer be possible. If the given flow were to be maintained, y_1 would have to increase so that E_1 would be greater. Alternatively, if y_1 were to remain fixed, q would be decreased. If we assume that q remains fixed, Fig. 11.15 illustrates the variation of the upstream and bump-crest depths with bump height. Note that once the flow over the crest becomes critical, it remains critical, with upstream depth adjusting to compensate

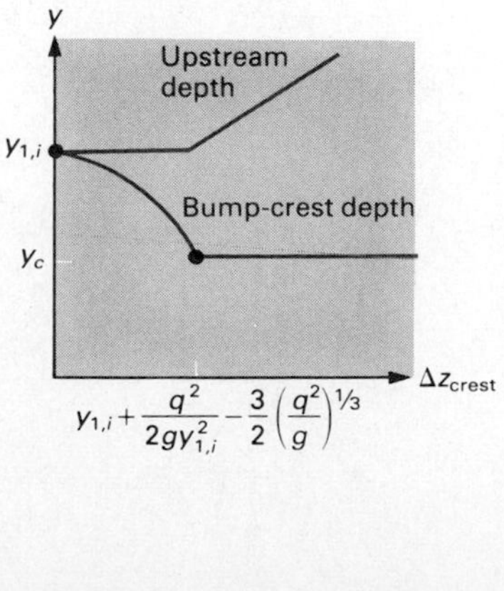

Figure 11.15 Variation of upstream depth and bump-crest depth with bump height for constant q.

for increases in bump height. This phenomenon illustrates an important characteristic of subcritical flow: The upstream flow adjusts to compensate for downstream conditions.

Flow elements that cause critical flow, such as sluice gates, bumps (dams), and changes in channel slope from mild to steep, are often used as control sections for open channel flows. A *control section* controls the flow upstream if it is subcritical and controls the flow downstream if it is supercritical.

EXAMPLE 11.6 Illustrates Frictionless Channel Flow with Variable Specific Energy

The sides of the irrigation canal of Example 11.5 are 8 ft high. A new proposal calls for the canal to deliver water at a rate of 3000 ft^3/sec. Can the canal handle the new flow rate? At a certain location, a telephone cable conduit spans the channel. The conduit is 5.5 ft above the canal bottom. Will the conduit be submerged at the new flow rate? If the conduit will be submerged, one proposal suggests lowering the water surface under the conduit either by excavating the canal bottom under the conduit or by filling in the bottom under the conduit. Which should be done? How much should the bottom be excavated or raised?

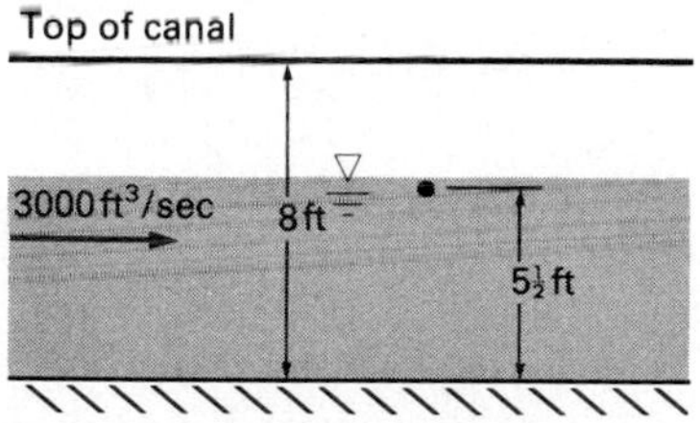

Figure E11.6a Conduit placed in irrigation canal.

SOLUTION

Given

Figure E11.6a

Canal 80 ft wide

Find

Water depth for 3000 ft^3/sec flow rate

Whether telephone conduit is submerged

Amount canal bottom should be excavated or raised so that telephone conduit is not submerged

Solution

We find the slope of the canal bottom by using Manning's formula, Eq. (11.14b), and the information in Example 11.5:

$$S_0 = \left(\frac{nV}{1.49R_h^{2/3}}\right)^2.$$

For an (assumed unfinished) concrete-lined canal, Table 11.1 gives $n = 0.014$. The hydraulic radius is

$$R_h = \frac{by}{b + 2y},$$

where b is the canal width (80 ft) and y the water depth (4 ft). Then

$$R_h = \frac{(80\text{ ft})(4\text{ ft})}{80\text{ ft} + 2(4\text{ ft})} = 3.64\text{ ft}.$$

The velocity V is

$$V = \frac{Q}{by} = \frac{1200\ \text{ft}^3/\text{sec}}{(80\ \text{ft})(4\ \text{ft})} = 3.75\ \text{ft/sec}$$

and

$$S_0^{1/2} = \frac{0.014(3.75\ \text{ft/sec})}{1.49(3.64\ \text{ft})^{2/3}} = 0.0149.$$

For the 3000 ft^3/sec flow rate, Manning's formula is

$$V = \frac{1.49}{n}(R_h^{2/3})S_0^{1/2}.$$

Now

$$V = \frac{Q}{by} \quad \text{and} \quad R_h = \frac{by}{b + 2y}.$$

Substituting into Manning's formula gives

$$\frac{Q}{by} = \frac{1.49}{n}\left(\frac{by}{b+2y}\right)^{2/3} S_0^{1/2}.$$

Solving for the depth, we have

$$\frac{1}{y}\left(\frac{b+2y}{y}\right)^{2/3} = \frac{1.49b^{5/3}}{nQ}(S_0^{1/2}).$$

The numerical values give

$$\frac{1}{y}\left(\frac{80\ \text{ft} + 2y}{y}\right)^{2/3} = \frac{1.49(80\ \text{ft})^{5/3}(0.0149)}{0.014(3000\ \text{ft}^3/\text{sec})} = 0.785.$$

Solving by trial and error gives

$$y = y_n = 7.12\ \text{ft};$$

therefore

The canal can handle the new flow rate and the conduit is submerged. **ANSWER**

We next investigate whether raising or lowering the canal bottom will lower the water surface below the conduit. The first step is to determine whether the flow is subcritical or supercritical. Calculating the Froude number,

$$V = \frac{Q}{by} = \frac{3000\ \text{ft}^3/\text{sec}}{(80\ \text{ft})(7.12\ \text{ft})} = 5.27\ \text{ft/sec},$$

and

$$\text{F} = \frac{V}{\sqrt{gy}} = \frac{(5.27\ \text{ft/sec})}{\sqrt{(32.2\ \text{ft/sec}^2)(7.12\ \text{ft})}} = 0.348.$$

The flow is subcritical. Assuming frictionless flow, Eqs. (11.23) and (11.25) give

$$\left(y_2 + \frac{q_2^2}{2gy_2^2}\right) = \left(y_1 + \frac{q_1^2}{2gy_1^2}\right) + z_{B_1} - z_{B_2}.$$

Choosing $z_{B_1} = 0$ and noting that

$$q = \frac{3000 \text{ ft}^3/\text{sec}}{80 \text{ ft}} = 37.5 \text{ ft}^2/\text{sec},$$

we have

$$y_2 + \frac{(37.5 \text{ ft}^2/\text{sec})^2}{2(32.2 \text{ ft/sec}^2)y_2^2} = 7.12 \text{ ft} + \frac{(37.5 \text{ ft}^2/\text{sec})^2}{2(32.2 \text{ ft/sec}^2)(7.12 \text{ ft})^2} - z_{B_2}$$

or

$$y_2 + \frac{21.84 \text{ ft}^3}{y_2^2} + z_{B_2} = 7.55 \text{ ft}. \quad \text{(E11.1)}$$

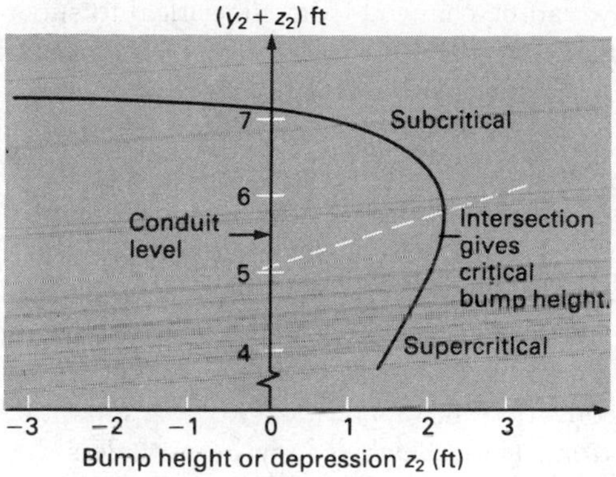

Figure E11.6b Subcritical and supercritical water level heights above $z_1 = 0$ level.

Positive values of z_{B_2} represent raising the canal bottom, and negative values of z_{B_2} represent excavating the canal bottom. The water surface height above the $z_{B_1} = 0$ level is $(y_2 + z_{B_2})$. Using trial and error, we solve Eq. (E11.1) for y_2 as a function of z_{B_2}. We then plot the values of $(y_2 + z_{B_2})$ for both subcritical and supercritical flows versus z_{B_2} as the solid curve in Fig. E11.6b. The curve shows that the water level is lowered below the conduit (5.5 ft) if the canal bottom is raised less than 2.24 ft and the flow is forced to change from subcritical to supercritical. We find the bump height, z_{B_2}, required for transition from subcritical to supercritical flow by first using Eqs. (11.6) and (11.25) with $F = 1$ to give

$$1 = \frac{V_2}{\sqrt{gy_2}} = \frac{q_2}{y_2\sqrt{gy_2}}, \quad \text{or} \quad q_2 = y_2\sqrt{gy_2}.$$

Equations (11.23) and (11.25) then give

$$y_2 + \frac{(y_2)^2 g y_2}{2gy_2^2} = y_1 + \frac{q_1^2}{2gy_1^2} + z_{B_1} - z_{B_2},$$

or

$$1.5y_2 = y_1 + \frac{q_1^2}{2gy_1^2} + z_{B_1} - z_{B_2}.$$

The numerical values give

$$y_2 = \frac{7.55 - z_{B_2}}{1.5}, \quad \text{or} \quad (y_2 + z_{B_2}) = 5.03 + \frac{z_{B_2}}{3}. \quad \text{(E11.2)}$$

Equation (E11.2) shows that the water surface height (above the $z_{B_1} = 0$ level) required for transition to supercritical flow depends on the value of the bump height z_{B_2}. In Fig. E11.6b, $(y_2 + z_{B_2})$ is plotted versus z_{B_2} as the dashed line. The intersection of the dashed line and the solid curve shows that transition occurs at a $(y_2 + z_{B_2})$ value of 5.80 ft. The corresponding value of z_{B_2} is 2.25 ft. Noting that the value of z_{B_2} corresponding to $(y_2 + z_{B_2}) = 5.5$ ft is 2.24 ft, we see that

The bump should look as sketched in Fig. E11.6c. **ANSWER**

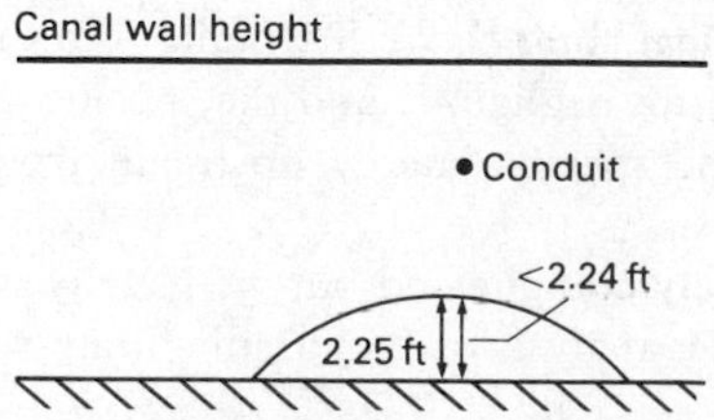

Figure E11.6c Bump schematic.

The maximum bump height is upstream of the conduit.

Discussion

Raising the channel bottom surface to lower the water surface below the conduit no doubt seems strange.

Could the water surface be lowered by changing the width of the canal at the conduit location? Should the canal be made wider or narrower to lower the water surface?

11.3.3 Hydraulic Jump

We have shown that open channel flow can be changed from subcritical to supercritical in a relatively efficient (low-loss) manner by sluice gates or changes in channel bottom geometry. In these cases, the flow state passes through critical with relatively small energy loss. We now consider transitions from supercritical to subcritical flow. Carrying out such transitions without significant energy loss is nearly impossible. The primary reason is that disturbances cannot propagate upstream in supercritical flow. Imagine trying to operate a sluice gate "backward," with the shallow supercritical flow entering and the deeper subcritical flow leaving. Intuition suggests that this is a highly unstable situation; any disturbance near the gate opening would cause the incoming liquid to "pile up" upstream from the gate.

Similarly, imagine attempting to convert from supercritical flow to subcritical flow using a bump in the channel bottom. If the height of the bump is less than $\Delta z_{B,\max}$, the flow remains supercritical over the crest and returns to its original supercritical state downstream from the bump. If the height of the bump is greater than $\Delta z_{B,\max}$, a physically impossible situation has been specified. In subcritical flow, the upstream depth adjusts to compensate; however, if the approach flow is supercritical, there is no mechanism by which the upstream flow can be "informed" about the too-high bump. Smooth passage through the critical state is possible only if the bump height is *exactly* equal to $\Delta z_{B,\max}$.

In practice, the transition between a supercritical flow and a subcritical flow is accomplished in a *hydraulic jump,* a reasonably short, extremely turbulent region of flow across which the state changes from supercritical to subcritical. Figures 6.15, 11.16, and 11.17 illustrate hydraulic jumps that occur in spillways. Like many other phenomena in fluid mechanics, the hydraulic jump is rather common. A hydraulic jump occurs in the kitchen sink as water from the faucet impinges on a flat surface and flows radially outward from the impingement point.* On a specific-energy diagram, a hydraulic jump is represented by a jump from the lower (supercritical) branch to the upper (subcritical) branch. A hydraulic jump is extremely turbulent, so losses in the jump are not negligible, and the specific energy downstream from the jump is less than the specific energy upstream from the jump.

The flow in a hydraulic jump is extremely complicated but we usually do not need to consider its details. A control volume analysis of a hydraulic jump reveals

* Try it in your sink.

Figure 11.16 Small hydraulic jump at outlet of a dam overflow. (Photograph by John Remington.)

Figure 11.17 Details of surface of the hydraulic jump shown in Fig. 11.16. (Photograph by John Remington.)

most of what we need to know about it. We have already used this idea once, in Example 4.5. Figure 11.18 shows a control volume for analysis of a hydraulic jump. Because jumps are usually short, we assume that the channel bottom is horizontal and that the flow is identical in all planes parallel to the paper. The continuity equation is

$$V_1 y_1(1) = V_2(y_2)(1),$$

where (1) is the unit width perpendicular to the paper. The momentum equation is

$$\sum F_x = \dot{M}_{x,\text{out}} - \dot{M}_{x,\text{in}}.$$

The forces are the result of hydrostatic pressure* at the ends of the control volume and shear stress at the channel bottom. Again, because a hydraulic jump is short, the bottom force is negligible and the momentum equation becomes

$$\frac{\gamma y_1^2}{2}(1) - \frac{\gamma y_2^2}{2}(1) = \rho y_2(1) V_2^2 - \rho y_1(1) V_1^2.$$

Solving for V_2 from the continuity equation, we have

$$V_2 = \frac{y_1}{y_2} V_1. \tag{11.40}$$

Substituting Eq. (11.40) into the momentum equation and solving for the depth ratio across the jump give

$$\frac{y_2}{y_1} = \frac{1}{2}\left[\sqrt{1 + 8\left(\frac{V_1^2}{g y_1}\right)} - 1\right],$$

* We assume that the streamlines at 1 and 2 are nearly straight and parallel.

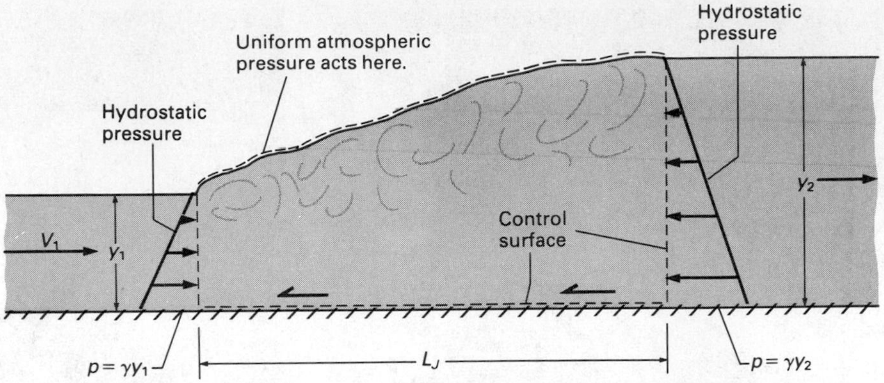

Figure 11.18 Control volume for analysis of hydraulic jump.

$$\frac{y_2}{y_1} = \frac{1}{2}(\sqrt{1 + 8\mathbf{F}_1^2} - 1). \tag{11.41}$$

The depths y_1 and y_2 on opposite sides of a hydraulic jump are called *conjugate depths*.

The energy loss is an important parameter of a hydraulic jump. The energy equation across a jump is

$$\left(\frac{V_1^2}{2g} + y_1\right) - \left(\frac{V_2^2}{2g} + y_2\right) = h_L.$$

Substituting Eqs. (11.40) and (11.41)—after some algebra—gives

$$h_L = \frac{(y_2 - y_1)^3}{4y_1y_2}. \tag{11.42}$$

Equation (11.41) works for $\mathbf{F}_1 > 1 (y_2/y_1 > 1)$ *and* $\mathbf{F}_1 < 1$ $(y_2/y_1 < 1)$; however, Eq. (11.42) shows that jumps with $\mathbf{F}_1 < 1$ have negative loss, which is a physical impossibility. Experiments have shown that Eqs. (11.41) and (11.42) are remarkably accurate for real jumps with $\mathbf{F}_1 > 1$ and $y_2 > y_1$.

Open channel designers often use hydraulic jumps as energy dissipators. A jump is usually located just downstream from a dam spillway (Figs. 6.15 and 11.16) to dissipate the kinetic energy of the flow. If this were not done, severe erosion of the downstream channel bed could result. Obviously, jumps must be located on a strong bed. Five classes of jumps, based on inlet Froude number, have been identified from experimental observation.

- $\mathbf{F}_1 = 1.0$ to 1.7: Standing wave, or undular, jump; dissipation of inlet kinetic energy less than 5 percent.
- $\mathbf{F}_1 = 1.7$ to 2.5: Smooth surface rise known as a weak jump; dissipation 5–15 percent.
- $\mathbf{F}_1 = 2.5$ to 4.5: Unstable, oscillating jump; each irregular pulsation creates a large wave that can travel far downstream, damaging earth banks and other structures; dissipation 15–45 percent.

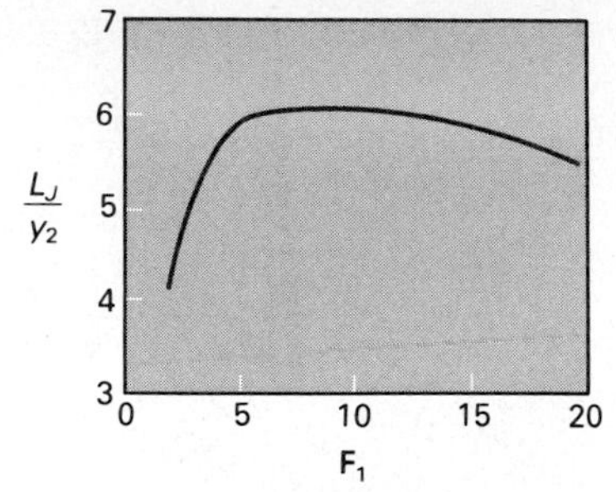

Figure 11.19 Length of hydraulic jump.

- $F_1 = 4.5$ to 9.0: Stable "steady" jump; best performance and action, insensitive to downstream conditions; dissipation 45–70 percent.
- $F_1 > 9.0$: Rough, somewhat intermittent, strong jump; dissipation 70–85 percent.

Our control volume analysis is incapable of predicting the length of a hydraulic jump. Dimensional analysis indicates that

$$\frac{L_J}{y_2} = f(F_1).$$

Figure 11.19 shows experimentally determined jump length as a function of Froude number.

EXAMPLE 11.7 Illustrates Hydraulic Jump Calculations

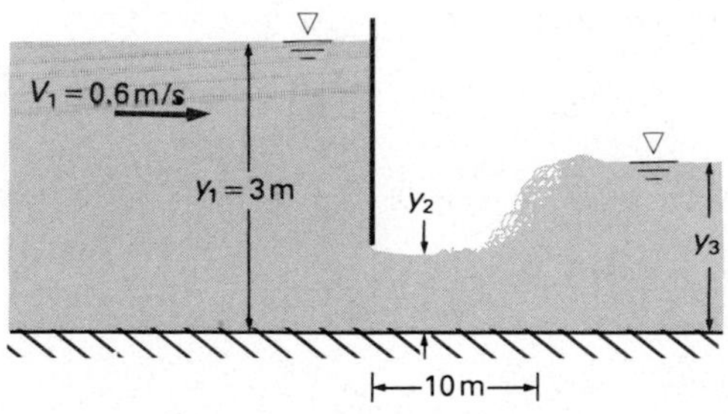

Figure E11.7a Flow under sluice gate, with hydraulic jump.

Figure E11.7a shows a hydraulic jump downstream from a sluice gate in a wide horizontal channel. Calculate y_2, y_3, and the percentage loss of mechanical energy. Sketch the flow to scale and include the energy grade line.

SOLUTION

Given

Flow under sluice gate with hydraulic jump
Figure E11.7a

Find

y_2, y_3, and percentage loss of mechanical energy in jump

Solution

We assume that the bottom of the channel is horizontal and that the flow is frictionless. Equation (11.23) gives

$$\frac{V_1^2}{2g} + y_1 = \frac{V_2^2}{2g} + y_2.$$

For a constant channel width, we have

$$V_1 y_1 = V_2 y_2,$$

so the specific energy equation is

$$\frac{V_1^2}{2g} + y_1 = \frac{1}{2g}\left(\frac{y_1 V_1}{y_2}\right)^2 + y_2.$$

The numerical values give

$$\frac{(0.6\ \text{m/s})^2}{2(9.807\ \text{m/s}^2)} + 3\ \text{m} = \frac{1}{2(9.807\ \text{m/s}^2)}\left[\frac{(3\ \text{m})(0.6\ \text{m/s})}{y_2}\right]^2 + y_2$$

or

$$3.018\ \text{m} = \frac{0.165\ \text{m}^3}{y_2^2} + y_2.$$

Because the critical flow occurs at the sluice gate, we solve for the supercritical (smaller) value of y_2. Trial and error give

$$y_2 = 0.244\ \text{m}.$$ **ANSWER**

We find y_3 by first finding the Froude number $\mathbf{F}_2$, using

$$V_2 = \frac{y_1 V_1}{y_2} = \frac{(3\ \text{m})(0.6\ \text{m/s})}{(0.244\ \text{m})} = 7.38\ \text{m/s}$$

and

$$\mathbf{F}_2 = \frac{V_2}{\sqrt{g y_2}} = \frac{7.38\ \text{m/s}}{\sqrt{(9.807\ \text{m/s}^2)(0.244\ \text{m})}} = 4.77.$$

Equation (11.41) then gives

$$y_3 = y_2[\tfrac{1}{2}(\sqrt{1 + 8\mathbf{F}_2^2} - 1)] = (0.244\ \text{m})[\tfrac{1}{2}(\sqrt{1 + 8(4.77)^2} - 1)];$$

$$y_3 = 1.528\ \text{m}.$$ **ANSWER**

The energy loss in the jump, Eq. (11.42), is

$$h_L = \frac{(y_3 - y_2)^3}{4 y_2 y_3}.$$

The percentage energy loss in the jump is

$$\Delta E = \frac{h_L}{V_2^2/(2g) + y_2}(100) = \frac{(y_3 - y_2)^3}{4 y_2 y_3 [V_2^2/(2g) + y_2]}(100).$$

The numerical values give

$$\Delta E = \frac{(1.528\ \text{m} - 0.244\ \text{m})^3(100)}{4(0.244\ \text{m})(1.528\ \text{m})[(7.38\ \text{m/s})^2/[2(9.807\ \text{m/s}^2)] + 0.244\ \text{m}]};$$

$$\Delta E = 47.0\%.$$ **ANSWER**

The scaled sketch of the flow is shown in Fig. E11.7b. We find the height of the sluice gate above the bottom of the channel by using Eq. (11.23), and assuming frictionless flow:

$$\frac{V_1^2}{2g} + y_1 = \frac{V_c^2}{2g} + y_c,$$

where Eq. (11.29) gives

$$V_c^2 = g y_c, \quad \text{so} \quad \frac{V_1^2}{2g} + y_1 = 1.5 y_c.$$

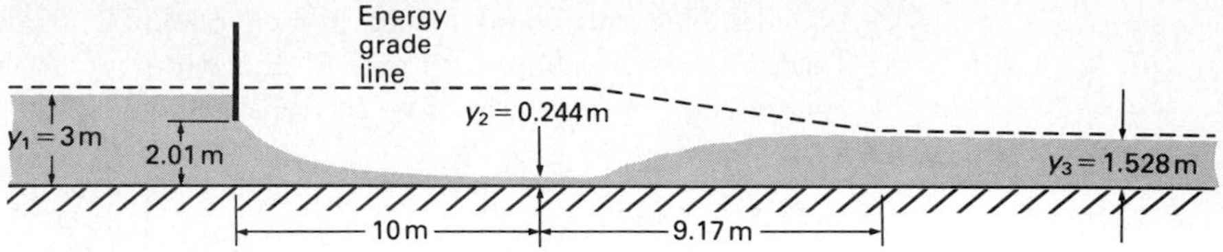

Figure E11.7b Scaled sketch of flow and energy grade line.

The numerical values give

$$3.018 \text{ m} = 1.5y_c, \quad \text{or} \quad y_c = 2.01 \text{ m}.$$

We find the length of the hydraulic jump by using $\mathbf{F}_2 = 4.77$ and Fig. 11.19:

$$L \approx 6y_3 = 6(1.528 \text{ m}) = 9.17 \text{ m}.$$

Discussion

Note that the energy grade line is level from point 1 to point 2, because we assumed frictionless flow and the floor of the channel is level.

11.4 GRADUALLY VARIED FLOW

Thus far our discussion of open channel flow has been limited to simplified cases such as uniform flow (which requires uniform bottom slope, channel cross section, and surface roughness) and frictionless flow, which is a rather idealized case. In this section, we discuss the analysis of varied flow with friction. We permit changes in bottom slope, channel shape, and surface condition and include friction losses; however, our analysis is not completely general because we restrict consideration to gradually varied flow (GVF). We assume that flow conditions vary slowly enough that the flow is one-dimensional (the independent variable is distance along the channel, x) and one-directional, with nearly parallel streamlines.

11.4.1 The Differential Equation of Gradually Varied Flow

The assumptions of one-dimensional and one-directional flow imply that the flow is described by an ordinary differential equation. Figure 11.20 illustrates the flow

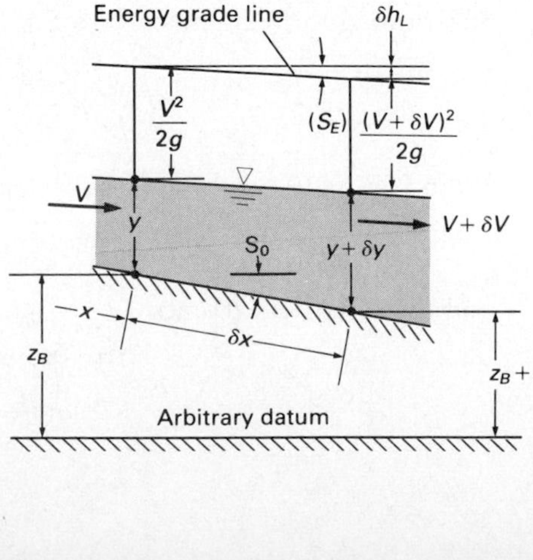

Figure 11.20 Gradually varied flow in small reach of channel.

between locations x and $x + \delta x$ in a channel. The channel bottom slope, S_0, may vary with x. The liquid velocity V and depth y vary with x. We first write the energy equation between cross sections x and $x + \delta x$:

$$\frac{V^2}{2g} + y + z_B = \frac{V^2}{2g} + \frac{d}{dx}\left(\frac{V^2}{2g}\right)\delta x + y + \left(\frac{dy}{dx}\right)\delta x + z_B + \left(\frac{dz_B}{dx}\right)\delta x + \delta h_L.$$

As usual, y is evaluated at the free surface, and kinetic energy correction factors are taken as 1.0.

The bottom slope (S_0) is

$$S_0 \approx -\frac{dz_B}{dx}.$$

We can write the head loss as

$$\delta h_L = \delta h_T = \left(\frac{dh_T}{dx}\right)\delta x = S_E\,\delta x, \tag{11.43}$$

where S_E is the slope of the energy grade line. Substituting S_0 and S_E into the energy equation, canceling terms where possible, and dividing by δx, we have

$$\frac{d}{dx}\left(\frac{V^2}{2g}\right) + \frac{dy}{dx} = S_0 - S_E. \tag{11.44}$$

Equation (11.44) involves derivatives of both velocity and depth. To eliminate the velocity derivative, we use the continuity equation for steady flow:

$$Q = VA = \text{Constant}.$$

Squaring and dividing by $2g$ give

$$\left(\frac{V^2}{2g}\right)A^2 = \frac{Q^2}{2g}.$$

Differentiating, we have

$$A^2\left(\frac{d}{dx}\right)\left(\frac{V^2}{2g}\right) + \left(\frac{V^2}{g}\right)A\left(\frac{dA}{dx}\right) = 0.$$

Substituting this expression into Eq. (11.44) gives

$$-\frac{V^2}{gA}\left(\frac{dA}{dx}\right) + \frac{dy}{dx} = S_0 - S_E. \tag{11.45}$$

From Fig. 11.10, we have (area changes with depth):

$$dA = b_0\,dy,$$

and the equation becomes

$$-\frac{V^2 b_0}{gA}\left(\frac{dy}{dx}\right) + \frac{dy}{dx} = S_0 - S_E.$$

Solving for dy/dx and using Eq. (11.34), we get

$$\frac{dy}{dx} = \frac{S_0 - S_E}{1 - (V^2 b_0/gA)} = \frac{S_0 - S_E}{1 - \mathrm{F}^2}. \tag{11.46}$$

Integration of Eq. (11.46) permits computation of liquid surface profiles; however, we can learn a great deal about gradually varied flow by examining Eq. (11.46) without integrating it. To simplify the discussion, we limit consideration to rectangular channels for which $b_0 = b =$ Constant and $A = by$.

Starting from an initial point on the liquid surface, the fluid depth increases or decreases accordingly as the right-hand side of Eq. (11.46) is positive or negative. The sign of the right-hand side must be determined by examination of both numerator and denominator. The denominator changes sign at $\mathrm{F} = 1$ so, for a given sign of the numerator, the change of depth downstream from an initial point is opposite for subcritical or supercritical flow. We can easily decide whether the flow at a given location is subcritical or supercritical by comparing the actual depth y to the critical depth y_c. If $y > y_c$, then $\mathrm{F} < 1$ and the flow is subcritical, so the denominator in Eq. (11.46) is positive; if $y < y_c$, then $\mathrm{F} > 1$ and the flow is supercritical, so the denominator is negative.

To examine the numerator, we first consider the total-head slope S_E. In terms of friction factor and hydraulic radius,

$$S_E = \frac{dh_T}{dx} = \frac{\delta h_L}{\delta x} = \frac{f}{8g}\left(\frac{V^2}{R_h}\right). \tag{11.47}$$

Using the Chezy coefficient C, we have

$$S_E = \frac{V^2}{C^2 R_h}, \tag{11.48}$$

or, in terms of Manning's formula,

$$S_E = \frac{n^2}{\xi^2}\left(\frac{V^2}{R_h^{4/3}}\right), \tag{11.49}$$

where $\xi = 1.0$ for SI units and 1.49 for BG or EE units.

The comparison of S_E and S_0 is often most easily accomplished by comparing the actual depth (y) to the normal depth (y_n) for the same flow rate. Using Eq. (11.49) and assuming a wide channel ($R_h = y$), we have for the flow rate

$$q = Vy = \left(\frac{\xi}{n}\right) y^{2/3} (S_E^{1/2}) y. \tag{11.50}$$

The normal depth for the actual bottom slope (S_0) is calculated from Eq. (11.14):

$$q = \left(\frac{\xi}{n}\right) y_n^{2/3} (S_0^{1/2}) y_n. \tag{11.51}$$

Equating the two expressions for q gives

$$\frac{S_E}{S_0} = \left(\frac{y_n}{y}\right)^{10/3}.$$

Table 11.2 Possibilities for surface slope from Eq. (11.46).

	$y > y_c$ (Denominator Positive)	$y < y_c$ (Denominator Negative)	$y = y_c$ (Denominator Zero)
$y > y_n$ (Numerator Positive)	$\frac{dy}{dx} > 0$ Depth increases	$\frac{dy}{dx} < 0$ Depth decreases	$\frac{dy}{dx}$ infinite Not GVF
$y < y_n$ (Numerator Negative)	$\frac{dy}{dx} < 0$ Depth decreases	$\frac{dy}{dx} > 0$ Depth increases	$\frac{dy}{dx}$ infinite Not GVF
$y = y_n$ (Numerator Zero)	$\frac{dy}{dx} = 0$ Uniform flow, depth constant	$\frac{dy}{dx} = 0$ Uniform flow, depth constant	$\frac{dy}{dx} = \frac{0}{0}$ Uniform critical flow, unstable

If the depth y is greater than the normal depth y_n, then S_E is less than S_0 and the numerator in Eq. (11.46) will be positive. Conversely, if $y < y_n$, then $S_E > S_0$ and the numerator will be negative.

The sign of dy/dx in Eq. (11.46) and, therefore, the direction of depth change in gradually varied flow are determined by the actual depth relative to both the normal depth and the critical depth. Table 11.2 summarizes the various possibilities.

11.4.2 Classification of Surface Profiles

There are twelve different types of liquid surface profiles in gradually varied flow.* These types are identified according to the bottom slope and the actual depth relative to the normal and critical depths. The slope designations are based on comparison of the bottom slope S_0 with the critical slope S_c (Eqs. 11.35 and 11.36) and with zero slope. Table 11.3 summarizes slope names and symbols.

Numbers are associated with the slope symbols according to Table 11.4. All "1" profiles and all "3" profiles have depth increasing. All "2" profiles have depth decreasing. As each slope symbol can be combined with a number, there seemingly

Table 11.3 Slope designations.

Slope Name	Condition	Class Symbol
Mild	$S_0 < S_c$	M
Steep	$S_0 > S_c$	S
Critical	$S_0 = S_c$	C
Horizontal	$S_0 = 0$	H
Adverse	$S_0 < 0$ (uphill)	A

Table 11.4 Depth designations.

Depth Condition	Number Designation
$y > y_n$, $y > y_c$	1
y between y_n and y_c	2
$y < y_n$, $y < y_c$	3

* Excluding uniform flow, which would make thirteen types.

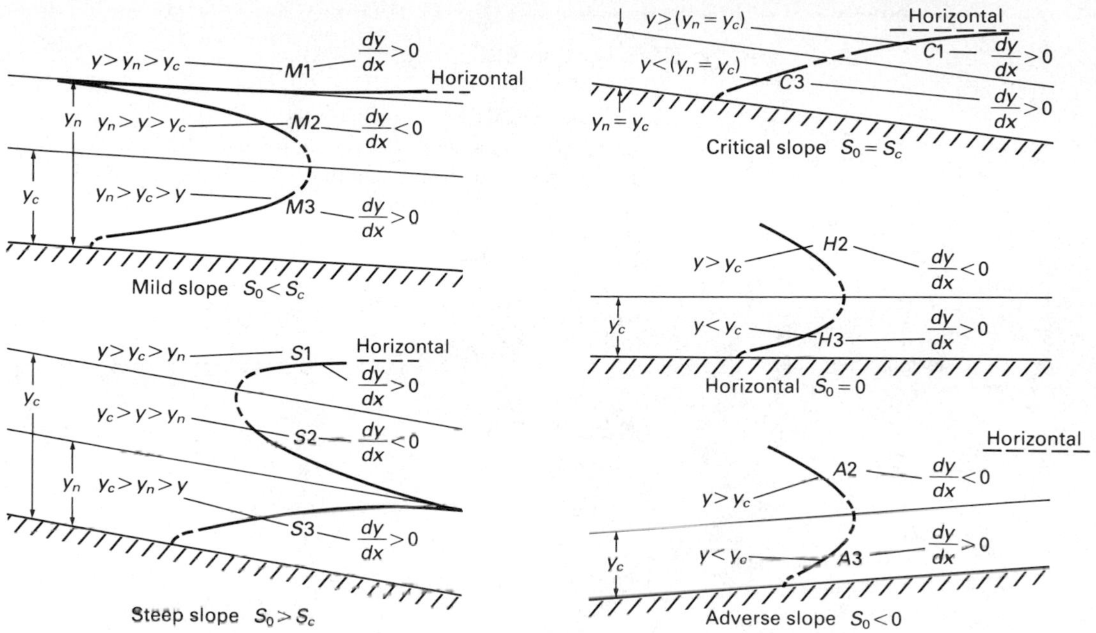

Figure 11.21 The twelve types of surface profiles in gradually varied flow.

would be fifteen profile types rather than twelve; however, there is no *C*2 profile because $y_n = y_c$ for critical slope, and there is no *H*1 or *A*1 because horizontal and adverse slope channels have no normal depth.

Figure 11.21 shows the twelve types of surface profiles. The horizontal (x) scale is greatly compressed. The sketches indicate the resulting profile shape in a reach of channel with constant slope, shape, and roughness. Note that all type 1 profiles approach a horizontal surface, and that all type 2 and 3 profiles approach the *lower* of critical depth or normal depth. Profiles at critical depth are shown as dotted lines normal to the channel bottom because of the singularity in dy/dx at $y = y_c$, $\mathrm{F} = 1$.

Figure 11.22 illustrates how the various profile shapes might actually occur. Note that all profiles associated with subcritical flow (*M*1, *M*2, *S*1, *C*1, *H*2, *A*2) extend *upstream* from a control section (such as a dam, sluice gate, or overfall), whereas all profiles associated with supercritical flow (*M*3, *S*2, *S*3, *C*3, *H*3, *A*3) extend *downstream* from a control section.

Figure 11.22 Actual occurrence of the twelve types of surface profiles.

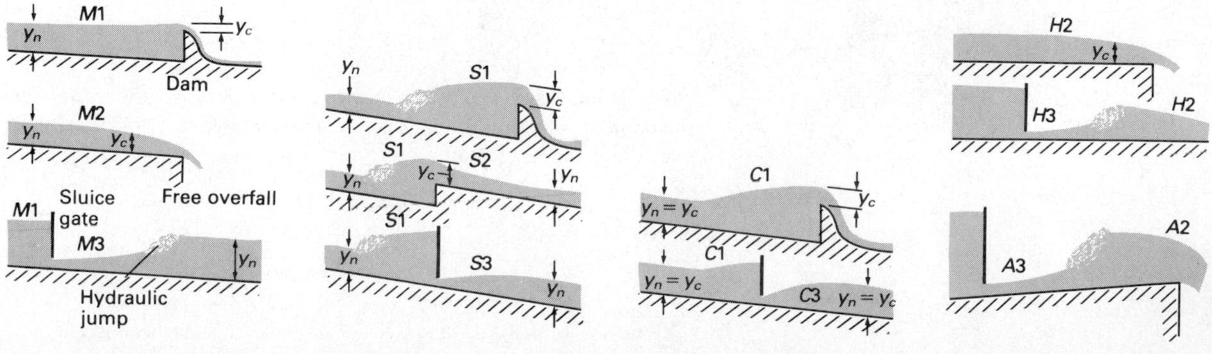

EXAMPLE 11.8 Illustrates the Classification of Surface Profiles

The channel shown in Fig. E11.8a is 6 m wide and carries 10 m^3/s of water. Critical flow occurs over the dam. Indicate the missing surface profile shape from the initial point to the dam. Sketch and label the types of surface profiles.

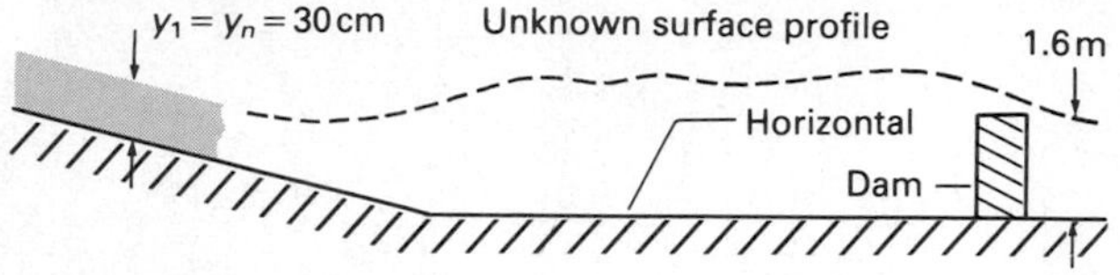

Figure E11.8a Given conditions of a gradually varied flow.

SOLUTION

Given

Figure E11.8a

10 m^3/s water flow rate

Critical flow over dam

Find

The missing surface profiles

Sketch and label the various surface profiles.

Solution

We check the state of the approach flow. The approach velocity* is

$$V_1 = \frac{Q}{by_1} = \frac{10 \text{ m}^3/\text{s}}{(6 \text{ m})(0.30 \text{ m})} = 5.56 \text{ m/s}.$$

The Froude number is

$$\mathbf{F}_1 = \frac{V_1}{\sqrt{gy_1}} = \frac{(5.56 \text{ m/s})}{\sqrt{(9.807 \text{ m/s}^2)(0.30 \text{ m})}} = 3.24.$$

The approach flow is supercritical.

We find the height H of the water surface a short distance upstream from the dam by assuming that the flow over this distance is uniform and frictionless. The specific energy equation, Eq. (11.23), gives

$$\frac{V_3^2}{2g} + y_3 = \frac{V_4^2}{2g} + y_4 + z_{B_4},$$

where point 3 is on the water surface a short distance upstream from the dam, $z_{B_3} = 0$, and point 4 is just above the dam. For critical flow over the dam, Eq. (11.6) gives

$$V_4^2 = c_4^2 = gy_4.$$

Then

$$Q = by_4V_4 = by_4\sqrt{gy_4},$$

* We assume small slope so that we can use depth y instead of flow thickness.

or

$$y_4 = \left(\frac{Q}{b\sqrt{g}}\right)^{2/3} = \left(\frac{10\ \text{m}^3/\text{s}}{6\ \text{m}\sqrt{9.807\ \text{m/s}^2}}\right)^{2/3} = 0.657\ \text{m}.$$

The other numerical values give

$$\frac{V_3^2}{2g} = \frac{Q^2}{2gb^2y_3^2} = \frac{(10\ \text{m}^3/\text{s})^2}{2(9.807\ \text{m/s}^2)(6\ \text{m})^2y_3^2} = \frac{0.1416\ \text{m}^3}{y_3^2}$$

and

$$\frac{V_4^2}{2g} = \frac{Q^2}{2gb^2y_4^4} = \frac{(10\ \text{m}^3/\text{s})^2}{2(9.807\ \text{m/s}^2)(6\ \text{m})^2(0.657\ \text{m})^2} = 0.328\ \text{m}.$$

The specific energy equation becomes

$$\frac{0.1416\ \text{m}^3}{y_3^2} + y_3 = 0.328\ \text{m} + 0.657\ \text{m} + 1.6\ \text{m} = 2.59\ \text{m}.$$

Solving for y_3 by trial and error gives $y_3 = 0.2458$ m and $y_3 = 2.569$ m. The first value is not possible, because the water would not flow over the dam. Therefore the approaching flow (at 1) is supercritical, and the flow just upstream from the dam is subcritical, with $y_3 = 2.569$; therefore a hydraulic jump must occur upstream from the dam. Assuming that the hydraulic jump occurs over a relatively short distance, we have from Eq. (11.41)

$$y_2 = \frac{y_1}{2}\left(\sqrt{1 + 8\text{F}_1^2} - 1\right) = \frac{0.30\ \text{m}}{2}\left(\sqrt{1 + 8(3.24)^2} - 1\right) = 1.23\ \text{m}.$$

Because

$$y_3 = 2.569\ \text{m} > 1.23\ \text{m} = y_2,$$

the hydraulic jump must occur on the incline, as shown in Fig. E11.8b, along with the water surface profile thus far. We must now classify the profile shapes in the reach from the hydraulic jump to the break in slope and from the break in slope to just upstream from the dam.

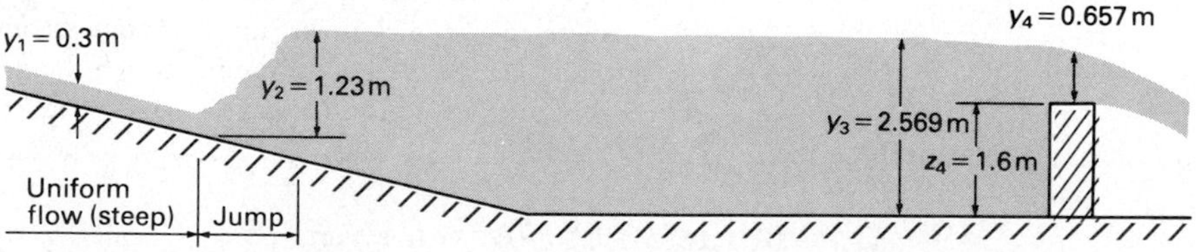

Figure E11.8b Initial sketch of surface profile segments.

Downstream from the jump, the flow rate q, bottom slope S_0, and channel surface roughness (Manning's n) are the same as upstream from the jump; therefore the normal depth, y_{n_2}, is

$$y_{n_2} = y_{n_1} = 0.3\ \text{m},$$

and the critical depth is

$$y_{c_2} = y_{c_1} = y_{c_4} = \left(\frac{q^2}{g}\right)^{1/3} = 0.657\ \text{m}.$$

Therefore $y_{n_2} < y_{c_2}$ and the slope is steep (S). Also, as

$$y_2 = 1.23 \text{ m} > y_{n_2} = 0.3 \text{ m},$$

the complete profile designation is $S1$.

Because the channel bottom is horizontal from the break in slope to the dam, the profile is an H profile. The critical depth at 3 is still

$$y_{c_3} = y_{c_4} = 0.657 \text{ m},$$

so, because

$$y_3 = 2.569 \text{ m} > y_{c_3} = 0.657 \text{ m},$$

the profile is type $H1$ ($dy/dx < 0$). Using this information, we sketch the profile in Fig. E11.8c. **ANSWER**

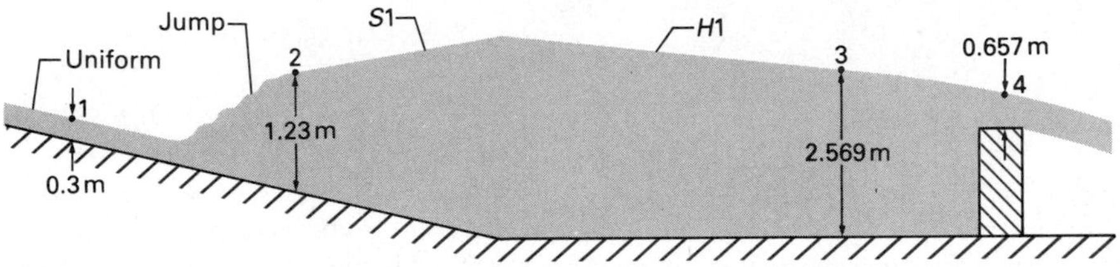

Figure E11.8c Sketch of complete surface profile.

Discussion

The flow in the vicinity of the dam is not closely one-dimensional and one-directional, so the values by y_3 and y_4 are only approximate (note, however, that the value of y_c is accurate). If the dam has a sharp crest, it acts as a full-span weir, a flow measurement and control device that we discuss in Section 11.5.1. A more accurate computation of y_3 can be based on the weir equations developed in that section.

Note that actual calculation of S_0 and S_c for the inclined reach of the channel was not necessary, because y_n could be calculated from the given data. The exact location of the hydraulic jump has not been determined. To determine it, we must assume a location for the jump and calculate the profiles upstream and downstream from it. The correct jump location is the location for which the momentum equation for the entire flow is satisfied.

11.4.3 Computation of Surface Profiles

Examining Eq. (11.46) and classifying surface profiles give us a general idea of the shape of a surface profile; however, we can obtain the actual shape only by integrating Eq. (11.46). We generally assume that Manning's formula is sufficiently accurate to evaluate the total head slope, so that S_E is related to y and V by Eq. (11.49). We presume that the bottom slope is known as a function of x. As both V and y vary with x, S_E, and F are variable. For an exact calculation, solving Eq. (11.46) simultaneously with the differential continuity equation is necessary. To do so usually requires a numerical solution.

A popular numerical solution is the following *direct step method*. The channel is divided into a number of finite-length reaches, and the right-hand side of Eq. (11.46) is evaluated with average values of the parameters over each reach. The calculations are most easily done by calculating x as a function of y, that is, by computing the length of channel necessary to cause a certain depth change. The approximate equation is obtained from Eq. (11.44):

$$\Delta x_i \approx \frac{(V_{i+1}^2/2g + y_{i+1}) - (V_i^2/2g + y_i)}{\overline{(S_0 - S_E)_i}}, \tag{11.52}$$

where Δx_i is the length required to change the depth from y_i to y_{i+1}, and V_i and V_{i+1} are related to y_i and y_{i+1} by

$$q = V_i y_i = V_{i+1} y_{i+1}.$$

Using Eq. (11.49), we have for the average value of $S_0 - S_E$:

$$\overline{(S_0 - S_E)_i} \approx \left(\frac{S_{0,i} + S_{0,i+1}}{2}\right) - \frac{\bar{n}^2}{\xi^2} \frac{(V_i^2 + V_{i+1}^2)/2}{(\bar{R}_h)^{4/3}}, \tag{11.53}$$

where $\bar{n}$ is the average value of Manning's n for Δx_i, and $\bar{R}_h$ is computed from the average values of y and b:

$$\bar{y} = \frac{y_i + y_{i+1}}{2} \quad \text{and} \quad \bar{b} = \frac{b_i + b_{i+1}}{2}. \tag{11.54}$$

A typical computation step follows:

- V_i, y_i, and $S_{0,i}$ are known.
- A value of y_{i+1} is chosen (usually $|\Delta y| < 0.1 y_i$).
- V_{i+1} is calculated from $V_{i+1} = V_i y_i / y_{i+1}$.
- $\bar{R}_h$ is calculated and $\bar{n}$ is selected.
- $S_{0,i+1}$ is estimated.
- $\overline{(S_0 - S_E)_i}$ is calculated from Eq. (11.53).
- Δx_i is calculated from Eq. (11.52).
- $S_{0,i+1}$ is checked.
- If S_0 changes rapidly with x, recomputing $\overline{(S_0 - S_E)_i}$ and obtaining a better value for Δx_i may be necessary.

By repeating this process for each reach, we can work out a complete surface profile. The entire surface profile does not have to be of a single type, because the stepwise calculation allows us to patch together various profile types. The calculations can be carried out by hand, although that is a rather tedious process. A FORTRAN or BASIC program, or an electronic spreadsheet, can be generated to perform the calculations. Use of computer support permits finer steps and, hence, greater accuracy.

EXAMPLE 11.9 Illustrates Computation of Surface Profiles in Gradually Varied Flow

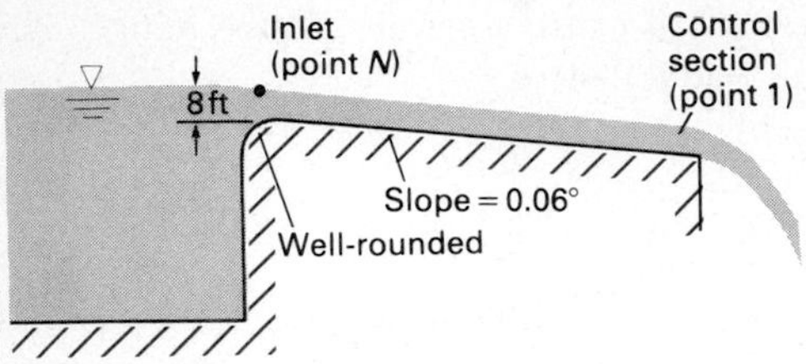

Figure E11.9 Gradually varied flow in long, rectangular channel.

The unfinished concrete channel ($n = 0.014$) shown in Fig. E11.9 is 12 ft wide and is laid on a 0.06° slope. The reservoir water depth is 8 ft above the bottom of the channel inlet. Calculate the flow rate in the channel if the channel length is 1000 ft.

SOLUTION

Given

Figure E11.9

12-ft-wide channel having 0.06° slope (tan 0.06° = 0.00105)

$n = 0.014$

Find

Flow rate for channel length of 1000 ft

Solution

For a channel length of 1000 ft, uniform flow probably is not established. We must therefore compute the gradually varied flow from the channel inlet to the free overfall. The *control section* for the flow is the critical point that occurs just upstream from the free overfall; we have to begin calculations there and compute the profile back upstream toward the reservoir. This would be a straightforward application of the direct-step method if we knew the critical depth or the flow rate. Unfortunately, we know neither; only the elevation of the reservoir surface above the channel inlet is given. To solve this problem, we must combine an iteration with the direct-step calculation of the free-surface profile. The iteration proceeds as follows:

- A value for the critical depth y_c is assumed. From the information given in Example 11.4, critical depth occurs about $5y_c$ upstream of the brink.
- The flow rate q is calculated from the critical depth.
- The depth at the channel inlet is computed by applying the energy equation to the frictionless flow between the reservoir and the channel inlet (labeled point N in Fig. E11.9a). The equation is

$$\frac{q^2}{2gy_N^2} + y_N = 8 \text{ ft}.$$

- The distance from the critical point to the channel inlet is computed using Eq. (11.52) N times, with

$$\Delta y = y_{i+1} - y_i = \frac{(y_N - y_c)}{N}.$$

The number of steps is arbitrary, with larger N implying more work and more accuracy. Note that Δx_i will be negative.

• The point where the depth equals y_N is computed from

$$x_N = 1000 - 5y_c + \textstyle\sum_1^N \Delta x_i.$$

If the computed value of x_N does not equal zero, then another value of y_c is assumed and the process is repeated.

The computational equations are

$$q = V_c y_c = \sqrt{g y_c^3} = 5.675 y_c^{3/2}, \tag{E11.3}$$

$$\frac{q^2}{2g y_N^2} + y_N = \frac{y_c^3}{2 y_N^2} + y_N = 8, \tag{E11.4}$$

$$\Delta y = y_{i+1} - y_i = \frac{y_N - y_c}{N}, \tag{E11.5}$$

$$V_{i+1} = \left(\frac{y_i}{y_{i+1}}\right) V_i, \tag{E11.6}$$

$$\bar{y} = \tfrac{1}{2}(y_{i+1} + y_i), \tag{E11.7}$$

$$\bar{V} = \tfrac{1}{2}(V_{i+1} + V_i), \tag{E11.8}$$

$$\bar{R}_h = \frac{b\bar{y}}{b + 2\bar{y}} = \frac{12\bar{y}}{12 + 2\bar{y}}, \tag{E11.9}$$

$$\overline{(S_0 - S_E)_i} = S_0 - \frac{\bar{n}^2}{\xi^2}\frac{\bar{V}^2}{\bar{R}_h^{4/3}}$$

$$= 0.00105 - 8.83 \times 10^{-5}\left(\frac{\bar{V}^2}{\bar{R}_h^{4/3}}\right), \tag{E11.10}$$

$$\Delta x_i = \frac{(V_{i+1}^2 - V_i^2)/64.4 + y_{i+1} - y_i}{\overline{(S_0 - S_E)_i}}. \tag{E11.11}$$

Equations (E11.5)–(E11.11) are applied N times. The calculated value of x at the channel inlet is

$$x_N = 1000 \text{ ft} - 5y_c + \sum_1^N \Delta x_i.$$

If x_N is not zero, we must assume a new value for y_c and return to Eq. (E11.3).

To illustrate the calculations, we choose $N = 5$ and guess

$$y_c^{(1)} = 5 \text{ ft}.$$

From Eq. (E11.3),

$$q^{(1)} = 63.44 \text{ ft}^2/\text{sec}.$$

Solving Eq. (E11.4) for y_N, we have

$$y_N = 6.538 \text{ ft}.$$

From Eq. (E11.5), we have

$$\Delta y = \frac{6.538 - 5}{5} = 0.308 \text{ ft}.$$

The results of applying Eqs. (E11.6)–(E11.11) are summarized in the first column of Table E11.9a.

Table E11.9a Summary of calculations five reaches.

	Channel Reach Number ($i = 1$ to $i = 5$)				
i	1 (control section)	2	3	4	5 (inlet)
y_i	5 ($= y_c$)	5.308	5.615	5.923	6.230
y_{i+1}	5.308	5.615	5.923	6.230	6.538
V_i	12.69 ($= V_c$)	11.95	11.30	10.71	10.18
V_{i+1}	11.95	11.30	10.71	10.18	9.70
$\bar{y}$	5.154	5.461	5.769	6.076	6.384
$\bar{V}$	12.32	11.63	11.01	10.45	9.94
$\bar{R}_h$	2.772	2.859	2.941	3.019	3.093
$\overline{(S_0 - S_E)_i}$	-2.39×10^{-3}	-1.89×10^{-3}	-1.49×10^{-3}	-1.16×10^{-3}	18.87×10^{-4}
Δx_i	-10.96	-37.63	-71.93	-117.33	-179.89
x_N	—	—	—	—	557.2

Table E11.9b Calculated value of channel inlet position for various values of y_c (channel divided into five reaches).

y_c	x_N
5.0	557.2
4.5	−3189.6
4.8	−12.6
4.81	26.60
4.803	−0.65

Table E11.9c Values of y_c for various numbers of reaches in channel.

N	y_c (for $x_N \approx 0$)
2	4.772
5	4.803
10	4.807
20	4.808
50	4.809
100	4.809

To compute the remaining four reaches, we wrote a BASIC computer program to evaluate Eqs. (E11.6)–(E11.11) and to calculate x_N. The results of this calculation are summarized in Table E11.9a. Note that the final depth in Table E11.9a is 6.538 ft, as calculated by Eq. (E11.4). This value is based on our assumed value for y_c. Because the calculated value of x_N (557.2 ft) is incorrect, our assumed value of y_c was incorrect and we must assume another value.

Table E11.9b summarizes calculations of x_N for various values of y_c, all calculated with $N = 5$. The value

$$y_c = 4.803 \text{ ft}$$

yields satisfactory results for $N = 5$.

To check the effect of the number of reaches, we use our BASIC program to repeat the calculations for $N = 2$, 10, 20, 50, and 100. Table E11.9c summarizes the results of these calculations. Clearly, to the correct number of significant figures,

$$y_c = 4.81 \text{ ft},$$

giving

$$q = 5.675(4.81)^{3/2} = 59.87 \text{ ft}^2/\text{sec}$$

and

$$Q = qb = (59.87 \text{ ft}^2/\text{sec})(12 \text{ ft}),$$

or

$$Q = 718.4 \text{ ft}^3/\text{sec}.$$

ANSWER

Discussion

Interestingly, the calculated value of y_c is within about 1 percent of the correct value, with the channel broken into only two reaches.

Note that the profile computations started at the critical flow control section and proceeded upstream.

11.5 MISCELLANEOUS TOPICS IN OPEN CHANNEL FLOW

We close this chapter on open channel flow by discussing flow measurement methods for open channel flow and an analogy between open channel flow and compressible flow.

11.5.1 Measurement Methods for Open Channel Flow

Measurement of liquid flow rate in an open channel can be accomplished by the traverse method or by use of a weir. In the traverse method, liquid velocity is measured at several points in the channel cross section.* The flow rate is the sum of each velocity multiplied by the appropriate area. Generally, the channel cross section is divided into vertical strips (Fig. 11.23), and the average velocity for each strip is determined. The flow rate for the strip is then

$$Q_{\text{strip}} = \bar{V}_{\text{strip}} A_{\text{strip}} = \bar{V}_{\text{strip}} y_{\text{strip}} \Delta b_{\text{strip}},$$

and the total flow rate is the sum of the individual strip flow rates. Many investigations have shown that the liquid velocity measured at a point 60 percent depth below the free surface is a close approximation to the average velocity. An even better approximation to the average velocity is the average of velocities measured at 20 percent depth and 80 percent depth. Thus in practice, only one or two velocity measurements per strip need be made.

A *weir* is an obstruction built across the channel that causes the upstream liquid to rise. The height of the surface upstream from the weir is a measure of the flow rate. Figure 11.24 shows several types of weirs. The rectangular and triangular (or V-notch) weirs are called *sharp-crested weirs*. Note that all weirs create a free "jet" of liquid (called the *nappe*) on the downstream side of the weir. At low flow rates, the nappe may be absent, with the liquid adhering to the downstream surface of the weir. If this condition occurs, the weir is not operating satisfactorily, and accurate flow measurement is difficult. Figure 11.25 shows the flow over a sharp-crested weir and illustrates the nomenclature of weir flow measurement.

The discharge over a full-span rectangular weir is calculated by

$$Q = \mathbf{C_d}\left(\frac{2\sqrt{2g}}{3}\right) L H^{3/2}, \tag{11.55}$$

where $\mathbf{C_d}$ is an empirically determined discharge coefficient that accounts for certain assumptions in the derivation of the equation, such as neglect of the approach velocity and the drawdown, as well as for effects of viscosity and surface tension. Dimensional analysis indicates that

$$\mathbf{C_d} = \mathbf{C_d}\left(\frac{H}{P}, \mathbf{R}, \mathbf{W}\right).$$

In most practical cases, Reynolds number and Weber number effects are negligible

* Velocities can be measured using a Pitot or Fechheimer probe (Figs. 4.16 and E4.15a) or a rotating cup current meter.

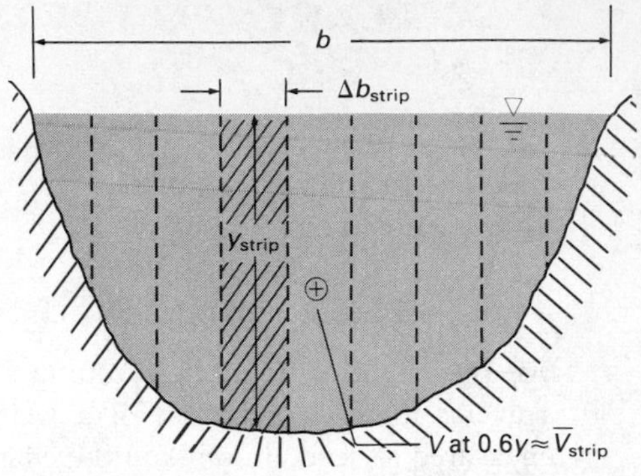

Figure 11.23 Flow rate can be measured by dividing channel into vertical strips and summing strip flows.

and $\mathbf{C_d}$ depends on H/P only. A widely used correlation is

$$C_d \approx 0.611 + 0.075\left(\frac{H}{P}\right). \tag{11.56}$$

The flow through a rectangular weir with end contractions is similar to flow through a full-span weir, except that the nappe contracts, making the effective flow area less than HL. Experiments have shown that if L is greater than $3H$, the amount of contraction of the nappe is approximately $0.1H$.

A triangular weir is useful for measuring relatively small flow rates. Figure 11.26 shows a triangular weir with vertex angle ϕ. The velocity at any depth is proportional to $\sqrt{2gh}$. The flow through area dA is*

$$dQ = \mathbf{C_d}\sqrt{2gh}\,dA.$$

The differential area is

$$dA = b\,dh.$$

From geometry,

$$b = 2(H - h)\tan\left(\frac{\phi}{2}\right),$$

and so

$$Q = \int dQ = 2\sqrt{2g}\,\mathbf{C_d}\tan\left(\frac{\phi}{2}\right)\int_0^H (H - h)\sqrt{h}\,dh.$$

Integrating gives

$$Q = \frac{8}{15}\mathbf{C_d}\sqrt{2g}\tan\left(\frac{\phi}{2}\right)H^{5/2} \tag{11.57}$$

for a triangular weir.

---

* Anticipating the need for an empirical discharge coefficient, we insert $\mathbf{C_d}$ in the equation; however, $\mathbf{C_d}$ has meaning only for the integrated flow rate equation.

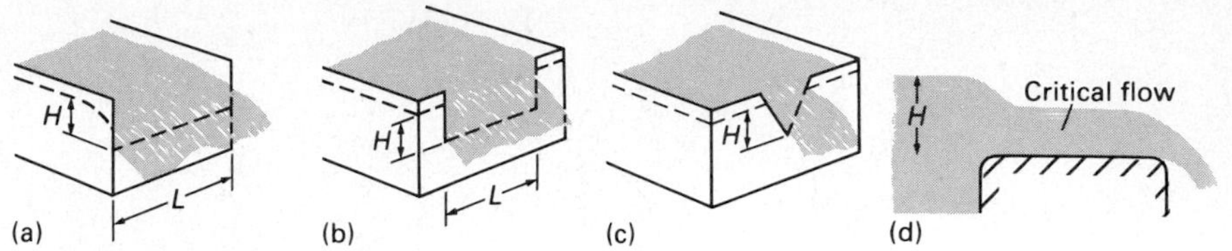

Figure 11.24 Various types of weirs for measuring flow rate in an open channel: (a) Full-span rectangular weir; (b) part-span rectangular weir with end contractions; (c) triangular or V-notch weir; (d) broad-crested weir.

The discharge coefficient is a function of vertex angle, Reynolds number, and Weber number. The two most commonly used vertex angles are 60° and 90°. For these angles and for H greater than about 0.2 m (0.6 ft) for $\phi = 90°$ and H greater than about 0.25 m (0.8 ft) for $\phi = 60°$, the value of $\mathbf{C_d}$ is approximately constant:

$$\mathbf{C_d} \approx 0.58. \tag{11.58}$$

A broad-crested weir is a large "bump" in the channel with a flat top surface (Fig. 11.27). A broad-crested weir should be high enough to cause critical flow to occur on the flat surface (see Section 11.3.2). The velocity over the weir crest is

$$V = V_c = \sqrt{gy_c},$$

and the flow rate is

$$Q = VA = \sqrt{g}y_c^{3/2}L. \tag{11.59}$$

The flow rate could be measured by measuring y_c; however, this is difficult to accomplish accurately because the flow is critical. Measuring the upstream depth H is easier. Applying the energy equation from a point on the upstream surface to a point on the crest and neglecting losses, we have

$$\frac{V_1^2}{2g} + H + P = \frac{V_c^2}{2g} + y_c + P.$$

Substituting $V_c = \sqrt{gy_c}$ gives

$$y_c = \frac{2}{3}H + \frac{V_1^2}{3g}.$$

If V_1 is small,

$$y_c \approx \frac{2}{3}H.$$

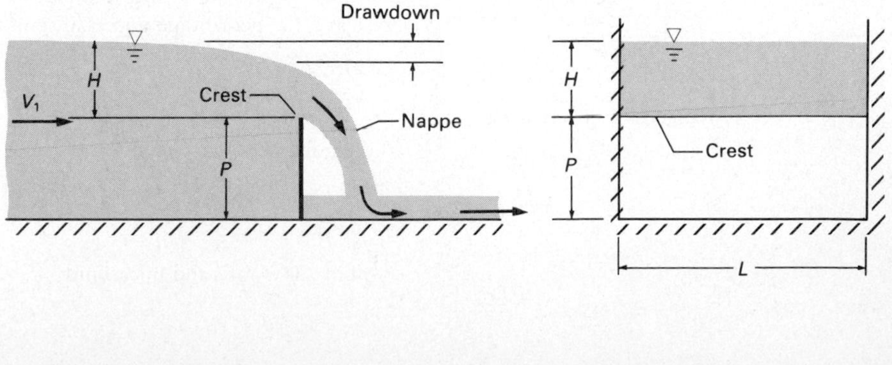

Figure 11.25 Flow over a full-span sharp-crested weir.

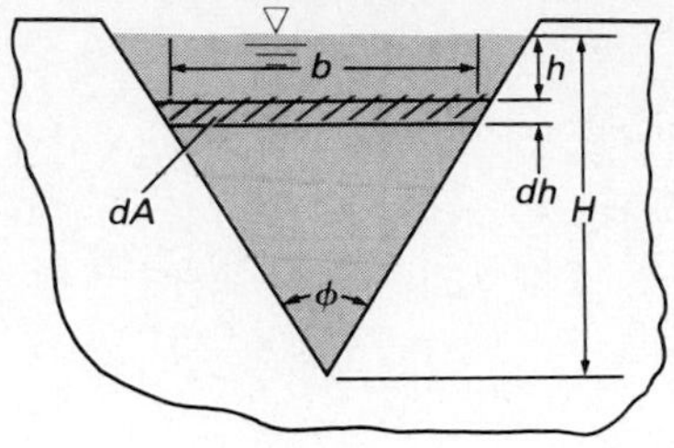

Figure 11.26 Illustration for analysis of triangular weir.

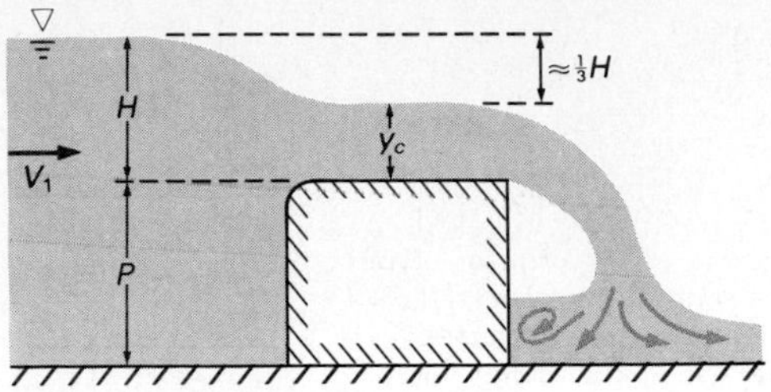

Figure 11.27 Flow over a broad-crested weir.

Substituting into Eq. (11.59) gives

$$Q \approx \frac{1}{\sqrt{3}}\left(\frac{2}{3}\right)\sqrt{2g}LH^{3/2}. \tag{11.60}$$

A more accurate equation replaces $1/\sqrt{3}$ with an experimentally determined discharge coefficient

$$Q = \mathbf{C_d}\left(\frac{2}{3}\right)\sqrt{2g}LH^{3/2}. \tag{11.61}$$

For flow with high Reynolds and Weber numbers, the discharge coefficient for a broad-crested weir is given by

$$\mathbf{C_d} \approx \frac{0.65}{[1 + (H/P)]^{1/2}}. \tag{11.62}$$

EXAMPLE 11.10 Illustrates Calculations for Weirs and Estimation of the Uncertainty in a Flow Measurement

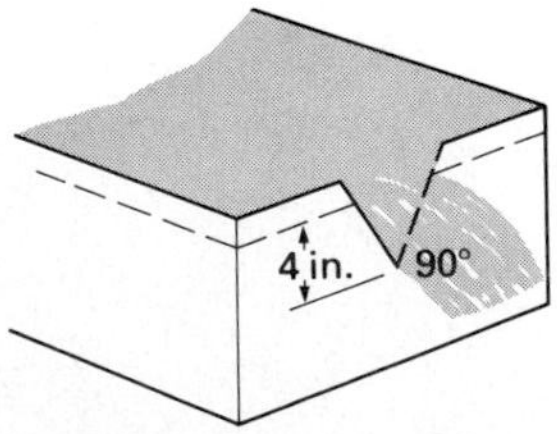

Figure E11.10 Triangular or V-notch weir.

An engineering laboratory experiment uses a triangular weir in an open channel to measure flow rate as shown in Fig. E11.10. The nominal weir angle is 90°. In a certain test, the head of water above the weir was 4 in. The uncertainty in the weir angle is $\pm 2°$ and the uncertainty in the depth measurement is ± 0.05 in. Compute the flow rate and the percentage uncertainty in the flow rate.

SOLUTION

Given

Triangular weir with angle of 90° $\pm 2°$

Head of water H above weir is 4 in. ± 0.05 in.

Find

Volume flow rate and uncertainty in volume flow rate

Solution

We find the volume flow rate through a triangular weir by using Eqs. (11.57) and (11.58):

$$Q = \frac{8}{15}\, C_d \sqrt{2g}\, \tan\left(\frac{\phi}{2}\right) H^{5/2}.$$

The numerical values give

$$Q = \frac{8}{15}\,(0.58)\sqrt{2(32.2\text{ ft/sec}^2)}\,\tan\left(\frac{90^\circ}{2}\right)[(4\text{ in.})(1\text{ ft}/12\text{ in.})]^{5/2},$$

or

$$Q = 0.159\text{ ft}^3/\text{sec}.$$ **ANSWER**

We can estimate the uncertainty in Q in one of three ways. In the first method, we calculate both the maximum possible value and the minimum possible value of Q:

$$\begin{aligned} Q_{max} &= \frac{8}{15}\,(0.58)\sqrt{2(32.2\text{ ft/sec}^2)}\,\tan\left(\frac{92^\circ}{2}\right) \\ &\quad \times [(4.05\text{ in.})(1\text{ ft}/12\text{ in.})]^{5/2} \\ &= 0.170\text{ ft}^3/\text{sec}, \end{aligned}$$

and

$$\begin{aligned} Q_{min} &= \frac{8}{15}\,(0.58)\sqrt{2(32.2\text{ ft/sec}^2)}\,\tan\left(\frac{88^\circ}{2}\right) \\ &\quad \times [(3.95\text{ in.})(1\text{ ft}/12\text{ in.})]^{5/2} \\ &= 0.149\text{ ft}^3/\text{sec}. \end{aligned}$$

The uncertainty U_Q in Q is

$$U_Q = (0.170 - 0.159)\text{ ft}^3/\text{sec} \quad \text{and} \quad U_Q = (0.149 - 0.159)\text{ ft}^3/\text{sec},$$

$$U_Q = +0.011\text{ ft}^3/\text{sec} \quad \text{or} \quad -0.010\text{ ft}^3/\text{sec}.$$ **ANSWER**

The second method is to estimate U_Q by taking the differential of Q:

$$U_Q \approx dQ = \left(\frac{\partial Q}{\partial \phi}\right)_H d\phi + \left(\frac{\partial Q}{\partial H}\right)_\phi dH.$$

Now

$$\left(\frac{\partial Q}{\partial \phi}\right)_H = \frac{4}{15}\, C_d\sqrt{2g}\,\sec^2\left(\frac{\phi}{2}\right) H^{5/2}$$

and

$$\left(\frac{\partial Q}{\partial H}\right)_\phi = \frac{5}{2}\left(\frac{8}{15}\right) C_d\sqrt{2g}\,\tan\left(\frac{\phi}{2}\right) H^{3/2},$$

so

$$dQ = \frac{4}{3}\, C_d\sqrt{2g}\left[\frac{1}{5}\sec^2\left(\frac{\phi}{2}\right) H^{5/2}\, d\phi + \tan\left(\frac{\phi}{2}\right) H^{3/2}\, dH\right].$$

Using

$$d\phi = 2^\circ\left(\frac{\pi \text{ rad}}{180^\circ}\right) = \frac{\pi}{90}\text{ rad}$$

and

$$dH = 0.05\text{ in.}\left(\frac{1\text{ ft}}{12\text{ in.}}\right) = 0.00417\text{ ft}$$

gives

$$dQ = \frac{4}{3}(0.58)\sqrt{2(32.2\text{ ft/sec}^2)}\left[\frac{\sec^2 45^\circ}{5}\left(\frac{4}{12}\text{ ft}\right)^{5/2}\left(\frac{\pi}{90}\text{ rad}\right)\right.$$
$$\left. + \tan 45^\circ\left(\frac{4}{12}\text{ ft}\right)^{3/2}(0.00417\text{ ft})\right] = 0.0105\text{ ft}^3/\text{sec.}$$

Using dQ as an estimate of the uncertainty in Q gives

$$U_Q \approx \pm dQ = \pm 0.0105\text{ ft}^3/\text{sec.}$$ **ANSWER**

The third method provides a more realistic estimate of the uncertainty. It recognizes that the maximum possible deviation in the weir angle ($\pm 2^\circ$) and the maximum possible deviation in the depth measurement (± 0.05 in.) probably do not occur simultaneously. Therefore the preceding estimates give the maximum possible uncertainty. A more realistic estimate of the uncertainty is the Pythagorean summation [5, 6] of the individual uncertainties:

$$U_Q \approx \pm\sqrt{\left[\left(\frac{\partial Q}{\partial \phi}\right)_H d\phi\right]^2 + \left[\left(\frac{\partial Q}{\partial H}\right)_\phi dH\right]^2},$$

or

$$U_Q \approx \pm\frac{4}{3}C_d\sqrt{2g}\left\{\left[\frac{1}{5}\sec^2\left(\frac{\phi}{2}\right)H^{5/2}\,d\phi\right]^2\right.$$
$$\left. + \left[\tan\left(\frac{\phi}{2}\right)H^{3/2}\,dH\right]^2\right\}^{1/2}.$$

The numerical values give

$$U_Q \approx \pm\frac{4}{3}(0.58)\sqrt{2(32.2\text{ ft/sec}^2)}$$
$$\times\left\{\left[\frac{\sec^2 45^\circ}{5}\left(\frac{4}{12}\text{ ft}\right)^{5/2}\left(\frac{\pi}{90}\text{ rad}\right)\right]^2\right.$$
$$\left. + \left[\tan 45^\circ\left(\frac{4}{12}\text{ ft}\right)^{3/2}(0.00417\text{ ft})\right]^2\right\}^{1/2},$$

$$U_Q \approx \pm 0.0075\text{ ft}^3/\text{sec.}$$ **ANSWER**

Discussion

Note that the first two methods provide an estimate of the maximum uncertainty and that the third method provides a more reasonable estimate of the actual uncertainty.

11.5.2 Analogy between Open Channel Flow and Compressible Flow

If you have studied Chapter 10, you might have noted the strong similarities between certain phenomena in open channel flow and compressible flow. Probably the strongest similarity is that between the hydraulic jump in open channel flow and the normal shock wave in compressible flow. Both are strongly dissipative phenomena that occur over a relatively short distance. Both are extremely complicated in detail but can be satisfactorily analyzed by the control volume approach. Both convert the flow from one state (supercritical, supersonic) to another (subcritical, subsonic).

Another obvious similarity is the phenomena of wave motions and their relevant dimensionless parameters. Weak free-surface gravity waves in open channel flow are similar to sound waves in compressible flow. The flow state is classified by the Froude number in open channel flow and Mach number in compressible flow. For Froude or Mach numbers of less than unity, disturbances at a point are propagated to all parts of the flow, but for Froude or Mach numbers of greater than unity, disturbances propagate downstream only.

A final illustration of the similarities between open channel flow and compressible flow is the comparison between the response of a frictionless open channel flow to a change of bottom elevation (see Section 11.3.2) and the response of an isentropic gas flow to area change (Section 10.4.1). We suggest that you study these sections in parallel and list the similarities.

Are these similarities simply coincidence, or is there a formal analogy between the two types of flow? If there is a formal analogy, we can interpret measurements or analysis in one type of flow in terms of the other type. We can demonstrate the formal analogy by considering the equations of continuity, momentum, and energy. Consider liquid flow in a rectangular open channel and gas flow in a closed passage. The continuity equation for channel flow is

$$Q = Vyb = \text{Constant},$$

and for gas flow is

$$\dot{m} = V\rho A = \text{Constant}.$$

Differentiating each logarithmically, we have

$$\frac{dV}{V} + \frac{dy}{y} + \frac{db}{b} = 0 \tag{11.63}$$

for channel flow and

$$\frac{dV}{V} + \frac{d\rho}{\rho} + \frac{dA}{A} = 0 \tag{11.64}$$

for gas flow.

For isentropic gas flow, the differential energy and momentum equations are identical. The energy/momentum equation is Eq. (10.63):

$$VdV + \frac{dp}{\rho} = 0.$$

Using

$$dp = \left.\frac{\partial p}{\partial \rho}\right)_s d\rho, \qquad c^2 = \left.\frac{\partial p}{\partial \rho}\right)_s, \quad \text{and} \quad \mathbf{M}^2 = \frac{V^2}{c^2},$$

we have for the energy/momentum equation for isentropic gas flow

$$\frac{d\rho}{\rho} = -\mathbf{M}^2 \frac{dV}{V}. \tag{11.65}$$

For open channel flow, the specific energy is

$$E = \frac{V^2}{2g} + y.$$

For frictionless flow in a horizontal channel, $E = \text{Constant}$, and

$$\frac{V\,dV}{g} + dy = 0.$$

Using

$$c^2 = gy \quad \text{and} \quad \mathbf{F}^2 = \frac{V^2}{c^2}$$

reduces the open channel energy equation to

$$\frac{dy}{y} = -\mathbf{F}^2\left(\frac{dV}{V}\right). \tag{11.66}$$

Comparing Eqs. (11.63) and (11.64) and Eqs. (11.65) and (11.66) suggests the following analogy between isentropic gas flow and frictionless flow in a horizontal rectangular channel:

- Mach number (gas flow) is analogous to Froude number (channel flow).
- Density (gas flow) is analogous to depth (channel flow).
- Area (gas flow) is analogous to channel width (channel flow).

The corresponding equations have identical form, so the analogy is complete and formally correct. This analogy is put to use in a water table in which a free-surface flow of water is used to simulate compressible flow of a gas [7]. A water table can be used for flow visualization, with hydraulic jumps representing shock waves and depth representing density. Depth measurements in a water table can be used to deduce densities in the analogous compressible flow. A water table is much cheaper to build and operate than a wind tunnel; therefore a water table may be called a "poor man's wind tunnel."

PROBLEMS

1. Do shallow waves propagate at the same speed in all fluids? Explain why or why not.

•**2.** Develop the vertical sluice gate equation,

$$\left(\frac{y_2^2}{y_1^2 - y_2^2}\right)\left(1 - \frac{y_2}{y_1}\right) = \frac{q^2}{2gy_1^3} = \frac{\mathbf{F}_1^2}{2},$$

for frictionless flow, where $\mathbf{F}_1$ is the upstream Froude number.

3. Water flows with an average velocity of 1.0 m/s and a normal depth of 0.5 m in a wide rectangular channel. Is the flow subcritical or supercritical?

4. A 10.0-m I.D. concrete pipe has Manning's $n = 0.015$. What is the value of Manning's n for a 25.0-m I.D. concrete pipe having the same value of absolute roughness?

5. Consider laminar flow down a wide rectangular channel making an angle θ with the horizontal. The fluid has kinematic viscosity ν and the volume flow rate per unit width is given by

$$q = \frac{gy^3 \sin\theta}{3\nu},$$

where y is the fluid depth perpendicular to the channel bottom. Find Manning's n for this flow.

6. A uniform flow of 110,000 ft³/sec is measured in a natural channel that is approximately rectangular in shape with a 2650-ft width and 17.5-ft depth. The water-surface elevation drops 0.37 ft per mile. Based on the computed Manning n, characterize the type of natural channel observed. Also compute the Froude number and determine whether the flow is subcritical or supercritical.

7. Water flows down a concrete street having a slope of 10° at the rate of 0.01 ft³/sec per foot of width. Find the water depth far down the street.

8. An unfinished concrete rectangular channel is 5 m wide and has a slope of 0.50°. The water is 0.5 m deep. Find the discharge rate for uniform flow.

9. A rectangular unfinished concrete channel with $b =$ 10.0 m carries 100 m³/s of water. The channel bottom slope is 0.0001. Find the normal depth.

10. A trapezoidal unfinished concrete channel with $b =$ 1.0 m and $\phi = 30°$ carries 100 m³/s of water. (See Fig. 11.6.) The channel bottom slope is 0.0001. Find the normal depth.

11. A prismatic fiberglass ($n = 0.012$) channel having a semicircular cross section is used as a water slide in an amusement park. The channel has a 1.0-m diameter, is 15 m long, and has a slope of 30°. The water flow rate down the slide is 1.0 m³/s. Find the normal depth.

12. A trapezoidal channel is made of wood planks (Manning's $n = 0.012$) and is used for a log slide. The bottom angle is 30° above the horizontal, the water flow rate is 5 ft³/sec, the angle ϕ in Fig. P11.12 is 60°, and the base b is 8.0 ft. Find the normal depth. Determine whether the flow is subcritical or supercritical.

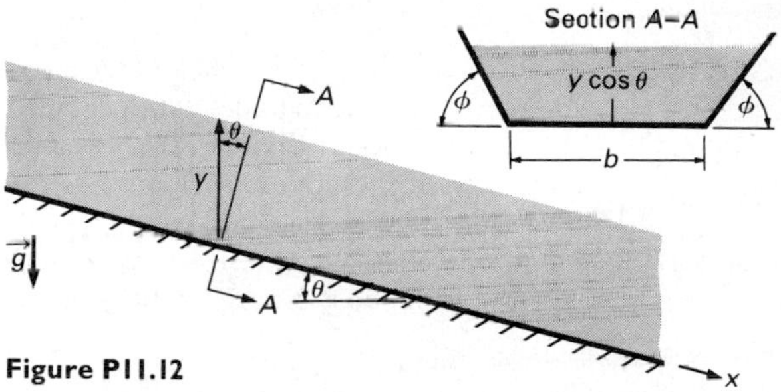

Figure P11.12

13. Find the depth at which water will flow at the rate of 6 ft³/sec inside a 40-in. I.D. concrete pipe having a slope of 0.0025.

14. A wide channel has a bottom slope $S_0 = 0.00015$, uniform laminar flow, and a normal depth $y_n = 0.125$ m. The fluid is water. Develop expressions for the water velocity parallel to the bottom and the discharge per unit width. Compare the form of this answer to that using Chezy's formula and solve for the Chezy coefficient.

15. Find the discharge per unit width for a wide channel having a bottom slope of 0.00015. The normal depth is 0.003 m. Assume laminar flow and justify the assumption. The fluid is 20°C water.

16. A 16-in. I.D. concrete pipe is flowing half full of water. The bottom slope is 0.0001. Find the flow rate.

17. The channel in Fig. P11.17 has two floodplains as shown. Find the discharge if the center channel is lined

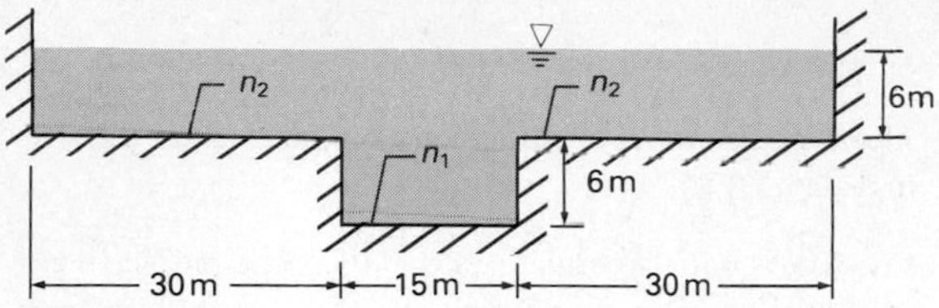

Figure P11.17

with brick and the two floodplains are lined with cobblestones. The slope S_0 is 0.00025.

18. A major river is divided into three parts or courses—the upper course, the middle course, and the lower course. The slope is 70 ft per mile in the upper course, 10 ft per mile in the middle course, and 1.0 ft per mile in the lower course. All three courses have rectangular cross sections. The upper course has a normal depth of 40.0 ft. The river width is 120 ft. Find the normal depths of the middle and lower courses.

19. The lower velocities of the middle course of the river in Problem 18 do not wear away the sides of the channel as rapidly as do the velocities in the upper course. Assume that the cross section of the middle course is a trapezoid with an angle ϕ (see Fig. 11.6) of 60°. Find the normal depth of the middle course if the base of the trapezoid b is 60 ft and the flow rate is 196,100 ft³/sec.

20. Find the uniform flow depth expected in a gravelly flood control channel during a 50-year storm flow of 1261 ft³/sec. The channel has a trapezoidal cross section with side slopes of 2:1, a bed width of 10 ft, and bed slope of 1 ft per 100 ft.

21. A large trapezoidal channel cut through stone has side slopes of 1:1 and a bed width of 291 ft. Find the uniform flow in the channel when the flow depth is 30.4 ft and the bed slopes 1 ft per mile.

•**22.** The spillway of a dam is 20.0 ft wide and has a flow rate of 5000 ft³/sec. The spillway makes an angle of 30° with the horizontal. Find the vertical water depth down the spillway. See Problem 78.

23. A rectangular brick-lined channel has a bottom slope of 0.0025 and is designed to carry a uniform water flow rate of 300 ft³/sec. Would the channel need fewer bricks if the channel were 2 ft wide, 6 ft wide, or 10 ft wide? Explain.

24. Show that the portion of a circular channel that will most efficiently transport a given volume flow rate Q is a semicircle.

25. Determine the values of ϕ and h for the triangular channel in Fig. P11.25 for the most efficient triangular channel (i.e., the channel of the smallest perimeter for a given flow area, slope, and flow rate).

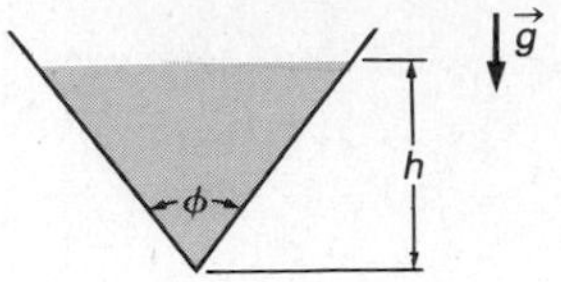

Figure P11.25

26. For $A = 1.0$ m² and $\phi = 90°$, plot P versus y using Eq. (11.20) and show that the optimum depth is $\sqrt{0.5}$ m.

27. Figure P11.27 shows a cross section of an aqueduct that carries water at 50 m³/s. The value of Manning's n is 0.015. Find the bottom slope.

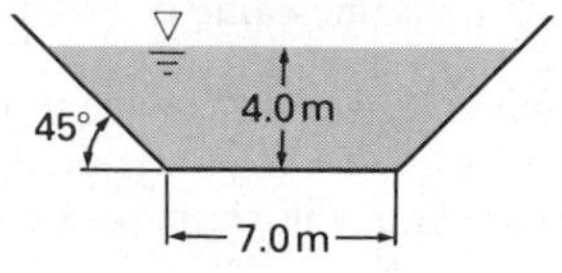

Figure P11.27

28. Consider a semicircular channel and a rectangular channel whose height is one-half the width. Find the diameter of the semicircular channel that will have the same discharge as the rectangular channel. Both channels are full and have the same slope and the same value of Manning's n.

29. The lowest river velocities occur in the lower course of the river in Problem 18 and permit silt to be deposited in the bed of the lower course. This silt raises the bottom of the river, and allows the river to overflow into the flood-plains. Assume that the cross section of the lower course is as shown in Fig. P11.29. Find the width b.

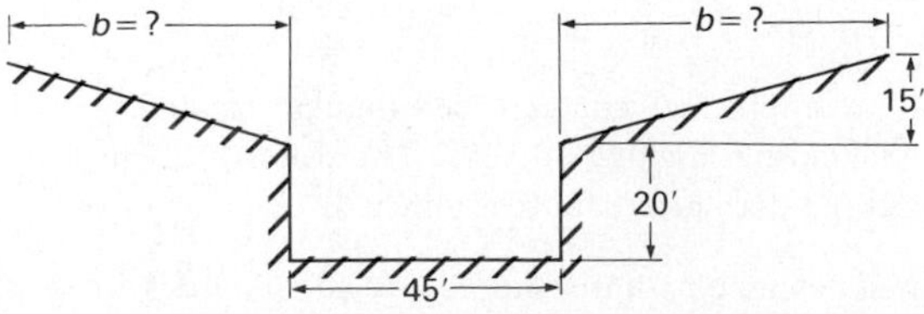

Figure P11.29

30. The rate of discharge through the canal in Fig. P11.30 is to be 10 m³/s. Find the width b and the bottom slope. The velocity will not exceed 5.0 m/s.

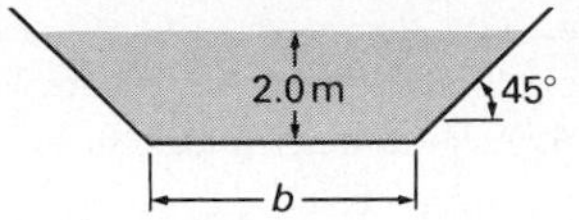

Figure P11.30

31. Consider a rectangular channel and a semicircular channel with the same bottom slope and Manning's n. Find the diameter of the semicircular channel that will have the same capacity as a rectangular channel measuring 3 m wide and 1.2 m deep.

32. Again, consider the river and floodplains in Fig. P11.29. The river has a Manning's n of 0.030. The floodplains are to be planted with light brush. Find the width b of the floodplains for the river to handle 30,000 ft^3/sec of water. Bottom slope = 0.0001.

33. A river has a rectangular cross section and measures 50 ft wide and 20 ft deep. Recent spring thaws have flooded the banks of the river. The flood water is estimated to be 10% to 15% of the total river flow rate. The bank of one side of the river can be cut back 5 ft. Will this widening of the river alleviate the flooding? If not, what other modifications should be made? The river's bottom slope is 0.00020.

34. Design a drainage ditch having trapezoidal cross section, side slopes of 2.5 : 1 in a clean, excavated earth bed, and a bed slope of 7.5 m/km. The ditch should be designed to carry the 25-year storm flow of 150.9 m^3/s without exceeding a 2-m flow depth. What is the average flow velocity during the storm?

35. Find the slope required for a 54-in.-diameter corrugated metal pipe to carry a uniform flow of 83 ft^3/sec when 50% full. What is the average flow velocity in the pipe?

36. Identify the type of existing surface in an excavated irrigation channel having a channel slope of 0.005 and uniform flow of 34 m^3/s. The cross section is trapezoidal, with side slopes of 3 : 1, a flow depth of 2 m, and top width of 13 m. What is the channel bed width?

37. Uniform flow in a sluggish channel having a nearly rectangular cross section that is 498 ft wide and 16.5 ft deep carries a flow of 8,250 ft^3/sec. Approximately how much does the water surface elevation drop along a river mile?

38. Find the diameter required for reinforced concrete pipe laid at a slope of 0.001 and required to carry a uniform flow of 19.3 ft^3/sec when the depth is 75% of the diameter.

39. Farmland along a small river is to be used as a floodplain to carry excess flow from the river during flooding. What width of property is required to carry an excess flow of 49.9 m^3/s at a depth not to exceed 1.1 m? Assume that the flow will be uniform and follow the natural slope of the farmland, which falls 0.36 m/km.

40. A developer needs to lay a storm-water pipe to drain the runoff from a nearly flat development site. The maximum permissible runoff allowed is 38 ft^3/sec, and the minimum velocity is 3 ft/sec to prevent settling of soil particles. For proper ground cover, the top of the pipe must be at least 2 ft below the ground at the upstream end. The pipe must tie into another pipe which is 6 ft 9 in. below the ground at a horizontal distance 1000 ft away. Should the developer use concrete or corrugated metal pipe? What pipe diameter should be used?

• **41.** A construction engineer is designing a channel to carry a water flow rate of 100 ft^3/sec down an angle of 0.0025° with the horizontal. The construction material is planed wood. The engineer is debating whether to use a 60° triangular channel or a 90° rectangular channel. Would less wood be used for the triangular channel or the rectangular channel?

• **42.** A city wants to lay out several athletic fields in the floodplain along both sides of a river. The main channel of the river is nearly rectangular and 14 ft deep, 350 ft wide, and consists of earth with some grass (n = 0.026). The floodplain consists of tall grass (n = 0.035) and extends 695 ft from the edge of the main channel to grassy vertical bluffs on each side. The channel bed drops about 2.64 ft per mile.

(a) Compute the uniform flow depth for the 50-year flood of 51,501 ft^3/sec.

(b) Compute the flow discharge and average flow velocities in the main channel and overbank reaches for the 50-year flood.

(c) If 200 yards are needed next to the bluffs on each side of the river for the athletic fields, at approximately what height must vertical levees be built to prevent the 50-year flood from flooding the athletic fields. List any assumptions made.

43. A wide river has a flow rate of 10 m^3/s per meter of width. Find the critical depth and the critical slope.

44. A trapezoidal channel with b = 10.0 m and ϕ = 30°

carries 100 m^3/s of water. (See Fig. 11.6.) Find the critical depth and the critical slope.

45. Water flows down a wide rectangular channel having Manning's $n = 0.015$ and bottom slope $= 0.0015$. Find the rate of discharge and normal depth for critical flow conditions.

46. A 90° triangular flume is built of planed wood and carries water at a rate of 0.25 m^3/s. Find the critical depth and the critical slope.

47. A 10-ft-wide rectangular channel is made of planed wood and carries 100 gal/min of water. The channel has a bottom slope of 0.001. Find the normal depth and the critical depth.

48. What is the minimum water depth necessary for a 40-ft-wide stream to handle 4000 ft^3/sec if the flow is not supercritical?

49. Compute the discharge that would produce a critical flow depth of 4 ft in a 2204-ft-wide rectangular channel.

50. Compute the width of a smooth contraction needed to produce a critical flow depth of 1.4 m in a rectangular channel having a discharge of 52 m^3/s.

51. Water flows from a reservoir to a free outfall through a clean, long, natural trapezoidal channel with a bed slope of 39.6 ft/mi. The cross section has a base width of 38 ft and side slopes of 2:1. The flow produces a flow depth of 11.75 ft at the outfall. Find the discharge in the channel and the elevation of the reservoir with respect to the outfall crest.

52. Compute the flow required to produce critical flow in a 48-in.-diameter pipe flowing half full.

53. Compute the critical flow depth in a 54-in.-diameter pipe having a discharge of 131.5 ft^3/sec.

54. What pipe diameter yields a critical flow depth that is 25% of the diameter for a discharge of 8.4 ft^3/sec.

55. Compute the discharge that produces a critical flow depth of 10 ft in a trapezoidal channel having side slopes 2:1 and a bed width of 268.3 ft.

56. Compute the critical flow depth for a discharge of 150.4 m^3/s in a trapezoidal channel having side slopes 3:1 and a bed width of 7.4 m.

57. A channel has a rectangular cross section, a width of 40 m, and a flow rate of 4000 m^3/s. The normal water depth is 20 m. The flow then encounters a 4.0-m-high dam. Find the water depth directly above the dam if the flow is critical. Assume frictionless flow.

58. A rectangular channel has a gradual contraction in width from 59 ft to 30 ft and a bed level drop of 6 in. below the upstream channel bed, which the increased velocity caused by the contraction scoured. If the upstream and downstream flow depths are both 10.1 ft, compute the discharge in the channel.

59. A rectangular channel has a gradual contraction in width from 62 ft to 35 ft and a bed level drop of 9 in. below the upstream channel bed, which the increased velocity caused by the contraction scoured. If the upstream and downstream flow velocities are 3.2 and 4.6 ft/sec, respectively, compute the discharge in the channel.

60. A flow of 1376 ft^3/sec passes under an open sluice gate in a rectangular channel having a gradual contraction in width from 80 ft to 52 ft. The channel bed has scoured to a level that is 9 in. below the upstream channel bed, owing to the increased velocity caused by the sluice gate and the contraction. If the upstream flow depth is 4 ft, compute the depth and velocity of the flow at the end of the contraction.

61. A flow of 873 ft^3/sec passes under a sluice gate in a rectangular channel having a gradual contraction in width from 80 ft to 52 ft. The channel bed has scoured to a level that is 6 in. below the upstream channel bed, owing to the increased velocity caused by the sluice gate and the contraction. If the upstream flow depth is 4 ft and the sluice gate opening is 1.5 ft, compute the depth, velocity, and Froude number of the flow at the end of the contraction.

• **62.** A hydraulic engineer wants to analyze a steady flow situation featuring a small timber bridge crossing. A single row of 1-ft-diameter wooden piers spans the 200-ft-wide river at a location where the channel bed is known to drop 1 ft. The upstream flow velocity is 11.4 ft/sec, and the velocity and depth downstream from the piers are 10.6 ft/sec and 5.9 ft, respectively. For this short transition reach, compute:

(a) The upstream flow depth and Froude number;

(b) The direction and magnitude of the resultant force on the fluid due to the piers and bed drop;

(c) The head loss in the transition;

(d) The number of 1-ft-diameter piers that would restrict this flow.

63. A channel has a rectangular cross section, a width of 40 m, and a flow rate of 4000 m^3/s. The water depth is 20 m. The channel bottom rises enough for the flow to become critical. The bottom then drops to the original level, and the flow encounters a 4.0-m-high dam. Find the water depth directly above the dam and the normal depth upstream of the dam.

64. Consider 100 ft^3/sec of water flowing down a rectangular channel measuring 10 ft wide. The normal depth is 3.00 ft. A 4.0-ft-diameter pier is located in the channel. Find the water depth as it flows past the pier. Assume frictionless flow.

65. Subcritical, uniform flow of water occurs in a 5.0-m-wide, rectangular, horizontal channel. The flow has a depth of 2.0 m and a flow rate of 5.0 m^3/s. The water flow encounters a 0.25-m rise in the channel bottom. Find the new normal depth. Is the flow above the rise subcritical, critical, or supercritical? Assume frictionless flow.

66. Subcritical uniform flow of water occurs in a 5.0-m-wide, rectangular, horizontal channel. The flow has a depth of 1.5 m and a flow rate of 15.0 m^3/s. Find the rise in the channel bottom for the Froude number above the rise to be 1.0. Assume frictionless flow.

67. Supercritical, uniform flow of water occurs in a 5.0-m-wide, rectangular, horizontal channel. The flow has a depth of 1.5 m and a flow rate of 45.0 m^3/s. The water flow encounters a 0.25-m rise in the channel bottom. Find the normal depth after the rise in the channel bottom. Is the flow after the rise subcritical, critical, or supercritical? Assume frictionless flow.

68. The flow in Problem 67 encounters a 0.25-m drop in the channel rather than a rise. Find the new normal depth. Is the flow after the drop subcritical, critical, or supercritical?

69. The higher velocities of the upper course of the river in Problem 18 carry sand and gravel along with the water and deposit them at the entrance of the middle course. This buildup is 120 ft high for a flow rate of 196, 100 ft^3/sec. Will a hydraulic jump occur at the entrance of the middle course?

70. A 5.0-m-wide channel has a slope of 0.004, a 8.0-m^3/s water flow rate, and a water depth 1.5 m after a hydraulic jump. Find the water depth before the jump.

71. A rectangular channel 3.0 m wide has a flow rate of 5.0 m^3/s with a normal depth of 0.50 m. The flow then encounters a dam that rises 0.25 m above the channel bottom. Will a hydraulic jump occur? Justify your answer.

72. Find the water depth over the dam and classify the flow in Problem 71.

73. An unfinished concrete channel 10 ft wide has a slope of 0.005 and carries a water flow rate of 120 ft^3/sec. The flow then encounters an obstruction that rises 0.50 ft above the channel bottom. Will a hydraulic jump occur? Justify your answer.

74. A 100-ft-wide unfinished concrete rectangular channel has a bottom slope of 0.0025 and a flow rate of 5000 ft^3/sec. An upstream sluice gate produces a minimum water depth of 1.5 ft in the channel. Will a hydraulic jump occur in the channel? Justify your answer. Find the power loss in a hydraulic jump if it should occur.

75. A 90° triangular flume has sides 2.0 m wide, a water flow rate of 1.0 m^3/s, and a depth of 0.50 m. Find the depth after a hydraulic jump and the power loss in the jump.

•**76.** A hydraulic engineer wants to analyze steady flow in a rectangular channel featuring a hydraulic jump immediately downstream from a sluice gate that is open to a vertical clearance of 3 ft. The flow depth upstream from the sluice gate is 8.7 ft, and the flow velocity beyond the sluice gate and prior to the hydraulic jump is 21.6 ft/sec. Assume that the flow upstream from the sluice gate is subcritical. Find:

(a) The discharge in the channel;

(b) The flow depth before and after the hydraulic jump;

(c) The flow velocities upstream from the sluice gate and beyond the hydraulic jump;

(d) The energy loss rate in the hydraulic jump;

(e) The force the sluice gate exerts on the fluid.
How does this compare with the force computed assuming a hydrostatic pressure distribution?

•**77.** Consider the flow down a prismatic channel having a trapezoidal cross section of base width b and top width $b + 2y \cos\theta \cot\phi$. The channel bottom makes an angle θ with the horizontal, and y is the vertical fluid depth. (See Fig. P11.77.) Show that

$$\frac{dy}{dx} = \frac{\tan\theta - (n^2Q^2/A^2R_h^{4/3})}{1 - \left[\dfrac{Q^2}{A^2 gy\cos^2\theta}\right]\left[\dfrac{b + 2y\cos\theta\cot\phi}{b + y\cos\theta\cot\phi}\right]}.$$

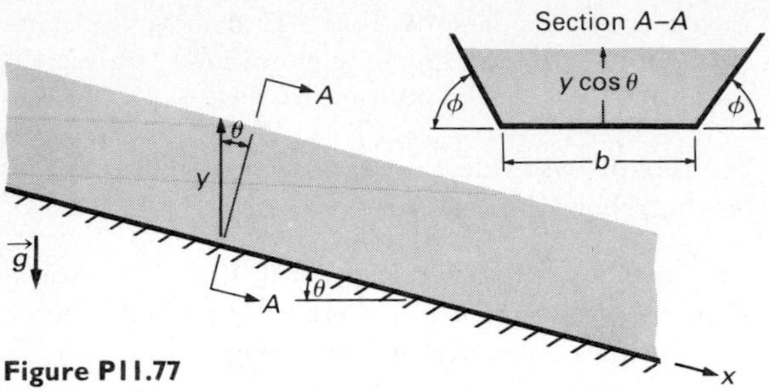

Figure P11.77

Note that the average fluid velocity is given by

$$V = \frac{Q}{A} = \frac{Q}{y \cos\theta(b + y \cos\theta \cot\phi)}.$$

Discuss the form of the equation for small values of $\theta (\theta \sim 1^\circ)$.

• **78.** Consider the flow down a prismatic channel having a rectangular cross section of width b. The channel bottom makes an angle θ with the horizontal. Show that

$$\frac{dy}{dx} = \frac{\tan\theta - (n^2Q^2)/(A^2R_h^{4/3})}{1 - Q^2/(A^2gy\cos^2\theta)},$$

where y is the vertical fluid depth, $A = by\cos\theta$, and Q is the volume flow rate. Discuss the form of the equation for small values of θ $(\theta \sim 1^\circ)$.

79. A rectangular channel is made of planed wood, has a bottom slope of 0.0005, and carries 55 ft^3/sec of water. At section 1 the depth is 2.5 ft, and at section 2 the depth is 2.0 ft. Which section is located upstream from the other? The channel width is 10 ft.

• **80.** A channel has a rectangular cross section, a slope $S_0 = 0.005$, a width of 50 m, and a water flow rate of 460 m^3/s. Find the uniform depth. A dam raises the water level 2.0 m. How far upstream from the high depth is the water level raised 0.5 m? Manning's $n = 0.014$.

• **81.** Consider an unfinished concrete ($n = 0.014$) drainage canal that has a trapezoidal cross section. The canal bottom is 15 ft wide and the angle ϕ is 60°. During a rainstorm the water in the canal rises to a level where a pump is activated to move water over a levee and into a lake. The canal has a bottom slope of 0.0008. The water flow rate is 500 ft^3/sec and the water depth at the pump inlet is 7.0 ft. Find the water depth $y(x)$ and classify the flow(s).

82. The river depth in most of the middle course in Problem 18 is 12.0 ft higher than at its entrance. Find the normal depth in the main section and the distance to attain 99% of the normal depth.

83. A rectangular earth channel is 10 ft wide and 16,000 ft long and has a bottom slope of 0.0015. The water depth is 2.0 ft high directly above a 3.0-ft-high dam at the end of the channel. The water flow rate is 300 ft^3/sec. Find the water depths at every 1000-ft interval upstream of the dam.

• **84.** A rectangular channel is 2.0 m wide and has a water flow rate of 3 m^3/s with a normal depth of 1.5 m. The channel makes a 90° bend in a horizontal plane. The radius of the inner side is 16.0 m and of the outer side 20.0 m. Find the difference in the water depths at the outer and inner sides.

• **85.** Bed elevations and channel widths are measured in a short, gradually converging rectangular channel with equally spaced sections:

Section	Channel width	Channel elevation
1	1102	100.00
2	875	102.00
3	568	100.66
4	553	100.00

Section 4 ends at a free overfall. For a discharge of 50,059 ft^3/sec, determine which section controls the flow, sketch the water surface profile throughout the channel, and show regions where the flow is subcritical and supercritical. Assume that no losses occur throughout the transition.

86. A gravelly, excavated earth channel is 100 ft wide and 15 ft deep, has a channel bed that drops 6.6 ft/mi, and ends at a free overfall. For a discharge of 10,000 ft^3/sec, estimate the distance upstream from the overfall that the depth becomes approximately uniform. Assume that the critical flow depth occurs at the overfall and compute the distance between the overfall and 95% of the uniform flow depth. How many steps, N, are required (using an equal depth increment) for the total upstream distance to vary less than 1% from the distance computed in the previous trial using $N - 1$ steps?

87. A 3-m deep, rectangular channel lined with unfinished concrete is used to convey water a distance of 300 m from a sluice gate to a free overfall. The channel is 6 m wide and has a bed slope that drops 1 m/km. The flow depth immediately downstream from the sluice gate is 0.4 m. Compute and plot the water surface profile between the sluice gate and free overfall for a discharge of 17 m^3/s.

88. Flow in a small laboratory flume over a triangular weir with a 90° vertex angle is used to demonstrate to high school students the principles of flow measurement. What head of water, H, is required to fill halfway a 430-gal tank in 30 sec?

89. Flow in a small laboratory flume over a full-span, 14.6-in.-wide rectangular weir is used to demonstrate to high school students the principles of flow measurement. How long will it take to fill halfway a 50-gal tank if the head of water, H, is 1.1 in. and the weir crest is 14 in. above the flume bottom?

REFERENCES

1. Chow, V. T., *Open Channel Hydraulics,* McGraw-Hill, New York, 1959.
2. Henderson, F. M., *Open Channel Flow,* Macmillan, New York, 1966.
3. Morris, H. M., and J. M. Wiggert, *Applied Hydraulics,* Ronald Press, New York, 1972.
4. White, F. M., *Fluid Mechanics,* McGraw-Hill, New York, 1979.
5. Kline, S. J. and S. McClintock, "Describing Uncertainties in Single Sample Experiments," *Mechanical Engineering,* January 1953.
6. "ASME Performance Test Code Supplement PTC 19.1—Measurement Uncertainty," American Society of Mechanical Engineers, 1985.
7. Thompson, P. A., *Compressible Fluid Dynamics,* Chapter 11, McGraw-Hill, New York, 1972.

Appendixes

Appendix A PHYSICAL PROPERTIES

Table A.1 Properties of the U.S. Standard Atmosphere (SI units)

Geometric Altitude z (m)	Temperature T (K)	Pressure p (Pa)	Density ρ (kg/m^3)	Gravitational Acceleration g (m/s^2)	Viscosity μ (N·s/m^2)	Kinematic Viscosity ν (m^2/s)
−5000	320.7	1.778 E5	1.931	9.822	1.942 E−5	1.006 E−5
−4000	314.2	1.596 E5	1.770	9.819	1.912 E−5	1.081 E−5
−3000	307.7	1.430 E5	1.619	9.816	1.882 E−5	1.163 E−5
−2000	301.2	1.278 E5	1.478	9.813	1.852 E−5	1.253 E−5
−1000	294.7	1.139 E5	1.347	9.810	1.821 E−5	1.352 E−5
0	288.2	1.013 E5	1.225	9.807	1.789 E−5	1.461 E−5
1000	281.7	8.988 E4	1.112	9.804	1.758 E−5	1.581 E−5
2000	275.2	7.950 E4	1.007	9.801	1.726 E−5	1.715 E−5
3000	268.7	7.012 E4	9.093 E−1	9.797	1.694 E−5	1.863 E−5
4000	262.2	6.166 E4	8.194 E−1	9.794	1.661 E−5	2.028 E−5
5000	255.7	5.405 E4	7.364 E−1	9.791	1.628 E−5	2.211 E−5
6000	249.2	4.722 E4	6.601 E−1	9.788	1.595 E−5	2.416 E−5
7000	242.7	4.111 E4	5.900 E−1	9.785	1.561 E−5	2.646 E−5
8000	236.2	3.565 E4	5.258 E−1	9.782	1.527 E−5	2.904 E−5
9000	229.7	3.080 E4	4.671 E−1	9.779	1.493 E−5	3.196 E−5
10000	223.3	2.650 E4	4.135 E−1	9.776	1.458 E−5	3.525 E−5
15000	216.7	1.211 E4	1.948 E−1	9.761	1.422 E−5	7.300 E−4
20000	216.7	5.529 E3	8.891 E−2	9.745	1.422 E−5	1.599 E−4
30000	226.5	1.197 E3	1.841 E−2	9.715	1.475 E−5	8.013 E−4
40000	250.4	2.871 E2	3.996 E−3	9.684	1.601 E−5	4.007 E−3
50000	270.7	7.978 E1	1.027 E−3	9.654	1.704 E−5	1.659 E−2
60000	255.8	2.246 E1	3.059 E−4	9.624	1.629 E−5	5.324 E−2
70000	219.7	5.520 E0	8.754 E−5	9.594	1.438 E−5	1.643 E−1
80000	180.7	1.037 E0	1.999 E−5	9.564	1.216 E−5	6.058 E−1
90000	180.7	1.644 E−1	3.170 E−6	9.535	1.216 E−5	3.837 E0

Table A.2 Properties of the U.S. Standard Atmosphere (EE & BG Units)

Geometric Altitude z (ft)	Temperature T (°R)	Pressure p (psia)	Density ρ (lbm/ft^3)	Density ρ (slug/ft^3)	Gravitational Acceleration g (ft/sec^2)	Viscosity μ (lb·sec/ft^2)	Kinematic Viscosity ν (ft^2/sec)
−15000	572.2	24.626	1.162 E−1	3.610 E−3	32.220	4.031 E−7	1.116 E−4
−10000	554.3	20.847	1.015 E−1	3.155 E−3	32.205	3.935 E−7	1.247 E−4
−5000	536.5	17.554	8.831 E−2	2.745 E−3	32.189	3.835 E−7	1.398 E−4
0	518.7	14.696	7.647 E−2	2.377 E−3	32.174	3.736 E−7	1.572 E−4
5000	500.8	12.054	6.590 E−2	2.048 E−3	32.159	3.636 E−7	1.776 E−4
10000	483.0	10.108	5.648 E−2	1.756 E−3	32.143	3.534 E−7	2.013 E−4
15000	465.2	8.297	4.814 E−2	1.496 E−3	32.128	3.431 E−7	2.293 E−4
20000	447.4	6.759	4.077 E−2	1.267 E−3	32.112	3.326 E−7	2.623 E−4
25000	429.6	5.461	3.431 E−2	1.066 E−3	32.097	3.217 E−7	3.017 E−4
30000	411.8	4.373	2.866 E−2	8.907 E−4	32.082	3.107 E−7	3.488 E−4
35000	394.1	3.468	2.375 E−2	7.382 E−4	32.066	2.995 E−7	4.057 E−4
40000	390.0	2.730	1.890 E−2	5.873 E−4	32.051	2.969 E−7	5.057 E−4
45000	390.0	2.149	1.487 E−2	4.623 E−4	32.036	2.969 E−7	6.423 E−4
50000	390.0	1.692	1.171 E−2	3.639 E−4	32.020	2.969 E−7	8.159 E−4
55000	390.0	1.332	9.219 E−3	2.865 E−4	32.005	2.969 E−7	1.036 E−3
60000	390.0	1.049	7.259 E−3	2.256 E−4	31.990	2.969 E−7	1.316 E−3
65000	390.0	0.826	5.716 E−3	1.777 E−4	31.974	2.969 E−7	1.671 E−3
70000	392.2	0.651	4.479 E−3	1.392 E−4	31.959	2.983 E−7	2.143 E−3
75000	395.0	0.514	3.511 E−3	1.091 E−4	31.944	3.001 E−7	2.750 E−3
80000	397.7	0.406	2.758 E−3	8.571 E−5	31.929	3.018 E−7	3.521 E−3
85000	400.4	0.322	2.170 E−3	6.743 E−5	31.913	3.035 E−7	4.501 E−3
90000	403.1	0.255	1.710 E−3	5.315 E−5	31.898	3.052 E−7	5.743 E−3
95000	405.8	0.203	1.350 E−3	4.196 E−5	31.883	3.070 E−7	7.316 E−3
100000	408.6	0.162	1.068 E−3	3.318 E−5	31.868	3.087 E−7	9.302 E−3
150000	479.1	0.020	1.112 E−4	3.456 E−6	31.716	3.512 E−7	1.016 E−1
200000	457.0	0.003	1.696 E−5	5.270 E−7	31.566	3.382 E−7	6.416 E−1
250000	351.8	0.000	2.263 E−6	7.034 E−8	31.42	2.721 E−7	3.868 E0
300000	332.9	0.000	1.488 E−7	4.625 E−9	31.27	2.593 E−7	5.608 E1

Table A.3 Physical properties of dry air and common gases at atmospheric pressure (SI units)

Gas	Temperature T (°C)	Density ρ (kg/m^3)	Absolute Viscosity μ (N·s/m^2)	Kinematic Viscosity ν (m^2/s)	Gas Constant R (N·m/kg·K)	Constant Pressure Specific Heat c_p (N·m/kg·K)	Constant Volume Specific Heat c_v (N·m/kg·K)	Specific Heat Ratio k
Air	0	1.290	1.71 E−5	1.33 E−5	287.0	1003.6	716.4	1.40
	50	1.090	1.95 E−5	1.79 E−5	287.0	—	—	1.40
	100	0.946	2.17 E−5	2.30 E−5	287.0	1010.3	723.1	1.40
	150	0.835	2.38 E−5	2.85 E−5	287.0	—	—	1.40
	200	0.746	2.57 E−5	3.45 E−5	287.0	1024.5	737.3	1.39
	250	0.675	2.75 E−5	4.08 E−5	287.0	—	—	1.39
	300	0.616	2.93 E−5	4.75 E−5	287.0	1044.6	757.8	1.38
	400	0.525	3.25 E−5	6.20 E−5	287.0	1068.5	781.3	1.37
	500	0.457	3.55 E−5	7.77 E−5	287.0	1092.3	805.1	1.36
Hydrogen	0	0.0900	8.4 E−6	9.33 E−5	4122.0	14194.9	10070.5	1.41
	100	0.0659	1.03 E−5	1.56 E−4	4122.0	14448.2	10323.8	1.40
	200	0.0520	1.21 E−5	2.33 E−4	4122.0	14504.3	10379.9	1.40
	300	0.0429	1.39 E−5	3.24 E−4	4122.0	14533.2	10408.8	1.40
	400	0.0365	1.54 E−5	4.22 E−4	4122.0	14508.9	10456.5	1.39
	500	0.0318	1.69 E−5	5.31 E−4	4122.0	14662.2	10537.8	1.39
Oxygen	0	1.430	1.92 E−5	1.34 E−5	259.8	914.8	654.8	1.40
	100	1.050	2.44 E−5	2.32 E−5	259.8	933.7	673.7	1.39
	200	0.824	2.90 E−5	3.52 E−5	259.8	963.0	703.0	1.37
	400	0.579	3.69 E−5	6.37 E−5	259.8	1023.7	763.7	1.34
	600	0.447	4.35 E−5	9.73 E−5	259.8	1068.9	808.9	1.32
	800	0.363	4.93 E−5	1.36 E−4	259.8	1099.9	840.3	1.31
Nitrogen	0	1.250	1.66 E−5	1.33 E−5	296.7	1039.2	742.3	1.40
	100	0.915	2.08 E−5	2.27 E−5	296.7	1042.1	745.3	1.40
	200	0.722	2.46 E−5	3.41 E−5	296.7	1051.7	755.3	1.39
	400	0.507	3.11 E−5	6.13 E−5	296.7	1091.5	794.7	1.37
	600	0.391	3.66 E−5	9.36 E−5	296.7	1139.2	842.4	1.35
	800	0.318	4.13 E−5	1.30 E−4	296.7	1181.5	884.7	1.34

Table A.3 continued.

Gas	Temperature T (°C)	Density ρ (kg/m^3)	Absolute Viscosity μ (N·s/m^2)	Kinematic Viscosity ν (m^2/s)	Gas Constant R (N·m/kg·K)	Constant Pressure Specific Heat c_p (N·m/kg·K)	Constant Volume Specific Heat c_v (N·m/kg·K)	Specific Heat Ratio k
Helium	0	0.178	1.86 E−5	1.04 E−4	2079.0	5229.0	3137.4	1.67
	100	0.131	2.29 E−5	1.75 E−4	2079.0	5229.0	3137.4	1.67
	200	0.103	2.70 E−5	2.62 E−4	2079.0	5229.0	3137.4	1.67
	300	0.0850	3.07 E−5	3.61 E−4	2079.0	5229.0	3137.4	1.67
	400	0.0724	3.42 E−5	4.72 E−4	2079.0	5229.0	3137.4	1.67
	600	0.0558	4.07 E−5	7.29 E−4	2079.0	5229.0	3137.4	1.67
	800	0.0454	4.65 E−5	1.02 E−3	2079.0	5229.0	3137.4	1.67
Carbon dioxide	0	1.96	1.38 E−5	7.04 E−6	188.8	814.8	625.9	1.30
	100	1.44	1.85 E−5	1.28 E−5	188.8	913.6	724.7	1.26
	200	1.13	2.68 E−5	2.37 E−5	188.8	992.7	803.9	1.23
	300	0.936	—	—	188.8	1056.7	867.9	1.22
	400	0.797	—	—	188.8	1110.3	921.1	1.21
	600	0.615	—	—	188.8	1192.0	1002.7	1.19

Table A.4 Physical properties of dry air and common gases at atmospheric pressure (EE & BG Units)

Gas	Temp. T (°F)	Density ρ (lbm/ft^3)	Density ρ (slug/ft^3)	Absolute Viscosity μ (lb·sec/ft^2)	Kinematic Viscosity ν (ft^2/sec)	Gas Const. R (ft·lb/lbm·°R)	Gas Const. R (ft·lb/slug·°R)	Const. Press Spec. Heat c_p (ft·lb/lbm·°R)	Const. Press Spec. Heat c_p (ft·lb/slug·°R)	Spec. Heat Ratio k
Air	0	0.0864	2.68 E−3	3.58 E−7	1.33 E−4	53.35	1716.	185.9	5981.	1.40
	32	0.081	2.50 E−3	3.61 E−7	1.44 E−4	53.35	1716.	186.7	6007.	1.40
	100	0.071	2.20 E−3	3.99 E−7	1.81 E−4	53.35	1716.	186.7	6007.	1.40
	200	0.0602	1.87 E−3	4.48 E−7	2.40 E−4	53.35	1716.	187.5	6033.	1.40
	400	0.0462	1.43 E−3	5.38 E−7	3.75 E−4	53.35	1716.	190.6	6132.	1.39
	600	0.0375	1.17 E−3	6.21 E−7	5.33 E−4	53.35	1716.	194.5	6258.	1.38
	800	0.0316	9.82 E−4	6.94 E−7	7.07 E−4	53.35	1716.	199.2	6409.	1.37
	1000	0.0272	8.45 E−4	7.63 E−7	9.07 E−4	53.35	1716.	203.8	6557.	1.36
Hydrogen	0	0.0059	1.80 E−4	1.76 E−7	9.61 E−4	766.0	24660.	2634.0	84750.	1.41
	100	0.0049	1.50 E−4	1.93 E−7	1.27 E−3	766.0	24660.	2661.0	85620.	1.40
	200	0.00415	1.29 E−4	2.14 E−7	1.66 E−3	766.0	24660.	2684.0	86360.	1.40
	400	0.0032	9.90 E−5	2.56 E−7	2.58 E−3	766.0	24660.	2692.0	86610.	1.40
	600	0.0026	8.10 E−5	2.95 E−7	3.65 E−3	766.0	24660.	2700.0	86870.	1.40
	800	0.0021	6.50 E−5	3.40 E−7	5.21 E−3	766.0	24660.	2707.0	87100.	1.39
	1000	0.00186	5.78 E−5	3.63 E−7	6.28 E−3	766.0	24660.	2707.0	87100.	1.39
Oxygen	0	0.0955	2.97 E−3	3.77 E−7	1.27 E−4	48.3	1553.	170.0	5470.	1.40
	100	0.0785	2.44 E−3	4.41 E−7	1.81 E−4	48.3	1553.	171.2	5508.	1.39
	200	0.0666	2.07 E−3	5.00 E−7	2.42 E−4	48.3	1553.	173.3	5576.	1.39
	400	0.0511	1.59 E−3	6.07 E−7	3.82 E−4	48.3	1553.	179.3	5769.	1.37
	600	0.0415	1.29 E−3	7.02 E−7	5.45 E−4	48.3	1553.	185.9	5981.	1.35
	800	0.0349	1.08 E−3	7.86 E−7	7.20 E−4	48.3	1553.	191.8	6171.	1.34
	1000	0.0301	9.04 E−4	8.63 E−7	9.23 E−4	48.3	1553.	196.7	6329.	1.33
Nitrogen	0	0.0840	2.61 E−3	3.28 E−7	1.26 E−4	55.2	1774.	192.8	6203.	1.40
	100	0.0690	2.14 E−3	3.80 E−7	1.77 E−4	55.2	1774.	193.3	6219.	1.40
	200	0.0585	1.82 E−3	4.29 E−7	2.36 E−4	55.2	1774.	193.7	6232.	1.40
	400	0.0449	1.40 E−3	5.16 E−7	3.70 E−4	55.2	1774.	195.7	6296.	1.39
	600	0.0364	1.13 E−3	5.95 E−7	5.26 E−4	55.2	1774.	199.5	6419.	1.38
	800	0.0306	9.51 E−4	6.66 E−7	7.01 E−4	55.2	1774.	204.1	6567.	1.37
	1000	0.0264	8.21 E−4	7.31 E−7	8.91 E−4	55.2	1774.	209.2	6731.	1.36

Table A.4 continued.

Gas	Temp. T (°F)	Density ρ (lbm/ft^3)	Density ρ (slug/ft^3)	Absolute Viscosity μ (lb·sec/ft^2)	Kinematic Viscosity ν (ft^2/sec)	Gas Const. R (ft·lb/lbm·°R)	Gas Const. R (ft·lb/slug·°R)	Const. Press Spec. Heat c_p (ft·lb/lbm·°R)	Const. Press Spec. Heat c_p (ft·lb/slug·°R)	Spec. Heat Ratio k
Helium	0	0.012	3.70 E−4	3.74 E−7	1.00 E−3	386.0	12420.	972.5	31290.	1.67
	100	0.0098	3.04 E−4	—	—	386.0	12420.	972.5	31290.	1.67
	200	0.0083	2.60 E−4	4.70 E−7	1.82 E−4	386.0	12420.	972.5	31290.	1.67
	400	0.0064	2.00 E−4	5.69 E−7	2.86 E−4	386.0	12420.	972.5	31290.	1.67
	600	0.0051	1.60 E−4	6.64 E−7	4.19 E−3	386.0	12420.	972.5	31290.	1.67
	800	0.0044	1.40 E−4	7.59 E−7	5.55 E−3	386.0	12420.	972.5	31290.	1.67
	1000	0.0037	1.20 E−4	8.54 E−7	7.42 E−3	386.0	12420.	972.5	31290.	1.67
Carbon dioxide	0	0.132	4.10 E−3	2.67 E−7	6.51 E−5	35.1	1129.	147.8	4755.	1.31
	100	0.108	3.86 E−3	3.26 E−7	9.72 E−5	35.1	1129.	157.9	5080.	1.29
	200	0.0915	2.84 E−3	3.74 E−7	1.32 E−4	35.1	1129.	169.6	5457.	1.26
	400	0.0702	2.18 E−3	4.73 E−7	2.17 E−4	35.1	1129.	185.2	5959.	1.23
	600	0.0570	1.77 E−3	5.62 E−7	3.17 E−4	35.1	1129.	198.7	6393.	1.21
	800	0.0480	1.49 E−3	6.38 E−7	4.28 E−4	35.1	1129.	208.8	6718.	1.20
	1000	0.0415	1.29 E−3	7.13 E−7	5.53 E−4	35.1	1129.	217.3	6991.	1.19

Table A.5 Physical properties of ordinary water and common liquids (SI units)

Liquid	Temperature T (°C)	Density ρ (kg/m^3)	Specific Gravity S	Absolute Viscosity μ (N·s/m^2)	Kinematic Viscosity ν (m^2/s)	Surface Tension σ (N/m)	Isothermal Bulk Modulus of Elasticity E_v (N/m^2)	Coefficient of Thermal Expansion α_T (K^{-1})
Water	0	1000	1.000	1.79 E−3	1.79 E−6	7.56 E−2	1.99 E9	6.80 E−5
	10	1000	1.000	1.31 E−3	1.31 E−6	7.42 E−2	2.12 E9	8.80 E−5
	20	998	0.998	1.00 E−3	1.00 E−6	7.28 E−2	2.21 E9	2.07 E−4
	30	996	0.996	7.98 E−4	8.01 E−7	7.12 E−2	2.26 E9	2.94 E−4
	40	992	0.992	6.53 E−4	6.58 E−7	6.96 E−2	2.29 E9	3.85 E−4
	50	988	0.988	5.47 E−4	5.48 E−7	6.79 E−2	2.29 E9	4.58 E−4
	60	983	0.983	4.67 E−4	4.75 E−7	6.62 E−2	2.28 E9	5.23 E−4
	70	978	0.978	4.04 E−4	4.13 E−7	6.64 E−2	2.24 E9	5.84 E−4
	80	972	0.972	3.55 E−4	3.65 E−7	6.26 E−2	2.20 E9	6.41 E−4
	90	965	0.965	3.15 E−4	3.26 E−7	—	2.14 E9	6.96 E−4
	100	958	0.958	2.82 E−4	2.94 E−7	5.89 E−2	2.07 E9	7.50 E−9
Mercury	0	13600	13.60	1.68 E−3	1.24 E−7	—	2.50 E10	—
	4	13590	13.59	—	—	—	—	—
	20	13550	13.55	1.55 E−3	1.14 E−7	37.5 E−2	2.50 E10	1.82 E−4
	40	13500	13.50	1.45 E−3	1.07 E−7	—	—	1.82 E−4
	60	13450	13.45	1.37 E−3	1.02 E−7	—	—	1.82 E−4
	80	13400	13.40	1.30 E−3	9.70 E−8	—	—	1.83 E−4
	100	13350	13.35	1.24 E−3	9.29 E−8	—	—	—
Ethylene glycol	0	—	—	5.70 E−2	—	—	—	—
	20	1110	1.11	1.99 E−2	1.79 E−5	—	—	—
	40	1110	1.10	9.13 E−3	8.30 E−6	—	—	—
	60	1090	1.09	4.95 E−3	4.54 E−6	—	—	—
	80	1070	1.07	3.02 E−3	2.82 E−6	—	—	—
	100	1060	1.06	1.99 E−3	1.88 E−6	—	—	—
Methyl alcohol (methanol)	0	810	0.810	8.17 E−4	1.01 E−6	2.45 E−2	9.35 E8	—
	10	801	0.801	—	—	2.26 E−2	8.78 E8	—
	20	792	0.792	5.84 E−4	7.37 E−7	—	8.23 E8	—
	30	783	0.783	5.10 E−4	6.51 E−7	—	7.72 E8	—
	40	774	0.774	4.50 E−4	5.81 E−7	—	7.23 E8	—
	50	765	0.765	3.96 E−4	5.18 E−7	—	6.78 E8	—

Table A.5 continued.

Liquid	Temperature T (°C)	Density ρ (kg/m^3)	Specific Gravity S	Absolute Viscosity μ (N·s/m^2)	Kinematic Viscosity ν (m^2/s)	Surface Tension σ (N/m)	Isothermal Bulk Modulus of Elasticity E_v (N/m^2)	Coefficient of Thermal Expansion α_T (K^{-1})
Ethyl	0	806	0.806	1.77 E−3	2.20 E−6	2.41 E−2	1.02 E9	—
alcohol	20	789	0.789	1.20 E−3	1.52 E−6	—	9.02 E8	—
(ethanol)	40	772	0.772	8.34 E−4	1.08 E−6	—	7.89 E8	—
	60	754	0.754	5.92 E−4	7.85 E−7	—	6.78 E8	—
Normal	0	718	0.718	7.06 E−4	9.83 E−7	—	1.00 E9	—
octane	16	—	—	5.74 E−4	—	—	—	—
	20	702	0.702	5.42 E−4	7.72 E−7	—	—	—
	25	—	—	—	—	—	8.35 E8	—
	40	686	0.686	4.33 E−4	6.31 E−7	—	7.48 E8	—
Benzene	0	900	0.900	9.12 E−4	1.01 E−6	3.02 E−2	1.23 E9	—
	20	879	0.879	6.52 E−4	7.42 E−7	2.76 E−2	1.06 E9	—
	40	858	0.857	5.03 E−4	5.86 E−7	—	9.10 E8	—
	60	836	0.836	3.92 E−4	4.69 E−7	—	7.78 E8	—
	80	815	0.815	3.29 E−4	4.04 E−7	—	6.48 E8	—
Kerosene	−18	841	0.841	7.06 E−3	8.40 E−6	—	—	—
	20	814	0.814	1.90 E−3	2.37 E−6	2.9 E−2	—	—
Lubricating	20	871	0.871	1.31 E−2	1.50 E−5	—	—	—
oil	40	858	0.858	6.81 E−3	7.94 E−6	—	—	—
	60	845	0.845	4.18 E−3	4.95 E−6	—	—	—
	80	832	0.832	2.83 E−3	3.40 E−6	—	—	—
	100	820	0.820	2.00 E−3	2.44 E−6	—	—	—
	120	809	0.809	1.54 E−3	1.90 E−6	—	—	—

Table A.6 Physical properties of ordinary water and common liquids (EE & BG Units)

Liquid	Temp. T (°F)	Density ρ (lbm/ft^3)	Density ρ (slug/ft^3)	Specific Gravity S	Absolute Viscosity μ (lb·sec/ft^2)	Kinematic Viscosity ν (ft^2/sec)	Surface Tension σ (lb/ft)	Isothermal Bulk Modulus of Elasticity E_v (lb/in^2)	Coefficient of Therm. Expansion α_T (°R^{-1})
Water	32	62.4	1.940	1.00	3.75 E−5	1.93 E−5	5.18 E−3	2.93 E5	2.03 E−3
	40	62.4	1.940	1.00	3.23 E−5	1.66 E−5	5.14 E−3	2.94 E5	—
	60	62.4	1.938	0.999	2.36 E−5	1.22 E−5	5.04 E−3	3.11 E5	—
	80	62.2	1.934	0.997	1.80 E−5	9.30 E−6	4.92 E−3	3.22 E5	—
	100	62.0	1.927	0.993	1.42 E−5	7.39 E−6	4.80 E−3	3.27 E5	1.7 E−3
	120	61.7	1.918	0.988	1.17 E−5	6.09 E−6	4.65 E−3	3.33 E5	—
	140	61.4	1.908	0.983	9.81 E−6	5.14 E−6	4.54 E−3	3.30 E5	—
	160	61.0	1.896	0.977	8.38 E−6	4.42 E−6	4.41 E−3	3.26 E5	—
	180	60.6	1.883	0.970	7.26 E−6	3.85 E−6	4.26 E−3	3.13 E5	—
	200	60.1	1.868	0.963	6.37 E−6	3.41 E−6	4.12 E−3	3.08 E5	1.52 E−3
	212	59.8	1.860	0.958	5.93 E−6	3.19 E−6	4.04 E−3	3.00 E5	—
Mercury	50	847	26.3	13.6	3.2 E−5	1.2 E−6	—	3.65 E6	1.0 E−4
	200	834	25.9	13.4	2.6 E−5	1.0 E−6	—	—	1.0 E−4
	300	826	25.7	13.2	2.3 E−5	9.0 E−7	—	—	—
	400	817	25.4	13.1	2.0 E−5	8.0 E−7	—	—	—
	600	802	24.9	12.8	1.7 E−5	7.0 E−7	—	—	—
Ethylene glycol	68	69.3	2.15	1.11	4.16 E−4	1.93 E−4	—	—	—
	104	68.7	2.14	1.10	1.91 E−4	8.93 E−5	—	—	—
	140	68.0	2.11	1.09	1.03 E−4	4.89 E−5	—	—	—
	176	66.8	2.08	1.07	6.31 E−5	3.04 E−5	—	—	—
	212	66.2	2.06	1.06	4.12 E−5	2.02 E−5	—	—	—
Methyl alcohol (methanol)	32	50.6	1.57	0.810	1.71 E−5	1.09 E−5	1.68 E−3	1.36 E5	—
	68	50.0	1.55	0.801	—	—	1.55 E−3	1.19 E5	—
	104	49.4	1.54	0.792	1.22 E−5	7.93 E−6	—	1.05 E5	—
	140	48.9	1.52	0.783	1.07 E−5	7.01 E−6	—	—	—
	176	48.3	1.50	0.774	9.40 E−6	6.25 E−6	—	—	—
	212	47.8	1.49	0.765	8.27 E−6	5.58 E−6	—	—	—

Table A.6 continued.

Liquid	Temp. T (°F)	Density ρ (lbm/ft^3)	Density ρ (slug/ft^3)	Specific Gravity S	Absolute Viscosity μ (lb·sec/ft^2)	Kinematic Viscosity ν (ft^2/sec)	Surface Tension σ (lb/ft)	Isothermal Bulk Modul. of Elastic. E_v (lb/in^2)	Coefficient of Therm. Expansion α_T (°R^{-1})
Ethyl alcohol (ethanol)	32	50.3	1.56	0.806	3.70 E−5	2.37 E−5	1.65 E−3	1.48 E5	—
	68	49.8	1.55	0.798	3.03 E−5	1.96 E−5	—	1.31 E5	—
	104	49.3	1.53	0.789	2.51 E−5	1.64 E−5	—	1.14 E5	—
	140	48.2	1.50	0.772	1.74 E−5	1.16 E−5	—	9.83 E4	—
	176	47.7	1.48	0.754	1.24 E−5	8.45 E−6	—	—	—
	212	47.1	1.46	0.745	—	—	—	—	
Normal octane	32	44.8	1.39	0.718	1.47 E−5	1.06 E−5	—	1.45 E5	—
	68	43.8	1.36	0.702	1.13 E−5	8.31 E−6	—	—	—
	104	42.8	1.33	0.686	9.04 E−6	6.79 E−6	—	1.08 E5	—
Benzene	32	56.2	1.75	0.900	1.90 E−5	1.09 E−5	2.07 E−3	1.78 E5	—
	68	54.9	1.71	0.879	1.36 E−5	7.99 E−6	1.89 E−3	1.53 E5	—
	104	53.6	1.67	0.858	1.05 E−5	6.31 E−6	—	1.32 E5	—
	140	52.2	1.62	0.836	8.19 E−6	5.05 E−6	—	1.13 E5	—
	176	50.9	1.58	0.815	6.87 E−6	4.35 E−6	—	9.40 E4	—
Kerosene	0	52.5	1.63	0.841	1.48 E−4	9.05 E−5	—	—	—
	77	50.8	1.58	0.814	3.97 E−5	2.55 E−5	—	—	—
Lubricating oil	68	54.4	1.69	0.871	2.74 E−4	1.61 E−4	—	—	—
	104	53.6	1.67	0.858	1.42 E−4	8.55 E−4	—	—	—
	140	52.6	1.63	0.845	8.73 E−5	5.33 E−5	—	—	—
	176	51.9	1.61	0.832	5.91 E−5	3.66 E−5	—	—	—
	212	51.2	1.59	0.820	4.18 E−5	2.63 E−5	—	—	—
	248	50.5	1.57	0.809	3.22 E−5	2.05 E−5	—	—	—

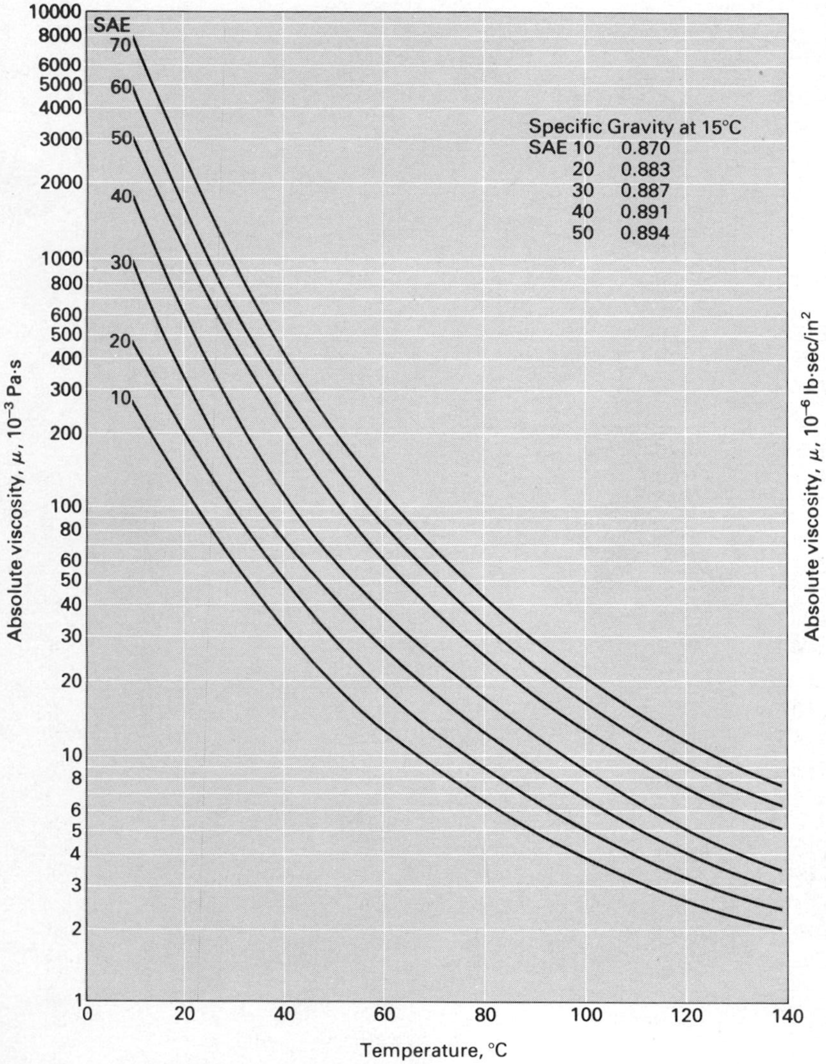

Figure A.1 Absolute Viscosity of SAE Oils (SI Units).

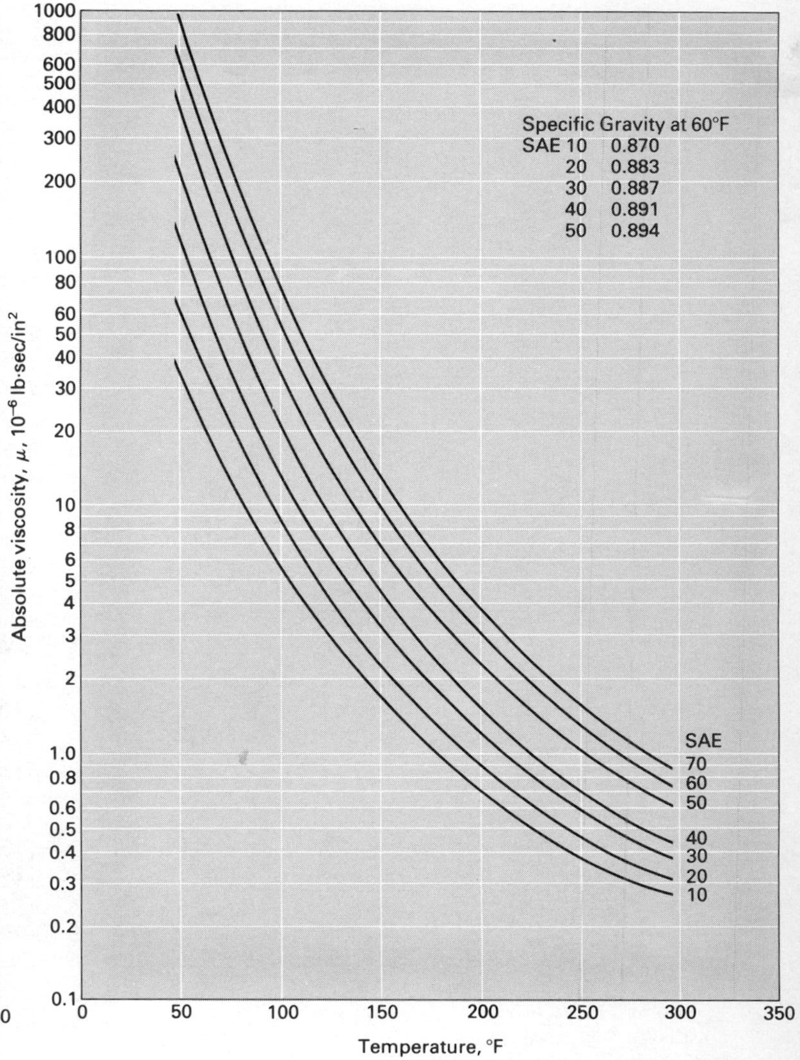

Figure A.2 Absolute Viscosity of SAE Oils (EE & BG Units).

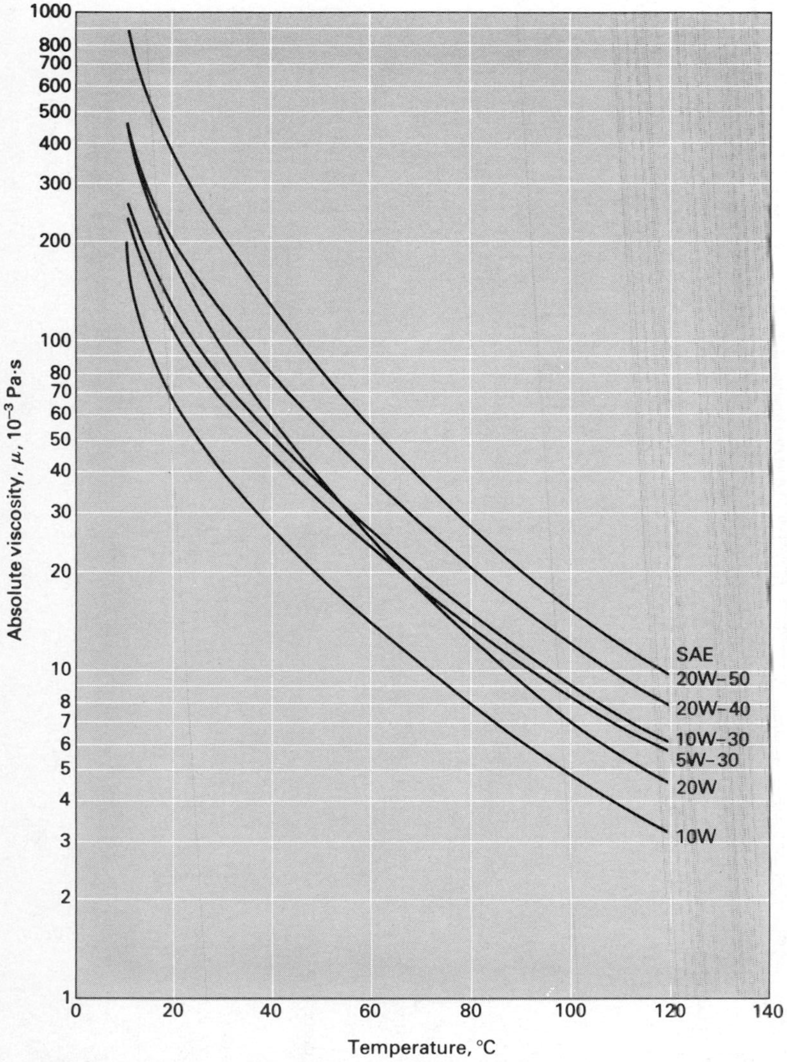

Figure A.3 Absolute Viscosity of SAE Multiviscosity Oils (SI Units).

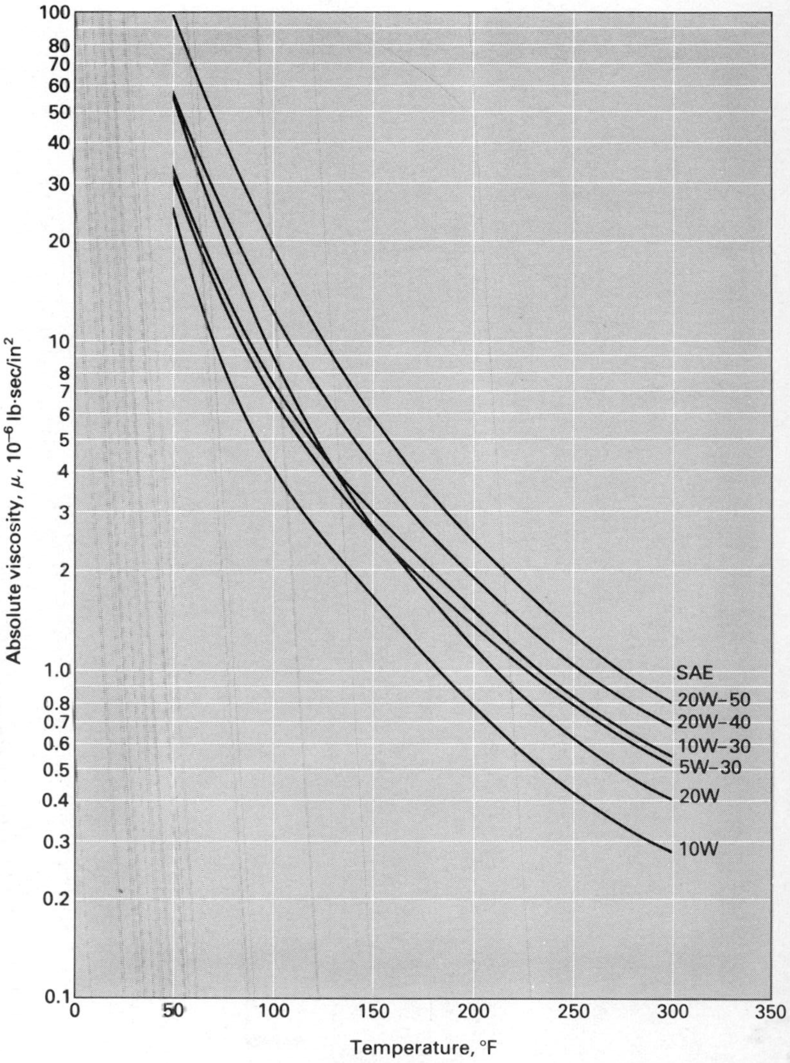

Figure A.4 Absolute Viscosity of SAE Multiviscosity Oils (EE & BG Units).

Appendix B PROPERTIES OF COMMON GEOMETRIC AREAS

Shape	Area	x_c	y_c	I_{xc}	I_{yc}	I_{xyc}
Rectangle 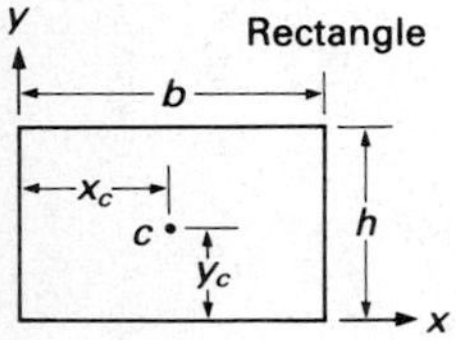	bh	$\frac{b}{2}$	$\frac{h}{2}$	$\frac{bh^3}{12}$ $I_x = \frac{bh^3}{3}$	$\frac{hb^3}{12}$ $I_y = \frac{hb^3}{3}$	0 $I_{xy} = \frac{bh}{3}$ $(h^2 + b^2)$
Any triangle, any side 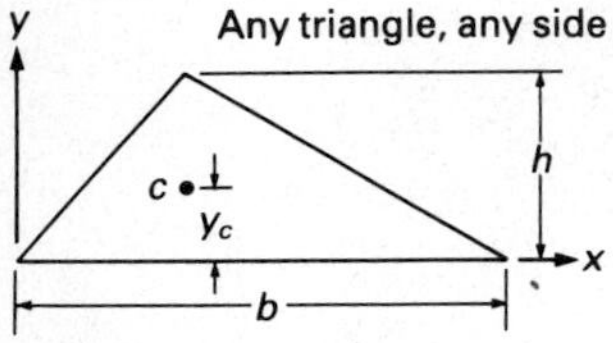	$\frac{bh}{2}$	—	$\frac{h}{3}$	$\frac{bh^3}{36}$ $I_x = \frac{bh^3}{12}$	—	—
Circle 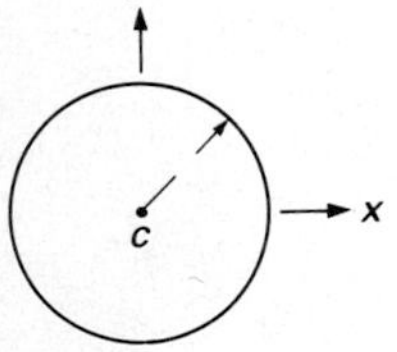	πr^2	0	0	$\frac{\pi r^4}{4}$	$\frac{\pi r^4}{4}$	0
Semicircle 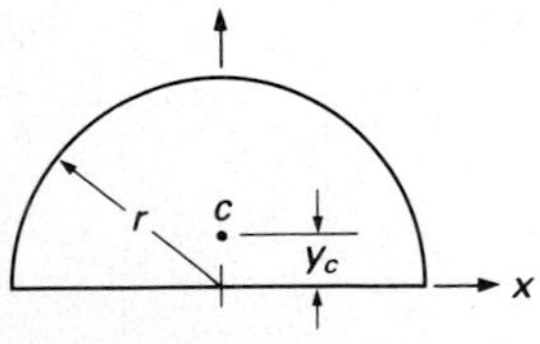	$\frac{\pi r^2}{2}$	0	$\frac{4r}{3\pi}$	$\frac{\pi r^4}{8}\left(1 - \frac{64}{9\pi^2}\right)$ $= 0.110r^4$	$\frac{\pi r^4}{8}$ $= 0.3927r^4$	0
Quarter circle 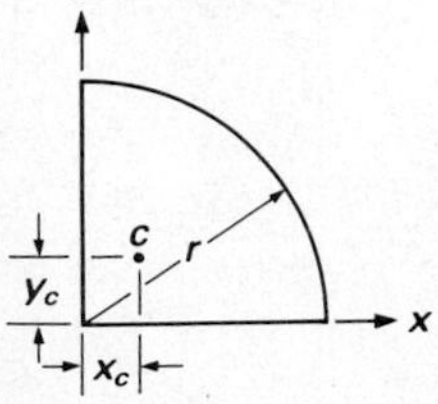	$\frac{\pi r^2}{4}$	$\frac{4r}{3\pi}$	$\frac{4r}{3\pi}$	$\frac{\pi r^4}{16}\left(1 - \frac{64}{9\pi^2}\right)$ $= 0.0549r^4$	$\frac{\pi r^4}{16}\left(1 - \frac{64}{9\pi^2}\right)$ $= 0.0549r^4$	—

Shape	Area	x_c	y_c	I_{xc}	I_{yc}	I_{xyc}
Circular segment 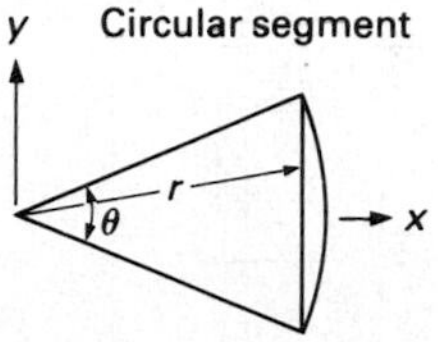	$\frac{r^2}{2}(\theta - \sin\theta)$	—	—	—	—	—
Circular sector 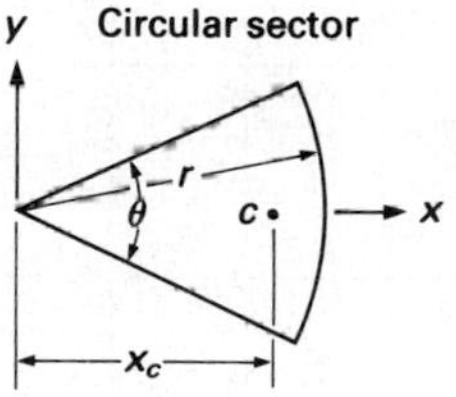	θr^2	$\frac{2r\sin\theta}{3\theta}$	0	$\frac{r^4}{16}\left(\theta - \frac{\sin 2\theta}{2}\right)$	—	0
Ellipse 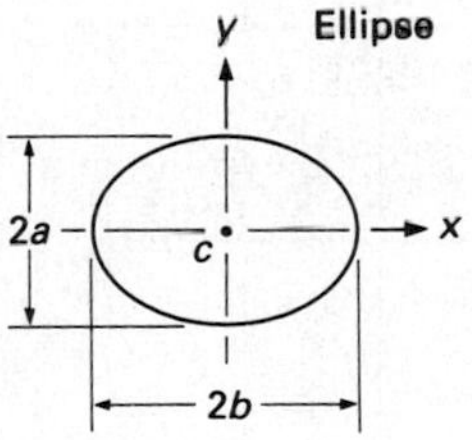	πab	0	0	$\frac{\pi ba^3}{4}$	$\frac{\pi ab^3}{4}$	0
Semiellipse 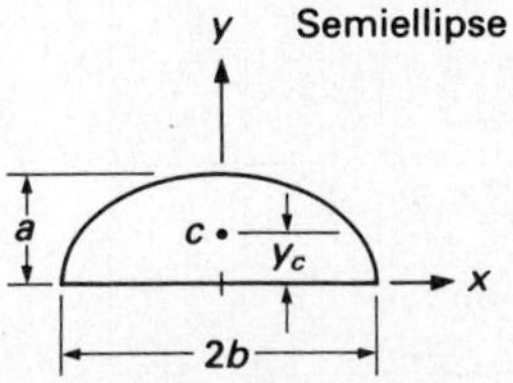	$\frac{\pi ab}{2}$	0	$\frac{4b}{3\pi}$	—	$\frac{\pi ab^3}{8}$	0

Appendix C CONVERSION FACTORS

Table C.2 Length conversion factors

	Angstrom (Å)	Micron (μ)	Mil	Millimeter (mm)	Centimeter (cm)	Inch (in.)	Foot (ft)	Yard (yd)	Meter (m)	Fathom (fath)	Rod (rd)	Statute mile (mi)	Int'l nautical mile (mi)
Angstrom (Å)													
Micron (μ)	$10^4 \frac{\text{Å}}{\mu}$												
Mil	$2.54\text{E}5 \frac{\text{Å}}{\text{mil}}$	$25.4 \frac{\mu}{\text{mil}}$											
Millimeter (mm)	$1.0\text{E}7 \frac{\text{Å}}{\text{mm}}$	$1000 \frac{\mu}{\text{mm}}$	$39.37 \frac{\text{mil}}{\text{mm}}$										
Centimeter (cm)	$1.0\text{E}8 \frac{\text{Å}}{\text{cm}}$	$1.0\text{E}4 \frac{\mu}{\text{cm}}$	$393.7 \frac{\text{mil}}{\text{cm}}$	$10 \frac{\text{mm}}{\text{cm}}$									
Inch (in.)	$2.54\text{E}8 \frac{\text{Å}}{\text{in.}}$	$2.54\text{E}4 \frac{\mu}{\text{in.}}$	$1000 \frac{\text{mil}}{\text{in.}}$	$25.4 \frac{\text{mm}}{\text{in.}}$	$2.54 \frac{\text{cm}}{\text{in.}}$								
Foot (ft)	$3.048\text{E}9 \frac{\text{Å}}{\text{ft}}$	$3.048\text{E}5 \frac{\mu}{\text{ft}}$	$1.2\text{E}5 \frac{\text{mil}}{\text{ft}}$	$304.8 \frac{\text{mm}}{\text{ft}}$	$30.48 \frac{\text{cm}}{\text{ft}}$	$12 \frac{\text{in.}}{\text{ft}}$							
Yard (yd)	$9.144\text{E}9 \frac{\text{Å}}{\text{yd}}$	$9.144\text{E}5 \frac{\mu}{\text{yd}}$	$3.6\text{E}5 \frac{\text{mil}}{\text{yd}}$	$914.4 \frac{\text{mm}}{\text{yd}}$	$9.144 \frac{\text{cm}}{\text{yd}}$	$36 \frac{\text{in.}}{\text{yd}}$	$3 \frac{\text{ft}}{\text{yd}}$						
Meter (m)	$1.0\text{E}10 \frac{\text{Å}}{\text{m}}$	$1.0\text{E}6 \frac{\mu}{\text{m}}$	$2.54\text{E}5 \frac{\text{mil}}{\text{m}}$	$1000 \frac{\text{mm}}{\text{m}}$	$100 \frac{\text{cm}}{\text{m}}$	$39.37 \frac{\text{in.}}{\text{m}}$	$3.281 \frac{\text{ft}}{\text{m}}$	$1.094 \frac{\text{yd}}{\text{m}}$					
Fathom (fath)	$1.829\text{E}10 \frac{\text{Å}}{\text{fath}}$	$1.829\text{E}6 \frac{\mu}{\text{fath}}$	$7.2\text{E}4 \frac{\text{mil}}{\text{fath}}$	$1.829\text{E}3 \frac{\text{mm}}{\text{fath}}$	$182.9 \frac{\text{cm}}{\text{fath}}$	$72 \frac{\text{in.}}{\text{fath}}$	$6 \frac{\text{ft}}{\text{fath}}$	$2 \frac{\text{yd}}{\text{fath}}$	$1.829 \frac{\text{m}}{\text{fath}}$				
Rod (rd)	$5.029\text{E}10 \frac{\text{Å}}{\text{rd}}$	$5.029\text{E}6 \frac{\mu}{\text{rd}}$	$1.98\text{E}5 \frac{\text{mil}}{\text{rd}}$	$5029 \frac{\text{mm}}{\text{rd}}$	$502.9 \frac{\text{cm}}{\text{rd}}$	$198 \frac{\text{in.}}{\text{rd}}$	$16.5 \frac{\text{ft}}{\text{rd}}$	$5.5 \frac{\text{yd}}{\text{rd}}$	$5.029 \frac{\text{m}}{\text{rd}}$	$2.75 \frac{\text{fath}}{\text{rd}}$			
Statute mile (mi)	$1.609\text{E}13 \frac{\text{Å}}{\text{mi}}$	$1.609\text{E}9 \frac{\mu}{\text{mi}}$	$6.336\text{E}7 \frac{\text{mil}}{\text{mi}}$	$1.609\text{E}6 \frac{\text{mm}}{\text{mi}}$	$1.609\text{E}5 \frac{\text{cm}}{\text{mi}}$	$6.336\text{E}4 \frac{\text{in.}}{\text{mi}}$	$5280 \frac{\text{ft}}{\text{mi}}$	$1760 \frac{\text{yd}}{\text{mi}}$	$1609 \frac{\text{m}}{\text{mi}}$	$879.7 \frac{\text{fath}}{\text{mi}}$	$320 \frac{\text{rd}}{\text{mi}}$		
Int'l nautical mile (mi)	$1.852\text{E}13 \frac{\text{Å}}{\text{mi}}$	$1.852\text{E}9 \frac{\mu}{\text{mi}}$	$7.291\text{E}7 \frac{\text{mil}}{\text{mi}}$	$1.852\text{E}6 \frac{\text{mm}}{\text{mi}}$	$1.852\text{E}5 \frac{\text{cm}}{\text{mi}}$	$7.291\text{E}4 \frac{\text{in.}}{\text{mi}}$	$6076.12 \frac{\text{ft}}{\text{mi}}$	$2026 \frac{\text{yd}}{\text{mi}}$	$1852 \frac{\text{m}}{\text{mi}}$	$1012.7 \frac{\text{fath}}{\text{mi}}$	$368.3 \frac{\text{rd}}{\text{mi}}$	$1.151 \frac{\text{st mi}}{\text{I.N. mi}}$	

Example: Convert 450.0 mils to centimeters.

Solution: Locate the *column* for the *smaller* unit (mil) and the *row* for the *larger* unit (cm). If unsure as to the smaller unit, try the second option if your first guess doesn't work. The intersection gives the conversion factor. In this case dividing by the conversion factor (equal to unity) cancels the mils, so

$$450.0 \text{ mil}\left(\frac{\text{cm}}{393.7 \text{ mil}}\right)$$
$$= 1.143 \text{ cm.}$$

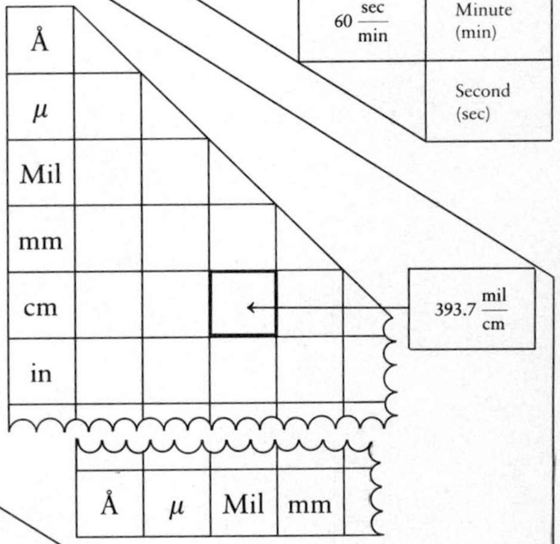

Table C.1 Time conversion factors

	Hour (hr)	Minute (min)	Second (sec)
Hour (hr)		$60 \frac{\text{min}}{\text{hr}}$	$3600 \frac{\text{sec}}{\text{hr}}$
Minute (min)			$60 \frac{\text{sec}}{\text{min}}$
Second (sec)			

Table C.3 Area conversion factors

	Square statute mile (mi^2)	Acre	Square meter (m^2)	Square yard (yd^2)	Square foot (ft^2)	Square inch (in^2)	Square centimeter (cm^2)
Square statute mile (mi^2)		$640\ \frac{\text{acre}}{\text{mi}^2}$	$2.589\text{E}6\ \frac{\text{m}^2}{\text{mi}^2}$	$3.098\text{E}6\ \frac{\text{yd}^2}{\text{mi}^2}$	$2.788\text{E}7\ \frac{\text{ft}^2}{\text{mi}^2}$	$4.014\text{E}9\ \frac{\text{in}^2}{\text{mi}^2}$	$2.589\text{E}10\ \frac{\text{cm}^2}{\text{mi}^2}$
Acre			$1233\ \frac{\text{m}^2}{\text{acre}}$	$4840\ \frac{\text{yd}^2}{\text{acre}}$	$4.356\text{E}4\ \frac{\text{ft}^2}{\text{acre}}$	$6.272\text{E}6\ \frac{\text{in}^2}{\text{acre}}$	$4.047\text{E}7\ \frac{\text{cm}^2}{\text{acre}}$
Square meter (m^2)				$1.196\ \frac{\text{yd}^2}{\text{m}^2}$	$10.76\ \frac{\text{ft}^2}{\text{m}^2}$	$1550\ \frac{\text{in}^2}{\text{m}^2}$	$1.0\text{E}4\ \frac{\text{cm}^2}{\text{m}^2}$
Square yard (yd^2)					$9\ \frac{\text{ft}^2}{\text{yd}^2}$	$1296\ \frac{\text{in}^2}{\text{yd}^2}$	$8361\ \frac{\text{cm}^2}{\text{yd}^2}$
Square foot (ft^2)						$144\ \frac{\text{in}^2}{\text{ft}^2}$	$929.0\ \frac{\text{cm}^2}{\text{ft}^2}$
Square inch (in^2)							$6.452\ \frac{\text{cm}^2}{\text{in}^2}$
Square centimeter (cm^2)							

Table C.4 Volume conversion factors

	Milliliter (ml) (cubic centimeter)	Cubic inch (in^3)	U.S. fluid ounce (fl oz)	U.S. liquid quart (qt)	Liter (l)	U.S. liquid gallon (gal)	Cubic foot (ft^3)	Cubic yard (yd^3)	Cubic meter (m^3)	Cubic statute mile (mi^3)
Milliliter (ml) (cubic centimeter)										
Cubic inch (in^3)	$16.39\ \frac{\text{ml}}{\text{in}^3}$									
U.S. fluid ounce (fl oz)	$29.574\ \frac{\text{ml}}{\text{fl oz}}$	$1.805\ \frac{\text{in}^3}{\text{fl oz}}$								
U.S. liquid quart (qt)	$946.4\ \frac{\text{ml}}{\text{qt}}$	$57.75\ \frac{\text{in}^3}{\text{qt}}$	$32\ \frac{\text{fl oz}}{\text{qt}}$							
Liter (l)	$1000\ \frac{\text{ml}}{\text{l}}$	$61.02\ \frac{\text{in}^3}{\text{l}}$	$33.81\ \frac{\text{fl oz}}{\text{l}}$	$1.057\ \frac{\text{qt}}{\text{l}}$						
U.S. liquid gallon (gal)	$3785.4\ \frac{\text{ml}}{\text{gal}}$	$231\ \frac{\text{in}^3}{\text{gal}}$	$128\ \frac{\text{fl oz}}{\text{gal}}$	$4\ \frac{\text{qt}}{\text{gal}}$	$3.7854\ \frac{\text{l}}{\text{gal}}$					
Cubic foot (ft^3)	$2.83\text{E}4\ \frac{\text{ml}}{\text{ft}^3}$	$1728\ \frac{\text{in}^3}{\text{ft}^3}$	$957.5\ \frac{\text{fl oz}}{\text{ft}^3}$	$29.92\ \frac{\text{qt}}{\text{ft}^3}$	$28.3\ \frac{\text{l}}{\text{ft}^3}$	$7.48\ \frac{\text{gal}}{\text{ft}^3}$				
Cubic yard (yd^3)	$7.646\text{E}5\ \frac{\text{ml}}{\text{yd}^3}$	$4.666\text{E}4\ \frac{\text{in}^3}{\text{yd}^3}$	$2.585\text{E}4\ \frac{\text{fl oz}}{\text{yd}^3}$	$807.9\ \frac{\text{qt}}{\text{yd}^3}$	$764.6\ \frac{\text{l}}{\text{yd}^3}$	$202.0\ \frac{\text{gal}}{\text{yd}^3}$	$27\ \frac{\text{ft}^3}{\text{yd}^3}$			
Cubic meter (m^3)	$1.0\text{E}6\ \frac{\text{ml}}{\text{m}^3}$	$6.102\text{E}4\ \frac{\text{in}^3}{\text{m}^3}$	$3.381\text{E}4\ \frac{\text{fl oz}}{\text{m}^3}$	$1.057\text{E}3\ \frac{\text{qt}}{\text{m}^3}$	$1000\ \frac{\text{l}}{\text{m}^3}$	$264.2\ \frac{\text{gal}}{\text{m}^3}$	$35.31\ \frac{\text{ft}^3}{\text{m}^3}$	$1.308\ \frac{\text{yd}^3}{\text{m}^3}$		
Cubic statute mile (mi^3)	$4.166\text{E}15\ \frac{\text{ml}}{\text{mi}^3}$	$2.544\text{E}14\ \frac{\text{in}^3}{\text{mi}^3}$	$1.409\text{E}14\ \frac{\text{fl oz}}{\text{mi}^3}$	$4.404\text{E}11\ \frac{\text{qt}}{\text{mi}^3}$	$4.166\text{E}12\ \frac{\text{l}}{\text{mi}^3}$	$1.101\text{E}12\ \frac{\text{gal}}{\text{mi}^3}$	$1.472\text{E}11\ \frac{\text{ft}^3}{\text{mi}^3}$	$5.452\text{E}9\ \frac{\text{yd}^3}{\text{mi}^3}$	$4.166\text{E}9\ \frac{\text{m}^3}{\text{mi}^3}$	

Table C.5 Mass conversion factors

	Slug	Kilogram (kg)	Pound (lbm) (avdp.)	Ounce (oz) (avdp.)	Gram (g)	Grain (gr)
Slug		14.594 $\frac{\text{kg}}{\text{slug}}$	32.174 $\frac{\text{lbm}}{\text{slug}}$	514.8 $\frac{\text{oz}}{\text{slug}}$	1.459E4 $\frac{\text{g}}{\text{slug}}$	2.252E5 $\frac{\text{gr}}{\text{slug}}$
Kilogram (kg)			2.2046 $\frac{\text{lbm}}{\text{kg}}$	35.27 $\frac{\text{oz}}{\text{kg}}$	1000 $\frac{\text{g}}{\text{kg}}$	1.543E4 $\frac{\text{gr}}{\text{kg}}$
Pound (lbm) (avdp.)				16 $\frac{\text{oz}}{\text{lbm}}$	453.6 $\frac{\text{g}}{\text{lbm}}$	7000 $\frac{\text{gr}}{\text{lbm}}$
Ounce (oz) (avdp.)					28.35 $\frac{\text{g}}{\text{oz}}$	437.5 $\frac{\text{gr}}{\text{oz}}$
Gram (g)						15.43 $\frac{\text{gr}}{\text{g}}$
Grain (gr)						

	Dyne	Poundal (pdl)	Newton (N)	Pound (lb) (avdp.)	Short ton	Metric ton	Long ton
Dyne							
Poundal (pdl)	1.383E4 $\frac{\text{dyne}}{\text{pdl}}$						
Newton (N)	E5 $\frac{\text{dyne}}{\text{N}}$	7.233 $\frac{\text{pdl}}{\text{N}}$					
Pound (lb) (avdp.)	4.448E5 $\frac{\text{dyne}}{\text{lb}}$	32.174 $\frac{\text{pdl}}{\text{lb}}$	4.448 $\frac{\text{N}}{\text{lb}}$				
Short ton	8.896E3 $\frac{\text{dyne}}{\text{s ton}}$	6.435E4 $\frac{\text{pdl}}{\text{s ton}}$	8896 $\frac{\text{N}}{\text{s ton}}$	2000 $\frac{\text{lb}}{\text{s ton}}$			
Metric ton	9.807E3 $\frac{\text{dyne}}{\text{m ton}}$	7.094E4 $\frac{\text{pdl}}{\text{m ton}}$	9808 $\frac{\text{N}}{\text{m ton}}$	2205 $\frac{\text{lb}}{\text{m ton}}$	1.102 $\frac{\text{s ton}}{\text{m ton}}$		
Long ton	9.964E3 $\frac{\text{dyne}}{\text{l ton}}$	7.207E4 $\frac{\text{pdl}}{\text{l ton}}$	9964 $\frac{\text{N}}{\text{l ton}}$	2240 $\frac{\text{lb}}{\text{l ton}}$	1.12 $\frac{\text{s ton}}{\text{l ton}}$	1.016 $\frac{\text{m ton}}{\text{l ton}}$	

Table C.6 Force conversion factors

Table C.7 Pressure conversion factors

	atm	Bar	psi	in. Hg (32°F)	ft H_2O (39.2°F)	in. H_2O (39.2°F)	mm Hg (0°C)	Torr	Pascal (N/m²)	dyne/cm²
atm		1.0132 bar/atm	14.696 psi/atm	29.92 in. Hg/atm	33.90 ft H_2O/atm	406.8 in. H_2O/atm	760 mm Hg/atm	760 Torr/atm	1.013E5 Pa/atm	1.032E6 dyne/cm²/atm
Bar			14.504 psi/bar	29.53 in. Hg/bar	33.46 ft H_2O/bar	401.5 in. H_2O/bar	750.1 mm Hg/bar	750.1 Torr/bar	1.0E5 Pa/bar	1.0E6 dyne/cm²/bar
psi				2.036 in. Hg/psi	2.307 ft H_2O/psi	27.68 in. H_2O/psi	51.71 mm Hg/psi	51.71 Torr/psi	6.895E3 Pa/psi	6.895E4 dyne/cm²/psi
in. Hg (32°F)					1.133 ft H_2O/in. Hg	13.60 in. H_2O/in. Hg	25.4 mm Hg/in. Hg	25.4 Torr/in. Hg	3386 Pa/in. Hg	3.386E4 dyne/cm²/in. Hg
ft H_2O (39.2°F)						12 in. H_2O/ft H_2O	22.42 mm Hg/ft H_2O	304.8 Torr/ft H_2O	2989 Pa/ft H_2O	2.989E4 dyne/cm²/ft H_2O
in. H_2O (39.2°F)							1.868 mm Hg/in. H_2O	1868 Torr/in. H_2O	249.1 Pa/in. H_2O	2491 dyne/cm²/in. H_2O
mm Hg (0°C)								1.00 Torr/mm Hg	133.3 Pa/mm Hg	1333 dyne/cm²/mm Hg
Torr									100 Pa/Torr	1000 dyne/cm²/Torr
Pascal (N/m²)										10 dyne/cm²/Pa
dyne/cm²										

	ft/min (fpm)	cm/s	m/min	in./sec	km/hr	ft/sec (fps)	statute mile/hr (mph)	Knots	m/s
ft/min (fpm)									
cm/s	1.969 fpm/cm/s								
m/min	3.281 fpm/m/min	1.667 cm/s/m/min							
in./sec	5 fpm/in./sec	2.54 cm/s/in./sec	1.524 m/min/in./sec						
km/hr	54.68 fpm/km/hr	27.78 cm/s/km/hr	16.67 m/min/km/hr	10.94 in./sec/km/hr					
ft/sec (fps)	60 fpm/fps	30.48 cm/s/fps	18.29 m/min/fps	12 in./sec/fps	1.097 km/hr/fps				
statute mile/hr (mph)	88 fpm/mph	44.7 cm/s/mph	26.82 m/min/mph	17.6 in./sec/mph	1.609 km/hr/mph	1.47 fps/mph			
Knots	101.3 fpm/knot	51.44 cm/s/knot	30.87 m/min/knot	20.26 in./sec/knot	1.852 km/hr/knot	1.688 fps/knot	1.151 mph/knot		
m/s	196.9 fpm/m/s	100 cm/s/m/s	60 m/min/m/s	39.37 in./sec/m/s	3.6 km/hr/m/s	3.281 fps/m/s	2.237 mph/m/s	1.944 knot/m/s	

Table C.8 Velocity conversion factors

Table C.9 Absolute viscosity conversion factors

$\frac{lb\cdot sec}{in^2}$	$\frac{lb\cdot sec}{ft^2}$	$\frac{lbm}{in.\cdot sec}$	$\frac{lbm}{ft\cdot sec}$	$\frac{N\cdot s}{m^2}$	$\frac{kg}{m\cdot s}$	Poise $\left(\frac{g}{cm\cdot s}\right)$	Centipoise	
	144 $\frac{lb\cdot sec/ft^2}{lb\cdot sec/in^2}$	386.1 $\frac{lbm/in.\cdot sec}{lb\cdot sec/in^2}$	4633 $\frac{lbm/ft\cdot sec}{lb\cdot sec/in^2}$	6895 $\frac{N\cdot s/m^2}{lb\cdot sec/in^2}$	6895 $\frac{kg/m\cdot s}{lb\cdot sec/in^2}$	6.895E4 $\frac{poise}{lb\cdot sec/in^2}$	6.895E6 $\frac{centipoise}{lb\cdot sec/in^2}$	$\frac{lb\cdot sec}{in^2}$
		2.681 $\frac{lbm/in.\cdot sec}{lb\cdot sec/ft^2}$	32.174 $\frac{lbm/ft\cdot sec}{lb\cdot sec/ft^2}$	47.88 $\frac{N\cdot s/m^2}{lb\cdot sec/ft^2}$	47.88 $\frac{kg/m\cdot s}{lb\cdot sec/ft^2}$	478.8 $\frac{poise}{lb\cdot sec/ft^2}$	4.788E4 $\frac{centipoise}{lb\cdot sec/ft^2}$	$\frac{lb\cdot sec}{ft^2}$
			12 $\frac{lbm/ft\cdot sec}{lbm/in.\cdot sec}$	17.86 $\frac{N\cdot s/m^2}{lbm/in.\cdot sec}$	17.86 $\frac{kg/m\cdot s}{lbm/in.\cdot sec}$	178.6 $\frac{poise}{lbm/in.\cdot sec}$	1.786E4 $\frac{centipoise}{lbm/in.\cdot sec}$	$\frac{lbm}{in.\cdot sec}$
				1.488 $\frac{N\cdot s/m^2}{lbm/ft\cdot sec}$	1.488 $\frac{kg/m\cdot s}{lbm/ft\cdot sec}$	14.88 $\frac{poise}{lbm/ft\cdot sec}$	1488 $\frac{centipoise}{lbm/ft\cdot sec}$	$\frac{lbm}{ft\cdot sec}$
					1 $\frac{kg/m\cdot s}{N\cdot s/m^2}$	10 $\frac{poise}{N\cdot s/m^2}$	1000 $\frac{centipoise}{N\cdot s/m^2}$	$\frac{N\cdot s}{m^2}$
						10 $\frac{poise}{kg/m\cdot s}$	1000 $\frac{centipoise}{kg/m\cdot s}$	$\frac{kg}{m\cdot s}$
							100 $\frac{centipoise}{poise}$	Poise $\left(\frac{g}{cm\cdot s}\right)$
								Centipoise

Centistoke					
Stoke (cm^2/s)	100 $\frac{centistoke}{stoke}$				
$\frac{in^2}{sec}$	645.2 $\frac{centistoke}{in^2/sec}$	6.452 $\frac{stoke}{in^2/sec}$			
$\frac{ft^2}{sec}$	9.29E4 $\frac{centistoke}{ft^2/sec}$	929 $\frac{stoke}{ft^2/sec}$	144 $\frac{in^2/sec}{ft^2/sec}$		
$\frac{m^2}{s}$	1.0E6 $\frac{centistoke}{m^2/s}$	10,000 $\frac{stoke}{m^2/s}$	929 $\frac{in^2/sec}{m^2/s}$	10.76 $\frac{ft^2/sec}{m^2/s}$	
	Centistoke	Stoke (cm^2/s)	$\frac{in^2}{sec}$	$\frac{ft^2}{sec}$	$\frac{m^2}{s}$

Table C.10 Kinematic viscosity conversion factors

Table C.11 Volume flow rate conversion factors

	$\frac{m^3}{s}$	$\frac{ft^3}{sec}$	$\frac{Liter}{s}$	$\frac{ft^3}{min}$	$\frac{gal}{min}$ (gpm)	$\frac{cm^3}{s}$
$\frac{m^3}{s}$		$35.61\ \frac{ft^3/sec}{m^3/s}$	$1000\ \frac{l/s}{m^3/s}$	$2119\ \frac{ft^3/min}{m^3/s}$	$15{,}850\ \frac{gpm}{m^3/s}$	$60E6\ \frac{cm^3/s}{m^3/s}$
$\frac{ft^3}{sec}$			$28.32\ \frac{l/s}{ft^3/sec}$	$60\ \frac{ft^3/min}{ft^3/sec}$	$449\ \frac{gpm}{ft^3/sec}$	$2.832E4\ \frac{cm^3/s}{ft^3/sec}$
$\frac{Liter}{s}$				$2.119\ \frac{ft^3/min}{l/s}$	$15.85\ \frac{gpm}{l/s}$	$1000\ \frac{cm^3/s}{l/s}$
$\frac{ft^3}{min}$					$7.481\ \frac{gpm}{ft^3/min}$	$471.9\ \frac{cm^3/s}{ft^3/min}$
$\frac{gal}{min}$ (gpm)						$63.09\ \frac{cm^3/s}{gpm}$
$\frac{cm^3}{s}$						

Table C.12 Power conversion factors

	$\frac{ft\cdot lb}{min}$	$\frac{g\cdot cal}{min}$	$\frac{Btu}{hr}$	Watt (W)	$\frac{ft\cdot lb}{sec}$	$\frac{g\cdot cal}{s}$	$\frac{Btu}{min}$	Horsepower (hp)	Kilowatt (kW)	$\frac{Btu}{sec}$
$\frac{ft\cdot lb}{min}$										
$\frac{g\cdot cal}{min}$	$3.086\ \frac{ft\cdot lb/min}{g\cdot cal/min}$									
$\frac{Btu}{hr}$	$12.96\ \frac{ft\cdot lb/min}{Btu/hr}$	$4.20\ \frac{g\cdot cal/min}{Btu/hr}$								
Watt (W)	$44.25\ \frac{ft\cdot lb/min}{W}$	$14.34\ \frac{g\cdot cal/min}{W}$	$3.4134\ \frac{Btu/hr}{W}$							
$\frac{ft\cdot lb}{sec}$	$60\ \frac{ft\cdot lb/min}{ft\cdot lb/sec}$	$19.44\ \frac{g\cdot cal/min}{ft\cdot lb/sec}$	$4.629\ \frac{Btu/hr}{ft\cdot lb/sec}$	$1.356\ \frac{W}{ft\cdot lb/sec}$						
$\frac{g\cdot cal}{s}$	$185.2\ \frac{ft\cdot lb/min}{g\cdot cal/s}$	$60\ \frac{g\cdot cal/min}{gm\cdot cal/s}$	$14.29\ \frac{Btu/hr}{g\cdot cal/s}$	$4.184\ \frac{W}{g\cdot cal/s}$	$3.086\ \frac{ft\cdot lb/sec}{g\cdot cal/s}$					
$\frac{Btu}{min}$	$778\ \frac{ft\cdot lb/min}{Btu/min}$	$252.0\ \frac{g\cdot cal/min}{Btu/min}$	$60\ \frac{Btu/hr}{Btu/min}$	$\frac{17.57\ W}{Btu/min}$	$12.96\ \frac{ft\cdot lb/sec}{Btu/min}$	$4.200\ \frac{g\cdot cal/s}{Btu/min}$				
Horsepower (hp)	$3.3E4\ \frac{ft\cdot lb/min}{hp}$	$1.069E4\ \frac{g\cdot cal/min}{hp}$	$2542\ \frac{Btu/hr}{hp}$	$745.7\ \frac{W}{hp}$	$550\ \frac{ft\cdot lb/sec}{hp}$	$6.416E5\ \frac{g\cdot cal/s}{hp}$	$42.44\ \frac{Btu/min}{hp}$			
Kilowatt (kW)	$4.425E4\ \frac{ft\cdot lb/min}{kW}$	$1.432E4\ \frac{g\cdot cal/min}{kW}$	$3413.4\ \frac{Btu/hr}{kW}$	$1000\ \frac{W}{kW}$	$737.6\ \frac{ft\cdot lb/sec}{kW}$	$238.7\ \frac{g\cdot cal/s}{kW}$	$56.83\ \frac{Btu/min}{kW}$	$1.341\ \frac{hp}{kW}$		
$\frac{Btu}{sec}$	$4.666E4\ \frac{ft\cdot lb/min}{Btu/sec}$	$1.512E4\ \frac{g\cdot cal/min}{Btu/sec}$	$3600\ \frac{Btu/hr}{Btu/sec}$	$1054\ \frac{W}{Btu/sec}$	$778\ \frac{ft\cdot lb/sec}{Btu/sec}$	$252.0\ \frac{g\cdot cal/s}{Btu/sec}$	$60\ \frac{Btu/min}{Btu/sec}$	$1.414\ \frac{hp}{Btu/sec}$	$1.054\ \frac{kW}{Btu/sec}$	

Table C.13 Energy conversion factors

	kWh	hp·hr	kcal	ft³·atm	Btu	Liter·atm	g·cal	ft·lb	Joule (J) (N·m)	in.·lb	ft·poundal	erg (dyne·cm)
kWh		1.341 hp·hr/kWh	860.6 kcal/kWh	1255 ft³·atm/kWh	3414 Btu/kWh	3.553E4 liter·atm/kWh	8.606E5 g·cal/kWh	2.655E6 ft·lb/kWh	3.6E6 J/kWh	3.186E7 in.·lb/kWh	8.543E7 ft·poundal/kWh	3.6E13 erg/kWh
hp·hr			641.6 kcal/hp·hr	935.9 ft³·atm/hp·hr	2546 Btu/hp·hr	2.649E4 liter·atm/hp·hr	6.416E5 g·cal/hp·hr	1.98E6 ft·lb/hp·hr	2.685E6 J/hp·hr	2.376E7 in.·lb/hp·hr	6.370E7 ft·poundal/hp·hr	2.684E13 erg/hp·hr
kcal				1.459 ft³·atm/kcal	3.966 Btu/kcal	41.29 liter·atm/kcal	1000 g·cal/kcal	3086 ft·lb/kcal	4184 J/kcal	3.703E4 in.·lb/kcal	9.929E4 ft·poundal/kcal	4.184E10 erg/kcal
ft³·atm					2.721 Btu/ft³·atm	28.32 liter·atm/ft³·atm	6855 g·cal/ft³·atm	2116 ft·lb/ft³·atm	2869 J/ft³·atm	2.539E4 in.·lb/ft³·atm	6.809E4 ft·poundal/ft³·atm	1.0133E9 erg/ft³·atm
Btu						10.41 liter·atm/Btu	252.0 g·cal/Btu	778 ft·lb/Btu	1054 J/Btu	9332 in.·lb/Btu	25020 ft·poundal/Btu	1.054E10 erg/Btu
Liter·atm							24.2 g·cal/liter·atm	74.74 ft·lb/liter·atm	101.3 J/liter·atm	896.8 in.·lb/liter·atm	2405 ft·poundal/liter·atm	3.578E7 erg/liter·atm
g·cal								3.086 ft·lb/g·cal	4.184 J/g·cal	37.03 in.·lb/g·cal	99.29 ft·poundal/g·cal	4.184E7 erg/g·cal
ft·lb									1.356 J/ft·lb	12 in.·lb/ft·lb	32.174 ft·poundal/ft·lb	1.356E7 erg/ft·lb
Joule (J) (N·m)										8.850 in.·lb/J	23.73 ft·poundal/J	1.0E7 erg/J
in.·lb											2.681 ft·poundal/in.·lb	1.130 erg/in.·lb
ft·poundal												4.214E5 erg/ft·poundal
erg (dyne·cm)												

Table C.14 Specific energy conversion factors

	erg/g	ft·lb/slug	Joule/kg	ft·lb/lbm	Btu/slug	Btu/lbm	g·cal/g	kcal/kg
erg/g								
ft·lb/slug	929.4 erg/g/ft·lb/slug							
Joule/kg	1.0E4 erg/g/J/kg	10.76 ft·lb/slug/J/kg						
ft·lb/lbm	2.989E4 erg/g/ft·lb/lbm	32.174 ft·lb/slug/ft·lb/lbm	2.990 J/kg/ft·lb/lbm					
Btu/slug	7.224E5 erg/g/Btu/slug	778 ft·lb/slug/Btu/slug	72.23 J/kg/Btu/slug	24.17 ft·lb/lbm/Btu/slug				
Btu/lbm	2.324E7 erg/g/Btu/lbm	2.5020E4 ft·lb/slug/Btu/lbm	2324 J/kg/Btu/lbm	778 ft·lb/lbm/Btu/lbm	32.174 Btu/slug/Btu/lbm			
g·cal/g	4.184E7 erg/g/g·cal/g	4.504E4 ft·lb/slug/g·cal/g	4184 J/kg/g·cal/g	1400 ft·lb/lbm/g·cal/g	57.90 Btu/slug/g·cal/g	1.800 Btu/lbm/g·cal/g		
kcal/kg	4.184E7 erg/g/kcal/kg	4.504E4 ft·lb/slug/kcal/kg	4184 J/kg/kcal/kg	1400 ft·lb/lbm/kcal/kg	57.90 Btu/slug/kcal/kg	1.800 Btu/lbm/kcal/kg	1.0 g·cal/g/kcal/kg	

Appendix D PIPE, TUBE, AND TUBING DIMENSIONS

Table D.1 Steel pipe (based on ASA Standards B36.10)

Nominal pipe size (in.)	Outside diameter (in.)	Schedule no.	Wall thickness (in.)	Inside diameter (in.)
$\frac{1}{8}$	0.405	40*	0.068	0.269
		80†	0.095	0.215
$\frac{1}{4}$	0.540	40*	0.088	0.364
		80†	0.119	0.302
$\frac{3}{8}$	0.675	40*	0.091	0.493
		80†	0.126	0.423
$\frac{1}{2}$	0.840	40*	0.109	0.622
		80†	0.147	0.546
		160	0.187	0.466
$\frac{3}{4}$	1.050	40*	0.113	0.824
		80†	0.154	0.742
		160	0.218	0.614
1	1.315	40*	0.133	1.049
		80†	0.179	0.957
		160	0.250	0.815
$1\frac{1}{4}$	1.660	40*	0.140	1.380
		80†	0.191	1.278
		160	0.250	1.160
$1\frac{1}{2}$	1.900	40*	0.145	1.610
		80†	0.200	1.500
		160	0.281	1.338
2	2.375	40*	0.154	2.067
		80†	0.218	1.939
		160	0.343	1.689
$2\frac{1}{2}$	2.875	40*	0.203	2.469
		80†	0.276	2.323
		160	0.375	2.125
3	3.500	40*	0.216	3.068
		80†	0.300	2.900
		160	0.437	2.626
$3\frac{1}{2}$	4.000	40*	0.226	3.548
		80†	0.318	3.364
4	4.500	40*	0.237	4.026
		80†	0.337	3.826
		120	0.437	3.626
		160	0.531	3.438
5	5.563	40*	0.258	5.047
		80†	0.375	4.813
		120	0.500	4.563
		160	0.625	4.313

(continued)

Nominal pipe size (in.)	Outside diameter (in.)	Schedule no.	Wall thickness (in.)	Inside diameter (in.)
6	6.625	40*	0.280	6.065
		80†	0.432	5.761
		120	0.562	5.501
		160	0.718	5.189
8	8.625	20	0.250	8.125
		30*	0.277	8.071
		40*	0.322	7.981
		60	0.406	7.813
		80†	0.500	7.625
		100	0.593	7.439
		120	0.718	7.189
		140	0.812	7.001
		160	0.906	6.813
10	10.75	20	0.250	10.250
		30*	0.307	10.136
		40*	0.365	10.020
		60†	0.500	9.750
		80	0.593	9.564
		100	0.718	9.314
		120	0.843	9.064
		140	1.000	8.750
		160	1.125	8.500
12	12.75	20	0.250	12.250
		30*	0.330	12.090
		40	0.406	11.938
		60	0.562	11.626
		80	0.687	11.376
		100	0.843	11.064
		120	1.000	10.750
		140	1.125	10.500
		160	1.312	10.126
14	14.0	10	0.250	13.500
		20	0.312	13.376
		30	0.375	13.250
		40	0.437	13.126
		60	0.593	12.814
		80	0.750	12.500
		100	0.937	12.126
		120	1.062	11.876
		140	1.250	11.500
		160	1.406	11.188

* Designates former "standard" sizes.
† Former "extra strong."

Table D.2 Copper tube and tubing

Type K Tube			Type L Tube		
Nominal Size (in.)	Outside Diameter (in.)	Inside Diameter (in.)	Nominal Size (in.)	Outside Diameter (in.)	Inside Diameter (in.)
1/4	0.375	0.305	1/4	0.375	0.315
3/8	0.500	0.402	3/8	0.500	0.430
1/2	0.625	0.527	1/2	0.625	0.545
5/8	0.750	0.652	5/8	0.750	0.660
3/4	0.875	0.745	3/4	0.875	0.785
1	1.125	0.995	1	1.125	1.025
1 1/4	1.375	1.245	1 1/4	1.375	1.265
1 1/2	1.625	1.481	1 1/2	1.625	1.505
2	2.125	1.959	2	2.125	1.985
2 1/2	2.625	2.435	2 1/2	2.625	2.465
3	3.125	2.907	3	3.125	2.945
3 1/2	3.625	3.385	3 1/2	3.625	3.425
4	4.125	3.857	4	4.125	3.905
5	5.125	4.805	5	5.125	4.875
6	6.125	5.741	6	6.125	5.845
8	8.125	7.583	8	8.125	7.725
10	10.125	9.449	10	10.125	9.625
12	12.125	11.315	12	12.125	11.565

Type M Tube			Type DWV Tube			Tubing	
Nominal Size (in.)	Outside Diameter (in.)	Inside Diameter (in.)	Nominal Size (in.)	Outside Diameter (in.)	Inside Diameter (in.)	Outside Diameter (in.)	Inside Diameter (in.)
3/8	0.500	0.450	1 1/4	1.375	1.295	1/8	0.065
1/2	0.625	0.569	1 1/2	1.625	1.541	3/16	0.128
3/4	0.875	0.811	2	2.125	1.041	1/4	0.190
1	1.125	1.055	3	3.125	3.035	5/16	0.238
1 1/4	1.375	1.291	4	4.125	4.009	3/8	0.311
1 1/2	1.625	1.527	5	5.125	4.981	1/2	0.436
2	2.125	2.009	6	6.125	5.959	5/8	0.555
2 1/2	2.625	2.495				3/4	0.680
3	3.125	2.981				7/8	0.785
3 1/2	3.625	3.459				1 1/8	1.025
4	4.125	3.935				1 3/8	1.265
5	5.125	4.907					
6	6.125	5.881					
8	8.125	7.785					
10	10.125	9.701					
12	12.125	11.617					

Appendix E TABLES OF COMPRESSIBLE FLOW FUNCTIONS FOR IDEAL GAS WITH $k = 1.4$

Table E.1 Isentropic flow functions [see Eqs. (10.30), (10.31), and (10.67)]

M	T/T_0	p/p_0	A/A^*
0.00	1.0000	1.0000	∞
0.02	0.9999	0.9997	28.942
0.04	0.9997	0.9989	14.481
0.06	0.9993	0.9975	9.666
0.08	0.9987	0.9955	7.262
0.10	0.9980	0.9930	5.822
0.12	0.9971	0.9900	4.864
0.14	0.9961	0.9864	4.182
0.16	0.9949	0.9823	3.673
0.18	0.9936	0.9777	3.278
0.20	0.9921	0.9725	2.964
0.22	0.9904	0.9669	2.708
0.24	0.9886	0.9607	2.496
0.26	0.9867	0.9541	2.317
0.28	0.9846	0.9470	2.166
0.30	0.9823	0.9395	2.035
0.32	0.9799	0.9315	1.922
0.34	0.9774	0.9231	1.823
0.36	0.9747	0.9143	1.736
0.38	0.9719	0.9052	1.659
0.40	0.9690	0.8956	1.590
0.42	0.9659	0.8857	1.529
0.44	0.9627	0.8755	1.474
0.46	0.9594	0.8650	1.425
0.48	0.9560	0.8541	1.380
0.50	0.9524	0.8430	1.340
0.52	0.9487	0.8317	1.303
0.54	0.9449	0.8201	1.270
0.56	0.9410	0.8082	1.240
0.58	0.9370	0.7962	1.213
0.60	0.9328	0.7840	1.188
0.62	0.9286	0.7716	1.166
0.64	0.9243	0.7591	1.145
0.66	0.9199	0.7465	1.127
0.68	0.9154	0.7338	1.110
0.70	0.9108	0.7209	1.094
0.72	0.9061	0.7080	1.081
0.74	0.9013	0.6951	1.068
0.76	0.8964	0.6821	1.057
0.78	0.8915	0.6691	1.047
0.80	0.8865	0.6560	1.038
0.82	0.8815	0.6430	1.030
0.84	0.8763	0.6300	1.024
0.86	0.8711	0.6170	1.018
0.88	0.8659	0.6041	1.013
0.90	0.8606	0.5913	1.009
0.92	0.8552	0.5785	1.006
0.94	0.8498	0.5658	1.003
0.96	0.8444	0.5532	1.001
0.98	0.8389	0.5407	1.000
1.00	0.8333	0.5283	1.000
1.02	0.8278	0.5160	1.000
1.04	0.8222	0.5309	1.001
1.06	0.8165	0.4919	1.003
1.08	0.8108	0.4801	1.005
1.10	0.8052	0.4684	1.008
1.12	0.7994	0.4568	1.011
1.14	0.7937	0.4455	1.015
1.16	0.7880	0.4343	1.020
1.18	0.7822	0.4232	1.025
1.20	0.7764	0.4124	1.030
1.22	0.7706	0.4017	1.037
1.24	0.7648	0.3912	1.043
1.26	0.7590	0.3809	1.050
1.28	0.7532	0.3708	1.058
1.30	0.7474	0.3609	1.066
1.32	0.7416	0.3512	1.075
1.34	0.7358	0.3417	1.084
1.36	0.7300	0.3323	1.094
1.38	0.7242	0.3232	1.104
1.40	0.7184	0.3142	1.115

M	T/T_0	p/p_0	A/A^*
1.42	0.7126	0.3055	1.126
1.44	0.7069	0.2969	1.138
1.46	0.7011	0.2886	1.150
1.48	0.6954	0.2804	1.163
1.50	0.6897	0.2724	1.176
1.52	0.6840	0.2646	1.190
1.54	0.6783	0.2570	1.204
1.56	0.6726	0.2496	1.219
1.58	0.6670	0.2423	1.234
1.60	0.6614	0.2353	1.250
1.62	0.6558	0.2284	1.267
1.64	0.6502	0.2217	1.284
1.66	0.6447	0.2152	1.301
1.68	0.6392	0.2088	1.319
1.70	0.6337	0.2026	1.338
1.72	0.6283	0.1966	1.357
1.74	0.6229	0.1907	1.376
1.76	0.6175	0.1850	1.397
1.78	0.6121	0.1794	1.418
1.80	0.6068	0.1740	1.439
1.82	0.6015	0.1688	1.461
1.84	0.5963	0.1637	1.484
1.86	0.5911	0.1587	1.507
1.88	0.5859	0.1539	1.531
1.90	0.5807	0.1492	1.555
1.92	0.5756	0.1447	1.580
1.94	0.5705	0.1403	1.606
1.96	0.5655	0.1360	1.633
1.98	0.5605	0.1318	1.660
2.00	0.5556	0.1278	1.688
2.02	0.5506	0.1239	1.716
2.04	0.5458	0.1201	1.745
2.06	0.5409	0.1164	1.775
2.08	0.5361	0.1128	1.806
2.10	0.5314	0.1094	1.837
2.12	0.5266	0.1060	1.869
2.14	0.5219	0.1027	1.902
2.16	0.5173	0.09956	1.935
2.18	0.5127	0.09650	1.970
2.20	0.5081	0.09352	2.005

M	T/T_0	p/p_0	A/A^*
2.22	0.5063	0.09064	2.041
2.24	0.4991	0.08784	2.078
2.26	0.4947	0.08514	2.115
2.28	0.4903	0.08252	2.154
2.30	0.4859	0.07997	2.193
2.32	0.4816	0.07751	2.233
2.34	0.4773	0.07513	2.274
2.36	0.4731	0.07281	2.316
2.38	0.4689	0.07057	2.359
2.40	0.4647	0.06840	2.403
2.42	0.4606	0.06630	2.448
2.44	0.4565	0.06426	2.494
2.46	0.4524	0.06229	2.540
2.48	0.4484	0.06038	2.588
2.50	0.4444	0.05853	2.637
2.52	0.4405	0.05674	2.687
2.54	0.4366	0.05500	2.737
2.56	0.4328	0.05332	2.789
2.58	0.4289	0.05169	2.842
2.60	0.4252	0.05012	2.896
2.62	0.4214	0.04859	2.951
2.64	0.4177	0.04711	3.007
2.66	0.4141	0.04568	3.065
2.68	0.4104	0.04429	3.123
2.70	0.4068	0.04295	3.183
2.72	0.4033	0.04166	3.244
2.74	0.3998	0.04039	3.306
2.76	0.3963	0.03917	3.370
2.78	0.3928	0.03800	3.434
2.80	0.3894	0.03685	3.500
2.82	0.3860	0.03574	3.567
2.84	0.3827	0.03467	3.636
2.86	0.3794	0.03363	3.706
2.88	0.3761	0.03262	3.777
2.90	0.3729	0.03165	3.850
2.92	0.3697	0.03071	3.924
2.94	0.3665	0.02980	3.999
2.96	0.3633	0.02891	4.076
2.98	0.3602	0.02805	4.155
3.00	0.3571	0.02722	4.235

(continued)

M	T/T_0	p/p_0	A/A^*	M	T/T_0	p/p_0	A/A^*
3.10	0.3422	0.02345	4.657	4.10	0.2293	0.005769	11.715
3.20	0.3281	0.02023	5.121	4.20	0.2208	0.005062	12.792
3.30	0.3147	0.01748	5.629	4.30	0.2129	0.004449	13.955
3.40	0.3019	0.01512	6.184	4.40	0.2053	0.003918	15.210
3.50	0.2899	0.01311	6.790	4.50	0.1980	0.003455	16.562
3.60	0.2784	0.01138	7.450	4.60	0.1911	0.003053	18.018
3.70	0.2675	0.009903	8.169	4.70	0.1846	0.002701	19.583
3.80	0.2572	0.008629	8.951	4.80	0.1783	0.002394	21.264
3.90	0.2474	0.007532	9.799	4.90	0.1724	0.002126	23.067
4.00	0.2381	0.006586	10.719	5.00	0.1667	0.001890	25.000

Table E.2 Normal shock functions [see Eqs. (10.42b), (10.43), (10.44), and (10.46)]

M_1	M_2	p_{0_2}/p_{0_1}	T_2/T_1	p_2/p_1
1.00	1.000	1.000	1.000	1.000
1.02	0.9805	1.000	1.013	1.047
1.04	0.9620	0.9999	1.026	1.095
1.06	0.9444	0.9998	1.039	1.144
1.08	0.9277	0.9994	1.052	1.194
1.10	0.9118	0.9989	1.065	1.245
1.12	0.8966	0.9982	1.078	1.297
1.14	0.8820	0.9973	1.090	1.350
1.16	0.8682	0.9961	1.103	1.403
1.18	0.8549	0.9946	1.115	1.458
1.20	0.8422	0.9928	1.128	1.513
1.22	0.8300	0.9907	1.141	1.570
1.24	0.8183	0.9884	1.153	1.627
1.26	0.8071	0.9857	1.166	1.686
1.28	0.7963	0.9827	1.178	1.745
1.30	0.7860	0.9794	1.191	1.805
1.32	0.7760	0.9757	1.204	1.866
1.34	0.7664	0.9718	1.216	1.928
1.36	0.7572	0.9676	1.229	1.991
1.38	0.7483	0.9630	1.242	2.055
1.40	0.7397	0.9582	1.255	2.120
1.42	0.7314	0.9531	1.268	2.186
1.44	0.7235	0.9477	1.281	2.253
1.46	0.7157	0.9420	1.294	2.320
1.48	0.7083	0.9360	1.307	2.389
1.50	0.7011	0.9298	1.320	2.458
1.52	0.6941	0.9233	1.334	2.529
1.54	0.6874	0.9166	1.347	2.600
1.56	0.6809	0.9097	1.361	2.673
1.58	0.6746	0.9026	1.374	2.746
1.60	0.6684	0.8952	1.388	2.820
1.62	0.6625	0.8876	1.402	2.895
1.64	0.6568	0.8799	1.416	2.971
1.66	0.6512	0.8720	1.430	3.048
1.68	0.6458	0.8640	1.444	3.126
1.70	0.6406	0.8557	1.458	3.205
1.72	0.6355	0.8474	1.473	3.285
1.74	0.6305	0.8389	1.487	3.366
1.76	0.6257	0.8302	1.502	3.447
1.78	0.6210	0.8215	1.517	3.530
1.80	0.6165	0.8127	1.532	3.613
1.82	0.6121	0.8038	1.547	3.698
1.84	0.6078	0.7947	1.562	3.783
1.86	0.6036	0.7857	1.577	3.870
1.88	0.5996	0.7766	1.592	3.957
1.90	0.5956	0.7674	1.608	4.045
1.92	0.5918	0.7581	1.624	4.134
1.94	0.5880	0.7488	1.639	4.224
1.96	0.5844	0.7395	1.655	4.315
1.98	0.5808	0.7302	1.671	4.407
2.00	0.5774	0.7209	1.687	4.500

M_1	M_2	p_{0_2}/p_{0_1}	T_2/T_1	p_2/p_1	M_1	M_2	p_{0_2}/p_{0_1}	T_2/T_1	p_2/p_1
2.02	0.5740	0.7115	1.704	4.594	2.72	0.4941	0.4166	2.364	8.465
2.04	0.5707	0.7022	1.720	4.689	2.74	0.4926	0.4097	2.386	8.592
2.06	0.5675	0.6928	1.737	4.784	2.76	0.4911	0.4028	2.407	8.721
2.08	0.5643	0.6835	1.754	4.881	2.78	0.4897	0.3961	2.429	8.850
2.10	0.5613	0.6742	1.770	4.978	2.80	0.4882	0.3895	2.451	8.980
2.12	0.5583	0.6649	1.787	5.077	2.82	0.4868	0.3829	2.473	9.111
2.14	0.5554	0.6557	1.805	5.176	2.84	0.4854	0.3765	2.496	9.243
2.16	0.5525	0.6464	1.822	5.277	2.86	0.4840	0.3701	2.518	9.376
2.18	0.5498	0.6373	1.839	5.378	2.88	0.4827	0.3639	2.540	9.510
2.20	0.5471	0.6281	1.857	5.480	2.90	0.4814	0.3577	2.563	9.645
2.22	0.5444	0.6191	1.875	5.583	2.92	0.4801	0.3517	2.586	9.781
2.24	0.5418	0.6100	1.892	5.687	2.94	0.4788	0.3457	2.609	9.918
2.26	0.5393	0.6011	1.910	5.792	2.96	0.4776	0.3398	2.632	10.06
2.28	0.5368	0.5921	1.929	5.898	2.98	0.4764	0.3340	2.656	10.19
2.30	0.5344	0.5833	1.947	6.005	3.00	0.4752	0.3283	2.679	10.33
2.32	0.5321	0.5745	1.965	6.113	3.10	0.4695	0.3012	2.799	11.05
2.34	0.5297	0.5658	1.984	6.222	3.20	0.4644	0.2762	2.922	11.78
2.36	0.5275	0.5572	2.002	6.331	3.30	0.4596	0.2533	3.049	12.54
2.38	0.5253	0.5486	2.021	6.442	3.40	0.4552	0.2322	3.180	13.32
2.40	0.5231	0.5402	2.040	6.553	3.50	0.4512	0.2130	3.315	14.13
2.42	0.5210	0.5318	2.059	6.666	3.60	0.4474	0.1953	3.454	14.95
2.44	0.5189	0.5234	2.079	6.779	3.70	0.4440	0.1792	3.596	15.81
2.46	0.5169	0.5152	2.098	6.894	3.80	0.4407	0.1645	3.743	16.68
2.48	0.5149	0.5071	2.118	7.009	3.90	0.4377	0.1510	3.893	17.58
2.50	0.5130	0.4990	2.137	7.125	4.00	0.4350	0.1388	4.047	18.50
2.52	0.5111	0.4910	2.157	7.242	4.10	0.4324	0.1276	4.205	19.45
2.54	0.5092	0.4832	2.177	7.360	4.20	0.4299	0.1173	4.367	20.41
2.56	0.5074	0.4754	2.198	7.479	4.30	0.4277	0.1080	4.532	21.41
2.58	0.5056	0.4677	2.218	7.599	4.40	0.4255	0.09948	4.702	22.42
2.60	0.5039	0.4601	2.238	7.720	4.50	0.4236	0.09170	4.875	23.46
2.62	0.5022	0.4526	2.259	7.842	4.60	0.4217	0.08459	5.052	24.52
2.64	0.5005	0.4452	2.280	7.965	4.70	0.4199	0.07809	5.233	25.61
2.66	0.4988	0.4379	2.301	8.088	4.80	0.4183	0.07214	5.418	26.71
2.68	0.4972	0.4307	2.322	8.213	4.90	0.4167	0.06670	5.607	27.85
2.70	0.4956	0.4236	2.343	8.338	5.00	0.4152	0.06172	5.800	29.00

Table E.3 Fanno flow functions [see Eqs. (10.91), (10.93), (10.94), and (10.95)]

M	$4C_f \frac{\ell^*}{D_h}$	p_0/p_0^*	T/T^*	p/p^*
0.00	∞	∞	1.200	∞
0.02	1778.450	28.94	1.200	54.77
0.04	440.352	14.48	1.200	27.38
0.06	193.031	9.666	1.199	18.25
0.08	106.718	7.262	1.199	13.68
0.10	66.922	5.822	1.198	10.94
0.12	45.408	4.864	1.197	9.116
0.14	32.511	4.182	1.195	7.809
0.16	24.198	3.673	1.194	6.829
0.18	18.543	3.278	1.192	6.066
0.20	14.533	2.964	1.191	5.456
0.22	11.596	2.708	1.189	4.955
0.24	9.387	2.496	1.186	4.538
0.26	7.688	2.317	1.184	4.185
0.28	6.357	2.166	1.182	3.882
0.30	5.299	2.035	1.179	3.619
0.32	4.447	1.922	1.176	3.389
0.34	3.752	1.823	1.173	3.185
0.36	3.180	1.736	1.170	3.004
0.38	2.706	1.659	1.166	2.842
0.40	2.309	1.590	1.163	2.696
0.42	1.974	1.529	1.159	2.563
0.44	1.692	1.474	1.155	2.443
0.46	1.451	1.425	1.151	2.333
0.48	1.245	1.380	1.147	2.231
0.50	1.069	1.340	1.143	2.138
0.52	0.9174	1.303	1.138	2.052
0.54	0.7866	1.270	1.134	1.972
0.56	0.6736	1.240	1.129	1.898
0.58	0.5757	1.213	1.124	1.828
0.60	0.4908	1.188	1.119	1.763
0.62	0.4172	1.166	1.114	1.703
0.64	0.3533	1.145	1.109	1.646
0.66	0.2979	1.127	1.104	1.592
0.68	0.2498	1.110	1.098	1.541
0.70	0.2081	1.094	1.093	1.493
0.72	0.1722	1.081	1.087	1.448
0.74	0.1411	1.068	1.082	1.405
0.76	0.1145	1.057	1.076	1.365
0.78	0.09167	1.047	1.070	1.326
0.80	0.07229	1.038	1.064	1.289
0.82	0.05593	1.030	1.058	1.254
0.84	0.04226	1.024	1.052	1.221
0.86	0.03097	1.018	1.045	1.189
0.88	0.02180	1.013	1.039	1.158
0.90	0.01451	1.0089	1.033	1.129
0.92	0.00891	1.0056	1.026	1.101
0.94	0.00481	1.0031	1.020	1.074
0.96	0.00206	1.0014	1.013	1.049
0.98	0.00049	1.0003	1.007	1.024
1.00	0.0000	1.0000	1.000	1.000
1.02	0.00046	1.0003	0.9933	0.9771
1.04	0.00177	1.0013	0.9866	0.9551
1.06	0.00389	1.0029	0.9798	0.9338
1.08	0.00658	1.0051	0.9730	0.9134
1.10	0.00993	1.0079	0.9662	0.8936
1.12	0.01382	1.011	0.9593	0.8745
1.14	0.01819	1.015	0.9524	0.8561
1.16	0.02298	1.020	0.9455	0.8383
1.18	0.02814	1.025	0.9386	0.8210
1.20	0.03364	1.030	0.9317	0.8044
1.22	0.03942	1.037	0.9247	0.7882
1.24	0.04547	1.043	0.9178	0.7726
1.26	0.05174	1.050	0.9108	0.7574
1.28	0.05820	1.058	0.9038	0.7427
1.30	0.06483	1.066	0.8969	0.7285
1.32	0.07161	1.075	0.8899	0.7147
1.34	0.07850	1.084	0.8829	0.7012
1.36	0.08550	1.094	0.8760	0.6882
1.38	0.09259	1.104	0.8690	0.6755
1.40	0.09974	1.115	0.8621	0.6632
1.42	0.1069	1.126	0.8551	0.6512
1.44	0.1142	1.138	0.8482	0.6396
1.46	0.1215	1.150	0.8413	0.6282
1.48	0.1288	1.163	0.8345	0.6172
1.50	0.1361	1.176	0.8276	0.6065
1.52	0.1434	1.190	0.8208	0.5960
1.54	0.1506	1.204	0.8139	0.5858
1.56	0.1579	1.219	0.8072	0.5759
1.58	0.1651	1.234	0.8004	0.5662
1.60	0.1724	1.250	0.7937	0.5568

M	$4C_f \frac{\ell^*}{D_h}$	p_0/p_0^*	T/T^*	p/p^*
1.62	0.1795	1.267	0.7870	0.5476
1.64	0.1867	1.284	0.7803	0.5386
1.66	0.1938	1.301	0.7736	0.5299
1.68	0.2008	1.319	0.7670	0.5213
1.70	0.2078	1.338	0.7605	0.5130
1.72	0.2147	1.357	0.7539	0.5048
1.74	0.2216	1.376	0.7474	0.4969
1.76	0.2284	1.397	0.7410	0.4891
1.78	0.2352	1.418	0.7345	0.4815
1.80	0.2419	1.439	0.7282	0.4741
1.82	0.2485	1.461	0.7218	0.4668
1.84	0.2551	1.484	0.7155	0.4597
1.86	0.2616	1.507	0.7093	0.4528
1.88	0.2680	1.531	0.7030	0.4460
1.90	0.2743	1.555	0.6969	0.4394
1.92	0.2806	1.580	0.6907	0.4329
1.94	0.2868	1.606	0.6847	0.4265
1.96	0.2930	1.633	0.6786	0.4203
1.98	0.2990	1.660	0.6726	0.4142
2.00	0.3050	1.688	0.6667	0.4083
2.02	0.3109	1.716	0.6608	0.4024
2.04	0.3168	1.745	0.6549	0.3967
2.06	0.3225	1.775	0.6491	0.3911
2.08	0.3282	1.806	0.6433	0.3856
2.10	0.3339	1.837	0.6376	0.3802
2.12	0.3394	1.869	0.6320	0.3750
2.14	0.3449	1.902	0.6263	0.3698
2.16	0.3503	1.935	0.6208	0.3648
2.18	0.3556	1.970	0.6152	0.3598
2.20	0.3609	2.005	0.6098	0.3549
2.22	0.3661	2.041	0.6043	0.3502
2.24	0.3712	2.078	0.5990	0.3455
2.26	0.3763	2.115	0.5936	0.3409
2.28	0.3813	2.154	0.5883	0.3364
2.30	0.3862	2.193	0.5831	0.3320
2.32	0.3911	2.233	0.5779	0.3277
2.34	0.3959	2.274	0.5728	0.3234
2.36	0.4006	2.316	0.5677	0.3193
2.38	0.4053	2.359	0.5626	0.3152
2.40	0.4099	2.403	0.5576	0.3111

M	$4C_f \frac{\ell^*}{D_h}$	p_0/p_0^*	T/T^*	p/p^*
2.42	0.4144	2.448	0.5527	0.3072
2.44	0.4189	2.494	0.5478	0.3033
2.46	0.4233	2.540	0.5429	0.2995
2.48	0.4277	2.588	0.5381	0.2958
2.50	0.4320	2.637	0.5333	0.2921
2.52	0.4362	2.687	0.5286	0.2885
2.54	0.4404	2.737	0.5239	0.2850
2.56	0.4445	2.789	0.5193	0.2815
2.58	0.4486	2.842	0.5147	0.2781
2.60	0.4526	2.896	0.5102	0.2747
2.62	0.4565	2.951	0.5057	0.2714
2.64	0.4604	3.007	0.5013	0.2682
2.66	0.4643	3.065	0.4969	0.2650
2.68	0.4681	3.123	0.4925	0.2619
2.70	0.4718	3.183	0.4882	0.2588
2.72	0.4755	3.244	0.4839	0.2558
2.74	0.4792	3.306	0.4797	0.2528
2.76	0.4827	3.370	0.4755	0.2499
2.78	0.4863	3.434	0.4714	0.2470
2.80	0.4898	3.500	0.4673	0.2441
2.82	0.4932	3.567	0.4632	0.2414
2.84	0.4966	3.636	0.4592	0.2386
2.86	0.5000	3.706	0.4553	0.2359
2.88	0.5033	3.777	0.4513	0.2333
2.90	0.5065	3.850	0.4474	0.2307
2.92	0.5097	3.924	0.4436	0.2281
2.94	0.5129	3.999	0.4398	0.2256
2.96	0.5160	4.076	0.4360	0.2231
2.98	0.5191	4.155	0.4323	0.2206
3.00	0.5222	4.235	0.4286	0.2182
3.50	0.5864	6.79	0.3478	0.1685
4.00	0.6331	10.72	0.2857	0.1336
4.50	0.6676	16.56	0.2376	0.1083
5.00	0.6938	25.00	0.2000	0.08944

Appendix F NUMERICAL TOOLS

This appendix presents a brief overview of some useful numerical methods and is intended to be a handy reference for problem solving. For those having no previous experience with numerical methods, the presentation is sufficient to convey the ideas behind the methods and to permit their use in problem solving. The methods presented were selected for simplicity of understanding and ease of implementation. A disk containing Fortran subroutines for these tools is available to instructors with the solutions manual. Note that many software packages such as MathCad®, TKSolver®, Lotus 123®, and Quattro® have built-in capability with respect to some or all of these problems.

Solving Nonlinear Algebraic Equations

Solving an algebraic equation for a single unknown is common to almost all fluid flow analyses. Any algebraic equation can be arranged in the form $f(x) = 0$; therefore this activity is often referred to as "finding the roots of $f(x)$" or "root finding." Many fluid flow analyses generate equations that cannot be solved explicitly for the unknown quantity in terms of known quantities. Such equations may involve transcendental functions ($\sin x$, e^x), or isolating the variable of interest may simply be impossible. In these cases numerical methods provide tools for completing the analysis.

Bisection. The bisection method is one of the simplest yet most robust* methods for finding the root of an equation. To begin, you must determine—either from knowledge of the physics of the problem or some other means—an upper bound (x_{ub}) and a lower bound (x_{lb}) that define the search interval within which the root must lie. Also, for this method to work, there must be only one root (or an odd number of roots) between x_{lb} and x_{ub}. For this condition, the sign of $f(x)$ is different at the two bounds. The bisection method proceeds by evaluating $f(x)$ at the midpoint of the search interval. If $f(x_{mp})$ has the same sign as $f(x_{lb})$, then x_{mp} replaces x_{lb} as the lower bound of the search interval. If not, then $f(x_{mp})$ must have the same sign as $f(x_{ub})$, and x_{mp} replaces x_{ub} as the upper bound. You continue this process until the search interval falls below a specified size. Because the root must lie between x_{lb} and x_{ub}, the difference between these values provides an upper bound on the error of the solution. This feature (an easily computed upper bound on the error) is not available for all methods and is a significant advantage of the bisection method.

The bisection method can be summarized as follows:

1. Define search by specifying: $f(x)$, x_{lb}, x_{ub}, and x_{tol}.
2. Compute $f(x_{mp})$, where $x_{mp} = \frac{1}{2}(x_{lb} + x_{ub})$.
3. If $f(x_{mp}) \cdot f(x_{lb}) > 0$, then $x_{lb} = x_{mp}$; else $x_{ub} = x_{mp}$.
4. If $(x_{ub} - x_{lb}) > x_{tol}$, then continue iteration by returning to step 2.
5. Root is at $x_{root} = \frac{1}{2}(x_{lb} + x_{ub})$.

---

* A numerical method is said to be *robust* if it works for many problems without requiring intervention to handle special cases.

The root cannot be farther than one-half the search interval from the midpoint of the interval, so this procedure produces a root with an error of no more than $\frac{1}{2}x_{tol}$. A final caution is: If $f(x)$ is discontinuous on the search interval, this procedure may return a false root.

Example 4.30 in the text illustrates an application of the bisection method.

Fixed-Point Iteration. Instead of seeking the location where a function takes on a value of zero, fixed-point iteration seeks the value of the independent variable for which the function is equal to the independent variable, or $x = g(x)$. To begin the procedure, solve $f(x) = 0$ for one of the x's to get $x = g(x)$. [*Note:* You may have several choices of which x to isolate.] Next, guess an initial value for x. Substituting this value into $g(x)$ produces a next iterate for x, which you again substitute into $g(x)$. If the procedure produces a sequence of iterates that approach a single value, the method has converged, and the equation is solved. If the sequence of iterates proceeds to unrealistic values, the method has diverged. The key question is: Under what conditions does fixed-point iteration converge? If $g(x)$ and $g'(x)$ are continuous on a search interval around the root and if $|g'(x)| < 1$ on the interval, fixed-point iteration converges for any initial guess on the interval. Consider the function displayed in Fig. F.1(a). An initial guess leads to a monotonic approach to the root. Figure F.1(b) shows a functional relationship where an oscillatory approach to the root is produced by the fixed-point iteration procedure. Figure F.1(c) shows an iteration that diverges because $|g'(x)| > 1$ on the search interval. Although $|g'(x)| < 1$ guarantees convergence, fixed-point iteration with $|g'(x)| > 1$ may converge; whether it does depends on the shape of the function and the initial guess. Note again that different choices for the x to isolate give different forms of $g(x)$ and therefore affect convergence.

The selection of a convergence criterion requires careful consideration. The most popular criterion is to terminate the procedure when the value between successive iterates changes by less than a specified value; that is, if

$$\left| \frac{x^{[k+1]} - x^{[k]}}{x^{[k+1]}} \right| < \text{tol},$$

terminate the procedure. This test has the useful property that the tolerance to be specified depends on the desired number of significant digits in the answer and not on the magnitude of the answer. As this test examines only the change in successive iterates, you should compute the value of $f(x_{root})$ as a further test of the accuracy of your solution.

In summary, fixed-point iteration involves the following steps:

1. Arrange the functional relationship in the form $x = g(x)$; select a first guess for the root, $x^{[1]}$ and specify a convergence tolerance, tol.
2. Compute $x^{[k+1]} = g(x^{[k]})$.
3. If

$$\left| \frac{x^{[k+1]} - x^{[k]}}{x^{[k+1]}} \right| > \text{tol},$$

iterate by returning to step 2.

4. The value of the unknown is $x^{[k+1]}$.

Note: If the calculations diverge, consider an alternative arrangement, if possible.

Compared to the bisection algorithm, each iterate requires fewer computations, but the method is less robust. References 1 and 2 provide a derivation of convergence

Figure F.1 Fixed-point iteration for (a) monotonic convergence, (b) oscillatory convergence, and (c) divergence.

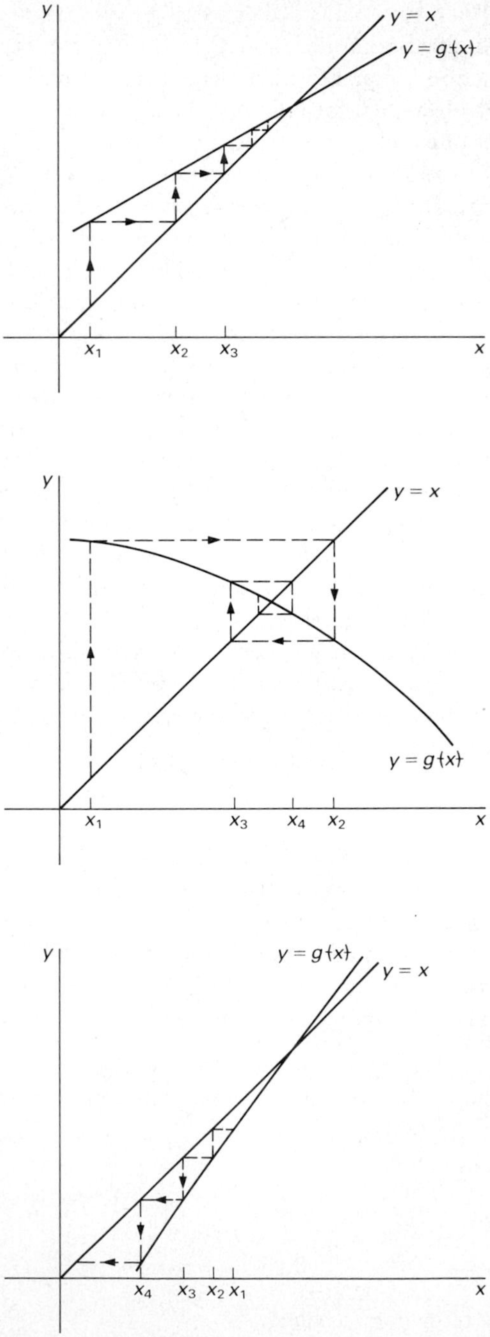

(a) Monotonic convergence

(b) Oscillatory convergence

(c) Divergence

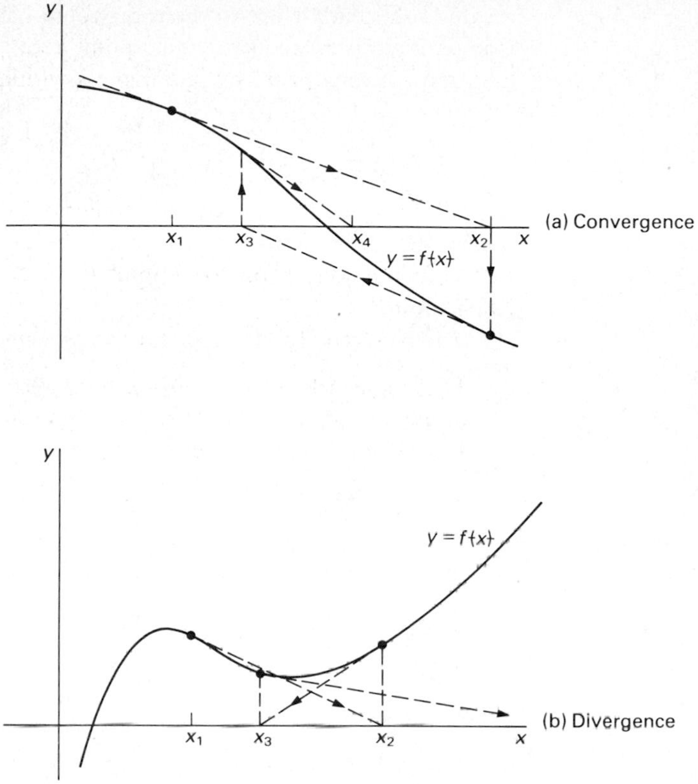

Figure F.2 Newton-Raphson iteration for (a) convergence and (b) divergence.

criteria for fixed-point iteration and present a procedure known as Aitken acceleration, which can be used to accelerate convergence to the root. Examples 7.3 and 7.15 in the text illustrate application of fixed-point iteration.

Newton–Raphson Iteration. The Newton–Raphson method is one of the most widely used methods for root finding, because it is strongly convergent and typically requires fewer iterations than the preceding methods. It is based on a linear approximation of the function in the neighborhood of the most recent iterate. Consider the function displayed in Fig. F.2(a). Starting from a first iterate $x^{[1]}$, you can compute $f(x^{[1]})$ and $f'(x^{[1]})$. The equation of a line passing through $(x^{[1]}, f(x^{[1]}))$ and tangent to $f(x)$ at that point can be written in the standard form $y_b = y_a + m(x_b - x_a)$, where m is the slope of the line. Because the line is tangent to the function at $(x_1, f(x_1))$, the slope of the line is simply the first derivative of the function at the point of tangency. The Newton–Raphson method selects the intersection of the tangent line with the x axis as the next iterate for the root:

$$x^{[k+1]} = x^{[k]} + \frac{y^{[k+1]} - y^{[k]}}{m} = x^{[k]} + \frac{0 - y^{[k]}}{m} = x^{[k]} - \frac{f(x^{[k]})}{f'(x^{[k]})}.$$

Clearly, for the problem depicted in Fig. F.2(a), this method will rapidly converge to the root. Figure F.2(b) depicts a problem for which the Newton–Raphson method diverges. Note that the question of convergence depends on both the shape of the function and the starting value for the iteration. Starting the procedure with a value sufficiently close to the root would have produced a convergent iteration. Comparing this procedure to fixed-point iteration shows [1] that Newton–Raphson iteration converges for any initial guess within a search interval for which

$$\left| \frac{f(x)f''(x)}{f'(x)} \right| < 1.$$

As with fixed-point iteration, this condition is sufficient to guarantee convergence but is not necessary for convergence. The discussion of convergence testing presented in the section on fixed-point iteration is directly applicable to Newton–Raphson iteration.

The Newton–Raphson method of root finding can be summarized as follows:

1. Arrange the relationship so that $f(x) = 0$; perform the calculus required to obtain an expression for $f'(x)$; select a starting value for $x^{[1]}$ and select a value for the convergence criterion, tol.
2. Compute the next iterate by
$$x^{[k+1]} = x^{[k]} - \frac{f(x^{[k]})}{f'(x^{[k]})}.$$
3. If
$$\left| \frac{x^{[k+1]} - x^{[k]}}{x^{[k+1]}} \right| > \text{tol},$$
iterate by returning to step 2.
4. The value of the unknown is $x^{[k+1]}$.

The closer the initial guess is to the actual root, the greater the probability that the iteration will converge. This initial guess may be based on an understanding of the physics of the problem or perhaps from some other, more robust, root-finding method.

Solving Ordinary Differential Equations

Ordinary differential equations (ODEs) typically arise from a need to find a variable when its rate-of-change (the derivative with respect to time) is a function of the other quantities in the problem and perhaps even of the variable itself. An introductory differential equations course typically focuses on finding analytic solutions to ODEs. Unfortunately, even simple-looking ODEs may not be easy to solve analytically if they are nonlinear. The numerical methods presented here can be used to solve any ODE, linear or nonlinear.

Although many methods are available for numerical solution of ODEs, only two are presented. Euler's method is the simplest and lays a foundation for understanding more complex methods. The Runge–Kutta method of order 4 is presented because it provides a good balance between computational effort and accuracy.

Euler's Method. Solving an ODE using Euler's method is probably what you would do if you thought about solving the problem graphically. Consider a first-order ODE arranged so that the derivative is isolated on the left-hand side of the

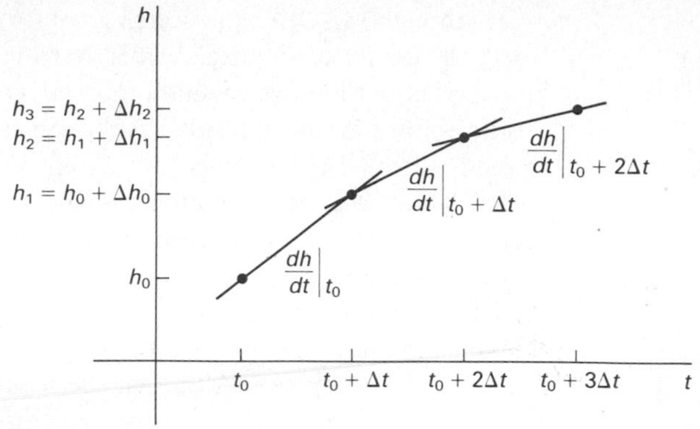

Figure F.3 Euler's method.

equation:

$$\frac{dh}{dt} = f(h, t); \qquad h(t = t_0) = h_0.$$

Note that in order to define the problem, you must specify a boundary or initial value, h_0. Because you know h_0 and t_0, the ODE provides a prescription for computing the derivative at the initial point. If you select a small interval, Δt, approximating $h(t)$ as a straight line with the slope computed at t_0 is reasonable. You can then compute an approximate value for $h(t_0 + \Delta t)$ by:*

$$h(t_0 + \Delta t) \approx h(t_0) + \left.\frac{dh}{dt}\right)_{t_0} \Delta t.$$

When you know the value of $h(t_0 + \Delta t)$, you can compute the derivative at $(t_0 + \Delta t, h(t_0 + \Delta t))$ and repeat the sequence of steps. Figure F.3 graphically depicts this process. Think of it as linear extrapolation from a known value to a new value using the ODE to compute the slope of the straight line used for extrapolation. Repeat the calculation sequence until you reach a termination criteria, such as h_{final} or t_{final}.

The steps required to implement Euler's method for solution of ODEs can be summarized as follows:

1. Arrange the ODE so that the derivative is isolated:
$$\frac{dh}{dt} = f(h, t).$$
2. Compute the value of the derivative at the point where h and t are known.
3. Use the derivative to compute a value for the function a small distance away:
$$h(t + \Delta t) = h(t) + \left.\frac{dh}{dt}\right)_t \Delta t$$
4. If the termination criteria is not yet satisfied, return to step 2 and repeat.

---

* This is precisely the value given for $h(t_0 + \Delta t)$ by a Taylor series expansion about t_0, neglecting the second-order and higher-order terms.

Although the ODE and starting values are dictated by the problem being analyzed, the choice of step size, Δt, is up to you. This method is based on a straight-line approximation over a small interval, so, logically, smaller values of Δt should produce more accurate results. Although this result often is true, taking this idea to the limit of extremely small time steps is undesirable. Because all computers use a finite number of digits, you can select such a small step size that the cumulative accuracy of the computer arithmetic degrades; the answer then actually becomes less accurate. The best approach is to perform a sequence of calculations to determine an appropriate step size. Each calculation in the sequence uses a successively smaller step until the answer, to the desired number of significant digits, ceases to change with step size. Although the sequence has converged, you have no guarantee of the accuracy of your answer. You know only that further decreasing the step size using this method will not improve your answer. Although examples can be constructed for which this method does not perform well, you will find it satisfactory for many of the ODEs resulting from flow analyses.

Although this development considered only a first-order ODE, you can easily extend this method to higher-order ODEs by rewriting them as a set of first-order ODEs. Consider the arbitrary third-order ODE:

$$f''' + Aff'' + Bf = C; \qquad f\{t_0\},\ f'\{t_0\},\ \text{and}\ f''\{t_0\}\ \text{specified.}$$

This equation can be reformulated as a set of first-order ODEs:

$$\begin{aligned} f_1 &= f'; \\ f_2 &= f'' = f_1'; \\ f_3 &= f''' = f_2' = C - Af_1f_2 - Bf. \end{aligned}$$

The solution procedure is: Compute $f'''\{t_0\}$ from the ODE; compute $f''\{t_0 + \Delta t\}$ using Euler's method; compute $f'\{t_0 + \Delta t\}$ using Euler's method; compute $f\{t_0 + \Delta t\}$ using Euler's method; compute $f'''\{t_0 + \Delta t\}$; and so on to termination.

Euler's method is both simple and robust. Reflection on the choices made during the calculation process reveals the path to alternative and more sophisticated methods. First, the method presented is sometimes referred to as the explicit Euler one-step method. The *explicit* comes from use of the slope at the beginning of the interval for the extrapolation across the entire interval. An *implicit* version of this method makes use of the slope at the end of the interval to construct the straight line. Although this method requires iteration, it is not difficult with a computer. If you "correct" the value of the end point "predicted" by the explicit slope, with a correction based on the slope at the end point, you have constructed what is known as a predictor–corrector algorithm. Another approach is to use information about the slope at a point (or points) inside the interval. This is precisely the idea on which the method presented in the next section is based. Examples 4.29 and 4.30 in the text illustrate Euler's method.

Runge–Kutta Method of Order 4. This method delivers a high degree of accuracy compared to the computational effort required. Like Euler's method, it is robust and relatively easy to use. The advantage provided by the Runge–Kutta method of order 4 (RK4), compared to Euler's method, is the ability to use larger steps and still obtain the same degree of accuracy. The simplest explanation of the method is based on the sequence of operations depicted in Fig. F.4. First, use Euler's method to obtain a first approximate value for y_{n+1}:

$$y_a = y_n + \left.\frac{dy}{dx}\right)_n \Delta x.$$

Figure F.4 RK4 method.

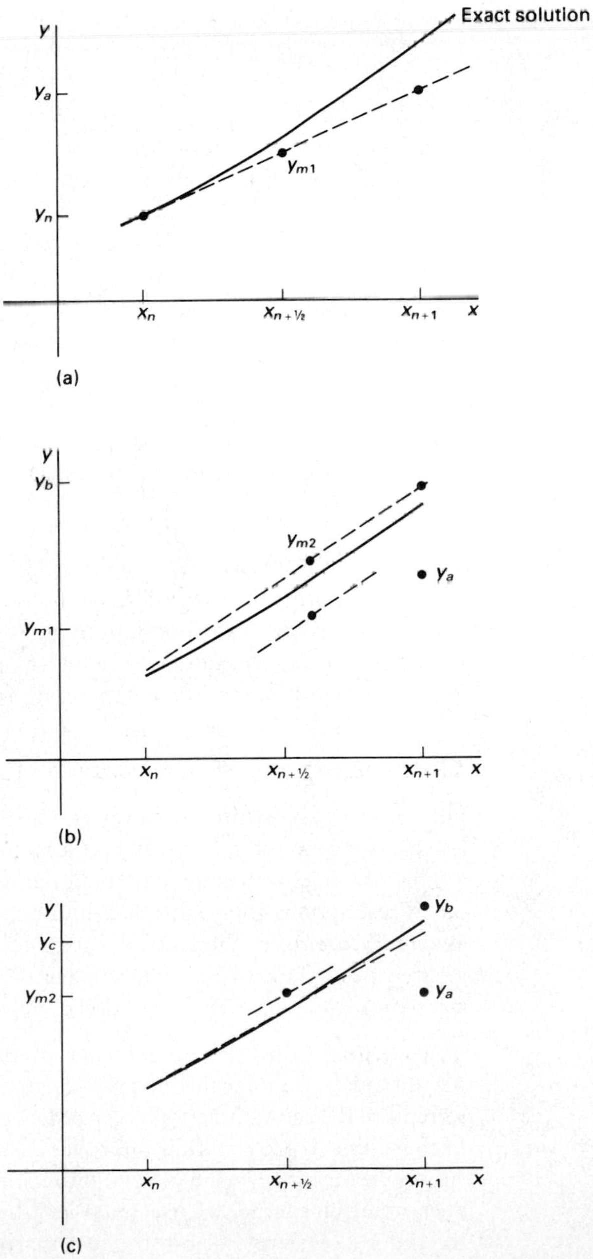

Next, obtain an approximate value for y at the midpoint of the interval, $y_{m1}(x_{n+1/2})$, using the same straight line. You can obtain a second approximate value for y_{n+1} by constructing a straight line through (x_n, y_n), evaluating the slope at y_{m1}:

$$y_b = y_n + \left.\frac{dy}{dx}\right)_{(x_{n+1/2},y_{m1})} \Delta x.$$

Repeat the procedure one more time. Use the second straight line to compute $y_{m2}(x_{n+1/2})$. Compute a third approximate value for y_{n+1} by using the derivative evaluated at $y_{m2}(x_{n+1/2})$:

$$y_c = y_n + \left.\frac{dy}{dx}\right)_{(x_{n+1},y_{m2})} \Delta x.$$

Construct the fourth approximation to y_{n+1} by using the slope at the end point predicted by the third approximation:

$$y_d = y_n + \left.\frac{dy}{dx}\right)_{(x_n,y_c)} \Delta x.$$

You now have four approximate values for y_{n+1}. A final "best" estimate is based on a weighted average of the four values:

$$y_{n+1} = y_n + \tfrac{1}{6}(y_a + 2y_b + 2y_c + y_d).$$

You can reduce the number of computations required to execute this calculation by reorganizing this expression to collect all the multiplications associated with Δx:

$$y_{n+1} = y_n + \Delta x\left(\tfrac{1}{6}f(x_n, y_n) + \tfrac{1}{3}f(x_{n+1/2}, y_{m1}) + \tfrac{1}{3}f(x_{n+1/2}, y_{m2}) + \tfrac{1}{6}f(x_{n+1}, y_d)\right).$$

You can easily modify this process to use a different number of slope evaluations, with the order of the method corresponding to the number of evaluations. Fewer evaluations require less computation per step, whereas more evaluations yield better accuracy but require more computations per step. The RK4 method represents a good trade-off between these competing factors.

Computing Integrals (Quadrature Formulas)

Fluid flow analysis often requires evaluation of an integral. When the anti-derivative cannot be found analytically (at least not to your knowledge) or the integrand is a table of values rather than a functional form, you must resort to numerical methods. Prescriptions for performing numerical integration are sometimes referred to as *quadrature rules*. Virtually all introductory calculus courses present basic quadrature rules to help explain the concept of integration, so you are probably already familiar with them even if you don't recognize the name.

Trapezoidal Rule. Consider computation of the integral of the function displayed in Fig. F.5 over the interval $[x_a, x_b]$. The basic idea of the trapezoidal rule is to replace the actual functional dependence $f(x)$ with a straight-line approximation. Because this approximation improves as the length of the interval decreases, you subdivide the range of integration into n intervals, using $n + 1$ points, as shown. To keep notation simple, we use equal-length subdivisions ($h = x_{i+1} - x_i$). The name for this rule is derived from the fact that, when $f(x)$ is replaced by a straight-line

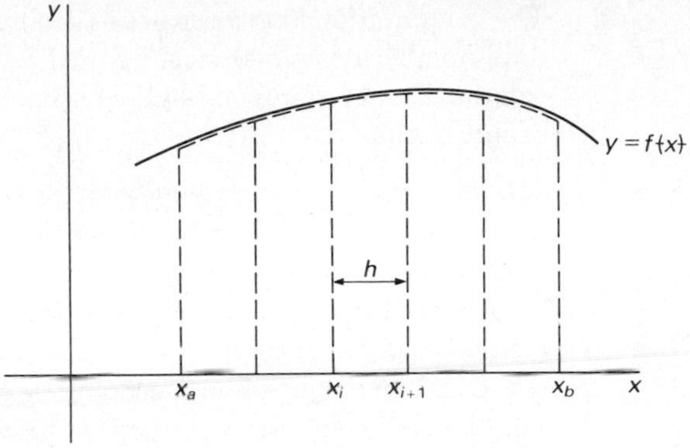

Figure F.5 Trapezoidal rule.

approximation over a subdivision, the value of the integral over that subdivision is simply the area of the trapezoid formed by the straight line, the x axis, and the vertical lines defining x_i and x_{i+1}:

$$\int_{x_i}^{x_{i+1}} f(x)\,dx \approx (x_{i+1} - x_i)\left\{\frac{1}{2}[f(x_i) + f(x_{i+1})]\right\} = \frac{h}{2}(f_i + f_{i+1}).$$

Repeating this computation for every subdivision of the interval and summing the values produces the value of the integral:

$$\int_a^b f(x)\,dx \approx \frac{h}{2}[(f_1 + f_2) + (f_2 + f_3) + (f_3 + f_4) + \cdots + (f_n + f_{n+1})]$$

$$= \frac{h}{2}(f_1 + 2f_2 + 2f_3 + \cdots 2f_n + f_{n+1}).$$

Increasing the number of points at which $f(x)$ is evaluated increases both the accuracy of the calculation and the number of computations required to complete it. You must use your knowledge of the problem to decide what is a sufficient number of subdivisions to use for your calculation. One common procedure is to perform a sequence of evaluations with successively smaller step sizes. When the value of the integral has converged to the desired number of significant digits, the sequence is terminated. Note that the method can be extended to nonuniformly sized intervals simply by summing the areas defined by the individual trapezoids.

Simpson's Rule. The main difference between the trapezoidal rule and Simpson's rule is how the function is approximated. The trapezoidal rule approximates the function by a straight line over an interval, but Simpson's rule approximates $f(x)$ with a parabolic arc over an interval. In the interest of brevity the derivation is omitted and consideration is limited to uniformly sized subdivisions. Note that uniquely specifying a second-degree polynomial requires evaluation of *three* constants and therefore *three* points on the curve must be used to compute each arc. The computations therefore use *pairs* of subdivisions. Summing over all pairs of subdivisions in the interval produces the expression for Simpson's quadrature rule:

$$\int_a^b f(x)\,dx \approx \frac{h}{3}(f_1 + 4f_2 + 2f_3 + 4f_4 + 2f_5 + \cdots + 2f_{n-1} + 4f_n + f_{n+1}).$$

The computation involves pairs of subdivisions, so an even number of subdivisions (odd number of points) must be used for the calculation. The accuracy of the calculation can be improved by increasing the number of subdivisons for the range of integration.

Solving Sets of Linear Algebraic Equations. Some fluid flow calculations require solution of simultaneous linear algebraic equations. Methods for solving sets of algebraic equations are categorized as either *iterative* or *direct*. Direct methods involve performing a fixed sequence of computations to arrive at a solution. Iterative methods involve computing a sequence of approximations to the solution, with the sequence converging to an acceptable approximation. Neither method is inherently superior. All computer arithmetic is performed with a finite number of digits and therefore all computer-based methods return (technically) approximate answers. Sometimes, iterative methods produce a solution with fewer computations than a direct method! Iterative methods are typically less demanding of computer memory. Despite the large number of options available for solving sets of algebraic equations, space limitations dictate limiting the presentations to one direct method and one iterative method.

Gauss Elimination. The Gauss elimination procedure is simply a codification of the procedure you probably learned in algebra for solving simultaneous equations by hand. The basic idea is to combine equations in a set so as to reduce to one the number of unknowns in a selected equation. This process is known as *elimination*. Once the value of this unknown is computed, it is used to solve for another, and then those to compute a third, until the entire set is solved. This process is known as *back-substitution*. The simplest way to present this procedure is by an example. In order to develop a robust procedure, we note potential difficulties and incorporate safeguards.

Consider the three algebraic equations:

$$2x_1 + 4x_2 - 2x_3 = 8; \tag{F.1}$$

$$x_1 + 3x_2 - 2x_3 = 3; \tag{F.2}$$

$$4x_1 - 2x_2 + 4x_3 = 18. \tag{F.3}$$

This set is more compactly expressed in the standard notation of linear algebra as $\mathbf{A}\boldsymbol{x} = \boldsymbol{b}$ where:

$$\mathbf{A} = A_{ij} = \begin{bmatrix} a_{11} & a_{12} & a_{13} \\ a_{21} & a_{22} & a_{23} \\ a_{31} & a_{32} & a_{33} \end{bmatrix} = \begin{bmatrix} 2 & 4 & -2 \\ 1 & 3 & -2 \\ 4 & -2 & 4 \end{bmatrix},$$

$$\boldsymbol{x} = x_i = \begin{Bmatrix} x_1 \\ x_2 \\ x_3 \end{Bmatrix},$$

and

$$\boldsymbol{b} = b_i = \begin{Bmatrix} b_1 \\ b_2 \\ b_3 \end{Bmatrix} = \begin{Bmatrix} 8 \\ 3 \\ 18 \end{Bmatrix}.$$

The **A** matrix is called the *coefficient matrix,* the $\mathbf{x}$ vector the *solution vector,* and the $\boldsymbol{b}$ vector the *vector of knowns.* Any of the equations can be multiplied by a constant without changing the solution, and equations can be combined by addition or subtraction without altering the solution. The goal of the elimination procedure is to combine the equations in such a way that all the lower triangular elements of **A** are zero. Multiplying Eq. F.1 by $-1/2$ and adding it to Eq. F.2 eliminates a_{21}:

$$\begin{bmatrix} 2 & 4 & -2 \\ 0 & 1 & -1 \\ 4 & -2 & 4 \end{bmatrix} \begin{Bmatrix} x_1 \\ x_2 \\ x_3 \end{Bmatrix} = \begin{Bmatrix} 8 \\ -1 \\ 18 \end{Bmatrix}.$$

Similarly, multiplying Eq. F.1 by -2 and adding it to Eq. F.3 eliminates a_{31}:

$$\begin{bmatrix} 2 & 4 & -2 \\ 0 & 1 & -1 \\ 0 & -10 & 8 \end{bmatrix} \begin{Bmatrix} x_1 \\ x_2 \\ x_3 \end{Bmatrix} = \begin{Bmatrix} 8 \\ -1 \\ 2 \end{Bmatrix}.$$

Except for the first row, all the first-column entries have been eliminated. Proceeding to work on the second column and multiplying Eq. F.2 by 10 and adding it to Eq. F.3, yields:

$$\begin{bmatrix} 2 & 4 & -2 \\ 0 & 1 & -1 \\ 0 & 0 & -2 \end{bmatrix} \begin{Bmatrix} x_1 \\ x_2 \\ x_3 \end{Bmatrix} = \begin{Bmatrix} 8 \\ -1 \\ -8 \end{Bmatrix}.$$

The coefficient matrix is now in *upper triangular form,* which means that the last equation in the set contains only a single unknown. Solving the third equation to obtain $x_3 = 4$ begins the back-substitution process. This value is used to solve the second equation to obtain $x_2 = -1 + 4 = 3$. Finally, back-substituting these values into the first equation gives $x_1 = 2$.

The steps followed in the preceding example reveal the essence of the Gauss elimination method.

1. Write the set of equations in matrix form: $\mathbf{A}\mathbf{x} = \boldsymbol{b}$.
2. Use linear combinations of the equations to reduce the coefficient matrix to upper triangular form.
   - Begin with first column.
   - To "eliminate" the entry for the first column of the ith equation, multiply the first equation by a factor of $-(a_{i1}/a_{11})$ and add it to the ith equation.
   - Repeat this procedure for the second through the Nth equations, so that the entire lower triangular portion of the matrix contains only zeroes. For the kth column, for all $i > k$, the kth equation is multiplied by $-(a_{ik}/a_{kk})$ and added to the ith equation to eliminate a_{ik}.
3. Perform back-substitution to obtain the solution vector.
   - Solve the Nth equation for the Nth unknown.
   - Back-substitute x_N into equation $N - 1$ to obtain x_{N-1}.

- Repeat this operation, proceeding back through the equation set until the entire solution vector is obtained.

Before ending this discussion, we need to identify three potential sources of difficulty and procedures for dealing with them. Because the elimination procedure requires multiplication by a factor of $-(a_{kj}/a_{kk})$, what happens if $a_{kk} = 0$? The best answer is to use a procedure known as *pivoting*. Before beginning the elimination associated with a_{kk}, examine all the a_{ik} on or below the diagonal to find the largest value. Interchange the equation with the largest value of a_{ik} with the current resident of the kth row. If producing a nonzero value for the diagonal element is not possible, the equation set is indeterminate, and a unique solution does not exist. In addition to eliminating the "zero on the diagonal" problem, pivoting improves the accuracy of your answer.

The second potential difficulty is excessive roundoff caused when one column of the coefficient matrix contains numbers of a different order of magnitude than the other columns. When this condition occurs, the finite-digit arithmetic performed by the computer can lead to serious roundoff errors. You can address this problem by scaling the equations. Before starting the elimination procedure, divide each row of the coefficient matrix by the largest element in the row.

If scaling fails to cure the roundoff problem, you have encountered the third potential difficulty: an ill-conditioned matrix. An ill-conditioned matrix suffers large roundoff errors despite scaling and pivoting. Applying Gauss elimination to an ill-conditioned matrix produces values for the unknowns that are not satisfactory solutions for the original equations. Solving equation sets that result in an ill-conditioned coefficient matrix requires special techniques that are beyond the scope of this discussion. Fortunately, although constructing an example of an ill-conditioned matrix is relatively easy, the analyses associated with this text rarely involve such a matrix.

Gauss–Seidel Iteration. Although direct solution methods are plentiful, iterative methods are sometimes preferable. If a matrix contains a large number of zeros, an iterative method may require fewer computations. An iterative method may sometimes reduce the roundoff problem encountered using a direct method. Finally, when implemented by hand, iterative methods are self-correcting. Although a computational error may lead to extra iterations, the final answer should be correct.

Iterative methods are similar to the fixed-point iteration procedure for finding the root of a nonlinear equation. In a fixed-point iteration, the equation is "split," so that instead of solving $f(x) = 0$, we solve $x = g(x)$. The solution is obtained by computing a sequence of iterates, $x^{[k+1]} = g(x^{[k]})$, which is terminated when the value of x converges to within a specified tolerance. A similar strategy can be pursued for solving a set of linear algebraic equations by "splitting" the coefficient matrix:

$$\mathbf{A}\boldsymbol{x} = (\mathbf{F} - \mathbf{G})\boldsymbol{x} = \boldsymbol{b}.$$

This equation is easily rearranged to accommodate an iterative procedure based on successive substitution:

$$\mathbf{F}\boldsymbol{x}^{[k+1]} = \mathbf{G}\boldsymbol{x}^{[k]} + \boldsymbol{b}.$$

The question is: How should **A** be split? First, a splitting that leads to a converging sequence of iterates is needed. Second, this system must be solved many times, so **F** should have a form that leads to an easy solution of the set. Consider splitting the coefficient matrix into three parts: a lower triangular matrix **L**, a diagonal matrix **D**, and an upper triangular matrix **U**. The equation set then becomes

$$(\mathbf{L} + \mathbf{D} + \mathbf{U})\boldsymbol{x} = \boldsymbol{b}.$$

Leaving only the diagonal part of the original coefficient matrix on the left-hand side makes solution of the resulting set easy, because each equation (row) contains only one unknown (nonzero element).† That is,

$$\mathbf{D}\boldsymbol{x}^{[k+1]} = \boldsymbol{b} - \mathbf{U}\boldsymbol{x}^{[k]} - \mathbf{L}\boldsymbol{x}^{[k]}.$$

This type of iteration is called Jacobi iteration. Rewriting this expression explicitly to show the required multiplications simplifies writing a computer implementation of this method:

$$x_i^{[k+1]} = \frac{b_i}{a_{ii}} - \sum_{j=1}^{i-1} \frac{a_{ij}}{a_{ii}} x_j^{[k]} - \sum_{j=i+1}^{N} \frac{a_{ij}}{a_{ii}} x_j^{[k]}.$$

Note that x_i does not appear on the right-hand side. Careful examination of the first sum reveals that, if execution proceeds from the first equation ($i = 1$) toward the last, all the $x_j^{[k+1]}$ have already been computed for $j < i - 1$. Therefore the new iterates can be used in this sum without complicating the solution. Intuition indicates that these higher-level iterates should be closer to the solution and that using them to compute the remaining unknowns should lead to a more rapidly converging method. This procedure can be expressed in matrix notation as

$$\mathbf{D}\boldsymbol{x}^{[k+1]} = \boldsymbol{b} - \mathbf{U}\boldsymbol{x}^{[k+1]} - \mathbf{L}\boldsymbol{x}^{[k]}.$$

This method is known as Gauss–Seidel iteration. It does, in general, converge more rapidly than Jacobi iteration.

To complete developing the Gauss–Seidel iteration method we need to identify a suitable convergence criterion for the iteration and provide some sense of the types of systems for which the iteration converges. As with fixed-point iteration for a single nonlinear equation, a common convergence criterion for Gauss–Seidel iteration is a maximum acceptable change in the value of successive iterates. That is, for every i, $i = 1 \ldots N$:

$$\left| \frac{x_i^{[k+1]} - x_i^{[k]}}{x_i^{[k+1]}} \right| < \text{tol}.$$

Gauss–Seidel iteration converges for any coefficient matrix that is diagonally dominant [2]. In a diagonally dominant matrix, the absolute value of the diagonal element is greater than the sum of the absolute values of all other elements in that row. Although this condition may seem restrictive, a large number of engineering problems generate mathematical statements that satisfy it. Further, diagonal dominance is sufficient to guarantee convergence but is not necessary for convergence.

---

† Note that all we are really doing is solving each of the equations for one of the x's in terms of the remaining x's.

Gauss–Seidel iteration for solving a set of linear algebraic equations can be summarized by the following steps:

1. Write the system of equations in standard form: $\mathbf{A}\boldsymbol{x} = \boldsymbol{b}$.
2. Specify a first guess of the solution vector $x_i^{[1]}$.
3. Compute a next iterate for the solution vector:

$$x_i^{[k+1]} = \frac{b_i}{a_{ii}} - \sum_{j=1}^{i-1} \frac{a_{ij}}{a_{ii}} x_j^{[k+1]} - \sum_{j=i+1}^{N} \frac{a_{ij}}{a_{ii}} x_j^{[k]}.$$

4. Unless the convergence criterion is satisfied, return to step 3 and compute a new iterate for the solution vector.

This method is the most popular iterative procedure for solving a set of simultaneous algebraic equations. It can be easily coded in the form of a computer program, and the iteration sequence is convergent for many engineering problems. Example 9.4 illustrates an application of Gauss–Seidel iteration using the Lotus 123® spreadsheet.

Curve Fitting

Engineering analyses often deal with data that are known only in a tabular format or on a point-by-point basis. An example is the relationship between two fluid properties, such as viscosity and temperature. An experimenter measures the viscosity at a finite number of temperatures in an effort to describe the functional dependence of viscosity on temperature. Transmittal or use of these data is awkward. Converting the data into a reasonably accurate analytic form usually is preferred. The goal of all curve-fitting procedures is to generate an analytic expression that represents the data as accurately as possible.

The beginning point in curve fitting is the choice of the form of the curve to be fit to the data. The most frequent choice is a polynomial ($y = a + bx + cx^2 + \cdots$); however infinitely many other choices are available (e.g., $y = a \cos x + be^x + cx^2 + \cdots$). The choice of curve form and degree is up to you. After choosing the form, you want to fit the "best" curve of that type to the data. This raises the question: "What is best?" The most common criterion requires that the *least-squares error* be minimized. Let (X, Y) be a data point and (x, y) be the corresponding value given by substituting x into the fitted functional expression: $y = f\{x\}$. In this method, the assumption is that all error is in the dependent variable and that the independent variable is known exactly. This condition can never be precisely true, but it is often a good approximation and is almost universally invoked. The error incurred by using the function to represent the relationship at point i is:

$$e_i = Y_i - y_i.$$

The sum of the square of the errors at all data points is

$$S = \sum_{i=1}^{N} (e_i)^2 = \sum_{i=1}^{N} [Y_i - f\{x_i\}]^2.$$

Minimizing this sum is the best way to fit a specified function to the data [1–4]. This condition alone does not guarantee a good fit; that also depends on whether the functional form chosen suitably represents the data. The least-squares procedure is an effective and efficient procedure for determining the constants to be used in the chosen function.

Although any function can be proposed for fitting data, we illustrate the method by fitting a straight-line approximation to a data set to develop the basics of least-squares curve fitting. We wish to obtain a least-squares fit of $y = a_1x + a_0$ to a set of N data points, (X_i, Y_i). We express the error incurred in the approximation as

$$S = \sum_{i=1}^{N} [Y_i - f(x_i)]^2 = \sum_{i=1}^{N} [Y_i - a_1x_i - a_0]^2.$$

In order to minimize S by appropriate choices of a_0 and a_1, we take the derivatives of S with respect to the unknowns (a_0 and a_1) and set them equal to zero:

$$\frac{\partial S}{\partial a_0} = \sum_{i=1}^{N} 2[Y_i - a_1x_i - a_0](-1) = 0;$$

$$\frac{\partial S}{\partial a_1} = \sum_{i=1}^{N} 2[Y_i - a_1x_i - a_0](-x_i) = 0.$$

Rearranging these equations into the standard form known as the "normal equations" yields

$$a_0N + a_1 \sum x_i = \sum Y_i;$$
$$a_0 \sum x_i + a_1 \sum x_i^2 = \sum x_iY_i.$$

We now have two equations for two unknowns and can solve the system by any standard method for solving a set of linear algebraic equations. The solution for a_0 and a_1 is unique and represents the best possible fit, in a least-squares sense, of a straight line to the data.

The procedure for least-squares curve fitting is easily extendable to higher-order polynomials. Fitting an Nth-order polynomial, $f(x) = a_0 + a_1x + a_2x^2 + \cdots + a_Nx^N$, generates the set of simultaneous equations:

$$\begin{bmatrix} N & \sum X_i & \sum X_i^2 & \sum X_i^3 & \cdots & \sum X_i^N \\ \sum X_i & \sum X_i^2 & \sum X_i^3 & \sum X_i^4 & \cdots & \sum X_i^{N+1} \\ \sum X_i^2 & \vdots & \vdots & \vdots & & \vdots \\ \vdots & \vdots & \vdots & \vdots & & \vdots \\ \sum X_i^N & \sum X_i^{N+1} & \sum X_i^{N+2} & \sum X_i^{N+3} & \cdots & \sum X_i^{2N} \end{bmatrix} \begin{Bmatrix} a_0 \\ a_1 \\ a_2 \\ \vdots \\ a_N \end{Bmatrix} = \begin{Bmatrix} \sum Y_i \\ \sum X_iY_i \\ \sum X_i^2Y_i \\ \vdots \\ \sum X_i^NY_i \end{Bmatrix}.$$

This system may at first appear solvable by standard methods, but the coefficient matrix is ill-conditioned. Fortunately, it is rarely desirable to use more than a fourth- or fifth-order polynomial to represent the data. For the small matrices associated with lower-order polynomials, the roundoff errors induced by ill-conditioning is not serious, and standard techniques provide reasonably accurate solutions.

The final topic in this section is least-squares fitting of functions other than polynomials. Sometimes, an understanding of the physics of a particular relationship suggests that the data be represented by a particular function, such as an exponential or a trignometric function. If the desired functional form can be reformulated into a polynomial form, the foregoing procedures can be employed. Consider the curve given by $y = ae^{bx}$. Taking the natural logarithm of both sides of this equation yields $\ln y = \ln a + bx$. The constants b and $\ln a$ can be found by using the procedure developed for straight lines. Although the result minimizes the least-squares error for the logarithmic equation instead of the original, the result is satisfactory.

Summary

Table F.1 presents a summary of the numerical tools developed in this appendix.

Table F.1

Roots of Equations

Bisection:

- Given $f(x) = 0$, x_{lb}, x_{ub}.
- $x^{[k+1]} = \frac{1}{2}(x_{\text{lb}} + x_{\text{ub}})$.
- If $f(x^{[k+1]})\, f(x_{\text{lb}}) > 0$, then $x_{\text{lb}} = x^{[k+1]}$; else $x_{\text{ub}} = x^{[k+1]}$.
- Compute next iterate until convergence.

Fixed-Point Iteration:

- Given $x = g(x)$, $x^{[1]}$.
- $x^{[k+1]} = g(x^{[k]})$.
- Compute next iterate to convergence.

Newton–Raphson Iteration:

- Given $f(x) = 0$, $f'(x)$, $x^{[1]}$.
- $x^{[k+1]} = x^{[k]} - \dfrac{f(x^{[k]})}{f'(x^{[k]})}$.
- Compute next iterate to convergence.

Solving ODEs

Euler's Method:

- Given $y' = f(x, y)$, $y(x_0) = y_0$.
- $y(x + \Delta x) = y(x) + \left.\dfrac{dy}{dx}\right|_x \Delta x = y(x) + f(x, y)\,\Delta x$.

Runge–Kutta Method of Order 4:

- Given $y' = f(x, y)$, $y(x_0) = y_0$
- $y_x + \Delta x = y_x + \Delta x(\frac{1}{6}f(x, y) + \frac{1}{3}f(x_m, y_{m1}) + \frac{1}{3}f(x_m, y_{m2}) + \frac{1}{6}f(x + \Delta x, y_e))$,
  where $x_m = x + \frac{1}{2}\Delta x$,
  $y_{m1} = y + \frac{1}{2}f(x, y)\,\Delta x$,
  $y_{m2} = y + \frac{1}{2}f(x_m, y_{m1})\,\Delta x$, and
  $y_e = y + f(x + \Delta x, y_{m2})\,\Delta x$.

Computing Integrals

Trapezoidal Rule:

- Given integrand $f(x)$ over range a to b.
- Subdivide range into N equal-sized intervals.
- $I = \int_a^b f(x)\,dx = \frac{\Delta x}{2}\sum_{i=1}^{N+1} c_i f(x_i), \qquad (c_i = 1, 2, 2, \ldots, 2, 2, 1).$

Simpson's Rule:

- Given integrand $f(x)$ over range a to b.
- Subdivide range into an even number of equal-sized subintervals.
- $I = \int_a^b f(x)\,dx = \frac{\Delta x}{3}\sum_{i=1}^{N+1} c_i f(x_i), \qquad (c_i = 1, 4, 2, 4, 2, \ldots, 2, 4, 1).$

Solving a System of Linear Algebraic Equations

Gauss–Seidel Iteration:

- Given $\mathbf{Ax} = \boldsymbol{b}$, $x_i^{[1]}$.
- $x_i^{[k+1]} = \frac{b_i}{a_{ii}} - \sum_{j=1}^{i-1} \frac{a_{ij}}{a_{ii}} x_j^{[k+1]} - \sum_{j=i+1}^{N} \frac{a_{ij}}{a_{ii}} x_j^{[k]}.$
- Compute next iterate to convergence.

Curve Fitting

Least-Squares Fit of a Polynomial:

- Given N sets of data points (X, Y), $y = \sum_{i=0}^{n} a_i x^i$; $\qquad (n \leq N)$.
- Solve $A_{ij}\, a_j = b_i$,

  where $A_{ij} = \sum_{k=1}^{N} X_k^{i+j-2}$, and

  $b_i = \sum_{k=1}^{N} X_k^{i-1} Y_k.$

REFERENCES

1. Gerald, C. F., and Wheatley, P. O., *Applied Numerical Analysis* (4th ed.), Addison-Wesley, Reading, Mass., 1989.
2. Johnson, L. W., and Riess, R. D., *Numerical Analysis* (2nd ed.), Addison-Wesley, Reading, Mass., 1982.
3. Hornbeck, R. W., *Numerical Methods,* Prentice-Hall, Englewood Cliffs, N.J., 1975.
4. Press, W. H., Flannery, B. P., Teukolsky, S. A., and Vetterling, W. T., *Numerical Recipes,* Cambridge University Press, Cambridge, England, 1986.

Appendix G LIST OF SYMBOLS AND DIMENSIONS

This appendix contains a list of the major symbols used in this book. Since there are many more concepts than characters available in both the Roman and Greek alphabets, some duplication is unavoidable. We have generally tried to follow the American National Standards Institute (ANSI) standard symbols; however, there is no ANSI standard for fluid mechanics per se. We have considered ANSI-Y10.2 (*Hydraulics*), ANSI-Y10.7 (*Aeronautical Sciences*), and ANSI-Y10.4 (*Heat and Thermodynamics*). If a particular symbol is used for a specific concept in one chapter and for a different concept in another chapter, we have noted the chapters in the list. We also include a listing of the appropriate dimension(s) of each concept in the M, L, T, θ-dimension system. A 1 in the dimensions column indicates a dimensionless quantity and a (—) in the dimensions column means that the idea of dimension is irrelevant to the particular concept.

SYMBOL	CONCEPT	DIMENSIONS
Roman Symbols		
A	Area (usually cross-sectional area)	L^2
A_{cv}	Surface area of control volume	L^2
A^*	Critical area	L^2
AR	Aspect ratio (Chapter 8)	1
a	Acceleration	LT^{-2}
a	Constant or exponent	Various
B	Atmospheric temperature lapse rate (Chapter 2)	θ^{-1}
B	Arbitrary system property (Chapters 3, 4)	Various
B	Constant in logarithmic velocity profile (Chapter 7)	1
b	Specific (per unit mass) value of B (Chapters 3, 4)	(Various) M^{-1}
b	Constant or exponent	Various
b	Width of open channel (Chapter 11)	L
C	Chezy coefficient (Chapter 11)	$L^{1/2}T^{-1}$
C	Constant	Various
c	Airfoil chord (Chapter 8)	L
c	Speed of sound (Chapter 10)	LT^{-1}
c	Speed of weak gravity wave (Chapter 11)	LT^{-1}
c_p	Specific heat at constant pressure	$L^2T^{-2}\theta^{-1}$
c_v	Specific heat at constant volume	$L^2T^{-2}\theta^{-1}$
$\mathbf{C_c}$	Contraction coefficient	1
$\mathbf{C_D}$	Drag coefficient	1
$\mathbf{C'_D}$	Viscous drag coefficient	1

SYMBOL	CONCEPT	DIMENSIONS
Roman Symbols		
$\mathbf{C_d}$	Discharge coefficient	1
$\mathbf{C_f}$	Skin friction coefficient	1
$\mathbf{C_L}$	Lift coefficient	1
$\mathbf{C_p}$	Pressure coefficient	1
$\mathbf{C_Q}$	Flow coefficient	1
$\mathbf{C_v}$	Velocity coefficient	1
D	Diameter	L
D_{eff}	Effective diameter	L
D_{eq}	Equivalent diameter	L
D_h	Hydraulic diameter	L
$\mathscr{D}$	Drag force	MLT^{-2}
D/Dt	Eulerian (material, substantial) derivative operator	T^{-1}
E	Specific energy (Chapter 11)	L
E_K	Kinetic energy	ML^2T^{-2}
E_v	Bulk modulus of elasticity	$ML^{-1}T^{-2}$
$E_{v,T}$	Isothermal bulk modulus	$ML^{-1}T^{-2}$
F	Force	MLT^{-2}
F_B	Buoyant force	MLT^{-2}
f	Darcy friction factor	1
$F\{\ \}$	Function	—
$f\{\ \}$	Function	—
$\mathbf{F}$	Froude number	1
g	Acceleration of gravity	LT^{-2}
g_c	Force unit/mass unit conversion factor	1
gh_L	Mechanical energy loss	L^2T^{-2}
$G\{\ \}$	Function	1
H	Boundary layer shape factor (Chapter 9)	1
H	Depth of liquid over weir (Chapter 11)	L
H.O.T.	Higher-order terms (in a series)	Various
h	Height	L
h	Depth of liquid below a free surface	L
h	Head	L
h_L	Head loss	L
h_p	Pump head	L
h_T	Total head	L
h_t	Turbine head	L

(continued)

SYMBOL	CONCEPT	DIMENSIONS
Roman Symbols		
$\tilde{h}$	Specific enthalpy	L^2T^{-2}
$\tilde{h}_0$	Stagnation enthalpy	L^2T^{-2}
I	Area moment of inertia (Chapter 2)	L^4
$\hat{i}$	Unit vector in x-direction	—
$\hat{j}$	Unit vector in y-direction	—
k	Number	—
k	Specific heat ratio	1
$\hat{k}$	Unit vector in z-direction	—
$\mathbf{K}$	Loss coefficient	1
L	Length of pipe or plate	L
L_e	Entrance length	L
L_{eq}	Equivalent length	L
$\dot{L}$	Total rate of mechanical energy loss	ML^2T^{-2}
ℓ	Length	L
ℓ	Mixing length in turbulent flow (Chapter 7)	L
ℓ^*	Choking (critical) length	L
$\mathscr{L}$	Lift force	MLT^{-2}
M	Momentum	MLT^{-1}
$\dot{M}$	Momentum flox	MLT^{-2}
m	Mass	M
$\dot{m}$	Mass flow rate	MT^{-1}
$\mathscr{M}$	Moment	ML^2T^{-2}
$\mathbf{M}$	Mach number	1
N	Rotational speed (rpm)	T^{-1}
N_s	Specific speed	$L^{3/4}T^{-3/2}$
n	Coordinate normal to streamline (Chapter 5)	L
n	Number	—
n	Exponent in power law velocity profile (Chapters 7, 9)	—
n	Manning's n (roughness factor) (Chapter 11)	1
$\hat{n}$	Unit vector normal (perpendicular) to area	—
p	Pressure	$ML^{-1}T^{-2}$
p_B	Back pressure	$ML^{-1}T^{-2}$
p_T	Total pressure	$ML^{-1}T^{-2}$
p_0	Stagnation pressure	$ML^{-1}T^{-2}$
p'	Modified pressure ($p' = p - \gamma h$) (Chapter 9)	$ML^{-1}T^{-2}$
Δp	Pressure difference	$ML^{-1}T^{-2}$

SYMBOL	CONCEPT	DIMENSIONS
Roman Symbols		
Δp_d	Pressure drop	$ML^{-1}T^{-2}$
Δp_L	Pressure loss	$ML^{-1}T^{-2}$
P	Perimeter	L
P	Height of weir	L
Q	Volume flow rate	L^3T^{-1}
$\dot{Q}$	Heat transfer rate	ML^2T^{-2}
q	Heat transfer per unit mass (Chapters 4, 10)	L^2T^{-2}
q	Volume flow per unit width of open channel (Chapter 11)	L^2T^{-1}
R	Radius of pipe, cylinder, or sphere	L
R	Radius of curvature (Chapters 5, 9)	L
R	Specific gas constant (Chapters 1, 10)	$L^2T^{-2}\theta^{-1}$
R_0	Universal gas constant	$ML^2T^{-2}\theta^{-1}$ (mole)$^{-1}$
R_h	Hydraulic radius	L
r	Radial coordinate (cylindrical coordinates)	L
r'	Radial coordinate (spherical coordinates)	L
r_{pc}	Critical pressure ratio	1
$\mathbf{R}$	Reynolds number	1
S	Specific gravity	1
S	Scale factor (Chapter 6)	1
S	Projected area for force coefficients (Chapters 8, 9, 10)	L^2
S_E	Slope of energy grade line (Chapter 11)	1
S_0	Slope of channel bottom (Chapter 11)	1
s	Streamline coordinate (Chapter 5)	L
s	Arc length (Chapters 8, 9)	L
$\hat{s}$	Unit vector in s-direction	—
$\tilde{s}$	Specific entropy	$L^2T^{-2}\theta^{-1}$
$\mathbf{S}$	Strouhal number	1
T	Temperature	θ
T	Averaging time (Chapter 5)	T
$\mathbf{T}$	Dimensionless wall shear stress (Chapter 9)	1
T_0	Stagnation temperature	θ
t	Time	T
$\mathcal{T}$	Torque	ML^2T^{-2}
U	Reference or given velocity	LT^{-1}
U	Velocity of coordinate system (Chapters 4, 6)	LT^{-1}
U	Velocity at edge of boundary layer (Chapter 9)	LT^{-1}

(continued)

SYMBOL	CONCEPT	DIMENSIONS
Roman Symbols		
u	Velocity in x-direction	LT^{-1}
u_τ	Friction velocity ($\sqrt{\tau_w/\rho}$)	LT^{-1}
u_{max}	Maximum velocity	LT^{-1}
$\tilde{u}$	Specific internal energy	L^2T^{-2}
V	Velocity	LT^{-1}
V_r	Relative velocity	LT^{-1}
$\forall$	Volume	L^3
$\forall_{cv}$	Volume of control volume	L^3
v	Velocity in y-direction	LT^{-1}
W	Weight	MLT^{-2}
W	Width	L
W	Work	ML^2T^{-2}
$\dot{W}$	Power	ML^2T^{-3}
$\dot{W}_s$	Sum of shaft power and shear power	ML^2T^{-3}
w	Velocity in z-direction	LT^{-1}
w_p	Pump work per unit mass	L^2T^{-2}
w_s	Sum of shaft work and shear work per unit mass	L^2T^2
w_t	Turbine work per unit mass	L^2T^{-2}
$\mathbf{W}$	Weber number	1
x	Cartesian or cylindrical coordinate, usually parallel to main flow direction	L
x	Arbitrary dimensional parameter (Chapter 6)	Various
y	Cartesian coordinate, usually perpendicular to main flow	L
y	Coordinate parallel to plane of area (Chapter 2)	L
y	Depth in open channel flow (Chapter 11)	L
y_n	Normal depth	L
z	Cartesian coordinate, usually pointing upward	L
z	Elevation above specified datum (Chapter 4)	L
Greek Symbols		
α	Kinetic energy correction factor	1
α	Velocity diagram angle	1
α	Angle of attack (Chapters 6, 8)	1
α_T	Coefficient of thermal expansion	θ^{-1}
β	Momentum correction factor	1

SYMBOL	CONCEPT	DIMENSIONS
Roman Symbols		
β	Velocity diagram angle	1
β	Diameter ratio of a flow meter (Chapter 7)	1
Γ	Circulation	L^2T^{-1}
γ	Specific weight	$ML^{-2}T^{-2}$
δ	Boundary layer thickness	L
δ^*	Boundary layer displacement thickness	L
Δ	Specific diameter (Chapter 7)	1
$\Delta(\ )$	Change of ( )	Various
$\delta(\ )$	Small change of ( )	Various
ε	Surface roughness	L
ζ	Vorticity	T^{-1}
η	Dimensionless coordinate in boundary layer (y/δ)	1
η	Efficiency	1
Θ	Boundary layer momentum thickness	L
θ	Angle, angle coordinate in cylindrical coordinate system	1
θ'	Angle coordinate in spherical coordinate system	1
κ	Constant in logarithmic velocity profile	—
Λ	Boundary layer pressure gradient parameter $(\delta^2/\nu)(dU/dx)$	1
λ	Strength of source in plane flow	L^2T^{-1}
λ	Boundary layer pressure gradient parameter $(\Theta^2/\nu)(dU/dx)$	1
λ'	Strength of distributed source	LT^{-1}
μ	Dynamic (absolute) viscosity	$ML^{-1}T^{-1}$
μ	Strength of plane doublet (Chapter 9)	L^3T^{-1}
ν	Kinematic viscosity	LT^{-2}
ξ	Dummy variable of integration	—
ξ	Unit-specific constant in Manning's formula (Chapter 11)	$L^{1/3}T^{-1}$
Π	Dimensionless group	1
ρ	Density	ML^{-3}
ρ_0	Stagnation density	ML^{-3}
σ	Surface tension (Chapters 1, 6)	MT^{-2}
σ	Normal stress (Chapter 5)	$ML^{-1}T^{-2}$
τ	Shear stress	$ML^{-1}T^{-2}$

(continued)

SYMBOL	CONCEPT	DIMENSIONS
Roman Symbols		
τ_{app}	Apparent stress in turbulent flow	$ML^{-1}T^{-2}$
ϕ	Angle, usually angle of shear deformation (Chapters 1, 3, 5)	1
Φ	Velocity potential (Chapter 9)	L^2T^{-1}
$\dot{\phi}$	Shear strain rate	T^{-1}
Ψ	Plane flow stream function	L^2T^{-1}
Ψ_s	Stokes stream function	L^3T^{-1}
$\vec{\Omega}$	Angular velocity of coordinate system (Chapter 4)	T^{-1}
$\mathbf{\Omega}$	Pump specific speed (Chapter 7)	1
ω	Angular velocity	T^{-1}
ω	Wave speed (Chapters 10, 11)	LT^{-1}
$\dot{\omega}_v$	Vortex-shedding frequency	T^{-1}

Special Symbols

$\{\ \}$	Function	
$\langle\ \rangle$	Time average in turbulent flow	
∇	Gradient operator	
$\oint$	Closed or cyclic line integral	

SUBSCRIPTS

Subscript	Meaning
abs	Absolute
atm	Atmospheric conditions
B	Bottom of open channel
c	Centroid (Chapter 2)
c	Critical (Chapter 11)
cv	Control volume
e	Exit
e	Edge of boundary layer (Chapter 9)
H	Horizontal
i	Initial (at $t = 0$)
i	Inlet
in	Entering control volume
m	Model
n	Normal or perpendicular, in n-direction
out	Leaving control volume
p	Center of pressure (Chapter 2)

SYMBOL	CONCEPT	DIMENSIONS
Roman Symbols		
p	Pertaining to specific fluid particle (Chapters 3, 5)	
p	Prototype (Chapter 6)	
p	At point p	
r	In radial direction or evaluated at location r	
ref	Reference quantity	
rev	Reversible process	
s	Isentropic process (Chapter 10)	
s	In s (streamline) direction	
sys	Pertaining to a system	
t	Throat (Chapter 10)	
t	At time t (Chapter 5)	
V	Vertical	
w	Wall	
x	In x-direction or evaluated at location x	
XYZ	Inertial reference frame	
xyz	Noninertial reference frame	
y	In y-direction or at location y	
z	In z-direction or at location z	
θ	In θ-direction	
0	Stagnation	
0	At point 0	
1	At point 1 or in plane 1, and so on	
∞	Far away, at infinity	

SUPERSCRIPTS

Superscript	Meaning
$\rightarrow$	Vector
	Time derivative, rate
—	Average, usually over cross-sectional area
′	Alternate, similar quantity
′	Fluctuating component in turbulent flow (Chapters 5, 7)
+	Dimensionless length or velocity in turbulent flow
*	Corresponding to sonic velocity, critical state
(1), [1] ①	Iteration 1, and so on

Appendix H DIFFERENTIAL EQUATIONS IN CYLINDRICAL (r, θ, x) AND SPHERICAL (r', ϕ, θ') COORDINATES

Continuity Equation in Cylindrical Coordinates

$$\frac{\partial \rho}{\partial t} + \frac{1}{r}\frac{\partial}{\partial r}(\rho r V_r) + \frac{1}{r}\frac{\partial}{\partial \theta}(\rho V_\theta) + \frac{\partial}{\partial x}(\rho V_x) = 0$$

Continuity Equation in Spherical Coordinates

$$\frac{\partial \rho}{\partial t} + \frac{1}{(r')^2}\frac{\partial}{\partial r'}[\rho (r')^2 V_r] + \frac{1}{r' \sin\theta'}\frac{\partial}{\partial \theta'}(\rho V_{\theta'} \sin\theta') + \frac{1}{r' \sin\theta'}\frac{\partial}{\partial \phi}(\rho V_\phi) = 0$$

Navier–Stokes Equations in Cylindrical Coordinates for a Fluid with Constant Viscosity

r component

$$\rho\left[\frac{\partial V_r}{\partial t} + V_r\frac{\partial V_r}{\partial r} + \frac{V_\theta}{r}\frac{\partial V_r}{\partial \theta} + V_x\frac{\partial V_r}{\partial x} - \frac{V_\theta^2}{r}\right]$$
$$= -\frac{\partial p}{\partial r} + \rho g_r + \mu\left[\frac{\partial^2 V_r}{\partial r^2} + \frac{1}{r}\frac{\partial V_r}{\partial r} + \frac{1}{r^2}\frac{\partial^2 V_r}{\partial \theta^2} + \frac{\partial^2 V_r}{\partial x^2} - \frac{V_r}{r^2} - \frac{2}{r^2}\frac{\partial V_\theta}{\partial \theta}\right]$$

θ component

$$\rho\left[\frac{\partial V_\theta}{\partial t} + V_r\frac{\partial V_\theta}{\partial r} + \frac{V_\theta}{r}\frac{\partial V_\theta}{\partial \theta} + V_x\frac{\partial V_\theta}{\partial x} + \frac{V_r V_\theta}{r}\right]$$
$$= -\frac{1}{r}\frac{\partial p}{\partial \theta} + \rho g_\theta + \mu\left[\frac{\partial^2 V_\theta}{\partial r^2} + \frac{1}{r}\frac{\partial V_\theta}{\partial r} + \frac{1}{r^2}\frac{\partial^2 V_\theta}{\partial \theta^2} + \frac{\partial^2 V_\theta}{\partial x^2} + \frac{2}{r^2}\frac{\partial V_r}{\partial \theta} - \frac{V_\theta}{r^2}\right]$$

x component

$$\rho\left[\frac{\partial V_x}{\partial t} + V_r\frac{\partial V_x}{\partial r} + \frac{V_\theta}{r}\frac{\partial V_x}{\partial \theta} + V_x\frac{\partial V_x}{\partial x}\right]$$
$$= -\frac{\partial p}{\partial x} + \rho g_x + \mu\left[\frac{\partial^2 V_x}{\partial r^2} + \frac{1}{r}\frac{\partial V_x}{\partial r} + \frac{1}{r^2}\frac{\partial^2 V_x}{\partial \theta^2} + \frac{\partial^2 V_x}{\partial x^2}\right]$$

Navier–Stokes Equations in Spherical Coordinates for a Fluid with Constant Viscosity

r' component

$$\rho\left[\frac{\partial V_{r'}}{\partial t} + V_{r'}\frac{\partial V_{r'}}{\partial r'} + \frac{V_{\theta'}}{r'}\frac{\partial V_{r'}}{\partial \theta'} + \frac{V_\phi}{r'\sin\theta'}\frac{\partial V_{r'}}{\partial \phi} - \frac{V_{\theta'}^2 + V_\phi^2}{r'}\right]$$

$$= -\frac{\partial p}{\partial r'} + \rho g_{r'} + \mu\left[\frac{1}{r'^2}\frac{\partial}{\partial r'}\left(r'^2\frac{\partial V_{r'}}{\partial r'}\right)\right.$$

$$+ \frac{1}{r'^2\sin\theta'}\frac{\partial}{\partial \theta'}\left(\sin\theta'\frac{\partial V_{r'}}{\partial \theta'}\right)$$

$$+ \frac{1}{r'^2\sin^2\theta'}\frac{\partial^2 V_{r'}}{\partial \phi^2} - \frac{2V_{r'}}{r'^2} - \frac{2}{r'^2}\frac{\partial V_{\theta'}}{\partial \theta'}$$

$$\left. - \frac{2V_{\theta'}\cos\theta'}{r'^2} - \frac{2}{r'^2\sin\theta'}\frac{\partial V_{\theta'}}{\partial \phi}\right]$$

θ' component

$$\rho\left[\frac{\partial V_{\theta'}}{\partial t} + V_{r'}\frac{\partial V_{\theta'}}{\partial r'} + \frac{V_{\theta'}}{r'}\frac{\partial V_{\theta'}}{\partial \theta'} + \frac{V_{\theta'}}{r'\sin\theta'}\frac{\partial V_{\theta'}}{\partial \phi}\right]$$

$$= -\frac{1}{r'}\frac{\partial p}{\partial \theta'} + \rho g_{\theta'} + \mu\left[\frac{1}{r'^2}\frac{\partial}{\partial r'}\left(r'^2\frac{\partial V_{\theta'}}{\partial r}\right)\right.$$

$$+ \frac{1}{r'^2\sin\theta'}\frac{\partial}{\partial \theta'}\left(\sin\theta'\frac{\partial V_{\theta'}}{\partial \theta'}\right)$$

$$+ \frac{1}{r'^2\sin^2\theta'}\frac{\partial^2 V_{\theta'}}{\partial \phi^2} + \frac{2}{r'^2}\frac{\partial V_{r'}}{\partial \theta'}$$

$$\left. - \frac{V_{\theta'}}{r'^2\sin^2\theta'} - \frac{2\cos\theta'}{r'^2\sin^2\theta'}\frac{\partial V_{\theta'}}{\partial \theta'}\right]$$

ϕ component

$$\rho\left[\frac{\partial V_\phi}{\partial t} + V_{r'}\frac{\partial V_\phi}{\partial r'} + \frac{V_{\theta'}}{r'}\frac{\partial V_\phi}{\partial \theta'} + \frac{V_\phi}{r'\sin\theta'}\frac{\partial V_\phi}{\partial \phi}\right]$$

$$= -\frac{1}{r'\sin\theta'}\frac{\partial p}{\partial \phi} + \rho g_\phi + \mu\left[\frac{1}{r'^2}\frac{\partial}{\partial r'}\left(r'^2\frac{\partial V_\phi}{\partial r'^2}\right)\right.$$

$$+ \frac{1}{r'^2\sin\theta'}\frac{\partial}{\partial \theta'}\left(\sin\theta'\frac{\partial V_\phi}{\partial \theta'}\right) + \frac{1}{r'^2\sin^2\theta'}\frac{\partial^2 V_\phi}{\partial \phi^2}$$

$$\left. - \frac{V_\phi}{r'^2\sin\theta'} + \frac{2}{r'^2\sin^2\theta'}\frac{\partial V_{r'}}{\partial \phi} + \frac{2\cos\theta'}{r'^2\sin^2\theta'}\frac{\partial V_{\theta'}}{\partial \phi}\right]$$

Euler's Equations

Obtain from Navier–Stokes equations by letting $\mu = 0$.

Appendix I GENERALIZED UPWIND TIME-ACCURATE SOLVER (GUTS)

The Generalized Upwind Time-Accurate Solver (GUTS) code provides a tool for modeling transient incompressible flows in simple geometries. The code is modular and commented so that it can serve as an instructional tool for those interested in learning some basic CFD concepts. It is an evolutionary descendant of the SOLA family of codes developed at the Los Alamos National Laboratory. We acknowledge the heritage of the code, but any and all errors are our responsibility and we appreciate any comments you may choose to forward for its improvement.

To use GUTS as a tool for analyzing flows, you need the GUTS disk (available from the publisher to instructors adopting this text) and any computer with a FORTRAN compiler. The disk contains the FORTRAN source code [GUTS.FOR], a dictionary file [GUTS.DIC], and a "readme" file [README.1ST]. You will need to compile the source code into executable code using the compiler available on your machine. With the exception of namelist input, all coding is in standard FORTRAN 77. The dictionary file contains a listing of all input variables, along with an explanation for each variable and a default value where appropriate. At the end of the file is a sample input file to test your compilation of the source code and to provide a model for your input files. You may use any editor to compose the input file provided it generates a pure ASCII file. The file system implemented in the code requires that the input file be named [INPFYL]. During execution all printed output is directed to a file named [PRTFYL]. When execution terminates, a restart file, [RDFFYL], is generated so that the analysis can be restarted from the point of termination. Finally, the README file contains any additional information you may require to execute the program.

The modular nature of GUTS is evident from the flow chart in Fig. I.1. The main program controls program flow to appropriately sequenced subroutines and tests for termination of execution. Each subroutine performs a task or group of tasks logically grouped by function. For example, ⟨SETUP⟩ performs (or calls subroutines to perform) all of the tasks required to initiate an analysis, including setting default values, opening file system, reading input file, generating computational mesh, and setting boundary conditions. Subroutine ⟨ADVANZ⟩ resets all the arrays to the new time-level values and computes a value for the next time-step size. The hydrodynamic solution is obtained by taking an explicit step in momentum using ⟨TILDE⟩ followed by iterative adjustment of the velocity in ⟨PRESIT⟩ to satisfy continuity at the new time level. The specified boundary conditions are enforced by calls to ⟨BC⟩. All printed output is generated in ⟨PRTOUT⟩. Tests for termination are performed in the main program and, if the analysis is to continue to the next time step, execution loops back up to ⟨ADVANZ⟩. If a termination test is satisfied, the appropriate output is written, the file system closed, and execution is terminated.

The models and methods used in GUTS were chosen to facilitate your review and exploration of the program. In implementing these models and methods in code, clarity of presentation was consistently selected over efficiency of execution.

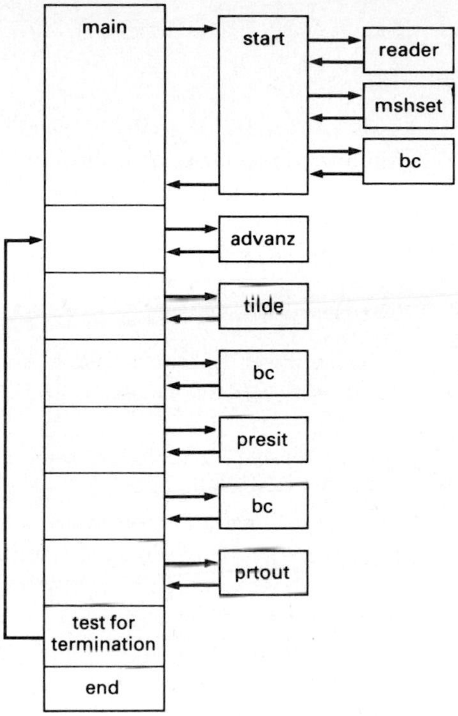

Figure I.1 GUTS Flow Chart

Compass notation (north, east, south, west) has been used with a naming convention designed to maximize the readability of the code. A "trailer" has been attached to each subroutine documenting its history and summarizing its function. Once you understand the overall functioning of the code, you can enhance its capabilities by adding additional features or use it as a testbed for examining more sophisticated models.

Appendix J DESIGN PROBLEMS

This Appendix describes seven design problems that are appropriate for junior- or senior-year engineering students. All have been assigned as part of an introductory fluid mechanics course at one or more institutions.

Automatic Gate

A wastewater treatment plant employs a process that involves several stages. Water in various stages of treatment is held in large, 15-ft deep, concrete tanks.

Two adjacent tanks are separated by a 1-ft-thick wall. Normally, the water from the two tanks is to be kept separate; however, if there is danger of a spill from overfilling one of the tanks, allowing the fuller tank to drain into the emptier tank is better than allowing a spill. Draining occurs through an opening in the separating wall, which normally is closed by a gate. The gate must open if a spill is imminent and the emptier tank has the capacity to accept the extra water.

Design the opening, the gate, and the actuating mechanism for opening and closing the gate. The following specifications must be met:

- The area of the opening is to be 4 ft^2.
- The gate must be open only if the level in the fuller tank is greater than 14 ft and the level in the emptier tank is less than 8 ft. (Note that either of the two tanks may be the fuller.)
- The gate must close when the levels in the two tanks are equalized.
- Energy/power/force for actuating the gate must be provided by the differential depths of the water. No external power sources may be used. Springs and counterweights may be used.

Water Jet Propulsion System

Figure J.1 is a schematic drawing of a water jet propulsion system for a boat. Water is ingested near the front of the boat and passes through a pump. The high-energy water from the pump is expelled from the rear of the boat at a high velocity to produce thrust.

Design a water jet propulsion system. Select duct size and configuration, including intake and exit nozzle. Also specify the pump requirements in terms of discharge, head, and rotational speed and estimate pump (impeller) size. The following specifications must be met:

- The system must produce a thrust of 2000 lb at a cruising speed of 25 mph.
- The system must be capable of a maximum static thrust of 3500 lb.
- The boat is 25 feet long, has a 7-foot beam, and has a draught of 3.5 feet.

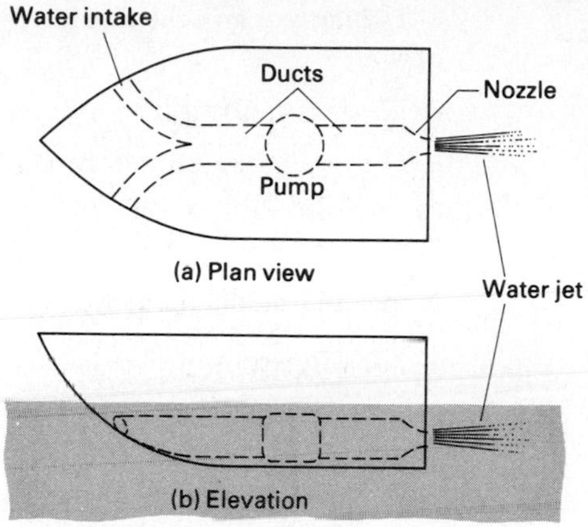

Figure J.1 Water jet propelled boat

The following constraints must be satisfied:

- The engine that drives the pump has a maximum speed of 5000 rpm.
- Water pressure cannot fall below 0.4 psia at any point in the system (to avoid vaporization of the water).

Models and useful information include the following:

- If you choose to use a gearbox between the pump and the engine, its efficiency can be estimated by

$$\text{Eff} = 0.6 + 0.35 \exp[-0.5(R - 1)],$$

where R is the speed ratio ($R > 1$).

- Mechanical energy losses in the duct system may not be negligible.

The design should give consideration to both the physical size of the system and the engine power requirement.

Garden Spray Jar

Figure J.2 shows a garden spray jar. The jar is filled with concentrated liquid chemical (fertilizer, insecticide). When the jar is connected to a water hose, the chemicals are drawn into the water stream and sprayed, diluted with water, onto a lawn or garden. Design a garden spray jar to meet the following specifications:

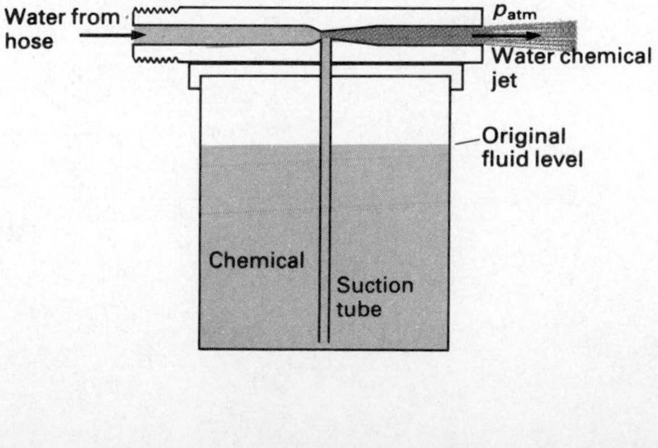

Figure J.2 Garden spray jar

- The jar is to mix chemicals with water at the ratio of 2 oz of chemicals to 1 gal of water.
- The jar will operate for at least 10 min without refilling.
- The preceding specifications must be met when operating with a typical "1/2-in." × 50-ft garden hose, with water supplied to the hose at typical city water pressure.

Notes and modeling assumptions include the following:

- The properties of the chemicals can be taken as identical to ethylene glycol.
- Do not assume ideal fluids.
- The design is not limited to the concept illustrated.

Air Hockey Set

An air hockey set illustrated in Fig. J.3 consists of four pieces. The first is a disc called the *puck*. It is made of hard rubber or plastic and is usually between 8 and 12 cm in diameter and between 1 and 2 cm thick. There are two *strikers*, which are slightly larger than the puck. A striker is a disc with a knob on one flat surface so that it can be held and manipulated by a player.

The most important and complex part of the set is the *air hockey table*. It has a hard, smooth surface with many small holes in it. Air is forced through the holes so that the puck can "float" or slide on a thin film of air, with little friction. The edges of the table are raised so that the puck cannot fly off. A slot is cut in each end of the table. The width of the slot is 2 to 3 times the puck diameter and its height is about 1.5 times the puck thickness. These slots are the goals. Players use the strikers to hit the puck. The object of the game is to hit the puck into the opponent's goal while defending your own goal.

Design an air hockey set for home use. The design should include:

- dimensions and material specification for all components;
- details of the air supply system;
- specifications for the air supply fan(s), including the number of fans, required air flow and pressure of each, and a reasonable estimate of each fan's efficiency or power requirement; and
- reasonable considerations for consumer safety.

Figure J.3 Air hockey set

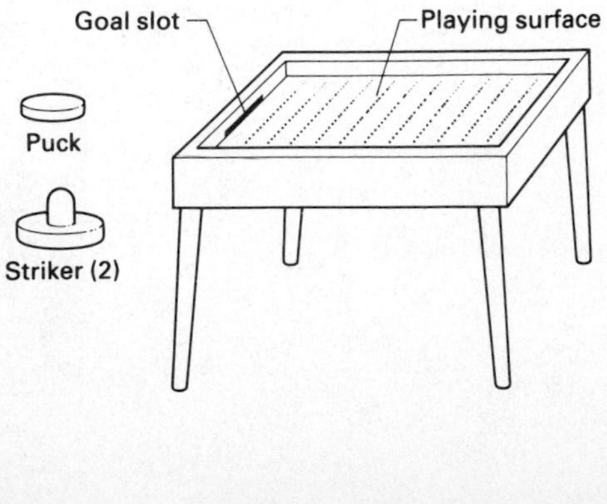

Air–Gas Jet Pump

A fan-driven jet pump is illustrated in Fig. J.4. The purpose of this device is to transfer a gas from the suction side ⓢ to the discharge side ⓓ against a total pressure difference $(p_{td} - p_{ts})$. A fan is used to produce a flow of primary air, $\dot{m}_j$. The primary air jet induces a secondary flow, $\dot{m}_s$, and the two streams continue to mix in the throat section ⓣ. Velocity pressure is recovered in the diffuser section.

Jet pumps are used to convey hot or corrosive gases as the secondary fluid, because this fluid does not come into contact with any moving parts (i.e., the fan).

The efficiency, η, of a jet pump can be defined as the ratio of the energy available from the secondary fluid to the energy expended by the primary fluid, or

$$\eta = \frac{\dot{m}_s(p_{td} - p_{ts})}{\dot{m}_j(p_{tj} - p_{td})}.$$

Efficiencies are inherently low, as much of the jet energy, measured by p_{tj}, is spent in turbulent mixing.

Design a jet pump to meet the following requirements:

- Primary fluid: standard air
- Secondary fluid: standard air
- Suction total pressure (p_{ts}): -2.0 in. H_2O
- Delivery total pressure (p_{td}): 2.0 in. H_2O
- Secondary inlet flow: 3000 ft^3/min
- Secondary suction pipe diameter: 3 ft
- Discharge pipe diameter: 3 ft

A jet pump with maximum efficiency is desired. Determine all relevant sizes, the required primary flow rate, and estimate the driving fan pressure and power. Suggest appropriate materials of construction.

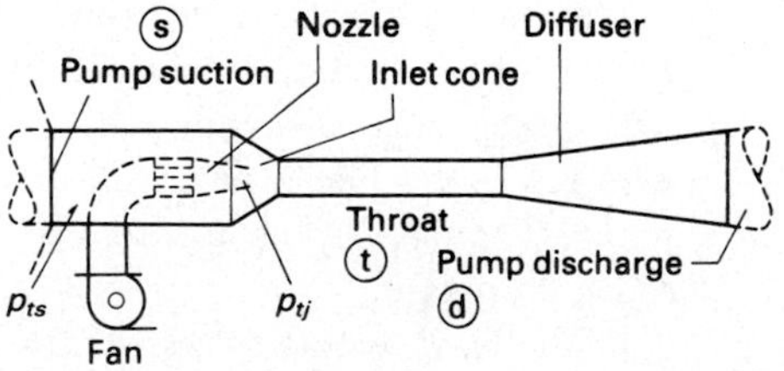

Figure J.4 Jet pump

Pipe System Sizing Software

The object of this project is the development of computer software that will do pipe sizing calculations for a single-branch, single-sized piping system. Such a system is illustrated schematically in Fig. J.5.

As a *minimum*, the software must have the following capabilities:

Input

- Fluid properties (density and viscosity)

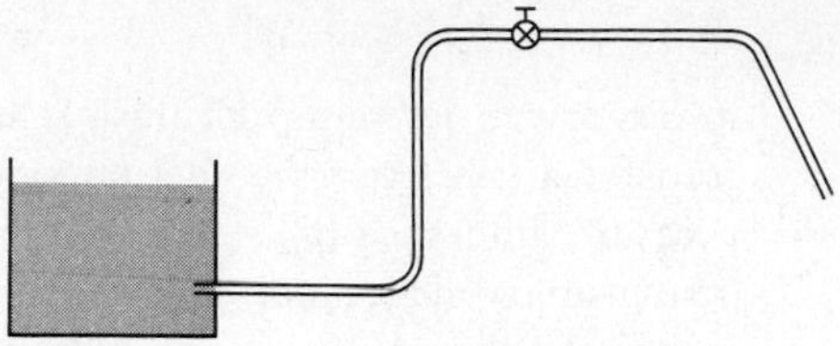

Figure J.5 Pipe system

- Pressure and elevation at the inlet and outlet
- Pipe material options (steel, cast iron, drawn tubing, or plastic)
- Flow rate
- System Description: pipe length, number of different types of components, and the number of each type present

Computational Capability

- The software must internally determine the friction factors for the pipe.
- For the minimum program, it is permissible to request keyboard input of local loss coefficients during the calculations; however, the request must identify the component by name or number and size (e.g., "Input the loss coefficient for a 6-inch globe valve" or "Input the loss coefficient for component number 4 in a 6-inch size."
- A better program would have some "standard" loss coefficient models built in. Typical components, such as globe valves or elbows, could then be specified by name or code number.

Output

The output must be the standard schedule 40 pipe size that will deliver the specified flow under the given available energy conditions.

Ram-Air Turbine Test Apparatus

A Navy aircraft uses an electronic countermeasures (ECM) radar jamming system. Power for the system is provided by a ram-air turbine (RAT). A ram-air turbine is a propeller-like device mounted on the front of the ECM pod, as shown in Fig. J.6. The RAT acts like a windmill driven by the relative wind when the airplane is flying. The RAT spins a generator that provides in-flight electrical power for the ECM hardware.

The Navy wants to test RATs on an aircraft carrier deck prior to flight. The basic concept requires directing a jet of air over the RAT on a stationary airplane to simulate the in-flight relative wind. An air jet 22 in. in diameter with a speed of 250 ft/sec is required. The maximum allowable air temperature is 170°F.

Power for the RAT test system can be provided by portable (cart-mounted) air compressors used to start jet engines. A jet-start compressor can provide 210 lb/min of air at a pressure of 77 psia and temperature of 515°F.

Design a system to provide the necessary air jet while using only the air from jet-start compressors as an energy source. The system should be as small as possible and use the minimum number of jet-start compressors.

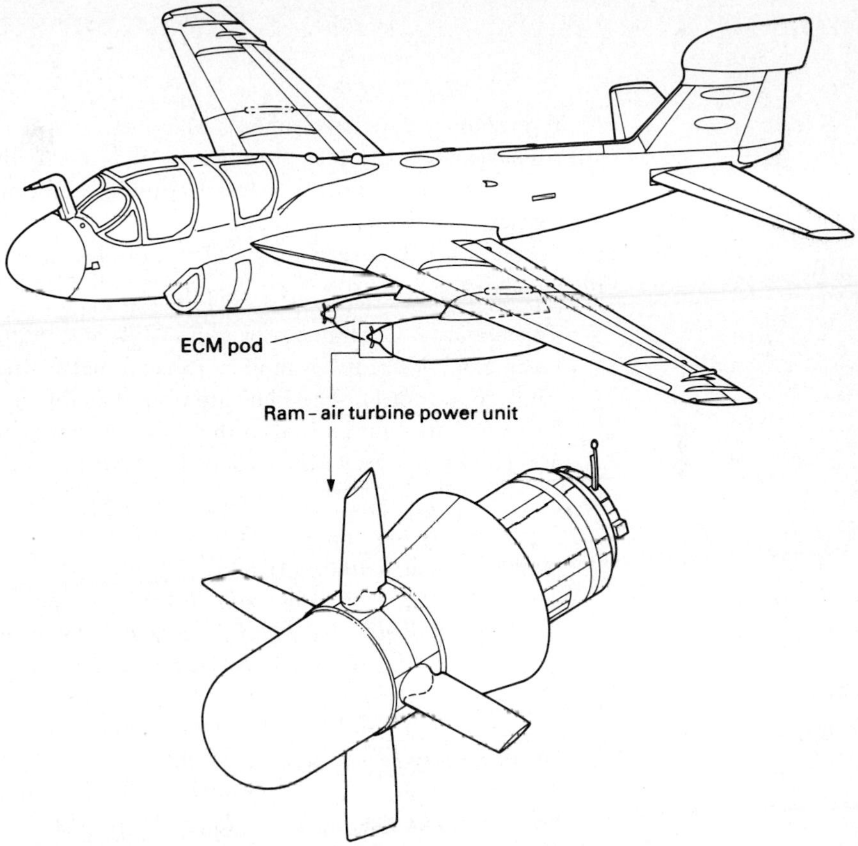

Figure J.6 Ram-air turbine (RAT) unit

Appendix K FLUID MECHANICS FILMS AND VIDEOS

Fluid mechanics is a visual subject. Many films are available to help engineers and students visualize complex flow phenomena. The films listed in this Appendix provide excellent support for class lecture and laboratory work in teaching fluid mechanics. The films are available for a modest rental fee or sometimes free from the agencies listed. Several of the films can also be obtained from public libraries or university film services.

A. The National Committee for Fluid Mechanics Films (NCFMF). Available from: Encyclopaedia Britannica Educational Corp., 425 N. Michigan Ave., Chicago, IL 60611. The films are also available in VHS video format. The films are summarized in the book *Illustrated Experiments in Fluid Mechanics, the NCFMF Book of Film Notes*, available from MIT Press.

1. *The Fluid Dynamics of Drag* (119 min)
2. *Vorticity* (44 min)
3. *Flow Visualization* (31 min)
4. *Deformation of Continuous Media* (38 min)
5. *Pressure Fields and Fluid Acceleration* (30 min)
6. *Surface Tension in Fluid Mechanics* (29 min)
7. *Waves in Fluids* (33 min)
8. *Secondary Flow* (30 min)
9. *Boundary Layer Control* (25 min)
10. *Channel Flow of a Compressible Fluid* (29 min)
11. *Low Reynolds Number Flows* (33 min)
12. *Flow Instabilities* (27 min)
13. *Cavitation* (27 min)
14. *Fundamentals of Boundary Layers* (24 min)
15. *Turbulence* (29 min)

B. University of Iowa Films. Available from University of Iowa, Media Library, Audiovisual Center, Iowa City, IA 52242.

1. *Introduction to the Study of Fluid Motion* (25 min)
2. *Fundamental Principles of Flow* (23 min)
3. *Fluid Motion in a Gravitational Field* (24 min)
4. *Characteristics of Laminar and Turbulent Flow* (26 min)
5. *Form Drag, Lift, and Propulsion* (24 min)
6. *Effects of Fluid Compressibility* (17 min)

C. St. Anthony Falls Laboratory Films. Available from St. Anthony Falls Hydraulic Laboratory, Mississippi River at 3rd Ave. S.E., Minneapolis, MN 55414.

1. *Hydraulic Model Studies* (20 min)
2. *Some Phenomena of Open Channel Flow* (33 min)
3. *Flow in Culverts* (20 min)
4. *Spillway Design, Guayabo Dam, El Salvador* (15 min)

5. *Surface Waves* (24 min)
6. *Fluid Mechanics: The Boundary Layer* (30 min)

D. Films available from Engineering Societies Library, 345 E. 47th St., New York, NY 10017.

1. *Turbulent Flow (An Introductory Lecture)* (35 min)
2. *Flow Separation and Formation of Vortices* (16 min)
3. *Hydraulic Analogy for Flow Studies* (24 min)
4. *Visual Cavitation Studies of Mixed Flow Pump Impellers* (22 min)

Index

E

F

G

N

O

P

Q

R

S

Table 8.4 Drag of various two-dimensional bodies.

Shape	Reynolds Number	C_D
Parallel flat plate →	$>10^5$	See below.
Normal flat plate →	$>10^3$	2.0
Circular cylinder →	All	See Fig. 8.10.
Square rod →	$>10^4$	2.0
Square rod →	$>10^4$	1.50
Equilateral triangular rod →	$>10^4$	Sharp edge forward: 1.40 Flat face forward: 2.0
C-section →	$>10^4$	2.30
C-section →	$>10^4$	1.20
Airfoils →	Various	(See Chapter 8, reference [5].)

Table 8.6 Drag of various axisymmetric objects.

Shape	Reynolds Number	C_D	
Sphere →	All	See Fig. 8.10.	
Cylinder/disc ($L = 0$) →	$>10^4$	$L/D = 0$	1.17
		$L/D = 0.5$	1.15
		$L/D = 1$	0.90
		$L/D = 2$	0.85
		$L/D = 4$	0.87
		$L/D = 8$	0.99
Hemispherical cup →	$>10^4$	1.40	
Hemispherical cup →	$>10^4$	0.40	
60° cone →	$>10^4$	0.50	
Parachute	$>10^5$	1.2	

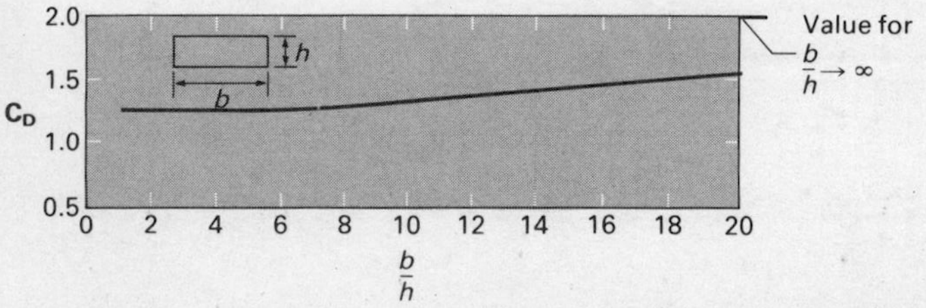

Drag coefficient for flow over a finite length flat plate, normal to the flow. Adapted from Fig. 8.9.

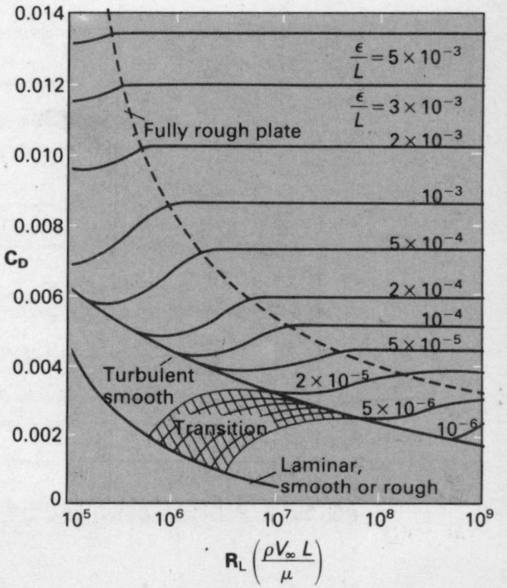

Drag coefficient for flow over a thin flat plate aligned with the flow. Adapted from Fig. 8.8.